Transforming the Future of Infrastructure through Smarter Information

Proceedings of the International Conference on Smart Infrastructure and Construction, 27–29 June 2016

Edited by
RJ Mair, K Soga, Y Jin, AK Parlikad and JM Schooling

Organising Committee:

Professor Lord Mair
Professor Kenichi Soga
Dr Ying Jin
Dr Ajith Parlikad
Dr Jennifer Schooling

Supported by:
Dr Krishna Kumar
Samantha Archetti
Tianlei Wu

Published by ICE Publishing, One Great George Street, Westminster, London SW1P 3AA.

Full details of ICE Publishing sales representatives and distributors can be found at:

www.icebookshop.com/bookshop_contact.asp

Other CSIC titles from ICE Publishing:

Bridge Monitoring: A practical guide
C. Middleton, P. Fidler, P. Vardanega. ISBN 978-0-7277-6059-3

Distributed Fibre Optic Strain Sensing for Monitoring Civil Infrastructure: A practical guide
C. Kechavarzi, K. Soga, M. Elshafie, R.J. Mair, N. de Battista, L. Pelecanos. ISBN 978-0-7277-6055-5

Whole-Life Value-Based Decision-Making in Asset Management
R. Srinivasan, A. Parlikad. ISBN 978-0-7277-6061-6

Wireless Sensor Networks for Civil Infrastructure Monitoring: A best practice guide
D. Rodenas-Herráiz, K. Soga, P. Fidler, N. de Battista. ISBN 978-0-7277-6151-4

A catalogue record for this book is available from the British Library

ISBN 978-0-7277-6127-9

© The authors and ICE Publishing 2016

ICE Publishing is a division of Thomas Telford Ltd, a wholly-owned subsidiary of the Institution of Civil Engineers (ICE).

All rights, including translation, reserved. Except as permitted by the Copyright, Designs and Patents Act 1988, no part of this publication may be reproduced, stored in a retrieval system or transmitted in any form or by any means, electronic, mechanical, photocopying or otherwise, without the prior written permission of the Publisher, ICE Publishing, One Great George Street, Westminster, London SW1P 3AA.

This book is published on the understanding that the author is solely responsible for the statements made and opinions expressed in it and that its publication does not necessarily imply that such statements and/or opinions are or reflect the views or opinions of the publishers. Whilst every effort has been made to ensure that the statements made and the opinions expressed in this publication provide a safe and accurate guide, no liability or responsibility can be accepted in this respect by the author or publishers.

Whilst every reasonable effort has been undertaken by the author and the publisher to acknowledge copyright on material reproduced, if there has been an oversight please contact the publisher and we will endeavour to correct this upon a reprint.

Commissioning Editor: Gavin Jamieson
Production Editor: Rebecca Norris
Market Development Executive: Elizabeth Hobson

Typeset by Manila Typesetting Company
Printed and bound in Great Britain by TJ International Ltd, Padstow

Contents

Fibre optics sensing - theory

Wireless sensor network systems

New sensor systems

Dynamic measurements and analysis

Data analysis - bridges

Data analysis - buildings

Data analysis - geotechnical structures

SECTION B: ASSETS

Asset information management

Deterioration modelling

Infrastructure resilience

Whole life cost and value

SECTION C: CITIES AND URBAN INFRASTRUCTURE

Cities

Preface

The International Conference on Smart Infrastructure and Construction (ICSIC), held in Cambridge from 27-29 June 2016, brought together world-leading academics and practitioners from the fields of infrastructure planning, asset management and infrastructure sensing and monitoring.

These Proceedings bring together the papers presented at the conference, providing inter-disciplinary insights into how smarter information leads to rapid, improved evidence-based decision making. The papers address many topics relating to smart infrastructure and construction, including the persistent barriers to industry integration of innovation and the development of novel, proactive solutions to infrastructure and construction challenges.

The papers were presented in three parallel streams focussed on three key aspects of Smart Infrastructure and Construction:

Sensors and data analysis
Recent innovations in sensor systems and development of new data analysis methods allow us to better understand the engineering performance of our infrastructure. Papers in this theme cover new developments in fibre optic sensing, wireless sensor networks and miniature low-power sensors, case studies using innovative sensor systems, and progress in data analysis methods. The findings from the work presented in the papers will lead to improvements in performance-based design, more efficient construction and a better-informed maintenance strategy.

Asset management
Smart sensing technologies offer immense potential to deliver a step-change in whole-life cost and value of infrastructure. Using the data generated by emerging technologies to make effective asset management decisions is critical for ensuring long-term value and sustainability of infrastructure. Papers in this theme discuss the short and long-term challenges in managing infrastructure, innovative models and tools for supporting investment and maintenance decisions, and how BIM can be used as an effective tool for asset management. Through an excellent collection of case studies, this theme offers insights into how the long-term resilience of infrastructure can be improved.

Cities and urban infrastructure
New technology and business models are emerging in urban infrastructure and service provision. Papers in this theme reveal new insights into the changing roles of infrastructure planning and construction in enhancing resilience and adaptability of the urban environment. Highlights include how new forms of data are transforming our understanding and management of infrastructure and development in cities. Papers from leading scholars and practitioners from engineering, geography, planning, urban design and architecture identify fresh opportunities for interdisciplinary engagement, with ideas for developing all-round and integrated solutions.

Professor Lord Mair, CBE FREng FRS,
Sir Kirby Laing Professor of Civil Engineering

Dr Jennifer Schooling
Director, Centre for Smart Infrastructure

SECTION A:
SENSORS AND DATA ANALYSIS

Proceedings of the International Conference on Smart Infrastructure and Construction
ISBN 978-0-7277-6127-9

© The authors and ICE Publishing: All rights reserved, 2016
doi:10.1680/tfitsi.61279.003

Distributed measurements with optical sensors in the EDF group: experience feedback and perspectives

Y-L. Beck*[1], F. Martinot[1], S. Desforges[2], E. Buchoud[3], J.M. Henault[3]

[1] *Electricité de France, Measures and Methods Division, Grenoble, France*
[2] *Electricité de France, Nuclear Buildings Surveillance, Lyon, France*
[3] *Electricité de France, Research and Development, Paris, France*
* *Corresponding Author*

ABSTRACT For the last decade, Distributed Optical Fiber Sensors (DOFS) have gradually played a prominent role in health monitoring of EDF structures. This paper first gives an overview of results obtained in leak detection applications in French hydraulic facilities. Then, it explains how this experience feedback has been transposed to nuclear application with recent experiments : EPR Flamanville 3 where a continuous surveillance is performed since 2013 and VERCORS, a 1/3 mock-up of a containment, where 2 km of OFS have been setup in 2014. Advantages, drawbacks and next challenges of this innovative technology are discussed from a user viewpoint.

1 OVERVIEW

EDF is progressing steadily on the quality of structural health monitoring with innovation. EDF has been involved for 15 years in the developpement of DOFS in earthen contrete structures. A use in concrete structures was considered in 2005, with the design of the EPR Flamanville 3 nuclear power plant. In this article, we first summarize EDF's operating experience (OPEX) on dikes with a focus on the Curban site. Then, we detail two large scale experiments in concrete at EPR Flamanville 3 and Vercors mock-up. New surface solutions are also designed to adress a broader range of applications. To finish, we discuss the advantages and next challenges for DOFS.

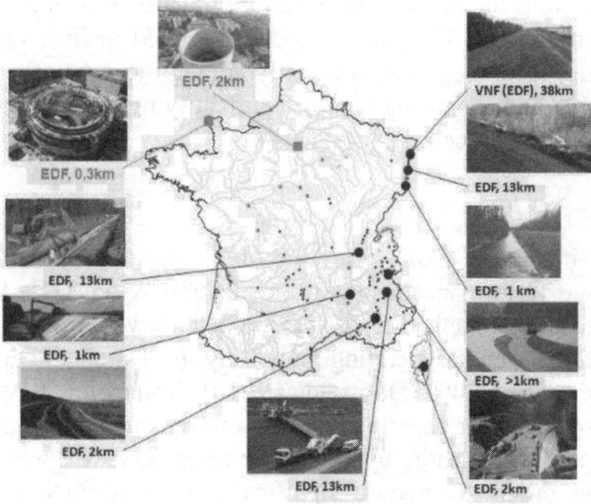

Figure 1: Overview of EDF structures equipped with DOFS

2 DOFS IN EARTHEN STRUCTURES

2.1 *Overview*

EDF operates over 580 km of channels for its hydroelectric facilities. The aging of these channels induce a more sensitivity to internal erosion processes.

Figure 2: View of the studied intake channel own by EDF on the left, of the Durance River, on the right.

Using the conventional monitoring systems, it is often difficult to detect internal erosion at an early stage despite the fact that this pathology is considered as the main cause of failure of earthen hydraulic structures. Thermal methods associated to fiber optics technologies are currently among the most effective and promising means for identifying the internal erosion processes.

2.2 *Early developments*

Since 2001, EDF is working on the development and the performance testing of acquisition systems. Early detection methods (passive and active) were tested in four large scale testing launch by the Dutch IjkDijk project in 2009 (Dornstädter 2010) (Koelewijn 2010). Indeed, four experimental dikes were constructed and put under charge by filling the reservoirs with water in order to provoke significant flows through the dikes and the measurements were recorded until failure. Optical fibers were installed to investigate the sensitivity analysis models to the position of the sensor. The use of passive fiber optic technology associated with daily analysis has been the most powerful monitoring device among the tested technologies.

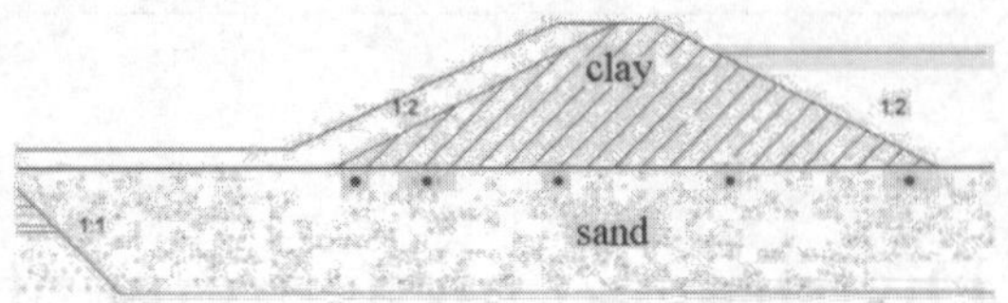

Figure 3 : Cross-sectional view of the dike showing the dike composition and placement of the fiber optics (red dot).

2.3 *Leakage detection in French rivers*

Eight EDF sites were instrumented by fiber optic sensors and monitored continuously. Various positions of fiber optics have been tested (eg. in facing upstream or at the downstream toe of the dike) in order to validate the existed and developed analysis models. The results of each analysis method are compared with visual inspections and / or conventional monitoring data (eg. flow measurements in drains or piezometric level) (Beck 2010)(Beck 2012).

2.4 *Leakage during impoundment at Curbans*

Recently, Curbans, a 5 km long and 43 to 50 m high intake channel, along the Durance River, was instrumented by 11km of fiber optics (Figure 3). Taking advantage of rehabilitation works of the bituminous concrete sealing and drainage system of the channel, a hybrid cable has been installed close to the drain in order to compare the results obtained by fiber optics with collected flow rate (Figure 5). The drainage system consists of 31 renovated drains which allow separating collected leaks from right bank and those from the left bank. The maximum length of each drain is 300m.

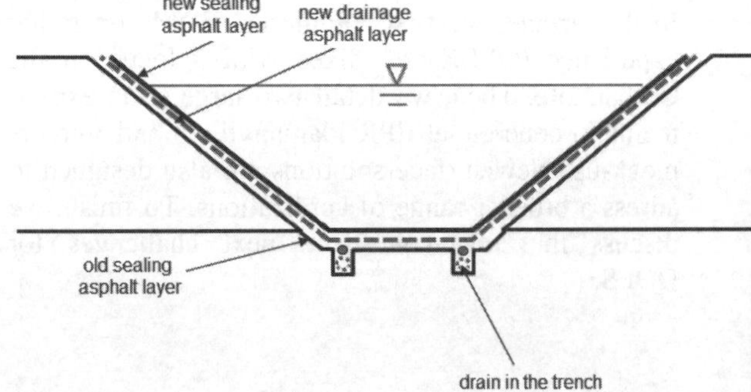

Figure 4: Cross-sectional view of upstream face of the Curbans' intake channel showing the sealing and drainage layer and placement of the fiber optics (red dot).

In order to heat the hybrid cable up to 10 W/m, the fiber optic installation has been split in four distinct cable sections of around 2.5 km long. At the end of rehabilitation works, one reference dT measurement has been carried with a 2 hours heating at 11 W/m before the filling of the channel. This reference dT is corresponding to a "no leak" state of the channel.

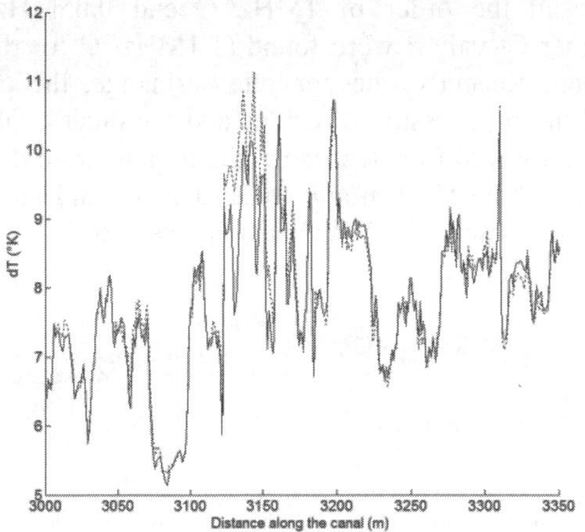

Figure 4: Example of results of leakage detection with active method (at 11 W/m). The dotted curve corresponds to the so called "reference dT", i.e. at dry state before channel filling, and the smooth curve is the measured during a filling phase of the channel.

The channel filling took place step by step. At each step where the water level of the channel was constant, an active measurement has been realized in the same condition than the reference measurement (2 hours, 11 W/m). As shown in Figure 4, a presence of leak is detected when reference measurements is well above over the analyzed heating measurement. In Figure 5, we identified a diffuse leakage zone from 0 m to 800 m and a concentrated leak from 3148 m to 3155 m.

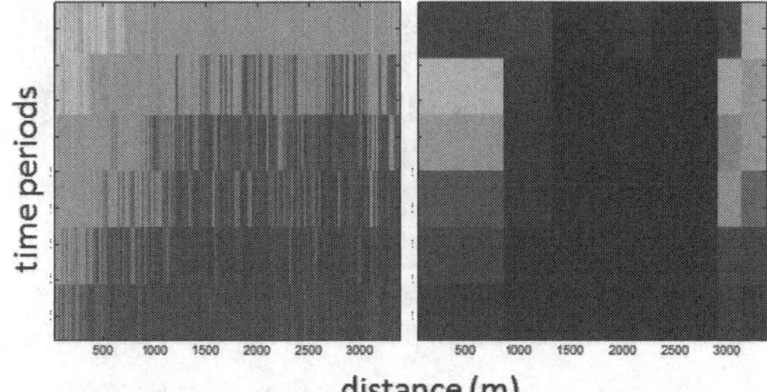

Figure 5 : Comparison between sensing parameters OF model (°C) and drainage flow rate (l/min)

3 DOFS IN CONCRETE STRUCTURES

EDF has developped DOFS based on Brillouin, Rayleigh and Raman measurements to improve the surveillance of mass concrete structures. We have not been alone to believe in these technologies since promising experiments have also been carried out by industrial partners like CEA, ANDRA, AREVA.

3.1 EPR Flamanville 3

Fiber optics sensors are located all along the top of inner containment's barrel, on a 165 m circumference, at a 24 m radius and a 45 m height.

Figure 6. EPR Containment

Three types of optical sensors (Figure 7) have been embedded in concrete. Among them, "DT" is a smooth flexible temperature / deformation sensor. "D" is a rigid strain sensor with a rough surface. "T" is a cable dedicated to temperature in "loose" configuration: the coating is not attached to the fiber. A redundancy in sensors (use of two cables for the DT sensor, use of multiple fibers for the T sensor and cable outputs (four measurement and four connection boxes) was also guaranteed. Eight vibrating wire strain gages (VWSG) and four PT100 probes are integrated in the structure for comparison.

It was decided to fix optical fibers before laying the outer ply reinforcement and to temporarily protect them with a cage and fireproof fabrics. Optical sensors were also installed 20 cm under prestressing tubes to limit the interaction with vibrating needles after concrete pouring. Surveillance during concrete pouring was planned to avoid interaction with a vibrating needle. All operations were performed with respect of the safety and time constraints of the containment erection thanks to a job rotation perfectly synchronized with builders.

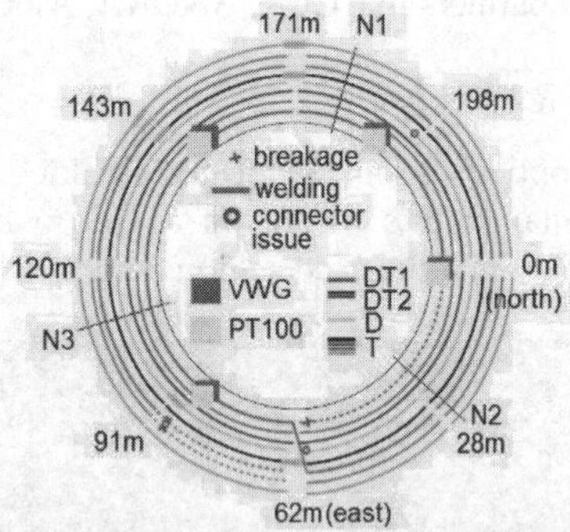

Figure 7. Schematics of the fiber optics instrumentation with corrections after concrete pouring

Sensors survived to mass concrete pouring which is an encouraging result for upcoming projects. However, 37 % (hypothesis of single access from 0 m in future) of optical lines were lost due to a use of precut cables lengths, connectors used instead of optical spliced to link each elementary sensor and quality defect on a sensor ending (Martinot 2014). Since June 2014, electronic devices have been remotely installed on a bungalow. A data acquisition is performed every hour.

The Raman temperature has followed seasonal temperatures measured in concrete (Figure 3).

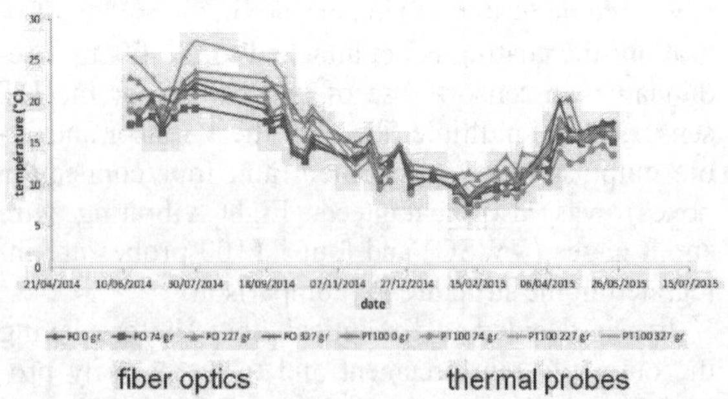

Figure 8. Comparison between Raman measurements and thermal probes

Raman data may be polluted at some points by the outside temperature because of frequent cable outputs (Martinot 2015).

Regarding strain, the relationship between Brillouin shifts, strain and temperature changes is $\Delta\gamma = Ct\Delta T + C_\varepsilon \Delta\varepsilon$ where Ct and C_ε are characteristical of the fiber type in the order of 1MHz/°C and 0.05MHz/µε. Higher Ct values were found (1.1MHz/°C for the D cable). Recently, the concrete shrinkage, the creep and the prestressing of cables and the outer containment allowed to detect more significant strain differences (about 150 µm/m between January and March). A good match with VWG data is obtained.

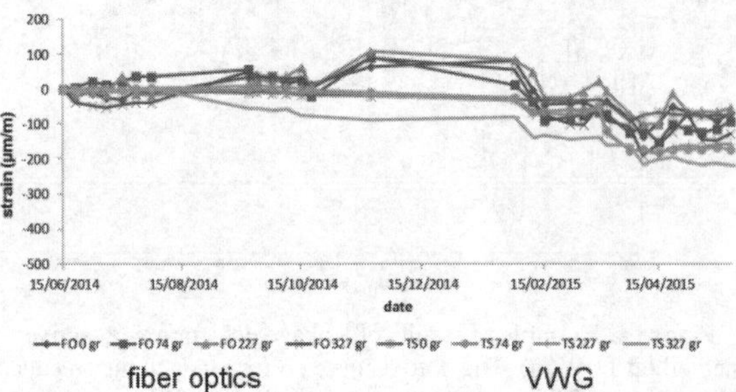

Figure 9. Comparison between Brillouin strain (corrected with PT100 temperature, Ct = 1.1MHz/°C, C_ε = 0.048MHz/µε) and VWG

As seen in Figure 10, the strain difference seems quite homogeneous on the circumference. The effect of notches is visible on the strain data before (existing effect on N2) and after prestressing (amplified effect on N1).

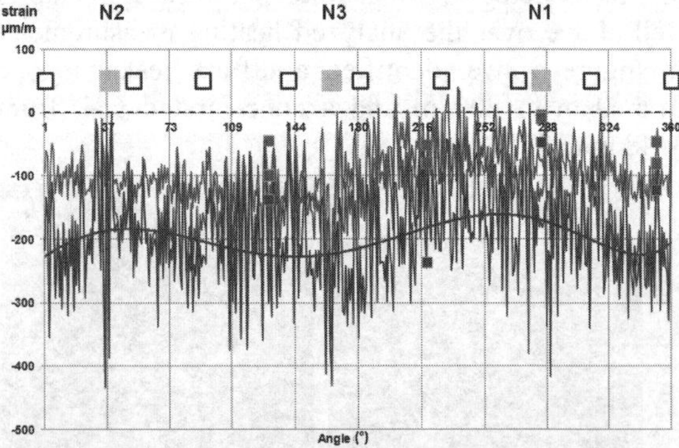

Figure 11. Prestressing effect after cable pulling on each notch

To conclude on theses analyses, results are very encouraging: optical fiber sensors seem to detect the strain due to concrete shrinkage, creep and prestressing. However, a consolidated thermal correction has to be proposed to truly quantify the phenomena.

3.2 *VERCORS*

The VeRCoRs mock-up (Figure 12) has been built by EDF to study the aging process of prestressed containment buidings of nuclear reactors (Figure 12. a.). Its scale corresponds to 1/3 of a double wall P'4 reactors. Its objective after 7 years of exploitation is to obtain an equivalent aging as in an industrial reactor operated during more than 60 years.

In order to measure the aging process, more than 700 traditional sensors have been installed measuring temperature, strain, humidity, displacement, into the concrete and onto some iron rebars. In addition, 2 km of optical fiber cables have been poured into the concrete (Figure 12b).

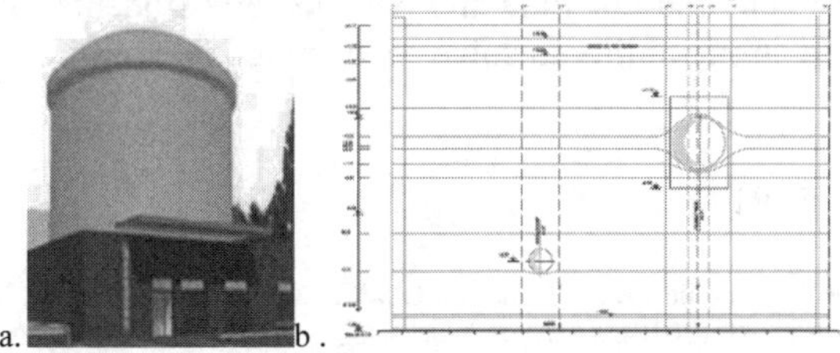

Figure 12: a. 3D representation of the mock-up [1]. b. Representation *z*- θ of the position of the optical fiber cables in the concrete in red.

As seen in Figure 13, the optical fiber sensors enable to measure distributed profiles of temperature and strain with a spatial resolution of 40 cm. Information have been collected during the several stages of the construction of the mock-up: pouring and hydration of concrete, early age and prestressing. Acquisition of data will be maintained to observe creep phenomenon, and effects during the pressurization tests that are required for leaktightness verification.

All these data will feed numerical models in order to predict the strain evolution during young age, the apparition of cracks and to understand the leak mechanism into concrete. A benchmark has been launched to compare the models of each participant to predict the aging process of the mock-up.

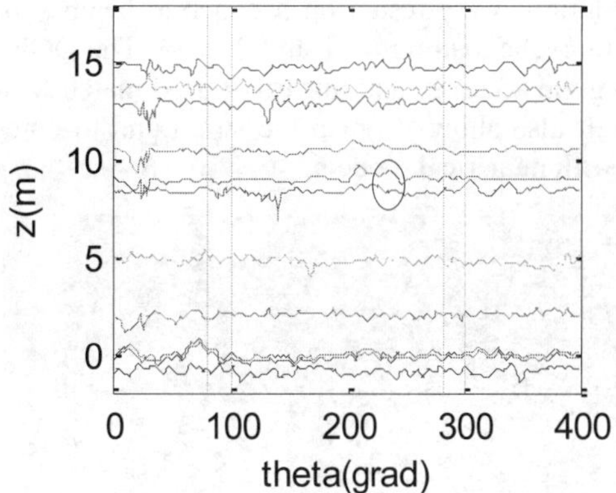

Figure 13 : Normalized strain profiles at the position of the optical fiber cables into the concrete. The circle represents the materials pass-by.

3.3 *New solutions for surface strain*

For existing structures (NPP, cooling towers) the sensors cannot be embedded in concrete anymore. If commercial solutions exist (scotch like) to measure the strain in surface, they appeared to not meet user's requirements for long term operation.

After numerical simulation aiming to optimize the strain energy transfers without exposing too much the sensor to cracks (Billon 2015), 50 configurations of cables, adhesive and integration methods were tested. Overall, the use of a soft resin to embed an "EPR-like" cable in a groove seems has given good results.

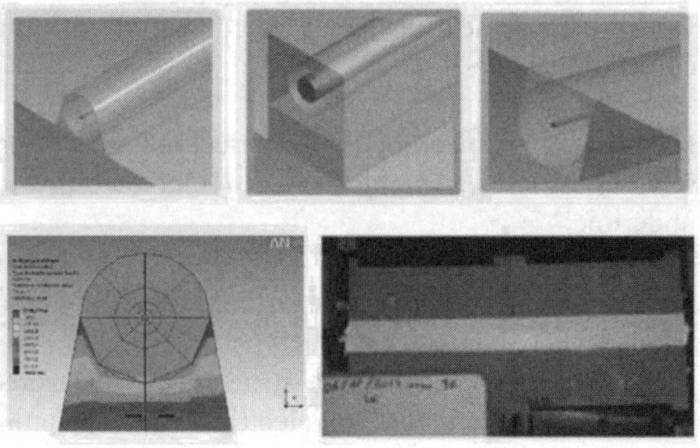

Figure 14. Opimization of sensing properties with models and test benches

Two solutions were tested on a concrete beam submitted to cyclic loading (at 1 and 2 tons). The DOFS show a good accuracy and linearity (good match with VWG). It also allows to map the strain and to compare it with numerical models.

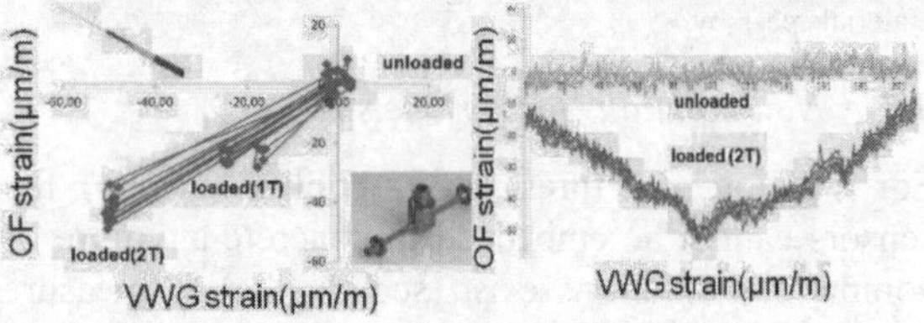

Figure 15. Optimization of sensing properties with models and test benches

4 CONCLUSION

The development of DOFS includes multiples theoretical and practical challenges. Experiments in living structures have allowed us to demonstrate the relevance of these new technologies compared with the conventional ones.

From a user's viewpoint, the DOF technologies bring obvious benefits. Among them, it is noteworthy to underline:

- a better representativity of structure's states: better estimates can be made regarding reliability and long term operation of a structure
- a cost reduction : on large structures, a measurement point is low-cost and electrical cables lengths are reduced
- a better versatility: a large range of applications can be addressed (multiple physical parameters, various integration methods) on multiple space scales.
- security in operation (light based systems)

Some advances remain however necessary:

- on site integration is still very specific to each site which may raise design costs. It has to be simpler and faster.
- optoelectronic devices remain expensive and bulky
- analysis methods have to be accepted by end-users with an appropriate training
- the massive data has to be exploited in a secured and efficient database. A web-based software may allow to share information more easily

ACKNOWLEDGEMENT

We would like to acknowledge our partners, mostly from IRSTEA, Tencate, géophyConsult, Sites SAS, Bouygues, Eiffage, Deltares and Grenoble University which contributed to the presented results in this paper. We also thank the colleagues from R&D department and our PhD and post-doctoral students for their strong ability to solve encountered problems.

REFERENCES

Beck Y.L., Khan A.A., Cunat P., Guidoux C., Artières O., Mars J. and Fry J.J. (2010), *Thermal Monitoring of Embankment Dams by Fiber Optics*, 8th ICOLD European Club Symposium DAM SAFETY, 5pp., 22-23 September, Innsbruck, Austria.

Beck Y-L., Cunat P., Reboud M., Courivaud J.-R., Fry J.-J. and Guidoux C. (2012), *Improvement of leakage monitoring in levees by the use of distributed fiber optics sensors*, International Conference on Scour and Erosion, 7pp. Paris, France

Billon A., Henault J. M., Quiertant et al. (2015). Qualification of a distributed optical fiber sensor bonded to the surface of a concrete structure: A methodology to obtain quantitative strain measurements. Smart Materials and Structures, 24(11).

Dornstädter J. 2010. *Detection of erosion flow path during 1:1 scale experiment using fibre optic Heat Pulse Method at Ijkdijk*, Annual Meeting of the European Working Group on Internal Erosion, 12 April 2010, Grenade, Spain.

Koelewijn A. 2010. *Performance of detection techniques at four full-scale seepage erosion tests*, Annual Meeting of the European Working Group on Internal Erosion, 12 April 2010, Spain.

Martinot, F., Martin G., Henault J.-M., Moreau G., Buchoud E., Beck Y.-L., Courtois A. (2015). "Distributed Fiber Optic Sensors in Concrete of a Nuclear Containment," *Proc. International Conference on Structural Health Monitoring of Intelligent Infrastructure,* Torino, Italy.

Martinot, F., Moreau G., Martin G., Henault J.-M., Thuault C., Beck Y.-L., Courtois A. (2014). "Optical Sensors in Concrete of the EPR Flamanville 3 Confinement Building," *Proc. 2nd International Conference on Technological Innovations in Nuclear Civil Engineering*, Paris, France.

Proceedings of the International Conference on
Smart Infrastructure and Construction
ISBN 978-0-7277-6127-9

© The authors and ICE Publishing: All rights reserved, 2016
doi:10.1680/tfitsi.61279.009

Distributed fibre optic sensors for the purposes of structural performance monitoring

A. Brault[1], N.A. Hoult*[1] , I. Trudeau[2] , T. Greenough[2], and B. Charnish[2]

[1] Queen's University Kingston, Canada
[2] Entuitive, Toronto, Canada
[*] Corresponding Author

ABSTRACT The structural design and construction industries have the potential to have a significant positive impact on the environment while at the same time reducing the cost of projects through optimized design. However, these benefits must not come at the expense of public safety or the serviceability of the structure. More comprehensive data is required to better understand the performance of structures designed using current procedures so that these procedures can be optimized. Until recently, the use of sensors during construction and operation of a building to acquire this data was difficult, expensive, and ultimately provided limited useful information. Distributed fibre optic sensors have the potential to overcome these difficulties including providing an extensive data set that can be used to more comprehensively evaluate the serviceability and ultimate limit states performance of a structure. This paper introduces the installation of a distributed fibre optic strain sensing system in a reinforced concrete commercial building. The system, based on measuring Rayleigh backscatter, enables strain to be measured with high accuracy over a gauge length that can be adjusted by the user along up to 70 m of fibre optic cable. The results of a live load test performed on a beam are introduced. The results were used to explore the beam's performance in terms of strain, deflection, flexural crack spacing, and support conditions. Distributed fibre optic sensors were found to show promise in this study, though future work is required.

1 INTRODUCTION

An ever increasing burden is being placed on the global economy and the environment by the need to build, repair and replace infrastructure. For the structural engineering community there is an opportunity to help alleviate this burden by reducing the use of materials and other resources through optimized approaches to design. Refined design could reduce the materials used in structures by 30%, reducing costs and CO_2 emissions by a similarly large amount (Allwood and Cullen, 2012). These significant savings are possible because steel and cement production accounts for approximately 50% of industrial CO_2 emissions (Allwood and Cullen, 2012) and in Canada alone the construction industry represents 6% of GDP ($73.8 billion) (Statistics Canada, 2011).

However, engineers are often limited in the design, construction, and analysis techniques they can use because of the need for conservatism in the absence of accurate information about structural behaviour. Conservative assumptions need to be made about the relative stiffness of elements as well as their support conditions such that each element is designed to support more load than it will ever carry. If a better understanding of how structural elements behave in-situ could be developed, future designs could be refined to reduce the use of materials. Distributed fibre optic sensors (FOS) can potentially provide critical data to remove this conservatism from design.

One of the challenges of installing sensors during the construction phase is that their installation has the potential to delay construction, resulting in increased costs and diminishing any benefits that optimized design might have offered. As a result, it makes sense to install the sensors during the period between casting and when finishing trades move into the floor so as to minimize the impact on schedule. This would require the fibres to be installed on the surface of the beams, which means that they need to be able to bridge cracks in the concrete surface while still providing useful readings. Additionally, the fibres could not be used to measure the dead load, early age creep and shrinkage response, but could be used to measure the live load response. The question then becomes, can effective comparisons be made between the measured and predicted live load response to better understand the structure's behaviour.

This paper details a pilot study on the installation and use of a distributed strain measurement system. The objectives of this research project were to:

1. Install fibre optic sensors on the face of a beam during the construction,
2. Measure the strains experienced by the beam under live loading, and
3. Use the measured strains to determine the beam's deflection and support conditions under live loading.

The next section of the paper will provide a brief background on fibre optic strain sensing followed by an introduction to the field monitoring site. The installation of the fibres will then be presented and the measurements discussed.

2 FIBRE OPTIC SENSING

Fibre optic strain measurement techniques can be divided into discrete and distributed measurement techniques. Discrete techniques, such as fibre Bragg gratings (FBGs), provide measurements at a point and as a result cannot be used to determine the support conditions or the deflected shape. Until recently most work in civil engineering was focused on the use of Brillouin backscatter (e.g. Mohamed et al., 2011) and long gauge FBGs (Schulz et al., 2001). Brillouin systems provide measurements over kilometres but have limitations in terms of the minimum strain (~35 $\mu\varepsilon$) and spatial resolution (~0.5 m) they can measure (Mohamed et al., 2011). Long gauge FBGs have high measurement accuracy but are not truly distributed as the number of gauges is limited.

Recently researchers have begun to use distributed Rayleigh backscatter based FOS systems, which have much higher strain (~1 $\mu\varepsilon$) and spatial (~5 mm) accuracy (Klar et al., 2014; and Villalba and Casas, 2012). Rayleigh based sensing enables the detection of localized deterioration in steel (Hoult et al., 2014), and can provide an understanding of the behaviour of bridge structures (Regier and Hoult, 2014). However, whilst researchers have mounted fibres to the surface of concrete beams and measured the strains in the past (e.g. Villalba and Casas, 2012; Regier and Hoult, 2014), these systems have never been installed in a reinforced concrete building to monitor strains.

3 FIELD MONITORING SITE

This study was performed during the construction of a reinforced concrete building in Ottawa, Canada. The building consists of 3 below grade floor slabs, 4 above grade floor slabs, and a roof slab with a floor-to-floor height of 5 m. One reinforced concrete beam situated on the 4th floor of the building was instrumented and tested under live loading. A simplified schematic of the entire level 4 floor slab is presented in Figure 1 along with a close-up of the instrumented beam, which illustrates the beam's location and span length.

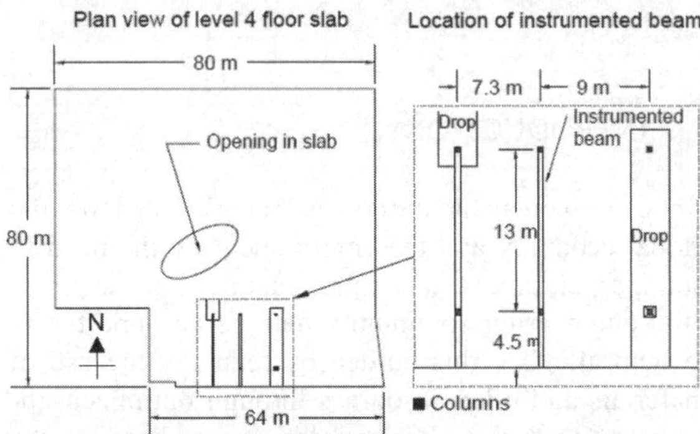

Figure 1. Simplified plan view schematic of the building's 4th floor slab with the location of the instrumented beam shown.

The beam is supported by 2 square reinforced concrete columns with side dimensions of 600 mm. The beam has a span of 13 m between the centres of its supports with a 4.5 m cantilever extending beyond the south column's centre. The reinforced concrete floor slab in the area of the beam has a thickness of 300 mm, and the beam itself has a height of 800 mm (excluding the floor slab thickness) and a width of 600 mm.

4 INSTRUMENTATION AND TEST SETUP

The beam was instrumented 3 months following the pouring of the beam when all the formwork, shores, and re-shores had been removed. The fibre optic cable used for instrumentation was a single-mode nylon-coated cable, which was bonded to the concrete surface of the beam with a two part epoxy (Loctite E-20HP Adhesive). Prior to bonding the fibre to the beam with epoxy, the concrete surface was sanded with a 150 grit flap wheel and cleaned with 99% isopropyl alcohol. Images of the fibre optic installation process are shown in Figure 2. The fibre was installed along the beam's length at two-separate heights, 75 mm from the bottom of the beam and 75 mm below the floor slab. The beam was instrumented with fibre optic sensors from the centre of the north column up to the beam's mid-span. The fibre optic sensor configuration is shown in Figure 3. Strain readings were taken using a Luna OBR 4600 fibre optic analyzer with a specified gauge length of 20 mm, a sensor spacing of 20 mm, and an accuracy of approximately 1 $\mu\varepsilon$.

In addition to the fibre optic sensors installed on the beam, a linear potentiometer was set-up 6.2 m from the centre of the north column as shown in Figure 3. The linear potentiometer captured vertical beam displacements with an accuracy of ± 0.024 mm.

The live load test was conducted using 6 scissor lifts consisting of 3 different types: lift type 1 (L1), lift type 2 (L2) and lift type 3 (L3) with a weight of 11.5 kN, 15.9 kN, and 15.6 kN, respectively. The scissor lifts were oriented perpendicular to the longitudinal axis of the beam, and each lift's vertical load was assumed to be divided equally into 2 separate loads applied through the centre of the wheels on either side of the lift's width. The magnitude, spacing, and orientation of each lift load is shown in Figure 3(b).

a) Concrete surface being sanded (left) and the prepared concrete surface cleaned with 99% isopropyl alcohol (right).

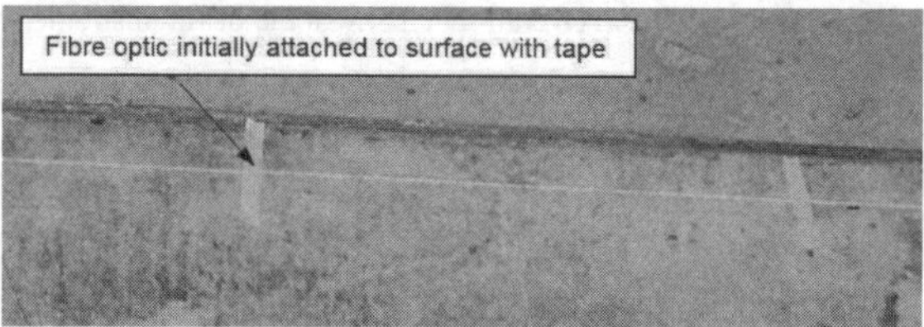

b) Fibre initially taped to concrete surface in the correct location

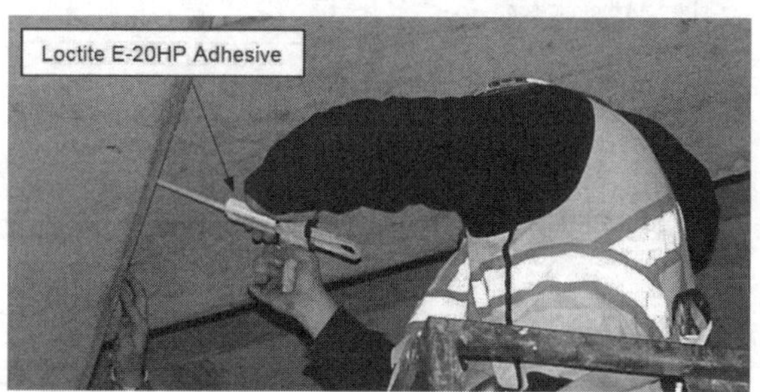

c) Fibre being bonded to concrete surface with Loctite E-20HP Adhesive.

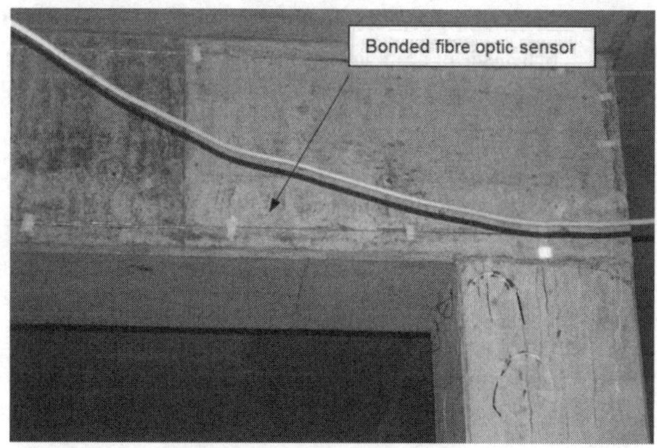

d) Fibre optic sensor bonded to concrete surface.

Figure 2. Images depicting fibre optic installation process.

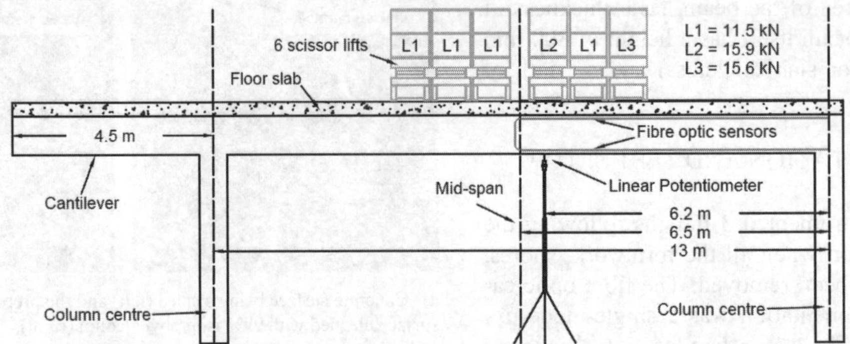

a) Overall schematic of the beam, instrumentation, and scissor lift live loading set up.

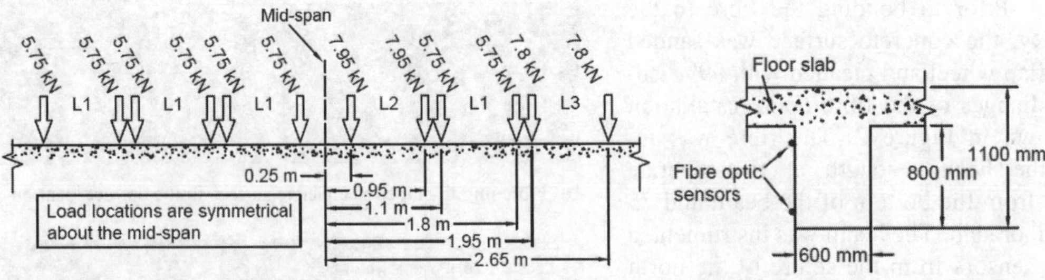

b) Applied live load magnitudes and spacing caused by scissor lifts.

c) Simplified cross section of the instrumented beam

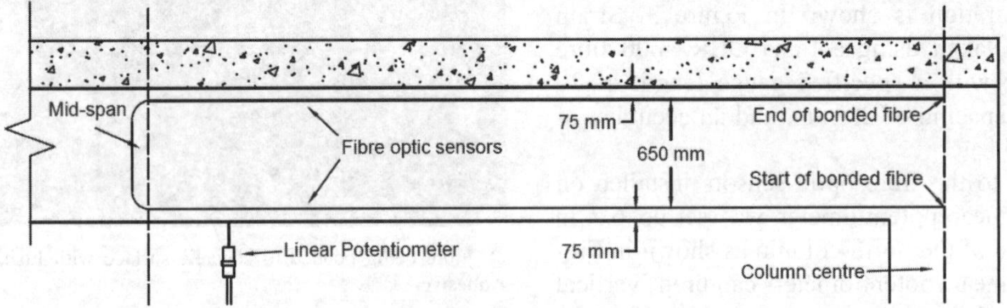

d) Close up of fibre instrumentation configuration.

Figure 3. Schematic of beam instumentation and live loading set up

A fibre optic reference reading was taken immediately prior to the loading of the beam. Because the duration of the entire testing process only took approximately 30 minutes, temperature fluctuations were minimal and were assumed to have a negligible effect on the fibre optic readings. This eliminated the need for temperature compensation of the fibre optic strain data. Displacement readings were taken at a rate of 1 Hz for the duration of the test.

5 RESULTS AND DISCUSSION

The top and bottom fibre optic strain readings taken when the beam was loaded with all 6 scissor lifts (as depicted in Figure 3) are presented in Figure 4. Strain results are shown for both fibres from the centre of the north column up until mid-span. It is important to note that the results in this section are solely from the live load test, and thus do not account for the dead load response.

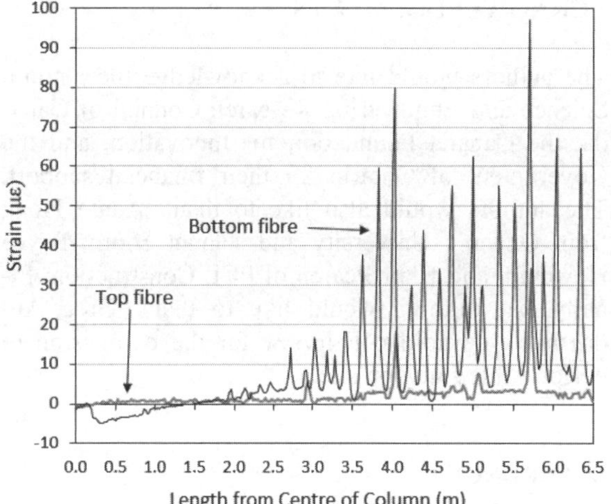

Figure 4. Fibre optic strain results from the instrumented beam loaded with all 6 scissor lifts.

In figure 4 it can be seen that the bottom fibre shows the largest tensile strains in the vicinity of the beam's mid-span. This is as expected since this is where the peak positive moment is developed under the live loading. Concrete cracking is also more evident near mid-span, which is seen in Figure 4 as positive spikes in the fibre optic data. This is also as expected since the bottom of the beam at mid-span experiences the highest tensile strains, leading to flexural cracking when the concrete's modulus of rupture is exceeded. The presence of flexural cracks is clear between 2.7 m and mid-span, with a fairly consistent crack spacing of approximately 200 mm.

Figure 4 shows that the top and bottom strain readings intersect at approximately 1.5 m, indicating an inflection point. The bottom fibre also measures compression between the column and the inflection point due to negative bending closer to the column, and suggesting that there is a degree of fixity at the north support.

The top fibre shows minimal strain differences along its length compared to the bottom fibre, suggesting that the height at which the top fibre was installed is close to the beam's neutral axis. The top fibre does however show large positive strain peaks at 4.0 m and 5.7 m, which are likely caused by flexural cracks extending up beyond the top fibre. This observation is corroborated by the tension spikes seen in the bottom fibre that are largest at 4.0 m and 5.7 m, suggesting that these are the largest flexural cracks.

Something to note is that the nylon-coated fibre optic cable used in this study measures average, rather than maximum, strains in the vicinity of cracks. This is due to slip under high strain gradients between the nylon coating, which is bonded to the concrete, and the fibre optic core where the strain readings are taken. This phenomenon surrounding surface strain measurements for reinforced concrete is explained in further detail elsewhere (Regier and Hoult, 2014). Regier and Hoult concluded that although individual strain readings may not be representative of local surface strains in the vicinity of a crack, the average strains can still be used to assess the structural behaviour. Also, by integrating the area under the strain spike, the approximate crack width can be determined. In this case an average crack width increase caused by live loading of 0.004 mm (at the height of the bottom fibre) was determined by integrating the strains in the vicinity of a crack. Using the strain measurements taken along the beam enables curvature, slope and displacement values along the beam's length to be determined, as discussed next.

Strain readings were measured at two separate known heights on the beam's cross section, thus curvature could be determined at 20 mm intervals along the beam's length. The curvature values can then be numerically integrated twice to calculate displacement readings if at least 2 boundary conditions are known. Thus, using an assumed vertical displacement at the centre of the north column of zero and the measured vertical displacement value from the linear potentiometer (-0.29 mm displacement at a length of 6.2 m), the displacements along the beam could be calculated through double integration. The displacements calculated from the centre of the north column to the beam's mid-span are presented in Figure 5.

The calculated displacement results match the expected behaviour, as the displacements are largest near midspan. The maximum displacement found was -0.292 mm located at 5.88 m along the beam, which is slightly north of mid-span. This is anticipated as the vertical loads (as seen in Figure 3(b)) applied north of mid-span were larger than those applied south of mid-span. The deflected shape seen in Figure 5 also illustrates negative curvature near the support, which is consistent with the compressive strains measured along the bottom of the beam near the support (Figure 4) resulting from some degree of fixity at the north column.

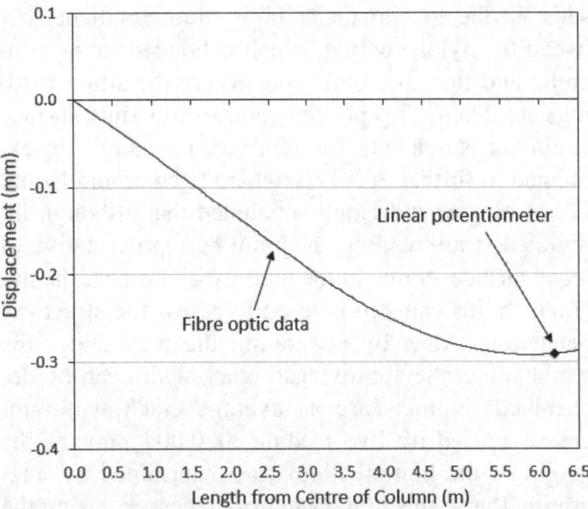

Figure 5. Calculated displacements from the fibre optic strain readings when the beam is loaded with all 6 scissor lifts.

6 CONCLUSIONS

The current research study was performed in order to determine if distributed fibre optic sensors can provide comprehensive data regarding the performance of reinforced concrete elements within a building. One beam was instrumented with distributed fibre optic sensors and a live load test was performed. The results showed that the behaviour of the beam could be captured and understood in far more detail than possible with discrete strain sensing systems. One single distributed fibre optic sensor was able to provide insight into the beam's continuous deflected shape, the exact point of inflection, the portion of the beam where flexural cracks are present, the spacing of flexural cracks, and the support conditions. Further research is required to measure the behaviour of several more reinforced concrete elements, and to compare these experimental values to predicted values using the building's design models.

ACKNOWLEDGEMENTS

The authors would like to acknowledge the Natural Science and Engineering Research Council of Canada, the Canada Foundation for Innovation, and the Government of Ontario for their financial support. The authors would also like to thank Adam Hoag from Queen's University and Tanner Blom, Bryce Howchin, and Blair Pearen of PCL Construction. Finally, the authors would like to thank Greg Andrushko of Cadillac Fairview for the permission to perform this study on site.

REFERENCES

Allwood, J. M. & Cullen, J. M. 2012, Sustainable materials: with both eyes open, UIT Cambridge Limited, Cambridge, UK.

Hoult, N. A. Ekim, O. & Regier, R. 2014, Damage/Deterioration Detection for Steel Structures Using Dist. Fiber Optic Strain Sensors, *J. Eng. Mech.*, **140**(12), 04014097.

Klar, A., Dromy, I. & Linker, R. 2014, Monitoring tunneling induced ground displacements using distributed fiber-optic sensing, *Tunnelling and Underground Space Technology*, **40**, 141-150.

Mohamad, H. Soga, K. Pellew, A. & Bennett, P. 2011, Performance monitoring of a secant-piled wall using distributed fiber optic strain sensing, *J. of Geotech. and Geoenvironmental Eng.*, **137**(12), 1236-1243.

Regier, R. & Hoult, N. A. 2014, Distributed Strain Behavior of a Reinforced Concrete Bridge: Case Study, *J. Bridge Eng.*, **19**(12), 05014007.

Schulz, W. L. Conte, J. P. & Udd, E. 2001, Long-gage fiber optic Bragg grating strain sensors to monitor civil structures, In SPIE's 8th Annual International Symposium on Smart Structures and Materials, 56-65.

Statistics Canada. 2011, Construction, *Government of Canada* <http://www.statcan.gc.ca/pub/11-402-x/2011000/chap/construction/construction-eng.htm> (Sept. 29th, 2015).

Villalba, S. & Casas, J. R. 2012, Application of optical fiber distributed sensing to health monitoring of concrete structures, *Mechanical Systems and Signal Processing,* **39**(1–2), 441–451

Proceedings of the International Conference on Smart Infrastructure and Construction
ISBN 978-0-7277-6127-9

© The authors and ICE Publishing: All rights reserved, 2016
doi:10.1680/tfitsi.61279.015

Development of self-sensing concrete sleepers for next-generation rail infrastructure

L.J. Butler*[1], N. Gibbons[1], H. Ping[2], J. Xu[3], P. Crowther[4] and M.Z.E.B. Elshafie[1]

[1] *University of Cambridge, Cambridge, United Kingdom*
[2] *Tongji University, Shanghai, China*
[3] *Harbin Institute of Technology, Harbin, China*
[4] *CEMEX UK, Birmingham, United Kingdom*
* *Corresponding Author*

ABSTRACT This study presents the early stage development of 'self-sensing' prestressed concrete (PC) sleepers that utilise integrated dynamic fibre optic strain sensors. One of the aims of this research is to demonstrate a future vision of rail infrastructure where self-sensing instrumentation can be directly incorporated during the manufacturing process of mass-produced structural elements (such as PC sleepers). In partnership with CEMEX UK, eight PC sleepers were instrumented in their prestressing facility with fibre optic sensors based on fibre Bragg gratings (FBGs). These sensors are capable of recording strains dynamically up to 1 kHz at strain resolutions of 10 microstrain which will allows the system to acquire strain data under passing trains. Sensor data was collected during the entire manufacturing process which included concrete casting, curing, detensioning as well as monitoring during the British standard (*BS EN 13230-2-2009*) rail seat bending qualification test. Using the installed sensor networks, the detensioning process and early-age qualification bend testing of the sleepers were captured in great detail. The recorded strain results confirmed the dynamic strain frequency capabilities of the sensors and showed good correlation with the expected structural response of the sleepers. These initial trials are very encouraging and will form the basis for a larger scale study on self-sensing PC sleepers involving laboratory testing, numerical modelling, and field monitoring. These next-generation sleepers have the ability to provide track designers and owners with invaluable real-time sleeper strain data under live train loads which could lead to refined design assumptions and better informed long-term maintenance planning.

1 INTRODUCTION AND BACKGROUND

Prestressed concrete railway sleepers have been in use for 73 years and make up the majority of railway sleepers in the UK (Taylor 1993). Railway sleepers are vital components in the rail system with the UK rail operator Network Rail predicting that more than 900 km of track will require annual sleeper replacement and it is estimated that approximately 70% of the sleepers in the UK rail network are concrete (Network Rail 2004). A number of specialised manufacturing facilities throughout the UK are capable of producing tens of thousands of sleepers per year to keep up with the demand within the national rail network. These factories are highly automated and robust quality control measures ensure superior levels of consistency of performance between various batches of sleepers. The railway sleeper carries the load transferred from the train axles and wheels, to the rail, and through the bearing pads and is supported on compacted ballast. Sleepers are only subjected to appreciable levels of stress under dynamic loads and their loading distribution is highly dependent on the condition of the ballast below. As a result, having knowledge of the response of sleepers under live train loading is essential to their design and the prediction of their long-term performance. To date, there have been a variety of studies that have attempted to measure the structural response of prestressed concrete sleepers using experimental and/or numerical simulations. Omondi et al. (2014) used acoustic emission methods to detect cracking under controlled laboratory conditions. They were able to detect the several main damage stages (cracking stages) using these techniques. In 2014, Loaec et al. (2014) discussed their study whereby FBG-based FOS were embedded in concrete to create 'smart' sleepers. This study arose based on a lack of consensus by rail experts in calculating the effective bending moments in sleepers over the long term in-situ response of the rail bed. They reported that based on static and dynamic laboratory tests, the results were repeatable and linear and may be used for in-situ monitoring of rail infrastructure however, no case studies have been reported. Kaewunruen and Remennikov (2009) tested prestressed concrete sleepers on ballast under controlled laboratory conditions using vibration monitoring and finite element model updating techniques. They used accelerometers to measure the induced vibrations from an impact hammer and used the dynamic characteristics (natural frequencies, mode shapes, etc.) to update their finite element model. In another study conducted in 2011, Kaewunruen and Remennikov attached external strain gauges to the top and bottom of a full-scale prestressed concrete sleeper and subjected it to a series of impact loads until failure. They were able to provide an empirical moment envelope which they found gave conservative dynamic design results. By considering the damage accumulation under repeated impact loading, they were able to provide predictions of residual design life of damaged sleepers. Although there have been many experimental studies, there have been a very limited number that have investigated the real-time response of prestressed concrete sleepers on ballast under live train loads. One such study by Sadeghi (2010) investigated the dynamic response of PC sleepers under real train loads and included measurements of actual wheel loads using an embedded load cell. Load cells were installed between the rail seat and sleeper and between the sleeper and the ballast. The results of this study indicated a parabolic pressure distribution under the sleeper but found that this shape was highly influenced by the speed of the passing train. That is, as the speed of train increased the more uniform the pressure distribution becomes. They concluded that for train speeds of 120 km/h and above, the assumption of a uniform pressure distribution is valid.

The main aim of this initial study is to validate the implementation of fibre optic sensor networks within the highly controlled process involved in producing prestressed concrete sleepers. By incorporating sensors directly into the sleepers during their manufacture, their internal strain state as well as a detailed record of their load history can be obtained. These initial trials will help develop confidence in installation techniques and sensor performance that will form the basis for an upcoming larger study involving full-scale laboratory testing and in-situ monitoring of prestressed concrete sleepers in real railway applications.

2 SLEEPER MANUFACTURING PROCESS

The prestressed concrete sleepers instrumented as part of this study were manufactured at CEMEX's Saltley factory in Birmingham on July 20th, 2015. The sleeper manufacturing process begins by cleaning and oiling the molds on the casting bed which provide the capacity for casting several hundred sleepers per day. The second stage of the process involves running the prestressing strands (six 7-wire 9.3 mm strands) out to the total length of the prestressing beds and placing them in the sleeper molds. The strands are then secured at one end of the bed (dead end) and are fed into the hydraulic jack head at the other end of the casting bed (live end). The strands are then prestressed up to 75% of their yield capacity. After prestressing, the concreting operations begin at one end of the bed and moves towards the opposite end at a rate of 1 m^3 every 3 mins. All concrete is produced on-site and is fed by an automated hopper system to the prestressing beds. In this way, the concrete quality is closely monitored during the entire casting process to maintain consistency between batches. After the concrete is cast, the top of the beds are covered in plastic sheeting to prevent excess moisture from evaporating. Once the concrete reaches initial set, the end plates are removed from the molds and finally sheeted with thermal blankets for curing. The testing beds are temperature controlled and thermocouples are cast into the concrete at several locations along the sleeper beds to monitor the development of curing temperatures and to provide a correlation with concrete maturity and compressive strength and to reduce the risk of delayed ettringite formation in the cement matrix. Approximately 18 hours after casting, the concrete will have achieved the required minimum compressive strength (35 MPa) for transferring the prestressing force. Prior to releasing the prestressing strands, a small additional tension force is introduced in order to loosen the jacking clamps. The molds on the prestressed beds at CEMEX's factory have been specially designed on rollers which allow for a 'soft' detensioning of the strands to avoid subjecting the very early age concrete to a sudden load change that can lead to premature cracking. Once all of the prestressing force has been released, the strands between each mold are cut using an automated saw. An automated demolder with a set of special clamps lifts eight sleepers (one mold) off the bed at one time and a crane transports the sleepers to the end of the factory for inspection and gauge readiness for dispatch. A minimum number of sleepers per casting cycle are tested according to BS EN 13230-2-2009 by means of a three-point bend test.

3 INSTRUMENTATION AND MONITORING

The sensor system was designed in order to be sufficiently robust, be capable of acquiring data at high frequency, provide adequate levels of strain resolution, and be readily easy to install and use during subsequent monitoring. Fibre-optic sensors based on fibre Bragg gratings (FBGs) were selected for this application as they offered the level of strain resolution and sampling frequency required. FBGs are similar to ordinary strain gauges as they are able to measure strain a specific point. However, unlike conventional metal strain gauges, FBGs have higher resolution, do not suffer from corrosion, are electromagnetically inert and can have multiple sensors integrated on one array (only one individual connection to the analyser is required). It should be noted however, that other fibre optic based sensors such as those that are capable of measuring distributed strain along an entire fibre (i.e. based on Brillouin Optical Time Domain Reflectometry) were not considered for this application as they are unable to measure strains dynamically. There are other systems based on Rayleigh backscatter that have been used successfully to monitor distributed strain changes dynamically (Kreger et al. 2007). However, the required analysers and hardware required to utilize this system were unavailable during the project planning stages.

To achieve adequate levels of robustness, the FBG sensors were manufactured using a draw tower process whereby the Bragg gratings are etched into the optical core as the glass is being drawn. To add additional robustness, the cables were also reinforced with a GFRP coating and had specially designed flexible splice protectors. The FBG analysers used for this study are capable of sampling at 1000 Hz with resolutions of approximately 10 με. A four channel FBG analyser manufactured by Micron Optics (sm130) was used to interrogate the array signals. One of the main considerations when using FBG-based fibre optic sensors is being able to properly compensate for the effects of temperature. To address this issue, the sensor arrays were designed to incorporate both strain (that measure total strain) and temperature compensation sensors. A total of six strain FBG sensors (three top; three bottom) and two FBG temperature compensating arrays (one top; one bottom) were used which required only two analyser channels (two arrays) per sleeper. Sensors were installed at locations at the rail seat section as well as at the mid-section. In total, eight G44 type prestressed concrete sleepers were instrumented as part of this study. Figure 1 illustrates the sensor arrangement for a typical G44 self-sensing sleeper.

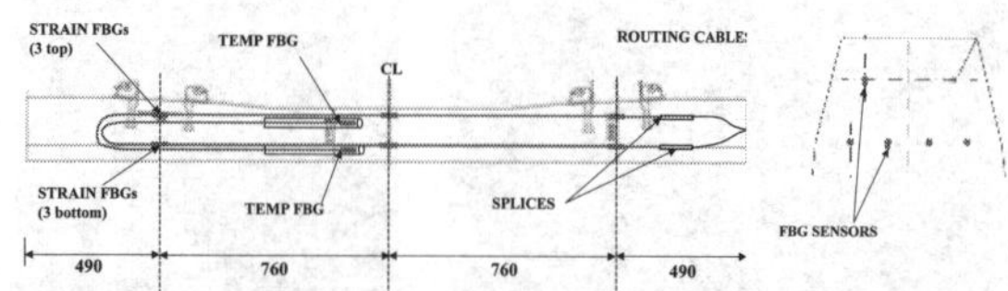

Figure 1. Sensor instrumentation and layout for typical G44 prestressed concrete sleeper (dimensions shown in mm).

The fibre optic cable arrays were installed once the prestressing strands had been fully tensioned. Sensor cables were affixed using a series of plastic cable ties spaced intermittently and specially manufactured exit collars were cast within the concrete to protect the cables where they exit the ends of the sleeper. Once the cables were installed, they were all plugged into the analysers and tested sequentially prior to concrete casting. Figure 2 depicts the installed sensors and exit locations of the cables. The two sleepers near the end of the casting bed were monitored continuously during concrete casting, curing and detensioning. Concrete curing took place over approximately 18 hours to ensure that the concrete had achieved adequate compressive strength prior to transferring the prestress force.

The fibre optic analyser outputs results in terms of wavelengths that can be converted to strain using equation 1.

$$\Delta\varepsilon_M = \frac{1}{k_\varepsilon}\left[\left(\frac{\Delta\lambda}{\lambda_0}\right)_S - k_T\frac{\left(\frac{\Delta\lambda}{\lambda_0}\right)_T}{k_{T_T}}\right] - \alpha_{conc}\frac{\left(\frac{\Delta\lambda}{\lambda_0}\right)_T}{k_{T_T}} \quad 1$$

Where,

$\Delta\varepsilon_M$ = change in mechanical strain

$(\Delta\lambda/\lambda_0)_S$ = change in relative wavelength of strain sensor

$(\Delta\lambda/\lambda_0)_T$ = change in relative wavelength of temperature compensating sensor

k_ε = gage factor (0.78)

k_T = experimentally derived constant for temperature compensating sensor (10 - 12×10^{-6})

k_{TT} = change of the refractive index of glass (between 8 and 9×10^{-6})

α_{conc} = linear coefficient of thermal expansion of concrete (between 10 and $12 \times 10^{-6}/°C$)

This equation accounts for several effects: total (raw) strain, strain measured from the loose tube temperature compensating sensors, and the thermal strain of the surrounding concrete. The end result is the mechanical strain produced by structural movements resulting from effects other than temperature ($\Delta\varepsilon_M$).

Figure 2. Instrumented sleepers and manufacturing process.

4 RESULTS AND DISCUSSION

The following section summarises some of the initial FBG sensor readings of this study specifically, pertaining to the detensioning and rail seat section bending qualification testing processes. In terms of the manufacturing process and materials properties, the instrumented sleepers were effectively identical to the other non-instrumented manufactured G44 type sleepers. Based on the compressive strength results provided by CEMEX UK, the concrete in the sleepers reached 42.7 MPa at transfer of prestressing and 93.6 MPa at 28 days after casting. Both the strength at transfer and at 28 days were well above the target values of 35 MPa and 60 MPa, respectively. The concrete mixtures used for the sleeper are designed to achieve rapid strength gain in order to facilitate 24 hour turnaround times on the prestressing beds.

4.1 Detensioning of Prestressing Strands

Sleepers were monitored continuously during the casting, curing and detensioning process. Results of the detensioning process along with the qualification testing are presented in this paper.

The entire detensioning process was captured using the installed sensors and is presented in Figure 3. It can be seen that there is a slight increase in positive strain in both top and bottom sensors due to the additional prestressing force that is introduced to aid in releasing the lock nuts at the jacking head prior to detensioning. While the strands are being detensioned, the strain sharply decreases (compression) and remains constant after the strands are cut. It is evident from the strain data that a differential strain exists between the top and bottom of the sleeper. The average strain change due to detensioning for the top and bottom sensors are approximately -179 με and -267 με, respectively. That is, the bottom of the section experiences a larger compressive strain change as compared to the top. This behaviour implies that a hogging bending moment is being introduced during the detensioning process and is consistent with the beam deflecting slightly upwards or cambering. The strain change exhibited in top sensor 2 (midspan section) was slightly larger than the strain recorded by sensors 1 and 3 (both at the rail seat section). The bottom sensors indicated that sensor 2 (midspan) experienced the smallest strain change as compared to sensors 1 and 3 (both at the rail seat section). Moreover, bottom sensors 1 and 3 measured very similar strain changes during the entire test. The overall level of noise in the recorded data was approximately 10 με which is consistent with the performance specifications of the FBGs and the analyser.

4.2 Qualification Testing

Following detensioning and cutting of the strands, the sleepers were lifted individually onto a forklift and brought to a specially designated storage area adjacent to CEMEX's materials testing laboratory. Two sleepers were selected to undergo qualification testing in accordance with BS EN 13230-2-2009. These tests were carried out at CEMEX's materials laboratory as part of their Saltley factory in Birmingham on July 21st, 2015; 24 hours after casting. A static test was performed at the rail seat section whereby an increasing concentrated load of 2 kN/s (120 kN/min) up until 223 kN (corresponding to the positive design bending moment at the rail seat section) and is held for 10 seconds (refer to Figure 4). If no cracking is observed during this test than the sleeper is deemed to have satisfied the design requirements.

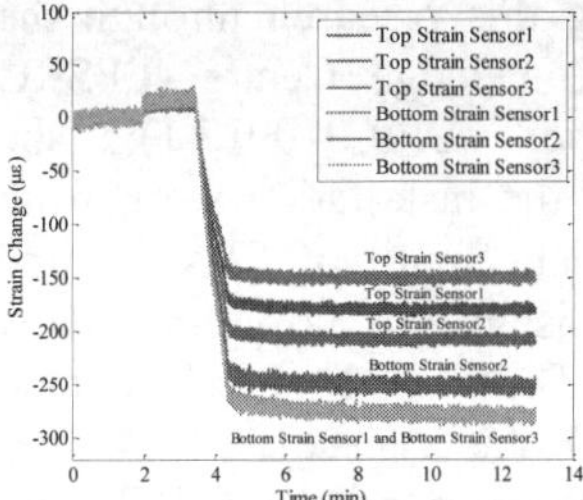

Figure 3. FBG sensor strain change readings due to detensioning of prestressing strands. Note that strain change readings are reported relative to the pre-detensioning strains.

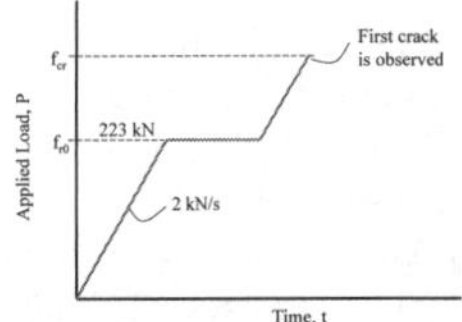

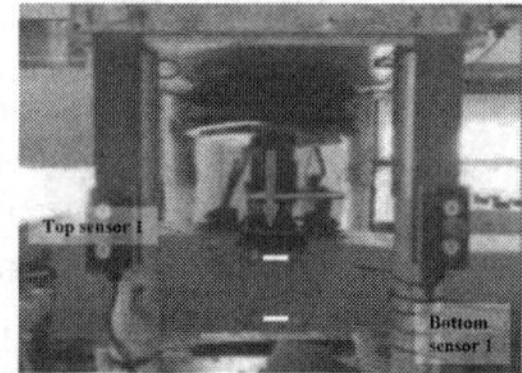

Figure 4. Static positive bending test at rail seat section on self-sensing sleeper.

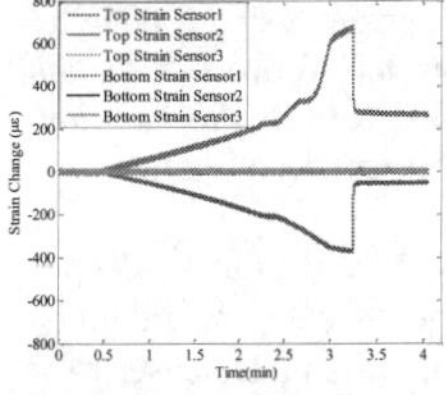

Figure 5. FBG sensor strain readings during static positive bending test at rail seat section. Note that strain readings are relative to strain state prior to testing (i.e., they do not include compressive strains due to detensioning).

The second part of the test involves continuing loading the sleeper up until the first crack is observed, noting the cracking force and then unloading. After testing, the sleepers are still fit for service and are returned to the storage yard. The tested self-sensing sleeper achieved the 223 kN target without cracking and the first crack was observed at a load corresponding to 290.9 kN.

The entire qualification testing process was monitored using the installed sensors and is presented in Figure 5. Note that strain sensors at location 1 (top and bottom) were at the rail seat location and subjected to the concentrated load. It is also important to note that the strain changes reported below are with respect to the state of strain prior to the start of the test. As would be expected, the strain change measured by sensors 2 and 3 (top and bottom), which are outside of the simply supported span of the bending test, remains fairly constant throughout the test. When the loading is applied at a constant rate (approximately 2 kN/s), the positive strain in bottom strain sensor 1 and the negative strain in top strain sensor 1 increased linearly at a strain rate of 114 με/min and -106 με/min, respectively. When the load reached 223 kN (at approximately 2.25 minutes into the test), the recorded strain changes in the top strain sensor 1 and bottom strain sensor 2 were 226 με and -205 με, respectively. This recorded behaviour confirms general beam theory whereby a section under positive bending experiences compression on the top and tension on the bottom of the section.

During the 10 second load holding time, the strain change values of the top and the bottom sensors remained constant and no cracking was observed thereby deeming the sleepers to have satisfied the design requirements. In the second loading part of the test, there was a flat section in the bottom strain sensor 1 when the strain change value reached 330 με. As the load increased, there was a gradual softening of the curve which may indicate the onset of cracking. This corresponded to top and bottom sensor strains of -352 με and 603 με, respectively, at the onset of non-linearity (at time ≈ 3.0 min).

As mentioned above, the strain change results presented in Figure 5 are reported relative to the strain state in the specimen prior to loading, that is, these relative strains do not include the negative strains induced by the detensioning of the prestressing strands reported in Figure 3. Prior to the start of the test, the average strain change relative to the pre-detensioning strain state are -101 με and -281 με for the top and bottom sensors, respectively. Therefore, to include the effects of detensioning, these compressive strains must be added to the results presented in Figure 5.

By including these results, the actual strain values of the top and bottom sensors become (-101 με - 352 με =) -453 με and (-281 με + 603 με =) 322 με, respectively. A net tensile strain at the bottom sensor of 322 με indicates that cracking in the concrete has occurred somewhere in the vicinity of the sensor. Likewise, if the detensioning strains are added to the post-bend testing strains, top = - 101 με - 51 με = -152 με and bottom = -281 με + 270 με = -11 με, then it indicates that similar to before the test began, the total section returns to a state of net compression. This corresponds well to visual observations of the crack closing following unloading and confirms that the sleeper is still suitable for service after testing. However, it should be noted that some additional positive strain (tension) has accumulated in the section due to the bend testing.

5 CONCLUDING REMARKS

Eight prestressed concrete sleepers were instrumented with a robust dynamic fibre optic strain sensor network during normal operating conditions at a sleeper manufacturing facility. The concrete casting, curing, and prestress detensioning processes were all able to be captured in great details with the use of this sensor system. In addition, the strain response of the sleepers during the BS EN 13230-2-2009 rail seat bending qualification test was monitored using this sensor system and the strain results were correlated to the loading changes. Results from this study have been encouraging as they have demonstrated that the sensing system chosen is capable of capturing the dynamic strain behaviour of the PC sleepers and the strains measured seem to be consistent with the expected structural response. By instrumenting and monitoring the sleepers during their production, the complete load (strain) history can be captured and used as a baseline for future monitoring.

Future research work will involve casting a second set of self-sensing prestressed concrete sleepers for full-scale dynamic laboratory testing on ballast. A number of time-dependent concrete properties will also be measured and used in conjunction with sensor results for comparison with design assumptions and for finite element model updating. Finally, several of the self-sensing sleepers will be installed on two new railway bridges that are currently being constructed in Staffordshire with their own fibre optic sensor networks. This completed study will provide track designers and owners with an unprecedented level of information on sleeper response under live train loads and will serve as a key demonstrator of using self-sensing prestressed concrete sleepers in future rail infrastructure.

ACKNOWLEDGEMENT

The authors gratefully acknowledge the EPSRC and Innovate UK for funding this research through the CSIC Innovation and Knowledge Centre (EPSRC grant reference number EP/L010917/1); the invaluable facilitation of the installation work from Andrew Carey and Stewart Smith of CEMEX and their team; the on-site assistance of Jason Shardelow of the CSIC and Jules Birks of Mott Macdonald (formerly of CSIC); and the assistance of Cedric Kechavarzi and Philip Keenan of the CSIC with sensor procurement and development.

REFERENCES

BSI. 2009. *BS EN 13230-2-2009 Railway applications-track-concrete sleepers and bearers,* British Standards Institution, Chiswick, United Kingdom.

Kaewunruen, S. & Remennikov, A.M. 2009. Application of vibration measurements and finite element model updating for structural health monitoring of ballasted railtrack sleepers with voids and pockets, *Excerpt from Mechanical Vibrations: Measurement, Effects and Control*, Ed. Robert C. Sapri, 621 – 641.

Kaewunruen, S. & Remennikov, A.M. 2011. Progressive failure of prestressed concrete sleepers under multiple high-intensity impact loads, *Engineering Structures* **31,** 2460 – 2473.

Kreger, S. T., Gifford, D.K., Mark E. Froggatt, M.E., Sang, A.K., Duncan, R.G., Wolfe, M.S., & Soller, B.J. 2007. High-resolution extended distance distributed fiber-optic sensing using Rayleigh backscatter. *Proc., SPIE—Sensor Systems and Networks: Phenomena, Technology, and Applications for NDE and Health Monitoring*, **6530**, Society of Photo-Optical Instrumentation Engineers (SPIE), Bellingham, WA.

Loaec, A. Petit, C. Lanticq, V & Lamour, V. 2014. Smart sleeper – measurement of bending moments in concrete sleepers laid on ballast tracks, *Transport Research Arena*, Paris La Defense, 14 – 17 April. 6 pgs.

Network Rail. 2004. Business Plan 2004 - Technical Plan, Section 9: Plans by Asset Type, 66 pgs.

Omondi, B. Agellis, D.G. Sol, H. & Sitters, C 2014. Acoustic emission behaviour of prestressed concrete sleepers under quasi-static homologation testing, *31st Conference of the European Working Group on Acoustic Emission (EWGAE)*, 3 – 5 December.

Sadeghi, J. 2010. Field investigation on dynamics of railway track pre-stressed concrete sleepers, *Advances in Structural Engineering* **13**(1), 139 – 151.

Taylor, H.P.J. 1993. The railway sleeper: 50 years of pretensioned, prestressed concrete, *The Structural Engineer* **71**(16/17), 281 – 295.

Proceedings of the International Conference on Smart Infrastructure and Construction
ISBN 978-0-7277-6127-9

© The authors and ICE Publishing: All rights reserved, 2016
doi:10.1680/tfitsi.61279.021

Distributed fibre optic sensors for measuring strain and temperature of cast-in-situ concrete test piles

N. de Battista*[1], C. Kechavarzi[1] , H. Seo[1] , K. Soga[1] and S. Pennington[2]

[1] *Centre for Smart Infrastructure and Construction, Cambridge University, Cambridge, UK*
[2] *Ove Arup & Partners Ltd., London, UK*
* *Corresponding Author*

ABSTRACT In this paper we present the use of distributed fibre optic sensor (DFOS) technology to measure the temperature and strain of reinforced concrete test piles during construction and during static load tests. Eight test piles were recently instrumented with DFOS, on three construction sites in London, by the Cambridge Centre for Smart Infrastructure and Construction (CSIC), in collaboration with Ove Arup & Partners Ltd. The concrete curing temperature profiles of the piles were used to detect the presence of significant defects in the piles. The load test strain profiles along the length of the piles were used to determine the load capacity of the piles and estimate the design parameters of the various soil strata, as well as the internal relative displacement of the piles under various loads. Being distributed in nature, DFOS give a much more detailed picture of the performance of a test pile, as compared to traditional embedded point sensors, such as vibrating wire strain gauges and extensometers. This is demonstrated with a sample of data obtained from one of the instrumented test piles.

1 INTRODUCTION

Full-scale testing of pile foundations is a well-established technique for validating foundation design before construction, as well as for quality control and continuous improvement in pile design and construction practices. Clause 7.5 of Eurocode 7 lays down requirements for pile load testing (British Standards Institution 2004), which accounts for a considerable proportion of the total value of the piling market in the UK (Federation of Piling Specialists 2006). It is imperative that pile testing translates into real value, both for specific projects and for the construction industry as a whole. This relies on the amount of useful knowledge that can be obtained from the pile testing process, as compared to the cost incurred.

This paper deals with the pile testing instrumentation of cast-in-situ reinforced concrete piles, constructed using bored or continuous flight auger (CFA) methods. We focus in particular on how distributed fibre optic sensors (DFOS) can be used to supplement traditional pile testing instrumentation, and potentially replace embedded point sensors in the future. The added value that DFOS technology can bring is demonstrated with extracts of monitoring data from recent pile tests carried out in London.

2 TESTING OF CAST-IN-SITU CONCRETE PILES

The fitness for purpose of a reinforced concrete pile depends on two aspects, both of which can be investigated with testing: pile integrity, which deals with the dimensional accuracy, structural quality and homogeneity of the pile; and pile stiffness, which defines the load-carrying capability of the pile for a range of displacements.

Pile integrity can be checked through site observations and inspection of pile records. It is also commonly inferred indirectly from the acoustic response of the pile to sonic waves introduced externally (e.g. low-strain integrity testing) or internally, via steel tubes cast into the pile (e.g. cross-hole sonic logging). Thermal integrity profiling (TIP) is becoming a more common method to infer integrity and uses the concrete curing temperature measured by three or more strings of closely-spaced temperature sensors embedded along the length of the pile (Piscsalko et al. 2015). These tests are intended to detect significant defects in the pile composition, such as soil inclusions, overbreak or voids.

Pile stiffness can be determined through load tests, either on working piles or on trial piles, with the latter typically being tested to failure at loads well above their safe working load. The loading can be static or dynamic and, while the former generally requires a larger working area and takes longer to perform, it gives a more direct measure of pile performance.

In static maintained trial load tests, a controlled load is applied and removed in stepped stages. The load is applied either from a bi-directional load cell embedded within the pile (compression testing only), or from a loading frame above the pile (compression, tension or lateral testing) (Figure 1). At each loading and unloading stage, the applied load and displacement at the pile head are measured, from which the load/settlement relationship of the pile can be derived.

In addition to the load and pile head displacement, it is also common practice to measure the strain at various levels within the pile during trial load testing. Traditionally this is done using embedded strain gauges, typically vibrating wire strain gauges (VWSGs) which are either welded directly to the reinforcement or mounted on lengths of rebar ("sister bars") that are tied to the reinforcement before it is inserted in the pile bore.

By measuring the change in strain along the pile for any constant load, it is possible to estimate pile-soil interface properties in the various geological strata that the pile penetrates, as long as the strain gauges are installed in sufficient number and in suitable locations. In practice, aspects such as cost, cabling congestion and data acquisition equipment limit the number of strain gauges that can practically be installed; they are typically placed several meters apart along the length of the pile. Therefore the reconstruction of the strain profile from the individual strain gauge measurements is not always reliable, as the individual point measurements can be influenced significantly by localised effects in the pile or the ground. On the other hand, localised effects which are not close to any of the strain gauges, but which might be important for the assessment of the pile performance, will not be detected by the instrumentation.

Figure 1. Static load testing frame for a trial test pile.

Another common measurement that is made during each stage of a trial load test is the relative displacement between the pile head and a limited number of locations along the length of the pile. This is traditionally measured by means of retrievable extensometers which are lowered down steel tubes embedded within the pile. From the extensometer readings, one can derive the shortening of the pile between measurement locations, for each load. However, as with the strain gauges, it is not possible to detect any localised effects that could occur between measurement points.

In summary, a complete trial pile load test would involve at least five independent sensor systems, each measuring a different action or reaction parameter: two external systems to measure load (e.g. load cell) and top deflection (e.g. displacement transducer); and three internal systems to infer integrity (e.g. temperature sensor strings for TIP testing), and to measure strain (e.g. VWSGs) and displacement (e.g. extensometers). In the remainder of this paper, distributed fibre optic sensing is proposed as a single alternative to integrity, strain and displacement sensors.

3 DISTRIBUTED FIBRE OPTIC SENSOR TECHNOLOGY

Until recently, practically all the sensor systems in use in the civil engineering and construction industry consisted of point sensors, where a measurement from one sensor represents a physical parameter acquired at a single point in space. With point sensors, it is only by increasing the number of individual sensors that one can obtain a spatially distributed measurement set. Indeed, all the sensor systems mentioned in the previous section, in relation to traditional pile test monitoring, conform to this mode of operation.

Over the past few years, we have seen the emergence of a new paradigm in instrumentation, namely "distributed sensing", where a single measuring device can record data from a large number of spatially distributed points. These systems tend to be easily scalable at minimal increase in system cost. Some examples of distributed sensing are 3D laser scanning, nano-composite sensing skins and distributed fibre optic sensors (DFOS).

DFOS systems use the principles of photonics in order to locate and quantify changes in the molecular structure of the glass along an optical fibre. In civil engineering applications, and particularly in relation to pile testing, this change is generally associated with a change in strain or temperature of the optical fibre, or a combination of both. In turn, this can be equated to a change in strain and / or temperature of the structure to which the optical fibre is bonded (e.g. in the case of steel members) or in which it is embedded (e.g. in the case of a concrete member).

Various DFOS measurement techniques exist, as described by Kersey (2011). In this paper we present the application of the Brillouin optical time domain reflectometry (BOTDR) technique for measuring strain and temperature along a single-mode optical fibre (Kurashima et al. 1993).

A BOTDR-based DFOS system consists of a BOTDR analyser and one or more lengths of fibre optic (FO) cables connected together in series to form a single circuit that is attached to or embedded in the structure to be monitored. The analyser transmits thousands of short optical pulses per second into one end of the FO circuit. As the light travels through the optical fibre, molecular density fluctuations inherent throughout the fibre's silica core cause a small fraction of this light to scatter and reflect back towards the analyser, from every location along the fibre. The analyser measures the frequency spectrum of the backscattered light caused by the light pulses and calculates the distance to where the scattering originated from, based on the speed of light within the fibre.

The frequency spectrum of the backscattered light consists of a number of components, one of which is the Brillouin frequency. The peak Brillouin frequency is shifted from the frequency of the input pulse by an amount that is proportional to the strain and temperature of the fibre, at the location where the backscatter originated from.

By recording the peak Brillouin frequencies of the backscattered light coming from closely spaced intervals along the FO circuit, and comparing them to those recorded from the same points at a previous time, it is possible to derive the change in strain and / or temperature that occurred at every measurement point along the fibre. The strain and temperature effects can be deconvoluted by taking measurements simultaneously from a loose-tube FO cable (temperature cable) and a tightly bonded FO cable (strain cable), installed adjacent to each other in the structure (Mohamad 2012).

At the time of writing, off-the-shelf BOTDR analysers could record data from an FO cable several tens of kilometres long, with a sampling resolution of 5 cm, spatial resolution of 50 cm, strain precision of ±30 με and temperature precision of around ±1 °C.

4 TESTING CAST-IN-SITU PILES WITH DISTRIBUTED FIBRE OPTIC SENSORS

Following several years of lab development and field trials, the BOTDR DFOS technique has recently been used successfully to monitor eight cast-in-situ reinforced concrete trial test piles, at three different sites in London. Four piles were constructed using the CFA technique and the other four were bored piles. The piles were between 25.5 m and 33.7 m long, and 0.9m in diameter. The stratigraphy at the test locations comprised Made Ground, Alluvium, River Terrace Deposits, the Lambeth Group and the Thanet Sand Formation. The Engineer was Arup and the DFOS instrumentation and monitoring was carried out by the Centre for Smart Infrastructure and Construction (CSIC) from the University of Cambridge.

4.1 *Instrumentation*

The DFOS cables were installed within the piles in U-shaped loops, consisting of a leg along one side of the pile, a wrap around the bottom of the pile, and a leg along the opposite side of the pile (Figure 2). Each loop consisted of two FO cables: a loose-tube temperature cable and a tightly-bonded strain cable, with the latter being pre-strained by hand before being fixed in place. The two cables were fixed to the pile reinforcement at the top and bottom of each leg and held loosely along the reinforcement with intermediate cable-ties.

Each pile was instrumented with at least one DFOS loop on the reinforcement cage. Two of the bored piles were also instrumented with a second DFOS loop at right angle to the first. Two of the CFA piles, which included a central bundle of reinforcement bars in addition to the reinforcement cage, also had a DFOS loop installed on this bundle.

Figure 2. Distributed fibre optic sensor cables being installed on the reinforcement cage of a CFA pile (top) and wrapped around the bottom of the cage (bottom).

As well as the embedded DFOS instrumentation, traditional pile testing sensors were used in all eight piles. Strings of temperature sensors at 30 cm intervals were used for assessing the pile integrity during concrete curing by the TIP method, while embedded VWSG pairs and retrievable extensometers were used to measure the strain and relative displacement at up to seven points within the piles during load testing. Seven of the piles were subjected to maintained loading in axial compression to a maximum load of between 8 and 25 MN, and one pile was subjected to maintained loading in axial tension to a maximum load of 5 MN.

4.2 *DFOS monitoring and results*

The temperature and strain in all the piles were recorded at a minimum of once every 30 minutes during concrete curing, starting from a few hours after concreting until several hours after the peak temperature was recorded. During the load tests, the temperature and strain were recorded at suitable intervals such that at least one measurement was taken during each loading and unloading cycle. In this section we present a small selection of the DFOS monitoring results, pertaining to one of the CFA compression piles. The reinforcement of this pile consisted of a 20 m-long cage and a 25 m-long central bundle.

Figure 3 shows the temperature profiles of this pile during concrete curing. The profiles show the temperature of the pile increasing steadily over its entire depth, until a maximum temperature of 25.8 °C was reached, 18 hours after the pile reinforcement was inserted in the concrete.

A section of this pile that coincided with the Alluvium and River Terrace Deposits reached consistently higher temperatures than the rest of the pile during curing. This is an indication of a possible overbreak at these strata, as the larger volume of concrete results in a slower dissipation of the heat of hydration. This profile matched the data recorded from the TIP sensors, which also indicated a possible overbreak in the same strata.

Figure 4 shows the strain profiles recorded in the centre of the CFA pile, during the second (final) load increment cycle. This cycle was carried out over 28

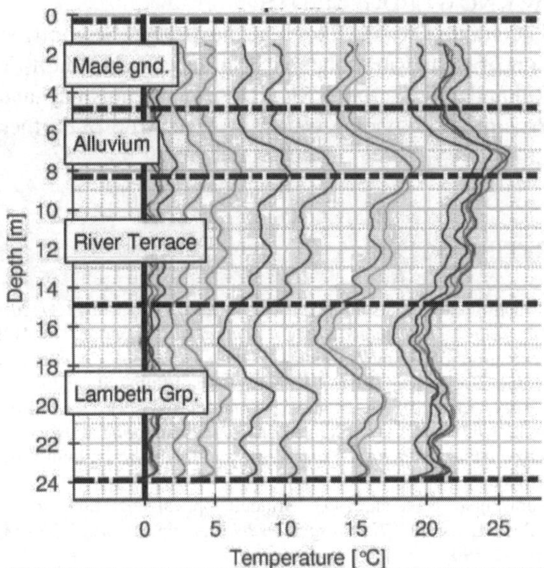

Figure 3. Temperature profile of a CFA pile, recorded at hourly intervals in the centre of the pile during concrete curing, from 4 to 18 hours after the pile was completed, when the concrete reached the peak temperature of 25.8 °C.

hours and consisted of seven load stages, from 7.7 MN (100% design verification load (DVL)) to 20 MN (100% DVL + 165% specified working load (SWL)). Each of these stages is represented by a single strain profile in the figure, with increasing compressive strain corresponding to increasing load. A maximum compressive strain of 772 με was recorded close to the top of the Alluvium stratum, under a load of 20 MN.

The strain profiles once again confirm the suspicion of an overbreak in the Alluvium and River Terrace Deposits strata. This is indicated by a sudden reduction of strain in the pile at these depths, when compared with the expected gradual decrease in strain. This is likely the result of a larger cross-section at these depths, hence a smaller strain for a given load.

By integrating the strain profile from top to bottom, it is possible to estimate the incremental internal displacement of the pile, relative to the pile head. This is shown in Figure 5, where the displacement profiles were derived from the strain profiles of Figure 4.

From the strain and displacement data, one is able to estimate a number of design parameters, such as the pile modulus at the pile head, the limiting shaft friction

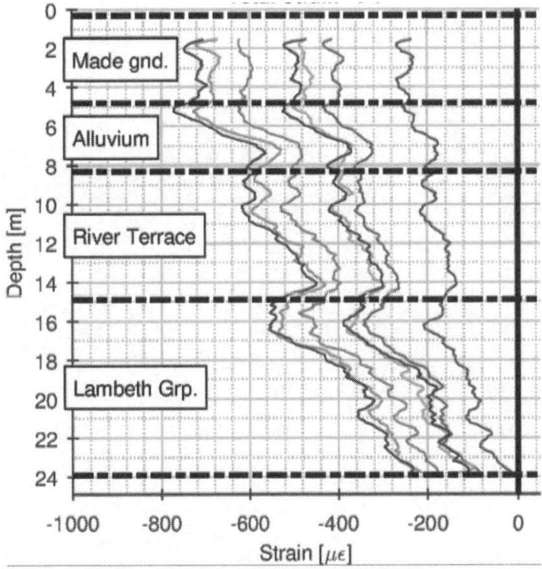

Figure 4. Strain profile of a CFA pile, as it was incrementally loaded vertically in compression, with loads of 7.7 MN, 11.4 MN, 13.3 MN, 15.2 MN, 17.1 MN, 18.9 MN and 20.0 MN (corresponding to the strain profiles going from right to left). Negative strain indicates compression.

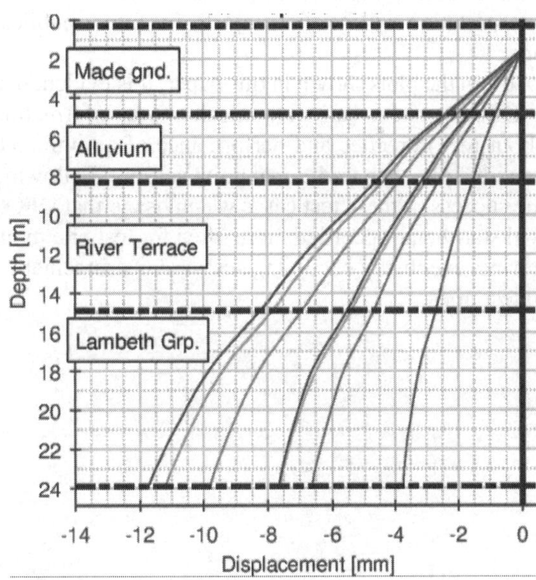

Figure 5. Displacement profile of a CFA pile, relative to the pile head, estimated from the recorded strain during vertical load testing as shown in Figure 4. Negative relative displacement indicates compression.

for the different soil strata and the pile base displacement.

The data from the conventional VWSG and internal displacement transducers within this pile agreed well with the data acquired from the DFOS system. This was also the case for the other seven piles. A quantitative comparison of the different data sets will be the subject of a future, more detailed paper on these case studies.

5 CONCLUSION

In this paper we have presented the application of distributed fibre optic sensor (DFOS) technology for monitoring the temperature and strain of reinforced concrete test piles during curing and load testing. Following years of development and field trials, this technology has been used in full-scale static pile testing to inform foundation designers about the integrity and load capacity of four bored and four CFA test piles in London.

A typical set of results from one of these test piles was used to illustrate the unprecedented level of detail that can be obtained from a single instrumentation system. A more in-depth presentation of results and their interpretation will be the subject of a future publication.

From the plots shown in this paper, it is evident that a DFOS system can give much more information about the properties and performance of a reinforced concrete pile than traditional point sensors. Following the success of these test cases we envisage that DFOS will start being specified more often for test pile monitoring, eventually becoming standard instrumentation in test piles.

ACKNOWLEDGEMENTS

The authors would like to acknowledge the contribution of (listed in alphabetical order): Peter Knott, Yi Rui and Zili Li from CSIC; Duncan Nicholson, Landi Proctor and Vivien Kwan from Ove Arup & Partners Ltd.

REFERENCES

British Standards Institution, 2004. *BS EN 1997-1:2004+A1:2013 Eurocode 7. Geotechnical design. General rules*, London, UK: BSI.

Federation of Piling Specialists, 2006. *Handbook on pile load testing*, Beckenham, Kent, UK: Federation of Piling Specialists.

Kersey, A.D., 2011. Distributed and multiplexed fiber optic sensors. In E. Udd & W. B. Spillman, Jr., eds. *Fiber optic sensors: An introduction for engineers and scientists*. Hoboken, NJ: John Wiley & Sons.

Kurashima, T. et al., 1993. Brillouin optical-fiber time domain reflectometry. *IEICE Transactions on Cummunications*, E76-B(4), pp.382–390.

Mohamad, H., 2012. Temperature and strain sensing techniques using Brillouin optical time domain reflectometry. In T. E. Matikas, ed. *Proceedings of SPIE 8346, Smart Sensor Phenomena, Technology, Networks, and Systems Integration*. San Diego, CA, p. 83461M.

Piscsalko, G.R. et al., 2015. Pile sensing device and method of using the same. US Patent 8,382,369 B2.

Proceedings of the International Conference on Smart Infrastructure and Construction
ISBN 978-0-7277-6127-9

© The authors and ICE Publishing: All rights reserved, 2016
doi:10.1680/tfitsi.61279.027

Distributed fibre optic long-term monitoring of concrete-lined tunnel section TT10 at CERN

V. Di Murro[*1,2], L. Pelecanos[1] , K. Soga[1] , C Kechavarzi[1] , R.F. Morton[2] and L. Scibile[2]

[1] *Centre for Smart Infrastructure and Construction (CSIC), University of Cambridge, United Kingdom.*
[2] *Centre for European Nuclear Research (CERN), Geneva, Switzerland*
**Corresponding author*

ABSTRACT. The Centre for European Nuclear Research (CERN) uses large and complex scientific instruments to study the basic constituents of matter by operating a network of underground particle accelerators and appurtenant tunnels. Several tensile cracks have been observed within a section of a concrete lined tunnel called TT10. As a result, there were concerns about the safety and structural health of CERN infrastructure and a long-term sensing plan was designed to monitor the behaviour of these tunnels. Additionally, two big challenges had to be met: (a) remote monitoring due to tunnel inaccessibility during the operation of the experiment and, (b) potential instrument malfunction due to high radioactivity. Therefore, there was a clear need for radiation-resistant monitoring instruments that could be operated remotely. Distributed Fibre Optic Sensing (DFOS) proved to be the most appropriate monitoring method as it appeared to satisfy the abovementioned requirements. Eight tunnel circumferential loops and one longitudinal were monitored with distributed fibre optic (FO) cables. In the progress readings taken so far, the circumferential tunnel loops showed some minor values of axial strains in the FO cable (peak absolute strain no more than 150$\mu\varepsilon$) and a somewhat consistent profile of the strains for all circumferential loops with high values of (minor tensile) strains at the sides and negative-compressive strains at the crown of the tunnel. This strain pattern would suggest a vertical tunnel elongation mechanism of deformation, i.e. there is compression at the tunnel crest and extension at the lateral two sides of the tunnel. Finally, the values of all observed axial strains seem to be insignificant suggesting no major movement or deformation of the tunnel lining over the relatively short monitoring period.

1 INTRODUCTION

The maintenance of tunnels requires a deeper understanding of their long-term behavior (Wongsaroj, 2005; Laver, 2010; Li, 2014).

The principal aim of this study is to investigate the long-term behaviour of an existing concrete-lined tunnel called TT10 at CERN, by adopting the new distributed fibre optic strain measurement technique developed at Cambridge University.

TT10 tunnel is a 700m long injection tunnel connecting the tunnel of the Proton Syncrotron (PS) ring to the main ring tunnel known as Super Synchrotron Protons (SPS), as shown in Figure 1.

Figure 1. The injection tunnel TT10.

The excavation of the tunnel was executed in December 1972 by using an alpine boring machine called road header (Figure 2). The cross-section of the tunnel has an internal diameter of 4.5m and has a horseshoe shape.

Figure 2. Excavation of TT10 tunnel.

2 INSTALLATION OF FIBRE OPTIC SENSORS

The main tunnel deformation mechanism observed in TT10 tunnel is localized compression and tension areas in the crown and in the shoulders of the tunnel respectively (Figure 3). Cracks on the floor of the tunnel were also observed in certain areas. In addition, ongoing tunnel deformation which is leading to lining failure can potentially affect the alignment and the integrity of the particle accelerator beam/line magnets.

For this reason a long-term sensing plan was designed to provide continuous monitoring for a long-term, in order to have a better understanding of the tunnel deformation. The monitoring method adopted makes use of innovative distributed fibre optic sensing (DFOS) system which showed to be the most suitable for a radioactive environment such as the one at CERN.

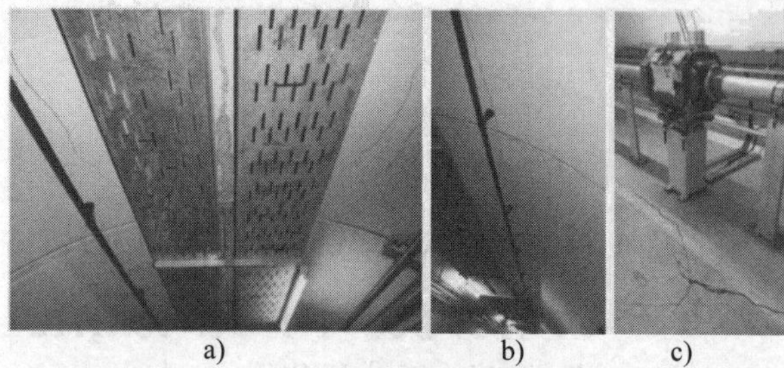

Figure 3. Observed cracks in TT10 tunnel lining: a) localised compression cracks on the crown of the tunnel; b) tension cracks on the shoulder of the tunnel and c) cracks on the tunnel floor.

2.1 *Method of monitoring*

The monitoring of CERN tunnels was carried out by using the distributed fibre optic sensing (DFOS) system, which has had applications in a wide range of engineering structures and the results have successfully been compared with other conventional instruments.

The principle of this technology is that a light is launched into a fibre optic cable from a Brillouin Optical Time Domain Analysis (BOTDA) analyzer. The Omnisens BOTDA analyser was used for CERN installation (Omnisens, 2013). When analysed in the frequency domain the backscattered light amplitude has a peak at some constant value of Brillouin frequency (Figure 4). If at any point the fibre optic cable is experiencing some strain, the value of Brillouin frequency for which the peak amplitude occurs is shifted. The frequency shift of the backscattered light is related to the strain and the temperature conditions of the cable through a linear relationship. The calculated strain can then be translated to other engineering quantities, depending on the type of the engineering problem (Hisham *et al.*, 2007; Hisham, 2008; Hisham *et al.*, 2010). Further studies will be carried out in the near future in order to translate the axial strain into displacements for CERN case study.

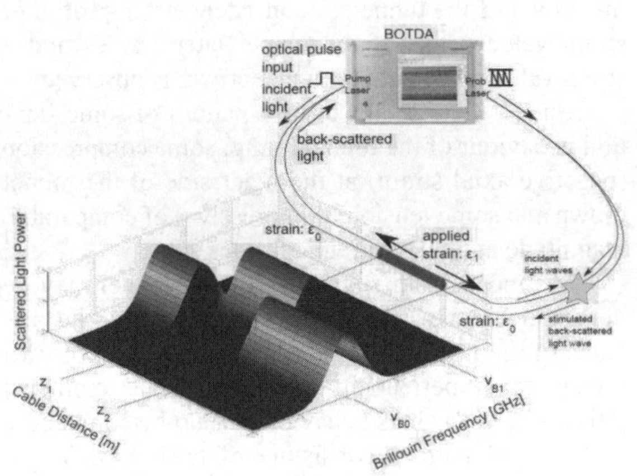

Figure 4. Principles of Distributed Fibre Optic Sensors.

2.2 *Monitoring scheme*

A section of TT10 tunnel of about 100 m length was chosen to be initially monitored. The FO cable runs in both the longitudinal and circumferential directions of the tunnel, forming several (eight) circumferential loops and one straight cable section in the longitudinal direction. The cable sections of interest are pre-strained (Figure 5) where a bold line refers to pre-strained section whereas dashed line refers to loose cable section. The pre-strained cable allows for accommodation of any compressive strain without cable buckling. The cable is attached on the tunnel lining using ten hook-and-pulley systems as shown in Figure 6. At the bottom of the tunnel the cable passes through a 5mm cut section in the floor concrete and then covered with glue and the section is then grouted.

At the end of the monitoring section the two ends of the FO cable are routed out of the tunnel section via a 50m vertical shaft (Figure 5) reaching a control room on surface and no further access is then required into the tunnel.

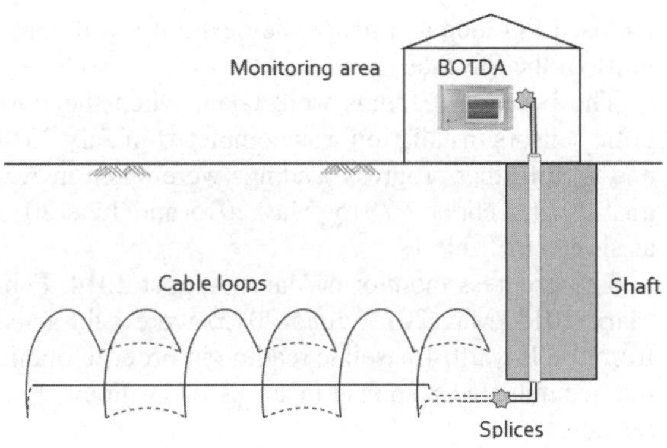

Figure 5. Installation scheme of FO sensors.

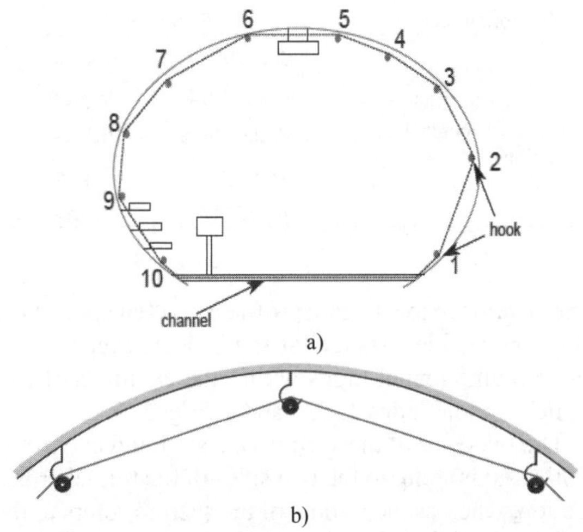

Figure 6. a) Cross section of FO installation. b) Method of attachment of FO cable: hook-and-pulley system.

The monitoring was planned to be taken approximately every 2-3 months for the first year and depending on the outcome of the readings the long-term monitoring plan would be designed.

3 CURRENT MONITORING DATA & RESULTS

The primary data obtained from the fibre optic sensors is the Brillouin Frequency, which is linearly proportional to the axial strain that the fibre optic cable is experiencing. The Brillouin frequency readings in the fibre optic cable provide a continuous signal through a closed FO loop and hence they exhibit a full continuity of the FO data.

The baseline readings were taken when the fibre optic sensors installation was completed in July 2014 and then further progress readings were taken in August 2014, February 2015, May 2015 and June 2015 as shown in Table 1.

The progress monitoring data (August 2014, February 2015, May 2015, June 2015) were subtracted from the July 2014 baseline reading in order to obtain the accumulated response in terms of Brillouin Frequency.

Table 1. Fibre optic readings: baseline and progress readings.

Monitoring section	Readings	
8 circumferential loops + 1 longitudinal section	July 2014	Baseline
	August 2014	Progress
	February 2015	Progress
	May 2015	Progress
	June 2015	Progress

The developed axial strain profile is plotted against the fibre optic cable distance for four selected representative circumferential loops within the examined TT10 tunnel section: loops 1, 2, 5 and 7 (Figure 7).

The two ends of the section (i.e. start and end of the plot's x-axis) refer to the two sides of the tunnel cross-section, whereas the middle of the section refers to the crown of the tunnel.

The graphs in Figure 7 show negligible strains developing within the first month (July 2014 – August 2014) after the deployment of the sensors, as the recorded values of axial strain do not seem to exceed about 20με for all the loops. Moreover, after seven months (July 2014 – February 2015) some notable values of axial strains can be seen in the FO cable. From the data collected after almost one year of monitoring (July 2014 – May 2015 and July 2014 – June 2015) small values of axial strains are experienced by the fibre optic cable, which do not go beyond around 100με for the mentioned loops.

The circumferential tunnel loops also seem to show a somewhat consistent profile of the strain s for all the examined loops. The lateral sides of the tunnel seem to experience larger strain values than those at the crown of the tunnel. A consistent strain profile of strain value peaks at the tunnel lateral sides and a strain value trough at the tunnel crown is observed.

Besides, there seems to be a pattern of some flexural behaviour of the tunnel lining: some compression (negative axial strain) at the inner side of the tunnel crown and some tension (compressive) of comparable magnitude at the lateral tunnel sides.

This mechanism suggests that the tunnel may be deforming in a vertical elongation shape as the distance of the lateral sides is decreasing and the crown seems to experiencing some hogging moments (Hisham, 2008). This behaviour would be compatible with a load pattern consisting of horizontal lateral stresses being larger than the corresponding vertical stresses, a case not uncommon in an over-consolidated soil environment (Figure 8).

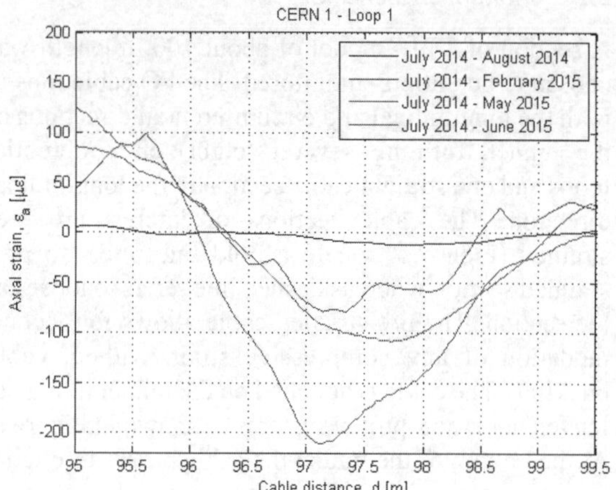

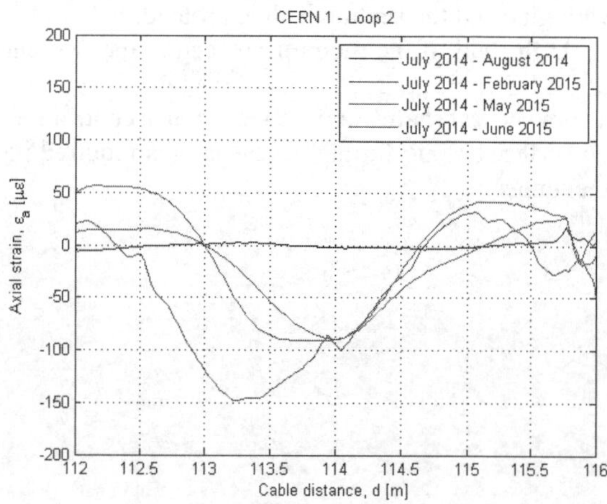

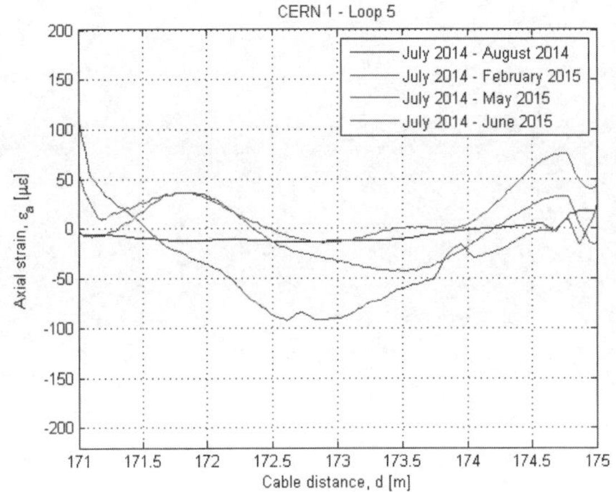

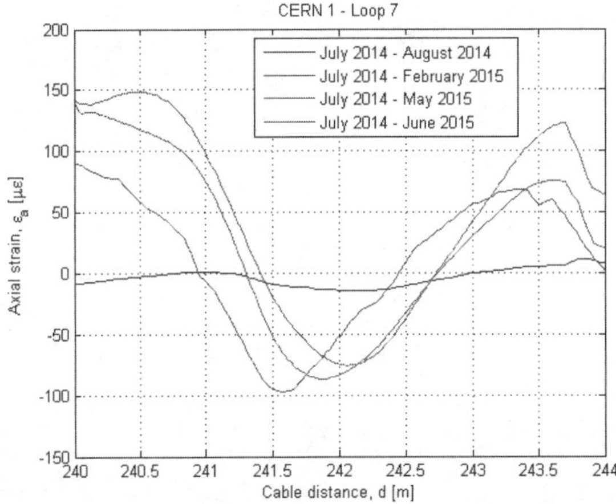

Figure 7. Fibre optic axial strain data: loops 1, 2, 5, 7.

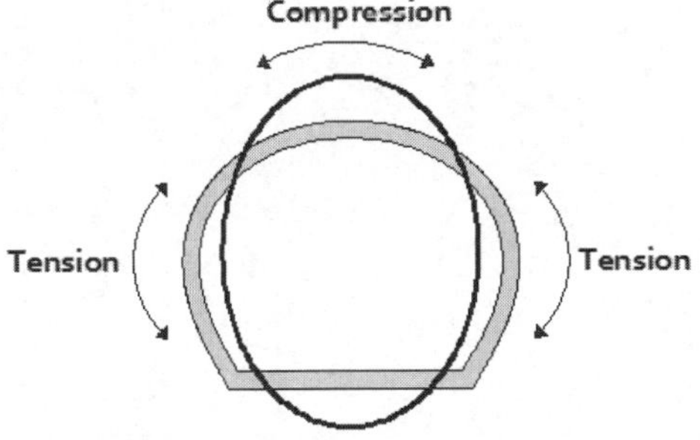

Figure 8. Tunnel deformation: elongation.

4 CONCLUSIONS

This paper describes the geomechanical investigation on the long-term behavior of CERN tunnels, by adopting an advanced monitoring system.

A novel technique of distributed strain sensing is introduced using Brillouin optical time-domain analysis to examine the performance of a concrete-lined tunnel called TT10. The fibre optic installation was deployed with a number of spatially distributed monitoring sections: 8 circumferential tunnel loops with a longitudinal section within the tunnel.

Initial baseline readings showed a good signal transfer and a clear trend in defining the various sections of interest. Five readings have been taken so far: the first set of baseline readings (July 2014) and four progress readings (August 2014, February 2015, May 2015 and June 2015).

The results from the circumferential tunnel loops suggest that the axial strains developed in the tunnel lining do not exceed peak absolute strain values of 150$\mu\varepsilon$.

The negative compressive strains at the crown of the tunnel were larger that the tensile strains at the sides of the tunnel. This strain pattern may have been caused by a vertical tunnel elongation mechanism of deformation, as the compression at the tunnel crest and extension at the lateral two sides of the tunnel are observed.

Finally, the values of all observed axial strains seem to be insignificant suggesting no real movement or deformation of the tunnel lining. This was expected as the purpose of this monitoring scheme is long-term tunnel deformations and no substantial strains would be anticipated over a period of ten months.

ACKNOWLEDGEMENT

The authors would like to thank the CSIC team (Centre for Smart Infrastructure and Construction) and the GS-SE group at CERN for transferring the knowledge regarding the fibre optic technique and for the assistance provided during the fibre optic installation process.

REFERENCES

Hisham, M., Bennett, P.J., Soga, K., et al. (2007). *Distributed Optical Fibre Strain Sensing in a secant Piled wall.* 7th International Symposium on Field Measurements in Geomechanics, ASCE 2007.

Hisham, M. (2008). *Distributed Optical Fibre Strain Sensing of Geotechnical Structures.* PhD thesis, Department of Engineering, Cambridge University, UK.

Hisham, M., Bennett, P.J., Soga, K., Mair, R.J., and Bowers, K.(2010). *Behaviour of an old masonry tunnel due to tunneling-induced ground settlement.* Geotechnique, 60(12), 927-938.

Laver, R. (2010). *Long term behavior of twin tunnels in London clay.* PhD thesis, Department of Engineering, Cambridge University, UK.

Li, Z. (2014). *Long term behavior of cast-iron tunnel cross passage in London Clay.* PhD thesis, Department of Engineering, Cambridge University, UK.

Omnisens (2013), *DITEST-STA-R Series: Fiber Optic distributed temperature & Strain monitoring system.* Omnisens, Morges, Switzerland.

Wongsaroj, J. (2005). *Three-dimensional finite element analysis of short and long term tunneling induced settlement in stiff clay.* PhD thesis, Department of Engineering, Cambridge University, UK.

Proceedings of the International Conference on Smart Infrastructure and Construction
ISBN 978-0-7277-6127-9

© The authors and ICE Publishing: All rights reserved, 2016
doi:10.1680/tfitsi.61279.033

Distributed monitoring of buried pipelines with Brillouin fiber optic sensors

X. Feng*, X.W. Zhang, W.J. Wu , X.Y. Li , X. Li and J. Zhou

[1] *Faculty of Infrastructure Engineering, Dalian University of Technology, Dalian, China*
* *Corresponding Author*

ABSTRACT The article presented here pertains to the distributed monitoring of the buried bell-and-spigot water pipelines. This study involved the monitoring scheme of distributed sensors and the experimental verification of a full-scale pipeline buried in a large soil trough. The sensor topology was proposed to fully describe the spatial deformation of the pipeline with Brillouin optical time domain analysis (BOTDA) sensors. Full-scale tests on buried pipeline were conducted in a large soil trough, simulating the non-uniform bedding under the pipeline. The longitudinal strains of the buried pipeline were monitored by the distributed Brillouin sensors for different load cases. The experimental results demonstrate that the distributed monitoring method based on Brillouin fiber optic sensors can detect the structural condition of the buried bell-and-spigot water pipelines.

1 INTRODUCTION

The pipelines are often referred to as "lifelines", as they carry water, oil and natural gas, essentially to the support of life. In practice, pipelines are usually classified as either continuous or jointed. The jointed cast iron pipes have been extensively used in the utility distribution systems around the world because the cast iron is less brittle and has higher load capacity. The bell-and-spigot joints are commonly adopted in the cast iron pipelines. Due to the structural deterioration, the damage in the pipe sections, as well as the defect in the pipe joints are inevitably happened during the life-cycle service. It is very difficult for the commonly-used inspection techniques to assess the structural conditions of the jointed bell-and-spigot pipelines because of the burial environment.

Recent developments of distributed fiber optic sensors based on Brillouin scattering effect promise to provide a cost-effective tool allowing monitoring of elongated structures such as pipelines (Ansari 2007; Bao 2009; Feng et al. 2013 & 2014). In the last decade, the Brillouin distributed fiber optic sensors have been applied to monitor the damage, leakage and failure of oil or gas pipelines. The distributed temperature sensors are used to detect the pipeline leakage by monitoring the environmental temperature distribution caused by the oil or gas leakage in the pipelines (Inaudi & Glisic 2010; Frings & Walk 2011). However, the temperature based approaches cannot monitor the structural deterioration. In order to detect the structural condition, the structural parameters of pipeline, such as strain, should be monitored in full extent. A number of studies based on distributed strain monitoring have been proposed for detecting the structural damages in pipeline (Zou et al. 2004; Ravet et al. 2006; Feng et al. 2015). According to the literature survey, the current studies mainly focus on the continuous pipelines, and rare works are reported for monitoring the structural condition of jointed pipelines. The Brillouin optical time domain reflectometry (BOTDR) based approach was adopted to monitor the jointed pipeline response to tunneling effect (Vorster et al. 2006). Glisic and Yao (2012) developed a method of health assessment based on distributed fiber optic sensors for buried bell-and-spigot pipeline. They demonstrated that the method is able to detect and localize the damage of pipeline caused by the earthquake induced ground movement. Mohamad et al. (2012) presented a method for monitoring twin tunnel interactions using distributed optical fiber strain measurements. Gue et al. (2015) used the BOTDR sensors to monitor an existing cast iron tunnel. They presented some practical guidance on the planning and installation of distributed fiber optic sensors. The aim of this paper is to investigate the performance of the Brillouin distributed fiber optic sensor to monitor the pipeline behavior under the non-uniform bendding conditions. The monitoring scheme was proposed for bell-and-spigot pipelines. And then the experimental verifications were conducted by a full scale testing in a large soil trough.

2 MONITORING SCHEME

Buried pipes have two dominant forms of behavior. The first involves circumferential response and load transfer around the pipe ring as it interacts with the surrounding soil. This has been studied at great length over the last century, and analytical and numerical solutions have supported the development of design equations considering the soil-pipe interaction of rigid and flexible pipes. The second dominant type of behavior involves variations in loading and changes in the soil and structural characteristics along the axis of the pipeline. This has received much less attention, and is little mentioned in pipe design procedures. Yet, longitudinal effects control the performance of the joints connecting different pipe segments, and it is being increasingly recognized that joint response often controls the long-term performance of the pipeline (Buco et al., 2006).

The bell-and-spigot joints are designed to permit rotation of one end of the segment relative to the next, and thereby releasing the longitudinal bending moments across the joints. The other key action of these joints is to transfer shear force between pipe ends, while remaining sealed against water movement into or out from the pipeline. The structural model of the bell-and-spigot pipelines is schematically plotted in Figure 1. In the model, the soil spring k_b describes the soil resistance due to the pipe-soil interaction, while the rotation springs k_r and the shear spring k_s represent the behavior of the joints, respectively. Thus the pipe segments behave as the Winkler elastic foundation beam. The joints between the segments are treated as the rotation and shear springs. To monitor the spatial deformation and damage of the pipeline, the parallel sensor topology was proposed in this study. For the installation on the pipeline, the parallel topology consists of three parallel sensors installed along the pipeline as shown in Figure 2. According to the strain analysis of the Winkler elastic foundation beam, three parallel longitudinal distributed sensors can fully describe the spatial deformation of the pipeline. Information on spatial deformation of the pipeline serves as an early warning about potentially high stress in the pipeline.

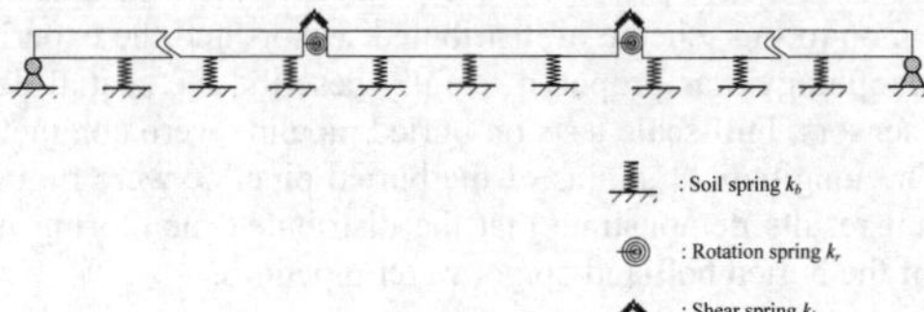

Figure 1. Schematic diagram of structural model for bell-and-spigot pipelines.

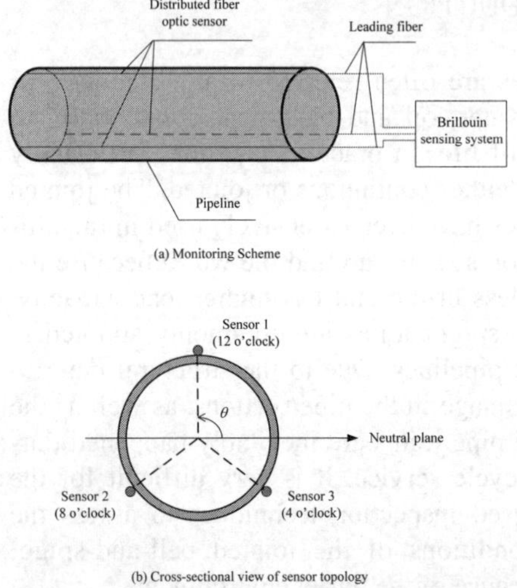

Figure 2. Parallel sensor topology for monitoring bell-and-spigot pipelines.

According to Figures 1 and 2, the bending behavior of the pipeline segments can be monitored by the parallel sensors. However, the distributed fiber optic sensors only measure the longitudinal strain, i.e. bending strain, of the pipelines. The rotations of the bell-and-spigot joints cannot be monitored by the parallel sensors due to the abrupt changes in the geometry of pipeline. In order to provide the continuous strain transfer, the small "bridges", which are made of steel, were proposed to be added at the locations of the joints, and then the distributed fiber optic sensors were installed on the "bridges". The "Omega-shape bridge" is shown in Figure 3 for monitoring the rotation of the joint. It is notable that the "bridge" configuration of the distributed sensor only qualitatively represents the deformation of the joint.

Figure 3. Monitoring scheme of bell-and-spigot joint.

3 EXPERIMENTAL VERIFCATIONS

To verify the proposed monitoring scheme with distributed fiber optic sensors, the full-scale experimental investigations were conducted in a large soil trough. In the test, the non-uniform bedding was simulated with different conditions. In practice, the non-uniform bedding is common to the buried pipelines. Construction conditions such as the use of unstable foundation materials, over-excavation under protruding bells or elsewhere, and non-uniform compaction of the bedding soil can all lead to sections where the pipe lack proper support. The non-uniform bedding influences the pipe behavior under the longitudinal bending, and often leads to the leakage of the bell-and-spigot joints.

3.1 Experimental setup

The standard DN200 ductile cast iron pipeline was selected as the experimental pipe. The nominal diameter of the pipe is 200 mm, and the length of the pipe is 6 m. Two pipeline segments were used in the experiments, and they were connected with one gasket bell-and-spigot joint. Three parallel distributed strain sensors were glued on the outer surface the pipeline with the epoxy resin. And the sensor "bridges" were installed according to Section 2.

A large soil trough was designed to simulate the burial conditions for the pipeline. The dimensions of the sol trough are 12 m × 1 m × 2 m. The soil trough was made of the steel frame and plates. The part of the bottom plates is movable. The plane size of the movable bottom plate is 1 m × 1 m. The movable bottom plates can drop in a controllable manner to simulate the non-uniform bedding conditions due to voids. The soil trough is shown in Figure 4, and the schematic diagram for simulating the non-uniform is plotted in Figure 5. In this figure, the numbers 1 to 6 represent the number of the bottom movable plates.

Figure 4. Soil trough for pipeline experiments.

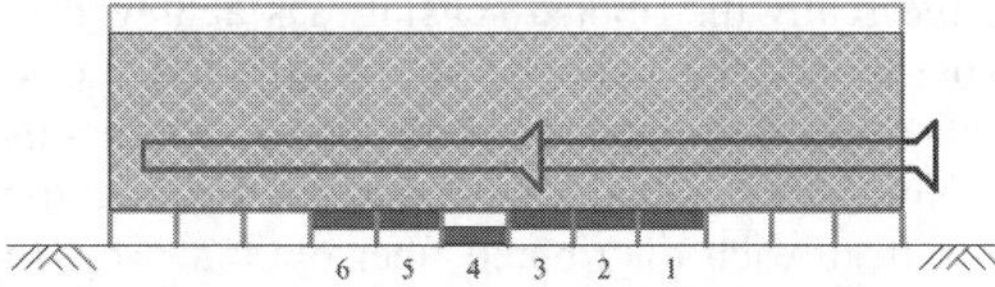

Figure 5. Schematic diagram for simulating the non-uniform bedding with the movable bottom plates.

The gravels were placed on the bottom of the soil trough, and the standard sands were paved to the height of 20 cm. The experimental pipeline was then installed. Two ends of the pipeline were supported to form the simply-supported boundary conditions. The soils were backfilled into the trough and compacted with jolt ramming squeezer.

3.2 *Experimental program*

As previously mentioned, the non-uniform bedding conditions were simulated by dropping the movable bottom plates of the soil trough. In the experiments, one or two plates were dropped in one load case. For the next load case, the plates dropped in the previous load cases were kept. The load cases are listed in Table 1. From the table, the void started from the middle location under the pipeline, and then it expanded to the both sides of the pipeline. With the void expansion, the effects of non-uniform bedding on the bending behavior of the pipeline were monitored by the distributed fiber optic sensors.

Table 1. Load cases of non-uniform bedding conditions in the experiments.

Load case	Sequence for dropping the bottom plates
1	Plate 4
2	Plate 3
3	Plate 5
4	Plates 2 and 6
5	Plate 1

The distributed strains were measured by a BOTDA interrogator NBX-6050A (Neubrex, Japan). For each load cases, two spatial resolutions, i.e. 10 cm and 20 cm, were selected to perform the measurements. In this article, only the results of 10 cm spatial resolution are presented since they can provide more accurate data. The distance resolution, i.e. sampling interval, is 5 cm in all measurements. Hence the really distributed sensing has achieved in the experiments. To calibrate the distributed measurements, the strain gauges were also used in this study. Nine strain gauges were mounted along the sensor path of each distributed fiber optic sensor. In total 27 strain gauges were placed on the test pipeline. The data were collected by NI compact DAQ. For each load case, the distributed fiber optic sensors as well as the strain gauges were interrogated simultaneously.

4 EXPERIMENTAL RESULTS AND ANALYSIS

The measured strains for each load case are plotted in Figures 6 to 10. In these figures, the measurements of three distributed fiber optic sensors were shown in the continuous curves, while the measured data of strain gauges were displayed by the scattered points. In each figure, "DS" denotes the distributed fiber optic sensor, and "SG" means the strain gauge. The numbers 1 to 3 represent the sensor locations shown in Figure 2(b).

As shown in Figures 6 to 10, the distributed measurements at the control points have good agreements with the data of strain gauges. There is a little difference between them because that the locations of strain gauges are not accurately same at the path of the distributed sensors. Moreover, the measurements of DS 1 are almost two times as those of DS 2 and DS 3. According to Figure 2(b), the top sensor has the double values of the two bottom sensors when the pipeline is subjected by the lateral bending load in the vertical place. Thus it is demonstrated that the proposed monitoring scheme can monitor the bending behavior of pipeline due to non-uniform bedding in the distributed manner.

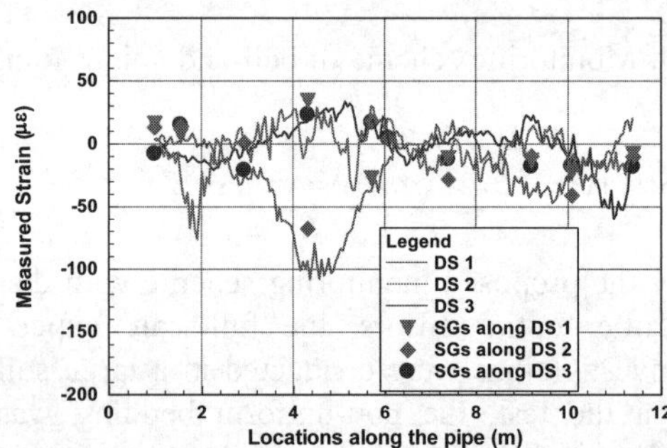

Figure 6. Measured strains along the pipeline for load case 1.

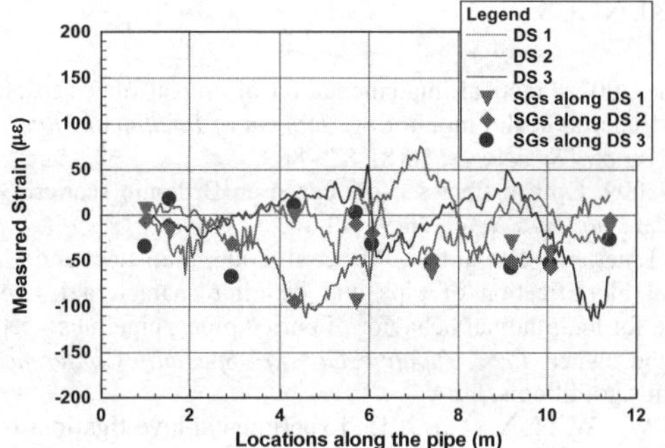

Figure 7. Measured strains along the pipeline for load case 2.

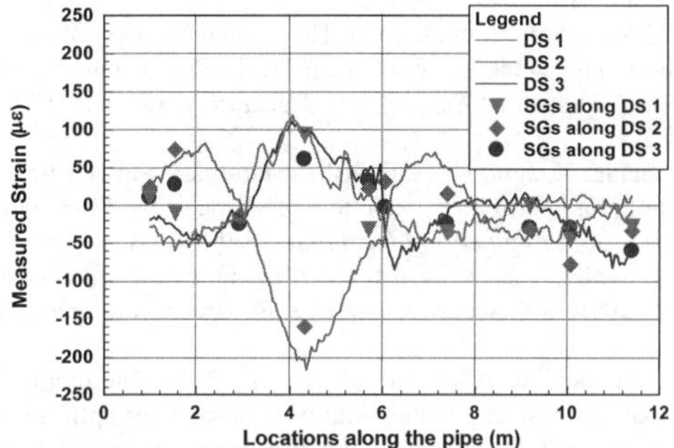

Figure 8. Measured strains along the pipeline for load case 3.

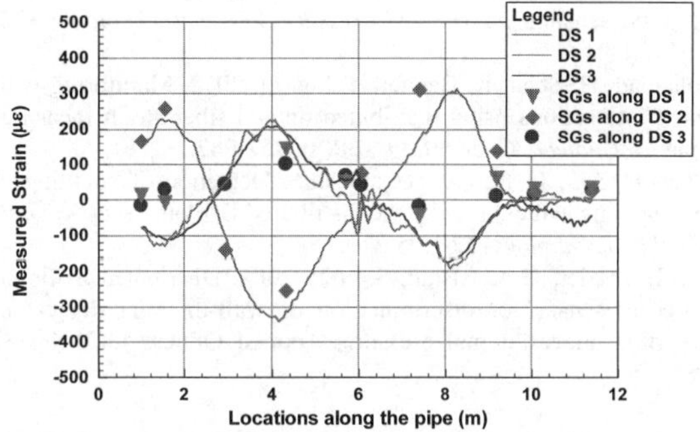

Figure 9. Measured strains along the pipeline for load case 4.

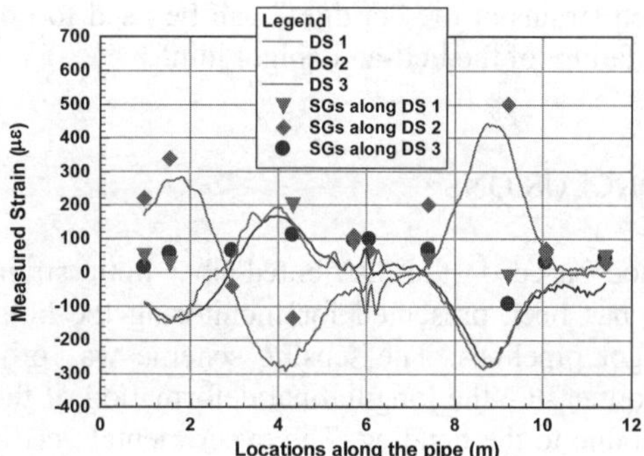

Figure 10. Measured strains along the pipeline for load case 5.

The location of bell-and-spigot joint is at the horizontal coordinate of 6 m in the experimental pipeline. From Figures 6 to 10, in the vicinities of the joints, the measured strains tend to zero. These observations indicate that the joint release the bending moment. The deformation of the experimental pipeline are caused by the non-uniform bedding condition due to the void.

In load case 1, the void with 1 m length formed in the left side to the bell-and-spigot joint. The bending behavior happened around the void in a finite range. In the other part of the pipeline, the longitudinal strains are almost zero. With the range of the void expanded, the bending behavior of the pipeline became complicated, i.e. the compressive and tensile strains appeared in the same sensor. The peak values of the measured strains become larger and larger with the increase of the range of the void. When the range of void tended to 6 m, the maximum strain on the top of the pipeline is about 450 με. But in each load cases, the measured strains in two ends of the pipeline are about zero. It is shown that the constrains of the pipeline keep simply-supported boundary conditions during the test.

In the first 3 load cases, the measured strains of the "bridges" at bell-and-spigot joint are less than 50με. However, in load case 4 and 5, the measured data at the "bridges" become larger. The abrupt variations qualitatively mean the condition of joint changes because of the large rotation. According to the excavation verification after the experiment, the rotation angle of the joint has exceeded the allowable value of the code. Therefore the abrupt change in the measured strains at the "bridges" can be used to detect the failure of the bell-and-spigot joint.

5 CONCLUSIONS

A method based on the distributed fiber optic strain sensors has been presented for monitoring the bell-and-spigot pipelines. The sensing scheme was proposed to measure the longitudinal deformation of the pipeline due to the bending. The experimental verifications were conducted on a full-scale ductile case iron pipeline in a large soil trough to monitor the structural responses under the non-uniform bedding condition due to void. The experimental results indicate that the proposed monitoring scheme can measure the bending behavior of the buried pipelines. The measured data at the joint can qualitatively reflect the changes in the condition of the bell-and-spigot joints. Thus the distributed fiber optic sensor based approach can detect the failure of the buried pipelines. However, further investigations, including comparison with analytical or numerical calculatioons, should be conducted under more complex loading conditions and for the quantitative assessment of the conditions of the pipeline joints.

ACKNOWLEDGEMENT

This research is based on work supported by the National Natural Science Foundation of China (No. 51378088) and the Fundamental Research Funds for the Central Universities (No. DUT14ZD202).

REFERENCES

Ansari, F. 2007. Practical implementation of optical fiber sensors in civil structural health monitoring, *Journal of Intelligent Materials, Systems and Structures*, **18**(8), 879-889.

Bao, X. 2009. Optical fiber sensors based on Brillouin scattering, *Optics and Photonics News*, **20**(9), 40-46.

Buco, J. Emeriault, F. Gauffre, P.L. et al. 2006. Statistical and 3D numerical identification of pipe and bedding characteristics responsible for longitudinal behavior of buried pipes, pipelines - service to the owner, *Proc. Pipeline Division Specialty Conference 2006*, Chicago, Illinois, USA.

Feng, X. Wu, W. Li, X. et al. 2015. Experimental Investigations on Detecting Lateral Buckling for Subsea Pipelines with Distributed Fiber Optic Sensors, *Smart Structures and Systems*, **15**(2), 245-258.

Voster, T.E.B. Soga, K. Mair, R.J. et al. 2006. The use of fiber optic sensors to monitor pipeline response to tunneling, *Proc. Geo-Congress 2006*, ASCE, Reston, VA, 1-6

Feng, X. Zhou, J. Sun, C. et al. 2013. Theoretical and experimental investigations into crack detection with BOTDR-distributed fiber optic sensors, *Journal of Engineering Mechanics*, **139**(12), 1797-1807.

Feng, X. Zhang, X. Sun, C. et al. 2014. Stationary wavelet transform method for distributed detection of damage by fiber-optic sensors, *Journal of Engineering Mechanics*, **140**(4), 1-11.

Frings, J. & Walk, T. 2011. Distributed fiber optic sensing enhances pipeline safety and security, *Oil Gas European Magazine*, **2**, 132-136.

Gue, C.Y. Wilcock, M. Alhaddad, M.M. et al. 2015. The monitoring of an existing cast iron tunnel with distributed fiber optic sensing (DFOS), *J. Civil Structural Health Monitoring*, **5**(5), 573-586.

Glisic, B. & Yao, Y. 2012. Fiber optic method for health assessment of pipelines subjected to earthquake-induced ground movement, Structural Health Monitoring, 11(6), 696-711.Inaudi, D. & Glisic, B. 2010. Long-range pipeline monitoring by distributed fiber optic sensing, *Journal of Pressure Vessel Technology*, **132**, 011701-1-9.

Mohamad. H. Soga, K. Bennett, P.J. et al. 2012. Monitoring twin tunnel interactions using distributed optical fiber strain measurements, *J. Geotech. Geoen. Eng.* **138**(8), 957-967.

Ravet, F. Zou, L. Bao, X. et al. 2006. Detection of buckling in steel pipeline and column by the distributed Brillouin sensor, *Optical Fiber Technology*, **12**, 305-311.

Zou, L. Ferrier, G.A. Afshar, S. et al. 2004. Distributed Brillouin scattering sensor for discrimination of wall-thinning defects in steel pipe under internal pressure, Applied Optics, 43(7), 1583-1588.

Proceedings of the International Conference on Smart Infrastructure and Construction
ISBN 978-0-7277-6127-9

© The authors and ICE Publishing: All rights reserved, 2016
doi:10.1680/tfitsi.61279.039

Study on the identification of dynamic behavior of beam structure by random access strain histories of Brillouin-based optical fiber sensor

M. Imai*[1] and S. Matsuura[2]

[1] *Kajima Technical Research Institute, Tokyo, Japan*
[2] *Yokogawa Electric Corporation, Tokyo, Japan*
* *Corresponding Author*

ABSTRACT As an accelerometer senses, temporal continuous measurement contains significant structural information to comprehend dynamic behavior. Recent development of optical fiber sensors can allow us to know distributed strain histories. Among the sensors, Brillouin optical correlation domain analysis (BOCDA) shows high signal-to-noise ratio, and thus it has a possibility to make its precision, spatial resolution, and sampling rate higher. These advantages straightforwardly contribute to the estimation of natural frequency and damping ratio of structures by dynamic strain history measurement. In BOCDA, on the other hand, the measured strain distribution profiles are not time-synchronized; therefore, its pitfalls interrupt to calculate a kind of its mode shape. To overcome this obstacle, time-position interpolation is executed to measure on a free oscillation plate test, and compared with synchronized strain gauge results. The results show that the interpolation method enables one to know strain distribution, which stands for vibrating mode, though the frequency of vibrating structure is limited.

1 INTRODUCTION

Dynamic behavior of civil engineering structures due to wind, earthquake, ambient vibrations, and/or live loading let us know valuable structural information. An accelerometer is the most common way to sense its dynamic structural behavior (Doebling et al. 1998), and the measurements can provide general information such as its natural frequency and damping ratio. By monitoring the mode frequency, it can be assumed whether its global structural system remains same or not. Also, mode shape given by multiple accelerometers can contribute to detect, localize and identify a local damage through a specialized data analysis (Fan & Qiao 2011).

For decades, optical fiber sensors (OFSs) have been developed and applied in various engineering fields. In dynamic measurement, fiber Bragg grating and Fabry-Perot interferometer have been popularly used (Glisic & Inaudi 2008). These successful OFSs can achieve high precision and high sampling rate, comparing well with the conventional electrical instruments. However, the measurement is spatially discrete, and one wire is typically necessary for one sensor. Spatially continuous strain profile can theoretically be given by Rayleigh-based (Parker et al. 2014, Kaplan et al. 2011) or Brillouin-based OFSs (Kishida & Li 2006, and Li et al. 2008).

Brillouin optical correlation domain analysis (BOCDA) is a Brillouin-based OFS. It uses high-power Brillouin scattering stimulated by continuous lightwaves; thus, it potentially shows high spatial resolution and short measurement time (Hotate & Ong 2002, Imai et al. 2010, and Manotham et al. 2012). Its sampling rate depends on electrical switching equipment. Sampling rate of the sensor based on correlation domain analysis is sure to be much higher in the near future, while electrical equipment is developed; however, data-processing issue is merging. In BOCDA, strain measurement is done point by point; therefore, the measured strain distribution is not time-synchronized. The time gap between measurement at one location and those at another location can be neglected in the case of static deformation, but cannot be neglected in the case of dynamic vibration. It has never become obvious so far, and recent development of BOCDA has exposed such a problem. To enhance the correlation-based Brillouin sensor, data-processing technique is experimentally investigated in this study.

2 STRAIN DISTRIBUTION HISTORIES

A conventional Brillouin-based strain sensor can statically measure strain distribution along the fiber at the specific moment, i.e., position vs. strain data set. Discrete sensor, e.g., electrical strain gauge, can dynamically measure strain history at the specific position, i.e., time vs. strain data set. It should be noted that this study tries to deal with strain distribution history, i.e., time vs. position vs. strain data set, thanks to the development of Brillouin-based sensor.

2.1 Brillouin Optical Correlation Domain Analysis

As a method of Brillouin-based sensors, time-domain method covers a long measurement range by an injected pulse lightwave, while BOCDA shows high spatial resolution by changing the frequency of two-encountered continuous lightwaves (Hotate & Ong 2002). Though their advantages and drawbacks are different, both the sensors are good enough to perform SHM in terms of strain distribution measurement. Having originated from telecommunication technology, time-domain method has successfully emerged to be a competent monitor of earthen structures or long structures, e.g. the deformation of a levee or the displacement of a pipeline. On the other hand, BOCDA developed to detect even a tiny strain change within the range of limited measurement length, thanks to its high spatial resolution.

2.2 Random Access Measurement

In BOCDA, strain is measured at one location by one location. Modulation frequency of light source determines measurement position; therefore, switching location takes little time without mechanical movement. Measurement time is almost equal to sweeping time for getting Brillouin spectrum at each location. Measurement position and traveling sequence from one location to another can be determined as we want. In other words, more resource for sampling rate can be allocated for a specific location, if you know the critical locations. Let us call it "random accessibility." This kind of feature can only be achieved by correlation domain method. In time domain method, the whole length of fiber has to be measured even if you only want to measure the specific region. Strain measurement with various densities is a kind of advantage of BOCDA-based sensor at the expense of time synchronization.

An example of random access measurement is illustrated in Figure 1. Assuming strain distribution change as curves shown in the figure, random access measurement, which is only one measurement at one moment, is denoted by a black-filled circle. After the time interval which denotes ΔT, BOCDA changes the location of measurement and measures strain at that location. Then, repetitive measurement is done at different locations as a set of routine. In the figure, there are six measurements for one routine. In the routine, the center of the span is measured twice, and the other four locations are measured only once. Again, the position and the order of measurement are fully controllable.

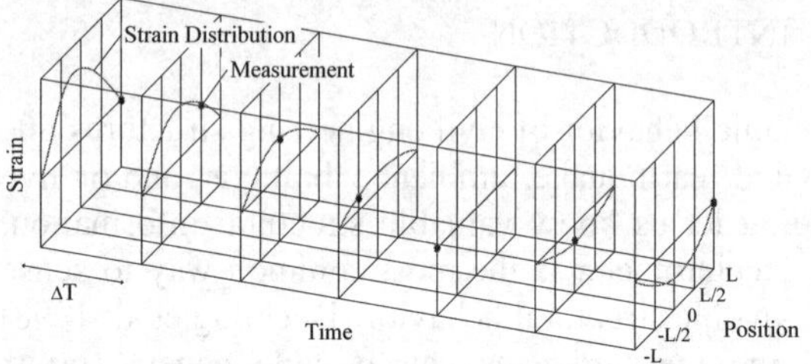

Figure 1. An example of random accessibility. Random access measurement is done at one location at one moment, that is, the location is traveling around to measure.

2.3 Time-position Interpolation

To comprehend the structural behavior at a specific moment, e.g., mode shape, which is sure to be one of the most important characters of dynamic structural behavior, time synchronization is sure to be required. Thus, post-processing method for non-sync measurement is proposed to achieve quasi-synchronization. Bi-cubic interpolation with plane triangulation is used for data processing in this study as written in the following formula, because it can be easily applied to get smooth continuous result. Consisting of 9 unit squares, the coefficients α_{ij} are solved by the given equations, e.g., given measurements, derivatives in x- and y-directions, and cross derivatives in xy-direction. This interpolation, in which the first and second derivatives are continuous, has been widely used for 2-D spatial interpolation (Lam 1983, and Watson 1992).

$$p(x, y) = \sum_{i=0}^{3} \sum_{j=0}^{3} a_{ij} x^i y^j$$

Treating time in the same way as position, the authors intend to apply the method for time in one-dimensional optical fiber instead of a two-dimensional one in a plane. Based on the discrete measurement, data is spatially and temporally interpolated, as shown in Figure 2. As a result, we can know interpolated strain distribution at any moment. It is clearly said that its availability depends on the relationship between measurement time interval and structural vibration frequency. Potentially, the phase shift cannot be perfectly compensated. The gap of synchronization can be neglected if its structure oscillates slowly, but it cannot in the case of transient state, especially. In this study, the basic concept of this proposed method for the first mode with less than 10 Hz is evaluated with the use of experimental measurement data which is currently 70 Hz in sampling rate.

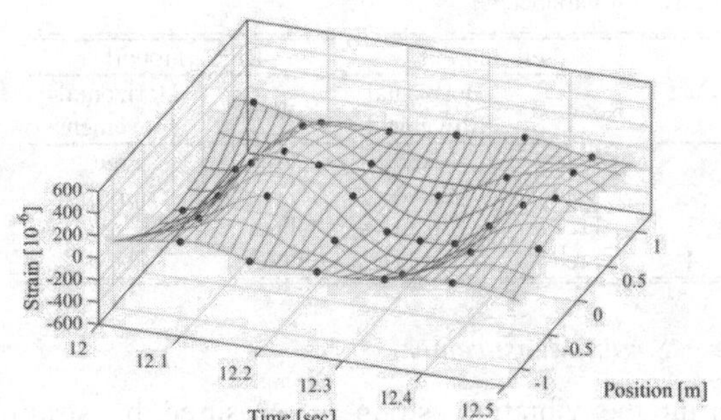

Figure 2. An example of interpolated results. Based on the measurement (black-filled circle), meshed data is interpolated in terms of time and position.

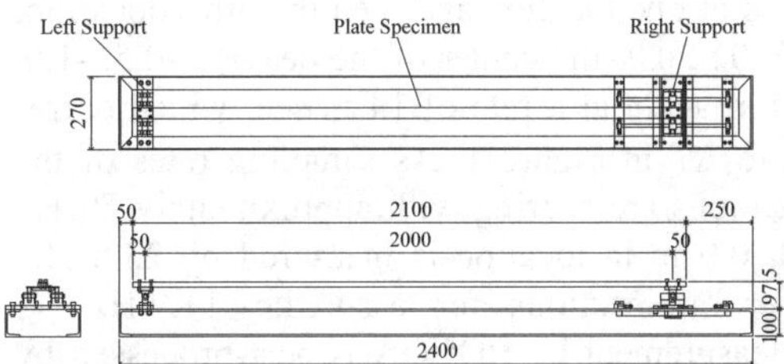

Figure 3. Plate specimen is fixed on two supports of the testing equipment.

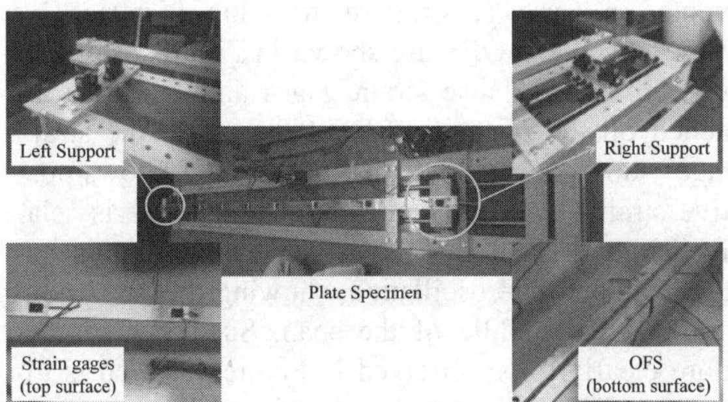

Figure 4. Experimental overview of plate specimen, left and right supports. Nine strain gauges and an optical fiber sensor are attached on the top and bottom surface of the plate, respectively.

3 EXPERIMENTAL STUDY

To evaluate the proposed data-processing method, random access measurement on vibrating plate specimen is executed and the processed results are compared with strain gauges in terms of its natural frequency, damping ratio, and strain distribution.

3.1 *Plate Specimen*

Figures 3 and 4 illustrate the testing equipment, which has a plate specimen made of duralumin on two supports as a single span beam. The constraint conditions of the right support, i.e., rotation and horizontal movement, can be changed, simulating an aging bridge support, whereas the left support is only allowed to be rotated. The length of the span is 2.0 m, and nine electrical strain gauges are attached on the top surface of the plate with 0.25 m-interval and a sensing optical fiber are attached on the bottom surface of the plate.

Four tests are totally conducted, as tabulated in Table 1. Changing the conditions of the right support, the plate is vibrated by pushing down on the middle of the span and then released instantly. Measurements are conducted while the plate shows gradually damped oscillation with its natural frequency.

Table 1. Test Parameters.

Case	Left Support		Right Support	
	Rotation	Horizontal Movement	Rotation	Horizontal Movement
1	Free	Fixed	Free	Free
2	Free	Fixed	Free	Spring
3	Free	Fixed	Free	Fixed
4	Free	Fixed	Fixed	Fixed

3.2 *Strain Measurement*

During the vibration, strain is measured by strain gauges at 200 Hz and BOCDA at 70 Hz, respectively. Of course, all nine strain histories by the attached strain gauges are synchronized. In BOCDA, strain is measured at one location and then the other location, i.e., 1.0, 0.5, 0.0 (the center of the beam) , -0.5, -1.0, and 0.0 m again at a rate of 14 m sec, which corresponds to ΔT in Figure 1. As sampling rates of the five locations are sharing with approximately 70 Hz in total, 0.0 m in location is measured by 23.8 Hz, and other four locations are measured by 11.9 Hz.

Measurement by BOCDA is post-processed by the above mentioned method, and then compared to results of strain gages. As an example of measurements, in Case 1, strain distribution histories by strain gages and OFS are shown in Figures 5 and 6, respectively. Because strain gages and OFS are attached on the opposite side of the plate, the strain gages shows negative strain whereas OFS shows positive strain. Even though OFS measurement is relatively scattered, it can be said that both figures behave damped free oscillation, showing the maximum strain at the middle of the span. Such behavior is more clearly to be observed in Figure 7 which is interpolated by 14 m sec in time, i.e., 70 Hz in sampling rate, and 0.25 m in position.

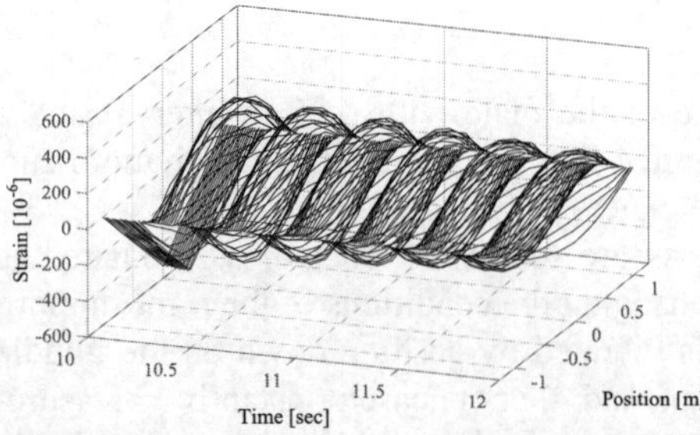

Figure 5. Strain distribution histories measured by nine strain gauges (0.25 m in interval) with 200 Hz sampling in Case1. The figure is re-sampled 100 Hz to be comprehensive.

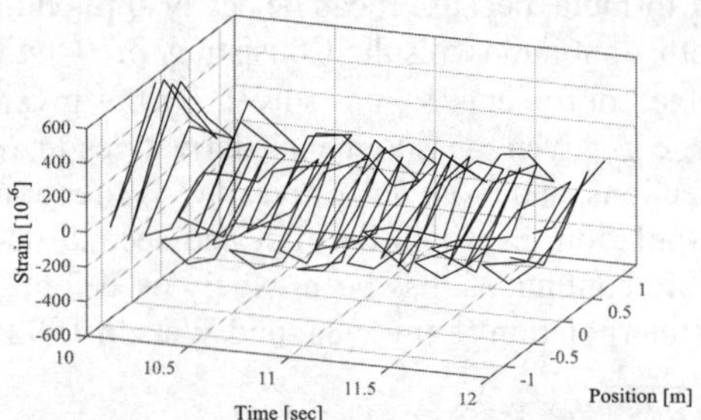

Figure 6. Strain distribution histories measured by OFS in Case1. Random access measurement is executed for five locations with 70 Hz sampling in total.

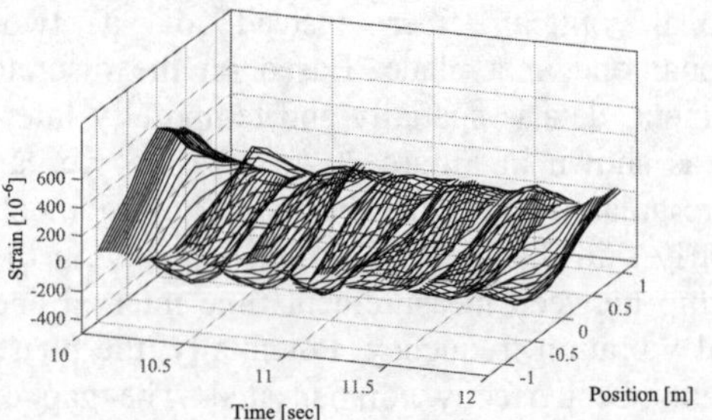

Figure 7. Strain distribution histories of interpolated OFS in Case1. Data is interpolated by nine locations (0.25 m in interval) and 70 Hz sampling.

3.3 *Natural Frequency*

Natural frequency is calculated by fast Fourier transform of strain histories at the center of the span for a period of 10 s just after the beginning of vibration. Thus, error in the natural frequency is dominantly determined by the sampling rate of the mid-span, not affected by strain distribution. As shown in Table 2, the results of calculated natural frequency by OFS are in good agreement with those by strain gauges for all four tests. The more the right support is constrained, the more the axial tensile force along the plate is applied; therefore the natural frequency is increasing from Case 1 to Case 4. It is difficult to discriminate Cases 2 and 3 by the natural frequency even if strain gauges are used; however, great variation of the structural condition can be successfully detected by OFS as a change in the natural frequency.

Table 2. Natural frequency.

Case	Strain Gauge	Interpolated OFS	Discrepancy
1	3.125 Hz	3.162 Hz	1.18 %
2	5.078	5.022	1.10
3	5.078	5.023	1.08
4	6.836	6.882	0.67

3.4 Damping Ratio

Similar to the natural frequency, the damping ratio is calculated by measurement at the center of the span for a period of 10 s just after the beginning of vibration. Interpolation process for strain distribution is no direct influence on identification of damping ratio as well. The results of calculated damping ratio are tabulated in Table 3, putting down with results by strain gauge and interpolated OFS. The energy loss by the movement of the right support seems to be extremely small; therefore, the damping ratio in Case 4 is smaller than that in other cases. It is clearly seen that the measurement discrepancy is increasing from Case 1 to Case 4 because of increasing vibrated frequency in comparison with the sampling rate; however, OFS can detect the tendency of damping ratio change as is the case with strain gauges.

Table 3. Damping ratio.

Case	Strain Gauge	Interpolated OFS	Discrepancy
1	5.69 %	5.57 %	2.11 %
2	4.13	4.32	4.60
3	4.19	3.64	13.12
4	3.86	3.27	15.28

3.5 Strain Distribution

Mode shape is one of the most important features of dynamic structural behavior, and it is much sensitive to being out of synchronization. It should be noted that a mode shape normally means a deflection curve at a specific time, but in this study, the strain distribution along the beam is investigated instead of a deflection curve. In fact, deflection curve must be corresponding to the results calculated by the double-integral of strain distribution.

Figures 8 represent measured strain distribution in Case 1. In the first oscillation, strain distributions at zero, π/3, 2π/3, and π in phase are drawn in the figures. There are three figures, i.e., (a) by strain gauges, (b) by OFS, and (c) by interpolated OFS. In (b), it is assumed that strain at certain location remains the same until the next measurement at that location is done. As a result, strain distribution is absolutely different from strain gauges. On the other hand, interpolated OFS (c) is much better, thanks to smaller time and position resolution as well.

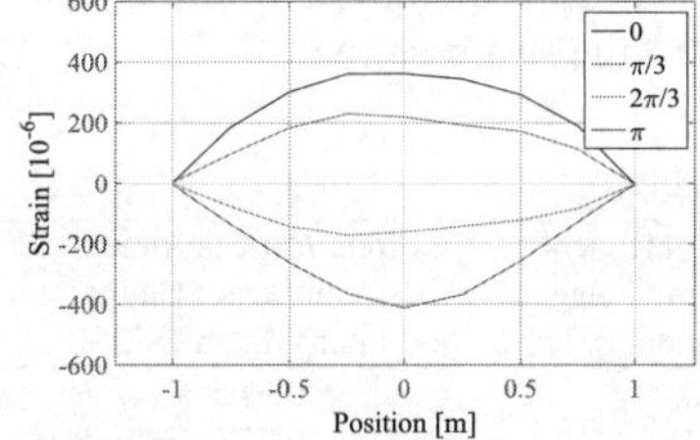

(a) Measured by strain gauges.

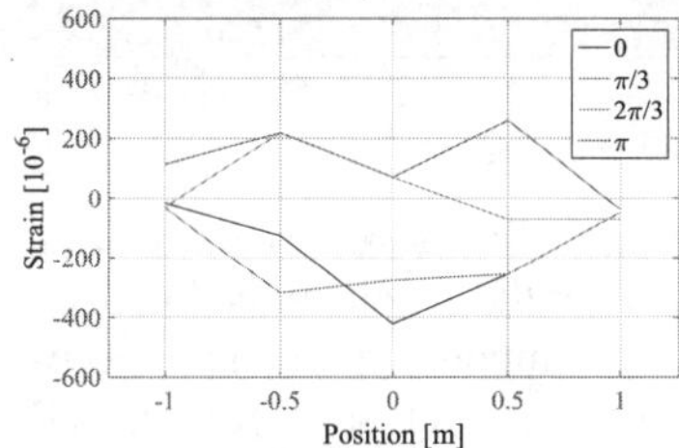

(b) Measured by OFS.

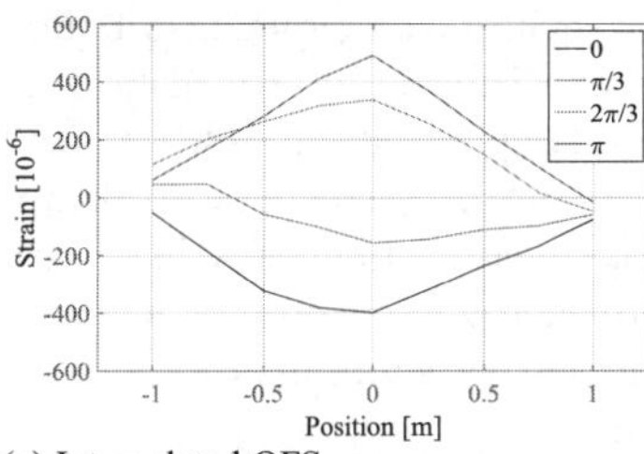

(c) Interpolated OFS.

Figure 8. Strain distribution of phase in zero, π/3, 2π/3, and π when the plate is oscillating at 3.125 Hz in Case 1.

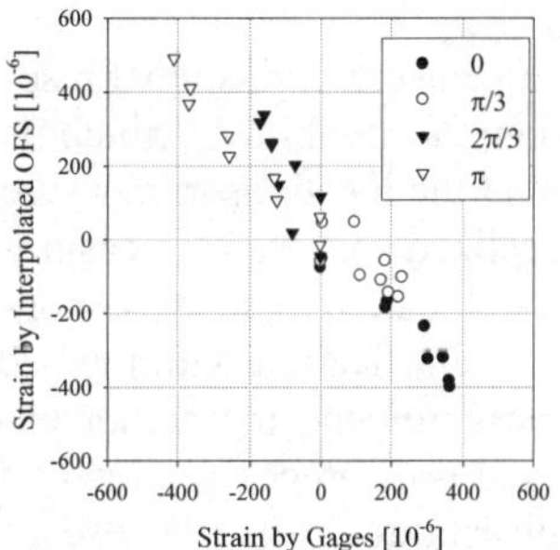

Figure 9. Strain comparison between strain gauges and interpolated OFS in Case1. Ideally, plots are placed in line, and that is correlation coefficient (R^2) is equal to 1.

To evaluate the results, correlation coefficient, i.e., R, between strain gauge and interpolated OFS is calculated using 36 data (9 locations and 4 moments) for each case, and R^2 (correlation coefficient) are tabulated in Table 4. If both were perfectly correlated, R^2 would be 1, and lower value in R^2 means lower similarities. When sampling rate is fast enough in comparison with the vibrating frequency, strain distribution behaves similar to the strain gauges.

Table 3. Strain distribution.

Case	Correlation Coefficient between Strain Gauge and OFS (raw data), R^2	Correlation Coefficient between Strain Gauge and Interpolated OFS, R^2
1	0.772	0.912
2	0.054	0.773
3	0.072	0.713
4	0.226	0.540

3.6 Results and Discussions

Strain on vibrating beam is measured by correlation-based OFS at 70 Hz, when the specimen is dominantly vibrating at its first mode. Its random access measurement is temporally and spatially discrete; therefore, it is post-processed by bi-cubic interpolation to get virtually synchronized strain distribution along the plate. It is experimentally proved that (1) natural frequency and damping ratio by OFS is well determined, (2) measurement discrepancy in damping ratio increases when natural frequency increases, and (3) time-position interpolation process successfully provides comprehensive strain distribution, especially for slower oscillation.

4 CONCLUSIONS

BOCDA has a potential to measure high spatial resolution with unique random accessibility, which is able to quickly travel around the locations to measure. For instance, it can be realized that more measurement is done at a critical location than at a non-critical one. For structural identification with the use of such random access measurement, the author experimentally investigates the post-processing method to determine its strain distribution with the use of measured strain-position histories, based on the test on the plate specimen. The method still has a limitation in the case of higher rate oscillation when compared with the measurement sampling rate; the study shows the fact that the comprehensive first mode vibrating with less than approximately several Hz can be estimated as a result of OFS measurement with position-time interpolation. It should be noted that the method has been validated in the steady state, and not in the transient state of vibration. Yet, this study shows the possibility of evaluating structural health in terms of natural frequency, damping ratio, and mode shape, which can be calculated by strain distribution.

ACKNOWLEDGEMENT

The authors gratefully acknowledge the continuous support being provided by Prof. Hotate of the University of Tokyo. This study was partially supported by Japan Science and Technology Agency, Adaptable and Seamless Technology Transfer Program, through a target-driven R&D: JST A-STEP, Contract No. AS2525002H.

REFERENCES

Doebling, S.W., Farrar C.R. & Prime. M.B. 1998, A summary review of vibration-based damage identification methods, *Shock and vibration digest*, **30(2)**, 91–105.

Fan, W. & Qiao, P. 2011, Vibration-based damage identification methods: a review and comparative study, *Structural Health Monitoring*, **10(1)**, 83–111.

Glisic, B. & Inaudi, D. 2008, *Fibre optic methods for structural health monitoring*, John Wiley and Sons, 21–26.

Hotate, K. & Ong, S.S.L. 2002, Distributed fiber Brillouin strain sensing by correlation-based, continuous-wave technique, cm-order spatial resolution and dynamic strain measurement, *Proc. Photonics Asia*, 299–310.

Imai, M., Nakano, R., Kono, T., Ichinomiya, T., Miura, S. & Mure, M. 2010, Crack detection application for fiber reinforced concrete using BOCDA-based optical fiber strain sensor, *Structural Engineering*, **136(8)**, 1001–1008.

Kaplan, A., Klute, S. M. & Heaney, A.2011, Distributed optical fiber sensing for wind blade strain monitoring and defect detection, *Proc. Int. Workshop on Structural Health Monitoring*, 13–15.

Kishida, K. & Li, C.H. 2006, Pulse pre-pump-BOTDA technology for new generation of distributed strain measuring system, *Structural health monitoring and intelligent infrastructure*, 471–477.

Lam, N.S.N. 1983, Spatial interpolation methods: a review, *The American Cartographer*, **10(2)**, 129–150.

Li, C.H., Tsuda, T., Sawa, T., Makita, A., Takano, H, Kishida, K, Wu, Z., Takeda, N. & Minakuchi, S., 2008, PPP-BOTDA method to achieve 10cm spatial resolution and 10Hz measuring speed in distributed sensing, *Technical report of IEICE*, OFT **108(245)**, 39–44.

Manotham, S., Kishi, M., He, Z. & Hotate, K. 2012, 1-cm spatial resolution with large dynamic range in strain distributed sensing by Brillouin optical correlation domain reflectometry based on intensity modulation, *Proc. Asia Pacific Optical Sensors Conference*, 835136.

Parker, T., Shatalin, S. & Farhadiroushan, M. 2014, Distributed Acoustic Sensing–a new tool for seismic applications, *First Break*, **32(2)**, 61–69.

Watson, D.F. 1992, *Contouring: a guide to the analysis and display of spatial data*, Elsevier, 101–161.

Proceedings of the International Conference on Smart Infrastructure and Construction
ISBN 978-0-7277-6127-9

© The authors and ICE Publishing: All rights reserved, 2016
doi:10.1680/tfitsi.61279.045

Condition monitoring of railway tracks and vehicles using fibre optic sensing techniques

W. Lienhart*[1], C. Wiesmeyr[2], R. Wagner[3], F. Klug[1], M. Litzenberger[2] and D. Maicz[3]

[1] *Institute of Engineering Geodesy and Measurement Systems, Graz University of Technology, Graz Austria*
[2] *AIT Austrian Institute of Technology GmbH, Vienna, Austria*
[3] *Hottinger Baldwin Messtechnik GmbH, Vienna, Austria*
* *Corresponding Author*

ABSTRACT Although the European railway network has been expanded considerably within the last decade, monitoring of railway tracks and railway vehicles is still limited to epoch wise test drives or to local continuous measurements. In this contribution we report about new approaches to continuously monitor railway tracks and vehicles over tenths of kilometres using distributed fibre optic sensing (DFOS) techniques. In a first approach fibre optic strain sensing cables are attached to the railway tracks and strain changes due to rail deformations are depicted by distributed Brillouin measurements (BOTDA, BOFDA). These measurements allow the early detection of possible damages of the railway facilities due natural causes like mudflow, avalanches, floods, landslides and can prevent secondary damage by fast and correct counter actions. In a second approach we use optical communication cables which are commonly already laid next to modern rail infrastructure to detect flat spots in railway wheels. If these spots become too large they can damage train tracks and in extreme cases even cause derailment. In this contribution we demonstrate that distributed acoustic sensing (DAS) allows continuous monitoring of trains and to extract an individual profile that indicates flat spots. Since already laid communication cables are used no additional infrastructure is required apart from the optical time-domain reflectometry (OTDR) instrument. In our field installations in the Austrian railway network we demonstrate that BOFDA and OTDR systems complement each other to monitor incidents and deformations and thus enable permanent condition monitoring of railway tracks and vehicles over very long distances.

1 INTRODUCTION

Within the last decade railway connections have become of major importance in the European transportation network. Outages due to rail track deformations or increased wear due to wheel flat spots are costly and can cause secondary damages e.g. due to derailment. Currently, the condition of railway tracks and vehicles is only monitored on a few distinct points of the railway network or by individual test drives using special measurement locomotives. For the early detection of possible damages it would be beneficial to monitor the whole network continuously with high spatial resolution.

Therefore, we investigated the applicability of distributed fibre optic sensing (DFOS) for monitoring of tenths of kilometres of railway tracks. Two different measurement principles were evaluated and their feasibility was verified in several experiments within the Austria railway network. These experiments focused on the detection of rail deformations and flat spots of railroad cars wheels.

2 DISTRIBUTED STRAIN SENSING

2.1 *Principle of Brillouin measurements*

Brillouin scattering is caused by acoustic-optical interaction which occurs when light is travelling in an optical fibre (e.g. Hayashi et al., 2014; Nöther, 2010; Horiguchi et al., 1989). In case of Brillouin Optical Time Domain Reflectometry (BOTDR) a high power laser pulse is sent into the fibre and the reflected signal caused by spontaneous Brillouin backscattering is recorded. In order to increase the measurement length to tenths of kilometres a loop configuration can be used. In this case a continuous pump laser is coupled into one end of the fibre and a pulse laser which emits light pulses (BOTDA, Brillouin Optical Time Domain Analyser) or amplitude modulated waves (BOFDA, Brillouin Optical Frequency Domain Analyser) is coupled into the other end of the fibre.

The sensing principle is based on the fact that the Brillouin frequency shift, changes linearly with temperature and strain (Zeni et al. 2015, Kurashima et al. 1990). This frequency shift can be converted into strain with conversion coefficients derived from sensor calibration. In practical applications the temperature sensitivity of the Brillouin frequency has to be compensated. This is usually done by integration of a loose fibre in the measurement loop which is only sensitive to temperature.

2.2 *Brillouin measurements of railway infrastructure*

Distributed strain measurements along railway tracks allow the early detection of possible damages of the railway facilities due to natural causes like extreme temperatures, mudflows, avalanches, floods, landslides and can prevent secondary damage by fast and correct counter actions,

In order to provide reliable results a rigid connection between the sensing fibre and the rail is required. Chuang et al. (2003) and Zeni et al. (2013) glued fibre optic cables to rails to measure strain. This approach delivers the highest strain resolution but fails in case of rail breakages. Rail breakages can occur for instance at low temperatures and may result in gaps with widths of several centimetres which can cause derailment of trains.

In terms of robustness, point wise anchoring of the fibre is preferable. The maximum gap width that can be covered depends on the used sensing cable and the spacing between the anchors. Larger distances between anchors enable the measurement of larger gaps without fibre breakage but reduce the sensitivity with respect to rail bending and buckling. Therefore, the selection of an appropriate anchor spacing is crucial for the performance of the measurement systems.

2.2.1 *Design of system components and laboratory investigations*

To clamp the fibre to the rail we developed a special cable clamp. This clamp consists of a base component which is rigidly connected to the rail. A cover plate with special grooves is then used to clamp the sensing cable to the base component. The geometry of the grooves is adjusted to the used sensing cable. For this application we choose the BRUsens V1 from Brugg Cables (Brugg Cables 2013). Laboratory investigations showed that large strains can be measured with this cable without fibre breakage.

In order to select the right anchor spacing we performed numerical studies and laboratory investigations. A rail breakage causes a longitudinal displacement of anchor points and if the fibre does not break this movement can be depicted reliably with a sensing cable stretched between the anchors. However, in case of lateral displacements between anchor points the sensitivity decreases considerably with increased anchor spacing. From the numerical simulations and the laboratory investigations an anchor spacing of 60 to 70cm was found to be most suitable. At this spacing rail breakages with gaps of several centimetres can be measured without fibre breakage and differential lateral displacements of more than 3mm can be depicted reliably.

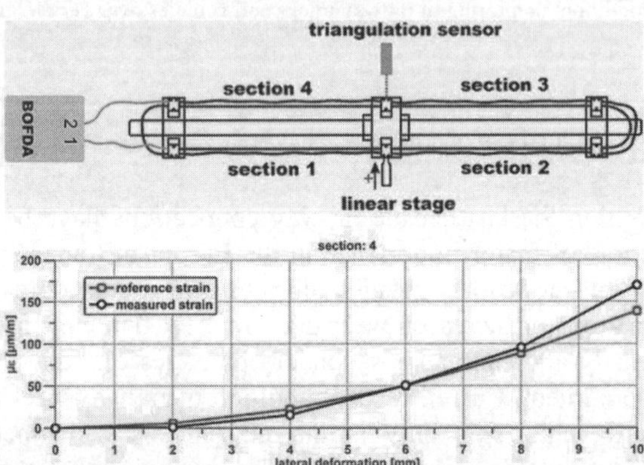

Figure 1. Test setup and comparison between fibre optic and reference measurements

Figure 1 shows the result of one experiment. In this experiment the sensing cable was mounted between 3 anchor points. The central anchor point was fixed on a translation stage. The applied lateral displacement was independently measured with a laser triangulation sensor. This reference displacement was converted into longitudinal strain and compared to the longitudinal strain measured with the BOFDA measurement instrument Fibris Terre fTB2505.

2.2.2 Field validation

The suitability of the proposed approach was verified in two field installations. In both field installations additional lose cables were used to compensate for temperature dependence. The first field installation focused on natural events where the rail track is locally deformed. Therefore one track of a rail track was assembled with a distributed fibre optic measurement cable on the inner and outer side (Figure 3, top). To simulate a natural event the rail track was laterally deformed at one position with a winch. The winch induced horizontal displacements to the railway sleepers. Figure 2 shows that the point force caused local bending with negative strain (shortening) on the inside track and positive strain (elongation) on the outside track.

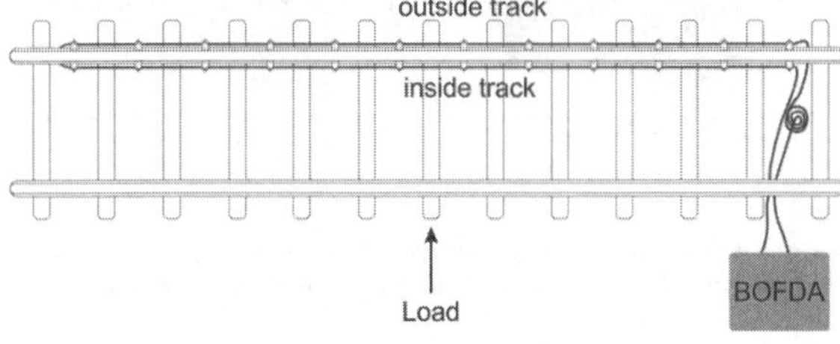

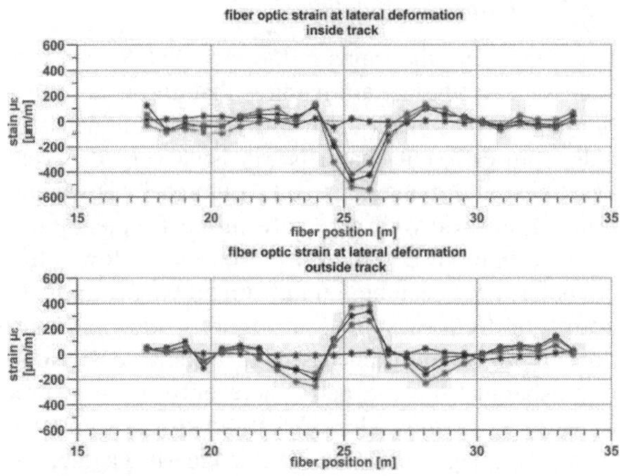

Figure 2. Experiment layout and measured shortening on the inside track and elongation on the outside track of the fibre optic sensing cable due to different point loads (blue: load step 1, red: load step 2, black: load step 3, purple: load step 4)

In the second field installation we measured the vertical deformation of a rail track. Therefore a static load, in terms of a locomotive (80t) and a passenger car (48t) were placed onto the assembled rail. The displacement of the anchors was independently measured using geodetic techniques. Figure 3 shows the results of the geodetic and fibre optic measurements during static loading with the passenger car. It can be seen that the position of the ring mount units can be identified by both measurement techniques. The vertical displacements at the position of the first ring unit are higher than expected which may have been caused by inadequate foundations of the train track in this section. The settlement of several millimetres strains the fibre up to 90μm which can be depicted reliably with the fibre optic measurement system. The vertical displacements of the second ring unit are less than 2mm and therefore the resulting measured strain is difficult to separate from measurement noise. Concluding it can be said although the position of individual axis may not be identifiable, problematic foundations can be identified.

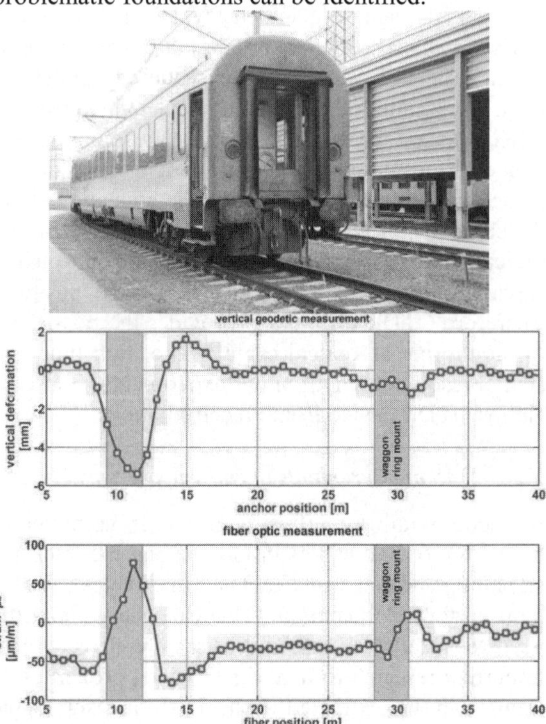

Figure 3. Measured vertical deformation (top) and measured strain (bottom) during static loading with a passenger carriage (weight 48t)

3 DISTRIBUTED ACOUSTIC SENSING

Distributed acoustic sensing (DAS) systems also use standard telecommunication cables as sensing element. DAS has been applied in a variety of applications where large longitudinal infrastructure objects were monitored for unusual activities. Examples are monitoring of oil pipelines for leakage and unauthorized interference as well as perimeter protection and the monitoring of road and railway infrastructure (Hill, 2015; Owen et al., 2012; Timofeev et al., 2015). Like with Brillouin sensing, lengths of several tenths of kilometres with spatial resolutions of better than 1 m can be realised. Power is only required at the DAS instrument but not along the sensing fibre. Therefore DAS requires less maintenance than other trespassing monitoring techniques such as microphones or cameras.

3.1 *Principle of DAS*

The optical sensing principle underlying DAS is optical time-domain reflectometry (OTDR). The OTDR device injects a sequence of light pulses into the optical cable and measures the backscattered light (Rayleigh backscatter). For an individual light pulse the reflections are measured with a certain sampling frequency. Considering the speed of light the reflected pulse travel time can be converted to a location along the cable length. Vibrations and local pressure on the optical fibre change its wave impedance causing Rayleigh backscatter from the respective position. Therefore OTDR technology provides the possibility of monitoring an optical cable with a high time and spatial resolution. In all of our experiments we used a Helios OTDR device from Fotech solutions.[1]

3.2 *DAS measurements of railway infrastructure*

In modern railway infrastructure it is common to have fibre optical cables installed along the track for signalling and data transfer. The same cables can also be used for DAS using an OTDR device without the need to install additional infrastructure. It is subject to current research to investigate which properties of trains can be extracted from DAS measurements (Timofeev et al., 2015).

[1] http://www.fotechsolutions.com/index.php/products/das

Our field test setup used an optical cable next to train tracks with a length of 1326m and measured with a spatial resolution of 0.68m. The temporal sampling rate is 5 kHz. Figure 4 shows the raw data, as well as a time-frequency image of a train recording as obtained from the OTDR device for two given positions along the track. The vibrations caused by the train are recorded between the seconds 5 – 20.

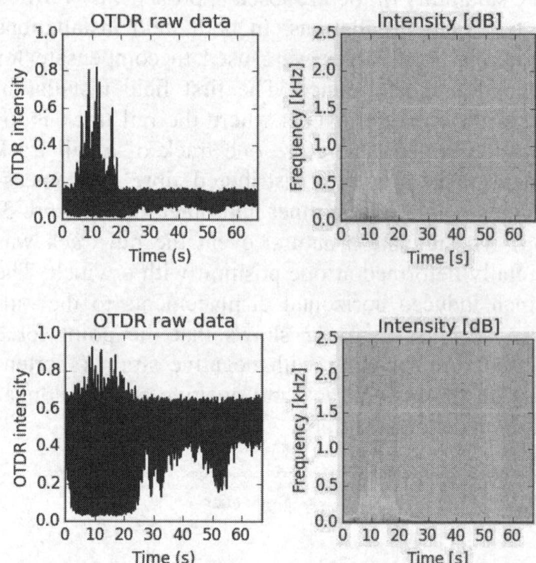

Figure 4. Raw data from a train recording for two given positions (left plots); Time-frequency image from raw data (right plots)

3.2.1 *Train identification*

For identifying the trains in the raw OTDR data it is necessary to find certain features that allow us to distinguish train signals from noise. The obtained signals from different channels can have very different baselines and behaviour, c.f. Figure 4. However, trains admit a certain structure in the frequency domain. While noise is mostly present in the lower frequencies, trains can be found from 50 Hz upwards. In both time-frequency plots in Figure 4 the time window in which the train passes is visually easily distinguishable from the part where only noise is present. By extracting selected features of the signal we obtain an image where the train is clearly visible, c.f. Figure 5.

The train can then be identified in each of the spatial channels as a number of connected pixels that are above an appropriate threshold. An illustration of the identification principle can be found in Figure 5. Please note that the optical fibre cable is installed with loops in some places leading to a different signal structure in the respective channels. However, this is a property of the track that is being monitored and can be configured upfront.

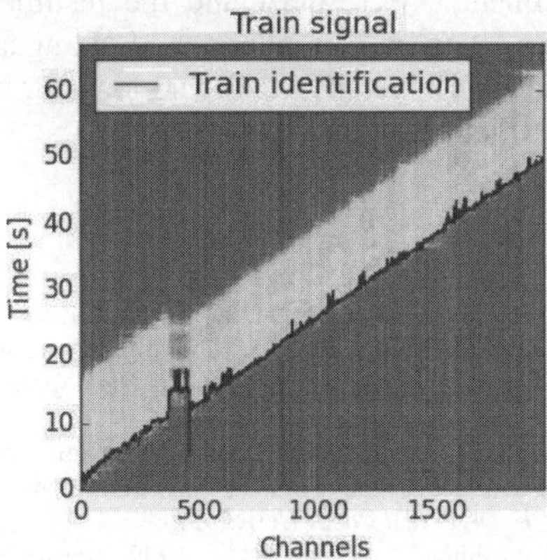

Figure 5. Train signal in time and spatial position; in the channels 370 to 460 the optical fibre is installed with a loop so the signal has a different structure and the train identification fails there

3.2.2 Flat spot detection

A flat spot is a fault in a railroad car's wheel that is typically produced during strong breaking when an axle stops rotating and is dragged along the rail ("wheel out-of-roundness"). If a flat spot is small the axle can still be used and it will be repaired when the car is serviced. Nevertheless it is important to identify faulty wheels as they lead to an increased wear of the whole infrastructure. If the out-of-roundness exceeds a certain limit the corresponding car has to be identified immediately and the faulty axle has to be replaced to avoid damage to the rails and in extreme cases even derailment.

Currently the standard is to have dedicated train monitoring measurement equipment that can identify flat spots. This equipment measures dynamic forces on the rail, resulting from the out-of-roundness of the wheel. In contrast to this technology, that can be applied only at certain points along the rail track, DAS continuously monitors faulty wheels with cables that are already part of the infrastructure. In this contribution the reference measurements for flat spots are done with an Argos® monitoring system[2], which allows determining faulty wheels with high accuracy. It measures dynamical as well as static forces exerted on the rail at one given point along the track (Maicz et al., 2012).

In OTDR signals flat spots change the characteristics of the train signal in several features. In Figure 6 we show a time-frequency plot for a train with two flat spots.

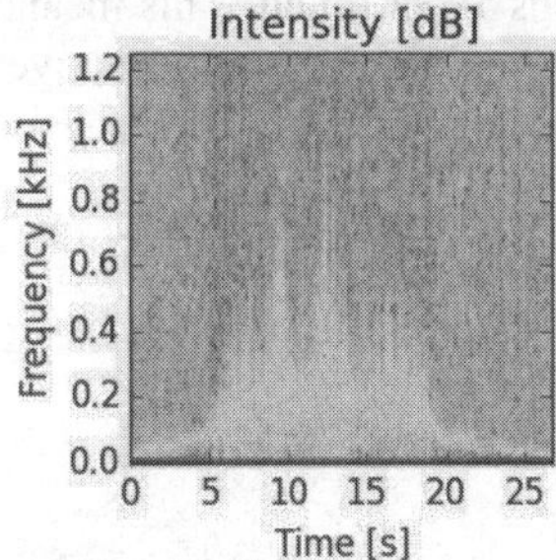

Figure 6. Time-frequency plot of OTDR signal from a train with two flat spots, one after a quarter of the length and the other one in the middle of the train.

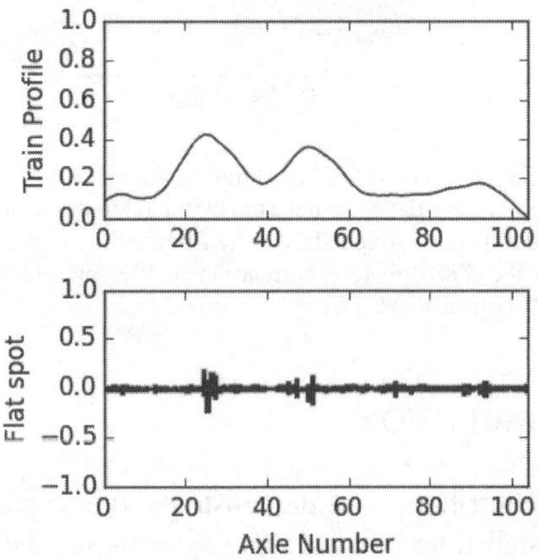

Figure 7. Top: train profile computed from the OTDR signal; Bottom: Argos® reference measurements

In Figure 7 we show the train profile for the train of Figure 6. In comparison to our computed profile we also show the reference measurement from the Argos® monitoring system. Every peak in the train profile corresponds to flat spots. Therefore, we eval-

[2] http://www.argos-systems.eu/?lang=en

uate for each axis how much the train profile at this position exceeds the average. This is used as a feature to classify the different axles as faulty/non-faulty.

In Figure 8 we show a scatter plot to illustrate the correlation between flat spots identified in the OTDR signals and the reference measurements for 39 recorded trains. Despite the different measurement principles, a high correlation between OTDR signal features and Argos® was found. This means that the flat spot detection via DAS is an attractive option with the major advantage of distributed sensing.

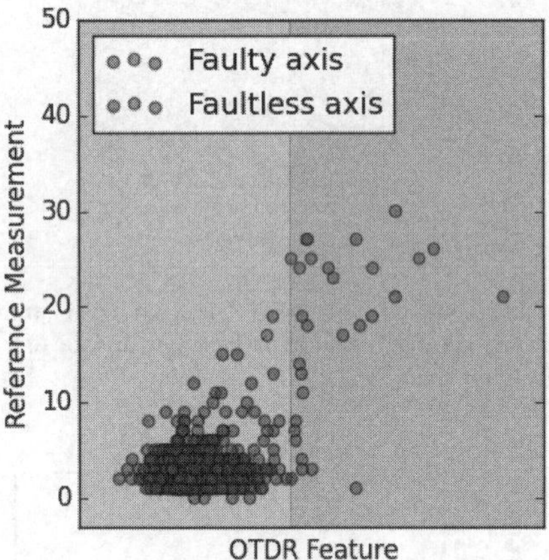

Figure 8. Scatter plot of OTDR feature vs reference measurement, red points indicate flat spots detected by the reference system, the green and red area distinguishes faulty from faultless axes determined by the OTDR feature; correlation coefficient: 0.668, sensitivity: 90%, specificity: 99%

4 CONCLUSION

In this contribution we demonstrated the feasibility of distributed fibre optic sensing systems for the condition monitoring of railway infrastructure. Brillouin measurements of sensing cables connected by anchors to the rails enable the measurement of rail breakages, rail buckling and displacements caused by natural events. We also showed the detection of out-of-roundness wheels in a field test installation along a railroad track using OTDR instruments in connection with existing fibre-optical cable infrastructure and confirmed the results with an independent point measurement system.

ACKNOWLEDGEMENT

We would like to acknowledge the funding support of the Austrian Research Promotion Agency (FFG project number 840448). Furthermore, we want to thank our project partners Austrian Federal Railways, R&D Department, NBG-FOSA and the Institute of Computer Languages Compilers and Languages Group of Vienna University of Technology for their valuable contributions.

REFERENCES

Brugg Cables, 2013. BRUsens Strain V1, *Datasheet Rev. 04*

Chuang, S.L. Hsu, A. & Young, E. 2003, Fiber optical sensors for high- speed rail applications, *Final Report for High-Speed Rail IDEA Project 19*

Hayashi, N. Mizuno Y. & Nakamura K. 2014, Alternative implementation of simplified Brillouin optical correlation-domain reflectometry. *IEEE, DOI: 10.1109/JPHOT.2014*

Horiguchi, T. Kurashima, T. & Tateda, M. 1989. Tensile Strain Dependence of Brillouin Frequency Shift in Silica Optical Fibers, *IEEE Photonics Technology Letters*, Vol 1, p. 107-108

Hill, D. 2015. Distributed Acoustic Sensing (DAS): Theory and Applications, *Frontiers in Optics*, Paper FTh4E—1

Kurashima, T. Horiguchi, T. & Tateda, M. 1990, Thermal Effects of Brillouin Gain Spectra in Single-Mode Fibers. *IEEE Photonics Technology Letters*, Vol. 2, p. 718-720

Maicz, D. Mittermayr, P. & Zottl, W. 2012. Wayside Train Monitoring: Erfahrungen mit dem Argos-System. *Der Eisenbahningenieur* 63 **9**, 152 – 156

Nöther, N. 2010, Distributed Fiber Sensors in River Embankments. Advancing and Implementing the Brillouin Optical Frequency Domain Analysis, *BAM-Dissertationsreihe*, Volume 64

Owen, A. Duckworth, G. & Worsley, J. 2012. OptaSense: fibre optic distributed acoustic sensing for border monitoring, *Intelligence and Security Informatics Conference (EISIC), 2012 European.*, 362 – 364

Timofeev, A. V. Egorov, D. V. & Denisov, V. M. 2015, The Rail Traffic Management with Usage of C-OTDR Monitoring Systems, *Int. J. of Computer, Electrical, Automation, Control and Information Engineering* **9**, 1492 – 1495

Zeni, L. Minardo, A. Porcaro, G. et al., 2013. Monitoring railways with optical fibers, *SPIE newsroom*

Zeni, L. Picarelli, L. Avolio, B. et al. 2015, Brillouin optical time-domain analysis for geotechnical monitoring, *Journal of Rock Mechanics and Geotechnical Engineering* 7, 458-462

Proceedings of the International Conference on Smart Infrastructure and Construction
ISBN 978-0-7277-6127-9

© The authors and ICE Publishing: All rights reserved, 2016
doi:10.1680/tfitsi.61279.051

Application of long-term structural monitoring data to bridge operations: a case study of a fiber-optic monitoring in a prestressed concrete bridge

M. Nishio*[1], J. Xu[1] and J. Mizutani[1]

[1] *Yokohama National University, Yokohama, Japan*
* *Corresponding Author*

ABSTRACT This paper presents the comparison procedure of the long-term structural monitoring data based on the time-series model estimation. The target is the fiber-optic strain monitoring data that has been acquired in a prestressed concrete bridge since December, 2011. The acquired data showed seasonal and daily trends due to significant environmental effects; i.e., air temperature, sunshine duration, and snow accumulation in winter. To extract features relate to any structural changes from the long-term data, authors propose to apply one of time-series models; seasonal ARIMA model. It was then shown that the distributions of estimated model coefficients, which were obtained by the sequential estimation procedure applied to the long-term data, could indicate the difference of strain behaviors that were due to the changes of thermal expansion behaviors through a year, which related to the existence of accumulated snow on the bridge in the case of target bridge. The application of Mahalanobis distance comparison was then verified using the feature vector of estimated seasonal ARIMA coefficients. The multipoint strain behavior could be compared by the one-dimensional measure with considering strong trends in the data including the daily one, through the period of Dec. 2011 to Sep. 2013.

1 INTRODUCTION

Although the structural monitoring systems, which are composed of various sensors, have been installed in many actual bridges, the strategies to analyze acquired long-term data have not been clearly introduced. Authors' group has also been working on a structural monitoring project in an existing concrete bridge since 2011. Here, a fiber-optic strain measurement system was installed, and long-term strain data have been acquired with one-hour time interval. It was then recognized that the acquired time-series data showed clear seasonal and daily trends mainly due to the environmental effects. The influences of environmental effects on the long-term structural responses were considered in many previous studies (e.g., Soyoz 2009, Cross 2013, and Zhou 2015), and most of them actually analyzed the influence of air temperature changes that were recognized as the most significant environmental effect. Zhou et al. 2015 indicated that the thermal expansion behavior also related to the detail temperature distributions within the whole structure by the analysis of data from a long-span bridge. This study actually gave recognition that the detail distribution of environmental effects (i.e., local effects) should also be considered in the analysis of the long-term structural responses.

In this paper, we firstly introduce the strategy to provide information about the structural conditions by applying the time-series analysis; the estimation of seasonal ARIMA (autoregressive integrated moving average) model, to acquired long-term strain data. There were actually some previous studies that also verified the applicability of time-series models to the monitoring data (e.g., Omenzetter 2005 and 2006). There, Omenzetter et al. applied the seasonal ARIMAX and seasonal ARIMA models to the strain monitoring data, and showed that the unusual events could be extracted by the comparisons of estimated model coefficients. In this study, we also adopted the seasonal ARIMA model because the data showed clear seasonal and daily trends. The coefficients of estimated model is expected to be used as the comparative feature to investigate structural condition changes through the long-term data. Here, plots of model coefficients obtained by the sequential model estimation are presented, and their applicability to the baseline distribution for condition assessment are verified. One of statistical comparison procedures; the Mahalanobis distance comparison, which is the measure for comparing multivariate feature vectors, is applied to investigate structural behavior changes through Dec. 2011 to Sep. 2013.

2 LONG-TERM STRAIN DATA IN THE TARGET BRIDGE

The target bridge in this study is an existing prestressed concrete (PC) bridge located in Hokuriku-area, Japan, where is actually the area of heavy snowfall in winter. The bridge was completed in 1972, and heavy corrosions were found in some PC cables in 2009. They were considered to be due to the snow melting agent that was used for the safety traffic operation in winter. The structural reinforcement work, which included the installation of external cables inside the box-girder, was conducted in 2011, and the continuous structural monitoring by using several sensors was started soon after the work.

2.1 *Target Bridge and Installed Sensor Network*

The bridge is a PC box-girder bridge with continuous four-spans, the total length of which is around 300m and the width is around 10m. One of the four spans with the length of 65m, which was between A1 and P1 indicated in Fig.1, was the target of the continuous monitoring in this study.

Twelve strain sensor units were attached on the inside web of the box-girder; P1/P2/P3-1/2/3/4 as shown in Fig.2. Each sensor unit consisted of two fiber-Bragg grating (FBG) sensors; one was completely attached on the structural surface in the longitudinal direction with the gage length of 50mm for the strain measurement, and the other was placed just next to the strain-FBG for the temperature correction. Three optical fiber lines; P1, P2, and P3, each of which consisted of four sensor units, were connected to a FBG interrogator. All equipments for the measurement system were set up inside the box-girder for the continuous data acquisition.

Figure 1. Full-view of the target bridge

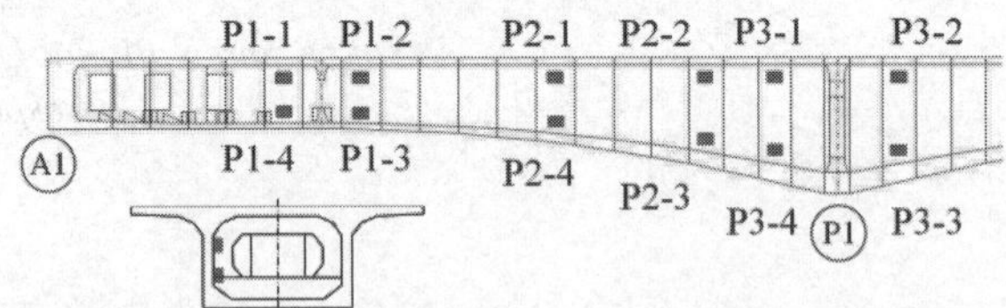

(a) Locations of sensor units

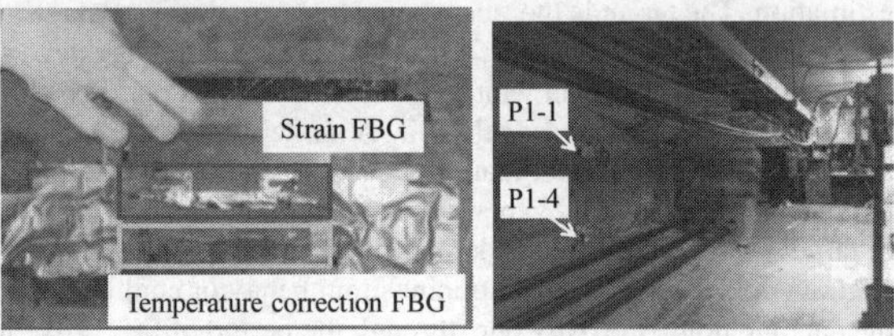

(b) Sensor installation

Figure 2. Installed fiber-optic strain sensor network

2.2 *Acquired data*

The continuous data acquisition was started in December 19th, 2011, which was the date soon after the reinforcement work. Each data acquisition was performed every one hour; here, the FBG wavelengths at all FBGs were recorded for two minutes in around 700Hz sampling rate. The raw data at each FBG with the length of around 84,000 points was then processed to an averaged wavelength, and the strain from the initial condition, which was the condition in the first data acquisition, was calculated with the temperature correction.

Figure 3 shows the obtained time-series strain data at two sensor units; P1-3 and P3-1, from December 19th, 2011 to September 29th, 2013. There were actually missing parts from December 2012 to March 2013, which were due to the maintenance of the measurement system. However, clear seasonal and daily trends were observed in all acquired time-series.

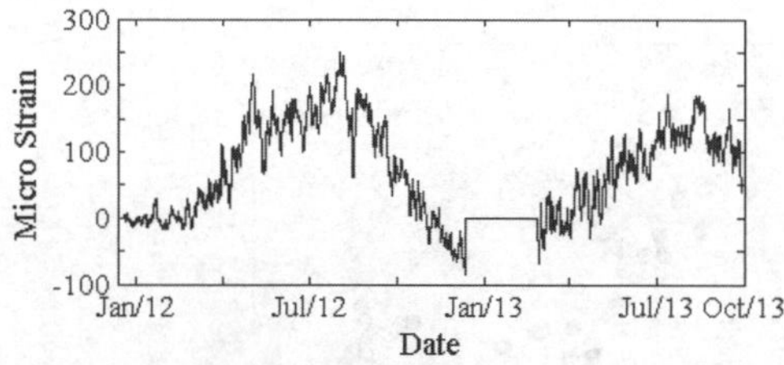

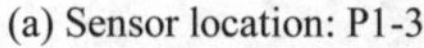
(a) Sensor location: P1-3

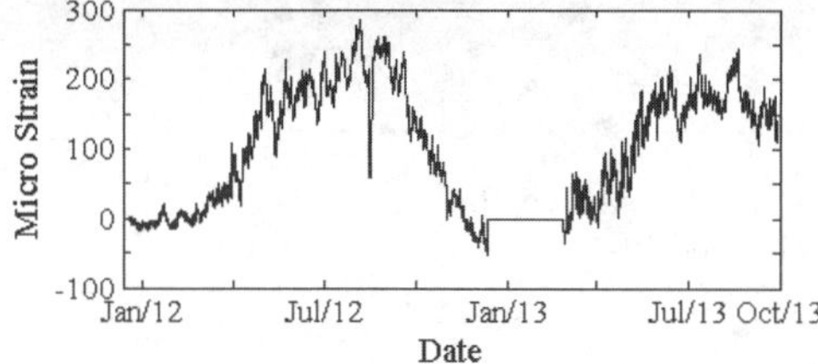

(b) Sensor location: P3-1

Figure 3. Installed fiber-optic strain sensor network

3 APPLICATION OF TIME-SERIES MODEL FOR ANALYZING LONG-TERM DATA

While the time-series models are used as a forecasting model in most cases, the coefficients of the estimated model are considered to be used as the feature for investigating the characteristic of the system. The seasonal ARIMA model can consider both the global trend and the periodic daily trend in the time-series that were also recognized in our strain data.

3.1 Theoretical Description of Seasonal ARIMA Model

In the seasonal ARIMA model, the non-stationary time-series can be transferred to the stationary one by taking differences several times for the ARMA model estimation. The definition of the seasonal ARIMA $(p, d, q)(P, D, Q)_S$ is written as:

$$\begin{aligned}&(1-\cdots-\varphi_p B^p)(1-\cdots-\beta_P B^{PS})(1-B)^d(1-B^S)^D y_t\\&=(1-\cdots-\psi_q B^q)(1-\cdots-\theta_Q B^{QS})\varepsilon_t\end{aligned}, \quad (1)$$

where y_t is t-th data variable, ε_t is the residual error with the white noise process, and φ, ψ, β, and θ are the AR, MA, seasonal AR, and seasonal MA coefficients with order of p, q, P, and Q, respectively. B is the lag-operator defined as:

$$B^k y_t = y_{t-k}; \quad (2)$$

therefore, d and D in Eq.(1) indicates the differential order for the non-periodic trend, and S is the seasonal differential order to consider the periodic trend in the time-series. In this study, the model estimations were conducted by using Matlab functions "arima" and "estimate" in the Econometrics toolbox; here, coefficients of the model were calculated by the maximum likelihood estimation.

3.2 Comparison Strategy of Estimated Model Coefficients

To compare the patterns of strain time-series continuously, the whole time-series strain data was divided into the three-day time-series. The seasonal ARIMA model estimation was then applied to each of all three-day time-series, and the distribution of estimated coefficients was statistically investigated to discuss the structural properties. One of issues in this strategy was: how to determine the model orders; i.e., p, d, q, P, D, Q, and S, in each estimation. Basically, it was firstly considered that the coefficients of the model with the same orders should be compared. Firstly, each of three-day time-series consisted of a weak global trend, which could be considered by d=1, and the 1-day periodic trend that could be removed by S=24. After checking the stationarity of the time-series with d=1 and S=24, the orders P and Q were then determined by the partial auto-correlation function (PACF) and the auto-correlation function (ACF), respectively.

In each seasonal ARIMA $(0,1,0)(1,0,1)_{24}$ estimation, the residual error term was checked by the Q-Q plot and the Ljung-Box test to determine whether the model was appropriately estimated or not. Only the coefficients that were judged as the ones from the appropriate model estimations were picked up for plotting the baseline distribution. The rate of adoption was more than 90% to all estimations.

3.3 Seasonal AR and MA Coefficients Plot

The plots of estimated seasonal AR and seasonal MA coefficients were then obtained for the strain data acquired from December 19th, 2011 to November 30th, 2012. The results in two of sensor locations; #P1-3 and #P3-1, are shown in Fig.4 (a) and (b), respectively. In both figures, the coefficients estimated from the data acquired in the period with snow coverage on the bridge are indicated by the black dots , and the coefficients indicated by the white markers are estimated from the data acquired in spring, summer, and autumn. It was then firstly observed that the distributions showed the linear regressions in all plots; each of them represented the same characteristic in the strain time-series. Meanwhile, the distributions of black dots in both figures seemed to show the different regression from the regressions of the white markers.

The two regressions in each plot were then examined statistically. First of all, each of two distributions; black dots and white markers (i.e., with and w/o snow accumulation), had a linear regression with the R2-statistic of over 0.8. The two regressions were then tested whether they were the same ones or not. The result, especially the result of Fig.4 (b), of the analysis of covariance (ANCOVA) then rejected the hypothesis: the two regressions were identical, with the p-value of $0.039 < 0.05$ under the 95% confidence interval. Actually, six of the eight sAR-sMA plots; the plots of the sensor locations #P1-1/2/3/4 and #P3-1/2/3/4, showed the significant differences between the two regressions.

This difference in regression was considered because the thermal expansion behavior of the bridge had been changed due to the nominal temperature distribution due to the snow accumulation in winter. The multipoint temperature measurement was then conducted through a whole year (Nishio et al., 2015). The significant seasonal trends including the difference between w/ and w/o accumulated snow were understood from the acquired data. The detail temperature distributions within the bridge were considered to affect the deformation patterns of structural responses.

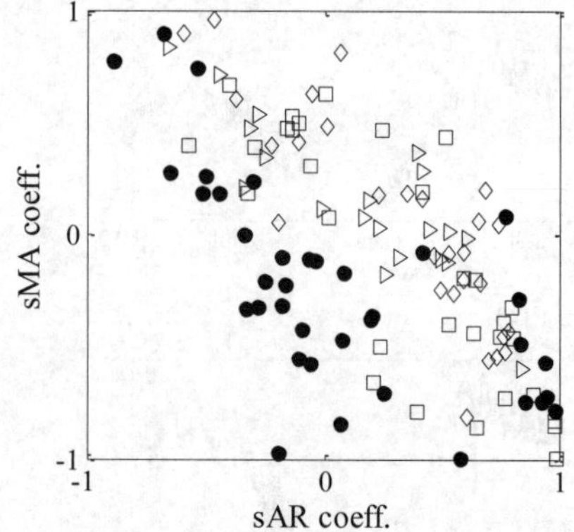

(a) Sensor location: P1-3

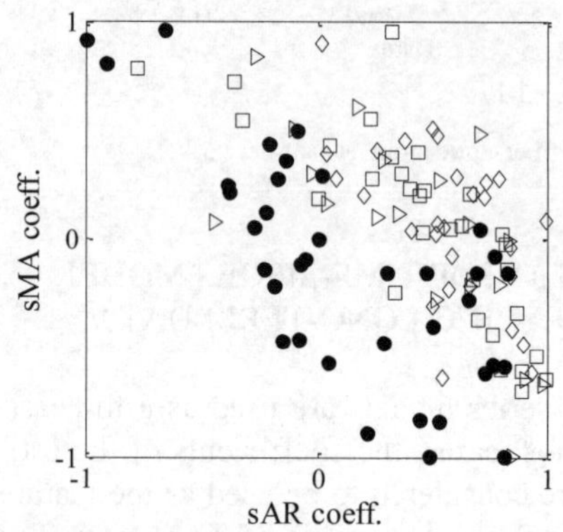

(b) Sensor location: P3-1

Figure 4. Seasonal AR and MA coefficients plots (●: 2011/12/19-2012/04/08 with snow accumulation, ◊: 2012/04/09-2012/06/30, □: 2012/07/01-2012/09/30, ▷: 2012/10/01-2012/11/30)

4 FEATURE COMPARISON BY THE MAHALANOBIS DISTANCE

The features from the long-term strain data in this study were multivariate; i.e., seasonal AR and MA coefficients from multi-sensing points. Therefore, the comparison and diagnosis strategy to provide information about structural behaviors/conditions. Here, we tried to apply the Mahalanobis distance comparison for this purpose.

4.1 *Basic Description about the Mahalanobis Distance Comparison*

The Mahalanobis distance is one of distance measures for multivariate statistics defined as:

$$D_k = \left(\mathbf{y}_k - \boldsymbol{\mu}_y\right)^T \mathbf{S}^{-1} \left(\mathbf{y}_k - \boldsymbol{\mu}_y\right). \tag{3}$$

D_k is the distance of each $\mathbf{y}_k$, which is the multi-dimensional feature vector, form their mean $\boldsymbol{\mu}_y$ normalized by the covariance matrix $\mathbf{S}$. For the feature comparison, the distribution of Mahalanobis distance of feature vectors from the basic condition (i=1-N), with the subscription B, is firstly derived using the mean $\boldsymbol{\mu}_B$ and the covariance matrix $\mathbf{S}_B$.

$$D_i^B = \left(\mathbf{y}_i^B - \boldsymbol{\mu}_B\right)^T \mathbf{S}_B^{-1} \left(\mathbf{y}_i^B - \boldsymbol{\mu}_B\right). \tag{4}$$

The distance of compared feature vector $\mathbf{y}_j^C$ (j=1-M) are then calculated using the mean $\boldsymbol{\mu}_B$ and the covariance matrix $\mathbf{S}_B$.

$$D_j^C = \left(\mathbf{y}_j^C - \boldsymbol{\mu}_B\right)^T \mathbf{S}_B^{-1} \left(\mathbf{y}_j^C - \boldsymbol{\mu}_B\right). \tag{5}$$

The multivariate feature vector can then be compared under one-dimensional measure.

4.2 *Comparison of Estimated Model Coefficients*

In the case of this study, the feature vector **y** was configured by the estimated sAR and sMA coefficients of seasonal ARIMA model at six sensor locations; P1-1/2/3/4 and P3-1/4; therefore, the length of the vector became twelve. Two base data sets were considered here; the vectors obtained from the data acquired in the seasons *with* and *without* snow accumulation in 2011-12. The compared data sets were the vectors calculated from the data in 2012-13.

Figure 5 shows the plots of Mahalanobis distance for investigating the possibilities of structural behavior changes through Dec. 2011- Sep. 2013. Notice that each Mahalanobis distance is from the model coefficients for the 3-days time-series, and some estimations, in which the residual showed autocorrelation, were removed. In addition, there were some missing parts in the strain data, which can be recognized in Fig.1; therefore, the 3-days period including those missing parts were also skipped in the seasonal ARIMA model estimation process. In the comparison of behaviors in the season with accumulated snow, Fig.5 (a), the mean and variance values of the distribution in 2012-13 get higher than that in 2011-12. On the other hand, the comparison in the case of "without accumulated snow" in Fig.5 (b) do not show significant changes between the two distributions. To discuss the structural changes by those plots, the comparison strategy here; setting two cases; *with* and *without* snow accumulation, must be more investigated including the study to clarify the effect of accumulated snow to the bridge behaviors in detail. Moreover, we have to consider what kind of information is helpful and applicable for the bridge owners before applying those statistical pattern recognition or machine learning techniques to the long-term data.

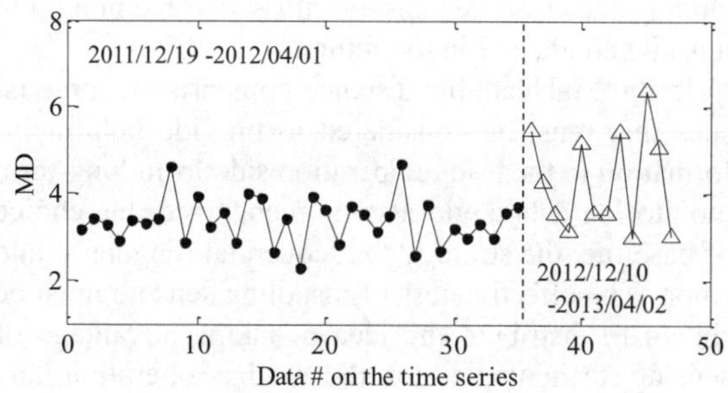

(a) Case of *with* snow accumulation

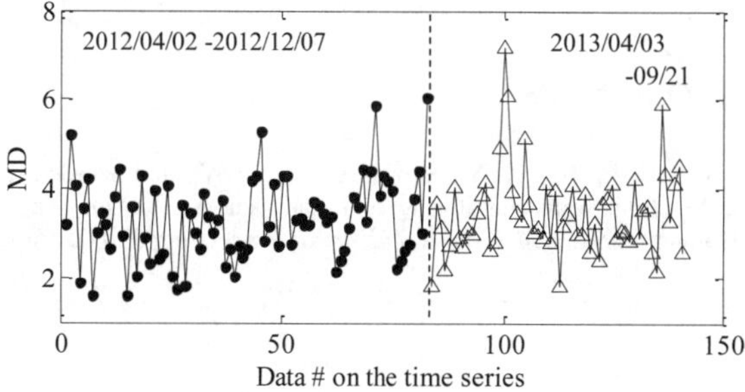

(b) Case of *without* snow accumulation

Figure 5. Comparison of extracted seasonal AR and MA vectors by Mahalanobis distances

5 CONCLUSION

In this paper, the application of seasonal ARIMA model estimation to the strain data acquired in the long-term monitoring were verified. Some conclusions are summarized below:

- The structural condition assessment strategy by applying the seasonal ARIMA model to the time-series strain data was presented, and the estimated coefficients plots showed the regressions.

- In the seasonal AR and seasonal MA coefficients plots, different regressions were recognized between the plots from the data acquired in the days with and without accumulated snow on the bridge.

- The application of Mahalanobis distance comparison was verified using the feature vector of estimated seasonal ARIMA coefficients. The multipoint strain behavior could be compared by the one-dimensional measure with considering strong trends in the data including the daily one. More detail consideration related to actual structural behaviors including those in the season with snow accumulation must be conducted in the future.

In the Mahalanobis distance comparison, some issues that must be considered to provide helpful information to the bridge operation side from long-term monitoring data could also be recognized; the choice of baseline, the strategy for sequential diagnosis, and so on. The effectiveness of installing sensors must be shown to distribute the idea of taking advantage of sensing technologies in the bridge operation and maintenance.

REFERENCES

Cross, E. J., Koo, K. Y., Brownjohn, J. M. W., and Worden, K. 2013. Long-term monitoring and data analysis of the Tamar Bridge, *Mechanical Systems and Signal Processing*, **35** (1): 16-34.

Nishio M., Xu J., Mizutani J. 2015. Structural Condition Assessment in a Prestressed Concrete Bridge based on Analysis of Environmental Effects on Long-term Strain Monitoring Data, *Proc. 10th International Workshop on Structural Health Monitoring*:

Omenzetter, P. and Brownjohn, J. M. W. 2005. A seasonal ARIMAX time series for strain-temperature relationship in an instrumented bridge, *Proc. 5th International Workshop on Structural Health Monitoring*:

Omenzetter, P. and Brownjohn, J. M. W. 2006. Application of time series analysis for bridge monitoring, *Smart Materials and Structures*, **15** (1): 129.

Soyoz, S. and Feng, M. Q. 2009. Long - Term Monitoring and Identification of Bridge Structural Parameters, *Computer-Aided Civil and Infrastructure Engineering*, **24** (2): 82-92.

Worden K., Manson G., Fieller N.R.J. 2000. Damage detection using outlier analysis, *Journal of Sound and Vibration*, 229 (3), 647-667.

Zhou, Y., Sun, L. and Peng, Z. 2015. Mechanisms of thermally induced deflection of a long-span cable-stayed bridge, *Smart Structures and Systems*, **15** (3): 505-522.

Proceedings of the International Conference on Smart Infrastructure and Construction
ISBN 978-0-7277-6127-9

© The authors and ICE Publishing: All rights reserved, 2016
doi:10.1680/tfitsi.61279.057

Distributed fibre-optic monitoring of tension piles under a basement excavation at the V&A Museum in London

L. Pelecanos*[1], K. Soga[1], S. Hardy[2], A. Blair[2], K. Carter[2] and D. Patel[2]

[1] *Centre for Smart Infrastructure and Construction, University of Cambridge*
[2] *ARUP, London*
* *Corresponding Author*

ABSTRACT A part of the Victoria & Albert Museum in London is currently being redeveloped and as part of this redevelopment an underground space of 2 levels was planned to be created for exhibitions and general storage. The deep excavation is supported by a base slab and a number of foundation "tension piles" which aim to carry the upward heave. Innovative distributed fibre optic (FO) sensors were installed within two tension piles prior to the excavation and regular readings were undertaken before and during basement excavation. This allowed the assessment of the induced stresses and strains within the instrumented piles due to the deep excavation. The monitoring data showed that the FO cables were able to pick up some tensile axial strains and forces due to concrete curing and basement excavation. During concrete curing the temperature within the pile drops whereas the axial strains increase, becoming more tensile. Finally, after the excavation the observed axial strain is tensile at the pile mid-depth and compressive at the pile top which is connected to the base slab.

1 INTRODUCTION

A section of the Victoria & Albert (V&A) Museum is currently being redeveloped. As part of this redevelopment, an underground space of 2 levels is planned to be created. The top level (from surface to a depth of 8m) is planned to be used for exhibitions, whereas the bottom level (depth: 8-12m) for storage.

The deep excavation is supported by a base slab and a number of foundation "tension piles" which aim to carry the upward heave load due to the deep excavation. The technical interest here was to examine what are the stresses and strains developed within the base slab and in the tension piles due to this excavation heave.

For this reason, distributed Fibre Optic (FO) sensors were installed in two tension piles, P6 and P7, (Figure 1) to monitor strains and potentially evaluate the effect of the excavation heave on the piles and the slab.

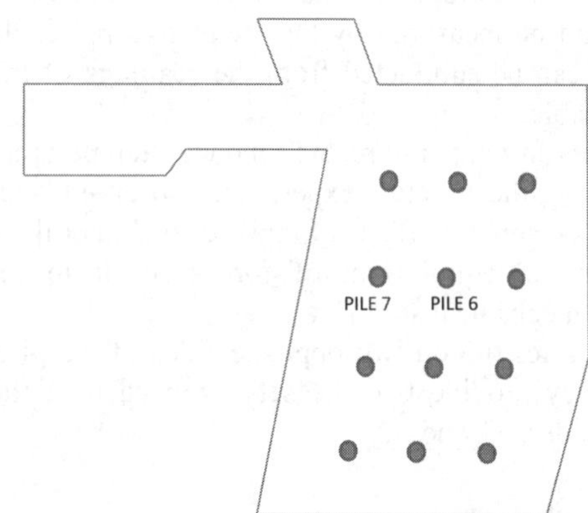

Figure 1. Location of piles monitored with FO cables.

2 FIBRE OPTIC MONITORING

2.1 Background

Distributed FO sensing techniques have been used widely over the last 10 years to monitor various types of civil infrastructure, such as tunnels (Mohamad et al., 2010, 2012; Cheung et al., 2010), piles (Klar et al., 2007), retaining walls (Mohamad et al., 2011; Schwamb et al., 2014; Schwamb & Soga, 2015) and slopes (Amatya et al., 2008). The principle of FO sensing is based on the strain-dependent change of light frequency within a bare optical fibre (Soga, 2014). A detailed explanation of the background theory and examples of recent applications may be found by Soga et al. (2015).

2.2 Monitoring scheme

The distributed fibre-optic sensing approach was adopted, using the Brillouin Optical Time-Domain Reflectometry (BOTDR) technique. Two types of FO cables were used: (a) a Fujikura reinforced “strain” cable and (b) an Excel loose tube “temperature” cable. The former (“strain”) cable records change in the Brillouin frequency, ΔVs, due to both mechanical load and temperature changes, whereas the latter (“temperature”) cable records frequency change, ΔVt, only due to temperature changes. Therefore, by using both cables together, the effects of temperature change can be measured by the “temperature” cable and then can be subtracted from the readings of the “strain” cable.

Changes in temperature, ΔT, induce “temperature strain”, ε_{temp}; the structure experiences observed “real strain”, ε_{real}, and the effects of applied load or soil reaction due to equilibrium of forces result in restrained “mechanical strain”, ε_{mech}.

Both cables run on two opposite sides of the pile, so that they provide two datasets from each of the two pile sides, S1 and S2.

2.3 Reading sets

Six sets of readings (R1-R6) were taken so far, as shown in **Error! Reference source not found.**. The first set of readings was taken immediately after the installation of the FO sensors when the concrete in the piles is expected to still be fresh. The second set was taken just before the excavation, and the concrete is therefore assumed to be cured and the last set of readings was taken after the first excavation, at a depth of about 8m. However, due to damage of the FO cables in P6, readings were taken only from P7 on that day. Three more readings were subsequently taken when the excavation was fully completed, at about 12m. For these last three readings, data for P6 is missing from one reading.

Table 1. FO reading sets.

Reading	Date	Comments
R1	22/10/14	After sensor installation
R2	23/01/15	Before excavation – cured concrete
R3	17/06/15	1st excavation – depth =8m
R4	30/07/15	2nd excavation – depth=12m
R5	11/08/15	2nd excavation – long-term 1st reading
R6	10/09/15	2nd excavation – long-term 2nd reading

3 MONITORING RESULTS

3.1 Pile response during concrete during

Figures 2 and 3 show the difference in the axial forces, F_a, from R1 and R2 for both opposite sides for the two piles. It is shown that in general positive values of F_a (tensile) are observed after concrete curing for both piles and their maximum value seems to reach around 4MN.

Besides, Figures 4 and 5 show the corresponding calculated vertical displacements, u_z, in the piles. The latter were obtained from integrating the real strain profiles with the depth and assuming zero displacement at the bottom of the pile. It is shown that minor displacements occur in both piles due to concrete curing which seems to be around 1-1.5mm for both piles. The values seem to be negative, i.e. moving downwards (remember, assuming a stationary point at the bottom of the pile), as a result of negative real strains from negative temperature difference during concrete curing.

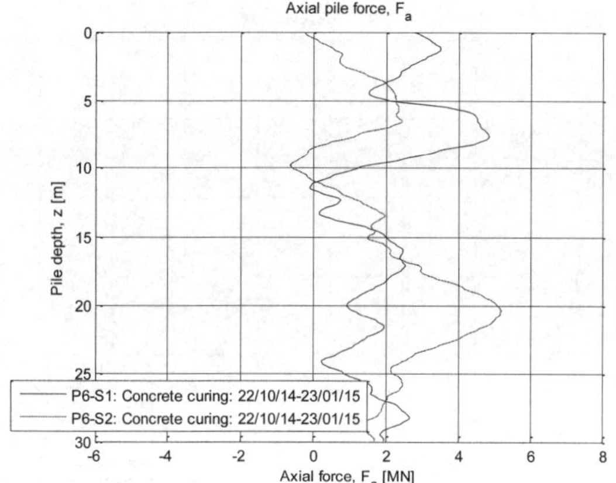

Figure 2. Axial force due to concrete curing in P6.

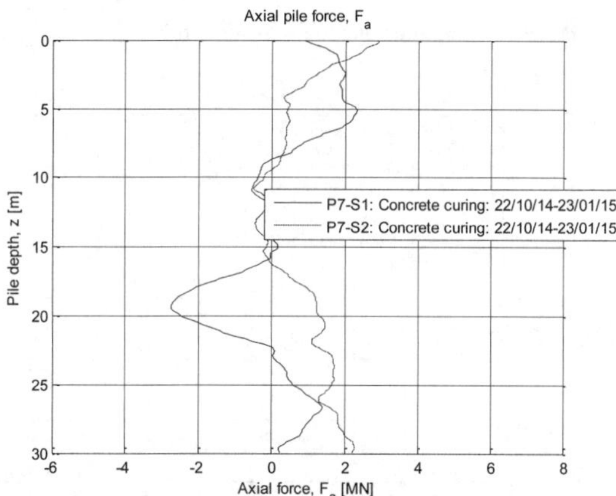

Figure 3. Axial force due to concrete curing in P7.

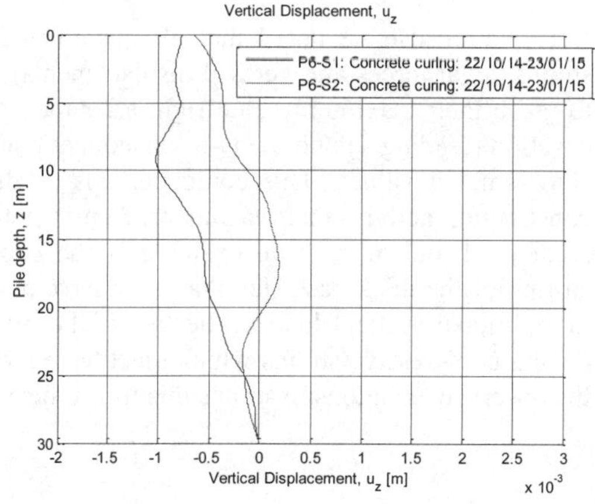

Figure 4. Vertical displacements due to concrete curing in P6.

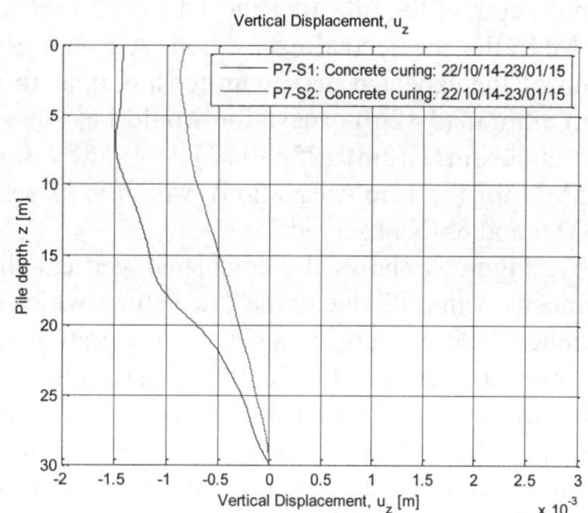

Figure 5. Vertical displacements due to concrete curing in P7.

3.2 Pile response during basement excavation

Here, representative monitoring results are shown only for P7 for brevity.

Figure 6 shows the axial mechanical strain, $\Delta\varepsilon$, profiles for P7 during basement excavation. Positive $\Delta\varepsilon$ (i.e. tensile) of about 100~150$\mu\varepsilon$ is observed after the 1st excavation for pile mid-depth. Negative $\Delta\varepsilon$ (compressive) of about 30~50$\mu\varepsilon$ is observed at the top of the pile, which could probably be due to the interaction (reaction) with the base slab. It is suggested that if soil heave is dragging the pile shaft upwards, then in a free-top pile, strain at the top should be close to zero. In contrast, a restraint-top pile (as in this case) should experience compression, i.e. negative strain.

Similarly, negative strain of about 50$\mu\varepsilon$ is observed at the bottom of the pile too. This could be due to the occurrence of some positive values of temperature strain. The latter could potentially be from the reaction of the soil due to seasonal (January – June) thermal pile expansion. However, in general the effects of temperature change during the excavation stages were found to be minimal as compared to that during concrete curing. After the 2nd excavation, the axial pile strains increase, with around 250$\mu\varepsilon$ at the pile mid-depth, -50$\mu\varepsilon$ at the pile base and around -150$\mu\varepsilon$ at the pile head.

Figure 7 shows the axial force profiles for the two piles during basement excavation. The maximum tensile axial force seems to be around 3MN around the mid-depth of the pile after the 1st excavation and 6-6.5MN after the second excavation. A rough estimation of the expected maximum tensile axial force due to a 8m (and 12m) excavation would be: $F = \gamma z A$ = (20 kN/m3)x(8m)x(25~30m2) = 4~5MN (and 6~7.5MN for the 12m excavation), which is close to the 3MN and 6MN observed.

Finally, Figure 8 shows the calculated vertical displacements within P7 due to the excavation, which as mentioned before, were obtained by integrating the real strain profiles with the depth and assuming zero displacement at the bottom of the pile. Considering P7 first, as shown from the latter figure, P7 seems to generally experience upward displacements of around 1.2mm and 2~2.5mm as a result of the 1st and 2nd excavations respectively. Negligible values of displacement are observed at the bottom of the pile, positive (upwards) further up the shaft and almost constant at the top 7m, perhaps due to the pile-top slab restraint.

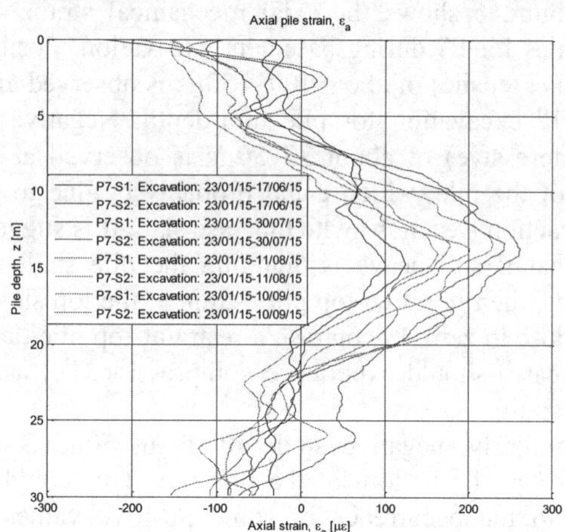

Figure 6. Axial mechanical strain due to excavation in P7.

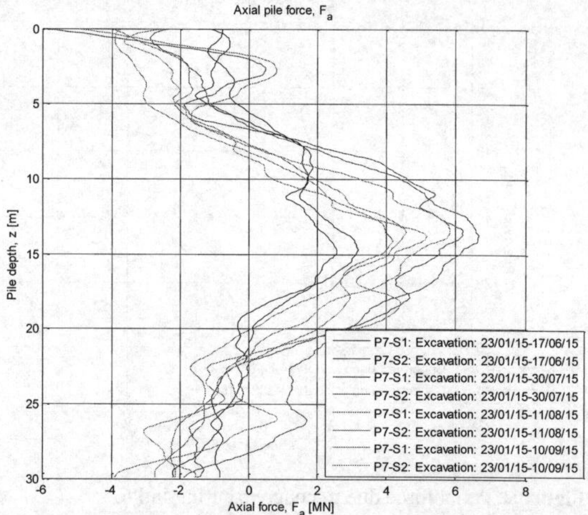

Figure 7. Axial force due to excavation in P7.

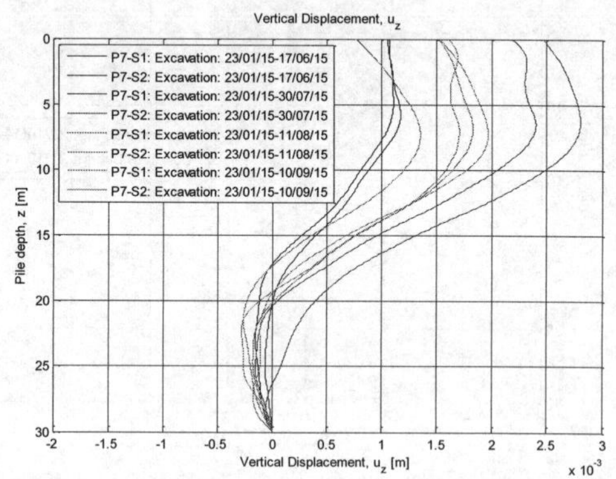

Figure 8. Vertical displacements due to excavation in P7.

It should also be noted that all quantities (axial strains, axial forces and vertical displacements) seem to reach their maximum value right after the 2nd excavation (reading 4) and the two subsequent readings show smaller values. This could perhaps be due to construction activity that was undertaken in between readings 4 and 5; such an example is the concrete casting of the thick base slab that must have resulted in additional vertical load at the top of the pile, inducing compressive strains which must have reduced the observed axial tensile strains due to soil heave.

4 CONCLUSIONS

This paper presents a case study of distributed fibre-optic monitoring of tension piles under a base slab for a basement excavation at the V&A museum in London.The following comments may be drawn from the current FO monitoring data:

- It was found that during concrete curing: the temperature (within the pile) drops and the axial strains and forces increase (i.e. becoming more tensile).
- Moreover, after the basement excavation: the temperature increases (which could be seasonal) and the observed axial strains (and therefore axial forces) are tensile at the pile mid-depth and compressive at the pile top which is connected to the base slab.
- Minor vertical displacements were calculated in the piles. During concrete curing, downward vertical displacements of about 1-1.5mm were observed in both piles. In contrast, after the 1st and 2nd excavation, minor upward displacements of about 1mm and 2.5mm were observed in P7, possibly as a result of base heave.
- Finally, the response of P7 immediately after the 2nd excavation showed the largest values of strains, forces and displacements, whereas the two subsequent readings showed smaller values. This could be due to construction activity (such as bottom base slab concrete casting) that took place in between these readings and that must have contributed to additional vertical load, inducing compressive strains resulting in a reduction of the tensile strains due to excavation heave.

ACKNOWLEDGEMENT

The Authors would like to acknowledge (a) the assistance of their colleagues at CSIC and ARUP and engineers from Wates, and (b) the financial support of EPSRC and Innovate UK.

REFERENCES

Amatya, B.L., Soga, K., Bennett, P.J., Uchimura, T., Ball, P., and Lung, R. 2008 Installation of optical fibre strain sensors on soil nails used for stabilising steep highway cut slope. In: Proceedings of 1st ISSMGE International Conference on Transportation Geotechnics, pp. 276-282.

Cheung, L. L., Soga, K., Bennett, P. J., Kobayashi, Y., Amatya, B., & Wright, P. (2010). Optical fibre measurement for tunnel lining monitoring. Proceedings of the ICE - Geotechnical Engineering, 163(3), 119-130.

Klar, A., Bennett, P., Soga, K., Mair, R. J., Tester, P., Fernie, R., St John, H. D., Torp-Peterson, G. (2006). Distributed strain measurement for pile foundations. Proceedings of the ICE - Geotechnical Engineering, 159(3), 135-144.

Mohamad, H., Bennett, P., Soga, K., Mair, R., & Bowers, K. (2010). Behaviour of an old masonry tunnel due to tunnelling-induced ground settlement. Geotechnique, 60(12), 927-938.

Mohamad, H., Soga, K., Bennett, P., Mair, R. J., & Lim, C. S. (2012). Monitoring Twin Tunnel Interaction Using Distributed Optical Fiber Strain Measurements. Journal of Geotechnical and Geoenvironmental Engineering, 138(8), 957-967.

Mohamad, H., Soga, K., Pellew, A., & Bennett, P. J. (2011). Performance Monitoring of a Secant-Piled Wall Using Distributed Fiber Optic Strain Sensing. Journal of Geotechnical and Geoenvironmental Engineering, ASCE, 137(12), 1236-1243.

Schwamb, T., & Soga, K. (2015). Numerical modelling of a deep circular excavation at Abbey Mills in London. Geotechnique, 65(7), 604-619.

Schwamb, T., Soga, K., Mair, R., Elshafie, M., R., S., Boquet, C., & Greenwood, J. (2014). Fibre optic monitoring of a deep circular excavation. Proceedings of the ICE - Geotechnical Engineering, 167(2), 144-154.

Soga, K. (2014). XII Croce Lecture: Understanding the real performance of geotechnical structures using an innovative fibre optic distributed strain measurement technology. Rivista Italiana di Geotechnica, 4, 7-48.

Soga, K., Kwan, V., Pelecanos, L., Rui, Y., Schwamb, T., Seo, H., & Wilcock, M. (2015). The role of distributed sensing in understanding the engineering performance of geotechnical structures. Proceedings of the XVI European Conference on Soil Mechanics and Geotechnical Engineering. Edinburgh.

Proceedings of the International Conference on Smart Infrastructure and Construction
ISBN 978-0-7277-6127-9

© The authors and ICE Publishing: All rights reserved, 2016
doi:10.1680/tfitsi.61279.063

Fibre-optic distributed acoustic and vibration sensing for monitoring of industrial plants and installations

P. Rohwetter*[1], R. Eisermann[1] , S. Großwig[2] and K. Krebber[1]

[1] *Federal Institute for Materials Research and Testing (BAM), Berlin, Germany*
[2] *GESO GmbH & Co. Projekt KG, Jena, Germany*
* *Corresponding Author*

ABSTRACT We propose the application of Distributed Acoustic Sensing (DAS) based on Rayleigh Coherent Optical Time-Domain Reflectometry (C-OTDR) to unconventional sensing tasks in industrial condition monitoring. As examples we present results on the way to fibre-optic remote sensing of dielectric damage processes in high voltage cable joints as well as to condition monitoring of passive rollers in large industrial belt conveyor systems.

1 INTRODUCTION

Distributed acoustic sensing (DAS) and Distributed Vibration Sensing (DVS) are collective terms for a class of fibre-optic sensing methods that, in spite of their relatively recent appearance, are already well established in various application fields. Of these, the most prominent ones are oil and gas exploration and production, as well as pipeline and perimeter security. Most implementations of DAS/DVS are based on the effect of elastic Rayleigh-backscattering in single-mode silica optical fibres (Shatalin et al. 1998) and enable the sensitive detection, localization and characterization of sound and vibration with ranges up to several tens of kilometres, and with spatial resolutions on the scale of several meters down to one meter. This opens the door to a number of potential applications for DAS in infrastructural monitoring.

Currently at BAM, fundamental research on several innovative uses of DAS/DVS of civil and industrial infrastructure monitoring is underway, including the monitoring of ageing bridges and of industrial pipeline systems. In this paper the focus is on two different types of infrastructure being investigated. These are critical insulations in high voltage cable systems and heavy duty belt conveyor systems, each presenting its own specific challenges to the application of DAS/DVS.

Power transmission in underground high voltage cables is most important in densely populated urban areas with lack of space for overhead lines. Insulations in cable joints and terminations are exposed to high local dielectric stress, making them especially vulnerable to degradative processes such as partial discharge (PD). Sensitive and specific monitoring of extended cable systems using conventional electrical PD detection techniques has been demonstrated (Gieselbrecht et al. 2012), however involving elaborate instrumentation. We propose DAS as a tool for quasi-distributed detection of critical PD by their acoustic emissions and present first results from laboratory experiments on real-scale high voltage joints, demonstrating the remote detection of PD through a single optical fibre with one-sided access (Rohwetter et al. 2015).

The other application considered here is the use of DAS/DVS for the condition monitoring of industrial belt conveyor systems. Such systems are often many kilometres long and are required to operate uninterruptedly for long periods of time. A large number of passive rollers, each having an individual expected lifetime well below the service life of the entire conveyor system, support the conveyor belt. Thus, ageing rollers have to be exchanged regularly. A roller failing unnoticedly may chafe, block, overheat, and eventually cause extensive damage. We demonstrate in a laboratory test that DAS may offer an elegant way of remote condition monitoring of individual rollers as a means to trigger timely replacement of rollers approaching failure.

Before turning to the applications, the sensing principle of C-OTDR-DAS is introduced.

2 SENSING PRINCIPLE

Optical single-mode silica fibres are drawn from a silica glass melt in a continuous process. During the solidification of the glass melt, nanometre-sized inhomogeneities of the refractive index are frozen into the material. Induced by the presence of these inhomogeneities, light propagating in the fibre experiences Rayleigh scattering, causing scattered radiation to propagate back towards the launching end of the sensing fibre. In coherent optical time-domain reflectometry (C-OTDR) DAS, short pulses of coherent laser light are used for probing the fibre. The coherence of the light causes a well-defined phase relationship between the individual backscattered partial fields. At any instant t, a superposition of a discrete number n_{scat} of such partial fields arrives at the detector located at the launching end of the fibre. The detected optical irradiance is proportional to the absolute square of the total optical field at the position of the detector,

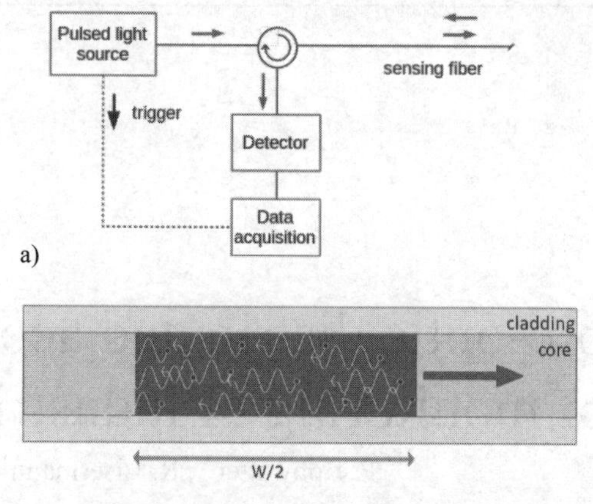

Figure 1. Sensing principle of Rayleigh C-OTDR-DAS: **a)** Simplified sketch of the optical setup.; **b)** Sketch of the scattering geometry. Partial backscattered fields (white wiggly lines) from discrete scattering centres (dots) covered by half the pulse length instantaneously arrive at the detector at the launching end of the fibre and interfere with each other.

$$I_{\text{det}} \propto E^*_{\text{det}} E_{\text{det}} = \sum_{j=1}^{n_{\text{scat}}} |a_j|^2 + 2 \sum_{j=2}^{n_{\text{scat}}} \sum_{k=1}^{j-1} a_j^* a_k \, e^{i(\phi_k - \phi_j)}, \quad (1)$$

where the collection of the contributing n_{scat} discrete scatterers changes as the probing pulse propagates down the fibre. In (Equation 1), a_i denote the relative amplitudes of the backscattered partial waves, and ϕ_i are the phases of the respective backscattered fields at the detector. These are related to the scatterer positions according to $\phi_j = 2 \times \frac{2\pi n_{\text{eff}}}{\lambda} x_j$, where λ is the vacuum wavelength of the laser light, c is vacuum light speed, and n_{eff} is the effective refractive index of the single mode fibre.

In (Figure 1a) a simplified sketch of a typical Rayleigh C-OTDR DAS setup is shown. A pulse of coherent laser light is launched into a single-mode optical fibre, acting as the sensing fibre, via a fibre-optic circulator. Rayleigh-backscattered light is detected by a fast photodetector, and the time-variant irradiance on the detector is sampled with high temporal resolution. (Figure 1b) is a cartoon of the scattered fields from a finite number of scatterers whose scattered fields interfere at the position of the detector. As the probing pulse propagates in the sensing fibre, the set of scatterers that contribute to the interference, according to (Equation 1), change. Thus, a Rayleigh backscatter-induced speckle interferogram unrolls on a "fast" timescale and is sampled from the photodetector output current with a rate of typically more than 100 MS/s.

In between the probing laser shots, acoustic or vibrational mechanical action upon the sensing fibre changes the phases ϕ_i of backscattered fields by modulating the geometrical distances between scatterers (that is, the coordinates x_i) as well as the effective refractive index n_{eff}, via the stress-optic effect (Rao 1997). The magnitude of the involved dynamic strains can be as small as a few times 10^{-9}. These cause a dynamic change on a shot-to-shot basis of the recorded time-dependent backscattered intensity. The resulting change of the speckle interferogram (sampled on the "fast" timescale), is sampled on the "slow" timescale defined by the probing laser firing period.

Apart from the direct detection Rayleigh C-OTDR scheme described thus far there exists a class of coherent Rayleigh reflectometry schemes that are known as "phase sensitive" or ϕ-OTDR (Posey et al. 2000). Those schemes are able to linearly resolve local changes of sensing fibre strain state, however at the cost of more complex instrumentation.

In the well-established field of security applications of Rayleigh DAS, for example intrusion detection or pipeline monitoring, the simpler direct detection C-OTDR scheme sketched in (Figure 1a) is common, due to its robustness and its lower cost. In direct detection C-OTDR, the phase information that is accessible to ϕ-OTDR is heavily distorted by the complexity, time-variability and nonlinearity of the dependence of $I_{\text{det}}(x, t_{\text{slow}})$ on local fibre strain $\epsilon(x, t_{\text{slow}})$. Nevertheless, in said applications much useful information can be recovered from analysing the recorded signal $I_{\text{det}}(x, t_{\text{slow}})$ in the frequency domain.

Our work introduced in the following two sections also relies on hardware that implements the simpler and more common direct detection C-OTDR scheme. In the case of acoustic partial discharge detection, specialized fibre-optic transducers and application-specific time-domain data analysis compensate for part of the inherent deficiencies of direct detection C-OTDR-DAS compared to ϕ-OTDR. In the second case, targeting fibre optic monitoring of passive conveyor rollers, our approach relies on specialized data analysis only, as the application requires the simplest possible procedures of sensing fibre application and mechanical robustness.

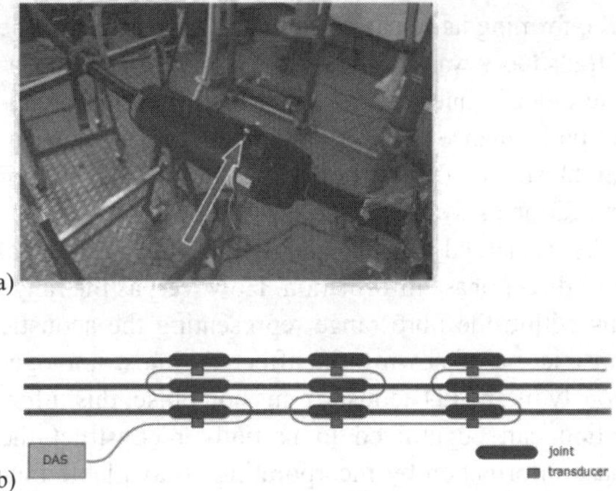

Figure 2. a) Installed and instrumented high voltage cable joint in a laboratory setup. fibre-optic acoustic transducer indicated; b) envisioned arrangement for quasi-distributed acoustic PD monitoring of cable joints.

3 QUASI-DISTRIBUTED PD DETECTION

In (Figure 2a) a laboratory setup of an instrumented high voltage joint is shown, representing one unit of the envisioned cable system monitored using C-OTDR sketched in (Figure 2b). The joint had an artificially induced defect simulating a common assembly fault. The transducer attached to the high voltage insulation is optimized for the conversion of weak short-pulsed body-sound to near-homogeneous dynamic radial strain of a coil wound out of approximately 30 m long special sensing fibre (fibrecore Ltd.,

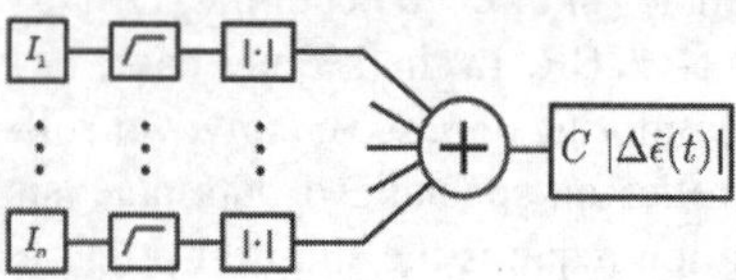

Figure 3. Flow diagram of Random Quadrature Demodulation

UK), forming its sensitive element. Whereas this type of transducer was originally developed for direct interferometric interrogation, in which it shows excellent performance (Rohwetter et al. 2014), its mechanical transduction properties are useful for C-OTDR interrogation as well.

As remarked earlier, the C-OTDR-DAS signal lacks direct phase information. However, as the range bins within the fibre range representing the acoustic transducer are known to be affected almost homogeneously by the PD acoustic emission pulse, this information can be utilized to partially reconstruct the phase information by incorporating all available binwise signals. The procedure for doing this was originally conceived by Pouet et al. (Pouet et al. 2006) for demodulating two-dimensional dynamic speckle patterns that arise in a Michelson interferometer with a diffuse reflector in one arm. The mathematical foundations of the application of the so-called Random Quadrature Demodulation to C-OTDR-DAS signals will be discussed in detail elsewhere (Rohwetter et al. 2016). Here we introduce it only briefly.

The procedure is easiest visualized in a flow diagram shown in (Figure 3). Signals from a consecutive set of range bins, covering the transducer section of the sensing fibre, correspond to different sets of scatterers entering (Equation 1). Each signal is high-pass filtered to isolate the quickly changing components from slow background variability caused by low frequency mechanical or thermal influences. Each high-frequency signal is then rectified before being summed to a single quantity. Analysis shows (Rohwetter et al. 2016) that the expected value of this quantity is proportional to the expected value of the modulus of the high frequency part of dynamic fibre strain:

$$\mathrm{RQD}(\mathrm{I}_1, I_2, \ldots, I_n) \propto |\Delta\tilde{\epsilon}(t)|\,, \qquad (3)$$

where $\mathrm{RQD}(\cdot)$ symbolizes the procedure described above and I_i are the single-bin C-OTDR signals. The loss of sign information is not important, for the primary sensing task is to reliably detect the presence of acoustic pulses. Nevertheless, the statistical variability of demodulated signal amplitude is a deficiency that has to be coped with.

PD in solid dielectrics under AC stress generally does not occur in singular, uncorrelated events, but rather appears in bursts with some degree of temporal correlation. This means that discharge events tend to recur during many AC cycles, often with a phase relationship relative to the AC voltage that is characteristic of the type of defect causing PD. This allows to construct AC-phase-resolved histograms such as shown in (Figure 4a), where the frequency of PD occurrence is indicated as a function of AC phase and discharge magnitude, and that often allow experts to assess the type and gravity of defect.

In (Figure 4b) a similar diagram is shown, whereby this time it is constructed from the magnitudes of acoustic pulses obtained from the amplitudes of spikes in the RQD-processed C-OTDR signal instead of discharge magnitudes. The remotely detected acoustic signature of the defect compares well with the corresponding histogram constructed from the electrical reference measurement shown in (Figure 4a), indicating an equivalent detection limit of about 1 nC discharge magnitude. In this way, C-OTDR-DAS may become a tool for the remote monitoring of insulations in high voltage assets.

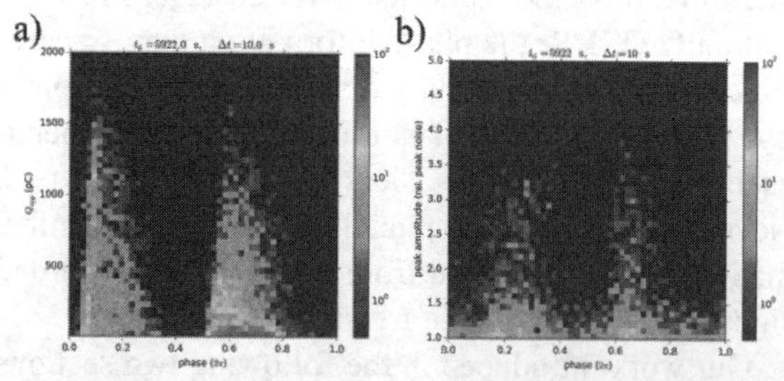

Figure 4. Comparison of AC-phase-resolved PD diagrams. a) electrical reference measurement; b) acoustic PD detection based on C-OTDR. PD above about 1 nC discharge magnitude are detected by the fibre-optic acoustic measurement channel.

4 ROLLER MONITORING

Roller monitoring poses a combination of challenges different from those encountered in quasi-distributed PD detection. Whereas PD acoustic emissions have a distinctly spiked shape in the time domain, sounds potentially carrying information can have quite diverse signature in roller monitoring.

We performed a series of experiments at the acoustic test stand of Rulmeca Germany GmbH, involving three different rollers of the same type of heavy duty passive roller for bulk conveyor use, selected out of a pool of ten. One roller was virtually factory-new, the two others were used and returned from the field after failing regular auditory inspection.

Five fibres out of 12 in a 300 m long standard telecomms optical fibre cable were spliced in series, totalling 1500 m of sensing fibre plus an additional 200 m of lead-in fibre. An approximately 100 m long section of cable was suspended on stands without immediate physical contact to the roller test stand and was guided parallel to the driving belt of the roller a total of six times in three vertically spaced levels. One pass of the cable was brought to close physical contact with the shaft of the respective roller. The various passes were intended to estimate the impact of air-acoustic coupling to the sensing fibre. In (Figure 5) roller "3" is shown mounted in the test stand. All of the setup was contained in an acoustically shielded chamber. The roller was driven by a belt circulating on two end drums outside the chamber, pressed against the roller from the top.

Experiments were performed at different typical belt velocities encountered in actual installations: 1 m/s, 3 m/s, 6 m/s, 10 m/s. C-OTDR probing pulse duration was varied in order to find the optimum balance between the partially complementary qualities of range resolution, sensitivity, and level of distortion. C-OTDR pulse repetition frequency was 44 kHz. Calibrated air-sound spectra were recorded for reference at a position approximately 1 m above the roller, using a sound level meter ("2270 Analyzer", Brüel & Kjær).

Exemplary time-dependent spectra obtained from C-OTDR time-domain data from the range bin closest to the point of contact between sensing cable and roller are shown in (Figure 6), comparing two of the three rollers at a belt velocity of 6 m/s and for a probing pulse duration of 30 ns. Spectra were numerically calculated from packets of 128 successive laser shots. Hence, the sampling rate of spectra was about 344 Hz,

Figure 5. Roller "3" in the test stand. Visible in the front: the segment of the sensing cable in close contact with the roller shaft.

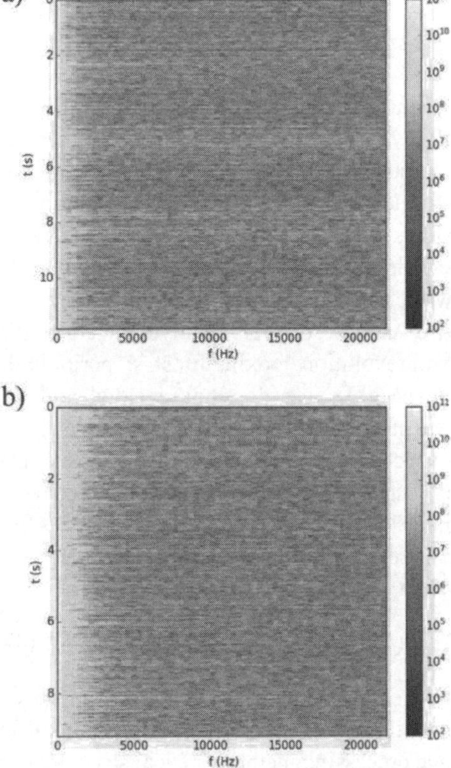

Figure 6. Exemplary time-dependent C-OTDR-DAS spectra recorded at a belt speed of 6 m/s, for three rollers with different health status: a) undamaged, b) moderate damage

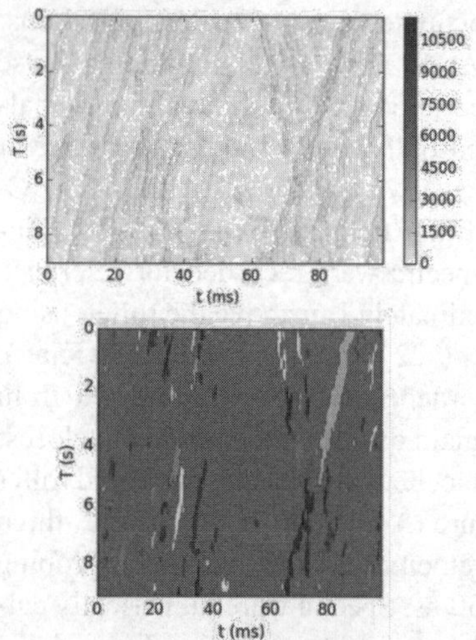

Figure 7. Signatures characteristic of a particular damaged roller obtained by a "wrapped time" plot and image processing. Top: raw plot, bottom: segmented plot with isolated and identified features

well in excess of the roller revolution frequency of 10 Hz.

It is obvious that the degree of damage reflects in the overall level of acoustic signal, whereas there appears to be no clear difference in spectral structure. The fibre-optically measured average acoustic levels of both rollers with no and moderate damage, respectively, even do not differ very strongly, although the damage was clearly audible to the human ear.

Likewise, in spite of their relatively well-defined frequency of revolution, looking for strict periodicities in the total spectral power does not reveal clear differences between the individual rollers. Instead, a combination of spectral and time-domain analysis seems to be a promising way to evaluate C-OTDR signals for roller monitoring. In (Figure 7) spectral intensity from a particular frequency band is colour-coded in a coordinate system

$$(t, T) = (t_\mathrm{m} \ \mathbf{mod}\ f_\mathrm{R}^{-1}, t_\mathrm{m}\ \mathbf{div}\ f_\mathrm{R}^{-1}) \ , \qquad (4)$$

with t_m the measurement time and f_R the frequency of revolution of the roller. The striped patterns, amplified using image processing methods, reveal structure contained in the signal relating to bearing fault that would not be revealed by pure frequency-domain analysis.

5 CONCLUSION

We demonstrated initial results on the application of fibre-optic C-OTDR-DAS to unconventional sensing tasks in industrial monitoring. Firstly, the remote detectability of partial discharge in high voltage cable joints was demonstrated. Next, the possibility to extract characteristic acoustic patterns of defective passive rollers was indicated by experimental results from an acoustic test stand. Further work on these subjects will include investigations related to sensing fibre application and data processing, as well as field trials.

ACKNOWLEDGEMENT

The authors thank Prof. R. Plath and A. Elben, from the department of High Voltage Engineering of Technical University of Berlin, Germany, and also U. Eisenhut from Rulmeca Germany GmbH, Aschersleben, Germany.

REFERENCES

Gieselbrecht, D. Kolturnowicz, W. Obralic, A. Ritz, T. Christensen, P. Schneider, B. Cohnen, K.H. 2012. Monitoring of 420 kV XLPE Cable System in Underground Tunnel, Proceedings of IEEE International Conference on Condition Monitoring and Diagnosis, Bali, Indonesia, 2012, 917 – 920

Posey, R. Johnson, G. A. Vohra, S. T. 2000. Rayleigh scattering based distributed sensing system for structural monitoring, Proceedings of the SPIE 4185, 678–681.

Pouet, B. Breugnot, S. Clémenceau, P. 2006, Robust Laser-Ultrasonic Interferometer Based on Random Quadrature Demodulation, AIP Conference Proceedings 820, 233–239.

Rao, Y. J. 1997. "In-fibre Bragg grating sensors," Measurement Science and Technology 8, 355–375.

Rohwetter, P. Lothongkam, C. Habel, W. Heidmann, G. Pepper, D 2014, Improved fibre optic acoustic sensors for partial discharge in elastomeric insulations, Proceedings of the SPIE 9157, 91571J.

Rohwetter, P. Eisermann, R. Krebber, K. 2015. Distributed acoustic sensing: Towards partial discharge monitoring, Proceedings of the SPIE 9634, 96341C-1.

Rohwetter, P. Eisermann, R. Krebber, K. 2016, Random Quadrature Demodulation for Direct Detection Single-Pulse Rayleigh C-OTDR, submitted to Journal of Lightwave Technology.

Shatalin, S.V. Treschikov V.N. Rogers, A.J. 1998. Interferometric optical time-domain reflectometry for distributed optical-fibre sensing, Applied Optics 37 5600–5604.

Proceedings of the International Conference on Smart Infrastructure and Construction
ISBN 978-0-7277-6127-9

© The authors and ICE Publishing: All rights reserved, 2016
doi:10.1680/tfitsi.61279.069

Optical fiber sensors for subsea and topside asset integrity monitoring applications

Victor Servette[1]* and Vincent Lamour[1]

[1]*Cementys, Houston, USA*
* *Corresponding Author*

ABSTRACT: Asset integrity monitoring is becoming a critical activity as Oil&Gas operators needs to secure their existing production assets and extend their service life. The last decade happened to be a turning point for sensing technologies as deeper subsea oil fields and intelligent oil wells required the extensive use of optical fiber sensors. As a result, optical fibers sensor technologies are now matured and off-the-shelf solutions for offshore and subsea asset integrity monitoring.

Fiber optics differs from other monitoring technologies by being totally passive: this means there is no need for subsea electronics or batteries power, making it an ideal choice for long term offshore and subsea asset integrity monitoring. Added to easy real-time communication, high multiplexing capabilities and distributed sensing, optical fiber brings new possibilities to existing traditional sensing systems.

Optical fibers are also extremely robust and, when deployed correctly, has a lifetime of more than 25 years, such as in the telecommunication industry.

Different optical fiber sensor technologies have been developed for Oil & Gas applications, creating a wide diversity of sensing capabilities, for both punctual (Fiber Bragg Gratings) and distributed sensing (Distributed Temperature Sensing, Distributed Strain Sensing, Distributed Acoustic Sensing, Distributed Pressure Sensing).

In this paper we present the latest optical fiber sensor monitoring solutions and applications for offshore and subsea asset integrity monitoring. The assets studied include Hulls and Topside Structures (Structural Health Monitoring), Risers (fatigue monitoring on Touch Down Zone and Hang-Off point), Mooring Chains and Turrets (load monitoring by load cells or load shackles), and subsea tiebacks and umbilicals (distributed monitoring of both temperature and strain for subsea leak detection and life of field monitoring).

1 INTRODUCTION

Optical fiber sensor is creating a revolution in the sensor market, by allowing new measurement capabilities. Its intrinsic properties make it very well adapted to the offshore and subsea environments:

- Passive sensors: the sensors do not need any electronics or batteries to send out measurements
- No inline power: well adapted for Class 1 Div 1 environments
- Good for static and dynamic measurements: 25 years without drift and up to 25 kHz for FBG
- Not affected by EM field: sensors can be implemented near high voltage or strong electromagnets
- Very high multiplexing possibilities: up to 50 000 sensors on a single fiber for distributed sensing
- Non-intrusive: a fiber is 125µm thick, only one fiber is needed
- Frequency measurements: high signal to noise ratio, insensitive to inline losses or de/reconnections

For comparison, a classic resistive strain gage may be sensitive to moisture ingress and EM fields, may needs to be recalibrated (offset drift) and typically requires 4 wires per gauge.

In this paper, we first present the two major technologies used for optical fiber sensor monitoring: Fiber Bragg Gratings "FBG" for discrete measurements and Optical Time Domain Reflectometry for distributed measurements. Then we present the different applications of optical fiber sensors in the Oil&Gas offshore and subsea industries.

2 TECHNOLOGIES

A wide range of optical fiber sensors were developed for different industries. In this paper, we focus on the two major technologies that has been used for the last decade in the Oil&Gas industry: Fiber Bragg Grating based sensors and Optical Time Domain Reflectometry based sensors.

2.1 *What is an optical fiber?*

In order to understand optical fiber sensors, one must first look at an optical fiber. An optical fiber is a flexible string made of extruded glass capable of transmitting light from one end to the other. It is extremely small (125μm, thinner than a hair) and totally passive: it does not transport energy other than light and does not need any power supply. The fiber core can be protected within a cable, making an optical fiber cable just as strong as a conventional electric cable.

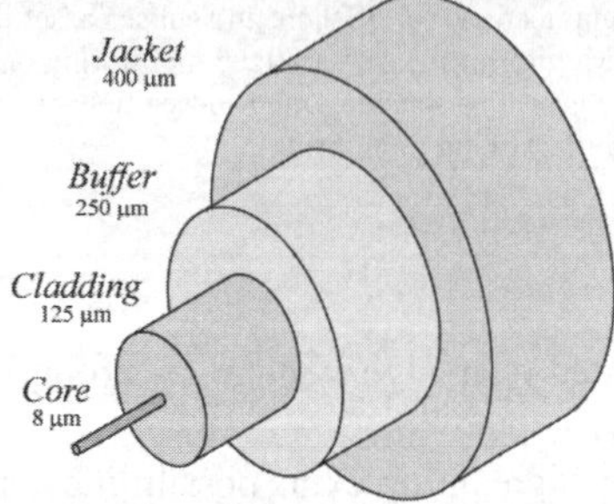

Figure 1: Cross section of an optical fiber cable

In order to confine the light, an optical fiber is made out of different layers of silica, with different optical indices (see Figure 1). The difference of index between the core and the cladding insures that the light will stay within the core of the fiber. The size of the core can change, making the fiber either single-mode (core of typically 9μm) or multi-mode (typically 50 or 62.5μm). The fiber is then protected by a buffer and inserted inside a cable, within different thickness and protection depending on the application.

2.2 *Fiber Bragg Grating sensors*

Fiber Bragg Grating (FBG) Sensor is the most common type of optical sensors, used for a wide variety of applications. A FBG is an optical element that can be implemented locally inside a fiber. It has the property of reflecting only one wavelength while transmitting all the others. The reflected wavelength depends on the properties of the fiber locally (strain and temperature).

It is composed of a small periodic variation of the optical index inside the core of the fiber. The frequency of the reflected wavelength is directly linked to the period of the grating (Figure 2), so one can easily see that a change in the period of the grating (due to either mechanical or thermal effects) will change the frequency of the reflected wavelength. They are similar to electrical strain gauges in their response, this is why they are often referred to as optical strain gauges.

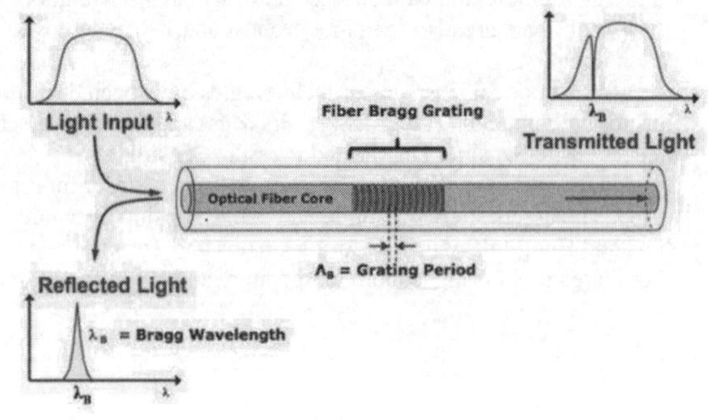

Figure 2: Fiber Bragg Gratings

2.3 *Distributed fiber optic sensors by optical time domain reflectometry measurements*

Distributed Sensing led a major breakthrough in the optical fiber sensing technologies: a single telecommunication fiber is able to provide measurements typically every foot for several miles! No punctual sensors are necessary, the fiber itself reacts to its environment and sends the information back to both ends, it acts both as sensor and as a communication device.

This technology is based on Optical Time Domain Reflectometry, meaning that it analyses the light backscattered from a fiber. A laser pushes a short single frequency light pulse through the fiber, that pulse is backscattered in the fiber due to different physical phenomena (Rayleigh, Raman, and Brillouin backscattering), this backscattered light is then analyzed. The location of the information is known thanks to the time of flight of the backscattered light (in the same way a radar computes distance).

The backscattered light is analyzed in frequency thanks to an optical spectrometer. The received spectrum contains interaction peaks due to different physical phenomena (see Figure 3):

- The Rayleigh peak has the same frequency as the laser pulse. Its intensity will vary with local vibrations and temperature
- The Brillouin peaks will see their frequency shifted due to both local temperature and strain changes
- The Raman peaks intensity will vary with temperature

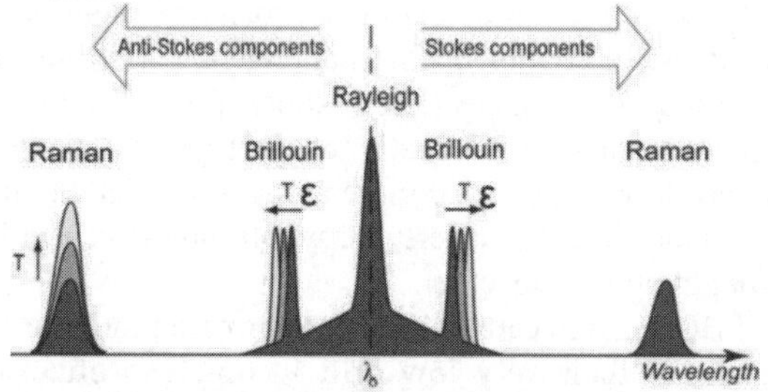

Figure 3: Typical spectrum of backscattered light

By analysing the different peaks, a single fiber can serve as a distributed sensor for temperature (DTS), strain (DSS), vibration (DVS) and acoustics (DAS).

3 APPLICATIONS

In this section, we will detail how optical fiber sensors can be used for Asset Integrity Monitoring in the Offshore industry. For the sake of clarity, we present the field applications from the subsea reservoir to the offshore platform.

3.1 Downhole applications

In the reservoir, several parameters are important to monitor in order to optimize the product flow and to enhance oil recovery. Temperature and pressure are the most common measured parameters. Today, the majority of pressure and temperature cells use embedded electronics and quartz technologies. They give good results but are their use may be limited by several factors: they cannot work permanently in high temperatures (>200°C or 390°F) and are difficult to multiplex on a single line.

For this reason, several companies developed pressure and temperature cells working on FBGs. These cells allow measurements in high temperatures with direct communication. The FBG is used as an optical strain gauge, attached to a sensitive element subject to pressure.

Distributed Fiber Optic Sensor measurements are also commonly performed in the downhole fields. Both temperature and acoustics can be monitored using an optical fiber cable deployed in the well by using DTS and DAS interrogators. The results were found to be extremely helpful during well completion and operation phases, this is why more than 10 000 wells (mostly onshore) have been equipped with distributed Fiber Optic Sensing measurements systems (Jacobs 2015).

The major advantage for using optical fiber sensors in subsea downhole applications relies on measuring remotely and permanently passive downhole sensors with very long existing umbilical lines (over 10 miles). The relatively high cost of optical wet mate connectors is still a limiting factor for Optical Fiber Sensor deployment on Subsea Wells.

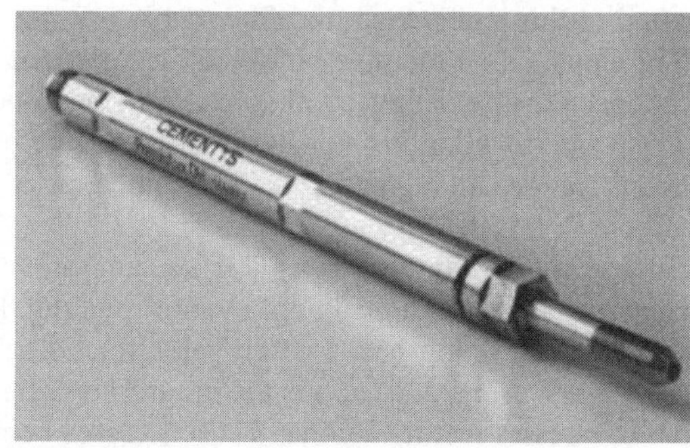

Figure 4: Example of an optical fiber downhole pressure/temperature gauge

3.2 *Subsea tree applications*

Subsea trees, BOP stacks are critical components, subjected to loads and vibrations from the riser system. A failure of critical components leads to safety and environmental issues. Structural Health Monitoring can be typically performed through 3 longitudinal strain sensors allowing vibration and stress analysis. If needed, a "wet mate" connector, allowing subsea connection and disconnection of the optical line, can be installed so the sensors can be interrogated by a ROV, for periodic survey readings.

Other Process sensors such as Pressure sensors, Flowmeters, thermometer or accelerometers can use the FBG technology to be integrated to existing optical telecommunication subsea networks without any power.

3.3 *Subsea tiebacks and flowlines*

Subsea tiebacks and flowlines are critical assets that are difficult to inspect. They need to be engineered as to make sure the pipe will not be obstructed by the transported product. The classic solution today is to double the line, to be able to send pigs for cleaning and be sure that one will be always available if the other gets clogged due, for example, to hydrate formation.

To remedy this issue, new pipes have been developed to increase the product temperature, either passively (insulated layers on the pipe) or actively (electrically trace-heated pipes). In both cases, it is important to measure the distribution of the temperature along the pipe: it will help production by reducing the risk of clogging and, in the case of heated lines, it will make sure the system is functioning correctly.

Distributed Temperature Sensing is very well adapted to this application: the fiber cable can easily be implemented on the pipe before its installation, to be connected to existing communication lines once laid. An interrogation box topside will then give out temperature data all along the line by connecting it to the fiber (DNV 2009).

Such a system can also be used to see and prevent the apparition of lateral buckling: by interrogating the fiber with a Brillouin based interrogator for Distributed Stress Sensing, one can see the progressive apparition of stresses inside the line. Those stresses being linked to the lateral buckling of the tieback, proactive measures can be put in place to make sure the design stress are never exceeded.

Figure 5: Example of a fiber optic cable inside a concrete weight coating on a subsea pipeline

3.4 *Corrosion*

Corrosion is omnipresent in the subsea and offshore environments. It is important to control it to ensure the durability and reliability of all the equipment installed. Today, several solutions exist to monitor the corrosion progress, for example cathodic protection monitoring or the use of coupons. The challenge is that those element require an inspection to know if they need to be replaced. This can be costly, especially if the coupon is hard to reach (for example a coupon installed on a production tubing, where the whole line needs to be brought up to inspect it).

FBG sensors can be used for corrosion monitoring thanks to their very low drift in time (no effect of moisture ingress or electromagnetic noise). If engineered and installed correctly, the instrumented coupon will relieve stress as its section is getting thinner due to corrosion, this stress relief can be monitored by an optical strain gauge. This type of measurements allows conditioned based maintenance approach, instead of a preventive maintenance strategy, and thus a lower maintenance cost as well as increasing the safety of the system.

3.5 *Risers*

Risers are equivalent to a field's highway: they link all the small oil routes at the seabed to structure topside. All the oil produced from a field will go through this complex structure (rigid or flexible risers), its health is thus critical. Today, most of the riser service lives are computed through cumulative fatigue models. Additionally, there are also sensors on the market to monitor them (mainly the hang of point), using either vibrating wire technology, strain gages or LVDT extensometers. They have been widely installed and give good results but have limitations: being electronic based sensors they need subsea power, meaning the batteries need to be changed by an ROV periodically for long term permanent applications. The technologies can also be limited in dynamic to a few Hertz, making the measurement of Vortex Induced Vibration difficult (DNV 2008).

Different companies are working on optical sensors for riser monitoring to avoid those limitation. An FBG based sensor installed at the hang-off point would not need any subsea power and electronics, and could give real time measurements up to 25kHz. As existing sensors, they could be clamped to an installed riser or fitted in the engineering design.

Even though they have a lot of advantages, solutions still need to be developed before an installation can take place: in order to function, the sensors need a continuous optical line to the platform. This can be done either by installing a dedicated umbilical, or by connecting the line to an existing optical line used for telecommunication. Both solutions have their pros and cons and need to be developed and properly installed.

Another critical part of the riser is the Touch Down Zone (TDZ): with tides, waves and other environmental factors, the location of the touch down point may vary up to several tens of meters. This can mean strong localized stresses and possible location of high fatigue damage accumulation. An FBG solution could be well adapted, thanks to its high multiplexing capabilities: an array of optical gauges can be deployed along the TDZ, allowing a complete stress analysis of the zone with low intrusion to this critical zone, all the sensors can be multiplexed on a single fiber. The array can be connected to the Subsea Umbilical Termination Assembly (SUTA) via a patch-cord and wet mate connectors once the riser is installed. There, an existing communication optical fiber umbilical line is used to link the sensors to the topside.

Finally, we can monitor the shape of the riser with time. Currents and temperature gradients can greatly affect the riser and create localized weak points. Distributed Strain Sensing (DSS) can be used to know the strains the all along the riser. These distributed strain values could be computed to displacement by integration and thus giving the shape. Here-again solutions need to be qualified before an installation can be done. The main issue would be the deployment of the optical fiber on an existing riser. Two solutions are thought-off, either by clamping a cable strait down the structure (piggy-back configuration), or twisting the cable regularly down the riser (helix configuration). The length of the riser could also be an issue: displacement being computed by integration of strain, the calculation error will increase rapidly without reference points.

3.6 Mooring chains and turrets

There are few sensors available on the market for mooring chain load or tilt monitoring applications. They can mainly be divided into three main categories: smart shackles, load cells or tilt sensors.

Smart shackles work by monitoring the stress inside a bended pin of the shackle to come back to load. They have to be installed on the last chain link and are responsible for the full load of the chain. Fiber Optic Shackles will give an absolute value of the load on the mooring chain without any moisture effect or electronic drift. This type of smart shackle has to be planned in advance before the chain installation.

The load cells differ from shackles by being installed on the chain stopper structure. A typical optical Load Cell will perform permanently in offshore environments without recalibration. Another possibility is to install optical strain gages on several links of the chain. Clamped to the link, they measure the variations in load with time.

Both smart shackles and load cells already exist on the market, but working with classic electrical gages. This means an important drift with time and a high dependency on environment (humidity, temperature…). These issues make the items unreliable after a few months of measurements. Transforming those items by including optical fiber gages would help this reliability issue. This is why several companies are working on the development and installation of such sensors.

A local measurement of tilt also gives valuable information on the load of the chain: most of the mooring systems use a catenary disposition, where the tilt of a known point is directly linked to the load of the line. This system can be either in 1 or 2 directions, to see the influences of currents. Those measurements can give an absolute value of the load, as well as information in load change with time. Once again, fiber optical technology has the advantage of not having any battery or electronics subsea: the sensor is totally passive and the measurements are taken real time thanks to a telecom optical cable connected to the platform.

In all applications, most sensors used in Offshore applications use classic electronic gages. This means that the measurements will deteriorate rapidly due to drift of the sensors and to their dependence on their environment (humidity, temperature, loses on the line) after only few months of operation.

This issue can be avoided by switching the electronic technology to optical FBG technology. The elements need only very little changes to be adapted to fiber optics.

3.7 *Structural health monitoring*

For long term safety reason, it is important to monitor the asset integrity of the topside structures. To do so, all the structures are linked to models assuring its good behavior. Sensors can be implemented on specific regions of the structure to get additional data and make sure the behavior is as it was predicted by the model.

But if the life of field of a platform needs to be extended, how can you be sure it is in good health? The model cannot be used anymore, and the lack of data concerning the structure makes it very difficult to ensure its integrity.

Optical sensors can be helpful in such a case: by installing several sensors at key positions, one can compute the frequency of resonance of the different modes of vibration, thanks to a vibration analysis. These sensors can be implemented at the end of the initial life expectancy of the structure, and make sure the frequencies do not vary with time: a change in frequency means a behavioral change of the structure, thus aging. This method doesn't give absolute information on the structure, but it is monitoring its aging.

Sensors using different technologies can be used for this application, but FBG based sensors are well adapted: being totally passive, they can be implemented on the structure without major changes. Other technologies might have trouble adapting to Class 1 Div 1 environments. All the sensors can be multiplexed on the same line, making the cable line non-intrusive. Lastly, they are long term sensors capable of measurements for several years without drift, and with high frequency measurements adapted to vibration analysis.

4 CONCLUSIONS

A multitude of industrial sectors will benefit from the new optical fiber sensor technologies. From the well to the topside structure, optical fiber sensor can bring complementary and previously impossible data to improve safety and production of Oil&Gas assets. Such long term monitoring solutions can provide near-term and long-term cost-saving benefits such as:

- Supervised lifetime extension for existing assets, which are candidates for replacement
- Asset Integrity Monitoring during installation, commissioning and early life, reducing Vessel Time when inspection is needed.
- Risk Based Management and Condition Based Maintenance during service life

Although the optical fiber technology is fully developed and already in place in other industries (civil engineering, nuclear, health sciences…), installation challenges slowed the implementations of optical fiber sensors in the Offshore industry. Most of those challenges are getting resolved (wetmate connectors, optical pass-through assemblies, optical umbilicals and jumpers…), meaning that the offshore industry will soon benefit from this Asset Integrity Monitoring enabling technologies.

REFERENCES

API 2014 Recommended Practice 2SIM: Structural Integrity Management of Fixed Offshore Structures, G2SIM01.

DNV 2009 Recommended Practice DNV-RP-F116 - Integrity Management of Submarine Pipeline Systems.

Trent Jacobs, 2015. Downhole Fiber-Optic Monitoring: An Evolving Technology, Journal of Petroleum Technology 66 (08), 44–53.

DNV 2008 Recommended Practice DNV-RP-F206 - Riser Integrity Management.

Proceedings of the International Conference on Smart Infrastructure and Construction
ISBN 978-0-7277-6127-9

© The authors and ICE Publishing: All rights reserved, 2016
doi:10.1680/tfitsi.61279.075

Fiber-optic sensors based on FBGs with increased sensitivity difference embedded in polymer composite material for separate strain and temperature measurements

V.V. Shishkin[*3], V.S. Terentyev[1], D.S. Kharenko[1], A.V. Dostovalov[1], A.A. Wolf[1], V.A. Simonov[1], M.Yu. Fedotov[2], A.M. Shienok[2], I.S. Shelemba[3] and S.A. Babin[1]

[1] *Institute of Automation and Electrometry, Siberian Branch of the Russian Academy of Sciences, Novosibirsk, Russia*
[2] *All-Russian Scientific Research Institute of Aviation Materials, Moscow, Russia*
[3] *Inversion Sensor Co. Ltd., Perm, Russia*
[*] *Corresponding Author*

ABSTRACT Sensors based on fiber Bragg gratings (FBGs) are known to be simultaneously sensitive to temperature and strain. To separate their contributions a sensor consisting of two FBGs with increased sensitivity deference is proposed. An array of such sensors embedded in carbon fiber-reinforced plastic has been studied. The measured temperature sensitivity of FBGs inscribed by femtosecond laser in boron doped fiber is twice higher than standard values. The accuracy of separate strain and temperature measurements of 45-60 με and 2.9-3.7 °C obtained in temperature range of 30-120 °C.

1 INTRODUCTION

Polymer composite materials (PCM) are widely used in aerospace industry, civil engineering and wind-power energetics due to their lightweight, strength, corrosion immunity and fire resistance. One of key questions for PCM construction details is their long-term reliability. That is why their condition should be monitored in real time to detect damages, delamination, wear and tear (during fabrication, transportation and exploitation). Sensors based on fiber Bragg gratings are ideally suited for embedding into composite materials since the fiber size is comparable with thickness of composite monolayer. In addition, they enable easy multiplexing and high accuracy with immunity to electromagnetic interference.

Main problems of embedded FBG sensors concern their simultaneous sensitivity to temperature and strain. Current techniques of their separation lead to significant increases in cost and complexity of interrogation equipment. Therefore simpler techniques are required, which will combine advantages of fiber sensors and applicability to embedding into composite materials with separation of temperature and strain contributions.

2 METHODOLOGY

To separate temperature and strain contributions, we explore the method based on using of two fibers with different characteristics (Sivanesan et al. 2002). We place in one measurement point two FBGs inscribed in different fibers thus providing different sensitivity to temperature and/or strain. The larger the sensitivity difference between sensors, the higher the measurement accuracy. As distinct from Sivanesan et al.

(2002), we introduce a polynomial being quadratic for temperature and linear for strain, which better approximates a shift of reflection spectrum $\Delta\lambda_{1,2}$ of FBGs:

$$\begin{cases}\Delta\lambda_1 = K_{1\varepsilon 1}\varepsilon + K_{1T1}\Delta T + K_{2T1}\Delta T^2 \\ \Delta\lambda_2 = K_{1\varepsilon 2}\varepsilon + K_{1T2}\Delta T + K_{2T2}\Delta T^2\end{cases} \quad (1)$$

where $\Delta T=T_1-T_0$, T_1 and T_0 are current and initial temperatures of the sample given that the dependence of the FBG spectral shift on temperature is known (Mahakud et al. 2013). To determine the coefficients of approximating polynomials $K_{1\varepsilon 1,2}$, $K_{1T1,2}$, $K_{2T1,2}$ we performed a calibration procedure, where the PCM sample is exposed to longitudinal strain load and to variable temperature. Effective coefficients $\bar{K}_{1\varepsilon 1,2}, \bar{K}_{1T1,2}, \bar{K}_{2T1,2}$ have been calculated for each FBG by least-square method together with root-mean-square error for wavelength shift $\delta\lambda_{RMS1,2}$. The accuracy of measurements using system (1) is estimated as:

$$\delta T = \pm\frac{\left[\bar{K}_{1\varepsilon 2}{}^2\delta\lambda_1^2+\bar{K}_{1\varepsilon 1}{}^2\delta\lambda_2^2\right]^{1/2}}{(\bar{K}_{1T1}+2\bar{K}_{2T1}\Delta T)\bar{K}_{1\varepsilon 2}-(\bar{K}_{1T2}+2\bar{K}_{2T2}\Delta T)\bar{K}_{1\varepsilon 1}} \quad (2)$$

$$\delta\varepsilon = \pm\frac{\left[(\bar{K}_{1T2}+2\bar{K}_{2T2}\Delta T)^2\delta\lambda_1^2+(\bar{K}_{1T1}+2\bar{K}_{2T1}\Delta T)^2\delta\lambda_2^2\right]^{1/2}}{(\bar{K}_{1T1}+2\bar{K}_{2T1}\Delta T)\bar{K}_{1\varepsilon 2}-(\bar{K}_{1T2}+2\bar{K}_{2T2}\Delta T)\bar{K}_{1\varepsilon 1}} \quad (3)$$

where $\delta\lambda_{1,2}^2 = \delta\lambda_{RMS1,2}^2 + \delta\lambda_{A1,2}^2$ takes into account the calibration accuracy and apparatus error $\delta\lambda_{A1,2}$.

To test the method the following pair of fibers has been chosen: boron doped fiber with acrylate coating PS1250-125/250 as a fiber with low temperature sensitivity, and another one with polyimide coating SM1500P-125/150 as a fiber with high temperature sensitivity. In each of these fibers 3 FBGs have been inscribed by a femtosecond laser. The fs inscription has been performed by point-by-point technique with femtosecond laser through the coating (Wolf et al. 2015), therefore mechanical characteristics have not been changed. It is known that the temperature sensitivity for FBGs inscribed by UV radiation in fibers PS1250 and SM1500P is different amounting to about 9 and 13 pm/°C correspondingly (Yoon et al. 2006). We supposed that the difference is enough for separate temperature and strain measurements with estimated accuracy about 20 °C and 200 με correspondingly.

For experimental tests we assembled the packages of prepreg based on epoxy fusion adhesive VSE-1212 and medium-tensile carbon fiber TohoTenax IMS65 of 500×500×2 mm size with layer reinforcing structure $[0]_n$ and formed slabs, from which the samples of carbon fiber-reinforced plastic (CFRP) 400×25×2 mm size have been fabricated by mechanical cutting (Figure 1). Two fibers, each of which comprises three FBGs of 2 mm length separated by L_1=40 mm, were placed close (h=1 mm) to each other between central prepreg layers in parallel with the long edge of sample in such a way to have the FBG sensor array in the central area of the sample (Figure 2). Fiber pigtails go out of the sample at angle of 15° through PTFE tubes safely securing the fiber with a sufficient flexibility. Molding of prepreg's packages was made in an autoclave with maximum temperature 180 °C and specific pressure of not more than 0.7 MPa.

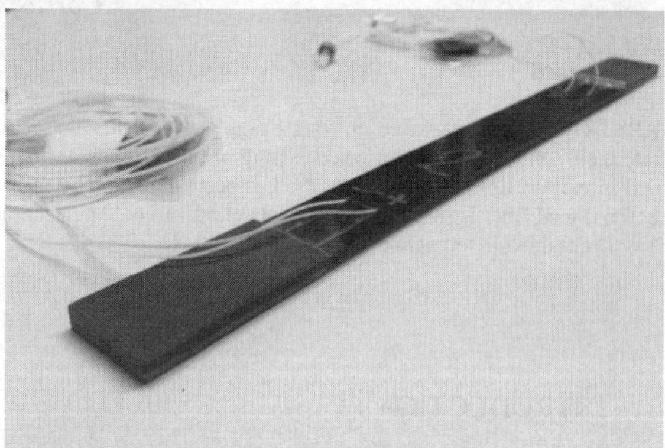

Figure 1. The CFRP sample with embedded fiber-optic sensors.

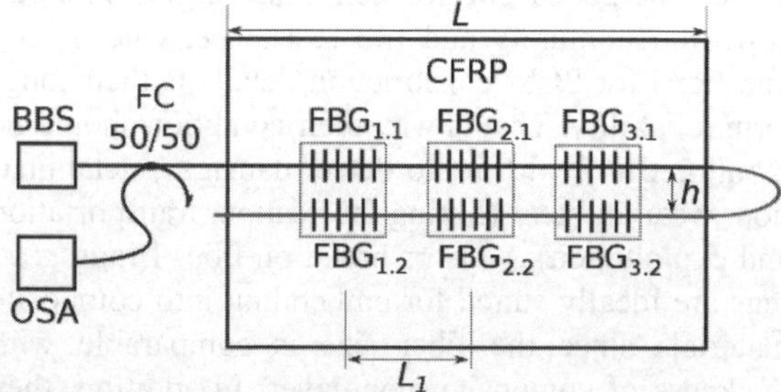

Figure 2. The topology of the fiber sensors in the sample.

3 EXPERIMENT AND RESULTS

Figure 3 show the measured spectra of the fiber sensors embedded into the CFRP samples. Since embedding into the sample do not influence sufficiently on the spectra, it appears possible to use standard interrogation techniques and algorithms for identification of spectral peak position. The FBGs in SM1500P fiber (red) have lower losses than that for FBGs in PS1250 fiber (black), which are probably of scattering nature since the intensity decreases from peak to peak. Such behavior may result from cladding modes influence on the FBG reflection.

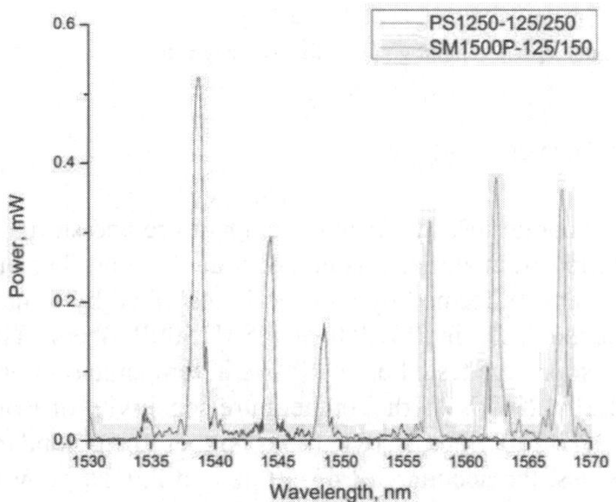

Figure 3. Reflection spectra of FBGs embedded into CFRP.

For the next step, coefficient of temperature and strain sensitivity for FBGs embedded into the CFRP sample have been determined with the use of testing machine Zwick/Roell Z-250 comprising temperature chamber and extensometer (Figure 4). During a test, the sample was heated to +30, +55, +80, +100 and +120°C. At each temperature one cycle of loading-unloading to 2.5 kN with a step of 0.5 kN was performed. The load value was limited by the grips we used. Peak wavelength shifts were measured by means of interrogator Astro A322 with resolution of 1 pm at 1 Hz sampling rate. Stretching of the sample was measured by extensometer with ~1 micron accuracy. The measured strain at longitudinal stretching amounts to 400 με. Besides, a resistance thermometer fixed at CFRP sample measured temperature with accuracy of 0.04 °C.

Figure 4. A sample of CFRP in testing machine.

As a result, wavelength shifts of the embedded FBGs have been measured at the tests. Figure 5 show that the FBGs in PS1250-125/250 fiber have higher temperature sensitivity than that for FBGs in SM1500P-125/150 fiber. The effect may be explained by the influence of cladding mode interference. Some influence may also have fs-induced modifications of the fiber. The picture also shows that FBGs in SM1500P-125/150 fiber have higher strain sensitivity. It may be explained by better adhesion of polyimide coating to polymer matrix of the sample. One can see that the readings for the FBGs in the same fiber are close whereas FBGs in different fibers diverge with increasing temperature that may be explained by different influence of cladding modes on the sensitivity coefficients in these fibers.

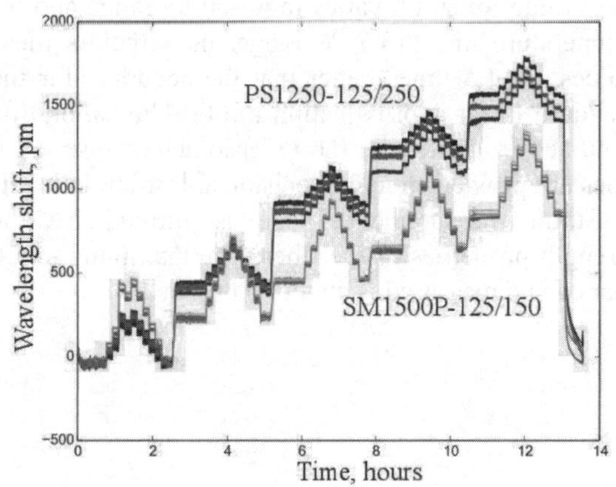

Figure 5. FBGs wavelength shifts at the test.

The measured sensitivity coefficients are $K_{1\varepsilon 1}$ = 0.43 pm/με, K_{1T1} = 22.3 pm/°C, K_{2T1} = -5.4 10^{-2} pm/°C^2 for FBGs in PS1250-125/250 fiber and $K_{1\varepsilon 2}$ = 1.13 pm/με, K_{1T2} = 17.0 pm/°C, K_{2T2} = -8.1 10^{-2} pm/°C^2 for FBGs in SM1500P-125/150 fiber. It is also useful to plot the whole field of wavelength shifts versus temperature and strain using the polynomial approximation (1) (Figure 6). In accordance with formulae (2) and (3), in case of the nonlinear polynomial approximation errors depend on the absolute values of temperature and strain.

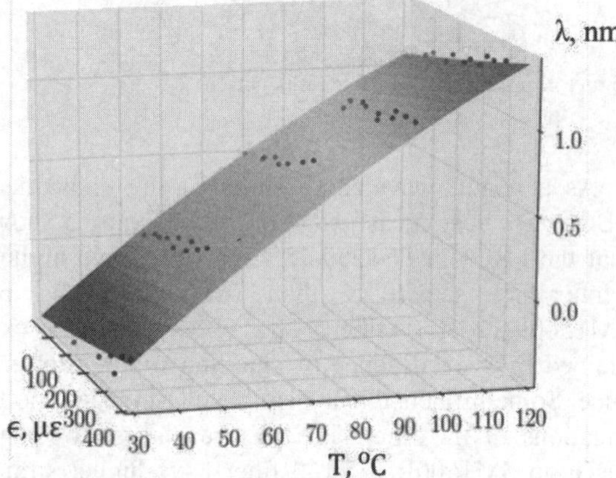

Figure 6. The field of wavelength shifts versus strain measurement temperature and strain.

As an example, in Figure 7 a contour graphic is presented demonstrating an accuracy of one sensor consisting of two FBGs in PS1250 and SM1500P fiber, with the above values of coefficients. The accuracy value for strain varies in 45-60 με range and for temperature in 2.9-3.7 °C range, nevertheless these values are 1.5 times better that the accuracy for the standard linear approximation of (1). The calibration accuracy is limited by the extensometer noise, as it does not provide enough precision at low strain level.

Strain measurement range is limited by the strength properties of the fiber. The maximum strain that can be measured is about 10 000 με.

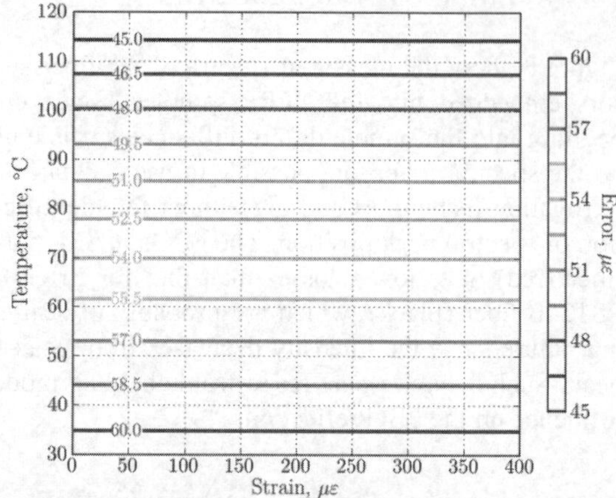

Figure 7. The contour graphics for accuracy value.

4 CONCLUSION

A separate measurement of temperature and strain in composite material was demonstrated using fiber optic sensors formed by two FBGs inscribed by femtosecond laser in PS1250 and SM1500P fibers. The sensors were tested up to 400 με in temperature range of 30-120 °C. As the temperature sensitivity of FBG in PS1250 fiber is two times higher than standard values, the accuracy of 45-60 με and 2.9-3.7 °C was obtained. To enhance the measurement accuracy a calibration at higher strain level should be performed.

Applications of fiber sensors embedded in a composite structure offers an opportunity for development of intellectual materials and constructions made of them, which are able to register and evaluate hazardous tension/hot conditions with separate measurement of strain and temperature response of the fiber sensors in real time. Adaptation of the described technique for the exploitation conditions of composite constructions will enable detailed information on the actual loads in the composite material under thermal and mechanical disturbances, provide a forecast of functionality of the construction parts, inform on the need for their repair or replacement.

Demonstrated technique of separate temperature and strain measurement have a potential in structural health monitoring of buildings and bridges loadbearing structures of which are reinforced by polymer composite materials.

ACKNOWLEDGEMENT

The authors acknowledge a technical assistance of A. F.Merzlyakov.

REFERENCES

Mahakud, R. Kumar, J. Prakash, Q. et al. 2013. Study of the nonuniform behavior of temperature sensitivity in bare and embedded fiber Bragg gratings: experimental results and analysis, *Applied Optics* **52**, 7570-7579.

Sivanesan, P. Sirkis, J.S. Murata, Y. et al. 2002. Optimal wavelength pair selection and accuracy analysis of dual fiber grating sensors for simultaneously measuring strain and temperature, *Optical Engineering* **41**, 2456–2463.

Wolf, A.A. Dostovalov, A.V. Babin, S.A. 2015. Direct writing of long-period and fiber Bragg gratings in specialty fibers by femtosecond laser pulses at the wavelength of 1026 nm, *CLEO/Europe-EQEC 2015: Conference Digest*, CM-4.2.

Yoon, H.J. Costantini, D.M. Limberger, H.G. et al. 2006. In situ strain and temperature monitoring of adaptive composite materials, *Journal of Intelligent Material Systems and Structures* **17**, 1059-1067.

Proceedings of the International Conference on Smart Infrastructure and Construction
ISBN 978-0-7277-6127-9

© The authors and ICE Publishing: All rights reserved, 2016
doi:10.1680/tfitsi.61279.081

Vibration and deformation monitoring of a long-span rigid-frame bridge with distributed long-gauge sensors

YD Tian[1] and J Zhang [2]
[1] *School of Civil Engineering, Southeast University, Nanjing, China*
[2] *Key Laboratory of Concrete and Prestressed Concrete Structure of Ministry of Education, Southeast University, Nanjing, China*

ABSTRACT Ambient vibration tests and analyses of a long-span rigid-frame bridge were conducted using distributed long-gauge fiber optic sensors. The concept of a long-gauge fiber optic sensor and its merit to reveal local and global structural features are presented. Monitoring of a long-span bridge was performed using long-gauge sensors, and a method to calculate structural deformation distribution from the measured long-gauge strains is proposed, in which shear deformation because of the Timoshenko effect is considered.Modal identification of the studied bridge using the measured long-gauge dynamic strains was also performed, from which strain and displacement mode shapes were identified. These results demonstrate the superiority of long-gauge fiber optic sensors and their successful application to the studied bridge.

1 INTRODUCTION

Bridge collapses such as the I-35W Bridge over the Mississippi River in 2007 have received tremendous public attention, which emphasizes the importance of effective management of civil infrastructure. Their maintenance is expensive and available budgets are limited, especially in developed countries. According to Federal Highway Administration data, of a total of ~590,000 bridges in the United States, 152,136 bridges are damaged or have functional problems. An available budget of $10.5 billion is insufficient to cover the annual required $17 billion maintenance costs (ASCE 2013). Similarly, it has been reported that 7.2 billion Euros will be required in Germany for civil infrastructure maintenance over the next 15 years, which is 70% higher than the available budget. Therefore, the importance of structural health monitoring (SHM) is becoming increasingly recognized by researchers and civil engineers, and is highly expected to be a promising tool to improve safety and maintain civil infrastructure efficiency.

SHM has been investigated widely and has been applied to a number of large-scale structures, including high-rise buildings and long-span bridges (Ko et al. 2002; Conte et al. 2008; Siringoringo and Fujino 2008; Pakzad and Fenves2009). SHM systems installed on civil structures are useful to monitor their in-service safety conditions; however, current SHM methods cannot satisfy engineering requirements completely, such as reliable structural performance evaluation and risk prediction (Brownjohn et al. 2009; Zhang et al 2012; ASCE 2013). The first challenging problem in the SHM field is to develop advanced sensing technologies for accurate and effective structural monitoring. The second challenging problem is to develop effective methods to process extensive monitoring data for structural identification and performance evaluation.

Application of accelerometers and displacement transducers in civil infrastructure monitoring has a long history (Abdel-Ghaffar and Scanlan 1985). Measurements from these types of sensors reveal structural dynamic characteristics such as frequency, damping and mode shapes; however, these parameters are global, are insensitive to minor structural damage, and it is difficult to use them to support decision making in structure maintenance and management. Strain is more sensitive to local damage and could be a good candidate for structural damage detection. However, measurements from traditional strain gauges are localized and cannot be used to monitor completely large-scale civil infrastructure because of their mostly short gauge lengths (1~2cm) (MacLeod 2010). The recently developed Fiber Bragg Grating (FBG) sensor is outstanding for strain monitoring with its special features of high precision, stable sensing capacity, and reliability (Ansari 2007), but it also has a short gauge length, which makes it is difficult to apply in large-scale structure monitoring. The emerging Brillouin Optical Time Domain Reflecting sensing system (BOTDR) and Pulse Pre-Pump-BOTDR technologies make distributed sensing with optical fibers possible (Murayama et al.2008). Unfortunately, they have not been applied widely in engineering practice because of their limited precision, space resolution, low speed, and high cost. Because of limitations in current sensor technologies described above, the SHM system of large-scale structures requires a large number of sensors for structural global and local information collection. For instance, over 1200 sensors were used in the SHM system of the Stonecutters Bridge, which cost $50 million. The expensive SHM system consists of complex sensors and related data-acquisition equipment, which is difficult to manage. Massive volumes of monitoring data from the SHM system are difficult to process for effective structural safety evaluation.

A long-gauge FBG sensor that can measure structural local and global information is described in this article, and its application on a long-span rigid-frame bridge is investigated. The long-gauge FBG sensor has a gauge length of 20 cm~2 m, and can be connected in series to make an FBG sensor array for area-distributed macro-strain measurement (Zhang et al. 2014). The concept and development of the long-gauge FBG sensor is presented in Section 2. The objective of this work is to show how to use the long-gauge FBG sensor for effective SHM. To achieve this goal, a long-span rigid-frame bridge and its monitoring using the long-gauge FBG sensors is described in Section 3. The method of using measured long-gauge dynamic strains for distributed deformation calculations and results from field test data is presented in Section 4. Section 5 provides strain modal identification results from field test data, and is followed by conclusions..

2 CONCEPTS OF LONG-GAUGE SENSING

Fully distributed monitoring of large-scale civil structures is difficult. Effective monitoring could be achieved by monitoring structural critical, easily damaged areas accurately using high-performance distributed sensing networks. As described previously, most sensors are point-type sensors, which are too global or local to reveal structural inherent characteristics. Recently developed distributed sensing technologies such as the BOTDR have limited properties and high costs. With the absence of suitable sensors for area-distributed monitoring, a concept of long-gauge sensing and its corresponding sensor have been developed. By designing FBG sensors with a long gauge, for instance, 20 cm to 2 m, the measured strain will be the average strain over the long-gauge length. It outputs static/dynamic structural strain responses with high precision. The measured long-gauge strain is related to structural rotations, thus it can reveal global (rotation measurements) and local (strain measurements) structural information. Based on the above described features of long-gauge sensing, together with basic FBG sensor characteristics such as high precision and resolution, long-gauge FBG sensors are ideal candidates for area-distributed monitoring of large-scale civil infrastructure.

3 BRIDGE DESCRIPTION AND SENSOR LAYOUT

SHM of a long-span rigid-frame bridge was performed using long-gauge FBG sensors. The bridge is a prestressed, concrete, continuous rigid-frame box girder bridge with a span length of 268 m (**Figure 1**). The bridge deck width is 34 m with two-way six lanes. The cross-section of the girder is a single box with a single chamber. The upper and bottom plate widths are 16.4 m and 7.5 m, respectively, and its height at the end of the girder is 15m. It has double-walled, rectangular, hollow reinforced concrete piers.

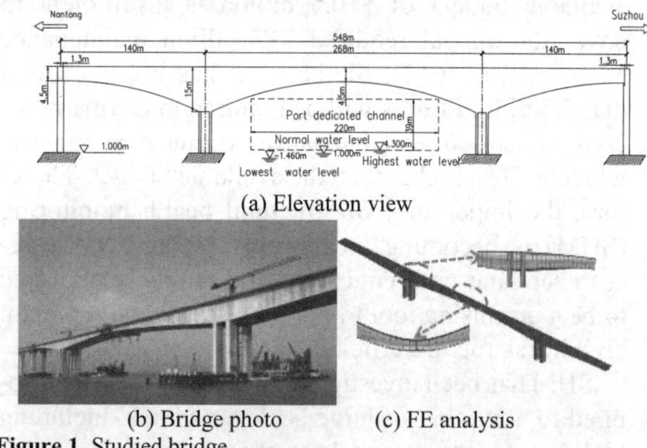

(a) Elevation view

(b) Bridge photo (c) FE analysis

Figure 1. Studied bridge

Finite element (FE) modeling and analysis of the bridge was performed using commercial ANSYS software to determine critical structural areas (Figure 1c). BEAM189 element was used in the FE modeling with 548 elements. The elasticity modulus, density, and Poisson ratio are 3.45×10^{10}N/m^2, 2500kg/m^3, and 0.2, respectively, which are based on the design drawing. The two bridge ends are simply supported. Figure 1c illustrates the calculated stress distribution of the structure under self-weight. Cracks and excessive deflections are two typical problems for general rigid-frame bridges. FE analysis results indicate that upper plates of the main girder at both ends of the middle span crack easily and the bottom plate at the center of the middle span is easy deflected excessively. Thus, these areas were considered critical, and the long-gauge FBG sensor network was distributed in these areas as shown in Figure 2, in which 52 long-gauge FBG sensors with a gauge length of 1 m were used. The sampling rate was set to 200 Hz. Typical strain responses measured during ambient vibration tests and their spectra are shown in Figure 3.

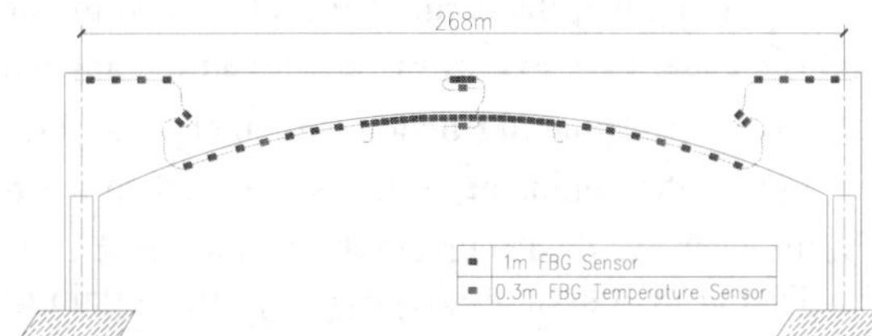

Figure 2. Sensor layout

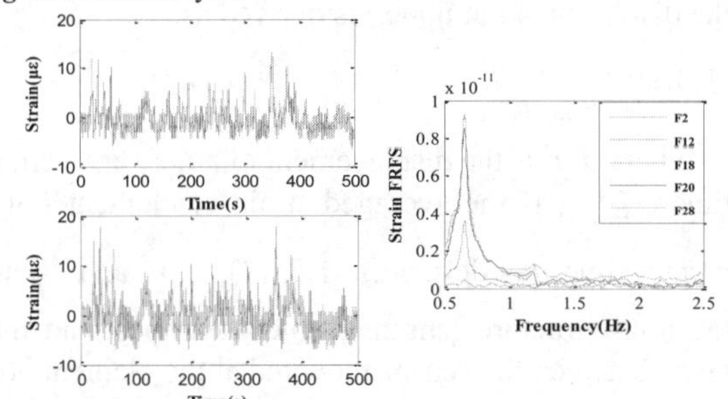

Figure 3. Typical long-gauge strain measurements and their spectra

4 DISTRIBUTEDDEFORMATION CALCULATION METHOD

In this section, a method to calculate the structural deformation distribution from measured long-gauge strains is presented. The studied bridge is a thin-walled box girder bridge with a length-depth ratio of 17.9, which induces the shear-deformation having an effect on vertical displacement to some extent.

4.1 Deformation calculation considering the shear strain

Deflection monitoring of long-span bridges is import for structural safety monitoring, but its measurement is challenging despite the development of many types of displacement-related sensors (Brown and Roberts 2006). Linear displacement transducers are accurate and have been used extensively in the laboratory. However, it is difficult to use them for bridge vertical deflection measurements because, in general, there are no fixed base points because of rivers or traffic on roads under bridges. A global positioning system can be used for bridge field tests; however, such a system is expensive and its accuracy needs improvement. Recently, equipment using high-resolution camera and radar technologies has been developed for deflection monitoring, but it is still not well accepted because of its high costs and limited performance. A convenient way to monitor deflection is to measure long-gauge strains and then calculate deflections. Shen and Wu (2010) studied deflection calculations using strain measurements, in which only pure bending deformation was considered. As mentioned above, the studied bridge is a thin-walled box girder bridge with a span-depth ratio of 17.9. Timoshenko beam theory states that shear stress and deflection cannot be neglected. Therefore, the question arises as to how to calculate the deflection of a thin-walled box girder bridge with a large span-depth ratio while considering shear deformation.

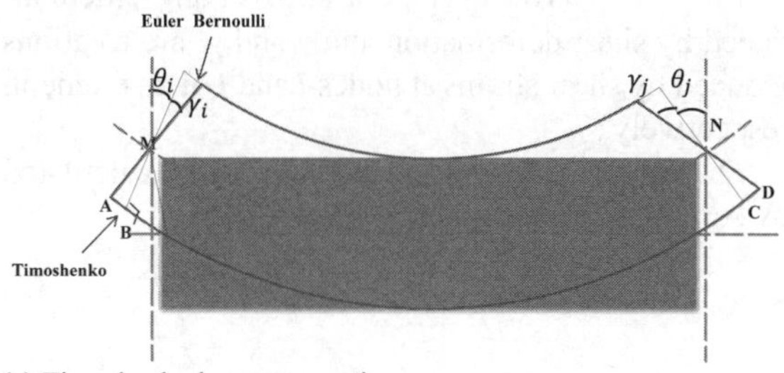

(a) Timoshenko beam conception

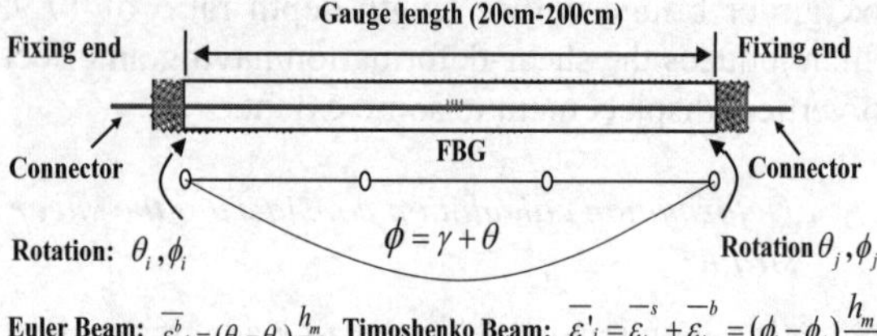

(b)Relationship between long-gauge strain and rotation

Figure 4. Difference between Timoshenko and Euler beams

For an Euler beam, if it is assumed that the beam element rotation in the clockwise direction is positive, the relationship between rotations and bending strain within an element is (Figure 4(b)):

$$\overline{\varepsilon^b}_i = (\theta_i - \theta_j)\frac{h_m}{L_m} \tag{1}$$

where $\overline{\varepsilon^b}_i$ is the average bending strain over an element, h_m is the distance between the sensor mounted at the beam bottom and the beam neutral axis, L_m is the element length, and θ_i and θ_j are rotations at nodes I and j of an element, respectively.

For a Timoshenko beam, the beam neutral axis is not perpendicular to the cross-section because of shear deformation. It is known that $\phi = \frac{dy}{dx} = \gamma + \theta$. The total rotation includes terms from the bending moment and from the shear force. The relationship between average shear strain and rotation induced by shear force of a Timoshenko beam is:

$$\overline{\varepsilon^s}_i = (\gamma_i - \gamma_j)\frac{h_m}{L_m} \tag{2}$$

where $\overline{\varepsilon^s}_i$ is the average strain over an element induced by shear deformation and γ_i and γ_j are rotations induced by shear strains at nodes I and j of an element, respectively.

Therefore, the total average strain can be calculated by summing Eq. (1) and (2) (**Figure 4(b)**):

$$\overline{\varepsilon'}_i = \overline{\varepsilon}_i^s + \overline{\varepsilon}_i^b = (\phi_i - \phi_j)\frac{h_m}{L_m} \tag{3}$$

where $\overline{\varepsilon'}_i$ is the total average strain induced by a bending moment and a shear force and ϕ_i and ϕ_j are the total rotation of an element at nodes i and j, respectively.

According to conjugate beam theory, the original beam deformation equals the moment of the conjugate beam . Based on fundamental knowledge of material mechanics, the deflection curve differential equation of the beam is $y'' = -\frac{M(x)}{EI}$ and $\overline{M}''(x) = \overline{q}(x)$. It is assumed that a conjugate beam corresponds to a real beam and that its length, x direction, and coordinates are the same. If the curvature $-M(x)/EI$ on the real beam is considered to be a load distribution $\overline{q}(x)$ on the conjugate beam, then its corresponding bending moment distribution $\overline{M}(x)$ is equivalent to the real beam's vertical displacement distribution $y(x)$. Hereafter, the over bar symbol ($\overline{\ }$) indicates that the parameter concerned refers to the conjugate beam. The relationships between strain $\varepsilon(x)$, curvature $k(x)$, and bending moment $M(x)$ are linear as shown below:

$$k(x) = \frac{\varepsilon(x)}{h_m} = \frac{M(x)}{EI} \tag{4}$$

where h_m is the distance from the beam's surface to the neutral axis.

Assuming that the conjugate beam's load distribution is equal to $\overline{q}(x) = \varepsilon(x)/h_m$, it can be concluded that the corresponding bending moment distribution $\overline{M}(x)$ of the conjugate beam is identical to the real beam's vertical displacement distribution $y(x)$.

Taking the curvature $\overline{q}(t) = \varepsilon(t)/h_m$ at a certain time as an external load imposed on the conjugate beam, the displacement at node j is derived as:

$$d_j(t) = \frac{L^2}{n^2}\left[\frac{j}{n}\sum_{i=1}^{n}\frac{\overline{\varepsilon}'_i(t)}{h_i}\left(n-i+\frac{1}{2}\right) - \sum_{i=1}^{j}\frac{\overline{\varepsilon}'_i(t)}{h_i}\left(j-i+\frac{1}{2}\right)\right] \tag{5}$$

where $d_i(t)$ is the displacement of node j at a certain time t; $\overline{\varepsilon}'_i(t)$ is the averaged strain of element i at a certain time t as defined in Eq. (3); L, n, and h_i are the total structure length, element number, and distance between the sensor mounted at the element bottom and the beam neutral axis of element i. Structural deformations can be calculated from the measured long-gauge strain even when shear deformation are considered.

4.2 Distributed deformation calculation of studied bridge

As described in the second section, long-gauge FBG sensors were deployed on the studied long-span rigid-frame bridge as shown in Figure 2. Figure 5 (a) illustrates typical long-gauge strain time histories near the central node of the middle span during ambient vibrations, which is collected by SM130 Interrogator. It should be noted that the temperature effect on measured stain is compensated by independent temperature-strain sensors. The largest strain of the time history is 17.45με. The deformation distributions of the middle span were calculated from the measured long-gauge strains using the proposed method. Figure 5 (b) plots the calculated deformation time history that corresponds to the measured strains as shown in Figure 5(a). The maximum value of the dynamic displacement time history is 12.69 mm. The deformation and span strain are small because the studied rigid-frame bridge is stiff. The deformation distributions of the studied bridge have been output from the measured long-gauge strains, which will provide valuable information to bridge engineers and owners for structural maintenance and management.

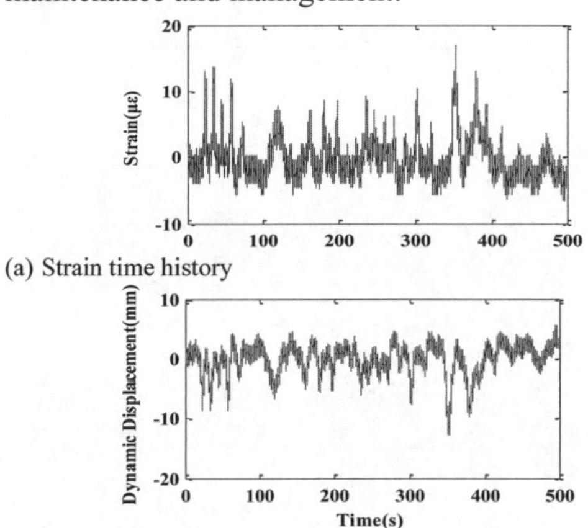

(b) Calculated displacement time history

Figure 5. Typical long-gauge strain and displacement time histories

5 LONG-GAUGE STRAIN MODAL IDENTIFICATION OF STUDIED BRIDGE

Besides deformation distribution calculations, structural modal parameters, including frequencies, damping, strain mode shapes, and displacement mode shapes, can be identified from long-gauge strain measurements.

The first and second frequencies of the studied bridge were found to be 0.66 Hz and 1.22 Hz, respectively, from the spectra of the measured long-gauge strains as shown in Figure 3. Figure 7(a) illustrates the identified strain mode shape in the first mode, in which the strain mode shape from the FE modal analysis is also plotted as a reference. Once the strain mode shape has been identified, the displacement mode shape can be calculated from the strain mode shape using the modified conjugate beam method proposed in the previous section. The calculated displacement mode shape is plotted in Figure 7(b), and is compared with the FE modal analysis results. The merit of using a long-gauge strain for modal analysis is that strain and displacement mode shapes are identified.

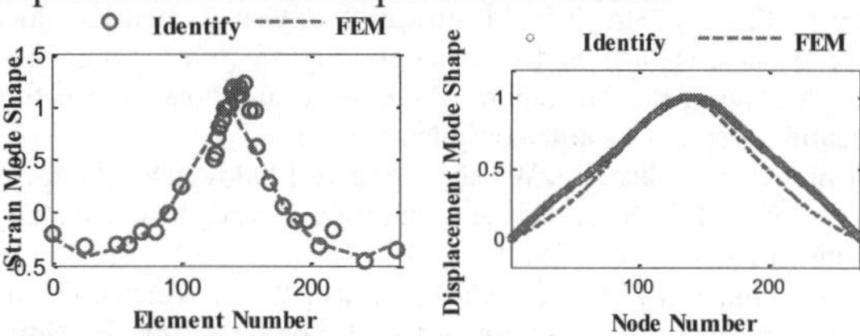

(a) Strain mode shape (b) Displacement mode shape

Figure 7. Identified strain and displacement mode shapes

6 CONCLUSIONS

Ambient vibration and deformation distribution identification of a long-span rigid-frame bridge has been investigated using a long-gauge FBG sensor. The following conclusions are drawn:

(1) The long-gauge FBG sensor can reveal local and global structural information. A monitoring system using the long-gauge FBG has been designed and applied in a studied bridge for ambient vibration testing.

(2) Deformation monitoring of long-span bridges is challenging. A method has been developed to calculate bridge deformation distribution from measured long-gauge strains. Shear deformation induced by the shear strain has also been considered. Deformation distribution of the studied long-span rigid-frame bridge has been calculated using field test data, and the peak value of the bridge deformation under general ambient loads is ~15 mm.

(3)Modal bridge parameters have been identified from long-gauge strain measurements to reveal structural dynamic properties. The displacement mode shapes have also been calculated from the identified strain mode shapes by using a modified conjugate beam method, which reflects that the long-gauge FBG sensor is multi-functional.

ACKNOWLEDGEMENT

This work was sponsored by the National High Technology Research and Development Program (863 Program) of China (2014AA110401).

REFERENCES

Abdel-Ghaffar A.M., and Scanlan, R.H.(1985). “Ambient Vibration Studies of Golden Gate Bridge: I Suspended Structure”. Journal of Engineering Mechanics, 111, 463-482.

Ansari, F. (2007).“Practical Implementation of Optical Fiber Sensors in Civil Structural Health Monitoring”. Journal of Intelligent Material Systems and Structures, 18(8), 879-889

ASCE. (2013). “Structural Identification of Constructed Facilities: Approaches, Methods and Technologies for Effective Practice of St-Id”. A State-of-the-Art Report. ASCE SEI Committee on Structural Identification of Constructed Systems.

Brown, C. J., Roberts, G. W., and Meng, X. (2006). “Developments in the use of GPS for bridge monitoring.” Proc. Inst. Civ. Eng. Bridge Eng.,159(3), 117–119.

Brownjohn, J. M. W., De Stafano, A., Xu, Y.-L., Wenzel, H., and Aktan, A. E. (2011). “Vibration-based monitoring of civil infrastructure: Challenges and successes.” Journal Civil Structure Health Monitoring, 1(3-4), 79–95.

Brownjohn, J.M.W., Magalhães, F., Caetano E., Cunha, A., Au, I., Lam, P. (2009).“Dynamic testing of the Humber Suspension Bridge”.International Conference of Experimental Vibration Analysis for Civil Engineering Structures. October, Wroclaw, Poland, 93-102.

Conte, J.P., He, X., Moaveni, B., Masri, S.F., Caffrey, J.P., Wahbeh, M., Tasbihgoo, F., Whang, D.H., and Elgamal, A.(2008).”Dynamic testing of Alfred Zampa Memorial Bridge”. Journal of Structural Engineering, 134, 1006-1015.

Catbas, F. Necati, Masoud Malekzadeh, and Tung Khuc(2013). "Movable Bridge Maintenance Monitoring." A Report on a Research Project Sponsored by Florida Department of Transportation Contract No. BDK78-977-10.

James M. Gere, Stephen P. Timoshenko,(1991).”Mechanics of Materials”. Springer US.

Ko, J.M., Sun, Z.G., Ni, Y.Q.(2002). “Multi-stage identification scheme for detecting damage in cable-stayed KapShuiMun Bridge”. Engineering Structures, 24, 857–868.

MacLeod, A.B.(2010).“Structural health monitoring of the traffic bridge in Saskatoon using strain gages”. Master of Science Thesis, University of Saskatchewan.

Malekzadeh, Masoud et al (2012). "Use of FBG sensors to detect damage from large amount of dynamic measurements." Topics on the Dynamics of Civil Structures, Volume 1, Proceedings of the 30th IMAC, A Conference on Structural Dynamics.

Malekzadeh, Masoud et al (2014). “An integrated approach for structural health monitoring using an in-house built fiber optic system and non-parametric data analysis”. Smart Structure and systems,14(5):917-942.

Malekzadeh, Masoud(2014). “Structural Health Monitoring Using Novel Sensing Technologies and Data Analysis Methods”. Doctor of University of Central Florida.

Murayama, H., Kageyama, K., Ohara, K., Uzawa, K., Kanai, M.(2008). “Novel measurement system with optic fiber sensor for strain distribution in welded tubular joints”. Proceedings of the ASME 27th international conference on offshore mechanics and arctic engineering.

Pakzad, S.N., and Fenves, G.L.(2009). “Statistical analysis of vibration modes of a suspension bridge using spatially dense wireless sensor network”. Journal of Structural Engineering, 135, 863-872.

Siringoringo, D.M., and Fujino, Y.(2008).“System identification of suspension bridge from ambient vibration response”. Engineering Structures, 30, 462-477.

Shen S., W u Z.S. (2010).“An improved conjugated beam method for structural deformation monitoring based on distributed optical fiber strain sensing technique”. China Civil Engineering Journal,43,63-70.

S. P. Timoshenko(1921). “On the correction for shear of the differential equation for transverse vibrations of prismatic bars”. Philosophical Magazine Series, 6:742–746.

Zhang, J., Prader, J., Moon, F., Grimmelsman, K.A., Aktan, E., Sayama, A. (2012).“Experimental Vibration Analysis for Structural Identification of a Long Span Suspension Bridge”, ASCE Journal of Engineering Mechanics, In Press.

Zhang, J., Hong, W., Tang, Y.S., Yang, C.Q., Wu, Q., Wu. Z.S. (2014).”Structural Health Monitoring of a Steel Stringer Bridge with Area Sensing”, Structural and Infrastructural Engineering.

Proceedings of the International Conference on Smart Infrastructure and Construction
ISBN 978-0-7277-6127-9

© The authors and ICE Publishing: All rights reserved, 2016
doi:10.1680/tfitsi.61279.087

The evaluation of bridges by distributed FBG sensors

K. Yang[*1], N. Tamehiro[2], K. Yamamoto[1] and R. Yonao[1]

[1] *Seismic Eng. Dept., KOZO KEIKAKU Eng. Inc. Tokyo, Japan*
[2] *Eng. Sales Dept., KOZO KEIKAKU Eng. Inc. Tokyo, Japan*
[*] *Corresponding Author*

ABSTRACT A distributed sensing system developed by authors for flexural structure using strain measurements from long-gage fiber Bragg grating (FBG) sensors for structural health monitoring (SHM) was introduced in this paper. By utilizing the sensing system, the following features of bridge can be evaluated: (1) strain distribution along the bridge; (2) bridge deflection distribution; (3) dynamic characteristics of bridge such as frequency and modal shape; (4) load level and (5) damage detection using modal macro-strain vector constructed by the strain measurement of all FBG sensors. To validate the effectiveness of the developed FBG sensing system, a series of field investigations on a reinforcement concrete (RC) bridge, a railway bridge and a pre-stressed concrete (PC) bridge were performed. The field observations showed good agreement between the results obtained from the sensing system and finite element method (FEM) analysis. This demonstrates that the sensing system could provide a multifunctional approach for SHM in the intelligent infrastructure.

1 INTRODUCTION

Regarding the on-going development of available fiber optic sensing techniques, FBG sensor in nature holds the excellent ability to provide a measurement having higher precision and measuring stability. A series of lab experiments and theoretical studies for the utilities of distributed long-gage FBG sensor system by authors has been performing since 2006 (Li et al. 2007, Yang et al. 2006).

On the bases of above studies, a series of field investigations on RC and PC bridges, and railway-bridge were conducted. The field observations showed good agreement between the results obtained from the FBG sensing system and FEM numerical results. This demonstrates that the sensing system provides a multifunctional approach for structural health monitoring in the intelligent infrastructure. It becomes possible to monitor and evaluate the under-use bridges much more exactly and scientifically than current monitoring method based on the veteran technician's judgment by his sense and experience.

2 A DISTRIBUTED SENSING SYSTEM

2.1 *Description of the sensing system*

The long-gage FBG sensor, whose effective sensing gauge length could be extended to several centimeters or meters through special packaging and broadening methods, as indicated in Figure 1, was developed by authors. Briefly speaking, it was applied to civil structures as a kind of distributed strain sensing technique by serially connecting a group of long-gage sensors together as shown in Figure 2.

By utilizing the sensing system, the following features of bridge could be evaluated: (1) strain distributions along the bridge; (2) bridge deflections based on the measured strain distributions; (3) dynamic characteristics such as frequency and modal shape of bridge from the time history of measured strains; (4) load level identification by measured strain distributions and (5) damage detection using modal macro-strain vector constructed by the strain measurement.

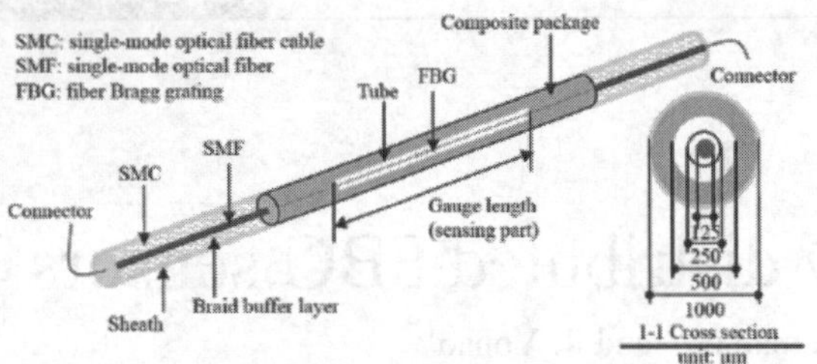

Figure 1. A packaged long-gage FBG sensor

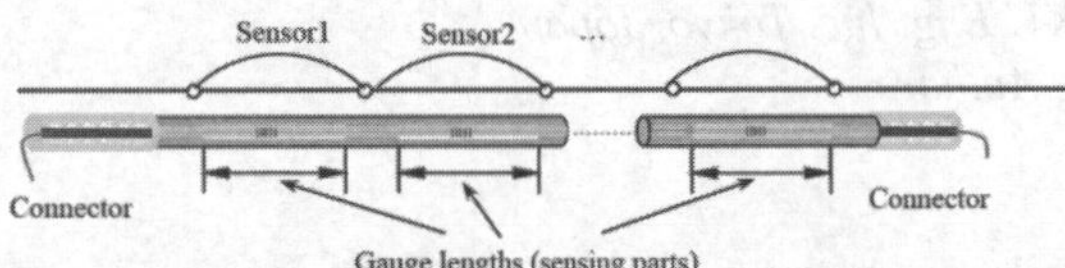

Figure 2. Distributed long-gage FBG sensing system

2.2 *Application of the sensing system*

By using the distributed sensing system that was introduced hereinabove, the time histories of the strains along the measured structure could be obtained. To clearly show the procedures of the application of this sensing system in practical engineering, following demonstrations are presented through a typical example using the sensing system on the whole span of a bridge, noted in Figure 3 which displays the time histories of the measured strains. The different colors represent different placed sensors.

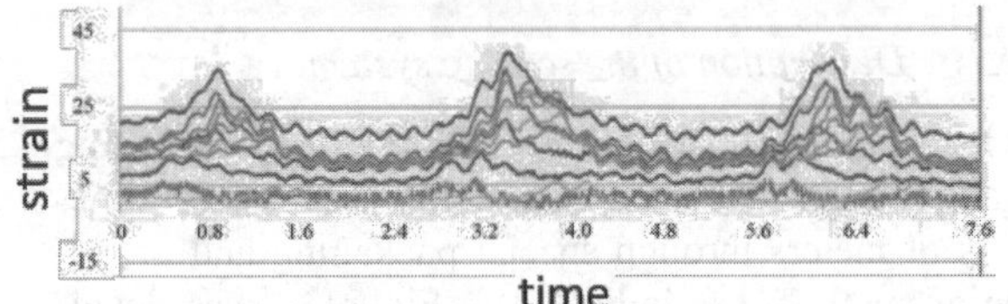

Figure 3. The time history of measured strains

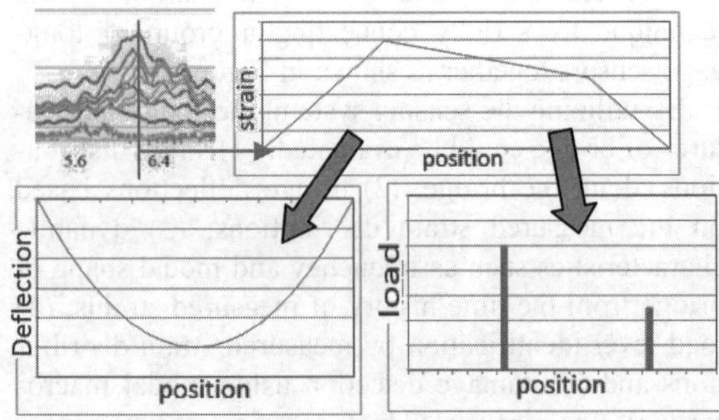

Figure 4: The profiles of applied load and deflection

On the one hand, the distribution of strain along the whole span of bridge could be described through choosing the strain values at the same time of the recorded time histories of strains. As shown in Figure 4, the distribution of strain along the bridge at certain time was obtained from the measured strains (Figure 3). According to it, the distribution of deflection can be determined through integrating the strain along the bridge (Figure 4); and furtherly the load level identification can be done by measured strains (Yang et al. 2007). Moreover, the traffic density passing through the bridge could be estimated as presented in Figure 5.

On the other hand, the natural frequency of bridge can be obtained by performing a Fourier transform from time history of measured strains. The Modal Macro Strain Vectors (MMSV), which are defined as the relative ratio of all components of the combination of macro strain magnitude of frequency response functions from all FBG sensors (Yang et al. 2014), are also calculated as shown in Figure 6.

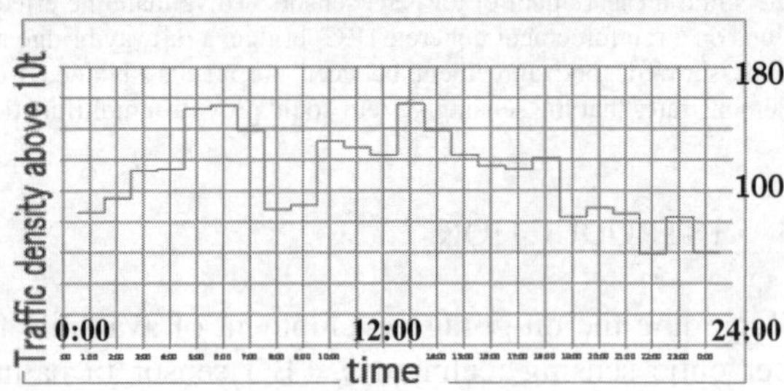

Figure 5: The statistical number of traffics above 10t

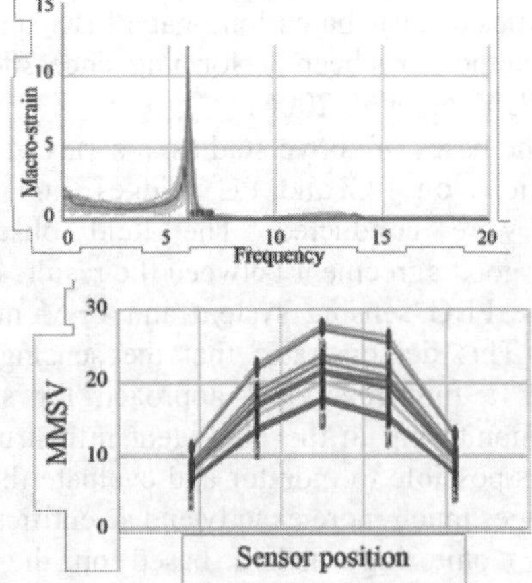

Figure 6: Natural frequency and macro strain modals

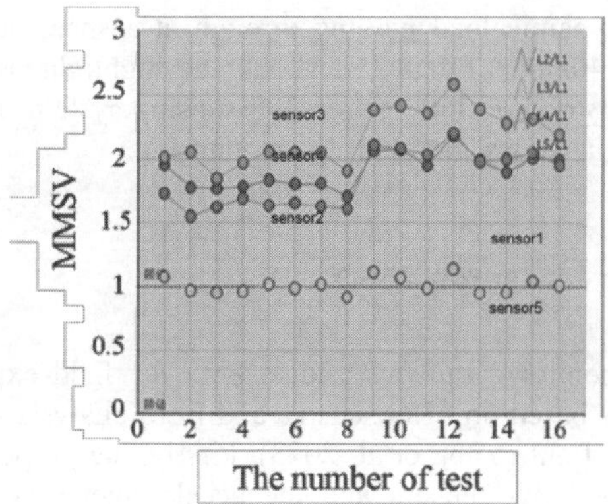

Figure 7: Damage detection

MMSV can be used as an index value for damage detection. The main idea is that the MMSV would keep constant as normalized mode shape of structure if there is no damage of structure, otherwise, the values of MMSV change abruptly even when a slight structure damage occur (Yang et al. 2014). Accordingly, the damage positions would be identified. Taking Figure 7 for example, when a crack occurs in the position near the sensor 3, the values of MMSV in sensor 3 varied most and the sensors 2 & 4 near sensor 3 also changed obviously. To the authors' knowledge, the MMSV based on the FBG sensing system could be a prominent index in damage detection of structure. The dynamic measurement was performed by setting zero start in every short record period to avoid bothersome temperature compensation.

3 THE EVALUATION OF STRUCTURES BY DISTRIBUTED FBG SENSORS

Based on the distributed sensing system, three field tests were carried out on the KAWANE bridge, a railway bridge and the Myoko bridge, to examine the validations and applications in engineering practice. The conditions and results of all tests are presented in this section.

3.1 The KAWANE bridge

The KAWANE bridge is located in Ibaraki, Japan, which is a RC bridge built in Oct. 1963.

To confirm the feasibility of the application of distributed FBG sensing system on this RC bridge, a bus was reserved as a load for the measurement, as shown in Figure 8. The bus totally weights 100kN. Specifically, the weight on the front and back wheels is 30kN and 70kN, respectively. When the bus stays at a fixed position on the bridge, it is served as a static load. However if the bus drives through the bridge at speeds of 10km/h, 30km/h or 40km/h, it was regarded as a dynamic load. A clear description of the static load positions is presented in Figure 9.

To demonstrate the capability of the distributed FBG sensors, both the field tests and the FEM analysis are performed and the detail configuration can be referred to Yang et al. (2009). 11 sensors with 1 meter gauge length are arranged along the bottom surface of a girder on the whole span of bridge, see Figure 8 and Figure 9.

Figure 8: Loading Bus

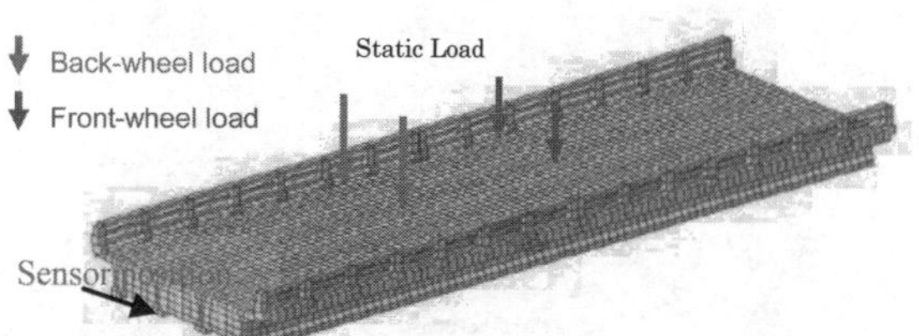

Figure 9: The position of static load on FEM model

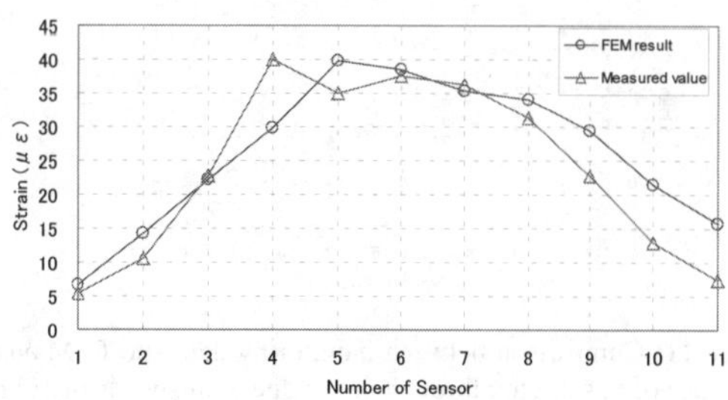

Figure 10: Comparison between monitoring and analysis in case of bus staying on the bridge

In the case of bus staying at the center of the bridge, the comparison between monitoring data and analysis results by FEM is shown in Figure 10. It can be observed that the strain distribution of monitoring data agrees well with that obtained from numerical analysis.

In the cases of bus driving through the bridge at speeds of 10km/h, 30km/h and 40km/h, the monitoring data is compared with the FEM analysis results, as demonstrated in Figure 11. The curves reveal that the strain obtained from the static FEM analysis in all tested cases show good agreements with that obtained by the FBG sensors, which then indicates that the static analysis by FEM could provide reasonable results for the dynamic loading when the speed is below 40km/h in this bridge.

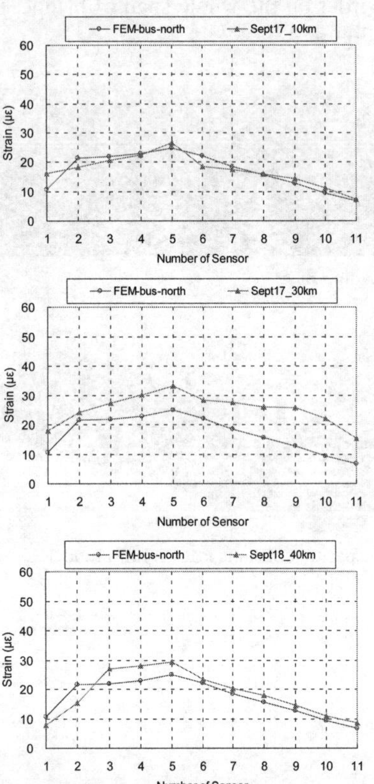

Figure 11: Comparison between monitoring data and FEM analysis in case of bus driving through the bridge with speeds of 10km/h, 30km/h and 40km/h

The obtained results of KAWANE bridge show that this bridge is on its good condition without detectable damages. The bridge conditions could be numerically simulated by static FEM analysis when the dynamic load passing through at a speed lower than 40km/h. Finally, it should be highlighted that the distributed long-gage FBG sensing system has a good accuracy in old bridge monitoring.

3.2 *The railway bridge*

To monitor a railway bridge, both the field experiments based on FBG sensors and FEM analysis were carried out (Yang et al. 2013). Firstly, the properties of the FBG sensors and the accelerometers which would likely to be installed in the railway bridge were chosen on the basis of the numerical analysis results which indicated that the frequency of the accelerometer should range from 0.5 to 50 Hz, and the measuring ranges should be from -1000 gal to +1000 gal to precisely and accurately record the data during this field experiment. Subsequently, the FEM model of bridge was improved up to coincide with the measured data. Two earthquakes over magnitude 3 occurred in monitoring period and the measured acceleration by the sensing system was compared with analytical result, together with a comparison of deflection. After the validation, according to the measured data from the field tests, FEM analysis was performed to examine the health conditions of the whole bridge.

Figure 12 shows the model for the FEM analysis and the measured earthquake and its acceleration response are demonstrated in Figure 13, which are the values obtained in the middle of the bridge perpendicular to the long axis.

Corresponding to the earthquake occurred in Aug. 30, 2013, the acceleration magnitude and its spectrum were recorded by the sensors and calculated by FEM analysis, which are displayed in Figure 14. Figure 15 notes that the comparison of the obtained deflections by FBG sensing system and FEM analysis. Observations of these figures reveal that the results obtained by distributed FBG sensing system agree well with those computed by FEM analysis. It then validates that the FBG sensors could effectively evaluate the properties of bridge during the dynamic loading.

Finally, the FEM analysis is conducted so as to investigate the overall conditions of the bridge according to the previous validated conditions. It would help to examine the bridge characteristics where the FBG sensors were not installed.

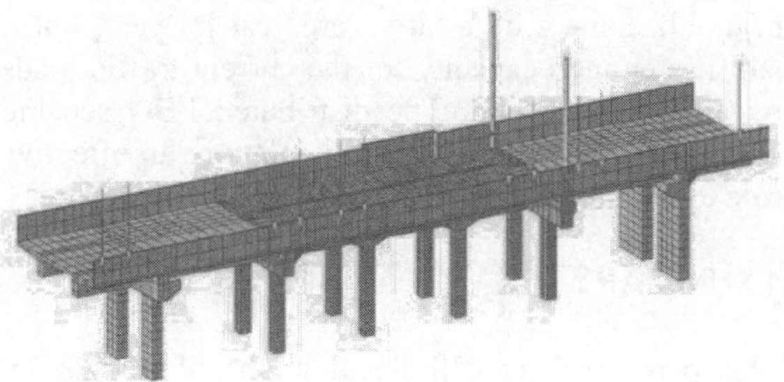

Figure 12: The FEM model of the railway bridge

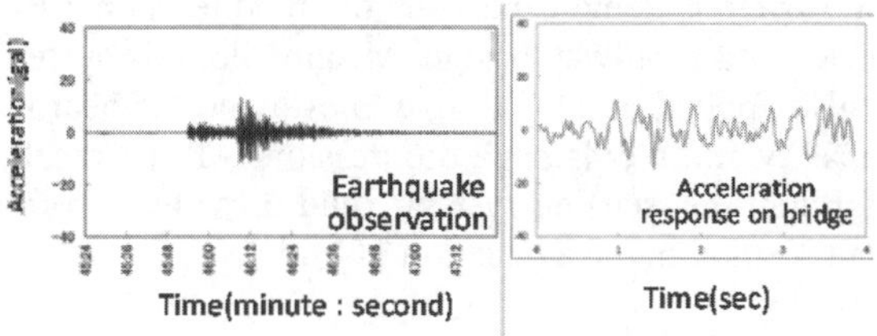

Figure 13: The measured earthquake and its acceleration response

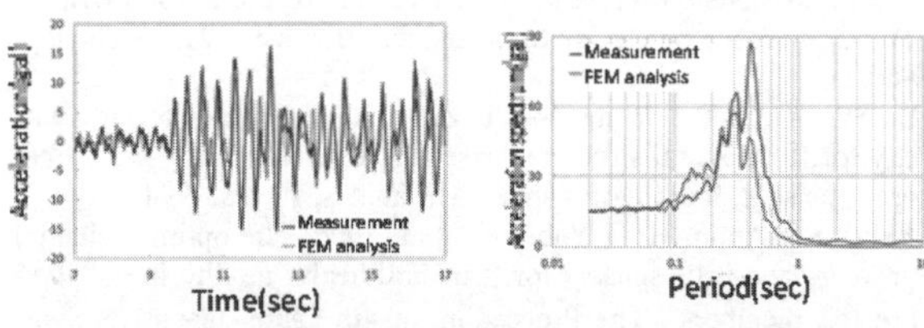

Figure 14: The comparison between the results from FBG sensing system & accelerometers and the analytical results in terms of acceleration magnitude and its spectrum

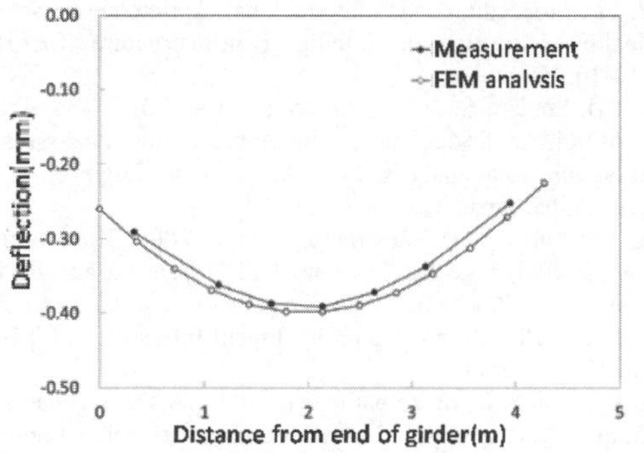

Figure 15: The comparison of the deflection by FBG sensing system and FEM analysis

3.3 The Myoko bridge

The last field test was carried on the Myoko bridge in Japan (Ishizawa et al. 2013). It is a 4 span PC continuous beam bridge built in 1972 as shown in Figure 16. In 2009, severe deteriorations and damages were found; therefore, the repairs were performed along the whole bridge. It was found that some PC steels were broken. Figure 17 indicates the view and the positions of some damages.

Although the Myoko bridge was subjected to sever damages, it is impossible to fully stop using it since it is located on the important national road. Because that the effectiveness of the distributed FBG sensing system has been validated, it was then practically employed in the Myoko bridge. There are two main purposes: 1) to estimate the bearing capacity of the bridge so as to manage the traffic loads passing through and 2) to observe the progress of the deteriorations in order to ensure whether the bridge can be used or not. Consequently, the configurations of the sensors in the first and fourth spans, which are bearing the most severe damages, are shown in Figures 18 and 19.

The experiments were performed by static load of 120t load in the middle of the tested span and dynamic load of a 20t truck driving through at a speed of 60km/h.

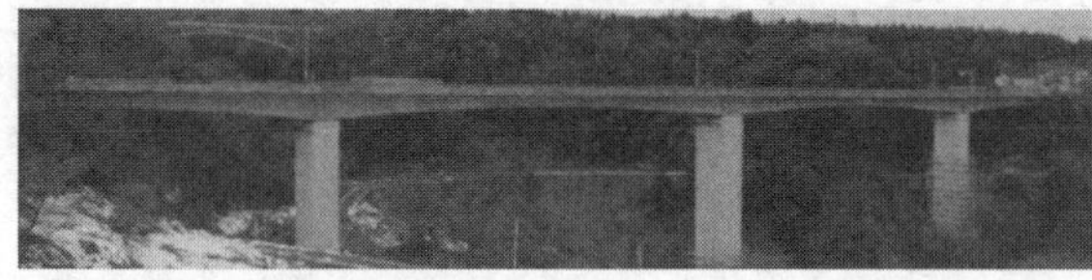

Figure 16: The overview of the Myoko bridge

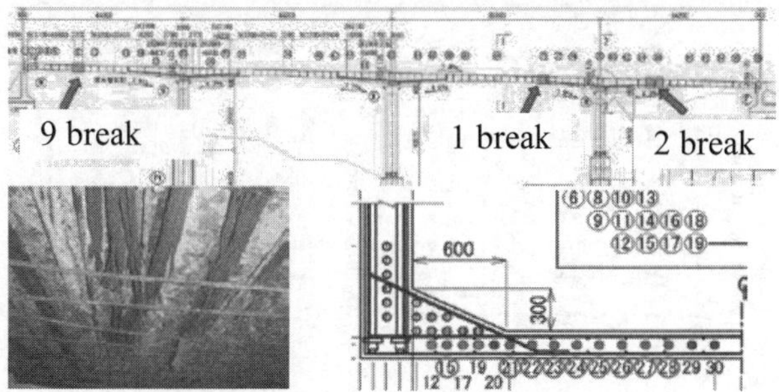

Figure 17: The view and locations of some damages

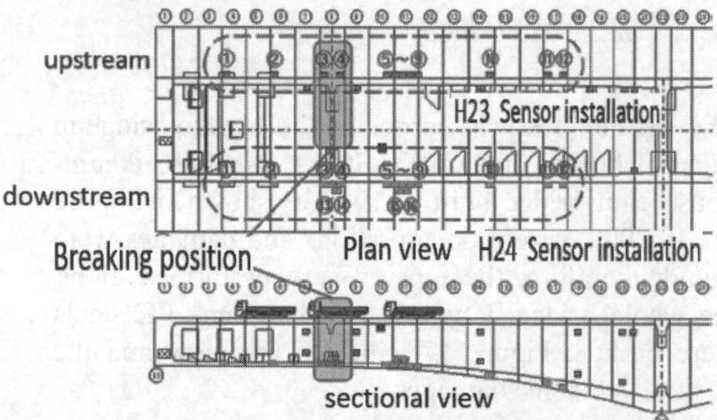

Figure 18: The sensor position of the first span

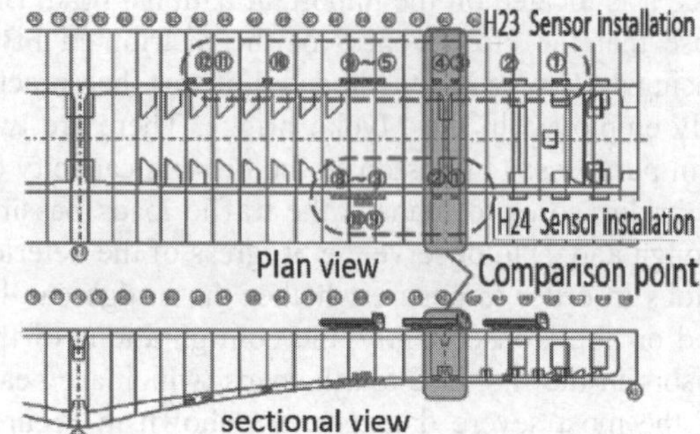

Figure 19: The sensor position of the fourth span

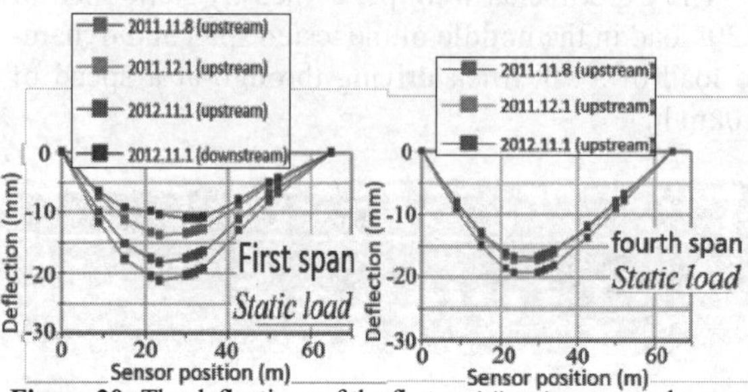

Figure 20: The deflections of the first and fourth spans under static load

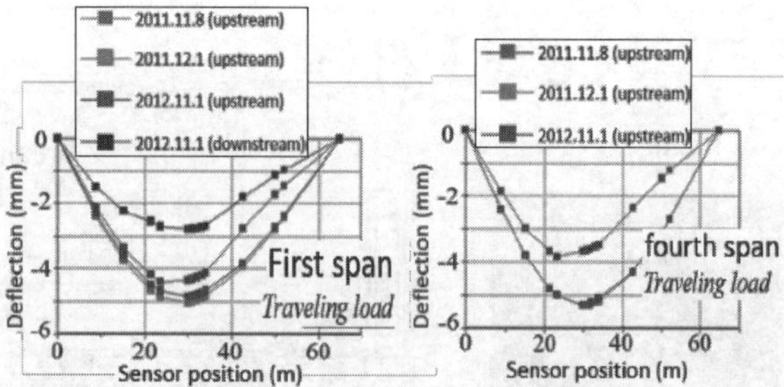

Figure 21: The deflections of the first and fourth spans under traveling load

Figures 20 and 21 show the deflections of first and fourth spans under static and traveling loads, respectively. The results reveal that the damages of the bridge become a little bit severe year by year, but it still has enough capacity for the current traffic loads by analytical criteria. The distributed FBG sensing system installed in the bridge is playing an effective role to monitor and evaluate the bridge in real-time.

4 SUMMARY

This paper presented the evaluations of bridges by distributed FBG sensing system, together with its description and application. The validation of the sensing system is demonstrated by the field tests on a RC bridge and a railway bridge. Meanwhile, this system is also applied in a PC bridge to estimate the bearing capacity and the damage progression. All the results highlight this sensing system could provide a prominent method in the evaluation of bridges.

REFERENCES

Ishizawa, K. Higuchi, R. & Maruyama, K. et al. 2013. The results and discussions on monitoring the structure of the Myoko Bridge. The Conference on Bridge Maintenance. Hokuriku, Japan. (In Japanese)

Li, SZ. Wu, ZS. & Watanabe, T. 2007. A Health monitoring strategy for RC flexural structures based on distributed long-gage fiber optic sensors, Journal of Applied Mechanics, 10, 983-994.

Yang, KJ. Araki, H. & Yabe, A. et al. 2006. The optimum length of long-gage FBG sensors for structural health monitoring of flexure RC members, The Proceeding of 4th China-Japan-US Symposium on Structural Control and Monitoring Oct.16-17, Hang-Zhou.

Yang, KJ. Araki, H. & Yabe, A. et al. 2007. Load identification of flexure RC structures using distributed FBG sensing system, P122, The Proceeding of 3th International Conference on Structural Health Monitoring on Intelligent Infrastructure (SHMII-3), Nov.14-16, Vancouver.

Yang, KJ. Suzuki, S. & Yamamoto, K. et al. 2013. The development of railway-bridge monitoring method and its assessment based on numerical analysis, JSCE Annual Civil Engineering Conference. Chiba. Japan. (In Japanese)

Yang, KJ. Yabe, A. & Yamamoto, K. et al. 2009. The Evaluation of KAWANE Bridge by Distributed FBG Sensors and by FEM Analysis, The Proceeding of 4th International Conference on Structural Health Monitoring on Intelligent Infrastructure (SHMII-4), July22-24, Zurich.

Yang KJ. Yonao R. & Yamamoto K. 2014. A Damage Detection Technique by Distributed Strain Measurements using Long-gage Fiber Bragg Grating Sensors, IABSE Symposium Madrid.

Proceedings of the International Conference on Smart Infrastructure and Construction
ISBN 978-0-7277-6127-9

© The authors and ICE Publishing: All rights reserved, 2016
doi:10.1680/tfitsi.61279.093

Applications and prospects for distributed sensing using polymer optical fibres

S. Liehr* and K. Krebber

Federal Institute for Materials Research and Testing (BAM), Berlin, Germany
* *Corresponding Author*

ABSTRACT One of the unique advantages of polymer optical fibres (POF) is that they can be used to measure very high strain values up to 100 % and beyond exceeding the strain limits of silica fibre-based sensor principles. In this paper the distributed strain measurement capabilities of POF based on backscatter change evaluation are summarized and distributed backscatter measurement technologies are introduced. Application examples in the structural health monitoring (SHM) field are presented: a promising approach is the integration into technical textiles for high-strain measurement in earthwork structures and crack detection in buildings. The potential of POF for future applications in SHM such as distributed relative humidity sensing is discussed.

1 INTRODUCTION

Optical fibre sensors (OFS) have gained increasing attention and commercial application during the last decade. One of the key advantages is their capability to measure continuously and spatially resolved (distributed) along the whole fibre length. Particularly distributed measurement of strain, temperature or vibrations opened new fields of application and even outperformed certain traditional measurement techniques. Most of the distributed OFS systems however, make use of and are limited to silica optical fibres. The unique mechanical advantages of polymer optical fibres (POF) further extend the scope of applications. In contrast to silica fibres, which are limited to maximum strain values of about 1 %, POF can be used to measure strain up to 100 % (Liehr 2015). This allows for high-strain applications such as crack detection or monitoring the dynamics of earthwork structures such as slopes, dykes or embankments.

The POF-specific backscatter effects that enable distributed strain measurement are introduced in section 2. In section 3 application examples with POF integrated into technical textiles and dynamic long-gauge measurement in a seismic shaking test are presented. Section 4 introduces the possibility for distributed humidity sensing using POF.

2 STRAIN MEASUREMENT TECHNIQUES

Two different POF types proved to be suitable for distributed backscatter measurement. However, their optical and mechanical properties, response to strain and the measurement techniques differ. Poly(methyl methacrylate) (PMMA) POF and low-loss perfluorinated (PF) POF are therefore presented in separate sub-sections. Both introduced measurement techniques are static (seconds to minutes measurement time).

2.1 *PMMA POF and OTDR*

The POF-specific local backscatter increase as a function of strain has first been observed in PMMA POF by Husdi et al. (2004) and has later been investigated for distributed sensing applications (Lenke 2007, Liehr 2009). Commercially available optical time domain reflectometers (OTDR) have initially been used to localize and quantify the relative backscatter changes – a short optical pulse is sent into one end of the POF and all backscattered power is sampled as a function of time or distance, respectively. Figure 1 shows exemplarily OTDR measurement traces of the same PMMA POF for different strain values applied on a 1.4 m long fibre section at 42 m.

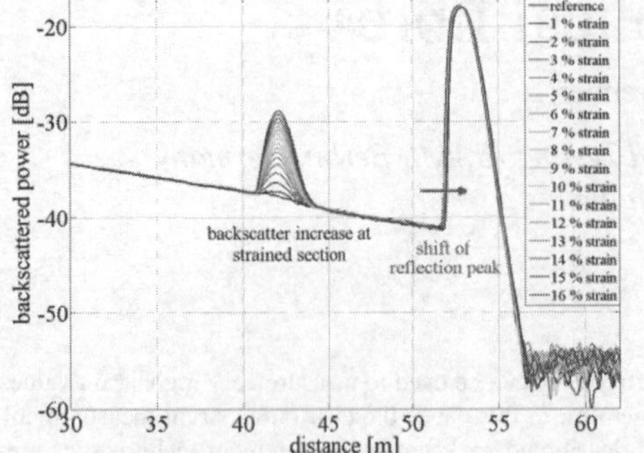

Figure 1. OTDR traces of strained POF (0% strain to 16% strain).

By evaluating relative backscatter changes, strained fibre sections can be located and strain values can be estimated. Spatial shifts of the reflection peak at the fibre end at 53 m can be used for total length change measurement. Figure 2 shows the relative backscatter increase as a function of applied strain up to 16 % strain and 45 %.

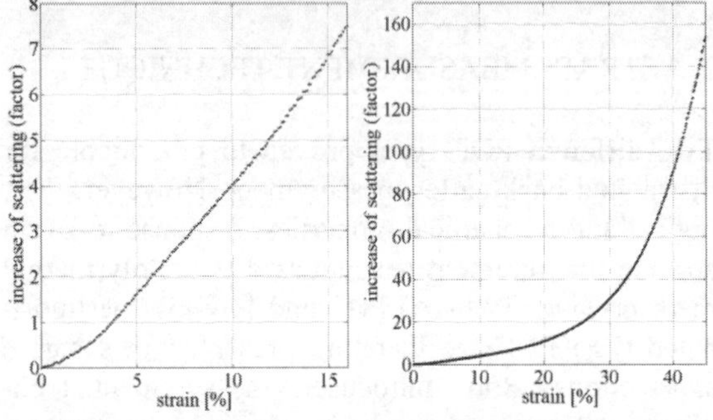

Figure 2. Relative backscatter power change (factor) as a function of strain up to 16 % strain (left) and 45 % strain (right).

It has to be noted that the backscatter increase signal has a temporal dependency decaying over time (Liehr 2009) which reduces the strain measurement precision. Standard PMMA POFs have a large 1 mm diameter core, are very robust and can endure strain exceeding 40 %, depending on the strain rate up to 80 %, and can, due to their relatively high attenuation, be used for sensor lengths up to 100-120 m.

2.2 *Low-loss POF and I-OFDR*

Perfluorinated (PF) POFs exhibit considerably reduced attenuation compared to PMMA POF (Liehr 2015) and allow for extending the measurement range up to several hundreds of metres. The maximum strain range of PF POF exceeds 100 %. These fibres are also compatible with low-cost standard telecom components regarding attenuation minimum around 1310 nm and core diameter (50 µm). Their gradient-index structure prevents modal dispersion and ensures maximum spatial resolution.

2.2.1 *I-OFDR setup*

We developed a PF-POF-compatible measurement setup based on incoherent optical frequency domain reflectometry (I-OFDR) that features by far increased signal stability and spatial resolution compared to OTDR (Liehr 2015), see Figure 3.

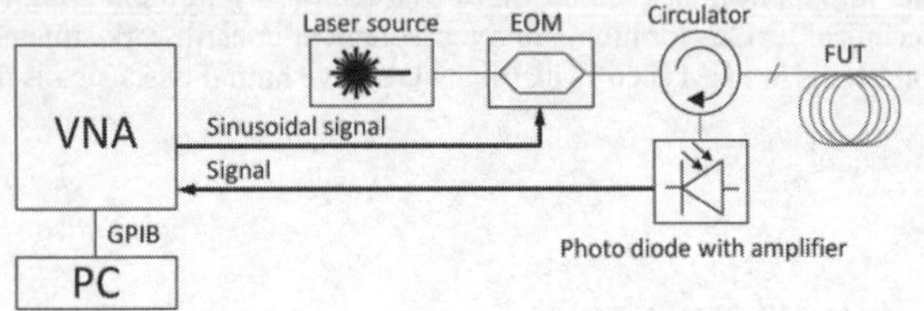

Figure 3. Schematic of the I-OFDR setup.

The measurement is conducted in the frequency domain: a vector network analyzer (VNA) generates a sinusoidal signal that amplitude-modulates a continuous-wave laser source via an electro-optic modulator (EOM). The optical circulator couples the modulated signal into the fibre under test (FUT). All backscattered power is detected by a photo detector and fed to the VNA. The VNA basically measures the complex frequency response of the FUT stepwise over a wide frequency range up to 2 GHz (corresponding to a spatial resolution of 3.4 cm). Applying the inverse fast Fourier transform (IFFT) to the frequency domain result yields the time domain response equivalent to an OTDR measurement. The I-OFDR approach shows significantly improved signal stability and spatial resolution compared to high-end OTDR devices (Liehr et al. 2010).

2.2.2 *Strain and length change measurement*

Compared to PMMA POF, the backscatter increase in PF POFs is less distinct and has a nonlinear dependency on strain. Although direct strain measurement is not possible, strained fibre sections can be located with cm-spatial resolution. A very useful characteristic of PF POF is the presence of strong randomly distributed scattering centres along the fibre. These scattering centres can be used to measure absolute length change Δl along the fibre. Figure 3 shows backscatter increase (several I-OFDR traces) due to different Δl applied to a 1 m long fibre section.

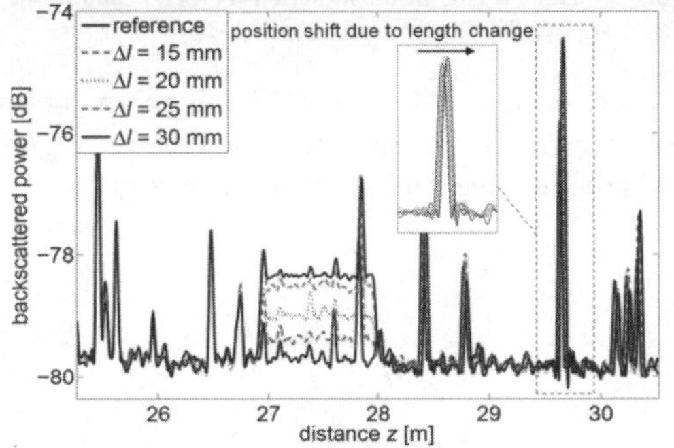

Figure 4. I-OFDR measurements of stepwise strained PF POF section between 27 m and 28 m, from Liehr (2015).

The applied strain results in a shift of the scattering centres towards greater distance. Due to the high signal stability of the I-OFDR approach and the high spatial resolution, relative length change measurement along the whole fibre can be conducted. By conducting step-wise cross-correlation of overlapping backscatter trace sections (with correlation length l_c) of a new measurement with a reference measurement, distributed length change results along the whole fibre length can be obtained. Figure 5 shows such measured length changes results along the same PF POF section as shown in Figure 4.

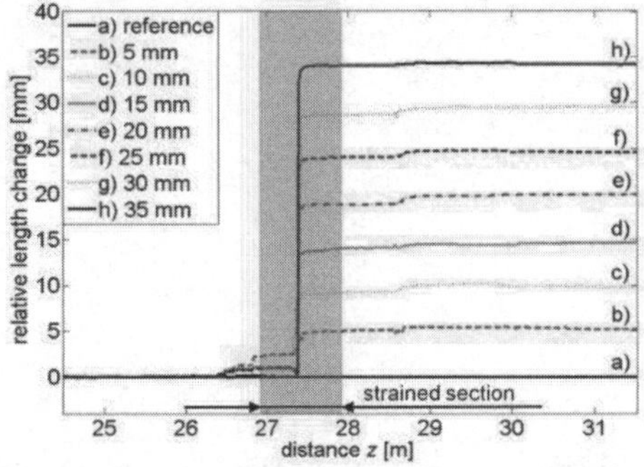

Figure 5. Distributed length change (cross correlation result with reference) for different length changes applied (correlation length $l_c = 2$ m), from Liehr (2015).

Figure 4 and Figure 5 show that, using I-OFDR and PF POF, strained fibre sections can be localized with cm-resolution and distributed length change can be measured with sub-mm resolution even at a relatively short correlation length of $l_c = 2$ m.

For certain applications, a precise distributed length change measurement may actually be the more interesting parameter than local strain results. Although PF POF can be strained up to 144 % (Liehr 2015), the elastic strain range is limited, depending on the overcladding and jacketing materials, to a few percent. Exceeding this limit, viscoelastic and plastic deformation occurs.

3 APPLICATION EXAMPLES

The small dimension of the sensor fibre and its geometric versatility eases application or integration into various structures. Application examples including POF integrated into technical textiles and bonded to a masonry structure are presented in this section.

3.1 *POF in technical textiles*

Today, technical textiles are used for various applications fulfilling different tasks from reinforcement to drainage. This already puts them into a location and orientation of interest to SHM monitoring schemes. The sensor integration into the textile during the production process has been achieved within the POLYTECT project (Liehr et al. 2009). The sensor fibres are low-cost and no additional working steps are required during textile integration. To prove the applicability of such "smart textiles", various laboratory and field tests have been conducted. Figure 6 shows a laboratory test setup for crack detection.

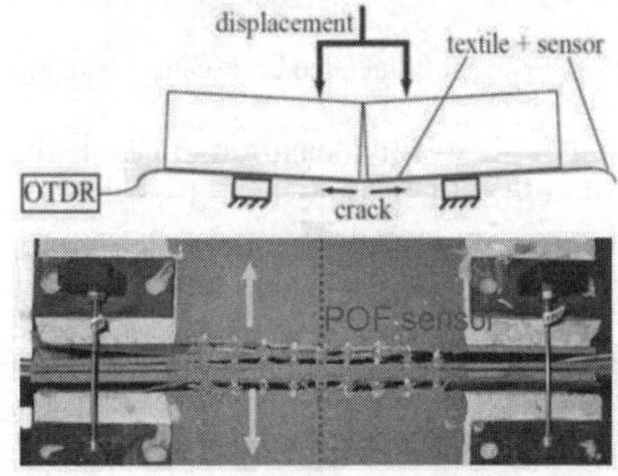

Figure 6. Test setup at the IfMB at Karlsruhe Institute of Technology for forced crack opening perpendicular to POF sensor textile.

A forced displacement on top of a two-stone sample results in a crack perpendicular to the PMMA POF sensor textile at the bottom of the sample. Cracks from 1 mm width could easily be detected with the OTDR. Figure 7 shows the development of the sensor signal (relative backscatter change) for increasing crack widths.

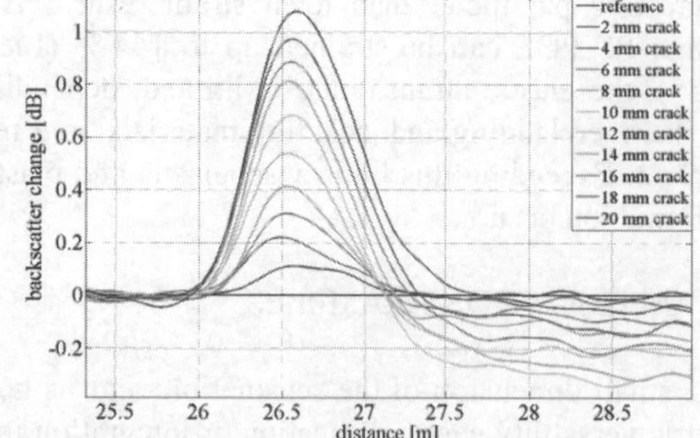

Figure 7. Backscatter change results for different crack widths.

Different textile types have been tested. Figure 8 shows exemplarily the integrated backscatter increase for two textile types. These results indicate that the position as well as the crack size could under known conditions be estimated from the sensor signal.

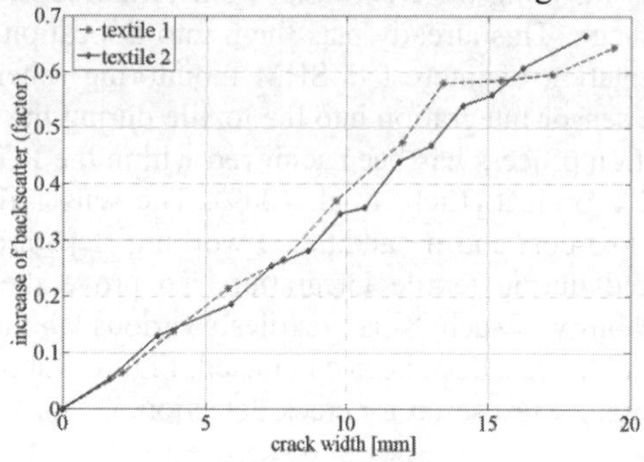

Figure 8. Sensor signal (integrated backscatter change).

Another high-strain measurement task that cannot be solved with silica fibre-based sensor systems is the measurement of strong deformations in earthwork structures such as a sliding slope. Figure 9 shows the installation of a PMMA POF sensor-equipped textile that has been installed perpendicular to the tear-off edge of a creeping slope at an open brown coal pit (Liehr et al. 2009).

Figure 9. Sensor textile installation (left) perpendicular to the tear-off edge (right), from Liehr et al. (2009).

After the installation, OTDR measurements have been conducted in irregular intervals. Figure 10 shows the relative backscatter change results.

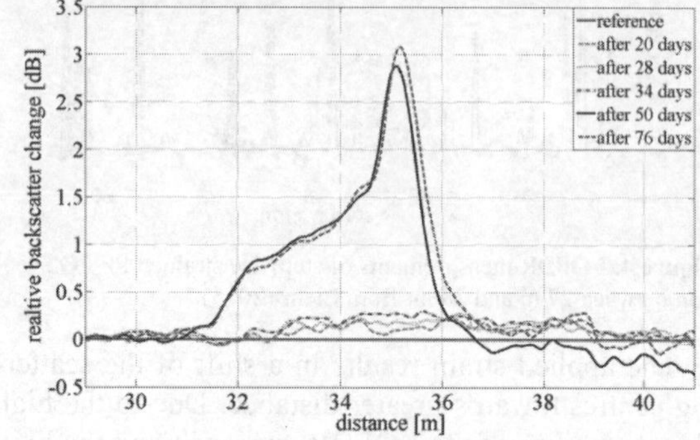

Figure 10. Change of backscatter signal relative to the reference.

The strong peak at 35 m indicates a very high local strain exceeding 10 % which has been caused by the rupture of the textile. These high strain values can only be endured by POF. An additional evaluation of the total length change of the sensor fibre from the position shift of the reflection peak at the fibre end (see Figure 1) is shown in Figure 11.

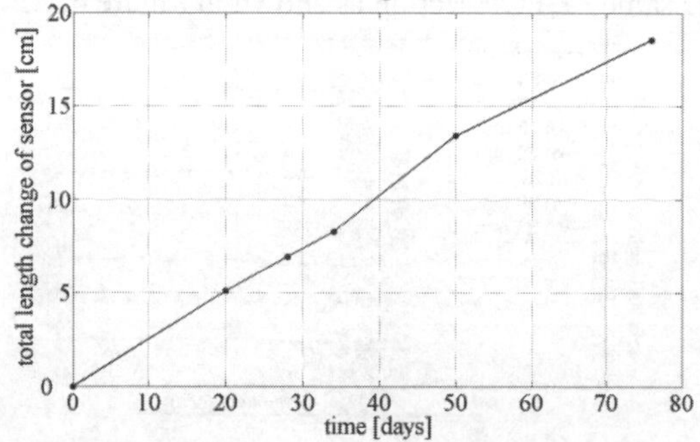

Figure 11. Total elongation obtained by peak shift evaluation.

This way, a relatively linear average creep velocity of about 2 mm per day has been determined. The use of PF POF and I-OFDR would considerably improve the spatial and length change resolution.

3.2 *Dynamic long-gauge measurement*

Based on the above described I-OFDR approach, we proposed to use reflection points in the sensor fibre as reference points for long-gauge strain measurement. By evaluating only a few modulation frequency results, it is possible to conduct precise length change measurement between reflections at an increased measurement repetition rate up to 2 kHz (Liehr 2012). µm-range resolution can be obtained at lower measurement repetition. Multiple sensor sections can be defined by individually placing reflections. Using this technique, the deformation of a masonry building on a seismic shaking table has been measured. Figure 12 shows the positions of 3 sensor sections fixed at the test object (1 PF POF, two silica fibres, each separated by reflections in the fibre).

Figure 12. Schematic of the installed sensor fibres (sensor 1 - PF POF; sensor 2 and 3 - silica fibre), from Liehr (2015).

Figure 13 shows the measured length changes Δl of all sensor sections during a seismic load from 27 s.

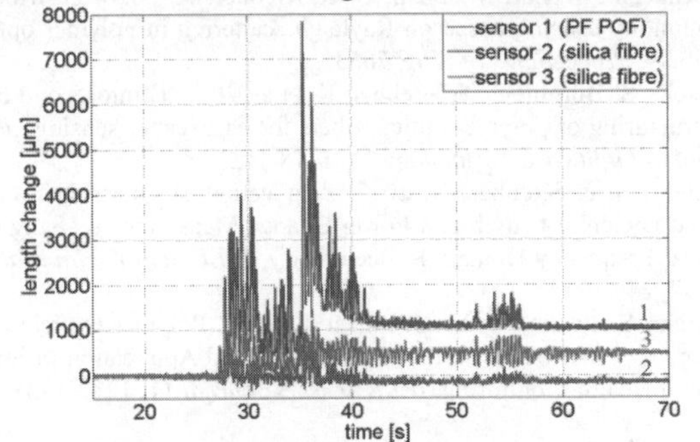

Figure 13. Length change results of all sensor fibres during the seismic load at a measurement repetition rate of 160 Hz.

The highest deformation occurred in the POF sensor 1 with a maximum elongation of 8 mm. It can be seen that new cracks along sensor 1 and sensor 3 caused a permanent deformation of 1 mm and 500 µm respectively. The advantage of this dynamic I-OFDR-based approach is that the gauge lengths can be chosen between cm and km and also multimode POF and silica fibres can be used for high-strain sensing applications. In addition to length changes also optical power changes at each reflection can be measured simultaneously. This allows for extending the scope of applications to other intensity-based fibre optic sensing principles (Liehr 2015).

4 DISTRIBUTED HUMIDITY MEASUREMENT

Distributed humidity sensing is a measurement task that cannot yet be accomplished for large-scale applications. POF have the potential to provide a solution for distances up to several hundred metres. Unlike silica fibres, POF are susceptible to relative humidity changes or the presence of moisture. We first investigated this effect and its dependencies in PMMA POF using OTDR (Lenke et al. 2010). PMMA absorbs water molecules to a concentration up to 3 %. This results in changes of fibre attenuation, backscattered power and refractive index of the fibre. Figure 14 shows exemplarily the measured relative backscatter change of a PMMA POF section (between 30 m and 40 m) which is placed in a climate chamber and is subjected to changes of relative humidity

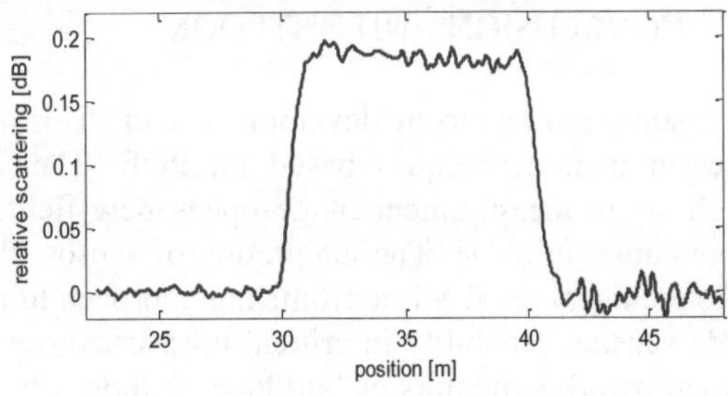

Figure 14. Relative backscatter change of a PMMA fibre section at lower relative humidity between 30 m and 40 m.

The section of relative backscatter increase can easily be localized. Cyclic measurements of backscatter changes and fibre attenuation show, that these effects are reversible and reproducible, Figure 15.

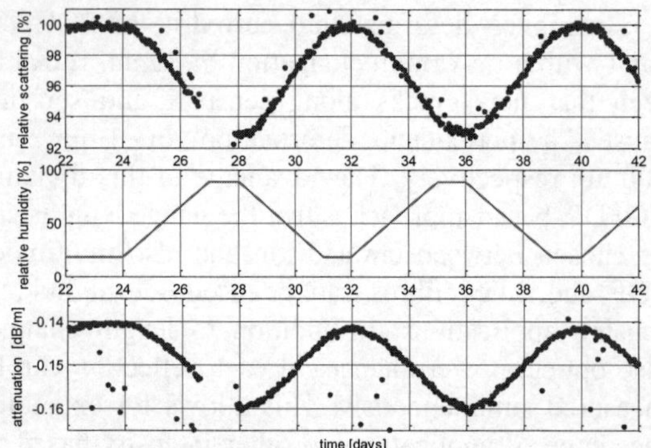

Figure 15. Relative backscatter change and fibre attenuation during cyclic relative humidity change in the climate chamber, from Lenke et al. (2010).

Relative humidity values along the length of the fibre can be derived from the backscatter change evaluation relative to a reference measurement. The response time to relative humidity changes in PMMA is in the order of one day at room temperature.

Recent results of the influence of relative humidity on PF POF using I-OFDR showed promising characteristics. Backscatter change as well as attenuation changes due to absorption have been measured (Liehr 2015). The absorption of OH groups has a spectral dependency which will be used to uniquely identify wet fibre sections and estimate relative humidity values along the sensor fibre. The use of low-loss PF POF and I-OFDR extends the measurement distance and spatial resolution compared to PMMA POF.

5 CONCLUSION AND OUTLOOK

We summarized recent developments in distributed measurement techniques based on POF. The very high strain measurement range opens new fields of application in SHM. The integration of sensor fibres into technical textiles is a promising approach to provide sensing capability in critical infrastructure such as earthwork structures or buildings. Laboratory and filed test results for strain measurement proved the applicability of smart POF textiles.

The development of the I-OFDR technique considerably improved the measurement resolution compared to existing OTDR techniques. A dynamic long-gauge strain measurement technique compatible with PF POF has been introduced.

Ongoing research on humidity influence promises sensor solutions for truly distributed humidity sensing applications. Further investigations show that POF are suitable also for quasi-distributed curvature and temperature sensing by creating scattering centres in the fibre core (Liehr et al. 2013). Research on fibre materials and production optimization is ongoing. Lower attenuation fibres and increased sensor lengths could be expected.

ACKNOWLEDGEMENT

This research has been carried out within the framework of the European project POLYTECT (NMP2-CT-2006-026789), within the BAM innovation program project "InnoPOF" and the EC's 7th Framework Programme (grant no. 227887 – SERIES Project). The authors are thankful for the financial support and the fruitful cooperation with all project partners.

REFERENCES

Husdi, I.R. Nakamura, K. & Ueha, S. 2004. Sensing characteristics of plastic optical fibres measured by optical time-domain reflectometry, *Measurement Science and Technology* **15**, 1553-1559.

Liehr, S. Wendt, M. & Krebber, K. 2010. Distributed strain measurement in perfluorinated polymer optical fibres using optical frequency domain reflectometry, *Measurement Science and Technology* **21**, 094023(6pp).

Lenke, P. Liehr, S. Krebber, K. et al. 2007. Distributed strain measurement with polymer optical fiber integrated in technical textiles using the optical time domain reflectometry technique, *Proceedings of POF Conference*, 21–24.

Liehr, S. 2015. Fibre Optic Sensing Techniques Based on Incoherent Optical Frequency Domain Reflectometry, Ph.D. dissertation, BAM Dissertationsreihe, Band 125.

Lenke, P. Wendt, M. Liehr, S. & Krebber, K. 2010. Distributed humidity sensing based on Rayleigh scattering in polymer optical fibers, *Proceedings of SPIE* **7653**.

Liehr, S. Burgmeier, J. Krebber, K. et al. 2013. Femtosecond laser structuring of polymer optical fibers for backscatter sensing, *Journal of Lightwave Technology* **31**, 1418-1425.

Liehr, S. & Krebber, K. 2012. Application of Quasi-Distributed and Dynamic Length and Power Change Measurement Using Optical Frequency Domain Reflectometry, *IEEE Sensors Journal* **12**, 237–245.

Liehr, S. Lenke, P. Wendt, M. et al. 2009. Polymer Optical Sensors for Distributed Strain Measurement and Application in Structural Health Monitoring, *IEEE Sensors Journal* **11**, 1330-1338.

Proceedings of the International Conference on Smart Infrastructure and Construction
ISBN 978-0-7277-6127-9

© The authors and ICE Publishing: All rights reserved, 2016
doi:10.1680/tfitsi.61279.099

Iterative filtering for Time-Frequency Localised Pulse Optimisation in STFT-BOTDR

Linqing Luo[1], Bo Li[1], Yifei Yu[1], Xiaomin Xu[1], Jize Yan[1, 2]*and Kenichi Soga[1]

[1] *Department of Engineering, University of Cambridge, Cambridge, UK*
[2] *Electronics and Computer Science, University of Southampton, Southampton, UK*
[*] *Corresponding Author*

ABSTRACT: Dynamic strain measurement in distributed fibre optic sensing (DFOS) is essential for structural health monitoring (SHM) of the strain changes induced by construction failure and other activities in infrastructure's life cycle. Among different DFOS systems, the Short Time Fourier Transform-Brillouin Optical Time Domain Reflectometry (STFT-BOTDR) takes the advantages of STFT obtaining full frequency spectrum to improve the performance of conventional BOTDR, providing an opportunity for dynamic sensing. The key parameters of distributed fibre optic sensors, spatial and frequency resolution, are strongly linked with the pulse's time-frequency localisation. In this paper, a set of Kaiser-Bessel functions is used to simulate different pulse shapes and compare their parameters in terms of Time-Frequency Localisation (TFL) and their Brillouin scattering spectrum. A method using iterative filtering algorithm to achieve the optimised pulse in terms of TFL is introduced to converge the Effective-pulse Width (TEW) in time-domain and Effective-pulse Linewidth (FEL) in the frequency domain, respectively, to the fundamental limitation. The optimised pulse can be fitted with 7th order Gaussian (super-Gaussian) shape and offer the best experimental performance compared to Rectangular pulse.

1 INTRODUCTION

Distributed fibre optic sensing, especially Brillouin Optical Time Domain Reflectometry (BOTDR), allows measurement of strain and temperature at any location along a single mode optical fibre up to a hundred kilometres (Bao and Chen 2011). Comparing with conventional sensing systems, this provides new opportunity for distributed and dynamic Structure Health Monitoring (SHM). However, the Time-Frequency Localisation (TFL) in the signal pulse affects the signal-to-noise ratio (SNR) and limits the resolutions, sensing distance and measurement speed, which increasing the difficulty for dynamic sensing (Luo et al. 2016).

Shape of the pulse is considered as having significant contribution on reforming the Brillouin scattering spectrum, hence the shape needs to be optimised to provide a good SNR of the spectrum to enhance the frequency resolution and remain good spatial resolution simultaneously. Different pulse shapes resulted in different spectrum bandwidth and different frequency error (Naruse and Tateda 2000). Lorentzian shape was considered to be better than Triangular shape in terms of Brillouin spectrum's peak power (Hao et al. 2013). However, previous research analysed the frequency domain information independently but omitted its iteration with time domain information which would contribute to the spatial resolution. Because of the TFL limitation of time-frequency analysis, improving in frequency resolution will sacrifice the spatial resolution. Therefore, a balanced and optimised pulse need to be introduced to improve the spatial resolution and frequency resolution, simultaneously.

In this paper, Kaiser-Bessel functions with different parameters of attenuation slope are used to simulate Gaussian, Hamming, Rectangular pulses as the input of the BOTDR system. The pulses are compared in a mathematical model to reveal their relation with the frequency resolution. An iterative filtering algorithm is introduced to optimise the pulse shape to enhance the system TFL. The simulation result shows that the Brillouin spectrum bandwidth can be improved by more iterations. The experimental result demonstrates that the ratio of peak frequency power and total power can be enhanced by using the optimised pulse generated by the iterative filtering algorithm comparing with a rectangular pulse in the STFT-BOTDR system.

2 PULSE EFFECT ON BRILLOUIN SPECTRUM

To evaluate the effect of the pulses on the BOTDR, the pulses are simulated by Kaiser-Bessel functions (Rabiner and Gold 1975) (Lewitt 1990). The different pulses are represented as below:

$$S_n = \begin{cases} \sqrt{\dfrac{\sum_{i=0}^{n} w[i]}{\sum_{i=0}^{M} w[i]}} & if\ 0 \le n \le M \\ \sqrt{\dfrac{\sum_{i=0}^{2M-1-n} w[i]}{\sum_{i=0}^{M} w[i]}} & if\ M \le n < 2M \\ 0 & otherwise \end{cases} \tag{1}$$

$$w[n] = \begin{cases} \dfrac{I_0\left(\pi\beta\sqrt{1-(\frac{2n}{N-1}-1)^2}\right)}{I_0(\pi\beta)}, & 0 \le n \le N-1 \\ 0 & otherwise, \end{cases} \tag{2}$$

Where N is the length of sequence. I_0 is the zeroth order modified Bessel function of the first kind, which is the solution of at zeroth order Bessel's differential equations. β is a non-negative parameter that decides the shape of the window, which represents the trade-off between the main-lobe width and the side lobe level.

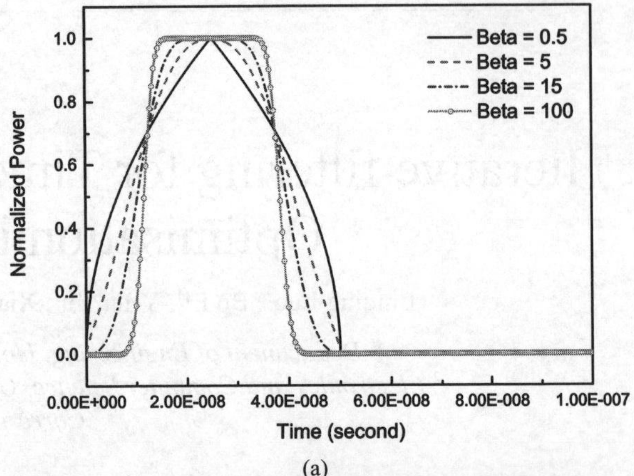

(a)

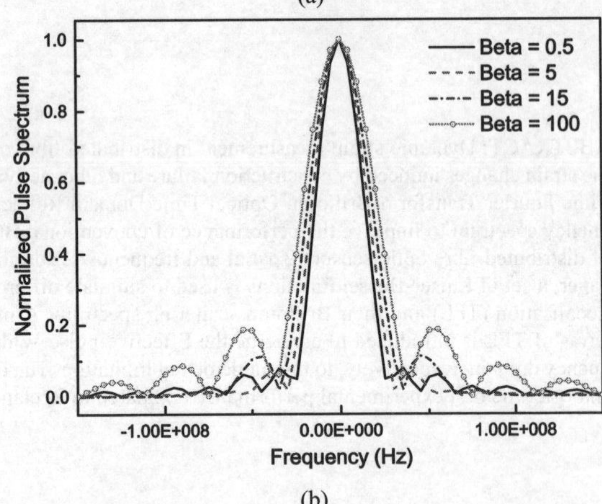

(b)

Figure 1. The representation of the Kaiser-Bessel shape pulses with different parameterisation beta (a)The time domain pulse shape and (b) frequency domain spectrum

Tuning β can generate different pulse shapes such as Gaussian, Hamming, Rectangular, etc. These pulse shapes have attenuating spectrum distribution, shown in Figure 1, where the β is chosen to be 0.5, 5, 15 and 100 with the same pulse width, and represent for shapes from Triangular to Rectangular, respectively. Assuming the noise effect can be omitted in the Brillouin scattering process, the pulses' effect on Brillouin scattering can be represented by a simplified mathematical model in Equation (3) (Naruse and Tateda 1999). After convoluting the pulses' spectrum and Brillouin scattering gain, the spectrum of Brillouin backscattered signals generated by different pulses are compared in

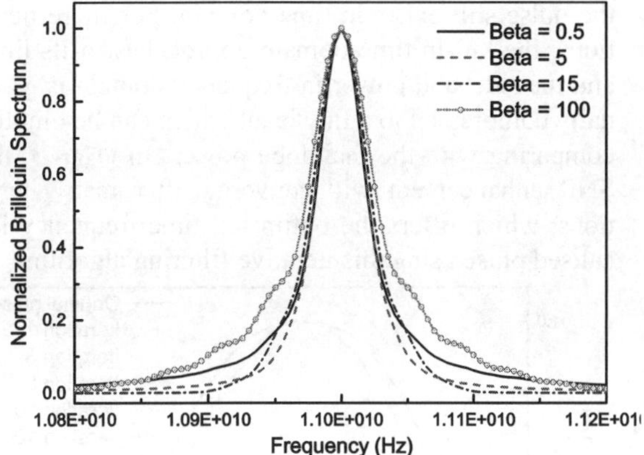

Figure *2.*

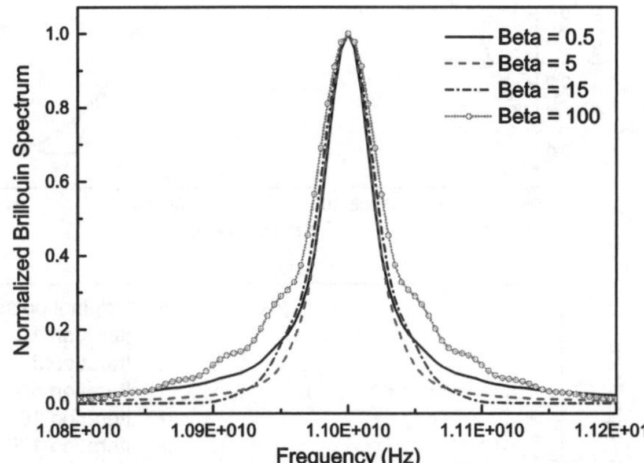

Figure 2. The Brillouin spectrum of different pulses with different parameterisation β of Kaiser-Bessel shape.

$$H(v) = \int_{-\infty}^{+\infty} P_p(f, f_0) \frac{h(\omega/2)^2}{[v - (f - S_B)^2] + (\omega/2)^2} df \quad (3)$$

where $P_p(f, f_0)$ is the power spectrum of the launched pulse. H(v) is the Lorentzian shape spectrum of the Brillouin backscattering light with peak frequency of $f - S_B$ and the full width at half maximum (FWHM) of ω. The term S_B expresses the frequency variation due to local acoustic waves, where the variation comes from the changes in the properties of the fibre, or the changes in strain or temperature.

Figure 2 shows that the ratio of the peak frequency power and the whole spectrum power increases while β decreases, indicating that the Brillouin spectrum expands when the pulse has larger linewidth. The **Figure** *3* shows the standard deviation of the peak frequency for different pulse shapes on a uniform optical fibre without strain and temperature variance but with varied averaging numbers. When β is equal to 5, the standard deviation is the smallest, which shows the best frequency resolution and the least measurement time among the four shapes using different β values. As a comparison, it needs 32 times of averaging for β equal to 5 to achieve a similar frequency resolution comparing with 256 times of averaging when β is equal to 15.

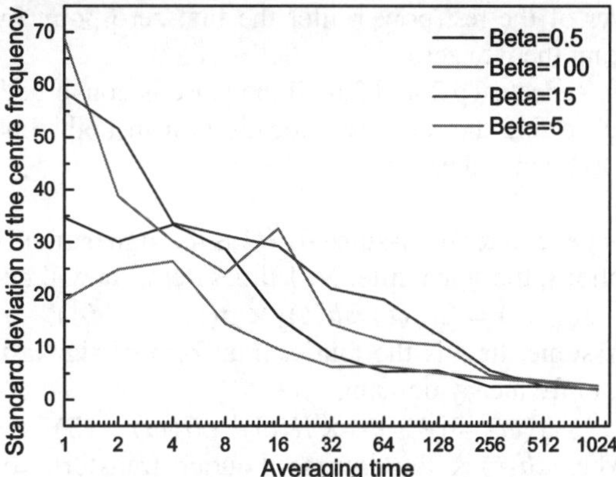

Figure 3. The standard deviation of the centre frequency on the uniform strain section when different pulses with different average numbers are applied to the BOTDR system

3 ITERATIVE FILTERING ALGORITHM

Due to the TFL, the improvement in frequency resolution sacrifices spatial resolution (Luo et al. 2016). Hence, an optimised pulse is needed to offer an optimised resolution in both time and frequency domain. As stated in earlier section, the sideband in frequency domain can increase the standard deviation of the peak frequency in BOTDR, i.e. reduce the frequency resolution. Similarly, in time domain, the increasingly length of the signal tail of the pulse shape will reduce the spatial resolution. Therefore, optimisation should focus on the reduction of the signal tails in time domain and the signal sidebands in frequency domain.

In this paper, an iterative filtering method to optimise the pulse shape and enhance the TFL is introduced by applying iterations to cut off the tails in time domain and sideband in frequency domain in consequence cycles to converge to the optimised pulse shape. The principle is:

1. Start with a rectangular pulse shape with a pre-defined frequency spectrum and time length.
2. Transfer the pulse into frequency domain and find its first positive frequency point that crosses zero at frequency δ. Then cut off the power of the rest harmonics in the sideband by making them to zero.
3. Transfer the modified frequency domain signal back to time domain. Find its first positive point which crosses zero along the signal at point τ and cut off the power of the rest power after the first zero point by making them to zero.
4. Repeat step 2 and 3 until the pulse is converged, which offers the optimised localisation in both time and frequency domain

In general terms, assume $a_N(t)$ is the signal after N iterations, the signal after N+1 times iteration will be

$$a_{N+1}(t) = [a_N(t) * h(t)] \times g(t) \quad (4)$$

Assume the A is the fourier transform of signal a, then in frequency domain,

$$A_{N+1}(\omega) = [A_N(\omega) \times H(\omega)] * G(\omega) \quad (5)$$

where $h(t)$ is the inverse Fourier transform of $H(\omega)$ and $G(\omega)$ is the Fourier transform for $g(t)$. $H(\omega)$ is the ideal low pass filter with cut-off frequency at δ expressed as

$$H(\omega) = U(\delta - \omega) \quad (6)$$

while $g(t)$ is the ideal rectangular window with close time at τ expressed as

$$g(t) = U(\tau - t) \quad (7)$$

The iteration method can also enhance the SNR of Brillouin scattering spectrum as shown in **Table** *1*. The SNR enhancement is different for different pulse shapes with same iteration number.

Table 1. The original SNR for pulses given different β and the SNR after the Modification

Pulse	Original SNR	Modified SNR
$\beta = 0$	23.308	25.495
$\beta = 5$	26.797	28.244
$\beta = 100$	20.257	29.086

The SNR increases when the iteration number is augmenting (**Figure** *4*). For one iteration, the frequency sideband reduction is an imperfect cut-off using a low pass filter and only the main lobe remains. The tail of the pulse still exists in time domain. For more iterations, the tail in time domain approaches to its limit and the sideband power in frequency domain is gradually compressed to a tiny level which can be omitted comparing with the mainlobe power. In **Figure** *5*, the SNR enhancement will converge after many iterations, which offers the optimised time-frequency localised pulse using this iterative filtering algorithm.

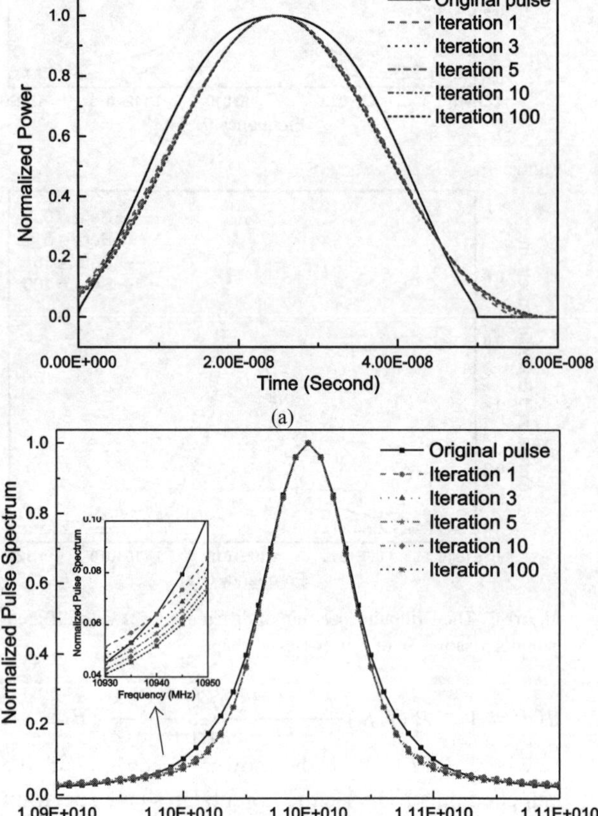

Figure 4. (a) The simulation result of pulse shape in time domain when pulse is original shape with β=5 and pulses are the modified shapes with different iteration numbers. (b) The simulated Brillouin scattering spectrum when these pulses are sent into BOTDR system.

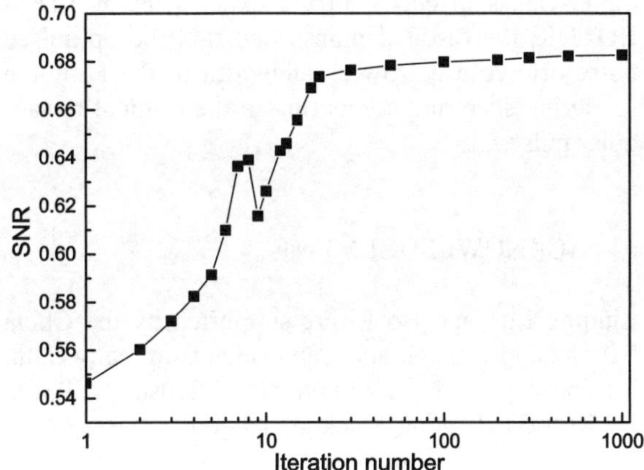

Figure 5. The SNR trend when iteration number increases

4 SUPERGAUSSIAN FITTING AND EXPERIMENTAL RESULTS

The initial pulse is a rectangular pulse with the pulse width of 50 ns and the period of 1 us. The iterative filtering method is then applied on the initial pulse to reduce its tail and sideband in time-frequency domain in iteration cycles. The optimised pulse shape is found out to a super-Gaussian curve, which can be fitted by 7th order Gaussian equations, whose coefficients are shown in **Table** *2*.

Table 2. The fitted 7th order super-Gaussian equation and the coefficients

Equation	f(x) = a1*exp(-((x-b1)/c1)^2) + a2*exp(-((x-b2)/c2)^2) + a3*exp(-((x-b3)/c3)^2) + a4*exp(-((x-b4)/c4)^2) + a5*exp(-((x-b5)/c5)^2) + a6*exp(-((x-b6)/c6)^2) + a7*exp(-((x-b7)/c7)^2)				
Coefficients:					
a1	0.33280	b1	0.04529	c1	0.05135
a2	0.00019	b2	-0.00397	c2	0.00696
a3	0.00000	b3	0.01891	c3	0.00000
a4	0.88100	b4	-0.01279	c4	0.06381
a5	-0.02095	b5	-0.02148	c5	0.03517
a6	0.12900	b6	-0.08062	c6	0.05222
a7	0.10550	b7	0.09070	c7	0.04861

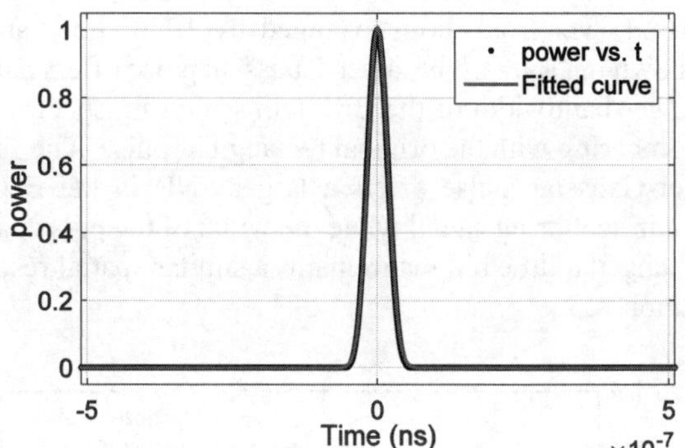

Figure 6. Modified pulse and its 7th order Super-Gaussian fitted curve

The Super-Gaussian pulse, shown in **Figure** *6*, is experimentally tested in a STFT-BOTDR setup shown in **Figure** *7*. An ultra-narrow-line-width laser with 1554.12nm wavelength was used as the light source. In branch A, after passing a coupler, a part of the continuous-wave (CW) light was modulated by an electro-optic modulator (EOM) with the 50 ns pulse generated by an Arbitrary Wave Generator (AWG, Agilent 33600). The pulsed light was then amplified by an Erbium-doped fibre amplifier (EDFA) and circulated into the sensing fibre to generate the Brillouin backscattered signal. This signal was heterodyned with the reference CW light in branch B and then down-converted to the radio frequency (RF) range using a wideband photo-detector. The signal was further down-converted to the intermediate frequency (IF) range, which was digitised in time domain and processed using the STFT signal processing algorithm to obtain the frequency peaks along the fibre under test.

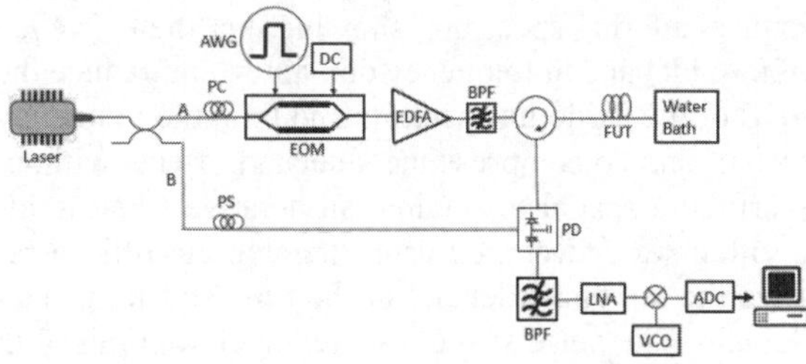

Figure 7. The experiment set up of the STFT-BOTDR

As a comparison, the super-Gaussian pulses (optimised from the 50ns rectangular pulse) is used as the input and sent into the STFT-BOTDR. The Brillouin scattering spectrums from the two pulses are compared, which are both averaged by 1000 times and shown in **Figure 8**. The super-Gaussian pulse offers narrower bandwidth of the Brillouin scattering spectrum comparing with the original rectangular pulse. The super-Gaussian pulse offers a larger SNR in the Brillouin scattering signal while the width of the pulse has changed a little bit, that remains a similar spatial resolution.

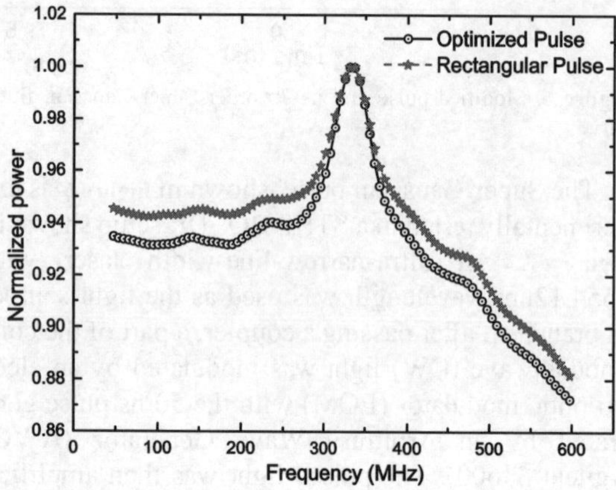

Figure 8. The normalised plot of experiment result of Brillouin spectrum generated by optimised pulse and rectangular pulse with same pulse width

5 CONCLUSION

In this paper, the Kaiser-Bessel functions with different attenuation slopes were used to simulate the Gaussian, Hamming, Rectangular, and other shapes. The pulse shapes were compared in terms of the Brillouin scattering spectrum, showing that the existence of the sideband in frequency domain would reduce the SNR of the Brillouin spectrum and limit the frequency resolution. To compress the sideband effects without sacrificing spatial resolution, an iterative filtering algorithm was developed using iterated cut-offs to reduce the tail and sideband in the time and frequency domain. The pulse shape was iteratively optimised in the time and frequency domain and therefore increases the ratio of the mainlobe power over the sideband power. The SNR was improved by optimised iterations. A rectangular pulse was modified as an illustration and the optimised pulse was fitted with a 7th order Super-Gaussian curve. In the experiment of STFT-BOTDR, the result demonstrated that the optimised pulse offered a narrower bandwidth of the Brillouin scattering spectrum comparing to the original rectangular pulse.

6 ACKNOWLEDGEMENT

Linqing Luo and Bo Li are supported by the China Scholarship Council and Cambridge Commonwealth, European, and International Trust. EPSRC EP/K000314/1 grant is acknowledged.

7 REFERENCES

Bao, X., and L. Chen. 2011. "Recent progress in Brillouin scattering based fiber sensors." *Sensors* 11 (4): 4152-4187.

Hao, Y., Q. Ye, Z. Pan, H. Cai, and R. Qu. 2013. "Analysis of spontaneous Brillouin scattering spectrum for different modulated pulse shape." *Optik* 124: 2417–2420.

Lewitt, R.M. 1990. "Multidimensional digital image representations using generalized Kaiser–Bessel window functions." *Journal of the Optical Society of America A*: 1834-1846.

Luo, L., B. Li, Y. Yu, X. Xu, K. Soga, and Y. Yan. 2016. "Time and Frequency Localized Pulse Shape fo Resolution Enhancement in STFT-BOTDR." *Journal of Sensors* 2016: 10.

Naruse, H., and M. Tateda. 2000. "Launched pulse-shape dependence of the power spectrum of the spontaneous Brillouin backscattered light in an optical fiber." *Applied Optics* 39 (34): 6376-6384.

Naruse, H., and M. Tateda. 1999. "Trade-off between the spatial and the frequency resolutions in measuring the power spectrum of the Brillouin backscattered light in an optical fiber." *Applied Optics* 38 (31): 6516-6521.

Rabiner, L.R., and B. Gold. 1975. *Theory and Application of Digital Signal Processing*. Englewood Cliffs, N.J.: Prentice-Hall.

Proceedings of the International Conference on Smart Infrastructure and Construction
ISBN 978-0-7277-6127-9

© The authors and ICE Publishing: All rights reserved, 2016
doi:10.1680/tfitsi.61279.105

Characterization on the distributed fibre optic sensors using a newly developed calibration system

Ying Mei[1], Xiaomin Xu[1], Jize Yan[1,2*] and Kenichi Soga[1]

[1]*Department of Engineering, Cambridge University, Cambridge, UK*
[2]*School of Electronics and Computer Science, University of Southampton, Southampton, UK*
[*] *Corresponding Author*

ABSTRACT : This paper presents the characterization of four tight-buffered strain cables and one loose-buffered temperature cable for the distributed fibre optic sensing using a multi-function calibration platform developed in the laboratory. The calibration platform comprises of two units: 1) an aluminum rig, which can automatically change the tested cable strain condition with the resolution of $1\mu m$ and the accuracy of $10\mu\varepsilon$; and 2) a water bath, which can uniformly change the submersed cable's temperature from 0-100°C with the resolution of ±0.1°C and the accuracy of ±0.05°C. Parameters that were tested to quantify the cable qualities are: the measurement accuracy, the measurement hysteresis during loading and unloading process, and the measurement's linearity of the measured Brillouin scattered frequency shift with regard to the input strain/temperature change. The quality of the five tested cables varied significantly and the sensing hysteresis was observed for several cables (e.g. temperature cable), which has to be considered in the future data analysis process when utilizing those sensing cables in the real site monitoring.

1 INTRODUCTION

Infrastructure monitoring is a very demanding application of the distributed fibre optic sensor (DFOS) in terms of distributed strain and temperature measurements. A typical DFOS system comprises two parts: the sensing cable and the analyser. When the analyser launches a light into the sensing cable, the majority of the light travels through the cable, but a small fraction is backscattered. By analysing the property of the backscattered light, cable's strain and temperature information can be obtained.

The key performance characteristics of DFOS could vary significantly using fibres with different cable tight buffering methods. There has been a number of calibration units reported in the past few years to evaluate the sensor performance. For example, a robust strain calibration device was developed in Zurich to calibrate the strain sensing cable parameters including cable longitudinal stiffness, yield strain, Brillouin conversional coefficient and fibre slippage (Iten 2011). A simple water tank was designed to compare the temperature coefficients for different types of sensing cables (Mohamad 2012) and a new 'dissimilar-fibre-splicing' method was proposed to calibrate sensor spatial resolution (Zhang et al. 2013). These previous calibration studies revealed that the sensing parameters can vary significantly for different fibre optic cables. It is therefore essential to calibrate their characteristic parameers prior to the real site applications.

This paper presents the performance characteristics of five fibre cables commonly used in infrastructure monitoring, using a newly developed strain and temperature calibration platform. The strain calibration unit consists of an aluminium rig and a micrometer mounted on it to automatically stretch the tested cable. The temperature calibration unit is a water bath which can accurately and precisely control the embedded cable's temperature. The cable performance was characterized in terms of the measurement accuracy, hysteresis and the linearity of the measured Brillouin scattered frequency shift with regard to the input strain/temperature change.

2 CALIBRATION UNITS

To quantify the sensing characteristics of different fibre optic cables, a calibration platform was established in the laboratory. This platform is made of an aluminium rig for strain calibration and a water bath for temperature calibration.

2.1 Sensing principle

The basic principle behind the distributed fibre sensor is that ambient parameters such as strain and temperature can influence the properties of the light signal travelling throughout an optics fibre. Therefore, when light is launched into a sensing cable, the majority of the light travels through the cable, but a small fraction is backscattered and the frequency shift of the Brillouin backscattered light is proportional to the cable strain/temperature.

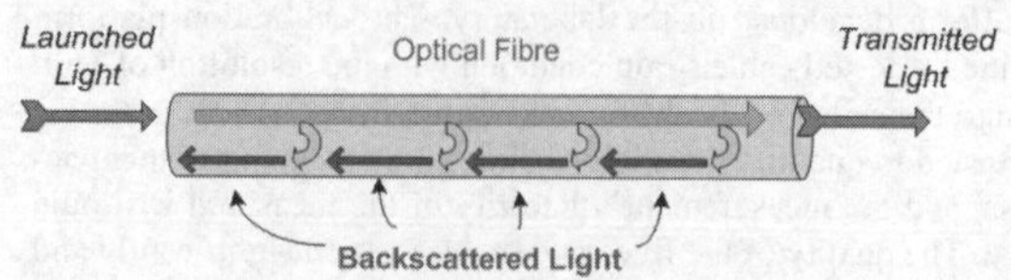

Figure 1. Brillouin backscattering on an optical fibre.

To convert the Brillouin frequency shift to strain/temperature measurement, coefficients (C_ε and C_T) are usually applied. However, for cables with different buffering methods, C_ε and C_T would be slightly different. For current available sensors, Brillouin C_ε vary in a range of 0.046-0.052MHz/ $\mu\varepsilon$ (Mohamad 2012).

Strain coefficient C_ε can be calculated as:

$$C_\varepsilon = \frac{\nu_B(\varepsilon) - \nu_B(\varepsilon_0)}{(\varepsilon - \varepsilon_0)}$$

Temperature coefficient C_T can be calculated as:

$$C_T = \frac{\nu_B(\varepsilon) - \nu_B(\varepsilon_0)}{(\varepsilon - \varepsilon_0)}$$

As has mentioned previously, the frequency shift of the backscattered Brillion spectrum is simultaneously sensitive to temperature and strain in the fibre. To distinguish them, strain sensing cable and temperature sensing cable are usually placed adjacent to each other for different measurements. The temperature cable consists of a loose buffered fibre to isolate strain effects and the strain cable contains a tight buffered fibre to efficiently transfer external strain to the fibre inside.

2.2 Strain calibration platform

The strain calibration set up (Figure2) consists of an electrical motorized linear stage to change the cable strain precisely and accurately and an aluminium base structure to mount the micro-motor. The calibration bench was put in a temperature controlled room to minimize the temperature effect on the Brillouin frequency shift readings.

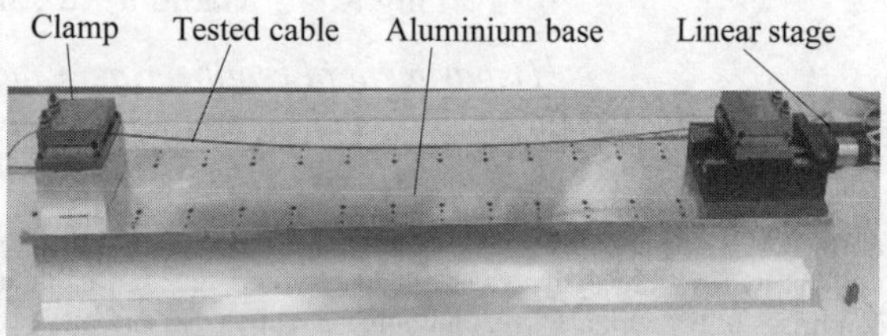

Figure 2. The strain calibration bench.

The micro-motor for cable stretching was selected as a M414 ball-screw type precision linear stage from PI Company considering its high performance in accurate motion control. Generally, this strain calibration bench can stretch 1.5m of the cable by a given displacement to a resolution of 1μm and to an accuracy of 10$\mu\varepsilon$.

Previous site strain measurement results showed that if the strain sensing cable's pre-strain is beyond 3000$\mu\varepsilon$, the glue used to attach cable to structures will start to degrade and the pre-strain will start to reduce. Therefore the sensing cable for this strain calibration experiment was pre-strained by less than 3000$\mu\varepsilon$.

2.3 Temperature calibration unit

A water bath was used as temperature calibration bench to allow a certain length of the fibre optic cable to be thermally isolated from the remainder of the cable and maintained at a desired temperature. The temperature controlling system used in this test was a C1G cooling system and a T100-ST 18 water bath from Grant Instruments with the dimension of 200mm (h) × 540mm (l) × 330mm (w).

This setup can uniformly change the cable's temperature from 0-100°C in a resolution of ±0.1°C with an accuracy of +/-0.05°C. The cable section was freely immerged into the water bath in the form of many loops as shown in Figure 3. There is a thermometer installed as well for temperature measurement reference.

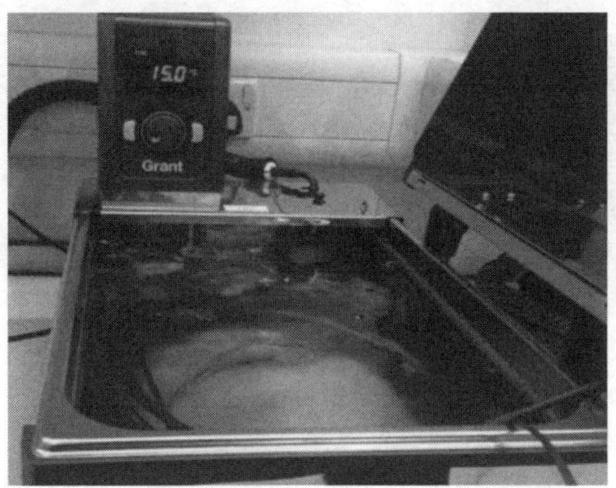

Figure 3. The temperature calibration bench.

3 RESULTS

The calibration process generally contains: 1) holding a certain section of the cable, 2) changing its strain/temperature condition accurately and precisely, 3) evaluating the input strain/temperature data recorded by strain sensor (or thermometer) with the frequency shift displayed on the analyser. Four types of strain sensing cable and one temperature sensing cable were tested in the laboratory. Cable's measurement hysteresis, accuracy and its linearity between input strain/temperature with regard to analyser measured strain/temperature were evaluated to quantify cable's sensing ability.

Cable	Fujikura reinforced Cable	Brugg Strain Cable	Tight buffer telecomm cable	NanZee Reinforced Strain Cable	Loose buffer telecomm cable
Details	Steel wire reinforced cable embedded with four single mode fibres	One up buffered optical fibre with robust outer sheath	Very compact tightly-jacketed cable	Tight buffered single mode fibre	Excel loose tube cable
Sketch	Nylon sheath; Steel wire; Coated fibres	Plastic protection layer; EPR; Soft plastic buffer; Coated fibre	PA protection layer; Coated fibre; Soft plastic buffer	Steel braids; Jacket; Coated fibre	Central buffer tube; Gel; Ripcord and Jacket; Coated fibres
Picture					

Table 1. The tested strain cables and temperature cable

3.1 *Tested cable*

Table 1 lists four strain sensing cables and one temperature sensing cable that were tested in the laboratory. The Fujikura Ltd. Designed strain sensing cable has four single-mode optical fibres embedded in the middle of the cable and two steel wires on sides to reinforce the cable. The ribbon shape of the cable makes it easier to attach the cable to structures. The Brugg Strain Cable has one single mode fibre tightly-buffered in a round tube. The jacket of this cable is so robust that it needs no reinforcement. The tight buffer telecomm cable is not particularly designed for distributed fibre optic sensing purpose. It has two fibres tightly buffered inside a squared protection layer. The NanZee Reinforced Strain Cable has four steel braids armouring around the optics fibre. Outside the steel braids there is a protection tube which makes this cable most robust one in the four tested strain cables. The loose buffer telecomm cable is used as temperature sensing cable. It has 8 coated single mode fibres floating within a hard tube which is filled with gel. The gel makes the external strain placed on the tube not able to be directly transferred to the fibre and therefore isolating strain effects. Due to the robust design, all tested sensing cables are ideal for harsh environment applications such as pipeline monitoring and burial in the solid.

3.2 Strain calibration results

During strain calibration, the tested strain cables were pre-strained to 1000$\mu\varepsilon$. Then, they were loaded 1000$\mu\varepsilon$ more and unloaded back to pre-strain condition in a step of 100$\mu\varepsilon$ using the strain calibration rig. At each loading stage, three measurements were recorded with Neubrex (NBX-5000) BOTDR analyser. The analyser was set with 0.5m spatial resolution and 0.05m sampling resolution. Strain recorded at cable pre-strained condition was used as a baseline reading.

The measured strain along the holding sections of four tested strain cables were averaged and plotted against the input strain change (Figure 4). The averaging was made over 60 data points (i.e. averaged over the length of 1m effective length with cables and readings were made three times). The error bars at each strain level was calculated as the standard deviation of all 60 data points.

The qualities of the tested four strain cables were then evaluated with four sensing parameters: C_ε, r^2, σ and ε_d. C_ε is the Brillouin frequency shift to strain conversional coefficient, r^2 is the coefficient determination of the best fitted line between measured strain and input strain value, σ represents measurement uncertainty which is the average of precision error calculated at different strain levels, and ε_d indicates slippage which is calculated as the difference between analyser measurement at pre-set cable condition and measurement after the cable is loaded and unloaded for one cycle.

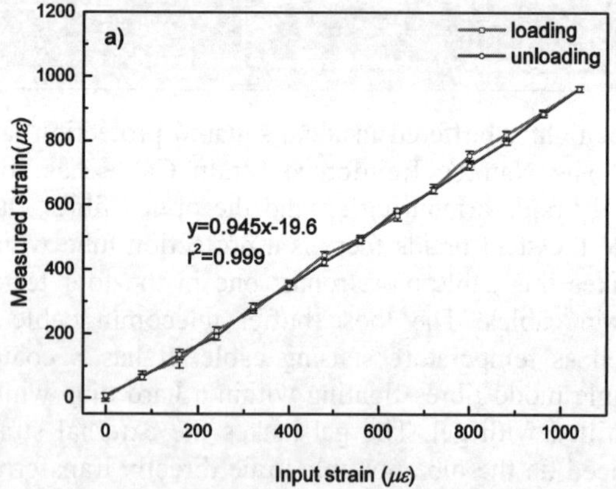

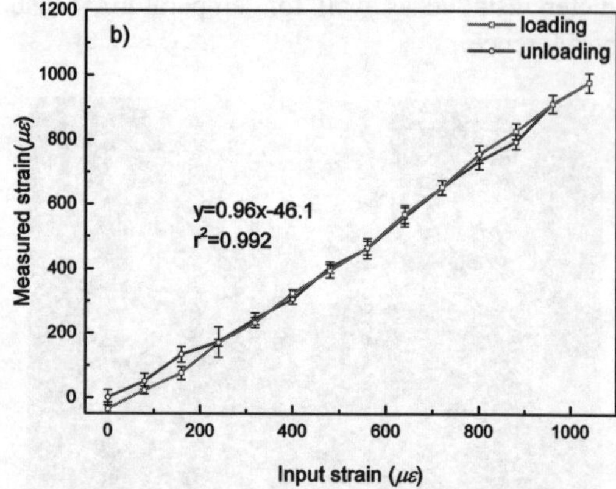

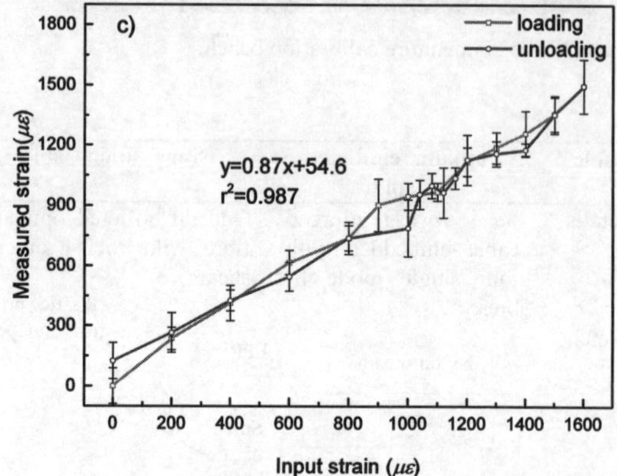

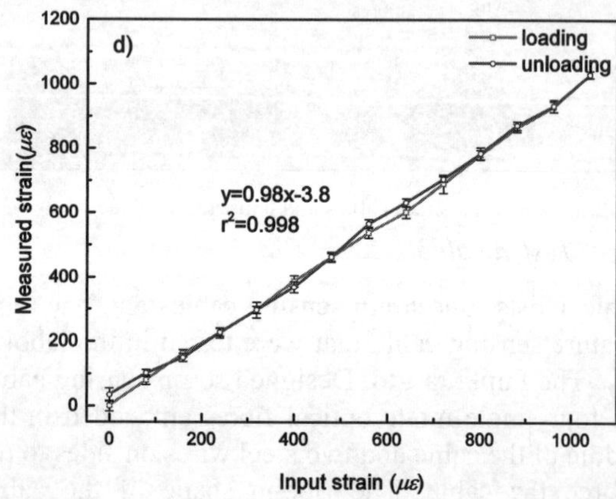

Figure 4. The analyser measured strain vs. input strain for a) Fujikura Reinforced Strain Cable, b) Brugg Strain Cable, c) Tight buffer telecomm Cable and d) NanZee Reinforced Strain Cable.

Table 2 compares the quality of the tested strain cables in regard of their sensing parameters. It can be seen that C_ε is in a range of 0.044-0.049 $MHz/\mu\varepsilon$ for four tested cables, which indicates C_ε can vary at least 13% for different types of strain cable. The table also illustrates Fujikura reinforced cable and NanZee reinforced cable have relatively higher measurement accuracy and better measured linearity compared to that of Brugg and tight buffer telecomm cables. The reason of that is probably due to the robust cable reinforcement causes the sensing fibre in a very stable condition during strain process. The small ε_d of the Fujikura cable implies that this cable has little slippage effects which can be proved in various case studies(Klar et al. 2006). In summary, the Fujikura Reinforced Strain Cable shows the best sensing performance during this strain calibration experiment.

Table 2. The comparison of the four tested strain sensing cable

Cable	Fujikura reinforced Cable	Brugg Cable	Tight buffer telecomm Cable	NZ reinforced Cable
C_ε ($MHz/\mu\varepsilon$)	0.047	0.048	0.044	0.049
r^2	0.999	0.992	0.987	0.998
σ ($\mu\varepsilon$)	11.1	22.9	82.9	15.5
ε_d ($\mu\varepsilon$)	6	22	125	34

3.3 *Temperature calibration result*

In temperature calibration, 12m of temperature cable and 7m of Fujikura Reinforced Strain Cable were spliced together and immerged in the water bath. Water taken from a tap was heated and cooled in a range of 20°C~45°C, at 5°C interval. At each temperature level, three repeated measurements were taken by NBX-5000 analyser after the temperature was stabilized. The baseline reading was taken at 20°C.

Figure 5 plots the temperature readings of the analyser at every 5cm along a) temperature sensing cable and b) strain sensing cable. The readings were calculated as an average of three repeated measurements at each temperature level. It can be clearly seen that temperature cable shows larger σ_T (measurement fluctuation) than strain cable does. For example, in the baseline condition, σ_T is 0.42°C for temperature cable and 0.19°C for strain cable. In the later experiment, σ_T does not change a lot and stays at around 0.49°C for strain cable. However for temperature cable, σ_T is building up during later heating/cooling process. It increases from 0.42°C to about 2.34°C after the whole experiment has finished. One explanation is that with temperature change, the property of cable gel becomes instable. When temperature is decreasing, the hardening of the cable gel induces external strain to the sensing fibre and thus causes measurement fluctuation. Because in practice temperature is not changing very spatially, an average result over distance may be used in the future for temperature compensation.

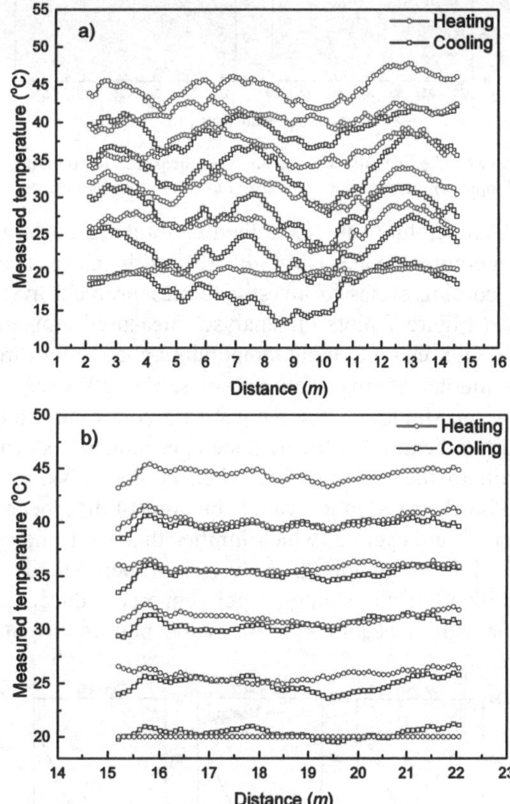

Figure 5. The average temperature measurement along the optic cable: a) temperature cable and b) strain cable.

Figure 6 presents the linear relationship of measured temperature plotted against temperature input. The strain cable shows no hysteresis while the temperature cable shows about 2.5°C hysteresis during the experiment. In addition, the measurement uncertainty for temperature sensing cable (2.22°C in average) is rather worse than that of strain sensing cable (0.75°C in average).

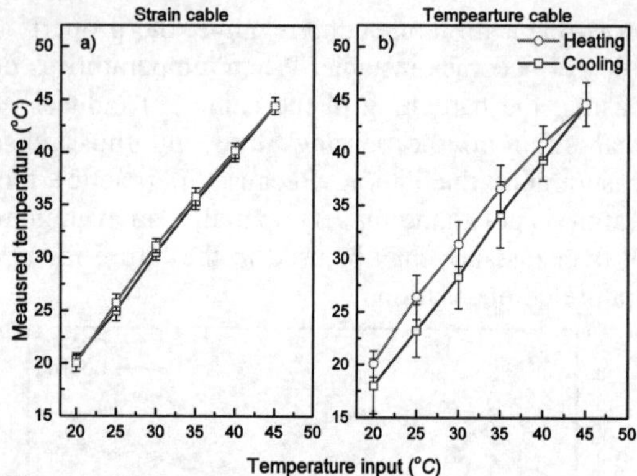

Figure 6. The linearity between analyzer measured temperature and input temperature for a) strain cable and b) temperature cable.

Temperature cable was then calibrated in a larger temperature range (5°C~85°C) with more heating/cooling cycles to investigate measurement hysteresis. Figure 7 plots the analyser measured temperature with regard to input temperature change for three continuous heating/cooling cycles. For all 3 cycles, σ_T is maximum at low temperature condition which again indicates that the hardened gel induces external strain to the sensing fibre when cooled. Hysteresis can be clearly seen in cycle 1, but almost disappear in cycle 2 and cycle 3, which implies that after one cycle of heating/cooling, cable gel property tends to stabilize and the changing gel property induced external strain becomes in the same pattern in short time.

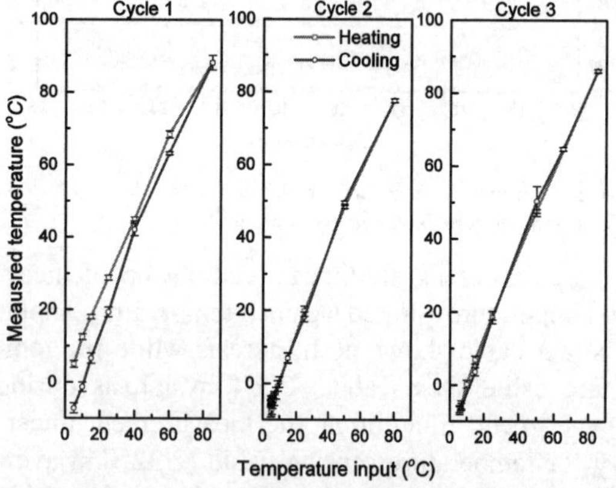

Figure 7. The linearity between measured temperature and input temperature for continuous three heating-cooling cycles.

4 CONCLUSION

In this paper, we presented the sensing characteristics of four commonly used tight buffered strain cables and a loose-buffered temperature cable, using a newly developed strain and temperature calibration platform. For the tight buffered strain sensing cables, it is found that the steel wire reinforced strain sensing cable is the best sensing cable, with the best measurement accuracy and the minimum slippage occurred during loading and unloading test. The poor linearity between strain change and measured Brillouin centre frequency was also observed in the other strain sensing cables. For temperature cable, much larger sensing hysteresis was observed comparing to that of strain cables. This is most likely due to the unstable property of the cable gel. However, the temperature hysteresis seems to be significantly reduced after one cycle of heating/cooling. The long-term temperature hysteresis needs to be further investigated in the future.

ACKNOWLEDGEMENT

This research is supported by the Cambridge Centre for Smart Infrastructure and Construction (CSIC) and Chinese Scholarship Council (CSC). The authors would like to thank Dr. Neil Houghton for designing and constructing the strain calibration bench. We would also like to thank Dr. Cedric Kechavarzi and Mr. Peter Knott for all their help during establishing the set up and conducting the experiment.

REFERENCES

Iten, M. 2011. "Novel Applications of Distributed Fiber- Optic Sensing in Geotechnical Engineering." ETH ZURICH.

Klar, A., P. J. Bennett, K. Soga, R. J. Mair, P. Tester, R. Fernie, H. D. St John, and G. Torp-Peterson. 2006. "Distributed Strain Measurement for Pile Foundations." In *Geotechnical Engineering*, **159**,135–144.

Mohamad, H. 2012. "Temperature and Strain Sensing Techniques Using Brillouin Optical Time Domain Reflectometry." In *SPIE*, **8346**,83461M.

Zhang, D., H. Cui, and B. Shi. 2013. "Spatial Resolution of DOFS and Its Calibration Methods." *Optics and Laser Technology* **51**, 335–340.

Proceedings of the International Conference on Smart Infrastructure and Construction
ISBN 978-0-7277-6127-9

© The authors and ICE Publishing: All rights reserved, 2016
doi:10.1680/tfitsi.61279.111

Industry challenges on resolution, linearity and optical budget of high-accuracy distributed Brillouin sensing

Nils Nöther* and Marko Krcmar

fibrisTerre Systems GmbH, Berlin, Germany
* *Corresponding Author*

ABSTRACT Distributed Brillouin sensing has become a state-of-the-art tool for strain and temperature monitoring in concrete and geotechnical applications throughout the civil construction industry. While the leading technologies, being BOTDA, BOFDA and BOTDR, are steadily advancing in terms of spatial resolution and measurement length, end-users in field installations often put the focus on softer parameters like linearity or optical budget when evaluating the performance of the technology.

This paper addresses the implications of high spatial resolution to the accuracy of relative and absolute strain and temperature data from the perspective of the Brillouin optical frequency domain analysis (BOFDA) technology, and outlines the need for a clear definition and a standardization scheme to make the terms dynamic range and optical budget comparable between different instruments and technologies. Data from field applications in concrete pile monitoring is used to discuss the above aspects.

1 DISTRIBUTED FIBER SENSING

For long-range measurements in geotechnical and industrial applications, distributed optical fiber sensors have become a tool of increasing importance throughout the past decade (Glisic and Inaudi, 2007).

Classic deformation monitoring (performed by strain gauges etc.) and temperature monitoring (Pt100 and alike) deliver data from fixed, single spots of a structure; quasi-distributed measurements (fiber Bragg gratings) provide a chain of discrete measurement points.

In contrast, an optical fiber connected to a device for distributed strain and temperature sensing (Brillouin DTSS) will provide a continuous profile of strain and temperature – spatially resolved down to less than 0.5 m – over a range of several tens of kilometers.

1.1 The Time Domain Approach

The most common approach to acquire spatially resolved information from an optical fiber is to measure the response of the fiber to an optical pulse that is injected from one end (Figure 2).

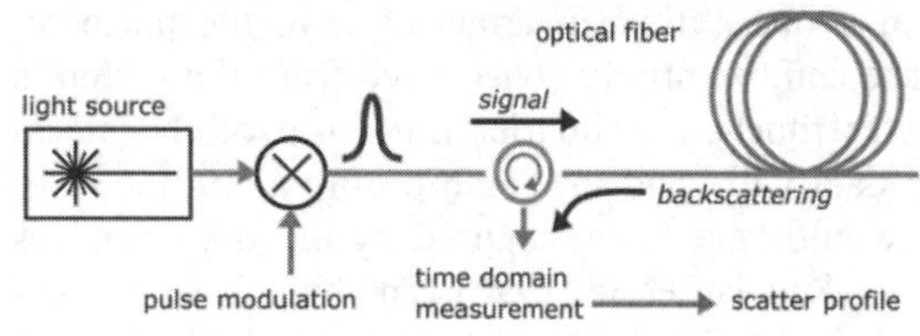

Figure 1. Basic principle of optical time domain reflectometry

The pulse runs along the fiber; everywhere it passes by, light is sent back to the injection end where it is measured over time. From this time of flight, a spatial profile of the fiber can be constructed, containing the distribution of the optical process that has generated the backwards-travelling light. For distributed strain and temperature sensing, this method has been implemented into the Brillouin Optical Time Domain Reflectometry (BOTDR); as a further advancement providing higher signal contrast and thus resolution and accuracy, many state-of-the-art instruments employ the Brillouin Optical Time Domain Analysis (BOTDA), see Nikles et al., 1996.

1.2 The Frequency Domain Approach

Measurements in the frequency domain provide the same result (a spatially resolved scattering profile of an optical fiber) while avoiding some implementation issues connected to the generation and reception of short optical pulses. Figure 3 shows the concept: Here, the pulses are replaced by sinusoidal waves of a tunable frequency (ca. 5 kHz to 100 MHz). By measuring the system's response to these frequencies (being the spectral components of the optical pulses in a harmonic representation), the transfer function of scattering and reflections along the fiber is recorded.

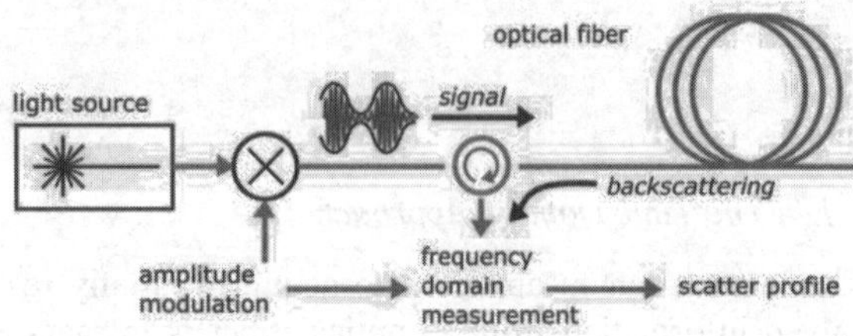

Figure 2. Basic principle of optical time domain reflectometry

As in a linear, time-invariant system, the pulse response can be entirely constructed from the system's transfer function, a Fourier transform of the measured data will yield the same profile along the fiber that would have been acquired by the use of pulses (Garus, Krebber et al., 1996). The benefit of this approach is the possibility of narrow-band filtering for each component of the pulse and its potential to be integrated into cost-efficient digital hardware; the implementation of this system for distributed Brillouin sensing is the Brillouin Optical Frequency Domain Analysis (BOFDA).

2 PERFORMANCE PARAMETERS

When comparing and classifying sensing systems with respect to real-life application requirements, a straight-forward approach is to apply performance parameters known from classical point-wise sensors to distributed sensing systems – such as precision, gauge-length, long-term stability etc.

However, the continuous nature of data resulting from distributed sensing systems requires a set of performance parameters that are fundamentally distinct from those characterizing classical point-wise sensors; and even among the different distributed technologies, different parameters or interpretations of such parameters will apply.

As an example, the precision and accuracy figures of distributed temperature measurements will fundamentally differ when comparing data recorded by Raman DTS systems to data recorded by Brillouin DTSS systems: Raman DTS employs the backscattering intensity (i.e., the intensity ratio between the Stokes and Anti-Stokes component of Raman-backscattered light), thereby providing absolute temperature data; the end-user will be interested in long-term effects altering this intensity distribution, thereby degrading precision in the long term.

In contrast, Brillouin DTSS data relies on a frequency information that is inherent to the physics of the fiber; once calibrated, the temperature response will remain stable; thus, the accuracy figure of the system becomes equal to the repeatability of the data. This inherent stability in turn puts high demand on the calibration of the Brillouin coefficients (connecting the Brillouin frequency reading to strain and temperature) and to separate mechanical strain from temperature impacts on the sensor (or, in this example of pure temperature readings, to provide strain-free implementation of the sensing fiber into the structure).

2.1 Spatial resolution

Among the parameters that are specific to distributed systems, the spatial resolution (along with the spatial sampling rate or spatial accuracy) is the one that is discussed most controversially – which often results in confusing and discouraging the end-customers which are confronted with a number of competing definitions.

From the COST299 guideline, the most application-oriented definition of spatial resolution is drawn:

"The spatial resolution is specified for a fiber by the minimum distance between two step transitions of the fiber's strain / temperature condition. It is directly related to the pulse length of the measuring instrument." (Thevenaz, Habel et al., 2007)

The first part of the definition provides a clear criteria to determine a system's spatial resolution (e.g. by fixing a fiber section on a linear strain stage). However, the second part (the relation to the pulse width) gives physical background that bears the danger of making an invalid reversal conclusion:

In linear time domain backscattering reflectometry (e.g. OTDR), the spatial resolution δz is connected to the optical pulse width by

$$\delta z = \frac{1}{2}\frac{c_0}{n}\Delta t_p$$

in which c_0 is the vacuum speed of light, n is the refractive group index of the fiber and Δt_p is the width of a presumed rectangular pulse.

This relation also applies to BOTDR systems and thereby makes the above definition of spatial resolution generally valid. However, the physics of Brillouin backscattering introduce limitations into this relation: A pulse width significantly shorter than 10 ns will introduce an uncertainty into the measurement value and degrade both spatial and strain/temperature precision (Thevenaz and Beugnot, 2009).

Therefore, characterizing an instrument's spatial resolution by merely providing the pulse width will not hold true under all circumstances; other impacts like fiber losses might also prevent the system from detecting a small event even though the nominal pulse width should theoretically allow the detection.

For completeness, the equivalent relation is found for frequency domain systems, where the governing parameter is not the pulse width, but the maximum modulation frequency f_{max}:

$$\delta z = \frac{1}{2}\frac{c_0}{n}\frac{1}{f_{max}}$$

In consequence, this issue can be seen as an example of parameter definitions that originated from other technologies or the physical background of the technology, but need an application-oriented revision with respect to what end users expect from data sheets and instrument comparisons. It stresses the obvious need for standardized parameter definitions as it has been done in the work of Itern, Spera, Jeyapalan et al. 2015.

2.2 *Dynamic range and loss budget*

Another set of parameters that is supposed to characterize the performance of a distributed sensing system is the systems dynamic range. Similar to the spatial resolution, the definition has been established for OTDR instruments, being the oldest and most widely spread distributed fiber sensing technology in the industry. The following figure shows the definition (specified by IEC 61746):

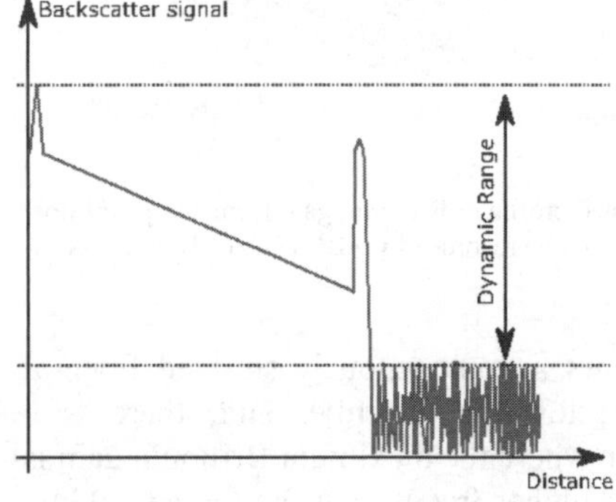

Figure 3. OTDR definition of dynamic range

Here, the dynamic range is defined as the difference (in a logarithmic backscattering graph) between the full scale level of the receiver and the upper limit of a range that contains at least 98% of all noise data points.

This definition allows the user of an OTDR to answer the central question: Will this instrument be able to detect a significant event even after a fiber section that induces optical losses of x dB?

Conclusively, an OTDR's dynamic range corresponds to its loss budget or attenuation budget – the maximum fiber attenuation to still allow quantitative distributed measurements.

For Brillouin DTSS systems, it is of course a straight-forward approach to adopt this exact definition, because it takes the components' specifications into account (laser power, detector's sensitivity) and provides a clear number that the end user might be used to. However, the system's dynamic range as defined for OTDR is not necessarily equivalent to the system's loss budget. Figure 4 explains why the loss budget is a far more complex figure for Brillouin DTSS systems than for OTDR.

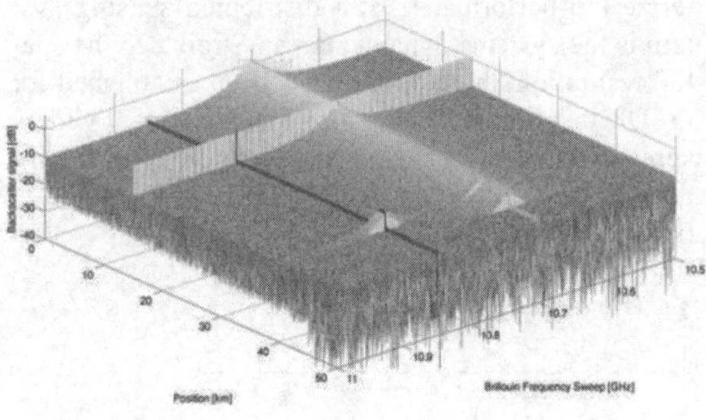

Figure 4. Distributed profile of Brillouin gain from an optical fiber (simulated, as it would be recorded by a BOTDA or BOFDA system)

The graph shows a homogenously strained fiber of 40 km length; towards the fiber end, there is a strained section where the maximum Brillouin gain is shifted towards higher frequencies, and at ca. 10 km, there is a strong Fresnel reflection (e.g., a bad connector or splice).

Brillouin gain, which corresponds to the major signal at the detector, decreases with fiber length and attenuation; at the fiber end, there is still a significant step down to the noise floor. Yet, for the detection of the strain section, this step is obviously not the only relevant criteria. This becomes clearer when looking at the two cross sections as they are indicated in Figure 4, see Figure 5.

The basic OTDR definition from Figure 3 still applies to characterize the system's performance in detecting high and low backscattering and reflection levels. The performance when clearly detecting small strain or temperature events, however, must take more parameters into account; especially, because the level of Brillouin backscattering received per section also depends on the overall length of fiber that has this exact Brillouin frequency shift. This origins from the non-linear nature of Brillouin scattering and is indicated by the lower level of backscattering at pos. 40 km in Figure 5 compared to the level of the homogenous fiber.

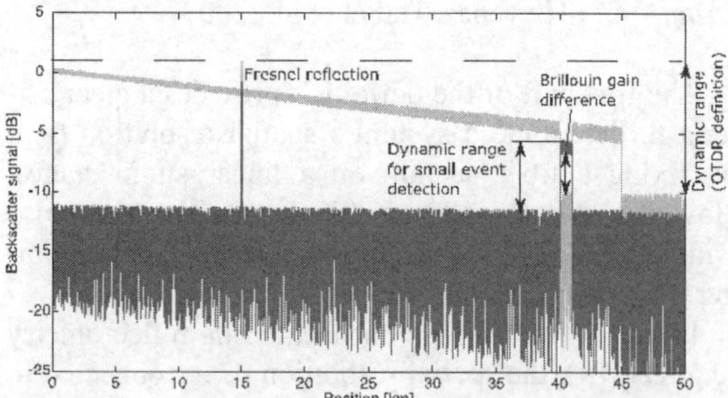

Figure 5. Cross-section at the marked positions in Figure 4

At this point, the loss or attenuation budget of a Brillouin DTSS system will substantially differ from its dynamic range and become an empirical parameter which is not meaningfully expressed by just adding up the components' sensitivity characteristics.

To provide an example for this problem by means of a BOFDA measurement, Figure 5 shows a section of a homogenous fiber of 475 m length, of which 0.5 m (at pos. 453.5 m) have been strained in a linear stage.

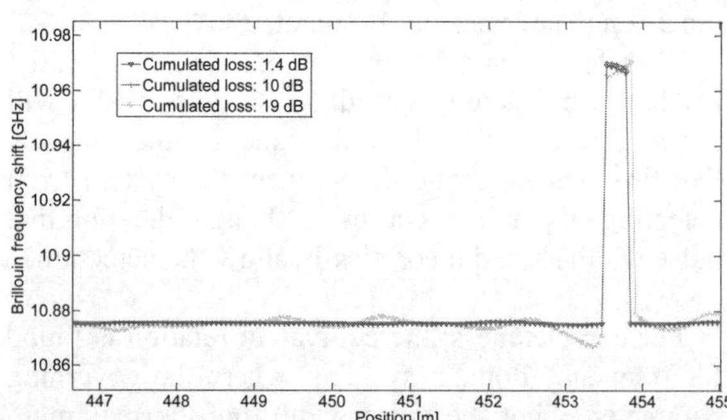

Figure 6. BOFDA measurements of a 0.5 m strained fiber section

Three measurements of the spatially resolved Brillouin frequency shift are shown.

The first one only comprises the inherent fiber attenuation and the losses at the connectors, cumulating to 1.4 dB. The second and third measurements were performed using an adjustable in-line attenuator, cumulating to attenuation values of 10 dB and 19 dB, respectively.

It becomes clear that the small event is resolved appropriately at all attenuation levels. However, the accuracy and repeatability obviously suffers from the induced attenuation.

To conclude: Similar to the spatial resolution definition, an application-oriented definition is needed for the loss budget of a Brillouin DTSS system, that not only takes into account the dynamic range of the instrument, but also gives a clear characteristic of the impact of optical losses on the detection reliability, the spatial resolution degradation and the data precision / accuracy figures.

While the COST299 guideline includes a generic definition of the dynamic range (Thevenaz, Habel et al., 2007), it lacks a definition of the optical loss or attenuation budget. This should be subject to further standardization works in close cooperation between the scientific community and the technology's end users from the industry.

3 BOFDA PILE TESTING

In order to provide an example of how different a spatially resolved strain measurement in the lab (as the one shown in Figure 6) can be from a distributed strain measurement performed within an industrial application, data from a static pile load testing campaign is shown.

Static load measurements for concrete piles using extensometers and parallel distributed optical strain sensor cables with Brillouin DTSS measurements were performed as sketched in Figure 7.

Such measurements, using Brillouin DTSS systems for static load testing of concrete piles, have been reported for various sensing configurations (Schwamb, Elshafie, and Ouyang 2011).

The present application comprises a concrete-poured pile of 5 m depth; reinforced, steel-armored fiber-optic sensing cables were fixed to the reinforcement cage before entering the cage into the ground. During pouring and curing of the concrete, the sensing cables were not damaged, so the overall optical loss remained within the limits of a few dB.

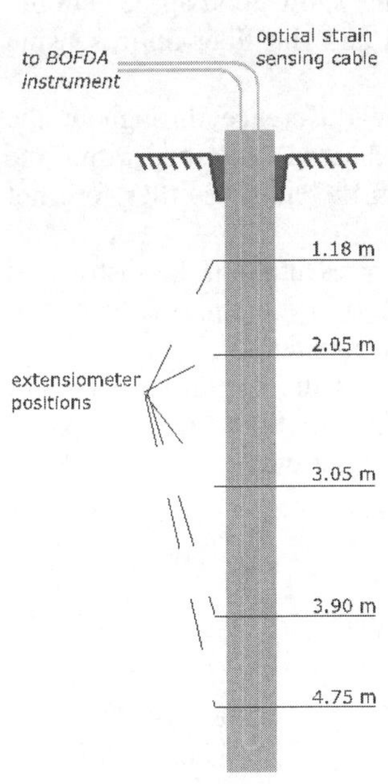

Figure 7. Measurement set-up for static pile load testing

During the tests, an increasing vertical load from 150 kN to 900 kN was induced onto the pile, while subsequently extensometer data as point-wise references, temperature data at the extensometer positions and distributed Brillouin strain data (using a BOFDA system) were recorded.

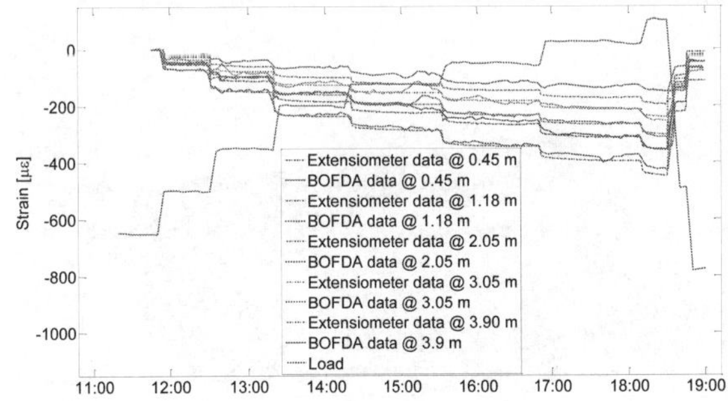

Figure 8. Strain evolution in extensometer and BOFDA data.

Figure 8 shows the evolution of strain for both the extensometers and the distributed Brillouin sensors (at the exact extensometer positions, compensated for the base line reading at 0 kN load) over time.

With the exemption of the lowest extensometer, all the measurement points show good agreement between the classical data and the fiber-optic sensing data.

The local temperature difference throughout the measurement campaign showed variations within the BOFDA uncertainty of 0.1°C and was therefore not accounted for.

Naturally, in such a representation, the distributed nature of Brillouin DTSS (representing the technology's true benefit) is not accounted for.

Figure 9 therefore shows the spatial evolution of strain at 3 selected points in time, again in good agreement to the extensometer data.

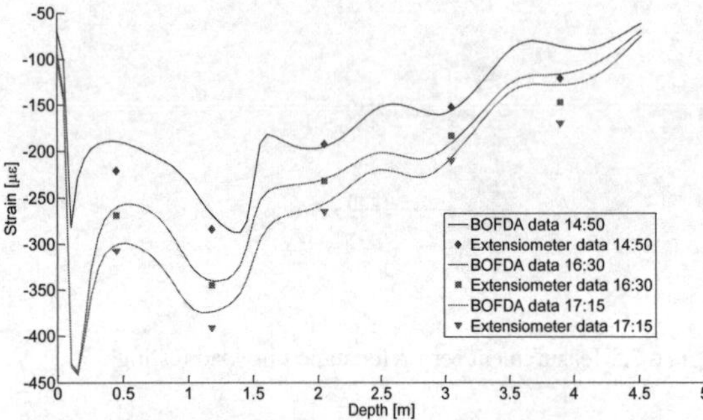

Figure 9. Strain evolution in extensometer and BOFDA data.

To combine the time-based view from Figure 8 with the spatially resolved view from Figure 9, a full 3D representation of the measurement campaign is shown in figure 10.

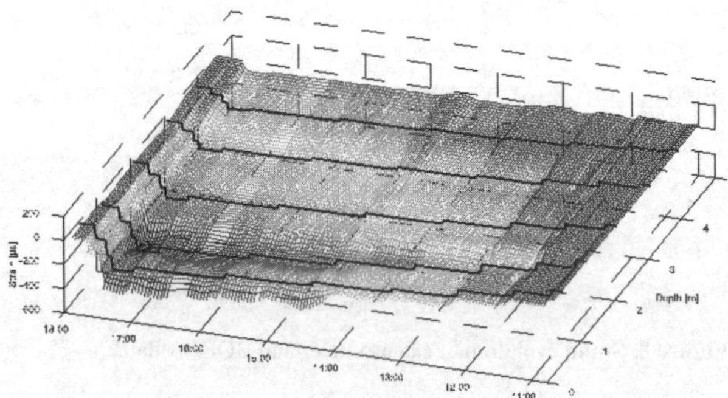

Figure 10. Strain evolution in extensometer and BOFDA data.

4 CONCLUSION

From the point of view of an end user of Brillouin DTSS systems, the definitions of spatial resolution and optical loss budget are often not sufficient to reliably characterize system performance in direct instrument comparisons. This field of lacking definitions has been outlined with examples from lab measurements and pile testing field data.

ACKNOWLEDGEMENT

We would like to thank Ryszard Rippel from Piletest Rippel for a fruitful collaboration on the presented studies. Furthermore, the great visionary work on new standards for fiber-optic sensing technologies of the group around Jey Jeyapalan within ASCE shall be mentioned.

REFERENCES

Glisic, B., Inaudi, D., 2007. Fibre Optic Methods for Structural Health Monitoring, John Wiley & Sons

Nikles, M., Thevenaz, L., and Robert, P.A., 1996. Simple distributed fiber sensor based on Brillouin gain spectrum analysis, Opt. Lett., vol. 21, p. 758.

Garus, D., Krebber, K., Schliep, F., et al. 1996. Distributed sensing technique based on Brillouin optical-fiber frequency-domain analysis, Opt. Lett., vol. 21, pp. 1402–1404.

Iten, M., Spera, Z., Jeyapalan, J., 2015, Benefits of Global Standards on the Use of Optical Fiber Sensing Systems for the Impact of Construction of New Utilities and Tunnels on Existing Utilities. Pipelines 2015: pp. 1655-1666.

Schwamb, T., Elshafie, M., Ouyang, Y., et al., 2011, A Monitoring of a secant piled wall using fibre optics sensors, 8th International Symposium in Field Measurements in Geomechanics,, Berlin, Germany

Thevenaz, L., Habel, W.R., 2007, Guideline for Use of Fibre Optic Sensors, COST Action 299 "FIDES"

Thevenaz, L., Beugnot, J.-C., 2009, General analytical model for distributed Brillouin sensors with sub-meter spatial resolution, SPIE vol. 7503, p. 75036A

Proceedings of the International Conference on Smart Infrastructure and Construction
ISBN 978-0-7277-6127-9

© The authors and ICE Publishing: All rights reserved, 2016
doi:10.1680/tfitsi.61279.117

Measurement error modification of linear non-uniform strain for BOTDA and a case study

J. Xu*[1], Y. Dong[1] and H. Li[1]
[1] *Harbin Institute of Technology, Harbin, China*
* *jinlongxu@hit.edu.cn*

ABSTRACT The distributed Brillouin optical fiber sensor, which is a powerful tool in civil structural health monitoring, can obtain continuous strain or temperature distribution with exceptional high spatial resolution and accuracy. But in practice, the strain distribution is rarely uniform, the non-uniform components can lead to inaccuracies in the measurement results. In this paper, we proposed a modification method for non-uniform strain for distributed Brillouin sensing by studying the characteristics of the Brillouin Gain Spectrum (BGS) under typical linear cases of non-uniform distributions. A novel BGS width change based approach was used to reduce the non-uniform strain measurement errors and reconstruct the accurate strain distributions. The proposed method was validated by the load test monitoring data of a wind turbine blade. The strain distribution contained obvious no-uniform strain components. Comparisons of the measured accurate strain distribution and the reconstructed results show that the new approach is capable of modifying non-uniform strain measurement and decreasing the measurement inaccuracies in the distributed BOTDA sensing.

1 INTRODUCTION

In recent decades, distributed fiber optic sensors, especially Brillouin scatter–based systems, have been widely researched and used in structural health monitoring (SHM) for their powerful capabilities of monitoring temperature and strain over extremely long distance with increasingly high accuracy (Bao et al. 2011). So in many field test applications, such as geotechnical engineering and bridge engineering, Brillouin sensing systems have been successfully used (Mohamad et al. 2011, Hoult et al. 2014). The classical scheme of BOTDA employs a stimulated Brillouin scattering (SBS) technique (Horiguchi et al. 1989). Two counter-propagating laser beams, i.e. a pump pulse and a CW probe wave, with a certain frequency difference near the Brillouin frequency shift, are injected from both end of the sensing fiber. The local acoustic wave can be excited by the coupling of the two laser beams, which results in an energy transfer from to the pump pulse to the CW probe wave. The relation between Brillouin gain of the CW and the frequency difference of the two laser beams is defined as the Brillouin gain spectrum (BGS) and the strain or temperature of the sensing fiber can be obtained by the Brillouin frequency shift of the BGS since it is linearly affected by temperature or strain change. The minimum measurable length that the BOTDA system can identify is defined as the spatial resolution which is determined by the pump pulse width.

In practice, the strain distribution is rarely uniform even within the spatial resolution length, the non-uniform components can lead to distortion in the Brillouin gain, resulting in inaccuracies in the measurement results (Naruse et al. 2003). Extraction of the accurate strain value from the distorted even multiple-peak BGS is a complicated task. In this paper, we proposed a non-uniform strain modification method for distributed Brillouin sensing by studying the characteristics of the Brillouin Gain Spectrum (BGS) under typical cases of non-uniform distributions. A novel BGS shape change based approach was used to optimize the non-uniform strain measurement errors and reconstruct the accurate strain distributions. Moreover, the proposed method was validated by a BOTDA strain monitoring data of a wind turbine blade load test which contained obvious no-uniform strain components.

2 THEORETICAL MODEL

The SBS process, i.e. the interaction between the pump pulse I_p and the CW probe I_{cw} can be described by the steady-state coupled intensity equations (Agrawal 1995):

$$\frac{d}{dz}I_p = -gI_{cw}I_p - \alpha I_p \quad (1)$$

$$\frac{d}{dz}I_{cw} = -gI_{cw}I_p + \alpha I_{cw} \quad (2)$$

where g is the gain coefficient defined as

$$g(\nu_B, \Delta\nu_B) = \frac{g_0}{\left(2\frac{\nu_B - \nu}{\Delta\nu_B{}^2}\right)^2 + 1} \quad (3)$$

z is the distance from the pump pulse end of the fiber, α is the fiber attenuation coefficient, g_0 is the center gain factor, $\Delta\nu_B$ is the Brillouin linewidth, ν is the frequency difference between the pump pulse and the CW probe, ν_B is the Brillouin frequency shift. The equations above only apply to the steady state which means the pump pulse width is larger than the phonon lifetime ($\approx 10ns$). When g is positive, the equations describe a Brillouin gain process which means the energy is transferred from the pump pulse to the CW probe. Conversely, when g is negative, the equations describe a Brillouin loss process which means the energy is transferred from the CW probe to the pump pulse.

Ref. [7] (Bao et al. 1995) solve these equations with a perturbation method and got the analytic solution of $I_{cw}(z)$. For the typical discrete non-uniform strain profile within the spatial resolution w, as illustrated in Figure 1, the Brillouin frequency shift within z and $z+\Delta l$ is ν_{Bs} and the Brillouin frequency shift within $z+\Delta l$ and $z+w$ is ν_B. Ref. [8] (Ravet et al. 2005) followed the Ref. [7] and got the analytic solution of $I_{cw}(z)$ by subdividing the integration interval into $[z, z+\Delta l]$ and $[z+\Delta l, z+w]$. The variation in the intensity of the CW probe at position z is given by

$$G = \frac{I_{cw}(z)}{I_{cw}(z+w)} =$$

$$\exp(-\alpha w)\exp\left\{\frac{\kappa_1}{\alpha}\exp(\beta_1)\left[\frac{\exp(-\beta_1 x_2)}{x_2} - \frac{\exp(-\beta_1 x_1)}{x_1} + \beta_1\left[E_1(\beta_1 x_1) - E_1(\beta_1 x_2)\right]\right\}\right. \quad (4)$$

$$\left. + \frac{\kappa_2}{\alpha}\exp(\beta_2)\left\{\frac{\exp(-\beta_2 x_3)}{x_3} - \frac{\exp(-\beta_2 x_2)}{x_2} + \beta_2\left[E_1(\beta_2 x_2) - E_1(\beta_2 x_3)\right]\right\}\right\}$$

Where $E_1(y)$ is the exponential integral, $\kappa_i = g_i I_p(0)$,

$$\beta_i = g_i I_{cw}(L)\frac{\exp(-\alpha L)}{\alpha}, i = 1,2,$$

and $x_1 = \exp(\alpha z)$, $x_2 = \exp[\alpha(z+\delta l)]$, $x_3 = \exp[\alpha(z+w)]$.

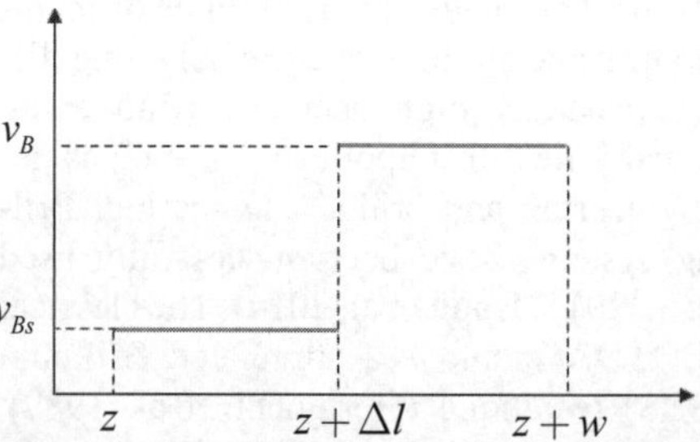

Figure 1. Typical discrete non-uniform strain profile within the spatial resolution w

For the typical linear non-uniform strain profile within the spatial resolution w, as illustrated in Figure 2, the Brillouin frequency shift change continuously from ν_{Bs} at z to ν_B at $z+w$. We can easily followed the Ref. [7] and Ref. [8] got the solution of $I_{cw}(z)$ by subdividing the integration interval into n sections within $[z,z+w]$. The variation in the intensity of the CW probe at position z is given by

$$G=\frac{I_{cw}(z)}{I_{cw}(z+w)}=$$

$$\exp(-\alpha w)\exp\left\{\sum_{i=1}^{n}\frac{\kappa_i}{\alpha}\exp(\beta_i)\left\{\frac{\exp(-\beta_i x_{i+1})}{x_{i+1}}\right.\right. \quad (5)$$

$$\left.\left.-\frac{\exp(-\beta_i x_i)}{x_i}+\beta_i\left[E_i(\beta_i x_i)-E_i(\beta_i x_{i+1})\right]\right\}\right\}$$

Where $E_1(y)$ is the exponential integral, $\kappa_i=g_i I_p(0)$,

$$\beta_i=g_i I_{cw}(L)\frac{\exp(-\alpha L)}{\alpha},i=1,2...n$$

and $x_i=\exp\left\{\alpha\left[z+(i-1)\frac{w}{n}\right]\right\}$.

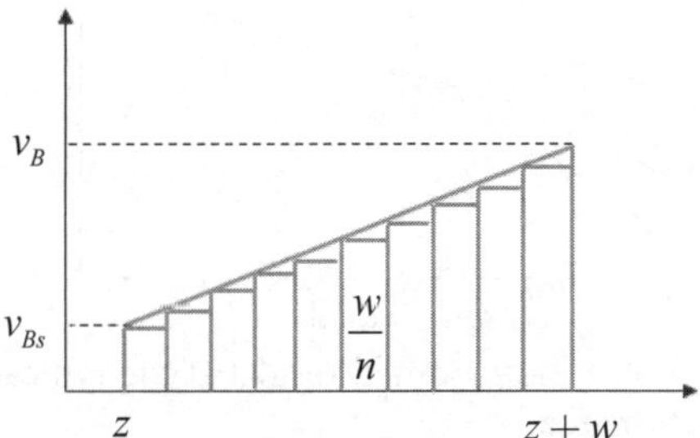

Figure 2. Typical linear non-uniform strain profile within the spatial resolution w

3 NON-UNIFORM STRAIN MODIFICATION METHOD

Here we discuss the typical linear non-uniform strain profile condition and its influence on the measurement errors. Figure 3 shows the BGS at location z for the fiber configuration in Figure 2. $\nu_{Bs}=10.80GHz$ is set as a constant and $\nu_B=\nu$ is set as a variable which changes from $10.80GHz$ to $10.86GHz$ with the step of $0.01GHz$. Then we calculated the BGS using the theoretical model in Equation (5) and got the BGS under each non-uniform case was shown in Figure 3. According to the strain measurement principles of BOTDA system, the measured strain value at location z is decided by the Brillouin frequency shift (BFS) of the BGS at location z. We can see that, with the increase of ν, the BGS at location z changes significantly. The BFS of the BGS is changed from 10.80GHz to 10.83GHz, which means the measured strain value is also changed. But the accurate BFS at location z is set as a constant ($\nu_{Bs}=10.80GHz$) and the accurate strain value is also a constant. So we can see that the non-uniform components can lead to inaccuracies in the measurement results of BOTDA system and the measured BFS value is larger than the accurate value. When $\nu<\nu_B$, we can easily get that the measured BFS value is smaller than the accurate value. The nonuniformity introduced BFS measurement error equals to the measured BFS of BGS at location z minus ν_{Bs}. Besides the BFS change, the width of the BGS is also changed with the increase of ν. The BGS width change equals to the measured width of BGS at location z minus the initial width of BGS when $v=\nu_{Bs}$. The nonuniformity introduced BFS measurement error and BGS width change under different cases of v are shown in Figure 4. We can see that the both the BFS measurement error and the width of the BGS increase with the increase of ν. Interestingly, these two parts almost share the same value at each case. So we try to modify the BFS measurement errors by using the BGS width change. The modified BFS value equals to the measured BFS minus the BGS width change and the modified results are shown in Figure 4. We can see that this modification method obviously decreases the measurement error. Note that in the case of $\nu<\nu_B$, the modified BFS value equals to the measured BFS plus the BGS width change.

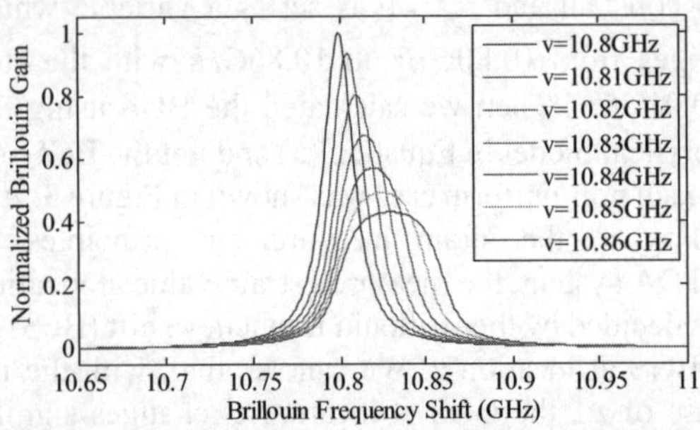

Figure 3. Normalized Brillouin Gain Spectrum at location Z corresponding to the typical linear non-uniform strain profile cases with $\nu_{Bs} = 10.80GHz$ and $\nu_B = \nu$ changes from $10.80GHz$ to $10.86GHz$ with the step of $0.01GHz$. The sensor settings are $P_p = 30mW$, $P_{cw} = 5mW$, $L = 1000m$, $z = 500m$, $w = 20m$.

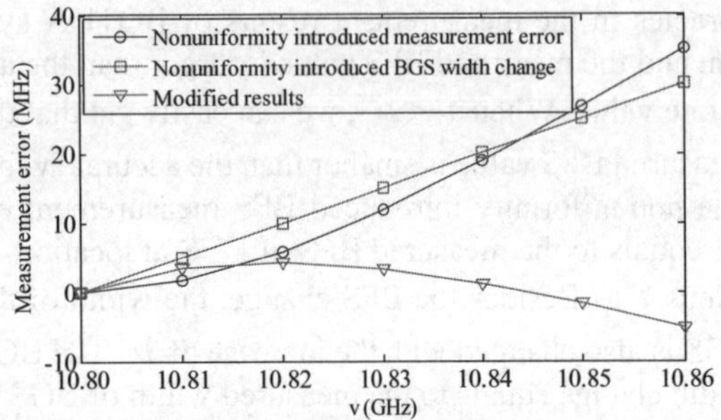

Figure 4. Nonuniformity introduced BFS measurement error and BGS width change under different cases of v and the modified results.

4 VALIDATION OF THE PROPOSED METHOD

We applied the proposed modification method in the BOTDA monitoring data of a wind turbine blade load test. In this case, the spatial resolution of the BOTDA is 50cm. Figure 5(a) shows the measurement errors between BOTDA measurement results and the accurate results. Here we defined measurement results by the higher spatial resolution (20cm) BOTDA system as the accurate results. We can see that the BFS distribution contains obvious and different types of no-uniform components. For the typical linear non-uniform conditions, when $\nu > \nu_B$, which means the BFS value linearly increases within the spatial resolution distance, the measured BFS value is larger than the accurate value. When $\nu < \nu_B$, the measured BFS value is smaller than the accurate value. The phenomenon is in good agreement with our theoretical analysis. Then we choose a section of the data as the test area which is the typical linear non-uniform distribution (shown in Figure 5(a)) and applied the proposed modification method. The initial width of BGS was obtained from the BGS at the strain free section of the sensing fiber. The modified results at the test area and its comparison with the BOTDA measurement results and the accurate results are shown in Figure 5(b). We can see that the proposed modification method obviously decreases the measurement error of the BOTDA results.

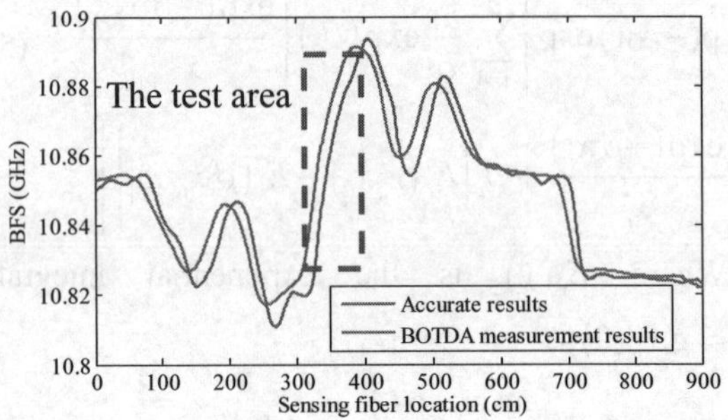

(a) Measurement errors of the BOTDA monitoring results in a wind turbine blade load test

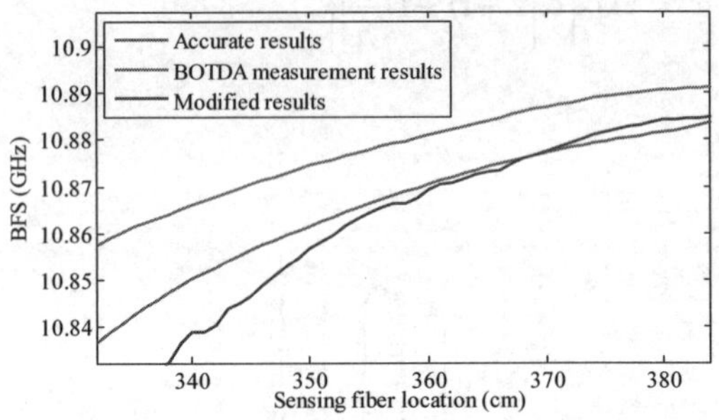

(b) Comparison of the modified results at the typical non-uniform test area

Figure 5. Validation test results of the proposed modification method

5 CONCLUSIONS

A non-uniform strain modification method for distributed BOTDA sensing has been proposed according to the theoretical analysis of the typical linear non-uniform distribution within the spatial resolution distance. With the BFS at z was set as 10.80GHz and the BFS at $z+w$ was changed from 10.80GHz to 10.86GHz, the BFS of the measured BGS was changed from 10.80GHz to 10.83GHz. The nonuniformity introduced BFS measurement error approximately equals to the BGS width change. The BGS width change can be used to modify the BFS measurement errors. The modified BFS value equals to the measured BFS minus the BGS width change in the case of $\nu > \nu_B$ and the modified BFS value equals to the measured BFS plus the BGS width change in the case of $\nu < \nu_B$. In the validation of the proposed method, the strain distribution of the wind turbine blade load test contained obvious no-uniform strain components. The phenomenon of the normal BOTDA measurement value was in good agreement with the theoretical analysis. Comparisons of the measured accurate strain distribution and the modified results show that the proposed method was capable of modifying the typical linear non-uniform strain measurement errors and increasing the measurement accuracies in the distributed BOTDA sensing.

ACKNOWLEDGEMENT

The financial support from the National High Technology Research and Development Program of China 863 Program 2014AA110401 and the National Key Technology R&D Program 2014BAG05B07 are gratefully appreciated.

REFERENCES

Bao, X. and Chen, L., 2011, Recent progress in Brillouin scattering based fiber sensors, Sensors. 11(4), 4152-4187.

Mohamad, H., Soga, K., Pellew, A., and Bennett, P. J., 2011, Performance monitoring of a secant-piled wall using distributed fiber optic strain sensing. J.Geotech. Geoenviron. Eng., 10.1061/(ASCE) GT.1943-5606.0000543, 1236–1243.

Regier and Hoult. 2014. Distributed Strain Behavior of a Reinforced Concrete Bridge: Case Study. J. Bridge Eng., 10.1061/(ASCE) BE.1943-5592.0000637, 05014007.

Horiguchi, T. and Tateda, M. 1989. Optical-fiber-attenuation Investigation Using Stimulated Brillouin Scattering Between a Pulse and a Continuous Wave, Opt. Lett. 14, 408–410.

H. Naruse, M. Tateda, H. Ohno and A. Shimada, 2003, Deformation of the Brillouin gain spectrum caused by parabolic strain distribution and resulting measurement error in the BOTDR strain measurement system, IEICE Trans. Electron. E86-C,2111–2121.

G. P. Agrawal, 1995.Nonlinear Fiber Optics, 2nd ed.

X. Bao, J. Dhliwayo, N. Heron, D. J. Webb, and D. A. Jackson, 1995, Experimental and theoretical studies on a distributed temperature sensor based on Brillouin scattering, J. Lightwave Technol. 13, 1340–1348.

Ravet, Fabien, Xiaoyi Bao, and Liang Chen., 2005, Simple approach to determining the minimum measurable stress length and stress measurement accuracy in distributed Brillouin sensing. Applied optics 44.25: 5304-5310.

Proceedings of the International Conference on Smart Infrastructure and Construction
ISBN 978-0-7277-6127-9

© The authors and ICE Publishing: All rights reserved, 2016
doi:10.1680/tfitsi.61279.123

Multiple windows algorithm for event detection in STFT-BOTDR

Yifei Yu[1], Linqing Luo[1], Bo Li[1], Jize Yan[1,2]* and Kenichi Soga[1]
1 Department of Engineering, University of Cambridge, Cambridge, UK
2 Electronics and Computer Science, University of Southampton, Southampton, UK
* *Corresponding Author*

ABSTRACT Short-time Fourier Transform (STFT) as a class of the time-frequency analysis have been utilised to deliver a fast-speed, accurate and low-variation BOTDR (Brillouin Optical Time Domain Reflectometer) system. The different window functions applied in the signal processing are important to improve the spatial resolution, frequency resolution, and variance of the BOTDR and to detect small event in the real application such as strain or temperature events smaller than the pulse length along the fibre. Window functions with long duration can achieve high frequency resolution, but be difficult to detect events less than the window length. While, small-length window offers better event detection performance than the long-duration window, but induces high variation along the fibre under test (FUT). For instance, the rectangular window offers a better frequency resolution; and the hamming window is better at the event detection. This paper introduces an optimised algorithm by utilising two separate windows and comparing the Brillouin frequency difference along the distance domain between the results generated by different windows to enhance the capability of small event detection, such as cracks and sharp temperature variation along the fibre

1. INTRODUCTION

Distributed optical fibre sensors based on Brillouin scattering have attracted significant interest for diagnosing deterioration and preventing disasters in the field of health monitoring for large civil structures, such as pipe lines, bridges, tunnels and dams in the past few decades (Bao and Chen 2011; Lee 2003). BOTDRs (Brillouin Optical Time Domain Reflectometers) have been utilised to measure the distributed strain and temperature along the entire optical fibre by monitoring the Brillouin Frequency Shift (BFS) of the scattering. In the BOTDR system, a laser probing pulse is launched into the sensing fibre under test (FUT) and its Spontaneous Brillouin Scattering (SpBS) spectrum along the fibre is measured at the same end of the fibre, which is convenient for the field applications (Horiguchi et al. 1995; Bergman, and Tur 2015). The conventional BOTDR system demands a time consuming frequency scanning of the entire SpBS spectrum and a great amount of averaging up to 2^{20} times to acquire the accurate Brillouin scattering spectrum (BSS) (Ohno et al. 2001; Wang et al. 2009).

Recently, a fast-speed Discrete Fourier Transform (DFT)-based BOTDR has applied the wideband detection architecture accelerated by a Digital Signal Processing (DSP) and/or Field Programmable Gate Array (FPGA) to replace the time-consuming frequency sweeping method, moreover, it takes only 1 s to measure the frequency-down-converted SpBS signal up to 500 MHz over the 1.5 km testing fibre (Ohno et al. 2001; Lu et al. 2008). This wideband detection has been demonstrated at a temperature resolution of 3 °C and a spatial resolution of 2 m for 6 km fibre (Lu, Dou, and Zhang 2008). Very recently, the zero-padding enhanced STFT-BOTDR has been demonstrated, which offers an analysis on the spectrum deformation of the BOTDR system to improve measured distance accuracy (Yu et al. 2015). In Short-Time Fourier Transform (STFT) BOTDR system, a short duration window is applied to achieve a better spatial resolution, which will worsen the frequency resolution of

SpBS. In contrast, a longer window would improve the frequency resolution but reduce the spatial resolution. In real applications, the frequency resolution of the SpBS determines the strain and temperature resolution, and the spatial resolution is essential for the strain or temperature events smaller than the pulse length detection.

This paper applied the zero-padding-enhanced STFT method to a temperature measurement of different length of fibre, for demonstrating the influence of the window type and duration to the BOTDR system to compare the spatial resolution, frequency resolution, variance and short-length event detection. The highest spatial resolution of the STFT-BOTDR is determined by the probe pulse length, which severely influenced by the number of data in each frame of STFT and linearly increases with increasing window size (Horiguchi et al. 1995; Wang et al. 2009). Accordingly, an event with length smaller than the pulse duration or the window length is not detectable, including the actual length and Brillouin frequency shift. In other words, a detectable event must contain a length larger than both of the pulse duration and window length in order to offer an undamaged Brillouin frequency shift and length. To overstep the uncertainty principle between frequency and time resolution of the STFT method (Flandrin, Lin, and Mclaughlin 2013), a combination of variant window types and durations algorithm has been proposed in this paper to deliver a comprehensive result. Based on the Brillouin frequency difference along the distance between two windows, the smaller-length event and Brillouin frequency transition region can be worked out. According to this indication, an optimised result from multiple windows can be achieved.

2. THEORY

Time-frequency distributions are designed to characterise the time-frequency content of signals. STFT is a linear time-frequency transform applied to evaluate the frequency and phase content of local sections of a signal as it changes over time (Hlawatsch and Francois 2008). The back-scattered Brillouin signal captured by the digitiser of a coherent heterodyne BOTDR system can be represented as discrete frequency domain information (Geng et al. 2007). In the case of a wideband receiver, the data to be transformed need to be split up into frames and each frame is Fourier transformed for observing localised spectrum of a signal. The discrete STFT can be represented by (Hlawatsch and Francois 2008):

$$STFT\{x[n]\}(m,\omega) \equiv \sum_{-\infty}^{\infty} x[n]w[n-m]e^{-j\omega n} \tag{1}$$

where x[n] is the signal to be transformed, and w[n] is the window function. The window duration and type are the two important variables in the STFT process. DFT requires the signal length to be finite in each frame, which means the N-point DFT operates on an N-elements data vector x[n] to produce an N-element result in a frequency domain X[K] (Durak and Arikan 2003).

$$X[k] = \sum_{n=0}^{N-1} x[n] e^{-j2\pi nk/N} \tag{2}$$

Since the frequency resolution is inversely proportional to the analysing window duration, it is difficult to ensure a good stationarity and good frequency resolution at the same time (Abdelghani and Fethi 2000). It is suggested to keep the duration of the sliding window as short as possible to ensure the stability, but reducing the frequency resolution of the spectrogram (Abdelghani and Fethi 2000), which is known as Heisenberg uncertainty principle (Busch, Heinonen, and Lahti 2007) and a time and frequency resolution tradeoff to be considered.

The various types of window can be utilised in signal with different characteristics. The rectangular window can be used to truncate the data sequence, while the other windows such as hamming, Gaussian and Blackman windows offer some data weighting (Harris 1978). The advantage of applying a window other than rectangular is to have lower sidelobes. However the disadvantage is a loss in frequency resolution (Hlawatsch and Francois 2008). If the actual signal length L is less than N, the data vector can be extended by "zero-padding" which accounts to putting zeros at the end of digitised time domain signal (Hlawatsch and Francois 2008). The advantages of zero padding and the corresponding peak-searching method applied in the STFT-BOTDR system are more immune to Brillouin spectrum deformation and asymmetry (Yu et al. 2015).

The minimum detectable frequency shift is a function of the Full Width at Half Maximum (FWHM) and Signal to Noise Ratio (SNR) in the BOTDR system(Horiguchi et al. 1995), which is given by:

$$\delta v_B = FWHM/\sqrt{2}(SNR)^{1/4} \quad (3)$$

3. EXPERIMENTAL RESULTS AND DISCUSSION

The FUT setup can be represented by Figure 1. The total FUT consists of nine sections of single mode fibre (SMF) controlled by a water bath during the experiment. The interval between the temperature controlled sections is 10 m. The temperature was set to be stable at 75 °C.

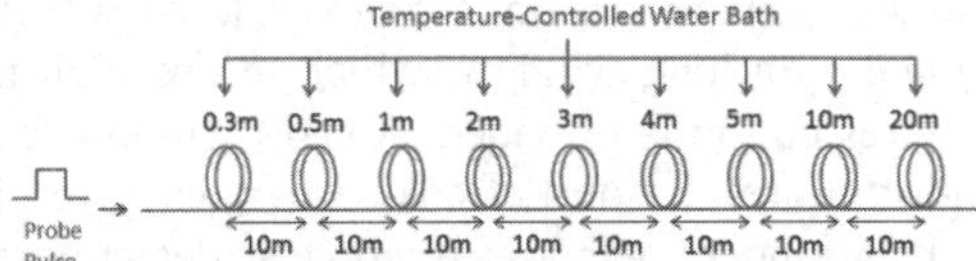

Figure 1. Experimental setup for the temperature-controlled unit of the STFT-based BOTDR system.

Figure 2 illustrates the schematic representation of the STFT-BOTDR system with temperature calibration unit. Branch A and B were split by a 90/10 coupler from a continuous-wave laser diode. The branch A was modulated by an Electro-Optical Modulator (EOM) with 34 ns modulating pulse, amplified by an Erbium-Doped Fibre Amplifier (EDFA). The amplified pulse light went through an optical filter followed by a circulator, and then launched into a 1.5 km FUT. The branch B is regarded as a reference light connected by a polarization scrambler (PS) and utilised as an optical local oscillator (OLO). The branch B mixed with the backscattering light via a coupler and a 26 GHz photodetector (PD) for coherent detection. The down-converted signal was captured by a 5 GS/s digitiser.

3.1 Window type

Five types of window have been applied to the STFT-BOTDR system, including Hamming, Gaussian, Rectangular, Blackman and Chebyshev window. With fixed window duration, readout and zero-padding, the measured Brillouin centre frequency along the distance axis is displayed in Figure 3. It is observed that the variations along the distance axis for five windows are different. The variation for the rectangular is 3.98 MHz, which is the lowest among these five windows. Moreover, the rectangular window brings the most rapid frequency transition region shown in Figure 4, which can be regarded as the shortest detectable event length. The spatial resolution for the rectangular window is 2m. Moreover, the FWHM measured for these five kinds of windows in the same position is illustrated in Figure 6. The rectangular window can offer lowest FWHM, so that the measured frequency resolution is the best in this window measurement. However, the rectangular window can not detect the event, of which the length is less than the pulse duation indicated in Figure 5.

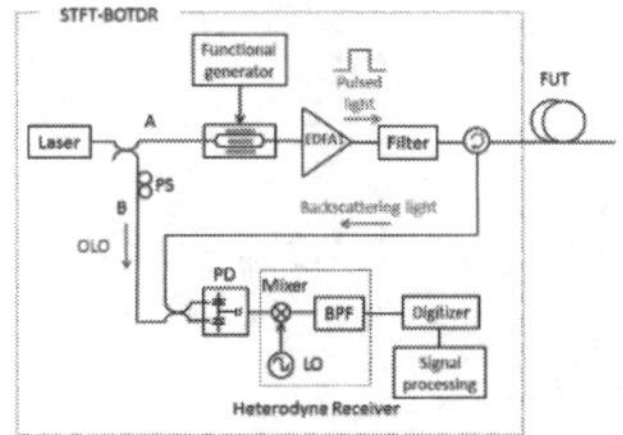

Figure 2. Experimental setup for STFT-BOTDR. PS, polarization scrambler; EDFA, erbium-doped fibre amplifier; EOM, electro-optic modulator; FUT, fibre under test; OLO, optical local oscillator.

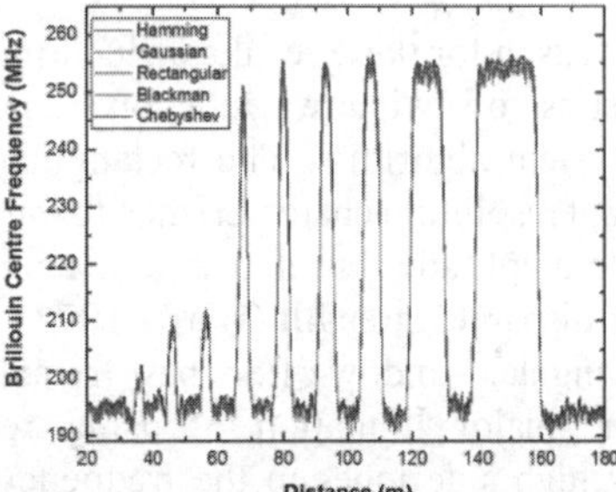

Figure 3. Experimental results of Brillouin centre frequency for 75°C temperature test with different window types.

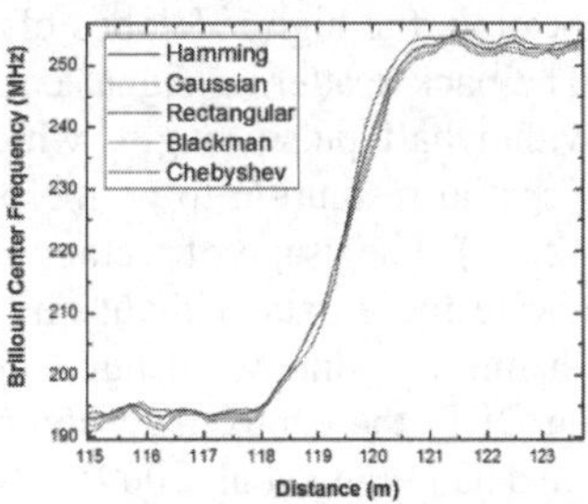

Figure 4. Experimental results of Brillouin centre frequency for 75°C temperature test with different window types in the frequency transition region.

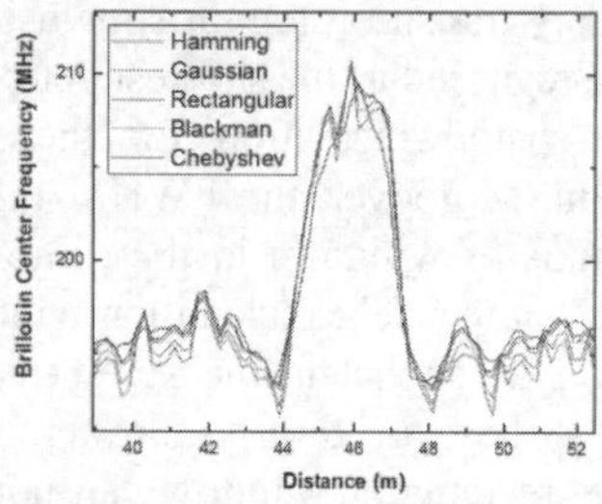

Figure 5. Experimental results of Brillouin centre frequency for 75°C temperature test with different window types in the 0.5m section.

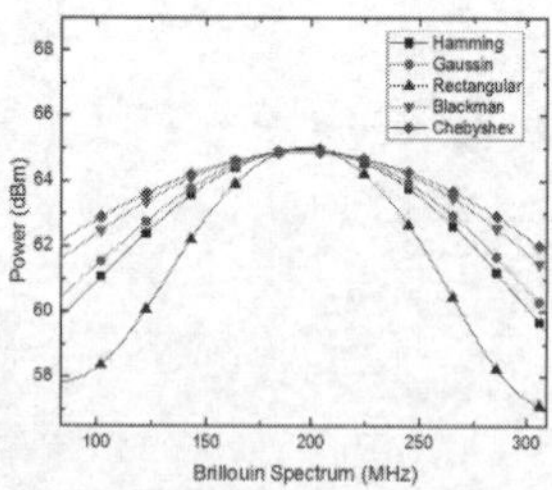

Figure 6. Experimental results of the Brillouin spectrum in a same position for the five types of windows.

The reason for these results is due to both of the characteristics of window functions and STFT-BOTDR system algorithm. The rectangular window has excellent resolution characteristics for sinusoids of comparable amplitude, but it is a poor choice for sinusoids of disparate strength (Harris 1978) . Furthermore, rectangular window offers best frequency resolution. The major limitation of using rectangular window is high side lobes in the frequency domain. But due to averaging process and Gaussian fitting, only the three peak points are utilised to find the Brillouin peak location, the high sidelobes effects can be minimised. The back scattering signal cannot be distinguished within half pulse length, which leads to limitation of spatial resolution to be the pulse length (Wang et al. 2009). The usage of rectangular window equals to truncate these data without any weighting. Because the hamming window changes the weight of each frame in DFT, the centre data contribute more than the fringe data (Wang et al. 2009). The hamming window shows good detection performance in the events less than the pulse length, because the window is optimised to minimise the nearest side lobe. In this case, it is observed that a window with high main lobe to side lobe ratio is capable to strain or temperature events smaller than the pulse length detection, such as hamming window. In the STFT-BOTDR system, a hamming window is good at less than pulse length detection, such as cracks; a rectangular window is recommended to be applied for low variation, high spatial resolution and frequency resolution.

3.2 Window size

Window size is crucial in the STFT-BOTDR signal processing. A long window brings high frequency resolution, but low time resolution. This equals to good strain/temperature resolution and bad spatial resolution in the BOTDR system. Moreover, low FWHM is generated by a long window, which means high frequency resolution is provided. A short window leads to high time resolution but coarse frequency resolution. This kind of window is good at detecting the event less than the pulse length. There are four durations of windows applied in the STFT process of the BOTDR system. It is observed that the lowest variation (2.48 MHz) along the distance is obtained by the window with 40 ns duration displayed in Figure 7. This relatively long window can also bring the shortest detectable event length (0.4 m) shown in Figure 8. Additionally, the FWHM for the 40 ns duration window is the lowest among these four windows illustrated in Figure 10, which indicates the highest frequency resolution. The long window is not good at 0.5 m event detection displayed in Figure 9. The shortest window (10 ns) is more sensitive to the event with length less than the duration of the probe pulse. The reason for the low variation along the distance for the long window is due to more data involved in a window, which is equivalent to more times of averaging. In this case, 40 ns window provides four times more averaging than the 10 ns window. In the rising area, because of the low FWHM, the double peaks effect is easier to observe in the long window signal processing. In the results of short window, the double peaks are merged. Accordingly, long window has advantage to provide high measured frequency resolution. In the 0.5 m section detection, the 40 ns window did not show strength. 10ns window is more sensitive to the short event detection, which is due to short duration of time data averaging. Long window can bring low variation, but at the same time it losses some information. If the length of the event is less than the window duration, the measured frequency shift is always less than the real value, which is similar to the pulse length demonstrated in Ref (H. Zhang and Wu 2008). Moreover, long window can bring low FWHM, which indicate high frequency resolution to the STFT-BOTDR system. As a result, a long window can offer low variation, high spatial resolution and frequency resolution; a short window is more sensitive to strain or temperature events smaller than the pulse length detection.

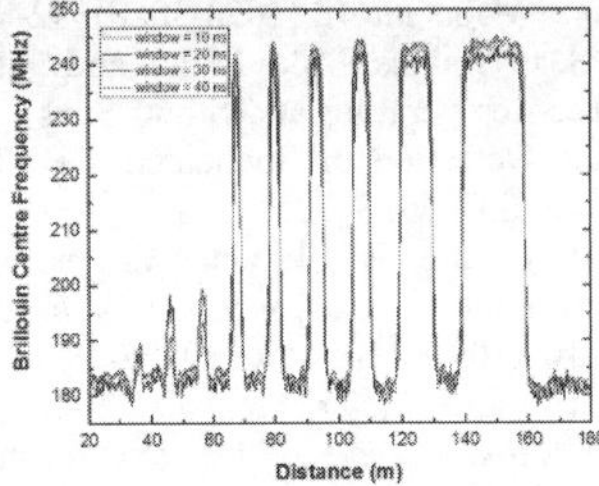

Figure 7. Experimental results of Brillouin centre frequency for 75°C temperature test with different window durations.

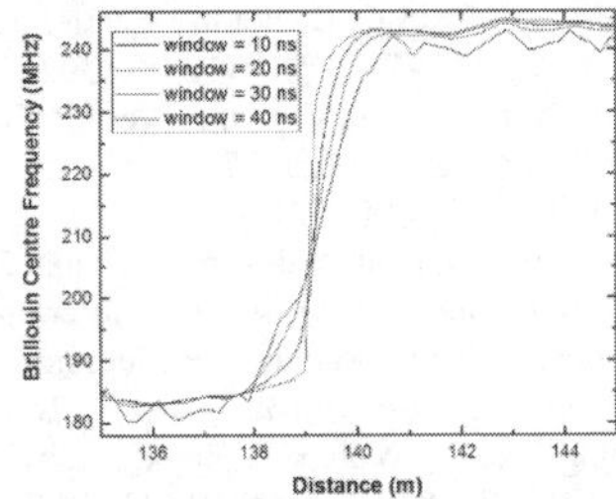

Figure 8. Experimental results of Brillouin centre frequency for 75°C temperature test with different window durations in the frequency transition region.

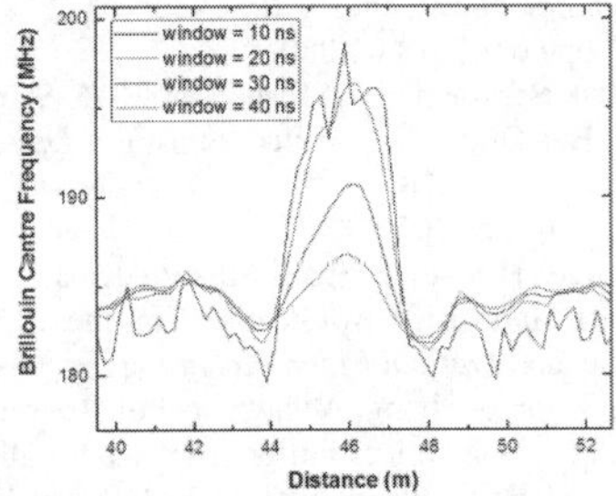

Figure 9. Experimental results of Brillouin centre frequency for 75°C temperature test with different window durations in the 0.5m section.

4. REFINEMENT ALGORITHM

The experimental results and physical explanation in section 3 demonstrate no window can be found for the STFT-BOTDR system to achieve high frequency resolution and small event detection simultaneously due to the time-frequency uncertainty principle. However, this paper utilised a combined window analysis algorithm to improve the system performance. In the signal processing, long rectangular window and short hamming window are applied. The rectangular window is longer than the probe pulse length; and the hamming window is shorter than the probe pulse length. For example, if the probe pulse is 34 ns, the rectangular window can be chosen to be 40 ns and the hamming window has a length of 20 ns. The difference between Brillouin frequency shift results generated by two windows are experimentally illustrated in Figure 11. In the temperature changing area (longer than the spatial resolution), the frequency difference has both positive and negative peaks shown in Figure 12. However, in the shorter temperature event (such a 0.5 m section), the frequency difference only have negative peaks displayed in Figure 13. Physically, the shorter windows can detect both the event longer or shorter than the spatial resolution, while the longer windows can only detect the event longer than the spatial resolution. Accordingly, the frequency difference between two kinds of window functions can be utilised as an indication whether the event section is longer or shorter than the spatial resolution. This window combination algorithm offers a rapid indication for small events happening along the fibre.

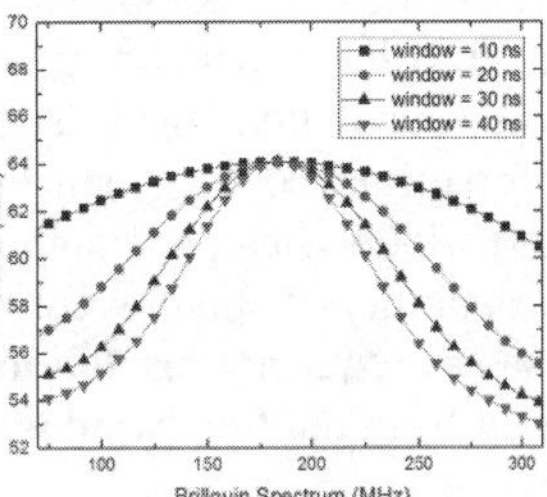

Figure 10. Experimental results of the Brillouin spectrum in a same position for the five durations of windows.

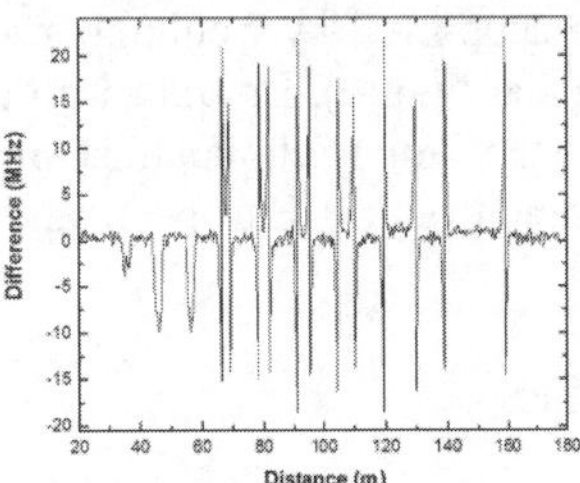

Figure 11. The difference between two Brillouin shifts for 40ns rectangular window windows and 10ns hamming window.

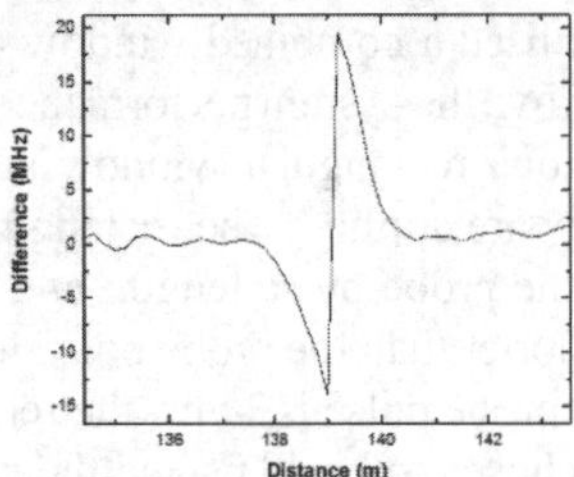

Figure 12. The difference between two Brillouin shift in the frequency changing area for 40ns rectangular window windows and 10ns hamming window.

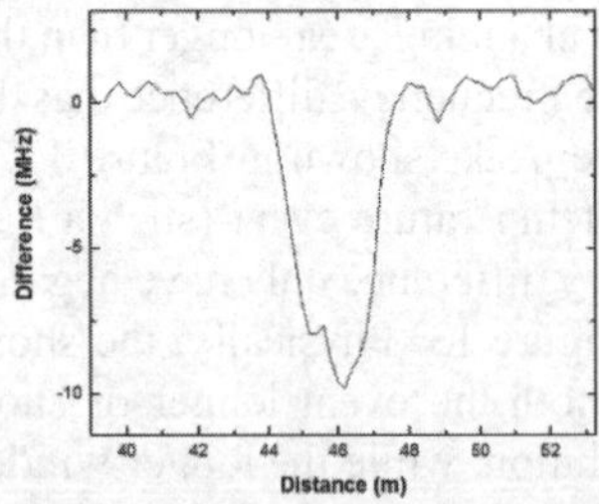

Figure 13. The difference between two Brillouin shift in the strain or temperature events smaller than the pulse length area for 40ns rectangular window windows and 10ns hamming window.

5. CONCLUSIONS

Five different types of window functions are utilised and experimentally compared in the STFT-BOTDR system in the paper. Long-window offers a better frequency resolution, while short-window offers a better performance for small event detection. Long and short windows can be combined together to enhance the capability of small event detection while keeping the same frequency resolution, indicating an optimised result from multiple windows analysis.

ACKNOWLEDGEMENTS

Yifei Yu is supported by Schlumberger Foundation (Faculty for the Future). Linqing Luo and Bo Li are supported by the China Scholarship Council and Cambridge Commonwealth, European and International Trust.

REFERENCES

Abdelghani, Djebbari, and Bereksi Reguig Fethi. 2000. Short-Time Fourier Transform Analysis of the Phonocardiogram Signal. *IEEE* **9** (7803-6542): 844–47.

Bao, Xiaoyi, and Liang Chen. 2011. Recent Progress in Brillouin Scattering Based Fiber Sensors. *Sensors (Basel, Switzerland)* **11** (4): 4152–87. doi:10.3390/s110404152.

Busch, Paul, Teiko Heinonen, and Pekka Lahti. 2007. Heisenberg's Uncertainty Principle. *Physics Reports* **452** (6): 155–76. doi:10.1016/j.physrep.2007.05.006.

Durak, Lütfiye, and Orhan Arikan. 2003. Short-Time Fourier Transform: Two Fundamental Properties and an Optimal Implementation. *IEEE Transactions on Signal Processing* **51** (5): 1231–42. doi:10.1109/TSP.2003.810293.

Flandrin, Patrick, Yu-ting Lin, and Stephen Mclaughlin. 2013. Time-Frequency Reassignment and Synchrosqueezing. *IEEE Signal Processing Magazine* **13** (1053-5888): 32–41.

Geng, Jihong, Sean Staines, Mike Blake, and Shibin Jiang. 2007. Distributed Fiber Temperature and Strain Sensor Using Coherent Radio-Frequency Detection of Spontaneous Brillouin Scattering. *Applied Optics* **46** (23): 5928–32.

Harris, F.J. 1978. On the Use of Windows for Harmonic Analysis with the Discrete Fourier Transform. *Proceedings of the IEEE* **66** (1): 51–83. doi:10.1109/PROC.1978.10837.

Hlawatsch, Franz, and Auger Francois. 2008. *Time-Frequency Analysis Concepts and Methods*. ISTE Ltd and John Wiley & Sons, Inc.

Horiguchi, Tsuneo, Kaoru Shimizu, Toshio Kurashima, Mitsuhiro Tateda, and Yahei Koyamada. 1995. Development of a Distributed Sensing Technique Using Brillouin Scattering. *Journal of Lightwave Technology* **13** (9512279): 1296–1302.

Lee, Byoungho. 2003. "Review of the Present Status of Optical Fiber Sensors." *Optical Fiber Technology* 9 (2): 57–79. doi:10.1016/S1068-5200(02)00527-8.

Lu, Yuangang, Rongrong Dou, and Xuping Zhang. 2008. Wideband Dectection of Spontaneous Brillouin Scattering Spectrum in Brillouin Optical Time-Domain Reflectometry. In *2008 International Conference on Optical Instruments and Technology*, edited by Xuping Zhang, Wojtek J. Bock, Xiaoyi Bao, and Ping Shum, 7158:715818–715818 – 7. doi:10.1117/12.807012.

Lu, Yuangang, Yuguo Yao, Xiaodong Zhao, Feng Wang, and Xuping Zhang. 2013. Influence of Non-Perfect Extinction Ratio of Electro-Optic Modulator on Signal-to-Noise Ratio of BOTDR. *Optics Communications*, no. **297** (June). Elsevier: 48–54. doi:10.1016/j.optcom.2013.01.080.

Motil, Avi, Arik Bergman, and Moshe Tur. 2015. State of the Art of Brillouin Fi Ber-Optic Distributed Sensing. *Optics and Laser Technology* **78**. Elsevier: 1–23. doi:10.1016/j.optlastec.2015.09.013.

Ohno, Hiroshige, Hiroshi Naruse, Mitsuru Kihara, and Akiyoshi Shimada. 2001. Industrial Applications of the BOTDR Optical Fiber Strain Sensor. *Optical Fiber Technology* **7**: 45–64.

Wang, Feng, Xuping Zhang, Yuangang Lu, Rongrong Dou, and Xiaoyi Bao. 2009. Spatial Resolution Analysis for Discrete Fourier Transform-Based Brillouin Optical Time Domain Reflectometry. *Measurement Science and Technology* **20** (2): 025202. doi:10.1088/0957-0233/20/2/025202.

Yu, Yifei, Linqing Luo, Li Bo, Linfeng Guo, Jize Yan, and Kenichi Soga. 2015. Double Peak-Induced Distance Error in Optical Time Domain Reflectometers Event Detection and the Recovery Method. *Applied Optics* **54** (28): 196–202.

Zhang, Hao, and Zhishen Wu. 2008. Performance Evaluation of BOTDR-Based Distributed Fiber Optic Sensors for Crack Monitoring. *Structural Health Monitoring* **7** (2): 143–56. doi:10.1177/1475921708089745.

Proceedings of the International Conference on Smart Infrastructure and Construction
ISBN 978-0-7277-6127-9

© The authors and ICE Publishing: All rights reserved, 2016
doi:10.1680/tfitsi.61279.129

Compressive sensing-based wireless sensors and sensor networks for structural health monitoring

Y. Bao* and H. Li

Key Lab of Structures Dynamic Behavior and Control of the Ministry of Education (Harbin Institute of Technology), Harbin, China

School of Civil Engineering, Harbin Institute of Technology, Harbin, China

* *Corresponding Author*

ABSTRACT Wireless sensor technology-based structural health monitoring (SHM) has been widely investigated recently. This paper presents the new developments and applications of compressive sensing (CS) for wireless sensors and sensor networks-based SHM in our research group. Frist, the group sprse optimization based compressive sensing for data sampling and recovery of wireless sensor network is introduced. Then, the lost data recovery for wireless sensors are presented. CS provides a data loss recovery technique, which can be embedded into smart wireless sensors and effectively increases wireless communication reliability without re-transmitting the data; the promise of this approach is to reduce communication and thus power savings. To embed into the smart sensor, a method called random demodulator is employed to provide memory and power efficient construction of the random sampling matrix. The program is embedded into the Imote2 smart sensor platform and tested in a series of sensing and communication experiments and field tests. Lastly, the fast moving wireless sensoing technique is presened. For the fast moving wireless data transmission, the Doppler effects are the main reason causing data packet loss. A field test on a cable-stayed bridge is performed to valid the ability of the CS-based robust wireless data transmission approach in obtaining high-quality data for the fast-moving wireless sensing technique.

1 INTRODUCTION

The structural health monitoring (SHM) technology has developed about one decades and a lot of civil infrastructures have been installed the SHM systems in the world (Ou & Li 2010; Mufti 2012). In the wired sensors based SHM system, the wired connection between sensors and data acquisition system reduces the reliability of SHM system, increasing the system cost, and cause great difficulties for the maintenance and replacement. As reported, the wired monitoring system on the Bill Emerson Memorial Bridge in Cape Girardeau, Missouri, USA, has cost more than $15,000 per sensor (Celebi 2006), and a large portion of that cost is the data transmission cables. Wireless sensors and wireless network have intelligent data processing capabilities with embedding algorithm and do not have cables that will great reduce the sensor installation cost. In comparison with the traditional wired sensor monitoring systems, wireless sensors and wireless network possess several advantages that make them an attractive alternative for monitoring large civil infrastructure.

Recently, wireless sensors have been used on many bridges for SHM purposes. The Jindo Bridge, a cable stayed bridge in Korea, has been installed 70 Imote2 smart wireless sensors for SHM (Jang et al. 2010). The Geumdang Bridge, a continuous beam bridge, which also has been installed 28 wireless sensors for SHM (Lynch et al. 2006). The Stork Bridge in Switzerland (Meyer et al. 2010), Ferriby Road Bridge in England (Hoult 2010), and Rock Island Arsenal Government Bridge in USA (Cho et al. 2015) are all have been installed the wireless sensors based-SHM system.

However, in the long-term monitoring of the structure, the large amounts of data acquisition and wireless transmission are likely to cause the instability of wireless sensor network. In addition, the radio interference, e.g., other devices operating on the same frequency, weather problems such as rain and lightning, poor installation and antenna orientation, large transmission distances, hardware problems, Doppler effects etc., will always cause the data packets lost in wireless sensor networks. Such data loss in wireless communication has been reported by several researchers for various applications (Pei et al. 2005; Kurata et al. 2004; Meyer et al. 2010).

Compressive sensing (CS) is new emerging theory that can reconstruct signals accurately from far fewer measurements than what is usually considered necessary (Donoho 2006; Candès 2006). The CS has been widely studied in SHM and wireless sensors and sensor networks. This paper presents the new developments and applications of CS for wireless sensors-based SHM in our research group.

2 COMPRESSIVE SENSING

CS provides a new sampling theory for simultaneous signal sensing and compression, which states that sparse or compressible signals can be exactly reconstructed from highly incomplete random sets of measurements. CS theory states that signals $x \in \mathbf{R}^n$ can be sensed using far less linear measurements,

$$y = \mathbf{\Phi} x + e \tag{1}$$

where $\mathbf{\Phi}$ is called a measurement matrix or sampling operator in an $m \times n$ matrix. As $\mathbf{\Phi}$ is an $m \times n$ matrix with $m << n$, the problem of recovering the signal x is ill-posed. But if x is known to be sparse in the basis $\mathbf{\Psi}$, i.e. many of the coefficients α are zero and the matrix $\mathbf{\Theta} = \mathbf{\Phi\Psi}$ obeys a so-called *restricted isometry property* (RIP), then α can be reconstructed exactly by solving the convex optimization program:

$$\hat{\alpha} = \arg\min \|\tilde{\alpha}\|_1 \text{ such that } \mathbf{\Theta}\tilde{\alpha} = y \tag{2}$$

Finally, the original data can be recovered approximately using α as follows:

$$\hat{x} = \mathbf{\Psi}\hat{\alpha} \tag{3}$$

Generally, $\mathbf{\Psi}$ can be any basis that provides a nearly sparse representation of x.

3 GROUP SPRASE OPTIMIZATION BASED COMPREESIVE SENSING FOR VIBRATION DATA

For most of vibration signals of civil infrastructures have sparse characteristic, namely, only a few modes contribute to the vibration of the structures. Therefore, the vibration data for most bridges and buildings are usually have sparse representation. Additionally, the measured vibration data by the sensors placed on different locations of structure almost has same sparse structure in the frequency domain. Such as the measured vibration data of a suspension bridge (Figure 1) by wireless sensors (Figure 2) shows that the multiple sensor test data have similar sparse property in frequency domain (Figure 3). Based on this group sparse property, the group sparse optimization method can be used to recovery the compressive sampling data. As shown in Figure 4, only using 10% random sampling data, the original data can be reconstructed using the group sparse optimization method. The modal identification results of Figure 5 shows that the mode shape can be exactly identified from the 10% random sampling reconstruction data.

Figure 1. A suspension bridge

(a)

(b)

Figure 2. Field test using wireless sensors: (a) wireless sensor node; (b) wireless base station

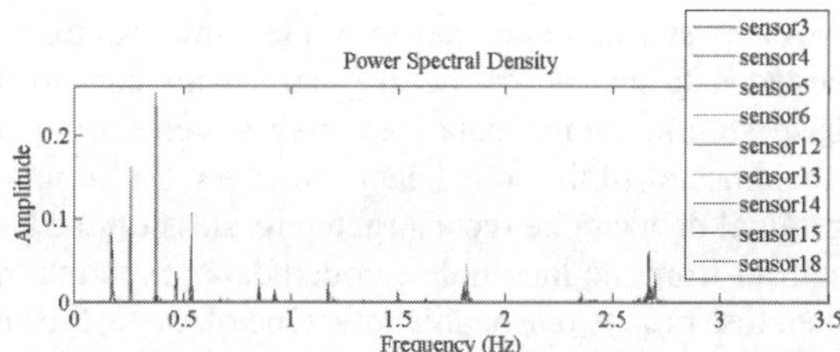

Figure 3. Power spectral density of the measured velocity data

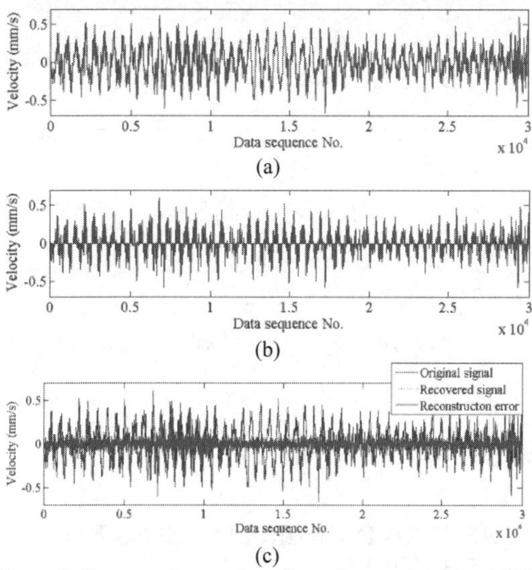

Figure 4. Compressive sensing of wireless sensor data: (a) Vibration velocity data measured by wireless sensor; (b) 10% random sampling velocity data; (c) Reconstruction from 10% random sampling data

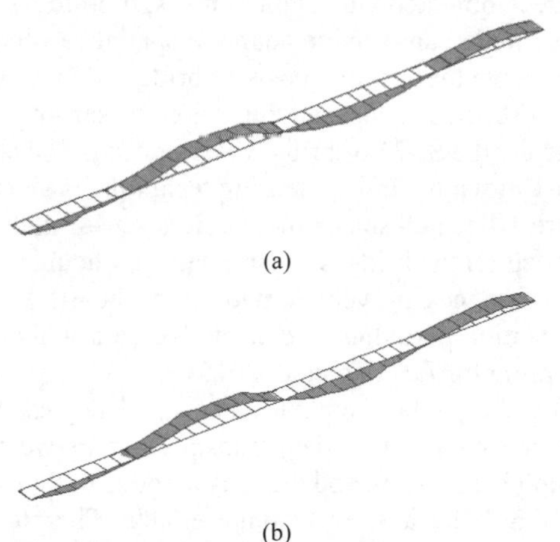

Figure 5. Identified modal shapes: (a) First modal shape identified from original data; (b) First modal shape identified from the data reconstructed by 10% random sampling data

4 COMPRESSIVE SENSING BASED LOST DATA RECOVERY FOR WIRELESS SENSORS

The CS theory can work for lost data recovery of wireless sensor in SHM based on two facts. One is that the vibration data of structure has sparsity in frequency domain, wavelet domain or other basis domain. Therefore, this vibration data of structure can be reconstructed from far few random measurements using CS theory (Bao et al. 2011; Bao et al. 2013; Bao et al. 2014). The second is that the randomly lost data can be equivalent to the compressed data. Then the lost data recovery is same as the data reconstruction from compressed measurements.

4.1 Embedding of CS-based data lost recovery algorithm into smart wireless sensor

This CS-based technique is implemented into the Imote2 smart sensor platform using the foundation of Illinois Structural Health Monitoring Project (ISHMP) Service Tool-suite as shown in Figure 6. To overcome the constraints of limited onboard resources of wireless sensor nodes, a method called random demodulator (RD) (Kirolos et al. 2006) is employed to provide memory and power efficient construction of the random sampling matrix. Adaptation of RD sampling matrix is made to accommodate data loss in wireless transmission and meet the objectives of the data recovery. The embedded program is tested in a series of sensing and communication experiments (Zou et al. 2014a).

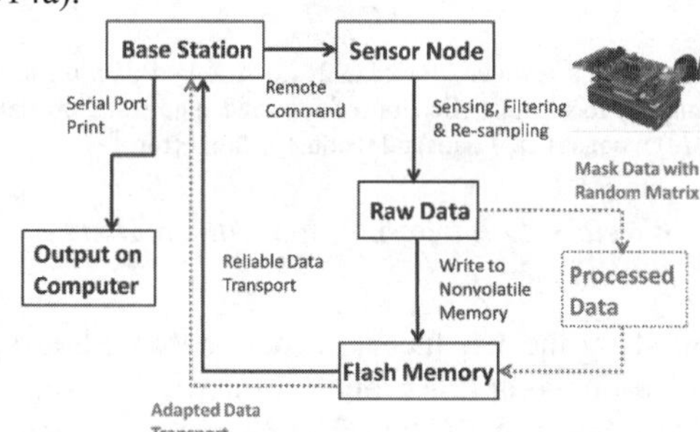

Figure 6. The procedures of the embedded CS-base algorithm

To demonstrate the performance of the embeded program, a series of sensing and communication experiments has been performed on the Songpu

Bridge in Harbin as shonwn in Figure 7. The Songpu Bridge is a single-tower cable-stayed bridge with a main span of 268 meters. The experiments are carried out along one sidewalk with sensors installed on the ground and antenna attached on top of the fence so that a direct communication path is assured for all tests.

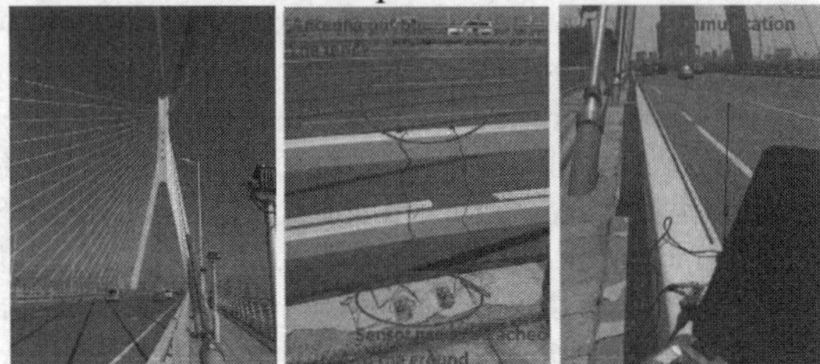

Figure 7. Field test on the bridge (Zou et al. 2014a)

The test results are shown in Figure 8. Figure 8 (a) shows the received transformed signal $\hat{y}$ with data loss on the base station. Figure 8 (b) is the locally magnifies $\hat{y}$ for the lost data points: random data loss. The recovered signal $\hat{x}$ with error and its spectrum are shown in (c), which shows the lost data is well recovered with small reconstruction error.

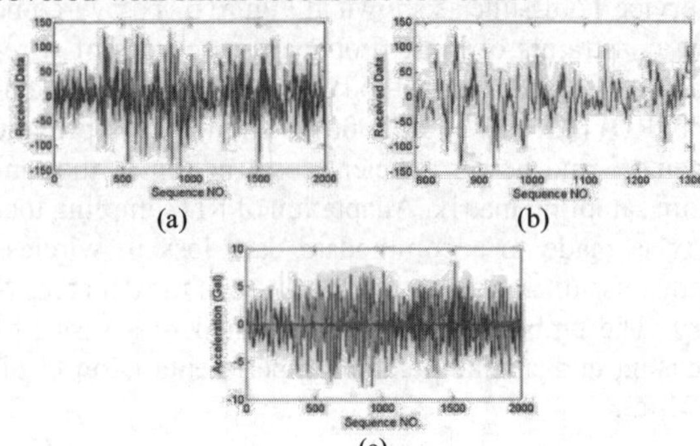

Figure 8. Data recovery for case 2: (a) received data on base station, (b) local magnification of received data: random data loss, (e) reconstructed data and reconstruction error [18]

4.2 *Wireless data transmission with random redundancy*

In spired by the CS theory, a new data coding and transmission method is proposed for wireless sensors in SHM (Zou et al. 2014b). The proposed method includes two coding stages, i.e., a source coding stage to compress the natural redundant information inherent in SHM signals and a redundant coding stage to inject artificial redundancy into wireless transmission to enhance the transmission reliability. After coding, a wireless sensor node transmits the same payload of coded data instead of the original sensor data to the base station. Some data loss may occur during the transmission of the coded data. However, the complete original data can be reconstructed lossless on the base station from the incomplete coded data given that the data loss ratio is reasonably low. One of the test results of a bridge cable vibration is shown in Figure 9, which shows the lost data can be exactly recovered.

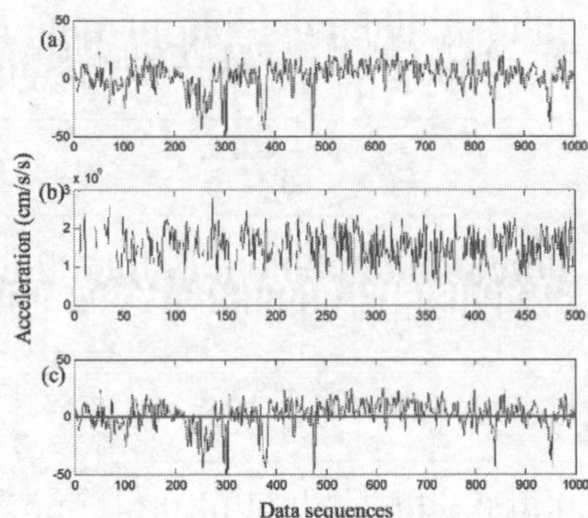

Figure 9. Lost data recovery resutls

5 FAST MOVING WIRELESS SENSING TECHNIQUES

Both the sensor nodes and base station of the traditional wireless sensors for SHM are fixed on the bridge. Compared with static sensors, mobile wireless sensor node can provide adaptive spatial resolutions. To achieve the fast diagnosis of bridge without interrupting traffic, a fast-moving wireless sensing technique for the SHM of bridges is proposed. The sketch of fast-moving wireless sensing technique is shown in Figure 10, which shows the wireless sensor nodes are installed on the bridges to automatically acquire data, and a fast-moving vehicle with an on-board wireless base station periodically collects the data without interrupting traffic (Bao et al. 2015).

However, in fast-moving states, the data packet loss rates during wireless data transmission between the moving base station and the sensor nodes will increase remarkably because of Doppler effects. Therefore, to solve the data-loss problem, the CS-based lost data recovery approach is used. Figure 11 shows the data recovery results of the data measured at a velocity of 30 km/h on a field test of bridge, which demonstrates the CS-based wireless data transmission can work well.

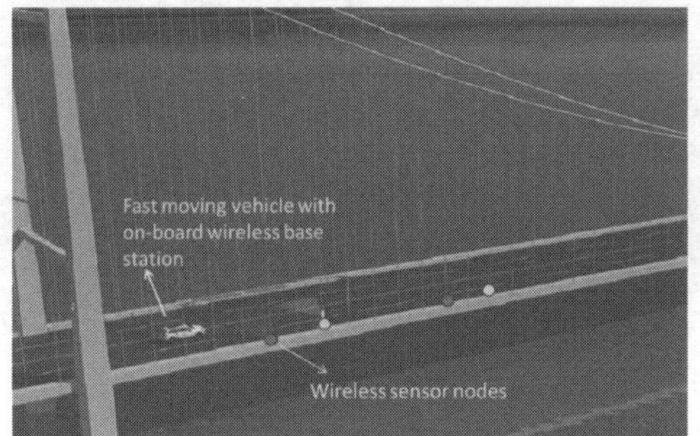

Figure 10. Sketch of fast-moving wireless sensing technique [20].

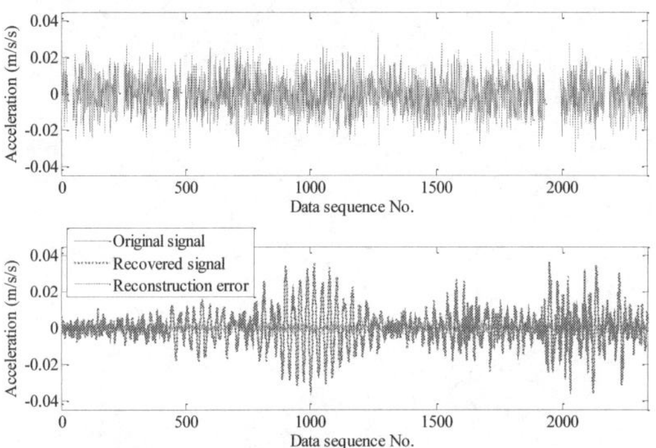

Figure 11. Data recovery results of the data measured at a velocity of 30 km/h: (a) original signal; (b) transformed signal $\hat{y}$ with 7.67% data packet loss; (c) recovered signal and the reconstruction error; and (d) the amplitude spectrum of the recovered signal (Bao et al. 2015).

6 CONCLUSIONS

This paper introduces the new developments and applications of compressive sensing (CS) for data sampling with group sparse optimization method and lost data recovery of wireless sensors-based SHM system in our research group. CS provides a good solution for data sampling and data loss recovery in wireless sensors and sensor network for SHM, which can be embedded into smart wireless sensors and effectively increases wireless communication reliability. Not only for the static wireless sensors, the CS-based data wireless transmission method also can be used to the fast moving wireless sensing techniques to recovery the lost data caused by Doppler effects. The investigation results of the CS based lost data recovery for wireless sensor showed a good performance of the approaches. More applications of CS for wireless sensors and networks are waiting to be explored.

ACKNOWLEDGEMENTS

This research is supported by the National Basic Research Program of China (Grant No. 2013CB036305), the NSFC (Grant No. 51378154, 51161120359), by Ministry of Science and Technology of China (Grant No: 2015DFG82080) and by Ningbo science and technology project (Grant No: 2015C110020).

REFERENCES

Bao, Y. Beck, J.L. & Li, H. 2011. Compressive sampling for accelerometer signals in structural health monitoring, *Structural Health Monitoring-An International Journal* **10**(3), 235-246

Bao, Y. Li, H. Sun, X. et al. 2013. Compressive sampling based data loss recovery for wireless sensor networks used in civil structural health monitoring, *Structural Health Monitoring-An International Journal* **12**(1):78-95.

Bao, Y. Li, H. & Ou, J. 2014. Emerging Data Technology in Structural Health Monitoring: Compressive Sensing Technology, *Journal of Civil Structural Health Monitoring* 2013, **4**(2): 77-90.

Bao, Y. Yu, Y. Li, H. et al. 2015. Compressive sensing-based lost data recovery of fast-moving wireless sensing for structural health monitoring, *Structural Control and Health Monitoring* **22**(3), 433-448.

Candès, E.J. 2006. Compressive Sampling, *Proceedings of the International Congress of Mathematicians*, Madrid, Spain: 1433-1452.

Celebi, M. 2006. Real-time seismic monitoring of the New Cape Girardeau Bridge and preliminary analyses of recorded data: an overview *Earthquake Spectra* **22**, 609-630.

Cho, S., Giles, R.K. & Spencer, B.F. 2015. System identification of a historic swing truss bridge using a wireless sensor network employing orientation correction, *Structural Control and Health Monitoring* **22**(2), 255-272.

Donoho, D. 2006. Compressed Sensing, *IEEE Transactions on Information Theory* **52**(4): 1289-1306.

Hoult, N.A., Fidler, P.R.A., Hill, P.G., et al. 2010. Long-term wireless structural health monitoring of the Ferriby Road Bridge, *Journal of Bridge Engineering*, ASCE **15**(2):153-159.

Jang, S., Jo, H., Cho, S., et al. 2010. Structural health monitoring of a cable-stayed bridge using smart sensor technology: deployment and evaluation, *Smart Structures and Systems* **6**(5-6), 439-459.

Kirolos, S. Laska, J. Wakin, M. et al. 2006. Analog-to-information conversion via random demodulation. *In Design, Applications, Integration and Software, 2006 IEEE Dallas/CAS Workshop on* (pp. 71-74). IEEE.

Kurata, N. Spencer, B.F. Jr. & Ruiz-Sandoval, M. 2004. Building Risk Monitoring Using Wireless Sensor Network, *Proceedings of the 13th world conference on earthquake engineering*, August 2-6, Vancouver, BC, Canada.

Lynch, J.P. Wang, Y. Loh, K.J. et al. 2006. Performance monitoring of the Geumdang Bridge using a dense network of high-resolution wireless sensors *Smart Materials and Structures* **15**, 1561–1575

Meyer, J., Bischoff, R., Feltrin, G. et al. 2010. Wireless sensor networks for long-term structural health monitoring. *Smart Structures and Systems* **6**(3), 263-275.

Meyer J. Bischoff R.B. Feltrin G. & Motavalli M. 2010. Wireless Sensor Network for Long-term Structural Health Monitoring, *Smart Structures and Systems* **6**(3), 263-275.

Mufti, A.A. 2002. Structural health monitoring of innovative Canadian civil engineering structures. *Structural Health Monitoring* **1**(1), 89-103.

Ou, J. & Li, H. 2010. Structural health monitoring in mainland China: review and future trends, *Structural Health Monitoring* **9**(3), 219-232.

Pei, J.S. Kapoor, C. Graves-Abe, T.L. et al. 2005. Critical Design Parameters and Operating Conditions of Wireless Sensor Units for Structural Health Monitoring, *in Proceedings of the 23rd International Modal Analysis Conference (IMAC XXIII)*, Orlando, FL, January 31-February 3.

Zou, Z. Bao, Y. Deng F. et al. 2014b. An approach of reliable data transmission with random redundancy for wireless sensors in structural health monitoring, *IEEE Sensor Journal* **15**(2), 809-818.

Zou, Z. Bao, Y. & Li, H. 2014a. Embedding Compressive Sensing based Data Loss Recovery Algorithm into Wireless Smart Sensors for Structural Health Monitoring, *IEEE Sensors Journal* **15**(2), 797-808.

Proceedings of the International Conference on Smart Infrastructure and Construction
ISBN 978-0-7277-6127-9

© The authors and ICE Publishing: All rights reserved, 2016
doi:10.1680/tfitsi.61279.135

Design of a wireless smart sensor node for fatigue assessment of welded joints

H. P. Ding*
Nanjing University, Nanjing, China
** Corresponding Author*

ABSTRACT Fatigue assessment of welded joints has been a significant way to predict the remaining life of large aging civil engineering structures and provide maintenance guidance for these structures during the past several years. To obtain accurate results, the authentic field-based data from structural health monitoring (SHM) system is usually preferred to be used in the assessment. However, for traditional wired SHM system, the high cost of installing and maintaining wires is restricted to these large structures such as the long-span bridges or tall buildings. This paper focuses on designing a wireless smart sensor node (WSSD) which is promising for estimating the fatigue life of welded joints. The WSSD is designed with high-performance micro-processor, large memory storage, high-precision signal conditioner, etc. The performance of the WSSD is validated using a cantilever beam alongside a wired strain gauge as the reference. During the evaluation, the accuracy, the resolution, the frequency response, and the performance in testing real structure's response of the WSSD are verified respectively under various excitations applied to the cantilever beam.

1. INTRODUCTION

Fatigue is a localized damage process of a component produced by repeated fluctuations of loadings which may be well below the structural resistance capacity. Due to unobvious cumulative damage and sudden fracture of a component, fatigue failures in civil engineering structures may result in heavy losses of life and property. In order to get the remaining life and maintenance guidance, fatigue assessment of components, especially welded joints, which are widely used in these metallic structures and more prone to get fatigued than a similar un-welded specimen due to pores, inclusions, undercuts, etc. (Radaj 1996), is critical.

Taking into account the wide use of welded joints in civil engineering structures, a large amount of research is conducted regarding fatigue assessment of welded joints (Schumacher & Nussbaumer 2006). The researchers have proposed several fatigue analysis methods such as the nominal stress method, the conventional hot spot stress method, and the Dong's method (Hobbacher 1996; Lotsberg & Sigurdsson 2006; Dong 2005). When these methods are adopted for fatigue assessment, it is necessary to obtain the stress spectra of welded details in critical locations. Although the stress spectra can be acquired from a theoretical stress analysis by assuming a loading and structural mode. However, the accuracy may be very much limited due to the difficulties and errors in both loading and structural model designs of complicated structures.

Recent advances in sensing, data acquisition, and communication have greatly promoted structural health monitoring (SHM) system capable of achieving the authentic field-based data for large complicated structures (Ko & Ni 2005). To obtain most reliable information for evaluating structural durability and reliability, a dense network of sensors installed throughout the structures are required for a robust SHM system. For such a system, the cost of installing wires to connect the sensors to a centralized monitoring station can run into thousands of dollars per sensing channel (Nagayama & Spencer 2007). Moreover, for some existing steel structures in operation, it is very difficult, if not impossible, to deploy a wired system for data acquisition and to supply power everywhere inside the structure. In recognition of the above problems, many researchers are actively exploring new technologies that can help advance the ability to economically realize SHM. In particular, wireless smart sensor node (WSSD) represents a significant improvement over traditional wired sensors. As a point of comparison, the WSSD can substantially reduce system costs to a few hundred dollars or less per sensing channel. Additionally, for fatigue assessment, due to portable design and local power supply of the WSSD, it is relatively easy to deploy wireless sensors at the most critical locations where fatigue cracks are expected to occur.

Mainly due to the above mentioned advantages, the WSSD has been investigated and implemented in various structures for SHM purpose during the past several years. For example, Ni et al. (2010) addressed the ambient vibration monitoring of the Guangzhou Television and Sightseeing Tower (GTST) using a prototype wireless sensing system. The authors demonstrated that wireless sensing technologies can be deployed reliably in monitoring the low-frequency and low-amplitude ambient vibration of super-tall structures such as the GTST. Kim et al. (2007) reported a large-scale deployment of wireless sensors for SHM on the Golden Gate Bridge. Ambient structural vibrations were reliably measured at low cost and without interference with the normal operation of the bridge.

There have been lots of applications using wireless sensing technologies for SHM as mentioned above, however, most of them are focused on how to use acceleration to detect damage. Limited number of reports have been addressed for fatigue assessment of welded joints using the WSSD in a real structure.

This paper aims to design a WSSD with high-performance micro-processor, large memory storage, high-precision signal conditioner, etc. Laboratory tests have been conducted to evaluate the performance of the WSSD.

2. WSSD DEVELOPMENT

WSSD differs from traditional wireless sensor node with several key features, such as high-performance integrated processor for data intensive applications and large memory storage for longer measurement. The WSSD typically includes a wireless communication platform, a signal conditioner, and a sensing unit. In addition, for measurement of weak signals, a high-performance amplifier with flexible gains is also required.

2.1 Wireless communication platform

As shown in Figure 1, the Imote2 (Crossbow 2007), which incorporates a low-power X-scale processor (Intel PXA271) and a low-power 802.15.4 radio transceiver (ChipCon CC2420), is chosen as the wireless communication platform for the WSSD.

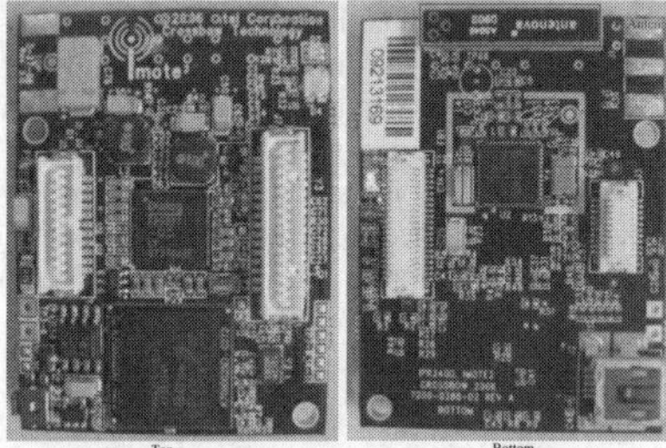

Figure 1. The Imote2 top and bottom view

The processor includes 256 KB of integrated SRAM, 32 MB of external SDRAM and 32 MB of FLASH memory making it extremely flexible in supporting large amount of data required for dynamic strain measurement and intensive computation. The radio offers a theoretical maximum transfer speed of 250 kbits/sec over the 2.4 GHz wireless communication band. The advantages mentioned above make the Imote2 well-suited for the applications of fatigue assessment.

2.2 Signal conditioner

The Imote2 does not have the digitizing capability which means raw signals (usually analog signals) cannot be processed directly by it. For fatigue assessment, accurate sensing of strain, from which the stress can be determined, is one of the most significant factors.

Therefore, a high-performance analog-to-digital converter (ADC) which greatly affects the measurement accuracy should be considered.

In this study, the Quickfilter signal conditioner (namely QF4A512) with user-selectable sampling rates is chosen mainly to serve as the ADC. The QF4A512 has a 4-channel 16-bit programmable ADC for analog differential or single ended inputs. As shown in Figure 2, each channel can be individually programmed for the gain, anti-aliasing filter cutoff frequency, analog to digital sampling frequency, and unique filter requirements (Quickfilter 2007).

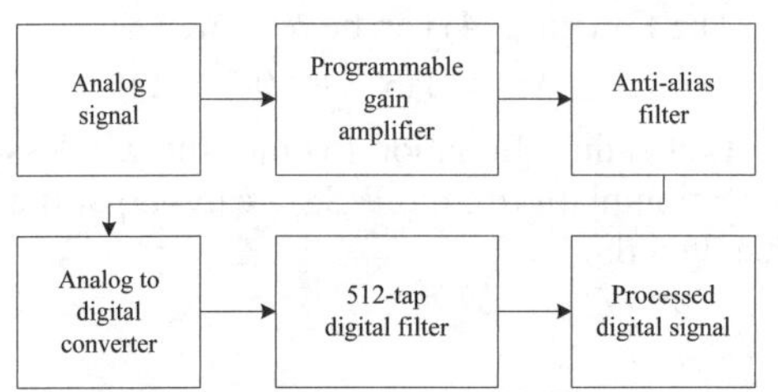

Figure 2. Functional block diagram of QF4A512

It should be noted that each channel of the QF4A512 has a programmable gain amplifier at gains of 1x, 2x, 4x, and 8x, however, it is not large and flexible enough for weak signals.

As the QF4A512 is adopted for converting analog signals to digital signals, it is important to get the accurate relationship between the input and the output. The best and easiest way to obtain the relationship is to carry out a laboratory calibration by using certain voltage values as the input. Figure 3 shows the fitted line between the input and QF4A512 output.

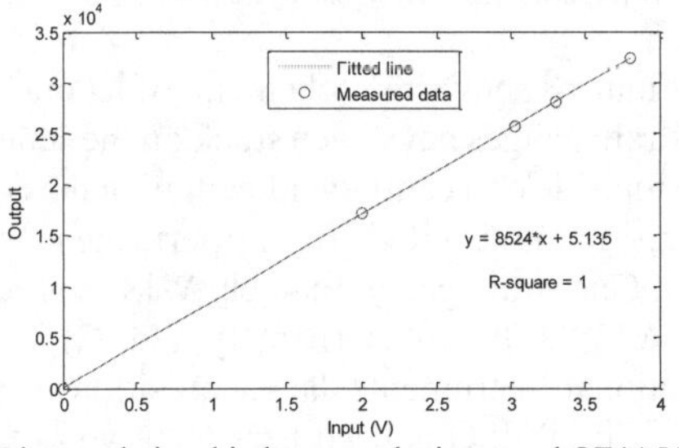

Figure 3. Linear relationship between the input and QF4A512 output

The relationship can be determined by,

$$y = 8524 * x + 5.135 \tag{1}$$

where x is the input voltage, y is the QF4A512 output.

2.3 Sensing unit and amplifier

The Imote2 does not have the sensing capability. In order to collect signals, external sensors should be integrated to the Imote2. Due to wide frequency range and inexpensiveness, the widely used 120 Ohm strain gauge is chosen as the sensor for the WSSD.

As a strain gauge is attached to an object by a suitable adhesive, the object deformation will result in the resistance change of the gauge. However, the resistance change is not easily detected by normal data acquisition system. In order to overcome the difficulty, a Wheatstone bridge is used in this study to convert resistance changes into voltages. It has to be considered that the voltage output of the Wheatstone bridge is usually quite small. Although the QF4A512 can provide a maximum gain of 8x as mentioned above, weak signals still can not be measured effectively with this gain.

The AD623 is an integrated single supply amplifier which offers superior user flexibility by allowing single gain set resistor programming, and conforming to the industry standard configuration. With an external adjustable resistor, the AD623 can be programmed for gains up to 1000. The relationship between the gain and the resistance of the resistor can be determined by the following formula,

$$G = 1 + \frac{100k\Omega}{R_G} \tag{2}$$

where G is the gain, R_G is the resistance of the external resistor.

Figure 4 shows how the strain gauge, the Wheatstone bridge and the AD623 are connected together. The circuit is designed to convert resistance change of strain gauge to voltage and amplify the weak voltage signals. In the Wheatstone bridge as shown in Figure 4, one of the resistors is replaced with a strain gauge, whereas the other three arms employ high-precision resistors with the same resistance as the strain gauge. Two adjustable resistors are used to configure the AD623. One serves as the adjuster for different gains while the other is used to bias the output voltage to the range corresponding to the input of the ADC.

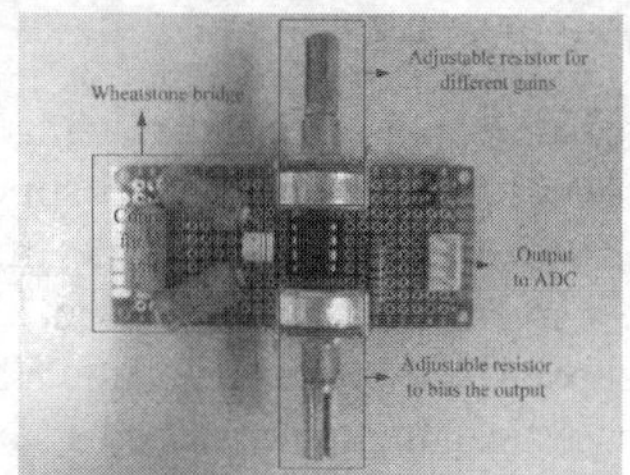

Figure 4. Circuit of sensing unit and amplifier

The output changes to ADC can be determined by the following formula (Lee et al. 2005),

$$\Delta V_o = \frac{E_{ex}}{4} * K * \Delta\varepsilon * G \tag{3}$$

where ΔV_o is the output changes to ADC, E_{ex} is the bridge excitation, K is the gauge factor, $\Delta\varepsilon$ is the strain changes, G is the gain.

2.4 WSSD design

Combining the parts mentioned above, the WSSD is designed with strong capabilities in catching weak strain signals, converting analog signals to digital signals, and data processing. Figure 5 shows the block diagram of the WSSD. A clear view of the WSSD can be found in Figure 6.

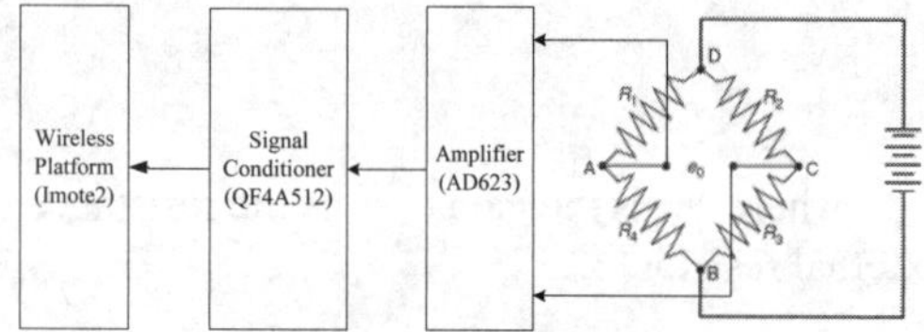

Figure 5. WSSD block diagram

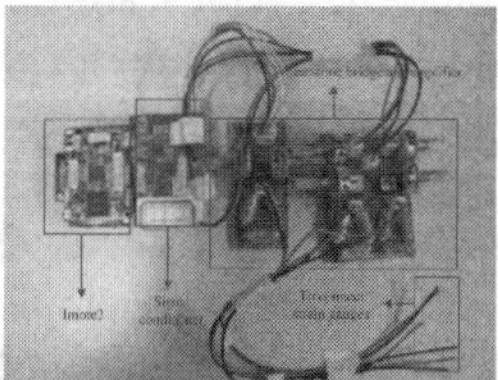

Figure 6. WSSD view

It is essential to know the meaning of the output value before using the WSSD to measure data. Therefore, the relationship between the input strain and the output of the WSSD is determined by combining Equations (1), (2) and (3), and can be expressed as the following formula,

$$\Delta y = 8524 * \frac{E_{ex}}{4} * K * \Delta\varepsilon * \left(1 + \frac{100k\Omega}{R_G}\right) + 5.135$$
$$\approx 2.892 * 10^6 * \Delta\varepsilon \tag{4}$$

where E_{ex}=3.2V, K=2.11, R_G is adjusted to 500 Ω, $\Delta\varepsilon$ is the input strain, and Δy is the output.

The Equation (4) demonstrates that the input of 1 $\mu\varepsilon$ corresponds to the output value of 2.892. The value can be also defined as the sensitivity of the WSSD with configured parameters. Higher sensitivity can be obtained with higher E_{ex} and smaller R_G.

The Equation (4) can be rewritten as,

$$\Delta\varepsilon = 345.8 * 10^{-9} * \Delta y \tag{5}$$

Embedding Equation (5) into the wireless communication platform, the WSSD can output the strain value directly.

3 EVALUATION OF WSSD

3.1 Experimental setup

To evaluate the performance of the developed WSSD, laboratory tests are carried out as shown in Figure 7.

Figure 7. Experimental setup: test system

Figure 8 shows the schematic of laboratory tests. Two strain gauges have been stuck on the same section of the top side of a cantilever beam with the controlled bending loads excited by a permanent magnetic shaker. One strain gauge (namely WlSG) is connected to the WSSD, the other (namely WdSG) is wired to the National Instruments data acquisition system (NI-DAQ). The WdSG is used here to serve as the reference to calibrate the WlSG. The whole wireless measurement system also includes a wireless base station which is used to receive the data from the WSSD.

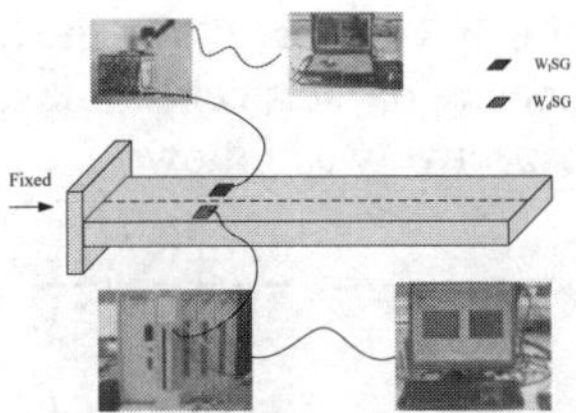

Figure 8. Schematic of laboratory tests

3.2 Typical characteristics tests

Several factors contribute to the performance of the developed WSSD. The first is the accuracy of the sensor which indicates the ability of a measurement to match the actual value of the quantity being measured. The second is the resolution of the senor which determines the smallest change the sensor can detect the quantity for its measurement. The final is the frequency response which depicts the quantitative measurement of the output spectrum of a system in response to a stimulus. It is a measure of magnitude of the output as a function of frequency, in comparison to the input. Considering the importance of accurate sensing of the WSSD in the fatigue assessment, the factors mentioned above would be evaluated respectively in the following sections.

3.2.1 Accuracy

With the reference WdSG, the accuracy of WlSG (representation of WSSD) is experimentally verified by using large random excitation applied to a cantilever beam. Strain data of the both system, WdSG and WlSG are recorded continuously at a sampling rate of 1000 Hz.

Figure 9 (a) shows the outputs of both the WdSG and WlSG in time history. A close-up view of the comparison has been shown in Figure 9 (b). From the result, one can tell that these measurements show good agreement.

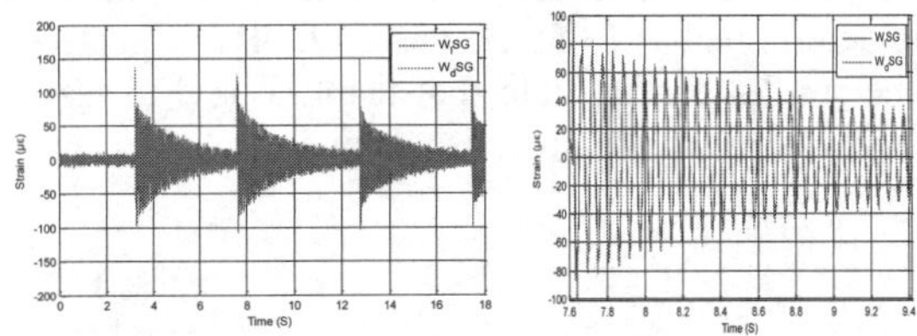

Figure 9. Comparison of both sensors outputs in time history: (a) zoom out (up); (b) zoom in (down)

In the frequency domain, the measured transfer function from the reference WdSG to the WlSG is shown in Figure 10. It should be noted that the transfer function should have a magnitude of 0 dB if the two compared signals are the same. Obviously, the signal from the WlSG is close to the reference signal from the WdSG over a wide frequency range. However, the performance of the WlSG is difficult to judge in the frequency range close to 0 Hz due to the reference sensor's inability to measure extremely low frequency excitation. Additionally, there exists a significant difference at about 50 Hz, which is due to the 50 Hz alternating current (AC) interference in the wired system.

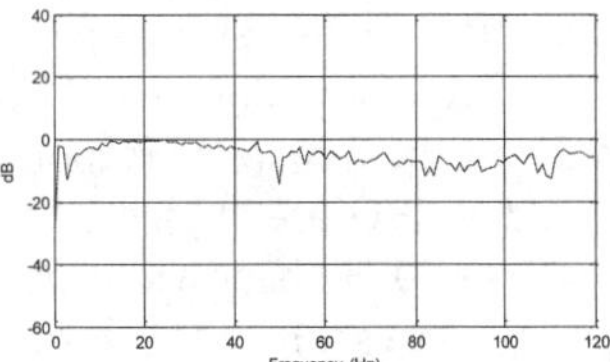

Figure 10. Transfer function from WdSG to WlSG

3.2.2 Resolution

The resolution is related to the precision with which the measurement is made. In order to obtain the resolution of the WlSG, a static load is conducted to the cantilever beam during the dynamic process. As shown in Figure 11, the resolution of the WlSG can be estimated to be 2 $\mu\varepsilon$ ($\pm 1 \mu\varepsilon$ noise), which is approximately ten times smaller than that of the WdSG. Moreover, by using a more precise amplifier, as well as higher resolution ADC, further improvement in resolution of the WlSG is considered possible.

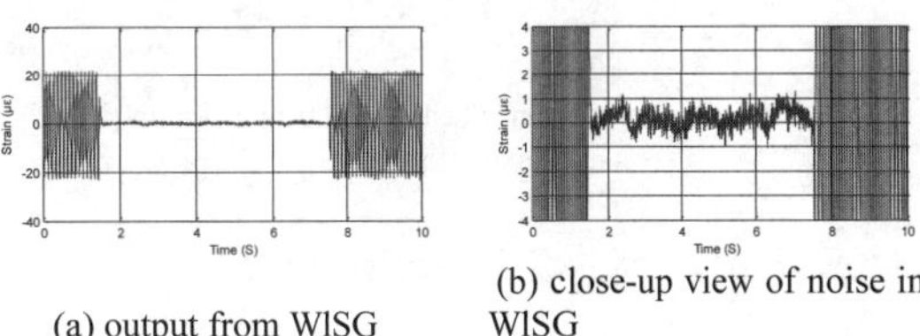

(a) output from WlSG

(b) close-up view of noise in WlSG

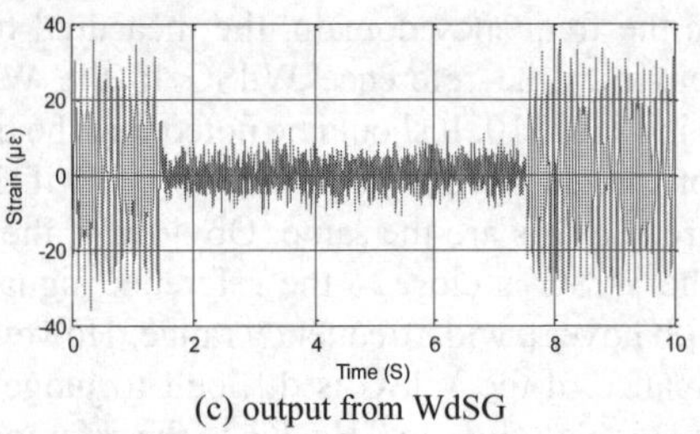

(c) output from WdSG

Figure 11. Resolution test:

3.2.3 Frequency response

White noise is a random signal with a flat power spectral density. Because of this property, many researchers used white noise to assess the sensor or system performance in frequency domain (Lynch 2005). To obtain and assess the frequency response of the WlSG, the WlSG and WdSG are subjected to a white noise excitation with a 50 Hz bandwidth. As shown in Figure 12, the responses of both sensors correlate well over a wide frequency range. However, a small discrepancy below 1 Hz can be found in the comparison. This is mainly due to two reasons. First, the shaker itself has poor performance in the frequency range below 5 Hz (Brüel & Kjær 2008). Second, the reference WdSG is not able to measure extremely small excitations. Nevertheless, the frequency response of the WlSG is considered sufficient for most fatigue assessment applications.

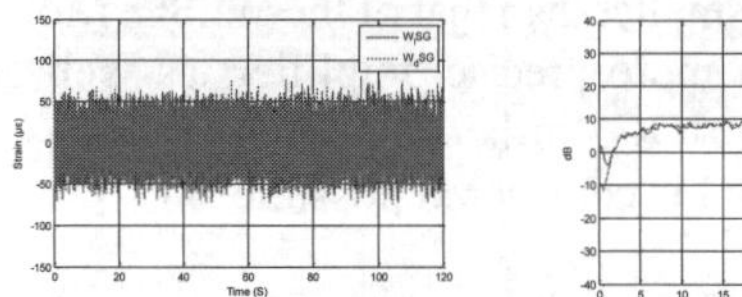

Figure 12. Frequency response test: (a) in time domain (b) in frequency domain

3.3 Natural frequencies measurement

To validate the performance of the WlSG in testing responses of real structures, the WlSG is used to measure the natural frequencies of the cantilever beam with the size of 290 mm long, 70 mm wide, and 2 mm thick. The results from the WdSG and theoretical calculation will be used to judge the performance of the WlSG.

According to Chopra (2007), the first and second natural frequencies of the steel cantilever beam can be calculated respectively as follows,

$$f_1 = \frac{3.526}{2\pi L^2}\sqrt{\frac{EI}{m}} \tag{6}$$

$$f_2 = \frac{22.03}{2\pi L^2}\sqrt{\frac{EI}{m}} \tag{7}$$

where E is the Young's modulus, L is the length of the cantilever beam, m is the mass per unit length, I is the cross-sectional moment of inertia and can be calculated from the equation,

$$I = \frac{1}{12}BH^3 \tag{8}$$

where B is the width, H is thickness of the plate.

Combining Equation (6), (7), and (8), the first and second natural frequencies are calculated as 17.85 Hz and 111.84 Hz respectively.

For the case where the cantilever beam is responding under large excitation, the measured frequencies from both the WlSG and WdSG can be found in Figure 13.

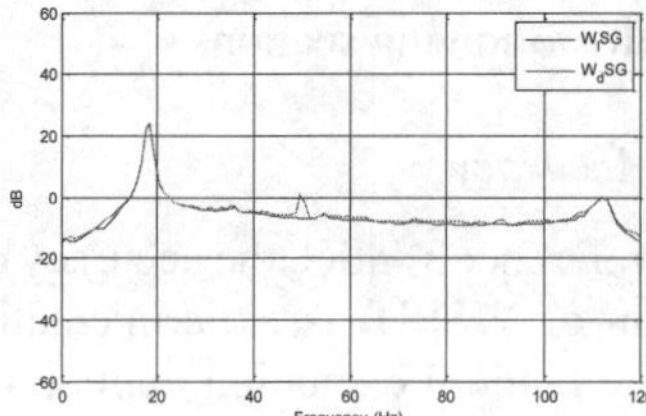

Figure 13. Natural frequency test of the cantilever beam

From the comparison in Figure 13, we can find that the two measurements show good agreement except for the frequency at 50 Hz. The discrepancy at 50 Hz is consistent with the transfer function shown in Figure10.

Additionally, as shown in Figure 13, the first and second natural frequencies of the cantilever beam can be determined as 18.55 Hz and 112.3 Hz respectively. The results demonstrate that using WlSG to test responses of real structures is promising.

4 SUMMARY AND DISCUSSION

This paper describes the WSSD which is developed to estimate the fatigue life of welded joints. In order to get accurate estimates, it is integrated with high-performance micro-processor, large memory storage, high-precision ADC, etc. Considering the importance of the accuracy, the resolution, and the frequency response of sensor, the performance of these parameters of the WSSD is evaluated using a cantilever beam alongside a wired strain gauge as the reference. In the evaluation, various excitations are applied to the cantilever beam to obtain the results compared with the reference. Firstly, according to the accuracies shown both in time and frequency domains, the results are quite promising in spite of some differences due to the AC interference in the wired system. Secondly, the obtained resolution of WSSD is considered to be sufficient for fatigue assessment applications. The noise in the WSSD has been found approximately ten times lower than that of the wired system. Finally, the wide frequency range of the WSSD is verified using the white noise signal as the excitation. After these evaluations, the WSSD is used to calculate the natural frequencies of the cantilever beam to verify the performance in testing the responses of real structures. The result obtained by WSSD shows good agreement with both the wired system and theoretically calculated results.

REFERENCES

Brüel & Kjær, Inc. 2008. Vibration test systems, Available from: <http://www.bksv.com>.

Chopra, A. K. 2007. Dynamics of structures: theory and applications to earthquake engineering, Prentice Hall, New Jersey.

Crossbow Technology, Inc. 2007. Imote2 Hardware Reference Manual, San Jose, CA.

Dong, P. 2005. Mesh-insensitive structural stress method for fatigue evaluation of welded structures, Battelle SS JIP training course material, Columbus, OH: Center for Welded Structures Research, Battelle Memorial Institute.

Hobbacher, A. 1996. Recommendations for fatigue strength of welded component, Cambridge, Abington Publishers.

Kim, S., Pakzad, S., Culler, D., Demmel, J., Fenves, G., Glaser, S. and Turon, M. 2007. Health monitoring of civil infrastructures using wireless sensor networks, Proceedings of the 6th International Conference on Information Processing in Sensor Networks, Cambridge, MA, April 25-27.

Ko, J. M. and Ni, Y. Q. 2005. Technology developments in structural health monitoring of large-scale bridges, *Engineering Structures*, Vol. 27, 1715-1725.

Lotsberg, I. and Sigurdsson, G. 2006. Hot spot stress S-N curve for fatigue analysis of plated structures, *Journal of Offshore Mechanics and Arctic Engineering*, 128(4), 330-336.

Lee, Y. L., Pan, J., Hathaway, R. and Barkey, M. 2005. Fatigue testing and analysis: theory and practice, *Elsevier Butterworth-Heinemann*, Burlington, MA.

Lynch, J. P. 2005. Design of a wireless active sensing unit for localized structural health monitoring, *Journal of Structural Control and Health Monitoring*, Vol. 12, 405-423.

Nagayama, T., and Spencer, B. F. Jr. 2007. Structural health monitoring using smart sensors, Report No. NSEL-001, Newmark Structural Engineering Laboratory, Department of Civil and Environmental Engineering, University of Illinois at Urbana-Champaign, Urbana, USA.

Ni, Y. Q., Li, B., Lam, K. H., Zhu, D., Wang, Y., Lynch, J. P. and Law, K. H. 2010. In-construction vibration monitoring of a supertall structure using a long-range wireless sensing system, *Smart Structures and Systems*, 7(2), 83-102.

Quickfilter Technologies, Inc. 2007. QF4A512 4-Channel Programmable Signal Conditioner., Allen, TX.

Radaj, D. 1996. Review of fatigue strength assessment of nonwelded and welded structures based on local parameters, *International Journal of Fatigue*, 18(3), 153-170.

Schumacher, A. and Nussbaumer, A. 2006. Experimental study on the fatigue behaviour of welded tubular K-joints for bridges, *Engineering Structures*, 28(5), 745-755.

Proceedings of the International Conference on Smart Infrastructure and Construction
ISBN 978-0-7277-6127-9

© The authors and ICE Publishing: All rights reserved, 2016
doi:10.1680/tfitsi.61279.143

Validation of wireless sensing technology densely instrumented on a full-scale concrete frame structure

X. Dong, X. Liu, T. Wright, Y. Wang* and R. DesRoches

School of Civil and Environmental Engineering, Georgia Institute of Technology, Atlanta, USA
**Corresponding author*

ABSTRACT The adoption of wireless sensing technologies in structural health monitoring (SHM) community has shown certain advantages over traditional cable-based system, such as simplified installation process and lower system cost. Recently, a new generation wireless sensing device, named *Martlet*, has been developed for SHM purpose. In order to obtain accurate high-resolution acceleration data and in the meantime reduce hardware cost, an accessory sensor board for *Martlet*, named integrated accelerometer *wing*, is developed. For performance validation, the *Martlet* wireless sensing system is installed on a full-scale two-story, two-bay concrete frame structure. Shakers from NEES@UCLA are used to provide dynamic excitations. In addition to *Martlet*, the performance of *Narada*, a precursor wireless sensing unit, is also validated in this test. The performance of the wireless sensing system is compared with a high-precision cabled sensing system through side-by-side comparison at critical locations of the frame. Results from this full-scale experiment demonstrate that the accuracy of the wireless sensor data is comparable to that of the counterpart cabled system. Furthermore, using the data collected by the densely deployed wireless and cabled sensors, detailed modal properties of the structure are identified for future model updating.

1 INTRODUCTION

In order to improve the safety assessment of engineering structures, structural health monitoring (SHM) technologies have been widely investigated for monitoring structural performance and identifying potential damage. In an SHM system, various types of data (such as acceleration, displacement, strain, and so on) are needed to evaluate the safety of a structure (Sohn, et al. 2003). In traditional SHM systems, coaxial cable is usually adopted for data acquisition because of its reliable performance. However, extensive cable lengths in a cabled SHM system may result in cost at a few thousand dollars per sensing channel (Çelebi 2002). Furthermore, in order to obtain more detailed structural information, it is required to install a large amount of sensors on the structure. Such dense instrumentation is usually not feasible due to the budget limit. In order to overcome the limitation of cabled SHM systems, significant efforts have been devoted to developing wireless SHM systems (Lynch & Loh 2006). Recently, a new generation wireless sensing unit, named *Martlet*, is developed for SHM purpose (Kane, et al. 2014). In order to obtain accurate high-resolution acceleration data and in the meantime reduce hardware cost, an accessory sensor board for *Martlet*, named integrated accelerometer *wing*, is developed (Dong, et al. 2014).

This paper evaluates the performance of *Martlet* with the integrated accelerometer *wing* through the experiments conducted on a full-scale two-story, two-bay concrete frame structure on Georgia Tech campus. In addition to *Martlet*, the performance of *Narada*, a precursor wireless unit, is also validated in this test (Swartz, et al. 2005, Zimmerman, et al. 2008). The frame is first installed with 42 cabled accelerometers (Kinemetrics models EpiSensor ES-T and ES-U) on columns and girders to capture the acceleration responses. To increase spatial resolution of the instrumentation, 23 *Narada* and 13 *Martlet* wireless units are interspersed between cabled accelerometers on the columns, girders and slabs. The performance of the wireless system is first validated by comparing to the high-precision cabled system. Furthermore, the wireless and cabled acceleration sensor data are utilized for extracting the modal properties of the frame structure.

The rest of the paper is organized as follows. The full-scale two-story, two-bay concrete frame structure is described first. Secondly, the wireless and cabled SHM systems are introduced, including the *Martlet* wireless sensing unit and the integrated accelerometer *wing*, followed by the instrumentation for both the wireless and cabled sensing systems on the test structure. In addition, measurement data from the wireless and cabled sensors are compared. Thirdly, structural modal properties, i.e. resonance frequencies and mode shapes, are extracted using both the wireless and cabled acceleration data. Finally, the paper is summarized with conclusions and future work.

Figure 1 Photo of test frames and two shakers on Frame 1 under test (viewing from southwest corner)

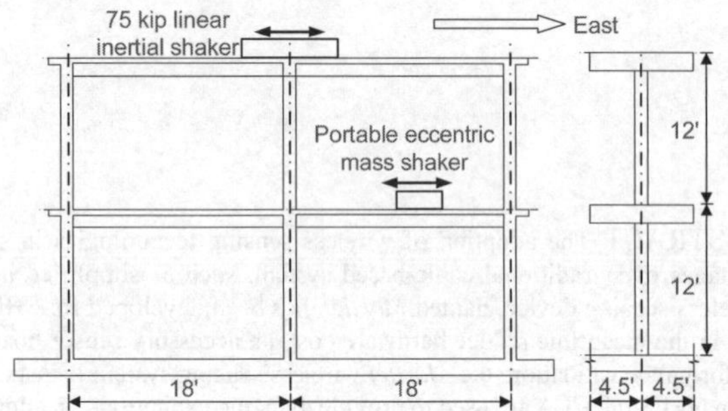

Figure 2 Elevation and side-view drawings

2 DESCRIPTION OF THE TEST STRUCTURE

The concrete test frames were built at full scale in the Structural Engineering and Materials Laboratory on Georgia Tech campus (Figure 1). Four identical frames with the same design were constructed. For experimental safety, two strong collapse prevention frames were constructed, one outside Frame 1 and the other one outside Frame 4. Although it is difficult to see in the photo, the total of six individual frames are separate from each other, with a gap between every two neighboring frames. Figure 2 shows the main dimensions of the test frame. Each test frame consists of two bays and two stories, and was meant to be representative of low-rise reinforced concrete office buildings in the central and eastern United States built in the 1950s-1970s. The frame design was carried out using simplified analysis techniques that would have been employed during the era when non-ductile reinforced concrete frames were built before modern seismic code was used. Frame 1 is an as-built bare frame as the reference structure, while different seismic retrofit measures are applied to the other three frames for seismic research. This paper reports on results from Frame 1 only.

Also shown in Figure 2 are the locations of the two shakers provided by NEES@UCLA. A 75-kip hydraulic linear inertial shaker is mounted on the second elevated slab, and a portable 980-lb eccentric mass shaker is installed on the first elevated slab of the frame. Through tens of test runs for each frame, the excitation amplitude generated by the linear shaker increases as the experiment carries on, gradually causing damage to the concrete frame. The eccentric shaker is used occasionally for low-amplitude sine sweeping.

3 WIRELESS AND CABLED ACCELERATION SENSING SYSTEMS

Although the modal analysis in this paper only involves acceleration data, a large number of other types of sensors (including strain gages, string pots, and LVDTs) were also instrumented on the test structure. This section first describes the *Martlet* wireless sensing device and the integrated accelerometer *wing*, as well as the cable-based system. The field testing setups of the wireless and cabled systems are then described, followed by the comparison between the wireless and cabled sensor data.

3.1 Martlet wireless sensing system and the integrated accelerometer wing

Martlet is a next-generation low-cost wireless sensing node developed for SHM applications (Kane, et al. 2014). The development of *Martlet* is a joint effort among three academic research labs at the University of Michigan, the Georgia Institute of Technology, and Michigan Technological University, respectively. The *Martlet* wireless node adopts a Texas Instruments Piccolo microcontroller as the core processor to execute onboard computation and data acquisition. The clock frequency of an earlier version (TMX320F28069) of the microcontroller can be programmed up to 80 MHz, and a more recent version (TMS320F28069) can support up to 90 MHz. The dimension of the *Martlet* node is 2.5 in by 2.25 in.

One distinct feature of the microcontroller on *Martlet* is the capability of high-speed data acquisition. The high-speed onboard microcontroller and the direct memory access (DMA) module on the microcontroller allows the *Martlet* node to collect sensor data at a sampling rate up to 3 MHz. The *Martlet* node adopts a 2.4 GHz radio for low-power wireless communication through IEEE 802.15.4 standard (Cooklev 2004). The extensible hardware design of the *Martlet* node enables various sensor boards to conveniently stack up through four wing connectors and work with the motherboard. The combination of the extensible design feature with onboard 9-channel 12-bit analog-to-digital conversion (ADC) allows the *Martlet* node to simultaneously sample analog signals from multiple sensors through different accessory sensor boards (termed "*wing*" boards). In addition to *Martlet*, the performance of *Narada*, a precursor wireless unit, is also validated in this test (Swartz, et al. 2005, Zimmerman, et al. 2008).

In order to obtain accurate acceleration measurement, and in the meantime reduce sensor cost, one solution is to integrate a low-cost microelectromechanical (MEMS) accelerometer and specialized signal conditioning circuit into a single *wing* board. The integrated accelerometer *wing* adopts a tri-axial MEMS accelerometer, the STMicroelectronics LIS344ALH model. A jumper on the board selects between ±2g and ±6g measurement scales. The noise density of the measurement is 25 $\mu g/\sqrt{HZ}$ along the x-axis and y-axis, and 50 $\mu g/\sqrt{HZ}$ along the z-axis.

The analog signals from the LIS344ALH accelerometer are directly fed into an onboard signal conditioner that performs mean shifting, low-pass filtering, and amplification (Dong, et al. 2014). A distinct feature of the integrated accelerometer *wing* is that the cutoff frequencies of low-pass filters and amplification gains are remotely programmable. This feature is achieved by adopting digital potentiometers (Digipots), whose resistance value can be programed on-the-fly by the *Martlet* microcontroller through an Inter-Integrated Circuit (I^2C) interface. In addition to the integrated accelerometer *wing*, another two types of commercial accelerometers (Silicon Designs 2012-002 and Crossbow CXL01LF1) are adopted in this experiment. In order to allow those two accelerometers work with *Martlet*, a smart ADC/DAC *wing* is stacked on top of the *Martlet* node (Kane, et al. 2014). During the experiment on Frame 1, two Silicon Designs accelerometers, two Crossbow accelerometers and fourteen integrated accelerometer *wing*s are connected with *Martlet* nodes. Meanwhile, twelve Silicon Designs accelerometers, four Crossbow accelerometers, and twenty-four integrated accelerometer *wing*s are connected with *Narada* nodes.

3.2 Cabled sensing system

Two accelerometer models from Kinemetrics (EpiSensor ES-T and ES-U) are adopted in the cable-based system. A 6-channel 24-bit Quanterra Q330 data logger is used to collect the cabled accelerometer data.

Table 1 provides the performance comparison of the Kinemetrics cabled accelerometers and the STMicroelectronics LIS344ALH accelerometer used in the integrated accelerometer *wing*. Available at a relatively low price, the STMicroelectronics accelerometer has a higher noise floor, yet is more convenient for wireless deployment due to the low requirement on power supply. During the experiment on Frame 1, nine EpiSensor ES-T and thirty-three EpiSensor ES-U accelerometers are deployed on the structure.

Table 1 Parameter of accelerometers used by the cabled and wireless systems

Specification	EpiSensor ES-T	EpiSensor ES-U	LIS344ALH
# of axis	3	1	3
Range	±4g	±4g	±2g / ±6g
Bandwidth	DC~200Hz	DC~200Hz	DC~1.8kHz
Noise floor	3.5 $ng/\sqrt{HZ}$	3.5 $ng/\sqrt{HZ}$	25 or 50 $\mu g/\sqrt{HZ}$
Power supply	20V	20V	2.4 ~ 3.6V

3.3 *Experimental setup of the first frame*

As shown in Figure 3 (a), the frame is first installed with 42 Kinemetrics cabled accelerometers (EpiSensor ES-T and ES-U) on columns and girders (60 data acquisition channels in total) to capture the acceleration responses. To increase spatial resolution of the instrumentation, 23 *Narada* and 13 *Martlet* wireless units with overall 66 acceleration channels are interspersed between cabled accelerometers on the columns, girders and slabs (Figure 3 (b)). As a result, there are a total of 126 acceleration channels installed on the structure, including both cabled and wireless channels.

As seen in Figure 3, some cabled and wireless accelerometers are deployed side-by-side at the same location, i.e. W43 and C33 (Figure 4), so that the data collected by the cabled and wireless data acquisition systems can be synchronized. In the meantime, the cabled and wireless data can be compared at the side-by-side locations. Most of the wireless units use a smaller-size 2 dBi omni-directional whip antenna. At locations with thicker concrete blocking between the wireless unit and wireless server, a directional antenna (6 dBi Intellinet 525138) is used by the unit.

3.4 *Vibration measurement on the first frame*

Figure 5(a) presents the example acceleration time histories recorded at beam and column joint at the first elevated slab (C13 and W13). The data is collected when the linear shaker operates on the second elevated slab with a scaled El Centro earthquake record. In this test, the acceleration record is scaled such that the corresponding maximum displacement is 4 inches. The amplification gain and cutoff frequency of the integrated accelerometer *wing* is set to be ×20 and 25 Hz, respectively. A 25 Hz low-pass digital filter has been applied to both the wireless and cabled data sets so that signal in the same frequency range is compared. Figure 5(b) shows the close-up comparison of acceleration time histories. Both plots demonstrate excellent agreement between the cabled and wireless system.

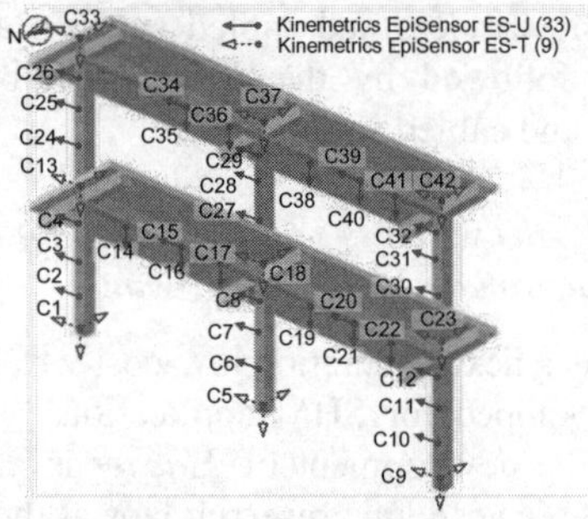

(a) Cabled sensing system

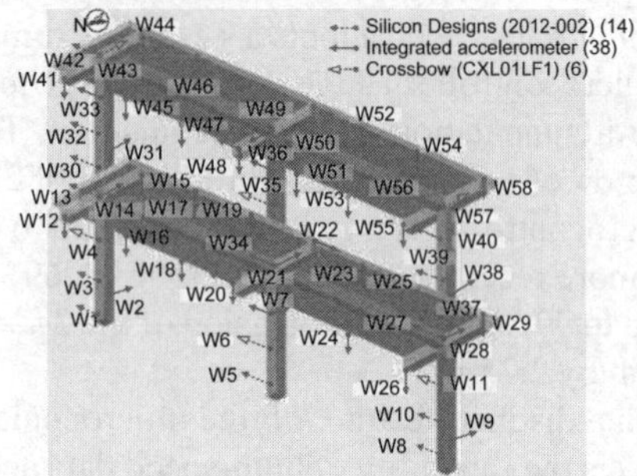

(b) Wireless sensing system

Figure 3 Accelerometer instrumentation on Frame 1

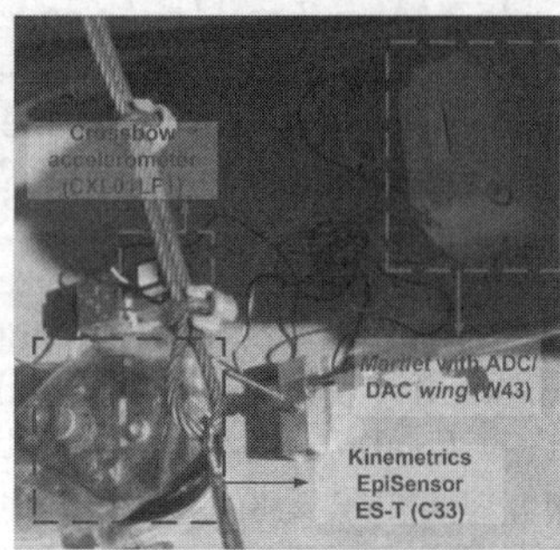

(a) Cabled and wireless accelerometer

Figure 4 Pictures of deployment on Frame 1

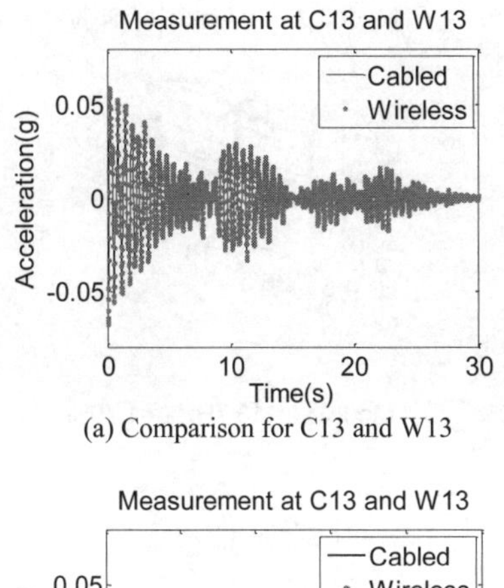

(a) Comparison for C13 and W13

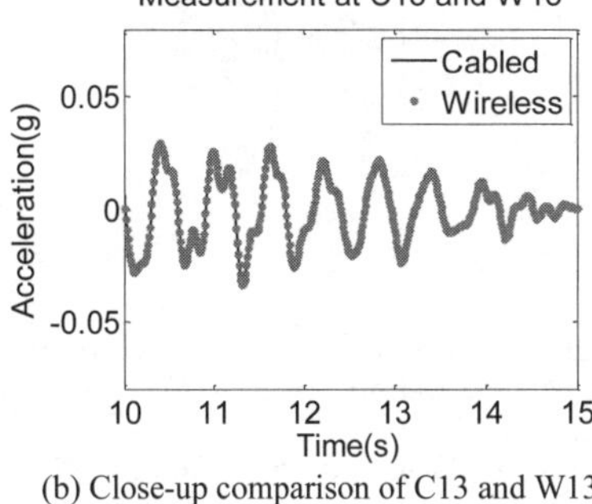

(b) Close-up comparison of C13 and W13

Figure 5 Comparison of acceleration time history plots between cabled and wireless system

4 STRUCTURAL MODAL PROPERTY ANALYSIS USING WIRELESS AND CABLED SENSOR DATA

Among the large number of tests performed with Frame 1, this paper reports modal analysis results using data from two tests when the linear inertial shaker was used. Using an accelerometer mounted on the moving mass of the linear inertial shaker, time history of the shaker force can be estimated as dynamic input to the structural system. With known input to the system, frequency response functions (FRFs) can be calculated for all response DOFs instrumented with cabled and wireless accelerometers. The commonly used eigensystem realization algorithm (ERA) is applied to extract the structural modal properties (Juang & Pappa 1985). Because the unidirectional inertial shaker force is along the horizontal in-plane direction, this paper focuses on the extraction of in-plane mode shapes of the frame. Extraction of out-of-plane modes using other sets of acceleration data will be reported in future studies.

The first set of data analyzed herein is when the linear shaker operates on the second elevated slab with a scaled El Centro record (the corresponding maximum displacement scaled to 1"). Figure 6 shows the first three in-plane modes of Frame 1 extracted from ERA method. Those three mode shapes are obtained by using data from the cabled system only (52 channels in total). Recall that Figure 3 shows all cabled channels are instrumented on the frame, and no cabled accelerometer is installed on the edge of the concrete slabs. The first mode shape shows all nodes moving along one direction, which is expected for this frame structure. In addition, Figure 7 shows modal analysis results from the test data when the maximum displacement amplitude of the El Centro excitation is increased to 4". Both cabled and wireless acceleration data, 95 channels in total, are used for modal analysis this time. The resonance frequencies of all identified modes show decrease, compared with the El Centro 1" results. With more sensor channels available for modal analysis, more detailed vibration modes are obtained with denser measurement nodes on the columns. The benefit of dense sensor nodes is particularly obvious for Mode 3 in Figure 7, where vertical vibration of the concrete slabs is captured by the wireless accelerometers.

From El Centro 1" test to the 4" test, the decrease in resonance frequencies is possibly due to structural damage caused by the shaker excitation. Difference observed between the two sets of mode shapes can be a promising indicator for locating and assessing the severity of structural damage.

5 CONCLUSIONS AND FUTURE WORK

This research evaluates the performance of a recently-developed wireless sensing unit, *Martlet*, with the integrated accelerometer *wing* during the experiment conducted on a full-scale two-story, two-bay concrete frame structure on Georgia Tech campus. In addition to *Martlet*, the performance of *Narada*, a precursor wireless unit, is also validated in this experiment. The acceleration time histories are compared between wireless and cabled systems, illustrating that the wireless data achieves comparable quality to the cabled data. In addition, modal properties of the structure are

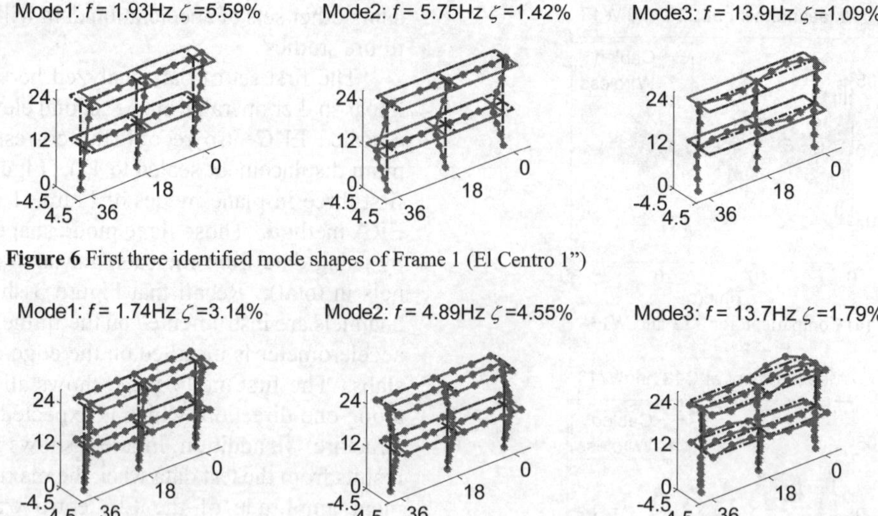

Figure 6 First three identified mode shapes of Frame 1 (El Centro 1")

Figure 7 First three identified mode shapes of Frame 1 (El Centro 4")

successfully obtained by using both the wireless and cabled acceleration data. The decrease in the resonance frequencies of the frame structure may indicate the existence of structural damage. The difference in the mode shapes may help to localize and assess the severity of the damage. In the future, data collected from different stages of the experiment will be analyzed so as to assess the growth of structural damage through the course of the experiment. In addition, more vibration modes may be obtained by analyzing structural response data from other tests.

ACKNOWLEDGEMENT

This research is partially sponsored by the National Science Foundation (#CMMI-1041607 and #CMMI-1150700), the Research and Innovative Technology Administration of US DOT (#DTRT12GUTC12), and Georgia DOT (#RP14-30). Any opinions, findings, and conclusions or recommendations expressed in this publication are those of the authors and do not necessarily reflect the view of the sponsors.

REFERENCES

Çelebi, M. 2002. "*Seismic Instrumentation of Buildings (with Emphasis on Federal Buildings)*," United States Geological Survey, Menlo Park, CA Report No. 0-7460-68170.

Cooklev, T., 2004. *Wireless Communication Standards : a Study of IEEE 802.11, 802.15, and 802.16*. New York: Standards Information Network IEEE Press.

Dong, X., D. Zhu, Y. Wang, *et al.* 2014 "Design and validation of acceleration measurement using the Martlet wireless sensing system," *Proceedings of the ASME 2014 Conference on Smart Materials, Adaptive Structures and Intelligent Systems (SMASIS)*, Newport, RI, USA.

Juang, J. N. & R. S. Pappa. 1985. "An eigensystem realization algorithm for modal parameter identification and modal reduction," *Journal of Guidance Control and Dynamics*, vol. 8, pp. 620-627.

Kane, M., D. Zhu, M. Hirose, *et al.* 2014 "Development of an extensible dual-core wireless sensing node for cyber-physical systems," *Proceedings of SPIE, Nondestructive Characterization for Composite Materials, Aerospace Engineering, Civil Infrastructure, and Homeland Security*, San Diego, California, USA.

Lynch, J. P. & K. J. Loh. 2006. "A summary review of wireless sensors and sensor networks for structural health monitoring," *The Shock and Vibration Digest*, vol. 38, pp. 91-128.

Sohn, H., C. R. Farrar, F. M. Hemez, *et al.* 2003. "*A Review of Structural Health Monitoring Literature: 1996-2001*," Los Alamos National Laboratory, Los Alamos, NM Report No. LA-13976-MS.

Swartz, R. A., D. Jung, J. P. Lynch, *et al.* 2005 "Design of a wireless sensor for scalable distributed in-network computation in a structural health monitoring system," *Proceedings of the 5th International Workshop on Structural Health Monitoring*, Stanford, CA.

Zimmerman, A. T., M. Shiraishi, R. A. Swartz, *et al.* 2008. "Automated modal parameter estimation by parallel processing within wireless monitoring systems," *Journal of Infrastructure Systems*, vol. 14, pp. 102-113.

Proceedings of the International Conference on Smart Infrastructure and Construction
ISBN 978-0-7277-6127-9

© The authors and ICE Publishing: All rights reserved, 2016
doi:10.1680/tfitsi.61279.149

Development of RTOS-based wireless SHM system: benefits in applications

Y.G. Fu*, K.A. Mechitov, V. Hoskere, B.F. Spencer, Jr

University of Illinois at Urbana-Champaign, Urbana, USA
* *Corresponding Author*

ABSTRACT The emergence of wireless smart sensor platforms with powerful computational capabilities has had broad impacts in the field of structural health monitoring (SHM), enabling numerous applications in civil infrastructure. However, as the demands of wireless SHM systems increase, certain properties of event-driven operating systems for these sensor nodes, such as the TinyOS operating system of the iMote2, begin to impose limitations on the development of SHM systems. The problematic characteristics include static resource allocation, single-application focus, lack of real-time scheduling support, and dependence on a non-standard programming language. To address these limitations, we consider the use of a real-time operating system (RTOS), commonly used for industrial control systems and similar applications, as an alternative solution in the development of Xnode, the next-generation smart sensor platform for civil engineering applications. In this paper, features of the RTOS environment are first analyzed systematically and compared with TinyOS to demonstrate how it addresses the major concerns of event-driven operating systems. A distributed data acquisition application from the Illinois SHM Services Toolsuite for the iMote2 is implemented as a demonstration of the RTOS-based framework and its advantages. Most importantly, benefits of the RTOS-based wireless SHM system are catalogued comprehensively, with a particular focus on flexible application framework and a more engineer-friendly environment.

1 INTRODUCTION

Structural health monitoring has gained increasing attention, as it helps to address the major concern about safety of aging infrastructure in our society. Since the 1990s, researchers have made continuous efforts to develop a series of wireless or smart sensor platforms to significantly facilitate damage detection diagnosis of world's structures (Spencer et al., 2004). To date, most of wireless platforms adopt event-driven operating systems to manage computing resources and host applications. Typically, TinyOS is among the most popular event-driven operating systems, which is designed for wireless platforms with extreme limited resources (Levis et al. 2005). In particular, iMote2, one of the most widely-used sensor platforms for data intensive applications such as SHM, is supported by TinyOS, and it has enabled the implementation of world's largest wireless smart sensor network (WSSN) on the Jindo Bridge in South Korea (Rice et al., 2010).

However, lessons learned from experience of iMote2 reveal several limitations of TinyOS (Rice & Spencer, 2009). The problematic characteristics include static resource allocation, single-application focus, lack of real-time scheduling support, and dependence on a non-standard programming language, etc. Of these limitations, concurrency model and programming efforts are considered as the two major concerns:

(i) There are only two level of executions in the concurrency model of TinyOS: tasks and events. Tasks are executed in a First In, First Out (FIFO) manner. Therefore, real-time applications are difficult to be realized, since critical tasks may be delayed by execution of previous tasks. Besides, uncertain delay of task executions is inevitable in some cases. As a result, the applications of TinyOS-based systems are limited.

(ii) Applications are developed in nesC, a dialect of C programming language, for TinyOS. The complex syntax and semantics in nesC makes it a challenge for engineers to develop applications. Besides, modeling nesC program requires involvement of hardware operations in most cases, which makes it complicated to update or debug code (Ammari, 2013). In addition, the single-application focus of TinyOS requires developers to consider every part of a system every time their code is updated. All of these dramatically increase programming efforts.

In order to address these concerns, we consider the use of a real-time operating system, commonly used for industrial control systems and similar applications, as an alternative solution in the development of Xnode, the next-generation smart sensor platform for civil engineering applications. In this paper, the RTOS environment is first analyzed and compared with TinyOS to demonstrate how it addresses the major concerns of event-driven operating systems. A distributed data acquisition application from the Illinois SHM Services Toolsuite for the iMote2 is implemented as a demonstration of the RTOS-based framework and its advantages. Most importantly, benefits of the RTOS-based wireless SHM system are catalogued comprehensively, with a particular focus on flexible application framework and engineer-friendly environment.

2 REAL-TIME OPERATING SYSTEMS

2.1 *Introduction to RTOS basics*

An RTOS is an operating system in which applications are run under specified time constraints with high reliability. Typical properties of an RTOS includes multi-tasking and preemptive scheduling. In general, an RTOS problem is divided into multiple tasks and handled by several processes. A scheduler is used to manage task sequencing at run time, based on specific scheduling algorithms.

There are many different RTOS that are used in the embedded systems market. FreeRTOS is selected as for development of Xnode due to its open source nature, portable C-language implementation, high degree of configurability, availability of free and commercial support options, and a large user community. Specifically, it has typical advantages as a class of RTOS, compared to event-driven operating systems, such as real-time scheduling and efficient inter-task communication. Also, it has specific advantages of small footprint, low overhead and fast execution, compared to some other classes of RTOS.

2.2 *Primary solution of FreeRTOS for Xnode*

FreeRTOS provides a priority-based preemptive scheduler, as well as inter-task communication and coordination tools (e.g., queues, mutexes and semaphores). It can efficiently address the limitations of event-driven operating systems on a low-power microprocessors. Primary solutions of FreeRTOS for Xnode include scheduling flexibility and programming efficiency, which are demonstrated through comparison between RTOS-based systems and TinyOS-based systems.

1) As shown in Figure 1, there are only two levels of concurrency in TinyOS, namely the task and interrupt contexts. Different tasks are managed within a single FIFO queue with no reordering or task priorities. Specifically, new tasks are put at the end of the task queue upon scheduling, and they cannot execute until all previous tasks are completed. In this way, some important tasks will be delayed by less critical tasks, for example a high-frequency sensor sampling task can be delayed by a periodic task that checks the battery level, introducing jitter into the sampling process. This is not desirable for real-time applications. By contrast, in FreeRTOS, important tasks can be assigned higher priorities. Tasks with lower priorities will be suspended to give way to important tasks and later be resumed after the important tasks are completed or blocked. This scheduling flexibility will significantly enable complex applications in wireless SHM systems, such as structural control, particularly with multiple applications coexisting on the same wireless network.

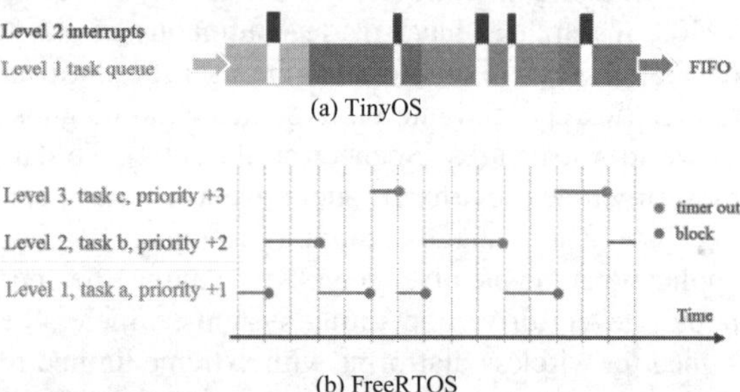

(a) TinyOS

(b) FreeRTOS

Figure 1. FIFO scheduling in TinyOS and priority-based scheduling in FreeRTOS.

2) Another major concern that FreeRTOS addresses is programming efficiency. FreeRTOS is predominantly written in standard C programming language, and hence developers can freely use many existing C

libraries, which implement useful functionality such as numerical algorithms and data compression, with much less porting efforts than that in TinyOS-based systems (written in nesC, which is a custom extension on the C language). Also there are several powerful tools written for the C language that can facilitate testing and debugging user-developed applications. In addition, the single-application focus of TinyOS, where application code, operating system code, libraries and device drivers are all compiled into a single image and share a single scheduler with no priority differentiation, which means that developers have to consider every part of a system and possible interactions of all components. By contrast, functionalities of a program in RTOS-based systems can be separated by different tasks and priority levels. Hence, developers just need to modify specific part of the code, rather than consider everything in a program. As a result, the cost and complexity of application development can be significantly reduced.

3 DEMONSTRATION: REMOTESENSING APPLICATION IN FREERTOS

3.1 RemoteSensing application in brief

The Illinois Structural Health Monitoring Project (ISHMP) Services Toolsuite, developed through collaboration between researchers in Smart Structures Technology Laboratory and Open Systems Laboratory at University of Illinois at Urbana-Champaign, is the software foundation for the iMote2-based remote sensing platform, which is built on TinyOS. RemoteSensing, a fundamental distributed data acquisition application in ISHMP Services Toolsuite (Rice & Spencer, 2009), is implemented in FreeRTOS as a demonstration of the RTOS-based framework and its advantages (Figure 2). Specifically, this application is used to help the coordination between gateway node and multiple sensor nodes to realize the remote sensing application.

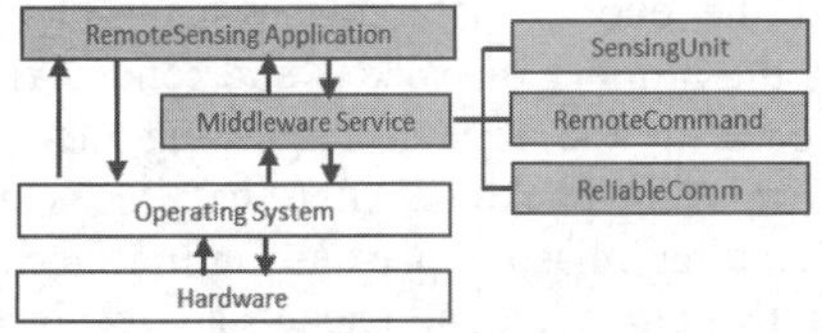

Figure 2. Structure of the RemoteSensing application, which is supported by three middleware services.

3.2 Sensing application framework

During RemoteSensing application, sensor nodes are required to start sensing when they receive commands from a gateway node, and to send sensor data back to the gateway node once the sensing is completed. To illustrate the flowchart of RemoteSensing in a single sensor node, a basic sensing application framework is developed in FreeRTOS (Figure 3). In the proposed framework, three tasks are defined to realize the sensing process, namely the Application Task, Sensing Task, and Radio Task. Commands are first received by the Radio Task and delivered to the Application Task. Then, the Application Task executes the commands and activates the Sensing Task to control sensors to generate sensor data periodically. Once sensing is completed, the Application Task gets sensor data from the Sensing Task and transfers it to the Radio Task for transmission back to the gateway node. During this process, inter-task communication and coordination is enabled by means of queues and semaphores. Meanwhile, different priorities are assigned for deciding which task should be executing at a particular time. Particularly, the Radio Task should be assigned a high priority, because any delay of the Radio Task may result in missed data packets that are transmitted from the gateway node.

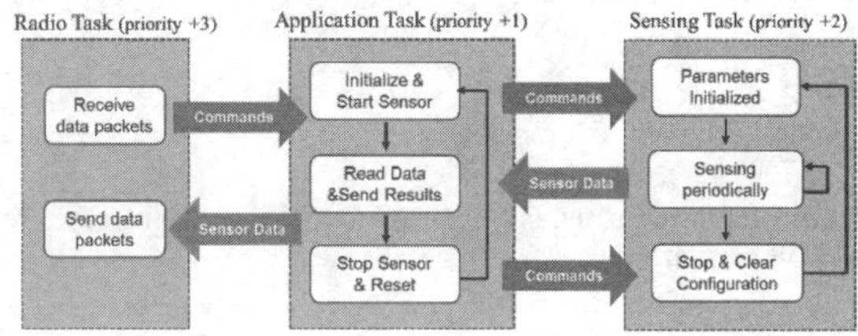

Figure 3. Flowchart of the sensing process in a sensor node in the FreeRTOS version.

3.3 SensingUnit Implementation

SensingUnit (SU) is a service component which governs the coordination between a gateway node and multiple sensor nodes during a sensing process (Sim & Spencer, 2009). It is used to transfer sensing information, initiate synchronized or unsynchronized sensing and transmit measured data. A series of states are defined using a state machine to facilitate the control of performance of different nodes in a sensor network. Basically, SU consists of two parts: SensingUnit GatewayNode (SUGW) and SensingUnit SensorNode

(SUSN), which are designated to serve a gateway node and a sensor node respectively.

Figure 4 shows the flowchart of SensingUnit in FreeRTOS, as well as its connection with the Application Task in a gateway node (GatewayNode_app) and with the Application Task in a sensor node (SensorNode_app). The main logic of SU in FreeRTOS remains the same as that in the TinyOS version, but interfaces of SU are modified significantly. Upon receipt of a command from GatewayNode_app and SensorNode_app, SUGW and SUSN will change its state to INIT in the initialization. Then GatewayNode_app will check each state of SUGW and trigger the transition between different states based on results of each iteration in a state machine. Sensing command will be transmitted from SUGW to SUSN by means of radio communication. Once the sensing process is completed in the sensor node, GatewayNode_app will directly send the data request command to SensorNode_app to retrieve the sensor data.

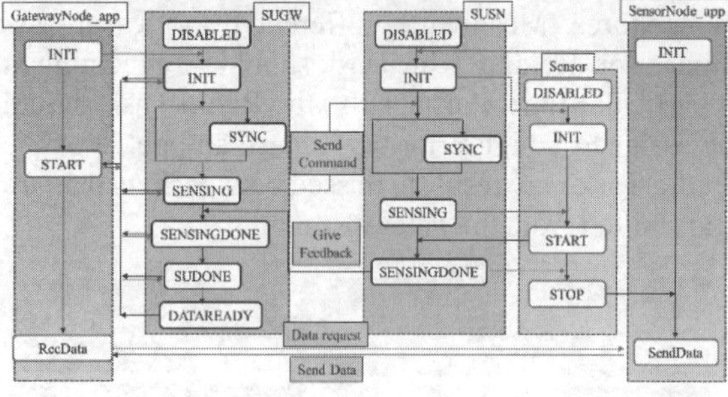

Figure 4. Flowchart and implementation of SensingUnit in the FreeRTOS version.

3.4 *RemoteCommand Implementation*

RemoteCommand is an efficient messenger for delivering commands, response or measured data between the gateway node and sensor nodes. It consists of several commands and event handlers, which are used in the designated order (Mechitov, 2012).

In the RemoteSensing application, RemoteCommand is applied to deliver sensing commands from SUGW to SUSN before sensing process, and it is also used to send data requests from the Application Task in the gateway node (Main_app_GW) to the Application Task in sensor nodes (Main_app_SN) after sensing is completed (Figure 5).

Figure 6 shows the flowchart of RemoteCommand in FreeRTOS. Main commands maintain the same interfaces as in the TinyOS version, but event handlers are replaced with an additional task with similar functionalities. At the beginning, sensor nodes will call registerCommand to specify a start function. Then executeCommand is called in a Gateway node to deliver commands. Once the commands are received by the RemoteCommand Task in sensor nodes (RecTask_SN), the corresponding start function is called in a designated way. After that, executionDone is called to send working results back. Upon RemoteCommand Task in the gateway node (RecTask_GW) receiving the data message, a callback function is executed and the next command is ready.

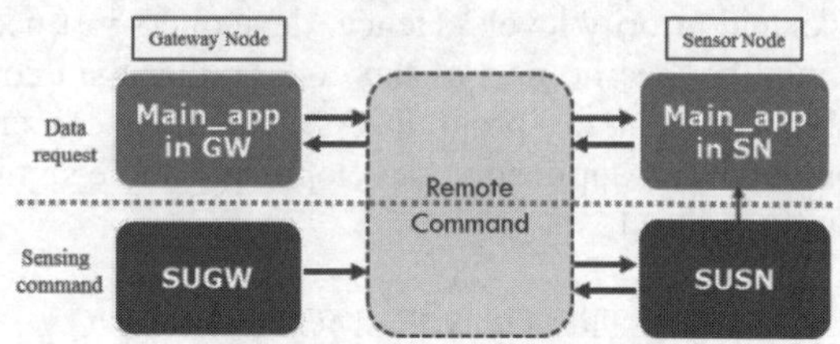

Figure 5. Implementation of RemoteCommand in RemoteSensing.

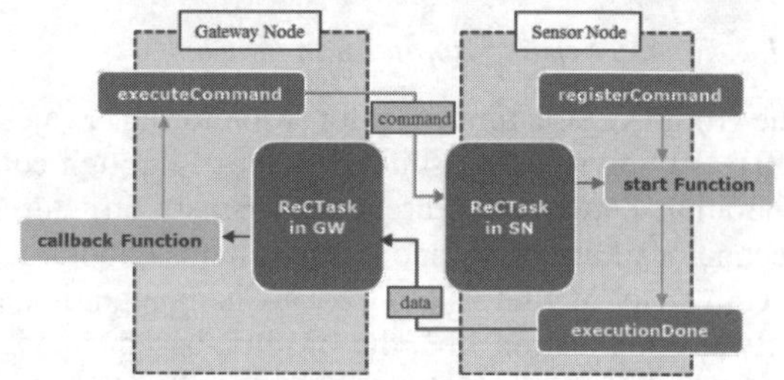

Figure 6. Flowchart of RemoteCommand in FreeRTOS.

3.5 *ReliableComm Implementation*

RelibleComm is a foundation service to address the concern of data loss in wireless systems, by means of acknowledgement-based reliable communication protocols (Nagayama & Spencer, 2007). The sender sends a message to the receiver through a series of data packets. At the end of data transmission, the receiver checks the number of packets and send acknowledgment back to the sender, requesting the sender to resend missing packets. After that, the sender will keep sending missing packets until it receives the acknowledgement that all packets are received.

Figure 7 shows the flowchart of ReliableComm in FreeRTOS. The Radio Task is implemented to receive data packets continuously in both the sender and the receiver. Two states, COMM and CHECK, are defined in a state machine to control the behavior of ReliableComm. Specifically, in the COMM state, data packets are transmitted between the sender and the receiver, while in the CHECK state, an acknowledgement is sent from the receiver to the sender, requesting missing packets. As shown in Figure 8, ReliableComm consists of several interfaces which are used to connect with RemoteCommand and the Radio driver.

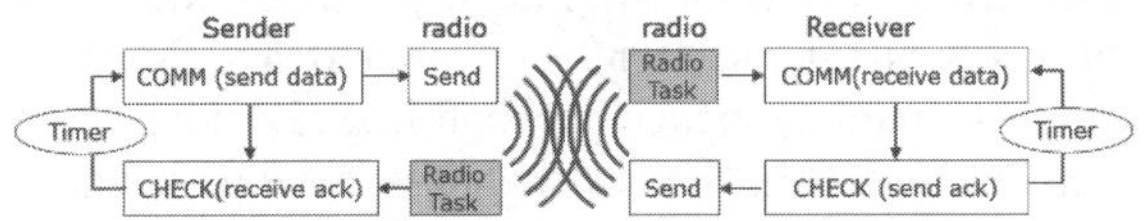

Figure 7. Flowchart of ReliableComm in FreeRTOS.

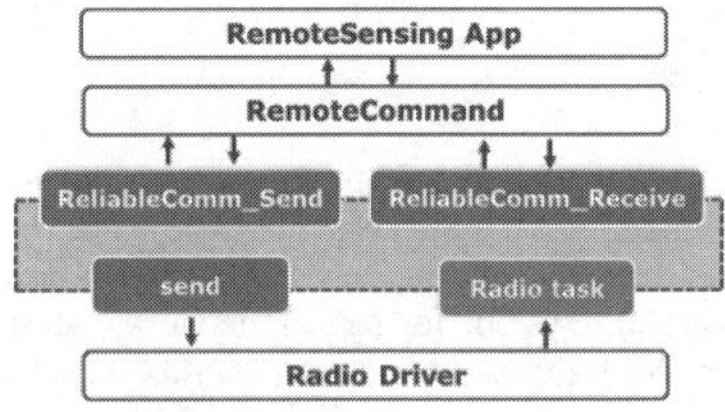

Figure 8. Implementation of ReliableComm in RemoteSensing

3.6 *ReliableComm Implementation*

A total of 4 types of tasks are defined for implementation of RemoteSensing in FreeRTOS, assigned with different priorities (Figure 9).

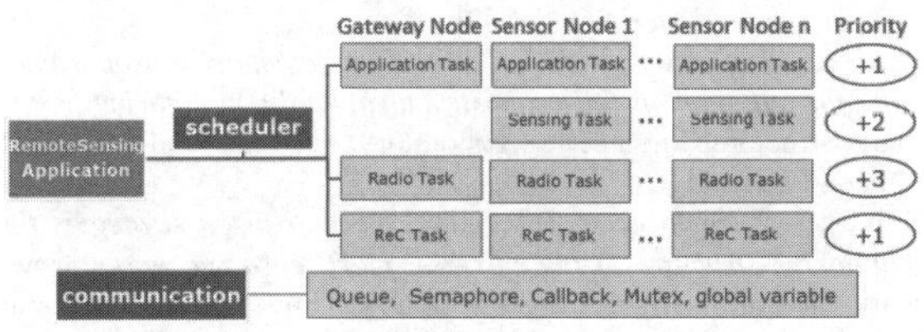

Figure 9. Structure of RemoteSensing in summary.

Specifically, the Radio Tasks have highest priorities to ensure the efficient communication between a gateway node and sensor nodes. Meanwhile, several tools are applied to enable inter-task communication, including queues, semaphores, callback functions, mutexs, etc. The diagram in Figure 10 demonstrates how the tasks in RemoteSensing are scheduled by FreeRTOS. Using FreeRTOS, different functionalities in RemoteSensing are isolated within separate tasks, and these tasks are executed efficiently, by means of a preemptive, priority-based scheduler.

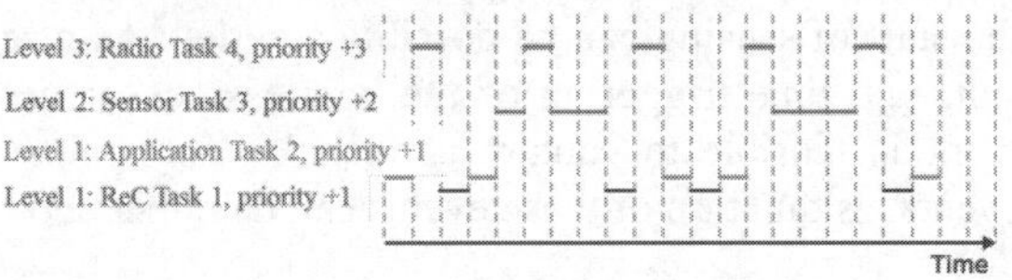

Figure 10. Tasks scheduled in the FreeRTOS version of RemoteSensing.

4 BENEFITS IN APPLICATIONS

4.1 *Flexible application framework*

One primary benefit of Xnode, a RTOS-based wireless SHM system, is the more flexible application framework. In moving to this new platform, the RTOS enables several benefits for application developers compared to the iMote2 (ISHMP/TinyOS-based) system, in part due to its real-time scheduler.

1) Over-the-air programming (OTA) can be realized by uploading a single module or the entire application image, which is a significant benefit for SHM applications. In general, wireless sensor nodes are densely deployed in a SHM project. If pre-installed damage detection algorithm is obsolete, it would be of great convenience to reprogram wirelessly for multiple sensor nodes, rather than to collect sensor nodes manually and update algorithms one by one.

2) Resource optimization is another attractive aspect. Developers can configure the system with dynamic resource allocation mechanism, hence resources can be allocated and deallocated at run-time. In this way, highly efficient operation is enabled for long-term structural health monitoring, especially for a large-scale wireless sensor network.

3) Real-time sensing and control applications are possible to be handled on the basis of the real-time scheduler. Specifically, different functionalities are defined by multiple tasks, and the task with the strictest deadline is assigned the highest priority, specified with maximum time spans. This is highly beneficial for real-time application development, such as structural control using wireless sensor nodes, in which time delay may result in serious detrimental consequences.

4) In addition, ability of SHM systems will be enhanced to capture extreme events, which is known as rare event detection. Because of strict timing control supported by RTOS-based system, uncertain delay in the start of sensing can be possibly avoided. As a result, real-time triggering of the wireless sensor network is feasible to capture earthquakes and impact events, as duration of those events can be rather short.

4.2 *Engineer-friendly environment*

Another attractive benefits of Xnode is the more engineer-friendly environment. In the field of SHM, most users are civil engineers who do not have a solid foundation of computer science knowledge or programming skills. This platform is attractive for those users, because of its more common programming language and stronger separation of concerns.

1) The C and RTOS-based system is easier for engineers to develop applications than TinyOS-based systems. Beginners will face a steep learning curve for programming on RTOS, using a standard programming language. However, in TinyOS, users have to understand the custom nesC extensions as well, which require deep understanding of the nesC/TinyOS concurrency model.

2) Some useful tools, such as an interactive debugger can help users to build their applications efficiently. In TinyOS-based systems, if users want to develop the RemoteSensing, they probably have to compile and install a test application in wireless sensor platforms every time when the code is updated or debugged. In particular, approaches to check if the code works well are limited (e.g., LED indicators are overly relied on in iMote2 development). However, development cost and complexity will be significantly reduced when developing the RemoteSensing application in RTOS-based systems.

3) RTOS-based systems provides an isolation of individual tasks and hence a separation of concerns. Application designers can focus on their specific parts without affecting the timing of other code. This is really beneficial for engineers. For example, algorithm development for damage detection is a main topic in the field of SHM, and implementation of algorithms in wireless SHM systems is always a major obstacle for non-programmers. In this case, RTOS-based systems will lower the barriers for people to enter application development in the field of SHM.

5 CONCLUSIONS

Primary limitations of event-driven operating systems for development of SHM systems are demonstrated, i.e., the concurrency model and inordinate programming effort. They are addressed by using FreeRTOS as alternative solution in the development of Xnode, due to its open source, portability, flexible scheduling and strong support for real time application. RemoteSensing application is implemented as a demonstration of RTOS-based framework and its advantages. Specifically, a basic sensing application framework is developed, and three middleware services (SensingUnit, RemoteCommand and Rliable-Comm) are implemented in FreeRTOS. Besides, great benefits from RTOS-based wireless systems are analyzed, with a particular focus on flexible application framework and engineer-friendly environment.

REFERENCES

Ammari, H.M. 2014. *The art of wireless wensor networks*. Springer, Berlin, Heidelberg.

Levis, P., Madden, S., Polastre, J., Szewczyk, R., et al. 2005. TinyOS: an operating system for sensor networks. *Ambient Intelligence*, Weber, W., Rabaey, J.M., Aarts, E., Eds. 115-148, Springer, Berlin, Heidelberg.

Mechitov, K.A. 2012. *A service-oriented architecture for dynamic macroprogramming of sensor networks*, Ph.D dissertation, University of Illinois at Urbana-Champaign.

Nagayama, T. & Spencer Jr, B.F. 2007. *Structural health monitoring using smart sensors*. Newmark Structural Engineering Laboratory. University of Illinois at Urbana-Champaign.

Rice, J.A., Mechitov, K., Sim, S.H., et al. 2010. Flexible smart sensor framework for autonomous structural health monitoring, *Smart Structures and Systems* **6**, 423-438.

Rice, J.A. & Spencer Jr, B.F. 2009. *Flexible smart sensor framework for autonomous full-scale structural health monitoring*. Newmark Structural Engineering Laboratory. University of Illinois at Urbana-Champaign.

Sim, S.H. & Spencer Jr, B.F. 2009. *Decentralized strategies for monitoring structures using wireless smart sensor networks*. Newmark Structural Engineering Laboratory. University of Illinois at Urbana-Champaign.

Spencer Jr., B.F., Ruiz-Sandoval, M & Kurata, N. 2004. Smart sensing technology: opportunities and challenges, *Structural Control and Health Monitoring* **11**, 349-368.

Proceedings of the International Conference on Smart Infrastructure and Construction
ISBN 978-0-7277-6127-9

© The authors and ICE Publishing: All rights reserved, 2016
doi:10.1680/tfitsi.61279.155

Low Power DSP with wireless monitoring for civil constructions

V. Kumar*[1], J. Yan[1,2], X. Xu[1], Y. Qian[1] and K. Soga[1]

[1] *Department of Engineering, University of Cambridge, Cambridge, UK*
[2] *Electronics and Computer Science, University of Southampton, Southampton, UK*

ABSTRACT Noise arising from the construction industry is a major source of noise pollution. Noise from construction sites affects not only the site itself, but also the surrounding area including neighbouring businesses and residents. The duration, complexity, schedule, location, method of construction and type of projects greatly affect the extent of noise impact. Although there are many noise regulation frameworks such as BS 5228, BS 7580 in the UK and 2002/49/EC across Europe to control and mitigate the impact of this construction noise, there is no standardized criteria for assessing construction noise impact. Hence there is a need to identify the extent and magnitude of the noise through using noise monitoring equipment. Such equipment should identify noise, quantify noise sources and differentiate various noise types such as piling, demolition, and hammering, reversing truck warning signals. The system should also provide a noise map of the locality and the surrounding area. This paper describes the development of a custom wireless sensor board for noise identification, monitoring and localisation using a low-power DSP and microcontroller. The system stores noise samples to a local SD memory card for future analysis and wirelessly transmits a summary of significant noise events in real time. This paper describes the result of an initial test of the system on a construction site in Cambridge. The noise event and their efficiency has been compared with high resolution, beamforming technology based SeeSV-S205 audio camera. Future work on the system will include further testing to develop better noise discrimination algorithms.

1 INTRODUCTION

The construction and demolition noise although lasts only for few days and fixed hours, it is one of the major source of noise pollution, affecting the residential and commercial premises (Shen et al. 2005). Several studies has been conducted in the past to show that residents, work offices get disturbed with many complaint logs as part of noise from these construction works affecting over a considerable amount of period. The adverse noise level creates harmful health effects such as concentration loss, mental, physiological and psychological dysfunction and has considerable socio-economic bearing. Some noise regulations across the UK (BS5228:2009, BS7580), Europe (2005/88/EC) are prevalent to decide the acceptable noise level. The acceptable noise level L_{Aeq} (Average equivalent of 'n' continuous sound sample for 15 minute) for industrial premises should be within 75 dB(A) while for offices, retail outlet it should be 70 dB(A). Similarly the residential properties has the acceptable noise level L_{Aeq} limit of 57.1 dB(A) between 0700 to 1900 hours Monday to Friday and 0700 to 1300 hours on Saturday, 53.1 dB(A) between 1900 to 2300 hours Monday to Friday while during night time between 2300 to 0700 hours Monday to Friday it should be within 48.2 dB(A) (Fahy & Walker 2004). In such cases it becomes important to predict, control and monitor the noise level to its acceptable limit without imposing unnecessary restrictions on contractors.

CSIC has developed a low power based Digital Signal Processor (DSP) solution which can provide event based noise recognition and monitoring for certain types of noise such as demolition, piling, vibration machine, reversing truck warning signal, public address system, and airplane and background noise using Frequency Selective method and Support Vector Machine (SVM) of Machine Learning approach. Although SVM algorithm has been applied for face, handwriting recognition and other general pattern classification, little effort has been done for noise recognition and discrimination at a construction site (Byun & Lee 2003).

2 APPROACH FOR NOISE DETECTION AND SEPARATION

Noise at a construction site can be detected using available sound level meter or microcontroller (MCU) based audio capturing system and the captured noise can be sent to any location for its processing and logging using various wireless methods such as Bluetooth, Bluetooth Low Energy (BLE), Zigbeee, WiFi or wired system with Ethernet. However the power consumption of the MCU based devices is typically higher due to the transmission of entire chunks of the data packet for its processing at far site. A portable battery to power up these devices contains only a limited amount of power. When the battery becomes exhausted, only option left is to change the battery or deploy other similar system. Due to the adverse location of the system in the construction site it becomes costly and time consuming to replace the system or even replace the battery. The low power DSP developed system performs audio processing at the device level locally using various DSP functions before the transmission of important events to a central server location thus enhancing the battery life of the system.

2.1 Frequency selective approach

Here noise is recorded and stored into SD card in time domain using a block of fixed size arrays and it is transformed in frequency domain with Discrete Fourier Transform (DFT) using sliding window mechanism. After generating the frequency response of the captured noise, the signal spectrum can be passed through a range of parallel frequency windows with narrow filter bands to separate various types of noise. The total number of noise events can be identified and recorded if the power of the signal within its duration is greater than a certain threshold. The extracted frequency response of the signal is again converted back into time domain using Inverse Discrete Fourier Transform (IDFT) for its future use. The complete process is shown in Figure 1.

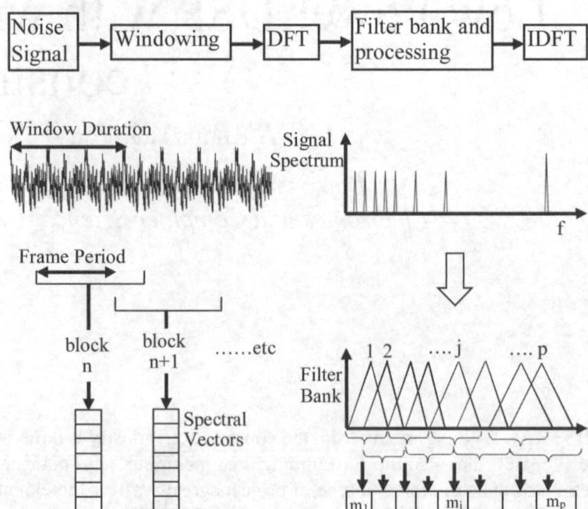

Figure 1. Frequency Selective method for noise identification

2.2 SVM approach

The SVM is one type of supervised machine learning algorithm to classify the noise data in binary format using a kernel method where positive label indicates data presence while negative label indicates the absence of data. The SVM based noise recognition algorithm is dependent on signal detection, parameter identification and calculation, model training, classification and decision module. Once data is recorded using microphone and parameters are identified for classification, decision module and parameters are extracted from the signal. Parameters may be dependent on energy stored in certain sub-bands of the noise signal (Zwan & Czyzewski 2008) as defined in (1) and sudden transient changes in the signal amplitude as defined in (2).

$$p = \frac{\sum_{n_{f1}}^{n_{f2}} P(n)}{\sum_{n_{f1}'}^{n_{f2}'} P(n)} = \frac{energy\ in\ band\ 1}{energy\ in\ band\ 2} \quad -(1)$$

$$tr_{length} = n_{|P(n)|<P_{thr}} - n_{|P(n)|=\max(p)} \quad -(2)$$

Here P(n) defines the parameter for the power spectrum of the signal which expresses the signal's energy accumulation in this particular frequency band while n_{f1}, n_{f2}, n_{f1}' and n_{f2}' are the indices of the respective band limits. The parameter P(n) calculation is based on transient calculation changes and hence it requires complete noise data to be stored in number of frames and accessed before the actual calculation can be started. There are many available techniques for noise data classification such as Hidden Markov Model (HMM), Gaussian Mixture Model (GMM), k-Nearest Neighbors (k-NN), Dynamic Time Warping (DTW), Artificial Neural Network (ANN) and SVM. The SVM (Cortes & Vapnik 1995) based noise classification methods use the parameters extracted from (1) and (2) where with sufficient examples of previous events database, classifiers are trained. These SVM based classifier techniques are primarily dependent on linear and nonlinear functions (Gaussian, Radial Basis Function, Sigmoidal Basis Function, Polynomial kernel function) for its kernel calculation (Keerthi & Lin 2003). Here the data vectors x_i with labels y_i is mapped to $Ø_n(x_i)$ in another dimensional space using an optimal hyperplane R^n so as to separate the data with positive and negative labels and sufficient margins while minimizing the data points falling in this margin. The hyperplane y_i for linear SVM (or nonlinear SVM) can be defined using (3) where x_i (or $Ø_n(x_i)$ for nonlinear SVM) and w represents data vectors and m-dimensional weighting vector of the hyperplane respectively while b represents bias point (Hearst et al. 1998). An additional cost factor, C can be added to improve the margin of the hyperplane. The margin can be found out using (3) and defined as 2/‖w‖.

$$Y_i = \begin{cases} f(w_i^T.x_i + b_i) > 1 \ \ if \ y_i = 1 \\ f(w_i^T.x_i + b_i) < -1 \ \ if \ y_i = -1 \end{cases} \quad -(3)$$

Here function f can be a linear or complex nonlinear function depending on the overlap of data points. If $y_i = 1$, data is accepted otherwise data is discarded.

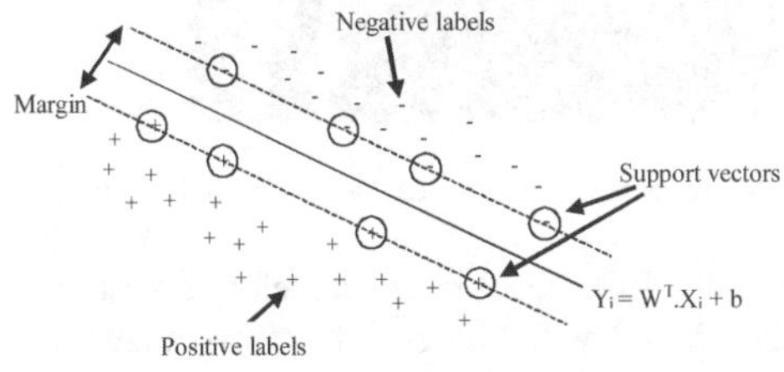

Figure 2. Linear SVM mechanism

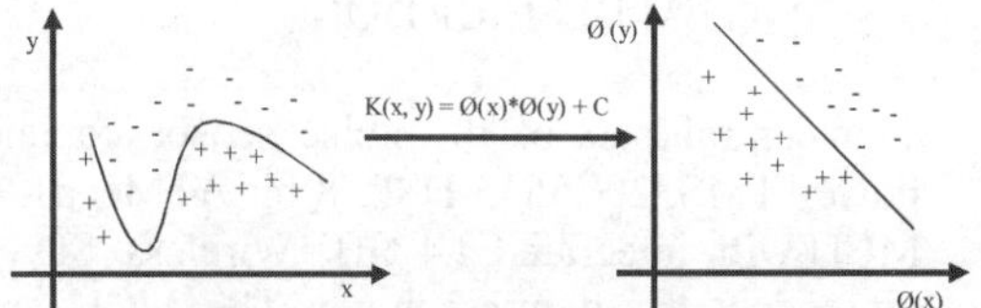

Figure 3. Non-linear SVM mechanism

As shown in Figure 2, the data can be easily separated using a linear equation while data being overlapped, a nth order polynomial as shown in Figure 3 can be defined to separate data points and using a suitable kernel function K(x, y) based on the previous training data sets with minimal training error, the polynomial space can be mapped to linearly separable data points.

2.3 *Acoustic camera SeeSV-S205*

The SeeSV-S205 as shown in Figure 4 is a noise capturing acoustic camera which can produce a noise hotspot representation of the area within its line-of-sight and records continuous audio stream in real-time for 180 second.

Figure 4. SeeSV-S205 Audio camera

The system has an in-built high resolution camera along with thirty numbers of highly sensitive digital MEMS microphones integrated with FPGA based high speed beamforming technology (Christensen & Hold 2004). The system has also pre-built software with real time frequency adjustment, auto image ranging, and linear/exponential image averaging etc. for real time audio and video processing. Once the file is captured and stored in Technical Data Management Streaming (TDMS) format it can be converted in Audio Video Interface (AVI) and Waveform Audio (WAV) file format for easier processing. A fixed audio threshold level from the WAV file is used to identify and record the noise event. These all format conversion and processing can be achieved using Matlab script or LabVIEW.

3 SYSTEM DESCRIPTION

The central part of the noise sensor contains low power TMS320C5515 DSP, 8 bit ATMega128RFA1 MCU with integrated 2.4 GHz Wireless, SD card, 32 bit codec driven microphone. The MCU and DSP communicates through configurable Universal Asynchronous Receiver/Transmitter (UART) interface. During its normal operation including the noise capture, its processing and storage operation in SD card, various blocks (such as core and LDO supply, IO and Memory interface, PLL, USB and Codec) of the DSP IC consumes about 14.03 mW/4.25 mA while various external interfaces (such as audio interface through codec, SD card, Clock and Memory) attached to DSP IC consumes about 264 mW/80 mA. Similarly MCU ATMega128RFA1 consumes 168 mW/51 mA. Hence the total power consumption of the DSP IC, the external interfaces and ATMega128RFA1 MCU is about 446.03 mW/135.25 mA. Since the source input for the noise sensor board is 5 V, the total power consumption at the source side is 446.03 mW/89.265 mA. The PCB as shown in Figure 5 has been designed to be powered either through a solar panel or battery. Considering the construction site where the construction activities typically last for about three to four weeks, the system has been powered using a 12 V, 33 Amp-hour battery providing continuous operation up to 37 days. If the power down mode of the system is utilized through software, the system can last for several months. As part of DSP applications development using Code Composer Studio (CCS), various functions to initialize UART communication, baud rate configuration, time and date stamp using an on-board Real Time Clock (RTC), SD card initialization, SD card read and write, noise capture using a microphone attached to the system, noise event transmission using its wireless radio and events logging to local storage and remote server has been achieved. When the DSP receives an audio signal on its microphone, it performs Fast Fourier Transform (FFT) over a frame consisting of 1024 fixed data points by means of sliding window mechanism as mentioned in section 2.1 and 2.2, and as per the difference of the amplitude of the frequency spectrum with the noise threshold (65 dB in our case), the DSP algorithms identify the type and magnitude of noise. The processed result is then sent to the MCU over UART for events generation as well as local storage in SD card and the inbuilt wireless radio within the MCU transmits the data to a gateway. The gateway then sends the data to a repository of a Raspberry PI based central server in real-time using 3G/4G modem card.

Top side of the PCB

Bottom side of the PCB

Figure 5. Wireless based MCU system with DSP

4 DEPLOYMENT AND RESULTS

The DSP based noise monitoring systems along with the acoustic camera SeeSV-S205 has been deployed for its construction activity as shown in Figure 6 at an upcoming James Dyson building in the Engineering department at University of Cambridge for the noise monitoring and logging. The Gateway was kept at a distance of about 30 m within the line of sight for its radio communication to the DSP system while the server was located within the department at a central location.

Figure 6. Noise sensor deployment at James Dyson Building

Since the result of the DSP based developed system has to be compared with the result of this acoustic camera, the window time of 180 second has been kept deliberately for the recording. In addition, the total number of the noise events within this 180 second time window has been manually counted for the efficiency calculation of the frequency selective and machine learning approach of the DSP based developed system and acoustic camera. Efficiency is defined in percentage as (the number of noise events recognized)/(total number of noise events counted manually)x100.

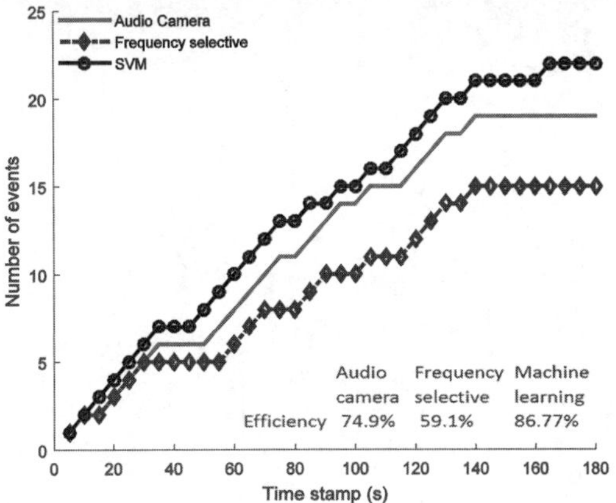

Figure 7. Real construction events at James Dyson Building

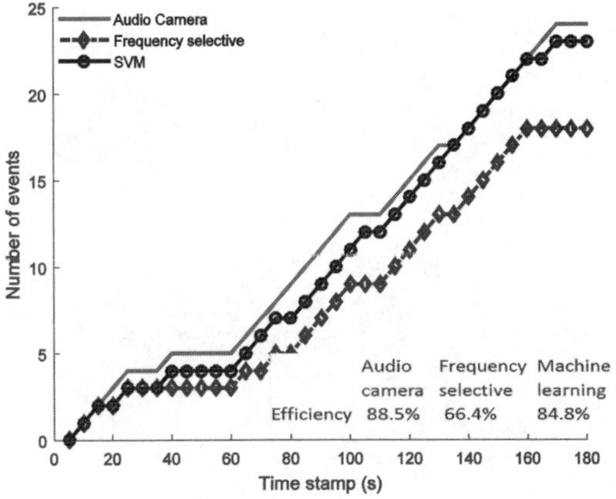

Figure 8. Controlled construction events at James Dyson Building

For the SVM approach, the developed SVM algorithm was trained with various construction noise video available at YouTube in order to enhance and optimize the algorithm for the real site deployment. The noise recording was obtained under two separate conditions, real and controlled construction events. During the real construction activity, the noise generation event within the time window was unknown and it depended on the type of construction works and noise generation while in controlled construction the noise activity was certain to happen within the time period.

5 DISCUSSION AND CONCLUSION

Here a low power DSP based and MCU integrated with wireless radio solution has been designed to capture the noise, store the raw data into SD card and send the noise event to a remote location using a Raspberry PI server. The processing of noise in the DSP card has been obtained using frequency selective and SVM methods. As shown in Figures 7 and 8, the number of noise events for a real and controlled construction activity using the Frequency Selective and SVM methods has been compared with an accurate audio camera SeeSV-S205. The SVM based machine learning algorithm shows a higher efficiency of 86.77 % and 84.8 % for real and controlled construction activities with the identification of the total number of noise events while the frequency selective approach shows a poor efficiency of 59.1 % and 66.4 %. Thus the total number of noise captured events from the developed SVM method in DSP card compares reasonably well with the audio camera SeeSV-S205. Future works include the refinement of these noise recognition algorithms for identifying and discriminating various types of noise such as Piling, demolition, truck reversal, vibration machine and public address system and deployment of the system at other upcoming construction sites.

ACKNOWLEDGEMENT

This research was funded and supported by Centre for Smart Infrastructure and Construction (CSIC) at the department of Engineering, University of Cambridge. We would specially like to thank Mr Paul Fidler from CSIC, Department of Engineering for his support in software development and system deployment. The authors would also like to thank Mr Jason Shardelow and Mr. Peter J knott for their deployment supports and trial of the system.

REFERENCES

Byun, H. & Lee, S. W. 2003. A survey on pattern recognition applications of support vector machines. *International Journal of Pattern Recognition and Artificial Intelligence*, 17(3), 459-486.

Christensen, J. J. & Hold, J. 2004. "Beamforming," *Bruel & Kjaer Technical Review*. 1-2004, 3-35.

Cortes, C. & Vapnik, V. 1995. Support-vector network. *Machine Learning Journal*, 20, 273-297.

Fahy, F. & Walker, J. 2004. Advanced applications in Acoustics, *Noise & Vibration*, London: Spon. Press, 187.

Hearst, M. A., Dumais, S. T., Osman, E., Platt, J. & Scholkopf, B., 1998. Support vector machines, in *Intelligent Systems and their Applications, IEEE*, 13(4), 18-28.

Keerthi, S. & Lin, C.-J. 2003. Asymptotic behaviors of support vector machines with Gaussian kernel, *Neural Comput.* 15(7), 1667-1689.

Shen, L.-Y., Lu, W.-S., Yao, H. & Wu, D.-H. 2005. A computer-based scoring method for measuring the environmental performance of construction activities. *Automation in Construction*, 14:297-309.

Zwan, P. & Czyzewski, A. 2008. Automatic sound recognition for security purposes. In: *Proc. 124th Audio Engineering Society Convention* Amsterdam.

Spectrum Digital, TMS320C5515 eZDSP USB Stick Technical Reference, 512845-0001 Rev A II 2010
http://support.spectrumdigital.com/boards/usbstk5515/reva/files/usbstk5515_TechRef_RevA.pdf

Atmel, "Atmega128rfa1: 8-bit AVR microcontroller with low power 2.4 GHz transceiver for Zigbee and IEEE 802.15.4," datasheet, 2001. [Online].
Available: http://www.atmel.com/Images/doc8266.pdf

Code Composer Studio (CCS), Texas Instrument, 2014

Proceedings of the International Conference on Smart Infrastructure and Construction
ISBN 978-0-7277-6127-9

© The authors and ICE Publishing: All rights reserved, 2016
doi:10.1680/tfitsi.61279.161

Wireless remote condition monitoring opens new monitoring applications in railway tunnel deformation and trackbed

S. Maddison and B. Smith

Senceive Ltd., London, United Kingdom

ABSTRACT Senceive is a developer and manufacturer of wireless sensor networks for remote geotechnical monitoring through the use of low cost, self-contained, self-configuring wireless sensors. These can be installed with a minimum of effort, and without the need for wiring. Considerable advances have been made for example in extending sensor node battery life to 10-15 years, increasing network robustness, data throughput and sensor precision. Additionally over the past two years applications have been extended into two new important and challenging asset classes: Rail Tunnels and Trackbed. In several highly challenging projects for London Underground and Network Rail, Senceive successfully met the asset holders' needs to monitor tunnel deformation during engineering works. The success of these projects has led to other long and short-term wireless tunnel monitoring projects for Senceive. Another growing area is for the monitoring of railway tracks, where Senceive's system has shown the ability to measure changes in both track cant and twist as well as longitudinal rate of change, with extremely high precision and stability. The client favoured the Senceive wireless over optical options, as the data was highly stable and repeatable, not prone to problems of dirt on reflectors, sight lines and far less susceptible to noise, spikes and outliers. Senceive believes these capabilities demonstrate significant steps in the acceptance of wireless monitoring of assets across the rail network, a trend that will continue to accelerate with the resulting economies of scale.

1 INTRODUCTION

Wireless sensor networks (WSN) are now becoming, or indeed have become, the established means of wireless Railway Condition Monitoring (Britton & Maddison 2008) with considerable benefits over the alternatives in terms of ease of deployment, speed of installation, and the ability to readily reconfigure and redeploy.

Prior to new applications for tunnel deformation and trackbed, deployments were typically on assets such as bridges, earthworks, retaining walls and viaducts.

Now with specially designed fixings, deployments can typically be achieved in a single shift or less, and data can be accessed remotely prior to leaving site. Fixings can be adjusted quickly and simply to level tilt meters, where necessary, and for steel lined tunnels employ powerful magnets for attachment. On the trackbed, sacrificial mounting plates are glued down using all weather adhesive in tens of seconds. This minimises track possessions and staff working time in hazardous locations. Furthermore WSN's ability to operate in remote locations enables the monitoring of assets that might well otherwise require regular manual inspection or surveying.

Senceive have made advances in the use of wireless for monitoring of railway tunnels and railway track. In tunnels, wireless has proven itself through its speed of installation and flexibility and lack of wires during intense engineering works, and offering a monitoring solution in confined spaces where there was no viable alternative. On track, for the monitoring of cant and twist, wireless now provides a good, if not better alternative to traditional automated optical monitoring through its high stability, high resolution of measurements and lack of cleaning, maintenance or configuration requirements, as attested by repeated use from satisfied users.

This paper describes these applications in greater detail, drawing on the experience of a number of deployments, namely the Bond Street to Baker Street tunnel relining and the Shell Centre redevelopment in London Underground, track lowering in the Box Tunnel for Network Rail, and at Pudding Mill Lane for track monitoring for Crossrail.

2 WIRELESS STRUCTURAL MONITORING SYSTEM

Acceptance of wireless has been and still is, in many instances, held back by some wireless solutions being unreliable, having noisy and unstable data, being difficult to install, showing insufficiently robustness or scalability, having short battery life and even the continued need for mains power for parts of the system.

Senceive has solved these problems, from integrating the lessons learned from some 10 years of challenging geo-technical deployments on bridges, earthworks and structures. These include practical issues with fixings, wireless propagation, deployment and configuration, sensor integration and energy consumption. These lessons were incorporated in a wide ranging and extensive development programme over the past 3 years, resulting in the launch of Senceive's 3rd generation step change FlatMesh system in 2014.

Whilst the system has always been a "true mesh", that is to say a non-hierarchical, self-organising mesh architecture, recent improvements impact all aspects of the system. The firmware installed on each wireless node enables a network of such nodes to communicate efficiently, providing network-wide data transmission, which ultimately provides an integrated monitoring system. Mesh networks have a "flat structure", whereby every node communicates with its neighbours on an equal status basis, without hierarchy and without the need for any specialised router or concentrator nodes, see Figure 1. This enables simple and quick installation, and robust, self-organising operation. Extensive development of proprietary algorithms within the node enable low power consumption, yielding an unprecedented 10-15 years of operational life, even at a reporting rate of 15 minutes, powering both high precision sensors and wireless communications from a single lithium 19Ah 'D' cell battery. In the event of a node failing or radio paths being blocked, data is automatically re-routed to ensure a seamless flow of information back to the user, with near 100% data transmission reliability.

Design challenges to integrate a wide range of different sensor types, and to operate them with extreme energy efficiency have been successfully surmounted, and system improvements continue to be made. A single node can now be fitted with three sensor interface modules, and each of these modules can support different sensor types and applications, each with one or more sensor channels. Developments continue to extend this range of sensors.

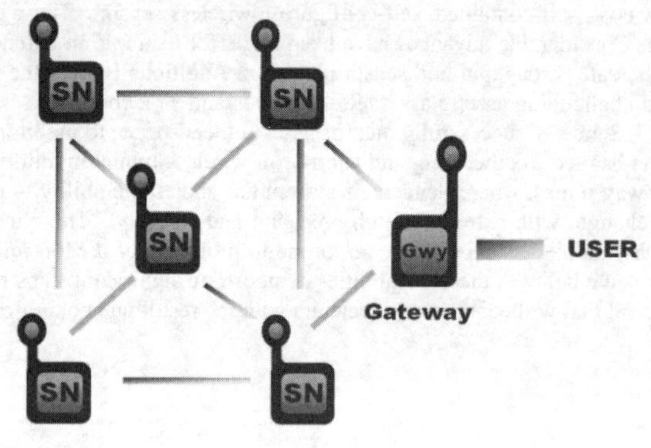

Figure 1. Key elements of FlatMesh system showing Sensor Nodes (SN)

3 TUNNEL MONITORING DURING ENGINEERING AND CONSTRUCTION WORKS

Several recent projects on London Underground (LU) and Network Rail demonstrate the effectiveness of wireless monitoring in Tunnels: Bond Street – Baker Street tunnel relining, the Shell Centre demolition and re-development, and the Box Tunnel electrification are discussed in more detail in this section.

3.1 *Bond Street – Baker Street Tunnel Relining*

This was a highly innovative and challenging 3 year project for tunnel lining replacement which was recently completed by LU / Tubelines. A 200m length of tunnel lining comprising concrete segments was progressively replaced with SGI segments, shift by shift in engineering hours, using two specially designed engineering trains. Senceive's wireless tilt sensors were deployed ahead of the works, leapfrogging ring by ring as the works progressed, and further rings of sensors were deployed behind on the newly completed rings, some following the work as it moved along the tunnel, and some being left behind long-term, to check on the tunnel stability post-works. Ahead of the works, high precision tilt sensors were mounted on lightweight restraining rings that had been installed to support the old concrete lining at its edge. These were used to detect any movement, deformation or potential instability, particularly in the immediate vicinity of replacement works. Sensors were similarly mounted on the new SGI rings to check for the long-term stability of the replacement tunnel lining.

This was an exciting and highly demanding environment where sensors were installed in the immediate proximity of heavy engineering activity (Figure 2), and the sensors were deployed and redeployed by the project's operatives. The system was further developed subsequent to the initial trials, enabling the data to be wirelessly accessed not only from the nearest platform at any time, as was originally required, but also by the engineers on the train during work shifts.

The system was an essential component of the project, as works were not permitted to proceed without accurate monitoring in place, both to provide assurance for the safety of the operations and personnel, and also to see that movements were consistent with those predicted. It was demonstrably easy to use, and the sensors showed extremely good long-term stability, in terms of the measurements taken and the data generated, as expected by the client.

As with earlier tunnel projects, such as the remedial grouting works at Bond Street (Wright et al. 2015), there was no viable alternative that would allow precise monitoring to take place in such close proximity to engineering activity in real time, and provide long-term monitoring during running hours as well.

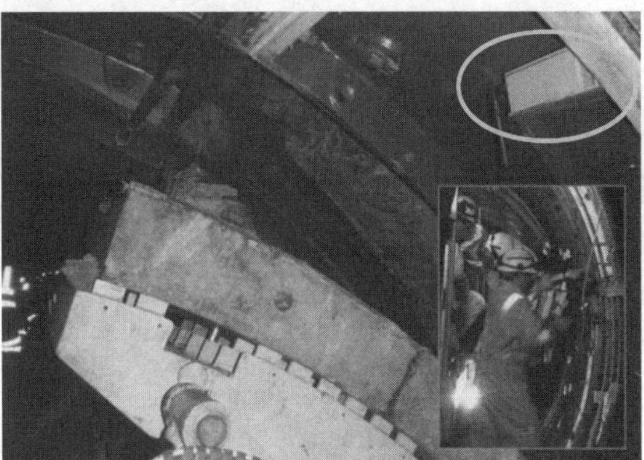

Figure 2. Wireless sensors in the immediate proximity of the tunnel lining replacement machine

3.2 *Shell Centre Redevelopment*

The major demolition and construction works of the Shell Centre will take place over a total of some five years from 2013. This lies immediately above the four London Underground Bakerloo and Northern Line running tunnels, some over 100 years old, and in places piles will be installed within a metre of the tunnel lining. Long-term stability monitoring of these tunnels is required, and to this end high precision tilt nodes are deployed around the tunnel rings, in order to determine deformation, as in the Bond Street to Baker Street relining, above.

Additionally the sensors are fitted along the middle of the track and on the crown (top) of the tunnel in order to determine tunnel and track heave and settlement longitudinally. In this particular case, the long-term stability of the wireless sensors is absolutely essential to determine the behaviour of the tunnel and ensure that it is within the predicted bounds of movement. There is a long history of movement and deformation in these tunnels (e.g. Burford 1988).

In this project, carried out in partnership with Getec, the specialist monitoring part of Keller, the wireless system had been selected over the use of optical methods based on automated total stations (ATS); the very limited space available in the tunnel envelope, coupled with the tunnel curves, meant that this any other method would simply not have been practicable. Although backed up by routine manual surveying, the latter is not only highly labour intensive, but of course only provides snapshots in comparison to the much more frequent measurements carried out automatically by the wireless system.

3.3 Box Tunnel Electrification

Part of the Great Western modernisation program includes electrification of key routes, including Brunel's famous 3km long Box Tunnel (Figure 3), between Bath and Chippenham on the Great Western Line, excavated through four distinct strata and two geological fault zones (Chambers & Chambers, 1839). It was completed in 1841 and subsequently partially lined due to structural instability. There is a complex and irregular geometry with 2km of brick lined sections of varying ages, 350m of unlined sections and 464m of free standing brick arches. In order to achieve clearances for new overhead electric lines, the track was lowered by up to 350mm. The key challenge was to implement an economical, resilient and precise wire free monitoring solution over 3km within a fully operational and congested construction site (Figure 4).

Figure 3. Monitoring system installation in the Box Tunnel

AECOM, responsible for the geological and structural survey, structural analysis, prediction and monitoring, worked closely with Senceive who implemented a wire and mains power free monitoring system incorporating 250 sensors and five gateways.

Monitoring was of critical importance to allow works to proceed, for reasons of safety. For this reason, it was essential that there were no single points of failure within the system. Senceive developed a mechanism of backing up each gateway, so that in the unlikely event of failure it would automatically switch to the standby gateway and ensure continuity of monitoring.

Figure 4. Box Tunnel work site, outside

With no available power or sunlight, an alternative was required to the normal solar powered gateway solution, and a design using very simple cyclical charging was developed. Lightweight rechargeable booster packs were held in the site office, and taken on site periodically at two week intervals. Plugged directly into the power socket of the gateway, the gateway's internal battery was recharged allowing operation for two weeks with a good margin of reserve. The booster packs were then removed at end of shift for recharging. Data was sent every 5 mins and helped verify predicted structural movements, identifying displacement trends before instability occurred, over the full 3km of the tunnel. This assisted project engineers to make decisions to control displacements to within set limits.

Despite extremely limited and constrained access, the system operated throughout the engineering works. Data was collated by five Cellular GSM/GPRS gateways, giving remote access in real time with SMS and email alerts for users in- and outside of the tunnel, free of wires and power, from anywhere with internet or telephone access.

4 RAILWAY TRACK CANT AND TWIST

Figure 5. Wireless sensors deployed on track sleepers to measure cant

The Docklands Light Railway (DLR) station at Pudding Mill Lane in East London was to be demolished and both the track and station be moved to make way for the new Crossrail tunnel that will surface next to it.

Morgan Sindall are responsible for the works, and are required to monitor the multiple Network Rail tracks as well as the DLR tracks at the Pudding Mill Lane site, to ensure that there is no adverse movement to the railway assets whilst the works are in progress, for a period of over 2 years. This was particularly important because of the large secant piling wall being driven into the existing embankment for the Crossrail cut and cover exit ramp. The piling into the embankment, and the subsequent excavation behind it to make way for the tunnel, had the potential of causing settlement in the embankment, with consequential effects on the track in terms of twist and longitudinal settlement.

Following earlier issues of stability and maintenance with an optical ATS system, they sought a new solution for measuring track tilt and twist. A pilot deployment was made using Senceive's high precision wireless tilt sensors, which were attached directly to the sleepers on the original DLR track. By measuring lateral tilt of a rigid sleeper, the relative change in cant from one rail to the other was computed.

The initial 6 month trial successfully showed outstanding stability, and resolution of track cant change to less than 0.1mm on a standard 1.435m track gauge. This was demonstrated by independent measurement using a proxy sleeper, as well as being validated by precise levelling of the tracks.

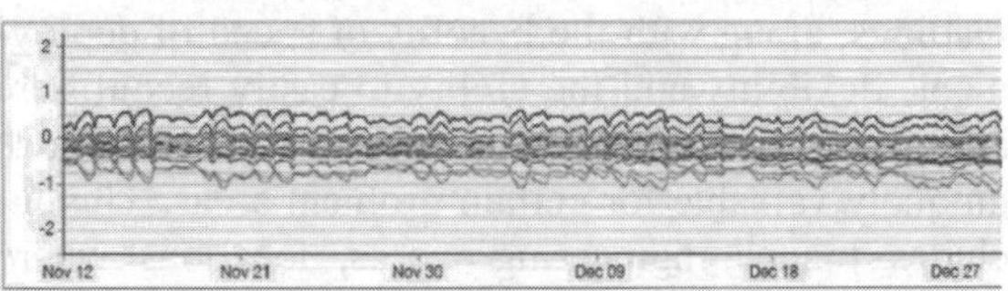

Figure 6. Data from busy rail line, measuring cant over 1.435m track gauge

This led to the design of a hybrid system, with Senceive's wireless tilt sensors mounted on the sleepers at intervals along the track, to measure tilt, twist and longitudinal settlement, cross correlated with a reduced system of optical measurements to detect slew.

This has resulted in the installation of a complete monitoring system incorporating circa 700 sensors extended over several hundred metres of parallel tracks (Figure 5), with a number of solar powered (and therefore wire and mains power free) cellular GSM/GPRS enabled gateways. These gather the data and forward it to a secure web server. The speed of installation has also proved to be a significant benefit, with the installation being conducted quickly and efficiently during limited track possessions. It has also been necessary to move the gateways at different phases of the construction work, with the construction and relocation of protection barriers. Moving the gateways was quickly and readily undertaken without any interruption of service or monitoring.

The cant change measurements are presented directly on a graphical representation. They are additionally processed to generate both relative and absolute twist measurements based on measured and designed geometrical characteristics of the track. A series of trigger levels are configured specific to different track sections which generate standard colour coded visible and email / SMS alerts.

5 CONCLUSION

Step change improvements in terms of reliability, 10-15 battery life, robustness and stability of sensor readings, along with the benefits of speed of deployment, flexibility and the ability to easily reconfigure and redeploy make Senceive's WSN solutions uniquely valuable for certain problem areas. This includes tunnel engineering works, locations where there is extremely limited space, and where there really is no viable monitoring alternative, including labour-intensive manual systems.

This is applicable not only to short-term deployments during engineering works for example, but also to long-term monitoring projects where operation for many years is required.

Moving forward, wireless is displacing or indeed often complementing other wired or optical solutions where they might previously have been seen as the first and only choice.

The prime benefit of wireless is of course that there are no wires. In the railway environment in particular this minimises personnel hazards, track access and possessions, and allows for the monitoring of difficult to access assets, where it might otherwise not even be cost effective to attempt automated monitoring. Senceive believes these capabilities demonstrate significant steps in the acceptance of wireless monitoring of assets across the rail network, a trend that will continue to accelerate with the resulting economies of scale.

ACKNOWLEDGEMENTS

Senceive would in particular like to thank the following for their support: London Underground / Tubelines; Network Rail; Costain; Morgan Sindall; C2HM Hill; Arup; Getec/Keller; Amey; TPS Consult/Carillion; SAL; Murphy Surveys; ITM; Soldata; SES; IPC and the Technology Strategy Board (Innovate UK).

REFERENCES

Britton M. & Maddison, M.S. 2008.Towards the reality of Intelligent Infrastructure with Wireless Meshed Sensors, 4th IET International Conference on Railway Condition Monitoring, IET London.

Burford D. 1988. Heave of tunnels beneath the Shell Centre, London 1959-1986. Géotechnique Volume 38 Issue 1, pp. 135-137

Chambers W., Chambers R. 1839.Great Western Railway – the Box Tunnel.Chamber's Edinburgh journal, 396 (Aug 31, 1839): 254-255.

Wright P., Skipper J., Maddison S., Brookfield J. & Moss N. 2015. An observational approach to railway tunnel voids grouting, Geotechnical Engineering 168 Issue GE3, pp 201-214.

Proceedings of the International Conference on Smart Infrastructure and Construction
ISBN 978-0-7277-6127-9

© The authors and ICE Publishing: All rights reserved, 2016
doi:10.1680/tfitsi.61279.167

Dynamic characteristics of a damaged pedestrian overpass measured by wireless sensor array

Y. Miyamori*[1], T. Kadota[2], Y. Zhang[3] and S. Mikami[1]

[1] *Kitami Institute of Technology, Kitami, Japan*
[2] *Oriental Consultants Co., Ltd. Tokyo, Japan*
[3] *Northeast Dianli University, Jilin, China*
* *Corresponding Author*

ABSTRACT In this study, a fundamental attempt for structural health monitoring was performed on a damaged pedestrian overpass by using wireless sensor network system. The pedestrian overpass had local damage on its girder. The vibration measurement was repeated before and after the repair work of the girder with 14 Imote2 wireless sensors. Tri-axial acceleration were measured by free damped vibration after human excitation. Natural frequencies, damping ratio and mode shapes of plural vibration mode were identified from measurement result. Its natural frequencies of each mode did not change by the repair work. Mode amplitudes were slightly changed in several vibration modes. A detailed finite element model was constructed to reproduce the bridge analytically and vibration characteristics were compared to the measurement results. In the model, every members including non-structural members were modeled. The structural damage was measured for repair designing and these damage was reproduced in the numerical model. Eigenvalue analysis and time history dynamic response analysis were conducted. A detailed FE model could reproduce the actual bridge. The error of natural frequencies were 4% at maximum up to 6th mode. However changes of natural frequencies and modal shape by repair work were not fully corresponded. In certain analytical case, dynamic response was changed at the damage location. However simulation results were not fully corresponded with the vibration experiment. The modeling error were need to be examined in the future study.

1 INTRODUCTION

As the deterioration of infrastructure is advancing rapidly in Japan, the maintenance strategy has been shifted to preventive maintenance from traditional breakdown maintenance. In such circumstances, it is expected that SHM will overcome the deficiencies of visual inspection by humans. Vibration-based SHM has been investigated as a way to assess damage by comparing the vibration characteristics between the past and present state of a structure (Boller et al. 2009; Doebling et al. 1996; Oshima 2000).

To practical application of SHM, it is necessary to establish a technique that is capable of detecting size and position of the damage from measurement result. In previous researches, various SHM techniques have been proposed and applied to actual bridges (Furukawa et al. 2007; Yoshioka et al. 2010). However, examples in girder type bridges are not enough although these girder bridges are the most common type in short or medium span length. Therefore, in this study, we conducted fundamental study of SHM by using smart acceleration sensors and high resolution FE model of a pedestrian overpass.

The pedestrian overpass had local damage on its girder. The vibration measurement was repeated before and after the repair work of the girder with Imote2 wireless sensors (Nagayama & Spencer, 2007; Rice & Spencer, 2009). Tri-axial acceleration were measured by free damped vibration after human excitation. Natural frequencies, damping ratio and mode shapes of plural vibration mode were identified from measurement result. A detailed finite element model was constructed to reproduce the bridge analytically and vibration characteristics were compared to the measurement results. In the model, every member including non-structural members were modeled. The structural damage, i.e. deformation and crack length, was measured for repair designing and these damage was reproduced in the numerical model. Eigenvalue analysis and time history dynamic response analysis were conducted. These results were compared and discussed.

2 BRIDGE VIBRATION MEASUREMENT

2.1 *The pedestrian overpass and its damage*

This pedestrian overpass as shown in Figure 1 is located in front of an elementary school for overpassing a national highway for students and other pedestrians. Its span length is 18.48m, width is 1.5m. The main superstructure is composed by the steel floor deck on two C-shaped steel girders. One girder had deformation and crack at the web as shown in Figure 2. This deformation was measured to encompass the damaged area at 100mm interval, and it was proved deformation in the out-of-plane direction about 160mm at most. A crack developed from edge of the lower flange toward the web, the crack length was about 200mm.

2.2 *Wireless sensor array and measurement setup*

Location of sensors and position of excitation is shown in figure 3. To identify the detailed vibration characteristic of the bridge, 14 sensors were arranged on both side in walkway of the span at equal interval. 8 MEMSIC SHM-H sensors were placed on middle of the span and 4 SHM-A sensors were located on the piers. In order to measure the three-axis acceleration component at all stations, the number of channels became 42chs. The vibrational characteristic was identified by using the damped free vibration waveform

Figure 1. The pedestrian overpass.

Figure 2. Damaged section (Upper: damaged, bottom: repaired).

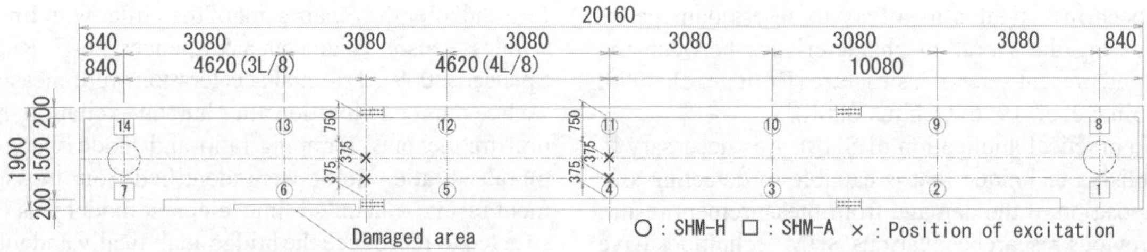

Figure 3. Plan view and sensor distribution

Table 1. Dynamic characteristics in the experiment.

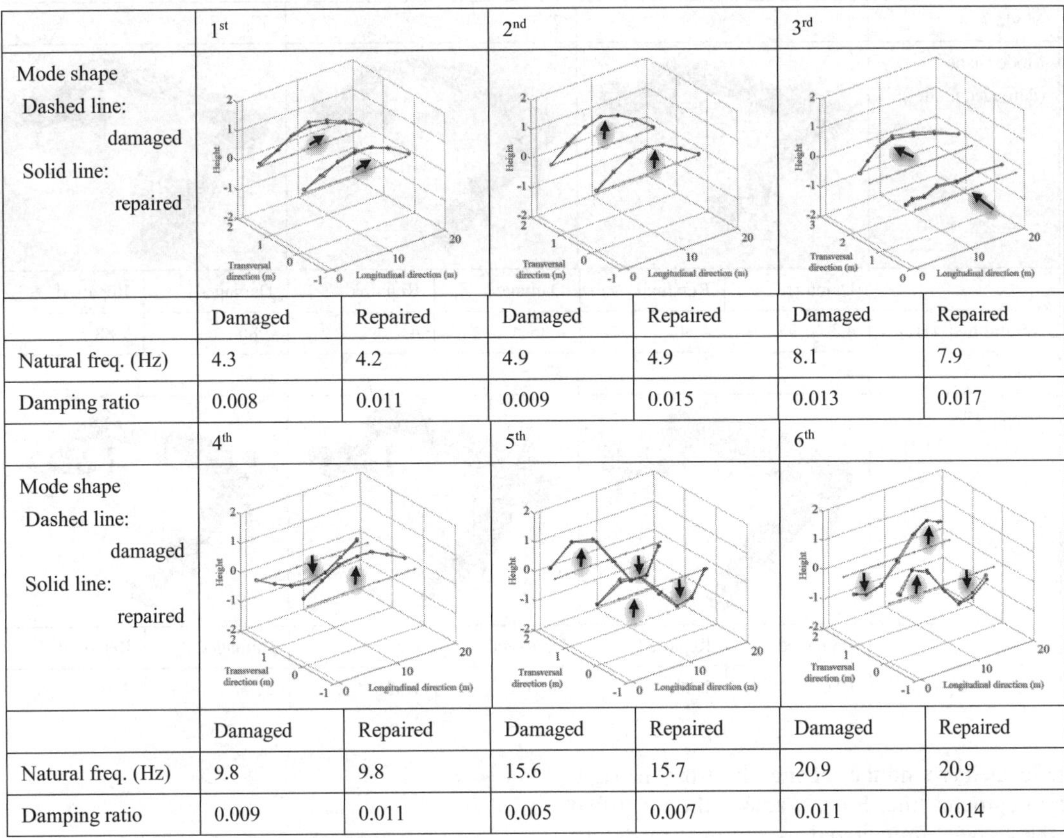

	1st		2nd		3rd	
Mode shape Dashed line: damaged Solid line: repaired						
	Damaged	Repaired	Damaged	Repaired	Damaged	Repaired
Natural freq. (Hz)	4.3	4.2	4.9	4.9	8.1	7.9
Damping ratio	0.008	0.011	0.009	0.015	0.013	0.017
	4th		5th		6th	
Mode shape Dashed line: damaged Solid line: repaired						
	Damaged	Repaired	Damaged	Repaired	Damaged	Repaired
Natural freq. (Hz)	9.8	9.8	15.6	15.7	20.9	20.9
Damping ratio	0.009	0.011	0.005	0.007	0.011	0.014

which caused by human jumping of a 60kg person. The sampling frequency of measurement was 280Hz, and the measurement time was 30 seconds per sample. In order to excite the 1st and 2nd bending modes and 1st and 2nd torsional modes, the person jumped at the position of the mark ×.

2.3 *Data analysis*

All measured data in 14 wireless sensor nodes were collected to the laptop PC via gateway node connected to the PC. The acceleration time history records were verified and their power spectra were calculated. The frequency resolution of the spectrum is about 0.1Hz. The natural frequencies of predominant modes were adopted from averaged value of peak of power spectra, because several acceleration data by the difference of the excitation position were obtained. In order to calculate the damping ratio from the logarithmic decrement, damped free vibration of each predominant mode was extracted by filtering of the IIR elliptic bandpass filter. Mode shapes were also obtained from relative displacement of filtered damped free vibration in each vibration mode.

3 DYNAMIC CHARACTERISTICS FROM MEASUREMENT

The vibration characteristics (natural frequency, damping ratio, mode shape) from the 1st to 6th mode were identified from the measurement and these results are shown in Table 1. Dashed line indicates mode shape in damaged state that is before repair work, solid line shows mode shape in intact state after repair. Also the plots in figures indicate measurement points, the

Table 2. Mode shape and natural frequency of the FE model.

Mode #	1st		2nd		3rd	
Mode shape (damaged model)						
	Damaged	Repaired	Damaged	Repaired	Damaged	Repaired
Natural freq. (Hz)	4.30	4.30	4.75	4.78	7.86	7.88
Mode #	4th		5th		6th	
Mode shape (damaged model)						
	Damaged	Repaired	Damaged	Repaired	Damaged	Repaired
Natural freq. (Hz)	10.07	10.07	16.16	16.35	21.58	21.61

damage located around the second plot from the right end of foreground line. From Table 1 the identified mode shapes are summarized as follows. In the 1st mode, the modal amplitude most predominates in longitudinal direction although vertical component was also caused by human excitation. The 2nd mode was vertical symmetric bending mode. The 3rd mode was vibration in transversal and vertical direction. Amplitudes of 2 girders were different directions because of the stairs side girder was restrained in horizontal direction. The 4th mode was torsional vibration mode. The mode had also an amplitude component in the transversal direction. The 5th mode was asymmetric bending mode with an amplitude in the transversal direction. The 6th mode was torsional mode. From above results, it was possible to identify a detailed vibration characteristics by a three-dimensional measurement.

Natural frequencies did not change so much according to existence of the damage. In the 1st and 3rd mode, natural frequencies decreased and it increased in the 4th mode by repair work. From basic relationship between structural stiffness and natural frequency,

Figure 4. FE model. (Upper: bird view, bottom: detail of upstairs)

natural frequencies should increase by recovery of structural stiffness. However the experimental results does not support this concept clearly because the damage was limited in local area.

Modal amplitudes in the 1st and 4th mode slightly changed around damage location. This changes might be caused by the damage. However modal amplitudes changed at other locations in other modes. Therefore numerical analysis simulated the local damage was conducted in the next chapter.

4 NUMERICAL EXAMINATION

4.1 *Detailed FE model and dynamic properties*

Analytical model was constructed by commercial FEM software Midas NFX as shown in Figure 4 (Kadota et al. 2015). In this model, major members which were C-shape main girders; deck plate; cylindrical piers and so on, were modeled in shell element. Pavement and adjustment mortar were modeled in solid element, steel balustrades and lateral stiffeners on the piers were modeled in bar element. Damaged area was modeled by moving nodes reproducing measured deformation in transversal direction. A crack of main girder web was modeled by using double contact. As a boundary condition, pier base was completely fixed. Connecting parts between main girder and stairs were modeled conveniently as a spring element of the six degrees of freedom.

Eigenvalue analysis was performed to the FE model and its results are summarized in Table 2. Modeling error is minimal in the fundamental mode. In higher modes error still re-mains about 3-4%. The error were larger than changes by the repair work.

In FE model, natural frequencies increase in 4 modes. There is no vibration mode which decreases natural frequency. This result corresponds to the basic theory. Therefore the experimental differences of dynamic characters by the repair is need to be considered with other effects than stiffness reduction due to the damage.

4.2 *Dynamic response analysis*

To reproduce the free vibration experiment after human excitation, dynamic response analyses were also conducted in the FE model. The Newmark β method

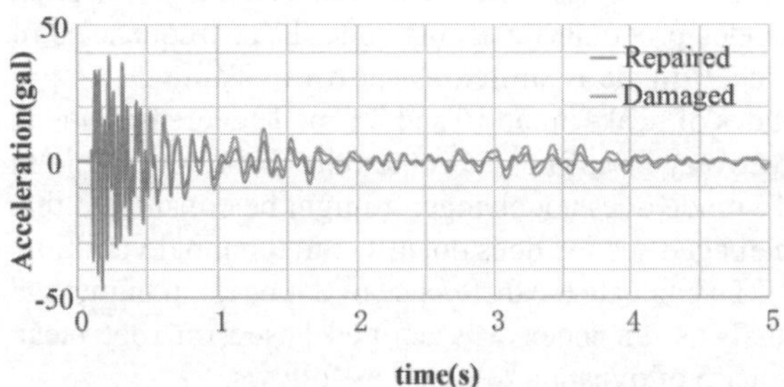

Figure 5. Time history acceleration at CH6 x-direction.

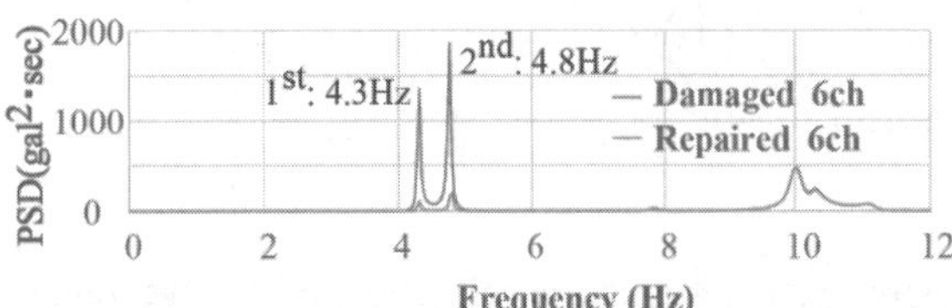

Figure 6. Power spectra of acceleration at CH6 x-direction.

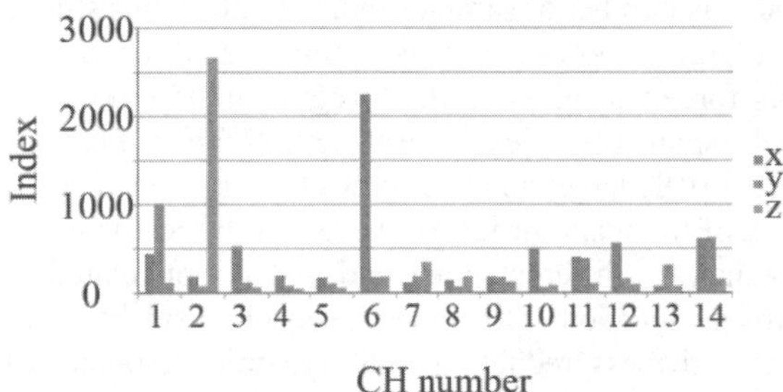

Figure 7. Index of change of acceleration response.

was adopted and time interval Δt was 1/280s considering to the experimental sampling frequency 280Hz. Calculation time was 24s.

Numerical simulations were conducted in 4 cases corresponding to excitation points as shown in Figure 3. In each case, an impulsive excitation force was given at the node of FE model near the excitation point. The magnitude of input was 7500N which was decided to agree with the maximum acceleration of the measurement.

Figures 5 and 6 are results in the case which excitation force was given at the foreground side in center of the span. Figure 5 is x-direction time history acceleration at the node corresponding to the sensor #6. This node located near the damaged point. Power spectrum of this acceleration response is shown in Figure 6.

Time history curve shows free vibration after impulsive excitation and the curve differs between 2 models after 1.5s. In the power spectra, 3 vibration modes predominate at this node. Peak frequencies of the 2nd mode is shifted as shown in Table 2. Amplitudes of peaks in the 1st and 2nd mode decreases due to recovery of stiffness. On the other hand amplitude in 4th mode does not change. It might be considered that damaged section does not affect to torsional vibration.

To evaluate the difference of 2 models in numerical analysis, an index was defined based on root mean square of dynamic response as follows.

$$Index = \max\left|\sqrt{\frac{\sum_{i=1}^{N}(x_i/y_i)^2}{N}}, \sqrt{\frac{\sum_{i=1}^{N}(y_i/x_i)^2}{N}}\right| \quad (1)$$

where, x is response acceleration of before repair model, y is response acceleration of after repair model and N is number of samples that is 6720 in this study.

Figure 7 shows the index in the case which excitation force was given at the foreground side in center of the span. The index value at sensor #6 in x-direction is relatively larger and this comes from change of stiffness of the main girder. However the largest value is obtained in different point and that is not related to damage location. Therefore the mechanism between local stiffness reduction and dynamic response of structure need to be examined in the future study.

5 CONCLUSION

In this study, a fundamental attempt for structural health monitoring was performed on a real damaged pedestrian overpass by using wireless sensor network.

The vibration measurement was repeated before and after repair work of the girder with 14 Imote2 wireless sensors. Natural frequencies, damping ratio and mode shapes of plural vibration mode were identified from measurement result. Its natural frequencies of each mode did not change so much by the repair work. Mode amplitudes were slightly changed in several vibration modes.

A detailed finite element model was constructed to reproduce the bridge analytically and vibration characteristics were compared to the measurement results. In the model, every member including non-structural members were modeled. The structural damage was also modeled. A detailed FE model could reproduce the actual bridge. The error of natural frequencies were 4% at maximum up to the 6th mode. However change of natural frequencies and modal shape by repair work were not fully corresponded. In certain analytical case, dynamic response was changed at the damage location. However simulation results were not fully corresponded with the vibration experiment. The measurement condition need to be considered carefully for such minor structural change. Modeling of structure and excitation condition were also need to be examined in the future study.

ACKNOWLEDGEMENTS

This research was supported by JSPS Grant-in-Aid for Scientific Research Grant Number 25870025. Vibration measurements in the pedestrian overpass were conducted with the supports from Hokkaido Development Bureau and students in Kitami Institute of Technology.

REFERENCES

Boller C., Chang F. & Fujino Y., 2009. *Encyclopedia of Structural Health Monitoring, Volume 1*. A John Wiley and Sons, Ltd.

Doebling S.W., Farrar C.R., Prime M.B. et al. 1996. Damage identification and health monitoring of structural and mechanical systems from changes in their vibration characteristics. A literature review, *Los Alamos National Laboratory Report*, **La-13070-Ms**.

Furukawa A., Otsuka, H., & Umebayashi, F. 2007. Verification of damage identification technique based on Fourier amplitude ratios due to unknown excitation force using a real bridge, *Journal of Structural Engineering, JSCE*, **53A**, 258-267. (in Japanese)

Nagayama T. & Spencer B.F., 2007. Structural Health Monitoring using Smart Sensors, *NSEL Report* **1**, https://www.ideals.illinois.edu/

Rice J.A. & Spencer B.F., Flexible smart sensor framework for autonomous full-scale structural health monitoring, *NSEL Report* **18**, https://www.ideals.illinois.edu/, 2009.

Kadota T, Miyamori Y, Watasaki R et al. 2015. Structural vibration characteristics of a pedestrian bridge by the 3D acceleration sensing of smart sensors and detailed 3D-FEM model, *Proceedings of the IABSE conference Nara 2015*, Paper No. IA-41.

Oshima T. ed., 2000. *Guidelines for bridge vibration monitoring. Subcommittee on bridge vibration monitoring*, Committee of structural engineering, Japan Society of Civil Engineers, Tokyo. (in Japanese)

Yoshioka, T., Itou, S., Yamaguchi, H. et al. 2010. Structural health monitoring of truss bridges based on damping change in diagonal member-coupled mode, *Journal of JSCE, Division A*, **66-3**, 516-534.

Proceedings of the International Conference on Smart Infrastructure and Construction
ISBN 978-0-7277-6127-9

© The authors and ICE Publishing: All rights reserved, 2016
doi:10.1680/tfitsi.61279.173

Wireless sensor network based pipeline failure detection system using non-intrusive relative pressure and differential temperature measurements

A.M. Sadeghioon*[1], N. Metje[1] , D.N. Chapman[1] and C.J. Anthony[2]

[1] *School of Civil Engineering, University of Birmingham, Birmingham, United Kingdom*
[2] *School of Mechanical Engineering, University of Birmingham, Birmingham, United Kingdom*
* *Corresponding Author*

ABSTRACT Aging infrastructure and a move towards sustainability have created new challenges for asset owners. On the other hand, advances in technology, such as low power electronics and sophisticated sensing methods, have made health monitoring of large infrastructure technologically feasible. However, despite the large body of research in this area infrastructure monitoring has not been widely adopted by asset owners of buried infrastructures. This is mainly due to complexity and high combined cost (production, installation and maintenance) of these systems. This paper describes the operation of a proposed non-intrusive relative pressure sensor, based on utilising low cost Force Sensitive Resistors (FSR). The paper also presents a novel method for detecting abnormal flow in pipes based on monitoring of the temperature differential between the pipe wall and its surroundings. The performance of these relative pressure sensors in conjunction with multiple temperature sensors has been assessed by deploying them for a period of 6 months on buried pipes at a water industry leak test training facility. The results showed that the proposed pressure sensors registered the expected systematic and daily pressure fluctuations in the network. In addition, the results showed that the calculated temperature differentials can be successfully used to detect abnormalities in the flow. This in conjunction with the relative pressure readings from the FSR based pressure sensors was shown to have the potential to be used to separate normal pressure variations from abnormal changes caused by pipe failure to provide a relatively low cost and easy to install monitoring system.

1 INTRODUCTION

Pipeline monitoring, specifically monitoring of water distribution networks is becoming increasingly crucial to achieve a sustainable future. Assessing the condition of these networks and detecting failures effectively can be challenging due to their large scale, age and complexity. Despite there being a large body of research conducted in the field of buried infrastructure monitoring (Liu and Kleiner 2013; Rajani and Kleiner 2004; Sinha and Knight 2004; Sinha et al. 2003) and a variety of commercial techniques (Bond et al. 2004), continuous pipeline monitoring has not been widely used on water pipelines.

The challenges associated with the underground environment of these pipelines (i.e. access to power) and the costs of installation and maintenance are the main factors in preventing these systems from being widely adopted by asset owners. Another common disadvantage of current pipeline monitoring systems is the invasive nature of their sensing techniques, which creates the requirement to access the medium transferred by the pipe (i.e. via tapping to measure pressure). This requirement can potentially cause issues with water quality and creates additional potential failure points on the system.

Various parameters can be measured in order to monitor the health of the buried water distribution networks; such as vibrations, flow, pressure, soil water content, strain, and corrosion (Sadeghioon 2015; BenSaleh et al. 2013). However, although parts of the water distribution networks are pressure managed, the internal pressure of water distribution pipes can fluctuate for a number of reasons. Some of the main reasons for a fluctuation of pressure in the pipes are: fluctuations in demand, intentional systematic pressure variations, pressure transients and pipe failure. Faults in the pipeline such as leaks, bursts and blockages will affect the internal pressure of the pipes (Misiunas 2005). Therefore, internal pressure is a suitable indirect indicator for assessing the operational and structural integrity of the pipes.

This paper describes the design of a novel non-intrusive relative pressure sensor based on utilising low cost Force Sensitive Resistors (FSR), which are attached in a concentric assembly external to the pipe. The performance of the sensor has been previously validated by comparison with commercial direct pressure sensors (Sadeghioon et al. 2014b). This paper presents details of how the performance of these novel relative pressure sensors, in conjunction with multiple temperature sensors (located on the pipe and in close proximity to the pipe), has been assessed by deploying them on buried pipes at a water industry leak test training facility. In addition, a novel method for detecting pipe failure based on monitoring the temperature difference between the pipe wall and its surroundings and relative pressure sensors is presented.

2 THEORY OF OPERATION

As the internal pressure of a pipe changes the internal stress applied to the pipe wall will also change, which results in the pipe expanding/contracting based on the amplitude of the change and the structure of the pipe (geometry and material). Although this expansion/contraction can be very small, the proposed sensing system indirectly uses this change in the radius to detect the changes of the internal pressure of the pipe.

The proposed relative pressure sensor assembly is composed of a stainless steel clip (concentric to the pipe), which tightly holds a Force Sensitive Resistor (FSR) on the exterior of the pipe. As the pipe expands/contracts the clip will also have to expand/contract with the pipe. Therefore, the contact force between the clip and the pipe will change. This change in the contact force is detected by the FSR. As the change in the contact force is caused by the change in the internal pressure of the pipe the output of the FSR sensor can be related to the change in the pressure of the pipe. Figure 1 shows a schematic of the relative pressures sensor assembly.

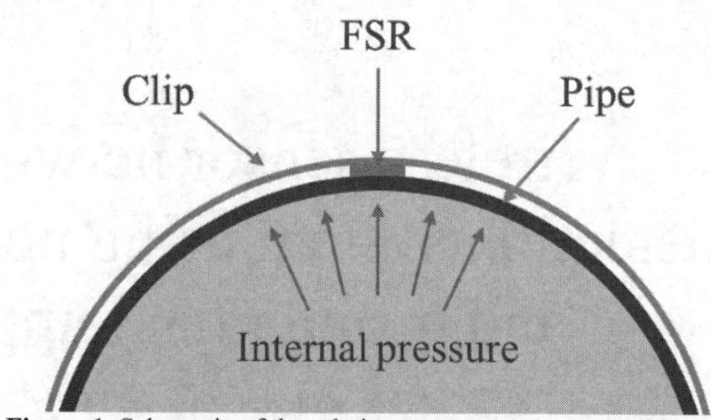

Figure 1. Schematic of the relative pressure sensor assembly

The relationship between the internal pressure of the pipe and the contact pressure between the pipe and the clip can be derived by modeling the pipe and the clip as two pressurized concentric shells with open ends. Using this method, Equation 1 can be used to estimate the contact pressure P_C between the pipe and the clip.

$$P_C = \frac{P.r_p^2.E_j.t_j}{(r_p^2.E_j.t_j)+(r_j^2.E_p.t_p)} \qquad \text{Equation 1}$$

Where P is the internal pressure in the pipe; r_j and r_p are the radii of the clip and the pipe; E_j and E_p are the respective material Young's modulii of elasticity of the clip and pipe and t_j and t_p are the thickness of the clip and pipe respectively. More detail on modelling of the proposed relative pressure sensor can be found in Sadeghioon et al. (2014b) and Sadeghioon (2015).

From Equation 1 the effect of the pipe geometry can be calculated for each specific pipe material. Figure 2 and Figure 3 show the effect of pipe geometry on the applied contact pressure on the sensor for medium density polyethylene (MDPE) and cast iron pipes (at 1 kPa) respectively.

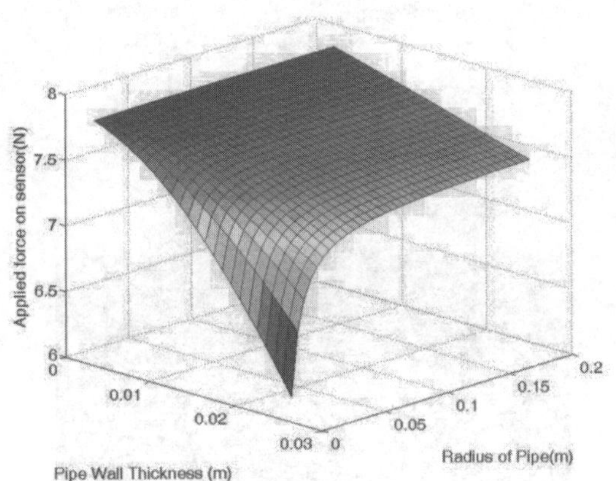

Figure 2. Effect of geometry on the applied force for MDPE pipes (Sadeghioon, 2015)

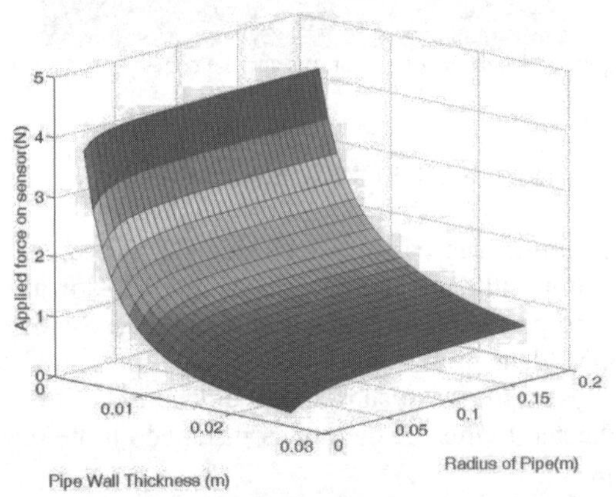

Figure 3. Effect of geometry on the applied force for MDPE pipes (Sadeghioon, 2015)

As shown in Figure 2 and Figure 3 the overall applied pressure on the FSR sensor for the MDPE pipe is significantly higher than the cast iron pipe. This is due to the fact the Young's modulus of the cast iron (Ep=100GPa) pipe is significantly higher than MDPE (Ep=1GPa) and therefore the expansion of the cast iron pipe is smaller than the MDPE pipe. It is also shown in Figure 2 and Figure 3 that the applied force on the FSR sensor decreases with an increase in the thickness of the pipe. However, the effect of an increase in the pipe wall thickness on the contact force is significantly larger for the cast iron pipe than the MDPE pipe. Figure 2 and Figure 3 also show that a decrease in the radius of the pipe results in a reduction in the contact force applied on the sensor. This is due to the fact that the change in the radius of pipes with smaller diameters and higher wall thickness for a given internal pressure is smaller than for pipes with larger diameter and lower wall thickness of the same material. Comparing the effect of the pipe material on the applied force on the sensor indicates that the proposed sensor in its current form is more suited for use on plastic pipes. However, the design parameters of the sensor assembly (i.e. the clip material and thickness) can be modified to compensate for this limitation to produce an increase in the resultant force on the sensor for metallic pipes.

Temperature is one of the other key parameters of any physical system. Pipe networks commonly have higher numbers of leaks/bursts during colder periods of the year (Kleiner and Rajani, 2002). Additionally, the temperature of the pipe is highly affected by the temperature and flow of the medium it carries. A change in the speed of the flow in the pipe results in the change in the residence time of the water in the pipe, which results in the heating/cooling effect of the medium on the pipe wall. This change, however, will have a significantly smaller effect on the temperature of the surrounding soil (i.e. 30cm away from pipe wall) due to the poor thermal conductivity of the soil compared to the pipe. It is proposed that the temperature differential between the pipe wall and the surrounding soil can be potentially used as an indicator of sudden changes in the flow within the pipe. This method when used in conjunction with the relative pressure readings from the FSR based pressure sensors has the potential to be used to separate normal pressure variations from abnormal changes caused by pipe failure.

3 TEST SETUP

The FSR based relative pressure sensors were deployed on plastic and metallic pipes at an industrial leak test facility to investigate their performance. The FSR sensors were laminated to protect them from environmental moisture and corrosion. The laminated sensors were attached to the pipes using stainless steel Jubilee clips.

The temperature of the pipe wall and its surrounding soil (approximately 30cm away from pipe) were also monitored in order to investigate the effect of flow on the temperature of the pipe. **Figure 4** shows the setup of the sensors deployed in these trials.

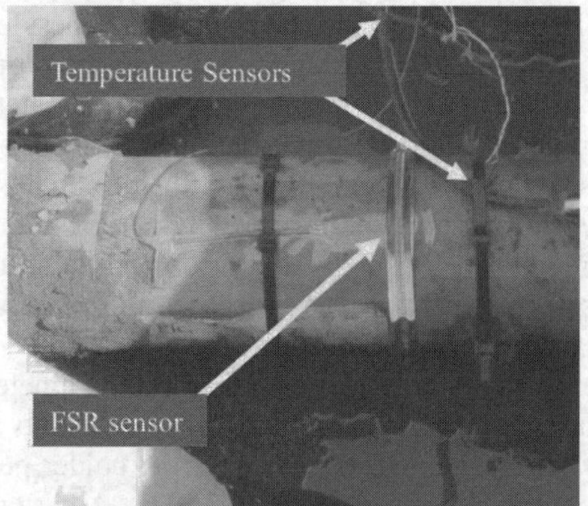

Figure 4. Setup of the sensors on the plastic pipe

Ultra low power wireless sensor nodes (Sadeghioon et al., 2014a) were also deployed during these trials in order to study their performance. These nodes were placed inside a hollow plastic post and were powered using a rechargeable battery. A 1W solar panel was also placed on the top of the post in order to charge the battery.

The nodes were connected to the sensors attached to the pipe under the ground via a multicore twisted pair cable.

Each node was programmed to measure all of the sensors attached to it and the voltage of its battery approximately every 17 minutes. These measurements were transmitted wirelessly to a mother node, which was placed inside a building close to the nodes. The received data was time stamped and stored on a laptop with internet connectivity (for remote access to the data). The sensors were monitored for a period of approximately 6 months. During this period multiple valve operations were carried out as part of personnel training taking place at the facility, during which various valves were opened and closed (creating temporary leaks). **Figure 5** illustrates the schematic of the setup of the node and the sensors attached to the pipe.

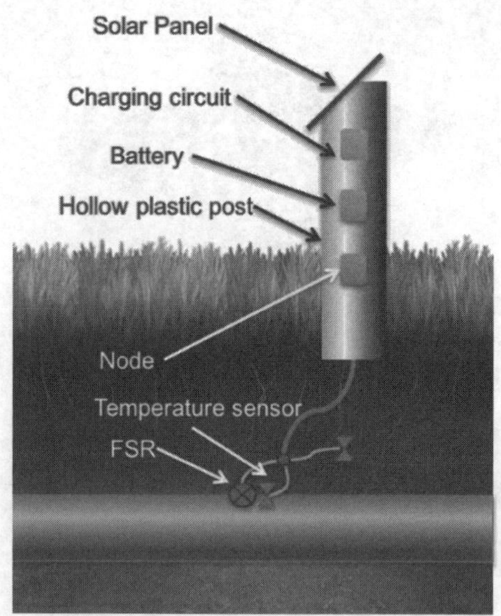

Figure 5. Schematic of the sensors and the nodes

4 RESULT AND DISCUSSIONS

The nodes successfully recorded the output of the sensors, although there were occasional downtimes due to corrosion of contacts. For the purpose of this paper, data from one of the nodes is presented and analysed. **Figure 6** shows a 20 day period of the data collected by one of the nodes installed on the plastic pipe.

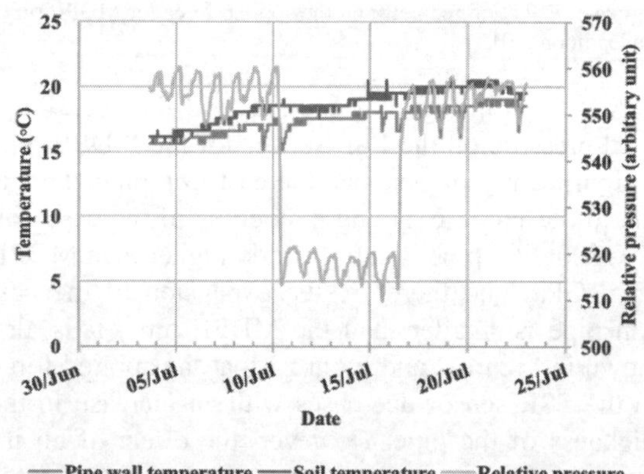

Figure 6. Relative pressure and temperature readings for a period of 20 days associated with the plastic pipe

As can be seen in **Figure 6** the FSR based relative pressure sensor successfully registered the daily pressure variations in the pipe system, which were caused by the fluctuations in the demand. In addition, **Figure 6** shows that systematic pressure changes were also registered by the FSR sensors (10th and 16th July).

During the period of monitoring various valves were opened and closed as a part of valve operation training. These operations created temporary leaks in the system. **Figure 7** illustrates relative pressure and calculated absolute temperature differential (between pipe and soil) for a 6 day period where, during the first three days and the last day, valve training was carried out at the test facility.

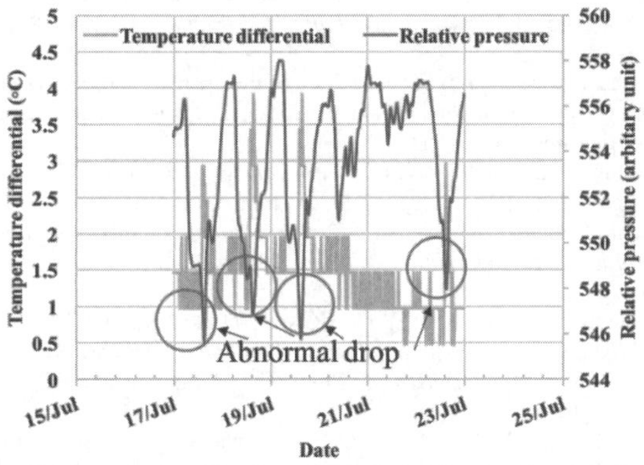

Figure 7. Temperature differential and relative pressure readings for a period of 5 days

It can be seen from **Figure 7** that the induced leaks are clearly detectable as abnormal pressure drops in the relative pressure readings. In addition, concurrent with these recorded abnormal pressure drops, the temperature of the pipe wall also dropped abnormally. However, this drop in the temperature of the pipe wall was not present in the temperature readings from the sensors placed within the soil adjacent to the pipe, leading to an abnormality in the temperature differential as seen in Figure 7. This sudden change in the temperature of the pipe wall was caused by the sudden increase in flow (caused by the leak) which resulted in an increase in the cooling effect of the water. However, due to poor thermal conductivity of the soil this change in temperature was not apparent in the soil temperature measurements.

It can be seen from **Figure 7** that the abnormal relative pressure drops and atypical temperature differential of the pipe only occurred during the days when leaks were induced in the pipe (during the valve training). This was very encouraging as it showed the potential for the relative pressure readings in combination with temperature differentials to be used for the detection of pipe failures and to differentiate them from normal variations. The main advantage of this method is the relative nature of the readings. This removes the need for time consuming and costly precision calibration processes, which can increase the cost of production and installation.

To further simplify the interpretation of the measurements for pipe failure detection, it is proposed that the absolute daily range of the relative pressure and the absolute daily range of the temperature differential is used to interpret the data. **Figure 8** and **Figure 9** show the daily range of the relative pressure and the absolute daily range of the temperature differential for the 5 day period presented in **Figure 7** respectively. The dotted lines represents the threshold for detection of abnormality.

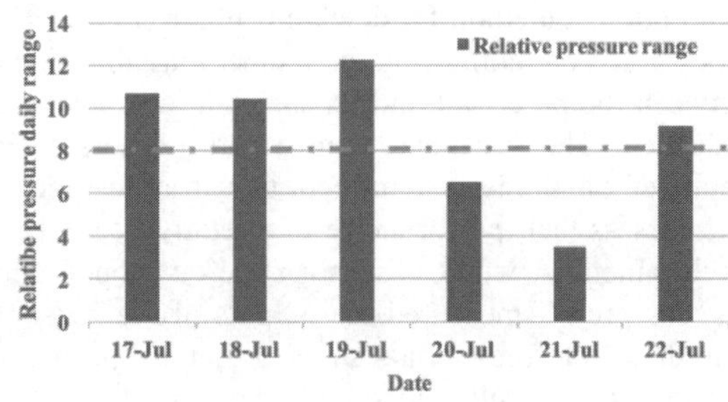

Figure 8. Daily range of the relative pressure

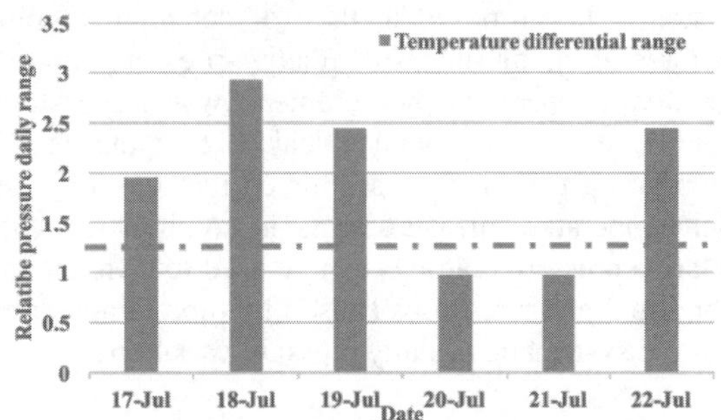

Figure 9. Absolute daily range of the temperature differential

As expected the absolute daily range of the relative pressure for the days when induced leaks occurred were higher than for the days without any training. Similarly, the range of the temperature differential (between the pipe wall and the surrounding soil) during the days when induced leaks occurred were also significantly higher than the other days. A fixed (as shown in **Figure 8** and **Figure 9**, dot-dash line) or adaptive thresholds could be used to flag abnormal events in the pipe based on historic behaviour of the system. Alternatively, learning algorithms could be trained using the processed data for each network based on the operational characteristics of that network. This can potentially provide better accuracy in detection of pipe failure events as operational parameters of each pipe network (i.e. each DMA, District Metered Area) is unique due to differences in consumptions and network architecture.

5 CONCLUSIONS

A FSR based relative pressure sensor has been presented, together with the theory behind its operation. In addition, a method for detecting abnormal flows in pipes based on monitoring the temperature difference between the pipe wall and its surroundings has been proposed. The FSR and temperature sensors were deployed on a pipe system in an industrial test facility to assess their performance in a relatively realistic environment. A wireless sensor node developed during this project measured the outputs of the sensors. The data from these trials showed that the proposed sensors successfully recorded the daily pressure changes in the pipes. In addition, these sensors successfully recorded systematic pressure changes in the system. The proposed method of detection of pipe failures using non-intrusive relative pressure and differential temperature measurements was successfully able to detect the induced leak events during the monitoring period. It is postulated that the differential temperature measurements in combination with relative pressure readings can be used to differentiate abnormal pressure drops caused by pipe failure from normal systematic or daily pressure variations.

ACKNOWLEDGEMENT

The authors gratefully acknowledge the support by the University of Birmingham and the UK Water Industry Research (UKWIR) who jointly funded this project.

REFERENCES

BenSaleh, M.S., Qasim, S.M., Obeid, A.M., et al. (2013) A review on wireless sensor network for water pipeline monitoring applications. *2013 International Conference on Collaboration Technologies and Systems (CTS)*, pp.128–131.

Bond, A., Mergelas, B. and Jones, C. (2004) Pinpointing leaks in water transmission mains. *Proceedings of ASCE Pipeline 2004*, pp. 1–10.

Kleiner, Y. and Rajani, B. (2002) Forecasting variations and trends in water-main breaks. *Journal of infrastructure systems*, 8 (4): 122–131.

Liu, Z. and Kleiner, Y. (2013) State of the art review of inspection technologies for condition assessment of water pipes. *Measurement*, 46 (1): 1–15.

Misiunas, D. (2005) *Failure Monitoring and Asset Condition Assessment in Water Supply Systems*. Lund University

Rajani, B. and Kleiner, Y. (2004) *Non-destructive inspection techniques to determine structural distress indicators in water mains*.

Sadeghioon, A.M. (2015) *DESIGN AND DEVELOPMENT OF WIRELESS UNDERGROUND SENSOR NETWORKS FOR PIPELINE MONITORING*. PhD dissertation, University of Birmingham.

Sadeghioon, A.M., Metje, N., Chapman, D., et al. (2014a) SmartPipes: Smart Wireless Sensor Networks for Leak Detection in Water Pipelines. *Journal of Sensor and Actuator Networks*, 3 (1): 64–78.

Sadeghioon, A.M., Walton, R., Chapman, D., et al. (2014b) Design and Development of a Nonintrusive Pressure Measurement System for Pipeline Monitoring. *Journal of Pipeline Systems Engineering and Practice*. 5(3), pp.3–6.

Sinha, S.K., Iyer, S.R. and Bhardwaj, M.C. (2003) "Non-contact ultrasonic sensor and state-of-the-art camera for automated pipe inspection." *In Sensors, 2003. Proceedings of IEEE.* **2003**. pp. 493–498.

Sinha, S.K. and Knight, M.A. (2004) Intelligent system for condition monitoring of underground pipelines. *Computer-Aided Civil and Infrastructure Engineering*, 19: 42–53.

Proceedings of the International Conference on Smart Infrastructure and Construction
ISBN 978-0-7277-6127-9

© The authors and ICE Publishing: All rights reserved, 2016
doi:10.1680/tfitsi.61279.179

Development and field experiment of routing-free multi-hop wireless sensor networks for structural monitoring

M. Suzuki[1*], K. Jinno[1], Y. Tashiro[1], Y. Katsumata[1], C. H. Liao[1], T. Nagayama[2], N. Makihata[3], M. Takahashi[3], M. Ieiri[3], H. Morikawa[1]

[1] *Research Center for Advanced Science and Technology/The University of Tokyo, Tokyo, Japan*
[2] *Department of Civil Engineering, the University of Tokyo, Tokyo, Japan*
[3] *Infrastructure solutions Division, JIP Techno Science Corporation, Tokyo, Japan*

[*] *Corresponding Author*

ABSTRACT To make wireless sensor networks practical tools for structural monitoring, there are two issues. The first is ensuring robustness of wireless communication. Multi-hop network technology is necessary to deploy sensors on a bridge because there are many obstacles on bridges, and the lengths of bridges are often longer than radio communication range. The second is reducing power consumption. Radio module and acceleration sensor are power hungry, so how to ensure battery life must be considered. The purpose of this paper is to solve these two issues while satisfying requirements of structural monitoring including time synchronization, 100% reliable transport, and high throughput. To this end, we are developing wireless structural monitoring system using Choco, which is a communication protocol which exploits concurrent transmission flooding (CTF). Our system can provide synchronization and robustness thanks to CTF. Furthermore, the system achieves low-power, 100% reliable, and high-throughput transport thanks to Choco's fine-grained scheduling. Through field deployment in an urban expressway, we show that applying CTF to structural monitoring is an effective approach.

1 INTRODUCTION

Many works have revealed that structural monitoring using Wireless Sensor Networks is promising (Kim, et al. 2007, Nagayama et al. 2010). The use of WSN for structural monitoring mitigates sensor installation cost and time significantly.

To make the deployment of wireless structural monitoring systems easier at lower cost, there are two issues. The first is robustness of wireless communication. Due to lack of robustness, prior proposed systems need deep knowledge of wireless communication and networking. For example, elaborated plans on RF communication and network topology balancing are important to deploy and operate systems normally.

The second is energy efficiency of wireless communication. Prior works assume that the dominant cause of power consumption is acceleration sensor, so they do not carefully study the power consumption of wireless communication. However, since power consumption of accelerometer has been rapidly decreased, power consumption of wireless communication accounts for a substantial fraction of total power consumption. Reducing the power consumption of wireless communication may provide a novel usage of structural monitoring.

However, it is difficult to construct a large-scale, low-power wireless network even with recently proposed communication protocols. The state-of-the-art networking protocols (Duquennoy et al. 2015, Ferrari et al. 2012) provide just 99.99% reliability for collecting even simple periodic traffic. Achieving low-power while satisfying three requirements of structural monitoring which are time synchronization, 100% reliability, and high-throughput communication, is still more difficult.

Recent works in wireless sensor networks have shown that Concurrent Transmission Flooding (CTF)

(Ferrari et al. 2011) is an efficient and practical communication primitive to construct robust and low-power wireless networks. CTF is a completely different approach for constructing multi-hop wireless sensor networks. When multiple nodes receive a packet, they immediately and concurrently forward the packet without inserting backoff. These concurrent transmissions do not generate severe destructive interference if the timing difference is below 0.5 us, and nodes can decode the packet successfully. Repeating this process, the packet can be transferred to all the nodes in a very short time. Furthermore, CTF is robust because it is flooding.

In this paper, we show our approach to structural monitoring using CTF. This paper makes three contributions:

- We present a wireless structural monitoring system using CTF. The system achieves robustness, low-power consumption while satisfying requirements of structural monitoring which are time synchronization, 100% end-to-end reliability, and high-throughput.
- We show an implementation of a low-jitter synchronized sampling mechanism while using CTF. CTF needs CPU occupation for several milliseconds, and this may produce large sampling jitter. To solve this issue, we exploit hardware support of timer compare function of a microcontroller, and mitigate the timing constraints. By this, the system achieves a sampling jitter below 50 us among sensor nodes.
- We show a large-scale field experiment of our structural monitoring system in an urban expressway. Although many CTF-based communication protocols have been proposed, field deployment has been rarely shown. To the best of our knowledge, our experiment is the first field deployment for structural monitoring using CTF.

The rest of this paper is as follows. Section 2 shows the use cases we consider, and describes requirements toward these use cases. Section 3 provides the design of our developed structural monitoring system. Section 4 shows the evaluation of the system, and Section 5 describes field experiment in an urban expressway. Finally, Section 6 concludes.

2 REQUIREMENT

There are tradeoffs among sensor node size, battery life, and sensor precision. To deploy sensor nodes easily, sensor node size is critical. Considering these tradeoffs, we consider three use cases below.

Use case 1: Short-term (to a month) monitoring using high-precision, power-hungry sensors.

Use case 2: Long-term (to several years) monitoring using high-precision sensors triggered by low-precision, low-power sensors.

Use case 3: Long-term (to several years) monitoring using low-precision, low-power sensors.

In all use cases, we assume that a monitoring system consists of several tens of sensor nodes, and the sampling rate of the sensors is higher than 100 Hz. In addition, measured data are collected wirelessly when the amplitude of acceleration exceeds a predefined threshold.

Prior works on structural monitoring (Kim, et al. 2007, Nagayama et al. 2010, and so on) consider only Use Case 1. Recently, power consumption of acceleration sensor significantly decreases. For example, ADXL362 manufactured by Analog Devices consumes just 2 uA when it samples acceleration at 100 Hz. Therefore, if the power consumption of wireless networking is kept very low, Use Case 2 and Use Case 3 become feasible.

Table 1. Use Cases.

Use Case	Sensor Type	Term	Measurement
1	High-Precision (1ug/√Hz)	Short-Term (- one month)	Always on
2	High-Precision (1ug/√Hz)	Long-Term (- several years)	Triggered
3	Low-Power (100 ug/√Hz)	Long-Term (- several years)	Always on

We aim to achieve these use cases with a single communication protocol. Most deployments needed specialized protocol development (Kim, et al. 2007, Nagayama et al. 2010). To make structural monitoring more common, practical tools for civil engineers, communication protocol development process should be removed. Toward this, we identify five requirements.

- Low-Power Consumption: To achieve Use Case 2 and 3, a wireless network should work in very low power consumption. To achieve battery life of several years with two D-cell batteries, power consumption of all operation including synchronization, sensor sampling, and storage should be lower than several 100 uA.
- Robustness: It should be possible for people who are not professional of wireless communication and network to deploy sensors easily.
- Synchronization: For analyzing measured acceleration waves, synchronization among wireless sensors is needed.
- 100% Reliable Transport: Data losses should not be allowed. If losses exist, the value of collected data decreases significantly.
- High Throughput: High throughput is required since acceleration sensors generate huge data.

In this paper, we show the design of our developed structural monitoring system which satisfies these five requirements by using CTF. Moreover, we show the practicality of the protocol through a field test. Several protocols which use CTF are proposed such as LWB (Ferrari et al. 2012), and P3 (Doddavenkatappa et al. 2015). However, these protocols have rarely been applied to real applications, especially to timing critical applications such as a structural monitoring system. To the best of our knowledge, our experiment is the first experiment which applies CTF for structural monitoring.

3 DESIGN

To construct a structural monitoring system using CTF, we use a wireless sensor network protocol called Choco. First, we describe Choco overview, and how Choco can satisfy five requirements described above. Next, we show how to achieve (several ten microseconds level) low-jitter sampling despite CTF's exclusively use of CPU for a several milliseconds. Finally, we show the sensors selected to develop sensor nodes.

3.1 Choco Overview

Choco is a slotted communication protocol, and all communications are performed using CTF like LWB. The sink node in Choco works as a master that coordinates communications of all the other nodes. The sink periodically transmits timing synchronization packets to synchronize the other nodes. Also, the sink transmits control packets to assign the slot owner. Only the node assigned as the slot owner is allowed to initiate a transmission, and the transmission is then forwarded by the non-owner nodes using CTF.

The scheduling of Choco is based on the backlog information from each node. Specifically, each node transmits backlog information in a packet header when it transmits a packet. The sink continues to schedule new data transmission slots for remaining packets or packet losses. Until there is no packet to be collected anymore, the sink indicates all the other nodes to sleep by transmitting sleep packets which contain a wake-up time. When the wake-up time comes, all nodes wake up, and the sink restarts the polling.

When a node which has no packets is requested to send a packet, it transmits a null packet, which consists of only a packet header to inform the sink of its zero backlog. When backlogs of all nodes are zero, the sink instructs all nodes to enter a sleep mode by sending sleep packets which include a wake-up time. On the wake-up time, the sink considers all backlogs unknown.

Choco can achieve synchronization, and robustness thanks to CTF. Compared to routing-based communication, CTF-based one is more robust because it can use many paths concurrently. Choco can reduce overhead of idle listening because every node can sleep long time when there are no packets to be collected. Choco also achieves 100% reliable transport because the sink node continues to reschedule lost packets until the packets are delivered. Moreover, Choco achieves high throughput by continuously scheduling slots to nodes who have packets to be collected.

3.2 Synchronized Sampling

To correlate data from different sensors, synchronization among sensor nodes is required. There are two approach to correlate data from different sensors. One is resampling after each node samples at different timing. The other is synchronizing even sampling timings of different nodes. From the perspective that removing data processing after measurement is important, we adopt the latter approach.

CTF and sensor sampling are timing-critical tasks. To achieve CTF, sensor nodes must forward packets just after the reception. If forwarding is deferred, the concurrent transmission becomes destructive. Also, sensor sampling must be done just after the sampling timing comes. If sampling is deferred, the quality of sensing becomes low. For example, if the CTF monopolize CPU for 1 ms, there are possibility that the sampling is deferred 1 ms. In Kim et al. 2007, to make the sampling accurate, the sensor nodes keep RFM completely off when sensing. This approach must stop synchronization, so time synchronization error accumulates. To achieve a high-quality structural monitoring system using CTF, sensor nodes must meet these two timing constraints.

To solve this issue, we used a hardware support of timer compare module, and choose sensors which have an external trigger input. Almost all modern microcontroller including MSP430, AVR, PIC, and Cortex-M have a compare mode. When a timer is configured as compare mode, the timer module toggles GPIO pins without the intervention of software. In addition, sensors which have an external trigger input can sample acceleration values by the change of trigger input.

3.3 *Sensors*

We selected a low-power sensor and a high-precision sensor from the perspective of precision, power consumption, and the existence of trigger function described above. Analog Devices ADXL362, and EPSON M-A351AU are selected as a low-power sensor, and a high-precision sensor, respectively. Both sensors have a trigger function. The specification is summarized as Table 2.

Table 2. Specification of selected sensors.

Sensor	Noise level	Power consumption
ADXL362	175ug/√Hz	10uA
M-A351AU	1ug/√Hz	20mA

4 EVALUATION

The measurement scheme is implemented on a sensor node called Di (see Figure 1). Di has TI MSP430F1611, TI CC2520, and Analog Devices ADXL362. We ported the source code of Glossy (Ferrari et al. 2011) to CC2520. Line-of-sight communication range of Di is about 400 m. The sampling frequency of acceleration sensing is set to be 100 Hz.

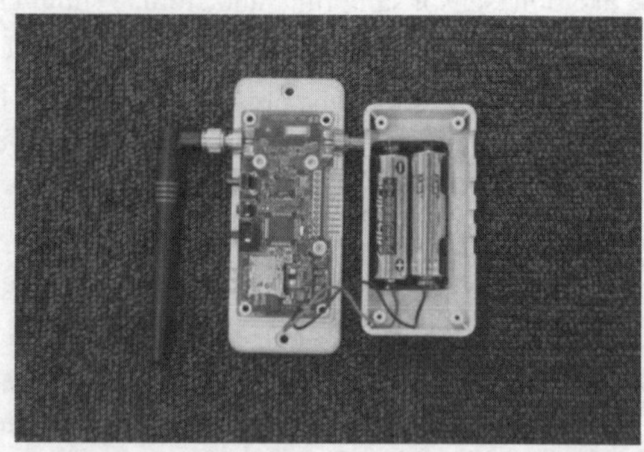

Figure 1. Developed sensor node called Di.

4.1 *Sampling jitter*

We measured sampling timing difference between two nodes which are placed nearby using a logic analyzer. Figure 2 shows the result. This result shows that delays due to task scheduling are completely removed. If there is a task scheduling delay, there must be sporadic large jitters (Kim et al. 2007).

Synchronization error is come from the resolution of clock. The resolution is about 30 us because the current implementation uses 32.768 kHz clock to generate sampling timing.

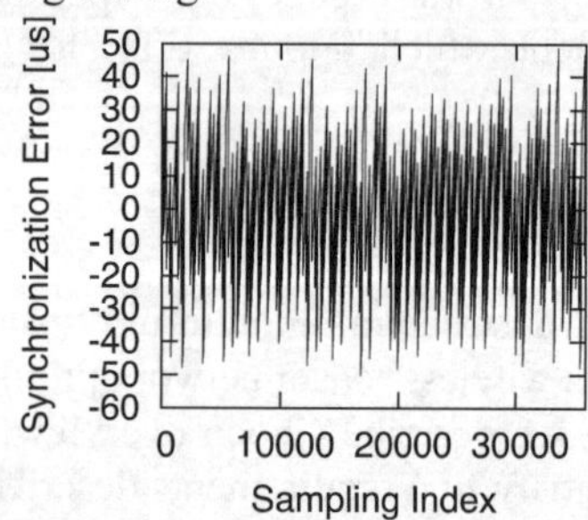

Figure 2. Sampling jitter

4.2 *Power Consumption*

We measured power consumption of a sensor node when the sensor node is executing time synchronization and sensing. Figure 3 shows the timeseries of power consumption.

In our implementation, sensor nodes can sleep almost all the time as shown in Figure 3. Each 5 ms, the power consumption rises to about 1 mA. This is due to CPU's activity of interrupt handling of timer compare. To generate a square wave whose cycle is 10 ms, the interrupt occurs each 5 ms. Every 31.25ms the power consumption rises up to 2 mA. This is due to the activity of CPU and SRAM for reading and writing the measurements of an acceleration sensor. From t=9.27, the power consumption rises to 20 mA. This is mainly due to the activity of RFM. When a sensor node does nothing, the power consumption is about 60 uA on average.

Table 3 summarizes the averaged power consumption per function.

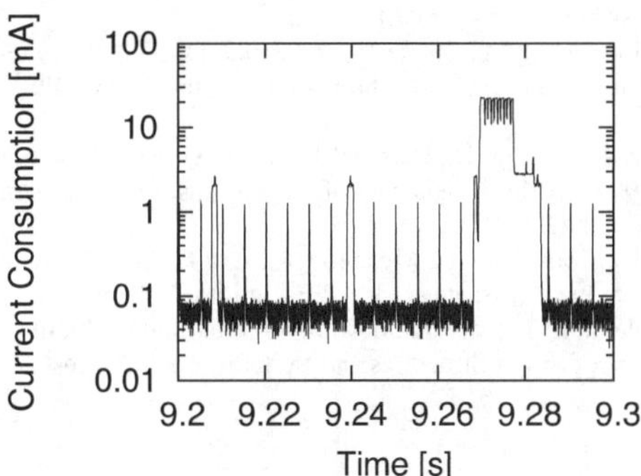

Figure 3. Timeseries of power consumption.

Table 3. Averaged power consumption per function.

Function	Average
Sampling	110 uA
Time Sync	5.6 uA
Data	23.3 uA
Idle	60 uA

5 INITIAL FIELD EXPERIMENT

5.1 Deployment Plan

We deployed 61 low-power sensor nodes over twelve viaducts. Some of the nodes are on the girders while the others are on the piers. Five sensors are deployed in each span like Figure 4 in an urban expressway bridges. Each span is about 30m, so the total length of this deployment is about 360 m.

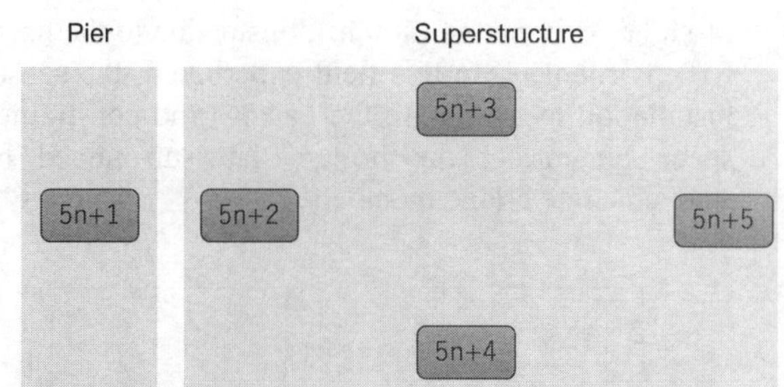

Figure 4. Schematic plan view of each span. 61 sensors are deployed over 12 spans (n=0, 1, 2, 3, …, 11).

5.2 Throughput

We conducted a data collection experiment. In this experiment, we measured acceleration over 40 seconds, and collected the data wirelessly. Figure 5 shows the timeseries of throughput averaged over five seconds. The average throughput is about 800 B/s. Data from 35 nodes can be successfully collected without any losses. The other nodes have stopped sending data before the completion. After the investigation of collected data, we found that a software bug of buffer management causes this problem. We can obtain data from these nodes without losses before they stop sending.

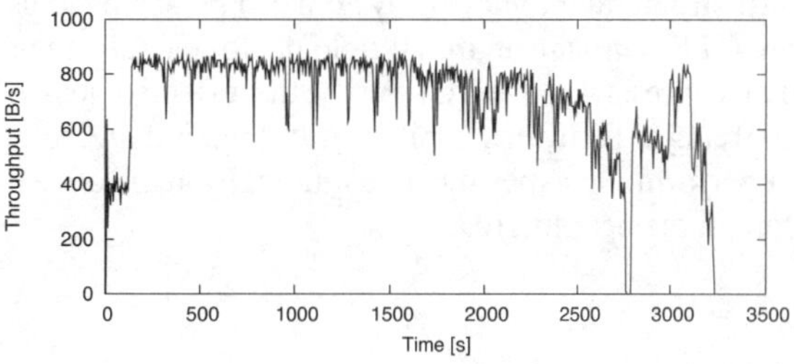

Figure 5. Time series of throughput.

5.3 Network Scale

We recorded the hop count of each received packet at the sink node. Figure 5 shows the average hop count from each node. The error bars show the standard deviation. The result shows that two spans can be reached within a single hop. The average hop count of the farthest node is 6.16.

The average single hop distance of 60 m is ensured in all areas. This indicates CTF does not show severe performance degradation in a network of this size. Though this distance is much shorter than that in line-of-sight environment (400 m), this is due to the harsh RF environment. In this field experiment, the sensor installation locations are the narrow and constrained space underneath the bridge slab surrounded by steel/concrete bridge members.

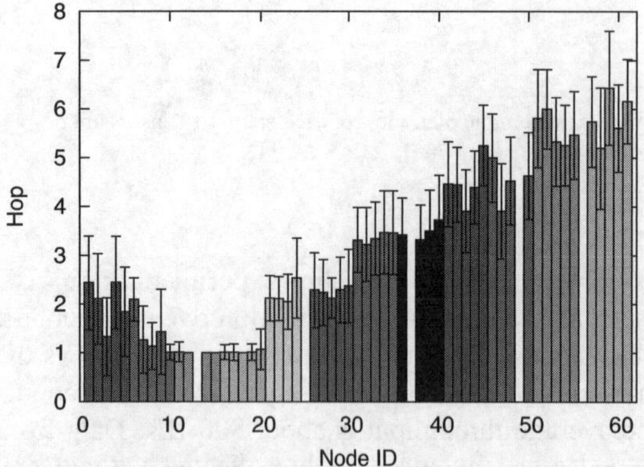

Figure 6. Average hop count of each node from the sink node.

6 CONCLUSION

In this paper, we have shown that a low-power and robust structural monitoring system can be achieved using CTF. Through an initial field deployment in an urban expressway, we can verify the effectiveness of applying CTF to structural monitoring. We are going to perform the experiments using high-precision sensors in the near future.

ACKNOWLEDGEMENT

We thank Metropolitan Expressway Company Limited for providing experiment fields and advice. This research is supported by Cross-ministerial Strategic Innovation Promotion Program "Infrastructure Maintenance, Renewal and Management Technology".

REFERENCES

Doddavenkatappa, M., and Mun C. C. 2015. P3: A practical packet pipeline using synchronous transmissions for wireless sensor networks. IPSN. IEEE.

Duquennoy S., Nahas B. A., Landsiedel O., and Watteyne T. 2015. Orchestra: Robust mesh networks through autonomously scheduled tsch. SenSys. ACM.

Ferrari F., Zimmerling M., Mottola L., and Thiele L. 2012. Low-power wireless bus. SenSys. ACM.

Ferrari F., Zimmerling M., Thiele L., and Saukh O. 2011. Efficient network flooding and time synchronization with glossy. IPSN. ACM.

Kim. S, Pakzad S., Culler D., Demmel J., Fenves G., Glaser S., and Turon M., 2007. Health Monitoring of Civil Infrastructures Using Wireless, IPSN, ACM.

Nagayama, T., Moinzadeh, P., Mechitov, K., Ushita, M., Makihata, N., Ieiri, S., Agha, G., Spencer, Jr., B.F., Fujino, Y., and Seo, J. 2010. Reliable multi-hop communication for structural health monitoring. Journal of Smart Structures and Systems. Techno-Press.

Proceedings of the International Conference on Smart Infrastructure and Construction
ISBN 978-0-7277-6127-9

© The authors and ICE Publishing: All rights reserved, 2016
doi:10.1680/tfitsi.61279.185

Multi-point monitoring of unstable slope with tilt sensors

T. Uchimura*[1], S. N. Tao[1], I. Towhata[2], L. Wang[3], L. Su[3], S. Nishie[3], H. Yamaguchi[3] and I. Seko[3]

[1] *University of Tokyo, Tokyo, Japan*
[2] *Kanto Gakuin Univerisity, Yokohama, Japan*
[3] *Chuo Kaihatsu Co., Tokyo, Japan*
* *Corresponding Author*

ABSTRACT A low-cost and simple monitoring method for early warning of rainfall-induced landslides has been proposed by the authors. Tilting angles in the surface layer of slope are mainly monitored in this method. A set of equipment has been developed for practical use, which is equipped with a MEMS (Micro Electro Mechanical Systems) tilt sensor and a volumetric water content sensor. In several case studies with this system, including a slope failure test conducted on a natural slope by applying artificial heavy rainfall, it detected distinct behaviors in the tilting angles in the pre-failure stages. Considering these behaviors, the authors has proposed to issue precaution at a tilting rate of 0.01 degrees per hour, and warning at that of 0.1 degree per hour, for conservative decision. The development of this system reduced the cost for slope monitoring reasonably. However, the number of available equipment for each slope is still limited due to financial restriction, and therefore, engineers need to select the position of sensors carefully for effective early warning. This paper introduce authors' recent attempt to develop more low-cost sensor network for slope monitoring. The cost for each sensor node was reduced by around onethird, and consequently, slopes can be monitored at many point, resulting in more meticulous observation of slope behaviors. The developed prototype of the proposed new sensor units were installed on an unstable slope site.

1 INTRODUCTION

There is a long history in prevention and mitigation of rainfall and/or scouring-induced landslides. Mechanical counter measures to prevent slope failure, like retaining walls and ground anchors, have been widely used. However, they are expensive and it is not realistic to apply such mechanical measures for all of these slopes with potential risk, because most of landslide occurs in small scale, but a large number of slopes. Therefore, careful monitoring of slope behaviors and consequent early warning is reasonable as alternatives.

The authors have proposed and developed an early warning system for slope failures, as one of feasible countermeasures (Figure 1) (Uchimura, et. al. 2015a). The system consists of minimum number of low-cost sensors on a slope, and the data is transmitted through wireless network. Thus, the system is low-cost and simple so that the residents in risk areas can handle it to protect themselves from slope disasters.

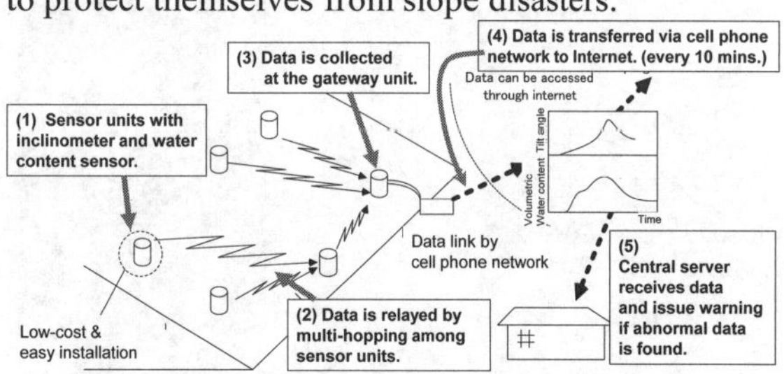

Figure 1. Proposed wireless monitoring and early warning system.

Figure 2 and Table 1 illustrates the sensor unit. It is equipped with a MEMS tilt sensor (nominal resolution = 0.00251 = 0.04 mm/m, SCA100T-D01 provided by MURATA) as well as a volumetric water content sensor (nominal resolution = 0.1%, EC-5 provided by DECAGON DEVICES). An additional temperature sensor is also used for temperature compensation for the tilt sensor. Each sensor unit is powered by 4 AA alkaline batteries and functions well in the field for a duration of more than 1 year.

The tilt sensor measures the tilt angle of a steel rod installed to a depth of typically 1 m in the unstable soil layer on the slope surface. The depth will be shorter if the unstable layer is thinner. The tilt sensor detects the average shear deformation of the slope surface layer. According to Osanai et al. (2009), most of landslides are shallow surface failures due to heavy rainfall, whose thickness is 1.2 m on average. Therefore, deformation of the whole surface layer can be detected with this arrangement in most cases. The tilt sensor is embedded into the ground after being attached to a steel rod to avoid large changes in temperature which could cause measurement errors.

The volumetric water content sensor is placed at a shallow position (typically a depth of 30 cm) in the slope. This sensor measures the dielectric constant of the soil, which is a soil parameter sensitive to the water contents. Soil mechanics theories state that the slope stability directly depends on the suction, or pore water pressure, rather than the water content. However, it is usually difficult to measure the suction of unsaturated soils, and it requires careful maintenance of the sensors. Thus, use of volumetric water content sensors is more suitable for simple monitoring.

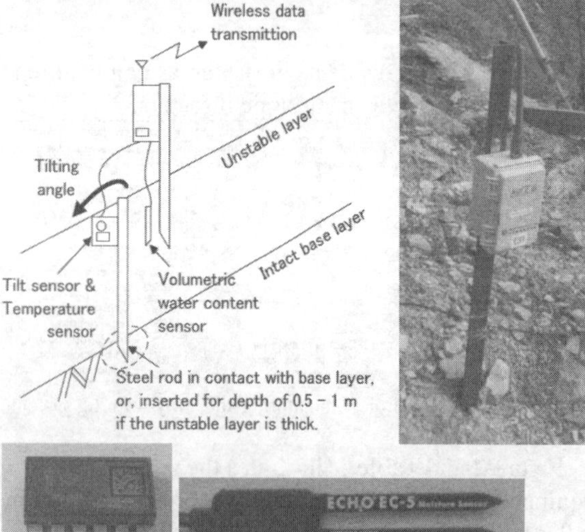

Figure 2. Sensor unit, tilt sensor (lower left) and volumetric water content sensor (lower right).

Table 1. Specifications of the developed sensor unit.

Sensor unit	
2 axis module (-30° ～+30°) 0.02°	
3 axis module (-90° ～+90°)	
Fall detection (The time 30° inclines)	
Water content meter (option)	
Easy form water content metor EC5-5	
Measurement accuracy : ±3%	
Wireless Communication	
Radio standard : ARIB STD-T67	
Transceiver frequency	: 429.2500～429.7375 MHz
Transmittable distance	about 600m (a non-obstacle) about 200m (inside grove)
Operating Temp.	-10℃～80℃
Power	C-type battery 4 pieces ※half year at 10min.sampling

The authors has installed the developed monitoring system at more than 80 sites in Japan, China, and Taiwan. The instability and/or failure of slopes was observed at some of those cases.

Figure 3 summarizes case studies on the tilting-rates observed in slope surface in pre-failure stages obtained at 10 natural slope sites under natural or artificial heavy rainfall. The vertical axis presents "the duration before failure or stabilization". That is, in cases the slope failed at the position of the tilt sensor, the duration is measured from the time when the corresponding tilting rate was observed to the time of failure. In cases the slope was did not fail and stabilized finally, the duration is measured from the time when the corresponding tilting rate was observed to the time of when the slope was stabilized. See Figure 4 for these definitions.

According to Figure 3, the order of tilting rate observed with slope deformation varied widely from 10^{-4} to 10 degree/hour depending on situations. The tilting rate tends to increase toward failure, and a shorter duration is remaining before failure when higher tilting rate is observed. The observed tilting rate was more than 0.1 degree/hour for all the cases in which the slope failed or nearly failed finally, while it was less than 0.01 degree/hour for all other cases.

Considering these behaviors, the authors has proposed to issue precaution at a tilting rate of 0.01 degrees per hour, and warning at that of 0.1 degree per hour, for conservative decision. Durations of 1 to 10 hours were remained before failure when at tilting rate of 0.1 degree/hour was observed.

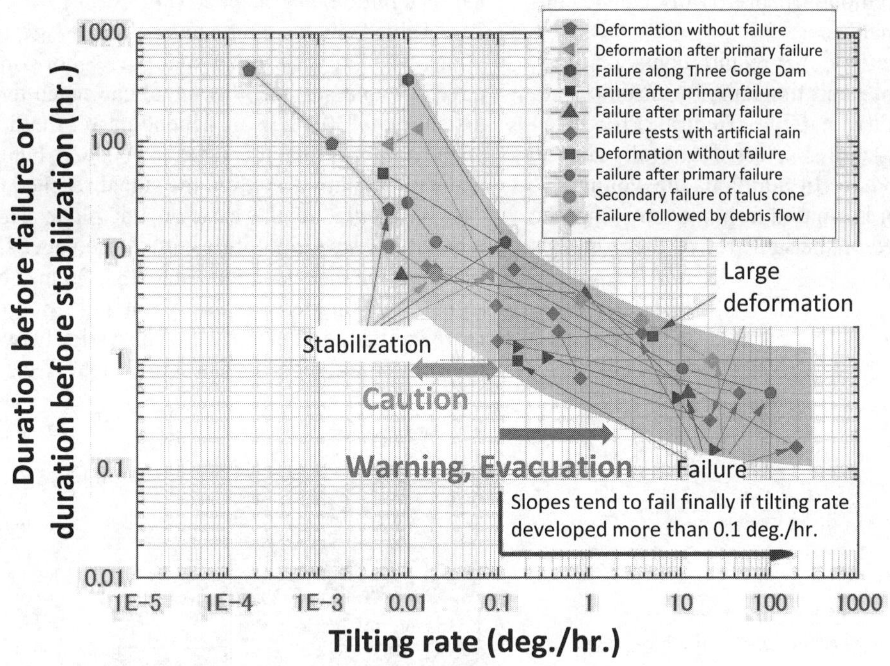

Figure 3. Summary of the tilting rate and duration observed on slope surface.

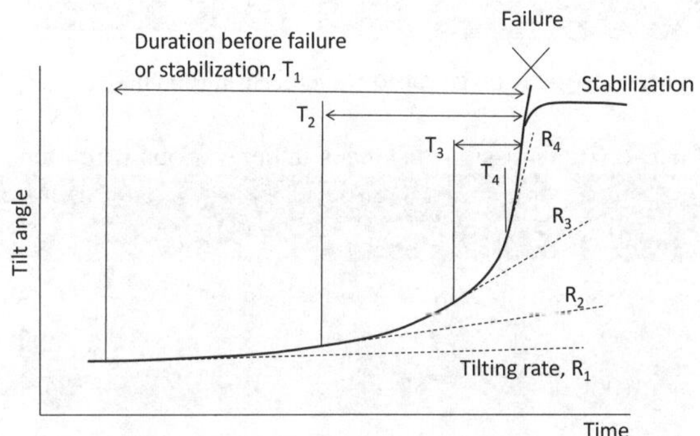

Figure 4. Definition of the tilting rate and the durations in Figure 3. (T_i: the duration before failure or stabilization; R_i: Tilting rate)

2 MULTI-POINT MONITORING OF UNSTABLE SLOPE

In order to improve the applicability of the developed monitoring and early warning system, the sensor equipment has been modified to be more low-cost, smaller in its size and weight, with simpler in installation and operation works (Uchimura et. al. 2015b). As a result, the total cost for the monitoring system is reduced, and larger number of sensors can be installed on the slope site. It helps us to figure out real-time situation of the slope in more details.

Figure 5 conceptualizes modified arrangement of the equipment. The proposed new sensor units are more low-cost and simple than the conventional units, but they have relatively shorter distance of radio transmission (around 30 m under worst condition). They are arranged densely on the high-risk areas among the site, and one conventional unit collect all the data of each area. Then, the data is transmitted through longer distance (300 - 600 m), and uploaded to the internet server. Thus, a meticulous sensor network can be constructed with reasonable cost.

In order to satisfy the requirements on cost, power consumption, radio transmission distance, accuracy and reliability, a CPU (Central Processing Unit) with radio-on-chip is selected carefully for the controller/radio module. In addition, the volumetric water content sensor is omitted from the unit. And, the tilt sensor is attached on the same rod as that supports the wireless module.

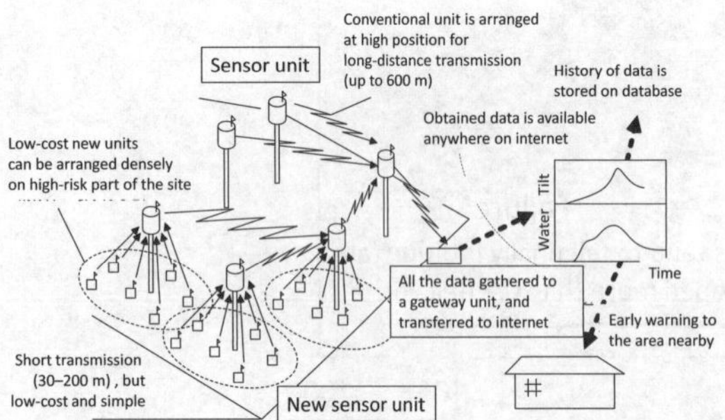

Figure 5. Modified arrangement of slope monitoring system.

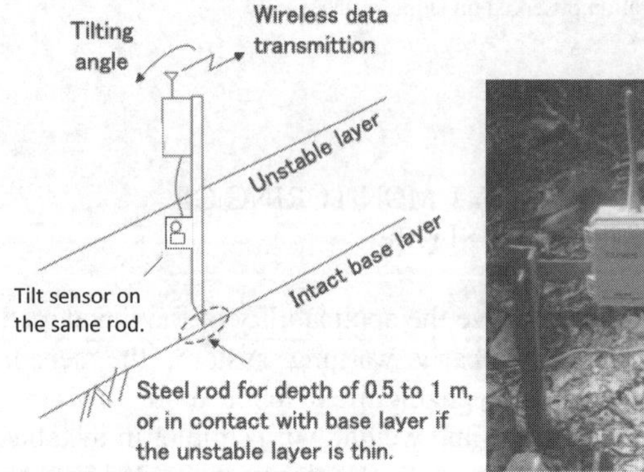

Figure 6. Modified sensor unit.

The conventional sensor unit has sufficiently long radio transmission distance, 300 m, or 600 m under good conditions. However, the proposed new sensor unit has shorter transmission distance, because it uses higher radio frequency (2.4 GHz band) than the conventional one (430 MHz band). The available distance was carefully evaluated for reasonable arrangement of sensors (Uchimura et. al. 2015b). The receiver was fix at a height of 1 m, while the transmitter was set at several heights and distances, and the radio signal recognition was checked (Figure 7).

Table 2 shows the results. When it is tested on a "flat plane", a wide open space with horizontal and paved ground surface, the signal can reach through a distance of 130 - 238 m depending on the height of antenna. Even when it is tested on a slope, but the area is open without vegetation, the signal reached through 215 m, similar to the case on flat plane. However, when it is tested in a forest with dense vegetation, the transmission distance reduced to 33 – 36 m. The radio signal with higher frequency intends to propagate more straight, and existence of obstacles between the transmitter and receiver has significant effect.

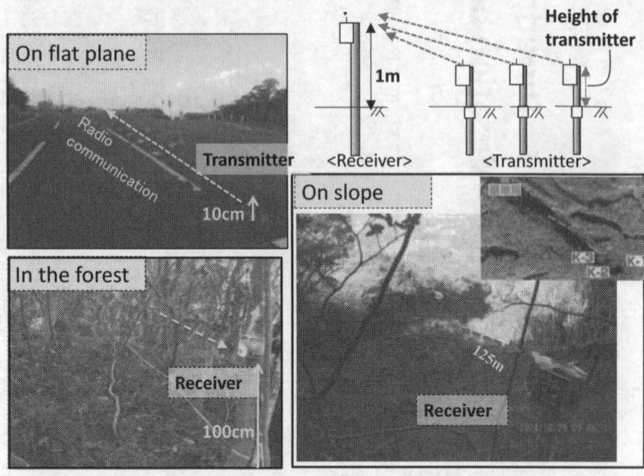

Figure 7. Evaluation of radio transmission distance.

Table 2. Transmission distances under various conditions.

Height of transmitter (cm)	Flat Plane (m)	Slope (m)	Forest (m)
10	130	215	33
50	204	–	33
100	238	–	34～36

3 TRIAL INSTALLATION OF MULTI-POINT MONITORING

As a trial of multi-point monitoring, a slope of mountainous region, Mansawa area, Japan, has been instrumented with 66 tilt sensors and monitored since March of 2015 (Figure 8). It is located near the boundary between Shizuoka and Yamanashi Prefectures, in south west direction of Mt. Fuji. The slope is formed by conglomerate, Sand stone, Mud stone layer of Miocene, and the surface of bedrock are weathered to be soft and sandy. Talus accumulation redeposited at the valley area.

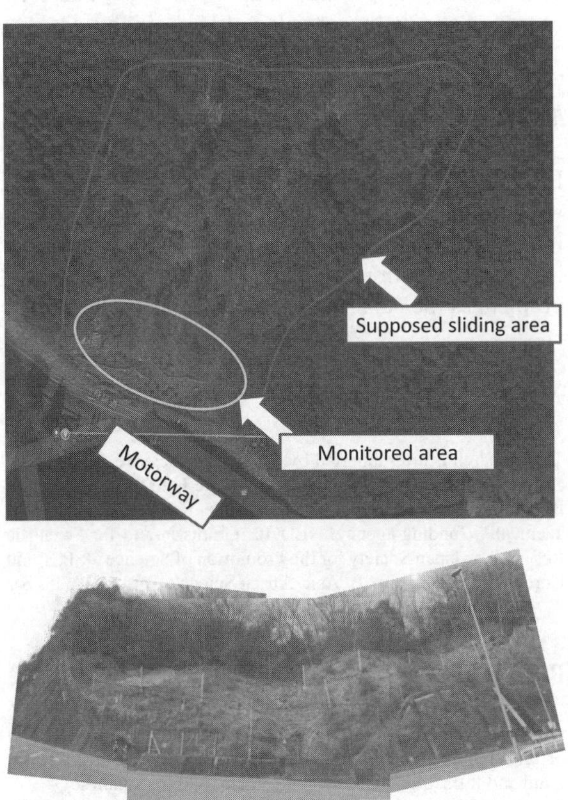

Figure 8. Photos of Mansawa slope monitoring site.

Figure 9 shows arrangement of sensors and other equipment for monitoring of Mansawa site. The 66 tilt sensors was installed in lattice arrangement with lateral spacing of 5 m. The newly developed sensor unit mentioned in previous chapter is also included. They are divided into 3 groups, left/middle/right zone, and one data receiver unit and one logger/gateway unit for internet collect all the data from respective group.

There were 8 heavy rainfall event during summer of 2015 as shown in Figure 10. And, the tilt angles accumulated due to each rain are summarized as Figure 11. The tilting rate averaged during each rainfall event is shown in Figure 12. Distribution of tilting behaviors are figured out by multi-point monitoring.

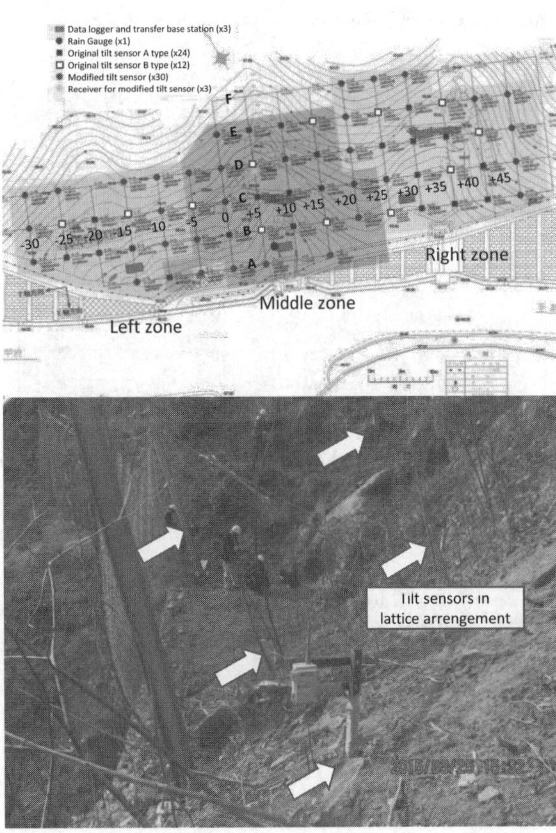

Figure 9. Equipment for Mansawa site.

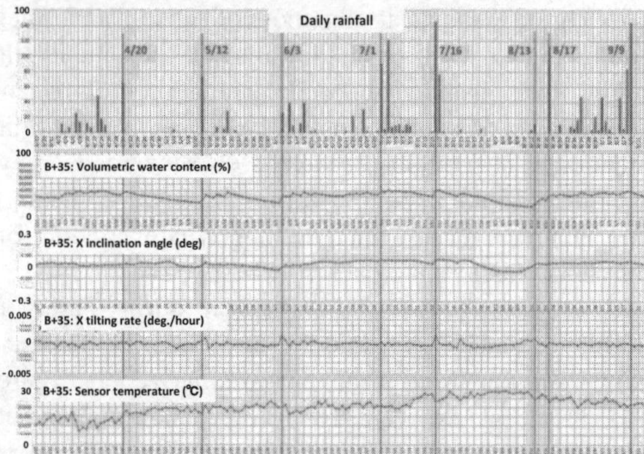

Figure 10. Observed behaviors in Mansawa due to rainfall.

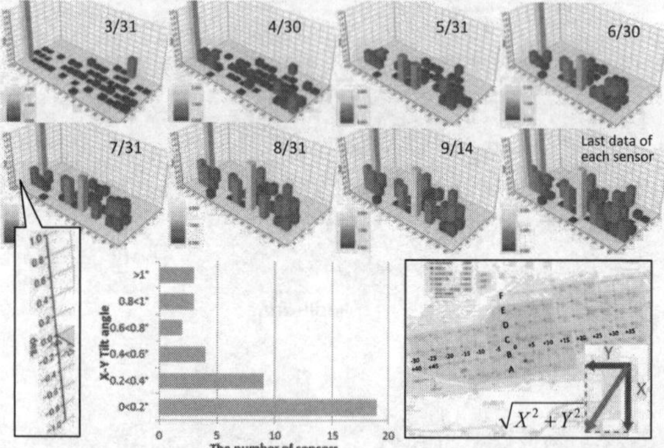

Figure 11. Distribution of accumulated tilt angles at each rain.

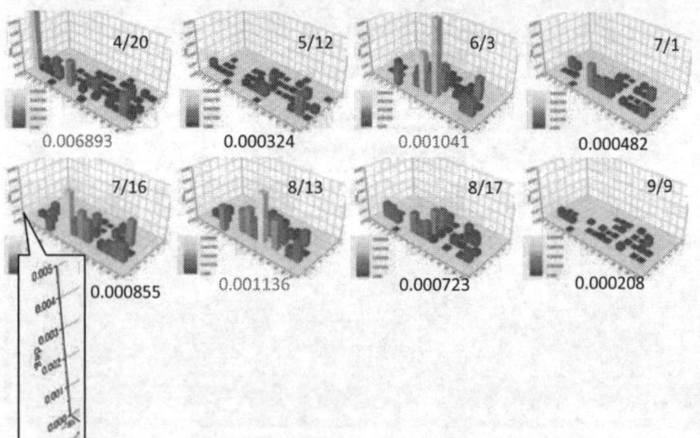

Figure 12. Distribution of tilting rate during each rain.

For practice, a criteria for issuing early warning have to be defined based on data from the large number of sensors. One of possible index for the criteria is simple sum of tilting rates from the sensors:

$$V_{alarm} = \sum_{n=1}^{N} \left(|V_n| * \frac{A_n}{A_0} * \partial_n \right)$$

Here, n is serial number of tilt sensors, V_n is tilting rate and A_n is the area of installation of the n-th sensor, and A_0 is the total area of monitored slope. δ_n gives weight for the n-th sensor considering geology, geography, and other factores. As the simplest example, values calculated with $\delta_n = 1$ for all the sensors are indicated in Figure 12. The rain at 4/20, 6/3, and 8/13 caused relatively higher value of V_{alarm} in this case.

4 CONCLUSION

Equipment for multi-point monitoring is newly developed, and tried at a slope site. Distribution of tilting behaviors are figured out by multi-point monitoring, and definition of warning criteria is essential problem as the next step of this research.

ACKNOWLEDGEMENT

This research is are supported by Council for Science, Technology and Innovation, “Cross-ministerial Strategic Innovation Promotion Program (SIP), Infrastructure Maintenance, Renovation, and Management”. (funding agency: NEDO), Grants-in-Aid for Scientific Research of Japan Society for the Promotion of Science (JSPS), and Core-to-Core Program “B. Asia-Africa Science Platforms” (JSPS).

REFERENCES

Osanai, N., Tomita, Y., Akiyama, K., Matsushita, T., 2009. Reality of cliff failure disaster. Technical Note of National Institute for Land and Infrastructure Management, No. 530 (in Japanese).

Uchimura, T., Towhata. I., Wang, L., Nishie S., Yamaguchi, H., Seko, I. and Qiao, J.-P. 2015a. Precaution and early warning of surface failure of slopes by using tilt sensors, Soils and Foundations, Vol.55, No.5, pp. 1087-1100.

Uchimura, T., Wang, L., Yamaguchi, H., Nishie, S., Eto, I., Tao, S., Lu, C.-W., Chang, J.-J. and Chen, C.-W. 2015b, Multi-point monitoring of unstable slope with low cost sensor network, Proc. of the 6th Sino-Japan Geotechnical Symposium 2015, Sapporo, Article number: JPN-35.

Proceedings of the International Conference on Smart Infrastructure and Construction
ISBN 978-0-7277-6127-9

© The authors and ICE Publishing: All rights reserved, 2016
doi:10.1680/tfitsi.61279.191

Wireless sensing on shield tunnels in Shanghai

F. Wang[1], H.W. Huang[*1], B. He[1], Y. Wu[2], H. Shao[3] and H.M. Wu[4]

[1] *Tongji University, Shanghai, China*
[2] *Wisen Innovation Co., Ltd., Jiangsu Wuxi, China*
[3] *Shanghai Rail Transit Maintenance Support Co., Ltd., Shanghai, China*
[4] *Shanghai Tunneling Engineering Co., Ltd., Shanghai, China*
[*] *Corresponding Author*

ABSTRACT Recent years more and more shield tunnels are being used in urban critical infrastructure in China, such as road and metro tunnels. Its safety during the construction and operation should be most crucial. Monitoring the tunnel behaviour including deformation, water leakage, crack and damage, etc. is an effective way to control the lifetime safety of shield tunnel. In this paper, a "smart" sensing system coupling the tilt sensor, seepage sensor, and crack sensor with MEMS-based and Zigbee-based mesh network and their application in road and metro shield tunnel are firstly proposed in China. All these sensors were developed and validated by indoor model test for shield tunnel segmental lining structure. MEMS tilt sensors were installed in 2 sections of the Shanghai Dalian road shield tunnel with 11m outside diameter and 480mm segment thickness. Tilt, seepage and crack sensors were installed in one interval shield tunnel in Shanghai metro with 6.2m outside diameter and 350mm segment thickness. The lateral convergence transferred from tilt sensor, area of seepage, opening and closing of joints can be obtained by WSN. Both measured data shows their stable and reliable. The merit of this system, compared to the conventional monitoring system, is its ability of real-time dynamic monitoring regardless of the running of road or metro tunnel.

1 INTRODUCTION

As the urban metro network rapid expanding, more and more structure issues also emerge, i.e., large convergence, differential settlement, crack and seepage. Most of these structure issues occur due to poor soil condition, poor workmanship, and adjacent construction and so on.

Structure Health Monitoring of the urban metro tunnel is necessary to understand the safety state of the operated tunnel. For an operated tunnel in a normal condition, the monitoring plan is the measurement of horizontal convergence by using the traditional monitoring system, e.g., total station (Bakker, et al., 1999). The frequency of the measurement is twice a year. In the meanwhile, the structural defects, such as leakage, joint open and segmental dislocation, are items of the traditional manual inspection plan for the metro tunnel with a frequency the same with that of convergence (Richards, 1998, Yuan, et al., 2012, Llanca, et al., 2013). However, for the urgent condition, the real time monitoring should be conducted for the early warning. In recent years, micro-electro-mechanical systems (MEMS) and wireless sensor network (WSN) are popular for structure health monitoring in civil engineering. As the technology evolving, MEMS sensors have become more integrated and less power-consuming which is ideal for its combination with WSN. Bennett et al. (2010) have applied several WSN trial systems in London, Barcelona and Prague subway tunnels. The battery powered sensor modules are small, completely wireless and easy to install. They automatically construct the network topology and acquire high quality live data 24 hours a day. Huang et al. (2012) discuss the WSN employed for the behavior study of shield tunnel structure. A wireless MEMS inclinometer prototype is developed for urban metro tunnel by Tongji University (He, et al., 2013), meanwhile a model for calculating the tunnel convergence was developed using this MEMS sensor (Huang, et al., 2013). A wireless crack sensor is developed for joint opening monitoring in shield tunnel (Wang and Huang, 2013). Seepage of shield tunnel is monitored by wireless sensor using Electrical Conductivity Method (Cheng and Huang, 2014).

This study presents a “smart” sensing system coupling the tilt sensor, seepage sensor, and crack sensor with MEMS-based and Zigbee-based mesh network and their application in road and metro shield tunnel are firstly proposed in China. Meanwhile the feasibility of WSN in metro shield tunnel is also discussed.

2 DEVELOPMENT OF WIRELESS SENSOR NETWORK FOR SHIELD TUNNEL

2.1 *Wireless sensor*

2.1.1 *Tilt sensor*

TJ-UWIS is a series of MEMS inclinometer prototypes developed by Tongji University. Three different specifications of TJ-UWIS sensors are provided for various purposes. There is single axis analog inclinometer (V1.3), dual axis analog inclinometer (V1.1) and single axis differential inclinometer (V1.2). The TJ-UWIS is compatible with Tongji wireless smart sensor network.

A series of tests are conducted in order to inspect the MEMS sensor's adaptability to various environments. All the sensors are tested under the same condition to compare their performances. Table 1 shows the performance indexes of TJ-UWIS.

Table 1. Performance index of TJ-UWIS

	Axis	Precision (°)	Resolution (°)	Range (°)
TJ_UWIS1.1	2	0.01	0.0025	-90~90
TJ_UWIS1.2	1	0.01	0.0025	-30~30
TJ_UWIS1.3	1	0.01	0.0013	-30~30

2.1.2 *Seepage sensor*

Electrical Conductivity Method is used to develop the seepage sensor. The concrete is almost nonconductive when concrete is dry. However, when concrete is soaked by water, its conductivity increases rapidly and resistance reduce greatly. Based on the concrete resistivity change characteristics, the tunnel leakage can be detected. Figure 1 shows the relationship between current and time under different flow rate.

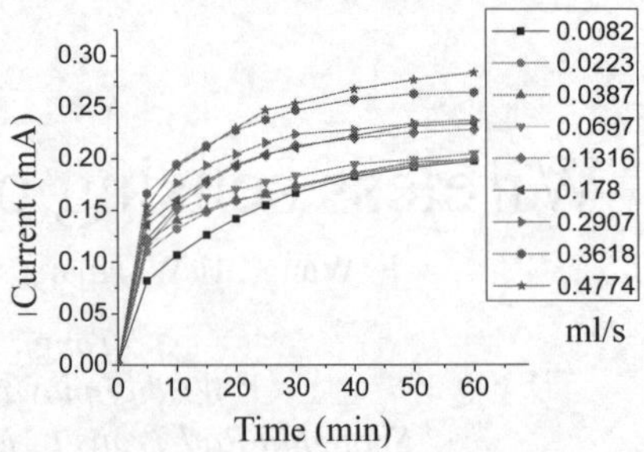

Figure 1. Current-time curve of different flow rate

2.1.3 *Crack sensor*

The structure of wireless crack sensor mainly consists of displacement transducer, temperature sensor, AD convertor, microcontroller and power module (Figure 2). The displacement transducer, LPDT (Linear Potentiometric Displacement Transducer) or LVDT (Linear Variable Differential Transformer), has small range and high precision. The temperature sensor is used for temperature compensation.

The microcontroller combines the excellent performance of a wireless transceiver (based on Zigbee protocol) with an industry-standard enhanced micro control unit. It has various operating modes, including ultralow power-consumption operating mode, making it have advantages in terms of power consumption and data rate. A series of tests are conducted in order to inspect the crack sensor's adaptability to various environments.

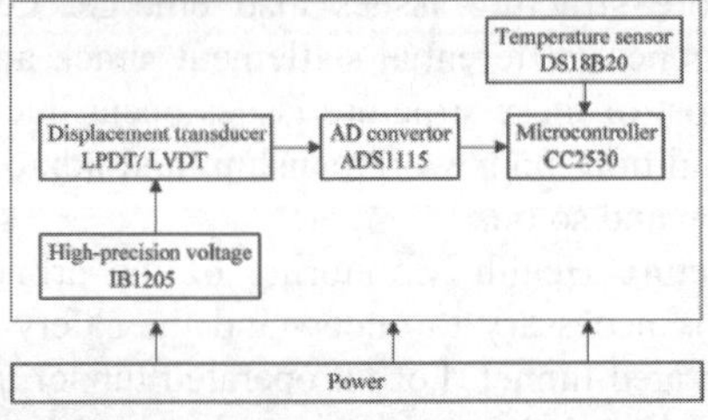

Figure 2. Structure of wireless crack sensor

2.2 *Wireless Sensor Network Deployment*

We deployed a WSN in a Shanghai shield tunnel. Figure 3 shows a model of the tunnel and the sensor nodes deployed. Here the base station is 20 meters away from the tunnel entrance, and is perpendicular to the tunnel mouth. An algorithm is proposed for the temporal and spatial compression and recovery of sensor data in the complex environment of a shield tunnel.

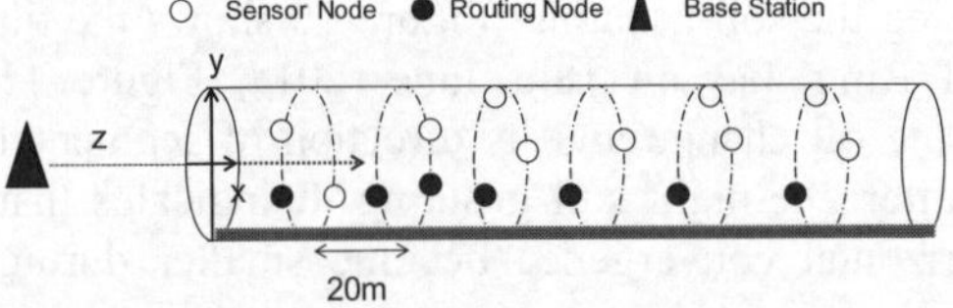

Figure 3. WSN deployment in a shield tunnel

3 MODEL TEST

Figure 4 shows the experimental facility in working condition. It includes two segments from a shield tunnel ring. The test-piece is fixed on two hinge supports which limit the linear displacement while rotation is allowed. Three horizontal hydraulic jacks are applied to simulate axial force in the ring. Vertical load is applied by a series of beams powered by another three jacks.

The convergence displacement of shield tunnel has a functional relation with its inclination. The mid-span displacement (Δ) can be calculated by following equation based on inclination I_1 and I_2 measured by Tilt sensors.

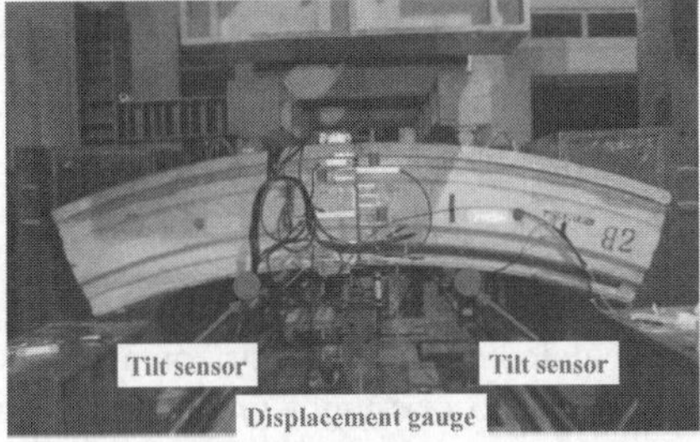

Figure 4. The experimental facility

$$\Delta = \frac{AC^2[\sin(\alpha_0) - \sin(\alpha_0 - I_1)] + BC^2[\sin(\beta_0) - \sin(\beta_0 + I_2)]}{AC + BC} \quad (1)$$

where α_0 = arctan (f/L_1), β_0 = arctan (f/L_2) as shown in Figure 5. The result will be compared with the real mid-span displacement measured by conventional instruments.

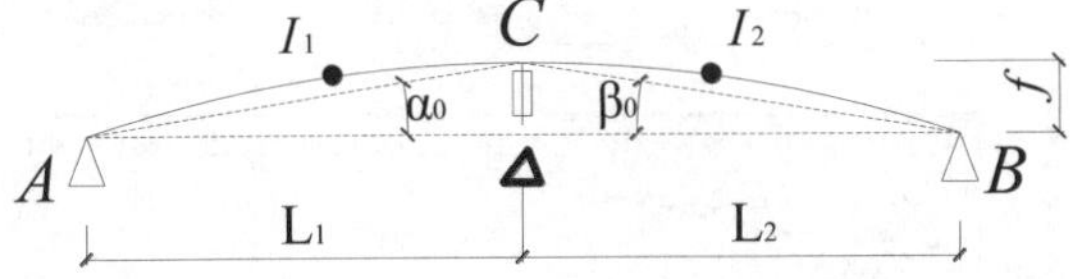

Figure 5. Sketch of the test-piece segment

Figure 6 shows the calculated displacement from the wireless inclinometer and the reference displacement from displacement gauges of each stage of load. The average error of the calculated displacement is about 7% which is acceptable. Although the variance of the calculated displacement is relatively large, indicating that the readings of the inclinometer are shifting constantly, they still spread evenly around the true value.

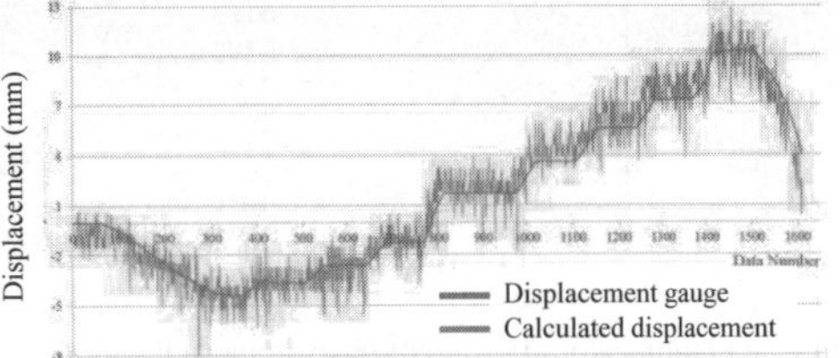

Figure 6. Calculated displacement by Tilt sensor

4 CONVERGENCE CALCULATION FOR SHIELD TUNNEL

The ring of shield tunnel is composed of several segments, shown in Figure 7.

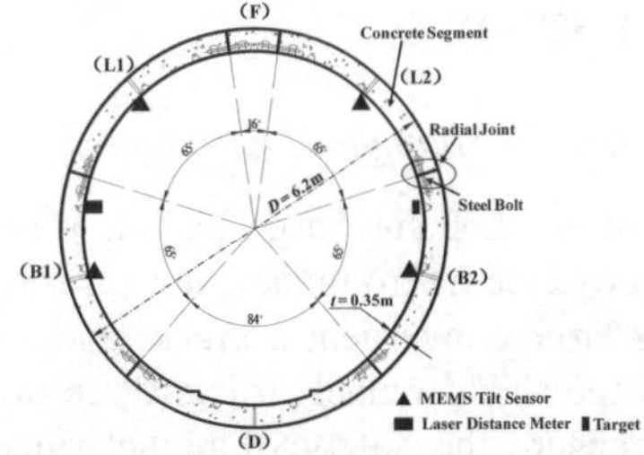

Figure 7. Cross section of metro shield tunnel

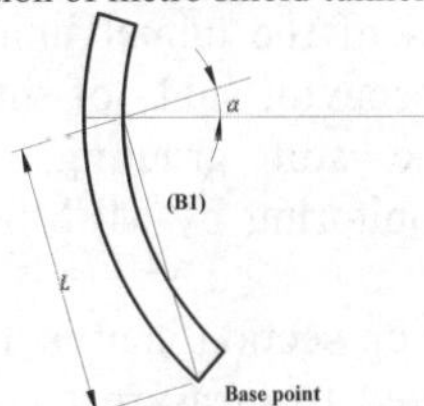

Figure 8. Geometry model of shield tunnel deformation

A simplified model is developed for calculating the horizontal convergence assuming that the segment is rigid, that is, tunnel deformation mainly consists of the joints rotation. Figure 8 presents the geometry model for the shield tunnel deformation. The horizontal convergence of tunnel can be calculated by Equation 2.

$$\Delta D' = L \times (\Delta\theta_1 - \Delta\theta_2) \times \cos\alpha \quad (2)$$

where $\Delta D'$ is the change of horizontal convergence, L is the distance from the base point to inner surface at the tunnel center point level, $\Delta\theta_1$ is change of the tilt sensor on segment B1, $\Delta\theta_2$ is change of the tilt sensor on segment B2, α is the angle shown in the Figure 8.

Figure 9. Calculation error at different measurement location

Considering the segmential lining is deformable, the calcultion error by Equation 2 is analyzed using K.M. Lee model under various joint stiffness, shown in Figure 9. Result indicates that the best installation point for tilt sensor is 120° (setting the crown of tunnel as 0°).

5 CASE STUDY

5.1 Case 1 for Shanghai Metro tunnel

Because of extreme surcharge loading on the ground surface above the metro tunnel, several serious issues including large convergence, crack, and seepage occurred in the shield tunnel structure are observed. In order to ensure the safety of tunnel operation, the convergence performance is recovered by soil grouting at two sides of the tunnel lining. WSN is employed for the behavior study of shield tunnel structure during the soil grouting. Applications of convergence monitoring by MEMS Tilt sensors are illustrated.

Two monitoring section, that is, ring 411 and ring 433, were selected to measure the convergence development during the soil grouting. The segmental lining for this metro tunnel has an outer diameter D of 6.2 m and a thickness of 0.35 m, as shown in Figure 7. One tilt sensor was installed on the inner surface of lining L1, B1, L2 and B2. Four MEMS tilt sensors are installed within one ring in total. Meanwhile one laser distance meter sensor was installed on B1 at the same level of tunnel center. Target for laser distance meter was installed on B2 at the same level.

Tilt sensors measure the rotation of segmental lining during the soil grouting. Figure 10 shows the tilt data of Ring 433 on 18th June, 2014. Figure 11 shows the tilt change and its direction of segmental lining after one-night soil grouting. It indicates that the horizontal convergence become smaller during the grouting.

Tilt sensors measure the rotation of segmental lining during the soil grouting. Figure 10 shows the tilt data of Ring 433 on 18th June. Figure 11 shows the tilt change and its direction of segmental lining after one-night soil grouting. It indicates that the horizontal convergence become smaller during the grouting.

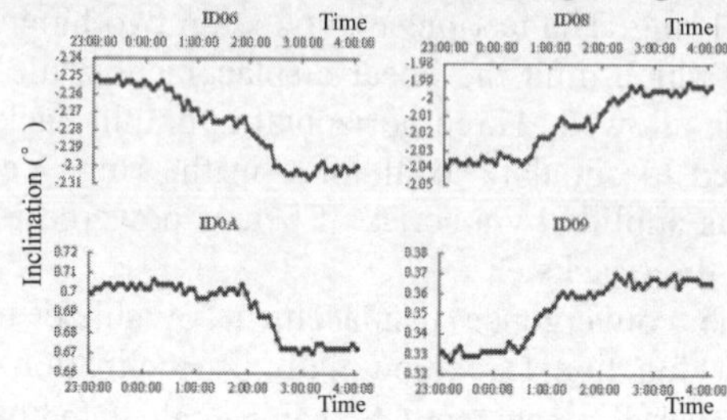

Figure 10. Tilt data of Ring 433 on 18th June

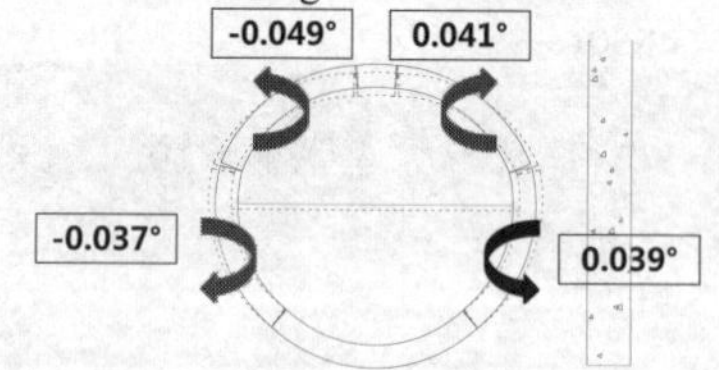

Figure 11. Schematic of segment rotation of Ring 433

The calculated change of horizontal convergence of Ring 433 due to one night grouting is -3.057 mm on 18th June. Laser distance meter and total station are conducted to verify the results. The change of horizontal convergence measured by laser distance meter and total station is -3 mm and -3.1 mm, respectively. For Ring 411, the calculated change of horizontal convergence is -2.735mm on 18th June, and the measured data by laser and total station is -2 mm and -3.7 mm, respectively.

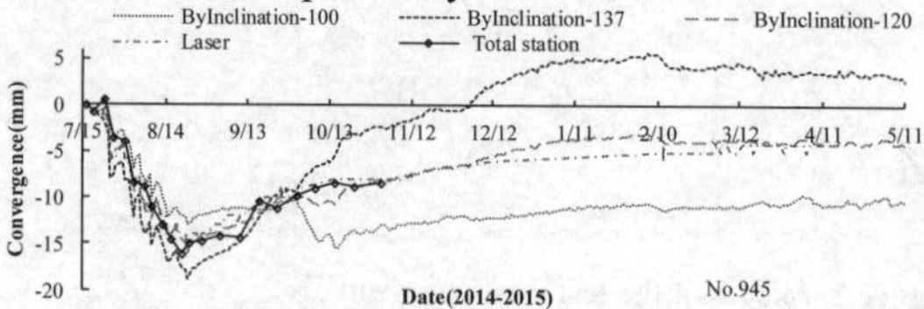

5.2 *Case 2 for Shanghai Metro tunnel*

There is a deep excavation for a business plaza nearby Shanghai metro, which would impact the tunnel safety. The nearest horizontal distance between the tunnel and deep excavation is 9.46 meters. The soil grouting is conducted for protecting the tunnel. To verify the best installation point, the tilt sensors are installed at 100° (setting the crown of tunnel as 0°), 120° and 137°.

Figure 12. Calculated convergence by tilts at different location

The convergence calculated by tilt sensor at 100° and 137° are much different from the convergence measured by laser displacement gauge and total station. However, the convergence calculated by tilt sensor at 120° agrees with the conventional method very well. The result proves that 120° is a best installation point of tilt sensor for convergence monitoring.

During grouting period, the convergence reduced about 16 mm. After grouting, the soil rebounded and the convergence kept stable at 5mm nearly. The grouting improved the soil condition and the grouting effect for convergence is about 5mm finally.

5.3 *Case 3 for Shanghai Metro tunnel*

Comprehensive demonstration for tilt sensor, seepage sensor and crack sensor is conducted in one section of Shanghai metro tunnel. The range of this demonstration is about 540 m from Ring 555 to Ring 1005. The distance between two monitoring sections is 12 m, 24 m and 36 m depending on the tunnel condition, shown in Figure 13.

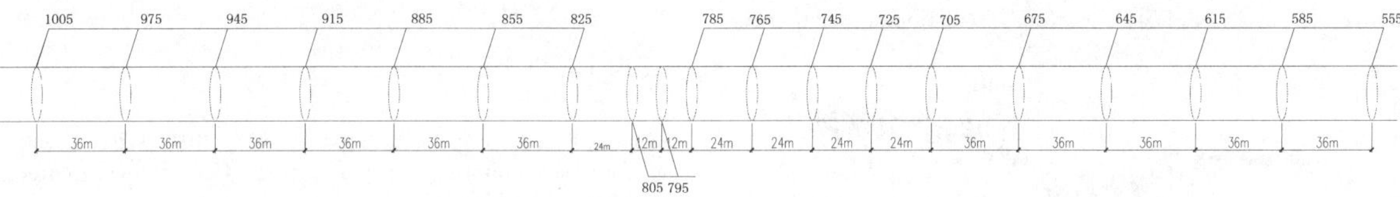

Figure 13. Monitoring sections along the shield tunnel

There are 30 tilt sensors, 4 seepage sensors and 8 crack sensors installed with two gateways to set up the wireless sensors network in total. Figure 14 shows the convergence change during one month. It is about 0.2 mm deformation in total and the convergence keep decrease during this period. The radial joint opening of Ring 765 is obtained by crack sensor. The displacement of joint is fluctuant around 0.25 mm, shown in Figure 15, which could be impacted by the vibration of running train. Circumferential joint and radial joint of Ring 765 is monitored for seepage. When the voltage of seepage sensor is above 0.1 V, there is seepage at the joint. Figure 16 indicates that there is seepage at the circumferential joint but not the radial joint.

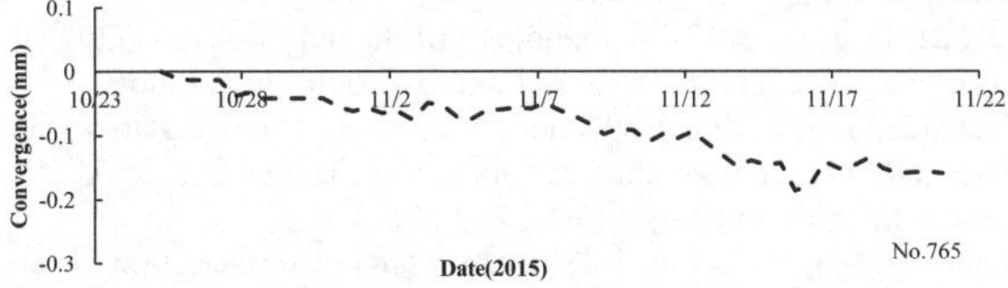

Figure 14. Convergence change of Ring 765

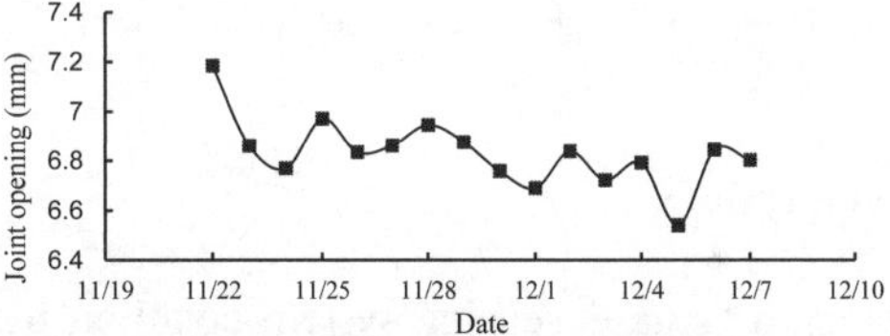

Figure 15. Opening of radial joint in Ring 765

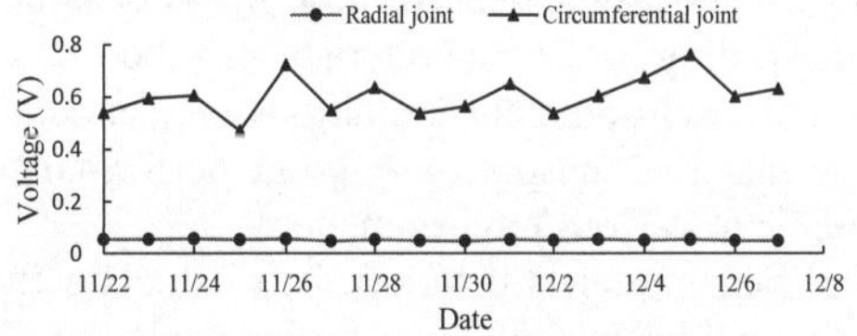

Figure 16. Seepage monitoring of Ring 765

5.4 *Case 4 for road tunnel*

MEMS tilt sensors are installed in 2 sections of the Shanghai Dalian road shield tunnel with 11m outside diameter and 480mm segment thickness. The tunnel ring consists of 8 segmental lining, shown in Figure 17. This tunnel is an underwater tunnel. Six tilt sensors are installed subject to the tunnel site condition.

Figure 18 shows that the total change of tunnel convergence is about 2 mm during the year. It is stable from January to June. However, the convergence increases obviously in summer. After summer it goes back partially. The temperature and water level increase obviously during summer in Shanghai. Therefore, the result indicates the convergence of this road tunnel increases along with the temperature and water level rising.

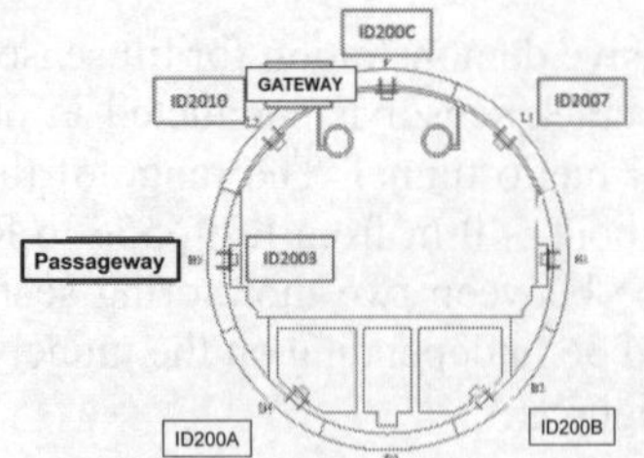

Figure 17. Cross section and tilt sensor installation of road tunnel

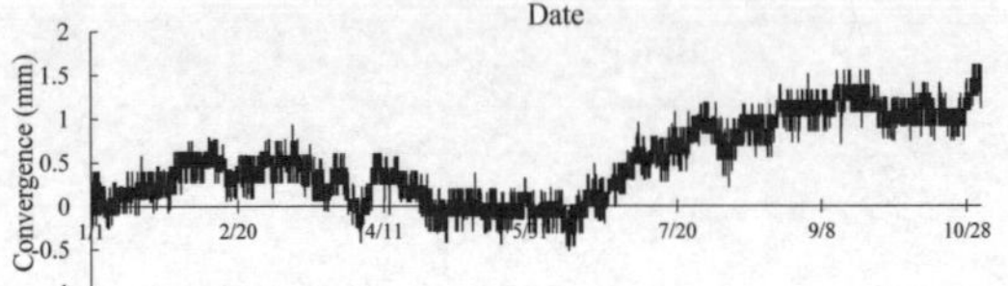

Figure 18. Convergence change of road tunnel during 1 year

6 CONCLUSION

In this paper, a "smart" sensing system coupling the tilt sensor, seepage sensor, and crack sensor with MEMS-based and Zigbee-based mesh network is developed, according to the characteristics of the tunnel structure and environmental conditions. A series of tests show that the sensors have good performance and adaptable to various environments.

A simplified model is developed to calculate the change of tunnel horizontal convergence using the tilt data. The best installation point is also recommended for Shanghai metro tunnel. According to the comparison result with conventional method, this model is good enough for field application. Furthermore, this method can obtain the real time performance of operated tunnel to ensure the safety of tunnel operation.

Four cases of the "smart" sensing system application in road and metro shield tunnel are firstly proposed in China. This system is improved significantly during these applications. So far, a huge site of 251 monitoring sections has been installed in Shanghai metro tunnel with length of 20 km. There are 518 tilt sensors and 20 gateways to compose the system. The intervals of monitoring section are 36 m, 60 m and 120 m, depending on the deformation history and tunnel condition. This system is still in testing phase and the testing data analysis is ongoing.

ACKNOWLEDGEMENT

This work was supported by the National Basic Research Program of China (2011CB013800), National Natural Science Foundation of China (51538009, 51278381, 51508403), Fundamental Research Funds for the Central Universities and the International Research Cooperation Project of Shanghai Science and Technology Committee (15220721600).

REFERENCES

Bennett, P. J. Kobayashi, Y. Soga, K. et al. 2010. Wireless sensor network for monitoring transport tunnels, Geotechnical Engineering **163(3)**, 147-156.

Cheng, S.F. & Huang, H.W. 2014. Monitoring Methods of Long-term Water Seepage in Shield Tunnel, Chinese Journal of Underground Space and Engineering **10(3)**, 733-738. (in Chinese)

He, B. Ji, Y. & Shen, R.J. 2013. Wireless inclinometer for monitoring deformation of underground tunnel, Optics and precision Engineering **21(6)**, 1464-1471.

He, B. Li, Y.G. & Huang H.W. et al. 2014. Spatial–temporal compression and recovery in a wireless sensor network in an underground tunnel environment, Knowledge and Information Systems **41(2)**, 449-465.

Huang, H.W. & Zhang, W. 2012. WSN application in urban subway shield tunnel, The Collection of the 11th cross-Strait Academic and Technical Symposium on Tunnel and Underground Engineering, Taiwan.(in Chinese)

Huang, H.W. Xu, R. & Zhang, W. 2013. Comparative performance test of an inclinometer wireless smart sensor prototype for subway tunnel, International Journal of Architecture, Engineering and Construction **2(1)**, 25-34.

Wang, K. & Huang, H.W. 2013. Smart crack sensor and its application in subway shield tunnel, Proceedings of the 6th International Conference on Structural Health Monitoring of Intelligent Infrastructure, Hongkong, China.

Llanca, D. et al. 2013. Methodology of diagnosis of urban tunnels in service, CRC Press-Taylor & Francis Group, Boca Raton.

Richards, J. A. 1998. Inspection, maintenance and repair of tunnels: international lessons and practice, Tunnelling & Underground Space Technology **13(4)**, 369-375.

Yuan, Y. Bai, Y. & Liu, J. 2012. Assessment service state of tunnel structure, Tunnelling & Underground Space Technology **27(1)**, 72-85.

Proceedings of the International Conference on Smart Infrastructure and Construction
ISBN 978-0-7277-6127-9

© The authors and ICE Publishing: All rights reserved, 2016
doi:10.1680/tfitsi.61279.197

Application of a wireless mobile system for real-time bridge scour monitoring

M.C. Wang[1], W.T. Chou[1], T.K. Lin[*2],K.C. Lu[2] and K.C. Chang[2]

[1]Department of Civil Engineering, National Chiao Tung University, Taiwan
[2]National Center for Research on Earthquake Engineering, Taipei, Taiwan
[*] Corresponding Author

ABSTRACT This paper proposes a wireless mobile system for real-time bridge scour monitoring. Through the multichannel wireless sensing system, the superstructure vibration of a bridge during flooding was first measured. A structural health monitoring algorithm combining the concept of array expression data and Naïve Bayes classification was then developed for the estimation of the corresponding scour depth. Moreover, by introducing a simple and intuitive software interface, the assessment result of the bridge scour depth can be rapidly displayed, allowing the users to easily interpret the current bridge safety status. To verify the performance of the proposed system, a series of full-bridge scour validation experiments was carried out. The result has shown that a reliable prediction of the bridge scour depth can be provided by the proposed system, which can be used for further evaluation of the overall safety condition before the critical damage or collapse of the bridge.

1 INTRODUCTION

Over the last few decades, structural health monitoring (SHM) has been an important field for structure engineering, and lots of researches have been conducted. With the support of numerous sensors such as temperatures, inclinometer, and accelerometers deployed on the structure, the health condition and the warning signal can be promptly announced by the developed diagnostic theories when the structure achieves its safety margin. (Cheung, Moe M. S. *et al.* 2007)

Due to the scour effect caused by flooding, bridge may face serious damage risk such as collapse, and related researches about bridge scour have been studied these years. For example, the characteristics of the soil layers of a bridge was analyzed to realize the depth change when the bridge is scoured (Ko, Y., *et al.* 2014).

Recently, accompanying with the advancement of wireless technology, the application of wireless sensors for SHM has been broadly studies. As a significant amount of research has been devoted to the implementation of real-time mobile SHM system, the integration of the wireless sensing technique and the SHM system has also been attempted. Since obtaining the real-time scour depth under the harsh environment during flooding is still a tough issue among the research of bridge scour, a wireless mobile system for real-time bridge scour monitoring is proposed in the study.

2. THE PROPOSED ALGORITHM FOR SCOUR DEPTH EVALUATION

The proposed algorithm is composed of three stages including the auto-regression–auto-regression with exogouos (AR–ARX) model, the Naïve Bayes (NB) classification, and the scour depth estimation module as shown in Figure1.

2.1 *AR–ARX Expression Array*

The DNA expression array extracted from human body was proven to be efficient in disease detection (A.D. Keller 2000). As the two-tier AR-ARX array has been successfully applied for SHM on small-scale structures (Sohn and Farrar, 2001), it is adopted to represent a characteristic array of the structure similar to the DNA expression array in the first part of the proposed algorithm.

The AR-ARX array can be established as follows. First, the bridge structure is considered to be a linear single-input-single-output (SISO) system, and the external excitation to the structure during flooding is assumed as random vibration. As a result, an AR model can be established where the coefficient calculated from the AR model is deposited in the first part of the expression array. To clearly illustrate the structure characteristic, an additional ARX model is established in the second stage. In the ARX model, the uncertain external input can be estimated by considering the error values in the first-stage AR model as the external input source. The detailed derivation of the equations is as follows.

For each time series measured from the superstructure, an AR model (Figure 2) with p AR term is first constructed as

$$x(\mathrm{t}) = \sum_{j=1}^{p} \phi_{xj} x(\mathrm{t}-\mathrm{j}) + \mathrm{e}_x(\mathrm{t}) \qquad (1)$$

where $x(\mathrm{t})$ is the structural response measured for time step t, and $e_x(\mathrm{t})$ is the error between the measurement and the prediction.

Under the assumption that the error $e_x(\mathrm{t})$ obtained from the AR model is mainly caused by the unknown external input, a double-tier ARX prediction model (Figure 3) is employed to reconstruct the input/output relationship between $e_x(\mathrm{t})$ and $x(\mathrm{t})$ as

$$x(\mathrm{t}) = \sum_{j=1}^{a} \alpha_j x(\mathrm{t}-\mathrm{j}) + \sum_{j=1}^{b} \beta_j \mathrm{e}_x(\mathrm{t}-\mathrm{j}) + \varepsilon_x(\mathrm{t}) \qquad (2)$$

Finally, the characteristic array database for the whole time history of the measured vibration can be established using the coefficient obtained in the AR-ARX model shown in Figure 4.

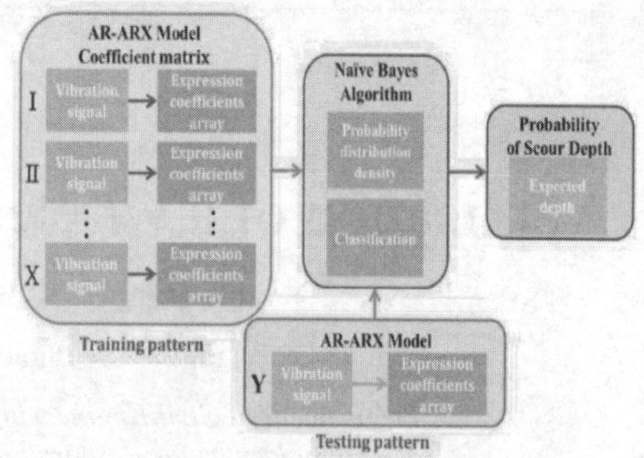

Figure 1. The proposed scour depth evaluation algorithm

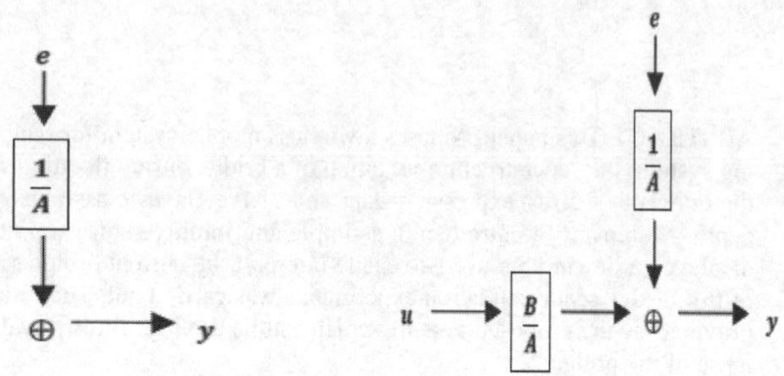

Figure 2. The AR model **Figure 3.** The ARX model

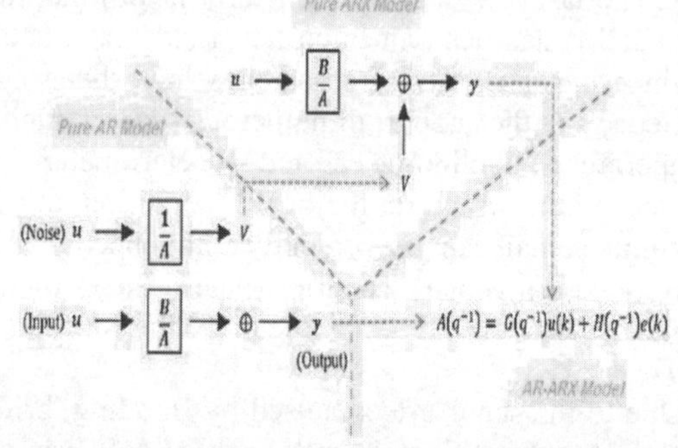

Figure 4. The two-tier AR-ARX model

2.2 *Naïve Bayes Classification*

In order to determine the real-time scour status, the Naïve Bayes classification is introduced in the second stage for the comparison between the training and testing patterns shown in Figure 5. The distribution of coefficients of the column vector is considered as a normal Gaussian distribution, and each coefficient of the row vector is independent. Therefore, the probability density distribution in the database can be condensed into two statistical parameters: the mean value and standard deviation.

In this study, the training pattern is obtained by analyzing the vibration response of the superstructure under different bridge scour conditions, and the time series of each condition corresponds to a specific AR–ARX characteristics array as shown in Figure 6. The application of NB classification is briefly introduced as follows:

Firstly, after processed by the AR-ARX with $n(\mathrm{n}=\mathrm{p}+\mathrm{a}+\mathrm{b})$ different components, an array X from an unknown scour condition can be expressed as

$$X_i = \{X_1, X_2, X_3 \cdots X_g \cdots X_n\} \quad (3)$$

To precisely detect the scour condition of the unknown array X, the training patterns M_i with i different scour levels are introduced. Similar to the unknown array X, every training pattern of a scour condition M_i is also composed of n components, given by

$$M_i^g = \{M_i^1, M_i^2, M_i^3 \cdots M_i^g \cdots M_i^n\} \quad (4)$$

where i represents the scour condition i

The scour condition of the unknown array X is then discriminated by the NB algorithm according to equation (5)

$$Class(\mathrm{x}) = \arg\max_i(\log p(\mathrm{x}|M_i)) \quad (5)$$

Moreover, the normal distribution can also be presumed for every component of M_i when the sample numbers are large enough

$$P_{\mu,\sigma^2}(\mathrm{x}) = \frac{1}{\sigma\sqrt{2\pi}} e^{\frac{(\mathrm{x}-\mu)^2}{2\sigma^2}} \quad (6)$$

By utilizing equations (5) and (6) with separate mean values μ and standard deviations α among the coefficients in the AR-ARX array, the corresponding scour condition of the bridge can be determined using equation (7). Moreover, the probability of the testing pattern can also be determined using the NB classification probability model.

$$Class(\mathrm{x}) = \arg\max_i \left\{ \sum_{coefficient} \left[-\log\sigma_i^g - 0.5\left(\frac{x^g - \mu_i^g}{\sigma_i^g}\right)^2 \right] \right\} \quad (7)$$

As the NB classification is mainly based on the statistical analysis, its reliability can be improved by enhancing the size of the training database.

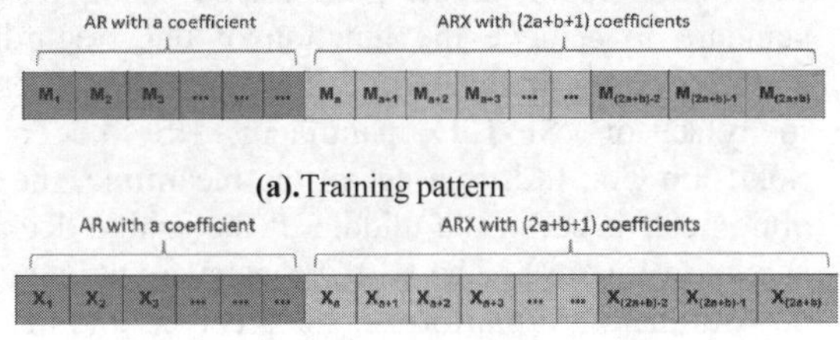

Figure 5. The array database for NB classification

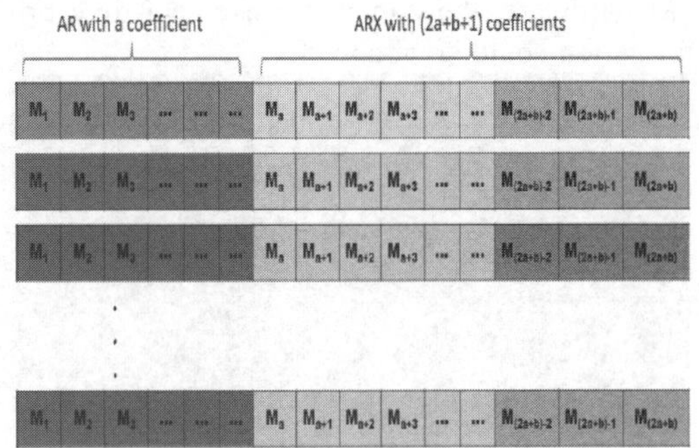

Figure 6. AR-ARX characteristics array

2.3 *Scour Depth Estimation Module*

After calculating the probability distribution of the testing pattern, the expected value of the scour depth can be obtained by multiplying the scour depth of every class with the corresponding probability, which can then be illustrated as

$$P_i = \alpha K_i \quad (8)$$

$$D_e = \sum_0^i P_i D_i \quad (9)$$

3 DEVELOPMENT OF THE WIRELESS BRIDGE SCOUR MONITORING SYSTEM

For its powerful and reliable processing capability, the NI CompactRIO-9074 is adopted as the SHM platform of the mobile bridge scour monitoring system. In addition, high efficiency and stability can be further provided by the integrated wireless sensors.

3.1 The SHM Hardware

The NTU-WSU wireless transmission unit (Figure 7) is adopted in this study, and the measured data are transmitted to the SHM platform by using RS232 standard to achieve the function of data recording and processing. In the part of sensor selection, the velocity sensor VSE-15D, manufactured by the Tokyo Sokushin Co., Ltd, is used. In the meantime, the inclinometer is set in the middle of the bridge deck as shown in Figure 8. The NI 9795 gateway is integrated with the SHM platform as the receiver, and the inclinometer is connected with the NI WSN-3202 module to reflect the inclination angle instantly when the bridge was under scour. Through the deployed SHM hardware, the comprehensive real-time bridge information could be obtained.

Figure 7. The NTU-WSU and SHM platform

Figure 8. The installed inclinometer on the bridge deck

3.2 SHM Software

The monitoring interface is developed on the basis of the software LabVIEW. As shown in Figure 9, the monitoring result is presented graphically. The warning levels and notification message for the safety condition of the bridge are provided in the top-left corner. Since long-time monitoring is a necessity of the proposed SHM system, the wireless connection quality is also included in the monitoring system. The battery status and connection quality between the wireless nodes are shown at the top of the figure. Moreover, the monitoring of the inclination angle of the bridge deck during the flooding is indicated at the top-right of the interface, which assists the determination of the structural characteristics over time.

The embedded depth of two bridge piers is illustrated in the main part of the interface, where S_1 and S_2 represent the conditions of the two individual piers P_1 and P_2. The time history of the vibration signal and the corresponding scour depth estimated from the proposed algorithm are displayed to reflect the real-time status of the bridge.

To verify the stability of the wireless module, the recorded data were preliminarily compared with a commonly used portable instrumentation device (SPC-51). The same velocity sensor VSE-15D was used to verify the consistency between the measured data of the wireless module and the signals obtained by the SPC-51. The vibration time history recorded from the test is shown in Figure 10. As identical results can be provided by the two systems, the NTU-WSU wireless module has been demonstrated to be reliable for the SHM system.

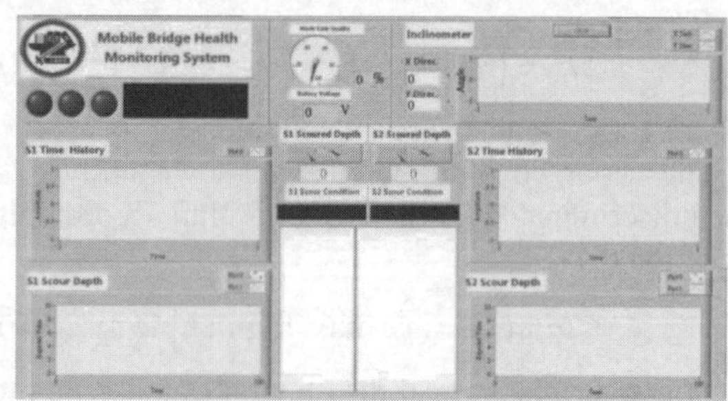

Figure 9. The interface of the monitoring system

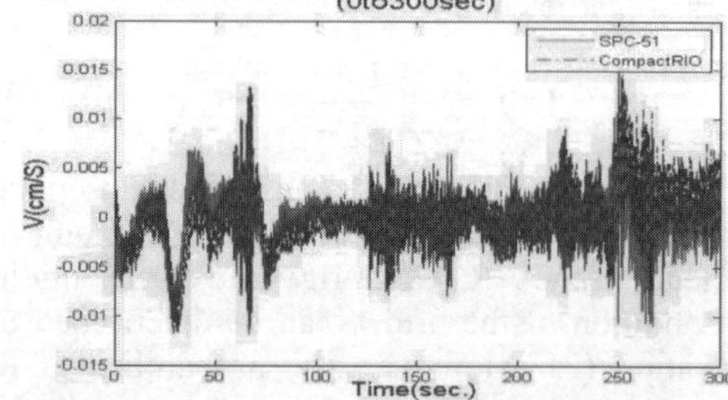

Figure 10. Comparison of time history between two systems

4 EXPERIMENTAL VERIFICATION OF THE BRIDGE SCOUR MONITORING SYSTEM

The velocity sensors were arranged parallel to the flow direction on the cap beam of the bridge column. The columns were partially buried in the sand, which was compacted around the test specimen to prevent an inadequate initial scour caused by the high porosity. The soil was gradually reduced through the scour process, which causes the boundary condition of the bridge column to change with time.

Figure 11 depicts the layout of the experiment. The inner camera was installed inside the hollow acrylic specimen for verifying the accuracy of the scour depth. A spare wired instrument was also deployed as for further comparison with the wireless system. The vibration response of the bridge at different scour depths was measured and recorded wirelessly, and the SHM platform system was applied to provide a real-time evaluation of the bridge scour depth. The estimated bridge scour depth was then presented through the display interface.

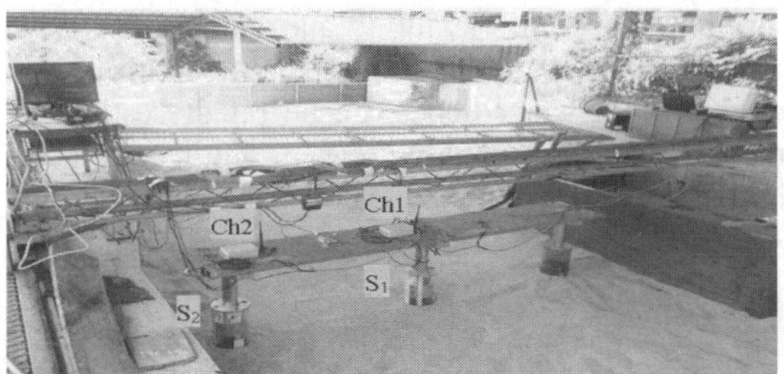

Figure 11. The condition of the experiment

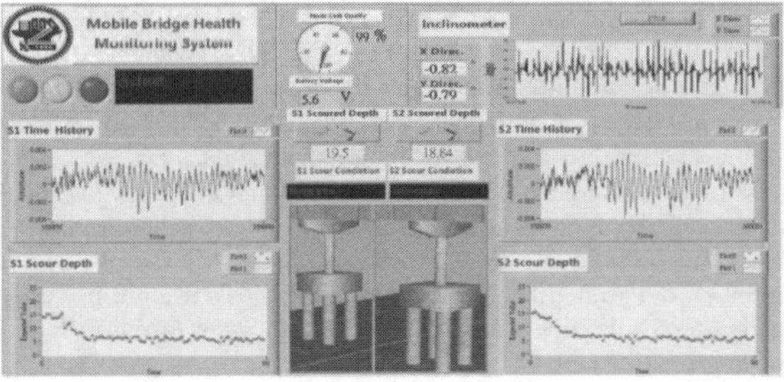

Figure 12. The experimental result

4.1 *Scour Depth Analysis and Comparison with Video Recording*

The experiment was lasted a total of 30 minutes, and the vibration signals were recorded with a sampling rate of 200 Hz when the water started to flume the bridge pier. Figure 12 shows the results of the bridge scour depth monitoring system after the experiment was completed. The vibration time history of sensors S_1 and S_2 are shown in Figures 13 and 14, respectively. The figures illustrate that the trend is similar between the two piers, and the times of the peak values of the vibration time history are highly consistent.

To verify the accuracy of the scour monitoring system, the video recording from the inner camera was compared with the estimated scour depth. As shown in Figure 15, S_1 indicates the evaluated embedded depth of the middle bridge pier, and S_2 is the estimated embedded depth of the left bridge pier. Moreover, the curves Camera1 and Camera2 represent the video recordings for S_1 and S_2, respectively.

For the record from the video recording, the embedded depth at the beginning was 25 cm, and it declined to 10 cm, which is 60% of the total scoured depth, in only 10 minutes. A smooth trend was then observed in the next 10 minutes. However, a bias was observed for the estimated scour depth by the proposed algorithm. The embedded depth was initially 15 cm and then converged with the video recording. When the experiment finished, the final embedded depth was detected as approximately 5 cm.

To modify the estimated embedded depths of S_1 and S_2 with the results from the video recording, a modification factor was applied

$$Modification\,factor = \frac{D_{inital} + D_{final}}{D_{inital} - D_{final}} \tag{10}$$

where D_{inital} is the initial depth before scour; D_{final} is the final depth when the experiment ends.

As shown in Figure 16, the overestimation phenomenon that occurred at the initial scour stage was improved after the modification factor was applied. In addition, a reliable scour depth trend, which is similar to the results of video recording, could be rapidly provided by the proposed monitoring system as only vibration of the superstructure is required.

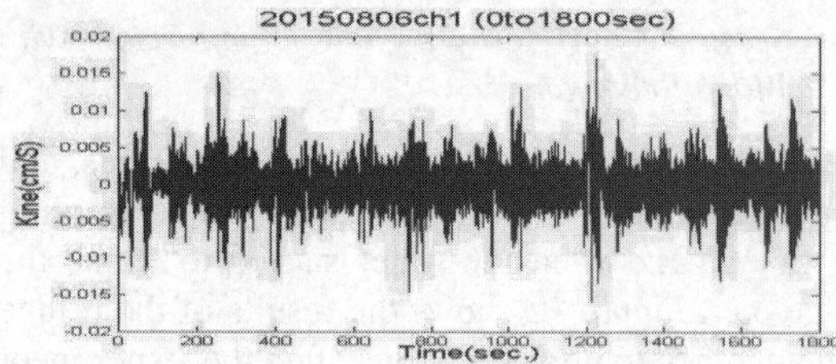

Figure 13. The time history of S_1 of the middle bridge pier

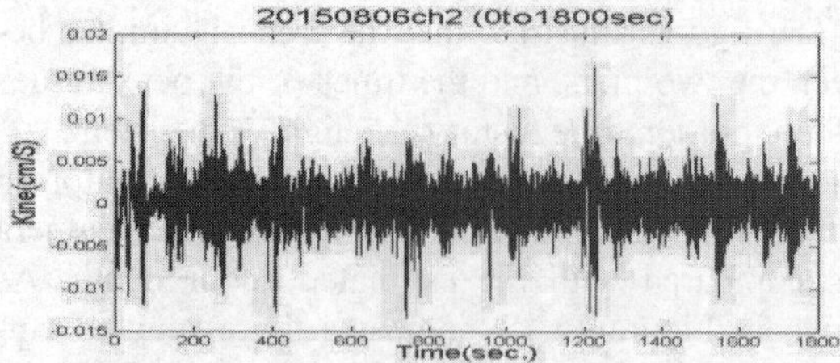

Figure 14. The time history of S_2 of the left bridge pier

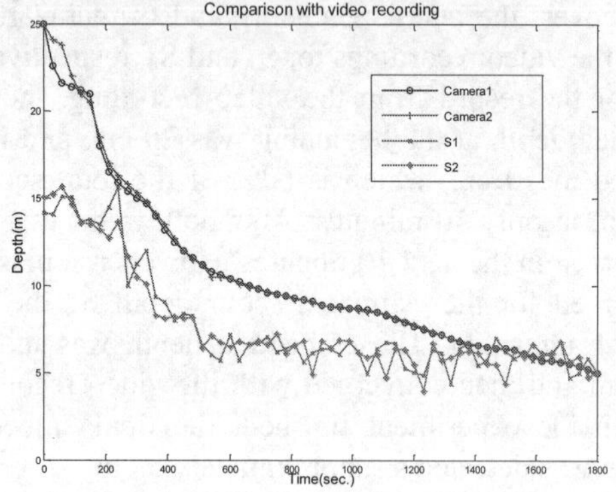

Figure 15. Comparison between the video recording and the proposed algorithm

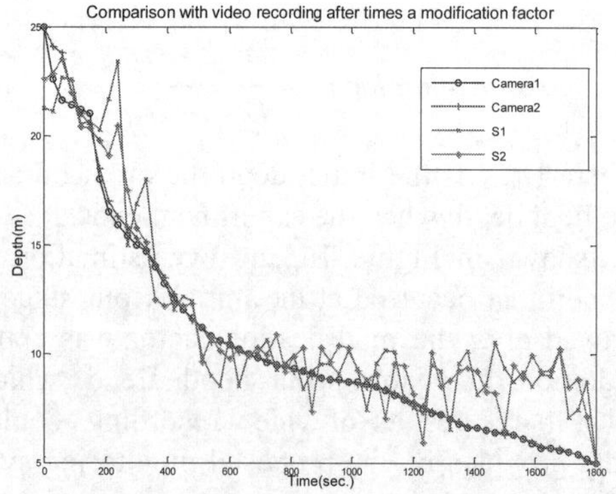

Figure 16. Comparison between the video recording and the proposed algorithm after modification

5. SUMMARY AND CONCLUSION

A wireless mobile system for real-time bridge scour monitoring is proposed in this study. A SHM algorithm combining the concept of array expression data and Naïve Bayes (NB) classification was developed to detect the scour depth. The array expression data were first generated from the measured vibration signal through the AR-ARX model. With the support of the NB classification, the approximate scour level can be rapidly determined, and the expected scour depth of every bridge pier can be calculated based on the corresponding probability of different scour levels.

The proposed real-time scour monitoring system was examined through a series of scour experiment. As demonstrated, the predicted scour depth of the entire scour process fits well with the recorded ones after considering an additional amplification factor. Subsequently, the rapid assessment of the bridge scour depth can be achieved, enabling users to easily interpret the current bridge safety state.

Compared with the existing bridge scour monitoring systems, the flexibility and mobility has been largely improved by the proposed system. Moreover, as the bridge scour depth can be rapidly estimated, the overall safety condition can be further evaluated before the critical damage or collapse of the bridge.

REFERENCES

Cheung, Moe M.S. Noruziaan. Bahman & Yang, C.Y. 2007. Health monitoring data in assessing critical behavior of bridges, *Structure and Infrastructure Engineering*, **v3**, **n4**, p325-342.

Ko, Y. Chiou, J. Tsai, Y. Chen, C. Wang, H. & Wang, C. 2014. Evaluation of flood-resistant capacity of scoured bridges, *Journal of Performance of Constructed Facilities* **28**, Special section: Performance of bridges under critical natural hazards, pp.61–75.

Keller A.D. Schummer M. Hood L. & Ruzzo W.K. 2000. Bayesian classification of DNA array expression data, *Technical Report UW-CSE*, 2000-08-01.

Sohn, H. & Farrar C.R. 2001. Damage diagnosis using time series analysis of vibration signals, *Journal of Smart Materials and Structures*, **10(3)**, pp.446-451.

Proceedings of the International Conference on Smart Infrastructure and Construction
ISBN 978-0-7277-6127-9

© The authors and ICE Publishing: All rights reserved, 2016
doi:10.1680/tfitsi.61279.203

Monitoring on the performance of temporary props using wireless strain sensing

X.M. Xu*[1], P.R.A. Fidler[1], D. Rodenas-Herraiz[2], W. Li[4], V. Kumar[1], J. Birks[1], J. Yan[1, 3] and K. Soga[1]

[1] *Department of Engineering, University of Cambridge, UK*
[2] *Computer Laboratory, University of Cambridge, UK*
[3] *School of Electronics and Computer Science, University of Southampton UK*
[4] *Aeroflex Incorporated, UK*

ABSTRACT Although temporary props have been extensively used in underground support systems, their actual performance is poorly understood, resulting in potentially conservative and over-engineered design. This paper presents the performance monitoring of 4 temporary props in an urban construction site using a newly developed wireless strain sensor node featuring a 24-bit ADC. For each prop, 6 strain gauges and 3 temperature sensors were directly attached onto the prop surface using super glue, and then connected to a wireless strain sensor node mounted in the middle span. Each sensor node transmitted both monitoring data and network diagnostic messages in near-real-time over an IPv6-based (6LoWPAN) wireless mesh sensor network. The data were also stored locally at each node on a micro SD card. Extensive testing and calibration was undertaken in the laboratory to ensure that the system functioned as expected. The prop loads are presented without correction for temperature effects and compared with the design loads. The monitoring data reveal the development of loads in temporary props during excavation, the formation of the basement and the extraction of the props. The network performance characteristics in terms of message reception ratio and network topology evolution are also highlighted and discussed.

1 INTRODUCTION

Temporary support systems in underground construction have become increasingly complex due to the increased complexity of underground infrastructure and surrounding ground conditions. This has potentially resulted in both conservative or unsafe designs (Bhalla et al. 2005). It is therefore essential to monitor the real performance of these supporting elements to ensure their satisfactory behaviour.

Wireless Sensor Networks (WSNs) are nowadays a mature technology, increasingly used for various large-scale applications including precision agriculture, environmental and infrastructure monitoring. Compared to the traditional sensor networks, the use of wireless technology has proven to offer distinctive advantages, such as flexible, faster and denser deployment of sensors in the field (Xu et al. 2015; Liu et al. 2015). This paper concerns the deployment of a WSN for performance monitoring of 4 temporary props in an urban construction site, for which a newly developed wireless strain sensor node was used.

2 WIRELESS STRAIN SENSOR NODE

A new wireless strain sensor node was developed by the Cambridge Centre for Smart Infrastructure and Construction (CSIC). Extensive testing and calibration was undertaken in the laboratory to ensure that the system functioned as expected.

2.1 *Wireless strain sensor node*

The CSIC SmartPlank version 2 sensor node is an 8-channel 24-bit ADC sensor board, as shown in Figure 1. The board supports 6 strain sensor analogue input channels, with a further 2 analogue channels specialised for use with load cells. The board also features three 1-wire connections for digital temperature sensors, such as the Maxim Dallas DS18B20, a real-time clock (RTC), power button, JTAG/ISP programming interface, a micro SD card socket for data logging purposes, and a multi-position switch for rudimentary in-field configuration. It provides a flexible and versatile platform to address the needs of a variety of applications. For example, depending on the application requirements, a quarter bridge or half-bridge can be easily reconfigured for foil strain gauges with either 120ohm or 340 ohm resistances.

Figure 1. Wireless strain sensor node developed in CSIC

The board is packaged in a robust (IP67) plastic housing.

2.2 Wireless strain sensor software

The application software running on the wireless strain sensor node was developed in Contiki OS (Dunkels et al. 2004). The program reads all 8 ADC sensors, the 3 digital temperature sensors and the time from the RTC. It then stores the readings on the micro SD card (if present) and transmits the sensor data via a UDP connection. Nodes use the Contiki OS standards-based IPv6 protocol stack (6LoWPAN/RPL) for link-local addressing and routing, and ContikiMAC at MAC layer for low-power operation. A more detailed description of the software can be found in Nawaz et al (2015).

2.3 Wireless strain sensor characterization

Sensor node testing and calibration was performed in a laboratory environment. The first test was to investigate the linearity and repeatability of strain gauges on all 6 channels. This was conducted using a 4-point bending test platform specially designed for sensor calibration. Figure 2 shows the incremental ADC readings from all 6 channels with 9 loading steps (up to 681$\mu\varepsilon$). Note that the variation of each channel's reading at each loading stage is less than 0.05$\mu\varepsilon$, and sub-microstrain measurement can be easily achieved, as indicated in the inset (ch1) of Figure 2.

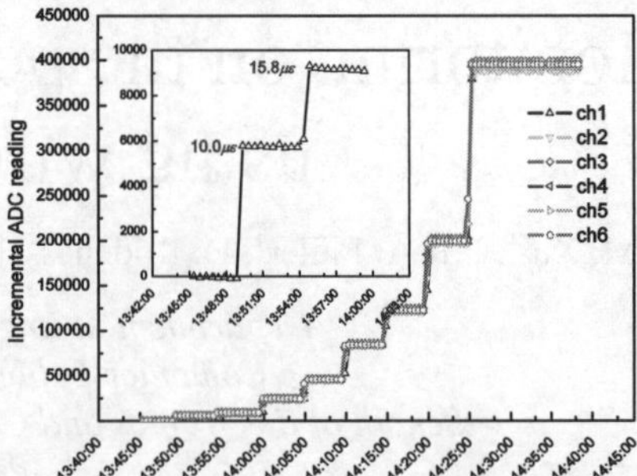

Figure 2. Calibration on the strain gauges using 4-point bending test platform.

The DS18B20 temperature sensors were tested in a water bath under heating and cooling cycles, with temperature readings from the sensors differing by no more than 0.3125 ºC from the water bath reference temperature. The load cell was tested using a direct shear apparatus (see Figure 3(a)), and the results is given in Figure 3(b). To further check the temperature effect, and robustness of the packaging, the sensor node was emerged in the water (5-45 ºC cycles), with all 6 strain gauges, 3 temperature sensors and 2 load cells connected, as shown in Figure 4.

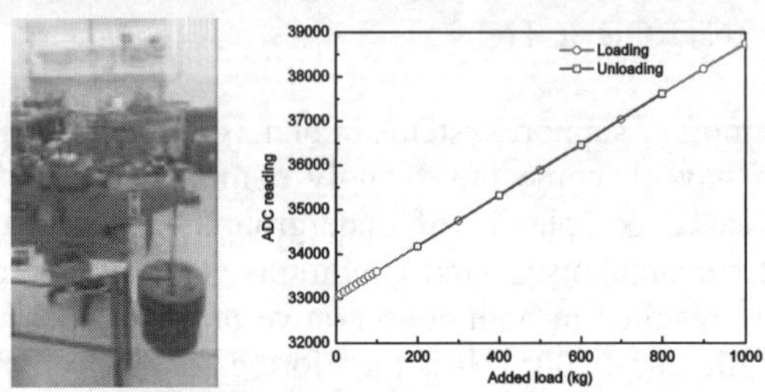

(*a*) Test apparatus (*b*) Test results

Figure 3. Calibration on the load cell using direct shear apparatus: (*a*) Test apparatus; (*b*) Test results.

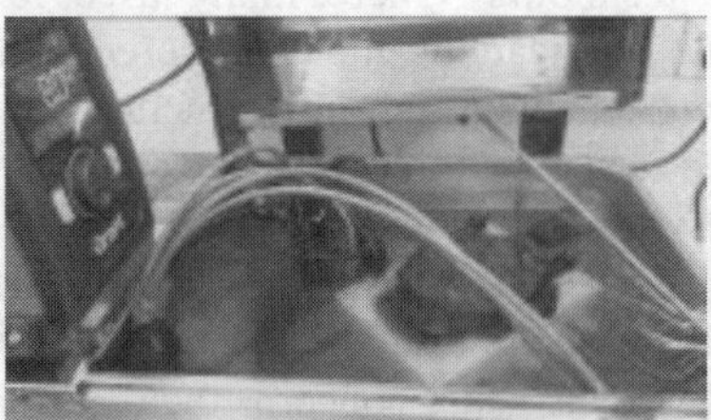

Figure 4. Robustness testing on the wireless strain sensor node.

3 APPLICATION

A wireless strain sensor network was deployed in an urban excavation site in Cambridge (UK), to provide an opportunity to understand the real performance of temporary supporting props in near-real-time.

3.1 Field overview

The Trinity Hall excavation site was situated on the eastern side of Thompson's Lane and 60m north of St. Clement's Church and Bridge Street, at the north-western end of Cambridge city centre. The proposed new student residence is to be a four storey building housing student flats with a single basement across the building footprint providing cycle parking and a common room. The approximately rectangular redevelopment site extended 24 m to 28 m eastwards from its 43 m long frontage onto the eastern side of Thompson's Lane. The site was bounded to the south by CATS library and Cambridge Spiritualist Church; to the north by Bishop Bateman Court and its rearward car park; and to the east by the rear boundaries of residential properties lining Portugal Place and Portugal Street.

The site stands at an approximate elevation of 7.7 m above sea level on land that slopes gently down towards the north. The investigation found a thick cover of made ground beneath the site associated with the historical raising of the site above the River Cam flood plain, together with the later construction of St. Clements Gardens. The foundations for the new four-storey residential block and basement will need to penetrate this made ground, the Alluvium and Terrace River Gravel, and could be based on the underlying Gault clay.

A number of monitoring technologies with which CSIC is familiar were used at the site in order to understand the real performance of various elements in the ground works during the basement excavation (as indicated in Figure 5). For example, an array of wireless MEMS inclinometers and accelerometers (together with humidity and temperature sensors) were deployed along the boundary wall of the CATS library to monitor the movement of the adjacent building during the sheet pile installation and subsequent basement excavation. The instrumentation was installed for a period of time prior to the works commencing. Fibre Bragg Grating (FBG) sensors were attached on a number of sheet piles along the southern and eastern walls of the basement to monitor the dynamic strain response of the sheet piles during installation, excavation and construction of the superstructure. For the temporary props both the newly developed wireless strain sensors and FBG sensors were used to monitor the load development in the temporary propping system during the basement excavation.

3.2 Wireless strain sensor field deployment

Four props in total, two centre props (props 1 and 2) and two corner props (props 3 and 4), were instrumented with wireless strain sensor, as indicated in Figure 6. These field deployments took place in stages as the excavation and associated archaeological work progressed. Due to the timing and space restrictions on site the wireless strain sensors were only deployed once the prop itself had been installed in the excavation. For example, the first wireless strain sensor was attached on prop 1 on 19th June 2015, while the last one was installed on prop 2 on 14th July 2015. FBG sensors were also installed on prop 1 and 4 for comparison, as indicated in Figure 5. Unfortunately however, these fibre-optic cables were damaged by mechanical diggers prior to the deployment of the wireless strain sensors. The wireless gateway and data logger was installed in February 2015, prior to the deployment of wireless MEMS tilt sensors on CATS library.

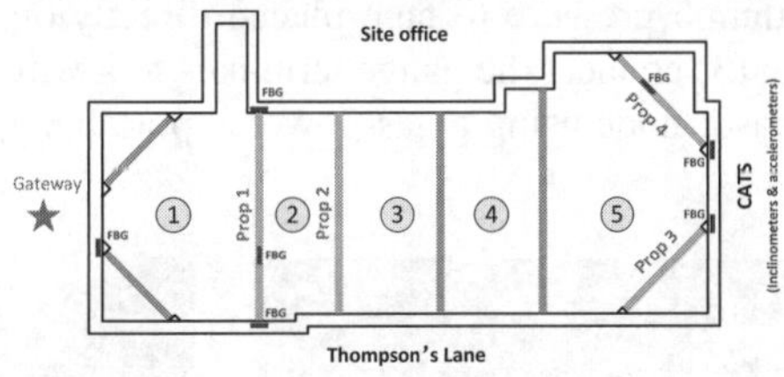

Figure 5. Layout of the instrumented temporary props, sheet piles and sensor network at Trinity Hall site

For each instrumented prop, one wireless strain sensor node was attached to 6 strain gauges and 3 temperature sensors. There were two kinds of props used in this project, namely tubular (props 1 and 3) and rhombic (props 2 and 4) ones. Gauges were attached at the 3, 6, 9 and 12 o'clock positions at the centre of the tubular props, as the most economical and practical option (Batten et al. 1999). The temperature sensors were attached next to the strain gauges to provide for temperature compensation of the measured strains. There was an exception to this configuration where one of the strain gauges on prop 1 (channel 3) was attached ¼ prop length away from the end of the prop at the Thompson's Lane side. The wireless sensor node was located in the middle of each prop. Detailed configuration of wireless strain sensors on each prop is described in Figure 6.

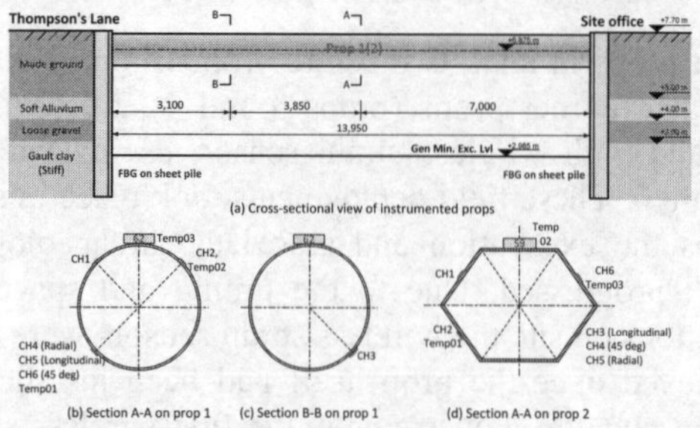

Figure 6. Configuration of wireless strain sensor on instrumented props and geological strata at Trinity Hall site

Foil strain gauges demand considerable care during field installation, due to their fragile nature. The attachment method is as follows: (1) remove the surface rust from the steel prop in the desired location with a battery-powered wire brush, and then with coarse and fine sand paper. Thoroughly clean the area with acetone; (2) apply super glue and align the gauge in the appropriate location. Immediately apply firm thumb pressure to tape placed directly over the gauge; (3) connect the gauge terminals to a wire from the sensor node using a gas-powered portable soldering kit.

(*a*) Instrumented props (*b*) Wireless strain sensor on prop

Figure 7. Field deployment of the wireless strain sensor network at the Trinity Hall excavation site.

In addition, foil strain gauges are very prone to deterioration due to water. They must be properly sealed if used in the underground structures or outdoor environments, where they are likely to encounter excessive moisture or erosion. For the first two props, the strain gauges were protected only with large quantities silicone sealant, as shown in Figure 7. This proved to be unsuccessful for water ingress protection, as evidenced by some of the sensor readings, which exhibited dramatic changes. For the later installations, a coating of M-Coat A was first applied over the entire gauge and terminal area. The installation of temperature sensors was relatively simple. These were embedded in silicone sealant to capture the temperature change of the prop itself, rather than that of the surrounding environment.

3.3 *Wireless strain sensor network*

Figure 8 presents the layout of the wireless strain sensor network at the Trinity Hall site (Triangle: strain sensor nodes on props; Square: tilt sensors on the wall of CATS library). Data messages are sent from each node at fifteen minutes intervals. Interestingly, it shows that sensor nodes were mainly routing message via the distant node 67 on prop 4, rather than using nearby nodes to forward messages.

Figure 9 shows the data message delivery ratio (MDR) computed from 4 wireless strain sensor nodes during the entire monitoring period. MDR for each node was obtained as the number of data messages successfully delivered to the gateway with respect to the total number of expected data transmissions. It can be observed from the figure that, the values of MDR for props 1 and 4 were above 80%, with their average PDRs of 99.7% and 97.5% from 11th July to 23th September 2015, respectively. The reduction of MDR in prop 1 between 7th July and 10th July 2015 was due to a transient fault with the gateway.

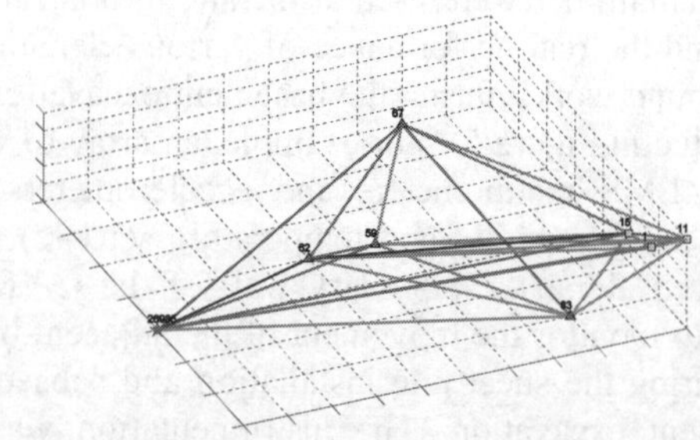

Figure 8. Network topology in Trinity Hall. Link colour represents the average number of connections made to the gateway per day

during the 110-day period. Grey lines indicate one or two connections; blue lines between 2 and 5 connections; green lines between 5 and 15 connections; and red lines more than 15 connections.

For props 2 and 3, it was observed that there were significant variations in MDR time history, with their average MDRs of 88.5% and 71.5%, respectively. The former may have been due to the ongoing excavation work, while the latter was probably due to the use of an internal chip antenna. It was witnessed that the external antenna on prop 2 was frequently disturbed by the digging bucket, as highlighted in Figure 7(a). Fortunately, all the data that failed to be delivered via the wireless network was later recovered from the local micro SD card storage.

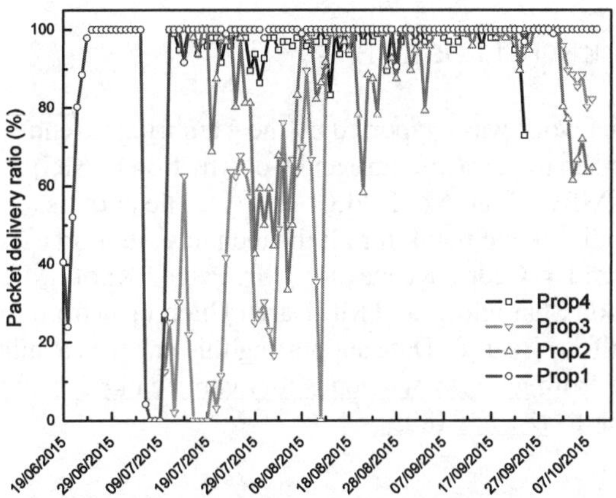

Figure 9. Packet delivery ratio at the gateway

3.4 *Monitoring results*

Figure 10 presents two examples of the measured incremental axial strain on prop 1 and 2 during the excavation period, with respect to the baseline readings taken immediately after the installation. Negative strains indicate compression. It can clearly be observed from Figure 10(a) that prop 1 mainly experienced slight tension, rather than compression. The data from the FBG measurements in the sheet piles at both ends of the prop also confirmed the prop performance. This is probably due to local reinforcement at each end of the prop. (Note that data received by via the WSN, shown in red, is incomplete; however the missing data, shown in grey, was subsequent retrieved from the SD card.)

Similar field performance was observed on prop 2, as indicated in Figure 10(b). The measured incremental strain increased to around $50\mu\varepsilon$ on 20th July, and reduced to approximate $-60\mu\varepsilon$ on 10th August. It again increased to about $80\mu\varepsilon$ on 17th August, and then gradually decreased to around $10\mu\varepsilon$ by the end of the monitoring period. Excavation levels are shown in Figure 10(b), for zones 2 and 3 (as indicated by the circled numerals in Figure 5). The excavation and backfilling was completed on 25th August 2015. Although the excavation level data is sparse, it is clear to see that the measured strain variations were in good alignment with the excavation levels.

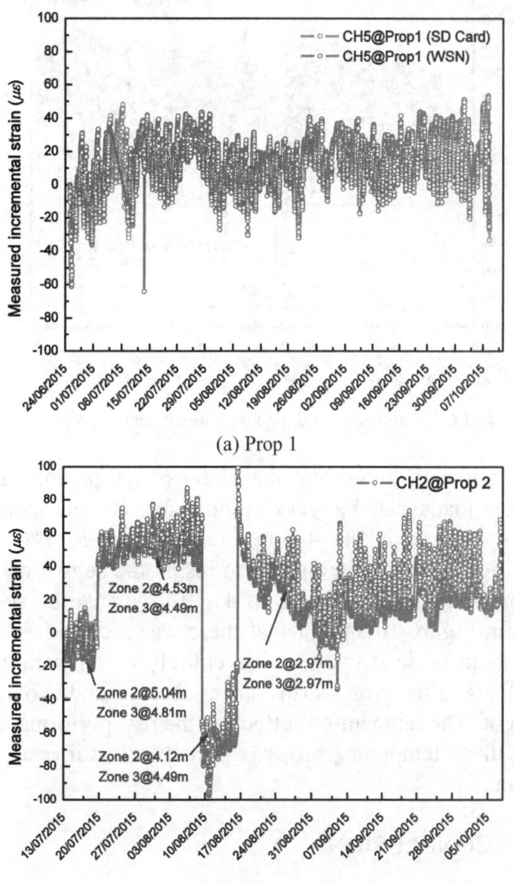

Figure 10. Examples of the measured incremental strain

Figure 11 plots the incremental axial loads on 4 instrumented props. The axial prop load was calculated using the measured incremental strain, Young's modulus for the steel (210GPa) and the cross-sectional area of steel props (0.021048m^2 and 0.016

m^2 for tubular and rhombic props, respectively). Although not temperature compensated, it is clear from the Figure that all the 4 props were not carrying much compression load in comparison to their design loads. Instead, somewhat surprisingly, tension loads were observed. These will be further investigated by looking into the data from the other channels on each prop. Nevertheless, this observation was confirmed by the FBG measurement data from the sheet piles at the ends of prop 1.

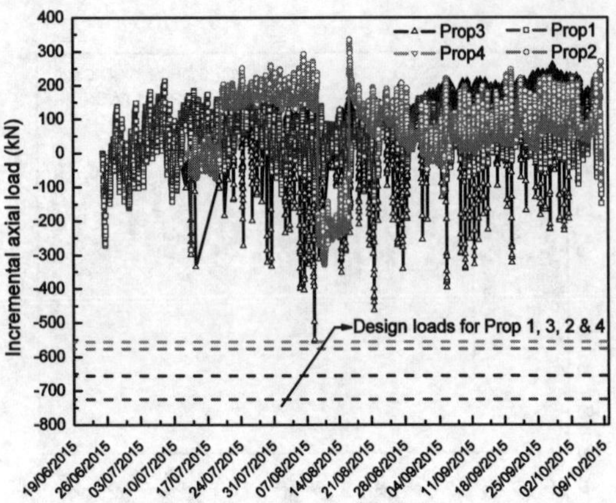

Figure 11. Incremental axial load on 4 instrumented props.

It is well know that the effect of temperature on prop loads can be very significant. The measured temperature on the 4 props varies from 3.75 ºC to 51.56 ºC, and the temperature inside the sensor node box varies from 4.06 ºC to 41.13 ºC. It can be seen from Figure 10 and 11 that there was a considerable amount of load cycling due entirely to temperature effects as the prop warms during the day and cools at night. The temperature effect on the real performance of these temporary props requires further investigation.

4 CONCLUSIONS

The paper presents the performance monitoring of 4 temporary props in an urban excavation site using a newly developed wireless strain sensor. Preliminary analysis on the sensing data from these props would seem to suggest that there is scope for more efficient design and construction in future schemes. The temperature effect on the real performance of temporary prop is to be further investigated.

The overall performance of the wireless sensor network in this construction site proved to be satisfactory, with average MDR of 88% over 110-day monitoring period. The small amount of data lost was recovered later from the on-board micro SD card storage.

The results of the lab calibration and field application of the new wireless sensor node shows very good performance. This presents the opportunity to build smarter temporary support systems, using props with integrated wireless strain and temperature sensors and load cells.

ACKNOWLEDGEMENT

This work was supported by the Cambridge Centre for Smart Infrastructure and Construction (CSIC) (EPSRC Grant No. EP/L010917/1). The authors would like to thank for all the technical supports from Dr. Cedric Kechavarzi, Mr. Peter J Knott, Mr. Jason Shardelow, and Mr. Daren Hitchings (from Balfour Beatty). Data supporting this paper is available from https://www.repository.cam.ac.uk/handle/1810/254805

REFERENCES

Batten, M. et al., 1999. Use of vibrating wire strain gauges to measure loads in tubular steel props supporting deep retaining walls. In *Proc. Instn of Civ. Engrs Geotechnical Engineering*. pp. 3–13.

Bhalla, S. et al., 2005. Structural health monitoring of underground facilities - Technological issues and challenges. *Tunnelling and Underground Space Technology*, 20(5), pp.487–500.

Dunkels, A., Grönvall, B. & Voigt, T., 2004. Contiki - A lightweight and flexible operating system for tiny networked sensors. *Proceedings - Conference on Local Computer Networks, LCN*, pp.455–462.

Liu, C., Teng, J. & Wu, N., 2015. A Wireless Strain Sensor Network for Structural Health Monitoring. *Shock and Vibration*, 740471, pp.1–13.

Nawaz, S. et al., 2015. Monitoring A Large Construction Site Using Wireless Sensor Networks. *Proceedings of the 6th ACM Workshop on Real World Wireless Sensor Networks*, pp.27–30. Available at: http://doi.acm.org/10.1145/2820990.2820997.

Xu, X. et al., 2015. SmartPlank monitoring on the real performance of timber structures in underground constructions. *Smart Structures and Systems*, 15(3), pp.769–785.

Proceedings of the International Conference on Smart Infrastructure and Construction
ISBN 978-0-7277-6127-9

© The authors and ICE Publishing: All rights reserved, 2016
doi:10.1680/tfitsi.61279.209

Wireless sensor monitoring of Paddington Station Box Corner

X.M. Xu*[1], S. Nawaz[2], P. Fidler[1], D. Rodenas-Herraiz[2], J. Yan[1,3] and K. Soga[1]

[1] *Department of Engineering, University of Cambridge, UK*
[2] *Computer Laboratory, University of Cambridge, UK*
[3] *School of Electronics and Computer Science, University of Southampton, UK*
* *Corresponding Author*

ABSTRACT This paper presents the real performance of three diaphragm wall panels on the southeast corner of Paddington Station Box during excavation, monitored using a wireless sensor network. In total, 15 LPDT displacement sensors, 12 tilt sensors, 13 relay nodes and a gateway were deployed at three different stages. Each wireless sensor node is programmed with Contiki OS using the in-built IPv6-based network layer (6LoWPAN/RPL) for link-local addressing and routing, and ContikiMAC at the medium access control (MAC) layer for radio duty cycling. Extensive testing and calibration was carried out in the laboratory to ensure that the system functioned as expected. Wireless tilt and displacement sensors were installed to measure the inclination, angular distortion and relative displacement of these corner panels at three different depths. The monitoring data reveal that the corner produced a stiffening effect on the station box, which might result in a breakdown of plane strain conditions. The network performance characteristics (e.g. message reception ratio and network topology status) and challenges are also highlighted and discussed.

1 INTRODUCTION

Spatial corner effects in deep excavations have been observed through traditional field instrumentation programs and numerical analyses (e.g. Ou et al. 1996; Tanner Blackburn & Finno 2007; Tan et al. 2014). The minimum D-wall deflections occurring near the pit corners during excavation were attributed to the three-dimensional stiffening effects caused by the higher stiffness at the corners. This would result in a breakdown of plane strain condition, which has been commonly adopted in engineering practice.

This paper presents real-time monitoring of the movement of three diaphragm wall corner panels in a very long and narrow station box, using a wireless sensor network (WSN). The site for the WSN deployment was an excavation for a new Crossrail station at Paddington, London, which took the form of an underground box (260m long, 25m wide and 23m deep). It is anticipated that these instrumentations would quantify the spatial corner effects, and to further improve the understanding of the performance near the corners of large deep excavations.

2 WIRELESS SENSING SOLUTIONS

The monitoring instrumentation installed onto the panels consists of displacement transducers and tilt sensors. The sensing information was transmitted wirelessly (via relays as required) to a gateway, which was connected to a mobile phone network. Extensive testing and calibration of the entire system were undertaken in the laboratory prior to on-site installation.

2.1 Wireless solutions

The tilt and displacement sensors used in the deployment were obtained from Wisen Innovation. These devices are internally based on the AVR ATmega1281 processor and the IEEE 802.15.4-compliant AT86RF231 radio. Fifteen sensors measured displacement using an LPDT while twelve measured tilt using Murata SCA100T MEMS inclinometers. For the thirteen relay nodes, Dresden Elektronik deRFmega128 modules were used. The gateway used a Memsic Iris mote acting as the root node and border router. This was attached to a Memsic MIB520 Gateway with data transferred over a USB connection and logged using a Raspberry Pi single board computer. Internet connectivity was provided by a 3G USB modem.

The application software running on the wireless sensor devices was developed in Contiki OS (Dunkels et al. 2004). Nodes use Contiki's standards-based IPv6 stack (6LoWPAN/RPL) for link-local addressing and routing, and ContikiMAC at MAC layer for low-power operation. A more detailed description of the software can be found in Nawaz et al (2015).

2.2 *Wireless sensor calibration*

All the wireless sensors were calibrated in the lab, using calibration platforms shown in Figurel, a laptop and a Sky gateway mote. The tilt sensor calibration platform can achieve 1/60 degree resolution, and that of the displacement sensor calibration platform can be as much as 0.01mm. For calibrartion the sensors were programmed with a version of the code which used a data transmission rate of 1 second per data message. For each tilt sensor, the calibration was performed in a range of -5 degree to 5 degree for both *X* and *Y* directions, with a minimum interval of 1/60 degree. The calibration range for each displacement sensor was made from -10 mm to 10 mm, 0.01 mm minimum interval. The calibration process was repeated up to 3 times for each sensor.

Figure 2 shows two examples of the calibration results from tilt sensor 18 and displacement sensor 04. It can be observed that the sensing data can be well characterized using the equations described in each figure. However, it was also found that there was significant discrepancy in the key characteristic parameter for each sensor. For example, the sensitivity coefficient of the displacement sensor used in this project varies from 0.03768 V/mm to 0.04256 V/mm, with a mean value of 0.0405075 V/mm. This is probably due to differences in the assembly and package of each sensor. It is therefore essential that all the sensors be individually calibrated prior to actual on-site installation.

(*a*) Tilt sensor calibration platform

(*b*) Displacement sensor calibration platform

Figure 1. Wireless sensor calibration platforms: (*a*) tilt sensor; (*b*) displacement sensor.

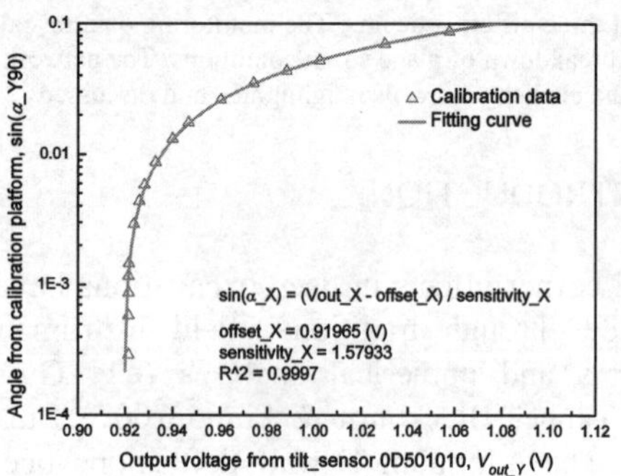

(*a*) Tilt sensor

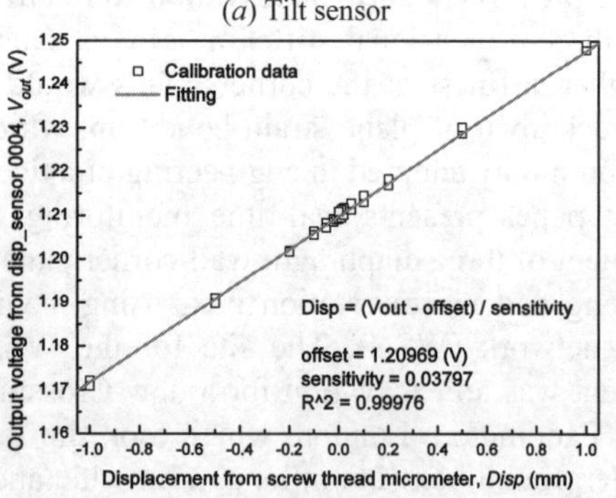

(*b*) Displacement sensor

Figure 2. Example results of the calibrated wireless sensors: (*a*) tilt sensor; (*b*) displacement sensor.

2.3 WSN Lab testing

Three lab tests in total were carried out in the laboratory to ensure that the WSN system was viable for deployment. For the first two tests, 15 wireless sensors and a gateway were tested, while for the third test 17 wireless sensors were tested. Satisfactory network performance was found for all these three tests, each lasting for around 2 weeks period.

3 FIELD DEPLOYMENT

A wireless sensor network was deployed in Paddington construction site in stages, including a gateway, 13 relays, 15 LPDT sensors and 12 tilt sensors.

3.1 Field overview

The Paddington Crossrail station is being built directly below Departures Road and Eastbourne Terrace, as marked with a red rectangle showing in Figure 3. The construction site is bounded by Eastbourne Terrace, Bishop's Bridge Road, Departures Road/Macmillan House and Praed Street. The inset of the Figure 3 also plots the three D-wall panels around the Southeast corner. Construction started in October 2011 and is due to be completed during 2017.

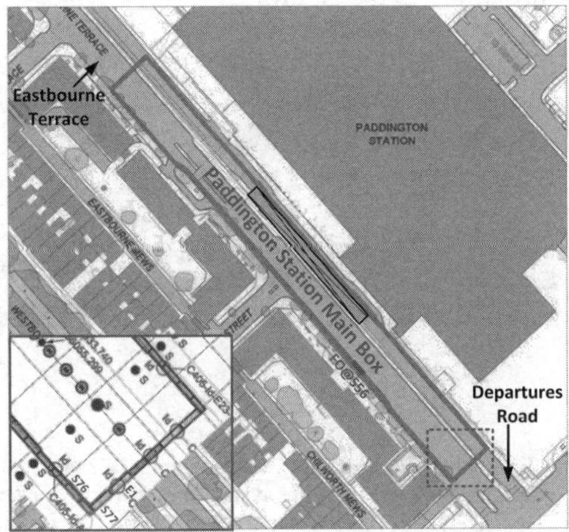

Figure 3. Paddington site main box site location.

The site is partially underlain by Pleistocene River Terrace Deposits (Lynch Hill Gravel), absent to the northwest of the site, over Eocene London Clay and Harwich Formation underlain by the Lambeth Group, Thanet Sand Formation and Cretaceous Upper Chalk. Recent Langley Silt is recorded above the River Terrace Deposits to the east of Paddington Station.

3.2 Sensor node locations

The parameters of particular interest in this monitoring scheme are the angular distortion and inclination of L-shaped corner panel (S77), as well as the relative movement of the panels immediately adjacent to it (S76 and E1). It was intended to instrument these three panels at four different levels (namely +119.0 m, +115.5 m, +113.1 m and +107.0 m), as the excavation proceeded.

At each installation level, five LPDT sensors and four tilt sensors were to be installed, including: (1) one LPDT sensor to span diagonally across the L-shaped panel S77; (2) two LPDT sensors to span across panels S76 and S77; (3) two LPDT sensors to span across panels S77 and E1; and (4) four biaxial tilt sensors on the three panels. The detailed layout of the sensors around the corner is illustrated in Figure 4.

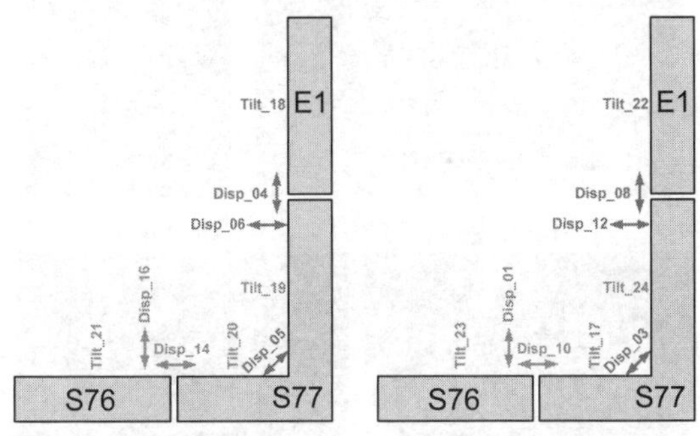

Figure 4. Sensor locations: (a) First stage at level +119.0m; (2) Second stage at level +115.50m.

3.3 Field deployment

Prior to field deployment, all the sensors were reprogrammed with a deployment version of the application software, which was also tested in the lab. Each node was also suitably labelled to inform operators on site of the monitoring undertaken.. To ease the installation of the LPDT sensors, a number of bespoke steel brackets were designed and manufactured at the University of Cambridge, including brackets for the diagonally mounted sensor spanning across the L-shaped panel, and others for mounting sensors across spacing between adjacent panels. Concrete blocks were prepared in the lab to test the installation process.

A gateway and four relays were firstly deployed in Paddington construction site on 22nd January 2014. The gateway was positioned outside the station main box and adjacent to the permanent opening (see Figure 5(a)), as it requires a power supply (110V) and good 3G signal coverage. One relay was placed on the top of panel N50 at Departures Road level, to ensure its good connectivity with both the gateway and other relays inside the main box. The other three relays were attached to panels (as indicated in Figure 5(b)) and plunge columns at Intermediate level. As the excavation progresses and slab casts, more relays were added at Concourse level. Note that the locations for attaching relays were very limited due to specific site requirements regarding the positioning of sensing instrumentation, with the D-wall panels and plunge columns only.

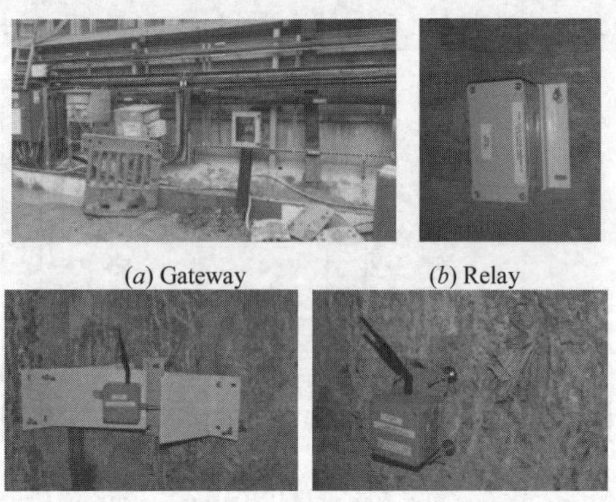

(*a*) Gateway (*b*) Relay (*c*) LPDT sensor (*d*) Tilt sensor

Figure 5. Field deployment of wireless sensor network at Paddington: (*a*) Gateway; (*b*) Relay; (*c*) LPDT sensor; (*d*) Tilt sensor.

The sensor installation at level +119.0m took place on 17th and 18th February 2014. Each sensor was attached onto a bracket using four screws, and the steel bracket was then mounted onto the D-wall using 4 concrete anchor bolts (bolt diameter 1/4 inches, minimum embedment 2 inches). The rest of sensors were installed on 14th March 2014 at level +115.5m and on 16th April 2014 at level +113.1m. Unfortunately, the installation at level +107.0m could not be realized due to the limited site access. The layout of the entire wireless sensor network is illustrated in Figure 6. All the sensors were removed from D-wall panels on 4th August 2014.

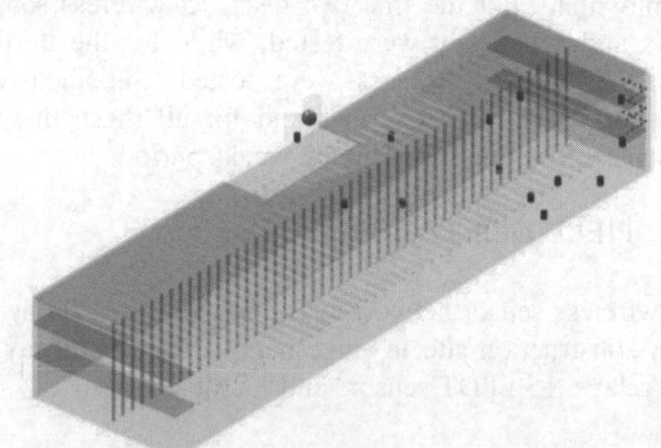

Figure 6. Model of Paddington station box and WSNs layout. Red sphere represents for gateway; blue cylinder for relays.

4 NETWORK DYNAMICS

Figure 7 presents the layout of the initial network topology, which is obtained from network diagnostic messages transmitted by all nodes in a periodical basis. Interestingly, it shows that sensor nodes were mainly routing messages via a far-off relay which was located on the opposite side of the station box in close proximity to the gateway.

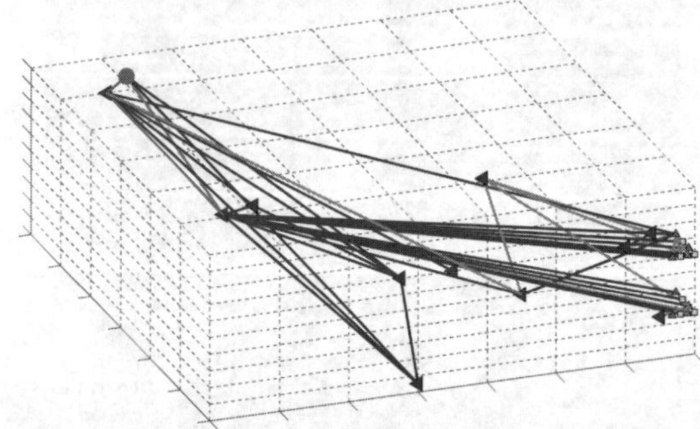

Figure 7. Initial network topology at Paddington (15th-19th March 2014). Link colour represents the average number of connections made to the gateway per day during the 5-day period. Grey line indicates one-two connections; blue line, between 2 and 20 connections; green line, between 20 and 200 connections; and red line, more than 200 connections.

Figure 8 shows the message delivery ratio (MDR) for 5 individual LPDT sensors and 4 tilt sensors during the entire monitoring period. The values of MDR for every node was obtained as the number of data messages successfully delivered to the gateway with respect to the total number of expected data transmissions. It can be observed from the figure that, the network experienced continuous connectivity problems that resulted in MDRs of below 10% in the first three months after deployment.

With the installation of two additional relays on 15th May 2014, an improvement in MDR for all sensor nodes (up to four times more) was observed (as shown in Figure 8). Unfortunately, this improvement only lasted for around 20 days, after which the MDRs dropped again. A more detailed description and explanation of the network dynamics can be found in Nawaz et al (2015).

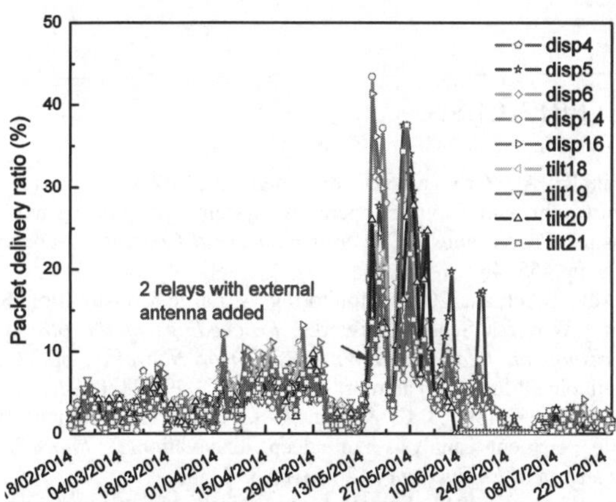

Figure 8. Packet delivery ratio at the gateway

5 MONITORING RESULTS

Although the WSN performance was not as good as expected, the received sensing data do provide sufficient information on the movements of three instrumented D-wall panels. For example, the measured displacement and inclination from four displacement sensors and two tilt sensors (as highlighted in Figure 5) are plotted in Figure 9. It can be observed from the figure that: (1) the maximum displacement for the L-shaped panel S77 was around 0.10mm (as indicated in Figure 9(a)), which corresponds to the angular distortion of about 1/2865 (0.02 degree) according to the sensor configurations. This might suggest that its extensive reinforcement may be unnecessarily, and significant cost savings may be possible; (2) the construction activity induced movement between panel E1 and S77 was up to 0.233 mm, as indicated in Figure 9(b); (3) the inclination on panel E1 was up to 0.10 degree. All the sensing data is to be further compared and analyzed with the readings from other instrumentations (e.g. FO sensing on panel S56 as highlighted in Figure 5, inclinometers, temporary prop loads, etc.), to gain some insights into the spatial corner effects of this long and narrow pit.

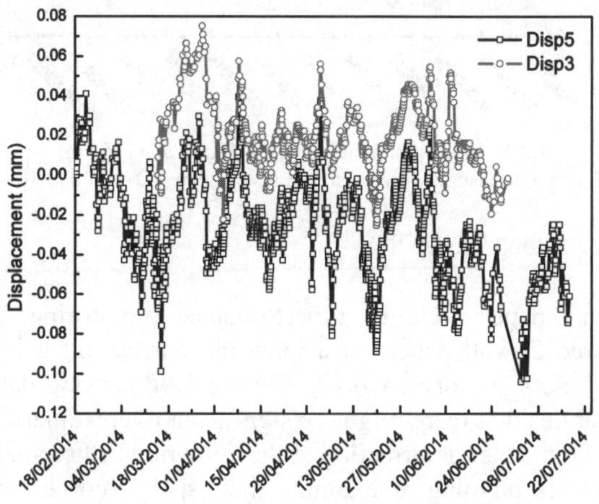

(*a*) Displacement in panel S77 (at +119.0m & +115.5m)

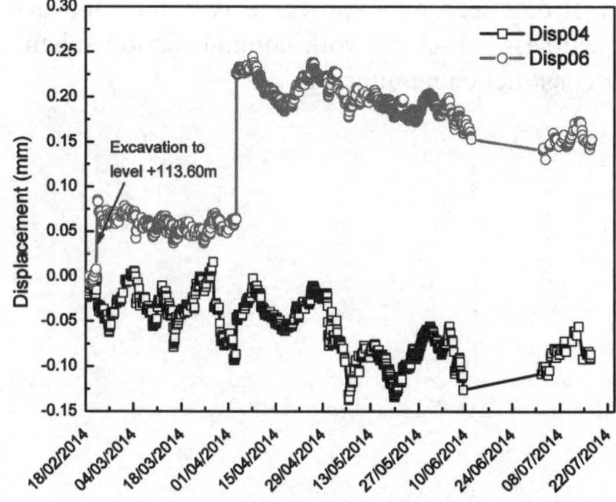

(*b*) Displacement between panels E1 and S77 (at +119.0m)

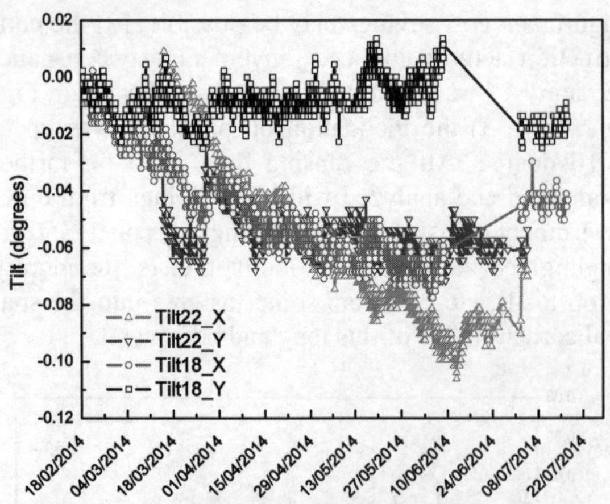

(*c*) Inclination of panel E1 (at +119.0m & +115.5m)

Figure 9. Measured movements on instrumented D-wall panels.

6 CONCLUSIONS

This paper presents a performance monitoring of three D-wall panels in a long and narrow pit using wireless sensor networks. The received sensing data implies that there might be significant overestimation on the panel deformation at the corners. Further analysis is ongoing to examine their spatial corner effects. The wireless network performance in this challenging environment was not satisfactory, and there is a strong need for improvements in the robustness of wireless sensor network communication schemes for construction monitoring.

ACKNOWLEDGEMENT

This research has been supported by Cambridge Centre for Smart Infrastructure and Construction (CSIC) (EPSRC Grant No. EP/K000314/1). We would like to thank Costain-Skanska Joint Venture (CSJV) and our industrial partner Crossrail for allowing access and instrumentation of the Paddington site. We would also like to thank Mr. Mohamad Alserdare from HS2 Ltd, Dr. Munenori Shibata from Japan Railway Technical Research Institute, Prof. Dan Zhang from Nanjing University, and Mr. Peter J Knott (CSIC) for their invaluable assistance with sensor network deployment. Data supporting this paper is available from https://www.repository.cam.ac.uk/handle/1810/254928

REFERENCES

Dunkels, A., Grönvall, B. & Voigt, T., 2004. Contiki - A lightweight and flexible operating system for tiny networked sensors. *Proceedings - Conference on Local Computer Networks, LCN*, pp.455–462.

Nawaz, S. et al., 2015. Monitoring A Large Construction Site Using Wireless Sensor Networks. *Proceedings of the 6th ACM Workshop on Real World Wireless Sensor Networks*, pp.27–30. Available at: http://doi.acm.org/10.1145/2820990.2820997.

Ou, C.-Y., Chiou, D.-C. & Wu, T.-S., 1996. Three-dimensional finite element analysis of deep excavations. *Journal of Geotechnical Engineering*, 122(5), pp.337–345.

Tan, Y. et al., 2014. Spatial corner effects of long and narrow multipropped deep excavations in shanghai soft clay. *Journal of Performance of Constructed Facilities*, 28(4). Available at: http://www.scopus.com/inward/record.url?eid=2-s2.0-84904470556&partnerID=40&md5=47e6e5b71e5f4241e2b33720251a9f1a.

Tanner Blackburn, J. & Finno, R.J., 2007. Three-Dimensional Responses Observed in an Internally Braced Excavation in Soft Clay. *Journal of Geotechnical and Geoenvironmental Engineering*, 133(11), pp.1364–1373.

Proceedings of the International Conference on Smart Infrastructure and Construction
ISBN 978-0-7277-6127-9

© The authors and ICE Publishing: All rights reserved, 2016
doi:10.1680/tfitsi.61279.215

Design of a wireless structural vibration control experimental system based on wireless sensor networks

Yan Yu[1], Xiaozhi Leng[1], Changping Yang[1], Luyu Li[2],Jinhe Guo[1] and Jinping Ou[2]

[1] *School of Electronic Science and Technology, Dalian University of Technology, Dalian, China*
[2] *School of Civil Engineering, Dalian University of Technology, Dalian, China*

ABSTRACT Major progress in developing structural regulators as an important method of vibration control have occurred for the last few decades. Large number of cables that are used in traditional wired control systems can be complicated and expensive. A wireless control system is designed and implemented in this study. Decentralized control strategies are employed in this control system. An optimal control algorithm based on Kalman estimator is embedded DSP controller in the control systems. To validate the performance of this control scheme, a three-story steel structure is developed with active mass dampers installed on each floor. Experimental results show that the wireless decentralized control hsa good control performance and various potential applications in industrial control systems. The designed experimental system may become a benchmark platform to validate the corresponding control algorithm.

1 INTRODUCTION

With the development of building technology, more ultra-large buildings are being constructed around the world. When the scale of the structure increases, centralized control systems with high instrumentation cost become difficult to reconfigure. These control systems could suffer from single point failure at the controller. However, decentralized control strategies may address these problems. Decentralized control was proposed in the 1970s to solve control problems of large-scale systems. At present, decentralized control systems are important in structural vibration control. Over the past few decades, studies have focused on decentralized control in structural engineering. For example, Sang-Myeong Kim conducted an experimental investigation of an active four-mount vibration isolation system and applied decentralized velocity feedback control. Yang Wang, R. Andrew Swartz, and J. P. Lynch et al. examined a three-stories, half-scale steel structure with magneto-rheological dampers installed on each floor and embedded closed-loop feedback control algorithms. Interactions between dynamically coupled subsystems are considered as unknown disturbances, and each controller is designed as a single-input, single-output subsystem. These decentralized control algorithms consider the interconnections between subsystems as unknown disturbances; thus, each decentralized controller aims to improve local control performance, which does not generally result in global optimal control. If the controller of a subsystem fails to operate properly during dynamic excitation, this subsystem may experience detrimental situations because neighboring subsystems focus on their own control performance and neglect the malfunctioning subsystem. In this decentralized control scheme, the controller of each subsystem requires local sensor data only, and each controller focuses on controlling the local subsystem.

2 CENTRALIZED AND DECENTRALIZED CONTROL ALGORITHMS

To For a lumped-mass model with n degrees of freedom, the system motion equation of a controlled structural system can be formulated as

$$M\ddot{X}(t)+C\dot{X}(t)+KX(t)=T_u u(t)+\dot{T}w(t) \quad (1)$$

where q(t) is the n ×1 displacement vector relative to the shake table; M, C, K are the n × n mass, damping, and stiffness matrices, respectively; w(t) and u(t) are the m1 × 1 external excitation and m2 ×1 control force vectors, respectively; and Tu and Tw are the control force and external excitation location matrices, respectively.

The state–space representation of the structural system can be written as

$$\dot{x}=Ax+Bu \quad (2)$$

where

$x=[x_1(t),x_2(2),\cdots,x_n(t),\dot{x}_1(t),\dot{x}_2(t),\cdots,\dot{x}_n(t)]^T$;

A is the system matrix given as $A=\begin{bmatrix}0 & I\\ -M^{-1}K & -M^{-1}C\end{bmatrix}$; and B is the control matrix given as $B=\begin{bmatrix}0\\ M^{-1}H\end{bmatrix}$.

The state–space vector x(t) can be related to the measurement vector y(t) as

$$y(t)=Cx(t)+Du(t) \quad (3)$$

where C and D are functions of the state variables selected to represent the measurements of the desired or available system. In this study, the relative displacements on all floors are measurable. However, the relative velocities cannot be quantified. Thus, C and D are expressed as

$$\begin{aligned}C&=\left[\text{eye}(3),\text{zeros}(3,3)\right]\\ D&=\text{zeros}(3,3)\end{aligned} \quad (4)$$

In this study, we assume a linear time invariant (LTI) system where stiffness, damping, and mass matrices do not change. Therefore, A, B, C, and D are assumed to be constant.

The linear quadratic regulator (LQR) control design balances the good system response and required control effort[19]. For the state–space representation of the structural control system defined in Equation (2), assuming static measurement feedback, we compute the optimal feedback control force u(t) using a gain matrix K as

$$u(t)=-Kx(t) \quad (5)$$

where the gain matrix K is designed to minimize cost function J. The classical linear quadratic regulator (LQR) control minimizes a quadratic cost function J as follows:

$$J(u)=\int_0^{\infty}(x^T(t)Q(t)x(t)+u^T(t)R(t)u(t))dt \quad (6)$$

where Q is a positive-definite Hermitian matrix and R is a positive-definite Hermitian matrix.

The LQR control design assumes full-state feedback. However, full-state information is rarely available to the controller. Typically, only a few measurements are available, e.g., acceleration and displacement. Thus, an estimator is used to reconstruct the state elements used for the control law, as shown in Figure 1.

In continuous time, the typical state–space plant model is represented as

$$\begin{aligned}\dot{x}&=Ax+Bu+Gw\\ y&=Cx+y\end{aligned} \quad (7)$$

where w is the process noise and v is the measurement noise. The noise variables are assumed to be unrelated white noise processes with a covariance defined by

$$E(w)=E(v)=0,\quad E(ww^T)=Q,\quad E(vv^T)=R \quad (8)$$

An estimate of the state is obtained from the resulting estimator given by

$$\dot{\hat{x}}=A\hat{x}+Bu+L(y-C\hat{x}) \quad (9)$$

where the open-loop estimate accounting for system dynamics or time update is corrected using the measurement weighed by the Kalman gain, L, as shown in Figure 1. If the estimated state is subtracted from the true state x, then the dynamics of the error in the estimate is represented by

$$\overline{x}=(A-LC)x \quad (10)$$

Equation 10 becomes zero if the estimator gain is selected properly and the system is observable. Note that D is assumed to be zero because input does not directly feed through to the measurements in our applications.

The steady-state Kalman gain, L, is given by

$$L = PC^T R^{-1} \tag{11}$$

Similar to the LQR solution, error covariance P is the positive definite solution to the algebraic Riccati equation

$$AP + PA^T \text{-} PC^T R^{-1} CP + GQG^T = 0 \tag{12}$$

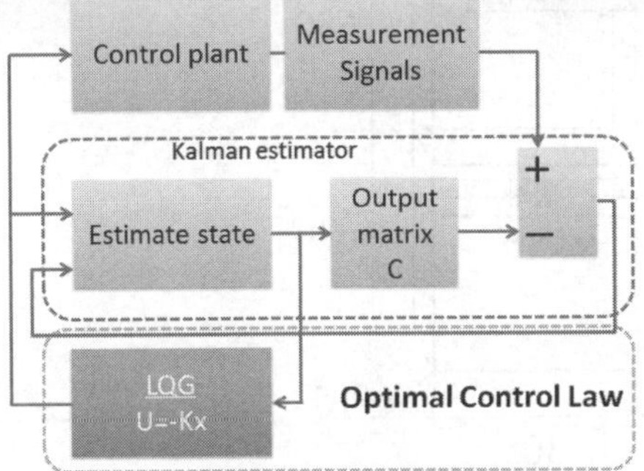

Figure 1. Block diagram of control algorithm

3 DESIGN OF WRIELESS DECENTRALIZED CONTROL SYSTRM

3.1 Overall framework of wireless control system

In this study, wireless centralized and decentralized control systems are designed. The wireless control system is composed of the following parts: (1) wireless Wi-Fi sensing unit, (2) controller, (3) laser displacement sensor, (4) three-story steel model, and (5) shake table. In the two systems, three laser displacement sensors are used as sensors. The three-story steel model is the experimental model, and active mass dampers (AMDs) are considered as the actuators. A displacement signal is sent to the wireless transmission units. The wireless receiving units obtain the displacement signal that is sent to the three DSP controllers. The controllers calculate the optimal control force signal. The control signal is sent to the power amplifier that can amplify the signal to -6 V–6 V. The enlarged control signal acts on the actuator, which can effectively suppress the vibration of the three-story steel model. The wireless centralized control system with the three-story steel model is illustrated in Figure 2.

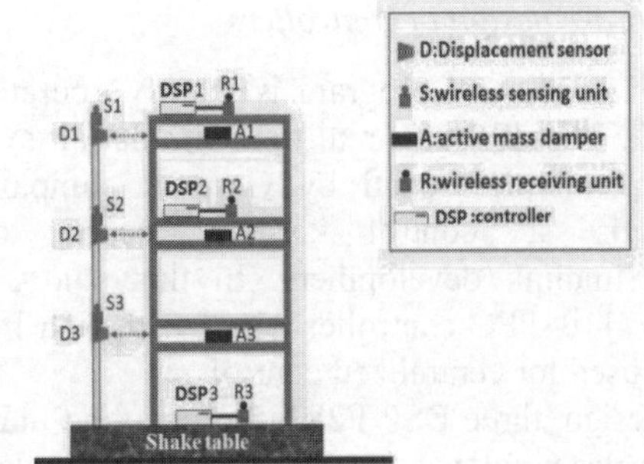

Figure 2. Wireless decentralized control system with three-story steel model

3.2 Hardware design of wireless sensing unit

As shown in Figure 3, the wireless sensing unit consists of three functional modules: sensor signal digitization, computational core, and wireless communication. The sensing interface converts analog sensor signals into digital data that are transferred to the low-power, 16-bit MSP430F1612 microcontroller through a high-speed serial peripheral interface port. The microcontroller communicates with a wireless transceiver through a universal asynchronous receiver and transmitter interface. The Wi-Fi module operates in the 2.4 GHz international ISM (industrial, science, and medical) band, and all the hardware components are internally referenced at 5 V.

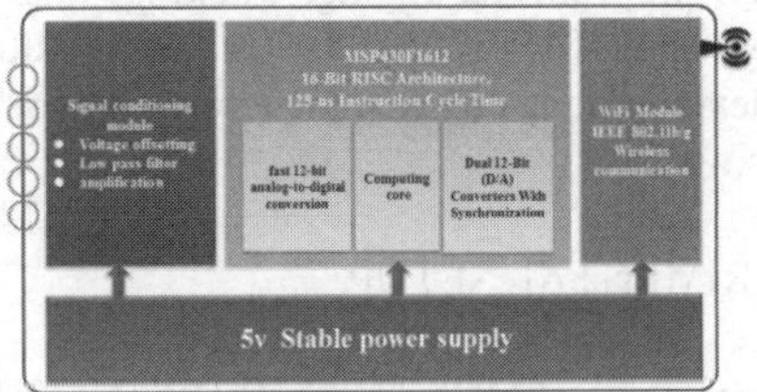

Figure 3. Block diagram of wireless Wi-Fi sensing unit

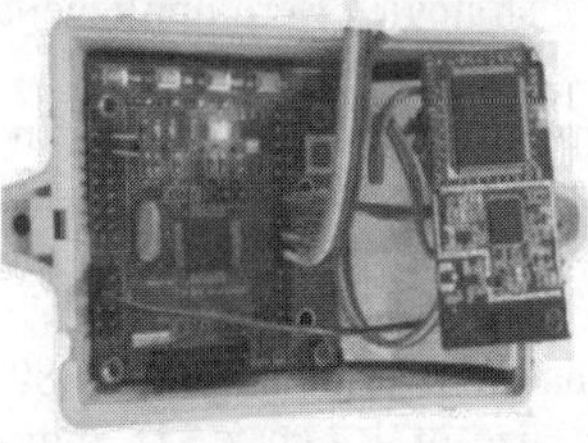

Figure 4. Hardware design of wireless sensing unit

3.3 *Hardware design of controllers*

The centralized control program is mainly operated by dSPACE. The dSPACE real-time simulation system, which was developed by German company dSPACE, is a control system based on MATLAB/Simulink development. In this study, a dSPACE DS1104PPC controller board with rich I/O interface is used for centralized control.

In this design, three DSP F28335 processing units are used for decentralized controllers, and the model-based development method is employed to develop the distributed control algorithm software. To shorten the development cycle, a decentralized code is generated by TargetLink. TargetLink is a software for automatic code generation that is based on a subset of Simulink/Stateflow models. TargetLink requires an existing MATLAB/Simulink model to work on and generates both ANSI-C and production codes that are optimized for specific processors.

The generated code is downloaded into DSP TMS320F28335, a high-precision processor with 32-bit floating-point processing units. The ADC result register is 16-bit, the digital quantity is in high 12 bits, and low 4 bits is invalid. Thus, the value in the ADC result should indicate a shift to the right 4 bits. As an editing tool for TMS320F28335, CCS can provide environment configuration, program editing, compiling, linking, program debugging, tracking, and analyzing. In general, TMS320F28335 can accelerate the software development process and improve work efficiency using CCS.

4 EXPERIMENTAL SETUP

In this study, three AMDs were used to control the response of a three-story structure subject to shake table excitation. Wireless centralized and decentralized control strategies were implemented for comparison.

The experimental system used in this study on wireless structural control is a three-story spring-steel structure fitted with three AMD control systems on each floor. The structure comprises several components and has an overall height of 1.89 meters. Spring steel has a size of 1 mm x 115 mm and inter-story height of 500 mm. The mass of each AMD is 1.984 kg. A custom-direct current motor drives the AMD with the control force that is amplified by the power amplifier.

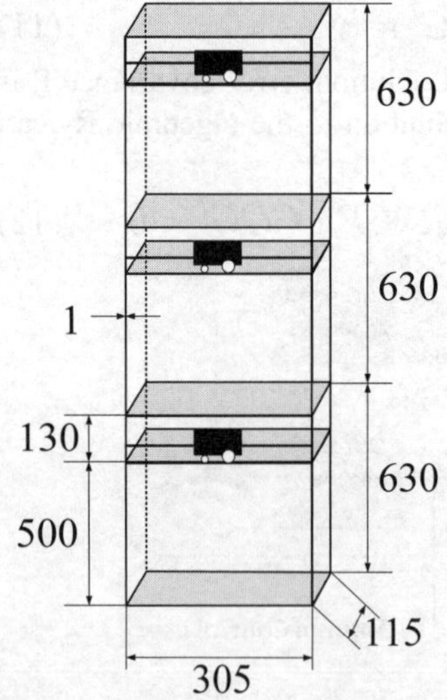

Figure 5. Properties of model

Table 1. Details of model parameters

Component	Properties
Total dimensions	305mm*115mm*189mm
Dimensions of spring steel	1mm*115mm*500mm
Mass of AMD	1.984kg

AMD is fitted with a DC motor. The input signals of the DC motor is the voltage signal that was generated by the controller. Thus, AMD can move along with the expected trajectory. The shake table provides an external excitation signal for the three-story model. The controller calculates the control signals according to the measured displacement signal and the LQR control algorithm. With respect to the DC motor and in the range of rated power, an approximately linear relationship between torque and current, T=K*I, was observed. Coefficient K is called constant torque. The current signal follows the control system, and the current signal changes to a corresponding voltage signal.

5 EXPERIMENTAL RESULTS AND ANALYSIS

5.1 Preliminary validation experiments

To verify the effectiveness of the control scheme, the model of an initial state was presented and made into a free vibration. The experimental results are presented in **Error! Reference source not found.**.

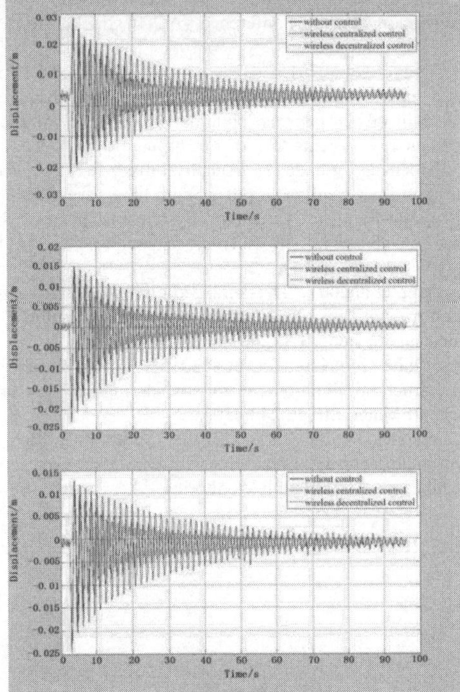

Figure 6. Displacements of three layers under an initial state

As shown in the experimental results, this process can significantly reduce the vibration amplitude model. Thus, the model quickly returns to a standstill. Experimental results show that the wireless control scheme that uses AMD is feasible and effective.

5.2 Seismic excitation situations

To accurately verify the general applicability of this method, the three-story model was given the seismic signals that were generated by the shaking table. The model obtains control performance by the displacement sensors. The experimental results are shown in succeeding pictures.

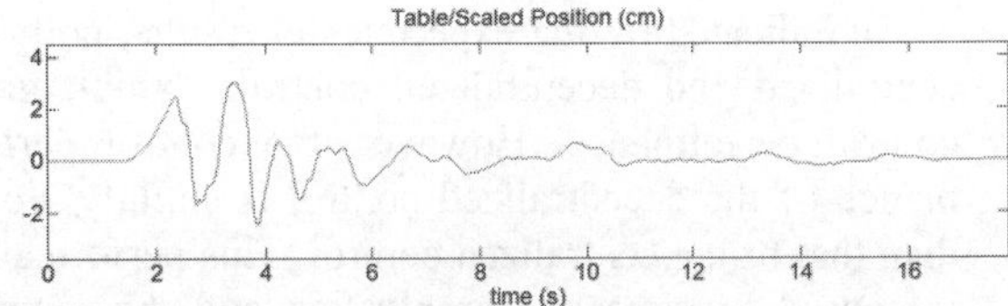

Figure 7. Displacement of shake table under Northbridge excitation

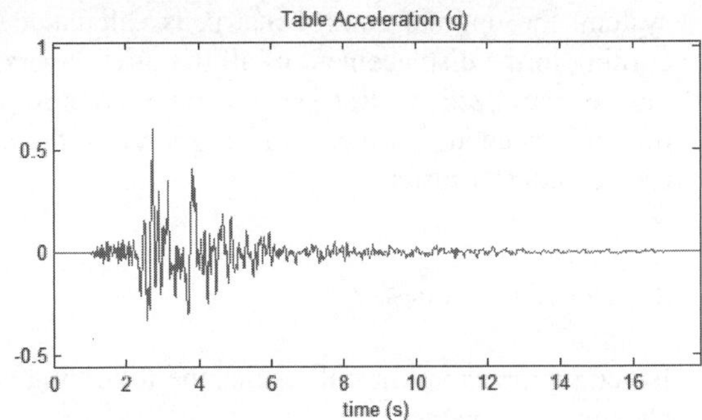

Figure 8. Acceleration of shake table under Northbridge excitation

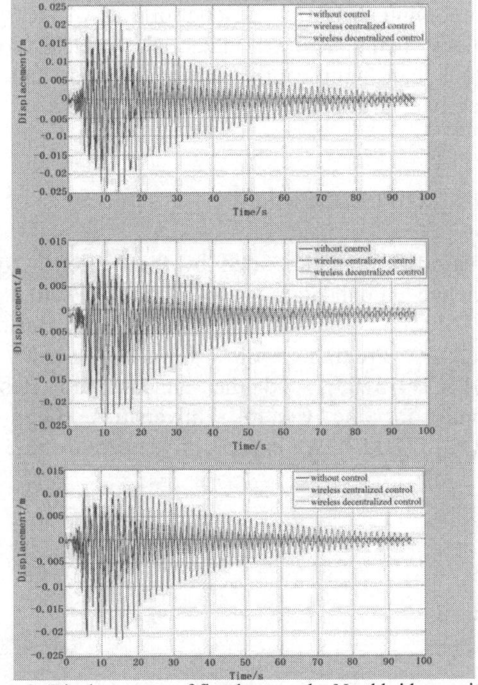

Figure 9. Displacements of first layer under Northbridge excitation

As indicated by the experimental results, both the centralized and decentralized controls exhibit good control performance. However, the control performance of the decentralized control is slightly worse than that of the centralized control. This performance is caused by time synchronization and the optimal gain matrix. In the decentralized control scheme, the input signal of the system is only the displacement of the current layer. However, in the centralized control system, the optimal control matrix is calculated according to the displacement of all the three layers. In the decentralized control system, time synchronization affects control, which can be ignored in the centralized control system.

6 CONCLUSIONS

Based on the experimental results, the following conclusions are obtained:

Decentralized control can effectively suppress vibration. More importantly, decentralized control can improve the stability of industrial control systems. Combining decentralized control and wireless technology helps achieve industrial wireless control, which can significantly reduce the cost and complexity of the industrial control system.

ACKNOWLEDGEMENT

This work was supported by the National Natural Science Foundation of China (Project No. 51161120359, 50921001), the National Basic Research Program of China (Grant No. 2011CB013705), and the Fundamental Research Funds for the Central Universities (Grant No. DUT15ZD117). We express our sincere thanks.

References

Lessard, Laurent.2012, "Decentralized LQG control of systems with a broadcast architecture." IEEE Conference on Decision and Control. Vol. 6241. 2012.

Mohammadi A, Plataniotis K N.2015, "Structure-Induced Complex Kalman Filter for Decentralized Sequential Bayesian Estimation."Signal Processing Letters, IEEE, 2015, 22(9): 1419-1423.

Palacios-Quinonero F, Rubio-Massegu J, Rossell J M, et al.2012, "Discrete-time static output-feedback semi-decentralized H∞ controller design: an application to structural vibration control."American Control Conference (ACC). IEEE, 2012: 6126-6131.

Qing, Xiangyun, et al.2015, "Decentralized unscented Kalman filter based on a consensus algorithm for multi-area dynamic state estimation in power systems." International Journal of Electrical Power & Energy Systems 65 (2015): 26-33.

Tong, Shaocheng, et al.2011, "Adaptive fuzzy decentralized control for large-scale nonlinear systems with time-varying delays and unknown high-frequency gain sign." Systems, Man, and Cybernetics, Part B: Cybernetics, IEEE Transactions on 41.2 (2011): 474-485.

Wang, Yang, et al.2007, "Decentralized civil structural control using real-time wireless sensing and embedded computing." Smart Structures and Systems 3.3 (2007): 321-340.

Yan Yu, Jinhe Guo, Luyu Li, et al.2015, "Experimental study of wireless structural vibration control considering different time delays." Smart Mater. Struct. 24 045005

Zhuoxiong Sun, Bo Li, Shirley J.Dyke1, et al.2015, "Benchmark problem in active structural control with wireless sensor network." Structural Control and Health Monitoring

Proceedings of the International Conference on Smart Infrastructure and Construction
ISBN 978-0-7277-6127-9

© The authors and ICE Publishing: All rights reserved, 2016
doi:10.1680/tfitsi.61279.221

Autonomous evaluation of human annoyance rate induced by subway trains using high-sensitivity wireless smart sensors

Wei Zhang[1*], Ke Sun[1], Huaping Ding[2], Robin E. Kim[3], Billie F. Spencer Jr.[4]

[1] *School of Earth Sciences and Engineering, Nanjing University, Nanjing, China*
[2] *School of Electronic Science and Engineering, Nanjing University, Nanjing, China*
[3] *Center for Integrated Smart Sensors, KAIST, Daejeon, South Korea*
[4] *Department of Civil and Environmental Engineering, University of Illinois at Urbana-Champaign, Urbana, USA*
* *Corresponding Author*

ABSTRACT The operation of subway trains induces secondary structure-borne vibrations in the nearby buildings. The vibration, along with the associated noise, can cause annoyance and adverse physical, physiological, and psychological effects on humans. Traditional tethered instruments restrict the rapid measurement and assessment on such vibration effect. This paper presents a novel approach for Wireless Smart Sensor (WSS)-based autonomous evaluation system for the subway train-induced human annoyance rate. The system was implemented on the MEMSIC's Imote2 platform, using a SHM-H high-sensitivity accelerometer board stacked on top. A new embedded application AnnoyanceRate, which quantitatively determines the adverse vibration impact on human comfort, was added into the Illinois Structural Health Monitoring Project Service Toolsuite. The system was verified in a large underground space, where a nearby subway station is a good source of ground excitation caused by the running subway trains. Using an on-board processor of the Imote2, each sensor calculated the distribution of vibration levels within the testing zone, and sent the distribution of human annoyance rate to the central server via radio to display the information. Also, the raw time-histories and frequency spectrum were retrieved from the WSS leaf nodes. Subsequently, spectral vibration levels in the one-third octave band, characterizing the vibrating influence of different frequency components on human bodies, was also calculated from each sensor node. Experimental validation demonstrates that the proposed system is efficient for autonomously evaluating the subway train-induced adverse effect on human comfort and the system holds the potential of greatly reducing the laboring of dynamic field testing.

1 INTRODUCTION

Train-induced vibration, along with the associated noise, can cause annoyance and adverse effects on humans, including physical, physiological, and psychological effects (Walker and Chan 1996, Lee and Griffin 2013). Thus, for those peoples who live or work in this specific vibrating environment for long periods of time, assuring that the human comfort is adequate enough is important. Dynamic field measurements can quantify the ambient vibration effects. The newly developed wireless smart sensors (WSSs) take advantage of ease of installation, wireless communication, on-board computation, battery power, locally data storage, relatively low cost, and small size (Jo et al. 2012). These features greatly facilitate the dynamic field measurement operations. A general-purpose accelerometer (SHM-A) sensor board for the Imote2 platform, an advanced WSS platform of Mote family developed by Intel research, was addressed by Rice and Spencer (2009). Subsequently, Jo et al. (2012) developed a high-sensitivity accelerometer (SHM-H) sensor board for measuring low-level ambient vibrations of civil structures. The SHM-H showed a satisfactory peak capture ability for model shapes under the low-level vibration (1-2 mg range), owing to its excellent capability of capturing low noise level property.

Traditionally, the Z vibration level, calculated from the dynamic field measurements, is used as the quantitative index to evaluate the subway train induced ambient vibration. However, the Z vibration level can only indirectly reflect the vibration impact on human bodies. Therefore, a more proper index relating human comfort should be considered for replacement.

In this paper, a Human Annoyance Rate (HAR) model was proposed to assess the adverse impact of subway train induced ambient vibrations. In addition, the model was embedded in the node of WSS so that the autonomous evaluation in site can be performed. Experimental validation was implemented in a large underground space adjacent to a subway station at Suzhou, China. The vertical ambient vibration responses were measured by WSSs; the decentralized on-board computation was executed and the results were directly sent to a central server wirelessly. Subsequently, the HAR distribution around the underground space floors was identified for human comfort assessment.

2 MODELS OF HUMAN ANNOYANCE RATE UNDER VIBRATION

The HAR is defined as the ratio of the number of people who feel the ambient vibration is unacceptable to the total number of people in statistics. Five calibration curves of HAR (A) to acceleration a_w are depicted in Figure 1. The five curves are correspond to restricted area, residence in the day-time, residence in the night-time, office and workshop. Note that the scale of A has been normalized to 1. In addition, as seen in the Figure, a threshold line, with the value of 0.07, was also plotted. The threshold value was so identified in that the HAR values of the 5 curves are all close to 0.07 when the a_w of the five circumstances takes their minimal value.

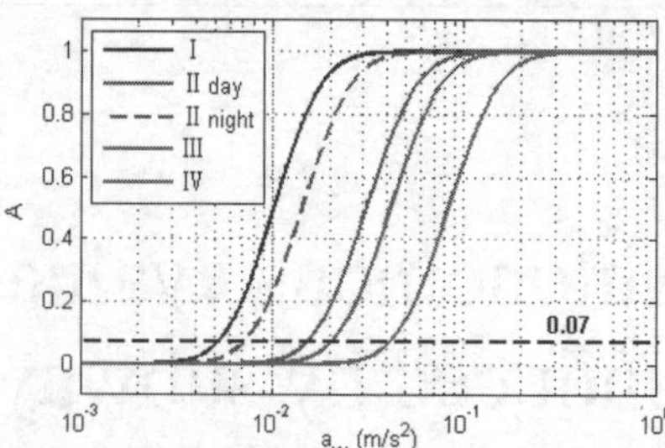

Figure 1. HAR calibration curves for the 5 circumstances

3 THE AUTONOMOUS EVALUATION SYSTEM

3.1 *Requirements and performances*

The former investigation reported that the frequency range of interest for subway induced vibrations in buildings is 1-80 Hz (ISO 2631-2 2003). Thus, to be able to capture the frequency contents, an wireless accelerometer needs a sampling rate greater than 160 Hz (Widrow 1956); meanwhile, the published data also revealed that a measuring accuracy of 1 mg, as well as a resolution of 0.05 mg, is necessary for capturing low-level ambient vibrations (Gupta et al. 2009, Degrande et al. 2006, Sanayei et al. 2013).

The SHM-H sensor board, developed from the University of Illinois at Urbana-Champaign, is a high-sensitivity accelerometer board addressing such rigorous requirements. The SHM-H sensor board provides flexible and highly accurate user-selected sampling rates by using a Quickfilter chip, QF4A512, which is a four-channel, 16-bit analog to digital converter (ADC), with programmable digital filters (Jo et al. 2012). Particularly, a low-noise and high-sensitivity MEMS accelerometer, Silicon Designs SD1221L-002, was embedded on the board to measure the acceleration in z-axis. The maximal resolution of the sensor is 0.043 mg, with a limit of the input range being ± 0.2 g (Jo et al. 2012). Also, the frequency response is 0~400 Hz, which satisfies the frequency range of interest. Other specifications also include a sensitivity of 2 V/g and a noise floor of 5 $ug/\sqrt{Hz}$.

The embedded software in the WSN node is developed based on the Illinois Structural Health Monitoring Project (ISHMP) Services Toolsuite, version 3.1.0, which implements key middleware functionality to provide high-fidelity measured data and to reliably transmit the data to the base station wirelessly across the sensor network, as well as a library of numerical algorithms. The open-source software, written in nesC language and designed to operate in TinyOS is available at http://shm.cs.uiuc.edu/software.html (Jang et al. 2010). An application for assessing the annoyance rate of ambient vibration using on-board computational functionalities has not been realized before. Thus, a new data processing application, named *HumanAnnoyanceRate*, has been developed for the on-board computation of the HAR, collaborating with the other existing components.

Unlike the centralized data processing strategy used in the traditional tethered systems, independent data processing strategy can be employed in the WSN network system. With all the necessary algorithms embedded in the computational core, distributed computing can be executed on each wireless leaf node, independently without having to communicate with each other. At first, the central server sends commands to activate the leaf nodes, establish a WSN, and set the measurement parameters. Then, after time synchronization, the nodes start collecting data. The measured data is firstly processed locally by the microprocessor on the Imote2. After that, the HAR results are wirelessly transmitted from the leaf nodes to the gateway node, which is connected to the central server. Figure 3 illustrates the schematic of the whole system.

4 EXPERIMENTAL VALIDATION

4.1 Background

The system was validated in the Xinghai Square, a two-story of large underground space adjoining Metro Line 1 in Xinghai station of Suzhou City, China. The underground space is comprised of two reinforced concrete frame structures, seamlessly standing on both sides of the subway station. The total cover area is 54511 m^2 and the maximal bury depth is 13.7 m underground. The rooftop of the underground space is made for the Sunken plaza. As seen, the distance, from the side wall of the Basements B1 and B2 to the central line of the subway rails, is only 2 m. Such a short distance can influence people in the underground space due to the subway train's operation.

4.2 Measurement layout

As discovered by amounts of investigations concerning subway train-induced ambient vibration, the amplitude of vibration level along z-direction is far greater than those along x and y directions in the horizontal plane (Degrande et al. 2006). Thus, the vibration along z direction, is normally employed to quantitatively characterize the major ambient vibration effect induced by subway trains.

One WSN leaf node, one WSN gateway node and a laptop were used to set up the field testing system referred above, as shown in Figure 2. The leaf node was mounted upon a steel U-shape base, which was firmly fixed onto the floor slab using epoxy. The central server, as well as the gateway node, was placed less than 50 m to both leaf nodes without any barrier in between. The sampling rate was set as 280 Hz and the cutoff frequency was 80 Hz.

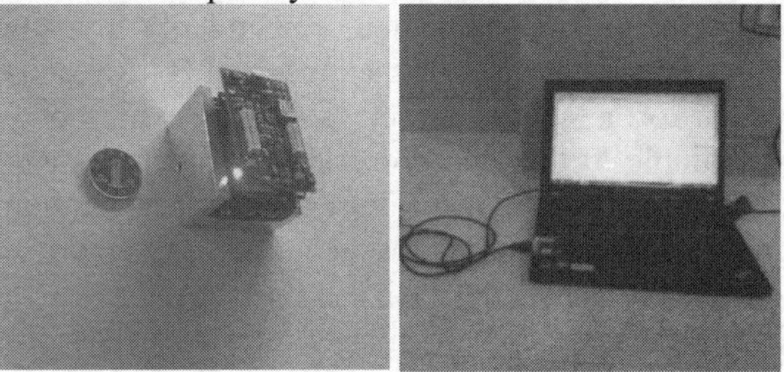

(a) Leaf node (b) Gateway node and central server

Figure 2. The testing system at site.

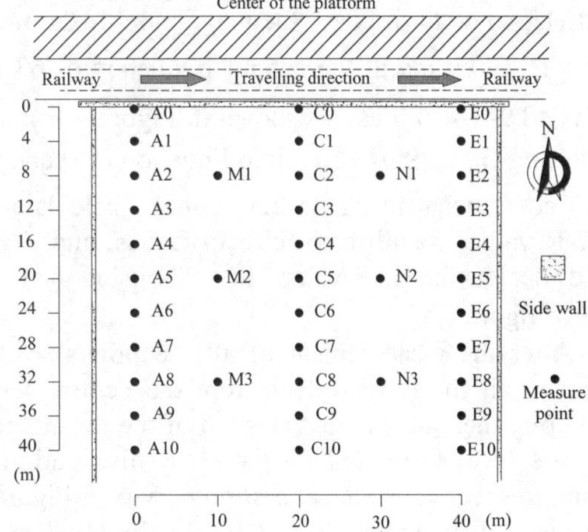

Figure 3. Measure point layout.

The measurement, following the procedure of ISO 5348 (1998), was performed in the basement 1 (B1).

The layout of the measure points are depicted in Figure 3. As shown, the testing area is square shaped, 40 m by 40 m. The north side wall is close to the subway rail. Totally 39 measurement points were identified along 5 vertical lines, as following: line A is close to the subway station entrance; line C is at the center of the platform; line E is close to the exit; line M and N are in the middle of lines A-C and C-E, respectively. The test duration was arranged at the low ebb of the subway operation, so that the weight variation caused by the passages on board can be minimized.

4.3 *Testing results*

The measurements were conducted during the period of train's arrival-stay-departure, lasting about 100 s. At each point, 5 sets of measurements were performed. Once the measurement is completed, the measurements are computed embeddedly by the MCU of the leaf node, and then the final HAR results are sent to the central server via the gateway node. An example of the data processing of measurement point E0 is illustrated in Figure 4. The acceleration signals of time domain was acquired, as shown in Figure 4 (a); the embedded PSD was performed to transmit the time domain signals into frequency domain ones, as shown in Figure 4 (b); the frequency-weighted calculation in one-third octave bands were implemented on the acceleration signals between 1-80 Hz, so as to discover the magnitude of the maximal a_w, existing at the central frequency of 63 Hz, was 28.94 $\times 10^{-3}$m/s^2, as shown in Figure 4 (c); substituting $a_w = 28.94 \times 10^{-3}$ into Equation (13) and use the coefficients in Table 2, we may calculate the HAR values for different circumstances, and Figure 4(d) depicts the HAR value of 0.2274 using the curve III of figure 1.

After the measurement of all the points, we retrieved all the HAR values from the central server and plot them as the contour map of the testing area. Figure 5 (a) to (e) exhibit the HAR distribution by using respective HAR calibration curves in Figure 1. In view of the threshold of 0.07 for the HAR in the vibrating circumstance, we can identify where the HAR exceeds its limit within the testing area. As seen from Figure 5 (a), most of the area's HAR value exceeds the limit of 0.07, suggesting that people will feel uncomfortable within this area if it is designed as a restricted one, such as the hospital or school room. Figure 5 (c) displays quite similar to 5 (a), but only the HAR of bottom part and the central upper part are less than the limit, meaning that the rest part of the area cannot be used as the residence at night. Likewise, Figure 5 (b) and (d) also exhibit quite similar; their maximal HARs, exceeding the limit, both occur at the top-right corner and the small top-left corner.

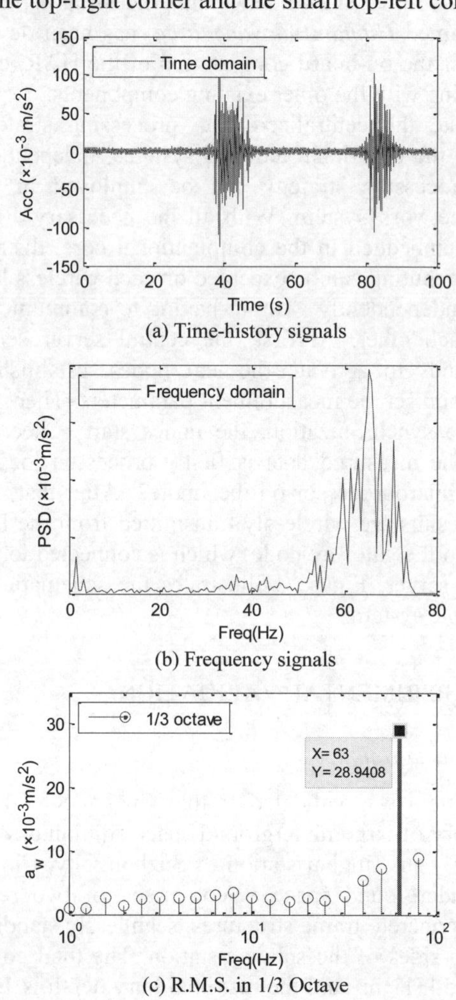

(a) Time-history signals

(b) Frequency signals

(c) R.M.S. in 1/3 Octave

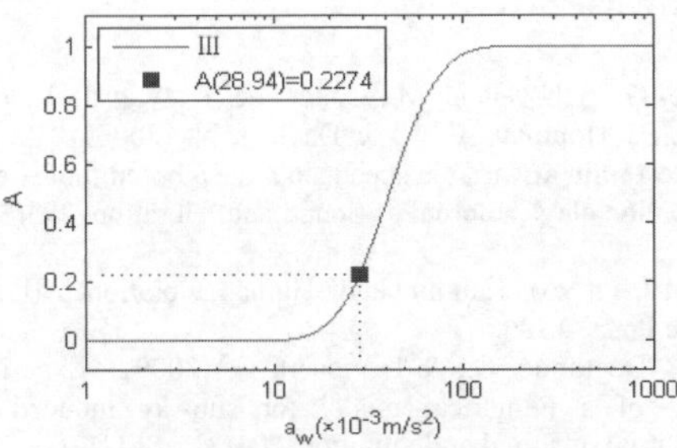

(d) HAR identification on calibration curve III

Figure 4. Data processing of measurement point E0

Other parts in both figures are all less than 0.07, suggesting they can meet the requirements of the residence at daytime or those of the office, respectively. Figure 5 (d) shows that all the HARs are less than the limit if the area is used as workshop purpose.

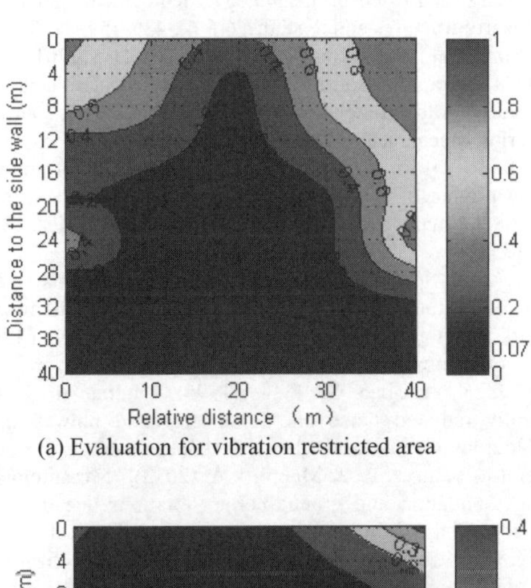

(a) Evaluation for vibration restricted area

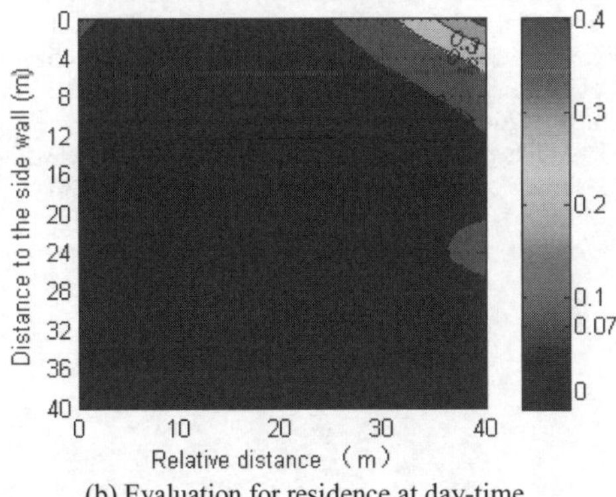

(b) Evaluation for residence at day-time

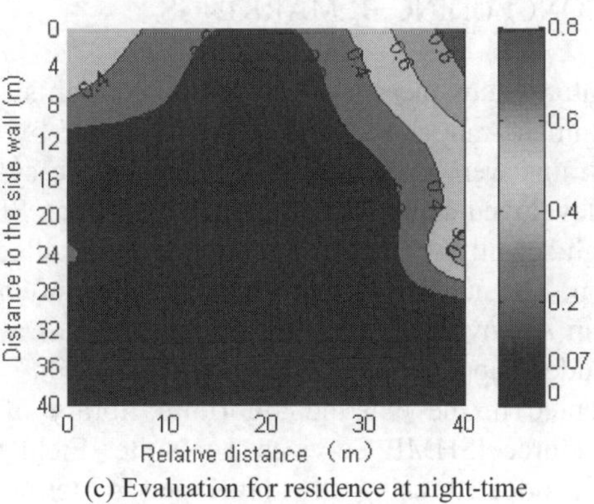

(c) Evaluation for residence at night-time

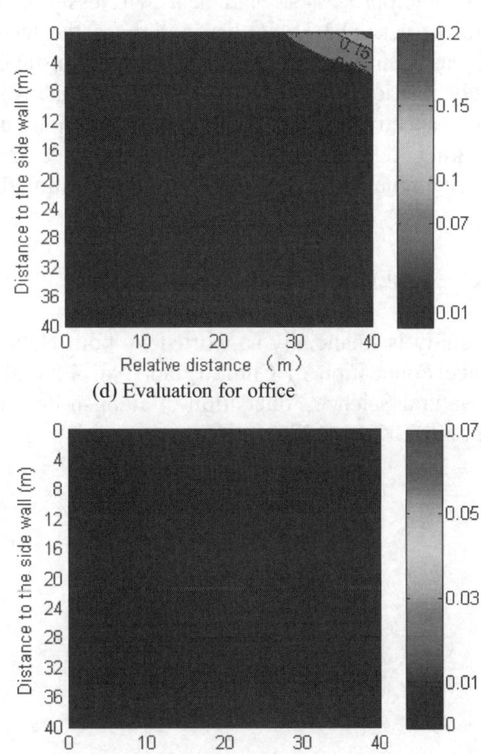

(d) Evaluation for office

(e) Evaluation for workshop

Figure 5. Human comfort evaluation for different circumstances

5 CONCLUDING REMARKINGS

An autonomous measurement and evaluation system for Human Annoyance Rate (HAR) induced by subway trains were presented in this paper; the system was developed using Imote2 platform, with an SHM-H high-sensitivity accelerometer board installed on the top. The numerical algorithms for calculating the Human Annoyance Rate were programmed as a new embedded application *HumanAnnoyanceRate*, implemented in the existing algorithm library of the open-source ISHMP Services Toolsuite. Field tests were conducted using the developed system on a large underground space adjoining a subway station. The HARs, were autonomously computed on the leaf nodes' microprocessors and sent wirelessly to the central server. The HAR distribution of the testing area was identified for different use circumstances, namely restricted area, residence, office and workshop, respectively. The potentials of autonomously evaluating the human comfort in subway train-induced ambient vibration were well demonstrated.

ACKNOWLEDGEMENT

This study is financially supported by both National Science Foundation of China (grant No. 40902076) and Natural Science Foundation of Jiangsu Province (grant No. BK20141224).

REFERENCES

Degrande, G., Schevenels, M., Chatterjee, P., Van de Velde, W., Hölscher, P., Hopman, V., ... & Dadkah, N. (2006), "Vibrations due to a test train at variable speeds in a deep bored tunnel embedded in London clay", Journal of Sound and Vibration, 293(3), 626-644.

Griffin M.J. (1996), Handbook of Human vibration [M], Lodon: Academic Press, 1996.

Gupta, S., Degrande, G., & Lombaert, G. (2009), "Experimental validation of a numerical model for subway induced vibrations", Journal of sound and vibration, 321(3), 786-812.

ISO2631, S. I. (1989), "Mechanical Vibration and Shock: Evaluation of Human Exposure to Whole-body Vibration. Part 2: Continuous and shock- inducedvibration in buildings (1-80Hz)", ISO Standard.

ISO2631, S. I. (2003), "Mechanical vibration and shock-evaluation of human exposure to whole body vibration Part 2: Continuous and shock induced vibration in buildings (1-80 Hz)", ISO Standard.

Jang, S., Jo, H., Cho, S., Mechitov, K., Rice, J. A., Sim, S. H., ... & Agha, G. (2010), "Structural health monitoring of a cable-stayed bridge using smart sensor technology: deployment and evaluation", Smart Structures and Systems, 6(5-6), 439-459.

Jo, H., Sim, S. H., Nagayama, T., & Spencer Jr, B. F. (2012), "Development and application of high-sensitivity wireless smart sensors for decentralized stochastic modal identification", Journal of Engineering Mechanics, 138(6), 683-694.

Lee, P. J., & Griffin, M. J. (2013), "Combined effect of noise and vibration produced by high-speed trains on annoyance in buildings", The Journal of the Acoustical Society of America, 133(4), 2126-2135.

Rice, J. A., & Spencer Jr, B. F. (2009), "Flexible smart sensor framework for autonomous full-scale structural health monitoring", Newmark Structural Engineering Laboratory. University of Illinois at Urbana-Champaign..

Walker, J. G., & Chan, M. F. K. (1996), "Human response to structurally radiated noise due to underground railway operations", Journal of sound and vibration, 193(1), 49-63.

Sanayei, M., Maurya, P., & Moore, J. A. (2013), "Measurement of building foundation and ground-borne vibrations due to surface trains and subways", Engineering Structures, 53, 102-111.

Widrow, B. (1956), "A study of rough amplitude quantization by means of Nyquist sampling theory", Circuit Theory, IRE Transactions on, 3(4), 266-276.

Yick, J., Mukherjee, B., & Ghosal, D. (2008), "Wireless sensor network survey", Computer networks, 52(12), 2292-2330.

Proceedings of the International Conference on Smart Infrastructure and Construction
ISBN 978-0-7277-6127-9

© The authors and ICE Publishing: All rights reserved, 2016
doi:10.1680/tfitsi.61279.227

Field monitoring of piling effects on a nearby masonry vault using distributed sensing

S. Acikgoz[1], L. Pelecanos*[1], G. Giardina[2] and K. Soga[1,2]

[1] *Centre for Smart Infrastructure and Construction, University of Cambridge*
[2] *Department of Engineering, University of Cambridge*
* *Corresponding Author*

ABSTRACT This paper presents a recent case study of monitoring the effects of piling on an adjacent old masonry vault in London. The monitoring scheme consists of 3 independent instrumentation sets that provide different types of information: (a) discrete total station point targets, (b) linear distributed fibre optic cable sensors and (c) surface distributed laser scanners. The availability of these sensors is able to shed some light on the actual response of the masonry structure through precise displacements and high-accuracy localised strains. The collected monitoring data show the location of cracks and provide indications for their opening magnitude. Relevant numerical analyses have also been conducted using (a) limit analysis mechanisms and (b) finite element deformation analysis which confirmed the observed field deformation mechanism and the presence of cracks within the structure. It is shown that such innovative sensing approaches can provide valuable detailed information about the real behaviour of structures that were not available before.

1 INTRODUCTION

Recent work during the London Bridge Station redevelopment involved construction of piles inside historic brick barrel vaults. The latter were monitored regularly by total stations in order to maintain safe operation of the transport systems throughout the course of the redevelopment.

Two additional spatially distributed sensing systems were used to provide more detailed monitoring data: (a) distributed Brillouin Optical Time Domain Reflectometry (BOTDR) and (b) laser scans.

This paper presents some of the monitoring data collected during piling within the masonry vault. Special emphasis is given on the relative merits of discrete total station prisms and distributed fibre optic (FO) and laser scan data, especially in detecting regions of localized cracks.

2 THE MASONRY STRUCTURE

Figure 1 shows a longitudinal section cut through the vault which also includes the geometric details of the arch and the local shallow soil stratigraphy. The arches were constructed on shallow concrete footings founded in made ground and soft alluvium.

During September to December 2013, 57 end-bearing piles were constructed under the arch. These are 450mm diameter CFA piles which terminate in London Clay.

3 MONITORING SYSTEMS

3.1 *Discrete-point total station prisms*

Total station devices use a laser beam and a precise servomotor and emit a modulated wave reflected from

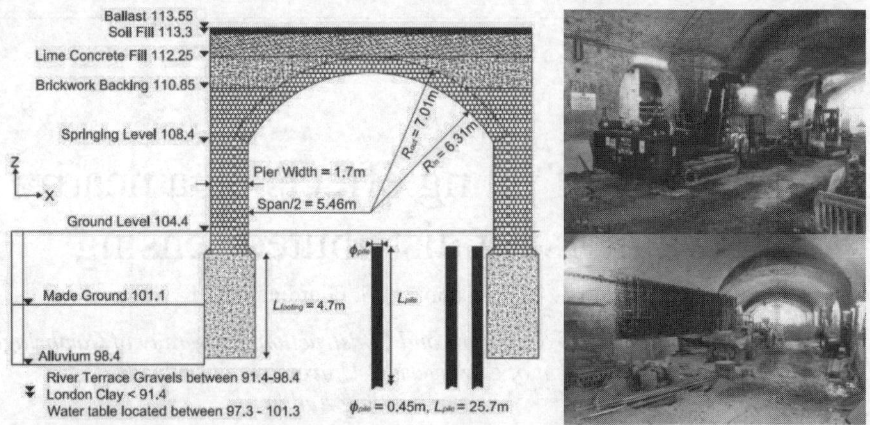

Figure 1. Longitudinal cross-section (left) and photos of the arch and piling works (right).

optical targets which is used by the device to determine the centre of the optical target and its relative location.

In this project a number of discrete-point prisms were installed in the masonry vault in an array of 3 prisms in each cross-section of the vault (Figure 2). The total station provides the exact location of the prisms and therefore using successive measurements, one can determine the relative displacements. Then by getting the difference between displacements of adjacent prisms and dividing this by their distance, one can obtain the averaged strain over that section.

3.2 *Distributed FO cables*

The distributed BOTDR technique uses optical fibre to monitor axial strains on civil infrastructure (Soga, 2014). A change in the Brillouin frequency of the back-scattered light within an optical fibre is linearly dependent on the applied strains. Therefore, the FO cables provide directly the axial strains within the monitored infrastructure (Soga et al., 2015).

Distributed FO sensing techniques have been used widely over the last 10 years to monitor various types of civil infrastructure, such as tunnels (Mohamad et al., 2010, 2012; Cheung et al., 2010), piles (Klar et al., 2007), retaining walls (Mohamad et al., 2011; Schwamb et al., 2014; Schwamb & Soga, 2015) and slopes (Amatya et al., 2008). The principle of FO sensing is based on the strain-dependent change of light frequency within a bare optical fibre (Soga, 2014). A detailed explanation of the background theory and examples of recent applications may be found by Soga et al. (2015).

In this masonry vault, a close loop of a FO cable was installed to monitor several points within the structure, and was installed as shown in Figure 2.

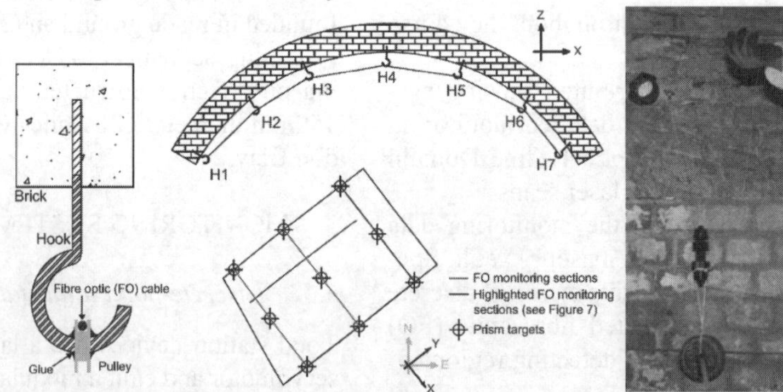

Figure 2. Attachment of the FO cables to masonry (left), arrangement of cables in arch (middle) and photos from the installation (right).

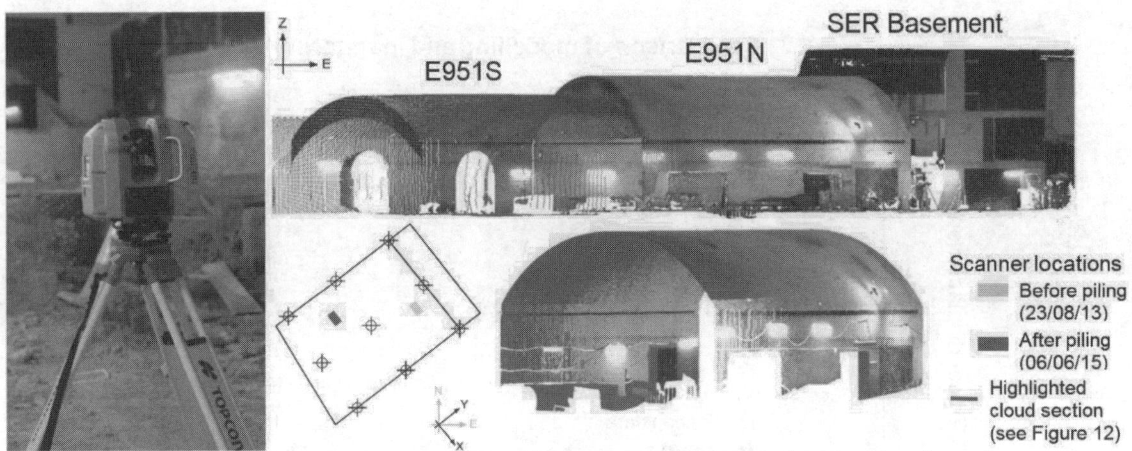

Figure 3. A laser scanner device (left), a colorized point cloud (top right) and locations of point clouds from laser scan surveys during piling (bottom right).

3.3 Distributed laser scans

Terrestrial laser scanners are geomatic devices equipped with a laser beam and precise servomotors, carrying out similar surveying operations to total stations. Several techniques exist to convert the laser scans to absolute structural displacements and these may be found elsewhere.

In this structure, laser scans were taken using firstly a FARO Focus 3D S20 laser scanner and later a Topcon GLS-2000 laser scanner. These provide a ranging error of 2mm and angle measurement error of 6 arc seconds (Figure 3). The data processing procedure took advantage of the open source software Cloud Compare.

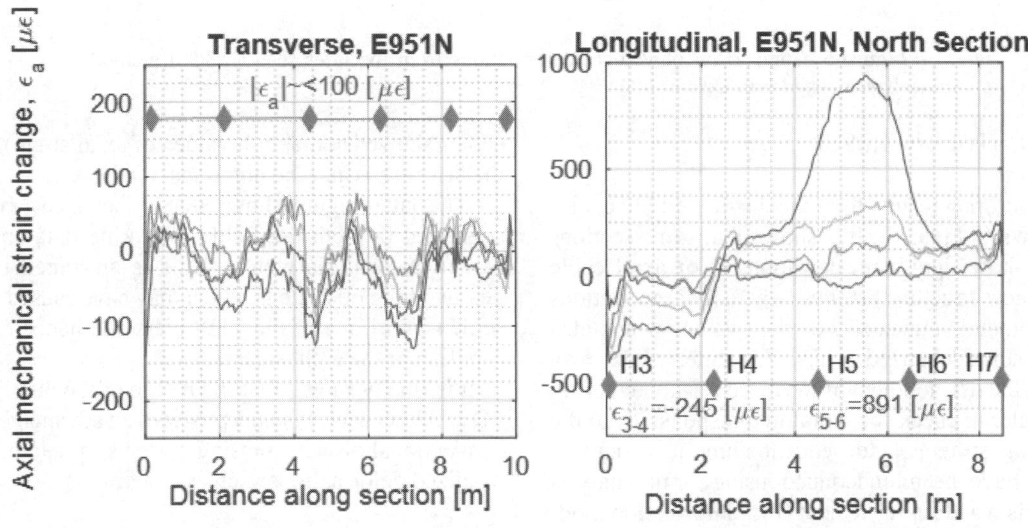

Figure 4. Change in axial mechanical strain during piling for selected sections

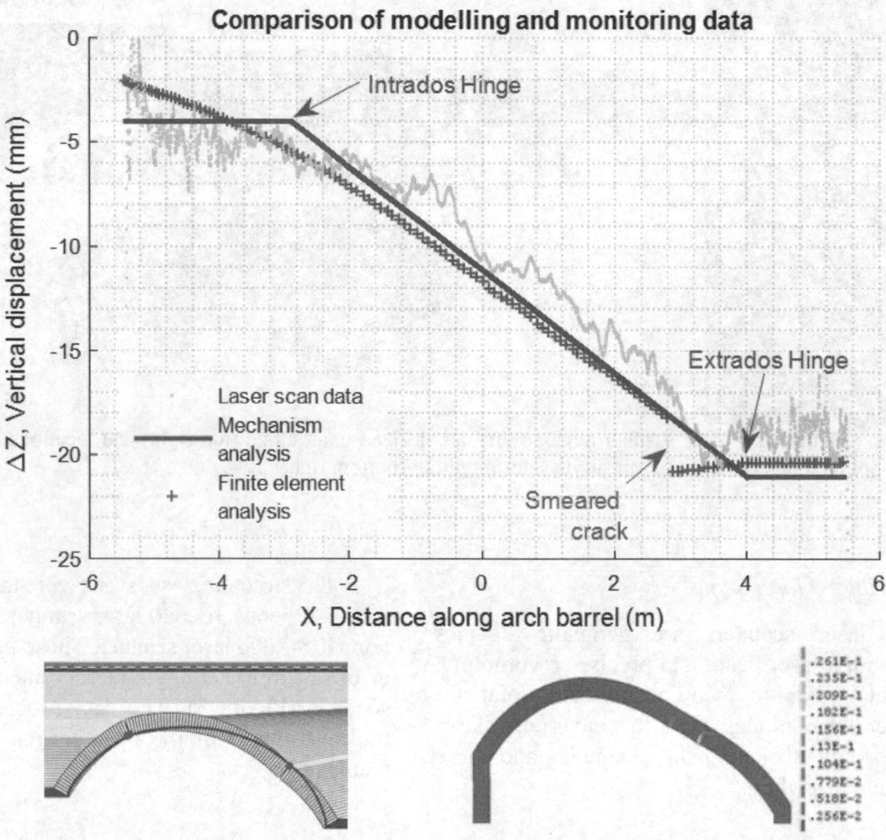

Figure 5. Comparison of vertical deformations from the laser scan (M3C2) method, mechanism and FE models.

4 MONITORING RESULTS

The monitoring results from the distributed FO cables are shown in Figure 4. It is shown that some sections do not show significant development of axial cable (and hence, structural) strains, whereas some sections show some high unexpected strain values which also exhibit some localized peaks. The latter peaks may suggest the development of localized tension cracks.

In order to check the monitored response from the 3 sensing systems, independent numerical analysis studies have been undertaken using limit analysis mechanism and finite element (FE) deformation models. Figure 5 shows the vertical displacements of one cross-section of the masonry arch (for which the FO cables showed some high values of axial strain) from the laser scans and the numerical analysis.

The latter figure shows that a localized crack is suggested by both the field monitoring data and the numerical analysis results. This is attributed to the nearby piling operations which may have caused some settlement of one of the vault piers and hence re-distribution of the arch stresses.

It is finally shown that the nature of the distributed FO and laser scan systems provides an advantage over conventional discrete sensing systems in determining localized deformations such as cracking.

5 CONCLUSIONS

This paper presents some monitoring data from a masonry vault at the London Bridge Station which experienced some movements due to nearby piling construction. Three independent monitoring systems have been deployed and the results have been analysed.

The analysis showed that the spatially distributed systems (fibre optics and laser scans) provide an advantage over conventional discrete systems, such as total station prisms, in detecting localized regions of high strains and possible cracking.

Relevant numerical analyses (both limit analysis mechanism and finite element deformation analyses) were able to match the observed trend of strains and displacements within the vault, therefore confirming the validity of the monitoring data.

ACKNOWLEDGEMENT

Funding for this project came from EPSRC and Innovate UK through the Cambridge Centre for Smart Infrastructure and Construction (CSIC) Innovation and Knowledge Centre (IKC). The contribution of industrial partners from Costain, Topcon, Network Rail and Soldata are also acknowledged. Finally, a number of colleagues from the University of Cambridge and CSIC have also contributed in the field work and the subsequent analysis.

REFERENCES

Amatya, B.L., Soga, K., Bennett, P.J., Uchimura, T., Ball, P., and Lung, R. 2008 Installation of optical fibre strain sensors on soil nails used for stabilising steep highway cut slope. In: Proceedings of 1st ISSMGE International Conference on Transportation Geotechnics, pp. 276-282.

Cheung, L. L., Soga, K., Bennett, P. J., Kobayashi, Y., Amatya, B., & Wright, P. (2010). Optical fibre measurement for tunnel lining monitoring. Proceedings of the ICE - Geotechnical Engineering, 163(3), 119-130.

Mohamad, H., Bennett, P., Soga, K., Mair, R., & Bowers, K. (2010). Behaviour of an old masonry tunnel due to tunnelling-induced ground settlement. Geotechnique, 60(12), 927-938.

Mohamad, H., Soga, K., Bennett, P., Mair, R. J., & Lim, C. S. (2012). Monitoring Twin Tunnel Interaction Using Distributed Optical Fiber Strain Measurements. Journal of Geotechnical and Geoenvironmental Engineering, 138(8), 957-967.

Mohamad, H., Soga, K., Pellew, A., & Bennett, P. J. (2011). Performance Monitoring of a Secant-Piled Wall Using Distributed Fiber Optic Strain Sensing. Journal of Geotechnical and Geoenvironmental Engineering, ASCE, 137(12), 1236-1243.

Klar, A., Bennett, P., Soga, K., Mair, R. J., Tester, P., Fernie, R., St John, H. D., Torp-Peterson, G. (2006). Distributed strain measurement for pile foundations. Proceedings of the ICE - Geotechnical Engineering, 159(3), 135-144.

Schwamb, T., & Soga, K. (2015). Numerical modelling of a deep circular excavation at Abbey Mills in London. Geotechnique, 65(7), 604-619.

Schwamb, T., Soga, K., Mair, R., Elshafie, M., R., S., Boquet, C., & Greenwood, J. (2014). Fibre optic monitoring of a deep circular excavation. Proceedings of the ICE - Geotechnical Engineering, 167(2), 144-154.

Soga, K. (2014). XII Croce Lecture: Understanding the real performance of geotechnical structures using an innovative fibre optic distributed strain measurement technology. Rivista Italiana di Geotechnica, 4, 7-48.

Soga, K., Kwan, V., Pelecanos, L., Rui, Y., Schwamb, T., Seo, H., & Wilcock, M. (2015). The role of distributed sensing in understanding the engineering performance of geotechnical structures. Proceedings of the XVI European Conference on Soil Mechanics and Geotechnical Engineering. Edinburgh.

Proceedings of the International Conference on Smart Infrastructure and Construction
ISBN 978-0-7277-6127-9

© The authors and ICE Publishing: All rights reserved, 2016
doi:10.1680/tfitsi.61279.233

Structural behavior sensing using small sized self-propelled inspection robot

A. Akutsu*, E. Sasaki and K.Takeya

Tokyo Institute of Technology, Tokyo, Japan
* *Corresponding Author*

ABSTRACT Application of maintenance methods to ensure the safety, extend of service life and reduction in the life cycle cost of civil engineering structures are important. In this paper, to perform structural behavior sensing of civil engineering structures which often include such areas that are very narrow or too dangerous for inspectors to reach, a new inspection method using a robot which can effectively approach target sections, and then perform both visual and detailed inspection with various devices installed in the robot has been proposed. A small sized self-propelled inspection robot was initially developed and this model found to be able to freely move on surfaces of various types of structures. Next, the ability to detect the deterioration such as cracks or corrosion of concrete or steel surfaces by attached cameras as visual inspection was examined. In addition, a function was installed to identify the location of the robot automatically by using same camera as visual inspection and to sense structural behavior by using an acceleration sensor directly attached to the robot. Finally, the robot's movement and performance of all installed functions were tested on an existing bridge and then attempted to increase their applicability.

1 INTRODUCTION

Maintenance, especially inspection and sensing, of structures has become more important topic in the civil engineering field due to the deterioration of structures after the high economic growth (Frangopol & Kim 2011) , (Nakashima & Nagai, 2013). It is important for structures to ensure the safety or lifelong duration and reduce their life cycle cost (Hinow & Mevissen 2011), and so it is necessary to perform easy and frequent inspection without using scaffolding, large equipment or many inspectors in addition to periodic inspection.

Recently, robotic inspection starts to gain attention as a part of the maintenance procedure. However civil engineering structures are made of various materials and have various surfaces, making the robot difficult to move across the surfaces. Moreover, they include many sections that are invisible, very narrow or too dangerous to access hence the robot is limited in size and information of its location also becomes important.

In this research, a small sized self- propelled inspection robot with the ability to move on various surfaces and angles was initially built. It was then further developed with the aim to achieve effective maintenance of structures without the need of scaffoldings or inspection passages. The proposed robot has a function to perform visual inspection and identify the location of the robot automatically by attached cameras, as well as to sense structural behavior. The efficiency of this robot's movement and all installed functions were tested on an existing bridge.

2 DEVELOPED INSPECTION ROBOT AND INSPECTION FUNCTION

2.1 *Requirements and developed inspection robot*

Wall climbing robots that have considered the aforementioned obstacles have already been developed in previous researches. There are many methods to attach robots to surfaces, such as by using vacuumed suction cups (Iwamoto, 2013), or electrostatic suction films (Wang, et al. 2014). The main purpose of these robots are to climb walls and so their ability to climb walls is high, yet their capacity is limited by the material of the walls. However, they can generally only perform visual inspection by camera, as if they have more devices attached, they become too large and heavy. Although there are flying robots also (Metni, & Hamel, 2007) they are limited as they are easily influenced by wind and are difficult to control. Considering these facts, the requirements of the inspection robot are determined as shown in Table 1. Functions to satisfy the requirements will be explained in following chapters.

In this research, the small sized self-propelled inspection robot are initially built which is satisfied the requirements and it has a simple movement system which is controlled by on and off signals shown in Figure 1. Moreover it attaches to surfaces using air pads put in a vacuum state, while moves by the expansion and contraction of the air cylinders. Polycarbonate is used for the body to reduce weight, and the size of the body is (W)153mm×(L)148mm×(H)60mm and weight is 328g.

This robot is controlled using cables, which can currently extend to 10 meters and have the capacity to be extended further. Therefore inspectors are able to control the robot from afar while viewing from the inspection cameras in real-time without putting themselves in any danger. Furthermore, the proposed robot can resist the influence of wind, even in high places, due to its small size and implemented vacuum system.

2.2 *Visual inspection function*

The essential advantage of using the proposed visual inspection function is that it can easily inspect areas where visual inspection is normally difficult. Moreover, it can significantly reduce the inspection cost by eliminating the scaffoldings or inspection passages. Proposed robot has CCD camera for visual inspection in front of the robot and it is able to show the surfaces of structures. In this research, the visual inspection function focused on two points; to the ability to recognize different colors and to check detailed surface conditions even when close to surfaces. Moreover the robot needs to be small sized due to inspect narrow areas. Hence, a small sized, light CCD camera with adequate quality to perform visual inspection was selected. An image taken by the CCD camera is shown in Figure 2. The image shows features of the surface such as holes and roughness and the color difference can also be accurately viewed. It is possible to install a scale to know the size of the cracks or holes (Figure 2(b)). As a result of numerous performance tests, it has

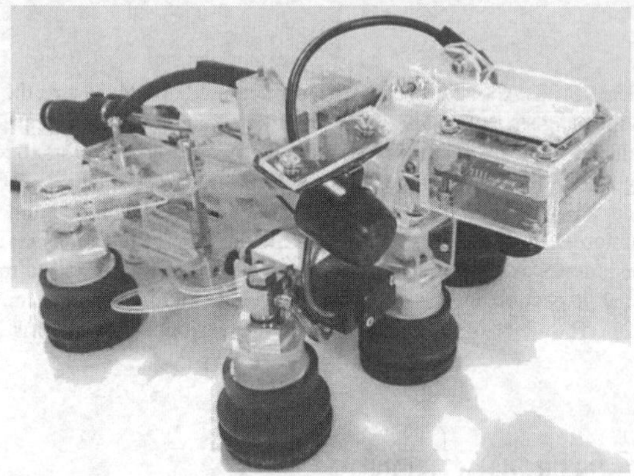

Figure 1. Proposed inspection robot

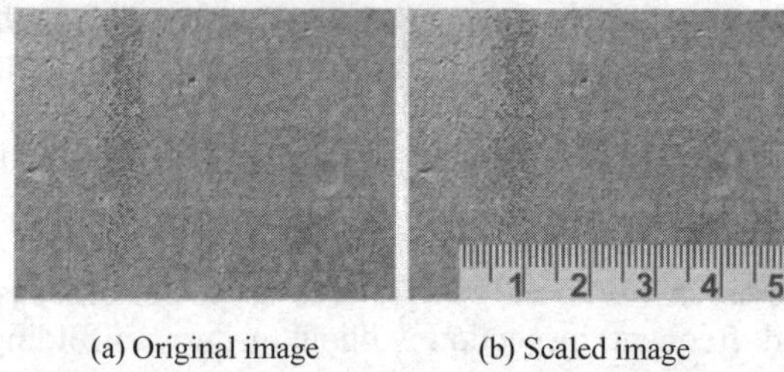

(a) Original image (b) Scaled image

Figure 2. Image taken by CCD camera on concrete wall

Table 1. Requirements of inspection robot

Requirements	Function for requirements
Ability to move on various surface conditions and roughness without special rails	Movement system with air pads and vacuum air
Small in size to inspect narrow sections	Miniaturization of robot body
Ability to perform visual inspection	Visual inspection function
An automatic identification system of robot location especially at invisible areas	Location identification function
Ability to sense structural behavior and changes	Function to sense structural behavior

been clarified that the CCD camera can take high quality images to perform effective visual inspection.

2.3 *Location identification function*

The inspection robot has a possibility to work in areas with low visibility. Therefore, it is important to obtain information of the robot location. In this research, the visual inspection camera was used to build the location identification function considering miniaturization and weight reduction of the robot. An advantage of using images for this function is that the progress of cracks or corrosion can be confirmed comparing with former obtained images. In this study the Phase-Only Correlation (POC) method (Takita, et al. 2003) that uses for image matching process, has been applied. In this method, only phase information is used to place emphasis on the edges of images and omit information regarding brightness. Applying POC method to two images then a POC function is obtained. When two images are similar, their POC function gives one distinct sharp peak. And then, when the two images are shifted, the position of the peak shifts from the center, and this shifted value from the central point shows how much the two images have changed in position.

A detailed outline of the location identification function is as follows. One picture is taken by the attached camera for each cycle of robotic movement, and then by using consecutive two images, the values of parallel translations in both x and y directions and a rotation angle can be obtained. The proposed robot can rotate approximately 10 degrees in one rotation cycle. Therefore, to determine the rotation angle, the image after rotation is rotated between -15 to +15 degrees. The rotation angle and the values of x and y are obtained by using the POC method at same time. By summating all obtained values the total parallel translation and rotation angle from the beginning to the end of the robot movement can be calculated.

2.4 *Mapping function*

Applying previous function, the location of the robot could be determined numerically. However, when the robot is working in invisible areas, it is difficult to image the real location. Therefore, a mapping function which is able to make a map of the robotic movement is developed and installed to help inspectors to recognize the location information visually. Moreover, using this function, it becomes possible to confirm the process of crack or corrosion conditions. Figure 3 shows an example of the location identification function and mapping function. This figure indicates that the two images moves 100pixel in x direction and 150pixel in minus y direction.

2.5 *Function to sense structural behaviors*

The proposed robot has not only cameras, but an acceleration sensor was also attached to sense structural behavior for detailed inspection. It is possible to obtain the acceleration data at the area where the robot is inspecting accordingly. In the real situation, it is difficult to replace acceleration sensor placed directly on structures, however using this robot, it becomes easy to move and replace its position. Furthermore, it is also possible to check the surface condition using attached

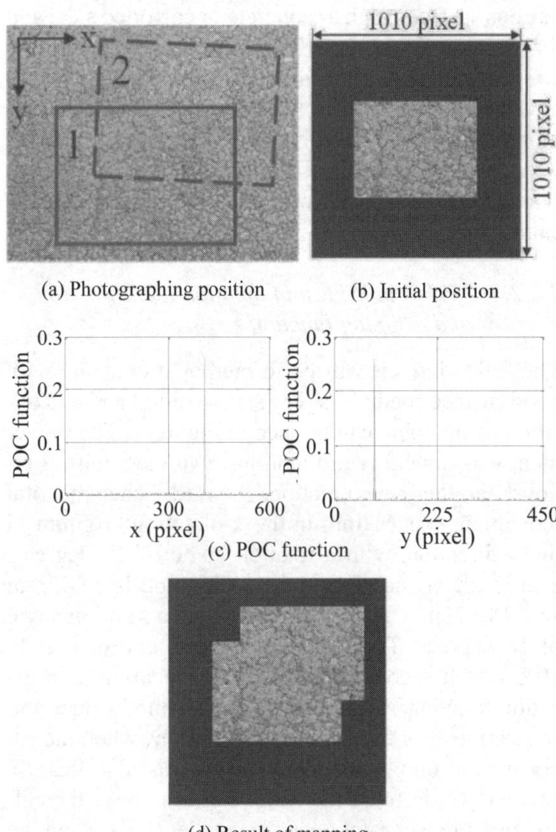

Figure 3. Location identification and mapping function

camera while the robot is moving, and it is also easy to measure the acceleration if necessary.

3 APPLICATION OF INSPECTION ROBOT AND ALL INSTALLED FUNCTION

In this research, a robot that is able to perform robotic inspection was developed. In this chapter, the performance of the robot's movement, visual inspection and location identification function were examined, using concrete walls and existing bridge as target structures.

3.1 Experiment on concrete walls

3.1.1 Robotic movement test

The proposed robot is able to move forward, retreat and rotation. In this test, all three movements were performed and evaluated on mainly outside concrete walls. It was confirmed that the robot could move on various surfaces such as concrete or corroded steel and that it can move up to 400 millimeters per second. As a result, it has been confirmed that the robot is able to advance and retreat well on various surfaces, and moreover, rotate smoothly on various surfaces. Moreover the performance of the robotic movement was scarcely affected by its weight or the elasticity of the cables.

3.1.2 Performance test of location identification and mapping function

The following test was performed on a concrete with good surface conditions. The start position and end position of the robot can be seen in Figure 4. The movement was three forward motions, two clockwise rotations, and then seven motions forward, hence the total movement was 61 mm in the x direction, 165mm in the y direction, with a rotation angle of 19 degrees. The result of analysis, in the x direction is 54.45mm and, 149.76mm in the y direction with a rotation angle of 18 degrees. The relative error was approximately 10% which is considered high. The location identification function is summing the all obtained values and so relative error may increase. However, when the robot moved only forward, the error was significantly smaller. To decrease the relative errors, make the calculation pitch of rotation angle smaller (0.01°), cut the distortion parts and increase sharpness and use the average of peak values of POC function if the difference between the peak and other peaks is less than 0.001.

The results of this test after modification is that the total movement was 63.9mm in the x direction and, 164.5mm in the y direction, with a rotation angle of 20.5 degrees. The relative error becomes half of the previous method and it was considered low and so, it has been confirmed that the location identification function performs well when used on real structures.

The mapping result, a map of the robot tracks, is shown in Figure 5. It is able to confirm that the modified analysis is tracking robot's movement better, hence demonstrate that the mapping function can track the movement of the robot effectively. It has thus been confirmed that the location identification function has the capacity to work in real inspection sites.

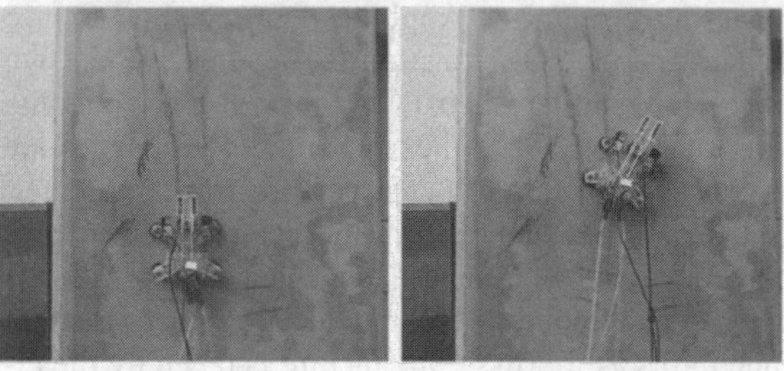

(a) Start position (b) End position

Figure 4. Performance test of Location identification and mapping function

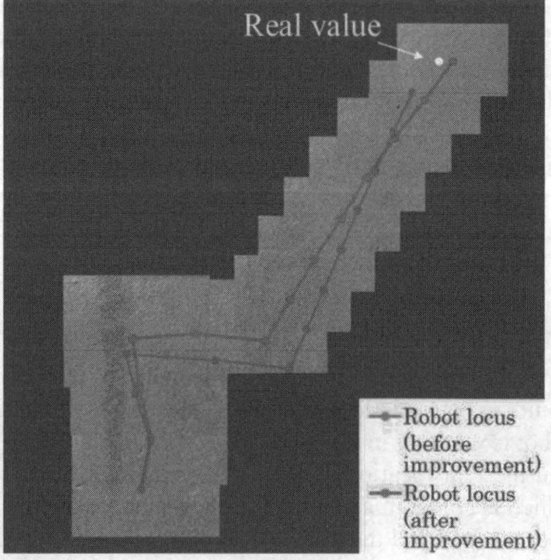

Figure 5. Result of mapping function

3.2 *Performance test of visual inspection and acceleration sensor on existing bridge*

The existing bridge used in this study is a simple steel through truss bridge. It is limited to five tons of weight due to corrosion. Live robotic inspection was performed to confirm whether the robot can work at sections that inspectors cannot reach, such as panel sections of truss members. In this test, the robot was tested on a lower chord and a vertical member of the truss as shown in Figure 6(a). Corrosion has significantly progressed in most areas of this bridge and some places even showed peeling of the corrosion surface. However this live testing confirmed that the robot could move and be applied in the real life scenarios.

The performance of two cameras were confirmed and Figure 6(b) shows two images, one from the camera for visual inspection, and one for looking in the forward direction. As shown, the visual inspection camera can take detailed images of surface conditions such as peeling, cracks, color and roughness of corrosion surface. Moreover, using the forward camera the inspector can check the conditions of not only close surfaces, but also the surface one step ahead of the robot. Thus, the inspectors are able to determine which sections need to be inspected, by judging the surface conditions in real time.

(a) Robot location

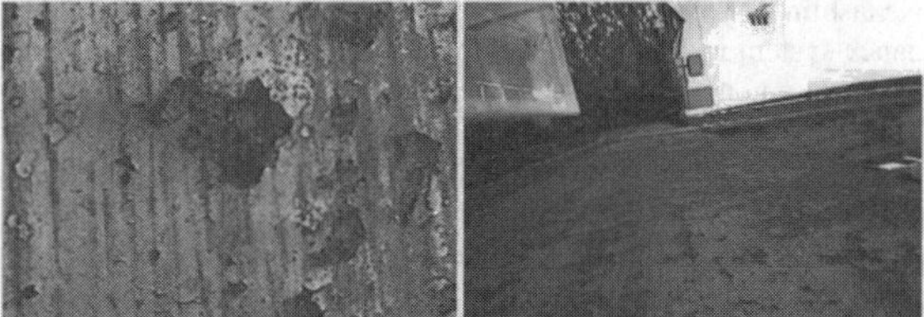

(b) Images from inspection and forward camera

Figure 6. Inspection at lower chord

A performance test of acceleration sensor was also carried out. In this test, the robot was attached next to an acceleration sensor that was attached directly to the bridge (Base acceleration sensor). Measurements were carried out for 5 minutes, with a sampling frequency

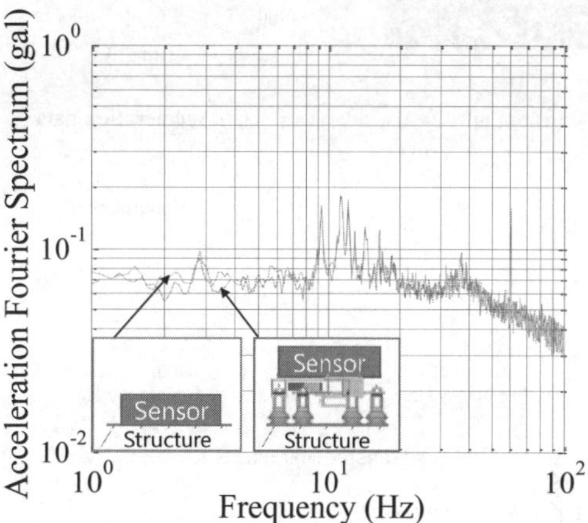

Figure 7. Acceleration Fourier Spectrum

of 1000Hz. Figure 7 shows the acceleration Fourier spectrum of both acceleration sensors. There were concerned that the damping of the air pads may influence the measurements, however, it can be seen that both acceleration Fourier spectrum lines correlate well especially at the important frequency band, excluding low frequency and high frequency for bridge vibration. As a result, it has been confirmed that it is also possible to recognize the natural frequency of bridges by using the acceleration sensor attached to the robot.

3.3 *Vibration test*

In previous experiment, it was confirmed that the both data from robot and acceleration sensor shows good match in frequency domain. However the raw acceleration data has gaps in amplitude values. It is considered that air pads have the possibility to affect raw vibration data due to nonlinearity of rubber behavior. Therefore it is important to confirm the influence of air pads to obtained vibration by robot.

In this test, the robot was also attached on a base plate next to the base acceleration sensor (Figure 8(a)) to know the characteristics of transfer of air pads. 10 waves which are sine and triangle wave of 1~5 Hz in 1Hz pitch were used. Figure 8(c) shows the gain and

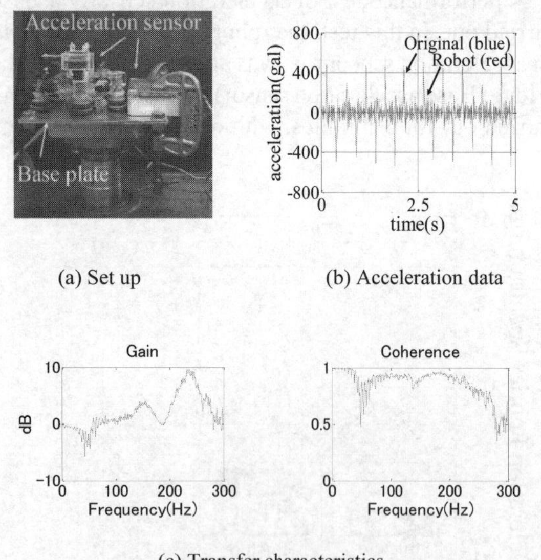

(a) Set up (b) Acceleration data

(c) Transfer characteristics

Figure 8. Vibration test

coherence wave form of transfer function in 2Hz triangle wave using both base acceleration data and robot acceleration data. All cases indicate similar results. It can be considered that disturbance of waveform at 50Hz has been affected by power supply noise. Moreover natural frequency of base plate was calculated 250Hz by eigenvalue analysis in FEM. Shell model was used for the analysis. Therefore the natural frequency of air pad can be considered 140Hz. Furthermore using half-power band width method, damping ratio has calculated to be 0.061. Modal analysis using solid model in FEM of air pads will be done to confirm the behavior and characteristics of air pad during vibrating in different view.

4 CONCLUSIONS

In this research, robotic inspection using a self-propelled robot without the need of scaffoldings or inspection passages for tunnels or road bridges has been proposed. This innovative inspection robot can perform robotic inspection of sections with various surface conditions, which has so far been difficult to perform. Numerous experiments were conducted on concrete walls, as well as an existing bridge, to examine the performance of the robots movement and the quality of the attached cameras for visual inspection. The results show that the proposed robot can be used on various surfaces such as concrete, corroded steel and even the under surfaces of structural members.

A location identification and mapping function is using images from the attached camera was also proposed. An experiment carried out on concrete walls and the results obtained indicate that the robot can identify its own current location by using both numerical values and a map of the member surface across which it walks. Another proposal was whether the robot could sense structural performance for detailed inspection, by attaching acceleration sensor. Using this function, the robot could successfully detect the natural frequency of the bridge. Also the vibration test was conducted. Then the natural frequency of air pad can be considered around 140 Hz and damping ratio has a possibility to be 0.061.

Using the proposed robotic inspection, inspectors are able to monitor structural conditions, even in such location as the under surface of a bridge, from a distant location. Moreover, the proposed robot makes inspections easy to perform for sections which currently require large robots or devices. There are future plans to attach small inspection devices, such as an eddy current probe, for detailed inspection. Thus in conclusion, the proposed inspection robot was highly effective is expected to become even more versatile.

REFERENCES

Fragonpol, D.M. & Kim, S. 2011. Chapter 5. Hinow, M. & Mevissen, M., 2011. Substation maintenance strategy adaptation for lifecycle cost reduction using genetic algorithm, IEEE Transactions on power delivery, 26(1), 197-204.

Iwamoto, T. 2013. Development of testing machine of concrete slab and vertical concrete wall Using impact echo method, Performance report of Grants in Aid for Scientific Research (in Japanese).

Karbari, V., and Lee, L.S. (Ed.) Service Life Estimation and Extension of Civil Engineering Structures: Service Life, Reliability and Maintenance of Civil Structures, Woodhead Publishing Ltd., Cambridge, U.K., 145-178.

Metni, N. & Hamel, T. 2007. A UAV for bridge inspection: Visual servoing control law with orientation limits, Automation in Construction, 17, 3-10.

Nakashima, M. & Nagai, K. 2013. An investigation of road maintenance system in Japan in developed Society, The 9th International Symposium on Social Management Systems.

Takita, K. Aoki, T. Sasaki,Y. Higushi,T. Et al. 2003. High-Accuracy subpixel image registration based on Phase-Only Correlation, IEICE TRANS. Fundamentals, E86-A(8), 1925-1934.

Wang, H. Yamamoto, A. & Higuchi, T. 2014. A crawler climbing robot integrating electroadhesion and electrostatic actuation, International Journal of Advanced Robotic Systems.

Proceedings of the International Conference on Smart Infrastructure and Construction
ISBN 978-0-7277-6127-9

© The authors and ICE Publishing: All rights reserved, 2016
doi:10.1680/tfitsi.61279.239

Ultrasonic sensor system for automatic depth measurement of surface opening cracks in concrete by means of a robotic arm

L. Belsito[1*], L. Masini[1], M. Sanmartin[1], K. Loupos[2], and A. Roncaglia[1]

[1] *Institute of Microelectronics and Microsystems/CNR, Bologna, Italy*
[2] *Institute of Communication & Computer Systems, Athens, Greece*
[*] *Corresponding Author*

ABSTRACT: The paper reports about the development of an automatic tool that can be used to measure the depth of cracks with millimetric precision on board of a structural inspection robot. The system is based on two piezoelectric ultrasonic transducers with central frequency of 50 kHz that are utilized to take Time-of-Flight (ToF) ultrasound measurements across the crack to determine its depth. In order to cope with the intrinsic inhomogeneity of concrete, the measurement method adopted in the automatic system is based on the simultaneous determination of ultrasound speed propagation and crack depth in one measurement series over the crack. To this purpose, the automatic system is designed to be able to position the two transducers in the required way on a concrete surface by using two stepper positioning systems mounted at 90° with respect to each other, able to displace the sensors with millimetric precision over the concrete wall, as needed in the measurement. The complete system, constituted by the ultrasonic sensors, the 2-dimensional positioning tool and a protection frame with steel supporting pins, can be displaced over the surface of a concrete tunnel by a robotic arm and placed across a crack in order to measure its depth automatically. The design, implementation, laboratory testing and early on-field testing of the system are presented, showing its capability to perform automatic crack depth measurements with millimetric precision.

1 INTRODUCTION

Recent progresses in robotics and computer vision have paved the way to the possibility of replacing humans with robots in inspection and structural evaluation of buildings and civil infrastructures. This method represents an alternative to the installation of wireless sensor networks able to take the needed measurements for an effective monitoring of the structure under investigation (Ferri at al. 2010, 2011), (Soga et al. 2014), (Stent et al. 2015).

In structural assessment, the depth of surface opening cracks on concrete structures is a very important and routinely monitored parameter in the current civil engineering practice. In concrete tunnels, such measurement is particularly relevant since the width and depth of surface opening cracks in the lining can be used in structural assessment models based on intra-dos measurements (Bhasin & Grimstad 1996).

Within ROBO-SPECT, a currently ongoing EC co-funded research project (FP7 - ICT – 611145), a semi-autonomous robotic system for tunnel structural inspection and evaluation is being developed (Loupos et al. 2014). Within the project, an automatic sensor system for ultrasonic depth measurements of surface opening cracks on concrete structures, suitable for use on a robotic arm, is being developed. In the present paper, the design, implementation and testing of such system are described.

2 CRACK DEPTH MEASUREMENT METHOD

The basic ultrasonic method relying on Time of Flight (ToF) measurements for the estimation of crack depth is schematically illustrated in Fig. 1. By generating an ultrasonic wave at one side of the crack, the time of flight measured at the other side can be used to estimate the depth, since the wave cannot propagate well through air and needs to reach the crack bottom to pass beyond it.

Consequently, by indicating with v the ultrasonic wave velocity in concrete and with τ the measured time of flight, we may write:

$$h^2 = v^2(\tau/2)^2 - d^2 \qquad (1)$$

where h is the depth of the crack and d the distance of the transmission and detection points of the ultrasonic waves from its centre, supposed to be the same in the two cases (Fig. 1).

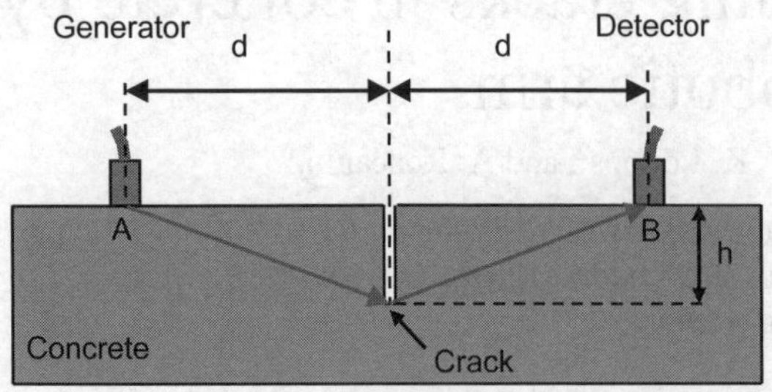

Figure 1. General method for crack depth measurement with Time Of Flight technique.

Eq. 1 simply derives from the application of Pythagoras' theorem to the geometric path represented in Fig. 1, which the ultrasonic waves have to go through after being generated from point A to reach point B in the figure.

Using this equation, the crack depth h can be quite straightforwardly determined from the ToF measurements since in that case the time of flight τ is directly determined from the ultrasonic measurement while the distance d can be measured with visual methods. However, in order to apply the formula, the ultrasound propagation velocity v has to be determined independently. In order to do this, the simplest method consists in displacing the transducers beside the crack in such a way that a direct path through the surface of concrete from the transmitter to the receiver is allowed for the ultrasonic waves. By measuring the ToF in this way and calculating the ratio between the distance and the measured time, the surface velocity can be determined. Such parameter can be identified with v in eq. 1 and used in subsequent ToF measurement experiments in order to determine the depth of a neighboring crack.

This method is straightforward, but can be prone to inaccuracies deriving from the intrinsic inhomogeneity of concrete, which is a composite material and can present a different structure from point to point. As a consequence of this characteristic, the ultrasonic velocity can be slightly different in different regions of the same concrete slab. For this reason, the method described above could give rise to measurement errors in case the inhomogeneities of the measured concrete sample are such that the surface velocity determined beside the crack is different than the one needed in eq. 1.

In order to overcome this problem, the alternative method illustrated in Fig. 2 can be adopted (Pinto et al. 2010). In this variant of the basic method, the same ToF measurement of Fig. 1 is performed several times displacing the transducers at various distances from the crack, indicated in the figure as d_1, d_2, d_3. For each of these measurements, eq. 1 holds, which in this peculiar case can be rewritten as:

$$d_i^2 = v^2\left(\tau_i - \frac{\tau_1}{2}\right)^2 - h^2 \qquad (2)$$

in which τ_i are the Time of Flight measurement results acquired at the various distances. Looking at eq. 2, it clearly appears that by plotting the squares of the distances versus those of the measured times of flight, the line intersecting the plotted points will yield as its y axis negative intercept the square of the crack depth.

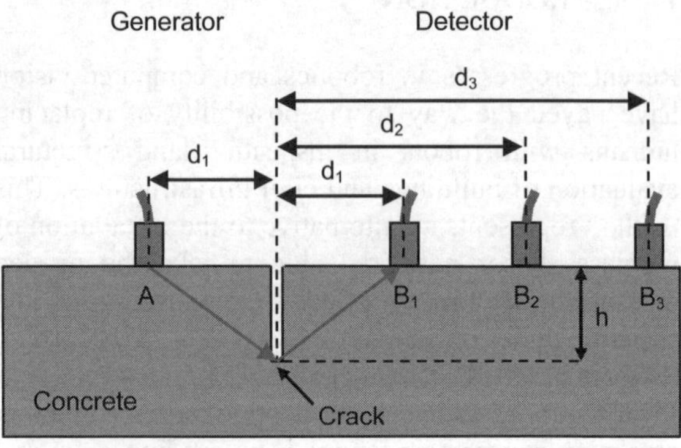

Figure 2. Alternative method for crack depth measurement with Time Of Flight technique including on-site determination of the ultrasonic propagation velocity.

In order to enable the application of this method to measurements performed with a robot, the design described in the following was adopted for the ultrasonic sensor system.

3 SYSTEM DESIGN

The block scheme of the ultrasonic sensor system is represented in Fig. 3.

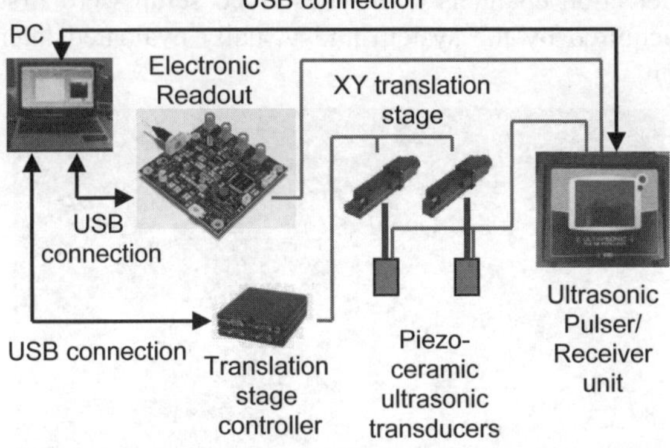

Figure 3. Block scheme of the ultrasonic sensor system with electrical connections.

The system is composed of a pair of piezo-ceramic transducers that can be used for both ultrasound generation and detection on concrete. Such transducers are operated by means of a Pulser/Receiver unit which generates the high-amplitude electric signals needed for ultrasonic generation in actuation mode and acquires the voltage produced by the piezoelectric transducers when used in sensing mode.

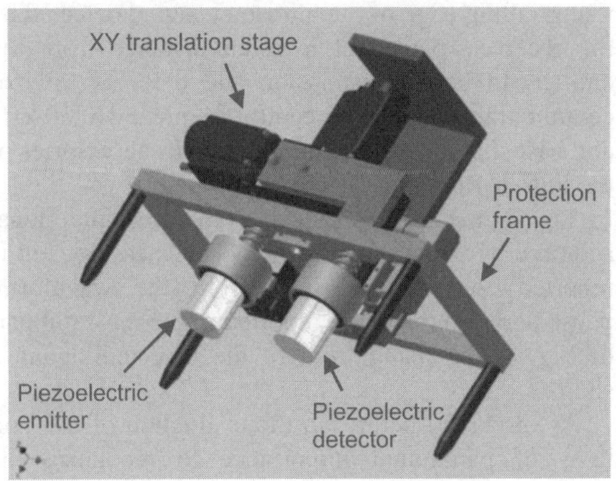

Figure 4. 3D model of the ultrasonic sensor system.

In order to be utilized in crack depth measurements on the robot, the mechanical design shown in Fig. 4 was adopted for the system. Since, according to the method illustrated in Fig. 2, during crack depth measurements one of the sensor needs to be displaced on the tunnel lining and contacted at increasing distances from the other one, an XY positioning tool was adopted in the design. Such tool was designed in order to be able to perform the needed displacements of the sensors on the tunnel lining without any intervention by the robotic arm, as shown in Fig. 4.

The XY stage was assembled by mounting two linear stepper motors from Standa at an angle of 90° and fixing them on the rigid structure of the prototype. The connection of this positioning system to the sensors was designed to enable the lifting and lowering of both sensors from the tunnel lining and vary the distance between them in the course of different ultrasonic measurements, as required by the method. The positioning of the sensors with the stage over the lining was also aided by introducing two spring connections between the transducers and the stage, so that the applied pressure could be absorbed by the system without risk of damage for the devices.

A four-point iron/aluminum support structure was designed around the positioning stages, in order to assist the placement of the system across the crack, as shown in Fig. 5, and provide the possibility of a steady operation on the tunnel lining.

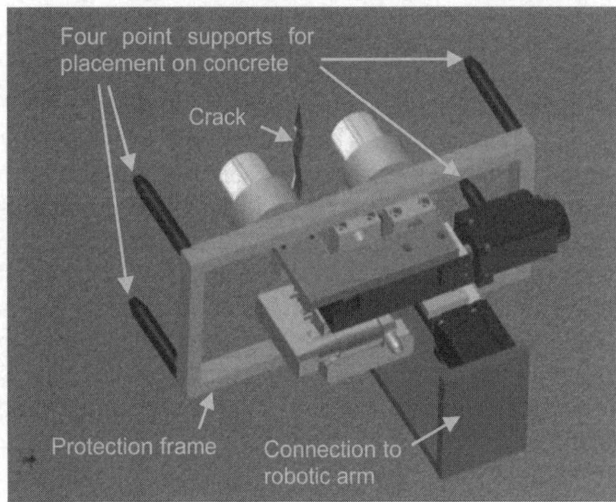

Figure 5. Use of the sensor system on the tunnel lining.

As can be observed from the 3d representation in the figure, the measurement is expected to be carried out after the four-point support has been positioned on the tunnel lining across the crack and maintained in place with the aid of the pressure applied by the robotic arm. After this positioning phase, all the measurements are performed by the ultrasonic module by operating the needed sensor displacements through the XY positioning stages, without any further intervention by the robotic arm.

The ultrasound sensor system prototype designed according to the guidelines reported above was all manufactured in massive aluminum with the exception of the rectangular protection frame, for which concave iron was utilized. A picture of the assembled system prototype is shown in Fig. 6.

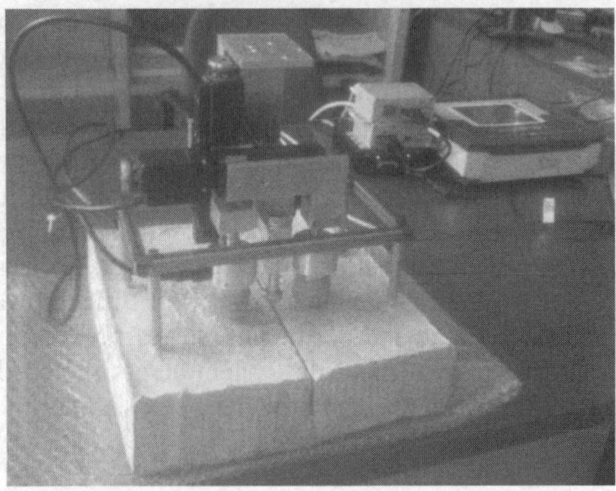

Figure 6. Ultrasonic sensor system prototype.

4 TESTING

In order to test the prototype, concrete samples with artificial cracks were produced using squared moulds by putting a Plexiglas slab vertically immersed into fresh concrete in order to create a narrow aperture in the specimen, suited for emulating the crack (Fig. 7). For these early tests, the apertures were created with roughly constant depth and prolonged enough to reach the lateral borders of the specimen and consequently be easy to measure in depth from the side of it.

First, in order to test the crack depth measurement method, preliminary ultrasonic surface velocity measurements on concrete specimens were performed by evaluating the ToF on the waveforms. In this operation, it was important to recognize the earliest variation of the detected waveform on the receiver channel, otherwise the ToF would be largely overestimated in the measurement. In these first experiments, the fronts of the waveforms on the detection channels of the described setup were first acquired by the system and visually evaluated later on.

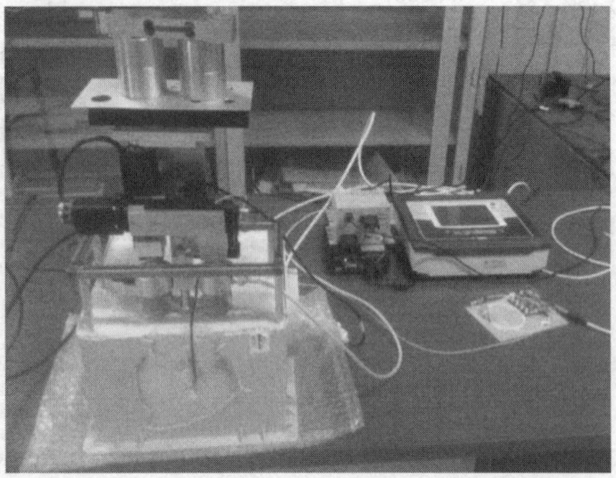

Figure 7. Laboratory setup for testing of the ultrasonic sensor system on artificially produced cracks on concrete specimens.

The ToF measurements were performed by evaluating the delay between the ultrasonic generation operated by a 58-E4900 Ultrasonic Pulser Analyser from CONTROLS and the first rising/falling edge of the ultrasonic signal detected by the receiver, placed at a given distance from the emitter. In both transmission and detection, piezo-ceramic transducers with central frequency of 50 kHz and base diameter of 3 cm, provided as accessories of the Pulser/ Receiver system, were used.

In Fig. 8, the typical shape of an ultrasonic signal acquired by the system during a ToF measurement is reported. As can be seen from the waveform, particularly on the detail reported in the top-right part of Fig. 8, the rising front of the detected signal is sharp.

As discussed before, in the evaluation of the ToF it is of paramount importance to recognize the earliest variation of such sharp waveform on the detection channel, otherwise the ToF may be overestimated in the measurement. In order to do that, different algorithms can be used in the Pulser/Receiver system. After an extensive experimental characterization of the method, the Akaike Information Criterion (Pan 2001) was adopted as the one giving the best performances in terms of accuracy in ToF and consequently in crack depth measurements. According to this method, a statistical correlation on a defined observation time between the signal and the noise baseline of the transducer was utilized to detect the rising front of the waveform. In this way, the front detection is usually located before the first visible edge of the signal, as shown in Fig. 8, whereas all the other fronts are not relevant in ToF detection.

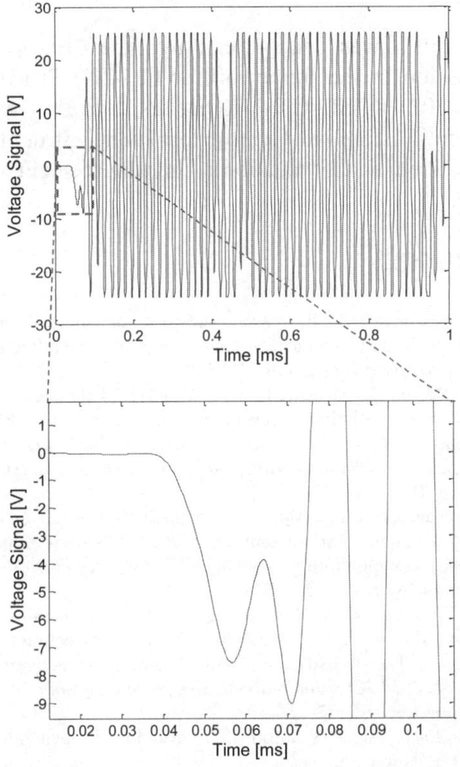

Figure 8. Typical shape of the ultrasonic waves acquired on the detection channel in ToF measurements on concrete.

Once the best algorithm for the automatic ToF estimation was determined, simultaneous crack depth and ultrasonic propagation velocity measurements on concrete specimens were carried out.

In Fig. 9, the result of a measurement performed on a 6 cm deep artificial crack is shown by way of example. As can be observed from the plot reported in the figure, the square of the crack-transducer distance shows a linear behavior when plotted versus the squared measured ToF, according to the model described by eq. 2 discussed before.

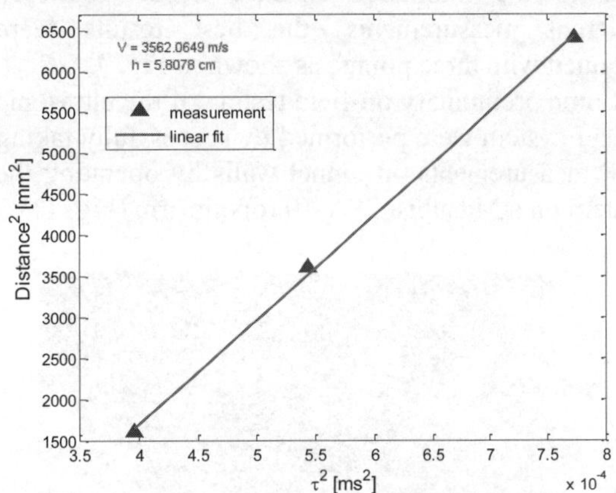

Figure 9. Result of crack depth measurement with ToF technique with in-place determination of ultrasound propagation velocity on concrete specimen with artificially produced cracks.

Using the intercept of the curve with the y-axis and its slope, a crack depth of roughly 5.8 cm and an ultrasonic propagation velocity around 3560 m/sec were estimated, with a measurement error on depth of about 5%.

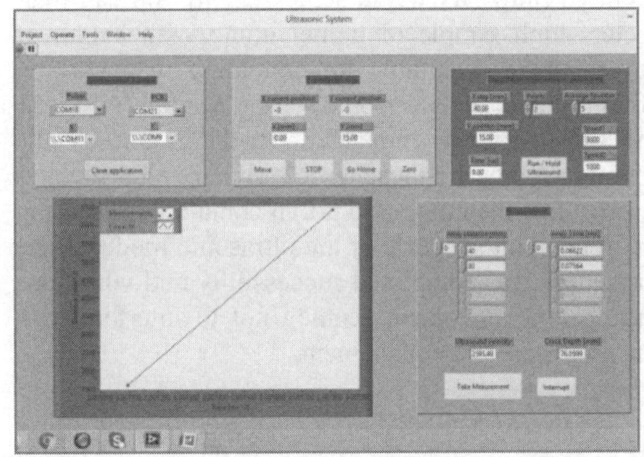

Figure 10. Labview user interface of the ultrasonic sensor system for ToF and crack depth measurements on concrete.

When performing such measurements with the automatic system, the number of ToF measurements used in the crack depth estimation procedure and their distances from the crack could be customized by the user by means of the PC Labiew interface shown in Fig. 10. Other customizable parameters where the speed of the displacement of the sensors, which could be set up to a maximum of 5 mm/sec. Due to the minimum distance needed between different measurements, the best results were obtained with three points, as shown in Fig. 9.

Some preliminary on-field testing of the ultrasonic sensor system were performed by successfully taking ToF measurements on tunnel walls by operating the system on a Mitsubishi PA-10 robotic arm (Fig. 11).

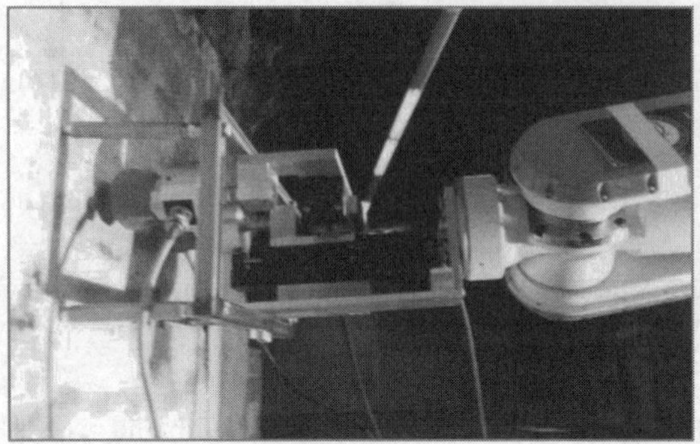

Figure 11. ToF measurement tests performed using the ultrasonic sensor system on a tunnel lining sample in VSH.

The on-field testing were performed in VSH facility (http://www.vsh-schiessen.ch) on a two-meters high sample of tunnel lining with length of several meters, suitable for emulating real working conditions for the ultrasonic sensor system. In Fig. 11 the sensor system contacted on the lining by the robotic arm during these tests is shown, after the arm approached the lining and got in contact with it using the four-point supports of the ultrasonic module. The procedure was completed successfully and vibrations during arm movement seemed not to interfere with the precision of the placement.

5 CONCLUSIONS

An automatic sensor system for crack depth measurements on concrete structures suitable for operation on a robotic arm has been presented. To this purpose, the system was equipped with two XY translation stages used to displace the ultrasonic sensors on the tunnel lining with no direct intervention by the robotic arm.

Laboratory tests on the prototype have been successfully performed on concrete specimens with artificially produced cracks, showing a relative accuracy in crack depth measurement around 5%.

Preliminary tests on-field using the system on a robotic arm have also been executed taking surface velocity measurements on tunnel lining samples in an underground infrastructure. Further on-field tests of the system on the robot developed within ROBO-SPECT project (FP7 ICT 611145) are currently ongoing.

ACKNOWLEDGEMENT

This work was supported by the European Commission in the framework of FP7-ICT-2013-10 project ROBO-SPECT (ROBotic System with Intelligent Vision and Control for Tunnel Structural INSPECTion and Evaluation), Grant agreement no: 611145.

REFERENCES

Bhasin, R. Grimstad, E. 1996. The use of stress-strength relationships in the assessment of tunnel stability, *Tunnelling and Underground Space Technology* **11**, 93–98.

Ferri, M. Belsito, L. Mancarella, F. et al. 2011. Fabrication and testing of a high resolution extensometer based on resonant MEMS strain sensors, *Tech. Digest of Transducers 2011 the 16th International Solid-State Sensors, Actuators and Microsystems Conference*, 1056 – 1059.

Ferri, M. Mancarella, F. Seshia, A. A. et al. 2010. Fabrication and packaging techniques for the application of MEMS strain sensors to wireless crack monitoring in ageing civil infrastructures, *Smart Structures and Systems* **6**, 225-238.

Loupos, K. Amditis, A., Stentoumis, C. et al. 2014. Robotic Intelligent Vision and Control for Tunnel Inspection and Evaluation – The ROBINSPECT EC Project, *Proceedings of ROSE 2014, 2014 IEEE International Symposium on Robotic and Sensors Environments*, 72-77.

Pan, W. 2001, Akaike's information criterion in generalized estimating equations, *Biometrics* **57**, 120–125.

Pinto, R. Medeiros, A. Padaratz, I. et al. 2010. Use of Ultrasound to Estimate Depth of Surface Opening Cracks in Concrete Structures, http://www.ndt.net/?id=9954.

Soga, K. Ledesma, A. Roncaglia, A. et al. 2012. Micro-measurement and monitoring system for ageing underground infrastructure (Underground M3), *Proceedings of the 7th International Symposium on Geotechnical Aspects of Underground Construction in Soft Ground*, 197-204.

Stent, S. A. I. Girerd, C. Long, P. J. G. et al. 2015. A Low-Cost Robotic System for the Efficient Visual Inspection of Tunnels *Proceedings of the International Symposium on Automation and Robotics in Construction* **32**, 1-8.

Proceedings of the International Conference on Smart Infrastructure and Construction
ISBN 978-0-7277-6127-9

© The authors and ICE Publishing: All rights reserved, 2016
doi:10.1680/tfitsi.61279.245

Bridge scour monitoring system development

Kuo-Chun Chang, Yung-Bin Lin, Meng-Huang Gu and Bo-Han Lee

National Center for Research on Earthquake Engineering, Taipei, Taiwan

ABSTRACT Induced by hydraulic deficiencies, scouring remains a major cause of bridge failure. Bridge scour failures have been reported frequently around the world. Collaborated with the interdisciplinary institute of the National applied Research Laboratories (NARLabs), we proposed a 6-hour forecasting technique by using the upstream rainfall to evaluate the flow level, flow velocity, and possible scour depth of the target bridges. This NARLabs Team also developed the sensory system with variety sensors such as arrayed scouring sensor for real-time scour depth response, arrayed forcing sensor for turbulent flow velocity measurement which different to the conventional surface flow velocity data, and bridge pier vibration sensor for bridge natural frequencies analysis during the flood period. This system has combined with the commercial video camera, surface flow velocity, and flow level to integrate into this bridge early warning monitoring system. All these sensors data have been connected to the cloud server with wireless sensor network for real-time signal process and synchronized to present to the public needed. This bridge multi-hazards early warning system has been successfully demonstrated in the field of Taiwan river basins.

1 INTRODUCTION

Collapse of bridges due to foundation scour is a worldwide problem. Scouring around bridge piers/abutments remains a major cause of bridge failure induced by hydraulic deficiencies. Scour excavates sediments around the foundations of a bridge causing a reduction in the safe capacity of the bridge. Between the 1960s and 1990s in the United States, more than 1,000 bridges had collapsed, among which 60% of the failures due to scour [1]. Similar problems also exist in many East Asian countries, especially in those areas subject to typhoon flood. For instance, in Taiwan there are over 20,000 bridges crossing rivers. During Typhoon Morakot in 2009, extreme amount of rainfall triggered fierce flooding throughout southern Taiwan to cause about 100 bridges collapsed in one single flood event [2]. Several bridge failure cases occurred during typhoon floods in Taiwan were shown in Figure 1. Causing reduction of safety capacity during flood, scour failure tends to occur suddenly without prior warning to the bridge pier structure. It is critically important to monitor the real-time scour depth changes in order to prevent catastrophic failure resulting in major operational disruption and life losses. A real-time bridge scour monitoring system should carry out the inspection of the scour depth recorded during times of flooding. The maximum scour may occur when the risk of scour is the highest flood discharge, at which the scour hole tend to be filled in as flood water recedes. The scour monitoring information is definitively helpful in assisting engineers to design safer and more cost-effective bridges.

Around bridge piers, the combined effects of the vortex system involving time-dependent flow pattern and sediment transport mechanism make scouring phenomenon extremely complex [3-7]. A number of formulas for estimating scour depth at bridge piers can be found in the literature. Most of the scour studies derived from experimental flume results mainly focus on the maximum scour depth by empirical regression equations. However, field data are limited due to observational difficulties [8-10]. Without sufficient measured data in the field, these empirical equations may not be accurate enough for field applications. A scour hole generally gets filled in as flood flow recedes. Bridge inspections after flood are not sufficient to fully determine the extent of scour damage at the highest flood discharge. Therefore, critical bridges subject to scour problems require safety monitoring.

Figure 1 Bridge failure cases during typhoon floods in Taiwan

A real-time scour monitoring system should improve bridge safety being cost-effective by guarding against premature or unnecessary maintenance [11]. Acquiring in situ scour-depth information using direct measurements is very difficult and dangerous in a typhoon flood which is usually accompanied by high-speed wind and torrential rainfall. Nevertheless, the development of measuring instruments with data acquisition systems generally faces the difficulties of surviving in monitoring of large-scale hydraulic and transportation structures in the flood event.

For monitoring bridge scour development, Prendergast and Gavin (2014) reported a broad range of instrumentation which were categorized as single use devices, pulse or radar devices, buried or driven rod systems, sound-wave devices, fiber-Bragg grating devices, and electrical conductivity devices. As discussed in the relevant literatures, most of these available scour monitoring techniques have been found to have limited applications. They are sensitive to many environmental and operational conditions in natural channels, such as temperature, salinity, turbidity, air entrainment, and debris. For instance, both sonar and radar can receive a substantial amount of noise caused by the turbidity of the flow, potentially making those systems unreliable for real-time monitoring of the scour process. The time-domain reflectometry (TDR) technique works by generating an electromagnetic pulse, which may attenuate and disperse its signal due to the environment of the transmission line. This drawback of the TDR technique reduces its ability to detect subtle changes in the scour process. A vibration-based sensor for scour measurement has been development and it would be more potential environmental and operational sensitivities compared to TDR and sonar devices. However, extreme flow condition with heavy debris impact would result in bridge failure in a flood. Hence, instrument development with data acquisition systems usually faces the difficulties of surviving in in the flood event.

A Wireless Sensor Network (WSN) system has been widely used in recent years as a promising technology in civil engineering and industrial applications that will greatly influence the methods of structural health monitoring. Sensornets have been used or proposed for numerous applications. They are able to provide spatial data over large areas for long periods of time and suitable for many monitoring applications. Smart and wirelessly networked sensors can collect and process a vast amount of data, from monitoring and control of structural damage, air quality, traffic conditions, to weather conditions and tidal flows. In the present study, a monitoring system that consists of a combination of vibration-based MEMS sensors and WSN is developed and utilized to obtain real-time measurements in the scour/deposition processes. These arrayed vibration-based sensors were packaged inside a stainless sphere with the proper protection of the full-filled resin, and they can measure free vibration signals when seismic activity occurs. By applying the wireless vibration-based sensors network system, the measured data of the water level, scour depth and depositional height are collected and analyzed.

Therefore, the bridge scour monitoring system faces the challenge of developing a real-time, reliable, and robust system that can be installed in a river-bed at a bridge pier (Figure 2). Moreover, it is well known that the established scour formula for estimating the maximum scour depth is related to the scour characteristics, including the flow depth, velocity, and sediment size. In reality, these limitations of the scour formula must be addressed before the formula can be properly applied. The recognition of any possible aggradation and degradation of the river-bed level in response to a channel disturbance is important for the prediction of channel bed variations. The scour process around the pier or abutment is essentially complex due to the three-dimensional flow patterns interacting with the sediments. However, most of the data obtained to develop the scour formula are collected from the laboratory instead of from the field. Thus, it is necessary to develop a real-time system for monitoring to observe and measure the scour-depth variations in the field. In this article, the authors propose a novel technique of scour monitoring that is more resilient to harsh environments with extreme flow conditions and is capable of measuring both scouring and deposition processes at a bridge pier. The research aim is to evaluate the applicability of scour depth estimation by numerical flow simulation for developing an early scour warning system at a highway bridge.

2 SENSOR & COMMUNICATION

2.1 MEMS Sensor

Measuring vibration is essential in detecting and diagnosing any deviation from normal conditions. The use of conventional accelerometers in vibration measurements is well known and accepted. The recent advances in wireless and embedded system technologies such as Micro-Electro Mechanical Systems (MEMS) sensors hold great promise. MEMS accelerometers are one of the simplest and most applicable micro-electromechanical systems. They have become indispensable in the automobile industry, and in computer and audio-video technology.

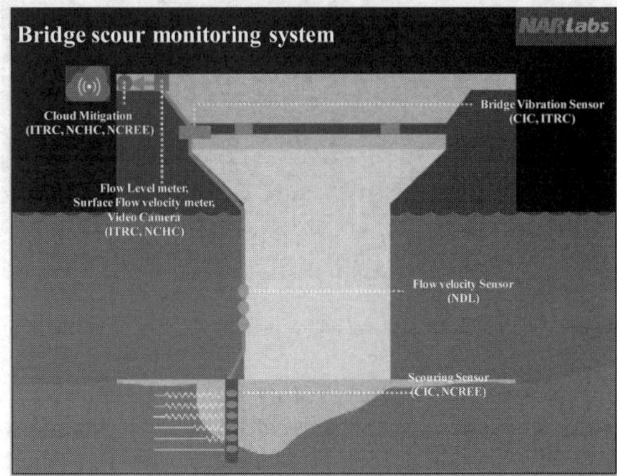

Figure 2 Bridge multi-hazards monitoring system

MEMS sensors are well recognized as the key building blocks for implementing disruptive applications in consumer devices. From game consoles to mobile phones, from laptops to white goods, consumer devices have already benefited in recent years from the use of low-g accelerometers for the implementation of motion-activated user interfaces and enhanced protection systems. Much has been written about the technology behind MEMS accelerometers, which are sensors capable of detecting linear accelerations.

In this paper the vibration-based sensor of LSM303DLH, STMicroelectronics, is used for the scouring indicator. The LSM303DLH is a system-in-package featuring a 3D digital linear acceleration sensor and a 3D digital magnetic sensor. The various sensing elements are manufactured using specialized micromachining processes, whereas the IC interfaces are realized using a CMOS technology that allows the design of a dedicated circuit that is trimmed to better match the sensing element characteristics.

Figure 3 Scouring sensor around the pier

As shown in Figure 3, this LSM303DLH accelerometer sensor was packaged inside a stainless sphere with the proper protection of the full-filled resin, and it can measure free vibration signals when seismic activity occurs. Appropriate packaging to resist harsh environments was the main long term robustness issue for bridge scouring measurements. Herein, this sensor was used for measuring the vibration signals as it emerged out of the riverbed and vibrated due to the flowing water. These packaged LSM303DLH accelerometers for bridge scouring measurement have been tested with different flowing velocities in the laboratory to ensure their performance and the WSN function before they were deployed in the field.

For early warning (Figure 2), this system also includes the conventional flow level meter and surface flow velocity meter for real-time recording the flow level and flow velocity variation.

2.2 *Communication System*

Wireless communication has spurred tremendous applications over the last few decades. In addition, wireless local area networks currently supplement or replace wired networks in many homes, businesses, and public safety issues. However, many technical challenges remain in designing robust wireless networks that deliver the performance necessary to support emerging applications, particularly for natural hazard monitoring. Commercially available Wi-Fi (802.11x) technologies offer potential capabilities to deliver secure mobile wireless networking that address these requirements. Although the potential of Wi-Fi technology is attractive, outdoor industrial environments pose numerous network design challenges, making successful WLAN implementations difficult and thus, many in place Wi-Fi networks fail to deliver expected performance.

Figure 4 Min-Chu Bridge monitoring system

As shown in Figure 2 and 4, the current access point technologies use two unlicensed spectrum bands. These are channels in the 2.4 and 5 GHz Industrial Scientific and Medical (ISM) bands. The 802.11 standard makes provisions for the use of several different modulation techniques to encode the transmitted data. These rules govern the transmission power and channel width and modulation. The most common of these is the 2.4 GHz band. This is used by the 802.11b/g and 11n devices. There are 11 channels but they are 20 MHz wide with a 1 MHz guard band on each side, so there are ultimately only three that don't overlap: channels 1, 6, and 11. These modulation techniques are used to enhance the probability of the receiver correctly receiving the data, thus reducing the need for retransmissions. As a result, many highly skilled man-hours and sophisticated testing equipment are required to install each Wi-Fi Access Point (AP) and ensure proper operation across its coverage radius along the highway and bridge. All the real-time data communicated and transferred to cloud and revealed the monitoring results on the internet.

3 FIELD DEPLOYMENT AND RESULTS

The scouring system was installed on the Min-Chu Bridge in the main channel of the Jhuoshu River where the greatest scour depth had occurred during the flood (Figure 4). All the sensor systems including the wireless station were fabricated in the laboratory factory. The signal and performance of all the sensors corresponding packaged waterproof and long-term durability were tested and analyzed before being used in the field.

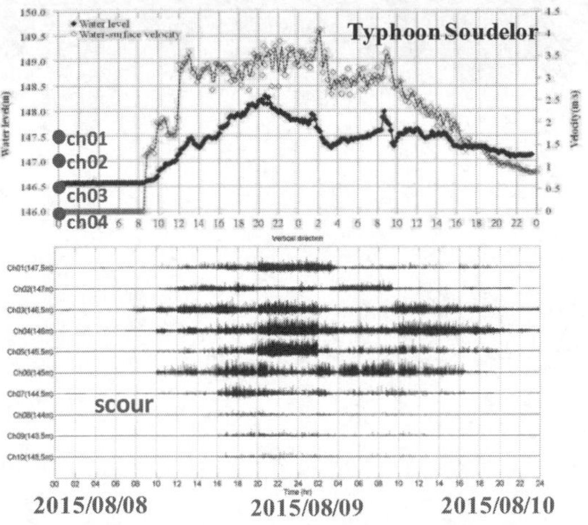

Figure 5 Min-Chu Bridge scour during Typhoon Soudelor

As shown in Figure 5, signal data of sensor monitoring were collected over the period of time from 8 through 10 Aug. during the Typhoon Soudelor. There were ten vibration-based sensors, named ch01 (located at 147.5m) to ch10 (located at 143.5 m) deployed around the pier. The distance between each sensor was 0.5m. Once the scour monitoring sensor emerges out of the riverbed, it will send vibration signal as it vibrates by the turbulent flow. Sensors ch01 and ch02 were initially exposed to air due to the previous flood event erosion, which were then able to detect flow turbulence to reveal signal fluctuations.

The flood stared at the time 08:00 on Aug. 8 from the flow level meter and it scoured obviously from the vibrated signal of sensor ch03. Along with the increase of water level and water velocity, the sensor measured the bridge scour from the signal of ch04 to ch 07. The noticeable fluctuations in the readings showed strong interactions between the sensor and the flow turbulence/sediment transport around the peak flow. However, after the first peak discharge in the period between 12:00 on Aug 8 and 02:00 on 9, the flood flow receded and the readings showed weak fluctuations indicating that ch07 was once again buried due to sediment deposition. At the meantime, Sensors ch03 to ch06 having readings were still in the flowing water. The flood of typhoon ended at time 22 of 10 Aug. The scour monitoring system reveals the deposition process from the time 21 of 9 Aug to time 22 of 10 Aug. From the ch07 sensor responses, it was recorded that there was a maximum scour depth of 2.5 m during this flood event.

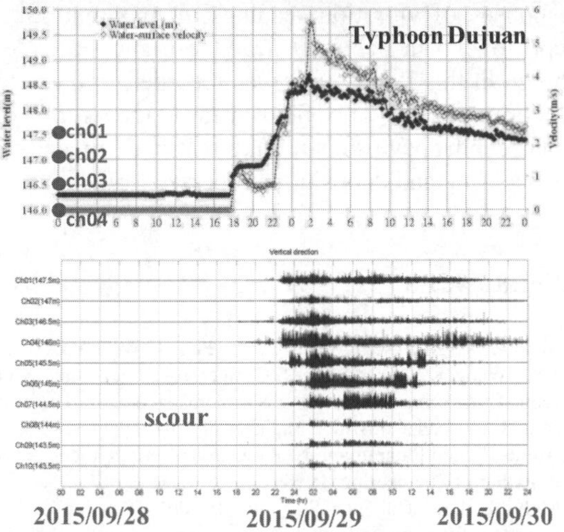

Figure 6 Min-Chu Bridge scour during Typhoon Dujuan

For Typhoon Dujuan, the scour began at time 23 of 29 Sep. as illustrated in Fig 6. The water level was 3m and the scour depth was 2.5m which corresponds to the Typhoon Soudelor. However, due to the larger flow velocity it revealed a little bit of scour depth potential from the scour sensor signal of ch08, ch09, and ch10.

4 SUMMARY

Scouring around bridge piers remains a major cause of bridge failure. Without prior warning to the bridge pier structure, scour failure tends to occur suddenly and result in major public disruption and life losses, particularly in a flood event. It is important to monitor in situ scour depth changes in order to prevent catastrophic failure. In the present study, the proposed novel technique of real-time bridge scour monitoring that is more resilient to harsh environments and is capable of measuring both scouring and deposition processes.

The system was applied and evaluated during the Typhoon Soudelor and Dujuan events, respectively. The flood event had a multiple-peak discharge and water stage hydrographs. The measured data in the scour and deposition processes were collected and analyzed. As a scour positioning indicator, the sensor exposed to running flow was able to detect flow turbulence to reveal signal fluctuations. From the sensor responses of vibration readings, a maximum scour depth of 2.5 m occurred near the highest peak discharge in this flood event was recorded in these two typhoon events. The field measurements of total scour depths have shown quite reasonable results for the scouring and deposition processes.

REFERENCES

1. Skibniewski M., Tserng H.P, Ju S.H., Feng C.W., Lin C.T., Han J.Y., Weng K.W. and Hsu S.C. Web-based real time bridge scour monitoring system for disaster management. The Baltic Journal of Road and Bridge Engineering, 2014, 9(1), 17–25.
2. Ahmed F. and Rajaratnam, N. Flow around bridge piers. Journal of hydraulic engineering, 1998, 124, 288-300.
3. Melville B.W., Coleman S.E. and Barkdoll B. Scour countermeasures for wing-wall abutments, Journal of Hydraulic Engineering, 2006, 132, 563-574.
4. Parker G. Surface-based bed load transport relation for gravel rivers. J. Hydraulic Research, 1990, 28(4), 417-436.
5. Chiew Y. M. Local scour and riprap stability at bridge piers in a degrading channel. Journal of Hydraulic Engineering, 2004, 130, 218-226.
6. Chang W.Y., Constantinescu G., Lien H.C., Tsai W.F., Lai J.S. and Loh C.H. Flow Structure around Bridge Piers of Varying Geometrical Complexity. Journal of Hydraulic Engineering, 2013, 139(8), 812–826.
7. Federico F., Silvagni G. and Volpi F. Scour vulnerability of river bridge piers. Journal of Geotechnical and Geoenvironmental Engineering, 2003, 129(10), 890-899.
8. Avent R.R. and Alawady M. Bridge scour and substructure deterioration: case study. Journal of Bridge Engineering, 2005, 10(3), 247-254.
9. Bolduc L.C., Gardoni P. and Briaud J.L. Probability of exceedance estimates for scour depth around bridge piers. Journal of Geotechnical and Geoenvironmental Engineering, 2008, 134(2), 175-184.
10. Lin Y.B., Lai J.S., Chang K.C., Chang W.Y., Lee F.Z. and Ten Y.C. Using MEMS sensors in the bridge monitoring system. Journal of the Chinese Institute of Engineering, 2010, 33, 25-35.
11. Lin Y.B., Lai J.S., Chang K.C. and Li L.S. Flood scour monitoring system using fiber Bragg grating sensors. Smart Materials and Structures, 2006, 15, 1950-1959.

Proceedings of the International Conference on Smart Infrastructure and Construction
ISBN 978-0-7277-6127-9

© The authors and ICE Publishing: All rights reserved, 2016
doi:10.1680/tfitsi.61279.251

A sensor system for real-time bridge scouring monitoring

Ssu-Ying Chen, Yi-Jie Hsieh, Chih-Chyau Yang*, Fu-Chen Cheng, Yu-Da Huang, Jin-Ju Chue, Chih-Ting Kuo, Chien-Ming Wu and Chun-Ming Huang

National Chip Implementation Center (CIC), Hsinchu, Taiwan
* *Corresponding Author*

ABSTRACT In Taiwan, many bridges have exceeded their 50-year life span. The strength of these old bridges is no longer affordable to the severe nature disasters. In other words, the bridges in Taiwan are likely to suffer from the damage. Since bridge pier scour is the main cause for bridge failure and collapse during the Typhoon in Taiwan, how to monitor the bridge scour becomes an important task. This paper presents a sensor system to detect the bridge scour in real time. The presented sensor system consists of under-water sensor nodes with the wired Ethernet communication protocol, a PoE switch and a data logger. The proposed under-water sensor node is implemented with two stacked octagon PCBs and enclosed in a steel hollow ball and then setup in the steel cage. Our developed under-water sensor node adopts the vibration sensing mechanism to detect the bridge scour by using the accelerometer sensor. The presented sensor system is now setup in Chih-Chang Bridge in the middle of Taiwan to monitor the bridge scour. The experimental results in the field during the Typhoon Dujuan show the presented sensor system can detect the bridge scour effectively with our proposed scour detection algorithm in real time.

Index Terms—accelerometer; under-water sensor node; rugged sensor system, bridge scour

1 INTRODUCTION

Bridges are the important pivots of traffic, the damage of bridges can cause the severe cost of human life and property. In Taiwan, many bridges have exceeded their 50-year life span. The strength of these old bridges is no longer affordable to the severe nature disasters. In other words, the bridges in Taiwan are likely to suffer from the damage. Scour is the main cause for bridge failure and collapse during the Typhoon in Taiwan [1]. The heavy rain brought by the Typhoon in July and August causes the bridge scour and makes the damage or collapse for bridges. In 2008, the Typhoon Sinlaku [2-3] caused the collapse of Hou-Feng Bridge, and thus six people died and 3 cars crushed due to the failure of bridge in this event. In 2009, the Typhoon Morakot [4] produced a record-breaking rainfall in Taiwan, therefore six people died and ten people injured due to the collapse of Shuang-Yuang Bridge. Thus, how to monitor the bridge health and real-time diagnose the bridge structure becomes an important task in Taiwan.

Bridge scour has been extensively studied in the world for more than a hundred years. Many methodologies and instruments have been employed to measure and monitor the local pier scour condition, such as bricks, sonar, radar sensors, Time-Domain Reflectometry (TDR), Fiber Bragg Grating (FBG) sensors and accelerometer sensors. The bricks sensors [5] are buried in the certain location of the sand before the rain season. After the floods, the bricks are digged out and the number of the remained bricks is calculated. Therefore, the bridge scour depth can be obtained. This method can be used only one-time and the scour condition cannot be real-time detected. The sonar and radar sensors [6] provide contactless measurement of streambed scouring near bridge pier and abutments, and usually used to show the final status of streambed after a flood. One of disadvantages of the sonar and radar is that they have limit for measuring status of streambed in real time as rush water contains sand, even rocks during a flood. The time domain reflectometry (TDR) [7][8] measures the reflections that results from a fast-rising step pulse travelling through a measurement cable. The depth of soil-water interface is determined by counting the round trip travel time of the pulse. However, the major drawback of TDR is that accuracy of TDR is dependent on the environment temperature and humidity. Monitoring the scour depth by the fiber Bragg grating (FBG) [7] is dependent on number of FBG elements. However, the cost of monitoring of the scour depth by FBG technique is higher than that of existing methods [7]. The costs of Radar and TDR are expensive due to high-speed hardware requirement. The frequency response with Fast Fourier Transform and the time domain response with the root mean square values of the accelerometer [9][10] are used to detect the scour. Since the accelerometer in [9] does not sense the vibration data by the flow directly, the result of scour depth may be inaccurate due to the unpredictable interferences in the complicated underwater environment. In addition, in order to obtain the frequency response result [10], it may consume the large computations to get the bridge scour condition.

This paper presents a sensor system for real-time bridge scouring monitoring. The presented sensor system consists of under-water sensor nodes with the wired Power over Ethernet communication protocol. The proposed under-water sensor node is implemented with two stacked octagon PCBs and enclosed in a steel hollow ball and then setup in the steel cage. The proposed architecture of the bridge scour monitoring system owns the scalability and flexibility for mass deployment. The presented rugged sensor system is now setup in Chih-Chang Bridge in Taiwan to monitor the bridge scour. The experimental results in the field show our sensor system can detect the bridge scour effectively with the proposed scour detection algorithm in real time.

In Section II, the developed bridge scour sensor system is introduced. In Section III, the algorithm for the bridge scour detection is presented. In Section IV, the experimental setups and results are illustrated. Finally, we conclude this paper in Section V.

2 THE DEVELOPED SENSOR SYSTEM

Figure 1 shows the architecture of proposed bridge scour sensor system. This system consists of under-water sensor nodes with wired Power over Ethernet communication protocol, a data logger, and a PoE switch. Each under-water sensor node has its own MAC/IP address, and it is connected to the PoE switch through the Power over Ethernet protocol. The data logger is connected to the PoE switch to retrieve the data of under-water sensor nodes through the Ethernet protocol.

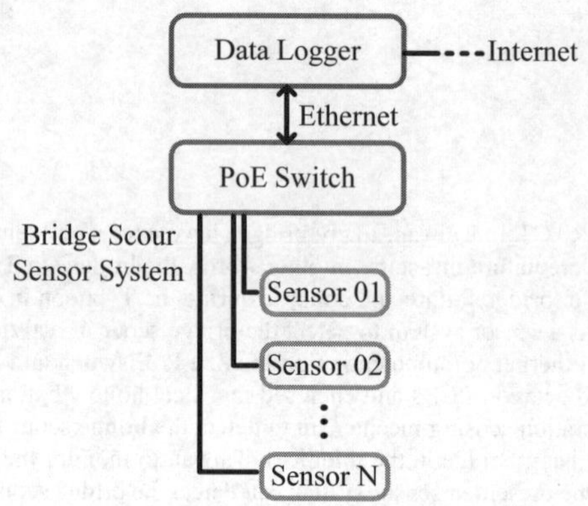

Figure 1. Shows the architecture of proposed bridge scour sensor system.

2.1 *Under-Water Sensor Node*

The main purpose of the under-water sensor is to monitor the scouring condition of the bridge pier. The proposed under-water sensor node is implemented with two stacked octagon PCBs enclosed with a plastic enclosure, and then setup in a steel hollow ball. Figure 2(a) shows the picture of two-stacked PCB design for the under-water sensor. The top PCB design is the power module, which operates as a DC-DC converter for generating 1.2~5V outputs from the 48V input to supply the power of under-water sensor node. An Ethernet PHY (TI, DP83640TVV) is used to send/receive Ethernet data via the PoE controller (TI, TPS2376DDAH). The bottom PCB design is the core module, which is composed of a Cortex-R4 micro-controller (MCU, TI, RM48L952) and an accelerometer sensor (STM, LSM303DLHC). After the MCU receives a Request packet, it retrieves the accelerometer sensor data through the I2C interface, and then sends the sensor data via the Ethernet response package to the data logger. Figure 2(b) shows that the two-stacked octagon PCB is setup in a plastic enclosure. The space between the PCBs and the plastic enclosure is filled up with the silicon to be waterproof. This plastic enclosure with the octagon PCB designs is further setup in a steel hollow ball to be rugged for the unpredicted under-water environment in the field. Figure 3(a) shows the rugged under-water sensor node with a steel hollow ball while Figure 3(b) shows the explosion figure of the under-water sensor node. With the proposed octagon PCB design, the placement area for PCB can achieve 40% increase compared with that using traditional rectangle PCB design. The diameter of the steel hollow ball with the octagon PCB design can have about the reduction of 15.5% compared with that using traditional rectangle PCB design, thus reduces 39.6% cost of under-water sensor nodes.

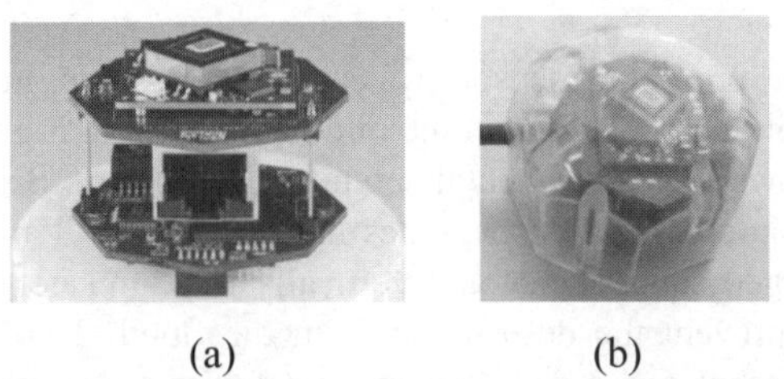

(a) (b)

Figure 2. The picture of two-stacked PCB design for the under-water sensor (a) without an enclosure (b) with an enclosure

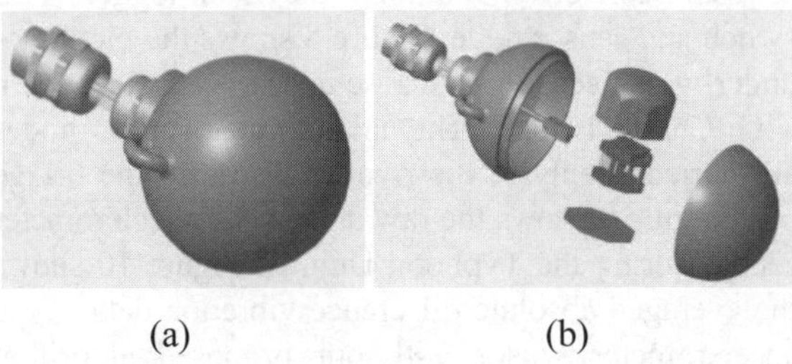

(a) (b)

Figure 3. The rugged under-water sensor node setup in a steel hollow ball. (a) the assembled under sensor node (b) the explosion figure

Our developed under-water sensor node adopts the vibration sensing mechanism to detect the bridge scour by using the accelerometer sensor. The accelerometer sensor owns the characteristics of low-cost, high sensitivity, small form factor, and low power compared with those in other instruments. The vibration data of accelerometer sensor can be detected easily no matter the river water is clean or mixed with sand. Moreover, it is easy to setup in the field without the direction alignment. The under-water sensor node with accelerometer sensor is buried into the sand of riverbed before the rainfall season. During the season of Typhoon, the heavy rain causes the river full of water and washes away the sand of the riverbed. The under-water sensor nodes are exposed and thus accelerometer sensor within the under-water sensor node is vibrated due to the river water flow.

2.2 *The Data Logger Program*

Figure 4 shows the GUI of data logger program for the bridge scour monitoring. In Figure 4, the Node IP indicates the IP address of corresponding under-water sensor node, Switch IP defines the IP address for the PoE switch, and Switch Port represents the port number of PoE switch that the current under-water sensor node is connected. While the under-water sensor node detects an unpredicted error, the data logger program can reset the power of the current sensor node via the corresponding port number of PoE switch. When the data logger program starts to receive the sensor node data, the Request packet with the payload "0x010300000003cccc@" is sent to the specified sensor node. Then, the data logger program starts to receive the sensor data from under-water sensor node. The data logger program stores the compressed sensor data to the disk with the path of folder being C:\temp\SensorNodeIPAddress\year-month\day\hour". The zipped file name is determined by minute and second such as minute-second.txt.gz.

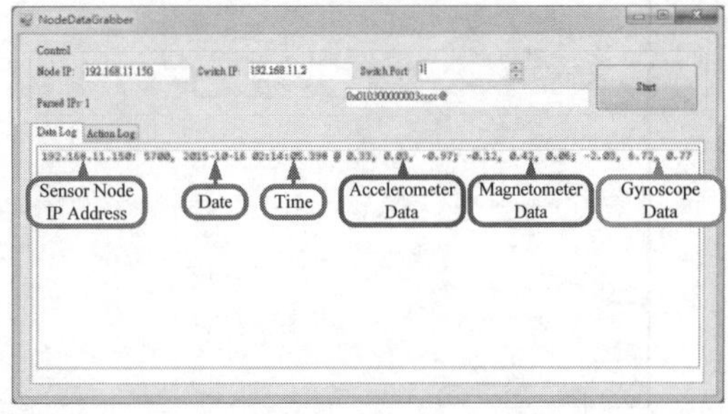

Figure 4. The data logger program for the bridge scour monitoring.

3 THE PROPOSED ALGORITHM FOR BRIDGE SCOUR MONITORING

In this paper, a sensor system with accelerometer sensors is presented to detect the bridge scour. The accelerometers are buried into the sand of riverbed in advance. During the season of typhoon, the heavy rain that comes with typhoon causes the river full of water. The sand of the riverbed is scoured and it causes the accelerometer sensors exposed and thus vibrated due to the river water flow. The main purpose of the under-water sensors is to monitor the scouring condition of the bridge pier and riverbed. The under-water sensors are arranged equidistance and vertically fixed on the steel cage. The under-water sensors are then buried deeply in the riverbed close to the bridge pier. In the normal condition, the sand of the riverbed can fully cover the under-water sensors, and the sensor nodes are in a steady state condition. When the water of the river becomes rapid due to the storm or heavy rain, it washes away part of the riverbed and the sensors originally buried in the sand become exposed and vibrated due to the scouring. The deeper the riverbed gets scoured, the more sensors are exposed. The vibration data of each sensor is real-time sent to the data logger through the Ethernet and the data logger program helps to identify the scouring condition.

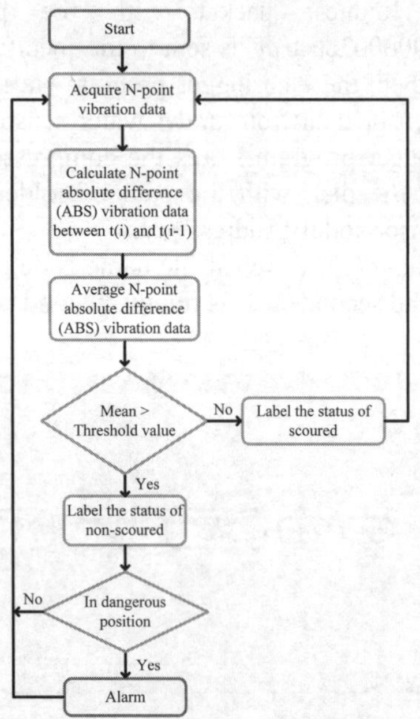

Figure 5. The scour detection loop for the scour detection.

Figure 5 illustrates the flow chart of the proposed algorithm to detect the bridge scour with accelerometers. The algorithm consists of scour detection loop for each sensor node. Each sensor node executes the scour detection loop. The data logger program acquires the N-point accelerometer values and calculates the absolute difference value between the accelerometer data at the current time frame and at the previous time frame in the detection loop. The absolute difference values in each detection loop are then averaged. If the averaged absolute difference value exceeds the threshold value, the sensor node is labeled as the status of scoured. Otherwise, the sensor node is labeled as the status of non-scoured. If the position of the scoured sensor node exceeds the position of alarm threshold, the alarm is triggered. Note that the threshold value and the alarm threshold position can be obtained from the experiment in the laboratory.

4 THE EXPERIMENTAL SETUPS AND RESULTS

The presented sensor system is now setup in the Chih-Chang Bridge in the middle of Taiwan. Figure 6 shows the experimental setup in Chih-Chang Bridge. 9 under-water sensor nodes labeled with "N" are arranged equidistance and vertically fixed in steel cage to prevent the destruction from the flood. Two PoE switches and a data logger are setup in a control box to acquire the sensor data, and send these sensor data to remote users via the internet. Figure 7 shows the physical connections among the data logger, PoE switch and sensor node. Figure 8 shows the picture of under-water sensor nodes setup in a steel cage in Chih-Chang Bridge. The under-water sensor nodes are buried deeply in the riverbed close to the bridge pier. Figure 9 shows the raw data of the accelerometer sensor during the Typhoon Dujuan. Figure 10 shows the averaged absolute difference vibration data of the accelerometer sensor with our proposed algorithm during the Typhoon Dujuan. The threshold absolute difference of vibration data is set to 0.002 according to the experiment results in the laboratory. At the time of ti, the Node k sensor starts to be scoured by the water and starts to be exposed from the sand. The value of averaged absolute difference vibration data for Node k sensor node starts to be larger than the threshold value from ti. ; Ti refers to the times {t1, t2, t3, t4, t5 } and Node k refers to the sensor nodes {Sensor01, Sensor02, Sensor03, Sensor04, Sensor05 }, respectively. By using the proposed algorithm shown in Figure 5, the bridge scour detection can be realized.

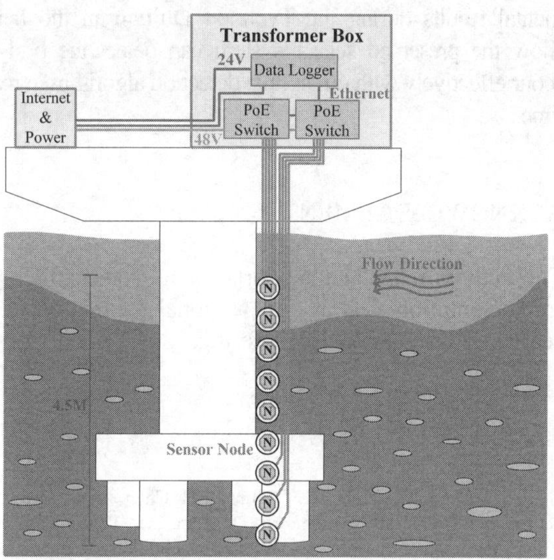

Figure 6. The experimental setup in Chih-Chang Bridge.

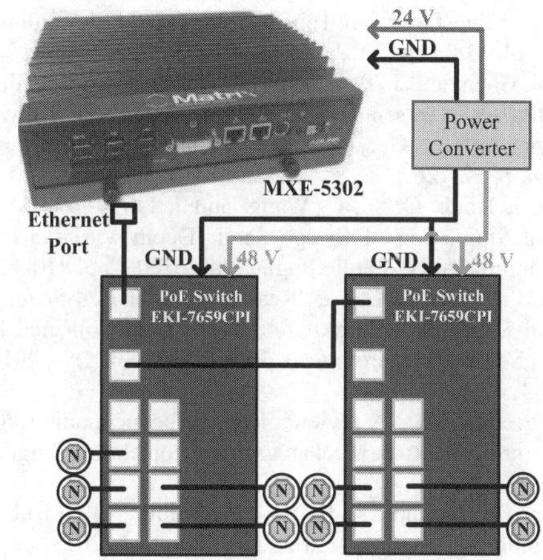

Figure 7. The physical connections among the data logger and PoE switch and sensor nodes.

Figure 8. The under-water sensor nodes setup in a steel cage.

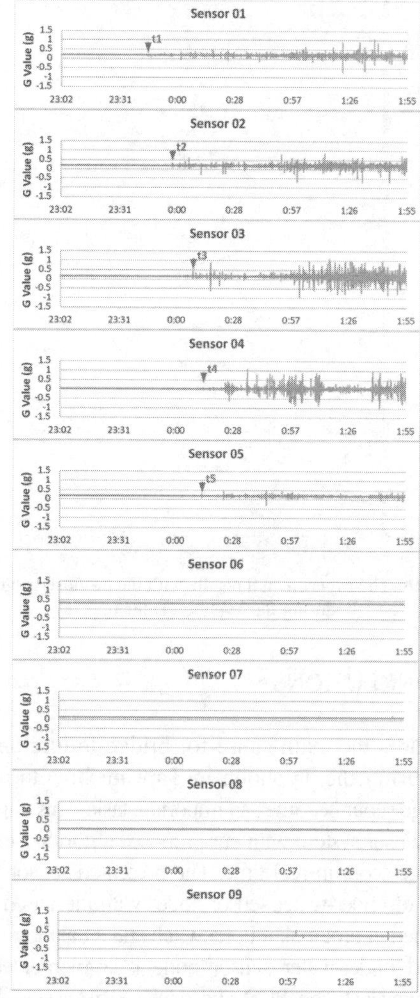

Figure 9. The raw data of accelerometer sensor during the Typhoon Dujuan.

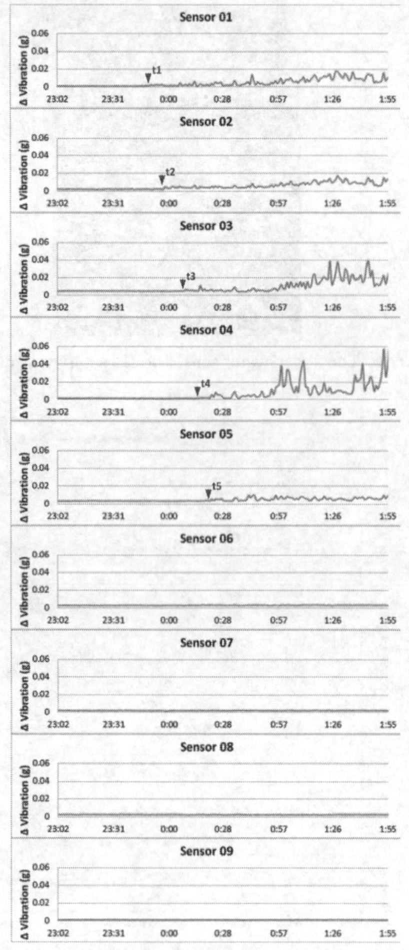

Figure 10. The experimental results with the proposed algorithm during the Typhoon Dujuan.

5 CONCLUSIONS

Scour is the main cause for bridge failure and collapse during the Typhoon in Taiwan, how to monitor the bridge scour becomes an important task. This paper presents a sensor system with our proposed algorithm to detect the bridge scour in real time. The presented sensor system consists of under-water sensor nodes with the wired Power over Ethernet protocol, a PoE switch and a data logger. The proposed under-water sensor node is implemented with two-stacked octagon PCB and enclosed in a steel hollow ball and then setup in the steel cage. The volume of the steel hollow ball with the octagon PCB design can achieve the large reduction compared with that using traditional rectangle PCB design, thus reduces the cost of underwater sensor nodes. Our developed under-water sensor node adopts the vibration sensing mechanism to detect the bridge scour by using the accelerometer sensor. The presented sensor system is now setup in Chih-Chang Bridge in the middle of Taiwan to monitor the bridge scour. The experimental results during the Typhoon Dujuan in the field show the presented sensor system can detect the bridge scour effectively with our simple detection algorithm in real time.

ACKNOWLEDGEMENT

The authors thank the supports of the National Chip Implementation Center and National Center of Research on Earthquake Engineering.

REFERENCES

[1] Y. C. Chen, T. S. Liao, K. C. Huang, H. Chen, “A low-power design of a bridge scour monitoring system,” in Proceedings of IEEE Instrumentation and Measurement Technology Conference, May, 2014, pp. 30-34.

[2] C. Y. Tsou, Z. Y. Feng, and M. Chigira, “Catastrophic landslide induced by Typhoon Morakot, Shiaolin, Taiwan,” Geomorphology, vol. 127, 2011, pp. 166-178.

[3] J. H. Hong, Y. M. Chiew, J. Y. Lu, J. S. Lai, Y. B. Lin, “Houfeng bridge failure in Taiwan,” Journal of Hydraulic Engineering, Vol. 138, Feb. 2012, pp. 186-198.

[4] C. Y. Chen, and J. T. Chiang, “Spatial and temporal distribution of disaster events in mountainous townships of Taiwan,” in Proceedings of IEEE Geoscience and Remote Sensing Symposium, July, 2014, pp. 21-26.

[5] J. Lu, J. Hong, C. Su, C. Wang, and J. Lai, “Field Measurements and Simulation of Bridge Scour Depth Variations during Floods,” Journal of Hydraulic Engineering, 2008, pp. 810–821.

[6] H. C. Yang and C. C. Su, “Real-time River Bed Scour Monitoring and Synchronous Maximum Depth Data Collected During Typhoon Soulik,” Hydrological Processes, vol. 29, 2015, pp. 1056-1063.

[7] L. J. Prendergast, “A review of bridge scour monitoring techniques,” Journal of Rock Mechanics and Geotechnical Engineering, 2014, pp. 138-149.

[8] J. Tao, X. Yu, and X. B. Yu, “Real-time TDR Field Bridge Scour Monitoring System,” Structures Congress, 2013, pp. 2996-3009.

[9] J. L. Briaud, S. Hurlebaus, K. Chang, C. Yao, H. Sharma, O. Yu, et al., “Realtime Monitoring of bridge scour using remote monitoring technology,” FHWA/TX-11/0-6060-1, Texas Transportation Institute, Austin, USA, 2011.

[10] M. Fisher, S. Atamturktur, A. Khan, “A Novel Vibration-based Monitoring Technique for Bridge Pier and Abutment Scour,” Structure Health Monitoring, vol. 12, 2013, pp. 114-125.

Proceedings of the International Conference on Smart Infrastructure and Construction
ISBN 978-0-7277-6127-9

© The authors and ICE Publishing: All rights reserved, 2016
doi:10.1680/tfitsi.61279.257

Understanding the advantages of satellite Earth Observation as a surveying tool for infrastructure monitoring

Maria de Farago[1], Geraint Cooksley[1], Mario Costantini[2] , Federico Minati[2], Francesco Trillo[2], Luca Paglia[2] and Jennifer Gates[3]

[1] Telespazio Vega UK, Luton, UK,
[2] e-Geos, Rome, Italy,
[3] Arup, London, UK

ABSTRACT New surveying sensors are being designed, others modified; new and old technologies implemented to best understand the movement behaviour of infrastructures from their initial concept to design, planning, construction and finally operations. A good geotechnical understanding of the area where the infrastructure will lay, its interactions with other structures, settlements and geotechnical processes induced during construction and operations will provide an accurate lifecycle risk assessment allowing decisions to be made. This level of information cannot be achieved by only one sensor but by the understanding of all surveying solutions and identifying when they are needed or if they are needed at all in some instances.

A good monitoring system should be designed for the whole lifecycle, should not repeat information unless redundancy and independent measurements are needed, should be planned accurately so the main areas of movement are identified and targeted with the most adequate sensors and should be organised in a comprehensive way so it's easy to find the formation when needed.

Of all monitoring techniques and methods in the Earth Observation industry, change detection and InSAR interferometry are the ones which have best responded to the necessities of infrastructure monitoring and this paper will demonstrate how these remote techniques have been successfully applied to specific civil engineering projects.

1 INTRODUCTION

InSAR interferometry and Coherent Change Detection (CCD) are the two earth observation techniques most used in the civil engineering industry to provide ongoing monitoring of surface motion and temporal surface changes respectively.

These technologies were developed in the early 90's when the first European Space Agency (ESA) satellites were launched (ERS and Envisat). These ESA satellites have covered the whole planet with extensive radar imagery coverage; Europe, North Africa and part of Middle East are the areas with the largest historical data stacks in the archives. These satellites covered a period of 19 years from 1991 to 2010 and it was not being until 2012 when ESA launched Sentinel-1a to continue C-band (~5 cm wavelength) radar satellite data acquisitions.

From 2007 new commercial satellites have been launched. The COSMO-SkyMed satellite constellation consists of 4 identical satellites providing a spatial resolution up to 1m/pixel and an unparalleled revisit frequency, allowing any area on earth to be imaged several times a day, and with up to 8 images per 32 day period in InSAR mode. This improvement in detection of features on the ground and being able to monitor them more frequently has dramatically improved the usefulness of InSAR surveying in the Civil Engineering sector proving to be very useful for the whole infrastructure lifecycle.

For the planning and design phase InSAR surveying provides a historical baseline analysis of movements on a site planned for future construction, assisting in decision making based on geotechnical retro-analysis and also to catalogue pre-existing motion areas to help clarify liabilities in case of construction induced movement.

During the construction phase, InSAR provides surface motion of the surrounding areas as well as complementing and/or validating in situ surveys; this larger area monitoring capability provides the insights to better understand and monitor geotechnical processes occurring on the site helping to mitigate risk during construction. The InSAR data can be combined well with dense in-situ sensors nearest to the works to provide a comprehensive monitoring from real-time and local extending into the wider area with frequent satellite derived updates.

For infrastructure maintenance, the extensive coverage of radar satellite imagery allows regional analysis of surface motion helping to plan maintenance strategy in the longer term. Areas sensitive to movement will be identified and measured and then catalogued in terms of priority to be inspected and/or monitored on the site. The technology can be used from regional studies of movement to measuring the details of a single structure such as a building. These regional maps of movement can be updated from monthly to annually providing with a long term surveying campaign to assist in risk planning management.

Coherent Change detection mapping allows the visualisation of the areas affected by changes which might need further investigation. Changes are sometimes related to geomorphological-geotechnical processes such as landslides, sinkholes, hydrogeological changes, etc; land use, areas changing from agricultural land to any other; evolution of built up areas to understand the extension of towns, cities and any other infrastructures; detection and monitoring of areas prone to flooding and understanding of recent made ground.

Comparing InSAR interferometry with in-situ ground displacement solutions (extensometers, levelling, automated prismatic levelling, ground based radar or laser scanning and GNSS-Global Navigation Satellite Systems), InSAR is unique in being a non-invasive remote sensing solution, covering very wide areas with no installation. In-situ sensors on the other hand provide real-time monitoring of the locations they are installed / visited in surveys. InSAR can exploit SAR data collected by satellite from 1992, while in situ methods can measure a baseline only from when the first survey or instrument is installed and data is collected on demand, whereas InSAR data allows retrospective and investigative analysis. InSAR can be used to guide the installation of in-situ monitoring (to focus on an identified motion area where continuous monitoring is needed) or define reference stable areas for benchmark locations or other sensors.

2 INSAR TECHNIQUE

The InSAR technique developed by Telespazio Group (e-Geos) is called PSP-IFSAR [1][2][3] and provides measurements of displacements – typically due to landslides and other slope failures, hydrogeological changes, tunnelling, oil/gas or mineral extraction, earthquakes and volcanic phenomena – of objects on the ground or the ground itself exhibiting radar backscattering properties stable over time, called Persistent Scatterers (PS). Typically, several PS are found in non-cultivated and scarcely vegetated areas, and in particular in man-made or natural structures like buildings, rocks, etc.

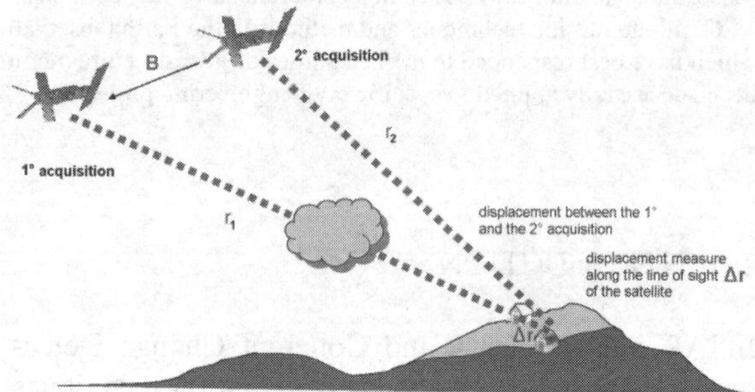

Figure 1. Principles of InSAR technology.

3 APPLICATIONS TO THE CIVIL ENGINEERING INDUSTRY

InSAR technology has being widely used for more than 10 years all over the world to monitor infrastructures from all civil engineering sectors: linear infrastructures such as railways, highways, tunnels, water pipelines; ports, airports; aquifers and water infrastructures.

This technology is specifically successful in man-made environments as infrastructures (such as buildings, roads, concrete and metallic structures etc) behave as high quality reflectors for radar satellites providing hundreds of thousands of potential measurement points depending on the size of the targets to monitor. This facilitates the persistent reflection of the radar signal by the same reflective points over time.

The direct benefits of InSAR technology will be analysed in two different case study contexts:
Firstly, an example in the Aquifer of Madrid where there have been demonstrated benefits on a regional scale for historical surface motion monitoring of the whole aquifer as well as an analysis of the cross comparison between satellite InSAR and in-situ instrumentation.

Secondly, surface motion studies in several locations in London including highlighting InSAR data in the context of other survey and in-situ instrumentation data for White Chapel Station in London as part of the construction of Crossrail where the benefits of using high resolution radar imagery is shown.

3.1 Example 1- Aquifer of Madrid. Regional surface motion PS-IFSAR and cross- comparison with in-situ techniques.

Canal de Isabel II (CYII), the Spanish public body which manages the Aquifer of Madrid commissioned an analysis of surface expressed movements related to the aquifer use for public water supply with PSP-IFSAR using ERS, ENVISAT and COSMO-SkyMed data. In this case CRs (Corner Reflectors- Figure 2) have been used to add purpose built and located InSAR measurements points on the location of several of the water extraction wells where the larger movements are expected.

Figure 2. CR installed on the location of one of the wells.

PSP-IFSAR results were also compared with measurements from other remote sensing technologies (GNSS), and in situ instrumentation (levelling, precision optics, extensometers and piezometers), Minati et al (2014).

Surface movement results from ERS and Envisat from 1992 to 2010 show alternation of subsidence as the ground water is drawn to be used during drought periods and heave as the phreatic levels start to recover. The deformation period coincides with water extraction and recovery cycles since 1992 with variations of +- 30milimeters. (**Figure 3**). The observed deformations show an almost elastic behaviour of the terrain (Minati et al, 2004).

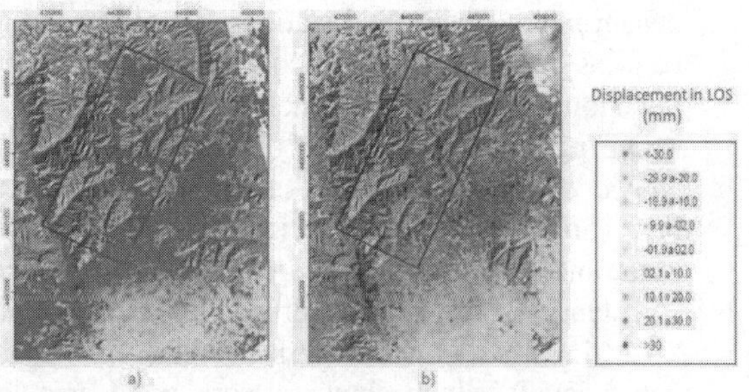

Figure 3. Subsidence and heave detected by InSAR due to water extraction from aquifer on drought season Apr 1992-Oct 1995 (left) and after recovery period Aug 2003 – May 2010 (right).

An examination of the historical piezometric levels reveals a strong spatial and temporal correlation between the displacement measured with ERS-Envisat IFSAR and the piezometric levels historically registered on the wells.

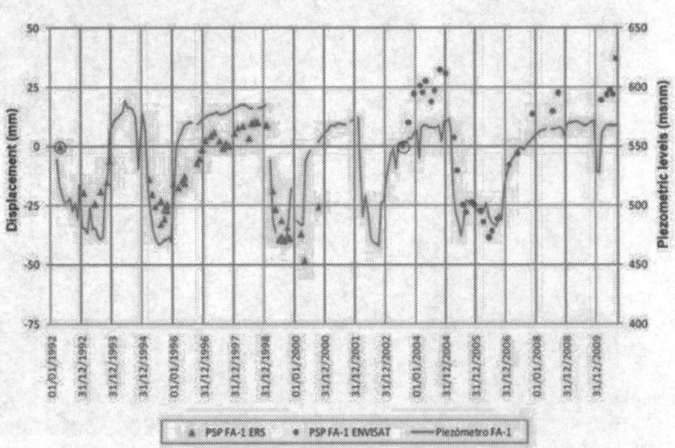

Figure 4. Time series showing displacement measured with PSP-IFSAR from ERS-Envisat and piezometric levels for one of the wells.

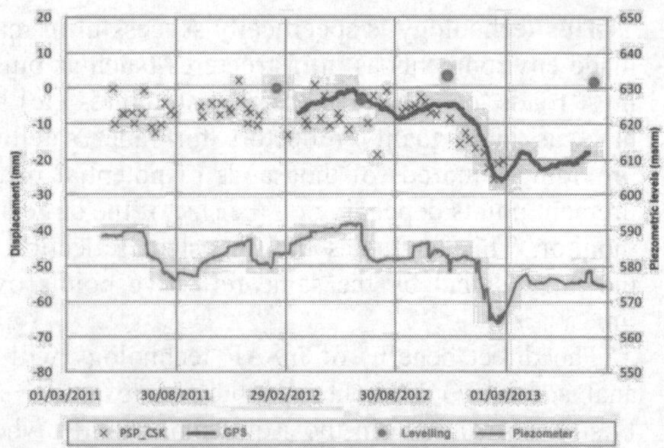

Figure 5. Comparison between in-situ measurements, GPS and COSMO-SkyMed PS IFSAR for the period of artificial recharge of the aquifer (2011-2013).

Data shows that for 10m of piezometric level drop, 4mm of subsidence has been detected in ERS and 7mm in Envisat within a radius of 1000m around the extraction well.

From 2011, and after the surface movement oscillations detected and measured with archived ERS and Envisat, CYII commissioned a surveying campaign using several technologies to effectively monitor surface movement in the area where a series of water injection tests were made to recharge the aquifer. The campaign lasted from 2011 to 2013 and 61 COSMO-SkyMed images where used together with GPS measurements from total stations located in the main wells as well as measurements from extensometers, precision levelling and tiltimeters.

During this period, 2011-2013, the artificial *recharge* of the aquifer gave a limited increase (less than 15m) in the piezometric levels, with localised effect and short duration therefore their influence on subsidence is almost inappreciable. At the end of 2012, there is a general sharp decrease of the piezometric levels which shows a very good correlation with a general and homogeneous deformation in the area of study. Deformations within this period show correlations with peizometric correlation of 3.5 mm per metre, similar to the correlations made for the historical period ERS.

None of the surveying studies for the water injection period show any significant movement that might have affected the surrounding infrastructures.

However, the historical studies ERS-Envisat (1992-2010) show that some infrastructures might have being affected due to the hydrological variation in the aquifer.

The PS-IFSAR COSMO-SkyMed results show subsidence measured on the infrastructures in the area of study, being sections of the M-50 the most affected.

Figure 6. Surface movement on M-50 highway and surrounding areas from COSMO-SkyMed (2011-2013).

The conclusions of this study are as follows:

- Results of PSP-IFSAR from 1992-2010 have shown strong spatial and temporal correlation with water extraction cycles and registered piezometric levels.
- Cross-comparison for the 2011-2013 between PSP-IFSAR and other in situ techniques have shown good correlation and precision in the measurements.

- PSP-IFSAR technology is adequate for terrain monitoring campaigns and provides detailed information of movements, thousands of measurements per km2 with millimetric precision and monthly and quarterly updates.
- Measurements from PSP-IFSAR complemented with GPS stations measurement provides continuous movement data and can be used as an absolute reference and calibration for the interferometric measurements.
- Subsidence and heave have been measured and associated with cycles of water extraction and recovery of the aquifer with maximum values of +- 30 millimetres in the areas surrounding the wells.
- Artificial recharge has produced very small variations in the phreatic level and its influence in surface movements is almost unobservable.

3.2 Example 2- London area. Regional surface motion PSP-IFSAR dataset.

A series of major underground projects are currently being undertaken in London with construction activities from Crossrail in particular, impacting the city over the last 5 years. This trend is set to continue with the green light having been given to the construction of Thames Tideway Tunnel, Northern Line extension, HS2 and potential future construction of Crossrail 2. With so much tunnelling activity, monitoring programs need to be thoroughly designed and implemented. In this context InSAR surveying with inherent wide area surveying capability can provide a synoptic measurement of ground motion.

With these projects in mind, a baseline ground motion study for the last 4 years was performed. 40 COSMO-SkyMed images from May 2011 to July 2015 were processed in order to locate existing areas of motion and to establish a very high resolution reference dataset to be used for future comparisons. Examination of figure 9 reveals that the most striking ground deformation pattern captured by the baseline study was that of displacement related to the Crossrail works, despite the analysis not being designed for its study. The InSAR data show at a regional scale the locations affected by ground movement, and for each measurement point the displacement is measured with millimetric precision. The availability of measurement points depends on how consistent is the radar reflectivity returned from on the ground, therefore the presence of measurement points cannot always be guaranteed. Data measurements are made on each pass of the satellite viewing the site therefore the frequency of displacement readings are not real time.

Figure 7. COSMO-SkyMed satellite footprints used for the study.

In-situ instrumentation technologies are used to provide a monitoring scheme designed to meet the spatial and temporal extents and magnitude of motion for a specific project. It is often challenging to have these established to allow a long baseline though once installed they provide highly accurate frequent and detailed capture of any ground motion.

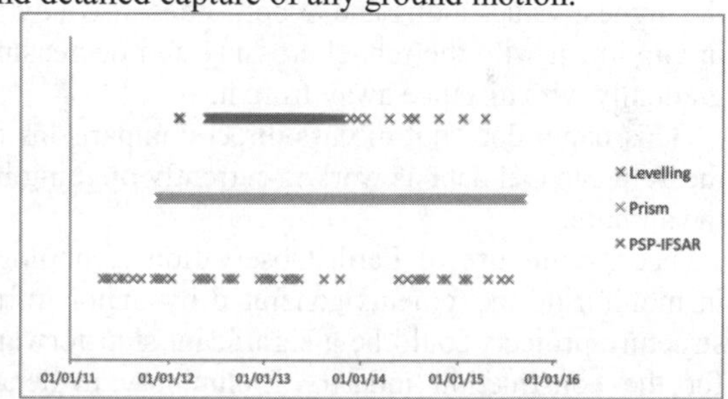

Figure 8. COSMO-SkyMed image dates used for the London baseline study together with Crossrail survey dates for Whitechapel Station.

The use of InSAR surveying alongside in-situ survey methods allows a large baseline dataset, detailed and real-time measurements with in-situ survey methods and ongoing wider area survey by InSAR.

Figure 9. Surface movement over central London showing surface settlements in areas of Crossrail works. COSMO-SkyMed (2011-2015).

InSAR can also be applied as a low cost long-term monitoring system beyond the normal retirement of in-situ surveying.

Figure 10 shows the location of the measurement points collected in the Whitechapel Station area from three surveying methodologies: prisms, levelling points and InSAR as well as their values of displacement. The InSAR results for average annual displacement given are compared to in-situ prism and levelling data, similarly analysed to calculate annually averaged displacements for their respective time periods. The figure indicates that the displacement of the three surveying methods coincide well showing the highest values along and at both sides of Brady St in alignment with the tunnel crossing and decreasing gradually with distance away from it.

This paper does not discuss direct comparisons of the Whitechapel data as work is currently ongoing by the authors.

The routine use of Earth Observation technology in monitoring movements generated by major infrastructure projects could be a significant step forward for the construction industry. Preliminary evidence from comparative studies indicates that in urban areas, InSAR data can be successfully benchmarked against conventional, in-situ monitoring techniques.

Given the scope of monitoring typically carried out around urban underground construction, the use of the InSAR technique offers the following potential opportunities:

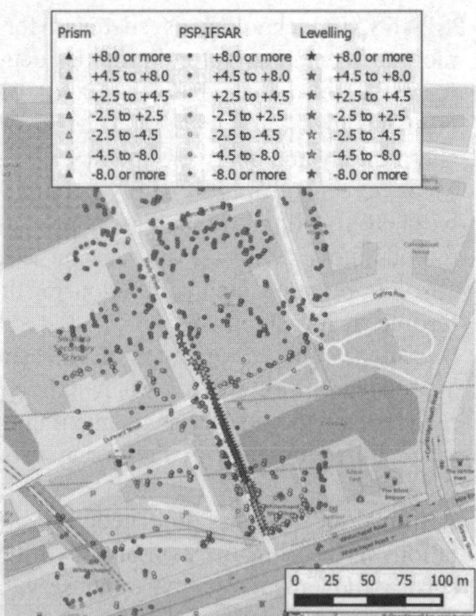

Figure 10. Whitechapel area showing the three data sets acquired for surface movement during the Crossrail excavation. Values are in mm/year. In-situ survey data set originates from Crossrail monitoring operations and COSMO-SkyMed data set (2011-2015) from Telespazio.

1. Prior to construction, a key advantage for a major project is the ability to request datasets for historic periods, allowing retrospective analysis of background movements, potentially reducing the period for in-situ background monitoring required by a client or by third parties;

2. during construction, understanding overall patterns of movement for a site allows monitoring teams to focus on critical areas and understand where monitoring should be reduced or extended of where the frequency of monitoring should be adjusted; and

3. commonly in the industry, major infrastructure clients and 3rd party infrastructure owners will expect ground movements to be monitored post-construction until ongoing movements diminish to a rate of less than say 2mm/year. The use of InSAR technology for meeting this requirement could significantly reduce ongoing monitoring costs for these projects during a period when the many of the project team will have demobilised.

Further studies into the cost benefits associated with adopting this technique in conjunction with traditional monitoring could help to further demonstrate the advantages of the technology in reducing overall instrumentation and monitoring costs associated with major projects.

4 CONCLUSIONS

25 years of development and implementation of InSAR technology in the Oil and Gas, Mining and Civil Engineering industries has positioned InSAR as a solid surveying technique. It can be used to provide retro-analysis of surface movement, complement other surveying practices, as a base to plan the surveying campaigns or to substitute more expensive in-situ surveying technologies in the final phases of post-works settlement monitoring.

The examples of the Madrid Aquifer and the London Regional surface motion study show that comparison of in-situ instrumentation measurements with remotely sensed InSAR allow not only validation of each other but also allow the advantages of each technique to be combined. Understanding their advantages and limitations will assist in designing the most technologically and economically efficient monitoring plan for the whole infrastructure lifecycle.

ACKNOWLEDGEMENT

Consent to use data from the Crossrail Project has been provided with thanks to Mike Black, enabling the comparison of traditional monitoring with InSAR technology and allowing the project's experience of the technique to be published.

REFERENCES

[1] M. Costantini, S. Falco, F. Malvarosa, and F. Minati, "A new method for identification and analysis of persistent scatterers in series of SAR images," in *Proc. Int. Geosci. Remote Sensing Symp. (IGARSS)*, Boston MA, USA, vol. 2, pp. 449–452, July 2008.

[2] M. Costantini, F. Malvarosa, and F. Minati, "A General Formulation for Redundant Integration of Finite Differences and Phase Unwrapping on a Sparse Multidimensional Domain," *IEEE Trans. on Geosci. Remote Sensing*, vol. 50, no. 3, pp. 758–768, Mar. 2012.

[3] M. Costantini, S. Falco, F. Malvarosa, F. Minati, F. Trillo, F. Vecchioli, "Persistent scatterer pair interferometry: approach and application to COSMO-SkyMed SAR data," IEEE Journal of Selected Topics in Applied Earth Observations and Remote Sensing, vol. 7, no. 7, July 2014.

[4] Minati F, Megías Farré E, Ibáñez Carranza J.C., 2014, Investigación sobre técnicas para la medición de subsidencias relacionadas con la explotación de acuíferos, *Cuadernos de I+D+i*, Canal Isabel II Gestión, Madrid

Proceedings of the International Conference on Smart Infrastructure and Construction
ISBN 978-0-7277-6127-9

© The authors and ICE Publishing: All rights reserved, 2016
doi:10.1680/tfitsi.61279.265

Fundamental investigation on applicability of Piezofilm Sensor in sensing low-frequency structural response of bridges

C.V. Dung* & E. Sasaki

Department of Civil Engineering, Tokyo Institute of Technology, Tokyo, Japan
* *Corresponding Author*

ABSTRACT Polyvinylidene Flouride (PVDF) is a film-type polymer that has been used as sensors and actuators in various applications due to its mechanical toughness, flexibility and low density. For the application of PVDF sensor in sensing structural response of bridges, the knowledge of PVDF sensor's performance in the low-frequency region is required. This study investigates the applicability of piezofilm sensor in sensing low-frequency structural response of bridges. Tensile tests using standard tensile specimens subjected to different tensile load levels and frequencies ranging from 0.1-20 Hz were conducted. The tensile test results indicate an excellent sensing performance of piezofilm sensor in the low-frequency region and at low-amplitude excitation. Transient excitations modulated from 1 Hz sinusoids with arbitrary amplitudes and durations were also generated. Piezofilm sensor also appears to exhibit a high consistency in measuring transient excitations with different arbitrary amplitudes and durations. It is shown that piezofilm sensor is capable of sensing low-frequency & low-amplitude response of bridges with a good accuracy.

1 INTRODUCTION

Polyvinylidene Fluoride (PVDF) is a thin film-type polymer that is mechanically tough, flexible, and low density. Since the discovery of piezoelectric effect in elongated and polarized films of polymers, particularly of PVDF, by Kawai (1969), fundamental properties of PVDF have been investigated in a great number of studies. PVDF is a semi-crystalline polymer consisting of long chain molecules with the repeat unit of CF_2CH_2, whose crystalline domains appear in four different forms that can be interconverted by the application of heat, electrical fields, and pressure (Lando et al 1966), (Sessler 1981), (Darasawa et al 1992). Thermal poling or corona poling orients the molecular dipoles in the crystalline parts and thus yields a permanent polarization. PVDF, when crystallized into its β-phase, exhibits piezoelectric property so that mechanical energy can be converted to electrical energy and vice versa. Therefore, PVDF has been used to manufacture sensors and actuators that can be utilized in various practical applications including shock impact and pressure sensors (Lee et al 1988) (Dutta & Kalafut 1990) (Bauer 2002) (Shirinov & Schomburg 2008), biomedical sensors (Wang et al 2006) (Kim et al 2009) (Chiu et al 2013), acoustic (Woodward & Chandra 1978) (Xu 2010) (Leon et al 2013) (Xu et al 2015), tactile sensors (Lee & Nicholls 1999) (Dargahi 2000) (Seminara et al 2011), active vibration control (Ma et al 2011), structural health monitoring of civil and aerospace structures (Matsumoto et al 2004) (Kurata et al 2013) (Bloomfield 1977), etc.

While many previous studies have focused on the use of piezofilm sensors (herewith PVDF sensors) in the out-of-plane mode in which mechanical stress is induced in the thickness (poled) direction, in-plane sensing mode of PVDF sensor for dynamic strain sensing has also been investigated in several studies. Lee & O'Sullivan (1991) developed a uniaxial strain rate gage that measures only strain rate along a specified direction by combining the effective surface electrodes, appropriate skew angle and the correct polarization profile. Wang & Wang (1997) presented a theoretical approach for feasibility analysis of the application of PVDF sensor to cantilever beam modal testing. Sirohi & Chopra (2001) investigated the behavior of piezoelectric elements including piezoceramic (PZT) and piezofilm(PVDF) as dynamic strain sensors of which the superior performance compared to conventional strain gages in terms of sensitivity and signal-to-noise ratio was demonstrated for the frequency range of 5-500 Hz. Correction factors to account for transverse strain and shear-lag effect due to the bond layer were analytically derived and experimentally validated by the same authors. Ma et al (2012) investigated the effects of PVDF sensor's area and the use of a charge amplifier on the measurement capability of a PVDF sensor attached to a cantilever beam that is subjected to impact loading. PVDF sensor proved to be capable of capturing most of the resonant frequencies (59.2 – 9533 Hz) from transient responses, and its sensitivity was demonstrated to be better than that of the conventional strain gage. Kotian et al (2013) presented an analytical investigation on the effects of stress-averaging for both in-plane sinusoidal stress waves and in-plane impact-induced stresses. It was concluded that error induced by stress averaging becomes more significant as sensor length increases, density of structure's material increases, and the magnitude of input stress increases, although the error induced by stress averaging is minimal for most practical applications. Furthermore, only very high frequencies (in orders of kHz) caused a significant reduction in PVDF sensor's output voltage due to stress-averaging.

Previous studies on in-plane mode strain sensing have so far focused on fundamental applications of PVDF sensors in sensing dynamic strain for small structures such as cantilever beam with high natural frequencies or tensile specimens subjected to high-frequency excitations. PVDF sensors with the aforementioned advantages can be a potential alternative for conventional strain gages in certain dynamic strain sensing applications in actual bridge structures. However, application of PVDF sensors in real bridge structures requires the knowledge of PVDF sensor's performance at low-frequency range. Therefore, this study aims to investigate the dynamic strain sensing capability of PVDF sensors at low-frequency region. An experimental investigation was conducted using tensile specimens subjected to sinusoidal tensile excitations of 0.1 – 20 Hz and to transient excitations with different durations and amplitudes.

2 SENSING PRINCIPLES

Consider the constitutive governing equation for the piezoelectric effect induced by one-dimensional mechanical deformation (IEEE, 1987):

$$D_3 = e_{31}S_1 + \varepsilon_{33}^S E_3$$

where D_3 is the electrical displacement component; e_{31} is the piezoelectric constant; S_1 is the longitudinal strain component; ε_{33}^S is the permittivity component at constant strain, and E_3 is the electrical field. Subscripts 1 and 3 indicate the longitudinal and poling direction, respectively (Figure 1). In case the external electrical field is absent ($E_3 = 0$), the relation is further reduced to:

$$D_3 = e_{31}S_1 = d_{31}c_{11}^E S_1 = d_{31}E_p S_1$$

where d_{31} is the piezoelectric constant, c_{11}^E is the elastic stiffness constant (i.e., Young's modulus, Ep) of the PVDF layer.

The effect of stress averaging on the magnitude of output voltage was investigated in Kotian et al (2013) who concluded that the error introduced by stress averaging is minimal for most typical cases although the effect of stress averaging becomes more significant as the sensor length increases, density of the structure's material increases, and the magnitude of input stress increases. For sinusoidal stress input, the error increases as the length of the PVDF sensor approaches the wavelength of the host structure.

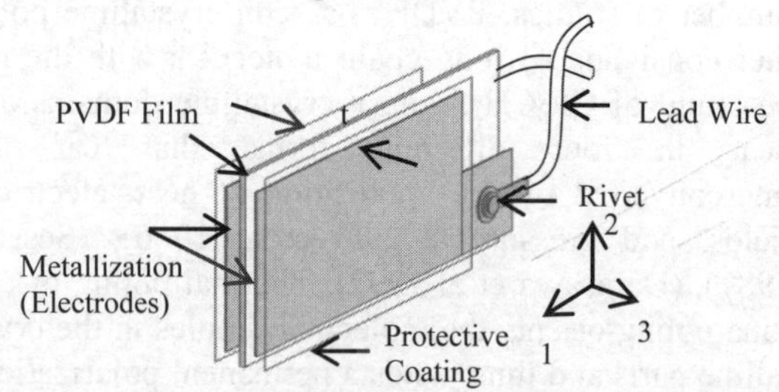

Figure 1. Schematic configuration of a PVDF sensor.

The generated electrical charge q and current i can be expressed as

$$q(t) = \int_A \boldsymbol{D}.\boldsymbol{n}dA = \int_A d_{31}E_p\,\varepsilon_1\boldsymbol{n}dA = d_{31}E_pl_pb_p\varepsilon_1$$

$$i(t) = \dot{q}(t) = d_{31}E_pl_pb_p\dot{\varepsilon}_1$$

where lp and bp are the length and width of the PVDF sensor, respectively.

For an ideal operational amplifier, the governing equation of the electrical circuit can be written as

$$\dot{V}_0(t) + V_0(t)\frac{1}{C_FR_F} = -\frac{d_{31}E_pl_pb_p\dot{\varepsilon}_1}{C_F}$$

For harmonic excitation, the solution to the above differential equation is

$$\bar{V}_0(t) = -(\frac{j\omega C_FR_F}{1+j\omega C_FR_F})\frac{d_{31}E_pl_pb_p\bar{\varepsilon}_1}{C_F}$$

where ω is the frequency of the harmonic excitation (rad/s); and the quantities with a bar represents their magnitudes.

3 TENSILE TESTS

Tensile test was conducted using standard tensile specimens whose configuration is illustrated in Figure 2. Eight frequencies ranging from 0.1 - 20 Hz were used to generate sinusoidal tensile load. Six nominal load ranges (i.e., six nominal strain amplitudes) (Table 1) were employed in the tensile test to evaluate the relationship between the output voltage obtained by the PVDF sensor attached on the tensile specimens and the applied tensile load. During the test, the output voltage signal from the PVDF sensor was transferred to the charge amplifier (MEAS, 2008) (herewith amplifier) before being sampled and recorded by the NI 4431-USB data acquisition. The lower cut-off frequency was selected 0.1 Hz which is the minimum value that could be set on the amplifier. Conventional foil strain gages were also attached at the midpoint on the side face of the specimen to obtain the "true" strain for verification. Both the applied load and displacement were also recorded. The sampling rate was set as 2000 samples/s. No filtering was applied for strain gages. Two PVDF sensors of 28- (PVDF28) and 52- (PVDF52) [MEAS, 2009] micrometer thickness were attached to both sides & at the midpoint of the specimen. The size of both PVDF sensors (DT4) was 22x171 mm. A sixty-second length of signal was recorded at each frequency and strain level. To investigate the effect of adhesive type, two types of adhesives including glue and tape were used to bond the gage to the surface of the host specimen. A commercial double-coated adhesive film tape and a glue-type adhesive that is frequently used to attach strain gage were selected. Smaller sensors (DT2) sized 16x73 mm were also used for this investigation.

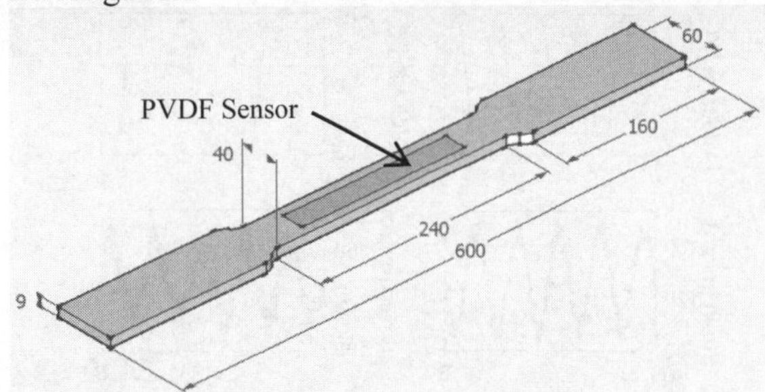

Figure 2. Tensile specimen configuration (unit: mm).

Table 1. Tensile load parameters.

Freq.(Hz)	0.1	0.2	0.5	1	2	5	10	20

ε_{amp} (με)	P_{amp} (kN)	P_{ave} (kN)	P_{min} (kN)	P_{max} (kN)
1	0.07	5	4.93	5.07
5	0.36	5	4.64	5.36
25	1.8	5	3.2	6.8
50	3.6	5	1.4	8.6
69.4	5	10	5	15
138.9	10	20	10	30

4 TEST RESULTS

4.1 Continuous Harmonic Excitations

Statistical correlation was used to evaluate the linear relationship between input load or displacement and the output voltage of the PVDF sensor. Correlation provides both direction and strength of the relationship and its scale -1 to 1 is independent of scale of variables. The Pearson correlation coefficient, r, is calculated by the ratio of covariance of the two variables x and y over the multiply of two standard deviations of the variables x and y.

$$r = \frac{cov(x,y)}{S_x S_y}$$

Figure 3 shows the actual applied load signal, the output strain measured by strain gages, and the output voltage of the PVDF sensors. It can be seen that the PVDF sensors exhibit an excellent signal-to-noise ratio and could even decently capture the strain signal of a micro strain amplitude.

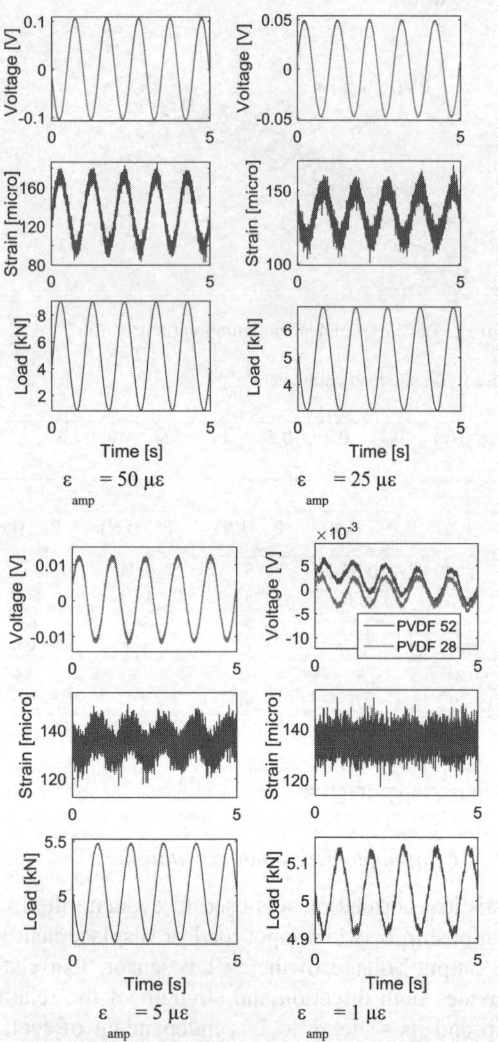

Figure 3. PVDF sensor's response to 04 different nominal strain amplitude levels at 1 Hz.

Figure 4 shows the results of correlation evaluation for all strain amplitude levels and all selected frequencies from 0.1 to 20 Hz. It can be seen that applied load and strain gage output hold a strong positive linear relationship. However, the correlation strength decreases as load amplitude decreases, which is due to the error introduced by noise in recorded strain gage data. Meanwhile a strong negative linear relationship between load and PVDF output was observed for all frequencies equal or above 1Hz. For frequencies lower than 1 Hz (i.e. 0.1, 0.2, and 0.5 Hz), a large variation in correlation coefficient was observed. This can be attributed to phase lag caused by filtering effect of the amplifier. Furthermore, PVDF28 appears to perform better than PVDF52 in lower strain amplitude levels and lower frequency.

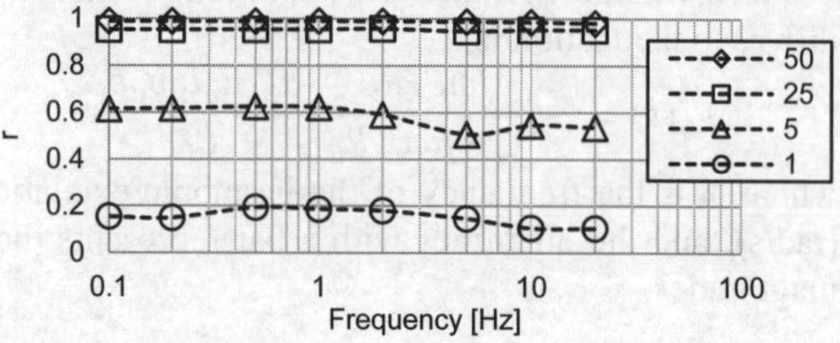

Figure 4. Correlation between load & strain gage.

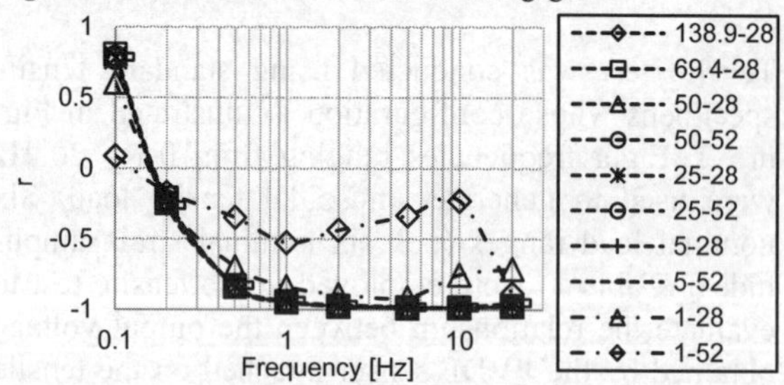

Figure 5. Correlation between load & PVDF output voltage.

Regarding the attachment method, the adhesive type may have considerable effect on phase lag and shear lag effect leading to loss in shear transfer from host structure to PVDF sensor. The results in Figure 6 indicate that the tape type results in lower correlation between load & output voltage than that for the glue type for frequencies lower than 2Hz. Also, the amplitude of output voltage signal for the tape type is lower than that for the glue type due to the shear lag effect. It is shown in Figure 6 that a PVDF sensor with larger size and attached by tape-type can perform as well as smaller-sized sensor attached to host structure by glue. From the view point of attachment method, the tape-type bonding is relatively easier to implement than the glue type for which the evenness of glue dispersion within a short hardening time is relatively more difficult to ensure than for the tape-type. Therefore, a larger PVDF attached by tape is more preferable than a smaller sensor bonded by glue.

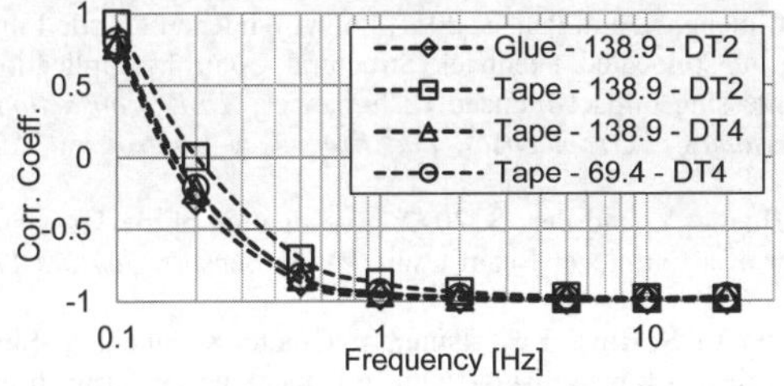

Figure 6. Correlation between load & PVDF sensor's output voltage with two different adhesive types.

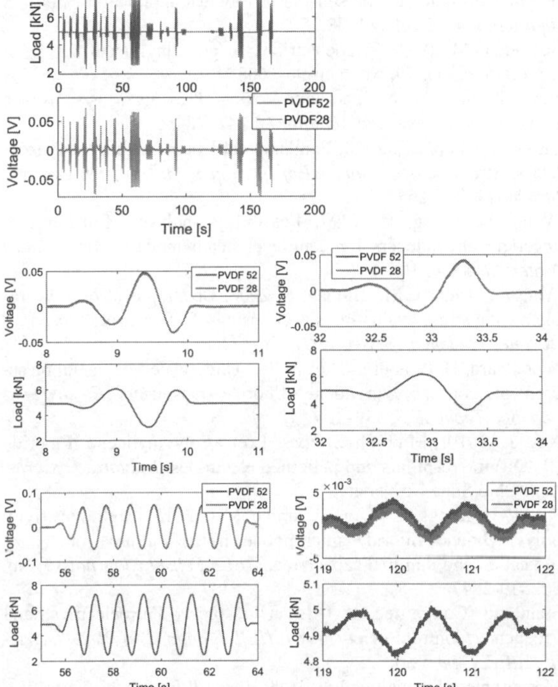

Figure 7. Response of PVDF sensors to transient excitation with different durations and amplitudes.

4.2 *Transient Excitations*

To investigate the performance of PVDF sensor in sensing transient excitations, arbitrary transient excitations modulated from 1-Hz sinusoids were generated with different amplitudes and durations. Figure 7 shows several actual transient applied load signals & PVDF sensor's output voltage out of total twelve events. The results indicate a strong, consistent correlation between load and PVDF output regardless of duration (number of cycles) and amplitude. The mean correlation coefficient of about -0.9255 & -0.923 for PVDF52 & PVDF28, respectively, appears to be similar to that for the continuous sinusoidal excitation at 1 Hz using 0.1 Hz of lower cutoff frequency for which the results were shown in figures 5 & 6. The response of the two PVDF sensors seems to be consistent despite of thickness difference. Furthermore, there appears to be a systematic effect of sensor thickness on correlation although this effect appears to be negligible (Figure 8).

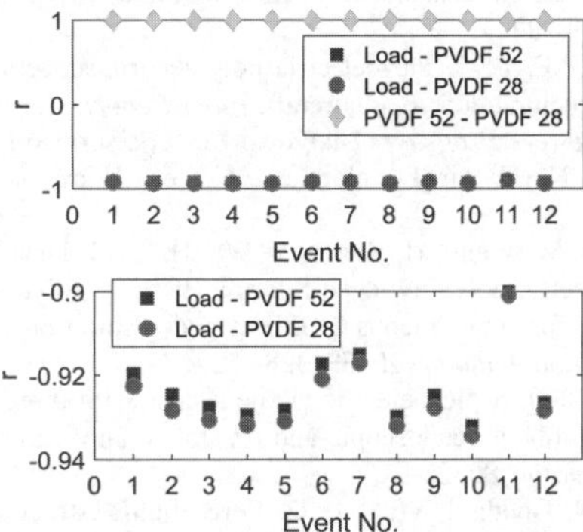

Figure 8. Correlation between load & PVDF sensor's response to transient excitations with different lengths and amplitudes.

5 CONCLUSIONS

Tensile tests using standard tensile specimens were conducted to evaluate the linear relationship between PVDF sensor's output voltage response and the input continuous harmonic & transient excitations with different amplitudes and durations in the frequency region of 0.1 Hz – 20 Hz. PVDF sensor has a high signal-to-noise ratio and can capture extremely low amplitude strain level. Phase lag is included in the PVDF sensor's response due to the high-pass filtering nature of PVDF sensor when connected to an external resistance and the effect of adhesive type. However, the effect of phase lag appears to be consistent throughout different strain levels, therefore, can be quantified and evaluated. The loading effect can be further navigated by future attempts to invent a charge amplifier that provides a cut-off frequency lower than 0.1 Hz, for example, 0.01 Hz. PVDF sensors with smaller thickness appear to be more preferable in sensing extremely low-amplitude excitation. A larger PVDF sensor attached to the host structure by adhesive tape can perform as well as a smaller sensor attached by glue. Overall, PVDF sensors can be a potential alternative to the conventional foil gages in sensing low-frequency & low-amplitude structural responses of steel bridges.

REFERENCES

Bauer, F. 2002. Ferroelectric PVDF polymer for high pressure and shock compression sensors. *11th International Symposium on Electrets*, 219-222.

Bloomfield, P.E. 1977. Piezoelectric polymer transducers for detection of structural defects in aircraft. *Final Report; Aeronautical Analytical Rework Program*, Analytical Rework/Service Life Project Office, Naval Air Development Center: Warminster, PA, USA.

Chiu, Y. Lin, W. Wang, H. Huang, S. Wu, H. 2013. Development of a piezoelectric polyvinylidene fluoride (PVDF) polymer-based sensor patch for simultaneous heartbeat and respiration monitoring. *Sensors and Actuators A* **189**, 328– 334.

Dargahi, J. 2000. A piezoelectric tactile sensor with three sensing elements for robotic, endoscopic and prosthetic applications. *Sensors and Actuators* **80**, 23–30

Darasawa, N. Goddard, W.A. 1992. Force fields, structures, and properties of poly(vinylidene fluoride) crystals. *Macromolecules* **25**, 7268–7281.

Dutta, P.K. Kalafut, J. 1990. Evaluation of PVDF Piezopolymer for Use as a Shock Gauge. *Special Report 90-23*, U.S. Army Corps of Engineers.

Headings, L.M. Kotian, K. and Dapino M.J. 2013. Speed of sound measurement in solids using polyvinylidene fluoride (PVDF) sensors. *Proceedings of the ASME 2013 Conference on Smart Materials, Adaptive Structures and Intelligent Systems,* SMASIS2013, Snowbird, Utah, USA.

Kawai, H. 1969. The piezoelectricity of poly(vinylidene fluoride), *Jpn. J. Appl. Phys.* **8**, 975–976.

Kim, K.J. Chang, Y.M. Yoon, S. and Kim, H.J. 2009. A Novel Piezoelectric PVDF Film-Based Physiological Sensing Belt for a Complementary Respiration and Heartbeat Monitoring System. *Integrated Ferroelectrics* **107**, 53–68.

Kotian, K. Headings, L.M. and Dapino, M.J. 2013. Stress Averaging in PVDF Sensors For In-Plane Sinusoidal and Impact-Induced Stresses. *IEEE Sensors Journal* **13**, no. 11, 4444-4451.

Kurata, M. Li, X. Fujita, K. He, L. and Yamaguchi, M. 2013. PVDF Piezo Film as Dynamic Strain Sensor for Local Damage Detection of Steel Frame Buildings. *Sensors and Smart Structures Technologies for Civil, Mechanical, and Aerospace Systems*, Proc. of SPIE Vol. 8692.

Lando, J.B. Olf, H.G. Peterlin, A. 1966. Nuclear magnetic resonance and x-ray determination of the structure of poly(vinylidene fluoride), *J. Polym. Sci.* Part A **4**, 941–951.

Lee, C.K. O'Sullivan, T.C. 1991. Piezoelectric strain rate gages. *J. Acoust. Soc. Am. 90* **2,** 945-953.

Lee., L.M. Graham, R.A. Bauer, F. Reed, R.P. 1988. Standardized Bauer PVDF piezoelectric polymer shock gauge. *Journal De Physique* **9**, No. 49, 651-657.

Lee, M.H. Nicholls, H.R. 1999. Tactile sensing for mechatronics—A state of the art survey. *Mechatronics* **9**, 1–31.

Ma, C. Chuang, K. and Pan, S. 2011. Polyvinylidene Fluoride Film Sensors in Collocated Feedback Structural Control: Application for Suppressing Impact-Induced Disturbances, *IEEE Transactions on Ultrasonics, Ferroelectrics, and Frequency Control*, no. 12, 2539-2554.

Ma, C. Huang, Y. and Pan, S. 2012. Investigation of the Transient Behavior of a Cantilever Beam Using PVDF Sensors. *Sensors* **12,** 2088-2117.

E. Matsumoto, S. Biwa, K. Katsumi, Y. Omoto, K. Iguchi, T. Shibata. Surface strain sensing with polymer piezoelectric film. *NDT&E International* 37 (2004) 57–64.

Seminara, L. Capurro, M. Cirill, P. Cannata, G. and Valle, M. 2011. Electromechanical characterization of piezoelectric PVDF polymer films for tactile sensors in robotics applications, *Sensors and Actuators A* **169**, 49–58.

Sessler, G.M. 1981. Piezoelectricity in polyvinylidenefluoride, *J. Acoust. Soc. Am.* **70**, No. 6, 1596-1608.

A.V. Shirinov W.K. Schomburg. 2008. Pressure sensor from a PVDF film. *Sensors and Actuators A* **142**, 48–55.

Sirohi, J. Chopra, I. 2000. Fundamental Understanding of Piezoelectric Strain Sensors. *Journal of Intelligent Material Systems and Structures*, 246-257

Wang, B. Wang, C. 1997. Feasibility analysis of using piezoceramic transducers for cantilever beam modal testing. *Smart Mater. Struct.* **6**, 106–116.

Wang, F. Tanaka, M. Chonan, S. 2006. Development of a wearable mental stress evaluation system using PVDF sensor, *Journal of Advanced Science* **18**, No. 1&2, 170-173.

Woodward, B. Chandra, R.C. 1978. Underwater acoustic measurements on polyvinylidene fluoride transducers. *Electrocompon. Sci. Technol.* **5**, 149–157.

Xu, J. 2010. Microphone based on Polyvinylidene Fluoride (PVDF) micro-pillars and patterned electrodes. *Doctoral Dissertation*, Ohio State University.

Xu, J. Headings, L.M. and Dapino M.J. 2015. High sensitivity polyvinylidene fluoride microphone based on area ratio amplification and minimal capacitance. *IEEE Sensors Journal* **15**, No. 5, 2839-2847.

Standards Committee of IEEE Ultrasonics, Ferroelectrics, and Frequency Control Society. 1987. *IEEE Standard on Piezoelectricity*, IEEE, New York.

Measurement Specialties Inc. 1999. *Piezo Film Sensors Technical Manual.* Hampton, VA, USA.

Measurement Specialties Inc. 2008. *Piezo Film Lab Amplifier*. Hampton, VA, USA.

Measurement Specialties Inc. 2009. *DT Series Elements with Lead Attachment*. Hampton, VA, USA.

Proceedings of the International Conference on Smart Infrastructure and Construction
ISBN 978-0-7277-6127-9

© The authors and ICE Publishing: All rights reserved, 2016
doi:10.1680/tfitsi.61279.271

Smartphones equipped with android application software for structural health monitoring

K. Fang[1, 2], J. D. Tian[1, 2], D.Y. Zhang[1, 2], H. Li[*1, 2]

[1] *Key Lab of Structures Dynamic Behavior and Control, Harbin Institute of Technology, Harbin, China*
[2] *School of Civil Engineering, Harbin Institute of Technology, Harbin, China*
[*] *Corresponding Author*

ABSTRACT Sensors with high accuracy, low cost, and easy integration into data acquisition system is one of the most important parts in structural health monitoring (SHM) system. In this paper, triaxial accelerometer and gyroscope embedded into the smartphones are utilized to monitor the vibration response of structure. However, in such an acquisition system, structural response data from different smartphones, compared to traditional wired accelerometers, is not always collected at the same time. So, nonsynchronous data processing technology is demonstrated comprehensively herein. First, the application based on Android operating system is described in detail. Modal identification method based on acceleration data which is collected asynchronously by different smartphones is introduced. Thus high performance SHM platform based on smartphones can be built. Then, numerical simulation study shows that modal identification method based on nonsynchronous data is feasible. At last, effectiveness and feasibility of SHM system based on the smartphones is proved in shaking table test on a three-story sheer frame building model.

1 INTRODUCTION

Structural Health Monitoring (SHM), mainly based on the data of structural dynamic response, is widely implemented on large-scale infrastructures. Dynamic response of structure, such as acceleration and strain, can be processed and analyzed to obtain the dynamic features of the large-scale structures, which can achieve the goal of real-time monitoring of structure eligibly (Carden et al. 2004). In order to measure the dynamic response of the structure accurately, the sensors should be properly selected and deployed on suitable position. Thus, wired accelerometers is widely used since its characteristics of stable data measurement in the early SHM. Further, signals measured by wired accelerometers needs to be amplified through signal amplifier (Yun et al. 2011). However, accelerometers, signal amplifying device and data acquisition equipment are connected via electric cables. To monitor the state of large structures, a large number of sensors and cables are needed. Therefore, the dynamic signal can be easily interference in the transmission path, not to mention the failure of the cables. Not only wasting a lot of manpower and costs but dynamic signals cannot be transmitted with high-quality. SHM with wired sensors is facing great challenges (Sandoval 2004). Mobile smart devices (e.g., smartphone) integrated with vibration sensors, CPU and large-capacity internal memory inside can support real-time data processing and storage and data transmission with wireless network and Internet. In addition, operating system for mobile smart devices with a wide range of users is open source which provides great convenience for the software developers. And smart devices are so cheap that can be deployed in structures in large quantity (Kotsakos et al. 2013). Compared with traditional wired sensor, smart devices communicate with each other and base station through wireless network, which saves a lot of cables and avoids low-quality data due to cable failure. However, there are also some issues in SHM based on smart devices, such as time synchronization, networking, and energy consumption *etc*. In order to monitor structural dynamic response accurately, smart devices record the dynamic data only when the response changed. This recording mode lead to the dynamic response data nonuniformly and asynchronously. Such data is obviously not able to meet the needs of SHM (Yu et al. 2013). To solve this problem, improved modal identification algorithm is proposed in this paper. Then feasibility of the algorithm is validated on a three-story shear frame model.

2 STRUCTURAL HEALTH MONITORING (SHM) SOFTWARE

An application software named "*Sensors*" was developed using java programming language based on the Android Operating System to monitor the data of structural location, acceleration, orientation and so on. The main interface when "*Sensors*" runs successfully is shown in Figure 1.

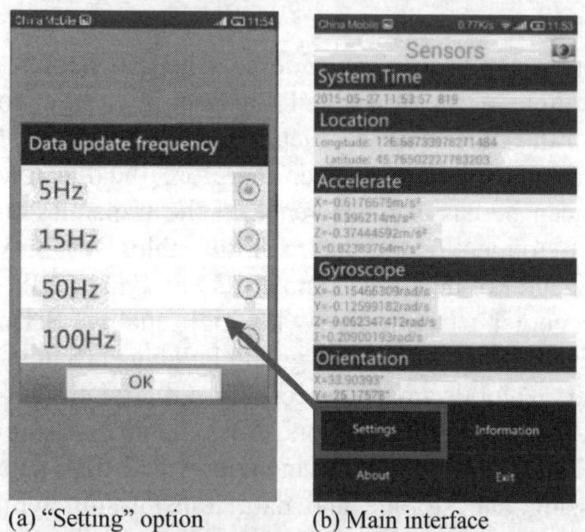

(a) "Setting" option (b) Main interface

Figure 1. Interactive interface

"*Sensors*" can be run successfully on any smart devices that installed Android 4.0 and above. "*Sensors*" can read the data collected by the embedded sensors such as camera, accelerometer, gyroscope, and GPS and show the data on the screen of smart device in real-time. Moreover, data recorded by "*Sensors*" is stored in files of CSV which includes the time and differences, and the data in the ROM memory of the smart device even can be send to the specified E-mail address through the Internet. In the main screen, click the "function" key to call out the main menu, which includes Setting, Information, About and Exit. The interface of "Setting" option, shown in Figure 1 (a), supports 4 types of sampling frequencies, such as 5 Hz, 15 Hz, 50 Hz, and 100 Hz to satisfy the dynamic data acquisition of different types of structures. "Information" shows the list of sensors the smart device uses. "About" includes the information of the developer and also allows the users to feedback questions and suggestions. "Exit" makes the software completely release the usage of the device memory.

3 STRUCTURAL MODAL IDENTIFICATION BASED ON ASYNCHRONOUS DATA

3.1 Derivation of mathematical formulas

For a structure with n degrees of freedom (DOF), the motion equation can be described as follows (Chopra 1995):

$$\boldsymbol{M}\ddot{\boldsymbol{x}}(t)+\boldsymbol{C}\dot{\boldsymbol{x}}(t)+\boldsymbol{K}\boldsymbol{x}(t)=\boldsymbol{F}(t) \tag{1}$$

where $\boldsymbol{M}$, $\boldsymbol{C}$, $\boldsymbol{K}$ are mass matrix, damping matrix, stiffness matrix of the structure, respectively. $\boldsymbol{F}(t)$ represents the excitation force. $\boldsymbol{x}(t)$ represents the displacement response of the structure.

The displacement response $\boldsymbol{x}(t)$ can be represented in modal coordinates as follows.

$$\boldsymbol{x}(t)=\sum_{i=1}^{N}\boldsymbol{\phi}_i\boldsymbol{q}_i(t) \tag{2}$$

where $\boldsymbol{\phi}_i$ denotes the i-th mode shape of the structure. $\boldsymbol{q}_i(t)$ denotes the response of the i-th modal coordinate system.

Therefore, the motion equation can be rewritten in modal coordinates as follows.

$$\ddot{\boldsymbol{q}}_j(t)+2\xi_j\omega_j\dot{\boldsymbol{q}}_j(t)+\omega_j^2\boldsymbol{q}_j(t)=\boldsymbol{f}_j(t)\ (j=1,\ldots,n) \tag{3}$$

where ξ_j and ω_j are the natural frequency and damping ratio of the j-th mode, $f_j(t)$ is the generalized force of the j-th mode.

Using Fourier transform, motion equation in modal coordinate system can be expressed in the frequency domain as follows:

$$\boldsymbol{Q}_j(\omega)=\boldsymbol{H}_j(\omega)\boldsymbol{F}_j(\omega) \tag{4}$$

where $H_j(\omega)$ is the transfer function of the j-th mode. $\boldsymbol{Q}_j(\omega)$ and $\boldsymbol{F}_j(\omega)$ are the Fourier transform of $\boldsymbol{q}_i(t)$ and $\boldsymbol{f}_j(t)$.

The displacement of the structure in frequency domain can be described as

$$\boldsymbol{x}(\omega)=\sum_{i=1}^{N}\boldsymbol{\phi}_i\boldsymbol{Q}_i(\omega) \tag{5}$$

where $\boldsymbol{x}(\omega)$ is the displacement of the structure in frequency domain.

In structural vibration analysis, the structural responses can be expressed only by the j-th mode of the structure when the frequency is very close to the j-th natural frequency of the structure in frequency domain. The response at the l-th DOF of the structure can be expressed as

$$\boldsymbol{x}_l(\omega)\big|_{\omega\to\omega_j}\approx\boldsymbol{\phi}_{lj}\boldsymbol{Q}_j(\omega) \tag{6}$$

where $\boldsymbol{x}_l(\omega)$ denotes the structural displacement at the l-th DOF; $\boldsymbol{\phi}_{lj}$ denotes the l-th element of the j-th mode shape of the structure.

Similarly, the excitation force can be represented as

$$\boldsymbol{F}(t)=\sum_{i=1}^{N}\boldsymbol{\phi}_i f_i(t) \tag{7}$$

Thus, the power spectral density function (PSDF) of the excitation force can be obtained as follow.

$$\begin{aligned}S_{F_j}(i\omega)&=E\left[F_j(i\omega)F_j^*(i\omega)\right]\\&=\phi_j^{\mathrm{T}}E\left[F(i\omega)F^*(i\omega)\right]\end{aligned} \tag{8}$$

Using the transfer function obtained from eq. (3), the PSDF of response is

$$S_{Q_j}(i\omega)=\left|H(i\omega)\right|^2 S_{F_j}(i\omega) \tag{9}$$

Then least squares method is introduced to minimize the eq. (10) to get determine the modal frequencies and damp ratios, respectively.

$$\underset{\omega_j,\delta_j}{\arg\min}\sum\left|\hat{S}_{Q_j}\left(i\omega,\omega_j,\delta_j\right)-S_{Q_j}\left(i\omega,\omega_j,\delta_j\right)\right|^2 \tag{10}$$

3.2 Identification of structural modal frequencies and shapes

According to the result derived in section 3.1, natural frequencies of the structure can be identified through peak picking method (Ren et al. 2004; Brincker et al. 2001) for the PSDF which is calculated by uniform and asynchronous data.

Substituting equation (5) into equation (6), the result can be as follows

$$\left.\frac{S_{X_l}(\omega)}{S_{X_m}(\omega)}\right|_{\omega\to\omega_j}\approx\frac{\phi_{lj}^{\;2}}{\phi_{mj}^{\;2}} \tag{11}$$

It is can be seen from the equation (8) that, when the frequency is close to the structural natural frequency ω_j, the PSDF ratio between two DOFs of the structure is approximately equal to the square of the ratio of corresponding DOFs of the l-th mode shape. Therefore, the modal shape of the structure can be evaluated by the PSDF.

In order to obtain modal information of structure, multi-point fitting method is used here. Half-power spectrum band range is chosen, which totally cover the peak. Assuming that $\phi_{l1}=1$, we can get

$$\begin{bmatrix}S_{X_l}(i\bar{\omega}_1) & \\ S_{X_l}(i\bar{\omega}_2) & 0\\ \vdots & \\ S_{X_l}(i\bar{\omega}_n) & \\ & S_{X_l}(i\bar{\omega}_1)\\ 0 & S_{X_l}(i\bar{\omega}_2)\\ & \vdots\\ & S_{X_l}(i\bar{\omega}_n)\end{bmatrix}\begin{bmatrix}\left|\phi_{mj}\right|^2\\ \left|\phi_{nj}\right|^2\end{bmatrix}=\begin{bmatrix}S_{X_m}(i\bar{\omega}_1)\\ S_{X_m}(i\bar{\omega}_2)\\ \vdots\\ S_{X_m}(i\bar{\omega}_n)\\ S_{X_n}(i\bar{\omega}_1)\\ S_{X_n}(i\bar{\omega}_2)\\ \vdots\\ S_{X_n}(i\bar{\omega}_n)\end{bmatrix} \tag{12}$$

where $\bar{\omega}_i$ means the frequency near the corresponding natural frequency ω_i.

Normalized mode shape can be obtained from eq. (12), which is more accurate than peak-picking method.

4. EXPERIMENTAL VERIFICATION

A three-story shear frame structure model of length 400 mm, width 150 mm, and story height 500 mm, is used here, on which smartphones and traditional wired accelerometers are deployed to monitor the accelerate signal of the structural model, as illustrated in Figure 2.

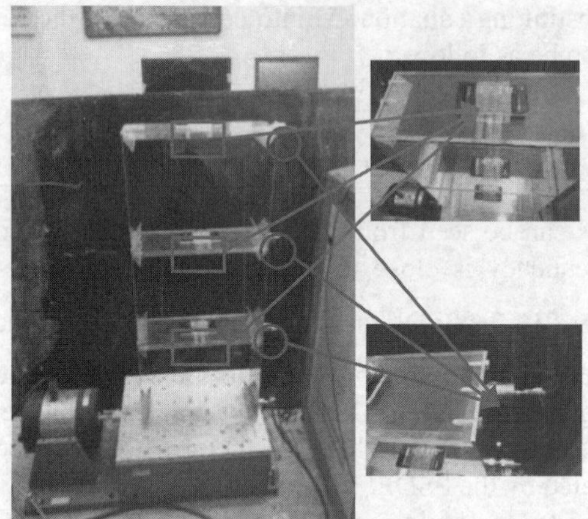

Figure 2. Three-story shear structure model

Slabs and columns of the frame structure model are simulated by plexiglass and aluminum plates, respectively and they are connected by bolts which are used to simulate the actual joints. The structure model is bolted in a small shaking table as an input signal source that can adjust the type of the excitations. Smartphones are tied on the center of the model with tapes while wired accelerometers of KD1100 produced by Yangzhou Kedong Co., Ltd. are also deployed on the short-side of each floor with magnets, as a reference for the measurement of the smartphones.

Using a structural model with three floors shown in Figure 2, there are 3 modes of the structure.

A band-pass Gaussian excitation with pass band frequency from 2 Hz to 20 Hz is used to excite the structure model. Acceleration data of 60s long with 100 Hz sampling rate are obtained and processed to identify the modal information of the structure model. Structural modal frequencies and mode shapes can be obtained from the power spectral density function (PSDF), PSDF and comparisons of the identified modal parameters are shown in Figure 3 and table 1 separately.

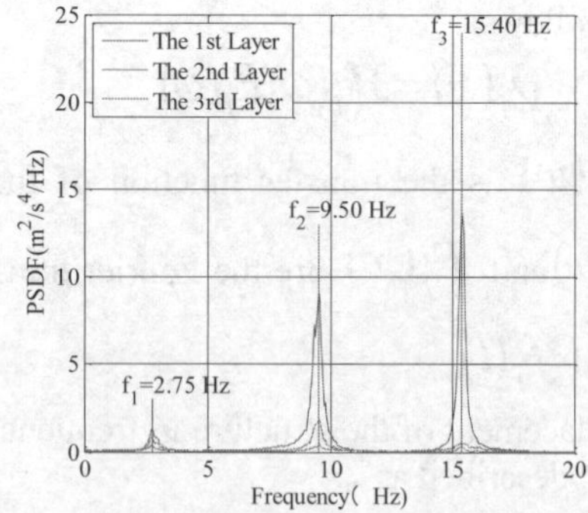

(a) PSDF obtained from data of smartphones

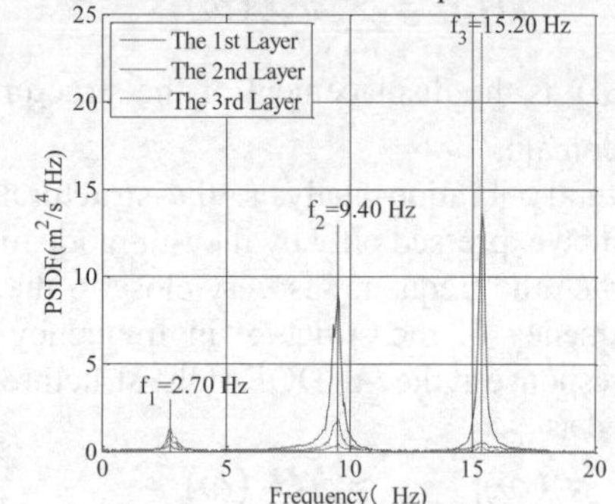

(b) PSDF obtained from data of accelerometers

Figure 3. Relationship of PSDF and frequency

Table 1. Comparisons of the different type of sensors

Sensor type	1st mode	2nd mode	3rd mode
Smartphone	2.75 Hz	9.50 Hz	15.40 Hz
Accelerometer	2.70 Hz	9.40 Hz	15.20 Hz
Difference	1.85%	1.06%	1.32%
Smartphones	[1 1.53 1.94]	[1 0.22 0.45]	[1 0.20 0.14]
Accelerometers	[1 1.17 1.53]	[1 0.34 0.68]	[1 0.65 0.36]
MAC	0.99	0.96	0.84

As shown in Table 1, smartphones can identify the modal frequency accurately, compared to wired accelerometers but large modal shape error occurs in high structure modes. This mainly because that the modal frequencies are determined by the location regardless of the magnitude of the peak of the PSDF.

But the modal shapes obtained by the multi-point fitting method are likely to be interfered by noise. Not only the location but the amplitude of the PSDF can affect the accurate of the modal shapes.

5 CONCLUSION

This paper provides a data acquisition scheme that an application software "*Sensors*" is developed in Android operating system based on smart devices in SHM systems. For the issue of time asynchronous when collecting data by smart devices, modal identification method based on asynchronous data is proposed and verified, which can accurately identify the modal information of structure.

A three-story shear structure model implemented with smartphones and wired accelerometers is introduced to illustrate the validity of SHM based on the application software "*Sensors*" and smartphones. Experimental results show that smartphones as a sensor for data acquisition is feasible to meet the needs of practical engineering compared to traditional wired sensors in SHM.

ACKNOWLEDGEMENT

The authors gratefully acknowledge the financial support of this work by National Basic Research Program of China (Grant No. 2013CB036305), by National Science Foundation of China (Grant No. 51378157, Grant No: 51161120359), by Ministry of Science and Technology of China (Grant No: 2015DFG82080), by Ningbo science and technology project (2015C110020), by the China Postdoctoral Foundation (Grant No. 2012M510967, 2014T70347) and by the Heilongjiang Postdoctoral Foundation (Grant No. LBH-Z12108).

REFERENCES

Brincker, R. Zhang, L.M. Andersen, P. 2001. Modal identification of output-only systems using frequency domain decomposition, *Smart materials and structures* **10**, 441–445.

Carden, E.P. & Fanning, P. 2004. Vibration Based Condition Monitoring: A Review, *Structural Health Monitoring* **3**, 355-377.

Chopra, A. K. 1995. *Dynamics of structures*, Prentice Hall, New Jersey, U.S.A.

Kotsakos, D. Sakkos, P. Kalogeraki, V. Gunopulos, D. 2013. SmartMonitor: using smart devices to perform structural health monitoring, *Proceedings of the VLDB Endowment* **6**, 1282-1285.

Ren, W. X. & Zong, Z. H. 2004. Output-only modal parameter identification of civil engineering structures, *Structural Engineering and Mechanics* **17**, 429-444.

Sandoval, S.R. 2004. Smart sensing technology: opportunities and challenges, *Structural Control and Health Monitoring* **11**, 349-368.

Yu, Y. Zhao, X.F. Wang, Y. Ou, J.P. 2013. A study on PVDF sensor using wireless experimental system for bridge structural local monitoring, *Telecommunication System* **52**, 2357-2366.

Yun, C.B. & Min, J.Y. 2011. Smart Sensing Monitoring and Damage Detection for Civil Infrastructures, *Journal of Civil Engineering* **15**, 1-14.

Proceedings of the International Conference on Smart Infrastructure and Construction
ISBN 978-0-7277-6127-9

© The authors and ICE Publishing: All rights reserved, 2016
doi:10.1680/tfitsi.61279.277

Smart Automation System dedicated to Infrastructure and Construction

Bruno Fazzari[1*], Angelo Stella[2], Giacomo Navarra[3] and Francesco Lo Iacono[3]

[1] *PEBRIT, Milan, Italy*
[2] *Bosch Rexroth Spa, Cernusco sul Naviglio (MI), Italy*
[3] *Faculty of Engineering and Architecture, "Kore" University of Enna, Enna, Italy*
[*] *Corresponding Author*

ABSTRACT This paper aims to show the State of Art for Smart Automation Systems dedicated to Infrastructures and Constructions (SASIC) The first part is devoted to exploit the topic of sensors: The paper highlights show some new sensor technologies might be adopted in these areas giving much more reliability to bridges and infrastructures. Bosch Rexroth produce a wide range of accelerometers dedicated to several applications. The concept is to develop dedicated smart sensors able to measure and communicate with a host computer and/or a supervisor system which will opportunely act. It would be possible to manufacture tailor made software able to fit with the new requirements every day more difficult to reach. Also for video cameras and accelerometers it would be possible to create enhanced devices able to give much more reliability to civil constructions. Having these kind of sensors available it would be possible to understand whether or not recovery energy through the movement of the bridge is feasible or it is not convenient. The Second part is dedicated to actuators: thanks to actuators developed by Bosch Rexroth, it would be possible to catch the movement and transforming it in electrical energy able to recharge the battery used by the sensors installed on the bridge. Of course Bosch Rexroth got a lot of devices able to transform the movement in electrical power able at least to charge the battery of the devices installed on the construction. In case the construction doesn't move, it would be possible to recover energy by solar cells positioned close to the sensors. The use of anthropomorphous robot or of an automatic portal could be a solution regarding measurements on the wear and tear of the construction materials. How this may be done? Thanks to the development of such advanced sensors and devices which Bosch Rexroth would be able to be active in this specific application fields. Due to these developments, it would be possible to measure the behaviour of a bridge or a construction after the installation and, periodically, remake these measures through this enhanced system having in real time the situation of the bridge or construction. A dedicated software might give indication on the actions which should be done in order to obtain the maximum reliability of the bridge or construction. Furthermore, being the bridge or the construction connected to a wide kind of sensors, it would be possible to acquire and store, like and aircraft black box, or transmit data even during an exceptional event like an earthquake.

1 INTRODUCTION

Italian buildings and infrastructural heritage are often ancient or characterized by a chronic lack of maintenance and, because of the seismic and hydro-geological risk, they are interested by high levels of vulnerability. In recent years, structural health monitoring (SHM) has emerged as a tool to support effective operation and maintenance of civil infrastructure. It mainly consists in a periodic observation of a structure or mechanical system by dynamic response measurements. The damage-sensitive features can be extracted from these measurements in order to determine the current state of system health. Under an extreme event, such as an earthquake or unanticipated blast loading, SHM could be used for rapid condition screening, to provide, in near real time, reliable information about the performance of the system during the event and about the subsequent integrity of the system. Moreover, from the viewpoint of the life-cycle management of civil structures, the economic and social impact of maintenance is crucial. Inspections and repairs entail huge direct and social costs, due to the interruption or reduction of the structure functionality. In order to define adequate and effective maintenance programs, owners of the buildings and infrastructures need information on the actual state of the structures before damages can occur with a consequent suspension of their functionality. Detecting damage at an early stage can also reduce the costs and recovery time associated with repair of critical damage.

Such information can be collected and elaborated by means of adequate monitoring systems and the numerous SHM techniques which have been developed (Doebling et al., 1996; Farrar & Worden 2013), but their diffusion is still limited by the high cost of sensors and devices needed.

During the last years Micro Electro Mechanical Systems (MEMS) technology and information technology have quickly improved, making available sensors with higher qualities, that can be used in measurement applications and on structural health monitoring (Furlong et al., 2005; Kant & Nagel, 2006; Deb & Blanton, 2006; Park & Gao, 2006; Cigada et al., 2007).

The large amount of data generated by a monitoring system can be handled by means of on-board processing which can be incorporated at the sensor level to allow a portion of the computation to be done locally. This approach results in the construction of the so called smart sensors able to reduce the amount of information to be transmitted over a monitoring network (Spencer et al., 2004; Ruiz-Sandoval et al., 2006).

In this paper it is shown how MEMS accelerometer can be used for monitoring of civil structures and infrastructures. Firstly, the monitoring system architecture based on MEMS sensors is presented in detail together with the technical properties of the proposed sensor. Then, results from several static and dynamic laboratory tests are reported, in order to define the sensor sensitivity and accuracy. Since MEMS sensors measurements are affected by environmental temperature, laboratory tests have also been performed to achieve the response of the sensor under a wide range of external operative temperatures. Finally, the characteristics of a new state-of-the-art sensor, produced by Bosch, are presented.

2 THE MONITORING SYSTEM

In this section, the main features of a monitoring system will be presented. The system has been developed by Wisenet Engineering S.r.l. and Laboratory of Experimental Dynamics of the "Kore" University of Enna (Navarra et al, 2015).

2.1 *The sensor board*

The sensor package constitute the main device of the monitoring system and it is comprised of two parts: the processor board and the sensor board (**Figure 1**).

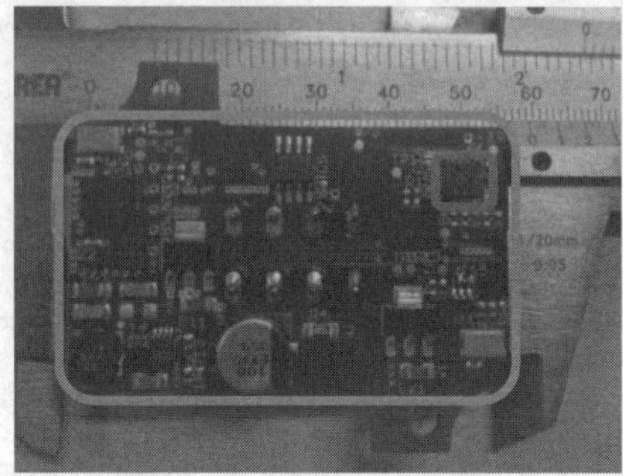

Figure 1. The Wisenet MEMS Sensor: the processor board inside blue rectangle and the MPU-6000 sensor board inside red rectangle.

The processor board have been developed by Wisenet Engineering S.r.l. and contains a microprocessor with 16K bytes flash (program) memory, 768 bytes SRAM and 256 bytes of data memory able to interpret commands from the master node and to elaborate data at the sensor level. The board mounts a clock sensor at 18M Hz to synchronize data acquisition. Analogical data are conditioned and converted on board and digital data are sent to the master node (PC) by means of a transceiver RS485. Each sensor can be identified univocally by setting a DIP switch available on the processor board. The board can be configured to be equipped with many kind of integrated circuits or sensors (gyroscope, accelerometer, temperature sensors, etc) and it communicates with the sensor board by means of Serial Peripheral Interface (SPI) protocol.

The sensor board consists of six axis MEMS sensor. The sensor includes a 3-axes MEMS accelerometer, a 3-axes MEMS gyroscope and a digital-output temperature sensor. In this work only the 3-axes accelerometer is considered. Static acceleration, for example due to gravity, can be also detected. Each sensor has a dedicated sigma-delta 16 bit ADC for providing digital outputs.

2.2 *The monitoring system architecture*

The architecture of the monitoring system consists mainly of:

- a general purpose PC to manage the flow of data coming from the network of sensor nodes;
- the master node which is connected to the personal computer by USB port and it has the main function of translating the physical USB protocol to the RS485 slave nodes protocol;
- the slave nodes or sensor nodes which represent the peripheral units of the system.

Each slave node is connected to the master node by a serial bus with a communication protocol based on the physical layer RS485 and logical protocol developed by Wisenet Engineering S.r.l. The RS485 standard ensures an optimal electromagnetic noise isolation while the logical protocol guarantees the optimization of the data flow on the bus, the coherence control of the data packs by means of a 16 bit cyclic redundancy check (CRC-16), the identification of the node ID information and the synchronization of data acquisition.

3 CHARACTERIZATION OF MEMS SENSORS

All the tests described in this section have been performed at the Laboratory of Experimental Dynamics of the Faculty of Engineering and Architecture of the "Kore" University of Enna. As detailed in the following, sensors have been tested by means of both static and dynamic tests in order to calculate floor noise, sensitivity, frequency response performance and sensitivity dependence on temperature.

3.1 *Static tests*

Data from accelerometers at rest are collected for several hours to observe the statistical properties of the sensor noise on the outputs and the effect of external temperature change on the sensor outputs. All tests reported herein have been conducted by using the sensor as an accelerometer configured with ±2 g range, 1k Hz sampling frequency and 260 Hz low pass filter.

From the analyses of the recorded acceleration data, the noise is fully characterized. The probability density function (PDF) have been computed and compared with a Gaussian curve to be sure that the noise itself was white and Gaussian. The standard deviation of the noisy measurement determines the characteristic of the sensor noise and, after the raw measurements are converted to physical units, the standard deviations, reported in Table 1, have been computed.

Table 1. Standard deviation of acceleration noise.

x-axis	$3.65\ 10^{-3}$ g
y-axis	$3.28\ 10^{-3}$ g
z-axis	$4.88\ 10^{-3}$ g

Scale factor calculation is one of the main steps to obtain the calibrated output variables from the raw sensor data. The procedure adopted to find the scale factors for each axis consisted in a series of static acquisitions of accelerations when, in turn, one of the axis of the accelerometer was close to the vertical position. The signals have been acquired for several positions of the sensor by varying the axis angle of very small quantities (in the order of 0.002°) to catch the real vertical alignment. The maximum value have been compared with the ground acceleration in order to compute the *scale factor sf* that can be used to convert the sensor output to the acceleration. Assuming a linear variation from -1 g to +1 g one can obtain the corrected output co_i for each axis according to:

$$co_i = m_{c,i} \cdot ro_i + b_{c,i} \qquad i = \mathrm{x, y, z} \tag{1}$$

where ro_i is the raw output from accelerometer axis, $m_{c,i}$ and $b_{c,i}$ are the coefficients of the linear transformation defined as:

$$\begin{cases} m_{c,i} = 2/\left(sf_i^+ - sf_i^-\right) \\ b_{c,i} = -\left(sf_i^+ + sf_i^-\right)/\left(sf_i^+ - sf_i^-\right) \end{cases} \qquad i = \mathrm{x, y, z} \tag{2}$$

Scale factors for the three axes, used to correct acceleration data of a specimen sensor, are summarized in Table 2 together with the temperatures $T_{c,i}$ at which tests have been performed.

Table 2. Scale factors and correction coefficients.

Axis	sf^+ [g]	sf^- [g]	$m_{c,i}$	$b_{c,i}$ [g]	$T_{c,i}$ [°C]
x	0.9910	-1.0110	0.9950	0.0059	26.9
y	1.0089	-0.9868	1.0022	-0.0111	27.0
z	1.0930	-0.9178	0.9946	-0.0871	26.4

3.2 *Thermal compensation*

The performance of the MEMS accelerometer sensor with changing temperature has been extensively studied by using a thermal chamber. The temperature has been varied from 8°C to 47°C in about 44 h.

Data from accelerometers and internal temperature sensor is collected and analyzed by keeping the sensor in the same position during temperature variation. Signals have been acquired and plotted respect to the temperature. It has been noticed that x and y axes are less affected by temperature changes than the z-axis, but the relationship between temperature and variation in sensitivity can always be assumed as a linear one.

Analogously to the previous subsection, the correction law for thermal compensation can be written in the form:

$$to_i = m_{t,i} \cdot T_i + b_{t,i} \qquad i = \text{x,y,z} \tag{3}$$

where to_i is the temperature corrected output, $m_{t,i}$ and $b_{t,i}$ are the temperature correction coefficients whose values are reported in Table 3.

Table 3. Thermal compensation coefficients.

Axis	$m_{t,i}$	$b_{t,i}$
x	3.537 10^{-4}	4.807 10^{-1}
y	-5.169 10^{-5}	7.720 10^{-3}
z	-2.348 10^{-3}	-3.817 10^{-1}

Once the static and thermal correction coefficients have been determined for each sensor and each axis, a general law that allow to compensate both errors can be derived from the combination of equations **Error! Reference source not found.** and **Error! Reference source not found.**, as follows:

$$co_i = m_{c,i} \cdot ro_i - m_i \cdot T_i + \hat{b}_i \qquad i = \text{x,y,z} \tag{4}$$

where co_i is the corrected output, ro_i is the raw output from the *i*-th accelerometer axis, $m_{c,i}$ is the correction coefficient defined in eq. (2), while m_i and b_i are the modified thermal correction and bias coefficients depending on the temperature $T_{c,i}$ at which the static calibration tests have been performed, as in the following:

$$\begin{aligned} m_i &= m_{c,i} \cdot m_{t,i} \\ b_i &= b_{c,i} + m_{c,i} \cdot m_{t,i} \cdot T_{c,i} \end{aligned} \tag{1}$$

3.3 *Inclinometer sensitivity analysis*

Multi position static tests have been done with the purpose of determining the angle sensitivity of the accelerometer sensor. Accelerometers have been mounted on an index table capable of measurable axis rotation with a resolution of 0.006°. The table is rotated from 0.0° to -1.4° in 35 steps so that it is possible to relate rotations and angle derived from sensor measurement channels.

The test has been repeated twice in order to measure angle variations on two axes per time. Sensor data have been corrected as explained in previous subsections and compared to the real angles measured on the reference table (Figure 2a-b).

Classical statistical error propagation analysis (Taylor, 1997) have been applied to the measurements and it has been found that the smallest measurable angles for each sensor axis are ±0.002°, ±0.002° and ±0.090° for the x, y and z-axis respectively. It is worth to note that the z-axis is subjected to a bigger error than the x and y-axis. This is mainly due to the fact that z-axis is characterized by a bigger value of noise as already remarked.

The measured angles present small deviations from the real table angle but these are smaller than the tolerances expected so it is possible to state that the presented sensor can be also used as an inclinometer for application in civil engineering structural monitoring.

3.4 *Dynamic tests*

In order to assess the efficacy of the proposed accelerometer for civil engineering applications, its performance was compared with a PCB model 393B04 high sensitivity piezoelectric seismic accelerometer.

The test procedure consisted in a head-to-head test on a APS 420 Long Stroke shaker in which both the reference and the MEMS accelerometers sense the same acceleration being fixed on the same rigid plate. In Figure 3 a picture of the dynamic test setup is reported.

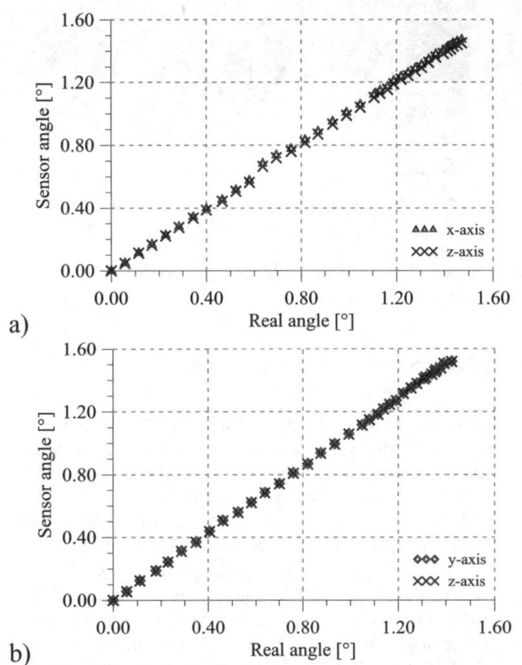

Figure 2. Measured angles from sensor compared to the real table angle: a) x and z axes; b) y and z axes.

Tests have been conducted by comparing, in turn, each of the three axes of the MEMS accelerometer in the frequency range 0.25÷100 Hz which is in good agreement with the civil engineering applications to whom the developed MEMS accelerometer will be addressed. The reference accelerometer data have been acquired by means of a 24-bit ADC National Instruments board model NI4499. Tests have been conducted by means of a self-developed software LabView® for the acquisition of time domain signals and post processing have been done by using self-developed Matlab® routines.

In order to verify real differences in dynamic performances between MEMS and reference sensors, sweep sine tests have been performed to calculate the Frequency Response Function (FRF) in both a low frequency range (from 0.25 Hz to 2.0 Hz) and a medium-high frequency range (from 1.0 Hz to 100 Hz).

Figure 3. Dynamical test setup for z-axis.

All tests consisted of 100 s time length signals acquired by using a sampling frequency equal to 1k Hz, and they have been repeated five times for each input frequency range and for each MEMS accelerometer axis set along the shaker moving axis. The FRFs have been computed referring to the 393B04 accelerometer.

Acceleration data from MEMS sensor have been calibrated by means of the procedure described in previous sections. In all cases the observed amplitude of the FRFs can be considered reasonably linear and constant from 0.5 Hz to 80 Hz; uncertainty grows for frequencies bigger than 80 Hz and lower than 0.5 Hz but for the latter it has to be noticed that the amplitude of the acceleration stimulus is low and recorded signals are very noisy.

However, although the dynamic response of the tested sensor is quite accurate, its use as an accelerometer for civil structures monitoring purposes is strongly limited by the floor noise. In the next section the main features of a new state-of-the-art MEMS sensors, produced by Bosch, are reported

4 MAIN FEATURES OF A STATE-OF-THE-ART MEMS SENSOR

In this section the main features of the XDK MEMS sensors, produced by Bosch, are reported. XDK is a hardware and software development platform for the realization of Internet of Things. It includes a full array of sensors and several devices to easily connect the XDK with other devices like PCs and mobile devices via USB, Bluetooth or WLAN.

Due to its housing with included battery in a small form factor of only 60 x 40 x 22 mm³, it can be retrofitted to objects of any size. Figure 4 depicts the sensor. XDK incorporates several built-in sensors as accelerometer s, gyroscopes, a magnetometer, a humidity sensor, a pressure sensor, a temperature sensor, an acoustic sensor and a digital light sensor.

In the same board are present also a 32-Bit microcontroller (ARM Cortex M3), 1MB Flash, 128 kB RAM, a radio device with Bluetooth and WLAN capabilities, a 560 mAh Li-Ion rechargeable battery, 3 programmable status LEDs, a Micro SD Card, and 2 programmable push buttons.

The XDK platform allow the use of the LWM2M communication protocol, the update of the firmware may be done "over the air" and extensive libraries are provided for the developer. XDK sensors can operate in a temperature range from 0 ºC to 60 °C and in a relative humidity range from 10% to 90%

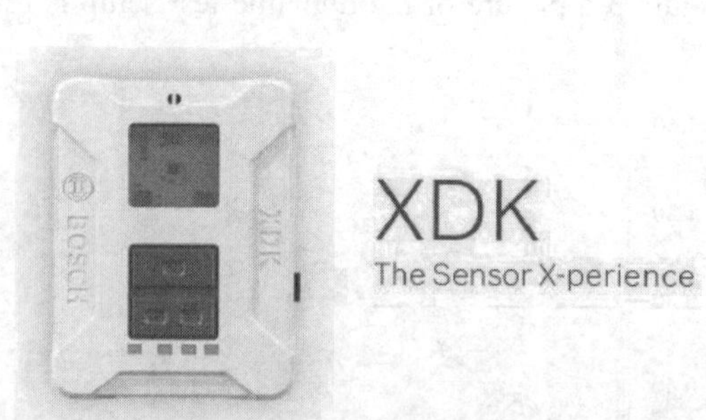

Figure 4. Bosch Micro Electronic Mechanical System.

CONCLUSION

In this paper it has been demonstrated that MEMS accelerometers can be used for monitoring of civil structures and infrastructures. In particular, a MPU- MEMS accelerometer has been integrated in a processor board developed by Wisenet Engineering S.r.l. in collaboration with the Laboratory of Experimental Dynamics of the University of Enna "Kore". The MEMS sensor have been calibrated and tested to certificate the possibility to be used in a wide range of civil engineering monitoring applications. The details of the static and thermal calibration procedure adopted to use the sensor both as an inclinometers, for static monitoring applications, and as a 3-axes accelerometer for structural dynamics applications have been reported.

Although the proposed system architecture and the programmable sensors can be used to implement reliable structural health monitoring systems, the capabilities of new state-of-the-art sensors, like the Bosch XDK, could greatly enhance the performance and the versatility of smart system giving to the system much more reliability and redundancy.

REFERENCES

Cigada, A., Lurati, M., Redaelli, M.& Vanali, M., 2007. Mechanical Performance and Metrological Characterization of MEMS Accelerometers and Application in Modal Analysis, in *Proc. IMAC-XXV: Conference & Exposition on Structural Dynamics*.

Deb N. & Blanton R.D., 2006. Built-in self-test of MEMS accelerometers, *Journal Microelectromech Syst*, **15**(1), 52-68.

Doebling S. W., Farrar, C. R., Prime, M. B. & Shevitz, D. W. 1996. Damage Identification and Health Monitoring of Structural and Mechanical Systems from Changes in Their Vibration Characteristics: A Literature Review, Los Alamos Nacional Laboratory.

Farrar, C. R. & Worden, K., 2013. *Structural Health Monitoring. A Machine Learning Perspective*. Wiley, Chichester, UK.

Furlong C., Kok R.& Ferguson C.F., 2005. Dynamic analysis and characterization of MEMS accelerometers by computational and opto-electromechanical methodologies, in *Proc. of the 2005 IEEE International Reliability Physics Symposium Proceedings*, Apr 17-21, San Jose, CA, USA.

Kant R.A. & Nagel D.J., 2006. Characteristics and performance of MEMS accelerometers, *AIP Conference Proceedings*, **368**(1), 166-176.

Navarra, G., Lo Iacono, F. & Oliva, M. 2015.Static and dynamic characterization of a low-cost MEMS sensor for structural monitoring purposes, *Proceeding of 22nd AIMETA Conference*, Genoa (Italy), 14-17 September, 287-296.

Park M. & Gao Y., 2006. Error analysis and stochastic modeling of low-cost MEMS accelerometer, *Journal of Intelligent and Robotic Systems: Theory and Applications*, **46**(1), 27-41.

Ruiz-Sandoval, M., Nagayama, T. & Spencer, Jr., B. F., Sensor Development Using Berkeley Mote Platform, *Journal of Earthquake Engineering*, **10**(2), 289–309.

Spencer, Jr., B. F., Ruiz-Sandoval, M. & Kurata, N., 2004. Smart sensing technology: Opportunities and challenges, *Journal of Structural Control and Health Monitoring*, 11, 349–368.

Taylor, J. R., 1997. *An Introduction to Error Analysis*, University Science Books.

Proceedings of the International Conference on Smart Infrastructure and Construction
ISBN 978-0-7277-6127-9

© The authors and ICE Publishing: All rights reserved, 2016
doi:10.1680/tfitsi.61279.283

Tri-band ground penetrating radar for subsurface structural condition assessments and utility mapping

D. Huston*, T. Xia, Y. Zhang, T. Fan, J. Razinger and D. Burns

University of Vermont, Burlington, USA
**Corresponding Author*

ABSTRACT This paper describes the operating principles and use of an impulse ground penetrating radar (GPR) system that operates with three center frequency bands to enable more agile sensing of structural conditions and utility mapping. The operation of a GPR is typically a tradeoff between low-frequency waves that penetrate well, but lack good spatial resolution with high-frequency waves that do not penetrate well but have better spatial resolution. A complication is that many GPRs use short-duration impulse-shaped signals with mixed time and frequency content. The electronic generation of impulse signals is hardware dependent. Most are capable of producing only a single type of electromagnetic wave with fixed frequency content. The result is limited sensing capability. Many structures and subsurface conditions are unpredictable and better sensed with a variety of different wavelengths. A workaround presented here is a system that combines three fixed frequency sensing bands with center frequencies of 400 MHz, 1.6 GHz and 2.3 GHz. This system can sense at three different depths and resolutions and provide a wider coverage of subsurface conditions. Results from tests aimed at sensing subsurface conditions in concrete structures, roadways and buried utility location are presented. In each of these three cases the system performs differently and better or worse for different frequency bands.

1 INTRODUCTION

Ground penetrating radar (GPR) is a nondestructive method of assessing subsurface conditions by launching, receiving and processing transient electromagnetic (EM) waves. Non-metallic structures, such as those made of reinforced concrete, pavement or soil can be well-suited for inspection with GPR. Since EM waves travel at the speed of light in a dielectric medium, the time, frequency, amplitude and spatial characteristics of these waves plays a large role in the sensitivity, penetration depth and spatial resolution

In general, it is desirable for the GPR signals to have the following characteristics: 1. *Short duration waveform transients for high-resolution down-range measurement* – The length of individual waveforms affects the spatial resolution in the downrange, i.e. depth, direction. This affects the ability to measure the thickness of thin layers and to identify small objects. 2. *Short duration waveform bundles to reduce overall measurement times* – A typical single GPR measurement involves launching and receiving a bundle of individual waveforms. Reducing the overall duration of the measurement waveform bundles reduces the overall site measurement times and allows for increasing the speed of the GPR (Xu, 2013). 3. *Wide frequency bandwidth for superior penetration depth and spatial resolution* – High-frequency waves impart superior spatial resolution, but tend to attenuate quickly with depth in lossy media, such as concrete or soil. For the same amount of loss per cycle low-frequency waves penetrate deeper. Frequency-dependent absorption can lead to frequency-dependent penetration depth. Wide bandwidth signals can penetrate through a wide range of depth and frequency-dependent media. 4. *Small amplitude* – GPRs inevitably emit electromagnetic waves into the environment. These radiated emissions can interfere with other instruments and are often the subject of government regulations. Reducing the launched signal amplitude reduces the radiated emissions. 5. *Large dynamic range* – Many structures amenable to GPR testing contain elements made of lossy dielectric media. This leads to return signals with small amplitudes, which prompt the need for GPR instruments to measure over wide amplitude ranges without distortion. Many test environments contain ambient electromagnetic noise that can contaminate the return signals, especially those with small amplitudes. Launching and receiving of waves with precise control and sensing over large dynamic ranges helps to alleviate these concerns. 6. *High cross-range spatial resolution* – The ideal EM plane wave has infinite lateral extent and no available cross-range resolution. EM waves that have a limited lateral spread have discernable cross range resolution that improves with the reduction in the launched wave spread. Diffraction effects place limits on the cross-range resolution using simple signal processing methods. More complicated sub-diffraction methods are possible. 7. *Polarization* – sometimes it is important to control the polarization of the launched waveforms and sensitivity to polarization on the received signals to discern polarization-dependent subsurface features, such as the direction of an array of reinforcing bars.

These desired characteristics conflict with one another. Building GPR systems requires compromises so that the properties of the test waves are best suited for a particular application. Most GPR systems are either impulse (I_GPR) or step-frequency (SF-GPR) (Huston, 2002).

I-GPRs launch short-duration impulse signals. Ideally the impulse is a delta-function with infinite frequency-domain bandwidth, but in reality is an impulse with finite duration, T_i, and finite frequency domain bandwidth, BW_i, with upper and lower frequency cutoffs f_u and f_l, respectively, Figure 1. Nominally, the time-frequency behavior is

$$BW_i = f_u - f_l \approx \frac{1}{T_i} \quad (1)$$

Hardware limitations, such as antennas, impose additional restrictions on the upper and lower ends of the frequency spectrum. Testing with I-GPR requires instruments capable of large amplitude dynamic ranges, fast data acquisition, and often the launching and receiving of bundles of multiple signals for sampling and averaging (Xia, 2013). Most commercial GPRs use the I-GPR technique, primarily due to the overall speed of the signal processing and overall system costs, but have the disadvantage of limited control over frequency domain behavior, especially the operating band.

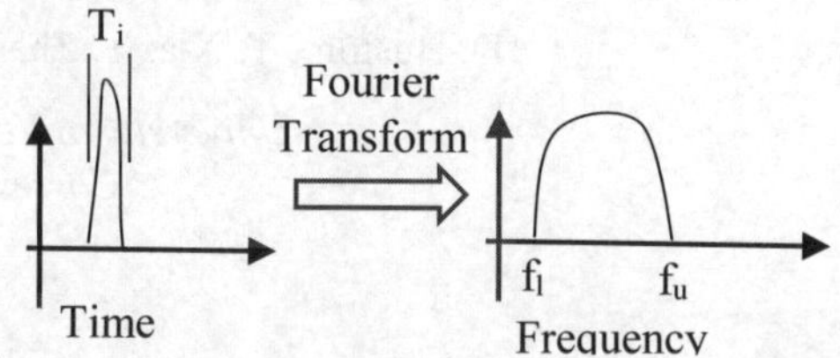

Figure 1 Time-frequency relation for I-GPR

SF-GPRs launch and receive sinusoidal waves. The technique uses bundles of waves with frequencies that step through a frequency band to provide wide bandwidth coverage. An inverse Fourier transform synthesizes a time domain impulse from a bundle of frequency domain data extracted from the bundle, Figure 2 and Figure 3.

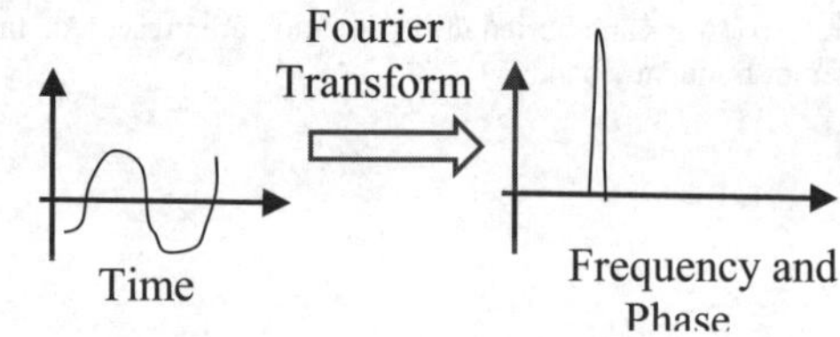

Figure 2 Time-frequency relation for single SF-GPR signal

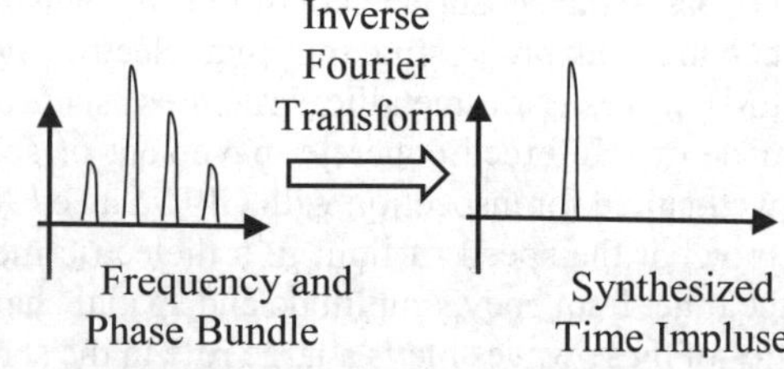

Figure 3 Frequency-time relation of signal bundles in SF-GPR synthesize time-domain impulse

SF-GPRs tend to be slower and more expensive than I-GPRs, but offer greater control over frequency-domain content. SF-GPRs tend to be favored by academic researchers and appear in certain high-performance GPRs.

A primary limitation of I-GPRs is that the circuitry and EM hardware tends to be fixed to a specific band of operation. Many testing situations can benefit from a wider bandwidth. In this context, the use of new tri-band I-GPR provides an enhanced frequency domain coverage. The concept is to combine three I-GPRs into a single instrument with each I-GPR covering a different part of the frequency spectrum, Figure 4.

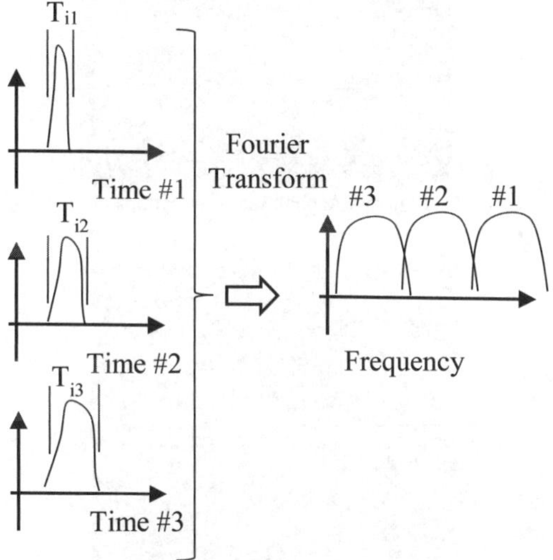

Figure 4 Concept of tri-band I-GPR

2 TRI-BAND I-GPR SYSTEM

An attempt at assembling a system with the characteristics of Figure 4 appears in Figure 5. Appearing in the figure is a 400 MHz and 1.6 GHz antenna for frequency bands #1 and #2, respectively. A supplemental 2.3 GHs antenna provided coverage of band #3. The primary components of the system came from commercial vendors and were adapted to the configuration. A key feature is to collect the data simultaneously and to register geometrically the relative positions of the various antennas and the structure under test.

Figure 6 Trailer with 400 MHz and 1.6GHz antennas

3 TEST RESULTS

The tri-band I-GPR provided measurement at a handful of structural and geotechnical sites. The following is a sampling of the results.

Figure 7 shows the results of testing the floor of a reinforced concrete building. This floor extends over earth retaining walls and an open room underneath. The corresponding B-scans show different levels of detail and penetration depth corresponding to the different bands.

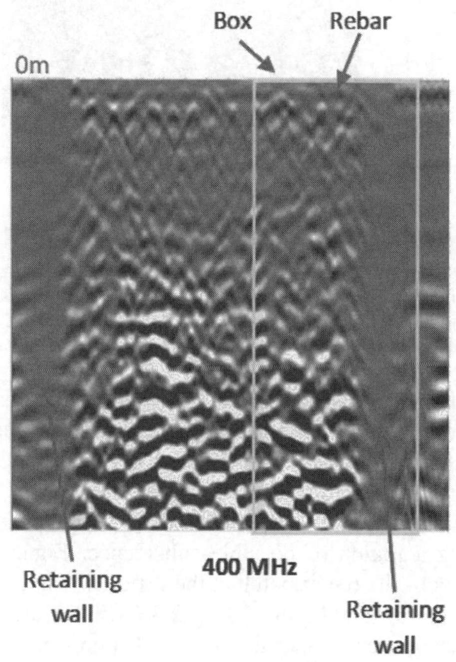

(a)

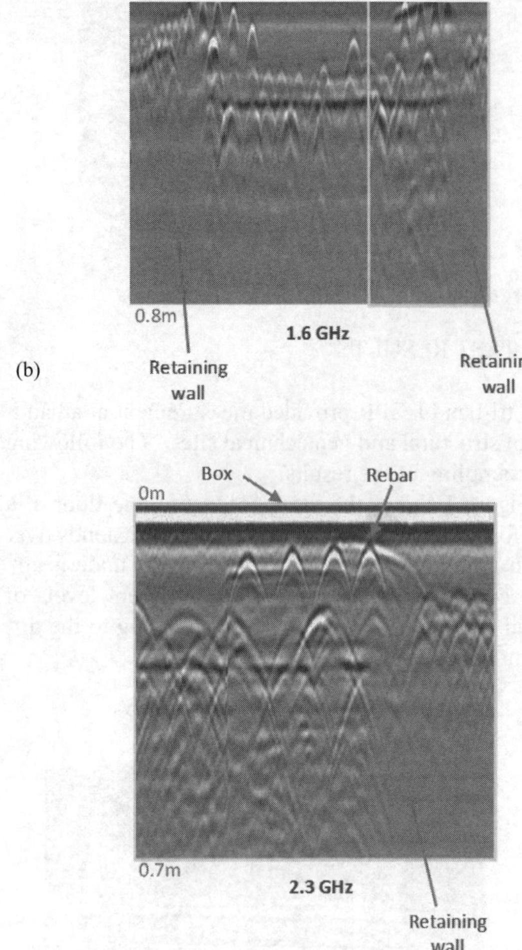

Figure 7 Tri-band I-GPR B-scan results from the floor of a reinforced concrete building: a. 400 MHz, b. 1.6 GHz, and c. 2.3 GHZ. The box corresponds to a co-located region in all three plots

The next set of tests examined the subsurface conditions of an asphalt bicycle and pedestrian roadway in Burlington, VT, USA. This is a region prone to poor drainage and frost heaving of clay layers in severe winters. From the topside, there is a large patch covering a region of possible subsidence, Figure 8. B-scans of the repair patch at the different bands appear in Figure 9 and Figure 10. At 2.3 GHz, features of the patch appear distinct, but deeper features are not distinguishable. At 1.6 GHz, the patch remains distinct and deeper features appear. At 400 MHz, the patch is visible, but not distinct, while deeper features, such as the pavement-soil interface layer, are visible.

Figure 8 Asphalt roadway with patch

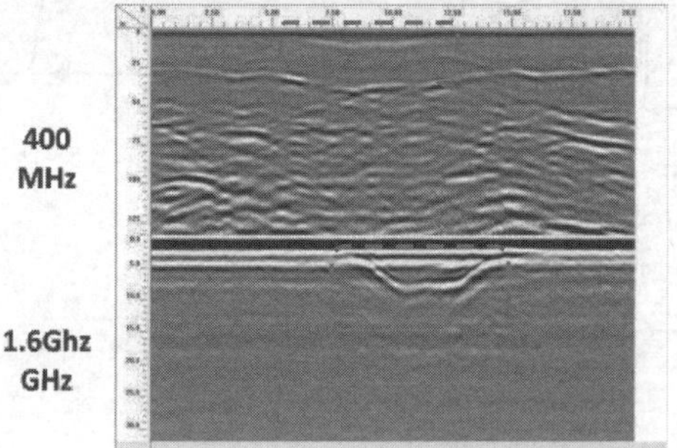

Figure 9 B-scans of repair patch at 400 MHz and 1.6 GHz

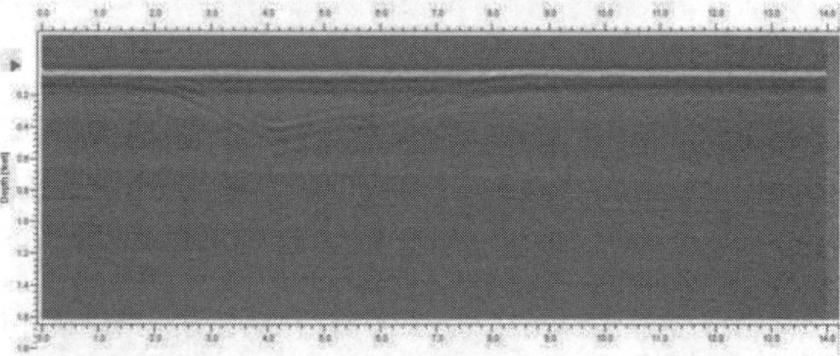

Figure 10 B-scan of repair patch at 2.3 GHz

A final set of data concerns the construction site for a new bus station in Burlington, VT. This site, being in a city with modern-era settlement spanning over 250 years has many unknown subsurface features, including the possibility of older archeological sites. A GPR scan of the site turned up many features with the 400 MHz band identifying deeper items, and 1.6 Hz and 2.3 GHz bands successful at near surface depths. As an example of the 400 MHz band performance, Figure 11 shows a B-scan of the site prior to excavation. Features including buried pipes appear as hyperbolas, while a distinct soil layer at a depth of about 2 m. appears. Subsequent excavation revealed a soil layer at the predicted depth, Figure 12. This soil layer is believed to the bed of the ancient Champlain Sea that arose during the receding glaciers in the last ice age, 10,000 B.C.E. (approx.)

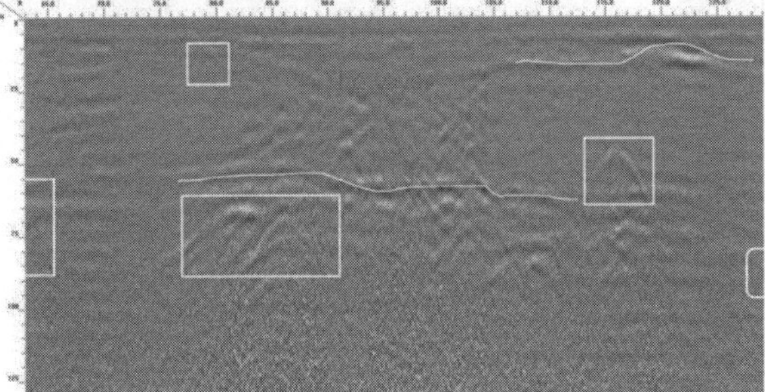

Figure 11 B-scan of bus station construction site, St. Paul St, Burlington, VT, USA. Detected features include buried pipes and a distinct soil layer at approx. 2 m.

Figure 12 Soil layer identified with 400 MHz I-GPR, believed to be bed of ancient Champlain Sea

4 CONCLUSIONS

This study presents the concept of multi-band I-GPR. The intent is to take advantage of the resolving capability of different frequency bands while retaining the speed, low-cost and ease of use of an I-GPR.

ACKNOWLEDGEMENT

This research was supported by the VT Agency of Transportation and the U. S. Army Research Laboratory and the U. S. Army Research Office under contract/grant number W911NF-13-1-0301.

REFERENCES

Huston, D.R. Fuhr, P.L. Maser, K, Weedon, W.H. 2002. Nondestructive Testing Of Reinforced Concrete Bridges Using Radar Imaging Techniques, Final Research Report NETC 94-2, New England Trans.Consortium, DOI: 10.13140/RG.2.1.4937.2002.

Xia, T. Venkatachalam, A.B. Huston, D. (2012) "A High Performance Low Ringing Ultra-Wideband Monocycle Pulse Generator" *IEEE Transactions on Instrumentation and Measurement*, **61**, 1, 261 - 266

Xu, X. Xia, T. Venkatachalam, A. Huston, D. (2013) The Development of a High Speed Ultrawideband Ground Penetrating Radar for Rebar Detection, *Journal of Engineering Mechanics*, **139**, 3, 272-285, DOI:10.1061/(ASCE)EM.1943-7889.0000458.

Proceedings of the International Conference on Smart Infrastructure and Construction
ISBN 978-0-7277-6127-9

© The authors and ICE Publishing: All rights reserved, 2016
doi:10.1680/tfitsi.61279.289

Sensing for smart infrastructure: prospective engineering applications

A. Klar*[1], E. Levenberg[1] , M. Tur[2] and A. Zadok[3]

[1] *Technion – Israel Institute of Technology, Haifa, Israel*
[2] *Tel-Aviv University , Tel-Aviv, Israel*
[3] *Bar-Ilan University, Ramat Gan, Israel*
** Corresponding Author*

ABSTRACT: Two parallel, yet complimentary, paths are being pursued by the scientific community with respect to the future of smart infrastructure. The first focuses on sensor technology and deals with advancing the capabilities and performance of the sensory gear. The second focuses on engineering applications and targets the development of interpretation models capable of transforming raw readings into information of engineering worth. This paper presents advancements made within various Israeli universities along these two paths. Firstly, innovations in the field of Brillouin distributed fiber optic sensing are discussed, together with presentation of prospective applications and future research directions. This is followed by an overview of recent advancements in the field of wireless embedded sensors, called Wisdom Stones, for civil engineering applications. It is concluded that expediting a smart infrastructure future requires a multi-disciplinary approach in which engineering needs are involved in the development of the sensing techniques.

1 INTRODUCTION

Smart infrastructure is an emerging field of study dealing with the development and large-scale embedment of hi-tech sensors in traditional civil engineering structures, and concurrently, with the collection and meaningful interpretation of the raw signals. Two parallel, yet complimentary, paths are being pursued by the scientific community with respect to the future of smart infrastructure. The first focuses on sensor technology and deals with advancing the capabilities as well as the performance-envelope of sensory gear. The second focuses on engineering applications and targets the development of interpretation models capable of transforming raw readings into information of engineering worth.

This paper overviews both recent and current research activities within the two paths taking place in different Israeli universities. It addresses advancements and applications in two technology categories: distributed fiber optic strain sensing, and buried wireless accelerometer networks.

2 DISTRIBUTED FIBER OPTIC STRAIN SENSING

Strain is a fundamental component of civil engineering design. It directly provides information on the stress levels in elastic systems and on the cumulative damage and fatigue in elasto-plastic systems. Most engineering design processes of civil engineering structures (e.g. buildings, foundation systems, embankments, tunnels, pavements, retaining walls, pipelines, etc.) involve either directly, or indirectly through stress analysis, limits on the allowable strain levels. It is therefore no wonder why measurement of the developed strain in civil engineering infrastructure is of the utmost importance for its effective design, construction, condition assessment, and maintenance.

Traditionally, strains were (and still are) evaluated locally using changes in electrical resistance of foil (bonded wire) strain gauges using Wheatstone bridge configuration, or by more advanced local devices such as vibrating wires, MEMS, or even fiber optic Bragg gratings. The interpretation of a local strain measurement for the purpose of understanding and analyzing civil engineering structures is, however, rather limited. A local stain measurement can neither be integrated to result in deformation, nor can it be differentiated to evaluate interaction forces. The development of distributed Brillouin fiber sensor technology, and its capabilities to provide spatial profiles of strains along conventional telecommunication fibers, has led to a re-evaluation of the manner in which strains can be used in civil engineering. First steps towards utilization of Brillouin based distributed sensing for civil engineering applications began approximately 10 years ago in the leading geotechnical group of Cambridge University, UK. The use of distributed sensing for research quickly spread to other civil engineering departments, among which are Technion, ETH, Princeton, and Queen's. The research work performed in these universities concentrated on schemes of installation and interpretation of the spatially distributed data for various civil engineering problems, among which are: piles foundations (Klar et al. 2006); evaluation of pipeline integrity to underneath excavation of a tunnel (Vorster et al. 2006); stressing and deformation of secant pile walls (Mohamad et al. 2007); landslide localization (Iten & Puzrin 2009); tunneling stressing (Mohamad et al. 2010); damage identification in concrete structures (Goldfeld & Klar 2012; Glisic & Inaudi 2012; Regier & Hoult 2014); and evaluation of tunneling induced ground deformations (Klar et al. 2014; Hauswirth et al. 2014). All of which are based on static, and low (1 m) sensing resolution application.

The following lines provides a brief overview of recent innovation in Brillouin optical sensing that may well revolutionize the manner in which distributed sensing is used in civil engineering.

2.1 *Principles of Brillouin scattering*

Brillouin scattering is a nonlinear process (Boyd 2008), in which acoustic phonons interact with a propagating light wave resulting in backscattering. Brillouin scattering may occur spontaneously due to interaction of thermally (and naturally) induced acoustic waves with an incident light wave, or intentionally by stimulating the interaction using a counter propagating light wave. The latter approach, of Stimulated Brillouin Scattering (SBS), has the advantage of stronger scattering, allowing for more precise measurements, as well as application of more advanced interrogation techniques. It has, however, the disadvantage of requiring access to the fiber from both ends (i.e. an optical loop).

In SBS, two optical waves (a 'pump' and a 'probe'), having a frequency difference of $\nu_a = \nu_{pump} - \nu_{probe}$, are counter-propagated along a fiber. The two waves interact so that the magnitude of one of them, the so-called probe wave, is being amplified at the expense of the other, i.e. the so-called pump wave. Their interference creates a moving optical intensity pattern with frequency ν_a and a wavenumber q equal to the sum of the probe and pump wavenumbers. This moving intensity wave initiates a density (acoustic) wave, which is of highest magnitude in case ν_a is equal to the Brillouin shift of the fiber, ν_B. The acoustic wave, in turn, couples a fraction of the incident pump wave into the probe, leading to the amplification of the latter. The stronger the acoustic wave, the larger the transfer of power between the two optical waves.

The value of ν_B in standard fibers and at the telecommunication optical wavelength range is about 11 GHz, but varies due to temperature or strain ($\Delta\nu_B/\Delta\epsilon = 0.05\ MHz/\mu\epsilon$, $\Delta\nu_B/\Delta T = 1\ MHz\ /°C$). By sweeping the frequency of either the pump or the probe wave an amplification (or gain) spectrum can be obtained, characterized by a peak value when ν_a is equal to ν_B. Since both temperature and strain affect the precise value of ν_B, evaluation of the changes in ν_B can infer the temperature or strain along the fiber. Such an interrogation is underlying Brillouin Optical Time Domain Analysis (BOTDA).

A comprehensive state of the art review of various interrogation techniques of Brillouin distributed sensing can be found in Motil et al. (2016). The following are a few innovative improvements, originating in Israel.

2.1.1 *Resolution advancements: random access distributed sensing*

Classical BOTDA is limited to a spatial resolution of roughly 1 m, due to constraints on the effective pulse width, T. Much higher resolution may be obtained with Brillion optical correlation domain analysis (B-OCDA) schemes, as proposed initially by Hotate & Hasegawa (2000). B-OCDA relies on the close relation between the magnitude of the stimulated acoustic field at a given location and the temporal cross-correlation between the pump and probe waves at that point. Judicious modulation of the two waves may restrict their correlation to discrete and narrow peaks, where SBS may build up. The interaction is strongly suppressed everywhere else.

The group of A. Zadok, at Bar Ilan University and coworkers, has adopted concepts from radar technology, and utilized them to establish a new paradigm of high-resolution B-OCDA. By using high rate, pseudo-random, phase coding of both the Brillouin pump and the probe waves, they were able to restrict the correlation between pump and probe to narrow peaks with arbitrary separation (Zadok et al. 2012). Fig. 1 shows the principle of random access distributed sensing. Fig. 1a illustrates the binary phase modulated Brillouin probe wave complex envelope, propagating in the positive z direction (top row). The sign of the optical field randomly alternates in between symbols through binary phase modulation. The bottom row of Fig. 1a illustrates SBS pump wave complex envelope, co-modulated by the same binary phase sequence, and propagating in the opposite direction. Fig. 1b demonstrates how the product between the pump envelope and the complex conjugate of the signal envelope generate a constant driving force, which prevails at discrete peak locations only (center), in which the two replicas of the modulation sequence are in correlation. Elsewhere, the driving force is oscillating about zero. Fig. 1c shows the magnitude of the resulting acoustic field, obtained by temporal integration over the driving force.

This technique allows for considerable improvement in the spatial resolution, to the level of 0.01 m. However, it is time consuming, and somewhat impractical for long fibers, as the coding needs to be repeated for each point of interest. Recently (London et al. 2014; Antman et al. 2014), Zadok's group overcame this time limitation by introducing a coded amplitude sequence, achieving a practical 0.02 m resolution over distance of 8.8 km and addressing all 440,000 resolution points (London et al. 2015).

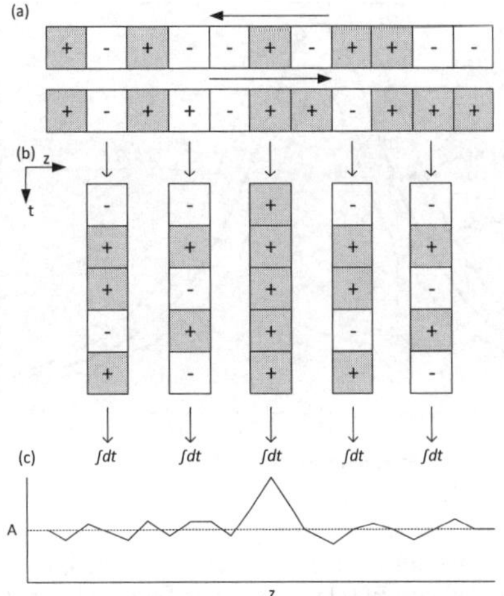

Figure 1. Principles of random access distributed sensing. (a) phase modulation of the probe and pump waves; (b) correlation between the coded waves; and (c) the induced acoustic field, allowing for high resolution Brillouin sensing.

2.1.2 *Dynamic sensing: slope assisted BODTA and Fast BOTDA*

Slope assisted BOTDA, suggested by the M. Tur's electro-optical group at Tel-Aviv University (Peled et al. 2011), allows for Brillouin based dynamic fiber optic sensing. Rather than evaluating the complete Brillouin gain spectrum (BGS), a single point (i.e. frequency) positioned along the linear, rising or falling, section can be interrogated, thus allowing conversion of amplitude changes to Brillouin frequency changes, as seen in Fig. 2.

The unique feature of Slope assisted BOTDA is that it allows evaluation of Brillouin frequency dynamic changes for an arbitrary initial Brillouin frequency shift (i.e. for a non-uniformed initially strained fiber). This is achieved by using a variable optical frequency probe wave, whose time evolution is tailored to such that wherever it meets the counter-propagating pump pulse, their frequency difference sits on the middle of the linear part of the local BGS. The Tel-Aviv group were able to demonstrate measurements of strain wave traveling at the speed of 4000 m/s (Peled et al. 2013). The approach can be extended to double (rising and falling) frequency points, to avoid drifting that might occur in the BGS evaluation (Motil et al. 2014).

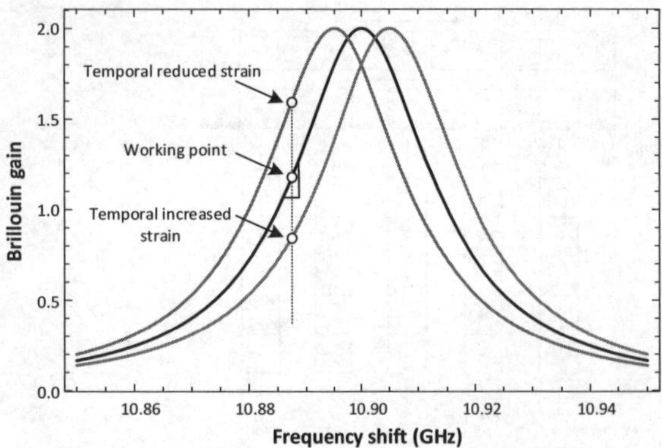

Figure 2. Slope assisted Brillouin sensing. A working point, positioned in the center of the linear raising section of the BGS, is selected based on a preliminary BOTDA scanning. Temporal changes in strain shift the BGS left of right. The amount of shift can be evaluated based on the gain change at the working point (consider a linear relation based on the slope).

Another promising approach for dynamic sensing was recently proposed by the Tel-Aviv group (Peled et al. 2012) and named Fast-BOTDA. In this approach an Arbitrary Waveform Generator (AWG) replaces the commonly used microwave synthesizer, allowing much faster switching speed in the order of nanoseconds over hundreds of MHz's. The Tel-Aviv group was able to demonstrate a sampling record of 11,300 full-fiber-BGS per second.

2.2 *Current and prospective applications*

Both the resolution and dynamic sensing improvements open new horizons for civil engineering applications. One clear horizon for high resolution distributed sensing is its application on a smaller scale for material characterization purposes. First steps towards utilization of high resolution distributed sensing for material characterization have been made at the Technion. It was demonstrated (Uchida et al. 2015) that a conjuncture helical envelope configuration of a single optical fiber can be used to evaluate the surface strain field, in both small and high strain level, of a specimen subjected to uniaxial loading. Uchida et al. (2015) validated the approach on cylindrical specimen of acrylic glass. Klar et al. (2016) explored the possibility of utilizing high resolution fiber optic distributed sensing for in situ geotechnical estimation of soil shear modulus distribution with depth. They demonstrated that a recursive analysis of an elastic problem together with a measured vertical strain can assist in evaluating the sought stiffness values, capturing variation along thin layers quite effectively. In these works the coherent Rayleigh backscatter analysis (Lu et al. 2010), another promising high resolution technique, was utilized for performing measurements.

Another, less intuitive horizon, arises from the high resolution sensing ability to evaluate small pattern changes due to stress and strain redistribution. That is, most of the distributed fibre optic applications in civil engineering focus on the strain levels and the shape of the profile itself, but disregard the fact that in statically indeterminate systems redistribution of stresses (and hence strains) occurs due to changes of stiffness (i.e. damage, plasticity, etc.). These small pattern changes can well be evaluated using high resolution sensing, and be a valuable input for a model-guided solution. In general terms, a model-guided analysis employs engineering models to perform "intelligent" signal processing of sensor measurements, extracting only the relevant information for the sought application. For example, Klar et al. (2014) demonstrated how the rate of change of peak strain location with tunnel advancement can reveal the value settlement trough length parameter without knowing the volume loss value.

One appealing prospective application, based on the ability of high resolution sensing to evaluate small pattern changes, is the evaluation of consolidation processes by monitoring the strain profile with depth. If one considers isochrones of a parabolic shape as an approximation for the 1-dimensional consolidation problem, he may well be able to track within time the consolidation front (supposedly be positioned at depth of $\sqrt{12c_v t}$, where c_v is the coefficient consolidation and t is time). So doing would enable an efficient determination of the degree of consolidation, as well as verification of the parameters involved in the design. Clearly, the parabolic isochrones approximation may not be sufficient, and specific, alternatively, characterizing points (potentially based on the exact Fourier series solution) may need to be established and tracked. Moreover, challenges related to fiber installation and the large strain magnitude involved in consolidation will have to be confronted. Nonetheless, this is a very appealing application that may be answered by a model-guided high resolution distributed sensing solution.

Pattern changes are not limited to static evaluation, and dynamic pattern changes may be a key component for the development of new exciting applications. For example, Eyov et al. (2013) have identified a new water-ground interface wave which develops at the sea bottom and travels roughly at the speed of sound. Apart from being a possible (and important) explanation to the manner in which energy is transferred from the sea to the shore, causing microseism, the special characteristics of the wave may be used to develop an early detection system for approaching tsunami waves (which induce this faster progressive interface wave). Dynamic distributed fibre optic sensing is an ideal candidate for the development of a suitable model-guided early warning system, as it will allow unique identification of this wave as it travels. Dynamic distributed strain sensing could also revolutionise the manner in which dynamic pile load testing is performed, revealing not only the total capacity of the pile, but how it is attributed to the soil layers along the pile.

With respect to condition assessment of bridges and buildings, dynamic sensing could potentially allow for full modal analysis, which is unprecedented. These are some of the prospective applications that may emerge from the above-mentioned innovations and improvements of distributed fiber optic sensing.

3 BURIED WIRELESS ACCELEROMETER NETWORKS

Displacements, similar to strain, also serve as fundamental component of civil infrastructure design. Observing and capturing the movement of 'points' within a system resulting from external loading can be extremely beneficial for all types of engineering analyses. The main technological obstacle in this connection is the need for an external reference that is unaffected by the loading, and is stable enough to enable the detection of minute displacements of the order of microns that are either fast occurring or slowly generated.

Inertial sensors (accelerometers) monitor velocity rates at their points of embedment, and can therefore offer information about the displacement field without the need for an external reference. Certain accelerometers can also detect the gravitational pull and hence provide self-inclination data. In recent years accelerometer technology had progressed significantly, offering superior measurement sensitivity, smaller physical sizes, broader range of operating temperatures, higher shock survivability, and improved power efficiency; at the same time, end-user unit costs continued to drop. Such an impressive set of attributes makes accelerometers ideal candidates for smart infrastructure applications.

3.1 Applications for wireless buried accelerometers

A myriad of engineering fields involve traveling loads with a wide array of characteristics. An obvious example is the transportation arena, but other arenas can come to mind such as agriculture/farming, border security and intrusion detection, and crowd management. In these fields, knowledge of load attributes is needed for usability analysis and management. The sought information in all abovementioned cases can be potentially provided with embedded clusters of wireless autonomous inertial sensors, and with proper interpretation of the load-generated acceleration traces.

The load identification problem has been researched for bridge structures, with a few applications in the geotechnical arena. Analyses were mainly founded on studying the high-frequency content in the measured vibrations. Such interpretation approach is deemed unsuitable for a smart infrastructure. By targeting fast occurring mechanical responses, such as propagating stress waves or acoustic signatures, extreme demands are placed on the measurement system, entailing high sampling rates, large data storage, and ultra-accurate synchronization levels among the sensors. These demands, collectively, prevent wireless technology, autonomous power sources, and deployment over wide areas.

Consequently, the interpretation approach advocated in recent years at the Technion was to focus the analysis on slow-occurring quasi-static information contained in the acceleration traces (Levenberg 2012; 2014; 2016; Levenberg et al. 2014). So doing sets the stage for wireless implementation because it considerably lessens the demands for power, storage, and signal acquisition. In general terms, the high frequency data contained in the raw acceleration traces is removed by means of smoothing, with the smoothing parameters determined via mechanical modeling of the problem being addressed (i.e. model-guided signal analysis).

It is worthy to note that, in principle, the load identification problem can be 'inverted' so that the sensor readings may be utilized for condition assessment. In this situation, the external loading is given/known and the sought information is the mechanical characteristics of the structure.

3.2 *Technological vision*

A vision for buried wireless inertial sensors, suited for smart infrastructure applications, has been exposed in Levenberg et al. (2014). The sensors nodes are provisionally called Wisdom Stones (WSs). Each WS is essentially an autonomous wireless high-end three-axis accelerometer encapsulated in a rock-like casing. The vision calls for the deployment of a dustfull of such stones during construction of new civil infrastructure. Once deployed, each unit would be wirelessly programmed to collect data at predefined dates or under certain conditions, and for predefined durations. The internally stored information could be wirelessly extracted for real-time or offline analysis.

From a technological standpoint, many of the desired WS features are unavailable. The three major limiters are: (i) long lasting, small, and cheap energy source for powering the units; (ii) long-range underground wireless communication that is power-efficient, does not necessitate large antennas, and can support high data transfer rates; and (iii) small-size, low-cost inertial sensors capable of covering a wide response range from micro-g's to several g's.

Figure 3. First generation prototype for a Wisdom Stone.

A first generation WS prototype was recently built at the Technion (see Figure 3). This unit tackled the measurement range issue by combining two different accelerometer technologies. The design and construction of a second generation WS is underway, targeting improvements in sensing abilities as well as wireless networking for synchronization and data transfer.

4 DISCUSSION AND CONCLUSIONS

This paper overviewed recent advancements in distributed fiber optic and wireless sensing carried out in Israel. Collectively, these efforts contribute to the field of embedded sensing on which the smart infrastructure vision in founded. In a broad sense, smart infrastructure is expected to provide two types of information in a reliable and timely manner: system usage and structural condition. Both types are needed for supporting reactive actions as well as long-term measures to improve design, enhance construction quality, and elevate operational efficiency.

The innovations mentioned in the paper can transform traditional infrastructure into smart infrastructure. Their spatial deployment facilitates data acquisition across different scales, from centimeters to kilometers, and allows the development of holistic awareness of infrastructure condition.

It was suggested that model-guided analysis is most suitable for instilling intelligence to data interpretation (as opposed to statistical analyses focused on identifying anomalies without a clear physical foundation). For example, measuring and understanding the patterns in which stresses and strains are redistributed in a statically indeterminate system may be part of the "intelligence" required by a smart infrastructure to identify damage at its early stages of development. It is unlikely that purely statistical methods would provide such insight, as these may not appear as anomalies.

It appears that most sensing developments are not guided by specific engineering needs, but rather pursue a general incentive for improvement. It is especially important that application needs drive and prioritize technological tradeoff decisions (e.g., acquisition rate versus sensing accuracy vs spatial resolution, etc.). At the same time, the civil engineering community focuses on application of recent sensing technology, without envisioning future (non-existing) sensors, thus does not provide a clear map of the sensing needs for smart infrastructures. We believe that expediting a smart infrastructure future demands multi-disciplinary approach in which the engineering needs are involved in (and drive) the development of the sensing techniques. One derivative of this insight is that academic institutions should encourage collaboration of researchers across disciplines, and consider modifying curricula to equip future engineers with a wider array of skills.

ACKNOWLEDGEMENT

The second author would like to acknowledge Netivei Israel - National Transport Infrastructure Company Ltd. for its financial support.

REFERENCES

Antman, Y., Elooz, D., Cohen, R., London, T., & Zadok, A. 2014 Brillouin Time-Domainand Correlation-Domain Analyses Combined, SENSORS, IEEE, 158–161.

Boyd, R.W. 2008 Nonlinear Optics, 3rd ed., Academic press, Waltham.

Eyov, E., Klar, A., Kadri, U., & Stiassnie, M. 2013. Progressive waves in a compressible-ocean with an elastic bottom. *Wave Motion*, **50**, 929–939.

Glisic, B. & Inaudi, D. 2012. Development of method for in-service crack detection based on distributed fiber optic sensors. *Structural Health Monitoring*, 11, 161–171.

Goldfeld, Y. & Klar, A. 2012. Damage Identification in Reinforced Concrete Beams Using Spatially Distributed Strain Measurements. *Journal of Structural Engineering*, 04013013.

Hauswirth, D., Puzrin, A. M., Carrera, A., Standing, J. R. & Wan, M. S. P. 2014. Use of fibre-optic sensors for simple assessment of ground surface displacements during tunneling. *Geotechnique*, 64(10), 837–842.

Hotate, K., and Hasegawa, T. 2000. Measurement of Brillouin gain spectrum distribution along an optical fiber using a correlation-based technique-proposal, experiment and simulation. IEICE T. Electron. E83-C(3), 405–412.

Iten, M., & Puzrin, A. M. 2009. BOTDA road-embedded strain sensing system for landslide boundary localization. In Proc. SPIE 7293, Smart Sensor Phenomena, Technology, Networks, and Systems 2009 (pp. 729312–729316).

Klar, A. Uchida, S., & Levenberg, E. 2016. In situ profiling of soil stiffness parameters using high resolution fiber optic distributed sensing. *ASCE Journal of Geotechnical and Geoenvironmental Engineering*. in print.

Klar, A., Bennett, P. J., Soga, K., Mair, R. J., Tester, P., Fernie, R., St John, H.D. & Torp-Peterson, G. 2006. Distributed strain measurement for pile foundations. *Proceedings of the Institution of Civil Engineers-Geotechnical Engineering*, 159(3), 135–144.

Klar, A., Dromy, I., & Linker, R. 2014 Monitoring tunneling induced ground displacements using distributed fiber-optic sensing. *Tunnelling and Underground Space Technology*, **40**, 141–150.

Levenberg, E. 2012. Inferring pavement properties using an embedded accelerometer. *International Journal of Transportation Science and Technology*, **1(3)**, 229–246.

Levenberg, E. 2014. Estimating vehicle speed with embedded inertial sensors. *Transportation Research Part C*, **46**, 300–308.

Levenberg, E. 2016. Backcalculation with an implanted inertial sensor. *Transportation Research Record: Journal of the Transportation Research Board*, **2525**, 3–12.

Levenberg, E., Shmuel, I., Orbach, M., & Mizrachi, B. 2014. Wireless pavement sensors for wide-area instrumentation. Proceedings of the 3rd International Conference on Transportation Infrastructure (ICTI): Sustainability, Eco-efficiency and Conservation in Transportation Infrastructure Asset Management, Losa & Papagiannakis (eds.), CRC Press, 307-319.

London, Y., Antman, Y., Cohen, R., Kimelfeld, N., Levanon, N. & Zadok, A. 2014. High- resolution long-range distributed Brillouin analysis using dual-layer phase and amplitude coding, *Opt. Express*, **22**, 27144.

London, Y., Antman, Y., Levanon, N., and Zadok, A. 2015. Brillouin analysis with 8.8 km range and 2 cm resolution. Paper 96340G, SPIE 9634, 24th International Conference on Optical Fibre Sensors (OFS-24), (Curitiba, Brazil, September 28, 2015).

Lu, Y., Zhu, T., Chen, L., and Bao, X., 2010. Distributed Vibration Sensor Based on Coherent Detection of Phase-OTDR. *Journal of Lightwave Technology* **28**(22), 3243–3249.

Mohamad, H., Bennett, P. J., Soga, K., Mair, R. J., & Bowers, K. 2010 Behaviour of an old masonry tunnel due to tunnelling-induced ground settlement, *Geotechnique*, **60**(12), 927–938.

Mohamad, H., Bennett, P. J., Soga, K., Klar, A., & Pellow, A. 2007. Distributed optical fiber strain sensing in a secant piled wall. In 7th International Symposium on Field Measurements in Geomechanics, ASCE, Boston, MA. September 24-27, 175, 1-12.

Motil, A., Bergman, A. & Tur, M. 2016. State of the art of Brillouin fiber-optic distributed sensing, *Optics and Laser Technology*, **78**(A), 81-103

Motil, A., Danon, O., Peled, Y. & Tur, M. 2014. Pump-Power-Independent Double Slope-Assisted Distributed and Fast Brillouin Fiber-Optic Sensor, IEEE Photonics *Technol. Lett.* **26**, 797–800.

Peled, Y., Motil, A., Yaron, L. & Tur, M. 2011 Slope-assisted fast distributed sensing in optical fibers with arbitrary Brillouin profile, *Optics Express*, **19**, 19845-19854.

Peled, Y., Motil, A., & Tur, M. 2012. Fast Brillouin optical time domain analysis for dynamic sensing, *Opt. Express*, **20**, 8584.

Peled, Y., Yaron, L., Motil, A., Tur, M. 2013. Distributed and dynamic monitoring of 4km/sec waves using a Brillouin fiber optic-strain sensor. Fifth Eur. Work Opt. Fibre Sensors. 879434879434.

Regier, R., & Hoult, N. 2014. Distributed Strain Behavior of a Reinforced Concrete Bridge: Case Study. *Journal of Bridge Engineering*, 05014007

Uchida, S., Levenberg, E., & Klar, A. 2015. On-specimen strain measurement with fiber optic distributed sensing. *Measurement*, **60**, 104–113.

Vorster, T. E. B., Soga, K., Mair, R. J., Bennett, P. J., Klar, A. & Choy, C. K. 2006 The use of fibre optic sensors to monitor pipeline response to tunneling. In ASCE Geocongress 2006, 1-6.

Zadok, A., Antman, Y., Primerov, N., Denisov, A., Sancho, J. & Thevenaz, L. 2012. Random-access distributed fiber sensing. *Laser & Photon. Rev.*, **6**: L1–L5. doi: 10.1002/lpor.201200013.

Proceedings of the International Conference on
Smart Infrastructure and Construction
ISBN 978-0-7277-6127-9

© The authors and ICE Publishing: All rights reserved, 2016
doi:10.1680/tfitsi.61279.297

Damage detection of a reinforced concrete column under simulated pseudo-dynamic loading using smart aggregates

Qingzhao Kong[1], Rachel Howser Robert[2], Pedro Silva[3], Gangbing Song[1,*], and Y.L. Mo[4]

[1] *Department of Mechanical Engineering, University of Houston, USA*
[2] *Energo Engineering, Houston, USA*
[3] *School of Engineering and Applied Science, George Washington University, Washington, USA*
[4] *Department of Civil and Environmental Engineering, University of Houston, Houston, USA*
[*] *Corresponding Author*

ABSTRACT Structural health monitoring of bridge columns in areas of high seismic activity is highly demanded. In this project, piezoceramic-based transducers known as smart aggregates (SA) were employed to detect the damage of a reinforced concrete (RC) bridge column subjected to pseudo-dynamic loading. The SA-based approach has been previously verified for static and dynamic loading but never pseudo-dynamic loading. An active sensing approach was developed to real-time evaluate the health status of the RC column during the loading procedure. The existence of cracks attenuated the stress wave transmission energy during the loading procedure and reduced the amplitudes of the signal received by SA sensors. To detect the crack occurrence and evaluate the damage severity, a wavelet packet-based structural damage index was developed. Experimental results show that the values of the damage index increase with the increasing of the cracks in the RC column. In addition to monitoring the general severity of the damage, the local structural damage indices monitored the cyclic crack open-close phenomenon subjected to the pseudo-dynamic loading.

1 INTRODUCTION

Reinforced concrete (RC) columns are often the primary elements of energy dissipation in bridges subjected to seismic loads. Failure of bridge columns can result in the collapse of bridge girders, which was the catastrophic scenario for the majority of bridges damaged in past earthquakes. After a seismic event, it is imperative to quickly assess the health status of an RC structure to provide vital structural safety information for decision makers. This is especially true for large-scale infrastructures where it is desirable to design an automated and distributed system to perform structural health monitoring in real-time or near real-time.

In this project, piezoceramic-based devices called smart aggregates (SA) were used as transducers for structural health monitoring of RC columns under a pseudo-dynamic loading procedure. In addition to the structural health monitoring function, the smart aggregate can be used to perform early-age concrete strength monitoring (Kong et al. 2013), impact detection (Song et al. 2007a), seismic detection (Gu et al. 2010), and soil freeze-thaw monitoring (Kong et al. 2014), proving that SA are multi-functional devices. In previous studies, SAs have been successfully implemented for structural health monitoring of various concrete structures under different loading cases including: a concrete bridge bent-cap under static loading (Song et al. 2007b), a concrete frame under static loading (Laskar et al. 2008; Zhao et al. 2006), a shear wall under reversed cyclic loading (Yan et al. 2009), columns under seismic loading (Liao et al. 2008), and columns under reversed cyclic loading (Moslehy et al. 2010; Hoswer et al. 2011). The purpose of this project was to further verify the effectiveness of the smart aggregate in health monitoring of a RC concrete column tested with pseudo-dynamic loads.

In this paper, SAs were embedded in the RC column and utilized to perform the structural health monitoring. A structural damage index was formed based on wavelet packet-based tools. The general increasing values of the developed structural damage indices for each implemented SAs successfully detect the damage severity during the test. In addition, the cyclic behavior of the local damage indices successfully monitored the crack opening and closing phenomenon subject to the pseudo-dynamic loading.

2 PRINCIPLES

Lead Zirconate Titanite (PZT) is one of the most popular materials among piezoelectric materials. Since PZT material is fragile and easily damaged, a smart aggregate (SA) was designed by embedding a waterproofed PZT patch with lead wires into a concrete block, as shown in Figure 1 (Song et al 2008).

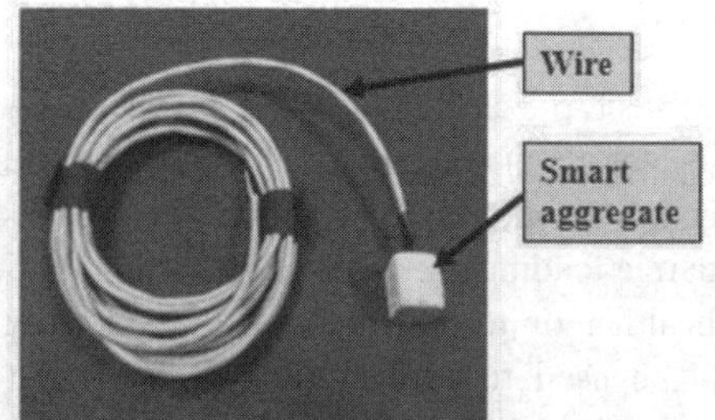

Figure 1. Structure of a smart aggregate

In the active sensing approach, one smart aggregate functions as an actuator to generate guided waves propagating in the concrete structure. Other distributed smart aggregates are used as sensors to detect the wave responses. If a crack appears in the structure, it acts as a stress relief in the wave propagation path. Therefore, the energy of the received signal from corresponding SA sensors will attenuate. The energy attenuation ratio is highly dependent on the severity of the cracks. In order to evaluate the crack severity, the wavelet packet-based structural damage index was adopted in this paper (Song et al 2008). The range of the structural damage index is from 0 to 1, which can be utilized to determine the damage severity of the structure. The value of 0 corresponds the structure in health status; the value of 1 means the structure is under severe damage condition or failure.

3 EXPERIMENTAL SETUP

A RC column instrumented with 10 SAs was fabricated. SAs were mounted with rebar before concrete casting. The dimension of the RC column and the location of SAs are shown in a structural drawing and a three-dimensional view, as shown in Figure 2.

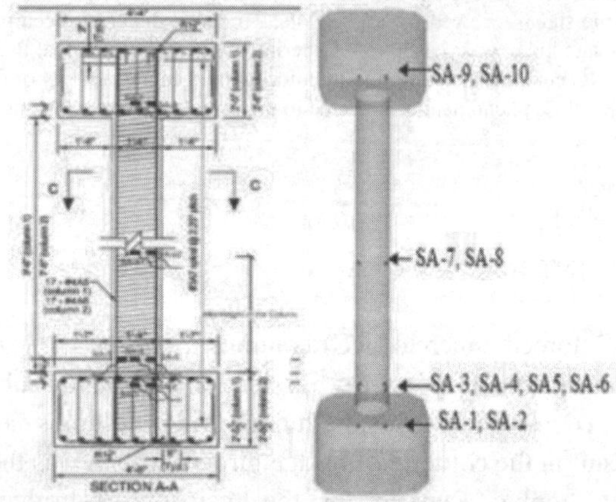

Figure 2. Dimension of the RC column and the location of SAs

A series of 330 kip actuators were placed at the top of the full-scale column to apply load to the specimen. The test concrete column was mounted to a strong floor using all-thread rods. Using the smart aggregate designation shown in **Error! Reference**

source not found., SA-1 was used as the actuator to generate repeated swept sine wave through a function generator. The amplitude, period, and the frequency range of the swept sine wave are 10V, 3s, and 100 Hz-10 kHz, respectively. SA-3 through SA-10 were used as sensors to detect the signal. The sensing signal was recorded by a NI-6353 data acquisition system. The sampling frequency of the data acquisition system is 100kS/s per channel.

4 LOADING PROTOCOLS

During the test, the concrete column was assumed to be a column in a curved bridge subjected to a simulated pseudo-dynamic earthquake record. After each step of the record, the stiffness of the column was calculated and the next step of the earthquake record was adjusted using a finite element analysis to account for any decrease in stiffness in the system. The magnitude of the acceleration was increased every 10 seconds. The acceleration record can be found in Figure 3.

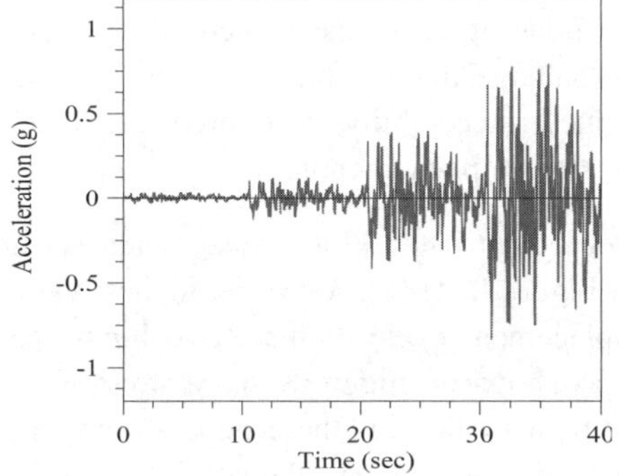

Figure 3. Loading history

5 EXPERIMENTAL RESULTS

5.1 Loading history

The test was stopped due to the actuators in the x-direction running out of stroke. At that point, the concrete column had yielded and was exhibiting spalling, but had not yet failed. Figure 4 shows the force-displacement and moment-curvature relationships in the x-, y- and z-directions for the concrete column, respectively. A photo of the yielded concrete column after test is shown in Figure 5.

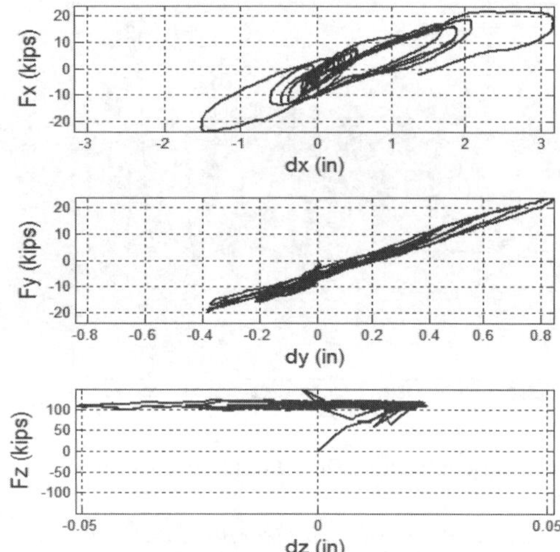

Figure 4. Force-displacement and moment-curvature relationships in the x-, y- and z-directions

Figure 5. A photo of the concrete column yielding after test

As shown in Figure 4, the x-direction had the largest displacement in the test. The displacement history in the x-direction at the top of the concrete column is shown in Figure 6. During the 38-hour test, 13 cycles of displacement in the x-direction were observed.

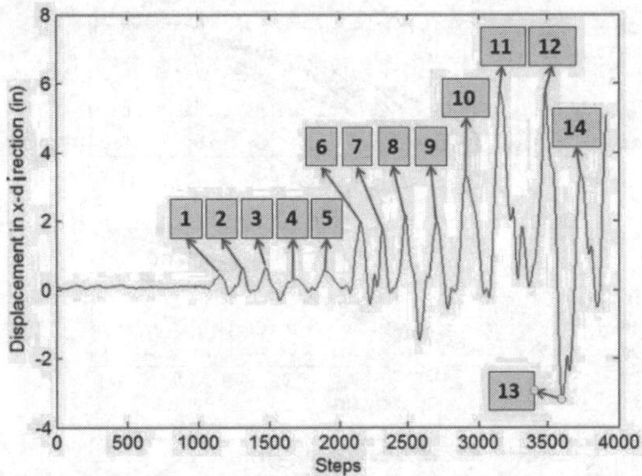

Figure 6. Displacement history in x-direction of the column

5.2 Wavelet packet-based structural damage indices

During the test, SA-6 and SA-9 failed to function. The possible reason for the dead sensors may be the wire broke during casting. Experimental results from the other sensors are shown below. Figure 7 shows the wave packet-based damage indices of all the functional SA sensors during the test. The damage indices of each SA sensor present generally increasing values, which correspond to the damage development of the RC column. Before step 1000, values of the damage indices are close to 0. As the load increased after step 1000, the values of damage indices increase. After step 3500, damage indices of all SA sensors report values close to 1, which means the RC column sustained a high level of damage. The results shown in the damage indices are in good agreement with the test record.

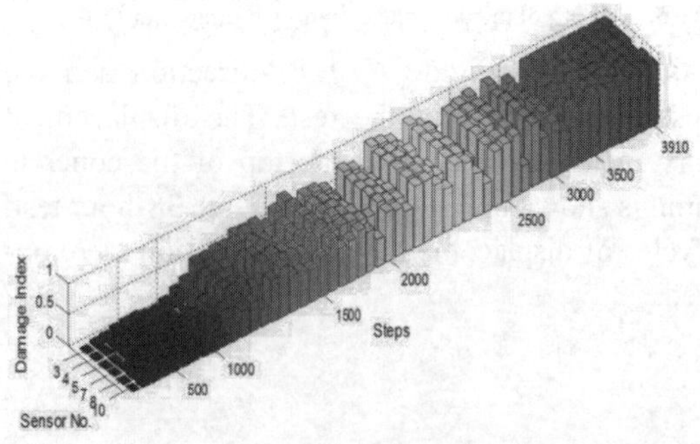

Figure 7. General damage indices throughout the test

Other than the general increasing trend shown in the damage indices, several local decreasing trends can also be found in the damage indices. Since the RC column was subjected to cyclic loading, the cracks in the RC column opened and closed correspondingly. The local increasing and decreasing values shown in the damage indices detected this opening and closing of the cracks during the test.

In the first 1000 steps, the displacement of the RC column was very small. Since no damage occurred during this time, the received signal of SAs did not change much. It can be considered that the RC column was in good health during the first 1000 steps.

From Step 1100 to Step 2100, a total of 5 extracted local damage indices, as shown in Figure 8 (a)-(e), exactly match the 5 displacement cycles. During each cycle shown in the displacement history, the RC column was pushed to one side and then pushed back. When load increased, cracks opened corresponding to damage increase. When load decreased, cracks closed corresponding to damage decrease. In each cycle of the damage indices, the increasing and decreasing values successfully monitored the crack opening and closing phenomenon.

From Step 2100 to Step 3300, damage indices can be found in Figure 8 (f)-(k). Since the highest value in each displacement cycle of the RC column was quite large, as shown in Figure 8, more stress wave energy was attenuated during these steps. Comparing to the previous damage indices, the latter damage indices present more clear cycles and higher damage levels. The highest damage level nearly reached 1.

After step 3300, the RC column almost reached the ultimate capability. Most of the stress wave energy was attenuated. SA sensors "lost contact" with the actuator. The local damage indices, as shown in Figure 8 (l), are not presenting the similar trends of increasing and decreasing values corresponding to displacement cycles as before. As the load cycled, the values of damage indices of all the SA sensors maintained a constant value near 1.

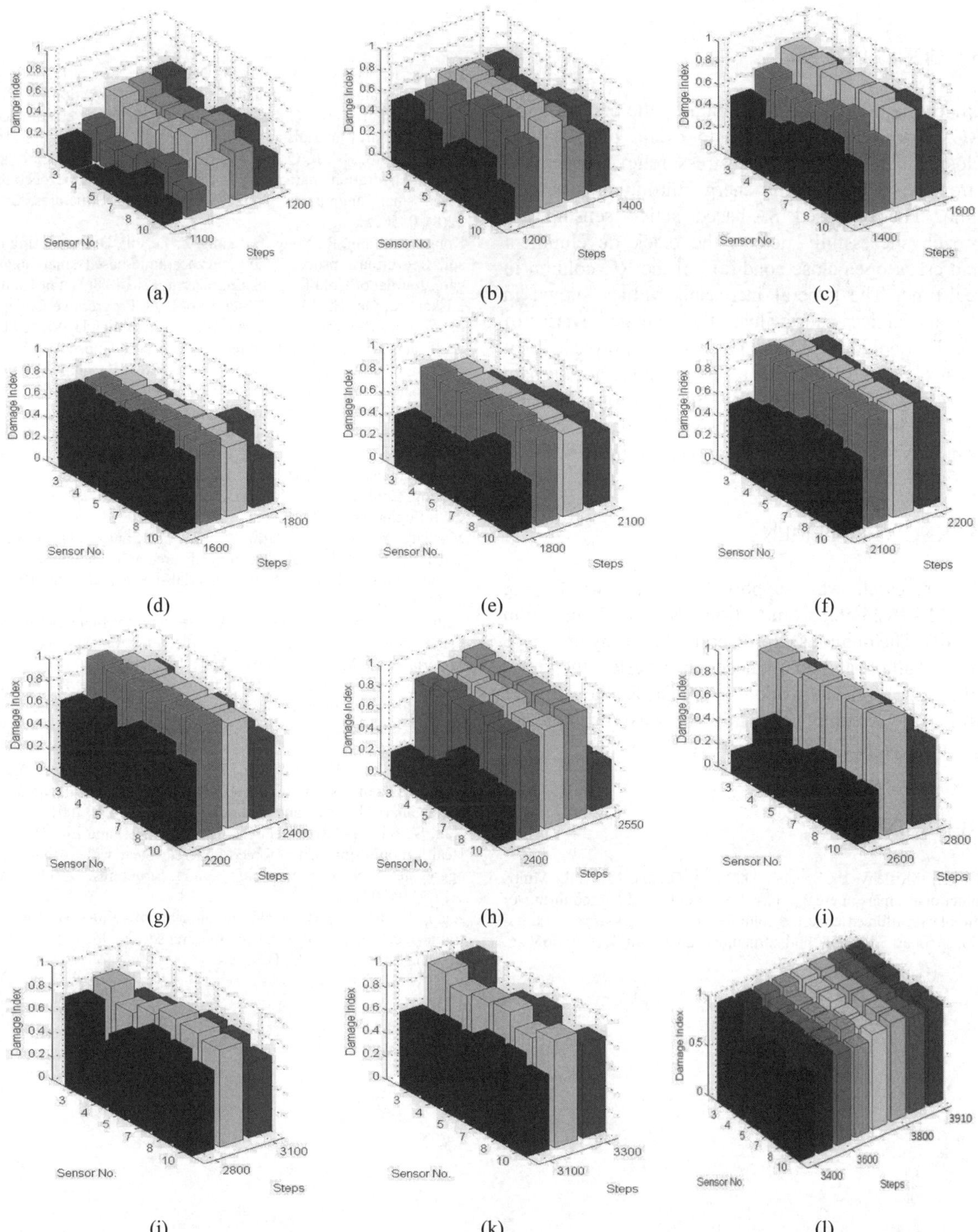

Figure 8. Local damage indices: (a) Cycle 1 (Step 1000-1200) (b) Cycle 2 (Step 1200-1400) (c) Cycle 3 (Step 1400-1600) (d) Cycle 4 (Step 1600-1800) (e) Cycle 5 (Step 1800-2100) (f) Cycle 6 (Step 2100-2200) (g) Cycle 7 (Step 2200-2400) (h) Cycle 8 (Step 2400-2550) (i) Cycle 9 (Step 2600-2800) (j) Cycle 10 (Step 2800-3100) (k) Cycle 11 (Step 3100-3300) (l) Cycle 12 (Step 3400-3910)

6 CONCLUSIONS

Under the pseudo-dynamic loading, the cracks in the RC column were subjected to cyclic opening and closing. Cracks, acting as stress relief, reduce the stress wave propagation energy throughout the column. The proposed SA-based active sensing approach successfully detects the crack development and crack open-close condition of the RC column in real-time. The general increasing values shown in damage indices can evaluate the damage severity of the RC column. When the values approach 1, it means that the RC column has reached the ultimate capability. The phenomenon of crack opening and closing during the cycles of pseudo-dynamic loading is clearly presented in the local damage indices.

ACKNOWLEDGEMENT

The research was supported by an award (No. CMMI- 0724190) from National Science Foundation (NSF). The opinions expressed in this study are those of the authors and do not necessarily reflect the views of the sponsor. The work done by Dr. Rachel Howser Roberts was completed while she was finishing her graduate degree at the University of Houston.

REFERENCES

Gu. H., Moslehy, Y., Sanders, D., Song, G., and Mo, Y.L. Multifunctional smart aggregate-based structural health monitoring of circular reinforced concrete columns subjected to seismic excitations. Smart Materials and Structures. 2010 Jun 1;19(6):065026.

Howser, R., Moslehy, Y., Gu, H., Dhonde, H., Mo, Y.L., and Ayoub, A., et al. Smart-aggregate-based damage detection of fiber-reinforced-polymer-strengthened columns under reversed cyclic loading. Smart Materials and Structures. 2011 Jul 1;20(7):075014.

Kong, Q., Hou, S., Ji, Q., Mo, Y. L., and Song, G. Very early age concrete hydration characterization monitoring using piezoceramic based smart aggregates. Smart Materials and Structures, 2013 22(8), 085025.

Kong, Q., Wang, R., Song, G., Yang, Z. J., Still, B. Monitoring the soil freeze-thaw process using piezoceramic-based smart aggregate. Journal of Cold Regions Engineering, 2014 28(2), 06014001.

Laskar, A., Gu, H., Mo, Y.L., and Song, G. Progressive Collapse of a 2-story Reinforced Concrete Frame. In: Binienda WK, editor. Earth and Space 2008: Engineering, Science, Construction, and Operations in Challenging Environments. Long Beach, CA: American Society of Civil Engineers; 2008. p. 1–9.

Liao, W.I., Gu, H., Olmi, C., Song, G., Mo, Y.L., and Loh, C.H. Structural Health Monitoring of a Concrete Column Subjected to Shake Table Excitations Using Smart Aggregates. Earth and Space 2008: Engineering, Science, Construction, and Operations in Challenging Environments. Long Beach, CA: American Society of Civil Engineers; 2008. p. 1–8.

Moslehy, Y., Gu, H., Belarbi, A., Mo, Y.L., and Song, G. Smart aggregate based damage detection of circular RC columns under cyclic combined loading. Smart Materials and Structures. 2010 Jun 1;19(6):065021.

Song, G., Olmi, C., and Gu, H. An overheight vehicle–bridge collision monitoring system using piezoelectric transducers. Smart Materials and Structures. 2007a Apr 1;16(2):462–8.

Song, G., Gu, H., Mo, Y.L, Hsu, T.T.C., and Dhonde, H. Concrete structural health monitoring using embedded piezoceramic transducers. Smart Materials and Structures. 2007b Aug 1;16(4):959–68.

Song, G., Gu, H., and Mo, Y. L. Smart aggregates: multi-functional sensors for concrete structures—a tutorial and a review. Smart Materials and Structures, 2008 17(3), 033001.

Yan, S., Sun, W., Song, G., Gu, H., Huo, L-S., and Liu, B., et al. Health monitoring of reinforced concrete shear walls using smart aggregates. Smart Materials and Structures. 2009 Apr 1;18(4):047001.

Zhao, X., and Li, H. Health monitoring of reinforced concrete frame-shear wall using piezoceramic transducer. Journal of Vibration and Shock. 2006;25(4):82

Proceedings of the International Conference on Smart Infrastructure and Construction
ISBN 978-0-7277-6127-9

© The authors and ICE Publishing: All rights reserved, 2016
doi:10.1680/tfitsi.61279.303

Detection of tensile force loss in a pre-stressing strand using coil impedance measurement

Jun Lee, Jimin Kim, Hoon Sohn*

Dept. of Civil and Environmental Engineering, KAIST, Daejeon, Korea
**Corresponding Author*

ABSTRACT In this paper, a new monitoring technique to detect tensile force loss of a pre-stressing strand is proposed using coil impedance measurement. First, a specially designed coil is installed at the outer surface of a wedge during construction stage. As a pre-stressing strand is gradually tensioned, the wedge is bounded between the strand and a barrel, and sequentially compressed. Magnetic permeability of the wedge is altered due to the magnetoelastic effect, and it induces impedance alteration of the coil. The advantages of the proposed technique are (1) simple installation, (2) low cost, (3) low power consumption and (4) high durability. The performance of the proposed technique is validated numerically and experimentally. Three dimensional model of a pre-stressing strand, a wedge and a barrel are developed, and variation of stress distribution is observed using commercial numerical simulation program, ABAQUS. Then, laboratory experiments are carried out to evaluate tensile force loss of a 3.3 m long and Φ 15.2 mm pre-stressing strand. The pre-stressing strand is tensioned by a universal tensile machine up to 180 kN, and decreased to 30 kN with 30 kN force resolution. Coil impedance is measured using an impedance analyzer, Agilent 4294A, for every force step. A monotonic relationship between the tensile force and the proposed damage index is successfully observed.

1 INTRODUCTION

Lately, prestressed concrete (PSC) bridges have been constructed widely as an alternative for reinforced concrete bridges since 1980s. Advent of pre-stressing method enables to overcome shortcomings of previous reinforced concrete bridges such as limited span length and cracking under service conditions. Due to its structural characteristics, the performance of a PSC bridge depends largely on the state of the pre-stressing strands (K.L. Rens, 1997 and G.A. Washer 2002).

In order to effectively inspect the tensile force of a pre-stressing strand, various techniques have been proposed. However, some disadvantages of the pervious techniques are (1) excessive power consumption, (2) troublesome installation on extremely narrow inner space of a structure, and (3) easily occurred sensor damage. For example, electromagnetic sensor requires excessive power to sufficiently magnetize a target pre-stressing strand so that satisfactory performance can be assured (Wang 2006). An ultrasonic technique requires careful attachment of piezo-electric sensors generating ultrasonic wave on the surface of a strand, which is vulnerable to a damage during construction procedures by inevitable contact with adjacent strands (Bartoli 2011). In recent, a specially designed pre-stressing strand is developed which contains an optical fiber with several fiber brag grating sensors at a core wire of the strand (Kim 2012). On the other hand, a tensile force inspection technique using eddy current sensor (ECS) was proposed so that the required power consumption and costs can be significantly reduced (Schoenekess 2007). Nevertheless, the location of ECS installation has been a bottleneck to be employed in in-situ environments. ECS should be closely contact on outer surface of a target pre-stressing strand to directly measure the electromagnetic variation of the strand.

In this study, a new monitoring technique through impedance alteration of coil is proposed to inspect the tensile force relaxation of a pre-stressing strand. The advantages of the proposed technique are (1) simple installation, (2) low cost, (3) low power consumption and (4) less vulnerability to be damaged by external component. The proposed technique detects variation of the magnetic fields of stressed wedge due to magnetoelastic effect using impedance alteration of the coil. The purposes of this study are as follows: (1) numerical and experimental investigation of the relationship between a tensile force and a magnetostriction on a wedge surface and, (2) development of coil impedance based tensile force monitoring (CI-TFM) system and its performance validation through a lab-scale experiment.

Section 2 introduces a magnetostriction for stress evaluation and proposes a tensile force monitoring based on coil impedance with a detailed explanation of its working principles. The experimental validations of CI-TFM system are provided in Section 3. Brief summary and future works are mentioned in Section 4.

2 WORKING PRINCIPLE

2.1 *Introduction of Magnetostriction*

Magnetostriction is defined as the deformation of any substance due to the presence of magnetic fields. It occurs in ferromagnetic materials including iron, cobalt, and nickel. Consider a magnetostrictive direction in the absence of any magnetic field. When a magnetic field is applied, the direction elongates into an ellipsoid with the symmetry axis along the direction of the applied magnetic field. This process is called Joule magnetostriction (direct effect). There is a limit to this induced strain, which is known as the saturation magnetostriction. The thought experiment above demonstrates the direct effect of magnetostriction. In the inverse effect, which is called magnetoelastic effect, mechanical stresses on a body affect its magnetic state. (J.A. HOMMEMA, 1997)

ECS is used to measure the changed magnetic field by mechanical stress, as shown in Figure 1. Under a given uni-axial stress σ' (green arrow), the flux density B for a given magnetizing field strength H is according to Le Chatelier's principle (Salach, J., 2010):

$$\left(\frac{d\lambda}{dH}\right)_{\sigma} = \left(\frac{dB}{d\sigma}\right)_{H} \tag{1}$$

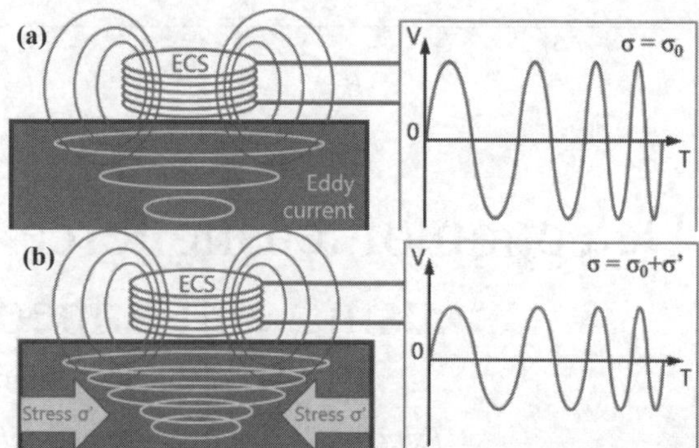

Figure 1. Basic configuration of ECS to evaluate a target structure subjected to different stress condition, (a) $\sigma=\sigma_0$, (b) $\sigma=\sigma_0+\sigma'$

Where, λ stands for a magnetostriction coefficient, σ is an applied stress, H is applied magnetic field strength and B is a stress induced flux density.

When stress alters magnetic flux density B, generated eddy current at material surface is different. Thus, an output voltage V_{out} of ECS is changed by an eddy current. By using this result, Z_c is defined:

$$Z_c(\omega) = \frac{V_{in}(\omega)}{I_{out}(\omega)} = i\omega L \frac{V_{in}(\omega)}{V_{out}(\omega)} \tag{2}$$

Where, $Z_c(\omega)$ is an impedance, $V_{in}(\omega)$ is an input voltage, $I_{out}(\omega)$ is an output current, $V_{out}(\omega)$ is an output voltage, ω is an angular frequency, L is a coil inductance, and i is a complex number.

2.2 *Schematic of Coil impedance based tensile force monitoring system*

In Figure 2, the variation of a pre-stressing strand force alters the stress of a wedge surface. Before tensile force is applied to the pre-stressing strand, the condition between the pre-stressing strand and the wedge can be considered as a simple contact. Then, as tensile force becomes to be applied gradually, the wedge is more and more fitted into an empty space of a barrel, which is a structural member to distribute the applied stress to a concrete structure, as shown in Figure 2 (b). In sequence, compressive force is provided along to a radial direction (green arrow) of the wedge. According to the magnetoelastic effect, magnetic field of the wedge is altered from H_0 to H_1, as tensile force is changed, respectively. Therefore, even though ECS induces magnetic field on the exposed surface, current density on the wedge surface varies with the pre-stressing strand force so that the output voltage is al-

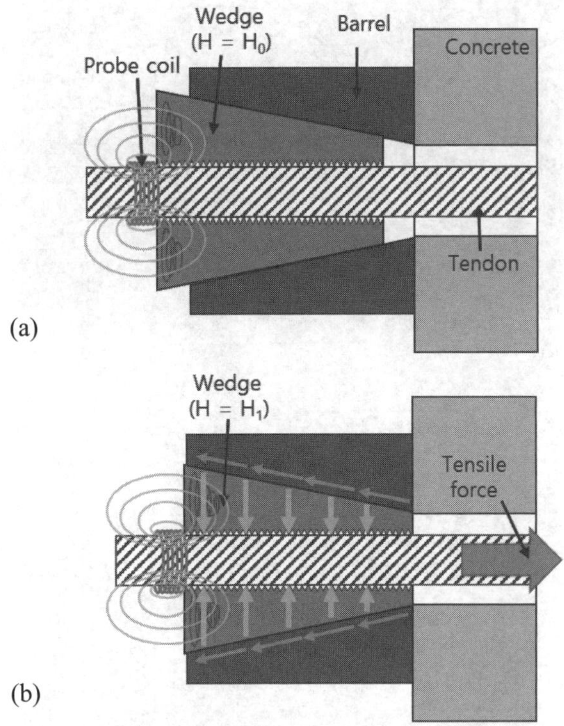

Figure 2. Stress distribution of the surface of the wedge (a) before and (b) after a tensile force is applied to a pre-stressing strand

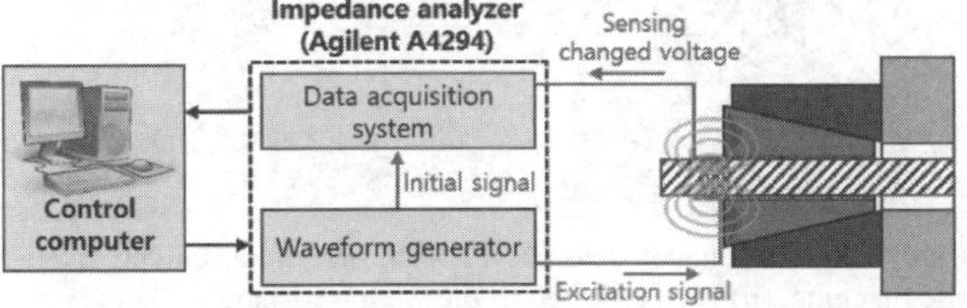

Figure 3. Schematics of the developed CI-TFM system

tered as well. The coil impedance is calculated by input voltage, changed output voltage and reactance of coil.

CI-TFM system is composed of an impedance analyzer, ECS, and a control computer as shown in in Figure 3. The entire process of tensile force monitoring follows the three steps: (1) ECS attached at the wedge surface generate the magnetic fields, (2) measurement of an induced coil impedance signal in ECS by impedance analyzer and (3) estimation of tensile force loss through calculation of the defined damage index (DI). Through these three steps, a strand force loss detection can be successfully carried out by simply attaching ECS on an outer surface of a wedge by measuring coil impedance.

Before calculating DI, the impedance under T_i, i^{th} tensile force, normalize by achieved impedance data at maximum tensile force, T_{max}. DI is defined by root-mean square level:

$$DI_i = \sqrt{\frac{1}{n}\sum_{k=1}^{n}\left|(N_i)_k\right|^2} \quad (3)$$

Where, N_i is the normalized value of the measured real part of impedance, $Re(Z_c)$, under T_i, and n is number of sampling point.

3 VALIDATION OF COIL IMPEDANCE MONITORING SYSEM

3.1 Numerical Analysis to Observe Stress Distribution of a Wedge

A numerical simulation is performed using a finite element analysis software, ABAQUS, to observe stress distribution of a wedge under the pre-stressing.

For simulating an actual pre-stressing strand system, a three dimensional model is developed. A wedge is modeled with three-pieces and a strand modeled a helically wound seven steel strands. The detailed mechanical properties of a barrel, a wedge and a strand such as young's modulus, Poisson's ratio and density are decided as shown in Table 1. The strand's geometric size is set Φ 15.2 mm × 300 mm, the barrel is set Φ 45 mm × 60 mm. A static frictional coefficient between the wedge and the strand is 0.9 and between the wedge and the barrel is 0.1. The end of the strand is forced 180 kN during 20 steps.

As shown in Figure 4 (a), the stress concentration is not occurred until the wedge is fitted to the barrel. When the wedge is fitted to the barrel in Figure 4 (b), the stress concentration at the wedge is changed from the bottom side of wedge and then the wedges are more tightly grabbed the strand. As the strand is more forced, large stress concentration is introduced in surface of the wedge as we expected. When the wedge is exactly fitted to the barrel, the large stress is mainly distributed at the wedge surface, which is connected

Table 1. Mechanical properties of a barrel, a strand, and a wedge used for numerical analysis using ABAQUS

Parameter	Barrel	Strand	Wedge
Young's modulus [GPa]	200	200	10
Poisson's ratio	0.35	0.35	0.33
Density [kg/m3]	7850	7850	7850

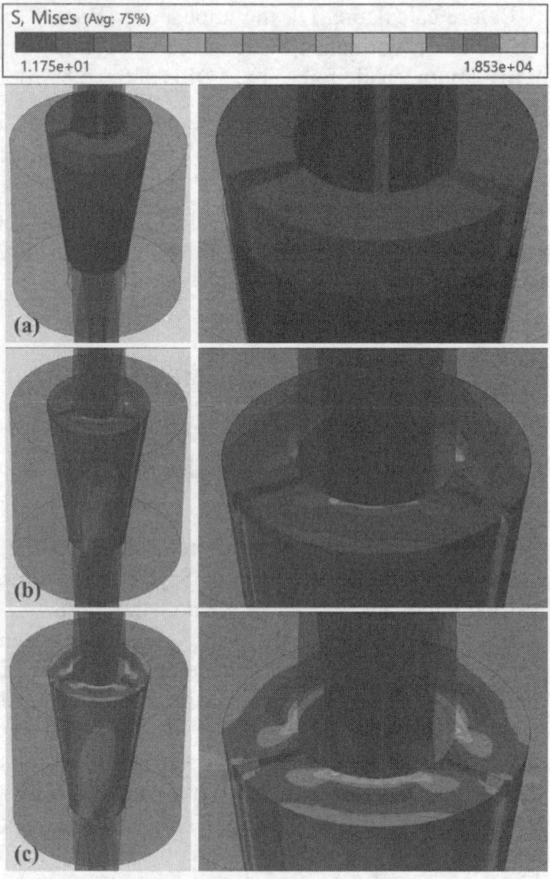

Figure 4. Numerical simulation for stress distribution of a wedge holding a pre-stressing strand using ABAQUS. (a) initial condition (T = 0 kN), (b) after 16th steps (T = 140 kN), (c) after 20th steps end (T = 180 kN)

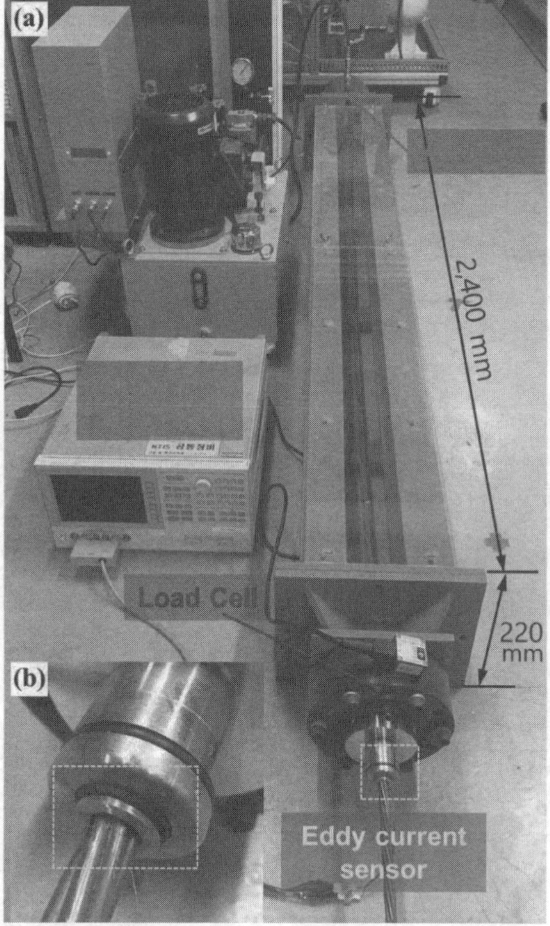

Figure 5. Experimental setup of CI-TFM system (a) universal tensile machine, (b) ECS attached to the wedge surface

with the strand surface, as shown in Figure 4 (c). The numerical simulation results offer a good explanation and possibility of inferring the tensile force of pre-stressing strand to monitoring the stress distribution of wedge surface.

3.2 Laboratory Experiment

Figure 5 (a) shows experimental setup to validate performance of the developed CI-TFM system. Universal tensile machine was manufactured considering a tensile force level for general design of a pre-stressing concrete. The test sequence is to acquire a baseline data under 180kN and then, gradually decreased a tensile force by 30 kN force step until it becomes 30 kN at the end.

In this experiment, the wedge produced by KTA Company is employed and its allowable pre-stressing stress is 1800 MPa.

ECS is designed as coin-shape and manufactured using a copper wire of Φ 0.1 mm in order to mount it through a pre-stressing strand. Inner and outer diameter of ECS are 17.2 mm and 26.5 mm, respectively. ECS continuously stacked-up until its electric resistance becomes 120 Ω respectively. Inductance of coil is 3.88 mH in air condition.

As shown in Figure 5 (b), ECS is attached to the wedge surface at the side of the load cell which is less vibration than hydraulic actuator. For measuring coil impedance, Agilent 4294a is used and sampling is 200 points / 1 MHz, as listed in Table. 2. For verifying the

Table 2. Experimental setup parameters value

Pre-stressing Strand Force Steps	30 kN ~ 180 kN (6 steps)
Strand Type	Φ15.2 mm x 3200 mm
Sensor Type	Single – Coin
Resistance of Sensor	120 Ω
Impedance Analyzer	Agilent 4294A
Frequency	0.4 MHz ~ 1 MHz
Point Number of Impedance	200 points / 1 MHz
Input Energy	0.16 mW · 15 sec ≒ 2.4 mJ
Number of Measuring	3 times (0hr, 4hr, 15hr)

repeatability of CI-TFM system, the experiment is repeated three times at 0 hour, 4 hour, and 15 hour after ECS exactly attached to wedge surface.

3.3 *Experimental Result*

First, coil impedance, Z_c, for each tensile force step is measured. According to the magnetoelastic effect, magnetic field on the surface of wedge, H_i, is altered by tensile force of the pre-stressing strand, which sequentially affects production of Z_c. In CI-TFM system, a $Re(Z_c)$ is more accurate monitoring than $Im(Z_c)$. Figure 6 shows the measured $Re(Z_c)$ from 180 kN to 30 kN, respectively. Because of variation of H_i by alteration of tensile force, the amplitude of $Re(Z_c)$ for each tensile force condition shows clear difference.

Second, the results of verifying the repeatability of CI-TFM system are shown in Figures 7 (a), (b) and (c). Data is normalized by each baseline data under 180 kN. Normalized impedance value under 150 kN and 120kN, when ECS attached after 15 hour as shown in Figure 7 (c), are slightly reversed and the resonance frequency and the maximum amplitude of each tensile test are different at each measuring time. These reason is related a mutual inductance between the wedge surface and ECS. The pieces of Wedge are able to rotate with strand as the center, and then the mutual inductance can be changed by location of the wedge. But a tendency to change tensile force is similar to each result.

For estimating the loss of the tensile force at pre-stressing strand, DI is employed which is previously defined through Eq. (3) in Section 2. Figures 8 (a), (b) and (c) show the calculated DI of each tensile force.

In all three cases, as the tensile force of the pre-stressing strand is gradually relaxed, DI becomes larger. The relation between DI and the tensile force of pre-stressing strand seems to be monotonic. As a result, the proposed CI-TFM system can be used in order to estimate how much force relaxation has been introduced from initial condition.

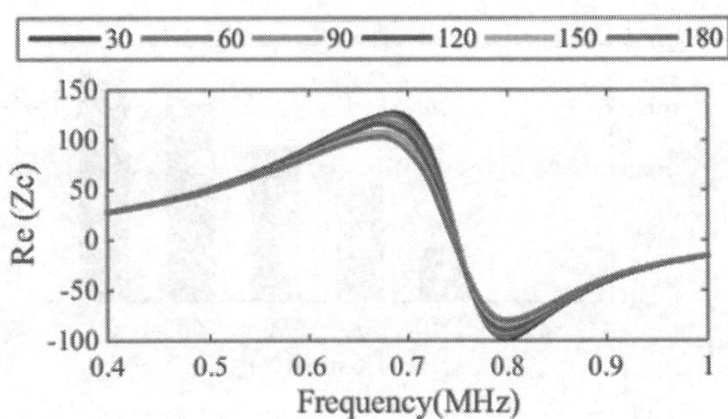

Figure 6. Comparison the measured $Re(Z_c)$ under each tensile steps.

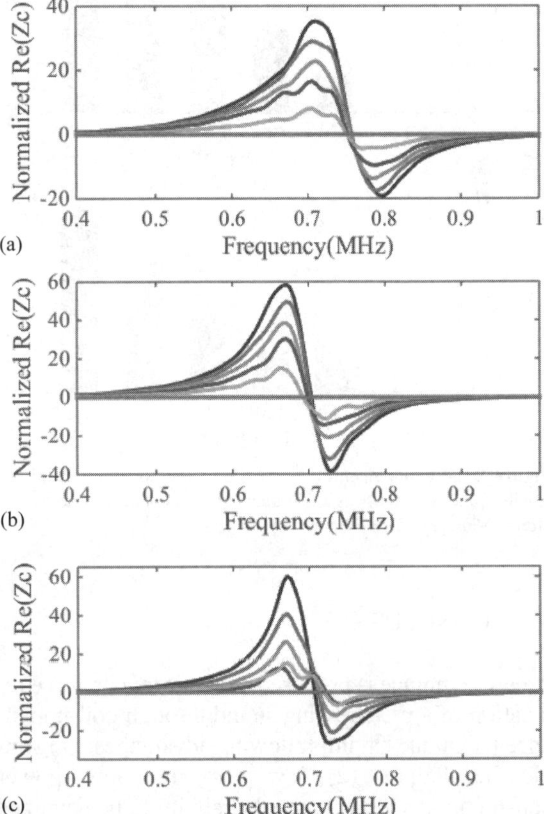

Figure 7. Comparison the normalized data measured $Re(Z_c)$ under each tensile step and time step. (a) after 0 hour, (b) after 4 hour, (c) after 15 hour

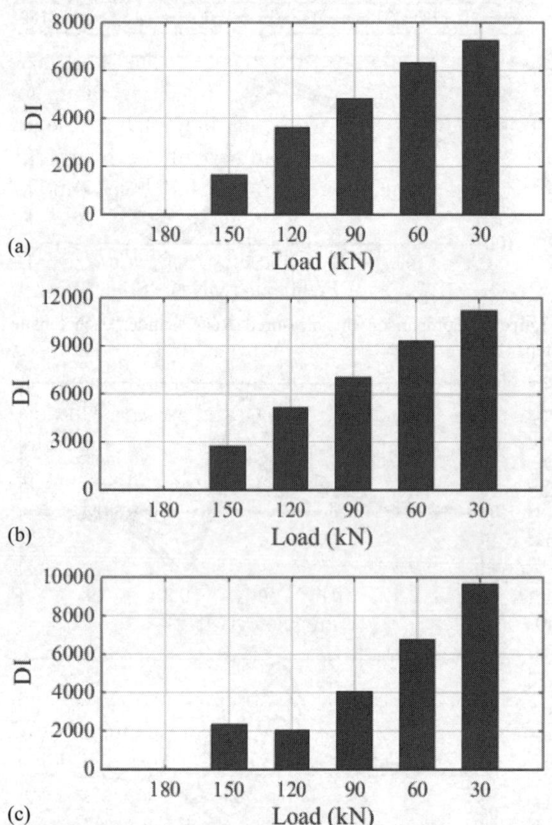

Figure 8. DI comparison under tensile force loss from an initial tensile force of 180 kN as ECS attached time (a) after 0 hour, (b) after 4 hour, (c) after 15 hour

4 CONCLUSION

A new technique is proposed to detect tensile force relaxation of a pre-stressing strand through coil impedance technique having following advantages: (1) simple installation, (2) low cost, (3) low power consumption and (4) less vulnerability to be damaged by external component. In this study, CI-TFM system is developed and its performance is experimentally validated using a 3.3 m long and Φ 15.2 mm pre-stressing strand which is tensioned by a specially designed tensile machine. As an initial condition, the pre-stressing strand was forced up to 180 kN and, ECS which is attached the surface of wedge was acquired which is used as a baseline data. Through calculation of the proposed DI for each tensile force with 30 kN decreasing step, tensile force of the pre-stressing strand was successfully estimated. A monotonic relationship between tensile force and the proposed DI was observed.

There are several issues to be considered for real applications in the future. First, CI-TFM system should be designed in sensor node, so that it is simply attached to a structure and carries out impedance measurement and data processing. In addition, to achieve the exact tensile force data, an improved algorithm of tensile force monitoring is needed.

ACKNOWLEDGEMENT

This study is supported by a grant from Smart Civil Infrastructure Research Program (13SCIPA01) funded by Ministry of Land, Infrastructure and Transport (MOLIT) of Korea government

REFERENCES

Bartoli, 2011, Use of interwire ultrasonic leakage to quantify loss of prestress in multiwire tendons, *Journal of Engineering Mechanics*, Vol. 137, No. 5, 324-333.

J.A. Hommema, 1997, Magnetomechanical Behavior of Terfenol-D Particulate Composites, *University of Illinois*

Jae-Min Kim, 2012, FBG Sensors Encapsulated into 7-Wire Steel Strand for Tension Monitoring of a Prestressing Tendon, *Advances in Structural Engineering*, Vol. 15, No. 6, 907-917.

K.L. Rens, T.J. wipf, & F.W. Klaiber, 1997, Review of nondestructive evaluation techniques of civil infrastructure, *Journal of performance constructed facilities*

Salach, J., Szewczyk, R., Bienkowski, A., & Frydrych, P., 2010, Methodology of testing the magnetoelastic characteristics of ring-shaped cores under uniform compressive and tensile stresses, *Journal of Electrical Engineering*, Volume 61, Issue 7, 93-95.

H.C. Schoenekess, W. Ricken, W.J. Becker, Method to Determine Tensile Stress Alterations in Prestressing Steel Strands by means of an Eddy current Technique, *IEEE Sensors journal*, volume 7, Issue 8, 1200-1205.

M.L. Wang, 2006, Application of Magnetoelastic stress sensors in large steel cables, *Smart Structures and Systems*, Vol. 2, No. 2, 155-169.

G. A. Washer, R. E. Green & R. B. Pond Jr., 2002, Velocity Constants for Ultrasonic Stress Measurement in Prestressing Tendons, *Research in Nondestructive Evaluation*, Volume 14, Issue 2

Proceedings of the International Conference on Smart Infrastructure and Construction
ISBN 978-0-7277-6127-9

© The authors and ICE Publishing: All rights reserved, 2016
doi:10.1680/tfitsi.61279.309

Integrated ROBOTIC solution for tunnel structural evaluation and characterization – ROBO-SPECT EC project

K.Loupos*[1], A.Amditis[1], A.Doulamis[1], P. Chrobocinski[2], J.Victores[3], M.Wietek[4], P.Panetsos[5], A.Roncaglia[6], S.Camarinopoulos[7], V.Kallidromitis[8], D.Bairaktaris[9], N.Komodakis[10] and R.Lopez[11]

[1] *Institute of Communication and Computer Systems, Athens, Greece*
[2] *AIRBUS DS, Elancourt, France*
[3] *Universidad Carlos Iii De Madrid, Madrid, Spain*
[4] *VSH Hagerbach Test Gallery Ltd, Flums, Switzerland*
[5] *Egnatia Odos Ae, Thessaloniki, Greece*
[6] *Consiglio Nazionale Delle Ricerche, Rome, Italy*
[7] *RISA SICHERHEITSANALYSEN GMBH, Berlin, Germany*
[8] *TECNIC SPA, Rome, Italy*
[9] *DBA EPE, Athens, Greece*
[10] *Ecole Nationale Des Ponts Et Chaussees, Paris, France*
[11] *ROBOTNIK AUTOMATION SLL, Valencia, Spain*
* *Corresponding Author*

ABSTRACT Recent developments in robotic systems, automation and computer vision and sensors have well prepared the ground towards automated robotic solutions for inspection of civil infrastructures and particularly tunnels ageing urgently requiring serious inspection and assessment. Nowadays, tunnel inspections are being performed manually and visually by human operators.

ROBO-SPECT is an EC co-funded research project (FP7 - ICT – 611145) that is driven by the tunnel inspection industry and that adapts and integrates recent research results in intelligent control in robotics, computer vision and active continuous learning and sensing, in an innovative, integrated, robotic system. ROBO-SPECT automatically scans the internal surface of tunnels for potential defects (spalling, delamination etc) and inspects and measures radial deformation in the cross-section, distance between parallel cracks, cracks and open joints that impact tunnel stability, with mm accuracies.

The robotic system (currently at a prototyping phase) consists of an intelligent robotic system supported by advanced control systems and an on-board high precision robotic tip, an advanced computer vision system based on high quality and stereo cameras for cracks/defects detection and 3D vision, an ultrasonic sensor able to detect crack width and depth with high accuracy. The system also includes a 3D laser scanner able to perform tunnel deformation measurements also with high accuracy. This publication describes the integration status and recent developments of the semi-autonomous robotic system as designed for concrete lining tunnels. The final system architecture, communication mechanisms and integration steps of the whole robotic system are also being presented.

1 INTRODUCTION

Structural health monitoring (SHM) can be defined as the diagnosis of the state of a structure and/or of its constituent materials and components or even the structure as a whole system. The parameters that may affect the structural integrity of any structure may range into a series of various parameters including ageing, loading, corrosion etc or even accidental actions. The assessment of structural integrity of existing civil structures in general is of primal importance in order to identify and determine its reliability levels on the ability to carry existing and future loads and fulfil its task having in mind human life, financial, maintenance and operational risks (Rucker 2006). The largest challenge over real-time monitoring systems is that all these infrastructures are in general unique. This raises a need for a system able to be adapted to different operational needs and structure types with different monitoring requirements (Loupos 2011-2).

1.1 Structural Health Monitoring Needs in Tunnels

Tunnels are critical type of structures where monitoring proves beneficial towards structural health identification. The SHM of a tunnel is usually focusing on the limits of deformations in terms of structure stability and thus investigating stresses, strains and deflections. Under the framework of tunnel inspection, we can highlight the inspection, assessment, maintenance and safe operation of the existing civil infrastructure (tunnels, bridges, roads, pipelines) of great challenges [ASCE 2009]. Civil infrastructures are significantly deteriorating and many are in need of inspection, damage assessment and repair suffering from ageing, environmental parameters, loading, usage and inadequate maintenance or repairs. The needs that ROBO-SPECT will be replying to are the following (Loupos 2014):

- High cost of new tunnel constructions (need for inspection, assessment and repair of existing);
- Transport demand is highly increasing and cannot cope with the rate of transport infrastructure and high tunnels uptime;
- Inspection and assessment should be speedy in order to minimize tunnel closures or partial closures;
- Engineering hours for tunnel inspection and assessment are severely limited;
- Currently tunnel inspections are predominantly performed through scheduled, periodic, tunnel-wide visual observations by inspectors who identify structural defects manually.
- Un-reliable classification of the liner conditions and lack of engineering analysis.

2 THE ROBO-SPECT EC PROJECT

ROBO-SPECT is a project co-funded by the European Commission under FP7-ICT (Robotics topic) that started in October 2013 and is coordinated by the Institute of Communication and Computer Systems (Athens, Greece). The objective of ROBO-SPECT is to provide an automated, faster and reliable tunnel inspection and assessment solution that can combine in one pass both inspection and detailed structural assessment that does not interfere with tunnel traffic. The robotic system will be evaluated at the research infrastructure of VSH in Switzerland, at London Underground and at the tunnels of Egnatia Motorway in Greece and the system is expected to:

- Increase the speed/reliability of inspections
- Provide assessment addition to inspection
- Minimize use of scarce tunnel inspectors and improve working conditions
- Decrease inspection and assessment cost
- Increase the safety of passengers
- Decrease inspection close time of tunnels

3 ROBO-SPECT TECHNICAL CONCEPT

The ROBO-SPECT system comprises of various distinct modules that combine and integrate the full robotic tunnel inspection system. The system modules can be summarised to the following main modules as can be seen in the diagram that follows:

- Mobile Vehicle
- Automated Crane
- Robotic Manipulator
- Sensign Systems (computer vision, ultrasonic etc)

Apart from these main modules, there are various subsystems that operate such as the sensorial systems (including computer vision, 3D laser profiler and ultrasonic sensors), navigation (including laser and landmarking) and communication systems. The above modules are supported by decision support system able to collect the tunnel collected data and perform the relevant processing to examine the tunnel structural condition as well as act as the user interface for the tunnel operators (Loupos 2013, Loupos 2014, Balaguer 2010). A ground control station (GCS) will also support by monitoring constantly the robot mission and being at short distance with the robotic system. A Control room incarnates either the safety control room of the organisation or the system in charge of the infrastructure monitoring along the time. Depending on the accessible communications, the Control Room can or not monitor in real time the robot mission and process the data to update the referential system and compute the tunnel integrity.

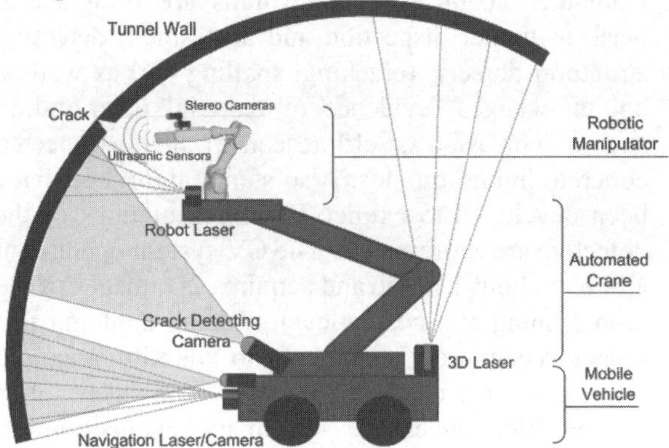

Figure 1. Conceptual design of the ROBO-SPECT system

As can be seen on the diagram above, the mobile vehicle is equipped with the crack detection cameras and the navigation systems and also carries the 3D laser profiler on board. An automated crane (boom) is positioned on the moving vehicle and is able to lift the sensors to the proper position around the tunnel surface in order to perform precise measurements. The sensorial systems in turn, (ultrasonic sensors and stereo cameras) have been positioned on the robotic manipulator that is fixed at the top of the automated crane giving the ability for very precise movements so that precise measurements close (or even in contact) to the tunnel surface can be performed.

4 ROBO-SPECT SYSTEM ARCHITECTURE, DESIGN AND INTEGRATION

4.1 ROBO-SPECT Overall System Architecture

The overall system architecture can be seen in the diagram below. A bottom down design approach has started from the identification of the three main modules of the overall system these being the control room, the ground control station and the actual robotic system.

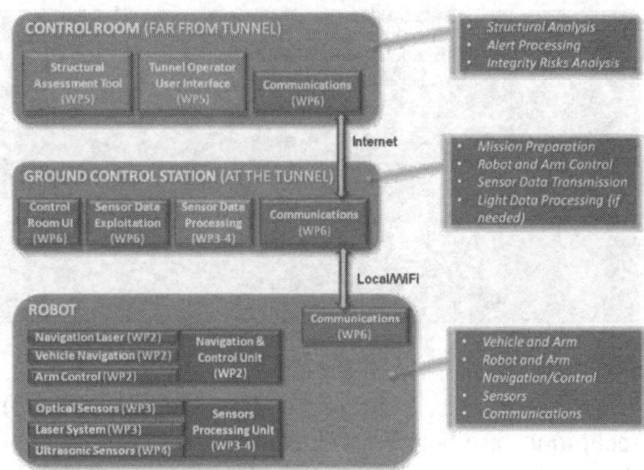

Figure 2. ROBO-SPECT High-level Architecture

The concept of operation, as can be seen in the diagram above, includes the robotic system that consists of the moving vehicle with its navigation systems and arm/boom control and the sensing components (optical sensors, 3D laser system and ultrasonic sensors). The ground control station (located at the tunnel level), includes the control room user interface and the sensorial data exploitation and processing and is responsible for the mission preparation, the robot and arm control, the sensor data transmission as well as a light data processing if needed. The control room includes the structural assessment tool and tunnel operator user interface and is the level where the structural analysis of the tunnel data will be performed and will include tunnel integriry analysis and alerts' processing.

4.2 Robotic Vehicle and Crane System

The ROBO-SPECT extended mobile robotic system is a wheeled robotic system able to extend an automated crane (boom) to the lengths commonly found in tunnels (up to 11 meters) sustaining a robot manipulator while being automated through the use of robotic controllers. This robotic system is composed by three subsystems: a mobile robot, an automated crane arm, and an industrial-quality robot manipulator. The manipulator is a 7-degree of freedom robotic system able to follow required trajectories avoiding obstacles. The extra degree of freedom ensures the positioning of the sensors attached in a variety of orientations.

Figure 4. ROBO-SPECT Integrated System

Component Based Software Engineering (CBSE) techniques are being developed as a set of low-level device drivers for each of the subsystems (i.e. mobile vehicle, crane and robotic arm) allow for the component's control integration. Currently, several robotic software architectures (YARP, ROS, OROCOS, etc.) for implementing CBSE exist and are interoperable. The dynamic and kinematic requirements of the robotic platform needed to reach the measurement area will be designed with special attention in keeping the vehicle stability as well as developing the platform modular enough to allow both road and railway navigation. A system global controller will be developed to improve the position and orientation errors at the end-effector of the robot and to improve the stability of the cameras and ultrasonic sensors. The three different subsystems (mobile robot, the automated crane arm, and robot manipulator) must fulfil a set of required behaviours conjunctly. Therefore a global controller will assure coherent and optimized trajectories. The global controller will have data of three very different natures as inputs: the 3D model data of the tunnel environment coming from the system sensors, associated uncertainties as well as additional semantic information regarding the state of the system and the required action/behaviour. The intelligent controller will be able to update its prior tunnel model of the environment continuously by using the 3D model stream as input while taking into account the uncertainties as confidence values of the given data. The semantic information will be treated as conditional clauses for generating trajectories that comply with the general system's requisites. The feedback will be used for the global controller to auto-tune its parameters.

4.3 Computer Vision Sensing

Computer vision (CV) algorithms are designed to perform tunnel inspection and assessment detecting structural defects (cracking, spalling etc) as well as colour changes (evidence of material deterioration such as corrosion or efflorescence) at the inspected concrete lining intrados. Also stereo algorithms have been developed to extract 3D representations of the defective areas of tunnels. The CV system operates at a rate of about 1 m/sec and acquires 2D images of the tunnel lining at a coarse level of detail applying fast object recognition techniques to identify areas of interest in the coarse 2D image and then, at a slower rate, concentrate the image acquisition on details of interest, thus allowing a higly detailed 3D geometric and radiometric documentation of the defects. Hierarchical computer vision schemes are applied to make the recognition process just-in-time, and thus significantly reduce the time and effort needed for visual inspections. The system will apply recent advances in active continuous learning to tunnels' inspection mechanisms, so as to achieve on-line understanding of the cracks while the system surveys the tunnels.

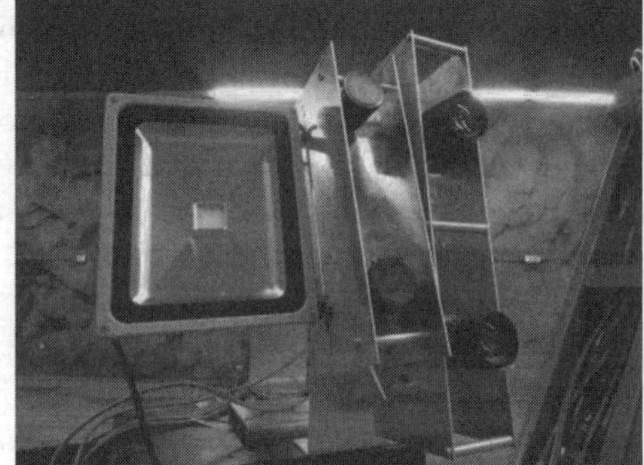

Figure 5. Stereo and Crack detection cameras on the robotic platform

All communication issues are treated through YARP (Yet Another Robotic Platform) system while a series of communication messages has been defined and tested to trigger the camera activation, image capturing, processing results, data transferring etc.

4.4 3D Laser Scanning

FARO LS SDK provides a compact and easy to use interface to control the scanner with its main functionality. All the basic operations like starting a scan, changing scan resolution and quality parameters, or determine the scan area through the horizontal and vertical angles. The developed application aims to remotely activate and control the laser scanner in a tunnel, in order to detect cracks of relatively big size, which exceeds the accuracy of the device. Creating a detailed full resolution 3D model of the inner surface of a tunnel is time-consuming and the requirements in data storage and computer memory for post-processing exceed the capabilities of an average computer. This is a significant drawback for processes that should be repeated within a year for safety reasons, such as the inspection of the tunnel static strength. Instead of several full 360°×305° scans, a number of narrow scans, with approximately 10° horizontal angle range are proposed, which result in a set of indicative cross-sectional data.

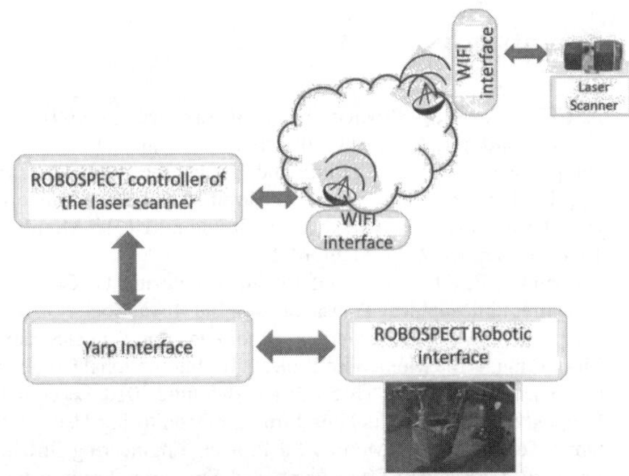

Figure 6. ROBO-SPECT controller regarding the operation of the laser scanner

Integration with the overall project is achieved through YARP (Yet Another Robot Platform), an open-source framework, which supports building a robot control system. YARP works as a middleware, consisting of libraries, protocols, and tools, designed to control and manage multiple devices and modules.

Each time YARP receives a message from one of the connected sensors, it sends a triggering signal to the CPU to execute the batch file, which in turn activates the laser scanner. In the YARP interface, some ports are allocated for the communication of different robotic components and sensors. In particular, the 3D laser scanner allocates a particular message assigned to a particular port in order to allow seamless communication among the different components of the ROBO-SPECT platform. The YAPR interface in the implementation of ROBO-SPECT robotic system controls the camera sensors, the 3D laser scanner, the frame grabber and other sensors installed. During the operation of YAPR a message is send over the communication buss channel of the robotic system. This message is passing though out the different modules of the robotic systems and sensors. Only the sensor that has been assigned to this port and to this interface are allowed to interact with the message.

4.5 Ultra-Sonic Sensor

The system includes a usual pair of piezo-ceramic transducers employed for ultrasound generation/detection on concrete operated by means of a Pulsonic Pulser/Receiver unit. Moreover, a custom fabricated fiber-optic ultrasound sensor is utilized in the setup, designed and fabricated and operated with the aid of a tunable infrared laser, an optical circulator and a photodiode. In order to be utilized in crack width and depth measurements on the robot, the mechanical design of the system has been updated in both types of crack measurement, the sensors need to be displaced on the tunnel lining independently of each other, a new sensor system composed by two XY positioning tools was designed, which will be able to perform all the needed displacements of the sensors on the tunnel lining without any intervention by the robotic arm, as shown in the figure below.

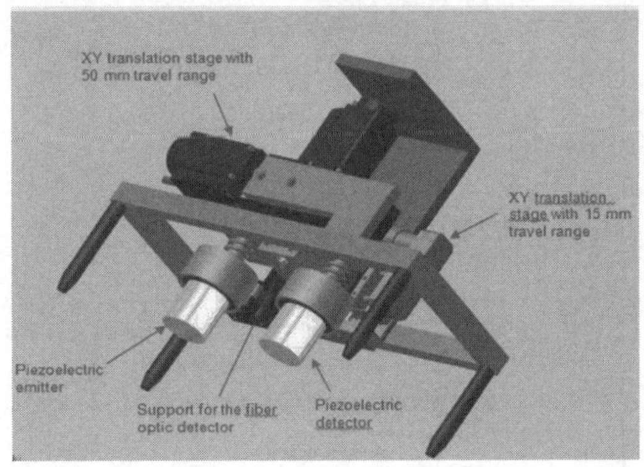

Figure 7. Ultrasonic sensor system suitable for operation on the robotic arm

5 OPERATIONAL DETAILS

In the flow diagram that follows, the operational sequence diagram of the system indicates the mission execution of the robotic system in a tunnel. The basic flow, includes a snapshot of the crack detection system that if positive drives the stereo capturing and 3D laser scanning of the particular tunnel slice. As a next step, the coordinates of the crack are passed to the crane system to direct the ultra-sonic sensor on the crack for a precise measurement. After the measurement is taken the robot returns to the 'home' position for the next "step".

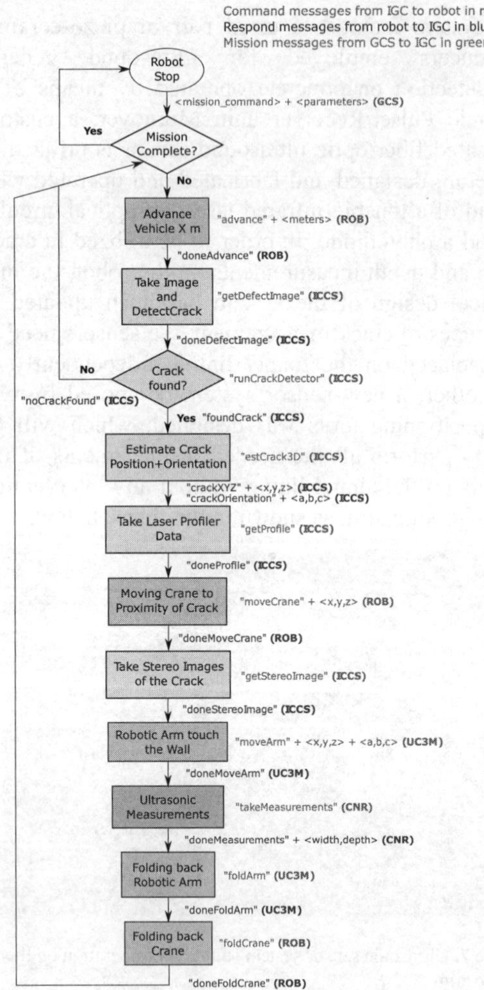

Figure 8. System Operation Flow Chart

6 CONCLUSIONS

The ROBO-SPECT project has just finished its 2nd year of activities and is now into the integration phase that will lead to the system evaluation and validation at actual tunnels. The final validation and evaluation tests are expected to be taking place in January-March 2016 and May-July 2016 in the actual tunnels of Egnatia Motorway (Greece) and London Underground (UK) respectively.

ACKNOWLEDGEMENTS

The research leading to these results has received funding from the EC FP7 project ROBO-SPECT (Contract N.611145). Authors would like to thank all partners within the ROBO-SPECT consortium.

REFERENCES

ASCE – American Society of Civil Engineers (2009). '2009 Report Card for American Infrastructure,' (online).

Bimpas Et Al - Bragg Grating And Botdr Fiber Optic Principles Applied For Real-Time Structural Monitoring - The Monico Project, 9th International Workshop On Structural Health Monitoring, 10-12 September, 2013, Stanford, Us.

Brownjohn Et Al – Structural Health Monitoring Of Civil Infrastructure, Philosophical Transactions of the Royal Society, 2015.

Loupos K. Et Al. - Fibre-Optic Technologies For Tunnel Structural Monitoring – The Monico Ec Project, 4th International Conference On Sensing Technology (ICST 2010), 3-5 June 2010, Lecce, Italy.

Loupos K. Et. Al - Real-Time Structure Monitoring Using Fibre-Optic Technologies - Monico Ec Project, Engineering Structural Integrity Assessment: From Plant And Structure Design, Maintenance To Disposal (Esia11), 24-25 May 2011a, Manchester, Uk.

Loupos K Et Al. - Application Of Fibre-Optic Technologies For Real-Time Structural Monitoring - The Monico Ec Project, 9th International Conference On Damage Assessment Of Structures (DAMAS 2011), 11-13 July 2011a, Oxford, Uk.

Loupos K. Et Al. - Fiber Sensors Based System For Tunnel Linings' Structural Health Monitoring", 2nd Conference On Smart Monitoring, Assessment And Rehabilitation Of Civil Structures, 9-11 September 2013 (Smar 2013), Istanbul, Turkey.

Loupos K Et Al. - Robotic Intelligent Vision And Control For Tunnel Inspection And Evaluation - The Robo-spect Ec Project, 2014 Ieee International Symposium On Robotic And Sensors Environments - Ieee International Symposium On Robotic And Sensors Environments (ROSE 2014), October, 2014, Romania.

Ross Et Al - In-service structural monitoring—a state of the art review. Struct. Eng.73, 23–31, 1995

Rücker Et Al., Federal Institute of Materials Research and Testing (BAM), Division VII.2 Buildings and Structures Unter den Eichen 87, 12205 Berlin, Germany - SAMCO Final Report 2006, F08a Guideline for the Assessment of Existing Structures.

Proceedings of the International Conference on Smart Infrastructure and Construction
ISBN 978-0-7277-6127-9

© The authors and ICE Publishing: All rights reserved, 2016
doi:10.1680/tfitsi.61279.315

Sensing capability enhancement of multifunctional carbon fiber-sprayed FRP composites fabricated with CNT-modified epoxy

I.W. Nam and H.K. Lee
Department of Civil and Environmental Engineering, Korea Advanced Institute of Science and Technology, South Korea

ABSTRACT The deterioration of infrastructures has been a major issue for numerous countries and has raised concerns for the importance of structural health monitoring. Fiber-reinforced polymer (FRP) composites have been widely used as a conventional retrofitting material, imparting enhanced load carrying capacity and ductility to the structures. The present study aims at developing a multifunctional FRP coating system which possesses stress/strain sensing capabilities by means of replacing ordinary epoxy resin with epoxy resin modified by carbon nanotube (CNT), which demonstrates enhancement in mechanical/electrical properties of composites. The FRP composite prepared in the present work was manufactured on the basis of the sprayed FRP fabrication method, which is one of the structural retrofitting methods, and two types of carbon fiber-sprayed FRP (CSFRP) were fabricated using ordinary epoxy resin or CNT-modified epoxy resin in an effort to compare their sensing performances. The sensing characteristics of the fabricated FRPs were evaluated by means of monitoring electrical resistance change induced in response to the applied stress. In particular, the sensing capability of the CSFRP manufactured with epoxy resin modified by CNT was more stable than that manufactured with ordinary epoxy resin. The preliminary result of the experiments demonstrated that the CSFRP facilitates resilient sensing capabilities as well as the retrofitting functions, which can effectively enhance potential and worth of FRP composites.

1 INTRODUCTION

The practical applications of retrofitting and strengthening of infrastructures are increasing throughout the world as the service life of the structures need to be extended (Hollaway and Teng 2008). Among various retrofitting materials, fiber-reinforced polymer (FRP) composites have been widely used as repair materials of infrastructures, imparting enhanced load carrying capacity and ductility to the structures (Hollaway and Teng 2008). The attempts to develop multifunctional FRP composites which additionally facilitate the evaluation of health conditions of structures have been conducted from late 1990s by means of incorporation of piezoresistive materials, of which the electrical resistance changes in response to external loading, into the FRP (Seo and Lee 1999; Park et al. 2002).

Seo et al. (1999) evaluated electrical resistance change of carbon fiber-reinforced polymer (CFRP) composites subjected to repeated tensile stress (Seo and Lee 1999). As the stress was applied repeatedly, the fatigue damage was accumulated in the composites, and thereby, their strength and stiffness decreased (Seo and Lee 1999). The authors also developed a damage detection model on the basis of stiffness reduction model, which enabled the prediction of mechanical behaviors of the composites under fatigue loads (Seo and Lee 1999). Park et al. (2002) adopted a concept of electrical ineffective length in an effort to predict electrical characteristics change of CFRP subjected to a tensile stress (Park et al. 2002). The formulated model nearly corresponded the result of tensile stress test to the extent that the electrical ineffective length was 5 mm (Park et al. 2002). Zhu et al. (2007) applied bending stress to CFRP composites, and measured the surface electrical resistance change on the top side (compressive zone) and bottom side (tensile zone) of the composites (Zhu et al. 2007). In addition, piezoresistive analysis model was used for comparison of experimental results with the calculated results. The calculated data and experimental results had similar behavior, and exhibited reversible electrical resistance change at the loaded and unloaded states (Zhu et al. 2007).

However, studies of developing multifunctional composites and assigning piezoresistive sensing capabilities to the sprayed FRP composites are limited. Therefore, the present study aims at developing multifunctional FRP composites which possesses stress/strain sensing capabilities of structures by means of replacing ordinary epoxy with epoxy resin modified by CNT, which enhances the mechanical/electrical properties of composites. FRP composites prepared in the present work were manufactured on the basis of the sprayed FRP fabrication method, and two types of carbon fiber-sprayed FRP (CSFRP) were fabricated using ordinary epoxy resin or CNT-modified epoxy resin in an effort to compare their sensing performances. The sensing characteristic of the fabricated FRPs were evaluated by means of monitoring electrical resistance change induced in response to the applied stress.

2 EXPERIMENTAL PROGRAM

2.1 *Materials and composite fabrications*

Carbon fiber, a proprietary product of Hyosung Inc., used in the present work was consisted of 6K of carbon filaments and showed 1.76 g/cm^3 of density. Carbon fiber was cut into 3 cm-length using a chopping device equipped in a spray gun. Multi-walled carbon nanotube, a proprietary product of Hanwha nanotech Inc. (CM-95), was used. Epoxy resin consisted of base resin YD-128 and a hardener TH-431, proprietary products of Kukdo chemical Inc., and a mix proportion of the epoxy to hardener was 5:3 by weight.

Following procedures were adopted in fabrications of the composites. The FRP composites were prepared on the basis of the sprayed FRP fabrication method. To set up a layer of fresh resin, the epoxy mixture (or CNT-modified epoxy mixture) was applied on a substrate using a paddle, and then, chopped carbon fibers were distributed on the layer of fresh resin using a spray gun with the chopping device. The sprayed carbon fibers were wet to epoxy resin with an aid of hand rollers. The aforementioned procedures were repeated for four times to form a CSFRP composite and the carbon fiber used in the composite fabrication amounted to 30 % of the total volume of the composite.

CNT dispersion process was additionally demanded in fabrication of CSFRP incorporating the CNT-modified epoxy. Three-roll milling (calindering) process was adopted as a CNT dispersion method. This takes advantage of shear forces generated by parallel rotating rollers (3 rollers) with a speed of 300 rpm (Nam et al. 2011). The CNT-added polymer materials were supplied into a gap (5 μm) of the rollers and the resultant mixtures were retrieved at the opposite gap of the rollers (Nam et al. 2011). Details of dispersion procedures can be found in Nam et al. (2011) and CNT content ratio reached to 1 % by total weight of the mixture. The fabricated composite samples were cured for three day at a room temperature before trimmed into dumbell-shaped samples in accordance with dimensions specified in ASTM D638.

2.2 *Method*

The methods of evaluating health conditions of structure materials (FRP, steel, etc.) thus far were detecting strain or damage extent with optical fiber, ultrasonic wave, memory alloy, sound wave, and so on (Loyola et al. 2011). However, some disadvantages were inherent in these methods, for instance, methods were limited in a local fraction or deliberated damage are inevitable for attaching the sensor (Loyola et al. 2011). The electrical resistance measurement method which can complement the aforementioned weaknesses is highlighted in the field of structural health monitoring (Seo and Lee 1999; Park et al. 2002; Zhu et al. 2007).

The electrical resistance measurement method is that carbonic or metallic materials are infused into the target structure and the electrodes are installed at the two ends of the target, measuring the electrical resistance (Seo and Lee 1999; Park et al. 2002; Zhu et al. 2007). The stress, strain and failure of the structure can be detected by monitoring the differences of the electrical resistance induced in response to the applied load. The electrical resistance measurement method can evaluate the entire area of the structure, does not require deliberated damage to attach a sensor, and economically inexpensive compared with above methods, hence an enormous number of studies are focusing on this method (Loyola et al. 2011).

In this study, the two wire method was used. Silver paste was applied to the two ends of the tensile specimen manufactured, as shown in Fig. 1. Metallic wires were attached on the two ends which act as electrodes. The other sides of wires were connected to the Digital Multimeter to measure the electrical resistance. The measured resistance was recorded in a connected computer in real time and hence, monitoring the variance of the electrical resistance.

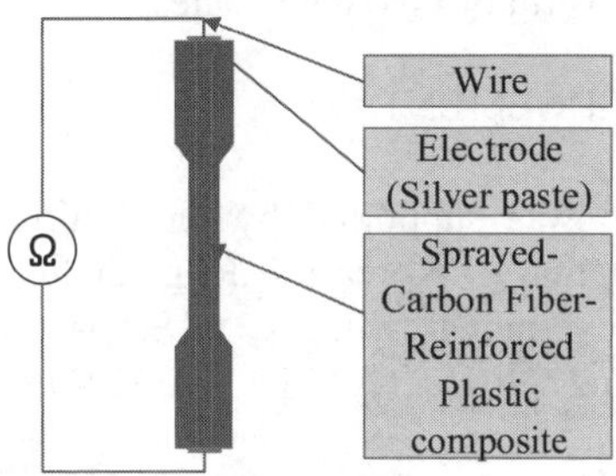

Figure 1. A schematic of the fabricated sample and resistance measurement

The electrical resistance change rate was calculated by the following equation (Kim et al. 2014).

$$\text{Electrical resistance change rate} = \frac{R_i - R_0}{R_0} \quad (1)$$

where, R_0 is the resistance value in the unloaded state, and R_i is the resistance values in the loaded state which varied with a lapse of time.

3 RESULTS AND DISCUSSION

The electrical resistance change rate of the CSFRP composite fabricated with the ordinary epoxy resin under repeated loading is shown in Fig. 2. The results showed that the electrical resistance tended to increase under loading and decrease under unloading. On the other hand, the fluctuations in the resistance were found during the repeated loading-unloading processes, stemming from the momentary deterioration or enhancement in fiber-fiber contact resistance at the processes. The momentary changes in the fiber contact resistance could have substantially affected overall electrical resistance of the samples, since the epoxy resin was present between fibers could not play a role as conductive bridges due to its intrinsically insulating characteristics.

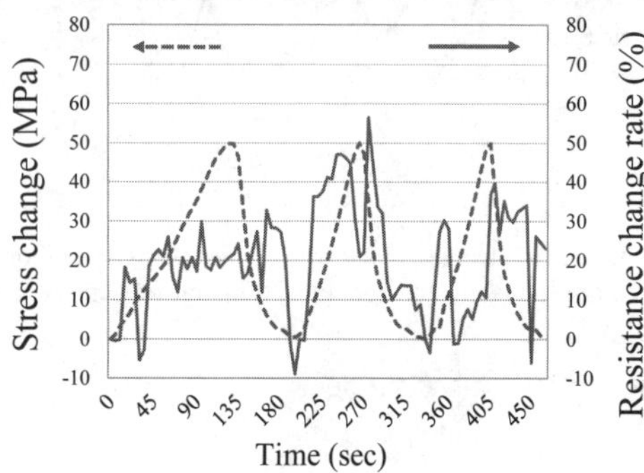

Figure 2. Stress change versus resistance change rate of the CSFRP composite fabricated with ordinary epoxy resin

The prediction of resistance values corresponding to the applied stress was feasible in the case of CSFRP composites fabricated with the CNT-modified epoxy resin (Fig. 3), whereas it was difficult in the case of CSFRP composites with the ordinary epoxy (Fig. 2). However, the resistance peak occurred posterior to the stress peak with delay of time, which can be attributed to characteristics of sprayed FRP composite comprised of chopped carbon fiber network. That is, once the carbon fiber network was loosened in the tensile loading process, this could entail a period of time to recover the network towards the original states.

In addition, tensile loading processes in the second and third cycles were corresponded to decreases of the resistance. This is likely due to the reduction of cross sectional area of the composite, resulted in denser CNT/carbon fiber networks and enhancement in their electrical contact conditions (Fig. 4). However, it was temporary phenomena. The increased resistance is attributed to the loosened network of carbon fibers (and/or CNTs) after a period of time corresponding to the decrease of resistance. The details of the examination on the sensing capabilities of the CSFRP composites will be presented.

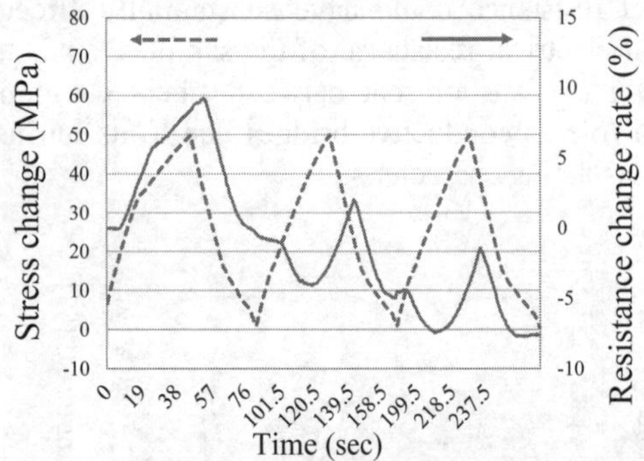

Figure 3. Stress change versus resistance change rate of the CSFRP composite fabricated with the CNT-modified epoxy resin

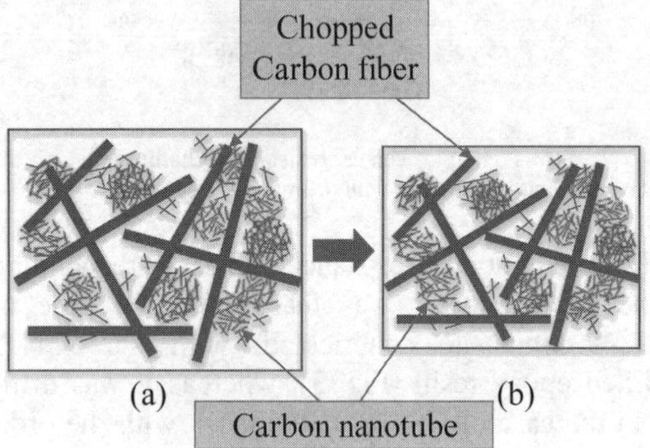

Figure 4. Schematics of cross sectional view of the CSFRP composites prior to loadings (a) and under the tensile loadings, which led to the reduction of cross sectional area of the composite.

4 CONCLUDING REMARKS

The present study aims at developing multifunctional FRP composites which possess stress/strain sensing capabilities of structures by means of replacing ordinary epoxy with epoxy resin modified by CNT. Two types of CSFRP were fabricated using ordinary epoxy resin or CNT-modified epoxy resin in an effort to compare their sensing performances. The sensing capability of the composites were evaluated by electrical resistance change results. The findings of the present study are summarized as follows.

1) An increase and decrease in the electrical resistance were observed from the two CSFRP composites subjected to repeated tensile loadings.
2) Fluctuations in the resistance were found in the ordinary CSFRP composite, stemming from momentary deterioration or enhancement in fiber-fiber contact resistance during the repeated loading and unloading processes.
3) The prediction of applied stress using the resistance values was more feasible in the CSFRP composites with the CNT-modified epoxy resin than those with the ordinary epoxy resin.

The preliminary results in the present experimental work demonstrated that the CSFRP composites can provide benefits in terms of multifunctionality - the resilient sensing capabilities and the retrofitting functions, which can effectively enhance potential and worth of FRP composites.

ACKNOWLEDGEMENT

This research was supported by the National Research Foundation of Korea (NRF) grant funded by the Korean government (Ministry of Science, ICT & Future Planning) (2015R1A2A1A10055694).

REFERENCES

Hollaway, L. & Teng, J.G. 2008. *Strengthening and rehabilitation of civil infrastructures using fibre-reinforced polymer (FRP) composites*, CRF press, Cambridge UK.

Kim, H. K., Park, I. S., & Lee, H. K. 2014. Improved piezoresistive sensitivity and stability of CNT/cement mortar composites with low water–binder ratio. *Composite Structures*, **116**, 713-719.

Loyola, B. R., Loh, K. J., & La Saponara, V. 2011, March. Static and dynamic strain monitoring of GFRP composites using carbon nanotube thin films. *In SPIE Smart Structures and Materials+ Nondestructive Evaluation and Health Monitoring* (pp. 798108-798108). International Society for Optics and Photonics.

Nam, I. W., Lee, H. K., & Jang, J. H. 2011. Electromagnetic interference shielding/absorbing characteristics of CNT-embedded epoxy composites. *Composites Part A: Applied Science and Manufacturing*, **42**, 1110-1118.

Park, J.B., Okabe, T., Takeda, N., Curtin, W.A. 2002. Electromechanical modeling of unidirectional CFRP composites under tensile loading condition, *Composites Part A: Applied Science and Manufacturing*, **33**, 267-275

Seo, D.C. & Lee, J.J. 1999. Damage detection of CFRP laminates using electrical resistance measurement and neural network, *Composite Structures*, **47**, 525-530.

Zhu, S. & Chung, D.D.L. 2007. Analytical model of piezoresistivity for strain sensing in carbon fiber polymer–matrix structural composite under flexure, *Carbon*, **45**, 1606-1613.

Proceedings of the International Conference on Smart Infrastructure and Construction
ISBN 978-0-7277-6127-9

© The authors and ICE Publishing: All rights reserved, 2016
doi:10.1680/tfitsi.61279.319

Nano-carbon cement based sensors for smart structures

G. Noiseux-Lauze, J. Orellana and G. Akhras*

Centre for Smart Materials & Structures
Royal Military College of Canada, Kingston, ON, Canada
**Corresponding Author*

ABSTRACT The purpose of Structural health monitoring (SHM) is to monitor the in-situ behaviour of a structure, assess its performance, detect damage, and determine the condition of the structure. The object of this research is to analyse one type of sensor for SHM. More specifically, the piezoristive property of nano-carbon reinforced cement based sensor is examined. The nano-additives, carbon nanotube (CNT) and carbon nanofiber (CNF), in a cement mixture produce a piezoresistive material that can reveal a variation of strain with the variation of its electrical resistivity. This study aims to assess the various mixtures to produce the best sensor as a replacement of conventional strain sensor. The sensing abilities of different cement based mixtures are assessed as cylindrical specimens are put through cyclic load tests. Different variables affecting the sensor effectiveness are examined such as the effect of nano-additives and aggregate content. Standard compressive tests are also conducted to assess the nano-additive effect on the material structural integrity. Two different type of carbon fibre, the carbon nano-fiber (CNF) and the carbon nanotube (CNT) were examined in two different matrixes: one with the cement paste and the other with cement mortar. The first sets of experiments with cement paste served to validate previous studies on the subject and helped established the process for the 2nd sets of experiments on cement mortar sensor. Results of the two sets of experiments are assessed and presented to highlight their sensing abilities.

1 BACKGROUND

Many civil infrastructures around the world are approaching the end of their useful life and many of them are in a state of utter disrepair. Particularly, many government structures are approaching or are already past their designed service life. With growing economic uncertainty and prevalent government expenditure rationalisation, significant efforts and innovative idea will be required to render the failing infrastructures back to a serviceable state and extend their service life while ensuring the safety of their users.

The lack of durability in construction materials and the inability to provide timely maintenance or retrofit increases the life cycle cost of any structure and leads very often to its early replacement. SHM can address this problem since many new and improved methods are being developed to expand damage detection, assess usage deterioration, and by the same token, increase the service life of the structures.

SHM is typically achieved by the continuous and autonomous monitoring of key structural parameters by embedding sensors such as electric-resistance strain gauges, optic sensors, and piezoelectric ceramic in strategic positions. Most of these sensors have considerable shortcomings such as low sensitivity, high cost, poor durability, and unfavorable compatibility particularly with concrete structures (Han *et al*, 2010). Furthermore, a high degree of knowledge of the type of damage that is to be expected and where this damage may occur is required. This can be difficult on certain structure (Akhras, 2010).

The current trend of SHM is based on the development of dense sensor networks capable of detecting very small defect that share some similarities with human nervous system, for integrated health management. The key requirements for the successful implementation of these dense networks are miniaturization and embedment (Rainieri *et al*, 2011). The evolution and advancement in nanotechnology over the past decade has generated a lot of interest in the field of construction materials and has made the development of dense sensor network more readily achievable. Civil infrastructures are the most expensive investment in any country and since concrete, in terms of volume used, is the most common and widely used material in the world (Han *et al*, 2011) the growing interest in improving its functionalities and properties is clearly justified.

Research to fabricate multifunctional self-sensing cement based material (including cement paste, cement mortar and concrete) using nano-particle is still in its early stage. Nevertheless, current research results show the significant potential of this technology. When concrete start to weaken with time, its structural integrity will deteriorate and its life cycle shortens. Any solution to prevent and avert the ageing phenomena is welcome. CNF and CNT concrete may have many structural benefits to deal with this issue: CNF seems to act usefully on the nonacraks and restricts their growth; they also increase some mechanical properties such as compressive strength and ductility. More interesting, they alter and improve the electric resistance of the mix which is central to the use of SHM (Akhras, 2010).

Self-sensing cement based material has favorable piezoresistivity (the ability of a material to change its conductivity when mechanically stressed), great durability and good compatibility when used in concrete structures. Some of the most promising technologies seem to lie with the use of CNT, CNF and nano carbon black (CB) as filler in the cement matrix. *Smart concrete* in its various forms not only enable the cementious material to have self-sensing ability but it also, as stated previously, improves its durability by inducing ductility, toughness and control crack growth.

2 OBJECTIVE AND WORK PLAN

Standard compressive strength tests on cylindrical cement-CNF composite specimen are performed in accordance with ASTM C39 (ASTM, 2012). In these test samples, no aggregate is added but only CNF, water, concrete admixture for the purpose of helping to disperse the CNF and the standard Portland cement type II. These specimens are instrumented with standard strain gauges and wired to a multimeter to record the variation of the specimens' resistivity under the compressive stress. As reported by Han (Han *et al*, 2011), the electrical resistivity of the specimen decreases reversibly under the compression provided that substantial cracking does not occur. The change in resistivity is later correlated with the strain data collected by the strain gauges and the impact of CNF concentration on the specimen piezorestive ability is analyzed.

The second testing phase will involve the cement-CNT composite and cement-CNF-CNT composite with the same process previously outlined. The third phase will include coarse aggregates in the cylinders with various concentrations of CNT and CNF. These future tests will determine if the combination of CNT and CNF effectively overcomes challenges associated with the formation of conductive network in concrete containing coarse aggregate. In other word, verify if having hybrid CNT and CNF in the concrete mix allows the piezoresistivity to perform in a comparable way to other successful research using only CNT in cement paste. This will also determine, as side objective, if any of the tested specimens could be calibrated and used as a load cell. The idea is to produce a concrete cylinder that is sensitive enough to act as a sensor.

3 EXPERIMENTAL TESTS

3.1 Mix Proportions

To perform all the preliminary tests, various proportions of CNF and CNT have been tried. Three samples of each mix were prepared and tested. The mix proportions are highlighted in Table 1.

Table 1. Mix proportion of the experimental tests

Test Specimen	CNT (% wt cement)	CNF (% wt cement)
0.1% CNF	0	0,1
0.2% CNF	0	0,2
0.1% CNT	0,1	0
0.2% CNT	0,2	0
Control	0	0
0.1% CNF /0.1% CNT	0,1	0,1
0.2% CNF /0.2% CNT	0,2	0,2

3.2 Specimen Preparation

Water was poured in a 1L graduated glass container and weighted IAW mix proportion on the scale. Superplasticizer (ADVA® Cast 575, a polycarboxylate-based) was weighted (2.5 wt% water) and manually mixed with water. Multi-walled Carbon Nanotube (MWNT), Nanocyl™ NC 7000l, with properties shown in Table 2, was added to the water.

Table 2: MWNT Properties

Property	Unit	Value
Average diameter	nm	9.5
Average length	microns	1.5
Surface area	m^2/g	250-300
Carbon purity	%	90

The Carbon nanofiber (CNF), Pyrograf®-III PR 24 XT-LHT, was weighted in accordance with the mix proportion and manually blended with the water solution. Properties of the CNF used are shown in Table 3.

Table 3: CNF Properties

Property	Unit	Value
Average diameter	nm	100
Average length	microns	50-200
Surface area	m^2/g	43
Dispersive surface energy	mJ/m^2	155

At higher CNT/CNF concentration, a magnetic stirrer was used to help with the dispersion. The solution was then put in the sound enclosure and the specimen was raise on a jack stand so that the ultrasonic liquid processor probe tip was submerged hallway in the solution (Figure 1). The Q 500 from Qsonica, LLC was used. This is a 500 Watt unit that was coupled with a 1" replaceable tip and a booster (2:1 gain). This setup allowed processing of as much as 1L at a time. The sound enclosure allowed to drastically reducing the high pitch noise heard during processing. This machine was installed in a dust hood for increased safety. The ultrasonic liquid processor was programed on a pulse setting (20 min on 20 min off) for 40 min at 90% power setting.

Other materials such as cement and silica fume were weighted and mix in the Hobart mixer for 5 minutes. The water/CNT/CNF solution was then gradually added in the Hobart mixer and mixed for another 5 minutes. The cement paste was then poured in the cylinder mould (D=75mm, H=150mm) and lightly vibrated (10 seconds) on a table to allow air bubble to escape.

Figure 1. Ultrasonic Liquid Processor

The specimens were left in the mould curing for a minimum of 5 days before being de-moulded. After a minimum of 14 days of curing, the specimen top and bottom surfaces were grinded smooth and perpendicular to their side. Two strain gages (Kyowa uniaxial, 20 mm, 120 ohms) were then glued with cyanoacrylate at each specimen mid-height on diametrically opposite side.

3.3 *Test Setup*

An apparatus to enable resistivity measurement was fabricated. It consists of two acrylic components that isolate the cylinder specimen from the steel test frame. As shown in Figure 2, cooper disk were glued and connected to a female adapter (shown in red). These female adapters are easily connected to the RCON™ impedance meter via standard test wires.

Figure 2. Test Setup

The impedance meter was connected to a computer via an USB cable and data was recorded with the supplied software. The recorded data are then easily exported to excel. Carbon conductive grease was used to enable a good connection between the cooper disks and the specimen grounded top and bottom surfaces. Both Strain gauges were connected to the HBM data acquisition system. Strains, load, stroke and time were recorded, synchronized with the resistivity measurements and merge in a single excel file. A cyclic compressing test was then programed on the computer. This test consisted of cyclic ranging from 1-20 kN with a loading rate of 0.2 kN/sec.

3.4 *Test Results and Discussion*

To evaluate the sensing ability of the specimen, the stress and strain were plotted against fractional change in resistivity (FCR) given by the following equation;

$$FCR = \frac{\rho_t - \rho_0}{\rho_0} \qquad [1]$$

where ρ_0 : Initial resistivity (prior to loading)
ρ_t : Resistivity at time t

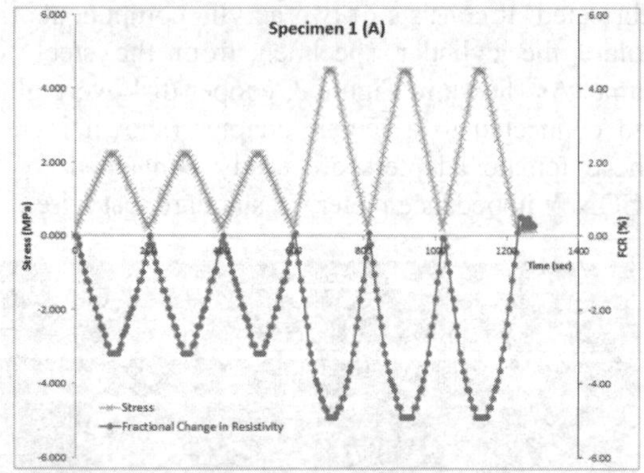

Figure 3. Cyclic Load Test for CNF 0.1%

Fig. 3 is an example of results obtains during testing. The blue lines represent the specimen stress in MPa ranging from 0.2 to 4.5 MPa resulting from load ranging from 1-10 kN and 1-20 kN respectively. The red lines represent the fractional changes in resistivity as previously defined. There is a clear and unambiguous correlation between the stress increase and the drop in resistivity. Furthermore, this test shows reversibility and consistencies as the specimen regains its initial resistivity at the end of each cycle. The tests were purposely done on low level of stress to illustrate the specimen sensitivity.

Fig. 4 is data collected from the same test as Fig. 3 and puts in relation the average of two strain gauges installed on the specimen during the experiment and the fractional change in resistivity. This graph provides a good indication of the recurrence of the relationship, as the narrow "bandwidth" is an indication of consistent results. On a fully calibrated model and after significant testing, it would be possible to accurately estimate the specimen strain by taking only resistivity reading.

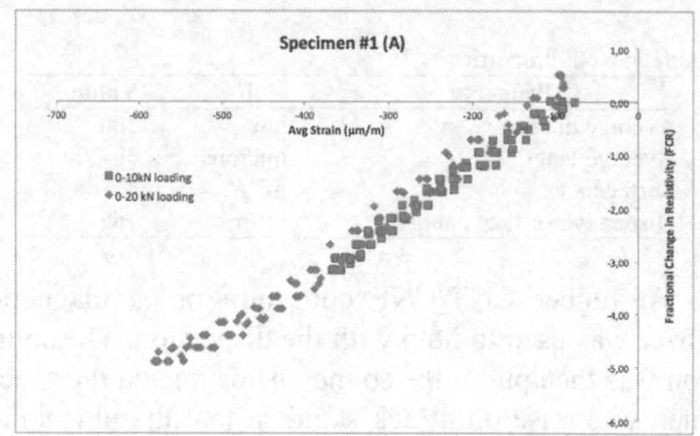

Figure 4. Fractional Change in Resistivity – Average Strain

A fundamental parameter of the strain gauge is its sensitivity to strain, expressed quantitatively as the gauge factor (GF). Gauge factor is defined as the ratio of fractional change in electrical resistance (R) to the fractional change in length (strain):

$$GF = \frac{\Delta R/R}{\Delta L/L} = \frac{(\rho_t - \rho_0)/\rho_0}{\varepsilon} = \frac{FCR}{\varepsilon} \qquad [2]$$

As it can be seen in Fig. 4, the relationship between strain and resistivity is slightly nonlinear (Azhari & Banthia, 2012). This particular test resulted in a gauge factor of approximately 70, which means it is much more sensitive than conventional foil strain gauges that possess a GF of approximately 2. This basically means that under a given strain, these cement-CNT/CNF based sensors will show larger change in resistivity than conventional strain gauges. If they could be made to be as reliable, the cement-CNT/CNF sensors would provide a higher strain data resolution due to their higher sensitivity.

While it is the change in the resistivity values that carries the most meaning for our application, absolute resistivity values remain useful. These values could indicate which ones of the combination would be better suited for a mixture containing aggregates. It is anticipated that the addition of aggregates will significantly increase specimen resistivity and challenge its ability to be used as a sensor. Table 4 shows the average resistivity of the specimens during cyclic loading.

Table 4. Average resistivity of specimens

Test Specimens (Cement Paste)	CNT (% wt cement)	CNF (% wt cement)	Resistivity (after 40 days) (ohm*m)
0.1% CNF	0	0.1	176.6
0.2% CNF	0	0.2	166.4
0.1% CNT	0.1	0	172.1
0.2% CNT	0.2	0	93.9
Control	0	0	346.1
0.1% CNF/0.1% CNT	0.1	0.1	96.7
0.2% CNF/0.2% CNT	0.2	0.2	19.7

4 ONGOING AND FUTURE WORK

The specimen in the second phase of testing contains relatively coarse aggregates with various concentrations of CNT and CNF. These tests will allow determining if the combination of CNT and CNF effectively overcomes challenges associated with the formation of conductive network in concrete containing aggregate. In other word, verify if having hybrid CNT and CNF in the concrete mix allows piezoresistivity comparable to results obtained using only CNT in cement paste. We will also examine if any of the tested specimens could be calibrated and used as a sensor.

5 CONCLUSION

The piezoresistive properties of CNF and or CNT-cement composites were investigated in this study. CNF/CNT-cement hybrid composite significantly increase the electrical conductivity of the specimens compared to CNT-cement or CNF-cement composites. Experimental results showed that the electrical resistance of the specimen changed in tandem with the strain levels. The effectiveness of the CNF and or CNF- cement sensor testing method and the whole fabrication procedure were confirmed. Results obtained during the first testing phase are coherent with other studies such as Azhari & Banthia, 2012, Tyson et al. 2011, Han et al. 2011 and Yu and Kwon 2009. The positive results in the first phase is instrumental in steering the second phase of research. As previously stated, these tests will determine if the combination of CNT and CNF effectively overcomes challenges associated with the formation of conductive network in concrete containing small to medium size aggregate. If successful, this would allow structural member to exhibit self-sensing capability.

ACKNOWLEDGEMENT

The authors would like to thank Canada's National Science and Engineering Research Council for its financial support.

REFERENCES

Akhras G. 2010, "Structural Health Monitoring for Municipal Buildings & Infrastructures." *Public Sector Digest*, 13-16.

Azhari F., Banthia N. 2012, "Cement-based sensors with carbon fibers and carbon nanotubes for piezoresistive sensing." *Cement and Concrete Composites*, 866–873.

Han B.G., Yu X., Kwon E., Ou J.P. 2010 "Piezoresistive MWNTs filled cement-based composites." *Sensor Letters*, **8** : 344–348.

Han B., Yu X., and Ou J. 2011, "Multifunctional and Smart Carbon Nanotube Reinforced Cement-Based Materials." In *Nanotechnology in Civil Infrastructure*, by K. Gopalakrishnan et al., 1-47. Springer-Verlag Berlin Heidelberg.

Rainieri C., Fabbrocino G., Song Y. and Shanov V. 2011, "CNT Composite for SHM: A Litterature Review." *Smart Materials, Structures & NDT in Aerospace.* Montreal: Smart Materials, Structures & NDT in Aerospace Conference.

Tyson B. M., Abu Al-Rub R. K., Yazdanbakhsh Y, Grasley Z., 2011, "Carbon Nanotubes and Carbon Nanofibers for Enhancing the Mechanical Properties of Nanocomposite Cementitious Materials." *Journal of Material in Civil Engineering* **23**, no. 7: 1028-1035.

Yu X. and Kwon E. 2009, "A carbon nanotube/cement composite with piezoresistive properties." *Smart Materials and Structures*.

ASTM International, 2012, "C39/C39M-12, Standard Test Method for Compressive Strength of Cylindrical Concrete Specimens,".

Proceedings of the International Conference on Smart Infrastructure and Construction
ISBN 978-0-7277-6127-9

© The authors and ICE Publishing: All rights reserved, 2016
doi:10.1680/tfitsi.61279.325

Surface deformation evaluation of sandwich plate with aluminum foam cores under high speed impact loading based on stereovision

Baohua Shan*[1,2], Boyi Zhang[1,2] and Yu Yan[2]

[1] *Country Key Lab of Structures Dynamic Behavior and Control (Harbin Institute of Technology), Harbin, China*
[2]*Harbin Institute of Technology, Harbin, China*
**Corresponding Author*

ABSTRACT In order to evaluate surface deformation of sandwich plates with aluminum foam cores under high speed impact loading, a surface deformation detection method based on the parallel stereovision model is proposed in this paper. Total five constraints are combined together to optimize the disparity map, and the maximum and minimum disparity constraint is achieved by adjusting the disparity range to remove false parallax. 3D point clouds of deformed surface are calculated according to the parallel stereovision model, and the deformed surface are gained by Delaunay triangulation accordingly. A stereovision surface deformation measurement system is integrated on the base of the above algorithm, the surface deformation measurement test of aluminum foam cores sandwich plates under high speed impact loading is performed with the integrated system. Experimental results show that the reconstructed deformed surface can accurately display the surface deformation of sandwich plates with aluminum foam cores under high speed impact loading, this verifies the effectivness and reliability of the proposed method.

1 INTRODUCTION

In the high-speed impact test, the depth and diameter of crater under high-speed bullet impact are the important characteristic parameters. These parameters can reflect damage condition of target plate subjected to bullet impact, and have a crucial influence on the flight security of aircraft and the impact resistance of structural component and material. As a result, researcher pay attention to this hot continuously (Pang *et al.* 2013; Chi *et al.* 2009).

At present, the scanner or camera is the most common tool for measuring the depth and diameter of crater under high-speed impact. The impact surface photo of crater is firstly obtained by scanner or camera. Then the wire-electrode cutting is used to cut open crater along the centerline. The cross-sectional photo of crater is correspondingly acquired by scanner or camera after grinding. PHOTOSHOP software is used to measure the diameter and depth of crater according to the method of reference (Murr *et al.* 2002). Finally, MATLAB is adopted to calculate the volume of crater by using the concept of rotating body (Zhou 2010).

The usage premise of the above method lies in a consideration that the deepest of crater is exactly located at the middle surface of crater and the crater volume is symmetrical. Therefore the above method measuring the diameter, depth and volume of crater is a kind of approximate method, and can't get the actual deformed surface of structure or material under high speed impact. To solve this problem, a stereovision-based surface deformation detection method is presented in this paper. A stereovision-based surface deformation measurement system is also integrated on the base of the proposed method, and a surface deformation evaluation test is conducted on the sandwich plate aluminum foam cores under high speed impact to testify the practicability and reliability.

2 THREE-DIMENSIONAL MEASUREMENT PRINCIPLE OF STEREOVISION

Based on the parallel stereovision model, this paper conducts 3D deformation measurement. The parallel stereovision model is the simplest and most typical binocular stereovision model, this model can measure 3D coordinates of measured point based on the principle of disparity (Zhang 2008). As shown in Figure 1, assuming that the imaging planes of two cameras are in the same plane, two cameras simultaneously observe the same point P in space. The image coordinates of point P in the left and right camera system are (X_{left}, Y_{left}) and (X_{righ}, Y_{right}), respectively. Because the imaging planes of two cameras are in the same plane, Y_{left} is equal to Y_{right}, and both are equal to Y.

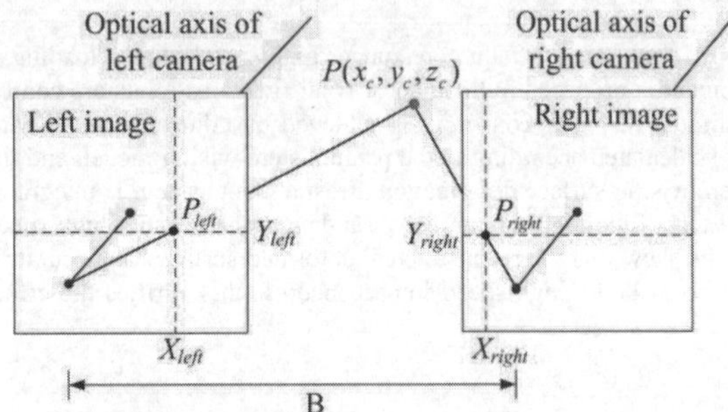

Figure 1. Parallel stereovision model

Supposing that 3D coordinate of point P in the left camera coordinate system is (x_c, y_c, z_c), and the disparity between the left and right image is $X_{left} - X_{right}$ Accordingly, the 3D coordinates of point P is expressed as follows (Wu 2008).

$$\begin{cases} x_c = \dfrac{B \cdot X_{left}}{Disparity} \\ y_c = \dfrac{B \cdot Y}{Disparity} \\ z_c = \dfrac{B \cdot f}{Disparity} \end{cases} \quad (1)$$

Where B is the substrate distance, f is the focal length.

In this paper, B and f can be acquired by epipolar rectification (Hartley & Zisserman 2003). As can be seen from Eq. (1), if B, f and $Disparity$ are all known, it is easy to calculate 3D coordinates of point P.

From Eq. (1), it can be seen that surface deformation evaluation needs to obtain the 3D coordinates of feature point. It is worth noting that any feature point on deformed surface are also feature points on the left and right images captured by cameras, respectively. According to the disparity of the same feature point on left and right images, this paper employs the binocular stereovision approach to reconstruct 3D coordinate of feature point through feature extraction and matching. Therefore, the main issue of surface deformation evaluation is how to identify the same feature point on the left and right images (Zhang 2000).

3 IMAGE PROCESSING

Reconstructing 3D deformation surface require the dense 3D point clouds. However, the disparity map which is obtained by feature matching is sparse and distributed unevenly, this means that only some sparse 3D points can be achieved (Giachetti 2000). Therefore, the area matching is firstly chosen to implement stereo matching of left and right images.

Then the Tsukuba image pair, which comes from the Middlebury database (Scharstein & Szeliski 2011), is adopted to compare the disparity maps gained by three correlation criteria such as SAD, SSD and ZNCC. Because the ZNCC correlation criterion has the most robust noise-proof performance and is insentive to the linear scale and offset in lighting on the left and right images (Pan 2009). The ZNCC correlation criterion is selected to conduct area matching. Moreover, three groups of window size, namely 7×7, 15×15 and 21×21, are employed to compute the disparity maps of the Tsukuba image pair. the disparity map of Tsukuba image pair is discontinuous and seems to be broken when the window size is 7×7. Although the dispairty map has good continuity, the edge and detail are blurry when a window size of 21×21 is utilized. Only a window size of 15×15 acquires the disparity map which has both the good continuity and outstanding detail. As a consequence, a window size of 15×15 is chosen to perform area matching in this paper.

Next, five constraints, namely the epipolar constraint, the uniqueness constraint, the sequence constraint, the left-right consistency constraint and the maximum and minimum disparity constraint, are combined together to optimize the disparity map.

Because area matching assumes that the disparity of image is continuous, the disparity should appear within a certain range in theory, and the corresponding probability distribution histogram of disparity should also mainly concentrated within some continuous interval. However, the false disparity may break way from this centralized range or rarely appear. Therefore, both the discontinuous disparity interval and the lower probability interval in probability distribution histogram can be regarded as the false disparities.

Among these five constraints, the first four constraints can be easily implemented. To achieve the fifth constraint, the false disparity can be eliminated by adjusting the disparity range, and the false disparity value within this interval can be replaced by the mean of its nearest neighbor points for optimizing the disparity map. Thereby, the maximum and minimum disparity constraint is implemented by adjusting the disparity range of image. The detailed process is given as below:

(i) Obtaining the disparity map processed by the first four constraints.

The first four constraints are employed to process the disparity map, and the false disparities are removed and replaced by 0. At the moment, the disparity range is $[S_{\min}, S_{\max}]$.

(ii) Acquiring the probability distribution histogram of disparity.

The statistical analysis is applied to the probability distribution of disparity matrix including 0 disparity value. The range of disparity matrix is divided into 20 equal portions, and the number of disparity value in each portion is counted. Accordingly, the disparity probability of each portion is gained by dividing the number of each portion by the total amount. To visually display the disparity distribution, the probability distribution histogram of disparity is illustrated in Figure 2.

(iii) Adjusting the disparity range.

To eliminate the discontinuous disparity, the probability threshold p is selected to make $p \in [p_{n\max}, p_{l\min}]$. Where $p_{n\max}$ is the maximum probability of the discontinuous disparity interval. $p_{l\min}$ is the minimum probability of the continuous disparity interval.

Assume that S is the disparity value corresponding to the probability threshold p. Then the disparity values which are less than p are removed away from the probability distribution histogram. As a result, the range of horizontal axis in the probability distribution histogram is narrowed down, that is the disparity range is changed into $[S, S_{\max}]$.

(iv) Optimizing the disparity map.

In the disparity map, 0 is used to replace the removed disparity value, then the interpolation method is adopted to substitute 0. This means that the disparity mean of eight nearest-neighbor points is utilized to approximately denote the disparity of central point in a 3×3 window. It should to be noted that the center of a 3×3 window is the position of 0 disparity value.

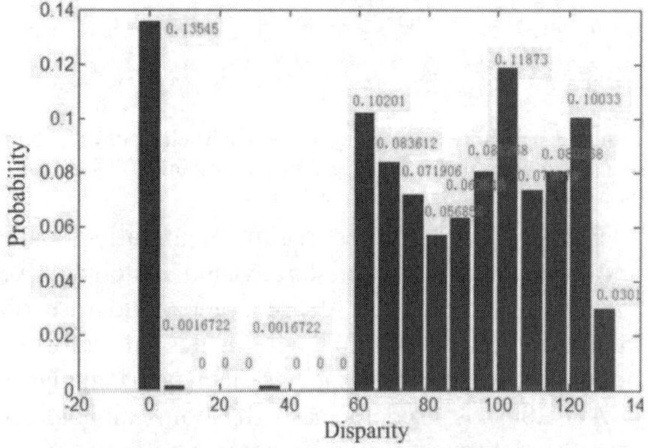

Figure 2. Histogram of disparity probability distribution

As shown in Figure 2, the disparity probability whose values are 0 is 0.13545. This means that the first four constraints eliminate 13.545% false disparity. The probability interval whose value is 0 indicates there no exists any disparity. As can be seen from Figure 2, the disparity is distributed within the range of [60, 130]. However, the probabilities of two discontinuous intervals are 0.0016722, so these two discontinuous intervals can be seen as the false disparities. Therefore, the probability threshold is supposed to be any value which belongs to [0.0016722,

0.0301]. Moreover, the disparities whose value is less than the threshold can be replaced by the mean of its eight nearest-neighbor points, and the disparity map is optimized accordingly.

The left and right images of 1st sandwich plate with aluminum foam cores under high speed impact loading are captured synchronously, and the rectangle zone which contains surface deformation is selected to conduct disparity optimization, the comparison result is given in Figure 3. As can be seen from Figure 3(a), there exist some discontinuous points in the original disparity map. Then the above optimization algorithms are used to eliminate the false disparities in Figure 3(a), and the disparity range is correspondingly adjusted. Figure 3(b) indicates that the disparity becomes continuous and the discontinuous points in Figure 3(a) are removed away.

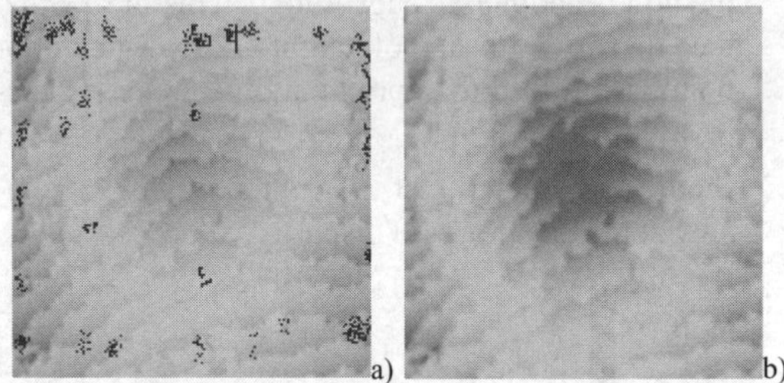

Figure 3. Disparity optimization of 1st sandwich plate with aluminum foam cores under high speed impact loading. a) Original disparity map; b) Optimized disparity map.

Finally, the deformed region is initially located and conducted per-pixel stereovision matching. According to Eq. (2), the dense point clouds are obtained by calculating 3D coordinates of deformed points on structural surface. Delaunay triangulation (GAO 2009) is used to mesh 3D point clouds, and then the deformed surface is correspondingly gained by illumination processing.

4 SURFACE DEFORMATION EVALUATION TEST OF SANDWICH PLATE WITH ALUMINUM FOAM CORES

4.1 Experimental setup

As shown in Figure 4, the integrated stereovision measurement system includes a computer, two CCD cameras, a tripod, surface deformation evaluation software. The cameras come with a 2/3 inch CCD and could record high-definition images with a pixel resolution of 1000×1000 at 60 fps. They are equipped with an optical zoom lens of F1.8-16 and focal length ranging between 12-30mm. In this work, the manufactured camera tripod can adjust horizontal distance, pitch angle and horizontal angle to satisfy the demand of stereovision detection. On the base of the above algorithm, the stereovision surface deformation evaluation software is compiled by MATLAB.

To test the precise of the integrated stereovision system, the distance measurement test is performed in lab. According to statistical analysis, experimental results indicate that the error of distance measurement at 95% probability is 0.03mm. Thereby, the accuracy of the integrated stereovision system in this paper is 0.03mm when the probability isn't greater than 95%.

Figure 4. Stereovision measurement system

4.2 Experimental setting

During test, the size of sandwich plate with aluminum foam cores is 200mm×200mm. The skin plate is made of steel, and the aluminum foam board is made of porous aluminum material 110um/ 1199. The sketch map of sandwich plate with aluminum foam cores is shown in Figure 5. Where t_f, t_c and t_b are the thickness of front plate, core plate and rear plate, respectively. Two sandwich plates with aluminum foam cores are adopted in test, both t_f and t_b of two plates are 0.8mm. However, t_c of two plates are different each other. t_c of the 1st plate is 5mm, and t_c of the 2nd plate is 10mm.

Because the skin plate is made of steel, the surface is smooth and has no texture, and reflects easily. Area matching algorithm isn't suitable for this kind of surface. Only the aluminum foam board is detected in this paper. The sandwich plate yields deformation under the impact of bullet, the photos of deformed aluminum foam boards are given in Figure 6. From Figure 6, there are some holes on two plates, and these holes are used to fix the aluminum foam board during the impact test.

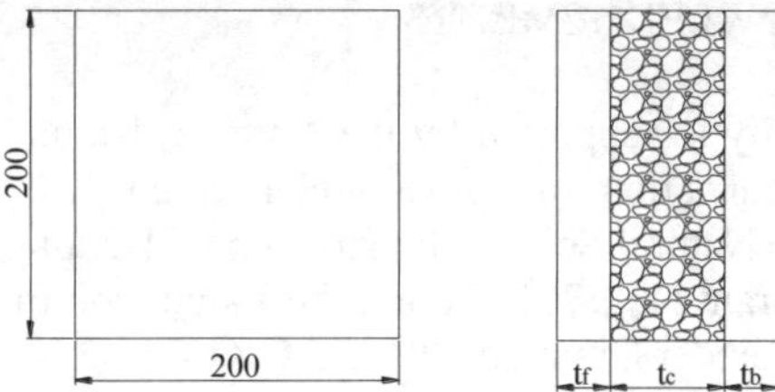

Figure 5. Sketch map of sandwich plate with aluminum foam cores

In test, two cameras are put away 110cm from the sandwich plate. The distance between two optical centers is 140mm, the angle of two optical axises is 10°, and the focal length of lens is 30mm. Firstly, the concave surface of 1st plate is measured, and the distortion calibration and epipolar calibration are used to process the left and right images of the concave surface of 1st plate.

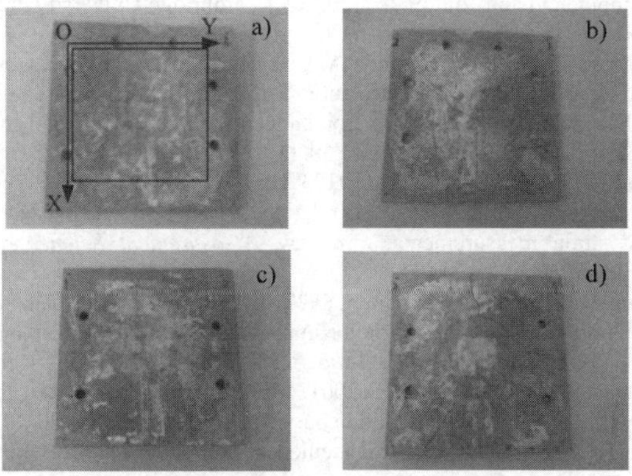

Figure 6. Photos of sandwich plates with aluminum foam cores: a) Concave surface of 1st plate. b) Convex surface of 1st plate. c) Concave surface of 2nd plate. d) Convex surface of 2nd plate.

During the impact test, the aluminum foam boards are fixed by bolts, and the displacements of screw holes can be seen as 0. Therefore, a certain point near to some screw hole can be chosen as the original point to establish the coordinate system. As shown in Figure 6(a), the coordinate transformation is conducted in the right-handed coordinate system composed of OX axis and OY axis. The rectangular zone of 605×580 in Figure 6(a) is selected to reconstruct 3D point clouds with the step size of 20 pixels, and total 899 points are obtained. The objective is to select undeformed surfac as a base plane for total points, then the deformed coordinate component in Z axis is the deflection value of concave surface.

4.3 Experimental results

The aluminum foam board is different from concrete material, whose surface is continuous and has no abrupt change. To get the more accurate result, 3D point clouds are processed by surface fitting. Then the local adding-weight linear regression method is utilized to perform data smoothing, and the curved surface is gained by triangulation of the fitted 3D point clouds. As shown in Figure 7(a), the change of deflection is displayed in different color, and the color bar denotes the deflection value.

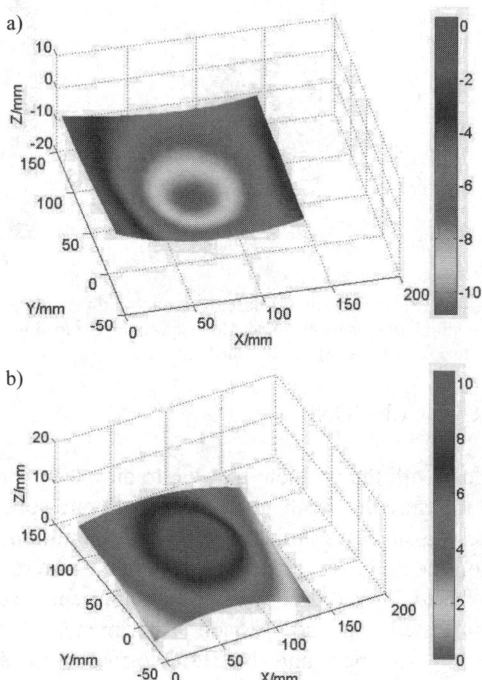

Figure 7. 3D deformation surface of 1st sandwich plate with aluminum foam cores under high speed impact loading. a) Concave surface. b) Convex surface.

The same method as detecting the concave surface of 1st sandwich plate is adopted to inspect the convex surface of 1st sandwich plate and the 2nd sandwich plate, and the measurement results are illustrated in Figure 7(b) and Figure 8.

a)

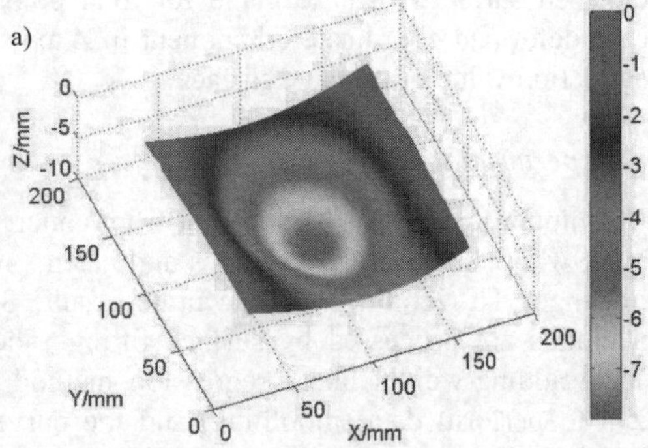

b)

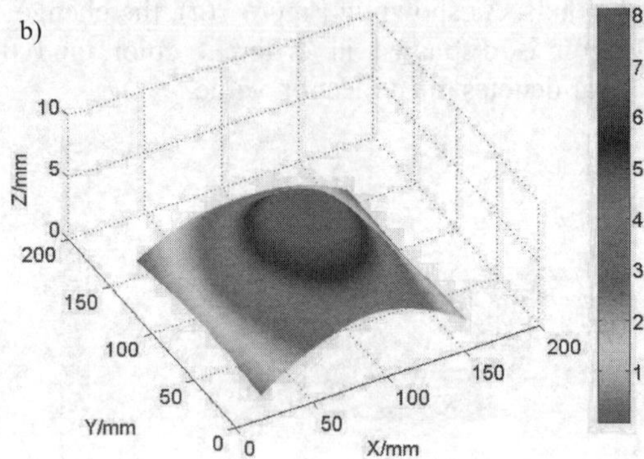

Figure 8. 3D deformation surface of second sandwich plate with aluminum foam cores under high speed impact loading. a) Concave surface. b) Convex surface.

5 CONCLUSIONS

To deal with the problem that the traditional method only measure the depth and diameter of crate approximately, a stereovision-based surface deformation detection method is proposed in this paper. A stereovision-based surface deformation measurement system is integrated. The comparison experiments of distance measurement and the 3D deformation surface measurement test of aluminum foam cores sandwich plates under high speed impact loading are performed with the integrated system. Experimental results show that the surface deformation measurement method based on the parallel stereovision model is feasible, and this verifies the reliability and effectiveness of the proposed method. The measured surface can accurately reflect the actual shape of sandwich plate with aluminum foam cores under high speed impact, and a new detection method is provide for measuring surface deformation of target plate under high speed impact in this paper.

ACKNOWLEDGEMENTS

This study is supported by the National Natural Science Foundation of China under Grant No.51478 148, the Natural Science Foundation of Heilongjiang under Grant No.E201434 and the foundation of Harbin City under Grant no. 2015RAQXJ028.

REFERENCES

Chi R. Q., Pang B.J., He M.J., Guan G. S., Yang Z. Q. & Zhu Y. 2009, Experimental investigation for deformation and fragmentation of spheres penetrating sheets at hypervelocity, *Explosion and shock waves*, 29(3), 231-236.

Gao H.W., Yu Y. & Liu X.Y. 2009, Research of 3D reconstruction experiment platform based on binocular vision, *Computer Engineering and Applications*, 45, 149-152.

Giachetti, A. 2000, Matching techniques to compute image motion, *Image and Vision Computing*, 18(3), 247-260.

Hartley R.I. & Zisserman A. 2003, Multiple View Geometry in Computer Vision. 2nd Ed. Cambridge, Cambridge University Press, UK.

Murr E., Trillo E.A., Bujanda A.A. & Martinez N. E. 2002, Comparison of residual microstructures associated with impact craters in Fcc stainless steel and Bcciron targets: the microtwin versus microband issue, *Acta Materialia*, 50(1), 121-131.

Pan, B., Qian, K.M., Xie, H.M. & Asundi, A. 2009, Two-dimensional digital image correlation for in-plane displacement and strain measurements: a review, *Measurement Science and Technology*, 20, 062001.

Pang B. J., Zheng W. & Chen Y. 2013, Dynamic impact behavior of aluminum foam with a Taylor impact test and a theoretical analysis, *Journal of vibration and shock*, 32(12), 154-158.

Scharstein D. & Szeliski R. 2001, Middlebury Stereo Vision Research Page, http//vision.middlebury.edu/stereo/data/scenes.

Wu F.C. 2008, Mathematical method in computer vision, Beijing, Science Press, 46-47.

Zhang G.J. 2008, Vision measurement, Beijing, Science Press, 134-136.

Zhang Y. J. 2000, Image understanding and computer vision, Beijing, Tsinghua University Press, 82-83.

Zhou D. L. 2010, Deformation and damage behaviors of solution treated AM60B magnesium alloy under high velocity impact, Ph. D. dissertation of Harbin Institute of Technology, 24-25.

Proceedings of the International Conference on Smart Infrastructure and Construction
ISBN 978-0-7277-6127-9

© The authors and ICE Publishing: All rights reserved, 2016
doi:10.1680/tfitsi.61279.331

A quantitative guided wave-based damage monitoring method for complex composite structures

Z. Wu *, K. Liu , S. Ma and Y. Zheng

State Key Laboratory of Structural Analysis for Industrial Equipment, Dalian University of Technology, China

* *Corresponding Author*

ABSTRACT Ultrasonic guided waves have emerged as one of the most prominent tools for structural health monitoring (SHM) due to their excellent propagation capability, through the thickness interrogation and ability to detect small damages within relatively large inspection areas. Accurate interpretation of captured guided wave signals is critical for the ultrasonic guided wave-based SHM. However, as the propagation mechanism of guided wave is quite complicated in composite structures due to geometric and structural complexity, it is quite difficult to extract the signal features for damage identification, such as the time of flight (ToF), the propagation mode and corresponding group velocity, accurately. To avoid direct interpretation of the ultrasonic guided wave signal and the effect of dispersion, the probability-based diagnostic imaging (PDI) has been studied intensively by many researchers. However, the PDI algorithm's influencing parameters, including the selection of certain damage index and frequency and the size of the effective elliptical distribution area are empirically determined. In addition, though good for localizing the defects when the related parameters is selected in an appropriate range, most of the methods lack the capability to quantitative evaluate the defects in composite. Therefore, the application of the method to quantitative monitor damages is limited in real world practices. In this study, the influences of the above-mentioned various factors on the diagnostic image pixel peak value are studied and discussed in detail. Then a novel quantitative guided wave-based damage monitoring method is developed, which is realized with probability by extracting the image pixel peak value, to improve the ability of damage quantification in composite. The validity of the approach is assessed by quantitative evaluating damages of different extent and at different locations on a stiffened composite panel.

1 INTRODUCTION

In recent years, composite structures have been widely applied in construction of components in mechanical, aerospace, shipbuilding, and other industries. However, laminated composite structures are prone to develop delamination during the manufacturing process as well as in-service events, thereby reduce the mechanical properties of a component and even lead to the sudden destruction of the entire structures. These hidden delaminations are not easily detectable by visual inspection. Thus, real-time Structural Health Monitoring (SHM) for complex composite structures becomes very important and urgently required (Gorgin et al., 2014).

Ultrasonic guided waves have emerged as one of the most prominent tools for structural health monitoring (SHM) due to their excellent propagation capability, through the thickness interrogation and ability to detect small damages within relatively large inspection areas (Gao et al., 2014). Accurate interpretation of captured guided wave signals is critical for the ultrasonic guided wave-based SHM. However, as the propagation mechanism of guided wave is quite complicated in composite structures due to geometric and structural complexity, it is quite difficult to extract the signal features for damage identification, such as the time of flight (ToF), the propagation mode and corresponding group velocity, accurately (Wang et al., 2010).

To avoid direct interpretation of the ultrasonic guided wave signal and the effect of dispersion, the probability-based diagnostic imaging (PDI) has been studied intensively by many researchers (Wang et al., 2009, Zhao et al., 2007). However, the PDI algorithm's influencing parameters, including the selection of certain damage index and frequency and the size of the effective elliptical distribution area are empirically determined. In addition, though good for localizing the defects when the related parameters is selected in an appropriate range, most of the methods lack the capability to quantitative evaluate the defects in composite. Therefore, the application of the method to quantitative monitor damages is limited in real world practices.

In this study, the influences of the above-mentioned various factors on the diagnostic image pixel peak value are studied and discussed in detail. Then a novel quantitative guided wave-based damage monitoring method is developed, which is realized with probability by extracting the image pixel peak value, to improve the ability of damage quantification in composite. The validity of the approach is assessed by quantitative evaluating damages of different extent and at different locations on a stiffened composite panel.

2 QUANTITAIVE DAMAGE MONITORING METHOD

In the PDI method, the monitoring area is meshed into uniformly distributed grids and the probability image is yielded with summing the signal difference features in elliptical patterns for various pitch-catch transducer pairs. Assuming that there are totally N sensing paths in a sensor network, the probability of damage occurring at certain grid (x, y) can be calculated as:

$$P(x,y)=\sum_{i=1}^{N} DI_i \cdot W_i[R_i(x,y)]. \qquad (1)$$

Here, DI_i is the damage index of the ith sensing path, $W_i[R_i(x, y)]$ is the weight distribution function of the ith sensing path. This weight is expected to increase with a decrease in the relative distance as a defect would cause the most significant signal change in the direct wave path, and that the signal change effect would decrease if the defect is away from the direct path of the sensor pair (Zhao et al., 2007, Ihn and Chang, 2004). $R_i(x, y)$ is defined as the relative distance from (x, y) to the ith sensing path, which is expressed as

$$R_i(x,y)=\frac{D_{a,i}(x,y)+D_{s,i}(x,y)}{D_i}-1, \qquad (2)$$

where D_i is the distance between the actuator and sensor for the ith sensing path, while $D_{a,i}(x, y)$ and $D_{s,i}(x, y)$ are respectively the distances between (x,y) and the actuator and sensor for the ith sensing path. The weight distribution function can be written as

$$W_i[R_i(x,y)]=\begin{cases}1-\dfrac{R_i(x,y)}{\beta}, & R_i(x,y)<\beta \\ 0, & R_i(x,y)\geq\beta\end{cases}. \qquad (3)$$

where β is a scaling parameter between interval (0,1), it is used to control the size of the effective elliptical distribution area in the PDI algorithm.

Through repeated experimental tests or theoretical modeling of different damage sizes at different locations, the relation between the peak values of probability maps versus damage sizes can be established (Qiu et al., 2013). First a damage with the same size is placed randomly in different locations. Then the peak value is obtained by the PDI algorithm in each location. Subsequently, a probability model of normal distribution is introduced with the expectation and standard deviation of peak values as μ and σ, respectively. At last a damage sample library is obtained by the probability models of different damage sizes.

As to an unknown damage, if the calculated PDI image peak amplitude is closer to the expectation μ of a damage model, the more likely the damage size gets closer to the model size. Thus the size probabilities that an unknown damage belongs to a certain size i in the damage sample library can be expressed as:

$$P_i=1-\left|F\left((2\mu_i-x);\mu_i,\sigma_i\right)-F\left(x;\mu_i,\sigma_i\right)\right|, \qquad (4)$$

where x is the image peak amplitude of the unknown damage, F is the cumulative distribution function of normal distribution, which can be calculated as

$$F(x;\mu,\sigma)=\frac{1}{\sigma\sqrt{2\pi}}\int_{-\infty}^{x}\exp\left(-\frac{(t-\mu)^2}{2\sigma^2}\right)dt. \tag{5}$$

3 EXPERIMENTAL DETAILS

Figure 1 shows a U7192 carbon fiber composite panel with four T shape stiffeners that was acquired from the COMAC for this study. Eight APC851 PZT sensors are mounted on the composite panel using the SMART Layer of Acellent Technologies Inc, as shown in figure 2. The dimensions of PZT sensors are 6.25 mm in diameter and 0.23 mm in thickness. The ScanGenie hardware of Acellent Technologies Inc is used for data acquisition. The sampling rate is set at 24 MHz. The Hann window modulated five-peak sine burst actuating signal with different scanning frequencies from 200 kHz to 400 kHz (25 kHz step) are drove to the actuators in this specimen.

As different damage indices (DIs) may have different sensitivity to the same damage, the selection of DIs is a key to distinguishing damage from benign changes such as noises and environmental variations. In this study, seven different damage indices including three kinds (SST, SSS, SSS6*dB*) which are based on the scatter signal, three kinds (SDT(Qing et al., 2006), SDS, SDS6*dB*) which are based on the signal difference, and SDCC (Zhao et al., 2007) which is based on the correlation coefficient are considered. The details of the DIs can be found in the Reference (Wu et al., 2015). Note that this study does not attempt to find an explicit criterion of the DI selection for a particular case of damage, but to prevent the different sensitivity of DIs from impacting on the analysis of the quantification results.

Previous studies (Wang et al., 2009, Zhao et al., 2007) have demonstrated that a small β reduces the affected zone, making the perception sharp and artifacts will be introduced; conversely, the resolution is lost if β is too large. As the scaling parameter is empirically determined on a case-by-case basis in previous studies, β are set from 0.1 to 0.9 in 0.2 increments, to investigate its effect on quantifying the damage in the composite panel.

To reduce the experiment cost, a kind of solid adhesive tape as thick as 3mm are bonded on the panel for artificial damage (Sohn et al., 2004; Qiu et al., 2013). Four damage sample cases of different positions (A, B, C and D) are established as illustrated in table 1 and figure 2. The damage levels are enlarged from D01 to D04 in each cases. The diameters of D01~D04 are 1.35 cm, 1.91 cm, 2.34 cm and 2.7 cm, respectively. Then the damage sample library can be obtained from the expectation and standard deviation of different damage sample levels. At last ten simulated defects with different extent at different locations are bonded on the stiffened composite panel to assess the effectiveness of the damage quantification algorithm, as shown in table 3.

Table 1. Details of different positions of artificial damage.

Damage	Location (mm)	Damage	Location (mm)
A	(284,278)	C	(284,354)
B	(284,430)	D	(360,278)

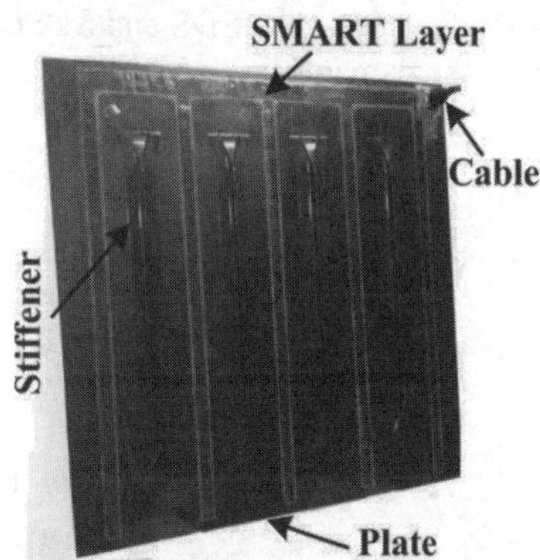

Figure 1. The stiffened composite panel.

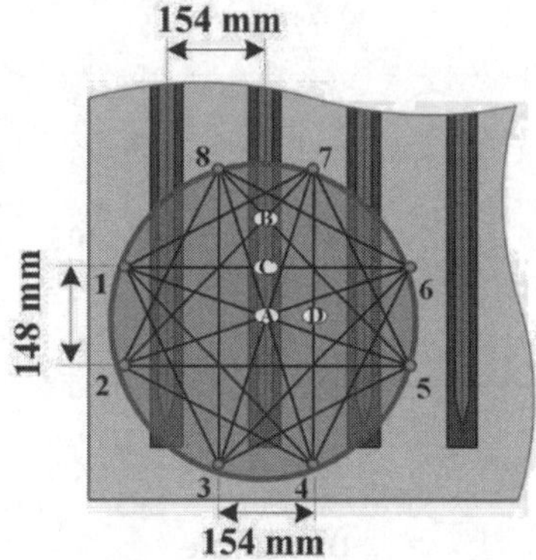

Figure 2. Diagram of the configuration of sensing paths, where the small circles indicate sensor locations, the large circles indicate the damage locations.

4 RESULTS AND DISCUSSION

It is feasible to evaluate the potential of the PDI algorithm for sizing defects by the enlarging damages from D01~D04. Figure 3 shows the monotonic relationship between the pixel peak value and damage size for both position A~D with SST damage index when β=0.1~0.9 . To consider the results of multiple frequencies, the pixel peak value in Figure 3 is calculated with averaged fusion. It is observed that the scaling parameter β has a small influence on sizing defects, though it shows a significant effect on localizing defects.

Moreover, the damage index shows a significant influence on sizing defects, as illustrated in Figure 4. Here the normalization pixel peak value is the average of both different scaling parameters and damage position. The results of SSS and SSS6*dB* show more sensitive to small damage as the SDS and SDS6*dB* show more linear to the damage size.

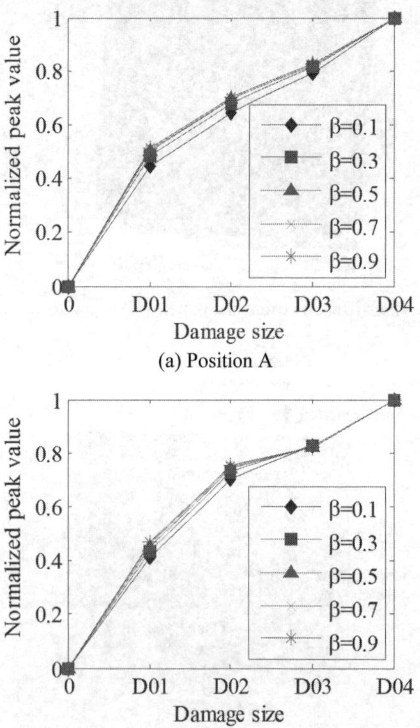

(a) Position A

(b) Position B

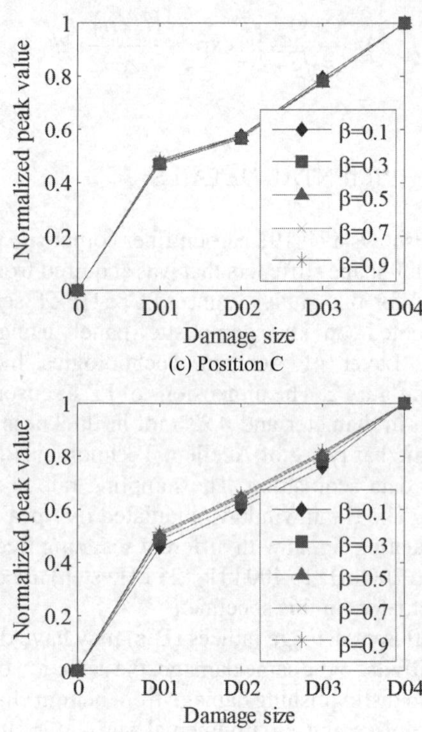

(c) Position C

(d) Position D

Figure 3. Pixel peak value (Normalization) versus damage size with SDS6*dB* damage index.

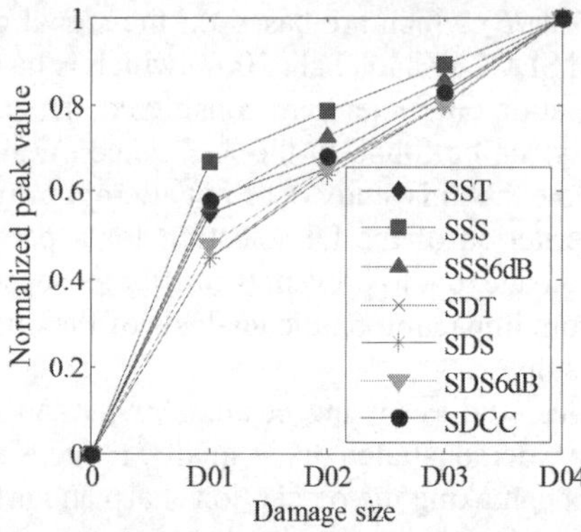

Figure 4. Pixel peak value (Normalization) versus damage size with different damage indices.

On the other hand, the image amplitude of damages with the same size D04 at different positions shows different value using the PDI algorithm, as shown in Figure 5. It is clearly illustrated that the center frequency of Lamb Wave affects not only the difference between various position damages, but also the amplitude of images. The variance of the image amplitude for D04 damage at position A~D can be calculated as shown in Figure 6. As the effect on the amplitude is compensated using normalized process as shown in Figure 6, it can be clearly seen that the effect of the frequency as well as the selected DI on the difference among damages of various positions.

In damage quantification, the amplitude variance is expected to be smaller. In other words, the more approximate amplitude is expected to be obtained with the same damage of various positions. So 275 kHz is selected for damage quantification since most of DIs are quite small in the frequency, as shown in Figure 6. The expectation and standard deviation of the PDI peak values at 275 kHz under different damage sizes are shown in table 2. Table 3 lists the estimation results from equation (4) using the damage sample library. The proposed algorithm showed great potential for damage quantification while the estimated sizes are matched with the actual sizes in all cases.

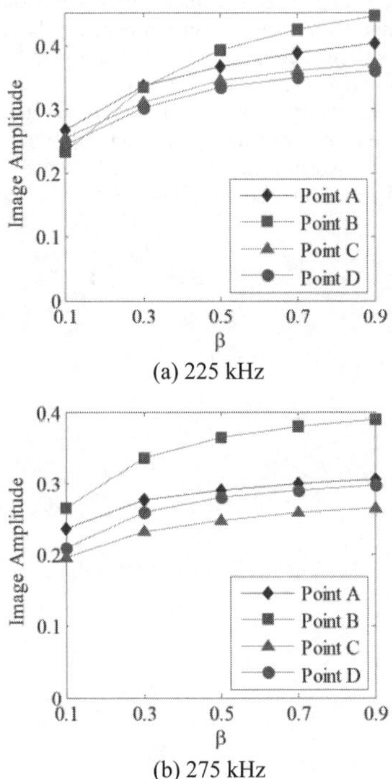

(a) 225 kHz

(b) 275 kHz

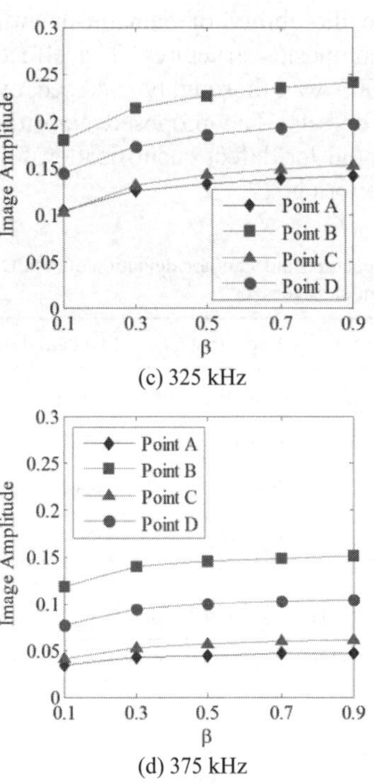

(c) 325 kHz

(d) 375 kHz

Figure 5. Image amplitudes of D04 damage with SST damage index.

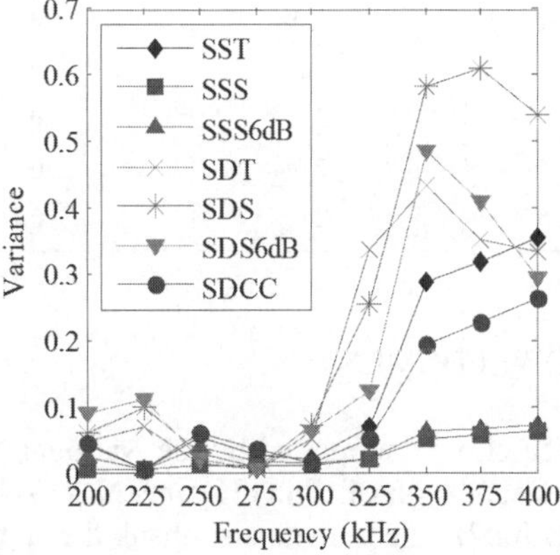

Figure 6. Variation of the difference between various position damages with frequency for different DI.

5 CONCLUSIONS

This study extended the previous work of probabilistic diagnostic imaging approach and developed a quantitative guided wave-based damage monitoring method to improve the ability of damage quantification in complex composite structures. The effectiveness of the algorithm was thoroughly assessed over several simulated defects. The proposed algorithm showed great potential for defect quantification for structural health monitoring.

Table 2. Expectation and standard deviation of the PDI peak values under different damage sizes.

Damage size (cm)	Expectation μ	Standard deviation σ
1.35	0.2210	0.0188
1.91	0.3071	0.0329
2.34	0.3773	0.0262
2.70	0.4702	0.0276

Table 3. Estimation results of damage size.

No.	Actual size (cm)	Estimation size (cm)	Error (cm)
1	1.4	1.2	0.2
2	1.4	1.4	0
3	1.7	1.8	0.1
4	1.7	1.9	0.2
5	2.0	2.0	0
6	2.0	2.0	0
7	2.3	2.2	0.1
8	2.3	2.4	0.1
9	2.5	2.6	0.1
10	2.5	2.4	0.1

ACKNOWLEDGMENT

This research was supported by the National Natural Science Foundation of China (Grants Nos. 91016024 and 51475067). The authors also thank the support of the Fundamental Research Funds for the Central Universities (NCET-11-0055).

REFERENCES

Gao, D. Wang, Y. Wu, Z. Rahim, G. & Bai, S. 2014. Design of a sensor network for structural health monitoring of a full-scale composite horizontal tail. *Smart Materials and Structures* **23,** 055011.

Gorgin, R. Wu, Z. Gao, D. & Wang, Y. 2014. Damage size characterization algorithm for active structural health monitoring using the A0 mode of Lamb waves. *Smart Materials & Structures* **23,** 623-626.

Ihn, J. B. & Chang, F. K. 2004. Detection and monitoring of hidden fatigue crack growth using a built-in piezoelectric sensor/actuator network: I. Diagnostics. *Smart Materials and Structures* **13,** 609-20.

Qing, X. P. Chan, H.-L. Beard, S. J. & Kumar, A. 2006. An active diagnostic system for structural health monitoring of rocket engines. *Journal of intelligent material systems and structures* **17,** 619-628.

Qiu, L. Liu, M. L. Qing, X. L. & Yuan, S. F. 2013. A quantitative multidamage monitoring method for large-scale complex composite. *Structural Health Monitoring-an International Journal* **12,** 183-196.

Wang, D. Ye, L. Lu, Y. & Su, Z. Q. 2009. Probability of the presence of damage estimated from an active sensor network in a composite panel of multiple stiffeners. *Composites Science and Technology* **69,** 2054-2063.

Wang, D. Ye, L. Su, Z. Q. Lu, Y. Li, F. C. & Meng, G. 2010. Probabilistic Damage Identification Based on Correlation Analysis Using Guided Wave Signals in Aluminum Plates. *Structural Health Monitoring-an International Journal* **9,** 133-144.

Wu, Z. Liu, K. Wang, Y. & Zheng, Y. 2015. Validation and evaluation of damage identification using probability-based diagnostic imaging on a stiffened composite panel. *Journal of Intelligent Material Systems and Structures* **26,** 2181-2195.

Zhao, X. Gao, H. Zhang, G. Ayhan, B. Yan, F. Kwan, C. & Rose, J. L. 2007. Active health monitoring of an aircraft wing with embedded piezoelectric sensor/actuator network: I. Defect detection, localization and growth monitoring. *Smart Materials & Structures* **16,** 1208-1217.

Proceedings of the International Conference on Smart Infrastructure and Construction
ISBN 978-0-7277-6127-9

© The authors and ICE Publishing: All rights reserved, 2016
doi:10.1680/tfitsi.61279.337

Monitoring system by vibration power generation

Y. Yoshida*[1], Y. Kobayashi[1] and T. Uchimura[2]

1 Railway Technical Research Institute, Tokyo, Japan
2 The University of Tokyo, Tokyo, Japan
** Corresponding Author*

ABSTRACT To solve a problem of power supply for structural health monitoring, we developed monitoring systems operated with power generation by vibration of a railway bridge. The greatest advantage of these monitoring systems is to reduce the initial cost of the system installation and the running cost for replacing batteries. In order to operate these monitoring systems by vibration power generation, we developed a technique to generate much power by vibration. Based on the technique, we installed the monitoring systems on the existing railway bridge, and confirmed their operation by vibration power generation.

1 INTRODUCTION

A power supply for a bridge health monitoring system often becomes a potential problem, since the commercial power is not always supplied at the bridge site. Even if the power supply can be used at the site, the wiring to each sensor or measurement equipment is likely to increase the initial cost of the system installation. In recent years, use of the battery-operated system, i.e. sensor network system (Nagayama et al. 2010), is becoming the mainstream. However, there is still the problem of the running cost required for replacing the battery every few years.

We have tried to use power generated by railway bridge vibration for operating the bridge health monitoring system. The power can be generated by a piezoelectric device attached to a bridge member. Using the power generated by vibration enables to eliminate the running cost for replacing the battery.

However, operating the monitoring system requires generating power by vibration efficiently, because the amount of the power generated by vibration is very small.

In this paper, we examined the technique to generate much power by vibration on an existing railway bridge. Based on the result, we installed two types of monitoring system operated by vibration power generation on the bridge and verified their operation under the real environment.

2 TECHNIQUE TO GENERATE POWER BY VIBRATION

2.1 *Vibration Power Generation*

A concept of vibration power generation appears in Figure 1. The piezoelectric device generates electric current according to stress (Imai et al. 2014). The stress is generated on the piezoelectric device attached to the bridge member vibrating when a train passes on the bridge. Vibration power generation with the piezoelectric device allows supplying power permanently, and does not require much cost for the installation.

Devices in the monitoring system, such as sensors or transmitters, are operated by power in the capacitor connected to the piezoelectric device through rectifier circuit. The power in the capacitor

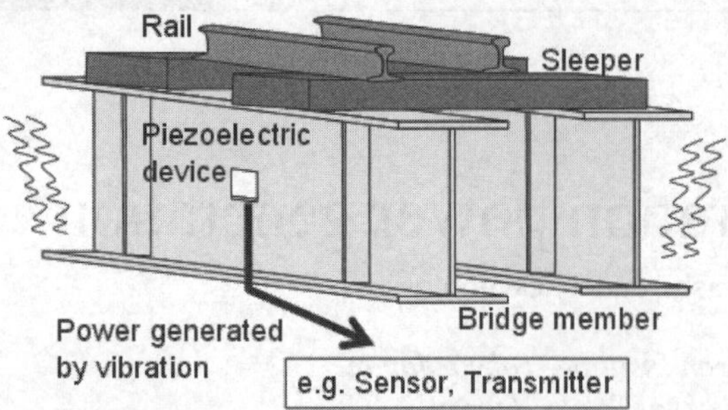

Figure 1. Concept of vibration power generation

Table 1. Variables in equation (1).

Variable	Factor
ω: frequency of stress σ_0: amplitude of stress	Bridge member
A: area s: Thickness d_α : piezoelectric coefficient ε : dielectric coefficient R_d: internal resistance	Piezoelectric device
R: load resistance	Circuit

increases a according to the current generated by the piezoelectric device Therefore, it is essential to generate the current efficiently by the piezoelectric device for operating the monitoring system.

2.2 *Factors Related to Vibration Power Generation*

The quantity of current generated by the piezoelectric device attached to the bridge member is shown in the equation (1). Table 1 summarizes the variables in the equation (1).

$$I = \frac{d_\alpha \sigma_0}{\frac{1}{j\omega A}\cdot\left(1+\frac{R}{R_d}\right)+\frac{\varepsilon}{s}R}\cdot e^{j\omega t} \qquad (1)$$

As Table 1 indicates, the variables are categorized into three factors: bridge member, piezoelectric device, circuit. We examined these factors to generate power efficiently by vibration on the railway bridge.

2.3 *Generating Power Efficiently*

In order to develop the technique to generate power efficiently by vibration, we compared the power generated by vibration under the different condition of the bridge member, the piezoelectric device and the

Table 2. Condition of factors

Factor		Parameter
Bridge member		flange, web, stiffener
Piezoelectric device	Area [mm^2]	20×20, 80×80
	Thickness [mm]	0.5, 1.0, 2.0
	Piezoelectric coefficient [pC/N]	210, 354
Circuit	Rectifier circuit	Full wave rectifier circuit Cockcroft-Walton circuit
	Capacitor	Ceramic capacitor (Capacitance:47μF)

circuit on an existing railway truss bridge. Table 2 shows the condition of these factors.

We attached the piezoelectric devices to the bridge members of different stress amplitude and frequency. The piezoelectric devices are different in their area, thickness and piezoelectric coefficient. As the circuit has to collect the power generated by vibration, we tested two types of the rectifier circuit.

We confirmed the power generated by vibration when the train passes on the bridge by measuring the voltage across the capacitor connected to the rectifier circuit. The stress on the piezoelectric device was measured to confirm the stress by vibration.

As an example of the measurement result, (a) the stress and (b) the capacitor voltage of the flange and the web is shown in Figure 2. The stress is generated on the piezoelectric device by vibration when the train passes on the bridge. The stress frequency of the web is much higher than that of the flange although the stress amplitude is almost the same. As a result, the capacitor voltage of the web increases, in contrast to that of the flange. According to equation (1), the reason for this is that the quantity of current increases with the stress frequency.

From the result, we found that the bridge member where a high frequency stress is generated, like the web, should be used. Other than this, we confirmed which piezoelectric device and circuit should be used to generate power efficiently by vibration.

3 MONITORING SYSTEM TO DETECT DAMAGE

Based on the findings in the previous chapter, we developed two types of the monitoring system operated by vibration power generation. One is the monitoring system to detect damage of bridges, and another is the monitoring system to predict deterioration of bridges.

In this chapter, we describe the concept of the monitoring system for detecting damage and the examination to confirm the operation in the real environment.

3.1 Concept

The concept of our developed monitoring system to detect damage is shown in Figure 3. This monitoring system senses damage by a sensor, and transmits the sensing results to the train which is passing on the bridge. The greatest advantage of this monitoring system is to reduce the cost of power supply for the monitoring system by vibration power generation. In addition, this monitoring system enables to collect the sensing results easily by receiving the sensing results on the running train. For example, this monitoring system allows an inspector with a receiver to collect the sensing results of multiple bridges equipped with the monitoring system on the running train at a stroke.

An example of the use of this monitoring system is to monitor fatigue crack of bridges. In this case, generation of fatigue crack at a structural weak point (i.e. welded part, notched part and so on) is detected by a sensor such as a conductive coating material with length of tens of centimeter drawn near the structural weak points. Checking the electrical conductivity of the conductive coating material, we can confirm that the fatigue crack, because the conductive coating material becomes non-conductive when it is broken by the crack.

3.2 Reducing Power Consumption

In order to operate this monitoring system by vibration power generation, it is necessary to reduce power consumption for collecting the sensing results. In a typical monitoring system for a large scale structure, multi-hop communication has been often em-

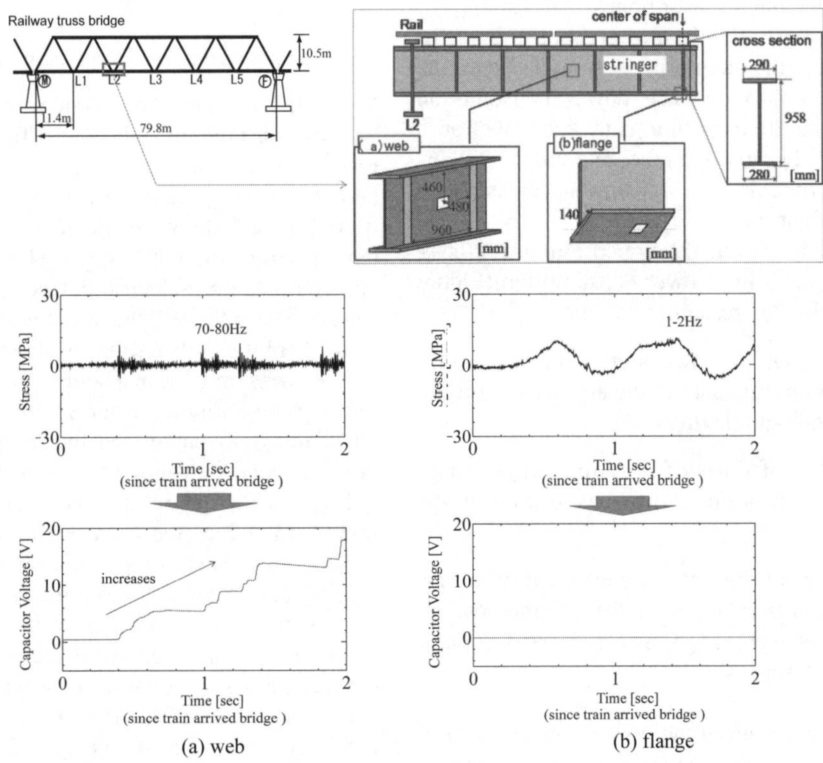

Figure 2. Stress, capacitor voltage

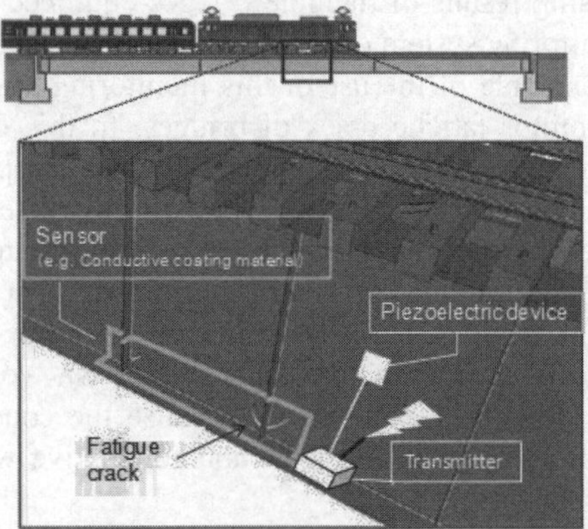

Figure 3. Concept of monitoring system to detect damage

ployed, which includes the process of aggregating the sensing results to a gateway by relaying between each sensor node. The process requires each sensor node to be in a standby state for receiving data of other sensor nodes. Additionally, each sensor node transmits data for a large number of times, since each sensor transmits data from the other sensor nodes.

Figure 4(a) illustrates the power consumption of the sensor node in the monitoring system based on the multi-hop communication. The power consumption for receiving and transmitting data accounts for a large proportion, because the receiving consumes current for a long time and the transmitting consumes a large amount of current.

On the other hand, our developed monitoring system is operated with low power consumption (Figure 4(b)) for the following reason.

- Receiving data is not necessary, since each sensor node transmits data to the approaching train directly without relaying.
- The number of times of transmission is small, since each sensor nodes only has to transmit its own data
- The amount of the current consumed by every transmission is small, since the communication distance between each sensor node and the approaching train is short.

In this way, we reduced the power consumption of the monitoring system to the level enough to be supplied by vibration power generation.

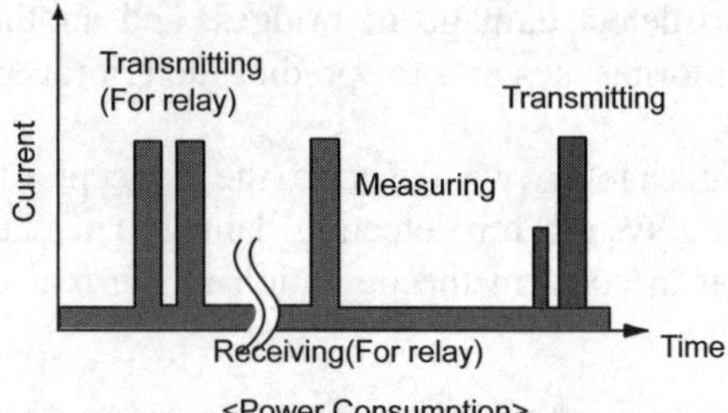

(a) Monitoring system with multi-hop network

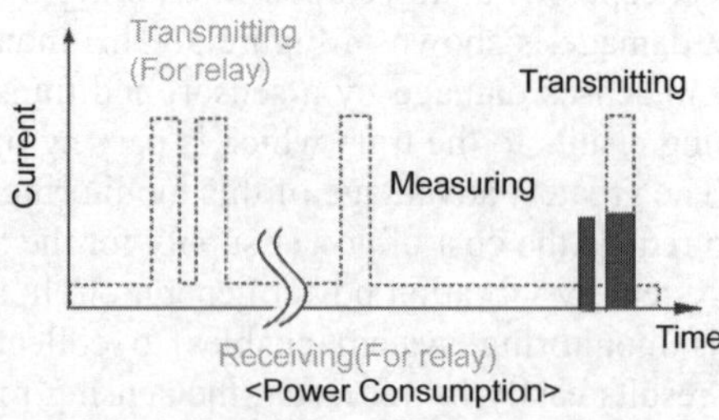

(b) Developed monitoring system

Figure 4. Power consumption of sensor node

3.3 *Transmitting and Receiving on Running Train*

We installed a prototype system on the existing railway truss bridge, in order to confirm that the sensing results are transmitted by vibration power generation, and the sensing results are received on the running train. Figure 5 shows the devices of the prototype system installed on the bridge. The prototype system consists of a piezoelectric device, a transmitter and a sensor. We attached the piezoelectric device to the web where the high-frequency stress is generated.

We measured the capacitor voltage of the prototype system. Figure 6 shows the capacitor voltage when the train passed on the bridge. The capacitor voltage increased with time, and dropped when the voltage reached about 5V. This result shows that the power was generated by vibration and consumed by sensing and transmitting. As a result of all the trials, sensing results were transmitted five times on an average with a ten rake train over the bridge.

We also tried to receive the sensing results by getting on the rearmost car of the train with a receiver in hand. As a result of the trial, we received the sensing results 3.7 times on an average. The reason why the number of receiving times was lower than that of the transmitting times is that the receiver on the rearmost

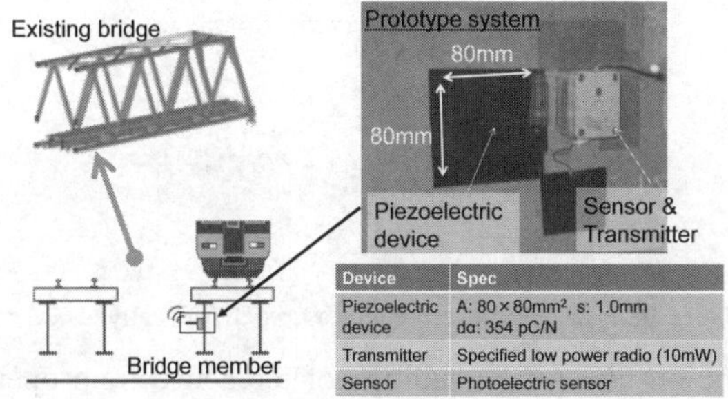

Figure 5. Prototype system to detect damage

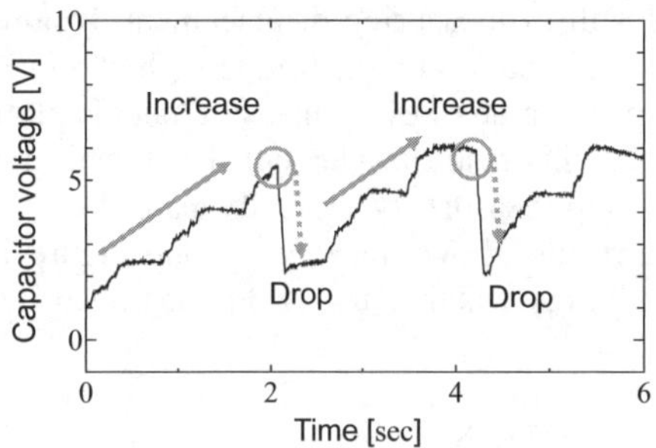

Figure 6. Capacitor voltage

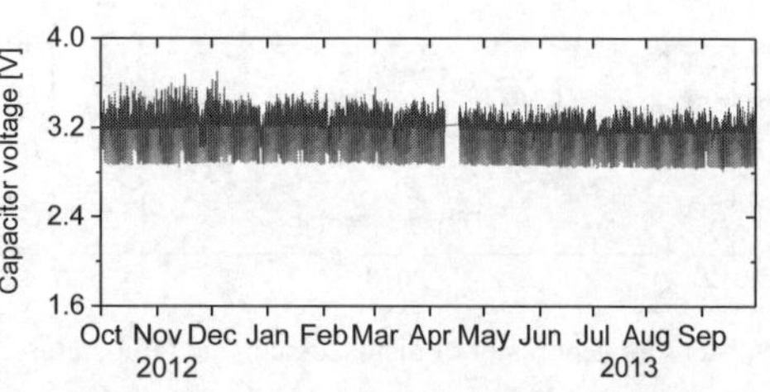

Figure 7. Voltage of electric double-layered capacitor

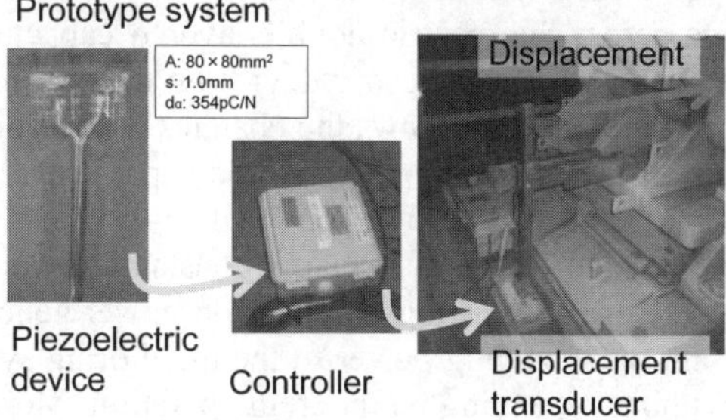

Figure 8. Prototype system to predict deterioration

car is too far from the transmitter to receive the sensing result transmitted shortly after arrival of the train to the bridge.

These results show that our developed monitoring system transmits the sensing results by vibration power generation only, and allows us to collect the sensing results on the running train.

4 MONITORING SYSTEM TO PREDICT DETERIORATION

4.1 Concept

We also developed the monitoring system to predict a tendency of bridge deterioration by vibration power generation. This monitoring system is useful to detect changes in performance of bridges by measuring as follows.

- Measuring at a constant time interval to monitor slow changes in displacement and stress of bridges (e.g. movable bearing displacement due to bridge deformation by temperature changes).
- Measuring dynamically to monitor changes when a train passes on the bridge (e.g. stress caused by train load)

4.2 Charging Power

It is necessary to supply power for the monitoring system constantly to measure at a constant time interval. Dynamic measurement requires to supply a large amount of power. The monitoring system mentioned in the previous chapter can't perform these measurement, because the power is generated only when the train passes over the bridge, and also this amount of the power is very small.

In the monitoring system detailed in this chapter, the above measurement becomes possible by charging an electric double-layered capacitor by vibration power generation. Charging the electric double-layered capacitor allows to measure at a constant time interval without depending on the train operation, and to measure dynamically.

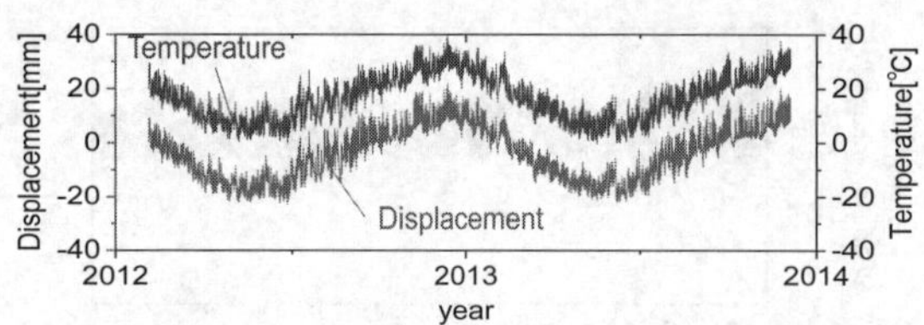

Figure 9. Measurement result of displacement and temperature

In order to confirm the power charged by vibration power generation, we constantly measured the voltage across the electric double-layered capacitor connected to the piezoelectric device on the existing truss bridge. Figure 7 shows the changes of the voltage across the electric double-layered capacitor. The result proves that the power charged in the electric double-layered capacitor remains constant for long term. Therefore, charging by vibration power generation allows supplying power to the monitoring system without depending on the train operation. Moreover, supplying the large amount of power is also enabled by accumulating power.

4.3 *Monitoring for Long-term*

In order to confirm that the monitoring system is operated with the power charged by vibration power generation for a long term, we installed a prototype system on the railway truss bridge. Figure 8 shows the devices of the prototype system. The prototype system aims to predict deterioration of the bridge movable bearing, and consists of a piezoelectric device, a controller, a displacement transducer, to measure as follows.

- Measuring displacement of the movable bearing and temperature every 10 minutes when a train does not passes on the bridge, in order to detect changes in the temperature deformation of the bridge
- Measuring displacement of the movable bearing dynamically after detecting an approaching train, in order to detect changes in the behavior of the bridge when the train passes

Figure 9 shows the displacement and the temperature measurement every 10 minutes. These results prove that the constant measurement becomes possible for more than two years by vibration power generation under the condition of the actual train operation, including several suspension due to weather or an accident.

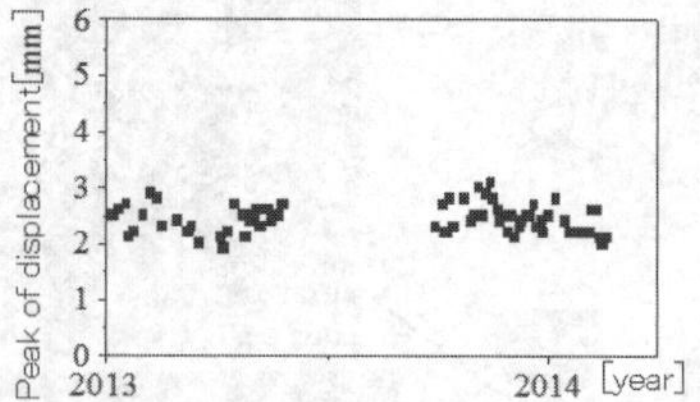

Figure 10. Peak of displacement measured dynamically

The result of measuring displacement dynamically when the train passes on the bridge is shown in Figure 10. The peak value of every measurement is plotted in Figure 10. This result proves our developed monitoring system to measure dynamically also.

These results shown in Figure 9 and Figure 10 are useful to judge whether the bridge has been deteriorated or not.

5 CONCLUSION

We developed two types of monitoring system operated by vibration power generation, based on the technique to generate power efficiently by vibration power generation. We confirmed that our developed monitoring systems are operated by vibration power generation on the existing railway bridge. Our developed monitoring system can be used for various purpose in addition to the above, such as t measurement of stress, acceleration and so on.

Part of this research has been conducted using a subsidy for railway technology development granted by the Ministry of Land, Infrastructure, Transport and Tourism.

REFERENCES

T.Imai,S.Fujimoto, M.Ichiki,2014.,Equivalent circuit model of impact-based piezoelectric energy harvester, Journal of Physics: Conference Series 557

T.Nagayama, P.Moinzadeh, K.Mechitov, M.Ushita, N.Makihata, M.Ieiri, G.Agha, B.F.Spencer. 2010.,Reliable multi-hop communication for structural health monitoring, *Smart Structure and Systems*,Vol.6, No.5-6,pp481-504

Proceedings of the International Conference on Smart Infrastructure and Construction
ISBN 978-0-7277-6127-9

© The authors and ICE Publishing: All rights reserved, 2016
doi:10.1680/tfitsi.61279.343

Smartphone based Cloud-SHM and its applications

Xuefeng Zhao[*1,3], Ruicong Han[1,3], Yan Yu[1,3], Hao Liu[1,3], Yanbing Ding[1,3], Mingchu Li[1,3] and Jinping Ou[1,2,3]

[1] *Dalian University of Technology, Dalian, China*

[2] *Harbin Institute of Technology, Harbin, China*

[3] *Research Center of Structural Smartphone Cloud Monitoring, Dalian University of Technology, Dalian, China*

[*] *Corresponding author*

ABSTRACT Smartphone, integrated with CPU, sensors, network, and storage capability, is developed rapidly in recent years. A cloud-structural health monitoring method based on smartphone was proposed, and a structural health monitoring system Orion-CC, which integrates functions of data acquisition, data analysis and data upload, was developed on smartphone to perform structural health monitoring without any other professional devices. And the feasibility of Orion-CC was proved by the cable force test compared with force balance acceleration sensor. An application D-Viewer was developed to measure the displacement, following a validation experiment. And then an application E-Explorer was developed to realize data communication without a network. A cloud-SHM data sharing website was built to make the data synchronization between smartphone and website, and realize data uploading and sharing, which can improve the efficiency of monitoring and big data integration, make the possibility for big data collection and quick structural safety evaluation.

1 INTRODUCTION

Structural health monitoring (SHM) as an effective measure to ensure the safety of large-scale engineering, has been developed rapidly these years. However, some problems should be considered further. First, the object of SHM is typically important infrastructure, but some structures with less significance may require SHM for safety purposes. Generally, an SHM system is relatively expensive. Additionally, SHM systems typically consist of a complex set of various types of sensors, devices and software, which requires professional installation and monitoring. Developing a low-cost and convenient monitoring technique will be helpful for the safety maintenance of less significant structures or some special construction procedure for field work.

Smartphone with an operating system and built-in sensors have already been applied to SHM research. The application of smartphones on SHM has also been validated preliminarily by authors, who proposed a new idea for SHM using smartphone's built-in sensors (Yu et al. 2012). Then Morgenthal demonstrated the possibilities and limitation of measuring mechanical oscillations using smart phones, and applied smart phones to measuring transient structural displacements (Morgenthal and Höpfner 2012; Höpfner et al. 2013). Reilly et al. developed an iShake system on smartphone for seismic monitoring (Reilly et al. 2013). Yi et al. developed a mobile-sensor data collector based on an Android smart phone to display sensor data and provide communication with sensors (Yi et al. 2013). Oraczewski et al. developed a smartphone-based platform that could be used for crack detection (Oraczewski et al. 2015). Feng and Ozer proposed a citizen sensor network according to the ubiquitous smartphones (Feng et al. 2015). Subsequently, they presented a SHM platform to integrate sensors, web and crowdsourcing (Ozer et al. 2015). The cable force test by internal sensor and displacement measurement by camera on smartphone has not developed by other group yet.

Authors' cable force test results indicated that the practicability and accuracy of this approach could satisfy engineering requirements (Zhao et al. 2015; Yu et al. 2015; Zhao et al. 2015). Authors also applied smartphone on offshore hoisting monitoring (Han et al. 2015).For more rapid disaster evaluation after earthquake, software was also developed to investigate seismic intensities (Zhao et al. 2015; Zhao et al. 2015).

In this study, Orion-CC, an application that integrates data collection and analysis, was developed to perform a cable force test without any other professional devices; an open website was built to collect monitoring data, make the possibility for big data collection and quick structural safety evaluation. An application D-Viewer, which can measure the displacement only by one smartphone's camera, has been launched to APP store for free download. And an application E-Explorer was developed for emergency rescue under earthquake without any network. By the validation of experiments, all of these applications have shown the feasibility of using smartphones for SHM.

2 FRAMEWORK OF CLOUD-SHM

The framework of Cloud-SHM based on smart phone is shown as Fig. 1. There are six levels, apperceive control level, network transmission level, information integration level, data processing level, decision level and information output level (Zhao et al. 2015; Zhao et al.2015).

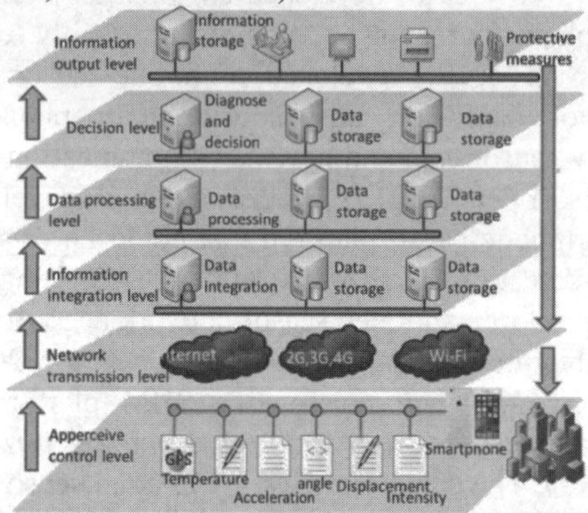

Figure 1. Framework of cloud-SHM

In apperceive control level, structural parameters can be obtained by built-in sensors, such as GPS information, temperature, acceleration, angle, displacement and so on. The interactive interface on smart phone can be applied to the seismic intensity investigation. The monitored data can be uploaded to server or website in network transmission level, and integrated in information integration level. Then the data can be processed to obtain the variation of structural parameters. According to the variation, the structural damage can be evaluated. Then the safety or damage information can be sent to smart phone user, based on the feedback, safety measures can be taken to provide guidance.

3 ORION-CC

3.1 Design of Orion-CC

Orion-CC was launched by our research group in the Apple App Store. It includes several functions as follows.

1. The collection of current location.
2. The establishment of monitoring project and the projects.
3. Data analysis. The frequency difference can be obtained by picking up frequency orders, shown as Fig.2(a). And the cable force can be calculated based on the cable parameters, which is shown as Fig. 2(b).

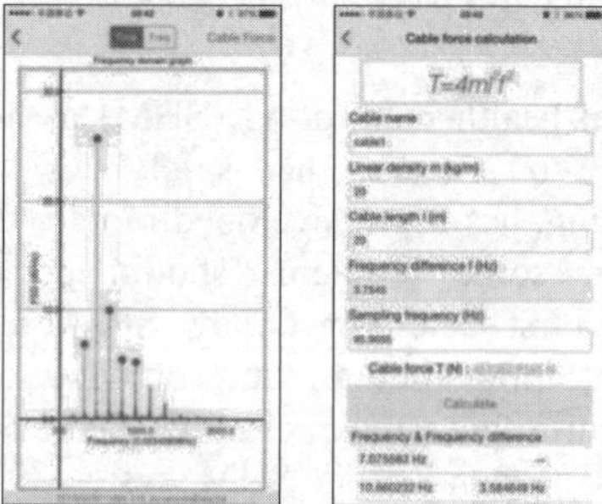

(a) Frequency spectrum (b) cable force calculation

Figure 2. Some interfaces of Orion-CC

3.2 Network function of Orion-CC

The collected data and cable force data can be uploaded to cloud structural health monitoring data sharing platform www.cloudshm.com. A same account is needed to register on website and smart phone. People can build monitoring project by logging with their own account, which is shown as Fig.3, then goes to the main interface as Fig.2 (a) to complete monitoring.

Fig. 3 Log-in interface with account

After the data collection and calculation, the project including GPS information, acceleration, cable force, etc. can be synchronized to website to realize the data sharing.

3.3 The application on cable force test

Vibration method is chosen to measure the cable force. In engineering applications, the frequency difference is often used to calculate cable force. The relationship between the cable force and frequency difference is shown as Eq. 1 (disregarding the effects of stiffness and sag) (Kim and Park 2007).

$$T=4ml^2(\Delta f)^2 \tag{1}$$

Where m is the linear mass density, l is the cable length and Δf is the frequency difference.

The comparison test between iPhone and force balance acceleration sensor was conducted on a bridge cable. The cable is 22.667 m long, and its linear mass density is 8.5 Kg/m. An iPhone with Orion-CC installed was employed to collect the data. The experiment picture is shown in Figure 4.

Figure 4. Test collection device

The frequency difference obtained by Orion-CC and force balance sensor was presented in Table 1.

Table 1. Frequency difference and cable force

	Frequency difference(Hz)	Cable force(KN)
Force balance sensor	2.327	94.59
iPhone	2.325	94.43
Error	0.086%	0.17%

As Table 1 illustrates, the error between the force balance acceleration sensor and iPhone is very small with a good consistency, which suggests that the smartphones installed with the Orion-CC software can serve well as a cable force measurement tool.

4 DATA SHARING PLATFORM

Except the data collection and transmission, Orion-CC also has functions of data processing and data uploading. The data collected by smart phone can be synchronized to the cloud-SHM data sharing platform www.cloudshm.com, which provides monitoring big data. Combining the strong public participation of the smartphone, the quick large-scale monitoring and evaluation under emergency condition becomes possible.

The cloud-SHM data sharing platform displays all uploaded project with GPS information, which is shown as Figure 5. People can get the information easily by checking the map.

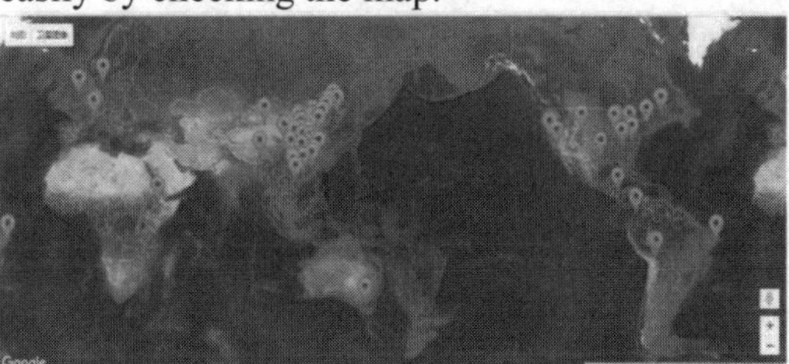

Figure 5. Projects on the platform

5 D-VIEWER

5.1 Fundamental

An application D-Viewer was developed for displacement measurement(Zhao et al. 2015). First, a laser device is mounted on the monitored object, making its optical path direction perpendicular to the movement direction of the monitored object. Then, the angle between the projection plate plane and the laser optical path is adjusted to 30 degrees. During experiment, use the camera of smartphone to shoot the laser spot on the projection plate. Once the monitored object moves from position 1 to position 2, the laser device will follow the movement, such that the laser spot on the projection plate will move proportionally. The ratio(K) of the actual size and the pixel size, as well as the original position of the laser spot of the video, are obtained through the APP first, after which the video is processed to obtain the pixel coordinates(X) of the laser spot centroid of each frame image. Then by formula L =X * K to acquire the actual coordinates of the laser spot centroid of each frame image, thereby visually reflecting the displacement of the monitored object. The experimental schematic diagram is shown in Figure 6.

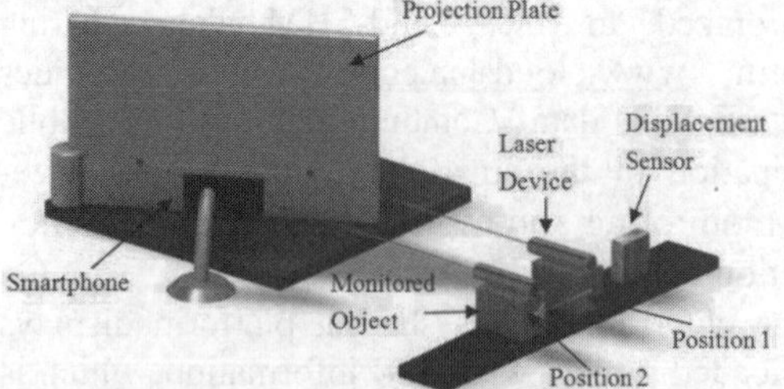

Figure 6. Experimental schematic diagram

5.2 Experiments and Results

A series of static and dynamic experiments were conducted by using samsung A5, and a laser displacement sensor was used to monitor the precise displacement of the monitored object. The results of dynamic experiment of vibration table simulate El Centro wave is shown in Figure 7.

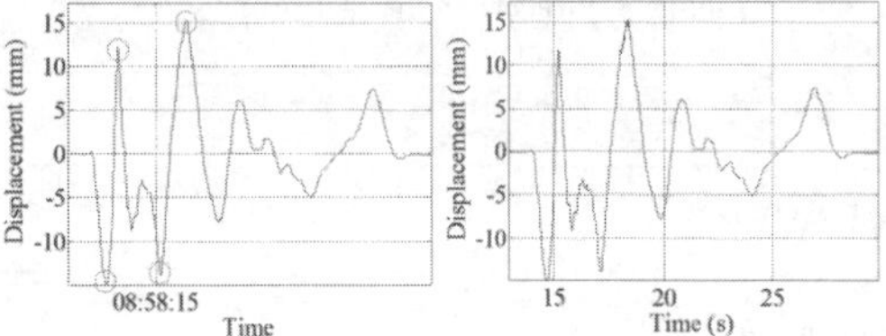

(a) SUMSANG A5 (b) Laser displacement sensor

Figure 7. Result of the A5 and the laser displacement sensor

Select four peak points displacement value as shown in Figure 2 to compare, the results are shown in Table 2.

Table 2. Comparison of the data of A5 and sensor

A5	-14.850	12.120	-13.860	15.380
Sensor	-15.070	11.650	-13.890	15.170
Difference	0.220	0.470	0.030	0.210
Error	1.46%	4.03%	0.22%	1.38%

The experimental results show that when the vibration table shakes with random vibration, the data difference between A5 and laser displacement sensor are very small, the displacement of vibration table can be accurately and completely monitored by the APP.

6 E-EXPLORER

Devastating earthquake can often cause the disaster area communication interrupt, traffic paralysis, etc (Peng et al 2015). It is difficult for the emergency rescue force to get the disaster area in time. Therefore, active local participation in the quake-hit areas to aid each other appeals extremely important. “E-Explorer” can send important information for personal survival, let rescue workers locate the positions of survivors trapped, creating an efficient self-help and mutual rescue platform for the earthquake-stricken people.

6.1 Communication and rescue principle

In the high seismic intensity area, the communication network will be cut off. Trapped survivors can't effectively convey help Information by means of usual communication way. "E-Explorer" communication module provides a new way of information transmission. This module allows two mobile devices to exchange information under the condition of no network connection. Through experimental verification, in the case of barrier-free, it can realize the communication within the scope of 30 to 40 m. In case there is an obstacle, the signal can pass through one or two thick walls blocked. So we can suppose that “E-Explorer” can still realize good communication with nearby people in the absence of any external network. Most important of all, each device equipped with “E-Explorer” software can be used as a node in the information transmission network, that is to say, everyone can communicate with the long-range goals by means of several number of nodes connected pairwise. Visual communication diagram are shown in figure 1 below. If the distance of two nodes is in the range of maximum communication distance, then there can build a "communication line" between them. Figure 8 shows the trapped survivors A through several nodes finally pass his help information to a relief worker B.

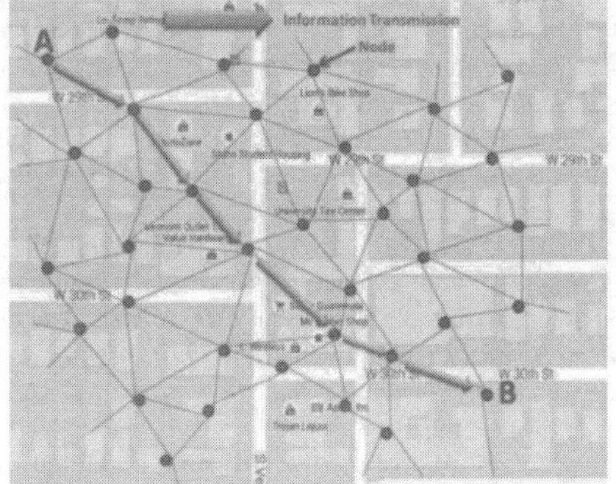

Figure 8. Personnel information transmission line schematic

6.2 Main functions and interfaces of E-Explorer

There are several typical interfaces to illustrate the function of E-Explorer. Figure 9(a) is the information inputting interface, people are encouraged to finish and submit them to make others know the state. After the submitting, the interface was shown as Figure 9(b), you can acquire the personal information of others.

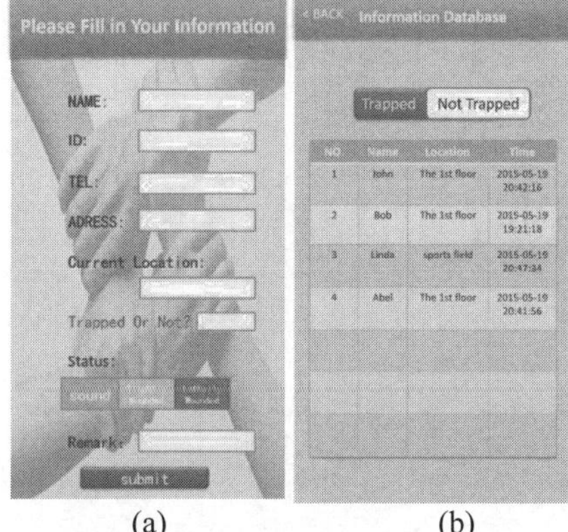

(a) (b)

Figure 9. information inputting interface

As long as you permit opening the blue tooth, the "E-Explore" can search the counterpart nearby automatically and to connect with it after matching successfully although without any network. Their connections will contribute to the information sharing between two databases, indicating that you can acquire the personal information of your partner and transmit the text message freely. Besides, when two users can't connect with each other when the distance is too far, if there is one user who have opened his "E-Explorer" between them, which can be seen as a node to connect them, they could also realize mutual connections and information transmission. If the distribution of users in the region is extremely dense, any pairwise users within the scope of connection distance can achieve communication. That is to say the "E-Explorer" constructs a temporary communication network system in the earthquake region.

7 CONCLUSION

In this paper, a Cloud-SHM method based on smart phone was proposed, the software Orion-CC was developed and applied to cable force test, the comparison between force balance sensor and iPhone proved the feasibility and convenience. The data sharing platform www.cloudshm.com was built to realize the data integration and safety evaluation. An application D-Viewer was developed to realize displacement measurement, which is validated by comparison test. Another application E-Explorer was developed to realize information transmission without any net, which is significant to rescue under extreme conditions. All the applications show the feasibility and potential of smartphone, which will provide more low-cost and convenient methods for SHM.

ACKNOWLEDGEMENT

This research was financially supported by key Projects in the National Science & Technology Pillar Program during the Twelfth Five-Year Plan Period (2011BAK02B01) and National Science Foundation of China (51221961).

REFERENCES

Feng, M. Fukuda, Y. Mizuta, M. & Ozer, E. 2015. Citizen Sensors for SHM: Use of Accelerometer Datafrom Smartphones. *Sensors*, **15**, 2980-2998.

Han, R. Zhao, X. & Yu, Y. 2015. Convenient posture monitoring on girder hoisting of Dalian Xinghaiwan cross-sea bridge based on smart phone. *Proceedings of the 7th International Conference on Structural Health Monitoring of Intelligent Infrastructure (SHMII 7 Torino 2015)*. Politecnico di Torino.

Höpfner, H. Morgenthal, G. Schirmer, M. Naujoks, M. Halang, C. 2013. On measuring mechanical oscillations using smartphone sensors: possibilities and limitation. *ACM SIGMOBILE Mobile Computing and Communications Review* **17(4)**, 29-41.

Kim, & Park, T. 2007. Estimation of cable tension force using the frequency-based system identification method. *Journal of Sound Vibration* **304(3)**, 660-676.

Morgenthal, G. & Höpfner H. 2012. The application of smartphones to measuring transient structural displacements. *Journal of Civil Structural Health Monitoring* **2**, 149-161.

Oraczewski, T. Staszewski, WJ. Uhl, T. 2015. Nonlinear acoustics for structural health monitoring using mobile, wireless and smartphone-based transducer platform. *Journal of Intelligent Material Systems and Structures,* DOI:1045389X15585902.

Ozer, E. Feng, MQ. & Feng, D. Citizen sensors for SHM: Towards a crowdsourcing platform. *Sensors*, **15**, 14591-14614.

Peng, D. Zhao, X. Zhao, Q. & Yu, Y. (2015). Smartphone based public participant emergency rescue information platform for earthquake zone-"E-Explorer". *Vibroengineering PROCEDIA 2015* **5**, 436-439.

Reilly, J. Shideh, D. Ervasti, M. Bray, JD. Glaser, SD. Bayen, AM. 2013. Mobile phones as seismologic sensors: automating data extraction for the ishake system, *IEEE transactions on automation science and engineering* **10**, 242-251.

Yi, WJ. Gilliland, S. Saniie, J. 2013. Wireless sensor network for structural health monitoring using System-on-Chip with Android smartphone. *SENSORS IEEE*, 1-4.

Yu, Y. Han, R. Zhao, X. et al. 2015. Initial Validation of Mobile-Structural Health Monitoring Method Using Smartphones. *International Journal of Distributed Sensor Networks* **2015**, 1-18.

Yu, Y. Zhao, X. Han, R. et al. 2015. Design and initial validation of external sensors board of smart phones for mobile structural health monitoring system. *Proceedings of the 7th International Conference on Structural Health Monitoring of Intelligent Infrastructure (SHMII 7 Torino 2015).* Politecnico di Torino.

Yu, Y. Zhao, X. & Ou, J. 2012. A new idea: Mobile structural health monitoring using Smart phones. *Intelligent Control and Information Processing (ICICIP), 2012 Third International Conference on. IEEE,* 714-716.

Zhao, X. Han, R. Ding, Y. et al. 2015. Portable and convenient cable force measurement using smartphone. *Journal of Civil Structural Health Monitoring* **5(4)**, 481-491.

Zhao, X. Liu, H. Yu, Y. et al. 2015. Convenient Displacement Monitoring Technique using Smartphone. *Vibroengineering PROCEDIA 2015* **5**, 579-584.

Zhao, X. Peng, D. Hu, W., et al. 2015. Quick seismic intensity map investigation and evaluation based on cloud monitoring method using smart mobile phone. *SPIE Smart Structures and Materials+ Nondestructive Evaluation and Health Monitoring. International Society for Optics and Photonics,* 94372A-94372A-7..

Zhao, X. Yu, Y. Li M. et al. 2015. Research on Cloud-SHM and its applications. *Proceedings of the 7th International Conference on Structural Health Monitoring of Intelligent Infrastructure (SHMII 7 Torino 2015).* Politecnico di Torino.

Zhao, X. Yu, Y. Hu, W. et al. 2015. Cable force monitoring system of cable stayed bridges using accelerometers inside mobile smart phone. SPIE Smart Structures and Materials+ Nondestructive Evaluation and Health Monitoring. International Society for Optics and Photonics, 94351H-94351H-7.

Zhao, X. Yu, Y. Li, M. & Ou, J. 2015. Cloud-Structural Health Monitoring based on smartphone. Vibroengineering PROCEDIA 2015 5, 241-246.

Zhao, X. Yu, Y. Li, M. & Ou, J. 2015. Research on Cloud-SHM and its applications. Proceedings of the 7th International Conference on Structural Health Monitoring of Intelligent Infrastructure (SHMII 7 Torino 2015). Politecnico di Torino.

Proceedings of the International Conference on Smart Infrastructure and Construction
ISBN 978-0-7277-6127-9

© The authors and ICE Publishing: All rights reserved, 2016
doi:10.1680/tfitsi.61279.349

Characterization of a traveling object with an underground cluster of accelerometers

O. Drori* and E. Levenberg

Faculty of Civil and Environmental Engineering, Technion - Israel Institute of Technology, Haifa, Israel
**Corresponding author*

ABSTRACT This paper tackled the problem of characterizing an object moving along a surface by means of a buried cluster of accelerometers. Sought characteristics include: path and speed of movement, number and spatial configuration of contact areas, and relative weight distribution across loaded zones. The suggested solution technique was based on solving an inverse problem. For this purpose the passing event was first simulated in a quasi-static mechanical model, and the unknowns were obtained from best matching measured and computed accelerations. The basic solution technique was demonstrated for synthetic acceleration traces generated under ideal conditions. A slightly modified solution technique was proposed for dealing with realistic/field data. Overall, the idea and solution approach are deemed workable, well suited for wireless implementation, and worthy of further development attention.

1 INTRODUCTION

This work is motivated by the desire to characterize a traveling object from readings obtained with a cluster of buried accelerometers. Sought after features include: (i) path and speed of movement; (ii) number and spatial configuration of contact areas; and (iii) relative weight distribution across loaded zones. These characteristics are not directly targeted by the sensors; they are identified from solving an inverse problem presuming known boundary conditions and mechanical model. This problem is extremely relevant to smart infrastructure wherein system usage information is to be deduced from data recorded by embedded sensors.

The problem of inferring the magnitude of a moving force with embedded sensors has been mostly researched in connection with bridge structures (e.g., Yu & Chan 2007). A typical application involves attaching strain gauges or accelerometers (or both) to beams and backcalculating the intensity of a force passing over the bridge. To achieve this goal, the so-called forward response of the structure to a moving and fluctuating load is typically solved with Bernoulli-Euler beam theory, employing a priori measured modal parameters. The passing event is then numerically simulated, attempting to best match the sensor readings. This latter step makes use of optimization algorithms. In the geotechnical or infrastructure arena, little work has been published on surface load identification. Most existing studies are based on recording and analyzing near surface high frequency ground vibrations, collected with microphones, geophones, or accelerometers (e.g., Itakura et al. 2000, Surinach et al. 2000, Hostettler et al. 2010).

The interpretation approach herein proposed does not deal with inertia effects such as propagating stress waves or acoustic signatures. While these carry pertinent information, they place high demands on the measurement system, with specific emphasis on elevated sampling rates and very accurate synchronization between sensors. Alternatively, the proposed method focuses the analysis on quasi-static responses. So doing sets the stage for wireless implementation because it considerably lessens power, storage, and signal acquisition demands (Liu & Tomizuka 2003). This approach builds and expands upon the work described in Levenberg (2014) wherein the synchronized readings of a pair of in-pavement accelerometers were employed to estimate the speed of passing vehicles.

The objective here is to present the general problem formulation along with a proposed solution technique. An application involving synthetic data is included, demonstrating the approach for ideal conditions. Preliminary experimental data involving field measurements are also presented and addressed.

2 PROBLEM FORMULATION

Figure 1 presents a plan view of an arbitrary object traveling in a straight line with speed V, approaching a cluster of accelerometers. A hexagonal boundary represents the object, encapsulating several zones, identified by L_j's and depicted with circular markers. The latter indicate contact areas through which forces generated by the moving body operate on the supporting medium. These forces are not necessarily timewise constant and do not necessarily operate normal to the surface.

A right handed Cartesian coordinate system is defined in the figure, positioned such that its x and y axes coincide with the surface of the medium while the z-axis points into it (not shown). The spatial arrangement of L_j's is assumed to be timewise constant. Accordingly, each L_j can be identified by its coordinate values x_j^L and y_j^L such that $x_j^L = V_x(t-\tau_j)$ and $y_j^L = V_y(t-\tau_j)+y_j^0$. The speeds V_x and V_y are the components of V relative to the coordinate axes such that $V^2 = V_x^2 + V_y^2$, t denotes time with an arbitrary origin, τ_j indicates the value of t when $x_j^L = 0$, and y_j^0 is the value of y_j^L when $t = \tau_j$.

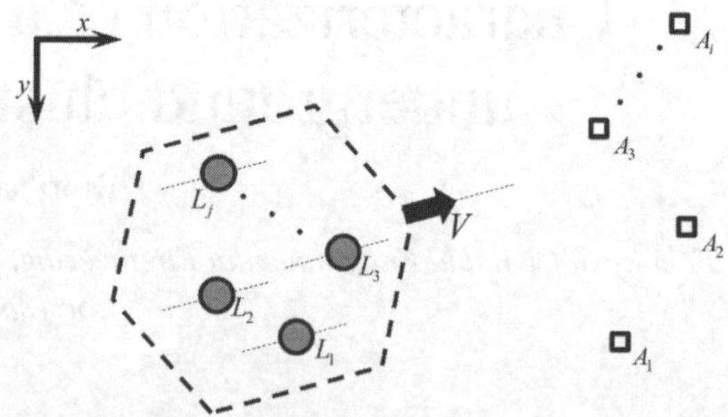

Figure 1. Plan view of a traveling object approaching a cluster of embedded accelerometers.

The sensors are illustrated in the figure by square markers, and identified as A_i's. They are stationary, prepositioned under the surface, each recognized by its coordinate values x_i, y_i, and z_i. Every sensor is capable of reporting accelerations in a certain orientation relative to the x, y, and z axes. Multichannel accelerometers, capable of measuring responses in several directions, are treated as separate, i.e., they are assigned different identifying indices. Accordingly, when the object passes in the vicinity of the cluster, it generates a single acceleration response trace $A_i(t)$ in each and every sensor. The resulting collection of traces embodies the mechanical properties of the supporting medium as well as all object characteristics. Accelerometers that are unable to detect the passing event are excluded from the analysis and not considered to be part of the cluster.

The problem being addressed is how to utilize cluster readings in order to estimate: (i) path and speed of object movement; (ii) number and spatial arrangement of L_j's; (iii) average force intensity, variation in magnitude, and orientation of each L_j; and (iv) contact properties of every L_j, e.g., area, shape, and stress distribution.

3 BASIC SOLUTION TECHNIQUE

Moving object characteristics are to be estimated by solving an inverse problem. For this purpose, the passing event is first simulated in a mechanical model with presumed boundary conditions and seed numerical values for all unknown parameters. Forward calculations are then executed to generate a set of projected/model quasi-static response traces $A_i^{\text{model}}(t)$ that correspond to the input (measured) set $A_i(t)$. Next, the sought unknowns are manipulated by an optimization algorithm until the discrepancy between all computed and measured traces is minimized. Finally, parameter values at the point of minimum are taken to be the estimated characteristics of the moving object.

For performing the minimization, the following scalar error term is suggested as an objective function:

$$Error = \frac{1}{IK}\sum_{i=1}^{I}\left(w_i \sum_{k=1}^{K}\left| A_i^{\text{model}}(t_k) - A_i(t_k) \right| \right) \quad (1)$$

wherein I is the total number of sensors involved in the computation, the t_k's are K preselected discrete times for performing the comparison, and the w_i's serve as weighing factors to counterbalance differences in signal strength across sensors. The most natural choice for the latter is $w_i = 1/\text{range}(A_i(\text{t}))$ in which $\text{range}(A_i(\text{t})) = \text{argmax}(A_i(\text{t})) - \text{argmin}(A_i(\text{t}))$ This choice also produces a unitless objective function.

4 APPLICATION FOR IDEAL CONDITIONS

2.1 *Generation of Synthetic Cases*

As a basic validation step, the proposed interpretation approach must be shown to yield correct results under perfect conditions in which the input acceleration traces are themselves model-generated. In this connection, three preliminary ideal cases were considered, involving a single loaded area traveling over an array of five virtual accelerometers; the cases differed only by the load travel path (see Figure 2). The underlying medium was chosen to be linear elastic isotropic and homogenous (weightless) half-space, characterized by a Young's modulus of $E = 100$ MPa and Poisson's ratio of $\nu = 0.3$.

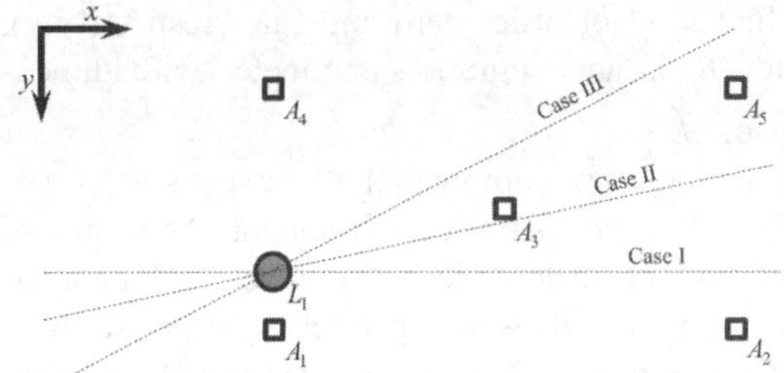

Figure 2. A single traveling load approaching a cluster of five accelerometers. Three optional trajectories are depicted.

All five sensors were recording accelerations in the vertical z-axis direction, and had a common embedment depth of $z = 300$ mm; their spatial coordinates are presented in Table 1. The load travel speed was $V = 10000$ mm/s, exerting a vertical force of constant intensity $P = 50$ kN, spread uniformly over a circular area with radius $a = 150$ mm.

Table 1. Coordinates of sensors in Figure 2. Depth coordinate was 300 mm for all; accelerations recorded in z-axis direction.

i	x_i *[mm]*	y_i *[mm]*
1	2000	2500
2	6000	2500
3	4000	1500
4	2000	500
5	6000	500

For generating model accelerations, a classic axisymmetric elasticity solution was employed as a computational kernel (Van Cauwelaert 2003):

$$w = \frac{qa(1+\nu)}{E}\int_{m=0}^{\infty}\frac{J_0(mr)J_1(ma)}{m}(2-2\nu+mz)e^{-mz}dm \quad (2)$$

wherein w is the vertical displacement in a half-space that is loaded over a circle with uniform vertical stress intensity q and radius a ($P = q\pi a^2$). The point of calculation is identified by r and z which designate, respectively, radial distance from the center of the load and depth below the surface. The terms $J_0(\cdot)$ and $J_1(\cdot)$ are Bessel functions of the first kind of order zero and one (respectively), while m is an integration parameter with dimensions of L^{-1}.

Equation 2 is computationally expensive, especially if high precision level is sought. As means of expediting the calculations the functional dependence of w on r was separately computed (with $z = 300$ mm) at 112 predetermined radial distances in the range of 0 to 100 m, and a cubic spline was defined to pass through the computed points for generating any intermediate values of choice.

Subsequently, synthetic (quasi-static) accelerations were generated by writing $r = r(t)$ in a manner that reflects the load movement relative to each virtual accelerometer, and then differentiating twice the resulting $w(t)$ with respect to time. The latter operation was performed numerically based on the central difference approximation formula employing a time interval of 0.6 milliseconds.

Case III acceleration traces are shown (as an example) in Figure 3 for a time period of one second commencing at $t = 0.9$ s. Five curves are included, each associated with a different sensor. In general terms these curves are similar in shape and are band-limited when viewed in the frequency domain (see Levenberg et al. 2014). As the load approaches a sensor, a positive (downward) acceleration is induced. At a certain point, a positive peak is attained, after which the acceleration is reduced to zero and then continues towards a negative (upward) peak. Because of the assumption of elasticity, the approaching shape is mirrored as the load continues on its path and moves away.

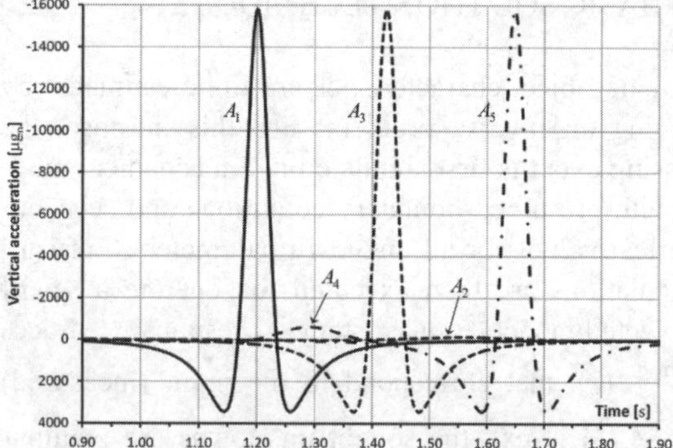

Figure 3. Synthetic accelerations for Case III (see Figure 2).

The timing of the peaks is linked to the speed of movement, while their relative magnitudes are linked to the travel path. In the figure it can be seen that similar acceleration traces were generated in A_1, A_3, and A_5. This means that the moving object passed near these three sensors with comparable offset. The signal is weaker for A_4 and weakest for A_2 indicating larger offsets. This information guides the convergence of the search algorithm to correctly estimate the sought characteristics.

2.2 *Interpretation of Synthetic Cases*

Model-generated acceleration traces were treated as inputs to the inverse problem. The medium for the analysis was taken as a linear elastic half-space with known E and ν. Also taken as known were the sensor coordinates x_i's, y_i's, z_i's. The unknown parameters in each case were: y_j^0, τ_j, P_j, V_x, and V_y (solved for $j = 1$). There was no attempt to backcalculate the contact area dimensions. Hence, Boussinesq's classic analytic solution was used as a computational kernel to economize run-time:

$$u_z = \frac{P}{2\pi E}\left(\frac{2(1-\nu^2)}{(r^2+z^2)^{0.5}} + \frac{(1+\nu)z^2}{(r^2+z^2)^{1.5}}\right) \quad (3)$$

wherein u_z is the vertical displacement in a half-space that is vertically loaded by a point force with intensity P. The cylindrical coordinates r and z

identify the calculation location relative to the point of force application.

As before, synthetic (quasi-static) accelerations were generated by writing $r = r(t)$ in a manner that reflects the load movement relative to each virtual accelerometer. Unlike before, to arrive at accelerations $u_z(t)$ was analytically differentiated twice with respect to time (resulting expression too long to include here). At this stage, the *Error* in Equation 1 was separately evaluated for each of the three cases, and an interior point algorithm was employed for minimization (Byrd et al. 1999). It was operated with the constraints: $\tau \geq 0$, and $P \geq 0$, and restarted manually with different seed values for the unknowns.

2.3 *Results for Synthetic Cases*

Final interpretation results are presented in Table 2, which also offers a side-by-side comparison with the true (input) parameter values. It can be seen that the method was able to converge to the correct answer in all three cases, giving the path and speed of movement, as well as the intensity of the load. Small discrepancies are due to systematic errors originating from utilizing an incorrect model (i.e., utilization of Equation 3 instead of Equation 2).

Table 2. Estimated load characteristics for synthetic cases.

	Case I		Case II		Case III	
Parameter	True	Estimated	True	Estimated	True	Estimated
τ [s]	1	1.000	1	1.000	1	1.000
y^0 [mm]	2000	2001	2400	2403	3000	3000
V_x [mm/s]	10000	10002	9806	9806	8944	8945
V_y [mm/s]	0	0	-1961	-1967	-4472	-4472
P [kN]	50	50.3	50	50.3	50	50.2
Error [%]	-	$1.93 \cdot 10^{-2}$	-	$8.23 \cdot 10^{-2}$	-	$2.30 \cdot 10^{-2}$

Interpretation of the synthetic cases involved additional examinations: (i) effects of incorrectly assuming Poisson's ratio (negligible); (ii) effects of letting the algorithm determine the number of traveling loads (not trivial, requires further investigation); and (iii) effect of using an incorrect Young's modulus (ratio P/E remains as constant).

The reliability of the optimal solution was studied next. To achieve this, a small perturbation about the optimum was applied to each parameter while others remained at optimum, and the corresponding changes in the *Error* term were observed. The outcome for Case 3 is shown in Figure 4 wherein the abscissa represents the parameter values normalized by their optimal levels and the ordinate depicts the objective function after normalization by its optimal value. The steepness of the resulting curves about their optimum serves to confirm that the solution is highly reliable. From comparing the outcome for different parameters, it is seen that the most trusted entity to identify is τ, while the least trusted parameter is P. It should be noted that the solution reliability expressed by Figure 4 also embodies the number and spatial arrangement of the sensors. Hence, such plot type can offer means for choosing optimal deployment topology.

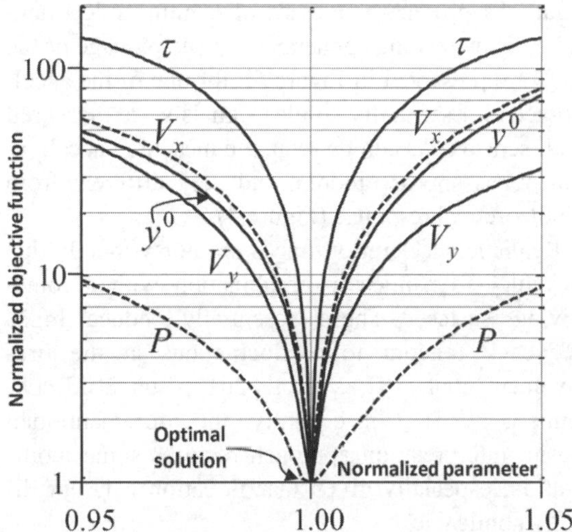

Figure 4. Sensitivity of objective function to changes in parameter values about their optimum (Case III).

5 FIELD EXPERIMENT

5.1 *Test setup*

A harvested and cleared wheat field was chosen to serve as test site. The soil comprised of dry alluvial clay extending to a large depth. Four uniaxial accelerometers were buried at a depth of 300 mm, each oriented vertically. Two types of accelerometers were employed: (i) piezoelectric (model KB12VD by Metra Mess). These sensors have a measuring range of ±0.6g$_n$, sensitivity of 10 v/g$_n$, and resonant frequency of 0.35 kHz; and (ii) MEMS technology (model VS9002 by Colibrys). These sensors have a measuring range of ±2g$_n$, sensitivity of 1 v/g$_n$, and resonant frequency of 1.3 kHz. All four accelerometers were connected to a data acquisition module (model 9234 by National Instruments), assuring synchronous data collection. Sampling frequency was about 1.7 kHz. The sensors were placed along a line that was at an angle of approximately 45 degrees with respect to the anticipated movement direction. Passing objects comprised of passenger cars, farming equipment, and heavy trucks.

5.2 *Measured Acceleration Traces*

Figure 5a presents a picture of a multi-axle truck-trailer. Raw readings generated by one passage of the truck are presented in Figure 5b for one of the accelerometers. Essentially similar signals were measured in all sensors. As can be seen, the measured acceleration traces appear random, and very different from the theoretical response (Figure 3).

Realistic acceleration traces are noisy mainly due to vehicle dynamics in combination with a rough driving surface. These essentially induce high-frequency random force fluctuations at the tire-pavement contact (Lak et al. 2011, Sun 2013, Levenberg 2015). Consequently, the aforementioned (basic) interpretation approach requires some modification, especially given that Equation 2 is not directly applicable.

5.3 *Analysis of Realistic Accelerations*

The approach for dealing with realistic accelerations is founded upon the premise that ideal quasi-static responses exist within the noisy traces, and can be exposed by application of smoothing techniques. To illustrate this point, Figure 5c presents a smoothed version of the signal shown in Figure 5b. As can be noticed, once high-frequency content is removed, the outcome closely resembles that expected under perfect conditions.

Accordingly, the suggested (modified) solution technique involves the following steps:

(i) Simulating the passing event in a quasi-static mechanical model with presumed boundary conditions and seed numerical values for all unknown parameters.

(ii) Generating a set of projected/model response traces $A_i^{\text{model}}(t_k)$ that correspond to the input (measured) set $A_i(t_k)$;

(iii) Plotting a frequency spectrum for every model trace by means of discrete Fourier transform, and identifying the highest meaningful frequency contained in each signal. This limit frequency is designated by f_i^L, i.e., $f_i^L = f^L\left(A_i^{\text{model}}(t_k)\right)$;

(iii) Smoothing each measured trace $A_i(t_k)$ to remove frequencies higher than f_i^L. The resulting outcome is designated by $A_i^{\text{smooth}}(t_k)$. A myriad of techniques may be utilized for this purpose;

(iv) Calculating an error term Err that quantifies the mismatch between smoothed accelerations and their corresponding model accelerations

$$Err = \frac{1}{IK}\sum_{i=1}^{I}\left(w_i^{\text{smooth}}\sum_{k=1}^{K}\left|A_i^{\text{model}}(t_k) - A_i^{\text{smooth}}(t_k)\right|\right) \tag{4}$$

wherein $w_i^{\text{smooth}} = 1/\text{range}\left(A_i^{\text{smooth}}(\text{t})\right)$; and finally,

(v) Manipulating the unknowns with an optimization algorithm (i.e., essentially repeating the above steps), until Err is minimized. Parameter values at the point of minimum are taken to be the moving object characteristics.

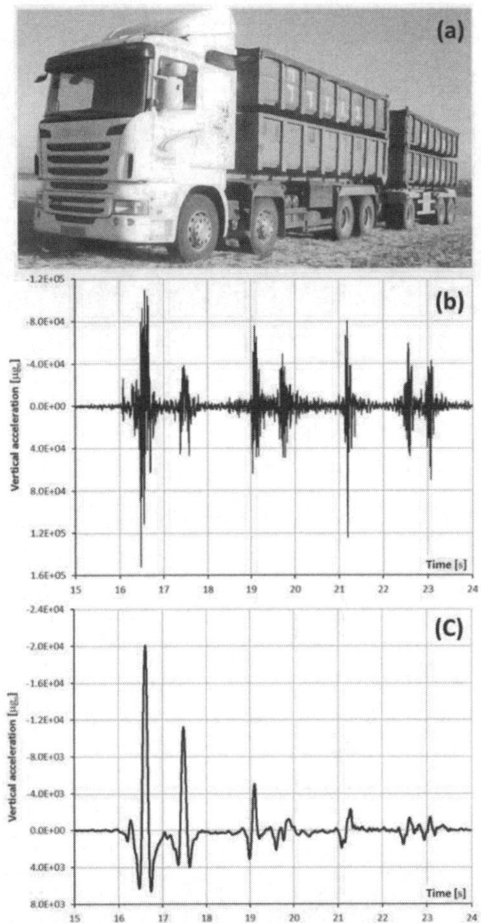

Figure 5. Field experiment: (a) picture of a passing track, (b) resulting raw readings from an underground accelerometer, and (c) smoothed acceleration traces.

6 CONCLUSION AND REMARKS

Buried accelerometers coupled to their surrounding medium record mechanical responses generated by a surface traveling object. Under ideal conditions acceleration traces have a smooth appearance, and a frequency spectrum that is essentially band-limited. Under realistic conditions, acceleration traces appear noisy, containing both quasi-static response and high-frequency signal content.

This work has shown that under ideal conditions it is possible to infer passing object characteristics from analysis of acceleration traces. The work also proposed a slightly modified solution technique for dealing with real measurements. In the latter case the raw signals are smoothed as part of the interpretation, with the smoothing effort itself being 'model guided.'

The approach was based on quasi-static mechanical modeling. As compared to classic analysis schemes that target propagating stress waves, the proposed interpretation technique poses much less stringent demands on sampling rates and synchronization level between the cluster members. This feature is particularly attractive when considering wireless implementation over wide areas with autonomous sensor nodes. It therefore coincides with the vision of smart infrastructure.

ACKNOWLEDGEMENT

This work was financially supported by Netivei Israel - National Transport Infrastructure Company Ltd. It is part of a larger research project aimed at developing new autonomous pavement sensors. Also acknowledged are Mr. Meir Ovadia and Mr. Arie Altberg for their assistance with the field experiment.

REFERENCES

Byrd, R.H., Hribar, M.E., and Nocedal, J. 1999. An interior point algorithm for large-scale nonlinear programming, *SIAM Journal on Optimization* **9(4)**, 877-900.

Hostettler, R., Birk, W., and Nordenvaad, L. 2010. On the feasibility of road vibrations-based vehicle property sensing, *Intelligent Transport Systems* **4(4)**, 356-364.

Itakura, Y., Fujii, N., and Sawada, T. 2000. Basic characteristics of ground vibration sensors for the detection of debris flow, *Physics and Chemistry of the Earth, Part B: Hydrology, Oceans and Atmosphere* **25(9)**, 717-720.

Lak, M.A., Degrande, G., and Lombaert, G. 2011. The effect of road unevenness on the dynamic vehicle response and groundborn vibrations due to road traffic, *Soil Dynamics and Earthquake Engineering* **31**, 1357-1377.

Levenberg, E. 2014. Estimating vehicle speed with embedded inertial sensors, *Transportation Research Part C* **46**, 300-308.

Levenberg, E. 2015. Backcalculation with an implanted inertial sensor, *Transportation Research Record: Journal of the Transportation Research Board* **2525**, 3-12.

Levenberg, E., Shmuel, I., Orbach, M., and Mizrachi, B. 2014. Wireless pavement sensors for wide-area instrumentation, *Proceedings of the 3rd International Conference on Transportation Infrastructure (ICTI): Sustainability, Eco-efficiency and Conservation in Transportation Infrastructure Asset Management*, Losa & Papagiannakis (eds.), CRC Press, 307-319.

Liu, S.C. and Tomizuka, M. 2003. Vision and strategy for sensors and smart structures technology research, *Proceedings of the 4th International Workshop on Structural Health Monitoring*, Chang, F-K (ed.), DES*tech* Publications Inc., 42-52.

Sun, L. 2013. An overview of a unified theory of dynamics of vehicle-pavement interaction under moving and stochastic load, *Journal of Modern Transportation* **21(3)**, 135-162.

Surinach, E., Sabot, F., Furdada, G., and Vilaplana, J.M. 2000. Study of seismic signals of artificially released snow avalanches for monitoring purposes, *Physics and Chemistry of the Earth, Part B: Hydrology, Oceans and Atmosphere* **25(9)**, 721-727.

Van Cauwelaert, F. 2003. *Pavement design and evaluation: the required mathematics and its applications*, Federation of the Belgian Cement Industry, Brussels, Belgium.

Yu, L. and Chan, T.H.T. 2007. Recent research on identification of moving loads on bridges, *Journal of Sound and Vibration* **305(1)**, 3-21.

Proceedings of the International Conference on Smart Infrastructure and Construction
ISBN 978-0-7277-6127-9

© The authors and ICE Publishing: All rights reserved, 2016
doi:10.1680/tfitsi.61279.357

An observability analysis for profile estimation through vehicle response measurement

T. Jothi Saravanan*, Zhao BoYu, Di Su and Tomonori Nagayama

Bridge and Structure Laboratory, Department of Civil Engineering, The University of Tokyo, Tokyo, Japan

* *Corresponding Author*

ABSTRACT To maintain road and railway infrastructure efficiently, the road or rail profile along the longitudinal direction need to be monitored frequently. Profile evaluation through vehicle response measurements potentially provides efficient solutions. However, the applicability of such profile evaluation for various type of vehicle models in terms of effective sensor installation locations is not clarified yet. Observability is a measure of whether the internal states of a system can be estimated through the measurement of its outputs. Hence, the observability rank condition (ORC) analysis of different time invariant linear vehicle models are conducted for the effective placement of the sensors in vehicles to estimate the profile. Since measurement of vehicle's absolute displacement is not practical, accelerations and angular velocities are assumed to be observed variables. Under this assumption, the profile is typically not observable. Therefore, second derivative of the profile is set as the variable to be identified so that non-static components of the profile is obtained as its double integral. The observability of this second derivative of profile is studied through the ORC analysis based on two approaches. The investigation of the ORC allows to determine the appropriate sensor types and their installation locations. This ORC analysis theoretically revealed a sensor type and placement strategy, which can be used as the guideline in the profile estimation through vehicle response measurement.

1 INTRODUCTION

To maintain road and railway infrastructure effectively, the road or rail profile along the longitudinal direction needs to be monitored frequently. Profile monitoring using vehicle response measurements potentially allowing for frequent profile evaluation has been proposed and studied. However, the applicability of such systems to various type of vehicle models in terms of effective sensor installation locations is not clarified yet. Hence, from the perspective of observability, sensor installation location's effect need to be investigated.

For a sensor setup to be effective, the important prerequisite is that interested states and parameters of the system should be observable. If the system is observable, the ability to precisely estimate the state of an observable system depends on the system driving noise (Franco et al. (2006)). However, when the system is not observable, an exact estimate of the state is not possible even if the noise level is insignificant (Goshen-Meskin and Bar-Itzhack (1992)). The observability of linear systems are vital structural properties which have close relationships with the state observers (Kalman (1963)). Particularly, observability can govern the stability of the Kalman filter. Detailed study on algebraic and geometric observability were carried out by Sedoglavic A. (2002), Anguelova M. (2004) and Pan and Wang (2005). Response-based road and rail profile estimation methods are developed by Doumiati et al (2011), Tsunanshima et al. (2014) and Zhao et al. (2016). However, the prerequisite on the types and locations of sensors are not theoretically explained for various type of vehicle model. In this paper, quarter car (QC) model (2 DOF), half car (HF) model (4 DOF) and 6 DOF vehicle model (train) are considered with appropriate effective sensor types, namely acceleration and gyro and their locations for profile estimation. Hence, its requirements are studied through the observability analysis.

2 OBSERVABILITY

2.1 *Observability rank condition (ORC) of linear system*

The state space model for the continuous time-invariant system is represented as,

$$\dot{x}(t) = Ax(t) + Bu(t) \tag{1}$$

$$y(t) = Cx(t) + Du(t) \tag{2}$$

where x is the state vector, y is the output vector, u is the input vector, A is the transition/state matrix, B is the input matrix, C is the output/measurement matrix, D is the feedback matrix.

A system is observable if, for any possible sequence of state vectors, the current state can be determined in finite time using only the outputs (Hermann and Krener (1977)). If a system is not observable, the current values of some of its states cannot be determined through output measurements. A continuous time-invariant linear state-space model with N states is observable if and only if, the rank of the observability matrix (O) is equal to N, where

$$O = \begin{bmatrix} C \\ CA \\ CA^2 \\ \vdots \\ CA^{N-1} \end{bmatrix} \tag{3}$$

2.2 *ORC of nonlinear system*

An analytic system is considered that of affine-input nonlinear system, which is written in the form (Hermann and Krener (1977), Chatzis et.al. (2015)):

$$\dot{x} = f(x) + \textstyle\sum_{j=1}^{l} g_j(x)U_j \tag{4}$$

$$y_i = h_i(x), i = 1, \ldots\ldots, n \tag{5}$$

where x is the state vector of size m, U_j $j=1,\ldots,l$ are independent, piecewise constant, inputs, h is the measurement matrix and n is the number of observation equations. A basic tool used in this algorithm is the Lie derivative which evaluates the change of a tensor field, $L_v(s(x))$ of scalar function s (x) along the flow of vector field, $v(x) = [v_1 \ldots\ldots v_v]^T$ and Lie derivative of a vector of k scalar functions, $\Omega = [s_1(\mathrm{x}), \ldots, s_k(\mathrm{x})]$ along the vector field $v(x)$ is given in Eq. (6) and (7) respectively.

$$L_v(s(x)) = \nabla s.v \tag{6}$$

$$L_v(\Omega) = d\Omega.v \tag{7}$$

The steps of the method can be outlined as:

1. Starting, k=0; $\Omega_0 = [h_1, h_2, \ldots, h_n]$, $\Delta\Omega_0 = \Omega_0$
2. $\Delta\Omega_{k+1} = [(L_f(\Delta\Omega_k))^T, \ (L_{g1}(\Delta\Omega_k))^T \ldots., \ (L_{gl}(\Delta\Omega_k))^T]^T$
3. $\Omega_{k+1} = \Omega_k \cup \Delta\Omega_{k+1}$
4. Calculate $d\Omega_{k+1}$ then $d\Omega_{k+1} = d\Omega_k \cup d\Omega_{k+1}$
5. If rank ($d\Omega_{k+1}$) = rank ($d\Omega_k$) or rank ($d\Omega_{k+1}$) = m or $k=m-2$ *end;* then $k=k+1$ and go to step 2.

2.3 *Separating the states of an unobservable system*

If the equations of the system are rational, which can be defined by the rational fraction, it is possible to distinguish between the observable and unobservable states of system. The observability algorithm for nonlinear system can directly separates the states into these categories by removing the i th column of the final matrix $d\Omega_k$, and the rank of the occurring matrix $\mathrm{d}\Omega_k^i$is computed. If the rank of $\mathrm{d}\Omega_k^i$ is less than the rank of $d\Omega_k$, the i th state is observable; otherwise it is not. Consequently, it is feasible to distinguish the states as falling within observable or unobservable sets. Similarly, it can be adopted for linear system also. Hence, the observable state variables alone can be separated from the unobservable system.

3 PROPOSED APPROCHES FOR PROFILE ESTIMATION

For estimating the profile (i.e., road or rail track profile along the longitudinal direction) using the vehicle (i.e., car or train), the location of sensors and their types are important. Although sensors to obtain displacement and rotational angle are expected to provide the profile component, they are not practical to use in the measurement field. Only acceleration and angular velocity are easily measurable in the field. Installation locations also have practical limitations. For high speed trains, axle-box accelerometers are used to obtain the profile directly (Li. Z et al. (2015)). However, for the normal commercial trains, axle-box accelerometers are not feasible to install. Car body and bogies are preferred as sensor installation locations. Under these limitations, profile estimation is generally difficult as explained in this paper. However, usually the profile components of very long wave length are not required, which ease the requirements. The second derivative of profile is first estimated through acceleration and angular velocity measurement and integrated twice with high pass filter to evaluate profile. Observability of the second derivative of the profile is studied herein.

The state space model for the continuous time-invariant system is represented as,

$$\dot{x}(t) = Ax(t) + Bu(t);\ y(t) = Hx(t) \tag{8}$$

where x is the system state vector, u is the input vector, y is the measurement vector, A is the state matrix, B is the input matrix and H is the measurement matrix. In this study the state vector is combined with the input vector. The state matrix can now be redefined by adding the input matrix to the original state matrix and increasing the size of the state matrix.

$$\tilde{x} = \begin{bmatrix} x \\ u \end{bmatrix} \tag{9}$$

The measurement matrix is appended by a null matrix because inputs are assumed unmeasured.

$$\widetilde{H} = [H \quad 0] \tag{10}$$

The two approaches for the estimation of profile as a part of the state vector are considered. One is to augment the state variables with the second derivative of the profile and estimate the second derivative. The profile is estimated as its double integration. The other is to alter state space model by adopting the first derivative of the state vector as new state vector. Thus, only the dynamic components are considered while the static components (i.e., displacement) are excluded from the state vector. The profile is estimated as the double integration of a state vector component. This process makes the second derivative of profile as an observable state even though the profile is not observable. The altered state space model is,

$$\ddot{\tilde{x}}(t) = A\dot{\tilde{x}}(t)\ ; \qquad \dot{y}(t) = \widetilde{H}\dot{\tilde{x}}(t) \tag{11}$$

where $\tilde{x}$ is augmented state vector and only the measurement matrix H, is modified while the transition matrix A, is unaltered.

4 ANALYSES OF VEHICLE MODELS

In order to obtain theoretically the appropriate locations and types of sensors to estimate the profile as the 'observable state', the four different vehicle models are considered for the analysis, namely QC model, HC model, 6 DOF model and 6 DOF with averaged geometry model (i.e., simplified 4 DOF vehicle model). Practical sensor types are accelerometers and gyros and their installation locations are car body and bogies. In these analyses the parameters of the vehicle models are considered known. The analyses are carried out using both the linear and non-linear observability check methods and are found to be consistent. The proposed approaches are investigated for all vehicle models.

All possible combination of measurements at car body and bogies with accelerometers and gyros are analyzed in terms of ORC and all cases where the profile or its derivatives are observable are extracted for each vehicle model. Note that only minimum combination of measurements are listed in the following sections.

4.1 QC model

A QC model (Figure 1) is a well-known model for simulating one-dimensional vehicle suspension performance. The dynamic equation of motion is,

$$\begin{bmatrix} m_1 & 0 \\ 0 & m_2 \end{bmatrix}\begin{Bmatrix} \ddot{z}_1 \\ \ddot{z}_2 \end{Bmatrix} + \begin{bmatrix} c_1 & -c_1 \\ -c_1 & c_1 + c_2 \end{bmatrix}\begin{Bmatrix} \dot{z}_1 \\ \dot{z}_2 \end{Bmatrix} + \begin{bmatrix} k_1 & -k_1 \\ -k_1 & k_1 + k_2 \end{bmatrix}\begin{Bmatrix} z_1 \\ z_2 \end{Bmatrix} = \begin{Bmatrix} 0 \\ k_2 u + c_2 \dot{u} \end{Bmatrix} \tag{12}$$

where *m, k, c, z* are the mass, elastic coefficient, damping coefficient and position of vehicle body respectively and *u* is the road profile.

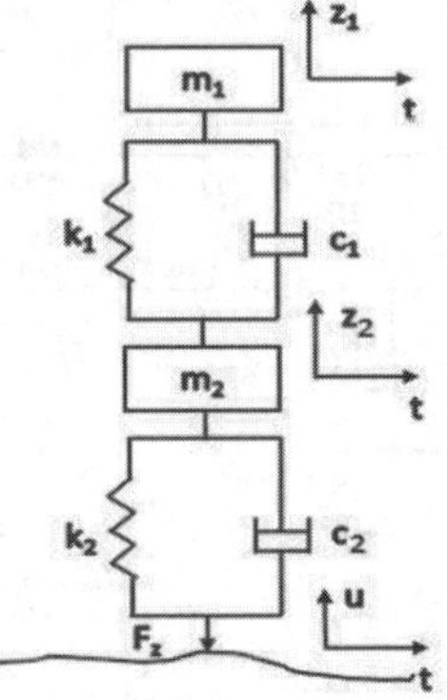

Figure 1. Quarter car model

The state vector is as follows:

$$\tilde{x} = (z_1\ \dot{z}_1\ z_2\ \dot{z}_2\ u\ \dot{u})^T \tag{13}$$

The results of ORC analysis of QC model is shown in Table 1. The numbers in the table represent the respective state variables in the state vector, while the bolded numbers represent the profile or its derivatives.

Table 1. ORC analysis results for QC model

Measurements	Observable states
$\ddot{z}_1$	Nil

The proposed two approaches are implemented with the following state vectors.
In approach (a), the state vector is,

$$\tilde{x} = (z_1\ \dot{z}_1\ z_2\ \dot{z}_2\ u\ \dot{u}\ \ddot{u})^T \tag{14}$$

In approach (b), the state vector is,

$$\dot{\tilde{x}} = (\ \dot{z}_1\ \ddot{z}_1\ \dot{z}_2\ \ddot{z}_2\ \dot{u}\ \ \ddot{u})^T \tag{15}$$

Table 2. ORC analysis for QC model using proposed approaches

Approach		Observable states
(a)	$\ddot{z}_1$	**7**
(b)	$\ddot{z}_1$	2, 4, **6**

Table 2 shows the observable states for the two approaches. Thus by measuring acceleration ($\ddot{z}_1$) only at the car body, the second derivative component of the profile is observable even though the system is unobservable.

4.2 *HC model*

The HC model (Figure 2) includes body pitch and bounce degrees of freedom. The dynamic equation of motion is derived using Lagrange function.

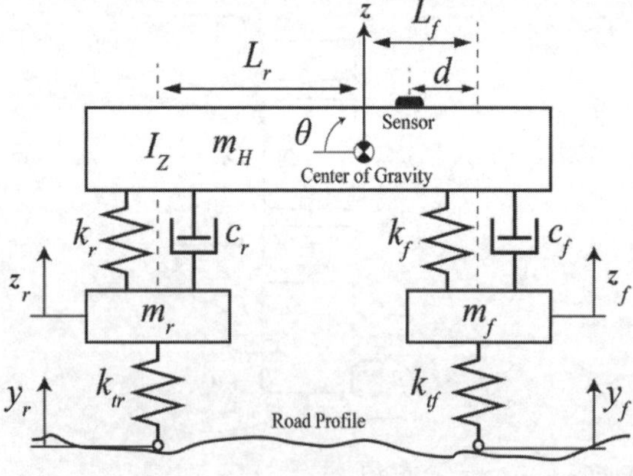

Figure 2. Half car model

$$M\ddot{x} + C\dot{x} + Kx = u \tag{16}$$

$$M = \begin{bmatrix} m_H & 0 & 0 & 0 \\ 0 & I_z & 0 & 0 \\ 0 & 0 & m_f & 0 \\ 0 & 0 & 0 & m_r \end{bmatrix} \tag{17}$$

$$K = \begin{bmatrix} k_f + k_r & L_r k_r - L_f k_f & -k_f & -k_r \\ L_r k_r - L_f k_f & L_f^2 k_f + L_r^2 k_r & L_f k_f & -L_r k_r \\ -k_f & L_f k_f & k_f + k_{tf} & 0 \\ -k_r & -L_r k_r & 0 & k_r + k_{tr} \end{bmatrix} \tag{18}$$

$$C = \begin{bmatrix} c_f + c_r & L_r c_r - L_f c_f & -c_f & -c_r \\ L_r c_r - L_f c_f & L_f^2 c_f + L_r^2 c_r & L_f c_f & -L_r c_r \\ -c_f & L_f c_f & c_f & 0 \\ -c_r & -L_r c_r & 0 & c_r \end{bmatrix} \tag{19}$$

The state vector and input vectors are as follows:

$$\tilde{x} = [z\ \ \theta\ \ z_f\ \ z_r\ \ \dot{z}\ \ \dot{\theta}\ \ \dot{z}_f\ \ \dot{z}_r\ \ y_f\ \ y_r\ \ \dot{y}_f\ \ \dot{y}_r]^T;$$
$$u = [0\ \ \ 0\ \ \ y_f k_{tf}\ \ \ y_r k_{tr}]^T \tag{20}$$

Table 3. ORC analysis results for HC model

Measurements	Observable states
$\ddot{z}\ \dot{\theta}$	6
$\ddot{z}_f\ \ddot{z}_r$	Nil

The results of ORC analysis of HC model is shown in Table 3. Because observation matrix depends on sensor location, the sensors are assumed to be located at the center of car body and bogies respectively. The proposed two approaches are executed with the following state vectors.

In approach (a), the new state vector is, $\tilde{x} =$

$$[z\ \ \theta\ \ z_f\ \ z_r\ \ \dot{z}\ \ \dot{\theta}\ \ \dot{z}_f\ \ \dot{z}_r\ \ y_f\ \ y_r\ \ \dot{y}_f\ \ \dot{y}_r\ \ \ddot{y}_f\ \ \ddot{y}_r]^T \tag{21}$$

In approach (b), the new state vector is,

$$\dot{\tilde{x}} = [\dot{z}\ \ \dot{\theta}\ \ \dot{z}_f\ \ \dot{z}_r\ \ \ddot{z}\ \ \ddot{\theta}\ \ \ddot{z}_f\ \ \ddot{z}_r\ \ \dot{y}_f\ \ \dot{y}_r\ \ \ddot{y}_f\ \ \ddot{y}_r]^T \tag{22}$$

Table 4. ORC analysis for HC model using proposed approaches

Approach		Observable states
(a)	$\ddot{z}\ \dot{\theta}$	6,**13,14**
	$\ddot{z}_f\ \ddot{z}_r$	**13,14**
(b)	$\ddot{z}\ \dot{\theta}$	2,5,6,7,8,**11,12**
	$\ddot{z}_f\ \ddot{z}_r$	5,6,7,8,**11,12**

Table 4 shows the observable states for the two approaches. The acceleration and angular velocity of car body and unsprung mass are the minimum combination of measurements, which results in observable profile derivatives. By implementing both the approaches, the second derivative component of the profile is observable.

4.3 6 DOF vehicle model

The 6 DOF vehicle model depicts the train model with car body and two bogies (Figure 3). For deriving the dynamic equation of motion for 6 DOF model Lagrange function is utilized.

$$M\ddot{x} + C\dot{x} + Kx = Dr + E\dot{r} \tag{23}$$

$$M = \begin{bmatrix} m_c & 0 & 0 & 0 & 0 & 0 \\ 0 & I_c & 0 & 0 & 0 & 0 \\ 0 & 0 & m_{t1} & 0 & 0 & 0 \\ 0 & 0 & 0 & I_{t1} & 0 & 0 \\ 0 & 0 & 0 & 0 & m_{t2} & 0 \\ 0 & 0 & 0 & 0 & 0 & I_{t2} \end{bmatrix} \tag{24}$$

$$K = \begin{bmatrix} 2k_s & 0 & -k_s & 0 & -k_s & 0 \\ 0 & 2k_s l_c^2 & -k_s l_c & 0 & k_s l_c & 0 \\ -k_s & -k_s l_c & 2k_p + k_s & 0 & 0 & 0 \\ 0 & 0 & 0 & 2k_p l_t^2 & 0 & 0 \\ -k_s & k_s l_c & 0 & 0 & 2k_p + k_s & 0 \\ 0 & 0 & 0 & 0 & 0 & 2k_p l_t^2 \end{bmatrix} \tag{25}$$

$$C = \begin{bmatrix} 2c_s & 0 & -c_s & 0 & -c_s & 0 \\ 0 & 2c_s l_c^2 & -c_s l_c & 0 & c_s l_c & 0 \\ -c_s & -c_s l_c & 2c_p + c_s & 0 & 0 & 0 \\ 0 & 0 & 0 & 2c_p l_t^2 & 0 & 0 \\ -c_s & c_s l_c & 0 & 0 & 2c_p + c_s & 0 \\ 0 & 0 & 0 & 0 & 0 & 2c_p l_t^2 \end{bmatrix} \tag{26}$$

$$D = \begin{bmatrix} 0 & 0 & 0 & 0 \\ 0 & 0 & 0 & 0 \\ c_p & c_p & 0 & 0 \\ c_p & -c_p & 0 & 0 \\ 0 & 0 & c_p & c_p \\ 0 & 0 & c_p & -c_p \end{bmatrix} \quad E = \begin{bmatrix} 0 & 0 & 0 & 0 \\ 0 & 0 & 0 & 0 \\ k_p & k_p & 0 & 0 \\ k_p & -k_p & 0 & 0 \\ 0 & 0 & k_p & k_p \\ 0 & 0 & k_p & -k_p \end{bmatrix} \tag{27}$$

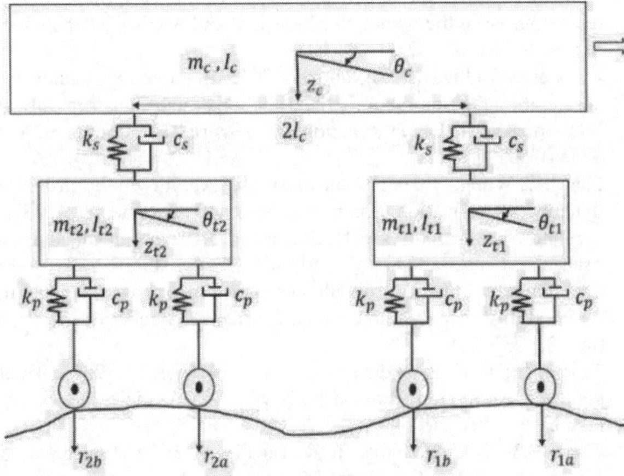

Figure 3. 6 DOF vehicle model

Let $x(t)$ be a state variable and $r(t)$ be a track geometry vector given as a forced displacement of four wheels. Sensors are assumed to be located at the center of car body and bogies respectively. The results of ORC analysis of 6 DOF model is shown in Table 5. The state vector is as follows,

$$\tilde{x} = [z_c\ \theta_c\ z_{t1}\ \theta_{t1}\ z_{t2}\ \theta_{t2}\ \dot{z}_c\ \dot{\theta}_c\ \dot{z}_{t1}\ \dot{\theta}_{t1}\ \dot{z}_{t2}\ \dot{\theta}_{t2}\ r_{1a}\ r_{1b} r_{2a} r_{2b}\ \dot{r}_{1a}\ \dot{r}_{1b} \dot{r}_{2a} \dot{r}_{2b}]^T \tag{28}$$

Table 5. ORC analysis results for 6 DOF train model

Measurements	Observable states
$\ddot{z}_c\ \dot{\theta}_c\ \dot{\theta}_{t1}\ \dot{\theta}_{t2}$	8,10,12
$\ddot{z}_{t1}\ \ddot{z}_{t2}\ \dot{\theta}_{t1}\ \dot{\theta}_{t2}$	10,12

Table 6. ORC analysis for 6 DOF model using proposed approaches

	Approach	Observable states
(a)	$\ddot{z}_c\ \dot{\theta}_c\ \dot{\theta}_{t1}\ \dot{\theta}_{t2}$	8,10,12, **21,22,23,24**
	$\ddot{z}_{t1}\ \ddot{z}_{t2}\ \dot{\theta}_{t1}\ \dot{\theta}_{t2}$	10,11, **21,22,23,24**
(b)	$\ddot{z}_c\ \dot{\theta}_c\ \dot{\theta}_{t1}\ \dot{\theta}_{t2}$	2,4,6,7,8,9,10,11,12,**17,18,19,20**
	$\ddot{z}_{t1}\ \ddot{z}_{t2}\ \dot{\theta}_{t1}\ \dot{\theta}_{t2}$	4,6,7,8,9,10,11,12,**17,18,19,20**

Table 6 shows the observable states for the two approaches. In both the approaches, by only measuring acceleration and angular velocity at car body and bogie the second derivative component of the profile is observable.

4.4 6 DOF vehicle model with averaged track geometry

The equation of motion for a 6 DOF vehicle model with averaged track geometry (Figure 4) is derived as described in the previous section (section 4.3). Only D and E matrix gets modified.

$$M\ddot{x} + C\dot{x} + Kx = Dr + E\dot{r} \tag{29}$$

$$D = \begin{bmatrix} 0 & 0 \\ 0 & 0 \\ 2c_p & 0 \\ 0 & 0 \\ 0 & 2c_p \\ 0 & 0 \end{bmatrix} \quad E = \begin{bmatrix} 0 & 0 \\ 0 & 0 \\ 2k_p & 0 \\ 0 & 0 \\ 0 & 2k_p \\ 0 & 0 \end{bmatrix} \tag{30}$$

In order to improve the results, state vector need to be reduced. Since input track geometry at front and rear axle are averaged by the bogie and then transmitted to the vehicle body, an averaged geometry between front and rear axles can be calculated.

An averaged geometry, $r_1 = \frac{r_{1a}+r_{1b}}{2}$; $r_2 = \frac{r_{2a}+r_{2b}}{2}$

$r^T(t) = [r_+, r_-]$ given as $r_+ = \frac{r_1+r_2}{2}$; $r_- = \frac{r_1-r_2}{2}$ (31)

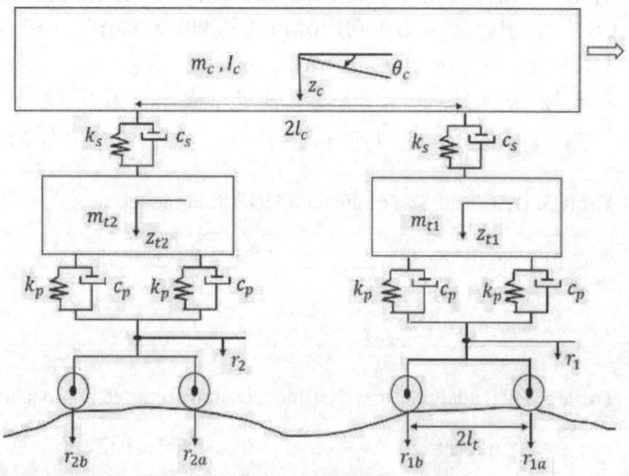

Figure 4. Simplified model with averaged geometry

The state vector is as follows, $\tilde{x} = [z_c\ \theta_c\ z_{t1}\ \theta_{t1}\ z_{t2}\ \theta_{t2}\ \dot{z}_c\ \dot{\theta}_c\ \dot{z}_{t1}\ \dot{\theta}_{t1}\ \dot{z}_{t2}\ \dot{\theta}_{t2}\ r_+\ r_-\ \dot{r}_+\ \dot{r}_-]^T$ (32)

However, the pitching motion of the bogie is not considered in this averaged 6 DOF model, thus it is a simplified model same as the HC model discussed earlier. Hence the state vector reduces to,

$$\tilde{x} = [z_c\ \theta_c\ z_{t1}\ z_{t2}\ \dot{z}_c\ \dot{\theta}_c\ \dot{z}_{t1}\ \dot{z}_{t2}\ r_+\ r_-\ \dot{r}_+\ \dot{r}_-]^T \quad (33)$$

The results and inferences are same as explained for HC model.

5 CONCLUSION

The observability analysis to theoretically obtain the appropriate sensor types and their placements for estimating profile (i.e., road or rail profile along the longitudinal direction) is presented. Acceleration and angular velocities are assumed to be observed variables. The second derivative of the profile is set as the variable to be identified, so that non-static components of the profile is obtained as its double integral. Four different types of vehicle models are considered with appropriate sensor types and their locations, for the numerical analyses and the results are presented. The two approaches to obtain the second derivative of profile as an observable state variable are examined. Approach (a) includes the second derivative of the profile in the state vector. Although the system is unobservable the augmented state variable is observable. Approach (b) alters state space model by taking the first derivative of the system equation. The second derivative component is observable. These analyses indicate that the profile can be estimated by an accelerometer and a gyro on car body or bogie. The profile is expected to be obtained by the double integration of the high pass-filtered second derivative. Profile estimation employing Kalman Filters based on these observability analyses is being studied. Also, the observability analysis will further be applied to 27-DOF train vehicle model and Full car model in the future.

REFERENCES

Anguelova, M. 2004, Non-linear observability and identifiability: general theory and a case study of a kinetic model for S.Cerevisiae, PhD thesis, School of Mathematical Sciences Chalmers University of technology and Göteborg University, Sweden

Chatzis, M.N. & Chatzi, E.N. & Smyth, A.W. 2015, On the observability and identifiability of nonlinear structural and mechanical systems, Structural Control and Health Monitoring, 22, 574-593

Doumiati, M. & Victorino, A. & Charara, A. & Lechner, D. 2011, Estimation of road profile for vehicle dynamics motion: experimental validation, American Control Conference, San Francisco, CA, USA, 5237-5242

Franco, G. & Betti, R. & Longman, R.W. 2006, On the uniqueness of solutions for the identification of linear structural systems. Journal of Applied Mechanics, 73(1), 153–162

Goshen-Meskin, D. & Bar-Itzhack, I. 1992, Observability analysis of piece-wise constant system- Part I: Theory, Aerospace and Electronic Systems, IEEE Transactions, 28(4), 1056-1067

Hermann, R. & Krener A. 1977, Non-linear controllability and observability, Automatic Control, IEEE Transactions, 22(5), 728–740

Kalman, R. 1963, Mathematical description of linear dynamical systems, Journal of the Society for Industrial and Applied Mathematics Series-A Control, 1(2), 152–192

Li, Z. & Molodova, M. & Nunez, A. 2015, Improvement in axle box acceleration measurement for the detection of light squats in railway infrastructure, IEEE Transaction on Industrial Electronics, 62 (7), 4385-4397

Pan, J. & Wang, R. 2005, Non-linear observability in the structural dynamic identification, Smart structures and materials 2005 : Sensors and smart structures technologies for civil, mechanical, and aerospace systems, San Diego, California, USA, 1045-1052

Sedoglavic A. 2002, A probabilistic algorithm to test local algebraic observability in polynomial time. Journal of Symbolic Computation, 33, 735-755.

Tsunanshima, H. & Naganuma, Y. & Kobayashi, T. 2014, Track geometry estimation from car-body vibration, Vehicle System Dynamics, 52, 207-219

Zhao, B.Y. & Nagayama, T. & Takada, S. & Takahashi, M. & Makihata, N. 2016, Development of response-based road profile estimation using multiple observables, EASEC-14, Structural Engineering and Construction, Ho Chi Minh, Vietnam.

Proceedings of the International Conference on Smart Infrastructure and Construction
ISBN 978-0-7277-6127-9

© The authors and ICE Publishing: All rights reserved, 2016
doi:10.1680/tfitsi.61279.363

Large-scale road surface evaluation using dynamic responses of commercial vehicles

N. Makihata*[1], B. Zhao[2], M. Toyoda[3], M. Takahashi[1], M. Ieiri[1] and T. Nagayama[2]

[1] *Infrastructure solutions Division, JIP Techno Science Corporation, Tokyo, Japan*
[2] *Department of Civil Engineering, the University of Tokyo, Tokyo, Japan*
[3] *Institute of Industrial Science, the University of Tokyo, Tokyo, Japan*
* *Corresponding Author*

ABSTRACT Response-based road condition evaluation is expected to provide road surface conditions effectively and efficiently. A large scale road surface condition evaluation using a large number of commercial vehicles is conducted based on Dynamic Response Intelligent Monitoring System (DRIMS) ; DRIMS estimate the International Roughness Index (IRI) using vehicle responses. A mobile device version of DRIMS, iDRIMS, is employed. The data is first analyzed to construct a Half-Car (HC) model of measurement vehicles. The parameters of the Half-Car (HC) model is identified through Unscented Kalman Filter(UKF) and Genetic Algorithm(GA). Then, IRI is estimated by analyzing vertical acceleration responses using the HC model. To verify the effectiveness of this method, the estimated IRI is compared with the reference IRI obtained by a road profiler. The comparison shows good agreement between iDRIMS and the reference and indicates the validity of large-scale road surface evaluation using general commercial vehicles. Finally, the data collection and analysis platform is built which succesfully collected and analyzed 6 months data from about 50 commercial vehicles.

1 INTRODUCTION

Quantitative and objective evaluation of road surface is important in order to maintain the road infrastructure cost-effectively and to ensure the safety of drivers. Local governments, which manage various types of roads from arterial roads to local streets, however, do not have practical ways to evaluate road conditions efficiently and objectively. Road profilers, typically used on expressways, are expensive and time-consuming.

In this paper an easy and efficient road monitoring and archiving platform involving a large number of commercial vehicles are proposed. Smartphone-based Dynamic Response Intelligent Monitoring System (iDRIMS) (Zhao et al. 2016), which can effectively evaluate road conditions, has the following advantages: (a) the algorithm can compensate the difference in vehicle dynamic characteristics and drive speed, (b) the difference in sensor installation location, which is often inevitable when sensors are installed on commercially used vehicles, can be compensated, and (c)the system embedded in smartphones allows for easy installation and operation while the accuracy of evaluation is reasonable.

This paper first presents the modification of road condition evaluation algorithms. A half-car (HC) model, which can represent both pitching and bouncing motion of the vehicle, is employed. The parameters of HC model are identified by UKF and GA when vehicle goes over a portable hump of known shape. Once the vehicle parameters are determined, vehicle transfer functions is estimated for various drive speeds at the specific sensor location. This transfer function is then multiplied to vehicle response to convert the measured accelerations to the standard quarter car (QC) responses, which is further converted to the International Roughness Index (IRI) (Sayers, et al. 1986).

Driving test on arterial roads and local streets are performed with commercial vehicles equipped with smartphones as the sensors. The IRI of target road is evaluated by a road profiler and compared with the iDRIMS outputs. The comparison shows that iDRIMS results in reasonable IRI estimations in most cases

Finally, the data collection and analysis platform is built and large scale road condition evaluation is performed. The iDRIMS measurement systems were installed to about 50 commercial vehicles, and the data is collected for about 6 months. The platform provides large-scale visualization of GPS trajectories, raw sensor data, such as acceleration, and analyzed IRI. The user can explore the data by zooming on maps and filtering with attribute values.

2 DYNAMIC RESPONSE INTELLIGENT MONITORING SYSTEM

2.1 *OVERVIEW OF DRIMS*

Dynamic response intelligent monitoring system (DRIMS) (Fujino, et al. 2005, Furukawa, et al. 2007, Asakawa et al. 2012, Nagayama et al. 2013), has been developed to evaluate road roughness by measuring vehicle responses and GPS information. The system is inexpensive and easily installed on an ordinary vehicle. Recently, a mobile device version of DRIMS, iDRIMS, was developed utilizing the sensors and GPS of a smartphone (see Figure 1).

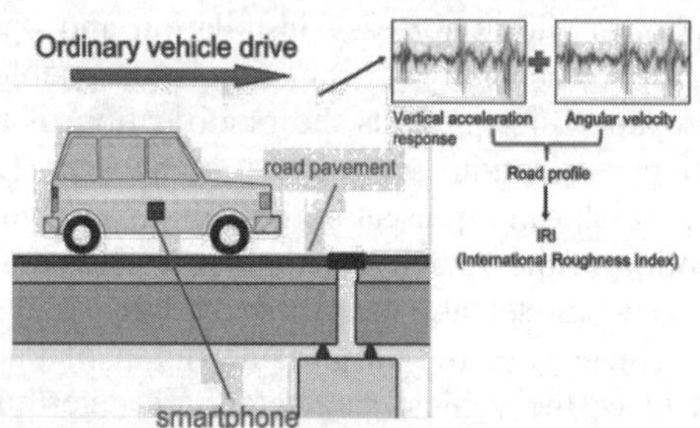

Figure 1. Scheme of the method of iDRIMS.

2.2 *MEASUREMENT EQUIPMENT*

As a measuring device, Apple's iPhone / iPod touch is adopted. The iOS application (iDRIMS measurement) was developed with the requirements on accurate sampling timing and easy manipulation assuming commercial vehicle drivers as the operators (see Figure 2).

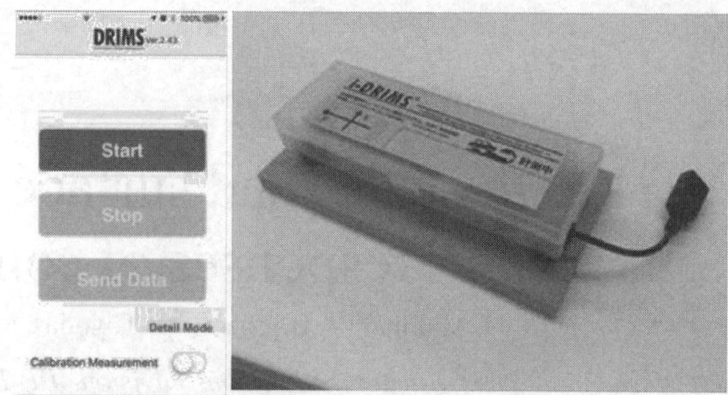

Figure 2. iDRIMS measurement application and the device.

2.3 *INTERNATIONAL ROUGHNESS INDEX*

International Roughness Index (IRI) was proposed by World Bank as an indicator of drive comfort. IRI is defined as:

$$IRI = \left\{\int_0^{L/V} \left|\dot{z}_s - \dot{z}_u\right| dt\right\} / L \tag{1}$$

where z_s is the vertical absolute displacement of sprung mass of the standard quarter car (QC) model (shown in Figure 3). z_u is the vertical absolute displacement of unsprung mass, L is the length of evaluation section, and V is the drive speed.

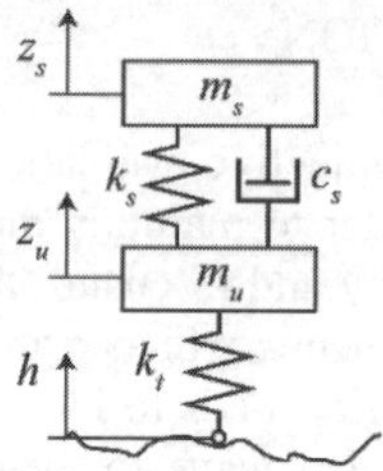

Figure 3. Standard quarter car model.

The process of IRI estimation is shown in Figure 4. The vertical acceleration of the measurement vehicle body, $\ddot{z}_m(t)$, is measured. The vertical acceleration RMS of sprung mass of standard QC model is estimated from $\ddot{Z}_m(\omega)$, the Fourier transform of $\ddot{z}_m(t)$, based on Parseval's theorem as follows:

$$RMS^2 = \frac{1}{T}\int_0^T \ddot{z}_s(t)dt = \int_0^\infty \left|\ddot{Z}_s(\omega)\right|^2 d\omega = \int_0^\infty \left|\ddot{Z}_m(\omega)\right|^2 \left|TF(\omega)\right|^2 d\omega \quad (2)$$

where $\ddot{z}_s(t)$ is the vertical acceleration of sprung mass of the QC model, $\ddot{Z}_s(\omega)$ is the Fourier transform of $\ddot{z}_s(t)$, $TF(\omega)$ is the amplitude ratio function. This amplitude ratio as a function of frequency ω is defined as follows:

$$\left|TF(\omega)\right| = \sqrt{\frac{PSD_{QC\_80km/h}(\omega)}{PSD_{car\_Xkm/h}(\omega)}} \quad (3)$$

where $PSD_{QC\_80km/h}(\omega)$ is the power spectral density of $\ddot{z}_s(t)$ at the standard drive speed of 80km/h and $PSD_{car\_Xkm/h}(\omega)$ is the power spectral density of $\ddot{z}_m(t)$ at the measurement vehicle's drive speed of X km/h.

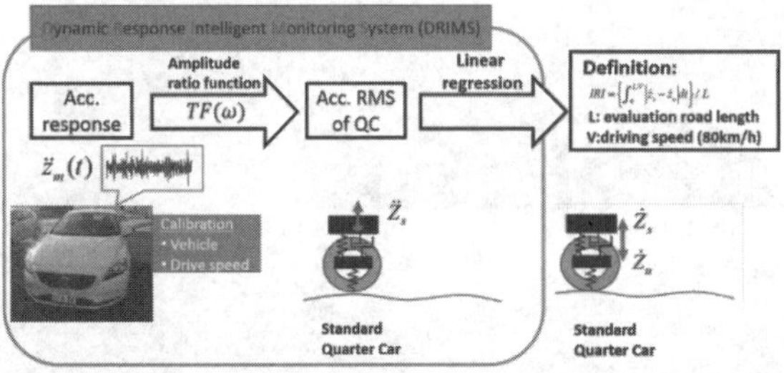

Figure 4. IRI estimation of iDRIMS.

Finally, IRI is estimated from the RMS by using the linear regression line between IRI and the RMS shown in Figure 5.

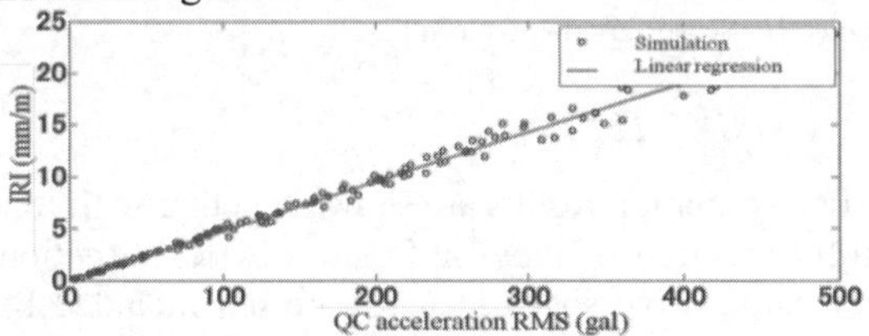

Figure 5. IRI and acceleration RMS of $\ddot{z}_s(t)$.

Note that $TF(\omega)$ plays an important role in compensating differences in vehicle dynamic characteristics and drive speeds. An effective and practical method to estimate $TF(\omega)$, hump calibration, is proposed.

2.4 HUMP CALIBRATION

Hump calibration is performed by measuring the vertical acceleration and pitching angular velocity of the vehicle body when the vehicle passes a portable hump of a specific shape at a specific speed. The hump employed has the height of 5cm and the width of 30cm (see Figure 6a). The speed is set at 10km/h. The measurement vehicle is modeled as a half car (HC) model (see Figure 6b), so that the bouncing, pitching, and axle modes can be reproduced. The HC model can account for the sensor installation location along the longitudinal direction. Unknown parameters are included as a part of the state-space vector and estimated through UKF and GA.

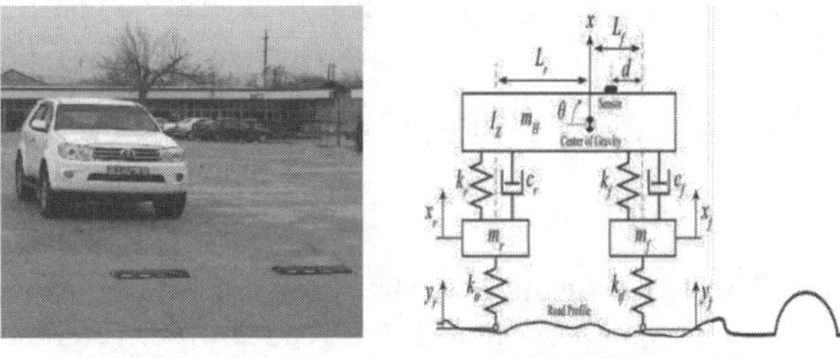

(a) Hump Test (b) Simulation model

Figure 6. Hump calibration experiment and simulation.

Once HC model is identified, vehicle motion simulation is performed with different drive speeds (from 20 to 110km/h with 10km/h interval) assuming a virtual road profile. Simulation is also performed using the standard QC model at the speed of 80km/h on the same profile. Then $TF(\omega)$ is calculated by Equation 2-3. This $TF(\omega)$ thus accounts for differences in vehicle dynamic characteristics and drive speeds. In order to check the repeatability of $TF(\omega)$ estimation, hump calibration is experimentally repeated three times and the results are shown in Figure 7. Good agreement is confirmed below 0.45 cycle/m including the bouncing and pithing mode. Therefore, a lowpass filter with the cutoff frequency of 0.45 cycle/m is employed. Even with this filter application, correlation coefficient between IRI and RMS of QC model is high as shown in Table 1.

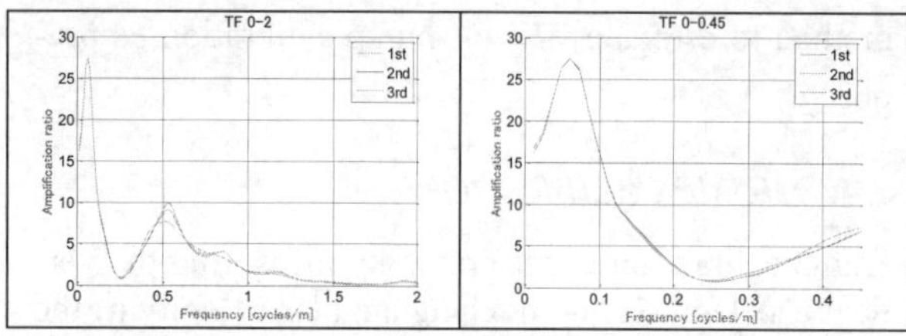

(a) $TF(\omega)$ at 0-2 cycle/m (b) $TF(\omega)$ at 0-0.45 cycle/m

Figure 7. Repeatability of three times $TF(\omega)$ estimation.

Table 1. Correlation between RMS and IRI.

Correlation function	Cutoff frequency	Correlation coefficient
IRI=0.0499×*RMS*	0.5 cycles/m	0.997
IRI=0.0511×*RMS*	0.45 cycles/m	0.996
IRI=0.0554×*RMS*	0.3 cycles/m	0.995
IRI=0.0654×*RMS*	0.1 cycles/m	0.981

3 ACCURACY VERFICATION

3.1 OVERVIEW

The IRI estimation accuracy of iDRIMS is examined by comparison with the road profiler reference data. The target road, consisting of 2 km of good condition section and 2 km of poor condition section, is shown in Figure 8.

Commercial vehicles are classified as van, small car, and light car. Three test vehicles representing these vehicle categories are selected as in Figure 9.

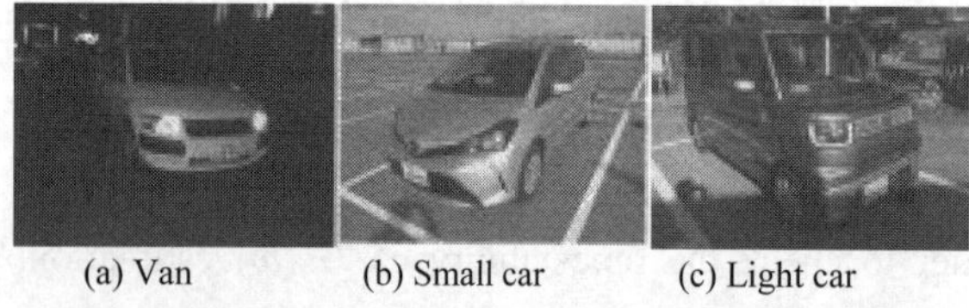

(a) Van (b) Small car (c) Light car

Figure 9. Test vehicles.

The sensor installation locations of these vehicles are oftentimes limited due to their usage in the commercial operation; the locations can vary from vehicle to vehicle. Therefore sensors are installed at various locations as in Figure 10 so that IRI estimation accuracy of these sensors are compared with each other.

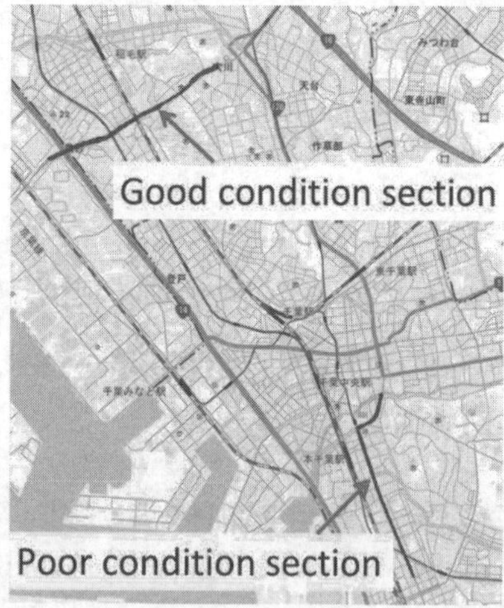

Figure 8. Test road sections

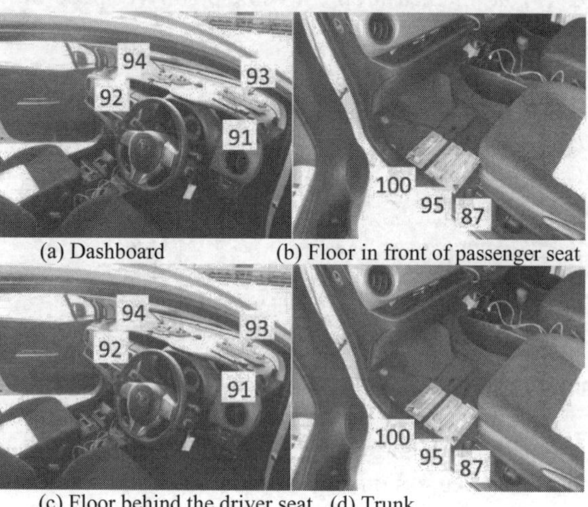

(a) Dashboard (b) Floor in front of passenger seat

(c) Floor behind the driver seat (d) Trunk

Figure 10. Sensor installation locations

3.2 DRIVE TEST

IRI estimation results are shown together with road profiler reference values in Figure 11. IRI at sections with large drive speed changes are not included because iDRIMS requires gradual drive speed change.

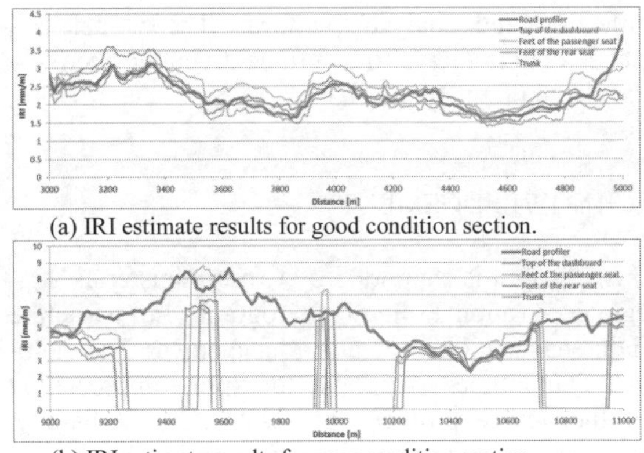

(a) IRI estimate results for good condition section.

(b) IRI estimate results for poor condition section.

Figure 11. IRI estimation results and road profiler reference data (small car).

The IRI error of iDRIMS is summarized in Table 2. For each test vehicle, the number of sensor units used is 14. The IRI error is the average value of two sections (i.e., the good condition section and poor condition section). The IRI error of iDRIMS installed at the floor behind the driver seat is larger than others. Note that most of iDRIMS currently installed on commercial vehicles are on the dashboard or on the floor in front of the passenger seat.

Because large drive speed change is often observed during commercial operation in particular at intersection and curves, algorithms to compensate the speed change is currently being studied.

Table 2 The IRI error of the three test vehicles

Installation position	IRI error [%]		
	Van	Small car	Light car
(a) Dashboard	22.0	14.9	13.5
(b) Floor in front of passenger seat	26.3	29.4	15.1
(c) Floor behind the driver seat	55.3	48.7	24.7
(d) Trunk	17.4	18.3	12.2

4 LARGE-SCALE ROAD SURFACE EVALUATION PLATFORM

4.1 OVERVIEW OF LARGE-SCALE ROAD SURFACE EVALUATION PLATFORM

A data collection and analysis platform is built for large-scale road surface evaluation based on iDRIMS. Figure 12 shows the overview of the platform. Commercial vehicles send their response data to the platform after their daily operations. The data is archived in the platform, and IRI is estimated for all trajectories by the method shown in Section 2. The platform provides large-scale visualization of estimated IRI, GPS trajectories, and raw sensor data (acceleration). The user can explore the data by zooming maps and filtering with attribute values.

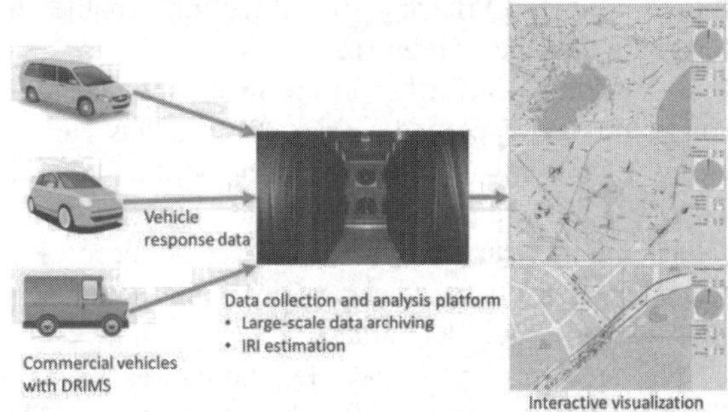

Figure 12. Overview of the large-scale road surface evaluation platform.

4.2 COLLECTION OF COMMERICIAL VEHICLE RESPONSE DATA

Using about 10 transportation vehicles and about 40 patrol vehicles of security company, the measurement is performed. The vehicle responses are measured during their commercial operations and collected. The measurement data of about 11,000 km was collected in three months in Chiba-city (see Figure 13); a large amount of vehicle response data has been efficiently collected in a short time.

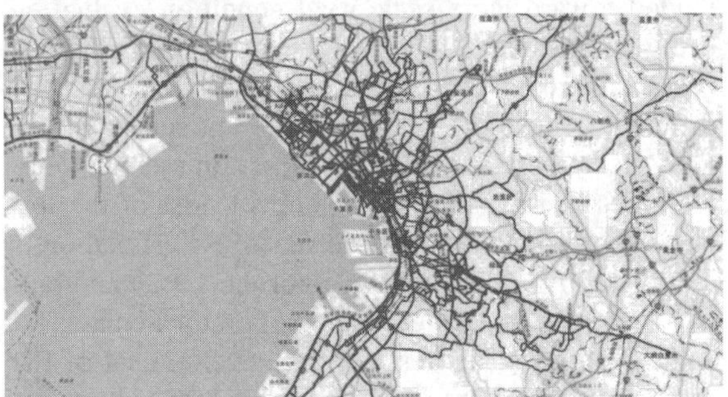

Figure 13. The measurement routes in Chiba-city.

4.3 *INTERACTIVE VISUAL ANALYSIS OF VEHICLE RESPONSE DATA*

The archived data is analyzed by interactive visualization (See the right hand side of Figure 12). It first shows rough IRI overview by a heat map view. When the user zoom into an interesting area, GPS trajectories with estimated IRI are shown. These trajectories can be filtered by IRI, velocity, and direction. Filtering by velocity is useful to remove unreliable segments. Filtering by direction enable to designate the lane of interest.

By zooming in further in the map, raw acceleration values are displayed on trajectories. This view is useful for finding out exact locations with bad conditions. By filtering segments with low z-acceleration values, the user can see where vehicles bounced. Filtering by high x and y-acceleration can choose segments with reliable IRI values.

The visualization is scalable, and can easily handle response data from 100 vehicles or more. The number of drawing primitives is efficiently controlled by zooming level. According to the designated area and zooming level, the visualizer extract required data objects from the data archive. To effective retrieval, the vehicle response data is indexed by their data types, areas, zooming levels, and time.

5 CONCLUSIONS

Smartphone-based Dynamic Response Intelligent Monitoring System, developed to evaluate road roughness condition using vehicle responses, is extended toward large-scale road condition evaluation using commercial vehicles. The system, iDRIMS, is first examined in terms of IRI estimation accuracy of different vehicle types and of different sensor installation locations considering typical commercial vehicle types and installation locations. Most of the test cases showed that the RMS error in IRI estimation is less than 30 %. Because the algorithms assume gradual drive speed change, IRI was not calculated at some sections of the test course; development of IRI estimation algorithm even under large speed change is needed. Finally, the vehicle response data measured for about 6 months by about 50 commercial vehicles is collected. The data is archived on the data collection and analysis platform. The vehicle response data is shown to be efficiently and effectively visualized on the platform.

ACKNOWLEDGEMENT

This research is supported by Cross-ministerial Strategic Innovation Promotion Program “Infrastructure Maintenance, Renewal and Management Technology”. The authors would like to express appreciation for the advice provided by Asahi-city, Chiba-city, Toyonaka-city and Sohgo security services co. ltd.

REFERENCES

Asakawa, H., et al. 2008. "Development and application of road monitoring system using dynamic response of vehicles." Proceedings of the 11th East Asia-Pacific Conference on Structural Engineering and Construction (EASEC-11).

Fujino, Yozo, et al. 2005. "Development of vehicle intelligent monitoring system (VIMS)." Smart Structures and Materials. International Society for Optics and Photonics.

Furukawa, T., et al. 2007. "Real-time diagnostic system for pavements using dynamic response of road patrol vehicles (VIMS)." Proc. 3rd International Conference on Structural Health Monitoring of Intelligent Infrastructure (SHMII-3).

Sayers, Michael W., Thomas D. Gillespie, and A. & V. Queiroz. 1986. The international road roughness experiment. Establishing correlation and a calibration standard for measurements. No. HS-039 586.

Sayers, Michael W., Thomas D. Gillespie, & William DO Paterson. 1986. Guidelines for conducting and calibrating road roughness measurements. No. Technical Paper 46.

Nagayama, Tomonori, et al. 2013. "Road condition evaluation using the vibration response of ordinary vehicles and synchronously recorded movies." SPIE Smart Structures and Materials+ Nondestructive Evaluation and Health Monitoring. International Society for Optics and Photonics.

Boyu Zhao, et al. 2016, "A method of response-based road profile estimation using multiple observables." Proceedings of the 11th East Asia-Pacific Conference on Structural Engineering and Construction (EASEC-14).

Proceedings of the International Conference on Smart Infrastructure and Construction
ISBN 978-0-7277-6127-9

© The authors and ICE Publishing: All rights reserved, 2016
doi:10.1680/tfitsi.61279.369

An experimental investigation on passive control for elevated water tanks

A. Roy[1], A. (D.) Ghosh[*1] and B. Basu[2]

[1]*Indian Institute of Engineering Science and Technology, Howrah, India*
[2] *Trinity College Dublin, Dublin-2, Ireland*
[*] *Corresponding Author*

ABSTRACT Elevated water tanks are vital elements of the social and economic infrastructure of a country. Besides catering to the water supply requirements of a community, their continued performance in the post-earthquake phase of earthquake-hit areas is also crucial from fire-fighting requirements. However, these structures are highly vulnerable to earthquake forces as they comprise huge water and tank masses concentrated at the top of tall supporting towers. The conventional technique of strengthening these structures involves a stiffer design which leads to increased seismic forces. It is thus pertinent to study the application of vibration control devices for the seismic protection of elevated water tanks. Amongst the various available control technologies, passive control is reliable in the post-earthquake scenario, where power outages are common, as it does not require external power and is also cost-effective. The Tuned Liquid Damper (TLD) is a viable option as it is well-established in the field of structural vibration control having several advantages. This paper investigates the passive vibration control of elevated water tank structures by the TLD through an experimental investigation employing Real-Time Hybrid Testing (RTHT) as the TLD is known to exhibit nonlinear behaviour. In this, the elevated water tank structure is numerically modelled while a full-scale TLD is considered with synchronized physical testing. The results from RTHT are compared with the established TLD model based on shallow water wave theory. Overall, the TLD is identified as an effective passive control device to mitigate the lateral vibrations of elevated water tanks.

1 INTRODUCTION

For fulfilling the vital needs of drinking water distribution and fire-fighting, elevated water tanks must remain unaffected in earthquake-hit areas in the post-earthquake phase. Moreover, the damage of such tanks may cause considerable economic loss. These structures are generally slender and the large water and tank mass concentrated on the supporting tower attract large seismic forces. Ghosh et al. (2013) have earlier conducted a detailed review of various reports on the damage and failure of elevated water tanks in past earthquakes. It was found that the tanks on shaft supports sustained greater damage and failure as compared to framed supports as the former do not have adequate redundancy and ductility. Importantly, during the actual occurrence of the earthquake, quite a few tanks were either empty or partially full. It is evident that the protection and safe performance of elevated water tanks during strong earthquakes is of vital concern and it is not always safe to assume that the tank is full for design purposes.

Normal strengthening practices for elevated water tanks causes an increase in the structural frequencies resulting in greater seismic base shear. So to avoid this, currently seismic vibration control schemes are being emphasized. Investigations on seismic isolation of elevated water tanks by several researchers (Shenton III and Hampton 1999, Shrimali and Jangid 2006, Shrimali 2007) reported its effectiveness. However, base isolation techniques involve considerable cost and difficult implementation. Currently, the incorporation of passive control devices is considered as an effective, reliable and comparatively economic solution for improving the dynamic response of structures. Retrofitting of elevated steel water tanks by viscous and friction dampers proved beneficial (Potty and Nambissan, 2008). A simple Tuned Mass Damper (TMD) was also proposed to mitigate the vibration of reinforced concrete elevated tanks under earthquake loads (Jaiswal, 2004). Another popular passive control device, the tuned liquid damper (TLD), is widely studied and is in application for structural protection under dynamic actions, especially for wind-excited tall buildings (Fujii et al., 1990, Kareem, 1990, Wakahara et al., 1992). The key advantages attributable to the TLD are affordable installation, operational and maintenance costs, ease of installation in case of existing structures, easy adjustment of natural frequency, effectiveness even against small-amplitude vibrations, function not restricted to unidirectional vibration, no need for low friction bearing surface etc.

This paper presents the experimental verification of a full-scale model of passive Tuned Liquid Damper (TLD) by employing Real-Time Hybrid Testing (RTHT) for lateral vibration control of elevated water tanks. This study is for flexible elevated water tank structures, such as those with annular shaft supports. The criterion of tuning the sloshing frequency of the TLD to the structural frequency necessitates dimensions of the TLD larger than those hitherto examined through experimentation and available in literature. The chief objective here is to validate the nonlinear model of the TLD based on the established shallow water wave theory for large container size by employing Real-Time Hybrid Testing (RTHT) for lateral vibration control of such flexible elevated water tank structures. As in case of the latter practically it would not be possible to test a full-scale structural model, RTHT is the standard option for such kind of experimental investigation. RTHT offers the unique advantage of the non-compulsion of using a material substructure of the elevated tank for performing the interaction of TLD-elevated water tank structure whereas the TLD, which is known to exhibit nonlinear behaviour, can be tested physically. During an earthquake excitation, to account for the realistic condition of a partially full tank which would precipitate sloshing of the water in the tank itself, the elevated water tank structure is modelled by a two-degree-of-freedom (2-DOF) system considering impulsive and sloshing modes. The structural response of the elevated tank system is evaluated numerically in real-time with the aid of the 2-DOF model formulated in Matlab/Simulink with base excitation and the sloshing force feedback. The water mass in the elevated water tank being a variable quantity, the natural frequency of the 2-DOF system characterizing the elevated water tank structure varies. Here, different tuning cases of the TLD frequency to the structural frequency corresponding to two different cases of tank-full conditions are considered, namely half-full and full tank conditions. The different tuning cases of the TLD demands different values of liquid depth of TLD to be examined. For this reason, three separate liquid depths in the damper container are explored. The results from RTHT are compared with the established TLD model based on shallow water wave theory. For this, the nonlinear model simulating sloshing motion of the water in the TLD as proposed by Sun et al. (1992) is utilized. Different base excitation characterizations are considered, namely harmonic, white noise and scaled recorded accelerograms.

2 REAL-TIME-HYBRID-TESTING (RTHT)

2.1 *General description of the hybrid model (test set-up)*

The RTHT reported in this paper was conducted using the MTS real-time hybrid testing system at Trinity College Dublin, Ireland. The test comprises of two components: testing of an experimental substructure of a TLD and simulation of numerical model of elevated water tank in the MATLAB/Simulink environment. The following are the chief constituents of the RTHT system:

1. A host PC running Matlab/Simulink
2. A real-time target PC with the shared common random access memory network (SCRAMNet)
3. A hybrid controller host PC running the graphical user interface to the MTS servo-controller
4. A MTS servo-controller hardware with SCRAMNet including a digital PID (Proportional Integral Derivative) actuator controller, signal conditioners, data acquisition system and interlock mechanisms
5. A 150 kN capacity hydraulic actuator (± 125 mm stroke) equipped with displacement and force sensors

2.2 *Numerical substructure: the 2-DOF model of elevated water tank*

The effect of sloshing of the water in the elevated water tank container is considered by modelling the tank structure as a 2-DOF system, one representing the first convective/sloshing mode of the water and the other denoting the lateral motion of the elevated water tank structure together with the remaining mass of the water assumed to move rigidly with the tank structure. The frequency of the latter mode is termed as the structural frequency and is considered as the tuning frequency for the TLD.

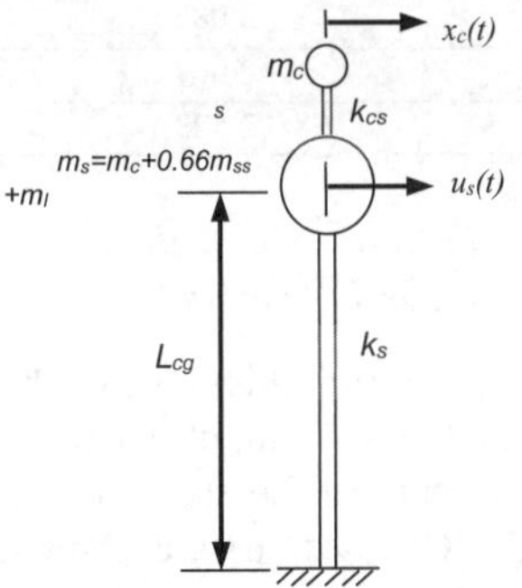

Figure 1. The 2-DOF elevated water tank model.

Here, the flexural stiffness of the shaft support, calculated from the deflection of the reinforced concrete staging acting as a cantilever, is represented by k_s (Figure 1). The associated mass, denoted by m_s, consists of two-thirds of the mass of the staging, the mass of the empty tank container and the water mass in excess of that in the first sloshing mode (Ibrahim *et al.*, 2001). This mass is considered to be lumped at the level of impulsive water mass. The damping coefficient related to m_s is denoted by c_s. ζ_s represents the damping ratio of the same. ω_s is equal to $\sqrt{(k_s/m_s)}$. T_s $[=2\pi/\omega_s]$ is the structural time period. The mass, stiffness and damping properties related to the sloshing liquid of the tank are m_{cs}, k_{cs} and c_{cs} respectively. The damping ratio and the natural frequency of the same are designated by ζ_{cs} and ω_{cs} respectively. The expressions for m_{cs} and ω_{cs} (Ibrahim *et al.*, 2001) are given by

$$m_{cs} = \left[\frac{d}{4.4h_w}\tanh\left(3.68\frac{h_w}{d}\right)\right]m_l \tag{1}$$

$$\omega_{cs} = \sqrt{\left\{3.68\frac{g}{d}\tanh\left(3.68\frac{h_w}{d}\right)\right\}} \tag{2}$$

where h_w is the water depth in the tank container, d is the tank container diameter, m_l is the total water mass within the tank container and g is the acceleration due to gravity. As per Livaoglu and Dogangun (2006), ζ_{cs} is chosen as 0.005.

The displacement of the tank structure relative to the base is denoted by $u_s(t)$. The displacement of the sloshing water mass relative to the structural displacement is designated by $x_c(t)$. Consideration of the dynamic equilibrium of the TLD-2-DOF structural system, subjected to horizontal ground acceleration $\ddot{z}(t)$, gives the following normalized equation of motion with respect to the structural mass.

$$\{\ddot{u}_s(t)+\ddot{z}(t)\}+2\zeta_s\omega_s\dot{u}_s(t)+\omega_s^2u_s(t)=2\zeta_{cs}\omega_{cs}\mu_{css}\dot{x}_c(t)+\omega_{cs}^2\mu_{css}x_c(t)+F/m_s \tag{3}$$

The normalized equation of motion of sloshing water mass in the tank container of 2-DOF system is

$$\{\ddot{x}_c(t)+\ddot{u}_s(t)+\ddot{z}(t)\}+2\zeta_{cs}\omega_{cs}\dot{x}_c(t)+\omega_{cs}^2x_c(t)=0 \tag{4}$$

where, $F(t)$ is the shear force developed at the base of the TLD due to water sloshing.

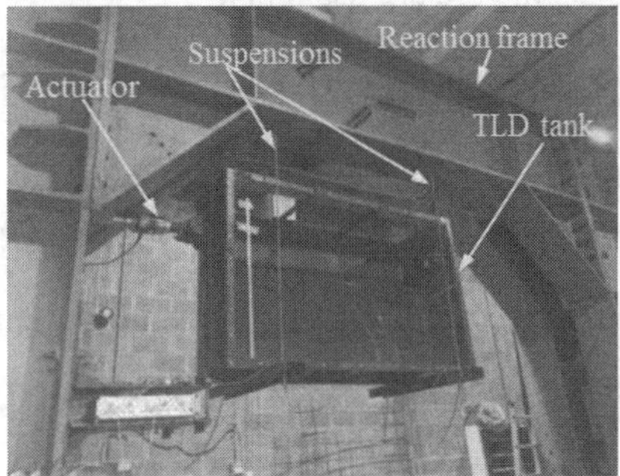

Figure 2. Photograph of experimental set-up.

2.3 *Experimental substructure test set-up*

A photograph of the test setup and the experimental substructure (i.e. the TLD) is presented in Figure 2. The setup is composed of a hydraulic actuator in the horizontal direction, a reaction frame and the data acquisition system. The MTS 244 actuator has a load capacity of 150 kN and maximum stroke of ±125 mm. It is attached to the left side of the TLD. For the measurement of the actuator displacement and the interactive force, a linear variable displacement transducer (LVDT) and a load cell are connected to the actuator. The full-scale model of the TLD is a rectangular tank having inner dimensions of 1.93 m (length) × 0.59 m (breadth) × 1.2 m (height). As the breadth of the TLD tank is much less compared to its length, the sloshing of the water is expected to be predominantly two-dimensional. With the purpose of minimizing the friction when the tank is compelled to relocate by the actuator, four steel cables are used to suspend the tank from the top of the reaction frame. Moreover, two capacitance wave gauges are mounted (with the sampling rate of 10 Hz/s) at the two end walls (left and right) of the tank to measure the water surface elevations at left and right walls.

3 TEST RESULTS

3.1 *Description of the example elevated water tank*

To examine the performance of the TLD for the vibration mitigation of the elevated water reservoir, an example reinforced concrete tank structure with annular shaft support is considered. This is a flexible structure for which the TLD is appropriate as a control device. The relevant data that are assumed to describe the structure are listed below:
Height of shaft staging = 30 m; Mean diameter of shaft = 2.5 m; Thickness of shaft wall = 0.125 m; Inner height of the tank container = 4 m; Mean diameter of the tank container = 10 m; Thickness of the container wall = 0.2 m; Thickness of the container bottom = 0.2 m; Thickness of the top cover of the container = 0.2 m; Height of water in the tank container = 3.8 m.

Since the water mass in the water tank is a variable quantity, the natural frequencies of the 2-DOF system representing the elevated water tank structure vary between the values corresponding to the full condition to that for the empty condition of the tank container. The structural frequency, as defined in Section 2.2, is found to be 2.823 rad/s, denoted by ω_{100} ($T_{s,100}$ = 2.226 s) when full, 2.969 rad/s, denoted by ω_{75} ($T_{s,75}$ = 2.116 s) when three-fourth full and 3.082 rad/s, denoted by ω_{50} ($T_{s,50}$ = 2.039 s) when half-full. The amount of water present in the tank is represented by a fraction of the tank full condition, denoted by α, and expressed as a percentage. The structural frequency at tank α full, is denoted by ω_{α}. The value of the structural damping of the elevated water tank is assumed to be 1% (Min et al., 2005).

3.2 *TLD Parameters*

With the fixed length of 1.93 m of the TLD tank, the resulting water mass is only dependent on the water depth. To have the fundamental linear frequency of the TLD to be equal to ω_{50}, ω_{75} and ω_{100}, three different water levels of TLD are examined: 421 mm, 382 mm and 337 mm (see Table1). The fundamental frequencies of the TLD for these depths are tabulated in Table 1.

Table 1. Fundamental frequencies of TLD for different cases.

Sl. No.	Water depth in TLD (mm)	Fundamental frequency (rad/s)
1	421	3.082
2	382	2.969
3	337	2.823

3.3 *Numerical results and comparison with simulation using nonlinear TLD model*

In this section, the displacement time histories of the structure for three different types of input acceleration to the numerical substructure of the elevated water tank (modelled as a 2-DOF system) are shown. Two different tank full conditions (i.e. two different values of α) are studied in each case: half-full and full tank cases. The comparison of the structural displacements obtained by RTHT and by the simulation study using the nonlinear model based on shallow water wave theory for three different TLD water levels are presented here.

First the elevated water tank structure, modelled as a 2-DOF system, is subjected to a harmonic base acceleration of amplitude 0.005 m/s^2. For all cases, the input frequency is equal to the structural frequency.

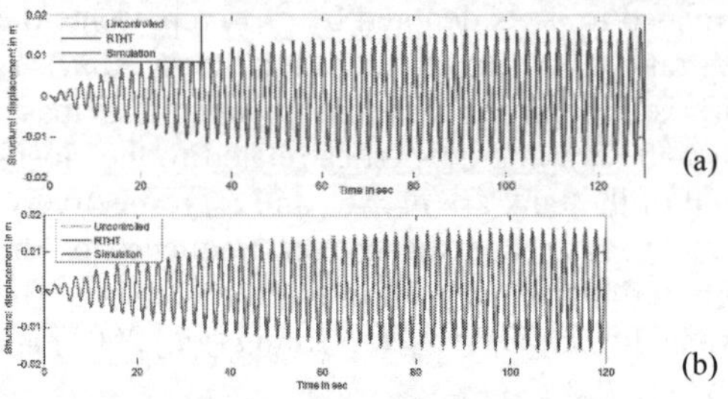

Figure 3. Displacement time history of elevated water tank structure to harmonic excitation for α=100% with different tuning of TLD - (a) ω_{TLD}=ω_{50} (b) ω_{TLD}=ω_{75}

Figure 3 shows the comparison of the time histories of the structural displacements obtained by RTHT and simulation study by the nonlinear model of TLD for α value of 100% with three different tuning conditions of TLD (namely ω_{TLD} = ω_{50}, ω_{75}). Here, the outcome from the simulation study is matching with the test result as can be seen from displacement time histories. But due to severe sloshing and splashing of water from the TLD tank, the test could not be carried out for the tuned case of TLD i.e. when $\omega_{TLD} = \omega_{100}$. In the other two tests, some reduction in the structural displacement is achieved.

For the half-full tank condition (α=50%), the comparison of the time histories of the structural displacements obtained by RTHT and the nonlinear model with the above mentioned two different values of ω_{TLD} is shown in Figure 4. In Figure 4(a), there is reasonable agreement between the results of experiment and simulation. The presence of beating is observed in the experimental time history. But in Figure 4(b), a mismatch is observed between simulated and experimental structural displacement time history. This is attributed to the slamming phenomenon that was observed during the experiment and which is not taken into account by the considered model of the TLD. Moreover, splashing of water out of the TLD tank was observed. However in Figure 4(a) and (b) good reduction in structural displacement could be obtained. The reductions achieved in peak structural displacement are given in Table 2. Here the reduction is evaluated by the ratio of the difference between uncontrolled and controlled structural displacement to the uncontrolled structural displacement, expressed as a percentage.

Table 2. Comparison of reductions obtained in simulation and RTHT for harmonic input.

α (%)	ω_{TLD}	Reduction in peak structural displacement (%) Simulation	RTHT
50	ω_{50}	30.25	36.82
50	ω_{75}	26.50	36.48
100	ω_{50}	20.55	3.74
100	ω_{75}	16.56	6.66

When the elevated water tank structure (modelled as a 2-DOF system) is subjected to a white noise base acceleration, Figures 5 and 6 present the comparison of the results obtained by RTHT and simulation study with three different tuning conditions of TLD (namely $\omega_{TLD} = \omega_{50}, \omega_{75}, \omega_{100}$) for values of α equal to 100% and 50% respectively. As can be seen from these figures, all the results obtained from RTHT match very well with those obtained from simulation. The reductions obtained in structural displacement are however not appreciable possibly due to very low mass ratio of 0.1%.

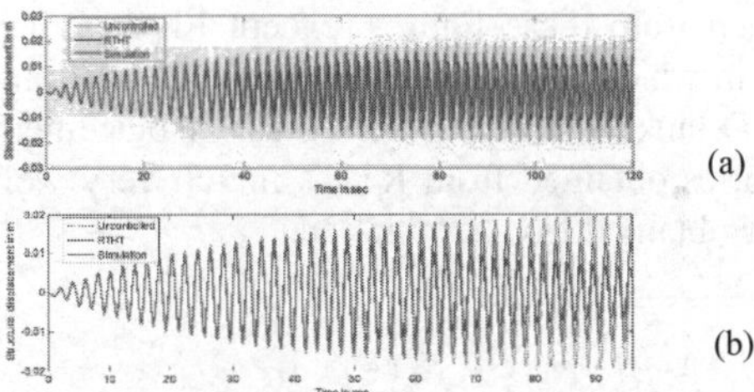

Figure 4. Displacement time history of elevated water tank structure to harmonic excitation for α=50% with different tuning of TLD - (a) $\omega_{TLD}=\omega_{50}$ (b) $\omega_{TLD}=\omega_{75}$

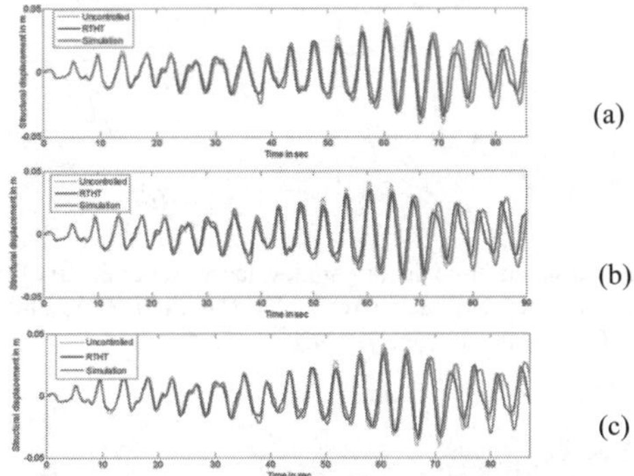

Figure 5. Displacement time history of elevated water tank structure to white noise excitation for α=100% with different tuning of TLD - (a) $\omega_{TLD}=\omega_{50}$ (b) $\omega_{TLD}=\omega_{75}$ (c) $\omega_{TLD}=\omega_{100}$

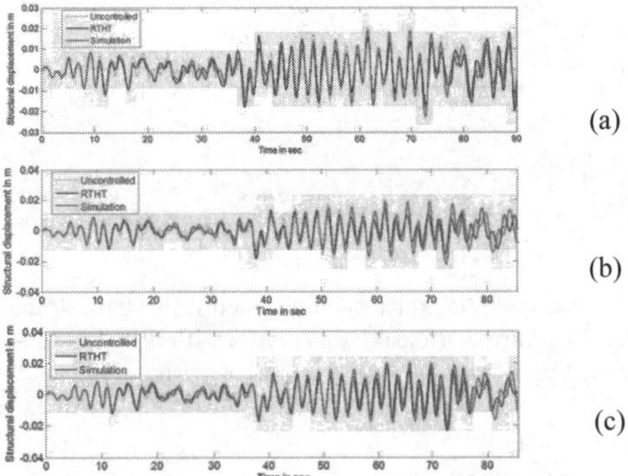

Figure 6. Displacement time history of elevated water tank structure to white noise excitation for α=50% with different tuning of TLD - (a) $\omega_{TLD}=\omega_{50}$ (b) $\omega_{TLD}=\omega_{75}$ (c) $\omega_{TLD}=\omega_{100}$

Next, the elevated water tank structure is subjected to the scaled down (by 0.1) recorded accelerogram of El Centro earthquake, 1940. As before, Figures7 and 8 show the comparison of the results obtained by RTHT and simulation study with the three different tuning conditions of TLD (namely $\omega_{TLD} = \omega_{50}, \omega_{75}, \omega_{100}$) for 100% and 50% α values of respectively. As can be seen from these figures, except Figures 7(a) and 8(a), in which case there was excessive sloshing in the TLD tank, in all the other cases, displacement time histories obtained from RTHT match very well with those obtained from simulation.

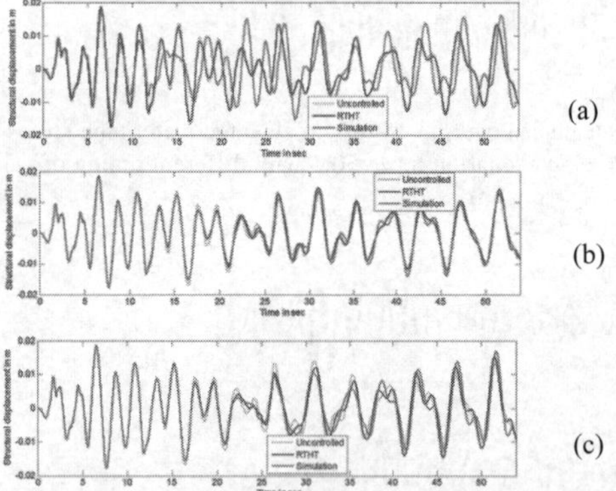

Figure 7. Displacement time history of elevated water tank structure to scaled El Centro earthquake for α=100% with different tuning of TLD - (a) $\omega_{TLD}=\omega_{50}$ (b) $\omega_{TLD}=\omega_{75}$ (c) $\omega_{TLD}=\omega_{100}$

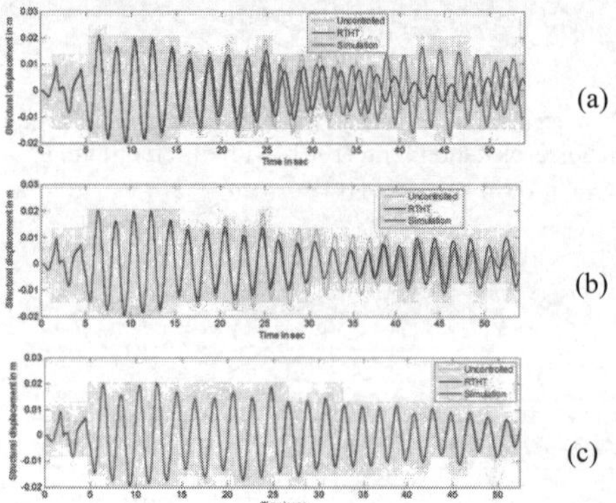

Figure 8. Displacement time history of elevated water tank structure to scaled El Centro earthquake for α=50% with different tuning of TLD - (a) $\omega_{TLD}=\omega_{50}$ (b) $\omega_{TLD}=\omega_{75}$ (c) $\omega_{TLD}=\omega_{100}$

4 CONCLUSIONS

There is an overall good agreement between the RTHT and simulated results by which the accuracy of both the methods are verified for large container size of TLD. Moreover, the TLD is identified as a promising passive control device for elevated water tanks.

ACKNOWLEDGEMENT

The financial support provided by the INSPIRE Programme, Department of Science and Technology (DST), Government of India (GOI) (Grant No. DST/INSPIRE Fellowship/2011/178) and the Science Foundation of Ireland (SFI) during this research is gratefully acknowledged by the first author.

REFERENCES

Fujii, K. Tamura, Y. Sato, T. and Wakahara, T. 1990. Wind-Induced Vibration of Tower and Practical Applications of Tuned Sloshing Damper, *Journal of Wind Engineering and Industrial Aerodynamics* **33**, 263-272.

Ghosh, A. (D.) Bhattacharyya, S. and Roy, A. 2013. On the seismic performance of elevated water tanks and their control using TLDs, *10th International Conference on Damage Assessment of Structures, (DAMAS),* Dublin, Ireland; Vol. **569-570** *of Key Engineering Materials*, 270-77.

Ibrahim, R. A. Pilipchuk V. N. and Ikeda T. 2001. Recent advances in liquid sloshing dynamics, *ASME Appl. Mech. Rev.* **54**(2), 133–199.

Jaiswal, O. R. 2004. Simple Tuned Mass Damper to Control Seismic Response of Elevated Tanks, *Proceedings of the 13th World Conference on Earthquake Engineering*, Vancouver, B. C. Canada, 1-6 August, Paper No. 2923.

Kareem, A. 1990. Reduction of Wind Induced Motion Utilizing a Tuned Sloshing Damper, *Journal of Wind Engineering and Industrial Aerodynamics* **36**, 725-737.

Livaoglu, R. and Dogangun, A. 2006. Simplified seismic analysis procedure for elevated tanks considering fluid-structure-soil interaction, *Journal of Fluids and Structures*, **22**, 421-439.

Min, K. W. Kim, H. S. Lee, S. H. Kim, H. and Ahn S. K. 2005. Performance evaluation of tuned liquid column dampers for response control of a 76-story benchmark building, *Engineering Structures* **27**(7), 1101-1112.

Potty, N. S. and Nambissan, S. 2008. Seismic Retrofit of Elevated Steel Water Tanks, *Proceedings of ICCBT (The International Conference on Construction and Building Technology)*, 2008– C-09, 99-108.

Shenton III, H. W. and Hampton, F. P. 1999. Seismic Response of Isolated elevated water tanks, *Journal of Structural Engineering, ASCE* **125**(9), 965–976.

Shrimali, M. K. and Jangid, R. S. 2006 Seismic performance of elevated liquid tanks isolated by sliding systems, *Proceedings of the First European conference on earthquake engineering and seismology*, Geneva, Switzerland, 3–8 September, Paper no. 1784.

Shrimali, M. K. 2007. Seismic Response of Elevated Liquid Storage Steel Tanks under Bi-direction Excitation, *Steel Structures* **7**, 239-251.

Sun, L. M., Fujino, Y., Pacheco, B. M. and Chaiseri, P. 1992. Modelling of Tuned Liquid Damper, *Journal of Wind Engineering and Industrial Aerodynamics* **41-44**, 1883-1894.

Wakahara, T. Ohyama, T. and Fujii, K. 1992. Suppression of wind-induced vibration of a tall building using Tuned Liquid Damper, *Journal of Wind Engineering and Industrial Aerodynamics* **43**(1-3), 1895-1906.

Proceedings of the International Conference on Smart Infrastructure and Construction
ISBN 978-0-7277-6127-9

© The authors and ICE Publishing: All rights reserved, 2016
doi:10.1680/tfitsi.61279.375

System for evaluating fatigue environment of steel girder bridge based on acceleration measurements

H. Sekiya*[1], C. Miki[1], O. Maruyama[1] & T. Kinomoto[2]

[1] *Tokyo City University, Tokyo, Japan*
[2] Metropolitan Expressway Company Limited, Tokyo, Japan
* *Corresponding Author*

ABSTRACT In bridge maintenance, particularly with regard to fatigue damage in steel bridges, it is important to continuously monitor the traffic load because the long-term cumulative effect of this load is the primary cause of fatigue damage. Bridge weigh-in-motion (BWIM), which was proposed as a conventional method of determining the traffic load, is based on performing inverse analysis using bridge response parameters, such as strain and displacement, under traffic loads. However, installation of the strain gauges used in this method requires the removal of paint from the surface of a steel bridge, and it is often difficult to use displacement gauges on bridges in service because they require a fixed reference point. This paper proposes a system for evaluating the fatigue environment of a steel girder bridge using microelectromechanical systems (MEMS) accelerometers. MEMS accelerometers are suitable for field measurements because they are cheap and compact and they can be easily attached magnetically to a painted metal surface. In addition, unlike conventional devices, such as laser displacement gauges, they do not require a fixed reference point. In this study, based on the free vibration of a bridge in service, the bridge displacement was determined from the measured acceleration. Based on the displacement response, a system for evaluating the fatigue environment that is able to determine the traffic load in each lane of a multilane bridge is proposed. Finally, measurements were performed on an actual bridge to verify the effectiveness of the proposed system.

1 INTRODUCTION

To perform appropriate bridge maintenance, particularly with regard to fatigue damage in steel bridges, it is important to continuously monitor the traffic load (Laman & Nowak 1996) because the long-term cumulative effect of this load is the primary cause of fatigue damage.

Axle load meters (Tokita et al. 2005) and bridge weigh-in-motion (BWIM) (Moses 1979) systems are conventional systems used to determine the live load. The BWIM method is based on performing inverse analysis using bridge response parameters, such as strain and displacement, under a live load. BWIM systems are easier and less expensive to install than systems based on axle load meters, which require road surface work. Furthermore, BWIM systems can be used to measure the weight of vehicles in motion on a bridge, meaning traffic flow does not need to be interrupted. However, installation of the strain gauges used in conventional BWIM requires the removal of paint from the surface of a steel bridge, and once the gauges are attached, they cannot be reused.

In recent years, accelerometers based on microelectromechanical systems (MEMS), which can measure responses to external forces, have been proposed for structural health monitoring (Gindy et al. 2007). Such devices are particularly suitable for field measurements because they are cheap and compact, they can

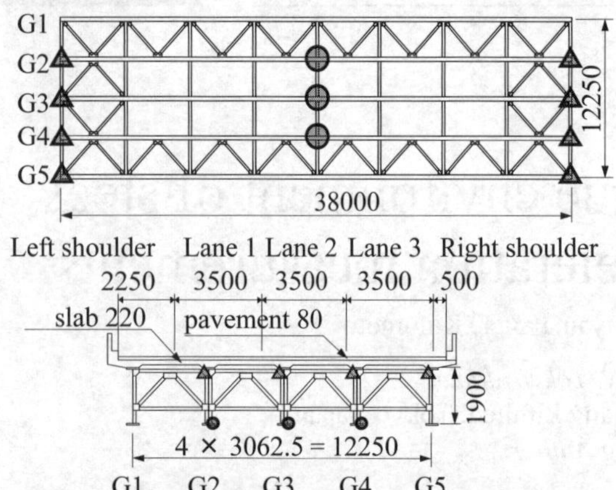

Figure 1. Test bridge and installation position of MEMS accelerometers and contact displacement gauges. Units: mm.

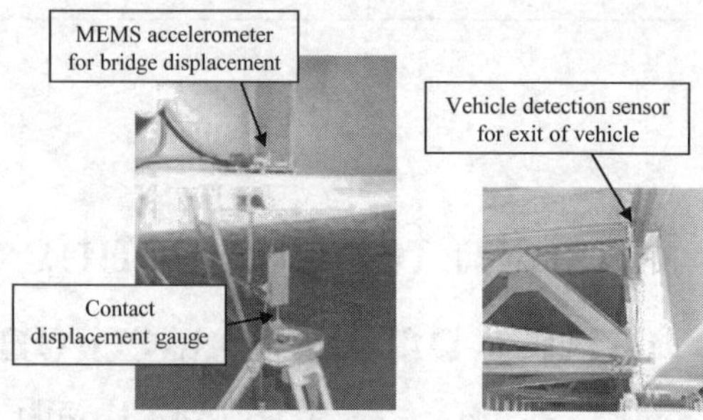

Figure 2. Installation of MEMS accelerometer and contact displacement gauge.

Table 1. Specifications of accelerometers and contact displacement gauge.

(a) Accelerometer used to determine bridge displacement (M-A550-AC, Seiko Epson)

Parameter [units]	Value
Acceleration range [m/s^2]	±49.0
Frequency bandwidth [Hz]	0.1–20
Sampling frequency [Hz]	100
Resolution [μm/s^2]	9.8
Noise density [μm/(s$^2\sqrt{\text{Hz}}$)]	7.8

(b) Accelerometer used to determine bridge displacement (M-G550-PC, Seiko Epson)

Parameter [units]	Value
Acceleration range [m/s^2]	±29.4
Frequency bandwidth [Hz]	0.1–148
Sampling frequency [Hz]	500
Resolution [μm/s^2]	1226
Noise density [μm/(s$^2\sqrt{\text{Hz}}$)]	981

(c) Contact displacement gauge (SDP-50C, Tokyo Sokki)

Parameter [units]	Value
Capacity [m]	0–25
Sensitivity [× 10^{-6} strain/mm]	500
Nonlinearity [mm]	0.1% RO
Frequency [Hz]	100

be easily attached magnetically to a painted metal surface, and their power consumption is very low (Lynch et al. 2004). MEMS accelerometers with wireless transmission capabilities can be easily attached and detached for repeated use (Park et al. 2013).

In the present study, based on the free vibration of a bridge in service, the bridge displacement response was determined using MEMS accelerometers. Because it is important to accurately detect vehicle entry and exit when using this system, a method of improving the sensitivity of the system response to passing vehicles is developed to achieve this. Based on the displacement response, a system for evaluating the fatigue environment that is able to determine the traffic load and the number of passing vehicles in each lane on a multilane bridge is proposed. Finally, measurements were performed on an actual bridge to verify the effectiveness of the proposed system.

2 FIELD MEASUREMENTS

Measurements of a bridge in service without traffic control were conducted using MEMS accelerometers and a contact displacement gauge. The test bridge is shown in Figure 1. The test bridge consists of five main girders with RC deck plates. The span length is 38 m, and the total width is 14.25 m. It is located in Tokyo and managed by the Metropolitan Expressway Co., Ltd.

The experimental setup of the accelerometers used to determine bridge displacement, those used to detect for vehicle entry and exit, and the contact displacement gauge is shown in Figure 2. The accelerometers used to determine bridge displacement were attached to the painted metal surface of the bridge using a magnet at the longitudinal center of the lower flange of the main girder. The accelerometers used to detect vehicle entry and exit were attached in the same way at the vertical stiffeners on both longitudinal edges of the main girder. The contact displacement gauge was fixed at the longitudinal center of the lower flange of the main girder to verify the accuracy of the displacement values obtained by the double integration of the measured accelerations. The specifications of the MEMS accelerometers and the contact displacement gauge are listed in Table 1.

3 DETERMINATION OF BRIDGE DISPLACEMENT FROM ACCELERATION

In theory, displacement can be calculated by double integration as

$$U(T)=\int_0^T[\int_0^T A(\tau)\mathrm{d}\tau]\mathrm{d}t+V_0T+U_0, \qquad (1)$$

where U is the displacement, A is the acceleration, V_0 is the initial velocity, U_0 is the initial displacement, t and τ are the time, and T is the total elapsed time. However, because bridges in service are always vibrating as a result of passing vehicles, it is difficult to determine the boundary conditions of the integration, i.e., the initial and terminal velocities and displacements of the bridge. In this study, the boundary conditions of the integration were estimated under the assumption that the bridge is vibrating before and after each vehicle passes. The estimated velocities and displacements are adapted to boundary conditions of the integration.

3.1 *Estimation of boundary conditions of integration*

The procedure for estimating the boundary conditions of integration, which are the velocity and displacement of free vibration, is as follows.

1. Separate the free and forced vibration regions by detecting vehicular entry and exit.
2. Transform the acceleration from the time domain to the frequency domain by Fourier transformation.
3. Remove frequencies below 1.0 Hz and above 20 Hz to eliminate the effect of forced displacement during vehicular passage.
4. Transform the acceleration from the frequency domain to the time domain by inverse Fourier transformation.

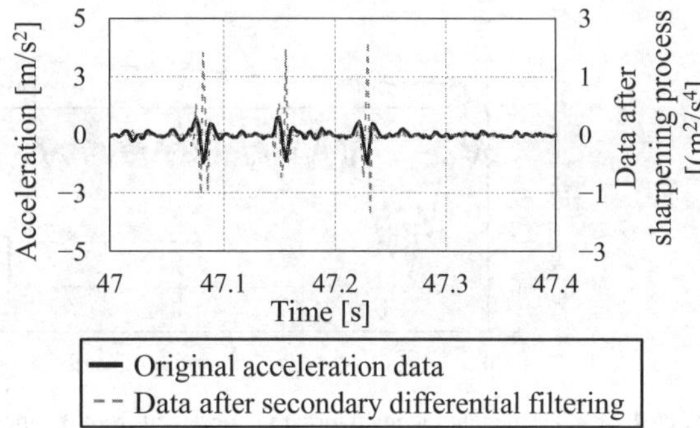

Figure 3. System response to vehicle entry after secondary differential filtering.

5. Take the integral of the acceleration with no components at frequencies below 1.0 Hz and above 20 Hz to obtain the velocity of free vibration.
6. Take the double integral of the acceleration with no components at frequencies below 1.0 Hz and above 20 Hz to obtain the displacement of free vibration.

3.2 *Improvement of system response to vehicle detection*

In this study, the time it takes for a vehicle to pass the full span of the bridge (passing vehicle time) was used as the integration time of the acceleration measured at the longitudinal center of the lower flange of the main girder. To accurately detect the bridge response to vehicle entry and exit, a secondary differential filter was adopted to square the measured acceleration at the vertical stiffener. The acceleration after applying the secondary differential filter is shown in Figure 3. The results demonstrate that using a secondary differential filter improves the response of the system to vehicle entry and exit in comparison with the original acceleration record.

3.3 *Determination of vehicle travel lane*

The vehicle traveling lane can be determined based on the system response to vehicle entry and exit at the vertical stiffeners of the longitudinal edges, which are labeled G2, G3, G4, and G5. The travel lane detection procedure is as follows.

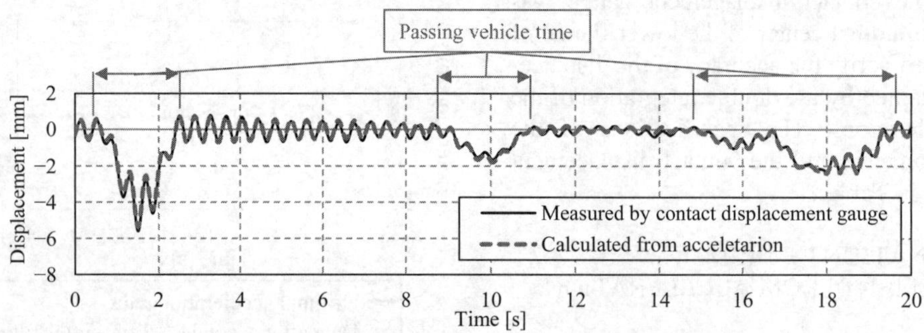

Figure 4. Displacement at the main girder G3 obtained from the measured acceleration.

1. At each vertical stiffener, system responses improved by secondary differential filtering that exceed the threshold are detected. A threshold of 0.005 was chosen based on trial and error.
2. It is determined whether there are girders that simultaneously show system responses exceeding the threshold. According to the combination of girders showing a response, the travel lane is estimated. The relationship between the girder response and the travel lane is shown in Table 2.

To verify the accuracy of this method, the travel lane determination results obtained using this method were compared with a video of cars passing over the bridge for 14 minutes. Table 3 shows the results of this comparison. Because small cars are not significant sources of fatigue damage to steel bridges (Fisher 1984), they were not considered in this study. The results show that 96% of passing vehicles were detected by our system and that the correct vehicle travel lane was estimated for 95% of all cars. Thus, the proposed method of vehicle detection and travel lane determination showed high accuracy.

3.4 *Displacement results from measured acceleration*

The displacement results at the main girder G3 obtained from the measured acceleration by using the estimated boundary conditions are shown in Figure 4. The vehicle entry and exit times, which were obtained from the vehicle response after secondary differential filtering, were used as the limits of the numerical integration. The calculated displacement was in good agreement with the reference displacement measured by the contact displacement gauge.

Table 2. Relationship between vehicle response at vertical stiffener and travel lane. ○ indicates that the system responded to vehicle entry and exit at the given stiffener.

Travel lane \ Girder	G2	G3	G4	G5
First travel lane	○	○		
Second travel lane		○	○	
Third travel lane			○	○

Table 3. Verification of accuracy of travel lane determination for each lane. Small cars were excluded from detection.

	Vehicle detection	Travel lane
Correct	**96%** (286)	**95%** (284)
Incorrect	4% (13)	5% (15)
Total	100% (299)	100% (299)

4 SYSTEM FOR EVALUATING FATIGUE ENVIRONMENT USING MEMS ACCELEROMETERS

In this section, a system for evaluating the fatigue environment of a steel girder bridge using the estimated displacement is proposed. The procedure used in this system is shown in Figure 5. The system comprises procedures for determining the displacement of the bridge from its acceleration and calculating the passing vehicle weight. The required data for this system are the accelerations recorded at the longitudinal center of the main girder and at the vertical stiffeners of both longitudinal edges. As with conventional BWIM using strain gauges, this system is composed of a small number of measuring points.

4.1 *Estimation of passing vehicle weight using estimated displacement*

Based on the BWIM method (Moses 1979), the vehicle axle weight was estimated by minimizing the sum of the squared errors between the displacement estimated from the acceleration and the sum over the influence lines, which increases with the weight of each axle. The sum of axle weights is equal to the vehicle weight.

4.2 *Calculation of influence line*

The accuracy of the estimated vehicle weight largely depends on the accuracy of the influence line. Therefore, in this study, the accuracy of the influence line obtained from the displacement estimated from the acceleration was validated by comparison with the influence line obtained from the displacement measured by the contact displacement gauge. The axle weights of a four-axle test truck were used to calculate the influence line; these weights are listed in Table 4. The influence line was calculated using a Fourier transform with a low-pass filter of 1 Hz, which was adopted to remove the dynamic response caused by traffic loads. The influence line results are shown in Figure 6. The influence line obtained using the MEMS accelerometers showed good agreement with the influence line obtained using the contact displacement gauge.

4.3 *Accuracy verification of BWIM*

To verify the accuracy of BWIM using MEMS accelerometers, a driving test was conducted three times on each lane of a bridge in service without traffic control. The results are listed in Table 5. Because it is thought that the accuracy of the displacement estimated from the acceleration decreases when multiple vehicles pass, such results were removed. The results in Table 5 showed good agreement with the total weight in Table 4 within an error of ±10%. The maximum error in this driving test was −9.5%. It is thought that an error of −9.5% was obtained because small vehicles could not be detected, which decreased the accuracy of the calculated displacement. Therefore, the accuracy of the BWIM results using the proposed system depends on the accuracy of the displacement calculated from the measured acceleration.

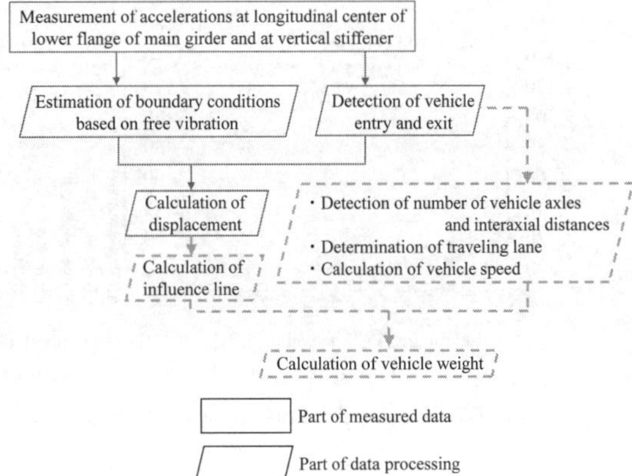

Figure 5. Procedure of the system.

Table 4. Specifications of test truck.

	Axle 1	Axle 2	Axle 3	Axle 4	Total
Weight [kN]	45.3	53.4	73.2	72.7	244.6

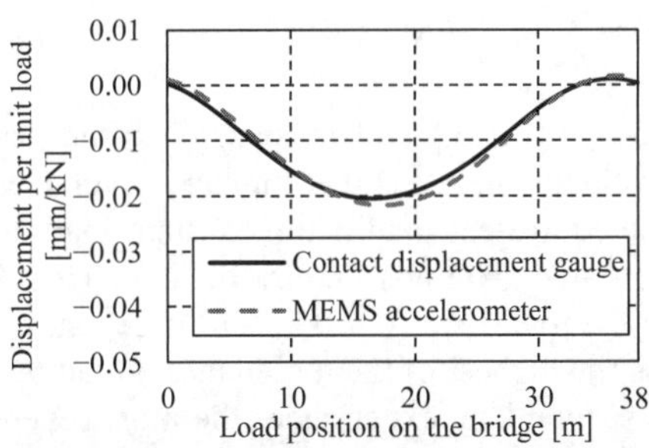

Figure 6. Influence lines obtained using contact displacement gauge and MEMS accelerometer.

Table 5. Results of BWIM for test truck.

No.	Lane	Estimated weight [kN]	Error [%]
1	1	221.4	−9.5
2	1	256.4	4.8
3	2	244.6	0.0
4	3	249.3	1.9
5	3	253.8	3.8

4.4 *Evaluation of fatigue environment for each lane*

The weights of all vehicles passing over lanes 1, 2, and 3 during a period of 3 h were calculated by determining the displacements of G2, G3, and G4, respectively. The results of BWIM using the proposed system for each lane for 3 h for a single vehicle influencing girder

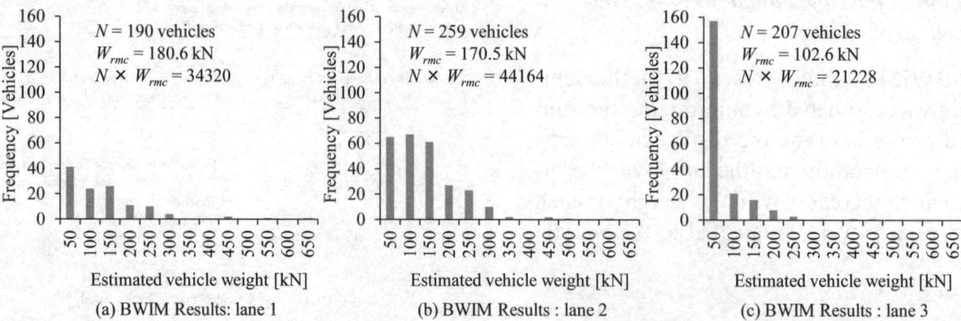

Figure 7. Results of BWIM using the proposed system for each lane for 3 h for a single vehicle influencing girder displacement

displacement are shown in Figure 7. The results of each lane in Figure 7 include the total number N of vehicles, the equivalent load W_{rmc}, and the sum of equivalent load $N \times W_{rmc}$. The equivalent load is defined as

$$W_{rmc} = \sqrt[3]{\sum_{i}^{N} W_i^3 \Big/ N}\,, \qquad (2)$$

where W_{rmc} is the equivalent load, W_i is the load of the ith vehicle, and N is the number of vehicles. In brief, the equivalent load is the constant load that is equivalent to a given varying traffic load. Therefore, because the total load $N \times W_{rmc}$ of the second travel lane was the highest of the three lanes, it can be said that the second lane experiences the most severe fatigue damage in this bridge. By applying this system to other bridges, the fatigue environments of many bridges can be compared.

5 CONCLUSIONS

In this study, a system using only MEMS accelerometers to evaluate the fatigue environment of a steel girder bridge was proposed. The weight of a test truck was accurately estimated using the proposed system. Furthermore, the number and weight of passing vehicles for a single vehicle influencing girder displacement in each lane were estimated.

ACKNOWLEDGEMENT

The field measurements in this study were obtained with support by Metropolitan Expressway Co., Ltd.; Shutoko Engineering Co., Ltd.; and Highway Technology Research Center. This study was supported by a grant from the Ministry of Education, Culture, Sports, Science and Technology (MEXT) (Grant-in-Aid for Scientific Research (KAKENHI)(A) 25249063).

REFERENCES

Fisher, J.W. 1984. *Fatigue and fracture in steel bridges*, 1st Edition, John Wiley & Sons, Inc., Hoboken, NJ, USA.

Gindy, M. Nassif, H. & Velde, J. 2007. Bridge displacement estimation from measured acceleration records, *Journal of the Transportation Research Board* **2028**, 136–145.

Laman, J.A. & Nowak, A.S. 1996. Fatigue-load models for girder bridges, *Journal of Structural Engineering (ASCE)* **122**, 726–733.

Lynch, J.P. Sundararajan, A. Law, K.H. et al. 2004. Embedding damage detection algorithms in a wireless sensing unit for operational power efficiency, *Smart Materials and Structures* **13**, 800–810.

Moses, F. 1979. Weigh-in-Motion system using instrumented bridges, *Transportation Engineering Journal of ASCE* **105**, 233–249.

Park, J.W. Sim, S.H. Jung, H.J. et al. 2013. Development of a wireless displacement measurement system using acceleration responses, *Sensors* **13**, 8377–8392.

Tokita, H. Nagai, M. & Miki, C. 2005. Evaluation of fatigue environment based on traffic data for the metropolitan expressway, *Journal of Structural Mechanics and Earthquake Engineering* **72**, 55–65.

Proceedings of the International Conference on Smart Infrastructure and Construction
ISBN 978-0-7277-6127-9

© The authors and ICE Publishing: All rights reserved, 2016
doi:10.1680/tfitsi.61279.381

Traffic volume estimation and load capacity evaluation using dynamic responses acquired in the structural health monitoring of a cable-stayed bridge

K. Wattana* and M. Nishio

Department of Civil Engineering, Yokohama National University, Yokohama, Japan
* *Corresponding Author*

ABSTRACT The effective condition assessments of existing structures will be realised if not only the structural responses but also the information about input loads can be understood from the structural health monitoring (SHM) data. This paper presents analysis of the response data acquired by a SHM system installed on an in-service cable-stayed bridge in Thailand (Bangkok). The SHM system consists of many kinds of sensors including accelerometers and temperature sensors. In addition, the vehicle counting system has been actually installed for the perpose of checking the traffic condition in this bridge. The relationships between the dynamic responses, and the temperature and the traffic volume of the bridge were investigated. The results revealed that the traffic volume was a dominant factor that influenced on variances of the responses. Then, the traffic effects were more investigated by using finite element (FE) models. The results showed that not only the traffic volumes, but also vehicle speeds had effects on the dynamic responses. In the case of the same traffic volumes, the response amplitudes decreased when the speeds of vehicles decreased. Furthermore, some of the response features that showed high correlations were then selected for constructing a linear regression model to estimate the total traffic volume per five minutes. The constructed model then showed the accurate fitting performance to the data, and it was also capable of predicting the traffic volume on the bridge. In addition, the predicted traffic volume could be used for identifying traffic conditions on the bridge.

1 INTRODUCTION

In most structural health monitoring (SHM) projects, the structural conditions are tried to be assessed only from their static and/or dynamic responses, i.e., outputs. However, those structural responses are actually caused by the inputs, which include both operational and environmental ones. Therefore, the behaviors of those inputs themselves must affect on the structural conditions. The effective condition assessments of existing structures will be realised if not only the structural responses but also the information about input loads can be understood from the SHM data. For instance, there is actually an approach to estimate traffic loads on bridges for rating fatigue loads; Weight-In-Motion (WIM). However, it basically requires the strict strain measurement for accurate estimations. Meanwhile, the load rating, which is another type of information about the live loads, can also be applicable for the stochastic discussion of structural reliability and that is for the planning of inspection and maintenance.

Traffic loads are the major live loads effecting on the bridge responses that are captured by the SHM system. The relationships between the traffic volume information and the SHM data, especially the modal parameters, have been addressed in several studies. Zhang et al. (2002) found that the resonant frequencies of the low-order global modes of a cable-stayed bridge varied about 1% due to changing traffic volume within a day. Kim et al. (2003) reported that the measured resonant frequencies could change up to 5.4% for short-span bridges but those changes could be hardly detected for the middle and long span bridges, because the mass of vehicles was relatively small in comparison to the mass of super-structures. In contrast, Cross et al. (2013) reported that traffic volume was a dominant factor in daily fluctuation of resonant frequencies for a long span bridge and suggested that the temperature variability effected on the resonant frequencies as well. In addition, Cornwell et al. (1999) reported that the resonant frequencies of a concrete deck bridge varied up to 6% over a day period due to the temperature differential across the deck. Moreover, the variability of temperature can contribute to the changes in stiffness of structures explained by the asphalt elastic modulus varying due to thermal effects (Peeters et. al 2001).

The idea of this study is that the traffic volume can be estimated by using some features that have appropriate correlations. First, the target bridge and its SHM system are detailed and the characteristics of the acquired traffic data and bridge responses are described. Then, the traffic effects were more investigated by using FE models. The next part, the estimation of traffic volume is introduced, and validation and its applications are presented. Finally, conclusions and discussions are given in the last part.

2 TARGET BRIDGE AND SHM SYSTEM

The target bridge in this study was the Bhumibol-Bridge-I, which is shown in Fig.1. This is a cable-stayed bridge crossing the Chaopraya River in Bangkok, Thailand. The bridge was opened on September 2006, and it is a part of the Industrial Ring Road; therefore, not only the passenger cars but also many trucks are crossing per day. The SHM system has been installed for ensuring the structural safety since April 2014 operated by the Department of Rural Road (DRR), Thailand. The acquired data are expected to be appropriately used for the effective operation and maintenance of this important bridge.

2.1 *Target bridge descriptions*

The target cable-stayed bridge consists of the composite deck in the main span with the length of 326 m and the post-tension concrete box girder decks in each of the 125 m-long side spans. The deck is supported by the two diamond-shaped towers with the height of 152 m and the four piers in the side span, both of which were constructed from the reinforced concrete. The connections between the deck and the towers are fixed.

Figure 1. Bhumibol I Bridge

2.2 *Installed SHM system*

The SHM system consists of several kinds of sensors and the data acquisition system. The deployed sensors were four tri-axial accelerometers ("AC"), and four temperature sensors ("TS"), each of which is installed on the main span deck and the stayed-cables as illustrated in Fig. 2. The data are acquired continuously with the sampling frequency of 50Hz in all sensors with synchronizations.To acquire the traffic volume information, the traffic counting system using the image processing technique was installed on the bridge. The system is known as a non-intrusive method of traffic-flow measurement using fifteen video cameras mounted on the road sign structures of the bridge. The acquired video data are fed into the processors that count and classify the passing vehicles (Allied Telesis, Inc, AT-MC 130XL model).

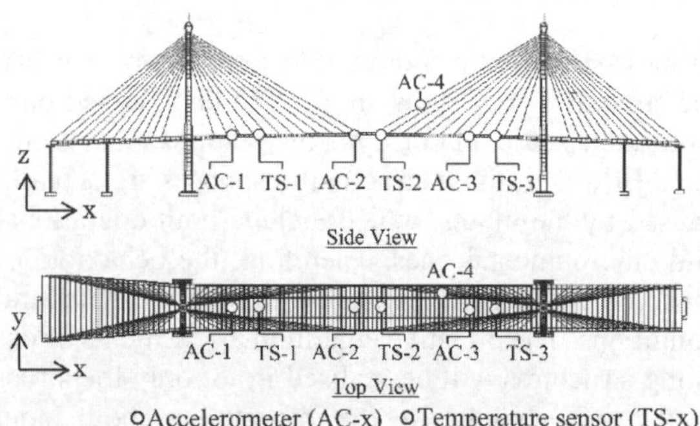

Figure 2. Sensor configuration of the SHM system

2.3 Acquired SHM data

2.3.1 Bridge responses and environmental data

Some data were selected from all SHM data for the verifications in this study; the accelerations (x, y, and z-directions) at the mid-span (AC-2), the accelerations of the stayed-cable in perpendicular direction (AC-4). The temperature data at the mid-span (TS-2) were also used as the significant environmental effects. Some of selected data; the vertical (z-direction) acceleration from AC-2, the accelerations of the cable AC-4 and the temperature in TS-2 for three days (from May 17th to 19th, 2014), are shown in Fig.3. It can be roughly observed that the amplitudes of all dynamic responses become relatively large during the daytime, and they decrease during the nighttime. In the temperature plot, the daily variance of each day is approximately 2 °C with the range of 29-32 °C. Notice that, in Bangkok, the daily and seasonal temperature variations are small; around 10 °C through a year. Therefore, the use of those three days data was considered to be appropriate for the investigation of the temperature effect in this study.

2.3.2 Traffic Data

The installed traffic counting system can classify the passing vehicles to several types of vehicles; the passenger car, and the trucks that are categorised to the single truck and the trailer. There are restrictions for the axle and gross weights of trucks; 5 tons for the single axle with single-tire, 20 tons for the tandem axles with dual-tires, and 25 and 45 tons for the restrictions of the gross weights of the single truck and the trailer, respectively. The passenger car is the vehicle with definition of the gross weights might vary within the range of 1 to 2 tons; therefore, the weight of 1.5 tons is used as the gross weight in the counting system. In order to represent the total traffic volume, the number of each type of vehicles was converted to the number of equivalent trucks with the gross weight of 25 tons by; multiplying by its gross weight and dividing by 25. Figure 4 shows the acquired traffic data; the plots of the number of equivalent trucks categorised to the passenger cars, the single trucks and the trailers. The figure firstly indicates that the single trucks and the trailers are the dominant vehicles passing on the bridge.

The number of the passenger cars shows high volumes only in the evening of three days, and in the morning of Monday. This is considered because the number of passenger cars can effect on the traffic-flow conditions on the bridge. For instance, the high number of passenger cars during the peak-hours can cause the dense traffic or traffic congestion. On the other hand, the number of the single trucks and trailers increases in the daytime and decreases in the nighttime because the trucks avoid the traffic congestion on the road networks during morning and evening peaks. In addition, the large drop in the number of equivalent trucks at around 8:00 am on Monday is due to the traffic congestion that always occurs in workdays. In summary, the traffic volumes have a daily trend and large fluctuations within a day.

3 FINITE ELEMENT MODELING FOR DYNAMIC RESPONSES

The dynamic responses obtained from the SHM system were described based on only the acquired data. However, we can use Finite Element (FE) models to predict the dynamic responses for more understanding the dynamic behaviors of the bridge. This chapter presents FE modeling of the target cable-stayed bridge and analytical dynamic responses, and compares the analytical responses with those from the measurement.

3.1 Finite element modeling and model validation

In this study, a three-dimensional finite element (FE) model of the Bhumibol I Bridge was developed by using Midas Civil software. The model has a total of 96 cable elements, 1,101 beam elements 385 plate elements and 1,127 nodes. The connections between the deck and the two towers and the four piers are fixed links. Table 1 summarises the resonant frequencies obtained from the power spectral density (PSD) of deck accelerations and the FE calculation. It indicates that the identified and calculated frequencies achieve a good agreement. Therefore, the FE model is validated and can be use as the base line FE model.

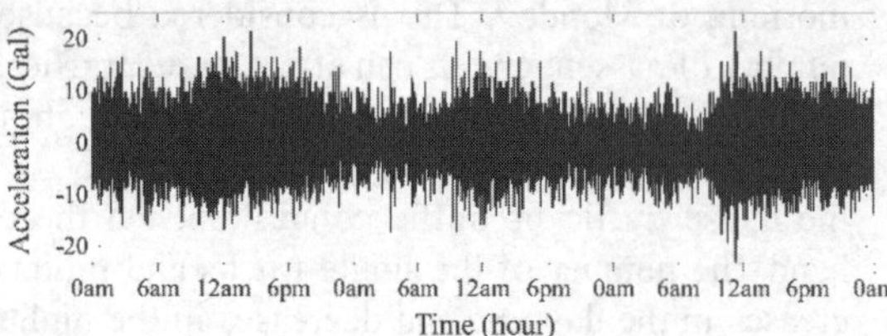

(a) Vertical acceleration of the deck at AC-2

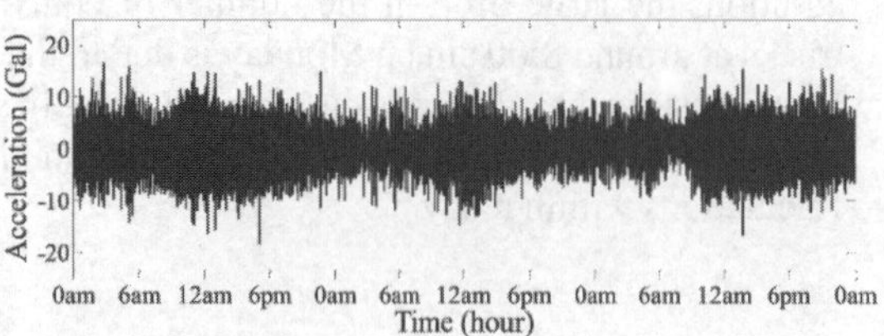

(b) Acceleration of the stayed-cable at AC-6

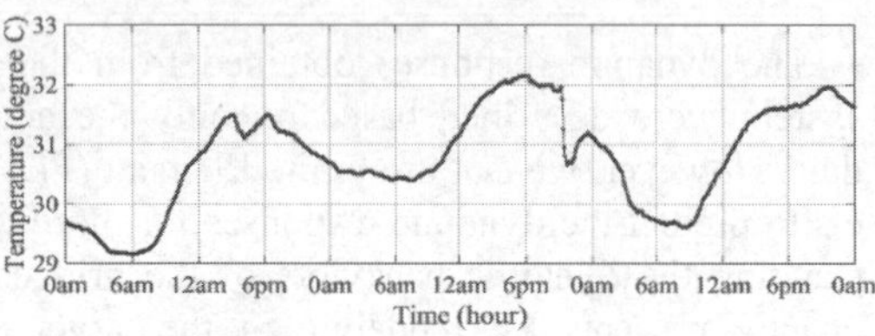

(c) The average temperature at TS-2

Figure 3. Acquired dynamic responses and temperature for three days (May 17th to 19th, 2014)

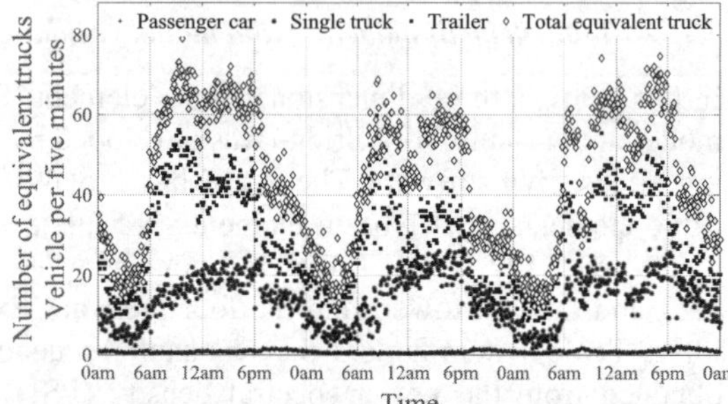

Figure 4. Monitored traffic volume for three days; Saturday, May 17, 2014, Sunday, May 18, 2014 and Monday, May 19, 2014

Table 1. Calaulated and Identified resonant frequencies.

Mode	FE	PSD	Difference (%)
1st	0.428	0.423	1.18
2nd	0.826	0.831	0.59
3rd	1.078	1.070	0.76
4th	1.331	1.324	0.54

3.2 *Dynamic responses obtained in FE models*

In the FE analysis, dynamic responses excited by the traffic loads were calculated every time increment of 0.02 second. The actual traffic volume obtained in the traffic counting system on Monday May, 19th was used to construct the load time history of the inputs for the FE model. The constructed input time history was assigned on each traffic lane node by the different arrival time which depended on the vehicle speeds.

To compare the calculated and measured accelerations of the deck at the location installed the AC-2 sensor, their RMSs of every 5 minutes interval were plotted together as shown in Fig. 5 (a). The RMSs of the calculated deck accelerations are a bit higher than those of the measured accelerations. However, they have the same trend that is large during the daytime and small during the nighttime. Moreover, both RMSs of the calculated and measured accelerations drop around 8.00 am at which the traffic congestion takes place. The drop of the RMSs of the calculated deck accelerations starts at around 7.00 am while the drop of the measured ones starts at around 6.00 am. The discrepancy might be caused by the slow traffic flow on the bridge during this period meanwhile the assigned speed of the vehicles in the FE models is constant. In order to investigate effects of the vehicular speeds on the accelerations obtained by the FE analysis, the assigned speed of 50 km/h was decreased to 40 km/h during 6.00 am to 7.00 am and to 30 km/h during 7.00 am to 8.00 am. Then the speed was increased to 40 km/h during 8.00 am to 8.30 am and then it was recovered to 50 km/h again. When the speeds of the vehicles decrease, the RMSs of the calculated deck accelerations decrease and they have more agreement with those of the measured ones as shown in Fig. 5 (b).

In summary, the dynamic responses obtained from the FE models showed more that not only the traffic volumes, but also vehicle speeds had effects on the dynamic responses. In the case of the same traffic volumes, the response amplitudes decreased when the speeds of vehicles decreased.

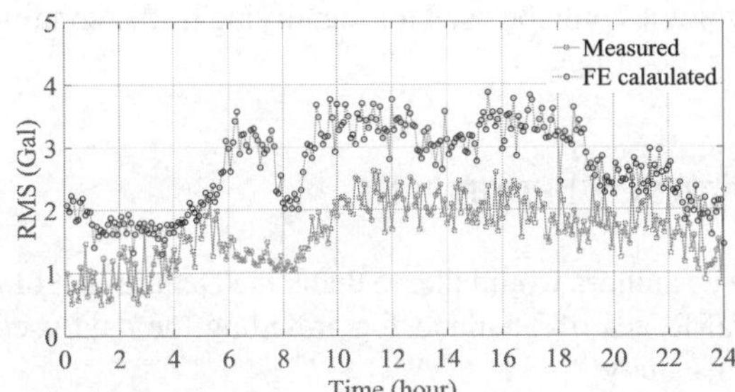

(a) Constant assigned speed in FE models

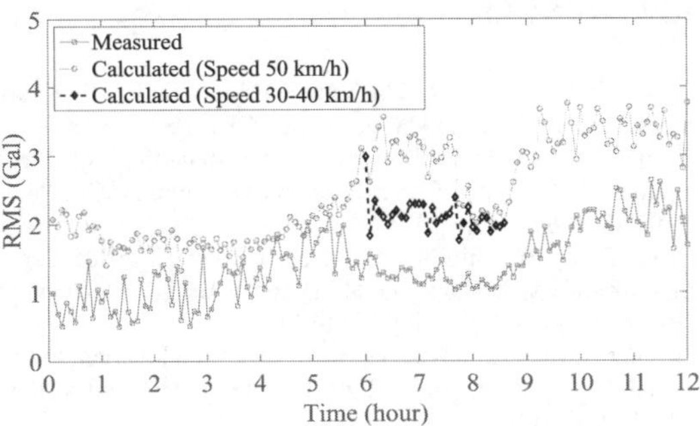

(b) Various assigned speed in FE models

Figure 5. RMSs of the calculated and measured accelerations of the deck on 19th May

4 TRAFFIC VOLUME ESTIMATION USING DYNAMIC RESPONSES

The correlation and regression analysis of the dynamic responses at each sensor and the temperature and traffic data to the peak frequencies in the target cable-stayed bridge has been conducted in our previous study (Wattana & Nishio 2015). Here, it was found that the total traffic volume had high correlations especially with the amplitude of accelerations of the deck and the stayed-cable, and the resonant frequencies of the lower modes. Then, the statistical modeling for estimating the traffic volume on the target cable-bridge bridge from the dynamic responses was constructed. The estimating model was defined as;

$$\hat{N}_{TR} = 420 + (151 \times \log RMS_{DECK}B4) + (69 \times \log RMS_{DECK}B2) + (113 \times \log RMS_{CA}B1), \quad (1)$$

where $\hat{N}_{TR}$ is predicted traffic volume (vehicle/5 min), $\log RMS_{DECK}B4$ and $\log RMS_{DECK}B2$ are logarithm of RMS of the vertical deck acceleration at mid span filtered by a band-pass filter with frequency range corresponding with the fourth and second modal frequencies, respectively (Gal), and $\log RMS_{CA}B1$ is logarithm of RMS of the cable acceleration filtered by a band-pass filter with frequency range corresponding with the first peak frequency (Gal).

The constructed model for estimating the traffic volume showed the accurate fitting performance to the one day trained data (May, 17th) , and it was also capable of predicting the traffic volume on the bridge for three-days data (May, 18th, 19th and 21st). Furthermore, it was considered that on the workdays (May, 19th and 21st), the traffic volumes have a same tendency and there is traffic congestion during the morning peak-hour (around 8.00 am) as shown in Fig. 6.

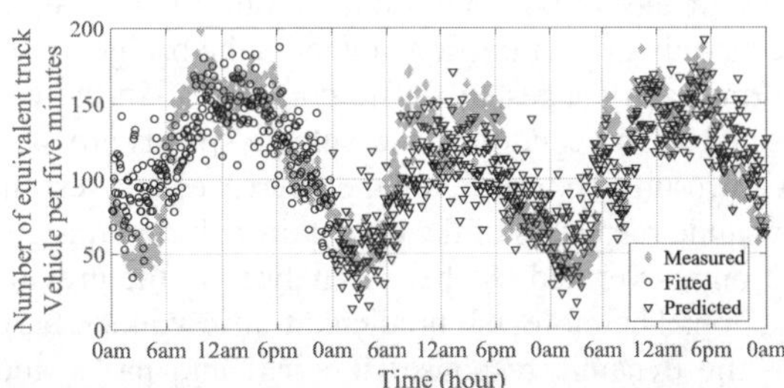

(a) Measured, fitted and predicted traffic volumes on Saturday, May 17, 2014, Sunday, May 18, 2014 and Monday, May 19, 2014.

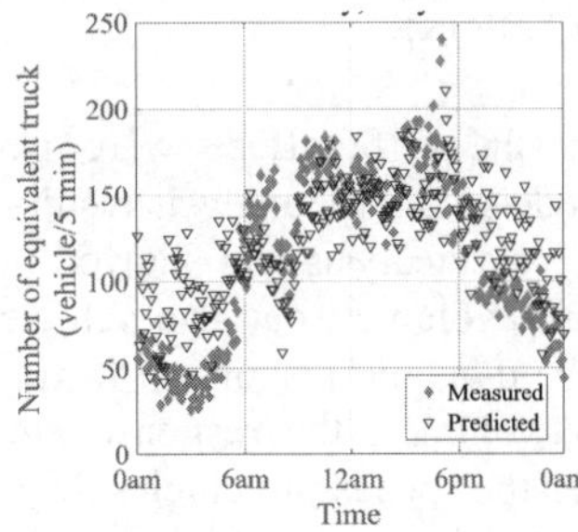

(b) Measured and predicted traffic volumes on Wednesday, May 21, 2014.

Figure 6. Measured, fitted and predicted traffic volumes.

Table 2. Actual and predicted traffic volume per day (Number of equivalent trucks per day).

Date	Actual	Predicted	Difference (%)
18th	28,844	25,895	10
19th	31,613	32,173	2
21st	32,964	34,949	6

Moreover, the actual and predicted traffic volume per day are shown in Table 2. It shows that the predicted traffic on the workdays (May, 19th and 21st) have a more accuracy (differences of 2% and 6%) than those on Sunday, May, 18th (10% difference). It might be caused by the speed of vehicles on Sunday that is relatively higher than those on workdays leading to more discrepancy with the data.

. It was then concluded that the constructed regression model for estimating the traffic volume only from the structural dynamic responses was applicable. In addition, the estimation model can recreate the daily fluctuations of the traffic volume especially during the time of the traffic congestion. In other words, it can accurately predict traffic volume on the bridge in both normal and congested traffic conditions. In the congested traffic condition, the vehicle speeds are slow. As a result, the traffic volume decreases and also the dynamic responses of the bridge drop during this condition, as verified by the FE analysis in the previous section. Therefore, the predicted traffic volume based on the dynamic responses is small and has a good agreement with the measured traffic volume.

5 CONCLUSIONS

In this paper, the traffic effects were investigated by using FE models. The FE analysis results agreed with those of the data analysis and showed more that not only the traffic volumes, but also vehicle speeds had effects on the dynamic responses. In the case of the same traffic volumes, the response amplitudes decreased when the speeds of vehicles decreased. Then, the statistical modeling for estimating the traffic volume on the target cable-bridge bridge from the dynamic responses was presented. The constructed model showed the accurate fitting performance to the data, and it was also capable of predicting the traffic volume on the bridge. In addition, the predicted traffic volume could be used for identifying traffic conditions on the bridge.

ACKNOWLEDGEMENT

The authors would like to thank the Department of Rural Roads of Thailand for providing the data used in this study.

REFERENCES

Cornwell P., Farrar C.R., Doebling S.W. & Sohn H. 1999, Environmental variability of modal properties. Exp. Tech, 45–48.

Cornell, C. A. 1967 “Bounds on the Reliability of Structural Systems,” Journal of the Structural Division, ASCE, 93 (ST1), 171-200, 1967.

Cross E. J., Brownjohn J.M.W. & Worden K. 2013. Long-term monitoring and data analysis of the Tamar Bridge. Mechanical Systems and Signal Processing 35, 16-34.

Hasofer A.M. & Lind N.C. 1974. An Exact and Invariant First Order Reliability Format, Journal of Structural Mechanics Division, ASCE, Vol. 100, No. 1.

Kim C., Jung D., Kim N., Kwon S. & Feng M. 2003. Effect of vehicle weight on natural frequencies of bridges measured from traffic-induced vibration. Earthquake Eng. Eng.,109-115.

Lutomirska M.. 2009. Live load model for long span bridge, Doctoral Dissertation.

Nowak, A.S. 1999. Calibration of LRFD Bridge Design Code, NCHRP Report 368, Transportation Research Board, Washington, D.C., USA.

Nowak, A.S. & Szerszen, M. M. 2001. Reliability-Based Calibration for Structural Concrete, Report UMCEE 01-04, Department of Civil and Environmental Engineering, University of Michigan, Ann Arbor, Mich., Nov.

Peeters B. & De Roeck G. 2000. Reference based stochastic subspace identification in civil engineering. Inverse Problems in Engineering, 8(1):47-74.

Peeters B., Maeck J. & De Roeck 2001. Vibration-based damage detection in civil engineering: excitation sources and temperature effects. Smart Mater. Struct., 10,518-527.

Wattana K. & Nishio M. 2015. A Analysis of multivariate SHM data of a cable-stayed bridge considering operational and environmental effects. Structural Health Monitoring of Intelligent Infrastructure, Proc. 7th international conference, Turin, Italy, July 1-3, 2015.

Zhang Q., Fan L. C. & Yuan W. C. 2002. Variability in dynamic properties of cable-stayed bridge. Earthquake Eng. Struct. Dyn., 2015-2021..

Proceedings of the International Conference on Smart Infrastructure and Construction
ISBN 978-0-7277-6127-9

© The authors and ICE Publishing: All rights reserved, 2016
doi:10.1680/tfitsi.61279.387

Seismic damage monitoring of high concrete dams with embedded PZT sensor network

J. Zhou*, Y. Zhang, and X. Feng

Faculty of Infrastructural Engineering, Dalian University of Technology, Dalian, China
**Corresponding Author*

ABSTRACT Seismic damage detection of concrete dams has always attracted much attention in hydraulic structure community. In this paper, amethod was proposed to perform seismic damage monitoring for concrete dams by using embedded PZT sensor network.As its importance in achieving the dam damage detection, the arrangement ofdistributedPZT sensor networkwas introduced in detail. A dam model system with a distributed PZT sensor networkwasusedasan object for verification. A shaking table is used to simulate the earthquake groundmotion for the testing object. Theseismic damage detection system is used not only for performing the seismic damage process monitoring by measuring the dynamic stress history but also for distributed detecting of the dam damaged region. By analyzing the sensor signals, the emergence and development of thestructuraldamagescan bemonitored timely.A damage index matrix is presented to evaluate the damage status of the dam in different paths. The experimental results demonstrate the timeliness andthe effectiveness of theproposed seismic damage detection method.

1 INTRODUCTION

Several high concrete dam is being or is about to be constructed in China to meet increases in energy demand (e.g., Jin Ping dam in the height of 305m and Ma Ji dam in the height of 300m). Most of them locate in a region of known high seismicity. Therefore, it is necessary to monitor the seismic health of these dams.

The use of strong-motion accelerometers and the on-site inspection are essential methods in seismic damage detection. Many researchers used this method to predict the damage location and severity for relatively simple structures, such as precast concrete frames (Belleri *et al.* 2014).Normally, the above method still face problems when applied to real large-scale complex structures, such as low signal to noise ratio, false frequency, system identification and model order estimation.

In recent years, piezoceramic (PZT) transducers have emerged as new tools for the health monitoring of concrete structures. Many researchers used this method to monitor the damage of concrete structures such as bridge bent caps (Song *et al.*2007). Although this monitoring approach can qualitatively evaluate the concrete structure damage status, the full damage process of concrete structures cannot be obtained. In a companion paper (Hou *et al.* 2012), it has proposed a method for not only evaluating the damage state but also obtaining the stress history of a concrete specimen. All the above-mentioned monitoring systems are successful in evaluation of the health status at many types of concrete structures, but they have not been utilized to monitor seismic damage of concrete dams, which is also important for timely safety assessment and dam reinforcement after the earthquake disaster.

In this paper, a novel seismic damage detection system was developed to perform seismic damage monitoring for concrete dams. The proposed method can be divided into two functions: (a) dam stress history monitoring based on direct piezoelectric effect and (b) assessment of damage by the variation in stress wave propagation. A dam model with embedded dual function PZT sensors was used as a testing object. A shaking table was used to simulate the earthquake ground motion for the testing object. Before the shaking table test, the sensitivity of sensors was calibrated by the loading test of concrete cylinder specimens. Experimental results have shown that the proposed seismic damage monitoring method for concrete dams is better than traditional methods with strong-motion accelerographs and on-site inspections. The proposed method has the potential to be applied to the seismic damage monitoring of real concrete dam.

2 SEISMIC DAMAGE MONITORING METHOD

2.1 Monitoring purpose

Since the collapse of the dam due to severe earthquakes will cause catastrophic consequences, the importance of seismic safety assessment of concrete dams is well recognized. Achieving the whole process monitoring of seismic damage of dams will have the following benefits. One is able to accurately evaluate the dam safety and timely provide information for the concrete dam reinforcement. The other is to help designers understand the failure mechanism of seismic damage of concrete dams and provide a reference for future optimized design of dams. To realize the above monitoring purposes, a seismic damage monitoring method for concrete dams, which based distributed PZT sensor network, is developed in this paper.

2.2 Monitoring principle

The adopted dual function PZT sensor is fabricated by embedding two pieces of PZTs and one conducting plate into a small brass case, the gap between them is filled with rubber material. The PZT sensor had a coin-like shape and diameter, thickness, and mass of 18.5 mm, 2.8 mm, and 5 g, respectively.

The adopted sensor can monitor the stress history of concrete dams by using direct piezoelectric effect during earthquake. The sensitivity can be related to the piezoelectric coefficient of the PZT patch by(Hou *et al.* 2012)

$$S=\sum_{i=1}^{3}A_3\alpha_i d_{3i} \tag{1}$$

where the coefficientsd_{31}, d_{32}, andd_{33} relate the normal stress in the 1, 2, and 3 directions respectively to a field along the poling direction, A_3 is the area of the adopted sensor in the 3 direction, and α_i is the stress ratio expressed as

$$\alpha_i=\frac{\sigma_i}{\sigma}, i=1-3 \tag{2}$$

whereσ_1, σ_2, andσ_3 are the normal stress in the 1, 2, and 3 direction respectively. owing to the constraint caused by the rubber is very small, the stress on the radial direction of the PZT can be neglect, so $\alpha_1=\alpha_2=0$.The adopted dual function PZT sensor contains two pieces of PZT patches, so the sensitivity of the sensors is double times than the sensors with one piece of PZT, Eqn1 can be written as

$$S=2A_3\alpha_3 d_{33} \tag{3}$$

The sensitivity can also be defined as the output voltage of dual function PZT sensor versus the applied stress in relation to

$$S=\frac{V}{\sigma} \tag{4}$$

where V is the generated voltage which can be obtained by related experimental equipment, σ is the applied stress on the adopted sensor which can be obtained by calculated. The strain rate of concrete dams due to the real earthquake always range from 10^{-4}/sto10^{-2}/s, three different loading were undertaken corresponding to the strain rate of 10^{-2}/s, 10^{-3}/s and 10^{-4}/s respectively. Calibration results showed that the sensors had constant sensitivity at three different strain rate. The average value of the sensors' sensitivity obtained from the calibration test was 2.47 V/MPa.

A pair of adopted sensors denoted as PZT 1 and PZT 2 respectively are embedded in the concrete dam. PZT 1 is used as actuator and PZT 2 is used as sensor. Stress wave with a range of frequencies is generated by the actuator PZT 1 and propagates to the sensor PZT 2.When the dam damage occurs between PZT 1and PZT 2,the stress wave will weaken at the damage area and the energy of received signal at PZT 2 will reduce. By comparing the intensity variation of the received signal at different frequencies, the damage in the concrete dam can be identified.

Root mean-square deviation (RMSD) of the energy of decomposed signal is a suitable damage index to monitor the state of the concrete structures (Song*et al.*2007; Laskar *et al.* 2009; Soh *et al.* 2000; Tseng and Naidu 2002). In this paper, RMSD between the energy vectors of the healthy state and the damaged state is also adopted as the dam age index to evaluate the damage status of concrete dams. The energy of the decomposed signal for damaged status is represented by $E_{i,j}$, where i is the time index and j is the frequency band. The energy of the decomposed signal for healthy status is represented by $E_{h,j}$. The damage index is defined as

$$I=\sqrt{\sum_{j=1}^{2^n}\left(E_{i,j}-E_{h,j}\right)^2 \Big/ \sum_{j=1}^{2^n} E_{h,j}^2} \quad (5)$$

where2^n is the number of the sensor signal decomposition at the frequency domain.

Based on historical experience, hundreds of aftershocks will rumble on since the main quake (e.g., A total of 233 aftershocks, with at least magnitude of 4.0, occurred after the Wenchuan 5-12 Earthquake). In order to demonstrate the damage location information and damage process information, a Path-History damage index matrix (PHDIM)$M_{p\times h}$ is defined as:

$$M_{P\times H}=\left[I_{i,j}\right]_{P\times H} \quad (6)$$

where the matrix element at the ith row and the jth column, $I_{i,j}$, is the damage index of the ith monitoring path at the jth excitation, p is the total number of monitoring paths and h is the total number of excitations. The damage status at different paths of the concrete dam at different excitations can be revealed by a three-dimensional damage index matrix plot. The PHDIM is useful in monitoring the seismic damage process to predict the failure of a concrete dam.

2.3 *Monitoring method*

The seismic damage detection system of concrete dams consists of two parts which connected by a local area network: on-site monitoring and the remote monitoring center, as shown in figure 1. The on-site monitoring of concrete dam includes three parts, that is, input signal subsystem, distributed PZT sensor network subsystem, and data acquisition subsystem. The remote monitoring center includes three parts, that is, data analysis subsystem, safety assessment subsystem and data storage subsystem.

Taking into account the actual complex and huge concrete dam, as well as the costs and other factors, only relatively few sensors are embedded into the limited positions. Therefore, a good distributed sensor network subsystem for monitoring concrete dams should use less measuring points to get enough in

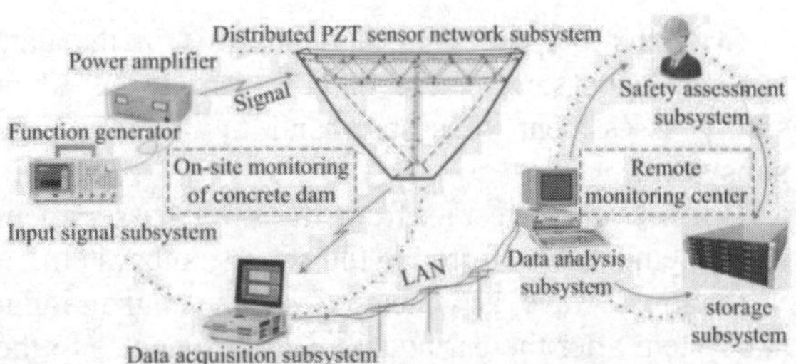

Figure 1 Seismic damage detection system of concrete dams.

formation. According to the conventional actual damage(e.g., Shapai dam appeared an opening several meters deep at the top of right-side contraction joint during Wenchuan Earthquake) and experimental studies for concrete dams (Zhou*et al.* 2000), it can be found that the upper area of the arch concrete dam is its weakest zone, especially the location of the structural joints. After the destruction of the arch direction of dam appeared, the crown cantilever of the dam will be the focus of monitoring areas. Based on the above considerations, a distributed PZT sensor network subsystem should include two parts: one part is the arch direction monitoring, and the other is the crown cantilever monitoring. The distributed PZT sensor network is a three-dimensional network, and can focus on monitoring the areas where are more prone to earthquake-induced damage for concrete dams. It should be noted that, due to the actual concrete dams often contain varying amounts and different forms of structural joints, the actual layout of distributed PZT sensor network should also consider the impact of structural joints.

2.4 *Monitoring process*

The seismic damage detection system monitored the stress history of dam by using direct piezoelectric effect during earthquake, and was used to detect seismic damage of the key areas by utilizing inverse piezoelectric effect after earthquake. The whole seismic damage detection process can be divided into the following steps:

(1) Collecting real-time stress history information through by distributed PZT sensor network subsystem during an earthquake.

(2) The above information is collected by the data acquisition subsystem and is sent to the remote monitoring center through the LAN.

(3) After preprocessing and analysis in the data analysis subsystem, these data is sent to the dam safety assessment subsystem and the data storage subsystem.

(4)Assessment results from the safety assessment subsystem are transferred to the storage subsystem.

(5)Stress wave is generated by the signal input subsystem after the earthquake and propagates to the distributed PZT sensor network subsystem.

(6)Utilizing the distributed PZT sensor network subsystem to monitor the concrete dam damage.

(7) Repeat steps 2 to4 to obtain the final result of the seismic damage detection for the concrete dam.

Steps 1 to 4 can detect the process of earthquake damage, while steps 5 to 7 can detect the damage areas of dam. The final result of the seismic damage detection can provide more timely, complete and comprehensive information for the dam repair job.

3 EXPERIMENTAL VERIFICATION

3.1 General description of the test

Shapai high arch dam was selected as the prototype dam, which has a height of 132 m, is located in the Wenchuan County, Sichuan Province, China. One contraction joint is arranged at each end of the arch, and two induced joints are arranged near arch crown. According to the elasticity-gravity similitude, a small-scale model dam was built on the shaking table. Figure 2 shows the upstream view of the model of the dam on the shaking table. Owing to the space limitations of the shaking table, the model dam was designed using a geometric ratio of 1:112, which produced a height of 1176 mm, crest length of 2215 mm, crest thickness of 85 mm, and bottom thickness of 257 mm. The model system also included the mountains that form the abutments and a partial foundation with a topographic feature near the dam. The system was fixed to the shaking table through a 120-mm thick concrete plate. For simplifying the test and reducing the effects of the uncertainties, the dam-reservoir interactions was ignored.

Fourteen dual function PZT sensors were embedded in the interior of the model when it was poured, as shown in Figure3. Only a single-layer sensor network was arranged in the streamwise direction owing to the size limitations of the dam model. Nine PZT sensors were embedded in the central axis of the

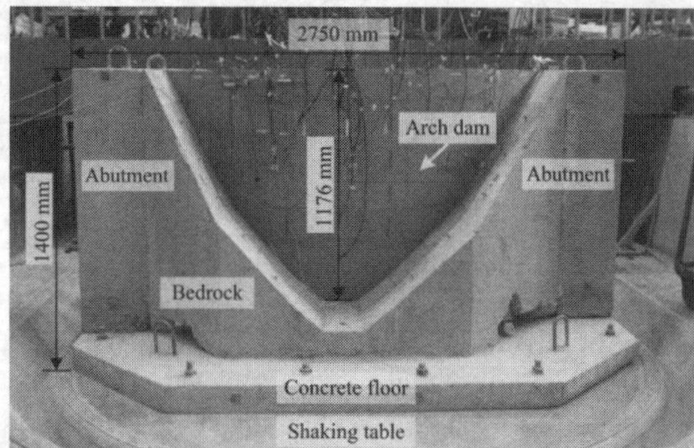

Figure 2. Upstream view of the dam model.

upper area of the dam in the arch direction and five were embedded in the central axis of the crown cantilever of the dam. It should be noted that the whole surface of the contraction joints is arranged precast concrete plates and here dam is completely separated, so that there cannot carry out active sensing but can perform the stress history monitoring. On the contrary, the surface of the induced joints is arranged discontinuously precast concrete plates, so that not only the stress history monitoring but the active sensing can be carried out. Total ten monitoring paths in the dam were considered as listed in **Error! Reference source not found.**. Frequency sweep in range of 1000~5000 Hz was conducted with amplitude of 90V. By analyzing the wave response, the damage location and the damage status for the dam structure can be predicted.

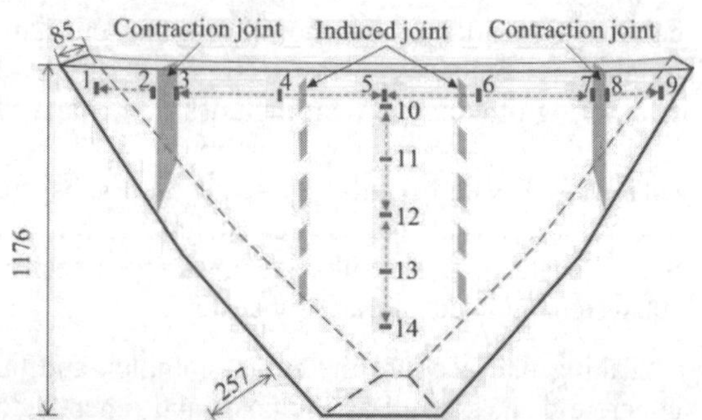

Figure 3. Distributed PZT sensor network embedded in concrete dam (mm).

Table 1. Monitoring paths for the active sensing

Path	A	B	C	D	E	F	G	H	I	J
Actuator	2	4	4	6	6	8	11	11	13	13
Sensor	1	3	5	5	7	9	10	12	12	14

The harmonic waves corresponding to the fundamental frequency were employed to excite the shaking table with six excitation levels, i.e. 0.1 g, 0.15 g, 0.2 g, 0.25 g, 0.3 g and 0.4 g. The advantage of the harmonic input is the repeatability of the model dam test. After each shaking, the distributed PZT sensor network was used to monitor the seismic damage of the model dam, and the next loading frequency was determined by the micro-amplitude white noise excitation. The amplitude of the harmonic wave was progressively increased until the model dam completely collapsed. The excitation amplitude took several seconds to reach the target value and then remained largely unchanged.

3.2 *General description of the test*

The distributed PZT sensor network was utilized for the real-time dynamic stress history monitoring of the concrete dam model. The stress history at different excitation levels in the arch direction and in the vertical direction are illustrated in Figure 4 and 5, respectively. In the upper area of the dam, the largest stress in the arch direction occurred at the position of PZT 5. The arch stress decreased from the position of PZT 5 to both contraction joints, and gradually increased from both contraction joints to the corresponding banks. Up to 0.15g excitation level, the stress near the arch crown gradually increased.In0.30-0.40g excitation level, the stress near the arch crown ceased to obviously increase, with the exception of near the abutments. This implies that the contraction and induced joints had fully opened and the upper area of the dam in the arch direction had been completely damaged. On the crown cantilever of the dam, the vertical stress gradually increased in 0.10-0.15g excitation level, maintained constant with a certain fluctuation in 0.20-0.25gexcitation level and reduced quickly in 0.30gexcitation level. The increase, instead of decrease, of the stress near the position of PZT10 in 0.30g excitation level demonstrated that the destruction in the arch direction affected the stress distribution of the crown cantilever and then the upper part of the crown cantilever will be destroyed.

The distributed PZT sensor network were also used to monitor the damage location and the damage status inside the dam during the interval between two excitation levels. The FFT spectrums for some

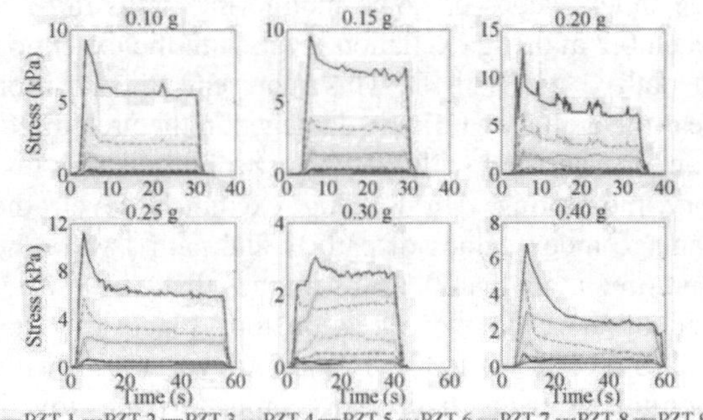

Figure 4. Stress history in the arch direction at different excitation levels.

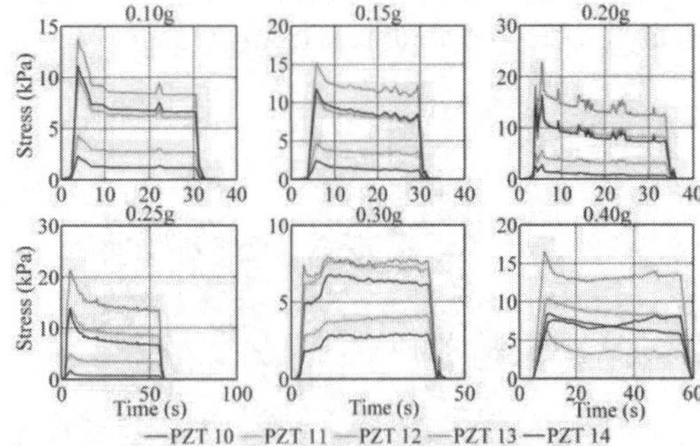

Figure 5. Stress history in the vertical direction at different excitation levels.

monitoring paths are shown in Figure 6. Based on the changes in the frequencies and amplitudes of the signals, the following conclusions are drawn. There was no great damage on the path A during the loading process, as shown in Figures 6(a). Minor damage occurred on the path C before 0.30g excitation level, after which there was serious damage, as shown in Figure 6(b). The path J was gradually destroyed under the different excitation levels, as shown in Figure 6(c).

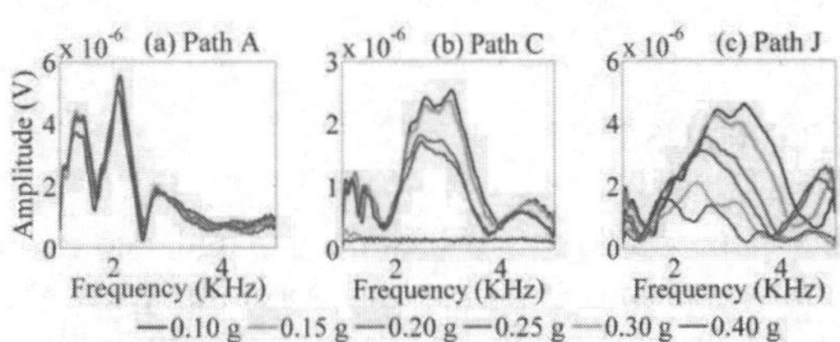

Figure 6. FFT spectrums for different monitoring paths: (a) path A; (b) path C; (c) path J.

Figure 7 is the three-dimensional plots of the PHDIM of this validation test. In figure 7, the damage index values of some monitoring paths did not reach 0.2 at 0.10g excitation level, with the exception of path C and path D. This represents the position near the arch crown firstly damaged in the arch direction, and collated with the values from the stress history monitoring. Until 0.20g excitation level, the damage index values of path H and path J were the first time more than 0.2.This implies that the destruction of the cantilever direction took place after the arch direction. At 0.30g excitation level, most of the monitoring paths had been completely destroyed, with the exception of path A and path F. This result collated with the values from the stress history monitoring and indicated that the contraction joints can effectively protect the abutment from damage.

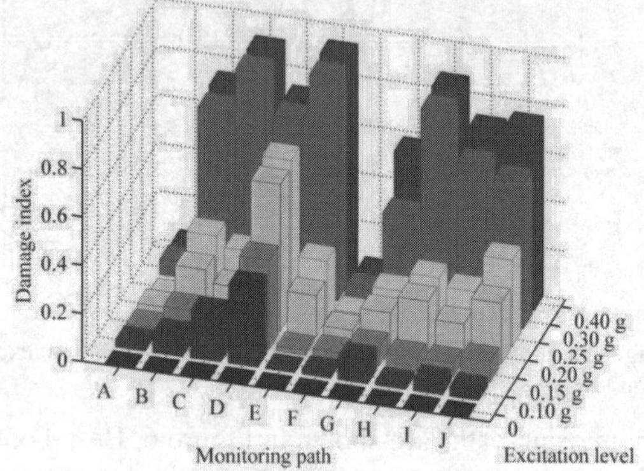

Figure 7. Path-History damage index matrix of the tested dam model.

By the above experimental results, it can be concluded that the dam damage occurred firstly in the arch direction. The presence of contraction joints and induced joints weakened the stiffness of the dam section and were therefore the first areas of the dam to damage. The destruction of the cantilever direction took place after the arch direction.

4 CONCLUSION

In this paper, a kind of seismic damage detection method was proposed. Following is a summary of the study and the conclusions drawn from the findings:

(1)This paper proposed a novel seismic damage detection system to perform seismic damage monitoring for concrete dams. Since the distributed PZT sensor network subsystem is an important part of the seismic damage detection system, the details of the arrangement were introduced.

(2)The dual function PZT sensor's performance is stable and reliable for the dynamic loading. The distributed PZT sensor network can be used not only for monitoring the dynamic stress history of dams during earthquake but also for performing active damage monitoring after earthquake. The results of stress history monitoring and active sensing can complement and verify each other.

(3) The proposed seismic damage detection method can effectively compensate for the lack of traditional methods with strong-motion accelerographs and on-site inspections, and help designers understand the failure mechanism of seismic damage of concrete dams. It has great application potential in seismic damage detection of concrete dams.

ACKNOWLEDGEMENT

This work was supported by the National Basic Research Program of China (Grant No. 2013CB035906) and the National Natural Science Foundation of China (Grant No.91215301).

REFERENCES

Belleri, A., Moaveni, B. and Retrepo, J.I. (2014). "Damage assessment through structural identification of a three-story large-scale precast concrete structure", Earthquake Engineering & Structural Dynamics, Vol. 43, No.1, pp. 61-76.

Song, G., Gu, H., Mo, Y.L., Hsu, T.T.C. and Dhonde, H. (2007). "Concrete structural health monitoring using embedded piezoceramic transducers", *Smart Materials and Structures*, Vol. 16, No. 4, pp. 959-968.

Hou, S., Zhang, H.B. and Ou, J.P. (2012). "PZT-based smart aggregate for compressive seismic stress A monitoring", *Smart Materials and Structures*, Vol. 21, No. 10, pp. 1-9.

Soh, C.H., Tseng, K.K-H., Bhalla, S. and Gupta, A.(2000). "Performance of smart piezoceramic patches in health monitoring of a RC bridge", *Smart Materials* , Vol. 9, No. 4, pp. 533-542.

Tseng K.K-H. and Naidu, A.S.K. (2002). "Non-parametric damaged detection and characterization using smart piezoceramic material", *Smart Materials and Structures*, Vol. 11, No. 3, pp. 317-329.

Zhou, J., Lin, G., Zhu, T., Jefferson, A.D. and Williams, F.W.(2000). "Experimental investigations into seismic failure of high arch dams", *Journal of Structural Engineering*, Vol. 126, No. 8, pp. 926-935.

Proceedings of the International Conference on Smart Infrastructure and Construction
ISBN 978-0-7277-6127-9

© The authors and ICE Publishing: All rights reserved, 2016
doi:10.1680/tfitsi.61279.393

Mechanical equivalent of logical inference: application to structural health monitoring problems

D. Bolognani*[1], C. Cappello[1] and D. Zonta[1,2]

[1] *University of Trento, Italy*
[2] *University of Strathclyde, UK*
* *Corresponding Author*

ABSTRACT Structural health monitoring requires engineers to understand the state of a structure from its observed response. When this information is uncertain, Bayesian probability theory provides a consistent framework for making inferences. However, structural engineers are often unenthusiastic about Bayesian logic and prefer to make inference using heuristics. Here we propose a quantitative method for logical inference based on a formal analogy between linear elastic mechanics and Bayesian inference with linear Gaussian variables. We start by discussing the estimation of a single parameter, by assuming that all of the uncertain quantities have a Gaussian distribution and that the relationship between the information and the parameter is linear. With these assumptions, the analogy is stated as follows: the expected value of the parameter corresponds to the position of a bar with one degree of freedom; uncertain information on the parameter is modelled as linear elastic springs in series or parallel. If we want to extend the analogy to *N* parameters, we prove that we must simply express the potential energy of the mechanical system, with *N* degrees of freedom, associated to the inference problem in question. We apply the mechanical equivalent to a real-life case study: the elongation of a cable belonging to Adige Bridge, a cable-stayed bridge North of Trento, Italy, in the case of two and three parameters to estimate.

1 INTRODUCTION

Structural engineers usually have a solid background in mechanics, yet not always a good relationship with probability theory. In most cases, this is not that critical because code-based design is practically probability-free, with serious probabilistic analysis typically being confined to the most recondite annexes of the codes (EN 1990:2002. It is different for those engineers who grapple with structural health monitoring (SHM), an activity where the objective is to estimate the state of a structure from an uncertain batch of observations, using uncertain models. A consistent framework for making inferences from uncertain information is Bayesian probability theory (Shon & Law 2005). Yet structural engineers are often unenthusiastic about Bayesian formal logic, finding its application complicated and burdensome, and they prefer to make inference by using heuristics. In this contribution, we wish to help structural engineers reconcile with probabilistic logic (Jaynes 2003) by suggesting a quantitative method for logical inference based on a formal analogy between mechanics and Bayesian probability. We will state the fundamentals of the analogy in the next section.

To start, we will limit the analogy to the case of linear Gaussian single-parameter estimation, which corresponds in the mechanical counterpart to mere linear elastic single-degree-of-freedom analysis: a cakewalk for structural engineers. In section 3, we apply this formal analogy to a classical inference problem: the estimation of the deformation of a cable belonging to a cable-stayed bridge, characterized by two independent parameters. We will carry out the simple problem of linear regression by solving the equivalent mechanical system of springs.

2 SINGLE- PARAMETER FORMULATION

In this section, we refer to the problem of logical inference of a single parameter based on uncertain information (Cappello et al. 2015). The goal is to estimate a parameter θ based on a set of uncertain information y_i. Further assumptions are that all the uncertain quantities have Gaussian distribution, and that the relationship between information and parameter is linear. When the problem is linear and Gaussian, in principle we can solve any logical inference problem using the following two fundamental rules.

First inference rule or inverse-variance weighting rule (Ku 1966). Given a set of n observations y_i of variance $\sigma_i{}^2$, the inverse of the variance $\sigma_\theta{}^2$ of the parameter is the sum of the inverse-variances of the observations, and the expected value of the parameter μ_θ is the inverse-variance weighted sum of the observations:

$$\frac{1}{\sigma_\theta^2} = \sum_{1=1}^{n} \frac{1}{\sigma_i^2}, \quad \mu_\theta = \frac{\sum_{1=1}^{n} \frac{y_i}{\sigma_i^2}}{\sum_{1=1}^{n} \frac{1}{\sigma_i^2}}. \tag{1a,b}$$

Second inference rule or linear propagation of uncertainties (Rabinovich 2005) (Kirkup 2006). The indirect measurement $y = x_1 + \cdots + x_m$ being the sum of m different arguments x_j of variance $\sigma_j{}^2$, the variance of the observations is the sum of the variance of the arguments and the mean value of the indirect observation is the sum of the arguments:

$$\sigma_y^2 = \sum_{j=1}^{m} \sigma_j^2, \quad \mu_y = \sum_{j=1}^{m} x_j. \tag{2a,b}$$

Before proceeding, it is also convenient, primarily to lighten notation, to introduce the quantity

$$w = \sigma^{-2} = \frac{1}{\sigma^2}. \tag{3}$$

The quantity w is compatible with the official definition of *accuracy* (ISO5725-6:1994) and the word itself intuitively connects to the practical meaning of w: the higher the *accuracy* w of an observation is, the more *accurate* our knowledge about the parameter becomes. Therefore, in the rest of the paper we will refer to the inverse-variance w simply as *accuracy*. Based on that, we can reword and reformulate the two basic inference rules.

First inference rule. Given a set of n observations y_i with accuracy w_i, the accuracy w_θ of the parameter estimation is the sum of the accuracy of the observations, and the mean value of the parameter μ_θ is the sum of the observations weighted with their accuracy:

$$w_\theta = \sum_{1=1}^{n} w_i, \quad \mu_\theta = \frac{\sum_{1=1}^{n} y_i w_i}{w_\theta}. \tag{4a,b}$$

Second inference rule. The indirect measurement $y = x_1 + \cdots + x_m$ being the sum of m different arguments x_j of accuracy $\sigma_j{}^2$, the inverse accuracy of the observation is the sum of the inverse accuracy of the arguments and the mean value of the indirect observation is the sum of the arguments:

$$\frac{1}{w_y} = \sum_{j=1}^{m} \frac{1}{w_j}, \quad \mu_y = \sum_{j=1}^{m} x_j. \tag{5a,b}$$

At this point, it is not difficult for a structural engineer to spot in (5a) the same form of the expression that provides the stiffness of a set of springs in parallel; and similarly, (5b) recalls of the stiffness expression of a set of springs in series. This opens a door to an analogy between the world of logic and the world of mechanics. Particularly, the analogy statements are the following.

- The value of the parameter is represented by the position of a rigid bar with one degree of freedom.
- An uncertain piece of information on the parameter is modeled as a linear elastic spring fixed at one edge and connected at the other to the bar, with stiffness equal to its accuracy and pre-stretch equal to its mean value.
- Multiple sources of uncertainties on the same information are modeled as serial springs, each with stiffness equal to its accuracy.
- The posterior mean value of the parameter corresponds to the position of the bar, in equilibrium.
- The posterior accuracy of the parameter corresponds to the resulting stiffness of the bar. The basic elements of the analogy are summarized in Table 1. Figure 1 shows the mechanical representation of simple linear Gaussian inference problems.

Table 1. Analogy between inference and mechanical models.

Symbol	Logical meaning	Mechanical meaning
w, σ^2	accuracy,inverse-variance	stiffness
σ^2	variance	flexibility
y	observation	pre-stretch
μ	expected value	equilibrium displacement

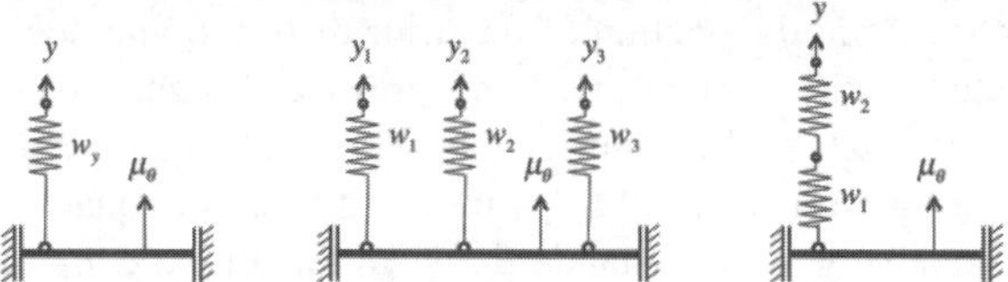

Figure 1. Mechanical analogy of simple linear Gaussian inference problems: parameter estimation based on one observation (a), three uncorrelated observations (b), one observation affected by three uncorrelated sources of uncertainty (c).

3 EXTENSION OF THE MECHANICAL ANALOGY TO N PARAMETERS

Now, we analyze a generic inference problem with N unknown parameters to estimate, represented by the vector $\boldsymbol{\theta} = (\theta_1,....,\theta_N)^T$: we imagine that each parameter is characterized by a prior mean value μ_{θ_i} and a prior standard deviation σ_{θ_i}; the latter is linked by the equation $w_{\theta_i} = \sigma_{\theta_i}^{-2}$ to the ith accuracy, which in our mechanical analogy represents the stiffness of the spring associated to each single parameter. The multivariate Gaussian distribution [9], linked to the N-dimensional vector $\boldsymbol{\theta}$, takes the form:

$$\mathrm{N}(\boldsymbol{\mu},\Sigma;\boldsymbol{\theta}) = \frac{1}{(2\pi)^{\frac{N}{2}}} \frac{1}{|\Sigma|^{\frac{1}{2}}} e^{\left\{-\frac{1}{2}(\boldsymbol{\theta}-\boldsymbol{\mu})^T \Sigma^{-1}(\boldsymbol{\theta}-\boldsymbol{\mu})\right\}}, \tag{6}$$

where $\boldsymbol{\mu}$ is the N-dimensional mean vector, containing the N values μ_{θ_i} associated to each parameter, Σ is the NxN covariance matrix, and $|\Sigma|$ denotes the determinant of Σ.

We have to pay attention to the structure of the Gaussian distribution. We can notice that the exponent is characterized by a quadratic form that corresponds to the potential energy $E_p(\boldsymbol{\theta})$ of a mechanical system with N degrees of freedom, related to the inference problem in question. It takes the following mathematical form:

$$E_p(\boldsymbol{\theta}) = -\mathrm{In}(\mathrm{N}(\boldsymbol{\mu},\Sigma;\boldsymbol{\theta})) = \frac{1}{2}(\boldsymbol{\mu}-\boldsymbol{\theta})^T \Sigma^{-1}(\boldsymbol{\mu}-\boldsymbol{\theta}). \tag{7}$$

We name here the inverse of the covariance matrix $\boldsymbol{\Lambda} = \Sigma^{-1}$, this is also known as *accuracy matrix* (Bishop 2006). Its diagonal terms represent the posterior stiffness $w_{\theta_i|\mathbf{y}}$ of each single parameter θ_i.

Now, to obtain the N diagonal elements to $\boldsymbol{\Lambda}$ we must get the second derivative of $E_p(\boldsymbol{\theta})$ with respect to each of the parameters θ_i; the elements out of diagonal are instead obtained by calculating the mixed derivatives of each parameter with respect to all other parameters. To obtain the covariance matrix we simply make the inverse of $\boldsymbol{\Lambda}$. The diagonal elements of Σ represent the posterior variance $\sigma^2_{\theta_i|\mathbf{y}}$ of each single parameter θ_i.

The posterior mean values $\mu_{\theta_i|\mathbf{y}}$ of each parameter θ_i correspond to those values that minimize the potential energy of our mechanical system. Therefore, to discover them, we have to resolve an algebraic system with N variables in which there are the partial derivatives of $E_p(\boldsymbol{\theta})$, each with respect to each parameter θ_i, set equal to zero.

4 A CASE STUDY: ELONGATION OF A CABLE BELONGING TO ADIGE BRIDGE

Adige Bridge (Cappello et al. 2015) was built in 2008, ten kilometers north of the city of Trento, Italy, and spans the Adige River. It is a two-span cable-stayed bridge with a steel-concrete composite deck 260 m long (Fig. 2). The composite deck is made from 4 "I"-section steel girders and a 25 cm cast-on-site concrete slab.

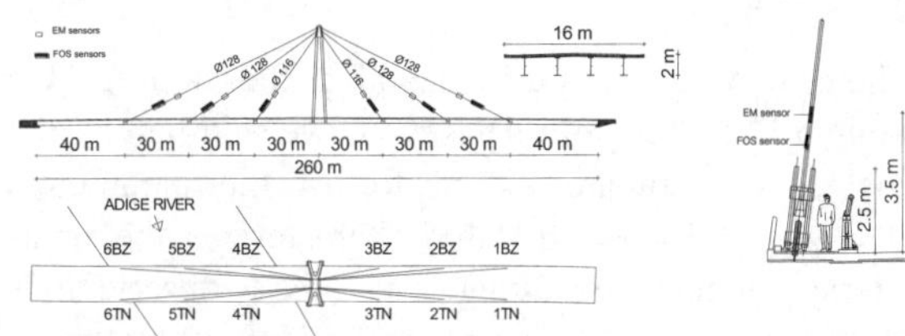

Figure 2. Longitudinal section of the bridge and sensor layout (upper left); plan view of the bridge (lower left); cross-section of the bridge (right).

The deck is also supported by 12 stay cables, 6 on each side, which have a diameter of 116 mm and 128 mm. Their operational design load varies from 5,000 kN to 8,000 kN. The cables are anchored to the bridge tower, consisting of four pylons and located in the middle of the bridge.

When the construction was completed, the Italian Autonomous Province of Trento, which owns and manages the bridge, decided to install a monitoring system to continuously record force and elongation of the stay cables. Elongations are recorded by 1 m long gauge sensors, placed on each of the 12 cables. These fiber-optical sensors (FOS) are based on Fiber Bragg Grating (FBG) technology (Measures 2001) and were supplied by Smartec SA. These sensors also record local temperature for thermal compensation.

4.1 Two parameters to estimate

As an example, we use data acquired from October 12, 2011, to November 25, 2012, for cable 1TN, purified of the effect of temperature. We consider only one sample a day, recorded between 4 AM and 6 AM, as we assumed the temperature in this period to be constant. We have discarded those days in which no samples were found in the time interval. Fig. 3 shows the data acquired, expressed in terms of difference of deformation and time:

$$\Delta y = y_i - y_1,\ \ \Delta t = t_i - t_1. \tag{8}$$

During the analysis, 411 deformation measurements were recorded with an uncertainty for each measurement equal to $w_y = 0.0016\,\mu\varepsilon^{-2}$, i.e. $\sigma_y = 25\,\mu\varepsilon$.

This is clearly a classical problem of the linear regression. We have to estimate the two parameters that best characterize the straight line fitting our time-dependent data set. The function employed is:

$$y = y_0 + \varphi \times t, \tag{9}$$

where y_0 y_0 is the intercept and φ the slope of the straight line fitting our dataset. As we said before, the goal is to estimate the vector of the parameters $\boldsymbol{\theta} = (y_0, \varphi)^T$ $\theta = (y_0, \varphi)^T$ that characterizes the parametric model resulting in the observations $\mathbf{y} = (y_1, y_2, ..., y_N)^T$, linearly dependent on the time t, as shown in Figure 4. We can represent the problem as a bar with two degrees of freedom: vertical translation and rotation. According to the parametric model defined in (9), we consider the slope of the bar linked to the parameter φ, its length to the time t and its distance from the ground floor to the parameter y_0.

Based on our experience, we assign to the two parameters φ and t two prior Gaussian distributions that give us the initial information about the state of the bar. We connect the left-hand end of the rigid bar to a vertical linear elastic spring with flexibility equal to the standard deviation of the prior distribution associated to the parameter y_0 and pre-stretch equal to its mean value. We connect the same end to a torsion spring with flexibility and imposed rotation equal respectively to the standard deviation and the mean value of the prior distribution associated to the parameter φ, as shown in Figure 4. Finally, we introduce the measurements as a system of linear springs, each with flexibility and pre-stretch equal respectively to the standard deviation and value associated to a single measurement. Each spring is placed at a distance from the torsion spring equal to the corresponding interval of time t_i.

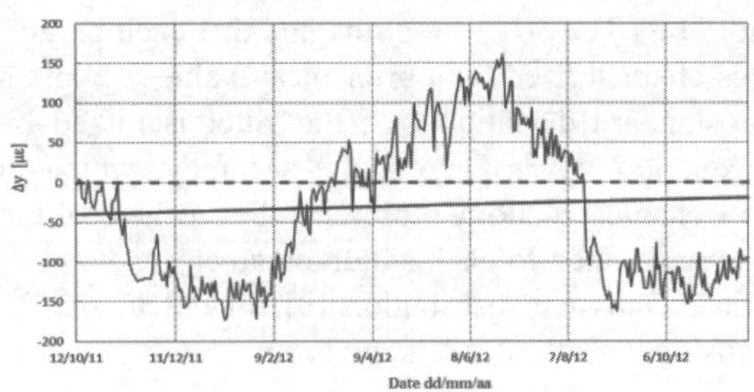

Figure 3. Relative strain of cable 1TN and interpolating lines.

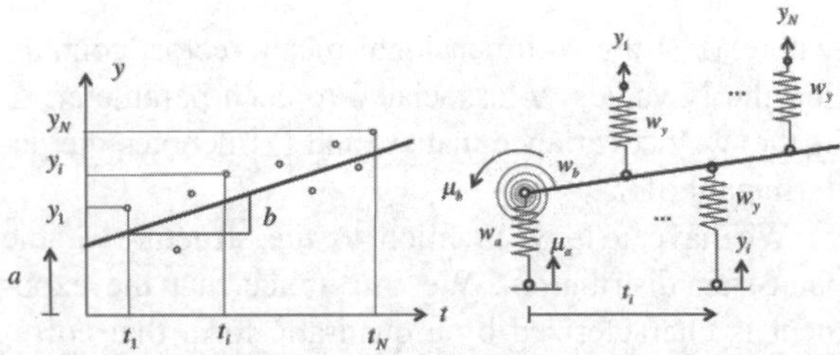

Figure 4. Representation of a linear regression problem with two parameters to estimate, in the world of Mechanics.

The elastic potential of the mechanical system of Figure 4 becomes:

$$\mathrm{E_p}(y_0,\varphi)=\frac{1}{2}w_{y_0}(y_0-\mu_{y_0})^2+\frac{1}{2}w_\varphi(\varphi-\mu_\varphi)^2+ +\frac{1}{2}w_y\sum_{i=1}^{N}[(y_0+\varphi t_i)-y_i]^2, \tag{10}$$

where $\Delta y_i = y_0+\varphi\cdot t_i - y_i$ represents the elongation suffered by the N springs linked to the observations, due to a generic translation y_0 and a generic rotation φ imposed on the system. The accuracy matrix is simply the Hessian matrix of (10):

$$\Lambda=\begin{bmatrix}\frac{\delta^2\mathrm{E_p}(y_0,\varphi)}{\delta y_0^2} & \frac{\delta^2\mathrm{E_p}(y_0,\varphi)}{\delta y_0\delta\varphi}\\ \frac{\delta^2\mathrm{E_p}(y_0,\varphi)}{\delta\varphi\delta y_0} & \frac{\delta^2\mathrm{E_p}(y_0,\varphi)}{\delta\varphi^2}\end{bmatrix}. \tag{11}$$

The inverse of the matrix (11) represents the covariance matrix Σ: the first term of its diagonal is the posterior variance associated to the parameter y_0 while the second term on the same diagonal is the posterior variance associated to the parameter φ.

To identify instead the values $\mu_{y_0|\mathrm{y}}$ and $\mu_{\varphi|\mathrm{y}}$ that represent the posterior mean values associated respectively to the parameters y_0 and φ, we must solve the system formed by the first derivative of (10) with respect to the parameter y_0 and the parameter φ, set equal to zero.

$$\begin{cases}\frac{\delta\mathrm{E_p}(\boldsymbol{\theta})}{\delta y_0}=w_{y_0}(y_0-\mu_{y_0})+w_y\sum_{i=1}^{N}[(y_0+\varphi t_i)-y_i]=0\\ \frac{\delta\mathrm{E_p}(\boldsymbol{\theta})}{\delta\varphi}=w_\varphi(\varphi-\mu_\varphi)+w_y\sum_{i=1}^{N}t_i[(y_0+\varphi t_i)-y_i]=0\end{cases} \tag{12}$$

The solutions of the system (12) give us the values of $\mu_{y_0|\mathrm{y}}$ and $\mu_{\varphi|\mathrm{y}}$ that represent the posterior mean values associated respectively to the parameters y_0 and φ and that minimize the potential $\mathrm{E_p}(y_0,\varphi)$ of our mechanical system. Now we can substitute the numerical values into the equations formulated above, and we obtain the final outcomes reported in Table 2, compared with the prior values of the parameters. Figure 3 reports the two straight lines interpolating our dataset.

4.2 *Three parameters to estimate*

We now extend the case of Adige Bridge, presented in the previous Section, by introducing the effect of temperature $\Delta\hat{T}$. Thus, we must estimate an additional parameter α and the model that fits our time dependent dataset becomes the following:

$$\Delta\hat{y}=y_0+\alpha\cdot\Delta\hat{T}+\varphi\cdot\Delta\hat{t}, \tag{13}$$

In Figure 6, we can note the N translation springs linked to the different measurements with stiffness $w_{LH}=\sigma_{LH}^{-2}=0.0016\,\mu\varepsilon^{-2}$ and the springs linked to the prior distribution: a translation spring associated to the parameter y_0, a rotational spring associated to α and a rotational spring associated to φ, whose their numerical values are the same as the case in Section 4.1. To determine the posterior standard deviation of the three parameters to estimate $(\mathrm{y}_0,\alpha,\varphi)$, we have to express the potential energy $\mathrm{E_p}(y_0,\alpha,\varphi)$ of the mechanical system represented in Figure 6, as a function of the three unknown parameter. We can now obtain the accuracy matrix $\boldsymbol{\Lambda}$ simply by calculating the Hessian Matrix associated to $\mathrm{E_p}(y_0,\alpha,\varphi)$, and the covariance matrix from the inverse of $\boldsymbol{\Lambda}$. To discover the values $\mu_{y_0|\mathrm{y}}$, $\mu_{\alpha|\mathrm{y}}$ and $\mu_{\varphi|\mathrm{y}}$, which represent the posterior mean values associated respectively to the parameters y_0, α and φ, we must solve the system formed by the first derivative of the potential energy with respect to the three parameters, set equal to zero.

Figure 7 shows the graphical representation, using the software Matlab, of the two surfaces fitting our data set. Finally Table 3 reports the numerical values obtained from the posterior distribution of the parameters.

Table 2. Prior and Posterior values of the parameters to estimate

Prior distributions			
Parameter y_0		Parameter φ	
$w_{y_0}\ [\mu\varepsilon^{-2}]$	0.0025	$w_\varphi[\mu\varepsilon^{-2}day^2]$	1
$\sigma_{y_0}[\mu\varepsilon]$	20.00	$\sigma_\varphi[\mu\varepsilon\ day^{-1}]$	1.0000
$\mu_{y_0}[\mu\varepsilon]$	0.00	$\mu_\varphi[\mu\varepsilon\ day^{-1}]$	0.0000
Posterior distributions			
Parameter y_0		Parameter φ	
$w_{y_0}\ [\mu\varepsilon^{-2}]$	0.6601	$w_\varphi[\mu\varepsilon^{-2}day^2]$	36893
$\sigma_{y_0}[\mu\varepsilon]$	2.44	$\sigma_\varphi[\mu\varepsilon\ day^{-1}]$	0.0103
$\mu_{y_0}[\mu\varepsilon]$	-49.07	$\mu_\varphi[\mu\varepsilon\ day^{-1}]$	0.0473

5 CONCLUSIONS

We have stated an analogy between the world of logic and the world of mechanics, allowing us to solve, using the methods of classical structural engineering, any complex inference parameter estimation problem, in which the values of the parameters have to be estimated based on multiple Gaussian-distributed uncertain observations. By simply expressing the potential energy of the mechanical system associated to our inference scheme, we are able, with a few trivial algebraic steps, to determine the posterior mean values and standard deviations of the parameters to estimate. With the aid of real-life structural health monitoring cases, we have showed how our approach allows structural engineers to solve simply general problems of linear regression. Although the examples shown in this paper are incidentally all structural engineering cases, the scope of application of the method is evidently the most general, and we seek to demonstrate in the future its applicability to inference problem arising from various disciplinary fields, including cognitive science, economics and law.

REFERENCES

Bishop, C.M. 2006. *Pattern Recognition and Machine Learning*, Cambridge: Cambridge University Press.

Cappello, C. Bolognani, D. & Zonta, D. 2015. Mechanical equivalent of logical inference from correlated uncertain information, Proc. 7th *International Conference on Structural Health Monitoring of Intelligent Infrastructure*, Torino.

Cappello, C. Zonta, D. Pozzi, M. Glisic, B. Zandonini, R. 2015. Impact of prior perception on bridge health diagnosis, *Journal of Civil Structural Health Monitoring* **5**, 509-525.

EN 1990:2002–Basis of structural design, Brussels: European Committee for Standardization, 2005.

Jaynes, E.T. 2003. *Probability Theory: The Logic of Science*, Cambridge: Cambridge University Press.

Kirkup, L. & Frenkel, R.B. 2006. *An Introduction to Uncertainty in Measurement*, Cambridge: Cambridge University Press.

Ku, H.H. 1966. Notes on the Use of Propagation of Error Formulas, *Journal of Research* **70(4)**.

Measures, R.M. 2001. *Structural Monitoring with fiber optic technology*. London: Academic Press.

Rabinovich, S.G. 2005. *Measurement Errors and Uncertainties: Theory and Practice*, New York: Springer.

Sohn, H. & Law, K-H. 2000. Bayesian Probabilistic Damage Detection of a Reinforced Concrete Bridge Column, *Earthquake Engineering and Structural Dynamics* **29(8)**.

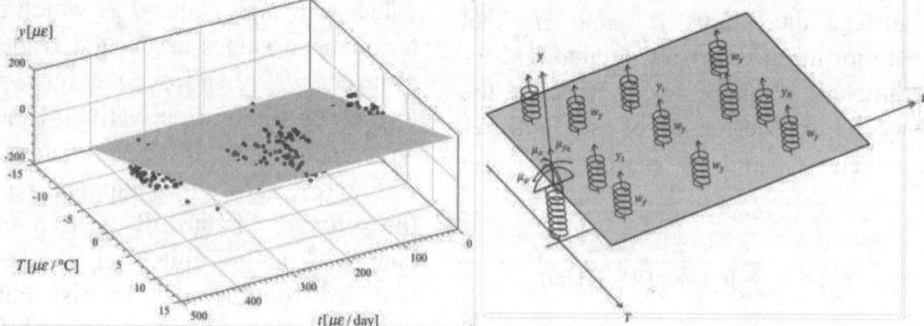

Figure 6. Representation of a linear regression problem with three parameters to estimate, in the world of Mechanics.

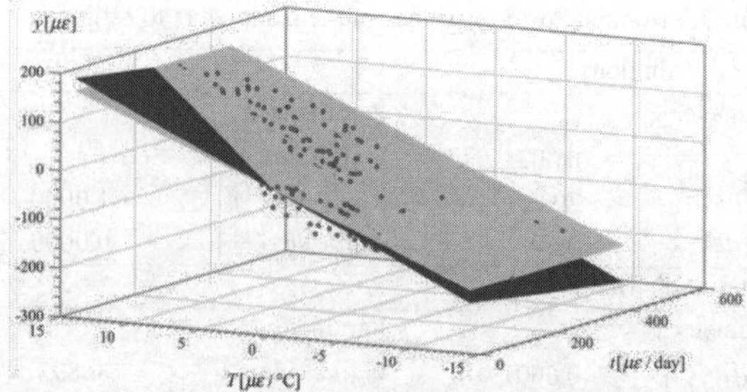

Figure 7. Representation of the two fitting surfaces associated to the prior parameters (gray) and to the posterior parameters (black).

Table 3. Posterior values of the three parameters to estimate

Posterior distributions			
Parameter y_0		Parameter φ	
w_{y_0} $[\mu\varepsilon^{-2}]$	0.6601	$w_{\varphi}[\mu\varepsilon^{-2} day^2]$	36893
$\sigma_{y_0}[\mu\varepsilon]$	2.54	$\sigma_{\varphi}[\mu\varepsilon\ day^{-1}]$	0.0106
$\mu_{y_0}[\mu\varepsilon]$	0.48	$\mu_{\varphi}[\mu\varepsilon\ day^{-1}]$	-0.1209
Parameter α			
w_{α} $[\mu\varepsilon^{-2}\ °C^2]$	27.88		
$\sigma_{\alpha}[\mu\varepsilon\ d\ °C^{-1}]$	0.20		
$\mu_{\alpha}[\mu\varepsilon\ d\ °C^{-1}]$	13.80		

Proceedings of the International Conference on Smart Infrastructure and Construction
ISBN 978-0-7277-6127-9

© The authors and ICE Publishing: All rights reserved, 2016
doi:10.1680/tfitsi.61279.399

Wind-induced variation of structural parameters of cable-stayed bridge

T.C. Huynh, S.H. Choi and J.T. Kim*

Pukyong National University, Busan, Republic of Korea
* *Corresponding Author*

ABSTRACT A structural identification of a cable-stayed bridge under strong wind conditions is performed using vibration responses measured by a wireless sensor system. Firstly, a cable-stayed bridge with the wireless monitoring system is described. Wireless vibration sensor nodes are utilized to measure accelerations from bridge deck and pylon. Secondly, dynamic responses of the cable-stayed bridge under the attack of two consecutive typhoons, Bolaven and Tembin, are analyzed under various wind speeds. Short-time Fourier transform analysis is selected to examine wind-induced variations of the bridge's responses based on the field measurements under the two typhoons. Finally, the structural identification of the bridge is performed under various wind velocities to examine the typhoon-induced variations of the bridge's structural properties. The variations of dynamic characteristics due to the typhoons are turned into the changes of the bridge's stiffness (or flexibility) via the fine-tuning process.

1 INTRODUCTION

Hwamyung Bridge, the longest cable-stayed bridge with the prestressed concrete box girder in Korea, has been monitored by adopting the smart sensing technologies. The long-term performance of the wireless monitoring system of the Hwamyung Bridge has been evaluated for the vibration measurement, the wireless communication, the capacity of solar-powered supply, and the survivability of sensors (Ho et al. 2012, Kim et al. 2014).

The aerodynamics (e.g., turbulence, vortex, buffeting, galloping) are often coupled with the dynamic responses of the cable-supported bridges. It has been reported that the natural frequencies of the cable-supported bridges decrease as the results of increasing the structural flexibility when the wind velocity increases (Fujino and Siringoringo 2013, Kim et al. 2014, Park et al. 2015). One important issue in the bridge aerodynamic is the non-stationary dynamic responses due to the vortex shedding during the typhoon. To extract the instantaneous changes in modal parameters induced by the typhoon, therefore, non-stationary random vibration responses should be dealt appropriately. This study has been motivated to examine the wind-induced variation of dynamic characteristics of Hwamyung Bridge by the short-term Fourier transform (STFT) and to analyze the corresponding change of the structural parameters (mainly focus on structural stiffness) of the bridge using the system identification.

2 VIBRATION MONITORING OF CABLE-STAYED BRIDGE UNDER TYPHOONS

Hwamyung Bridge, completed in 2011, is located on the Nakdong River to connect Gimhae-si with Hwamyeong-dong (Korea). The bridge has a total length of

1,414 m, a width of 17.9 to 27.8 m, and a design speed of 80 km/h. As a main structure of Hwamyung Bridge, the two-pylon cable-stayed bridge has three spans including a 270 m long mid-span and two 115 m long side-spans, and total 72 multi-strand type stay cables (see Figure 1).

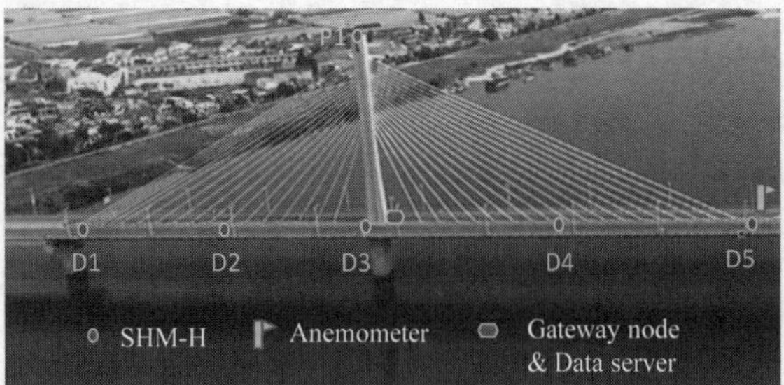

Figure 1. Hwamyung cable-stayed bridge with wireless monitoring system

For vibration monitoring of the bridge, seven sensor nodes (Imote2/SHM-H, Jo et al. 2010) including 6 leaf nodes and 1 gateway node were installed on the bridge (see Figure 1). Among the leaf nodes, 6 Imote2/SHM-H sensors were placed at five locations of the deck (i.e., D1~D5) and at the top of the west pylon (i.e., P1). The base station including a gateway node and a PC was installed at the nearest pylon P1. For the for on-site wind velocity and wind direction measurements, an ultrasonic wind sensor (Model 85000/RM Young Co.) was installed at the middle of the bridge. The vibration signals were hourly measured in a duration of 10 minutes with sampling frequency of 25 Hz.

By the end of August 2012, two consecutive typhoons named Bolaven and Tembin passed through the Korean peninsula and affected the site of the Hwamyung Bridge. During the attack of the typhoons, the maximum wind speeds were recorded over 20 m/s on-site. Among the 6 deck/pylon sensor nodes except the gateway node, unfortunately, only a few sensor nodes including two deck nodes (D2 and D5) had made successful monitoring and wireless communication performances during the typhoon attacks. It was observed that the decks D2 and D5 experienced the maximum acceleration response about 0.008g and 0.013g, respectively when the wind speed was about 20 m/s. Figure 2 shows the relationship between the deck acceleration in term of root mean square (RMS) and the mean wind velocity during the two typhoons. The acceleration response were obtained for five different wind speeds: 3.72 m/s, 5.05 m/s, 11.70 m/s, 17.92 m/s, and 19.36 m/s. A similar general trend in which the deck acceleration increases as the wind speed increases can be observed for the decks D2 and D5. As compared to the deck D2, the deck D5 experienced much higher acceleration during the typhoons. It is understandable since the deck node D5 was installed at the center of mid-span which could exhibit significant flexibility during strong wind conditions.

Figure 3 shows the stochastic subspace identification (SSI)'s stabilization chart and the typical power spectral density (PSD) response of the deck. It can be seen that the PSD resonant peaks were consistently identical to the stable modes indicated by the SSI method. From the stabilization chart, natural frequencies of four vertical modes were extracted as V1: 0.419 Hz, V2: 0.692 Hz, V4: 1.048 Hz, and V6: 1.401 Hz. Note that the order of modes (including the noise modes) follow the previous results on the bridge (before the pavement on the road way) reported by Ho et al. (2012a). After completing the asphalt pavement in Feb 2012, the natural frequency of the bridge decreased due to the increment of deck's mass (Nguyen et al. 2013), as observed in Figure 4.

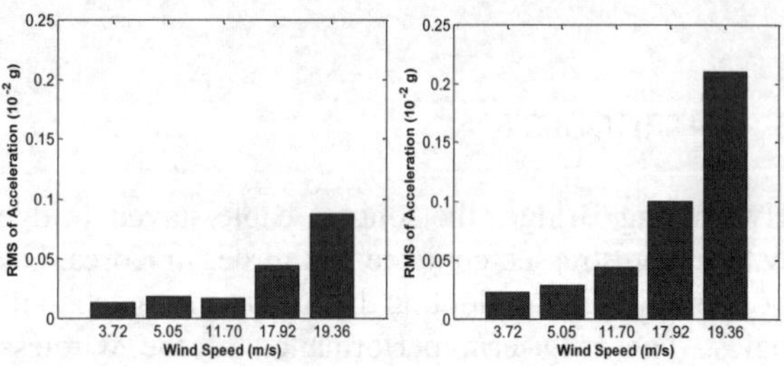

Figure 2. Relationship between mean wind velocity and RMS of deck acceleration during typhoons

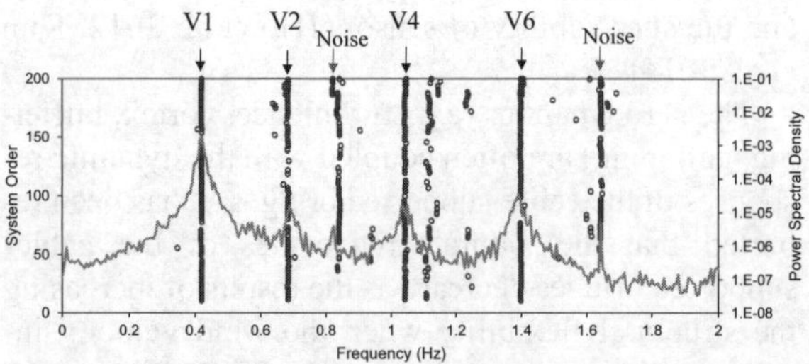

Figure 3. SSI's stability chart and PSD response for modal identification of the deck

Figure 4 shows identified mode shapes of the four vertical modes (i.e., Modes V1, V2, V4 and V6) of the deck measured at wind speed 3.72 m/s. Modal amplitudes of the four modes were extracted from only two sensor nodes (D2 and D5), with the reference of the published mode shapes extracted from all five locations (e.g., D1 ~ D5) by Ho et al. (2012a). That was because other sensor nodes were power-off due to the typhoon attack. It is observed that the normalized modal amplitudes show consistent results as compared to the mode shapes by Ho et al. (2012).

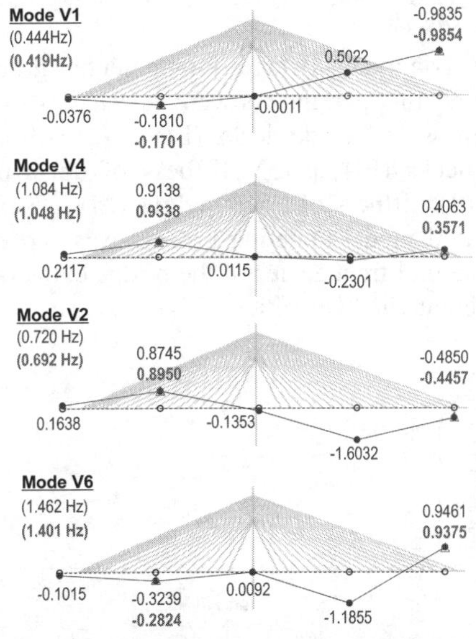

Figure 4. Measured mode shapes of the deck at wind speed 3.72 m/s. Blue color indicates results by Ho et al. (2012)

3 VARIATION OF STRUCTURAL PROPERTIES UNDER TYPHOONS

3.1 Wind-induced variation of modal parameters

As stated previously, the non-stationary random vibration responses should be examined by time-varying analysis. Due to the simplicity by using fast Fourier transform (FFT), the STFT (Quatieri 2001) was selected for the time-varying analysis in this study. The STFT analyses were performed for the deck vibration signals (D2 and D5) under three wind speeds: 3.72 m/s, 11.70 m/s, and 19.36 m/s. Figure 5 shows the time-frequency STFT of the deck D5 acceleration signals at wind speed 19.6 m/s. For the four vertical modes (i.e., Modes V1, V2, V4 and V6), natural frequencies are plotted as functions of time. It is found that the STFT analyses produce natural frequencies fluctuating along with the short period. It is observed from the table that the natural frequencies decreased as the wind velocity increased (see Table 1).

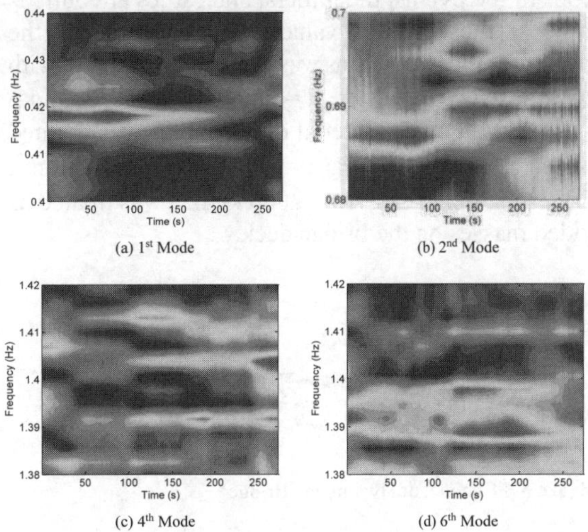

Figure 5. STFT analysis of the deck D5 vibration responses under wind velocity 19.36 m/s

Table 1. Extracted natural frequencies (Hz) for three wind speeds

Wind speed (m/s)	Mode V1	Mode V2	Mode V3	Mode V6
3.72	0.4242	0.6958	1.0559	1.4099
11.7	0.4181	0.6805	1.0437	1.3824
19.36	0.4181	0.6836	1.0315	1.3794

3.2 Wind-induced variation of deck's stiffness

The structural identification of the Hwamyung Bridge is performed for various wind velocities under the two consecutive typhoons, Tembin and Bolaven. First, a FE model of the bridge is established. Next, the modal strain energy (MSE)-based method (Stubbs and Kim 1996, Kim et al. 2013, Huynh et al. 2013) is employed to fine-tune the FE model of the as-built structure with respect to typhoon-induced variation of dynamic characteristics.

3.2.1 Finite Element Modeling of Bridge

As an analytical model of the bridge, a finite element (FE) model was established using SAP2000 (see Figure 6). The two pylons and the girders were modeled by frame elements while the stay cables were modeled by cable elements. In the FE modeling, the dimensions of the pylons, the girders, and cables are built as same as the designed values. The material of the girder and the two pylons was defined as concrete with the following properties: E = 28.6 GPa, $\upsilon = 0.2$, and $\rho = 2500$ kg/m^3. The material of the cables was defined as steel with E = 195 GPa, $\upsilon = 0.3$, and $\rho = 7850$ kg/m^3. The asphalt pavement ($\rho = 2000$ kg/m^3) was treated as added masses on the bridge deck.

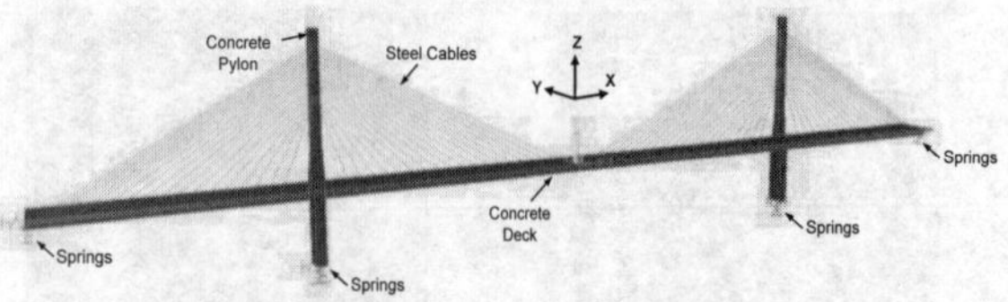

Figure 6. FE model of Hwamyung Bridge

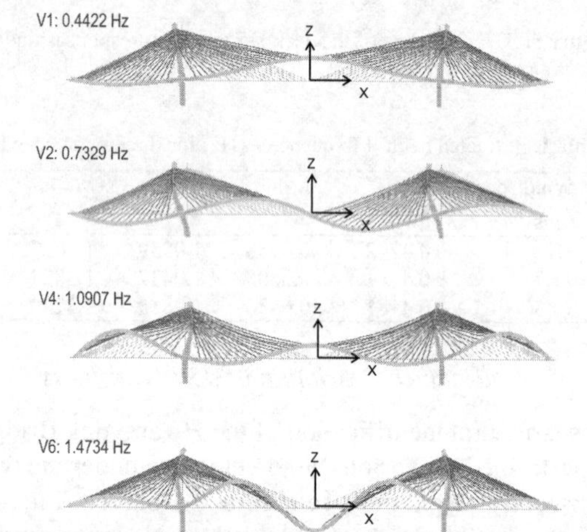

Figure 7. Mode shapes and natural frequencies (Hz) of initial FE model

The supports of the two pylons and the two girder ends were modeled by three directional springs (see Figure 6). Since the stiffness of the supporting springs depend on the foundation conditions which is beyond the scope of this study, they were assumed as $k_{d\text{-}x}$ = 5E+8 N/m, $k_{d\text{-}y}$ = 5E+8 N/m, $k_{d\text{-}z}$ = 5E+8 N/m for the deck supports, and $k_{p\text{-}x}$ = 10E+8 N/m, $k_{p\text{-}y}$ = 10E+8 N/m, $k_{p\text{-}z}$ = 10E+8 N/m for the supports of the pylons and the girders. To complete the FE model, the tension force of each stay cable in the FE model was assigned using the tension force measured from lift-off test (Kim et al. 2013). Figure 7 shows the four numerical modal parameters generated by the initial FE model of the bridge. It is found that these four calculated modal parameters were respectively identical to the four experimental vertical modes (i.e., Modes V1, V2, V4, and V6) of the bridge (see Figure 4).

3.2.2 Structural identification for various wind speeds

For the Hwamyung Bridge, four potential groups of model-updating parameters were selected as follows: (1) stiffness of outside-decks $(EI)_{o\text{-}d}$; (2) stiffness of middle-decks $(EI)_{m\text{-}d}$; (3) stiffness of upper-pylons $(EI)_{u\text{-}p}$; (4) stiffness of lower-pylons $(EI)_{l\text{-}p}$. Next, the MSE-based model updating process was carried out for the natural frequencies of the bridge extracted under different wind velocities.

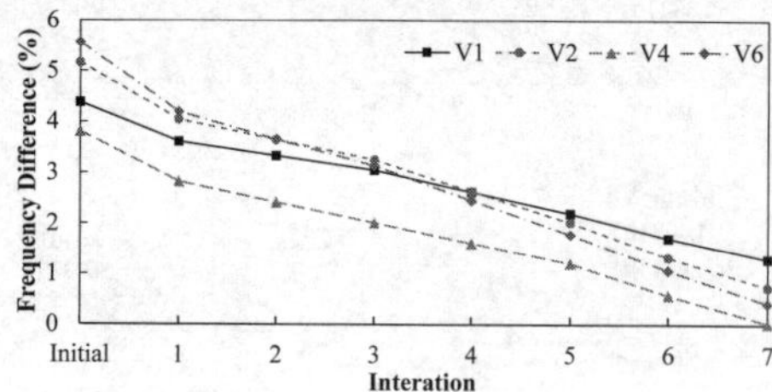

Figure 8. Convergences of four natural frequencies between FE model and experiment (wind speed 3.72 m/s)

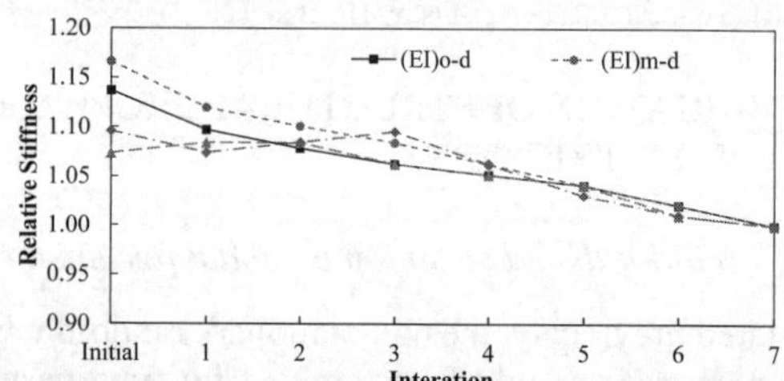

Figure 9. Relative change in four model-updating parameters of FE model (wind speed 3.72 m/s)

Figure 8 shows the typical errors of the natural frequencies of the updated model, with compared to the experimental natural frequencies (wind speed 3.72 m/s). It is found that the natural frequencies of Modes

V2, V4 and V6 matched very well with the experimental frequencies. The frequency difference converged to 0.03-1.3 (%) after the final iteration. The maximum difference between the numerical and the experimental values reduced from 5.56% to 1.3% after updating process.

The relative changes in the model-updating parameters of the FE model during the updating process are shown in Figure 9. Compared to the initial parameters of the initial FE model, the stiffness of the decks was decreased by 14.6~16% while the stiffness of the upper-pylons and the stiffness of the lower-pylons were reduced by 6.8~8.7%. This is reasonable since the stiffness of the decks is more sensitive to the change in natural frequencies than those of the pylons as described previously.

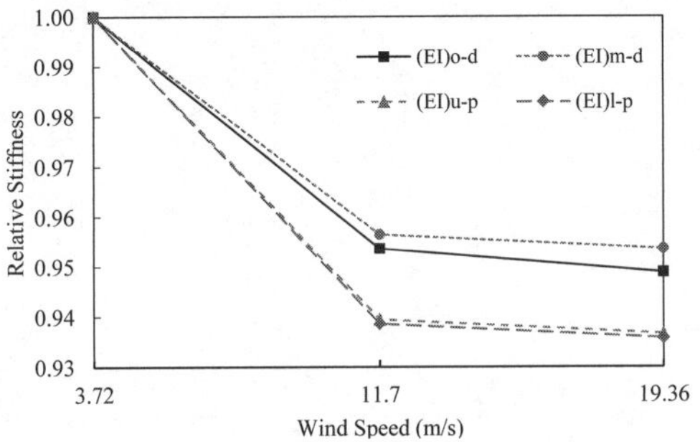

Figure 10. Updated structural parameters versus wind speeds for SSI and STFT analyses

Updated modal-updating parameters of the bridge under three wind speeds are shown in Figure 10. It is proved that the stiffness of the decks and pylons decreased when the wind speed increased. In other word, the increment of the wind speed led to the increment of the flexibility of the bridge resulting to the decrement of the natural frequencies and the relaxation of cables, which were recently reported by Fujino and Siringoringo (2013), and Kim et al. (2014). It also observed that the stiffness of the decks and pylons was rapidly reduced as the wind speed increased from 3.72 m/s to 11.7 m/s, but slightly decreased when the wind speed increased from 11.7 m/s to 19.36 m/s.

5. CONCLUSION

In this paper, a structural identification of a cable-stayed bridge under back-to-back typhoons was performed using vibration responses measured by a wireless sensor system. Dynamic responses of the cable-stayed bridge under the attack of two consecutive typhoons, Bolaven and Tembin, were analyzed under various wind speeds. Short-time Fourier transform analysis were selected to examine wind-induced variations of the bridge's responses based on the field measurements under the two typhoons. From the system identification results of the cable-stayed bridge under various wind speed, it was proved that the stiffness of the decks and pylons decreased when the wind speed increased. In other word, the increment of the wind speed led to the increment of the flexibility of the bridge resulting to the decrement of the natural frequencies.

ACKNOWLEDGEMENT

This research was supported by a research grant (Code 12 Technology Innovation E09) from Construction Technology Innovation Program funded by Ministry of Land, Transportation and Maritime Affairs (MLTM) of Korean government. The graduate students involved in the research were also supported by the Brain Korea 21 Plus program of Korean Government.

REFERENCES

Fujino, Y. and Siringoringo, D.M. 2013. Lessons learned from structural monitoring of long-span bridges and a tall base-isolated building, *Proceedings of SHMII-6*, Hongkong.

Ho, D.D., Lee, P.Y., Nguyen, K. D. et al. 2012. Solar-powered multi-scale sensor node on Imote2 platform for hybrid SHM in cable-stayed bridge, *Smart Struc. Syst.*, **9**(2), 145-164.

Huynh, T.C., Lee, S.Y., Kim, J.T. et al. 2013. Simplified planar model for damage estimation of interlocked caisson system, *Smart Struct. Syst.*, **12**(3-4), 441-463.

Jo, H., Rice, J.A., Spencer, B.F. et al. 2010. Development of a high-sensitivity accelerometer board for structural health monitoring, *Proceedings of the SPIE*, San Diego, March.

Kim, J.T., Ho, D.D., Nguyen, K.D. et al. 2013. System identification of a cable-stayed bridge using vibration responses measured by a wireless sensor network, *Smart Struct. Syst.*, **11**(5), 533-573.

Kim, J.T., Huynh, T.C., and Lee, S.Y. (2014), "Wireless structural health monitoring of stay cables under two consecutive typhoons", *Struct. Monit. Maint.*, **1**(1), pp. 47-67.
Park, J.H, Huynh, T.C., and Kim, J.T. 2015. Wireless monitoring of typhoon-induced variation of dynamic characteristics of a cable-stayed bridge, *Wind and Structures*, **20**(2), 293-314.
Quatieri, T.F. 2001. Discrete-time speech signal processing: principles and practice, Prentice Hall Press, Upper Saddle River, NJ.
Stubbs, N., and Kim, J.T. 1996. Damage localization in structures without baseline modal parameters, *AIAA Journal*, **34**(8), 1644-1649.

Proceedings of the International Conference on Smart Infrastructure and Construction
ISBN 978-0-7277-6127-9

© The authors and ICE Publishing: All rights reserved, 2016
doi:10.1680/tfitsi.61279.405

Finite element model updating based on a Kriging surrogate model for bridge performance evaluation

H.Y. Jung, J.H. Lee, S.S. Jin and H.J. Jung*
Dept. of Civil and Environmental Engineering, KAIST, Daejeon, Korea
* *Corresponding Author*

ABSTRACT In order to evaluate the performance of a bridge structure, finite element (FE) model updating methods have been frequently used in these days. In conventional FE model updating approaches, a precise initial FE model is used for obtaining accurate results. However, a precise FE model causes the considerable computational burden such as excessive calculation time. This paper presents a new FE model updating strategy based on a surrogate model, which is able to significantly reduce the calculation time without sacrificing the accuracy. In this study, a Kriging model (or response surface model) is selected as a surrogate model. The sequential sampling strategy is introduced for more efficient construction of an accurate Kriging surrogate model. The performance and efficiency of the proposed FE model updating approach are verified with the field-test data obtained from an in-service bridge structure in Korea. It is demonstrated that the proposed FE model updating method has better performance in the calculation time than the conventional one.

1 INTRODUCTION

As for civil infrastructure such as bridges and buildings, it is necessary to continuously evaluate and reinforce the structural performance for maintaining their serviceability and prevent structural failure. To this end, a finite element (FE) model of a target structure is needed, but the current design and assessment procedures using an FE model do not have any quantitative linkage to actual existing structures (Catbas et al. 2013). It is, thus, necessary to link an FE model with the corresponding existing structure. This process is called FE model updating, which is based on an inverse problem of identifying structural parameters by updating an initial FE model based on experimental data.

During the process of FE model updating, the precision of an initial model plays a crucial role in securing the accuracy and reliability of an updated FE model (i.e., the minimization of the systematic error in the FE model). Definitely a high-fidelity (or sophisticated) FE model is able to give a more accurate and reliable result, while simple modeling approaches may result in a considerable systematic error mainly due to modeling simplifications, the omission of structural components, and FE discretization errors. However, the high-fidelity model consumes excessive calculation time, resulting in the considerable computational burden to users.

In order to alleviate the computational burden of a high-fidelity FE model as an initial model in a FE model updating process, a surrogate model has recently received considerable attention as one of the promising alternatives. Surrogate modeling is a computational modeling technique which emulates a simulation model in the form of a mathematical and/or statistical approximation using the input and output of an FE analysis. A surrogate model is also called a response surface model. The model updating process can be replaced with the response surface (Jones 2001). It is expected that the updating parameters for an FE model can be obtained in a relatively short calculation time with a surrogate model without sacrificing the accuracy.

In this paper, a new FE model updating strategy based on a Kriging surrogate model is proposed for bridge performance evaluation, and the sequential sampling strategy is introduced in the process for more efficient construction of an accurate Kriging surrogate model. The performance and efficiency of the proposed FE model updating approach are verified with the field-test data obtained from an in-service bridge structure in Korea.

2 THEORY

At this section, the brief description of the Kriging surrogate model and the sequential sampling strategy is made. A more detailed description can be found in Jin and Jung (2015a, 2015b).

The Kriging model is one of the surrogate models originated from Geostatistics (Krige 1994). It is a way of modeling a function as a realization of the Gaussian process, i.e., a stochastic process with a mean μ and a variance σ^2. It is based on a spatial correlation among the values of the function. The Kriging basis with k dimensions (i.e., the correlation function) can be expressed as

$$\psi^{ij} = exp\left(-\sum_{p=1}^{k} \theta_p \left\|x_p^i - x_p^j\right\|\right) = \text{corr}\left[y(x^i), y(x^j)\right] \quad (1)$$

where the subscript "p" denotes the dimension of a sample x, the superscripts "i" and "j" indicate the i-th and j-th samples, respectively, and $\left\|x_p^i - x_p^j\right\|$ is the Euclidean distance between the two samples in a parameter space with 2-norm. The Kriging basis contains the parameters corresponding to each dimension (i.e., $\boldsymbol{\theta}_p = [\theta_1, \theta_2, \ldots, \theta_p]$) which determine how fast the correlation decays in each dimension (i.e., input).

The correlation matrix of all the samples can be constructed as

$$\boldsymbol{\Psi} = corr[\boldsymbol{Y}, \boldsymbol{Y}] = \begin{bmatrix} corr[y(x^1), y(x^1)] & \cdots & corr[y(x^1), y(x^n)] \\ \vdots & \ddots & \vdots \\ corr[y(x^n), y(x^1)] & \cdots & corr[y(x^n), y(x^n)] \end{bmatrix} \quad (2)$$

The covariance matrix can be derived from the correlation matrix (ψ) as

$$COV(\boldsymbol{Y}, \boldsymbol{Y}) = \boldsymbol{\sigma}^2 corr(\boldsymbol{Y}, \boldsymbol{Y}) = \boldsymbol{\sigma}^2 \boldsymbol{\Psi} \quad (3)$$

The variance σ^2 determines the overall dispersion relative to the mean of the Kriging model.

In order to construct the Gaussian process, it is necessary to estimate the parameters such as μ and σ^2. The parameters can be estimated by maximum likelihood estimation (MLE), and its log-likelihood function is given as

$$ln(L) \approx -\frac{n}{2} ln(\sigma^2) - \frac{1}{2} ln|\Psi| - \frac{(\boldsymbol{Y}-\boldsymbol{1}\mu)^T \Psi^{-1} (\boldsymbol{Y}-\boldsymbol{1}\mu)}{2\sigma^2} \quad (4)$$

where $\boldsymbol{1}$ is an n-by-1 unit vector and $\boldsymbol{Y}$ is the true function value of the observed samples.

By taking the derivatives of Eq. (4) with respect to μ and σ^2, repsectively, and setting these to zeros, MLEs of μ and σ^2 are expressed as

$$\hat{\mu} = \frac{\boldsymbol{1}^T \psi^{-1} \boldsymbol{Y}}{\boldsymbol{1}^T \psi^{-1} \boldsymbol{1}} \quad (5)$$

$$\hat{\sigma}^2 = \frac{(\boldsymbol{Y}-\boldsymbol{1}\hat{\mu})^T \psi^{-1} (\boldsymbol{Y}-\boldsymbol{1}\hat{\mu})}{n} \quad (6)$$

MLEs for μ and σ^2 are sequentially computed from the correlation matrix (ψ), so the only remaining parameter to be determined is θ_p in Eq. (1). In order to find the optimal θ_p, thus, the optimization method is applied by maximizing Eq. (4) under the observed samples.

After some manipulation, the augmented log-liklihood can be obtained as

$$ln(L) = \left(\frac{-1}{\hat{\sigma}^2(1 - \hat{\boldsymbol{\psi}}^T \boldsymbol{\Psi}^{-1} \hat{\boldsymbol{\psi}})}\right)(\hat{y} - \hat{\mu})^2 + \left(\frac{\hat{\boldsymbol{\psi}}^T \boldsymbol{\Psi}^{-1} (\boldsymbol{Y} - \boldsymbol{1}\hat{\mu})}{\hat{\sigma}^2(1 - \hat{\boldsymbol{\psi}}^T \boldsymbol{\Psi}^{-1} \hat{\boldsymbol{\psi}})}\right)(\hat{y} - \hat{\mu}) \quad (7)$$

where $\hat{\mu}$ and σ^2 are the MLEs from Eqs. (5) and (6), $\widehat{\boldsymbol{\psi}}$ is a correlation vector between the new (x_{new}) and observed samples (X), i.e., $\widehat{\boldsymbol{\psi}} = corr[y(x_{new}), \boldsymbol{Y}]$. The prediction ($\hat{y}$) is obtained by maximizing the log-likelihood function in Eq. (7). Taking the derivative of Eq. (7) with respect to $\hat{y}$ and setting this to zero, the prediction ($\hat{y}$) is obtained as

$$\hat{y}(x_{new}) = \widehat{\boldsymbol{\mu}} + \widehat{\boldsymbol{\psi}}^T \boldsymbol{\Psi}^{-1}(\boldsymbol{Y} - \mathbf{1}\widehat{\boldsymbol{\mu}}) \quad (8)$$

By evaluating the prediction values, the Kriging surrogate model can be constructed in the linear combination form.

The sequential sampling strategy is an adaptive sampling approach and the sampling continues until the target accuracy of the Kriging model is achieved. Figure 1 describes the basic concept of the sequential sampling strategy. In the strategy, a new sample is added to improve hte accuracy of the suroogate model and reduce the variance. The new sample has high likelihood of improvement based on statistical interpretation of prediction. In this study, the sequential sampling strategy based on expected improvement (i.e., EI(x)) is introduced and the expected improvement approach is the criteria to evaluate how much improvement of current best value is expected if a new sample is obtained (Jones 2001). Figure 2 represents the flowchart of the proposed Kriging surrogate model with the sequential sampling strategy. As shown in the figure, there are two stopping criteria to determine whether the sequential surrogate modeling stops or not as follows:

$$R_i^2 \geq R_{Th}^2 \quad (9)$$

$$\Delta RMSE_i \leq \Delta RMSE_{Th} \quad (10)$$

where R^2 is the R-squared value representing a measure of how well the observed outputs are reproduced by the prediction model and $RMSE$ is the root-mean-square error meaning the squared difference between the observed and predicted outputs in teh validation sample, and the subscript "i" indicates the i-th state of an infill sample and "Th" indicates the treshold value.

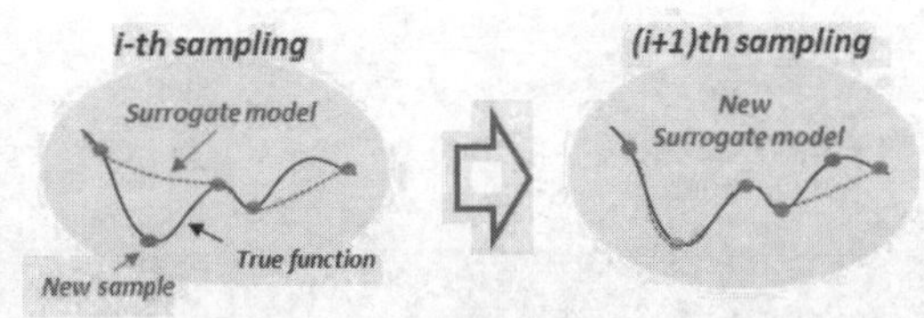

Figure 1. Construction of the surrogate model with the sequential sampling strategy

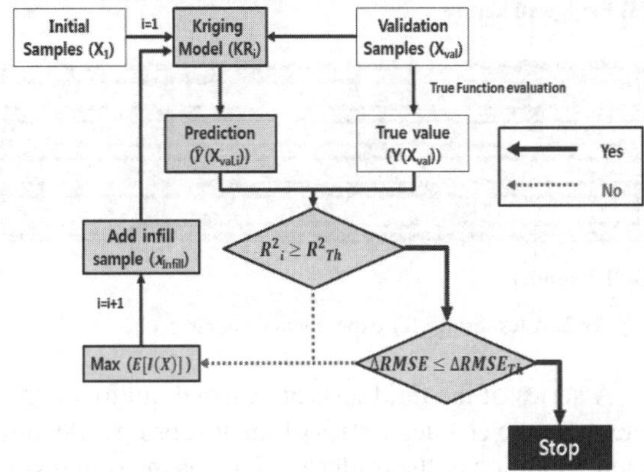

Figure 2. Flowchart of the Kriging surrogate model with the sequential sampling method.

3 EXPERIMENTAL VALIDATION

In order to validate a model updating efficiency of the proposed approach consisting of the Kriging surrogate model and the sequential sampling strategy, an experimental validation is carried out. A series of the experiments are performed with a full-scale bridge structure, which is a testbed bridge on the highway in Korea. It has a simple span, five main girders, five cross girders and four diaphragms. First, an FE model is constructed with 768 frame elements and 1728 shell elements by using ANSYS APDL. The boundary conditions of the FE model are introduced to both ends as a hinge and a roller support, respectively. Figure 3 shows a full-scale in-service bridge structure and the plan view of its FE model.

(a) Bridge structure

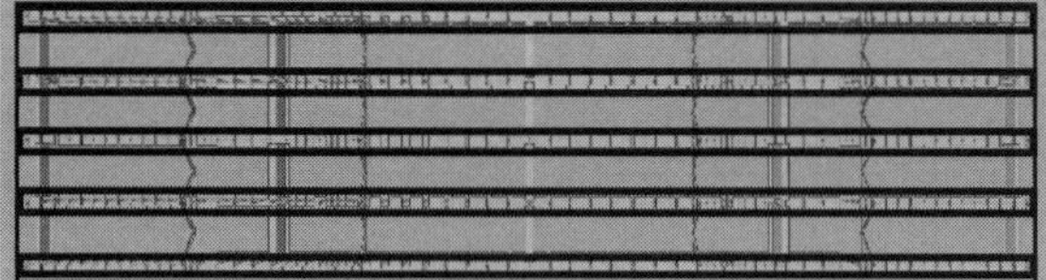

(b) FE model

Figure 3. A test bridge for experimental validation.

A series of the field tests are carried out to identify the dynamic characteristics of the test bridge. An ambient vibration of the bridge structure is measured with 15 accelerometers during 2 hours. The sampling frequency of 100 Hz and the low pass filter are used. The stochastic subspace identification (SSI) method is employed to identify the natural frequencies and their corresponding mode shapes. Figure 4 shows the sensor layout plan, and figure 5 shows the graphical description of the lowest six natural frequencies and the corresponding modes shapes extracted from the test results. As shown in the figure, the first mode (i.e., 3.909 Hz) and the third mode (i.e., 10.759 Hz) are vertical bending modes, while the second mode (i.e., 4.727 Hz) and the fourth mode (i.e., 13.494 Hz) are torsional bending modes. The remaining fifth and sixth modes (i.e., 14.950 Hz and 19.173 Hz) are lateral bending modes.

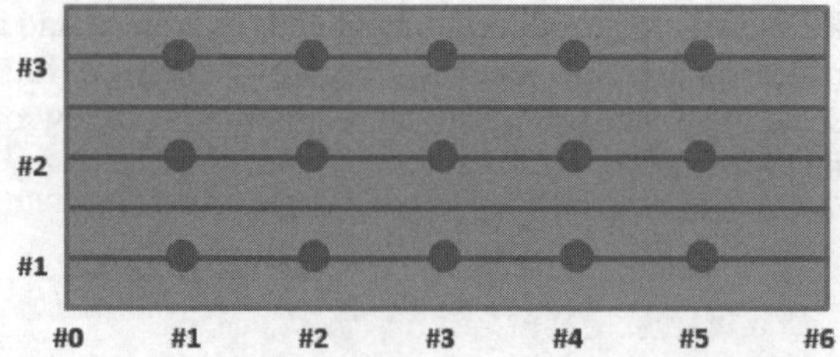

Figure 4. Sensor layout plan.

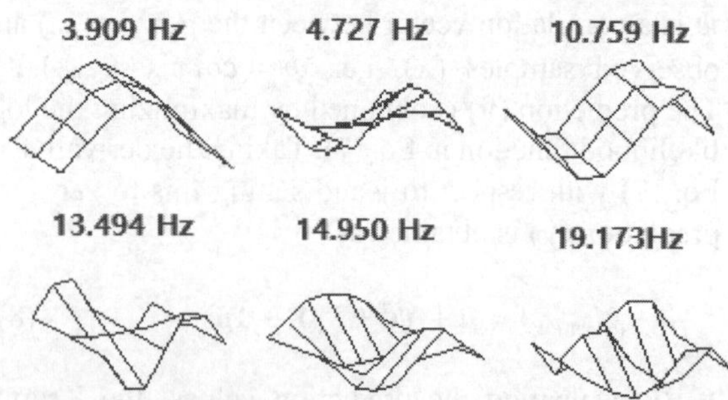

Figure 5. The lowest six natural frequencies and the corresponding mode shapes extracted from the test results.

The Young's moduli of the five main girders and three groups of cross girders, which mainly affect the behavior of the structure are selected as the updating parameters for a Kriging surrogate model. Entirely, the lowest six natural frequencies are selected as target outputs. The 1st and 2nd modes of vertical bending, torsional bending and lateral bending are all used.

In order to construct an initial Kriging surrogate model, 30 initial samples are generated by Latin hypercube sampling (LHS) first. Moreover, 500 validation samples are additionally generated to evaluate the accuracy of the Kriging surrogate model. By using the sequential sampling strategy, the Kriging surrogate model is updated. The threshold values of R-square value (i.e., R^2_{Th}) and root mean square error (i.e., $\Delta RMSE_{Th}$) are 0.99 and 0.005, respectively, for the stopping criteria.

In order to validate the performance of the Kriging surrogate model, the additional 300 samples are randomly selected. For the target outputs of the 300 additional samples, the prediction values from the Kriging surrogate model are compared with those from the FE model analyses. The R^2 and RMSE values are shown in figure 6. As shown in figure 6, the R^2 values in all the cases are larger than 0.99, and the RMSE values are less than 0.005. It demonstrates that the constructed Kriging surrogated model accurately reproduces the target modal properties.

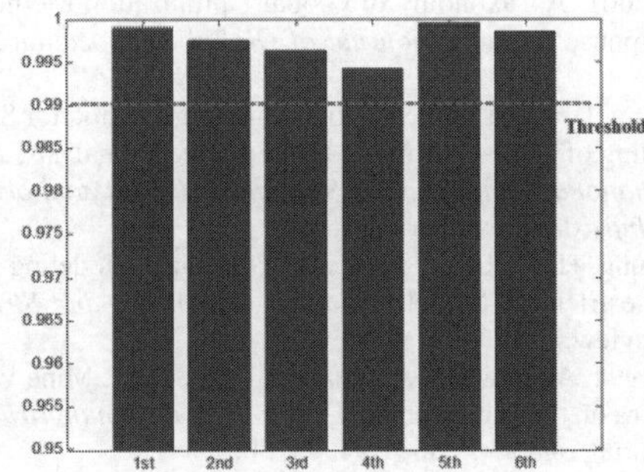

(a) R^2 values of the additional 300 samples

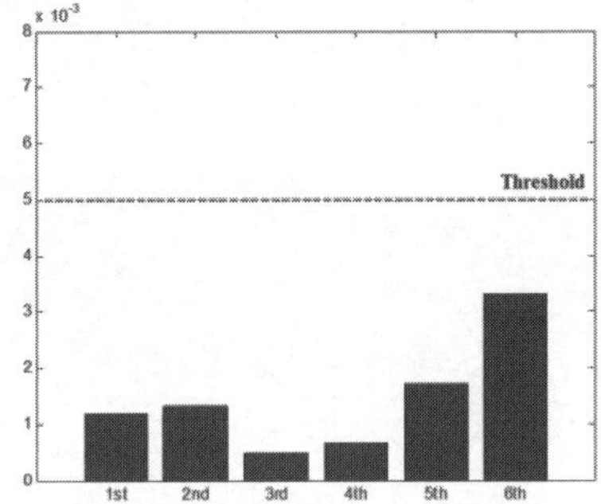

(b) RMSE of the additional 300 samples

Figure 6. R^2 and RMSE of the additional 300 samples

In this study, the genetic algorithm (GA) and the Nelder-Mead simplex algorithm are used for FE model updating. The GA can be used to assess the accuracy of the Kriging surrogate model indirectly in the model updating process. In addition, the Nelder-Mead simplex algorithm is used to improve the solution from the GA. ANSYS APDL and MATLAB are linked to perform modal analysis iteratively with optimization tools of MATLAB.

Table 1 shows the FE model updating results using the Kriging surrogate model. In table 1, the updated values of the target outputs are compared with the experimental test results and initial FE model values. As shown in the table, all the updated values are closed to the experimental results. It means that the model updating accuracy of the Kriging surrogate model is improved significantly. However, the error level of the Kriging model, from f_2 to f_4 is relative high. This relatively high error level is associated with the updating parameters. For constructing the Kriging surrogate model, three groups of cross girders are assigned to the updating parameter. In order to obtain the accurate updating results, updating parameters of the cross girder have to be considered individually. Also, the boundary conditions of the FE model have to be included in the updating parameters to improve the analysis results.

Table 1. FE model updating results

Target output	Experimental Result (Hz)	Initial FE model		Kriging model	
		Value(Hz)	Error(%)	Value(Hz)	Error(%)
f_1	3.909	3.109	20.5	3.992	2.1
f_2	4.727	3.457	26.9	4.476	5.3
f_3	10.759	9.025	16.1	11.309	5.1
f_4	13.494	10.262	23.9	12.516	7.2
f_5	14.950	12.220	18.4	14.958	0.05
f_6	19.173	15.226	20.6	18.528	3.3

The total computation time for the Kriging surrogate model-based FE model updating process is about 14 hours. However, the conventional FE model updating method consumes about 330 hours. It is, thus, clearly show that the proposed Kriging surrogate model-based FE model updating technique could substitute the conventional model updating technique to effectively address the computation cost issue.

4. CONCLUSION

In this study, a new FE model updating method consisting of the Kriging surrogate model and the sequential sampling strategy is presented. A brief description of the theoretical background on the Kriging surrogate model and the sequential sampling strategy is first made. Then, the experimental validation of the proposed method is performed by using a field-test data obtained from a full-scale in-service bridge structure in Korea. It is demonstrated that the proposed FE model updating method based on the Kriging surrogate model is able to accurately reproduce the target output (i.e., modal properties). In addition, the proposed Kriging surrogate-based FE model updating method shows a drastic reduction of calculation time for FE model updating compared with the conventional approach (i.e., 14 hours vs. 330 hours). It is, therefore, concluded that the proposed FE model updating approach could be a substitute for the time-consuming conventional FE model updating approach.

ACKNOWLEDGEMENT

This research was supported by a grant (13SCIPA01) from Smart Civil Infrastructure Research Program funded by Ministry of Land, Infrastructure and Transport (MOLIT) of Korea government and Korea Agency for Infrastructure Technology Advancement (KAIA).

REFERENCES

Catbas, F. N., Kijewski-Correa, T. L. & Aktan, A. E., eds. 2013. Structural Identification (St-Id) of Constructed Systems: Approaches, Methods and Technologies for Effective Practice of St-Id, ASCE, Structural Engineering Institute, Reston, VA.

Jones, D. R. 2001. A Taxonomy of Global Optimization Methods Based on Response Surfaces, *Journal of Global Optimization* **21**, 345-383.

Jin, S. S. & Jung, H. J. 2015a. Sequential Sampling Method for Surrogate Modelling of Time-consuming Finite Element Analysis, *The Seventh International Conference on Structural Health Monitoring of Intelligent Infrastructure*, Torino, Italy.

Jin, S. S. & Jung, H. J. 2015b. Sequential Surrogate Modeling for Efficient Finite Element Model Updating. *Computers and Structures*. under review.

Krige, D. G. 1994. A Statistical Approach to Some Basic Mine Valuation Problems on the Witwatersrand. *Journal of the South African Institute of Mining and Metallurgy* **94**, 95-111.

Proceedings of the International Conference on Smart Infrastructure and Construction
ISBN 978-0-7277-6127-9

© The authors and ICE Publishing: All rights reserved, 2016
doi:10.1680/tfitsi.61279.411

Smart structural health monitoring system for damage identification in bridges using relative wavelet entropy

M. Moravvej, M. El-Badry* and P. Joulani

Department of Civil Engineering, University of Calgary, Calgary, Canada
* *Corresponding Author*

ABSTRACT A smart structural health monitoring system for damage identification in bridges is proposed. The system is based on the fact that structural damage induces abrupt changes and singularities in measured acceleration signals of bridges and that discrete wavelet transform is a powerful tool to decompose the measured signals in order to detect these singularities. In addition, relative wavelet entropy is used to quantify the degree of damage-induced disorder in the signals. The main advantages of the system are that (1) there is no need to obtain the dynamic data of the undamaged state of bridges (reference-free); (2) there is no need to control or even measure the input excitations (response-only); (3) it is capable of evaluating both the global dynamic properties of bridges and the local structural condition of their elements; and (4) it can be utilised for identification of different types of damage in different types of structure. To illustrate robustness of the system, experiments were conducted on two structural systems: a reinforced concrete beam strengthened with steel-reinforced polymer (SRP) sheets, and a precast hybrid truss girder made of glass fibre-reinforced polymer (GFRP) tubes filled with concrete reinforced and connected to pretensioned top and bottom concrete chords by double-headed GFRP bars. Different damage scenarios including concrete cracking and debonding of the SRP sheets in the strengthened beam, and rupture of the GFRP tubes in the precast truss elements were investigated. The results demonstrate ability of the system to detect, localise, and estimate severity of the damage in the elements tested.

1 INTRODUCTION

Civil infrastructures are designed and built to be safe against failure, and to perform satisfactorily over their service life. Over the past few decades, however, these infrastructures, particularly concrete bridges, have been deteriorating at an alarming rate due to aging, inadequate maintenance, adverse environmental conditions, and constantly growing transportation demand. Corrosion of reinforcing steel is the major source of deterioration of concrete bridges. Cracking of concrete reduces the structural stiffness and expedites corrosion of the steel. Therefore, structural health monitoring (SHM) of in-service bridges has gained increasing attention due to its potential to facilitate more economical maintenance and management of such crucial infrastructure. However, robust damage identification techniques (DITs) are still needed for detection of imminent damage in bridges at the earliest possible time in order to avoid catastrophic failures.

Despite all the advances in bridge monitoring techniques, the following problems have been historically difficult to solve: (1) in most in-situ cases, data from the intact (undamaged) state of in-service bridges are not available; therefore, it is not possible to simply compare before and after damage states to evaluate the current condition of bridges and their elements (Fan & Qiao 2010); (2) measuring input excitations of bridges for evaluating their global dynamic properties is not practical since it requires the bridges' normal operations be interrupted (Farrar et al. 1999); (3) local methods provide only a localised knowledge of the structure's condition and require the vicinity of damage to be known in advance and be accessible for testing (Doebling et al. 1996); (4) most of the current DITs are designed for specific types of structure and are limited to identifying a particular type of damage (Carden & Fanning 2004); and (5) bridges experience varying operational and environmental conditions which lead to changes in their measured dynamic responses. These changes can be wrongly interpreted as an indication of damage (Farrar & Worden 2012).

To overcome the difficulties associated with the traditional systems, a smart structural health monitoring (SSHM) system is proposed in this paper. The proposed system is called smart since it utilises a DIT that requires only the output acceleration responses of in-service bridges to identify structural damage and to evaluate the global dynamic properties of bridges.

The proposed DIT works based on the following: (1) structural damage induces disturbances in the measured acceleration signals of structures; see Figure 1; (2) these disturbances are detectable by means of discrete wavelet transforms (DWTs); and (3) the detected disturbances can be quantified using spectral entropy (Shannon 1948; Powell & Percival 1979; Rosso et al. 2001).

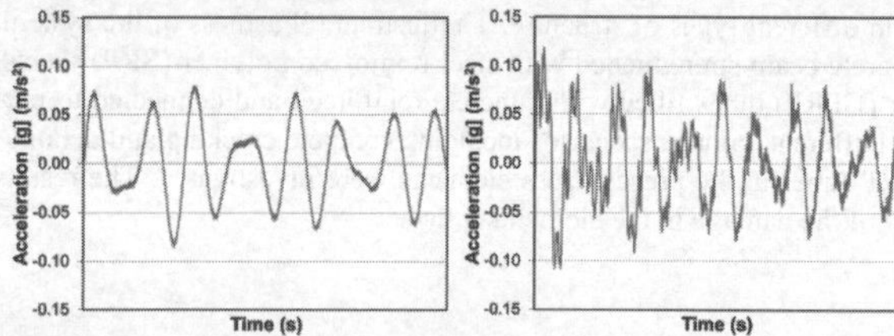

Figure 1. Acceleration signals obtained at (a) a slightly damaged location and (b) a severely damaged location.

The main advantages of the proposed SSHM system are that it (1) is reference-free since there is no need to obtain the dynamic data of the undamaged state of bridges; (2) is response-free since there is no need to control or even measure the input excitations; (3) is capable of evaluating both the global dynamic properties of bridges and the local structural condition of their elements; and (4) can be utilised for identification of different types of damage in different types of structure.

The efficacy of the proposed SSHM system in identification of different types of damage is examined experimentally by testing two structural systems: (1) a reinforced concrete (RC) beam strengthened with steel-reinforced polymer (SRP) sheets; and (2) a precast hybrid truss girder made of glass fibre-reinforced polymer (GFRP) tubes filled with concrete reinforced and connected to pretensioned top and bottom concrete chords by double-headed GFRP bars.

2 THEORETICAL BACKGROUND

The following subsections provide the mathematical definitions of wavelet transform, wavelet energy, and relative wavelet entropy. It is also explained how these quantities can provide useful information of the signals in a simple way and how damage can be identified through proper use of these quantities.

2.1 *Wavelet transform (WT)*

In general, WT is a convertor of a signal into a different mathematical form in order to emphasise its specific properties that are of our interest (Gao & Yan 2011). Continuous wavelet transform (CWT) is defined as the product of a continuous signal, $f(t)$, and a basic wavelet function, $\psi(t)$. The result of this product is wavelet coefficients, defined by Eq. (1), which show how well a wavelet function correlates with the signal.

$$C(s,\tau)=\frac{1}{\sqrt{s}}\int_{-\infty}^{+\infty} f(t)\,\psi^{*}(\frac{t-\tau}{s})\,dt \tag{1}$$

where $\psi^{*}(t)$ is the complex conjugate of the wavelet, S is a scale factor, and τ is a shifting factor.

In practice, an acceleration signal is sampled at discrete time intervals through a data acquisition system. By adopting the values of 2^j and $2^j k$ for the scale and shifting factors, respectively, the corresponding wavelet family can be expressed by Eq. (2) and the corresponding wavelet coefficients, $C_j(k)$, can be obtained accordingly from Eq. (1).

$$\psi_{j,k}(t)=2^{-\frac{j}{2}}\psi\left(2^{-j}t-k\right) \tag{2}$$

2.2 *Wavelet energy*

The wavelet coefficients can be used as a direct estimation of the wavelet energy. In this context, the energy of the signal at each scale, E_j, and the energy of the signal at each sampled time, $E(k)$, are defined, respectively, as:

$$E_j=\sum_k |C_j(k)|^2 \tag{3}$$

$$E(k)=\sum_j |C_j(k)|^2 \tag{4}$$

Consequently, the total wavelet energy, E_{total}, and the wavelet energy ratio, p_j, can be obtained by:

$$E_{total} = \sum_j \sum_k |C_j(k)|^2 = \sum_j E_j \tag{5}$$

$$p_j = \frac{E_j}{E_{total}} \tag{6}$$

The wavelet energy ratio vector, $\{p_j\}$, represents the energy distribution of the signal over different frequency bandwidths and provides a suitable tool for detecting and characterising singular features in the signal in order to detect the structural damage.

2.3 *Wavelet entropy (WE)*

The wavelet entropy can quantify the degree of disorder in a measured signal and is defined as:

$$S_{WT}(p) = -\sum_j p_j \ln[p_j] \tag{7}$$

If damage occurs at a location, the degree of disorder of the acceleration signals at that location will increase because of complex energy dissipation mechanisms and increase in nonlinearity due to gaps and frictions (Lee et al. 2014). Consequently, the probabilistic distribution of wavelet energies increases. As a result, the wavelet entropy value of the signal will also increase which can be utilised as an effective quantitative measure of the damage severity.

2.4 *Relative wavelet entropy (RWE)*

RWE describes the degree of dissimilarity between two probability distributions of two sets of signals and can be defined as:

$$S_{WT}(p|q) = \sum_j p_j \ln\left[\frac{p_j}{q_j}\right] \tag{8}$$

RWE will be equal to zero only if the wavelet energy ratio vectors $\{p_j\}$ and $\{q_j\}$ are exactly the same. For the application of RWE in damage detection, these two sets of signals must be chosen in such a way that the degree of dissimilarity between them represents the severity of possible targeted damage.

3 EXPERIMENTAL PROGRAM

The robustness of the proposed SSHM system in damage identification is investigated experimentally in two structural systems as discussed below.

3.1 *SRP-strengthened RC beam*

The beam was 3150 mm long with a 150 mm × 300 mm rectangular cross section. In addition to the internal reinforcement, one layer of SRP sheet was externally bonded to the bottom surface over the entire span of the beam. One layer of U-shape SRP strip was also bonded adjacent to each support to work as an anchorage system. The beam had already been severely damaged due to approximately 650,000 fatigue cycles between 24 kN and 72 kN and a post-fatigue static test from zero to failure. More details on the fatigue and post-fatigue tests are presented in Lahamy et al. (2015).

Figure 2 shows the damaged state of the beam after the fatigue and post-fatigue tests and before performing a series of impact tests. The SRP sheet is severely debonded in the left half of the beam following significant concrete cracking near the mid-span. The SRP sheet remained perfectly bonded to the concrete in the right half of the beam.

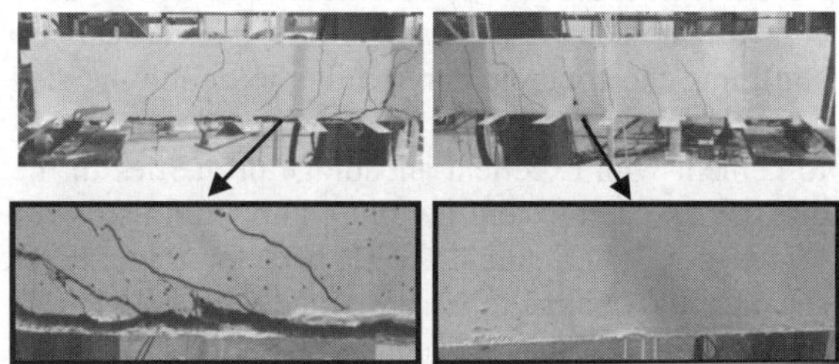

Figure 2. Damaged state of the beam.

To identify the debonding profile of the SRP sheet, one accelerometer was attached to the top concrete surface of the beam, and another accelerometer was attached to the SRP sheet at the beam bottom surface as shown in Figure 3. The location of the pair of accelerometers was successively changed from measurement point 2 to 12.

Another series of impact tests was conducted to identify the cracking profile. In these tests, one accelerometer was attached to the top surface of the beam at Location 12 (L12), where the beam seems undamaged. The location of this accelerometer was kept unchanged during the tests (see Figure 4). Another accelerometer was successively moved from measurement point 2 to 11.

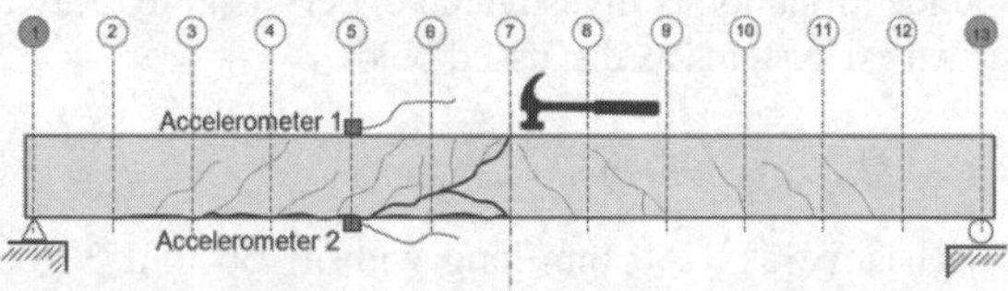

Figure 3. Typical test setup for identifying debonding profile.

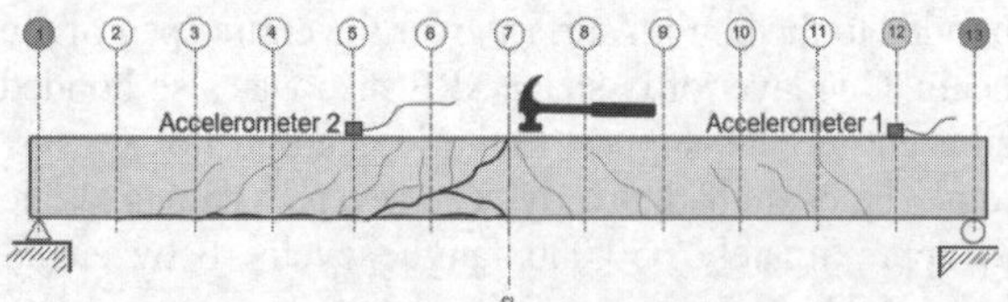

Figure 4. Typical test setup for identifying cracking profile.

3.2 Hybrid FRP-concrete truss girder

The truss girder system consists of pretensioned top and bottom concrete chords connected by precast truss elements made of GFRP tubes filled with concrete reinforced and connected to the chords by double-headed GFRP bars. The vertical truss elements are predominantly in compression and the diagonal truss elements are mainly in tension. The GFRP tubes enhance the compressive strength of the verticals by confining the concrete core while the double-headed GFRP bars serve as non-corrosive internal reinforcement with excellent anchorage properties in the diagonals. The top and bottom chords are also reinforced with longitudinal GFRP bars for flexural resistance and control of cracking and with GFRP stirrups to provide shear resistance (El-Badry 2007).

A 3-m long two-panel truss girder, of which the dimensions and reinforcement are shown in Figure 5, was fabricated and tested under a vertical point load applied at mid-span. The load increased monotonically from zero to failure and produced various types of damage in the different elements of the girder (Joulani et al. 2016). Accelerometers were attached to the vertical elements at three locations along their heights as shown in Figure 5, establishing a 3×3 matrix of locations; Eq. (9). The elements of each column of the matrix represent the locations of the sensors on each vertical tube. A series of impact tests was conducted before and after the static loading test using only two accelerometers at a time to cover all the nine measurement points.

$$[\text{Location}]_{3\times3} = \begin{bmatrix} L11 & L12 & L13 \\ L21 & L22 & L23 \\ L31 & L32 & L33 \end{bmatrix} \quad (9)$$

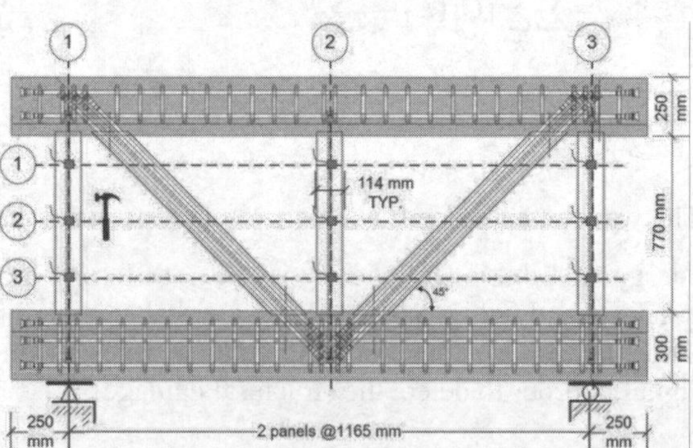

Figure 5. Typical dimensions and reinforcement of a two-panel truss girder and the sensor arrangement

4 RESULTS AND DISCUSSION

In addition to selecting appropriate locations for measuring sets of vibrational signals, selecting a mother wavelet that fits the pattern of damage more properly will result in detecting the damage content more accurately. In the present work, the measured signals have been analysed using the Daubechies wavelet family, particularly db1 (also known as Haar wavelet). This family of wavelet has been employed for two main reasons: (1) among several alternatives, successful use of Daubechies wavelets in damage identification has been reported by many researchers (Ren and Sun, 2008; Qiao et al., 2012; Xiang & Liang, 2012; Solis et al., 2013); and (2) as shown in Figure 6, the acceleration signals obtained at damaged locations contained disturbances, which are correlating with wavelet function db1.

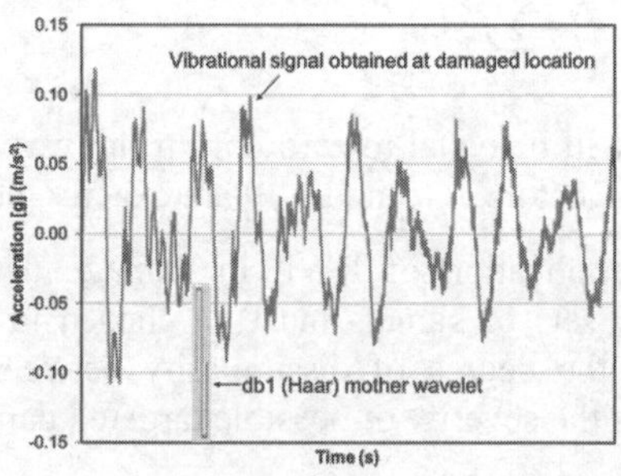

Figure 6. Similarity of Haar wavelet and disturbances in signals

4.1 *SRP-strengthened RC beam*

Figure 7 illustrates the debonding severity and concrete cracking profiles of the SRP-strengthened RC beam. The values on the graphs represent the percentage of a specific type of damage, i.e. debonding or cracking, which has been calculated by normalising corresponding RWE indices. To identify the debonding, the RWE index at each location was calculated by analysing the acceleration signal obtained at the top concrete surface through eight levels of wavelet decomposition (using db1) relative to the signal obtained at the beam bottom surface. The RWE indices corresponding to the cracking profile were calculated by analysing the acceleration signal obtained at the top concrete surface relative to the signal obtained at L12 for each impact test.

As can be seen, the debonding severity graph is perfectly matching the actual debonding profile of the beam, where the SRP sheet is still bonded to the concrete over the right half of the beam (zero percent damage content) and is significantly debonded over the left half of the beam, particularly around L6 with 41 percent damage content. In addition, the concrete cracking profile shows that the concrete cracking is distributed over the entire length of the beam and is more sever around L5, L6, and L7, as is the case in the actual cracking pattern of the beam.

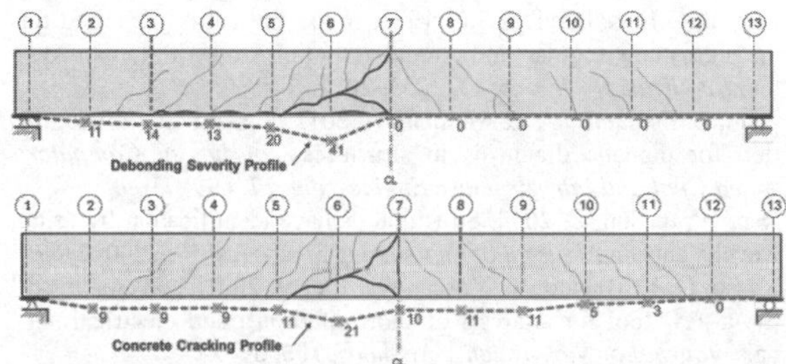

Figure 7. Debonding severity and concrete cracking profiles of the SRP-strengthened RC beam

4.2 *Hybrid FRP-concrete truss girder*

The wavelet analysis of the acceleration signals recorded by sensors during the impact tests on the vertical truss elements results in a 3×3 matrix of wavelet energy ratio vectors, as given in Eq. (10). Each cell in the matrix is a vector of 10 elements (i.e. $j=10$ levels of wavelet decomposition) representing the energy distribution of the signals over ten frequency bandwidths. Comparing any two wavelet energy ratio vectors in Eq. (10) using Eq. (8) describes the degree of dissimilarity between the two energy distributions, which are utilised here to identify possible damage. Performing this analysis for the entire matrix results in a 3×3 RWE matrix of each location relative to the eight other locations. This matrix identifies both the location and severity of damage. When a particular location is affected by damage, its RWEs will be higher compared to others.

The matrix of RWEs obtained from the impact tests is given in Eq. (11), in which the locations affected by the damage, and the estimated damage severity, are recorded. The values are then normalised, scaled to 100, and depicted in Figure 8. It can be seen from the figure that the central vertical element of the truss is the most critical followed by the right element. This finding can be validated using the maximum strains in the GFRP tubes induced by the static load, as given in Eqs. (12) and (13). The results of the RWE analysis also agree with the physical damage caused by the static loading test, in which rupture of the GFRP tubes occurred only in the central vertical truss element between its bottom and mid. height (see Figure 9), while the two other GFRP tubes showed no sign of rupture.

$$[P_{10}]_{3\times3} = \begin{bmatrix} \{P11\} & \{P12\} & \{P13\} \\ \{P21\} & \{P22\} & \{P23\} \\ \{P31\} & \{P32\} & \{P33\} \end{bmatrix} \quad (10)$$

$$[RWE]_{3\times3} = \begin{bmatrix} 22.8 & 51.4 & 36.9 \\ 27.3 & 59.5 & 35.4 \\ 27.1 & 64.8 & 35.3 \end{bmatrix} \quad (11)$$

$$[\text{Axial strain}] = \begin{bmatrix} -483 & \text{NA} & -644 \\ -667 & -12282 & -1318 \\ -514 & \text{NA} & -2553 \end{bmatrix} \times 10^{-6} \quad (12)$$

$$[\text{Hoop strain}] = \begin{bmatrix} 1880 & \text{NA} & 1497 \\ 281 & 6543 & 179 \\ 613 & \text{NA} & 1113 \end{bmatrix} \times 10^{-6} \quad (13)$$

5 CONCLUSIONS

A smart structural health monitoring (SSHM) system using a damage identification technique (DIT) based on relative wavelet entropy (RWE) has been presented and experimentally verified. The DIT is both response-only and reference-free and the instrumentations are very simple, which make the technique a practical means for damage identification in existing bridges. The main conclusions drawn from the present study are:

1. RWE is a sensitive damage feature, which can be used to detect and localise different types of damage in different types of structures, as well as to estimate damage severity.

2. Application of the SSHM system to the SRP-strengthened RC beam resulted in debonding and cracking profiles perfectly matching the actual damaged state of the beam.

3. The SSHM system could identify the damage content in the concrete-filled GFRP tubes of the truss girder and could help in decision-making regarding maintenance of the girder. Results of the damage identification analysis were verified by the strain gauge data and visual inspection of the actual damage of the tubes.

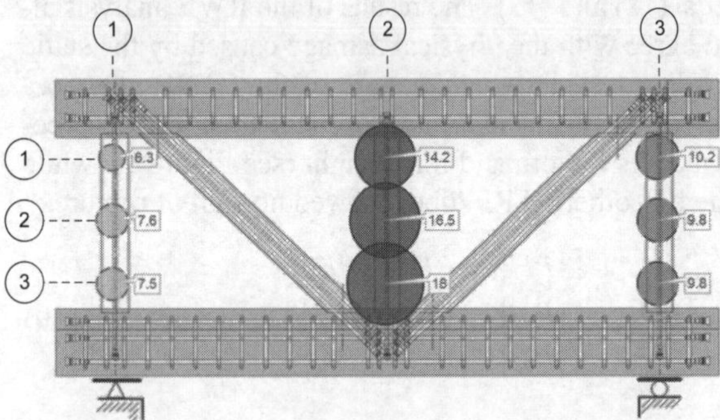

Figure 8. Location and severity of damage in the truss elements

Figure 9. Vertical truss elements after static test

ACKNOWLEDGEMENT

The financial support received from the Natural Sciences and Engineering Research Council of Canada (NSERC) is gratefully acknowledged.

REFERENCES

Carden, E.P. & Fanning P. 2004. Vibration based condition monitoring: a review, *Structural Health Monitoring,* **3**, 355–377.

Doebling, S.W., Farrar, C.R., Prime, M.B. et al. 1996. Damage identification and health monitoring of structural and mechanical systems from changes in their vibration characteristics: a literature review, *Los Alamos National Laboratory report*.

El-Badry, M. 2007. An innovative hybrid FRP-concrete system for short and medium-span bridges, Proceedings of *the COBRAE 2007 Conference on "Benefits of Composites in Civil Engineering,"* Stuttgart, Germany, 16 pp.

Fan, W. & Qiao, P. 2010. Vibration-based damage identification methods: a review and comparative study, *Structural Health Monitoring,* **10**, 83–111.

Farrar, C.R., Duffey, T.A., Cornwell, P.J. et al. 1999. Excitation methods for bridge structures, *Proceedings of the 17th International Modal Analysis Conference*, Kissimmee, FL, 1063–1068.

Farrar, C.R. & Worden, K. 2012. *Structural Health Monitoring: A Machine Learning Perspective*, John Wiley & Sons, Ltd, USA.

Gao, R. & Yan, R. 2011. *Wavelets: Theory and Applications for Manufacturing*, Springer Science and Business Media, USA.

Joulani, P., El-Badry, M. & Moravvej, M. 2016. Static Load Behaviour of Hybrid FRP-Concrete Two-Panel Truss Girders Reinforced with Double-Headed Glass FRP Bars, Proceedings of *the CSCE 5th International Structural Specialty Conference, ISSC-V*, London, Ontario, Canada, 10 pp.

Lahamy, H., Lichti D., El-Badry, et al. 2015. Evaluating the capability of time-of-flight cameras for accurately imaging a cyclically loaded beam, *Proceedings of the Conference on Videometrics, Range Imaging, and Applications, part of SPIE Optical Metrology*, Munich, Germany, 11 pp.

Lee, S.G., Yun, G.J. & Shang, S. 2014. Reference-free damage detection for truss bridge structures by continuous relative wavelet entropy method, *Structural Health Monitoring,* **13**, 307–320.

Powell, G.E. & Percival, I.C. 1979. A spectral entropy method for distinguishing regular and irregular motion of Hamiltonian systems, *Journal of Physics,* **12**, 2053–2071.

Qiao, L. Esmaeily, A. & Melhem, H. 2012. Signal pattern recognition for damage diagnosis in structures, *Journal of Computer-Aided Civil and Infrastructure Engineering*, **27**, 699–710.

Ren, W. & Sun, Z. 2008. Structural damage identification by using wavelet entropy, *Journal of Engineering Structures,* **30**, 2840–2849.

Rosso, O.A., Blanco, S., Yordanova, J. et al. 2001. Wavelet entropy: a new tool for analysis of short duration brain electrical signals, *Journal of Neuroscience Methods,* **105**, 65–75.

Shannon, C.E. 1948. A mathematical theory of communication: I and II, *Bell System Technical Journal*, **27**, 379–443.

Solis, M., Algaba, M. & Galvin, P. 2013. Continuous wavelet analysis of mode shapes differences for damage detection, *Journal of Mechanical Systems and Signal Processing*, **40**, 645–666.

Xiang, J. & Liang, M. 2012. Wavelet-based detection of beam cracks using modal shape and frequency measurements, *Journal of Computer-Aided Civil and Infrastructure Engineering*, **27**, 439–454.

Proceedings of the International Conference on Smart Infrastructure and Construction
ISBN 978-0-7277-6127-9

© The authors and ICE Publishing: All rights reserved, 2016
doi:10.1680/tfitsi.61279.417

Clarification of cross-sectional vibration characteristics for damage identification of belt conveyor support structure

S. Rana*[1], T. Nagayama[1], K. Hisazumi[2] and T. Tominaga[2]

[1]*Department of Civil Engineering, The University of Tokyo, Japan*
[2]*Nippon Steel and Sumitomo Metal Corporation, Japan*
* *Corresponding Author*

ABSTRACT Belt conveyors, widely used in various industries worldwide, are often exposed to corrosive environment. Decades after construction, many of the support structure of belt conveyors have severe degradation, which may cause structural failure and functional stop of associated industries. Damage identification of the support structure is therefore important to ensure the safety and reliability. However, application of existing global vibration based techniques to these structures is difficult due to unavailability of baseline condition, the possible presence of multiple corroded members in a single structure, and the effect of non-structural components. In this paper, cross-sectional vibration characteristics are clarified on a numerical model of support structure. Numerical analysis reveals that there exist some eigenmodes, named as cross-sectional mode (CSM), in which the main member vibrates strongly in cross-sectional direction. When a part of the continuous main member is damaged, the cross-sectional mode becomes localized only to the damaged panel. In addition, the local vibration frequency decreases significantly from CSM. This mode is named as localized cross-sectional mode (LCSM). The sensitivity analysis of CSM and LCSM reveals that the presence of damage on a panel affects CSM frequencies and LCSM frequencies of other panels marginally. For multiple damages, multiple LCSMs corresponding to the damage severities and locations exist; multiple damages can be independently analyzed. The damage identification is therefore feasible by comparing the frequencies of CSM and LCSM, which is available from current state of the structure without the need for before-after comparison. The existence and observability of CSM are experimentally confirmed.

1 INTRODUCTION

Belt conveyors, one of the important industrial infrastructures, are used worldwide to carry various types of materials. Some of them are of small scale while others are of large scale and used in mining fields and factories of brick, asphalt, cement, concrete, and steel. Large belt conveyors usually consist of non-structural members, such as belt, machinery parts, and walkways, and the structural members, that is support structures (Figure 1). They are most often situated in corrosive environment, which results in corrosion damages in support structures over decades. Sometimes, damages are severe enough to cause sudden collapse of the structures. These accident may result in human injury and death and also cancellation of operation of the factories. Therefore, to maintain the safety and reliability of these structures, structural damage identification is necessary.

Vibration-based damage detection methods assess damage by using dynamic characteristics of the structures. Change in structural parameters due to damage is considered to alter the dynamic properties consequently. Thus the structural condition is assessed by monitoring the changes in dynamic properties. Over the last few decades, these vibration-based damage detection methods have been intensively studied in civil, mechanical, and aerospace engineering communities (Doebling et al. 1996, Salawu 1997, Mottershead & Friswell 1993, Chang et al. 2003, Carden & Fanning

2004, Humer et al. 2006, Brownjohn 2007, Fan & Qiao 2011, Gunes & Gunes 2013).

Figure 1. Belt conveyor structure

Although a large number of researchers have been studying global vibration-based damage detection methods, the application of these methods is mostly limited to simulations and laboratory studies (Humar et al. 2006). Some essential limitations associated with the practical application of those global vibration based method are low sensitivity of dynamic characteristics to damage, sensitivity to environmental factors, and incomplete nature of vibration characteristic measurement.

In addition to these challenges, there are other difficulties which limits the application of these methods to belt conveyor support structures. For example, in most cases, baseline condition of the whole structure is not available, which is required by existing methods. Second, multiple damages are present in a single belt conveyor support structure. Existing methods can detect only a single or a few damages at most. Third, the contributions of non-structural members to global vibration modes are significant as their dimensions are not small as compared to the support structure. In addition, these members are replaced more frequently than structural members which also changes the vibration properties of the structures over time even if corrosion or other damage does not take place.

To address these issues, Honarbakhsh et al. (2015) proposed damage identification techniques using periodic and isolated local vibration modes. However, these techniques are applicable only to the secondary members of the support structure; damage assessment of main members is difficult because periodic and isolated local vibration modes do not exist for these members. This paper presents the characteristics of cross-sectional vibration modes of the main members from a perspective of damage identification.

2 FINITE ELEMENT MODELLING

The support structure of belt conveyors is usually a steel truss structure, which consists of the main members and the secondary members (Figure 2). Main members are continuous along the longitudinal direction while the secondary members are connected to the main members. Main members are usually much stiffer than the secondary members. The structure is made from different types of steel sections.

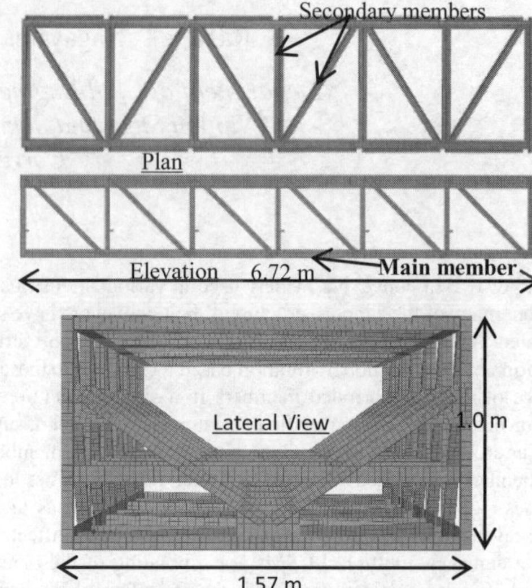

Figure 2. Finite element model of the structure

To investigate the vibration characteristics of the support structure in the cross-sectional direction, the structure is modeled with shell elements using the commercial FEM software-ABAQUS (2015). Figure 2 shows the outline of the structure. The main member is 6.72 m long and each panel is 1.08 m long; a panel is defined as a portion of the truss between the neighboring vertical secondary members. The thickness of the main member is 6 mm. Secondary members are connected to the continuous main members. The connections between the main members and secondary members are modelled by fastener connection.

3 VIBRATION CHARACTERISTICS OF THE BELT CONVEYOR SUPPORT STRUCTURE

Vibration modes of the structure are obtained through eigenvalue analysis. Typical global vibration modes are vertical bending mode, and lateral torsional bending mode (Figure 3). The modal frequencies of these modes are tabulated in Table 1. In the global vibration modes almost all members are under vibration.

Figure 3. Global Vibration modes

There are other modes in which the modal amplitude of the main members is large. In these modes, the vibration of the main members is coupled with secondary members. These modes occur in higher frequency range as compared to the lower order global vibration modes. Figure 4 shows a vibration mode shape while the modal frequencies of some of these modes are tabulated in Table 1.

Figure 4. Coupled local bending mode of the main member

3.1 Cross-sectional Vibration Modes

The eigenvalue analysis of the support structure reveals that there exist some modes in which only the main member vibrates locally in cross-sectional direction. These vibration modes are defined as cross-sectional modes (CSM) of the main members (Rana et al. 2014). As the main member is continuous in the longitudinal direction, the vibration is not localized to a single panel. Longitudinal member portions of other panels also vibrate (Figure 5). In the cross-sectional modes, both web and flange vibrate in phase in out of plane direction; there exist a nodal line along the longitudinal direction of the member where vibration amplitude is much smaller than the free edge (Figure 5). These modes occur usually at high frequency range as compared to the global modes (Table 1). There exist a group of cross-sectional modes sharing similar mode shapes within a small frequency range. One of the 1st CSM of the main member obtained by the eigenvalue analysis is shown in Figure 5. Similarly, the 2nd and the 3rd CSM's are also observed. The cross-sectional modal frequencies are tabulated in Table 1.

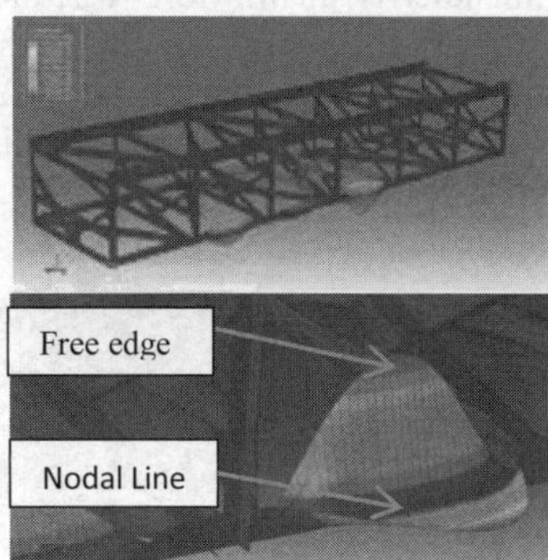

Figure 5. Cross-sectional vibration mode (CSM) of main member

Table 1. Frequency of Global modes and Cross-sectional modes

Mode	Frequency
1st vertical bending	42.41 Hz
1st lateral torsional	44.38 Hz
1st lateral bending	62.62 Hz
2nd vertical bending	100.95 Hz
1st CSM of main member	133 – 137 Hz
2nd CSM of main member	281 – 287 Hz
3rd CSM of main member	450 – 456 Hz
Coupled local bending mode of main member	217.7 Hz
Coupled local bending mode of main member	237.4 Hz

3.2 Localized Cross-sectional Vibration Modes

Consider a typical damaged support structure containing damage in the main member at certain location of the continuous member as shown in Figure 6. The panel containing the damage is defined as the damaged panel. In the FE model of the structure, damage is simulated by changing the thickness of a certain portion of the member. In one panel, the thickness of a 0.54 m portion of the main member is reduced to 3.6 mm, resulting in 20% average cross-sectional loss (Figure 6).

Figure 6. Damage simulated in the main member

The eigenvalue analysis is performed on the damaged structure. Modes exist in which vibration becomes confined only to the longitudinal member in the damaged panel in the cross-sectional direction only. In the modes, the vibration is localized in the damaged member. The vibration of the other part of the continuous main member is small. Moreover, the damaged member in the panel does not vibrate in the CSM of the neighboring undamaged panels (i.e., CSM of the main member). The mode is named localized cross-sectional mode (LCSM). The 1st and 2nd LCSMs are shown in Figure 7.

Figure 7. LCSM: (a) 1st mode, (b) 2nd mode

The frequency of the LCSM is also significantly lower than the CSM of the main member. The 1st CSM of the main member is found around 135.4 Hz. The 1st LCSM of the damaged member is around 115.0 Hz, resulting in 20.4 Hz frequency change. The 2nd CSM is around 284.7 Hz. The 2nd LCSM is around 231.2 Hz resulting in 53.5 Hz frequency change. The 3rd mode changed from around 450.1 Hz to 343.2 Hz resulting in 106.9 Hz frequency change. Hence the LCSMs are sensitive to damages of the main member. Note that the damaged member also vibrates in other high frequency modes, where the longitudinal member deforms in a bending manner as shown in Figure 8. These modes do not have high sensitivities to the damage. Thus, these modes are not applicable in detecting and locating the damage.

Figure 8. Local bending mode of main member

Frequencies of global modes and other modes where all members have large mode shape amplitude are affected by the structural properties of all members and provide insignificant information about local damage while LCSM frequency mainly reflect the local properties of the corresponding member. Taking advantage of mode localization of the damaged panel, the existence and location of damage can be determined through identification of the LCSM frequency in comparison with that of CSM.

For the structure containing multiple damages as shown in Figure 9, multiple LCSMs corresponding to the damage severities and locations exist (Figure 10). Each LCSM is independent of damages on other members. Thus, multiple damages can be identified independently.

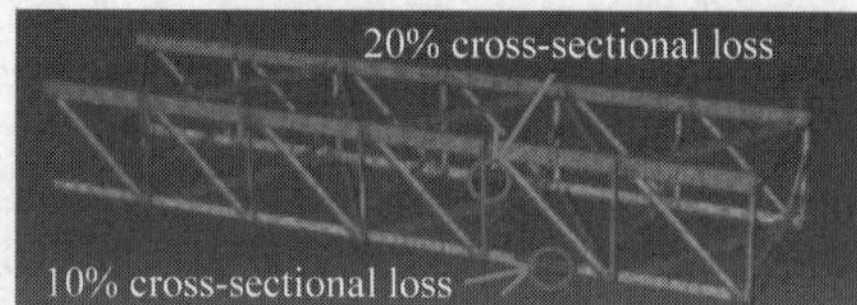

Figure 9. Structure containing multiple damages

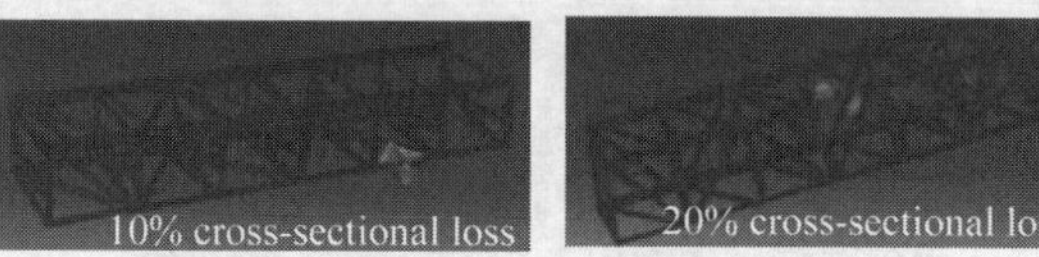

Figure 10. Distinct LCSM for multiple damages

4 SENSITIVITY ANALYSIS OF CSM

The sensitivity analysis of CSM frequency under the presence of damages in the main member is performed for several cases. CSM from the undamaged member is observed by varying the damage level in the single damaged member in the structure (Figure 11(a)). The damage is simulated as before by changing the thickness. Figure 12 shows the sensitivity of 1st CSM frequency for this case. In all the 1st, 2nd and 3rd CSM frequencies, the difference is found less than 1%.

Moreover, CSM from the undamaged member is observed by varying the damage level in the multiple damage members (Figure 11(b)). Figure 13 shows the sensitivity of 1st CSM frequency for the multiple damage case. In all the 1st, 2nd and 3rd CSM frequencies, the difference is found less than 1% in this case also. Therefore, CSM of the main member is considered practically unaffected by the presence of damage in the damaged panels.

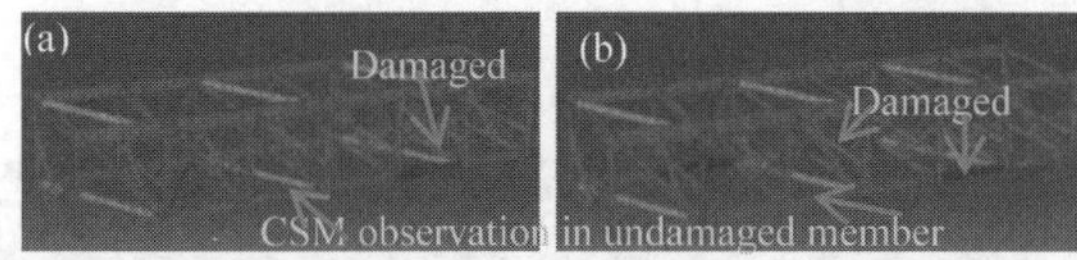

Figure 11. CSM sensitivity observation. a) for single damage, b) for multiple damage

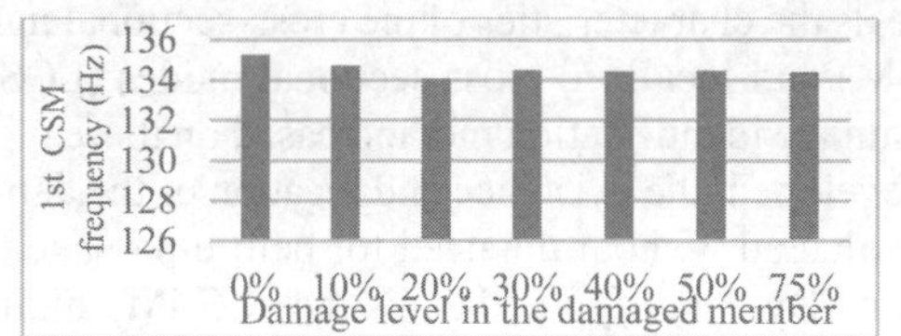

Figure 12. CSM sensitivity in the single damage case

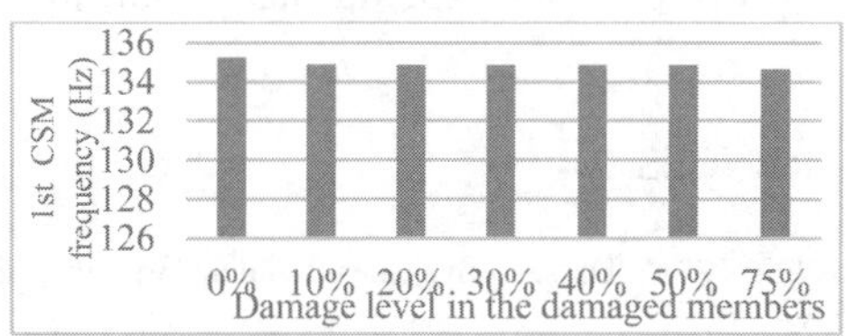

Figure 13. CSM sensitivity in the multiple damages case

5 SENSITIVITY ANALYSIS OF LCSM

The sensitivity analysis of LCSM frequency under the presence of other damages in the main member is performed for several cases. LCSM of a damaged member is examined when the damage level of the damaged member in the other panel is changed (Figure 14(a)). The member corresponding to the LCSM is damaged by 30 % while the damage level on the other panel is varied. The damage is simulated by changing the thickness of the member. Figure 15 shows the sensitivity of 1st LCSM frequency for the case. All the 1st, 2nd and 3rd LCSM frequencies are very close and the difference is found less than 0.5%.

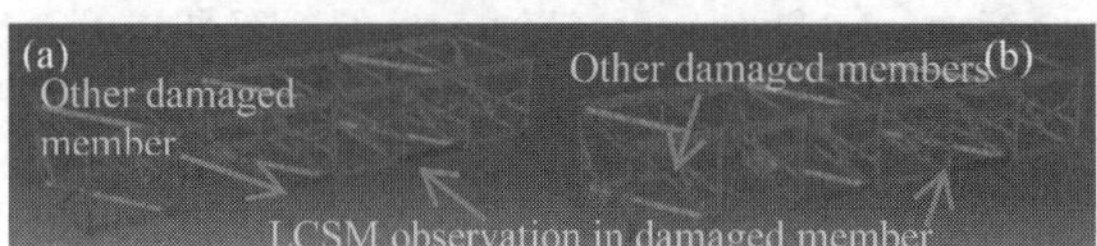

Figure 14. LCSM sensitivity observation. a) for single damage, b) for multiple damage

Moreover, LCSM of the damaged member is observed by varying the damage level in the multiple damage members (Figure 14(b)). Figure 16 shows the sensitivity of 1st LCSM frequency for the multiple damage case. The difference is found less than 1%. Therefore, LCSM is considered practically independent of damage presence in other panels and dependent on the local properties of the member.

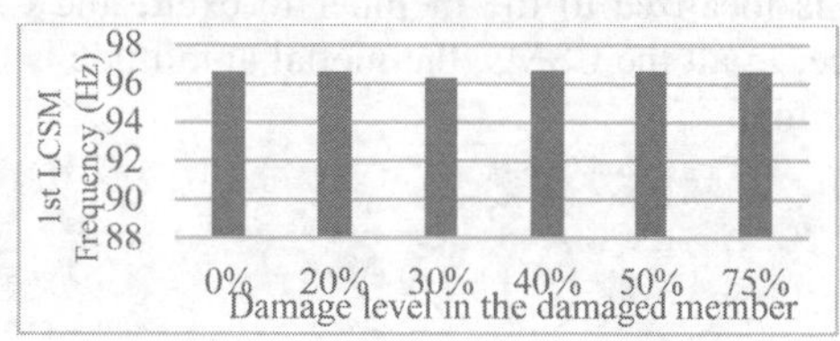

Figure 15. LCSM sensitivity in the single damage case

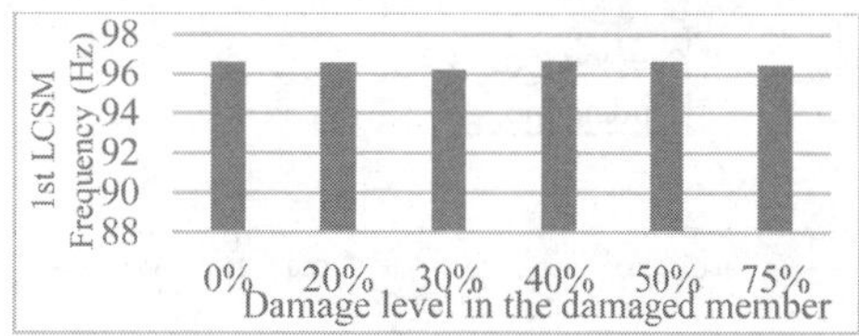

Figure 16. LCSM sensitivity in the multiple damages case

6 EXPERIMENTAL VALIDATION

To ensure the observability of the CSM of the main member, an experiment has been performed on a decommissioned belt conveyor support structure. A non-contact scanning type Laser Doppler Vibrometer (LDV) is used to measure the velocity responses of the structure (Figure 17). The cross-sectional mode has large mode shape amplitude along the free edge of the main member. Measurements are performed on a typical member along the free top edge of the web in out of plane direction. An impact hammer with a load cell is used to excite the member. The sampling frequency is 3.125 kHz. Measurement is repeated at least 3 times for averaging. The power spectrum is calculated from the impulse response after employing exponential window (Figure 18).

Figure 17. Experimental setup

For the 1st CSM, a strong peak appears at 152.0 Hz. Similarly, for the 2nd CSM and 3rd CSM, strong peaks appear at 303.3 Hz and 474.8 Hz respectively (see Figure 18). A significant amount of excitation energy is localized to the member to excite the CSM. Hence, in all the CSMs, the modal amplitude is relatively high.

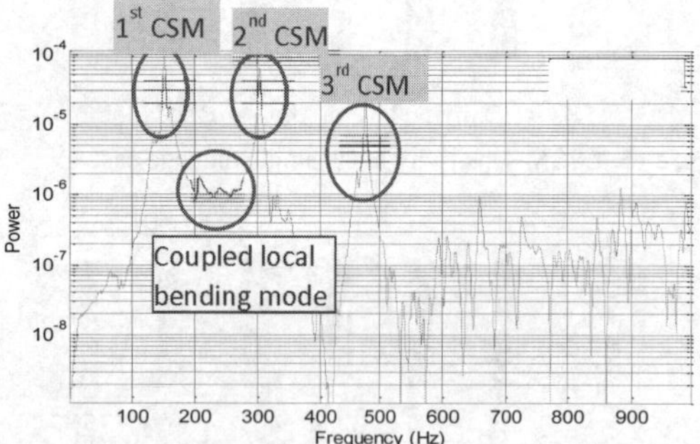

Figure 18. Frequency Spectrum

To confirm the observability of the CSM, measurement on several points both along the longitudinal (X) and vertical direction (Y) of a member is performed to obtain the CSM shapes. Figure 19 shows one typical frequency response function (FRF) obtained from scanning type LDV.

Figure 19. Experimentally measured FRF

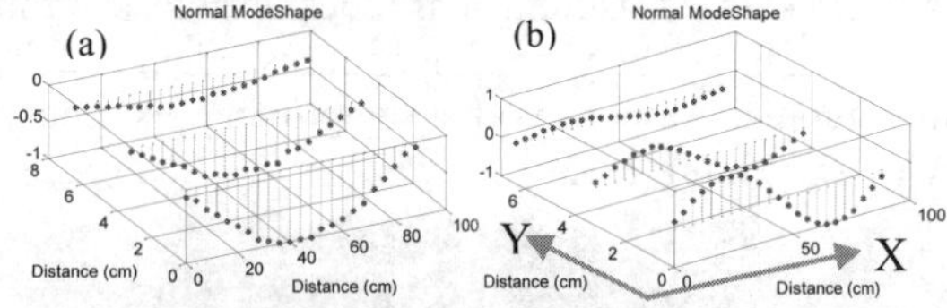

Figure 20. Mode Shape; (a)1st CSM shape, (b) 2nd CSM shape

CSM shapes (Figure 20) are determined by applying Eigen-system realization algorithm (ERA). The y = 0 axis is the free edge of the member. The free edge vibration is comparatively larger than the others. This large vibration at the edge agrees with the mode shapes obtained from numerical analysis.

7 CONCLUSION

Dynamic characteristics of cross-sectional vibration modes of main longitudinal members of belt conveyor support structure are clarified from a damage identification perspective. Eigenvalue analysis of a FE model reveals the characteristics of the cross-sectional modes (CSM) and localized cross-sectional modes (LCSM). A damage identification method based on these modes is developed; the existence and location of the damage is evaluated without the need for before-damage condition. By identifying the distinct LCSM, multiple damages are also identifiable. The existence and observability of CSM are experimentally confirmed.

REFERENCES

Abaqus FEA. 2015. Dassault Systèmes Simulia Corp., Providence, RI, USA.

Brownjohn, J. M. W. 2007. Structural health monitoring of civil infrastructure. Philosophical Trans. of the Royal Society A: Mathematical, Physical and Engineering Sciences, **365**(1851), 589-622.

Carden, E.P. & Fanning, P. 2004. Vibration based condition monitoring: A review. Structural Health Monitoring, 3, 355–377.

Chang P.C., Flatau, A., & Liu, S. C. 2003. Review Paper: Health Monitoring of Civil Infrastructure. Structural Health Monitoring, **2**(3), 257–267.

Doebling S.W., Farar, C.R., Prime, M.B. & Shevitz, D.W. 1996. Damage identification and health monitoring of structural and mechanical systems from changes in their vibration characteristics. A Literature Review. Los Alamos National Laboratory report La-13070-MS.

Fan, W. & Qiao, P. 2011. Vibration-based Damage Identification Methods: A Review and Comparative Study. Structural Health Monitoring, **10**, 83-111.

Gunes, B. & Gunes, O. 2013. Structural health monitoring and damage assessment Part I: A critical review of approaches and methods. International Journal of Physical Sciences, **8**(34), 1694-1702.

Honarbakhhsh, A., Nagayama, T., Rana, S., Tominaga, T., Hisazumi, K., & Kanno, R. (2015). Damage identification of belt conveyor support structure using periodic and isolated local vibration modes. Smart Structures and Systems, **15**(3), 787-806.

Humar, J., Bagchi, A., & Xu, H. 2006. Performance of Vibration-based Techniques for the Identification of Structural Damage. Structural Health Monitoring, **5**(3), 215–241.

Mottershead J.E. & Friswell M.I. 1993. Model updating in structural dynamics: A survey. Journal of Sound and Vibration, **167**(2), 347-75.

Rana, S., Nagayama, T., Hisazumi, K. & Tominaga, T.2014.Damage Assessment of Support Structure of Belt Conveyor by Cross-sectional Local Vibration Modes. Proceedings of 10th International Workshop on Advanced Smart Materials and Smart Structures Technology, Taipei, Taiwan.

Salawu, O.S. 1997. Detection of structural damage through changes in frequency: A review. Engineering Structures, **19**, 718–723.

Proceedings of the International Conference on Smart Infrastructure and Construction
ISBN 978-0-7277-6127-9

© The authors and ICE Publishing: All rights reserved, 2016
doi:10.1680/tfitsi.61279.423

Detection of uniform structural temperature distributions for a temperature based method of structural health monitoring

J. Reilly[1], H. Abdel-Jaber[1], M. Yarnold[2] and B. Glisic[1]

[1] Princeton University, Princeton, USA
[2] Tennessee Tech University, Cookeville, USA

ABSTRACT. Structural Health Monitoring aims to characterize the state of a structure based on some combinations of sensing technologies and analysis of acquired data. This process normally seeks to find changes in the mechanical response of a structure, primarily bridges in this work, based on strain or displacement data. Temperature acts as a distraction in this type of application of SHM, clouding the mechanical input-output behavior of the structure. Thermal effects on most structures, however, play too large of a role to be ignored or taken out of consideration. Combining the output strain and displacement data with input temperature data has the potential to robustly characterize a structure. To begin implementing this Temperature Driven approach, time periods with a uniform temperature distribution on the structure need to be identified. These time periods will reduce any effects of thermal gradients on the output strain and displacement measurement, leaving a much simpler input-output relationship. An initial algorithm was developed to recognize these uniform temperature distribution from years of data collection. This algorithm can modify the strictness of the required level of temperature uniformity, as perfect uniformity can be impossible to find in some cases. This algorithm has been validated on a real structure, the Streicker Bridge on Princeton University's campus, and the initial findings will be used to upgrade and improve the algorithm.

1 INTRODUCTION

Temperature effects on bridges are often eliminated from analysis of Structural Health Monitoring (SHM) data, but can induce comparable stresses to daily traffic loads (Yarnold 2013). An SHM method using temperature as the driving input may give a much more in depth analysis of the structure. The aim of this paper is to create a method to use temperature as a main forcing function, characterizing this input to an output strain and displacement.

Current research is beginning to realize the important role temperature can play in SHM methods. Previous work seeks to assess if the addition of temperature consideration can enhance a current model's ability to monitor a structure (Laory 2011). Temperature also becomes the sole focus of other work. Numerical models have been used to relate the structural behavior of a bridge to the temperature distribution, providing a way to quantify the thermal response and incorporate this response into assessing total behavior of a structure (Kromanis 2014). Few cases consider temperature as a driving force in analysis. Temperature has been incorporated into Structural Identification, to characterize a temperature input to strain and displacement output for the purpose of finite element model calibration. Temperature has also been incorporated as a forcing function with normal structural health monitoring methods, continually monitoring a structure to track key parameters with respect to a baseline response (Yarnold 2013).

The method proposed in this paper provides the ability for deeper, more accurate analysis of this relationship between temperature as input and strain and displacement as output. Thermal gradients in a bridge can produce un-desirable non-linearities in the thermal response. A non-uniform temperature distribution in the cross section of a bridge would create vertical and horizontal curvatures, but also deplanation of the cross-sections, changing at each location with the differing temperature. These changing curvatures and deplanations can obscure the identification of the effects of temperature on the thermal response. Identifying time periods with uniform temperature distributions will allow for a direct comparison of the total structural response for different temperatures. This comparison will identify the thermal effects on the structure, allowing for variations in strain or displacement to be identified and incorporated into a thermal signature for the bridge. This thermal signature will provide an in depth characterization of the behavior of the bridge (Yarnold 2013).

Data from a real structure, the Streicker Bridge at Princeton University's campus, are used to ease the description of the method. Close to 100 fiber-optic strain and temperature sensors were embedded in the concrete deck during construction of the Streicker Bridge in 2009. The sensors are based on Fiber-Bragg Grating (FBG) principle. Besides the principal sensor network, an additional set of redundant FBG temperature sensors was installed at certain locations. While principal and redundant temperature sensors both have FBG as the internal sensing element, they were produced by different manufacturers and function based on different temperature-transfer mechanism. Detailed information regarding the SHM system at the bridge is given in (Sigurdardottir 2015).

As the first step of the method, a preliminary long-term data validation was performed. With valid data, uniform temperature distributions across the structure are identified to start examining the temperature-strain relationship. Comments on data validation and an initial algorithm for identification of uniform temperature distributions are presented in this paper.

2 COMMENTS ON LONG TERM DATA VALIDATION

An initial stage in long-term SHM analysis should include a data validation, in order to ensure proper data interpretation. To examine the long-term stability of the temperature sensors, the temperature readings were compared to ambient temperature measurements taken at Trenton Airport, approximately ten miles away. Principal-sensor readings at each location were then compared to the 'Trenton data', as well as to redundant-sensor readings. Two examples of comparison are shown in Figure 1 and Figure 2.

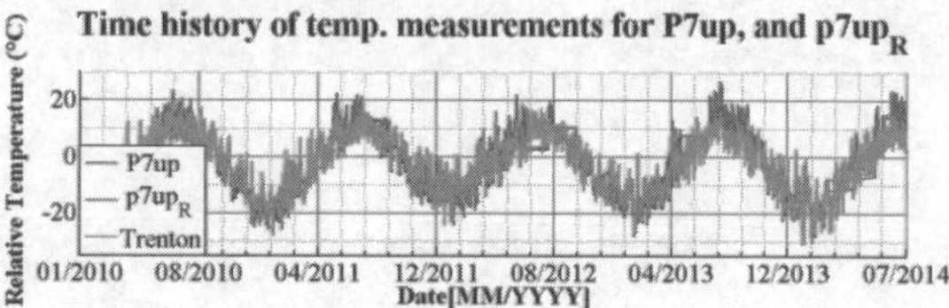

Figure 1. Time History P7up and P7up$_R$ with Trenton

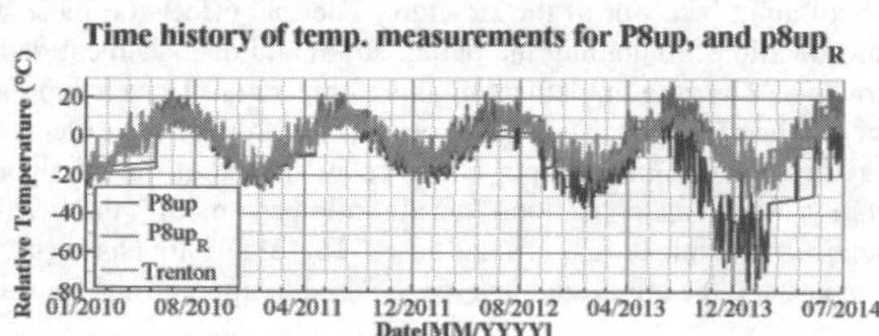

Figure 2. Time History P8up and P8up$_R$ with Trenton Data

Figure 1 shows the temperature data for principal and redundant temperature sensors at the same location (Pier 7) as well as ambient temperature data from Trenton, while Figure 2 shows the same for Pier 8. Both graphs show temperature values on the y-axis relative to a reference temperature at the beginning of monitoring. Pier 7 shows a general agreement between the principal and redundant sensors with Trenton. For Pier 8, general agreement was observed for the first three years between the two sensors and the Trenton data, until some initial discrepancies occurred in the winter of 2013 and then large differences in the winter of 2014. The principal temperature sensor begins to read temperatures much lower than the redundant sensor and the Trenton data, which still remain in good agreement. Very similar behavior was exhibited by eight sensors in total, prompting a failure analysis and an exploration of data regeneration. These two parts of the analysis exceed the scope of the paper and only a summary of results is presented here.

Failure analysis demonstrated that one possible explanation for this winter discrepancy comes from the mechanics of the principal temperature sensor. The sensing FBG is placed in a loose tube within the sensor packaging, but the splice protector was not attached to the packaging in installation. The movement of the splice protector could induce some strain in the FBG, which could affect the reading and introduce error.

In order to continue working with the temperature data from the Streicker Bridge, the data from the failed sensors needed to be regenerated. This regeneration process relied upon the initial years of data where all the temperature sensors were reading reliably. Nearby properly functioning principal sensors were used to define a basic relationships between the functioning sensors and the malfunctioning sensor. The relationships would then be used to regenerate data for the malfunctioning sensor in the years after the failure. To evaluate this process, the failed sensor was compared to Trenton data for its years during the failure before and after the regeneration, shown in Figure 3 and Figure 4. Temperature in these figures consists of values relative to an initial reference temperature.

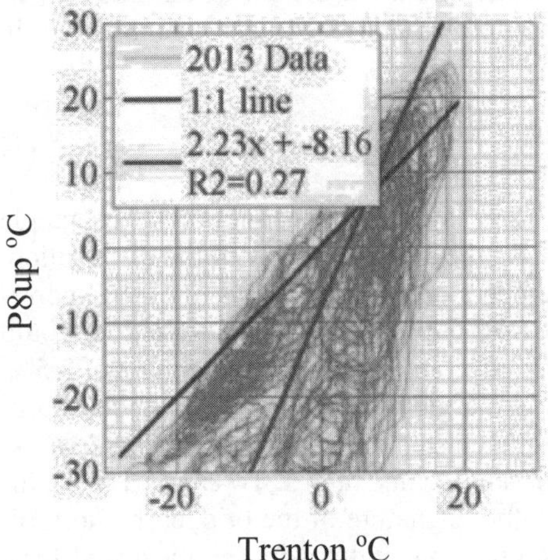

Figure 3. P8up Failed with Trenton

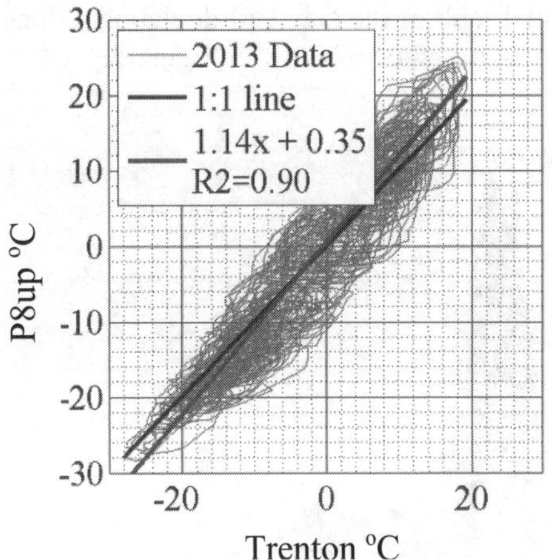

Figure 4. P8up Regenerated with Trenton

The preliminary sensor regeneration successfully provides reasonable data for further work with the temperature readings for the eight failed sensors in the bridge. This data would allow for further analysis to be completed. Process of validation and regeneration of data is still in course.

3 UNIFORM TEMPERATURE DISTRIBUTION

To begin characterizing the temperature-strain relationship on the bridge, the data was sorted to find time periods with a uniform temperature distribution across the bridge. A uniform temperature distribution will eliminate most of the difficult to model non-linear effects from a temperature gradient through the cross section or along the span of the bridge. Comparing two time periods with different but uniform temperature distributions will provide a clear first description of how the changes in temperature affect the strain in the bridge. As the data validation is still in process, only data from 2010, free of any sensor failures, is analyzed.

3.1 Standard Deviation Metric

Looking for times with absolutely uniform distributions may yield close to zero time periods, so some degree of non-uniformity must be tolerated. One metric explored for a uniform temperature distribution was the standard deviation of the readings at each time step. An initial look at the standard deviation values for each time step is shown below in Figure 5.

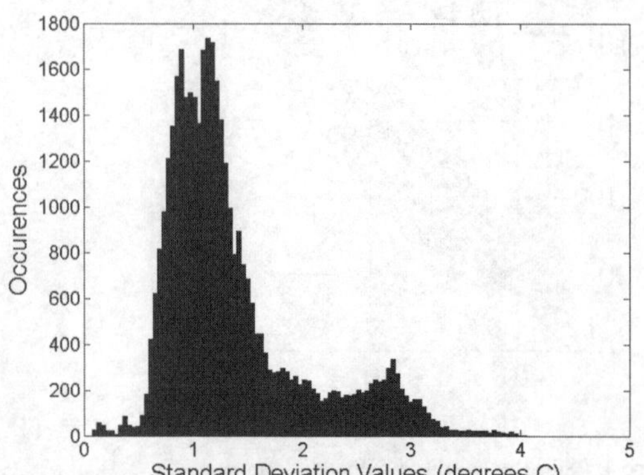

Figure 5. Standard Deviation Histogram, 100 Bins

Figure 5 shows a histogram of the total number of times a time step had each value of standard deviation, using 100 bins to span the range of values. Most of the temperature distributions have a standard deviation at just over 1°C. The algorithm needs to be tailored so that enough time periods with uniform distribution are identified to start characterizing the temperature signature of the bridge, without sacrificing too much of the uniformity of the distribution. Cutoff levels of σ = 0.25, 0.5, and 1 °C were made to begin examining appropriate bounds. Initially a cutoff of σ = 0.25 as a maximum standard deviation was used to identify time periods of uniform temperature distributions. This cutoff yielded very few results, with only two unique time periods. The resulting amount of time periods for this cutoff was not nearly sufficient to begin constructing a temperature signature for the structure. Next, a cutoff of σ = 0. 5 maximum standard deviation was examined. Figure 6 below shows the time periods where standard deviations under 0.5 °C occur.

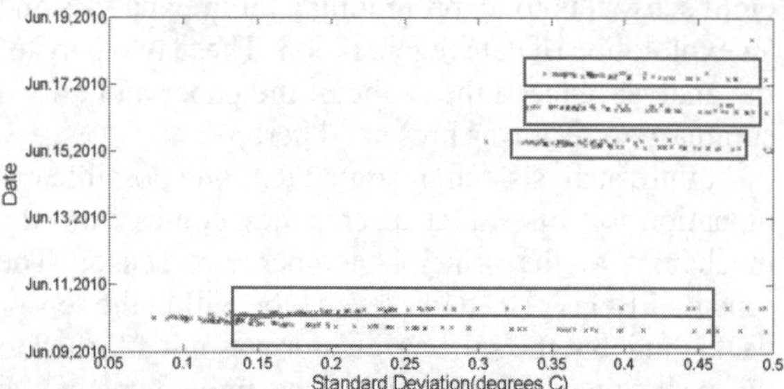

Figure 6. Standard Deviations under 0.5 °C

The y-axis here refers to the date of the corresponding standard deviation. It is important to identify the date for each occurrence of these standard deviations. Figure 6 shows that for this cutoff of 0.5 °C, all of the time periods in 2010 under this criteria occur in June. While there are many instances of these close to uniform distributions, they all fall under 5 distinct time periods, seen in red in Figure 6. The sensors take measurements every 5 minutes, so each cluster represents a uniform temperature distribution over a period of a few hours. At the max standard deviation here of σ = 0.5°C, 95% of the readings would fall within ±1°C (2σ) of the mean. 2σ here is only an approximation of 95% as the distribution is assumed to not be normal. In a simplified unrestrained thermal expansion, shown in Equation 1, a 1°C Δt and a coefficient of thermal expansions (α) of approximately ±12 με/°C would result in 12 με.

$$\varepsilon = \alpha \Delta t \tag{1}$$

An error of 12 με is within the bounds of an acceptable strain added by the non-uniformity of the distribution. Seasonal temperatures can vary by 40-50 °C, corresponding to 480-600 με. This 12 με variation would then correspond to an error on the magnitude of 2.5%, an acceptable margin of error. While this bound provides 5 unique clusters, they all occur in June, at essentially the same temperatures. To create a more robust temperature signature of the bridge, the uniform temperature gradients will need to be identified for a variety of temperatures, hopefully from all four seasons.

In an effort to identify a greater variety of time periods for a uniform distribution, Figure 7 below shows a similar plot for standard deviations under 1°C.

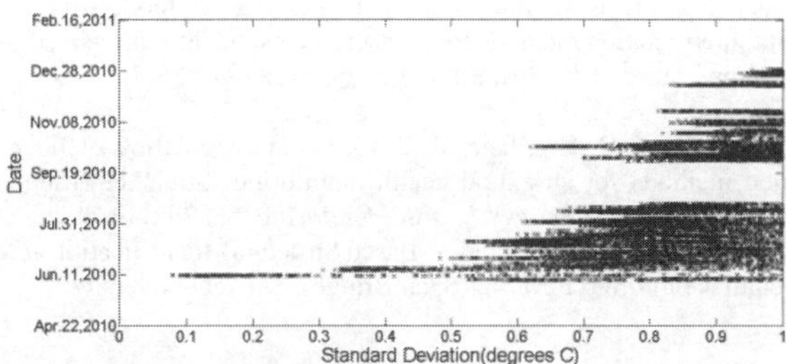

Figure 7. Standard Deviations under 1°C

Doubling the standard deviation cutoff more than doubles the number of time periods found, and more importantly returns a variety of unique time periods. Not only are more time periods found, but a much greater variety of temperatures are located, ranging from June to December 2010. This bigger boundary for the standard deviation unquestionably returns a much better set of distributions to begin building a temperature driven model, but needs to be verified to not adversely affect any strain readings beyond linear thermal effects. The large gap around September corresponds to a time period where the Streicker Bridge was not being actively monitored, rather than a time period with no uniform temperature distributions. With a maximum standard deviation of 1°C, 95% of readings would fall within 2σ or ±2 °C. Using (1), this temperature distribution would result in a strain error of about 24± με. While this error in strain would be worse than the 0.5 °C cutoff, it is still acceptable, as it would correspond to about 5% of the strain resulting from normal seasonal thermal changes of 40-50 °C. A magnitude of 24με is also not likely to be indicative of a serious damage to the structure.

Each of these unique time periods for Figure 7 consists of approximately 20 – 60 distinct time steps. These time steps generally occur over the course of a few hours, as the bridge has settled into a uniform temperature distribution. The time steps will each have corresponding temperature and strain readings. Each temperature reading from the time period can be averaged, and then correlated to the average strain over the time period. This averaging over the hours of uniform temperature will help increase the precision and decrease the uncertainty in the characterization of the strain-temperature relationship. This will be conducted in future work.

4 CONCLUSIONS

The temperature based method for SHM will use temperature as a driving force in an input-output model. In order to begin exploring this temperature to strain and displacement relationship, the sensor data has to be validated.

The raw temperature sensor data from the Streicker Bridge was not adequate to begin performing in depth analysis evaluating the temperature-strain relationship. The data validation process identified eight failed sensors, and then regenerated the failed sensors' data from nearby sensors. Sensor validation and regeneration is currently still ongoing for the Streicker Bridge.

To begin identifying uniform temperature distributions on the bridge, the standard deviation of each time step of measuring was found as an initial method. Using only the sensor failure-free 2010 data, the temperature distribution at each time-step of measurement was found. An appropriate cutoff for a maximum permissible standard deviation was examined by looking at 3 discrete cutoff values. Maximum values of σ = 0.25 and 0.5 °C were found to be inadequate to identify enough unique time periods to begin characterizing the temperature-strain-displacement relationship. σ = 0.5 °C as a boundary provided a moderate amount of unique time periods (5 time periods) but lacked diversity in temperature at these time periods. It was necessary to consider all time steps where the standard deviation of the temperature distribution was less than σ = 1.0 °C. This cutoff provided a large amount of unique temperature distributions, spanning a variety of seasons / temperatures. This boundary was deemed allowable in terms of the strain effects caused by the non-uniformity of the distributions. These non-uniformities can induce non-linearities and unwanted curvatures in the structure, complicating the thermal effects on the strain and displacement. Identifying these uniform temperature distributions allows analysis to begin on the linear thermal effects on the structure.

ACKNOWLEDGEMENT

This research has been supported by National Science Foundation Grant CMMI- 1434455. Any opinions, findings, and conclusions or recommendations expressed in this material are those of the authors and do not necessarily reflect the views of the National Science Foundation.

REFERENCES

Kromanis, R. 2014. Predicting Thermal Response of Bridges Using Regression Models Derived from Measurement, *Computers and Structures* **136**, 64-77.

Laory, I. Trinh, T. & Smith, I. 2011. Evaluating two model-free data interpretation methods for measurements that are influenced by temperature. *Advanced Engineering Informatics* **25(3)**, 495-506.

Sigurdardottir, D. H. Glisic, B. 2015. On-site validation of fiber-optic methods for structural health monitoring: Streicker Bridge. *Journal of Civil Structural Health Monitoring.* **5**, 529-549

Yarnold, M. 2013. Temperature-Based Structural-Identification and Health Monitoring for Long-Span Bridges, Drexel University.

Proceedings of the International Conference on Smart Infrastructure and Construction
ISBN 978-0-7277-6127-9

© The authors and ICE Publishing: All rights reserved, 2016
doi:10.1680/tfitsi.61279.429

A Wavelet-energy based damage identification method for steel bridges

H. Wang[1], and M. Noori[*2] and J. Zhang[1]

[1] *International Institute for Urban Systems Engineering (IIUSE), Southeast University, Nanjing, China*
[2] *California Polytechnic State University, San Luis Obispo, USA& IIUSE, Southeast University, Nanjing, China*
**Corresponding Author*

ABSTRACT Strain is sensitive to damage, especially in steel structures. However, traditional strain gauges do not fit bridge damage identification because they only provide the strain information of the point where they are installed. While traditional strain gauges suffer from some drawbacks, long-gauge FBG strain sensor is capable of providing the strain information of a certain range and all the damage information within the range can be reflected by the strain information provided by FBG sensors. The wavelet transform is a signal processing method to analyze the signals and is capable of providing multiple levels of details and approximations of the signal. In this paper a wavelet packet transform-based damage identification is proposed for the steel bridge damage identifications. The strain data obtained via long-gauge FBG strain sensors are transformed into a modified wavelet packet energy rage index first to identify the location and severity of damage. The results of numerical simulations show that the proposed damage index is a good candidate and is capable of identifying both the location and severity of damage under noise effect.

1 INTRODUCTION

During the service life of structures such as long-span bridges, gantry cranes and frame structures, damages are often observed· Hou and Noori (2000). Those damages may be caused by various factors such as excessive response, accumulative crack growth, or impact by a foreign object. If not monitored properly, damages finally result in structural failure, whose consequences can be fatal. This leads to the necessity of establishing non-destructive techniques for damage detection that are both accurate and practical.

Non-destructive damage identification methods can be classified as local or global damage identification, Fan and Wei (2011). Local damage identification techniques, such as ultrasonic and X- ray methods, require that a priori vicinity of damage is known and readily accessible for testing, which is hard to be guaranteed in the case of some long-span bridges. An ideal damage identification method typically consists of two major components: a network of sensors for collecting response measurements, and a data analysis algorithm for interpretation of the measurements in terms of the physical condition of the structures, Hou and Noori (2000).

Over the past decade, dynamic testing and analysis have been revolutionized by the major advances in the field of structural dynamics and mechanical vibration measurements, Adewuyi and Phillips (2011). Specifically, the advent of Fiber Bragg Gratings (FBG) sensing has provided flexibility and practicality for cases where the traditional sensors are either incapable or not feasible to make appropriate and reliable measurements. FBG sensors are light, compact, flexible and immune to electromagnetics interference, compared with traditional sensors. Moreover, FBG sensors can form large-seale distributed sensing networks with multiple channels easily. FBG sensors have become the most popular fiber-optic sensing system for civil engineering applications, Majumder and Mousumi (2008). They provide real-time quasi distributed sensing whether embedded within or attached on the surface of a structure.

The wavelet transform (WT) is a relatively new way to analyze the signals, which can be viewed as an extension of the traditional STFT transform with adjustable window location and size. Its ability to examine local data with a "zoom lens having an adjustable focus" provides multiple levels of details and approximations of the original signal. Therefore, with wavelet packet transform (WPT), signals can be easily decomposed into separate frequency bands and then analyzed easily.

Damage detection methods using Wavelet packet transform (WPT) have been paid attention to in many fields, such as in civil, aerospace and mechanical engineering. Sami, Yildirim and Poyraz (2008) utilized the wavelet packet component energy as input in-neural network models for damage assessment, which is proved useful for damage detection. Hasan Han, Ren and Sun (2005), utilized WPT and probabilistic modelling for failure prediction. However, few researchers have carried out damage detection using strain signals.

In this article, a new damage index that combines Wavelet Packet Energy and Strain-Time area is proposed for damage identification. Several scenarios are taken into consideration to verify the feasibility of the proposed damage index.

2 WAVELET PACKET TRANSFROM

Wavelet analysis is a signal processing method that relies on the introduction of an appropriate basis and a characterization of the signal by the distribution of amplitude in the basis, Ren and Sun (2005). If the wavelet is required to form a proper orthogonal basis, it has the advantage that an arbitrary function can be uniquely decomposed and the decomposition can be inverted. The wavelet is a smooth and quickly vanishing oscillating function with good localization in both frequency and time. A wavelet family $\Psi_{a,b}(t)$is the set of elementary functions generated by dilations and translations of a unique admissible mother wavelet: Ψ(t)

$$\Psi_{a,b}(t)=\frac{1}{\sqrt{a}}\Psi\left(\frac{t-b}{a}\right) \tag{1}$$

Where $b \in R, a > 0$ are the scale and translation parameters, respectively, and t is time (or location, if wavelet transform is utilized in spatial distributed signal). As the scale parameter a increases, the wavelet becomes wider. So, each parameter shows the signal at different scales and with variable time (or space localization). The wavelet transform (in its continuous or discrete version) correlates the function $f(t)$ with $\Psi_{a,b}(t)$. The continuous wavelet transform (CWT) is the sum over all time of the signal multiplied by a scaled and shifted version of a mother wavelet

$$C(a,b)=\frac{1}{\sqrt{a}}\int_{-\infty}^{+\infty} f(t)\Psi\left(\frac{t-b}{a}\right)dt \tag{2}$$

The results of the transform are wavelet coefficients, which determine how the wavelet function signal expresses the signal.

One possible drawback of the CWT is that in high frequency region the frequency resolution is quitter poor. WPT is one extension of the WT that provides complete level-by-level decomposition. The wavelet packets are alternative bases formed by linear combinations of the usual wavelet functions, Yuan and Zhou (2001). As a result, the WPT enables the extraction of features from signals that combine both stationary and non-stationary characteristics with arbitrary time-frequency resolution. After j levels of decomposition (Figure 1) a signal $f_j^i(t)$ can be expressed as:

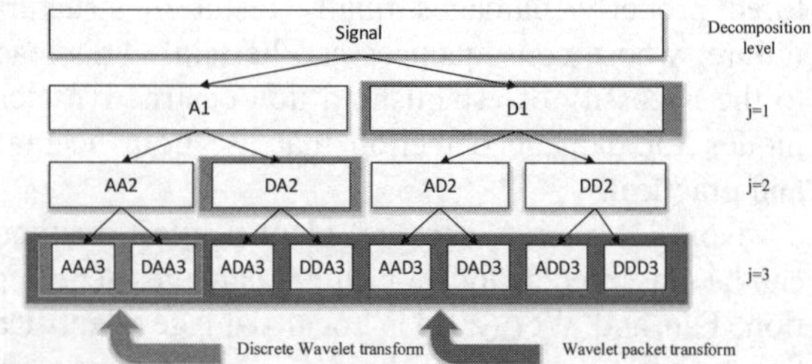

Figure 1 3 level Wavelet Transform and packet transform

$$f_j^i(t) = \sum_{k=-\infty}^{\infty} c_{j,k}^i \psi_{j,k}^i(t) \tag{3}$$

3 MODIFIED WAVELET PACKET ENERGY RATE

The wavelet packet energy representation can provide a more robust signal feature for classification than directly using the wavelet packet coefficients, Yen and Lin (2000) and Sun and Chang (2002). The wavelet packet energy E_{n_j} at j level of node n is defined as:

$$E_{n_j} = \int_{-\infty}^{+\infty} f^2(t)dt = \sum_{m=1}^{2^j}\sum_{n=1}^{2^j}\int_{-\infty}^{+\infty} f_j^m(t) f_j^n(t)dt = \sum_{i=1}^{2^j} E_{n_j^i} \tag{4}$$

where the wavelet packet component energy $E_{n_j^i}$ can be considered to be the energy stored in the component signal $f_j^i(t)$:

$$E_{n_j^i} = \int_{-\infty}^{+\infty} f_j^i(t)^2 dt \tag{5}$$

Strain is more sensitive to damage than other raw data, such as deflection, velocity and acceleration. In certain cases, damage can be directly identified via raw strain data. For structural dynamic responses, the envelope area of strain-time curvature (EASC) is proposed as a damage index:

$$S_n = \int_{-\infty}^{+\infty} f_n(t)dt \tag{6}$$

To quantify the damage from wavelet packet component energies, the modified wavelet packet energy rate (MWPER) is proposed herein to detect both the location and the severity of the damage. MWPER of node n is defined as:

$$D_n = \sum_i^{2^j}\left[\left[\left(E_{n_j^i}\right)_a - \left(E_{n_j^i}\right)_b\right]\times\left[(S_n)_a - (S_n)_b\right]\right] \tag{6}$$

where subscripts a and b stand for damaged, and undamaged status, respectively. Wavelet packet energy rate is capable of utilizing frequency domain information, and the envelope area of strain-time curvature extracts the amplitude information from strain time-history data. Thus, MWPER can utilize both the frequency and amplitude information of the raw data.

4 VALIDATION VIA NUMERICAL MODELS

To illustrate the proposed wavelet identification, the bridges are simplified as a fixed-ends beam. Properties of the beams are as follows: mass density p = 7800kg/m^3, elastic modulus E = 210GPa, length L = 4m, cross section area A = 0.384 x 10“^{3}m^2, and the moment of inertia of the cross section I_x = 0.148 x 10^{6}m^4, I_y = 0.148 x 10^{6}m^4. Geometrical Properties of the beam are marked in Figure 2. Different damage scenarios are considered and listed below:

Table 1. Damage Scenarios

No.	Damage Location (Ld/L)	Hammar Point	Stiffness Loss	No.	Damage Location (Ld/L)	Hammar Point	Stiffness Loss
1	0. 25	0. 75	0.0	8	0. 25	0. 50	9.5
2	0. 25	0. 75	1.3	9	0. 25	0. 25	9.5
3	0. 25	0. 75	5.4	10	0. 25	0. 75	19.5
4	0. 25	0. 75	9.5	11	0. 38	0. 75	9.5
5	0. 25	0. 75	19.5	12	0. 50	0. 75	9.5
6	0. 25	0. 75	32.4	13	0. 63	0. 75	9.5
7	0. 25	0. 75	38.0	14	0. 50+0. 25	0. 75	9.5

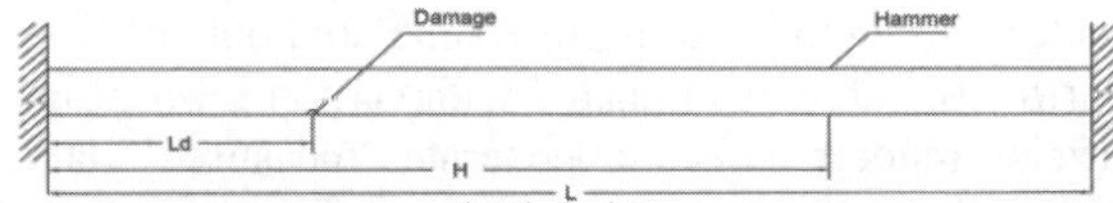

Figure 2. Fix ends beam

4.1 Comparing MWPER Damage Index and Conventional Methods

In this section Scenario No.4 is used to present the effectiveness of MWPER by comparing MWPER, wavelet packet energy rate (WPER) and Envelope Area of Strain-time Curvature (EASC). A single damage with 9.5% stiffness loss is considered and the damage is assumed to be at the 20th element, located lm away from the left support. All the 3 damage indexes are normalized to make the peak values the same (Figure 3).

All 3 damage indexes are capable of indicating the existence of damage in scenario No. 4. While the absolute values of WPER and EASC are much larger than MWPERR on elements, which indicates that WPER and EASC are not so sensitive to small damage and may lead to a false indication of damage in some cases of damage severity. MWPER is stable with a disturbance less than 0.05 on intact elements.

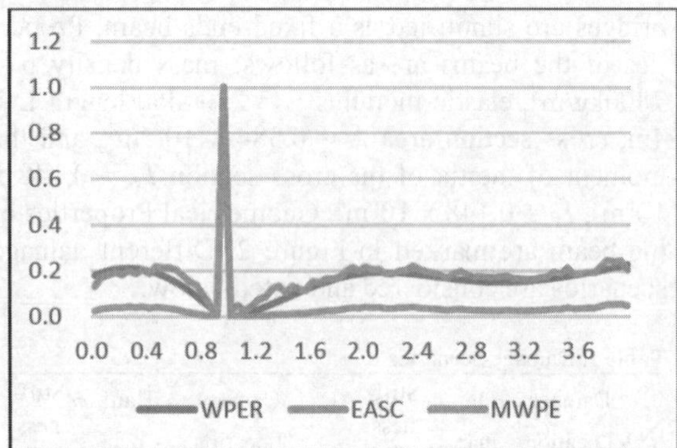

Figure 3. Comparison between MWPER, WPER and EASC

4.2 *Verification of Noise Effect*

All results presented in this article are based on strain time-history data obtained from finite element computations of the response, and hence contain no experimental noise. For real cases, experimental noise is inevitable. To evaluate the robustness of MWPER under measurement noise, the simulated data of Scenario No.4 are contaminated with certain level of artificial random noise to generate 'measured' data. Normally distributed random noises whose amplitude is 2%, 5% and 10%, respectively, of the root-mean-square (RMS) value of strain data are added to the strain time-history data. MWPERs in Figure 4 are normalized to make the comparison clear. With the increase of the noise level, disturbance on intact elements increase, especially elements nearby the ends of the beam. However, under a noise level of 10%, the damage can still be identified clearly.

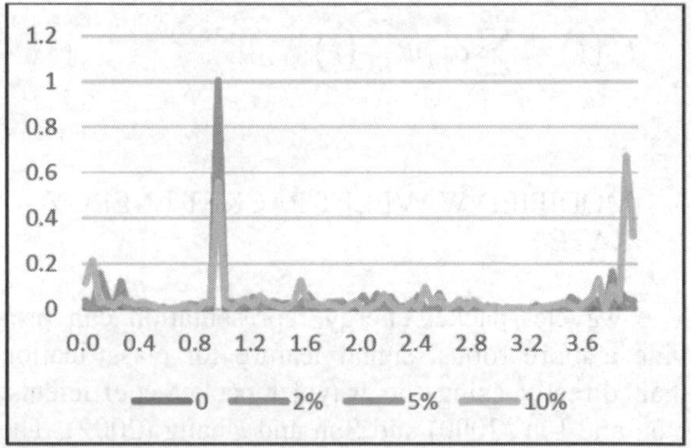

Figure 4. MWPER under measurement noise

4.3 *Effect of Stiffness Loss Levels*

As stated in the previous sections, MWPER is capable of identifying the location of a damage under a certain noise level, which means the 2nd level of damage identification. To accomplish the 3rd level of damage identification, MWPERs of different stiffness loss levels (Scenario No. 1-7) are discussed to show MWPER's ability to quantify the damage. Figure 5 shows the MWPERs on the 20th element with different levels of stiffness loss are different, while MWPERs have a positive correlation with stiffness loss level. With more stiffness loss level data of finite element data, MWPER is capable of identifying stiffness loss level.

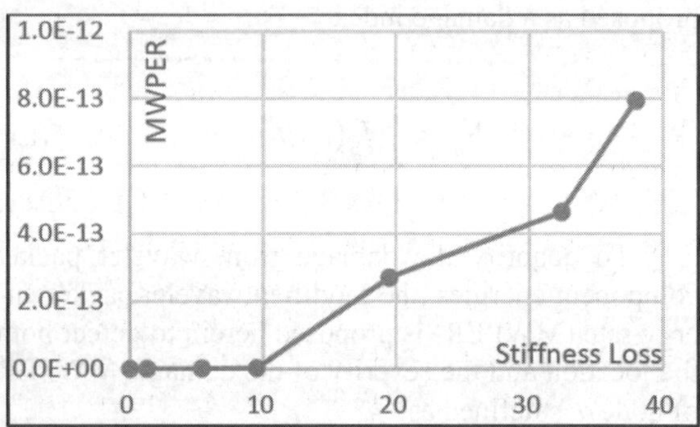

Figure 5. MWPERs of different stiffness loss levels

4.4 *Effects of Damage Location*

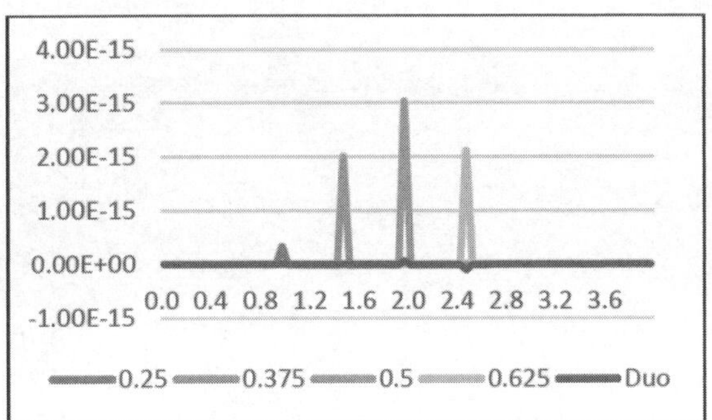

Figure 6. MWPERs of different damage locations

5 CONCLUSIONS

In this article, a wavelet based method, MWPER, isn proposed for bridge damage identification based on FBG sensing. It is applied to 14 different damage scenarios to verify its capacity. It is shown that MWPER is capable of identifying the location of damages on a fixed-end beam under a noise level of 10%. Both stiffness loss and damage location affect MWPER. Under certain excitations, MWPER is capable of quantifying the stiffness loss level. More research needs to be done to quantify stiffness loss under both effects. Instead of hammer impact excitations, running car excitations will be

ACKNOWLEDGEMENT

This research was supported by the International Institute for Urban Systems Engineering of Southeast University, China and by a grant provided by 1000 Program for the Recruitment of Global Experts, and by a special grant, Shuangchuang, provided by Jiangsu Province, China. These supports are gratefully acknowledged.

REFERENCES

Adewuyi, A., Philips, A., & Wu, Z. 2011. Vibration - based damage localization in flexural structures using normalized modal macro-strain techniques from limited measurements. *Computer - Aided Civil and Infrastructure Engineering* **26.3**, 154-172.

Ekici, S., Selcuk Y., & Poyraz. M. 2008. Energy and entropy-based feature extraction for locating fault on transmission lines by using neural network and wavelet packet decomposition. *Expert Systems with Applications* **34.4**, 2937-2944.

Fan, W., & Pizhong, Q. 2011. Vibration-based damage identification methods: a review and comparative study. *Structural Health Monitoring* **10.1**, 83-111.

Han, J.G., Ren, W.X., & Sun, Z.S. 2005. Wavelet packet based damage identification of beam structures. International *Journal of Solids and Structures* **42.26**, 6610-6627.

Hou, Z., Noori, M., & Amand, R. 2000. Wavelet-based approach for structural damage detection. *Journal of Engineering Mechanics* **126.7** (2000): 677-683.

Lau, K.T., Yuan L, and Zhou L.M., 2001. Strain monitoring in FRP laminates and concrete beams using FBG sensors, *Journal of Composite Structures,* **51(1)**, 9-20.

Majumder, A., & Mousumi, E. 2008. Fibre Bragg gratings in structural health monitoring—Present status and applications. *Sensors and Actuators A: Physical* **147.1**, 150-164.

Ren, W.X., Sun, Z.S., 2008. Structural damage identification by using wavelet entropy. Engineering Structures **30.10**, 2840-2849.

Sun Z, Chang C.C. 2002. Structural damage assessment based on Wavelet Packet Transform, *Journal of Structural Engineering*, **128(10)**, 354-1361.

Yen, G.G., & Lin, K.C. 2000. Wavelet packet feature extraction for vibration monitoring, *Journal of Industrial Electronics, IEEE Transactions* on, **47(3)**, 650-667.

Proceedings of the International Conference on Smart Infrastructure and Construction
ISBN 978-0-7277-6127-9

© The authors and ICE Publishing: All rights reserved, 2016
doi:10.1680/tfitsi.61279.435

Effect of vertical excitation on horizontal performance of sloped rolling-type seismic isolation bearings

S.J. Wang[*1], C.H. Yu[2], W.C. Lin[1], J.S. Hwang[3] and K.C. Chang[2]

[1] *National Center for Research on Earthquake Engineering, Taipei, Taiwan*
[2] *National Taiwan University, Taipei, Taiwan*
[3] *National Taiwan University of Science and Technology, Taipei, Taiwan*
[*] *Corresponding Author*

ABSTRACT Past researches have demonstrated that the adoption of sloped rolling-type seismic isolation bearings can meet the rigorous performance-based design requirements for the protected object, especially for high-precision equipment. It is because that the isolation bearing has an excellent and steady horizontal acceleration control performance as well as an efficient inherent gravity-based self-centering capability. In addition, the built-in friction damping mechanism can facilitate the isolation bearing to suppress excessive horizontal displacement responses and to stop rolling motion respectively during and after excitation. This study, first, reviewed a generalized analytical model for the sloped rolling-type isolation bearing in which the two V-shaped surfaces in contact with cylindrical rollers are designed with arbitrary sloping angles. A series of numerical studies using the developed analysis program were performed to investigate the influence arising from vertical excitation on the horizontal control performance of the isolation bearing. The numerical results indicated that if the vertical excitation is large sufficiently, it will quite affect the horizontal control performance of the isolation bearing. Under the circumstances, adopting the exact generalized analytical model, rather than the simplified one, can obtain more precise and conservative analysis and design results for the isolation bearing.

1 INTRODUCTION

Rolling-based metallic seismic isolation bearings in which either a ball rolls on concave (Zhou et al. 1998, ISO-Base^{TM}) or conical (Kasalanati et al. 1997) surfaces or a rod rolls on curved (Jangid & Londhe 1998, Mahmood & Amirhossein 2011, CRS) or sloped (Tsai et al. 2007, Lee et al. 2010, Ou et al, 2010, Wang et al. 2014) surfaces (or rails) have been numerically and experimentally demonstrated for their excellent horizontal seismic isolation performance. Since the rolling friction force and the restoring force due to gravity are very limited and much smaller than the input seismic force, the rolling motion can be activated immediately once an earthquake occurs, and the horizontal acceleration transmitted to the protected object can be significantly reduced during an earthquake. Besides, such isolation bearings have an efficient inherent gravity-based self-centering capability (without permanent dislocation) after excitation.

Among various types of such isolation bearings, the sloped rolling-type seismic isolation bearing attracted much attention recently. In addition to possessing the advantages aforementioned, the design of constant sloping surfaces in contact with cylindrical rollers can ensure that the transmitted horizontal acceleration remains essentially constant regardless of the excitation intensity. Thus, the isolation bearing does not have a fixed vibration natural period, and can offer maximum horizontal decoupling between the protected object and input excitation. It is worth noting that the zero post-elastic stiffness performance, i.e. the constant transmitted horizontal acceleration performance, is the most attractive feature.

This feature can make the isolation bearing easily meet the rigorous performance-based design requirements for structural and nonstructural systems if the maximum transmitted acceleration response is selected as the seismic performance criterion. Furthermore, to have a better displacement control and more effectively stop rolling motion respectively during and after excitation, the built-in damping mechanism provided by additional sliding friction was employed (Lee et al. 2010, Ou et al, 2010, Wang et al. 2014).

The sloped rolling-type seismic isolation bearings discussed in Lee's [2010[a], 2010[b]] and Wang's [2014] studies were basically composed of three bearing plates (denoted as upper, intermediate, and lower bearing plates hereafter) and cylindrical rollers. In Lee's study, the upper and lower bearing plates were designed with flat surfaces in contact with rollers, while the intermediate bearing plate was designed with a V-shaped surface. In Wang's study, three bearing plates were designed with V-shaped surfaces in contact with rollers, and the sloping angles were identical. Because of different designs of rolling surfaces, the equations of motion derived in Lee's and Wang's studies were essentially different and cannot be exchanged. Therefore, Lin (2015) proposed a generalized analytical model for the sloped rolling-type seismic isolation bearing in which the rollers move between two V-shaped surfaces designed with arbitrary sloping angles. The generalized analytical model was theoretically and experimentally verified to be capable of describing the dynamic behavior of the two different isolation bearings respectively discussed in Lee's and Wang's studies.

In this study, the generalized analytical model is adopted to numerically discuss the influence arising from vertical excitation on the transmitted horizontal acceleration response of the sloped rolling-type seismic isolation bearing. According to the quantitative discussion results, designers should pay more attention to the horizontal performance of the isolation bearing when the acceleration excitation possesses a considerable vertical component.

2 GENERALIZED EQUATIONS OF MOTION

A simplified model, a cylindrical roller sandwiched between two V-shaped surfaces designed with different sloping angles θ_1 and θ_2, to represent the dynamic behavior of a sloped rolling-type seismic isolation bearing in one principle horizontal direction and in the vertical direction is illustrated in Figure 1(a), in which M, m_1, and m_2 are the seismic reactive masses of the protected object, superior bearing plate, and roller, respectively; r is the radius of the roller; and θ_1 and θ_2 are the sloping angles of the superior and inferior bearing plates, respectively. The free body diagrams when $\mathrm{sgn}(x_1)=\mathrm{sgn}(x_2)=1$ and $\mathrm{sgn}(\dot{x}_1)=\mathrm{sgn}(\dot{x}_2)=1$ are shown in Figure 1(b), in which $\ddot{x}_g\,(\ddot{z}_g)$ is the horizontal (vertical) acceleration excitation; $x_1\,(z_1)$, $\dot{x}_1\,(\dot{z}_1)$, and $\ddot{x}_1\,(\ddot{z}_1)$ are the horizontal (vertical) displacement, velocity, and acceleration responses of the protected object and superior bearing plate relative to the origin O, respectively; $x_2\,(z_2)$, $\dot{x}_2\,(\dot{z}_2)$, and $\ddot{x}_2\,(\ddot{z}_2)$ are the horizontal (vertical) displacement, velocity, and acceleration responses of the roller relative to the origin O, respectively; the positive directions of x and z are correspondingly defined to be rightward and upward in the figure; g is the acceleration of gravity; I is the moment of inertia of the roller; α is the angular acceleration of the roller (the positive rotation is defined to be clockwise in the figure); $f_1\,(f_2)$ and $N_1\,(N_2)$ are the rolling friction force and normal force acting between the superior bearing plate and roller (between the roller and inferior bearing plate), respectively; and F_D is the built-in friction damping force acting parallel to the slope of the bearing plates.

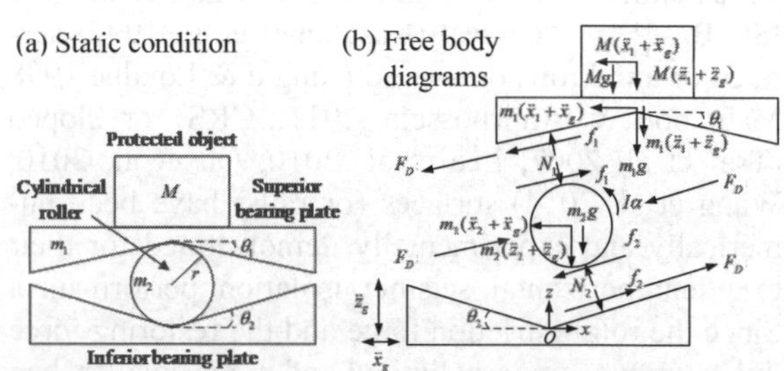

Figure 1. Simplified model.

By taking the free body diagrams of the superior bearing plate and roller, and considering the compatibility conditions, a total of nine variables, α, $\ddot{x}_1$, $\ddot{x}_2$, $\ddot{z}_1$, $\ddot{z}_2$, N_1, N_2, f_1, and f_2, can be solved. With reasonable neglect of the term of $m_2/(M+m_1)$ due to the fact that m_2 is in general much smaller than $(M+m_1)$, the exact mathematical form of the solution for $\ddot{x}_1$ is given in Equation (1)

$$\begin{aligned}\ddot{x}_1 = &\frac{-(\cos\theta_1+\cos\theta_2)}{2(M+m_1)\left[1+\cos(\theta_1-\theta_2)\right]}\{2F_D\,\mathrm{sgn}(\dot{x}_1)\\ &+(M+m_1)\left[\ddot{x}_g(\cos\theta_1+\cos\theta_2)\right.\\ &\left.+(g+\ddot{z}_g)(\sin\theta_1+\sin\theta_2)\mathrm{sgn}(x_1)\right]\}\end{aligned} \tag{1}$$

Hereafter, Equation (1) is generally called the exact generalized equations of motion. The following two assumptions for further simplicity are taken into consideration: (1) assuming the sloping angles θ_1 and θ_2 are small enough, the higher order terms of θ_1 and θ_2 are negligible; and (2) assuming the term of $\ddot{z}_g$ is neglected. Thus, $\ddot{x}_1$ can be approximated by

$$\begin{aligned}\ddot{x}_1 = &\frac{-(\cos\theta_1+\cos\theta_2)}{4(M+m_1)}\left[2F_D\,\mathrm{sgn}(\dot{x}_1)\right.\\ &\left.+(M+m_1)g(\sin\theta_1+\sin\theta_2)\mathrm{sgn}(x_1)\right]-\ddot{x}_g\end{aligned} \tag{2}$$

Hereafter, Equation (2) is generally called the simplified generalized equations of motion. By using the exact and simplified generalized equations of motion respectively given in Equations (1) and (2), a generalized analytical model can be provided to analyze and design the sloped rolling-type seismic isolation bearing.

3 ANALYSIS PROGRAM

The Visual Basic of Application (VBA) is adopted to develop the analysis program in this study. The Newmark-β method with constant average acceleration is used for solving the exact and simplified generalized equations of motion at each time step. Note that during two or more consecutive time steps in numerical integration, it is possible to have no relative motion between the superior and inferior bearing plates of the simplified model shown in Figure 1. This phenomenon is defined as the relative static state hereafter. Thus, let $\ddot{x}_1$ in Equations (1) and (2) equal to zero, the critical characteristic strengths due to the built-in friction damping force, $F_{D,C}$, can be respectively calculated as

$$\begin{aligned}F_{D,C} = &\left|\frac{1}{2}(M+m_1)\left[\ddot{x}_g(\cos\theta_1+\cos\theta_2)\right.\right.\\ &\left.\left.+\mathrm{sgn}(x_1)(g+\ddot{z}_g)(\sin\theta_1+\sin\theta_2)\right]\right|\end{aligned} \tag{3}$$

$$\begin{aligned}F_{D,C} = &\left|\frac{1}{2}(M+m_1)\left[2\ddot{x}_g\frac{1+\cos\theta_1\cos\theta_2}{\cos\theta_1+\cos\theta_2}\right.\right.\\ &\left.\left.+\mathrm{sgn}(x_1)g(\sin\theta_1+\sin\theta_2)\right]\right|\end{aligned} \tag{4}$$

It can be seen that $F_{D,C}$ is dependent on $\ddot{x}_g$ and $\ddot{z}_g$ (only $\ddot{x}_g$ for the simplified generalized equation of motion). If $F_{D,C}$ is calculated to be smaller than the design built-in friction damping force F_D at a specific time step (e.g. at t_i), the relative static state occurs. Under this circumstance, at the time step t_i, $\ddot{x}_1$ and $\dot{x}_1$ should be modified to be zero, and x_1 should remain the same as that at the previous time step t_{i-1}. In order to have a better acceleration control performance for the sloped rolling-type seismic isolation bearing, Wang (2014) suggested employing a pounding prevention mechanism—an arc rolling range with a fixed curvature radius which is larger than the roller radius. Therefore, in this analysis program, the arc rolling range is rationally assumed to be composed of many infinitesimal segments designed with a fixed and continuously varied sloping angle.

4 NUMERICAL STUDY

4.1 *Design Cases*

A simplified model composed of a cylindrical roller with a sectional radius (r) of 15*mm*, a superior bearing plate with a sloping angle of θ_1, and an inferior bearing plate with a sloping angle of θ_2, as shown in Figure 1(a), is employed. Twenty-five design cases are numerically studied, in which a commonly used combination of θ_1 and θ_2 respectively varying from 2 to 6 degrees with an increment of 1 degree is designed, i.e. (θ_1, θ_2)=(2°~6°, 2°~6°). To control the maximum transmitted horizontal acceleration response not greater than 0.15*g* according to the simplified generalized equation of motion given in Equation (2), and to have an acceptable self-centering performance, the built-in friction damping forces F_D are designed to be the same for the twenty-five cases and are dominated by the design case of (θ_1, θ_2)=(6°, 6°), i.e. F_D=301*N*. The protected object above the isolation bearings is assumed to be an important facility, and the total seismic reactive mass of the protected facility and superior bearing plate, $(M+m_1)$, is designed to be 1000*N-sec²/m*. The protected facility is ideally regarded as a single degree-of-freedom system in this numerical study for simplicity. Considering the pounding prevention mechanism suggested in Wang's study (2014), an arc rolling range of 21.6*mm* is designed at the intersection of two inclines of V-shaped surfaces. Therefore, the curvature radiuses (*R*) are designed to be 618*mm*, 206*mm*, 155*mm*, 124*mm*, and 104*mm* corresponding to the sloping angles varying from 2 to 6 degrees, respectively.

As detailed in Table 1, three recorded earthquake histories, denoted as ElCentro, Kobe, and ChiChi hereafter, as well as three generated artificial acceleration histories compatible with the required response spectra (RRS) specified in AC156 (2007) and IEEE Std 693™-2005 (2006), denoted as AC156-1, AC156-2, and IEEE hereafter, are adopted as the unilateral and biaxial acceleration inputs with different peak acceleration (PA) scales, i.e. 25%, 50%, 75%, and 100% of the original PA value.

Table 1. Acceleration input program

Acc. input	Description	Excitation direction		100% PA (*g*)
El Centro	IMPVALL/I-ELC180 IMPVALL/I-ELC-UP Imperial Valley, U.S., 1940/05/19	Unilateral	X	0.31
		Biaxial	X	0.31
			Z	0.21
Kobe	KOBE/KJM000 KOBE/KJM-UP Kobe, Japan, 1995/01/16	Unilateral	X	0.82
		Biaxial	X	0.82
			Z	0.34
ChiChi	CHICHI/CHY028-N CHICHI/CHY028-V Chi-Chi, Taiwan, 1999/09/21	Unilateral	X	0.76
		Biaxial	X	0.76
			Z	0.34
AC 156-1	RRS in AC156 3rd floor (8.75m in elevation) of a 7-story building (24m in height) at Taipei City	Unilateral	X	0.5
		Biaxial	X	0.5
			Z	0.25
AC 156-2	RRS in AC156 3rd floor (8m in elevation) of a 3-story building (12m in height) at Nantou County	Unilateral	X	1.00
		Biaxial	X	1.00
			Z	0.5
IEEE	RRS in IEEE Std 693™-2005 for high performance level	Unilateral	X	1.00
		Biaxial	X	1.00
			Z	0.80

4.2 *Numerical Comparison*

The hysteresis loops of different design cases (θ_1 and θ_2 respectively varying from 2 to 6 degrees with an increment of 2 degrees) obtained from the exact and simplified generalized equations of motion under unilateral and biaxial inputs of 100%-Kobe are presented in Figure 2, in which BE and UE denote the numerical predictions by using the exact generalized equation of motion given in Equation (1) under biaxial and unilateral excitation, respectively, and US denotes those by using the simplified one given in Equation (2) under unilateral excitation. By comparing the BE and UE results, the influence of vertical excitation on the horizontal acceleration performance can be discussed. It is found that when the roller moves between two sloped surfaces, the BE result reflects many obvious fluctuations rather than a perfect constant. However, when within the fixed curvature range (or the arc rolling range), the difference of the BE and UE results is not very significant. For further discussion, two indices, R^2 and ER, are defined as

$$R^2 = 1 - \frac{\sum_{i=1}^{m}\left[(A_{BE})_i - (A_{UE})_i\right]^2}{\sum_{i=1}^{m}\left[(A_{BE})_i - (A_{BE})_{mean}\right]^2} \tag{5}$$

$$ER = \frac{\max(|A_{BE}|) - \max(|A_{UE}|)}{\max(|A_{UE}|)} \times 100\% \tag{6}$$

where A_{UE} and A_{BE} are the horizontal acceleration responses under the conditions of UE and BE, respectively; R^2 is the coefficient of determination to quantitatively evaluate the difference of A_{BE} and

A_{UE} (i.e. with and without considering the vertical excitation); ER is the error ratio to quantitatively evaluate the maximum underestimate in horizontal acceleration responses if the vertical excitation is not considered; the subscript i represents the analysis data point at time t_i; m represents the total number of analysis data points; and the subscript *mean* represents the mean value of total analysis data points.

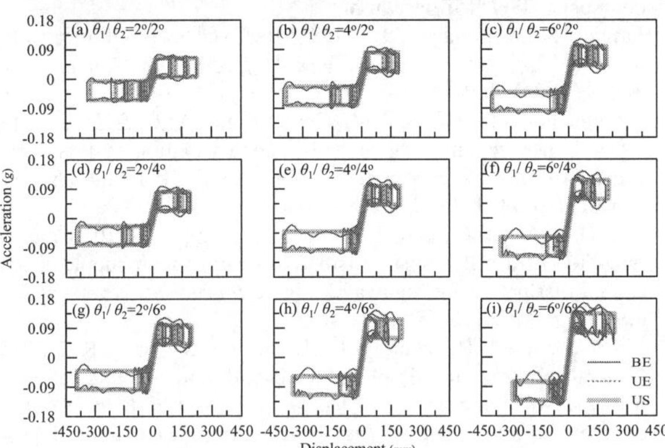

Figure 2. Comparison of hysteresis loops under 100%-Kobe.

The variations of R^2 and ER with different vertical PA values under all the biaxial acceleration inputs are shown in Figures 3 and 4, respectively. The variations of ER with different vertical PA values for the design cases of (θ_1, θ_2)=(2°, 2°), (3°, 3°), (4°, 4°), (5°, 5°), and (6°, 6°) under all the biaxial acceleration inputs are shown in Figure 5. To focus on the dynamic behavior of the sloped rolling-type seismic isolation bearing when the roller moves between two sloped surfaces, as well as to avoid erroneous judgment due to small excitation and/or responses, the calculation results of the roller moving within the fixed curvature range (or the arc rolling range) are excluded in the figures and not discussed in the following.

It can be seen from Figure 3 that a large number of R^2 values are farther from 1. In this numerical study, the minimum R^2 value is 0.675 for the design case of (θ_1, θ_2)=(5°, 3°) under biaxial-100%-Kobe, which might be an unacceptable value in engineering practice. As observed from Figure 4, the difference between the maximum horizontal acceleration predictions by using the exact generalized equation of motion under unilateral and biaxial excitation, i.e. the ER value, might be somewhat noticeable. It is apparent that a larger vertical PA value will lead to a more significant effect vertical excitation on the ER value. As observed from Equation (1), the term of $\ddot{z}_g$ is multiplied by a parameter which is related to θ_1 and θ_2, i.e. $(\cos\theta_1+\cos\theta_2)(\sin\theta_1+\sin\theta_2)/[1+\cos(\theta_1-\theta_2)]$. Therefore, to more clearly understand the tendency of variations of ER with various vertical PA values, the calculation results of five design cases, (θ_1, θ_2)=(2°, 2°), (3°, 3°), (4°, 4°), (5°, 5°), and (6°, 6°), under all the biaxial acceleration inputs are marked with different symbols in Figure 5. It is very evident from the figure that the larger the vertical PA value, the more significant effect of vertical excitation on the ER value is. In this numerical study, the maximum ER value is 36.59% for the design case of (θ_1, θ_2)=(2°, 2°) under biaxial-100%-IEEE, which might not be very acceptable in engineering practice either.

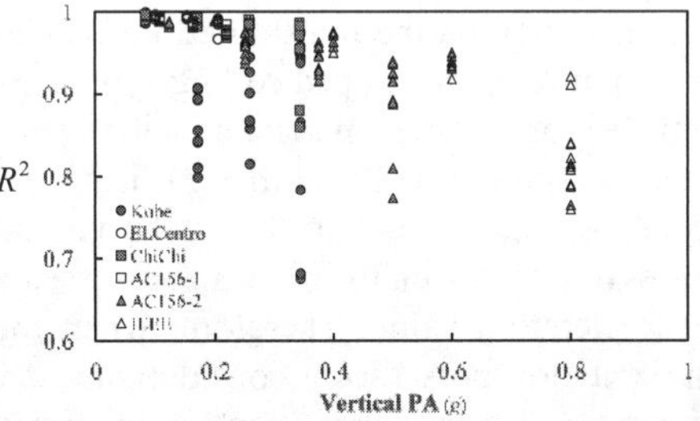

Figure 3. Variations of R^2 with different vertical PA scales.

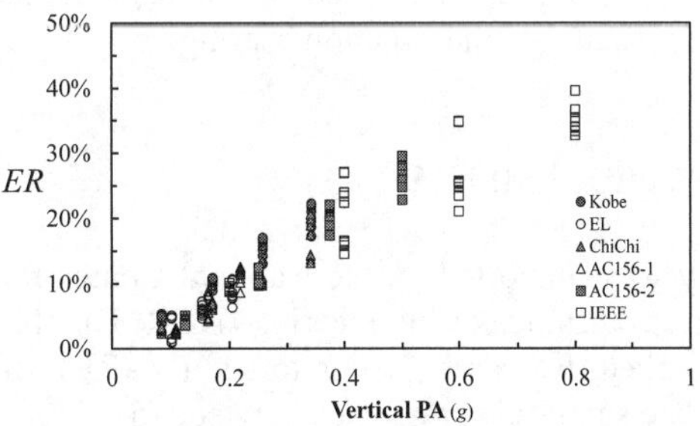

Figure 4. Variations of ER with different vertical PA scales.

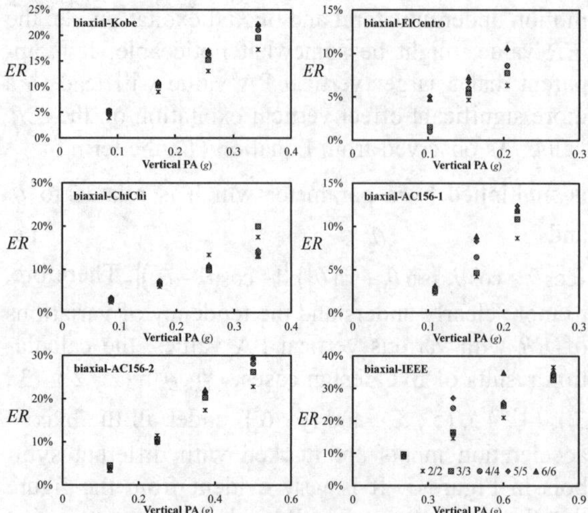

Figure 5. Variations of *ER* with different vertical PA scales for $(\theta_1, \theta_2) = (2°, 2°)$, $(3°, 3°)$, $(4°, 4°)$, $(5°, 5°)$, and $(6°, 6°)$.

5 CONCLUSIONS

In this study, a series of numerical studies are performed to quantitatively discuss the influence arising from vertical excitation on the transmitted horizontal acceleration response of the sloped rolling-type seismic isolation bearing. The numerical result reveals that the vertical excitation will lead to significant underestimate in the maximum transmitted horizontal acceleration response, especially when it possesses a larger peak acceleration value. Therefore, when the acceleration excitation possesses a considerable vertical component, adopting the exact generalized equations of motion, rather than the simplified ones, can obtain more precise and conservative analysis and design results for the isolation bearing.

ACKNOWLEDGEMENTS

The study was supported by the National Center for Research on Earthquake Engineering (NCREE), National Applied Research Laboratories (NARL) of Taiwan. The support is greatly acknowledged.

REFERENCES

AC156 2007. Acceptance criteria for seismic qualification by shake-table testing of nonstructural components and systems, ICC Evaluation Service Inc..

CRS, cosine curved rail system. http://www.oiles.co.jp/en/menshin/building/units/crs.html.

IEEE Std 693[TM]-2005 2006. IEEE recommended practice for seismic design of substations, Institute of Electrical and Electronics Engineers (IEEE) Power Engineering Society, NY.

ISO-Base[TM] seismic isolation platform. http://www.worksafetech.com/ pages/ISO %5FBase.html.

Jangid, R.S. & Londhe Y.B. 1998. Effectiveness of elliptical rolling rods for base isolation, *Journal of Structural Engineering, ASCE*, **124**(4), 469-472.

Kasalanati, A., Reinhorn, A.M., Constantinou, M.C. & Sanders, D. 1997. Experimental study of ball-in-cone isolation system, *Proceedings of the ASCE Structures Congress XV*, Portland, Oregon, April 13-16, 1191–1195.

Lee, G.C., Ou, Y.C., Niu, T., Song, J. & Liang, Z. 2010[a]. Characterization of a roller seismic isolation bearing with supplemental energy dissipation for highway bridges, *Journal of Structural Engineering, ASCE*, **136**(5), 502-510.

Lin, W.C., Yu, C.H., Wang, S.J., Hwang, J.S. & Chang, K.C. 2015. Generalized exact and simplified analytical models for sloped rolling-type isolation bearings. *Proceedings of the 14th World Conference on Seismic Isolation, Energy Dissipation and Active Vibration Control of Structures (14WCSI)*, San Diego, U.S.A., September 9-11.

Mahmood, H. & Amirhossein, S. 2011. Using orthogonal pairs of rollers on concave beds (OPRCB) as a base isolation system - part I: analytical, experimental and numerical studies of OPRCB isolators, *The Structural Design of Tall and Special Buildings*, 20(8), 928-950.

Ou, Y.C., Song, J. & Lee, G.C. 2010[b]. A parametric study of seismic behavior of roller seismic isolation bearings for highway bridges, *Earthquake Engineering and Structural Dynamics*, 39(5), 541-559.

Tsai, M.H., Wu, S.Y., Chang, K.C. & Lee, G.C. 2007. Shaking table tests of a scaled bridge model with rolling type seismic isolation bearings, *Engineering Structures*, 29(9), 694-702.

Wang, S.J., Hwang, J.S., Chang, K.C., Shiau, C.Y., Lin, W.C., Tsai, M.S., Hong, J.X. & Yang, Y.H. 2014. Sloped multi-roller isolation devices for seismic protection of equipment and facilities, *Earthquake Engineering and Structural Dynamics*, 43(10), 1443-1461.

Zhou, Q., Lu, X., Wang, Q., Feng, D. & Yao, Q. 1998. Dynamic analysis on structures base-isolated by a ball system with restoring property. *Earthquake Engineering and Structural Dynamics*, 27(8), 773-791.

Proceedings of the International Conference on Smart Infrastructure and Construction
ISBN 978-0-7277-6127-9

© The authors and ICE Publishing: All rights reserved, 2016
doi:10.1680/tfitsi.61279.441

Analytical simulation on seismic mitigation of raised floor systems using sloped rolling-type isolation devices and magnetorheological dampers

Pei-Ching Chen* and Shiang-Jung Wang

National Center for Research on Earthquake Engineering, Taipei, Taiwan
** Corresponding Author*

ABSTRACT Sloped rolling-type isolation device has been demonstrated as one of the most effective approaches to mitigate seismic damages of equipment and facilities above the raised floor system for high-tech fabrication. It provides excellent horizontal decoupling between the protected facilities and input excitation since its dominant vibration period is absent. Therefore, the dynamic response of the isolated object becomes less dependent on the frequency content of input excitation. Although supplemental damping can advance the isolation device to suppress excessive displacement responses during strong excitation, it also increases the absolute accelerations above the raised floor system and results in fabrication shutdown or even failure of the protected facilities. In this study, analytical simulation of a conceptual smart isolation system for raised floor systems are performed to investigate its feasibility and effectiveness. This system consists of four sloped-rolling type isolation devices and one magnetorheological damper in each horizontal direction. In addition to passive control, four control algorithms are designed and applied to regulate the input voltage for the magnetorheological damper including fuzzy-logic control, switching control, LQR control, and model-based control. Four spectrum compatible artificial ground motions are adopted in the simulation. Different structural periods are considered in the simulation in order to investigate the efficiency and adaptiveness of the smart isolation system. Simulation results from various types of linear and nonlinear control strategies are discussed and summarized. It reveals that the smart base isolation system can effectively protect the equipment and facilities above the raised floor system with low power consumption and high reliability.

1 INTRODUCTION

Raised floor systems are elevated above the building slabs in order to create hidden space for wiring mechanical and electrical cables as well as allocating facilities and sensors. Raised floors are widely used in modern buildings, especially for high-tech fabrication laboratories, data centers, and telecommunications applications. Since raised floors are installed at a certain height from the slab, additional structural support and lighting are required. Recent reports indicate that the equipment placed on the raised floor systems could be severe damaged under small and medium earthquakes even though the buildings and structures remain intact. For instances, servers, computers, networks and telecommunications of high-tech companies may be malfunctioned which could lead to enormous pecuniary loss, generally several times the cost of facilities after an earthquake. Therefore, it has become a significant issue to enhance the seismic performance of the raised floor systems for the past decade.

Seismic isolation technology has been regarded as one of the most effective approaches to mitigate seismic risks and potential damages of high-tech equipment and facilities. In particular, sloped rolling-type isolation device with multi-roller and built-in damping mechanisms has been demonstrated to effectively reduce the transmitted acceleration above the raised floor system (Wang et al. 2014). The device is composed of three bearing plates with constant sloping angles and two pairs of orthogonal rollers as shown in Fig. 1. In each horizontal direction, the rolling mechanism of two rollers provides the in-plane seismic isolation capability. Main characteristics of this isolation device include: (a) the acceleration transmitted to the protected object remains constant regardless of the earthquake intensity, and (b) the isolation device is self-centering after earthquake excitations. In addition, an embedded adjustable linear spring modules can be designed to generate required normal forces, increasing the sliding friction forces between the side plates and bearing plates. This built-in damping module has an advantage of suppressing excessive displacement responses during severe excitations. However, it is predicted that the floor acceleration at high-rise buildings could lead to significant displacement of the sloped rolling-type isolation devices due to the dominant long-period responses. As a result, the isolation devices incorporated into active or semi-active control, which are generally referred as smart isolation system (Ramallo et al. 2002; Lu and Lin 2008) could provide a solution for mitigating the seismic responses of raised floor systems subjected to long-period excitations.

In this paper, preliminary analytical simulation on seismic mitigation of raised floor systems was conducted. The sloped rolling- type isolation devices were selected to isolate the excitation input to the raised floor systems. In addition, magnetorheological (MR) dampers were adopted to provide the semi-active controllability for the isolation devices. Four simple control algorithms were designed and applied to the smart isolation system to investigate the seismic performance on both acceleration and displacement transmitted to the raised floor.

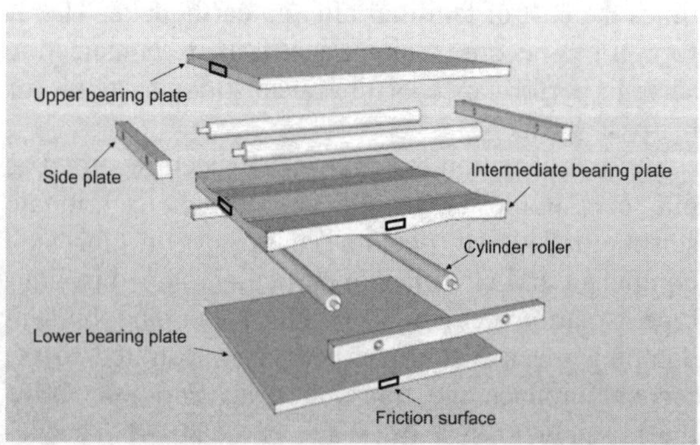

Figure 1. Illustration of the sloped rolling-type isolation device.

2 ANALYTICAL MODEL

The analytical simulation in this preliminary study contains three parts: (a) four sloped rolling-type isolation devices, (b) equipment placed on the raised floor, and (c) a MR damper to provide controllability. The schematic setup of this analytical study is shown in Fig. 2. It is noted that a rigid mass was used to represent the equipment and the acceleration on the raised floor was assumed identical to that on the top of the isolators.

MATLAB/Simulink was used to perform the analytical simulation. All the simulations were conducted using a sampling rate of 2000Hz. The ode5 solver of MATLAB/Simulink using the Dormand-Prince Formula was adopted in the simulation. It was assumed that each raised floor system is supported by four sloped rolling-type isolation devices. The mass of the equipment placed on the raised floor was 2420 N-s^2/m which indicates that each isolation device takes the mass of 605 N-s^2/m.

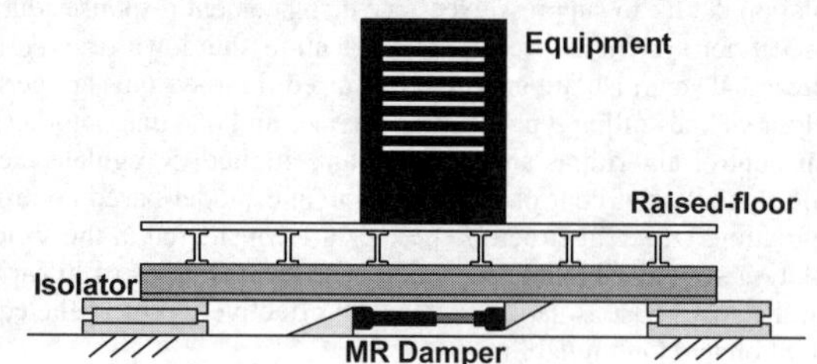

Figure 2. Schematic setup of the analytical study.

2.1 *Rolling-type isolation device*

The multi-roller bearing developed by Wang et al. was adopted as the sloped rolling-type isolation device in this paper. Each single roller of the device is allocated between a flat surface and a V-shaped surface with a sloping angle of θ in the principle horizontal direction. By neglecting the effect of the vertical ground acceleration, the simplified equation of motion in the horizontal direction when the roller moves apart from the fixed curvature can be represented as

$$\ddot{x} + \ddot{x}_g = -\frac{1}{2} g \sin\theta \,\mathrm{sgn}(x) - \frac{[\mu_r N + F_d]}{M + m} \mathrm{sgn}(\dot{x}) \tag{1}$$

where x, $\dot{x}$, and $\ddot{x}$ are the relative displacement, relative velocity, and relative acceleration on the isolation device; the parameters g and $\ddot{x}_g$ represent the gravity and ground acceleration, respectively; M and m are the mass of the protected object and the upper plate, respectively; μ_r is the ratio of the rolling resistant coefficient between the roller and the bottom plate; F_d is the inherent damping force of the device; and N is the normal force acting between the upper plate and the roller. When the roller moves within the fixed curvature range, the simplified equation of motion in the horizontal direction can be expressed as

$$\ddot{x} + \ddot{x}_g = -\frac{g}{4R}x - \frac{[\mu_r N + F_d]}{M + m}\mathrm{sgn}(\dot{x}) \tag{2}$$

where R is the fixed curvature radius between two inclines of the V-shaped surface of the bottom plate.

Only one horizontal direction was considered in the analytical simulation. The mass of the upper plate was 48.88 N-s^2/m. The sloping angle of the V-shaped surfaces was 6 degrees. The curvature radius of the arc was 100 mm. The inherent damping force was assumed 10 N. The rolling resistant coefficient between the roller and the upper plate, μ_r, was assigned as 0.002.

2.2 *Magnetorheological damper*

It is known that a MR damper is filled with MR fluid and can be controlled by a magnetic field by varying the input power of the electromagnet. Therefore, a MR damper was adopted to provide the isolation device with controllability in the analytical simulation. The mechanical model proposed by Spencer et al. was used to simulate the dynamic response of the MR damper as shown in Fig. 3. The governing equation of the model can be expressed as

$$c_1\dot{y} = \alpha z + k_0(x - y) + c_0(\dot{x} - \dot{y}) \tag{3}$$

where the evolutionary variable z is

$$\dot{z} = -\gamma|\dot{x} - \dot{y}|z|z|^{n-1} - \beta(\dot{x} - \dot{y})|z|^n + A(\dot{x} - \dot{y}) \tag{4}$$

After solving Eqs. (3) and (4), the total force F can be written as

$$F = c_1\dot{y} + k_1(x - x_0) \tag{5}$$

The parameters for the MR damper in the analytical simulation were identical to those identified by Spencer et al., but slightly scaled to the force level satisfying the condition for the isolation device. The maximum allowable input voltage to the MR damper was 2.25 V. The responses of the modified MR damper subjected to 1-Hz, 50-mm sinusoidal displacement with a variety of input voltages are shown in Fig. 4. It appears that the modified MR damper can provide a maximum force close to 1200 N which is approximately 5% the weight of the equipment placed on the raised floor.

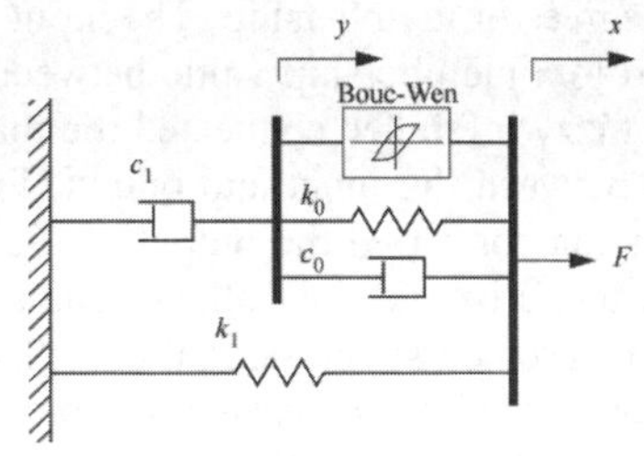

Figure 3. Mechanical model of the MR damper (Spencer et al. 1997).

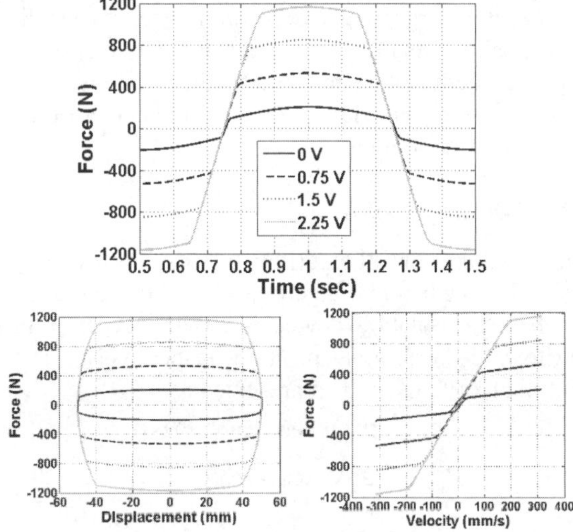

Figure 4. Responses of the MR damper model.

3 CONTROLLER DESIGN

In addition to the passive control (maximum input voltage was retained) for the MR damper, four control algorithms were designed and applied to the isolated raised floor system to investigate the seismic responses including fuzzy-logic control, switching control, model-based control, and linear-quadratic regulator.

3.1 *Fuzzy-logic control*

Fuzzy-logic control method (Zadeh 1965) was implemented to control the MR damper for the isolation device in this study. The control rules were established by engineering intuition and experience. In the fuzzification, crisp values were converted into fuzzy sets and degrees of membership. The input space can be mapped to a membership value between 0 and 1. Then, the fuzzy inference connected the mapping relationship between the input and output. Finally, the defuzzification converted the output of the inference rules back to a crisp control voltage. The control targets were focused on suppressing the displacement of the isolation as well as decreasing the transmitted acceleration. As a result, the triangular membership functions can be defined through trial and error as shown in Fig. 5. The fuzzy rule base is shown in Table 1 where NL, NM, NS, ZR, PS, PM, and PL represent negative large, negative medium, negative small, zero, positive small, positive medium, and positive large, respectively. Mamdani's method was used for fuzzy inference. For defuzzification, the center of area method was adopted.

3.2 *Switching control*

The switching control algorithm adopted in this study is simple. The input voltage V_{MR} was set 0 V and 2.25 V when the isolation device is moving away from and toward the original position, respectively. The control algorithm can be represented as

$$\begin{aligned} V_{MR} &= 0\text{V when } \operatorname{sgn}(x \cdot \dot{x}) > 0 \\ V_{MR} &= 2.25\text{V when } \operatorname{sgn}(x \cdot \dot{x}) < 0 \end{aligned} \tag{6}$$

3.3 *Model-based control*

One of the simple approaches to design the controller for the MR damper can be obtained by the equation of motion. The equations of motion of the isolated system when the roller moves within the sloped surface may be expressed as

$$\ddot{x} + \ddot{x}_g = -\frac{1}{2} g \sin\theta \operatorname{sgn}(x) - \frac{[\mu_r N + F_d]}{M+m} \operatorname{sgn}(\dot{x}) + \frac{F_{MR}}{M+m} \tag{7}$$

where F_{MR} is the MR damper force. The MR damper force can be assigned to cancel the disturbance force, i.e.

$$F_{MR} = \frac{1}{2}(M+m) g \sin\theta \operatorname{sgn}(x) + (\mu_r N + F_d)\operatorname{sgn}(\dot{x}) \tag{8}$$

Identical design procedure can be applied when the roller moves within the curvature range.

3.4 *Linear-quadratic regulator*

The linear-quadratic regulator, one of the most common control methods in semi-active structural control application, was designed and used in the numerical study. The state-space form of the equation of motion can be represented as

$$\begin{aligned} \dot{\mathbf{z}}(t) &= \mathbf{A}\mathbf{z}(t) + \mathbf{B}u \\ \mathbf{y}(t) &= \mathbf{C}\mathbf{z}(t) + \mathbf{D}u \end{aligned} \tag{9}$$

where $\mathbf{z}$ is the state vector defined as $\mathbf{z} = [x \ \ \dot{x}]^T$ and u is the control force. The system matrix **A**, and the control force input matrix **B** are

$$\mathbf{A} = \begin{bmatrix} 0 & 1 \\ 0 & 0 \end{bmatrix}; \quad \mathbf{B} = \begin{bmatrix} 0 \\ \dfrac{-1}{M+m} \end{bmatrix} \tag{10}$$

The relative displacement and absolute acceleration of the isolation device were selected as the output signals. The output matrix **C**, and the control output matrix **D** are

$$\mathbf{C} = \begin{bmatrix} 1 & 0 \\ 0 & 0 \end{bmatrix}; \quad \mathbf{D} = \begin{bmatrix} 0 \\ \dfrac{-1}{M+m} \end{bmatrix} \tag{11}$$

In order to effectively reduce the relative displacement and absolute acceleration of the protected object, the state-feedback control force is obtained by minimizing the quadratic cost function

$$J = \int_0^\infty \left(\mathbf{y}^T \mathbf{Q} \mathbf{y} + R u^2\right) dt \tag{12}$$

where **Q** and R are the state weighting matrix and the control cost, respectively. The state feedback gain **K** can be obtained by solving the Ricatti equation. Then, the control force u can be calculated by letting u=-**Kz**. Finally, the desired force for the MR damper can be calculated by

$$F_{MR} = u - \frac{1}{2}(M+m) g \sin\theta \operatorname{sgn}(x) - (\mu_r N + F_d)\operatorname{sgn}(\dot{x}) \tag{13}$$

Identical design procedure can be applied when the roller moves within the curvature range.

Table 1. The fuzzy rule base.

Control Voltage		Displacement				
		NL	NS	ZR	PS	PL
Velocity	NL	ZR	ZR	ZR	PM	PL
	NS	ZR	ZR	ZR	PS	PM
	ZR	ZR	ZR	ZR	ZR	ZR
	PS	NM	NS	ZR	ZR	ZR
	PL	NL	NM	ZR	ZR	ZR

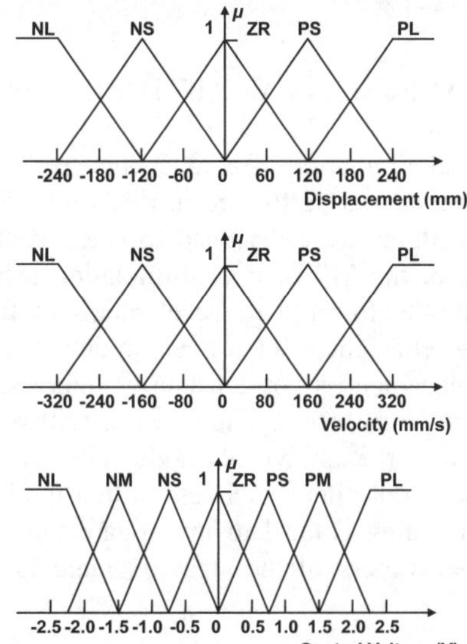

Figure 5. Membership functions for fuzzy-logic control.

4 SIMULATION RESULTS

The raised floor system was assumed to be placed in the building located at the Neihu Science Park in Taipei city. Four spectrum compatible artificial ground motions were adopted in the simulation, namely TAP021EW, TAP021NS, TAP041EW, and TAP041NS. The peak ground acceleration of each ground motion was 0.24 g. In order to realize the effect of the structural period on the seismic responses of the smart isolation system, a 5% damped, single-degree-of-freedom (SDOF) model with four different structural periods (T= 0.5 s, 1.0 s, 1.5 s, and 2.0 s) was used to represent the amplified slab acceleration which was input to the raised floor system. The analytical simulation scheme is illustrated in Fig. 6.

The fuzzy-logic and switching controllers mentioned previously are able to obtain the control voltage directly while the rest two controllers need to convert the desired control force into the control voltage for the MR damper. The algorithm for selecting the control voltage can be depicted in Fig. 7 which is referred as clipped control strategy (Dyke et al. 1996). However, a transition region using linear interpolation to determine the control voltage was adopted to prevent the MR damper force from chattering due to the clipping process in the simulation.

Table 2 to Table 4 show the maximum acceleration, velocity, and displacement of the isolated raised floor system located at the top of an SDOF structure subjected to the four spectrum compatible artificial ground motions. It is evident that the acceleration transmitted to the equipment of each control case is smaller than that of the passive case for all the four ground motions despite of the difference of structural period. However, the displacement is enlarged for some cases which may not be acceptable because large displacement indicates that additional clearance is required between the raised floor and the structural wall. Figure 8 shows the displacement versus acceleration relationship of the raised floor subjected to TAP021EW (T=1.5 s). It represents the case that the four controllers perform better than the passive case on both acceleration and displacement. Consequently, it is possible to design a controller for the smart isolation system to reduce both acceleration and displacement.

Table2. The maximum acceleration of the simulation results (m/s^2).

Ground Motion	Structural Period (sec)	Control Algorithm				
		Passive	Fuzzy	Switching	Model	LQR
TAP021EW (PGA=0.24g)	0.5	1.037	0.667	0.667	0.666	0.666
	1	1.039	0.667	0.678	0.669	0.666
	1.5	1.045	0.676	0.685	0.676	0.673
	2	1.045	0.677	0.685	0.677	0.673
TAP021NS (PGA=0.24g)	0.5	1.038	0.670	0.674	0.669	0.668
	1	1.041	0.672	0.682	0.670	0.672
	1.5	1.046	0.668	0.685	0.674	0.673
	2	1.044	0.671	0.680	0.670	0.671
TAP041EW (PGA=0.24g)	0.5	1.038	0.668	0.667	0.666	0.667
	1	1.039	0.669	0.672	0.669	0.668
	1.5	1.044	0.676	0.684	0.673	0.672
	2	1.039	0.672	0.677	0.669	0.671
TAP041NS (PGA=0.24g)	0.5	1.038	0.667	0.668	0.668	0.668
	1	1.039	0.670	0.680	0.671	0.672
	1.5	1.046	0.676	0.685	0.674	0.673
	2	1.039	0.672	0.683	0.672	0.672

Table 3. The maximum velocity of the simulation results (m/s).

Ground Motion	Structural Period (sec)	Control Algorithm				
		Passive	Fuzzy	Switching	Model	LQR
TAP021EW (PGA=0.24g)	0.5	0.815	0.999	0.852	0.870	0.879
	1	1.115	0.923	1.063	1.066	0.914
	1.5	1.665	1.637	1.605	1.605	1.605
	2	1.618	1.602	1.563	1.576	1.575
TAP021NS (PGA=0.24g)	0.5	0.966	1.214	0.983	0.995	0.991
	1	1.229	1.292	1.241	1.232	1.232
	1.5	1.742	1.647	1.667	1.660	1.652
	2	1.598	1.332	1.360	1.348	1.387
TAP041EW (PGA=0.24g)	0.5	0.957	0.950	0.739	0.768	0.896
	1	1.020	0.939	0.938	0.927	0.919
	1.5	1.403	1.475	1.422	1.445	1.434
	2	1.250	1.399	1.380	1.397	1.405
TAP041NS (PGA=0.24g)	0.5	0.885	0.949	0.830	0.840	0.823
	1	1.227	1.252	1.215	1.219	1.220
	1.5	1.692	1.561	1.588	1.565	1.542
	2	1.425	1.549	1.495	1.495	1.489

Table 4. The maximum displacement of the simulation result (m).

Ground Motion	Structural Period (sec)	Control Algorithm				
		Passive	Fuzzy	Switching	Model	LQR
TAP021EW (PGA=0.24g)	0.5	0.345	0.617	0.466	0.480	0.484
	1	0.438	0.557	0.520	0.511	0.446
	1.5	0.886	0.754	0.748	0.748	0.749
	2	1.122	1.217	1.143	1.132	1.131
TAP021NS (PGA=0.24g)	0.5	0.372	0.844	0.649	0.670	0.685
	1	0.443	0.777	0.669	0.697	0.707
	1.5	0.705	1.002	0.809	0.867	0.903
	2	0.972	0.965	0.944	0.962	0.991
TAP041EW (PGA=0.24g)	0.5	0.238	0.476	0.367	0.419	0.422
	1	0.260	0.605	0.386	0.400	0.411
	1.5	0.364	0.677	0.551	0.566	0.573
	2	0.866	1.040	1.040	1.045	1.050
TAP041NS (PGA=0.24g)	0.5	0.402	0.582	0.546	0.564	0.547
	1	0.492	0.712	0.623	0.622	0.606
	1.5	0.714	0.763	0.734	0.791	0.798
	2	1.048	1.089	0.981	0.952	0.926

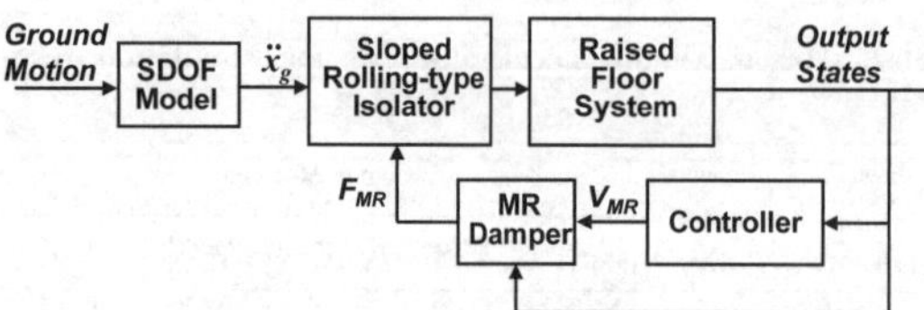

Figure 6. The analytical simulation scheme.

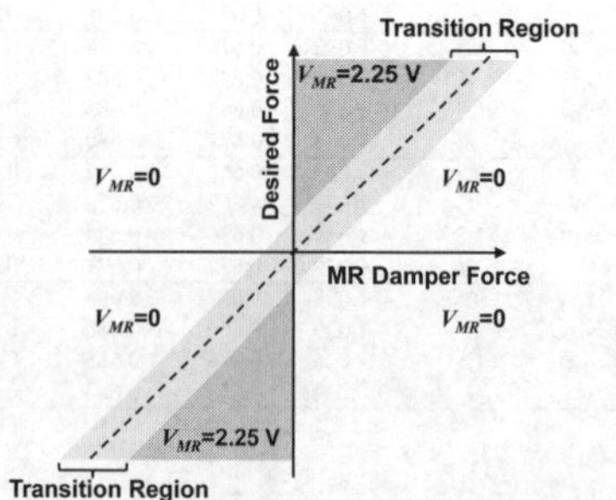

Figure 7. Illustration of algorithm for selecting the control voltage

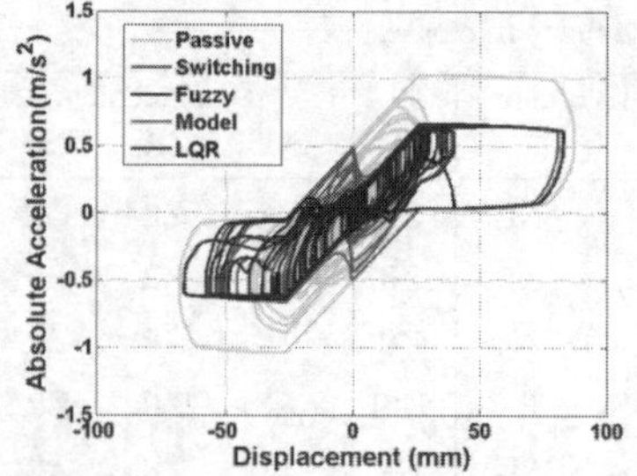

Figure 8. Responses of the raised floor system subjected to TAP021EW (T=1.5 s).

5 CONCLUSIONS AND FUTURE WORK

Analytical simulation of a conceptual isolated raised floor system was performed in this study. Four control algorithms were designed to calculate the input voltage for the MR damper. Simulation results indicate that the transmitted acceleration to the raised floor system can be effectively reduced. However, the displacement is not constantly suppressed for all the four control cases. Future work will be focused on refining the analytical model and synthesis of nonlinear controllers such as sliding-mode control and scheduling control to investigate the feasibility and effectiveness of the smart isolated raised floor system.

REFERENCES

Dyke, S.J. Spencer, B.F. Sain. M.K. & Carlson, J.D. 1996. Modeling and control of magnetorheological dampers for seismic response reduction, *Smart Materials and Structures* **5(5)**, 565–575

Lu, L.Y. & Lin, G.L. 2008. Predictive control of smart isolation system for precision equipment subjected to near-fault Earthquakes, *Engineering Structures* **30(11)**, 3045-3064.

Ramallo, J.C. Johnson, E.A. & Spencer, B.F. 2002. Smart base isolation systems, *Journal of Engineering Mechanics* **128(10)**, 1088-1099.

Spencer, B.F. Dyke, S.J. Sain. M.K. & Carlson, J.D. 1997. Phenomenological model of a magnetorheological damper, *Journal of Engineering Mechanics*, **123(3)**, 230-238.

Wang, S.J. Hwang, J.S. Chang, K.C. Shiau, C.Y. Lin, W.C. Tsai, M.S. Hong, J.X. & Yang, Y.H. 2014. Sloped multi-roller isolation devices for seismic protection of equipment and facilities, *Earthquake Engineering and Structural Dynamics* **43(10)**, 1443-1461.

Zadeh, L.A. 1965. Fuzzy sets, *Information and Control* **8**,338–353.

Proceedings of the International Conference on Smart Infrastructure and Construction
ISBN 978-0-7277-6127-9

© The authors and ICE Publishing: All rights reserved, 2016
doi:10.1680/tfitsi.61279.447

A new and efficient method for the determination of Output Frequency Response Function of nonlinear vibration systems

N.N.L. Nik Ibrahim* and Z.Q. Lang*

The University of Sheffield, Sheffield, United Kingdom
* *Corresponding Author*

ABSTRACT In this paper, a new method exploits the technique of Associated Linear Equations (ALEs) is used in determining the new concept of Output Frequency Response Function (OFRF) that was proposed recently. The OFRF reveals a significant link between the system output frequency response and the parameters that define the system nonlinearity and can therefore facilitate a more systematic design. Simulation studies demonstrate the effectiveness of the new method and the OFRF based design of a building structure vibration isolation system has then been used to demonstrate how the new method can be applied to implement a design for application in earthquake engineering.

1 INTRODUCTION

Most vibration systems are inherently nonlinear. Therefore a proper design of the system nonlinearity has great significance for achieving desired vibration control performance. In this context, an appropriate design of nonlinear damping and /or stiffness is often needed. Traditionally, harmonic balance and averaging etc methods have been used to analyse the effects of nonlinear damping and/or stiffness on the system vibration responses. Then, the nonlinear damping and/or stiffness parameters are designed based on the results of analysis.

Recently, a new concept known as the Output Frequency Response Function (OFRF) has been proposed (Lang et al. 2007).The OFRF reveals a significant link between the system output frequency response and the parameters that define the system nonlinearity and can therefore facilitate a more systematic design. The implementation of the OFRF based design requires (1) determining the OFRF of the system to be designed and (2) using an OFRF based optimization procedure to find an optimal solution to the design. However, currently available methods (Lang et al. 2007)(Jing 2012) for the OFRF determination have a complexity issue requiring a significant number of numerical simulations to generate the system responses under different values of the system design parameters.

In order to resolve this problem, in this paper, a new method for the determination of the OFRF of a wide class of nonlinear systems including nonlinear vibration isolators is proposed. The method exploits the technique of Associated Linear Equations (ALEs) (Vazquez et al. 2006) and can significantly reduce the number of numerical simulations needed for the determination of the system OFRF. This result enable the OFRF based design of nonlinear systems especially nonlinear vibration isolators to be more easily implemented so as to significantly facilitate the application of nonlinear designs in engineering practice.

Simulation studies have first been performed on a single degree of freedom model to verify the effectiveness and demonstrate the advantage of the new method. The OFRF based design of a building structure vibration isolation system has then been used to demonstrate how the new method can be applied to implement a design for application in earthquake engineering.

2 DETERMINATION OF OFRF USING ALE

2.1 Nonlinear Differential Equation (NDE) Model

Consider the nonlinear systems which can be described by a differential equation of a polynomial form and known as Nonlinear Differential Equation (NDE) model

$$\sum_{m=1}^{M}\sum_{p=0,p+q=m}^{m}\sum_{l_1,\cdots,l_{p+q}}^{L} c_{pq}(l_1,\dots,l_{p+q})\prod_{i=1}^{p} D^{l_i}y(t) \times \prod_{i=p+1}^{p+q} D^{l_i}u(t) = 0 \quad (1)$$

where M and L are maximum degree of nonlinearity in terms of $y(t)$ and $u(t)$, and the maximum order of derivative while the operator D is defined by

$$D^l x(t) = \frac{d^l x(t)}{dt^l} \quad (2)$$

A specific case of system (1) can, for example, be a second order nonlinear passive engine mount whose dimensionless motion governing equation is given by

$$m\ddot{y} + (c_1 + c_2 y^2)\dot{y} + (k_1 + k_2 y^2)y = -m\ddot{x}_1 \quad (3)$$

Equation (3) is obviously a specific instance of equation (1) with
$c_{1,0}(2) = m,\ c_{1,0}(1) = c_1, c_{1,0}(0) = k_1,$
$c_{3,0}(0,0,1) = c_2,\ c_{3,0}(0,0,0) = k_2$ and
$c_{0,1}(2) = -m.$

2.2 OFRF

For nonlinear systems that can be described by model (1), and satisfy the following assumptions:

- The system is stable at zero equilibrium
- The systems can equivalently be described by the Volterra series model with its maximum order $N \geq M$ over a regime around the equilibrium,

there exists an explicit analytical relationship between the system output frequency response and the model parameters which define the system nonlinearity such that (Lang et al. 2007)

$$\hat{Y}(j\omega) = \sum_{j_1=0}^{m_1}\sum_{j_2=0}^{m_2}\cdots\sum_{j_{S_N}=0}^{m_{S_N}} \gamma_{(j_1,\cdots,j_{S_N})}(\omega)x_1^{j_1}\dots x_{S_N}^{j_{S_N}} \quad (4)$$

where x_i, $i = 1,\dots,S_N$ are the parameters which define the system nonlinearity; m_i is the maximum power of x_i, $i = 1,\dots,S_N$. $x_1^{j_1}\dots x_{S_N}^{j_{S_N}}$ represent the coefficients of the term $\gamma_{(j_1,\cdots,j_{S_N})}(\omega)$which is a set of all monomials involved in the representation of $\hat{Y}(j\omega)$.

The set of all monomials involved in the representation of $\hat{Y}(j\omega)$, denote as M_{S_N} , and can be determined priori by the following algorithm (Peng & Lang 2008):

1. Set $S_N \geq 0$ and $N = 1,2,\dots,S_N$.
2. $M_1 = [1]$. Calculate M_N by using

$$M_N = \left[\bigcup_{l_1,\dots,l_N=0}^{L}[c_{0N}(l_1,\dots,l_p)]\right] \bigcup\left[\bigcup_{q=1}^{N-1}\bigcup_{p=1}^{N-q}\bigcup_{l_1,\dots,l_N=0}^{L}[c_{0N}(l_1,\dots,l_p)]\otimes M_{N-q,p}\right] \bigcup\left[\bigcup_{p=2}^{N}\bigcup_{l_1,\dots,l_p=0}^{L}[c_{p0}(l_1,\dots,l_p)]\otimes M_{N,p}\right] \quad (5)$$

where $\otimes$ is the Kronecker product,

$$M_{Np} = \bigcup_{i=1}^{N-p+1}(M_i \otimes M_{N-i,p-1})$$

and

$$M_{N1} = M_N \quad (6)$$

3. Lastly, the set of the parametric characteristics in equation (8) can be expressed as

$$M_{S_N} = \bigcup_{N=1}^{S_N} M_N \quad (7)$$

This algorithm provides an effective way to determine all the monomials involved in the OFRF (4). After knowing these monomials, an OFRF representation of the system can be determined where every monomial is paired with an OFRF "coefficient". The OFRF representation can be written as

$$\hat{Y}(j\omega) = \sum_{N=1}^{S_N} M_{Nr} P_{Nr} \quad (8)$$

where $n = 0,1,\ldots,S_N$ and $r = 0,1,\ldots,b$. b is the maximum number of elements in M_N.

2.3 Associated Linear Equations (ALEs)

Any system that possesses a Volterra series representation can be described by a series of associated linear equations (ALEs) (Vazquez et al. 2006). From these ALEs, analysis of the output of the nonlinear system can be done order by order.

The ALEs for every N can be determined by the following algorithm:

1. Set $N = 1,2,\cdots,S_N$, where S_N is the maximum order of the system nonlinearity that is taken into account.
2. $J_0 = J_1 = 0$. The ALEs for every nth-order can be written as

$$m\ddot{y}_N(t) + c\dot{y}_N(t) + ky_N(t) = a_N x^N + J_N - J_{N-1} \quad (9)$$

where

$$J_N = \sum_{n=1}^{N}\left(\sum_{j_1=1}^{n-l+1} \cdots \sum_{j_i=1}^{n-l+i+j_1-\cdots-j_{i-1}} \cdots \sum_{j_l=0}^{n-j_1-\cdots-j_i-\cdots-j_{l-1}} y_{j_1}(t) y_{j_i}(t) y_{j_1}(t)\right)$$

3. The estimation of the output signal and the output spectrum for the system up to $S_N th$-order thus can be written as

$$\hat{y}(t) = \sum_{N=1}^{S_N} \hat{y}_N(t) \quad (10)$$

and

$$\hat{Y}(j\omega) = \sum_{N=1}^{S_N} \hat{Y}_N(j\omega) \quad (11)$$

Using the algorithm discussed above, the ALEs for the system described as Equation (3) up to 5th-order are

$$m\ddot{y}_1(t) + c\dot{y}_1(t) + ky_1(t) = -m\ddot{x}_1$$

$$m\ddot{y}_3(t) + c\dot{y}_3(t) + ky_3(t) = -c_2 y_1{}^2(t)\dot{y}_1{}^2(t) - k_2 y_1{}^3(t)$$

$$m\ddot{y}_5(t) + c\dot{y}_5(t) + ky_5(t) = -c_2 y_1(t) y_3(t) \dot{y}_1(t) \dot{y}_3(t) - 2k_2 y_1{}^2(t) y_3(t) \quad (12)$$

Lastly, the estimation of the output signal and the output spectrum can be written as

$$\hat{y}(t) = \hat{y}_1(t) + \hat{y}_3(t) + \hat{y}_5(t) \quad (13)$$

and

$$\hat{Y}(j\omega) = \hat{Y}_1(j\omega) + \hat{Y}_3(j\omega) + \hat{Y}_5(j\omega) \quad (14)$$

2.4 Determination of OFRF by using ALEs

Then, to determine the OFRF of the nonlinear system, the OFRF "coefficients" need to be determined per Nth-order as

$$\hat{Y}(j\omega) = \sum_{N=1}^{S_N} M_{Nr} P_{Nr} = \sum_{N=1}^{S_N} \hat{Y}_N(j\omega) \quad (15)$$

It is good to take note that the number of simulations needed per Nth-order is r simulations.

Consider system that has been defined by Equation (3). Using the algorithm discussed in Section 2.3, all monomials involved in the representation of the system output frequency response up to the 5th order that were determined by using the 3 step procedure are

$$M_1 = [1]$$
$$M_3 = [c_2, k_2]$$
$$M_5 = [c_2{}^2, c_2 k_2, k_2{}^2] \quad (16)$$

Then, to determine the OFRF representation of the system, every monomial are paired with an OFRF "coefficient", thus the result of the OFRF representation will be

$$\hat{Y}(j\omega) = \hat{P}_1(j\omega) + c_2\hat{P}_{31}(j\omega) + k_2\hat{P}_{32}(j\omega) + c_2{}^2\hat{P}_{51}(j\omega) + c_2 k_2 \hat{P}_{52}(j\omega) + k_2{}^2\hat{P}_{53}(j\omega) \quad (17)$$

where r for third and fifth order is 2 and 3 respectively. This means that to solve the 3rd-order and 5th-order, 2 and 3 simulations are needed respectively.

Next, as this system will be solved up to 5th-order, the ALEs that is needed will be up to fifth order

$$\hat{Y}(j\omega) = \hat{Y}_1(j\omega) + \hat{Y}_3(j\omega) + \hat{Y}_5(j\omega) \quad (18)$$

and the solution for the OFRF "coefficients" can be determined as below

$$\hat{P}_1(j\omega) = \hat{Y}_{11}(j\omega)$$

$$\begin{bmatrix} \hat{P}_{31}(j\omega) \\ \hat{P}_{32}(j\omega) \end{bmatrix} = \begin{bmatrix} c_{21} & k_{21} \\ c_{22} & k_{22} \end{bmatrix}^{-1} \begin{bmatrix} \hat{Y}_{31}(j\omega) \\ \hat{Y}_{32}(j\omega) \end{bmatrix}$$

$$\begin{bmatrix} \hat{P}_{51}(j\omega) \\ \hat{P}_{52}(j\omega) \\ \hat{P}_{53}(j\omega) \end{bmatrix} = \begin{bmatrix} c_{21}{}^2 & c_{21}k_{21} & k_{21}{}^2 \\ c_{22}{}^2 & c_{22}k_{22} & k_{22}{}^2 \\ c_{23}{}^2 & c_{23}k_{23} & k_{23}{}^2 \end{bmatrix}^{-1} \begin{bmatrix} \hat{Y}_{51}(j\omega) \\ \hat{Y}_{52}(j\omega) \\ \hat{Y}_{53}(j\omega) \end{bmatrix} \quad (19)$$

3 SIMULATION STUDIES AND DISCUSSIONS

3.1 *The OFRF based analysis of a Nonlinear single degree of freedom system*

Consider a nonlinear single degree of freedom system as shown in Figure 1 (Peng & Lang 2008) whose motion governing equation is given by

$$m\ddot{y} + (c_1 + c_2y^2)\dot{y} + (k_1 + k_2y^2)y = -m\ddot{x}_1 \quad (20)$$

where $y = x_2 - x_1$ is the movement of mass m relative , to the base.

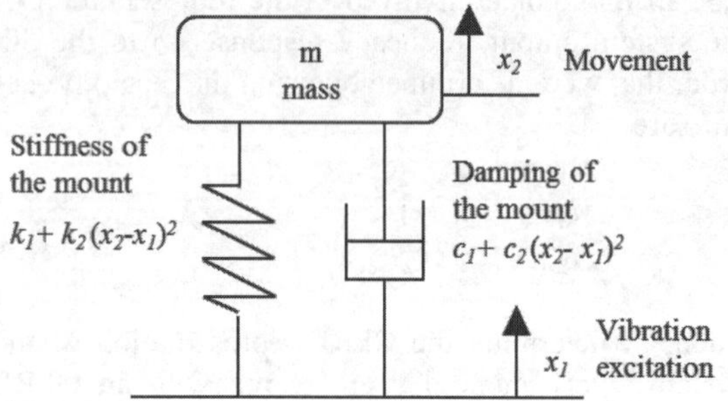

Figure 1. Schematic of a nonlinear single degree of freedom system

In this analysis, the input base excitation, $\ddot{x}_1$ is the acceleration data recorded during the main shock of the 2011 Great East Japan Earthquake and m is the total mass of the building as considered in (Dan & Kohiyama 2013).

Based on the results in Section 2.4, 3 simulations are needed to solve all the OFRF "coefficients" in order to determine the OFRF of the system. Table 1shows the three different values of c_2 and k_2 used in the three simulations.

Table 1. Value of c_2 and k_2 used in the three simulations

Simulation	c_2	k_2
1	15×10^6	2.00×10^9
2	17×10^6	2.25×10^9
3	20×10^6	2.50×10^9

This yields three sets of results in both the time and frequency domain, namely $\boldsymbol{y_{1i}, y_{3i}, y_{5i}}$ and $\hat{\boldsymbol{Y}}_{1i}(\boldsymbol{j\omega}), \hat{\boldsymbol{Y}}_{3i}(\boldsymbol{j\omega}), \hat{\boldsymbol{Y}}_{5i}(\boldsymbol{j\omega})$, i=1,2,3. Figure 2 shows a comparison of the simulated system frequency domain output in the case of $\mathbf{c_2 = 17 \times 10^6}$ and $\boldsymbol{k_2 = 2.25 \times 10^9}$ with the summation of the ALE solutions in the frequency domain $\hat{\boldsymbol{Y}}_{12}(\boldsymbol{j\omega}) + \hat{\boldsymbol{Y}}_{32}(\boldsymbol{j\omega}) + \hat{\boldsymbol{Y}}_{52}(\boldsymbol{j\omega})$, indicating again a good accuracy of the ALE solution but in the frequency domain.

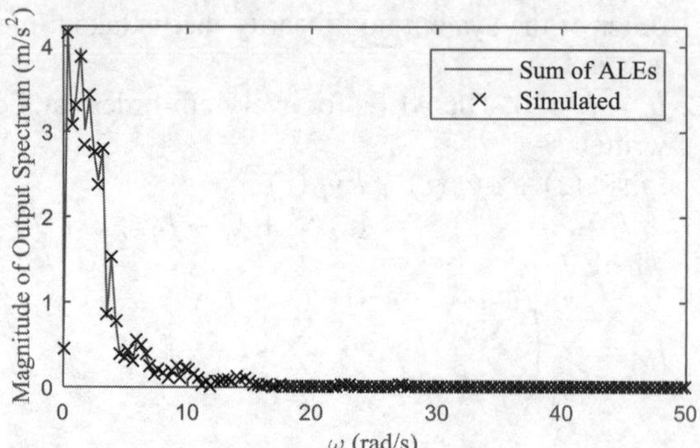

Figure 2. Comparison between the simulated output spectrum and the total output spectrum when $\mathbf{c_2 = 17 \times 10^6}$ and $\boldsymbol{k_2 = 2.25 \times 10^9}$.

From the solutions to the ALEs, the OFRF "coefficients" are then determined using equation (19). It is worth pointing out that the above approach only needs three simulations under the different choices of system parameters as shown in Table 1 to determine the system OFRF. In contrast, at least six simulations are needed when the original OFRF analysis in (Lang et al. 2007) is used for the same purpose.

In Figure 3, a comparison of the simulated system output spectrum when $c_2 = 18 \times 10^6$ and $k_2 = 2.30 \times 10^9$ with the spectrum evaluated using the OFRF showed that the OFRF as determined above can well represent the system output frequency responses.

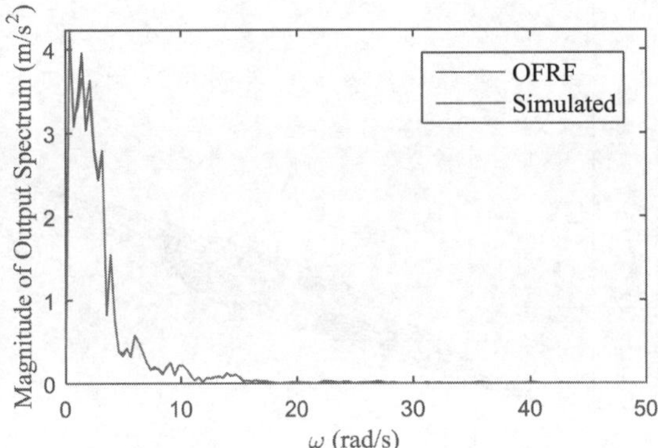

Figure 3. Comparison of the amplitude of $\hat{Y}(j\omega)$ and $Y(j\omega)$ when $c_2 = 18 \times 10^6$ and $k_2 = 2.30 \times 10^9$.

From the OFRF as determined above, the design of the system parameters c_2 and k_2 can be performed efficiently as the OFRF shows the relationship between the parameters and the output frequency response. For example, Figure 4 shows the OFRF based relationship between the parameters, c_2 and k_2 and magnitude of the output spectrum at 2 rad/s frequency.

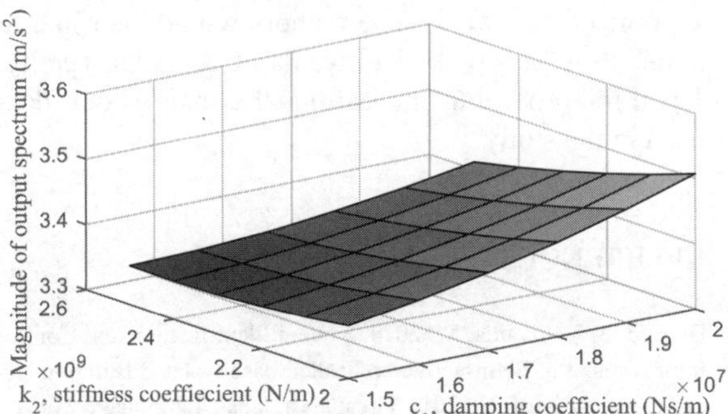

Figure 4. The relationship between the parameters, $\mathbf{c_2}$ and $\boldsymbol{k_2}$ and magnitude of the output spectrum at frequency= $\boldsymbol{2\ rad/s}$.

3.2 *Implementation of the OFRF into the system identification of the building*

This method also works in single input multi output system. In this subsection, how OFRF can be used in the system identification of the building will be discussed. The equation of motion of the building system of the building at the Keio University is given as

$$\boldsymbol{M\ddot{x}} + \boldsymbol{C\dot{x}} + \boldsymbol{Kx} = \boldsymbol{Eu} + \boldsymbol{F\ddot{z}} \qquad (21)$$

with

$$\boldsymbol{x} = [x_1 \quad x_2 \quad \cdots \quad x_{10}]^T$$

$$\boldsymbol{M} = \text{diag}\,(m_1, m_2, \dots, m_{10})$$

$$\boldsymbol{C} = \begin{bmatrix} c_1 + c_2 & -c_2 & \cdots & 0 & 0 \\ -c_2 & c_2 + c_3 & & 0 & 0 \\ \vdots & & \ddots & & \vdots \\ 0 & 0 & & c_9 + c_{10} & -c_{10} \\ 0 & 0 & \cdots & -c_{10} & c_{10} \end{bmatrix}$$

$$\boldsymbol{K} = \begin{bmatrix} k_1 + k_2 & -k_2 & \cdots & 0 & 0 \\ -k & k_2 + k_3 & & 0 & 0 \\ \vdots & & \ddots & & \vdots \\ 0 & 0 & & k_9 + k_{10} & -k_{10} \\ 0 & 0 & \cdots & -k_{10} & k_{10} \end{bmatrix}$$

$$\boldsymbol{E} = [1 \quad 0 \quad \cdots \quad 0]^T$$

$$\boldsymbol{F} = [-m_1 \quad -m_2 \quad \cdots \quad -m_{10}]^T$$

where $\boldsymbol{M}$ is the mass matrices, $\boldsymbol{C}$ is the damping matrices, $\boldsymbol{K}$ is the stiffness matrices, $\boldsymbol{x}$ is the displacement vector (relative motion between each floor and the ground), $\boldsymbol{u}$ is the control force and $\boldsymbol{\ddot{z}}$ is the ground acceleration.

The structural parameters of the building are as shown in Table 2.

Table 2. Structural parameters of the building.

Floor	Mass, m $\times$ 10^6kg	Stiffness, k $\times$ 10^6N/m	Damping, c $\times$ 10^6Ns/m
RF	2.4999	999.6	8.0487
7F	2.0664	1156.4	9.3110
6F	2.0371	1381.8	11.126
5F	2.0369	1568.0	12.625
4F	2.0500	1813.0	14.598
3F	2.0331	1803.2	14.520
2F	1.8264	1979.6	15.940
1F	2.4906	2763.6	22.252
B1F	3.4382	2273.6	18.306
B2F	4.9814	66.836	0

For this investigation, the control force of the system is described as

$$u = (C_1 + C_2{x_1}^2)\dot{x}_1 + (K_1 + K_2{x_1}^2)x_1 \qquad (22)$$

which was the damping and stiffness functions of the passive engine mount. Then, the OFRF of the building system was determined. Figure 5 shows the comparison between the magnitude of simulated output spectrum with the spectrum evaluated using the OFRF when $C_2 = 3 \times 10^5$ and $K_2 = 7 \times 10^5$ for the 7F floor.

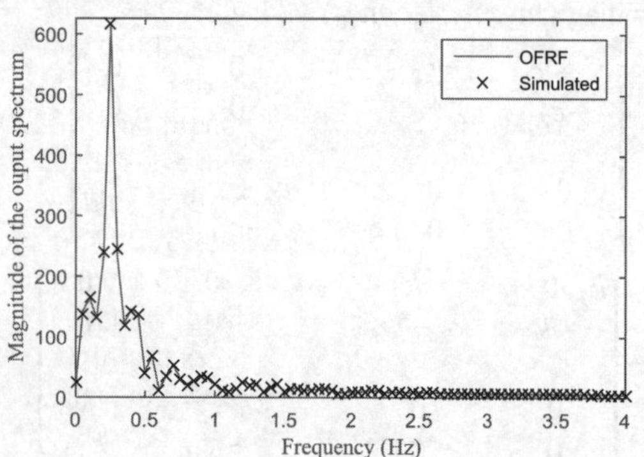

Figure 5. Comparison between the amplitude of $\hat{Y}(j\omega)$ and $Y(j\omega)$ when $C_2 = 3 \times 10^5$ and $K_2 = 7 \times 10^5$ for the 7F floor.

From Figure 5, it can be seen that the result from the OFRF gave an excellent agreement to the simulation data. This method also works for other floors. This show that the method of determining OFRF using ALEs work. Lastly, Figure 6 shows the OFRF based relationship between the parameters, C_2 and K_2 and magnitude of the output spectrum which is useful in designing process.

4 CONCLUSIONS

A new method of determining output frequency response function (OFRF) using Associated Linear Equations (ALEs) was proposed in this paper. This allows OFRF, which reveals a significant link between the system output frequency response and the parameters that define the system nonlinearity to be determined with significantly less number of numerical simulations compared to previous works. The design of NDE system can be done efficiently using OFRF. Then, the OFRF based design of a building structure vibration isolation system has then been used to demonstrate how the new method can be applied to implement a design for application in earthquake engineering.

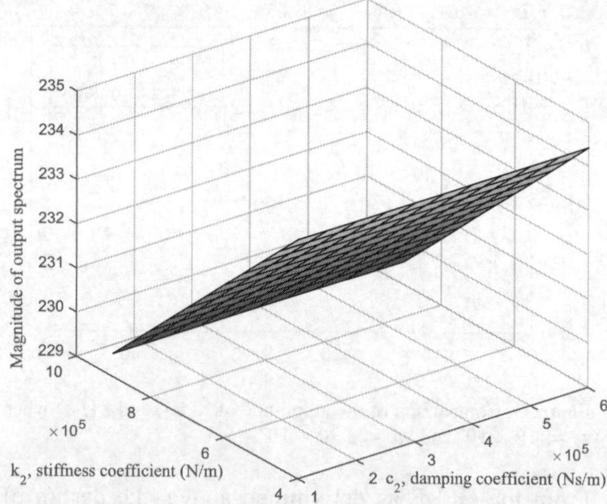

Figure 6. The relationship between the parameters, c_2 and k_2 and magnitude of the output spectrum at frequency $= 0.3\ Hz$ for the 7F floor.

ACKNOWLEDGEMENT

Authors would like to thank the Royal Society for support of this work. The authors would also like to thank Prof Masayuki Kohiyama at Keio University, Japan for providing the earthquake main shock data used in the study.

REFERENCES

Dan, M. & Kohiyama, M. 2014. System Identification and Control Improvement of Semi-active-controlled-base-isolated Building Using the records of the 2011 Great East Japan Earthquake. *Safety, Reliability, Risk and Life-Cycle Performance of Structures and Infrastructures.* 3841 -3847

Jing, X. 2012 Nonlinear Characteristic Output Spectrum for Nonlinear Analysis and Design. *IEEE/ASME Transactions on Mechatronics.* **19**. 1-13.

Lang, Z.Q. Billings, S.A. Yue, R. & et al. 2007. Output frequency response function of nonlinear Volterra systems. *Automatica.* **43**, 805-816.

Peng, Z.K. & Lang, Z.Q. 2008. The Effects of Nonlinearity on the Output Frequency Response of a Passive Engine Mount. *Sound and Vibration.* **318**. 313-328.

Vazquez Feijoo, J.A. Worden, K. & Stanway, R. 2006. Analysis of Time-invariant Systems in the Time and Frequency Domain by Associated Linear Equations(ALEs). *Mechanical Systems and Signal Processing*. **20**. 896-919.

Proceedings of the International Conference on Smart Infrastructure and Construction
ISBN 978-0-7277-6127-9

© The authors and ICE Publishing: All rights reserved, 2016
doi:10.1680/tfitsi.61279.453

Shaking table test of seismic bracings used in piping systems

X.Q. Nin, J.W. Dai*, D.Z. Wang, Y.Q. Yang and W. Bai

Key Laboratory of Earthquake Engineering and Engineering Vibration,
Institute of Engineering Mechanics, China Earthquake Administration, Harbin, China
* *Corresponding Author*

ABSTRACT The recent earthquake damage has highlighted the importance of performance-based design and functional recoverability of nonstructural components. Directing at seismic bracings used in piping systems, this paper firstly introduces the construction and design procedure of seismic bracings. The seismic code requirements and relevant seismic research of bracings are also reviewed. The shaking table test of seismic bracings in a 6-meter full-size piping system is conducted. In addition with theoretical analysis, the seismic performance and failure mode of seismic bracings are studied. The results indicate that failure of bracings in piping systems is mostly due to the bolt looseness in connection components, and the longitudinal response of pipes is basically identical to the floor response. Based on the results, both technical improvement measures to enhance weak connections and design optimization suggestions considering cost are put forwarded. The use of real-time monitoring in bracings may be an effective way in condition-based maintenance.

1 INTRODUCTION

For modern architecture, the post-earthquake function loss is mostly embodied in the nonstructural system. The low recoverability of buildings due to the low vulnerability of nonstructural components has been widely observed. The good operation of piping systems in buildings is the foundation to ensure the whole function implementation. Seismic bracings are important parts to connect structure and nonstructural components like piping systems, which should have enough stiffness and strength to surfer earthquake actions and to make piping systems restored quickly, in order to achieve the goal of building function recoverability. However, the specialized study and information about seismic bracings are very little, and applicable codes and guidance greatly depend on past experience, engineering judgment and intuition, rather than test and analysis results. Therefore, it becomes necessary to study the seismic response characteristics of seismic bracings.

1.1 Type and design of seismic bracing

The types of seismic bracing mainly include lateral and longitudinal seismic bracing, single tube and door-shaped seismic bracing (Figure 1), which is primarily composed of U-steel, seismic connecting component, limit fastener, expansion bolt and other parts (Figure 2). The design of seismic bracing is divided into four steps: firstly, determine the location and orientation of bracing; secondly, determine the design load requirements; thirdly, select the correct shape, size and maximum length of bracing; finally, choose the appropriate type and specification of fastener to fix bracing on the building structure.

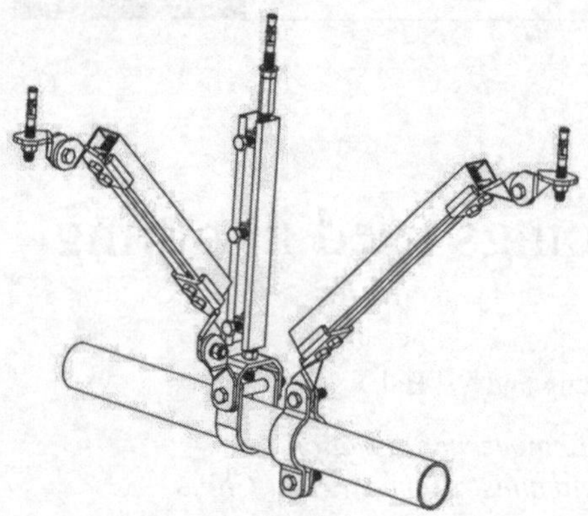

(a) Single tube seismic bracing

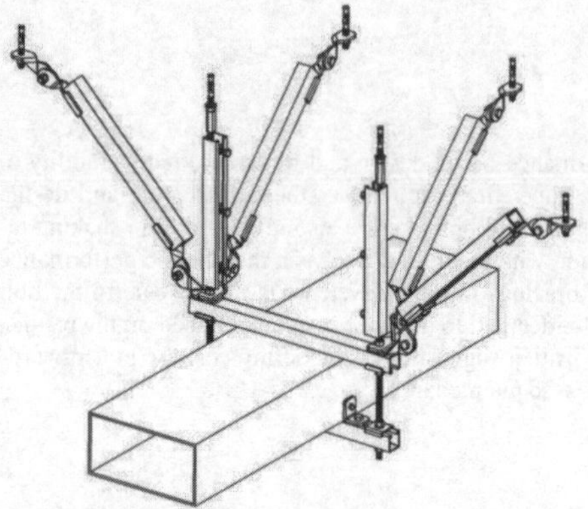

(b) Door-shaped seismic bracing

Figure 1. Types of seismic bracing.

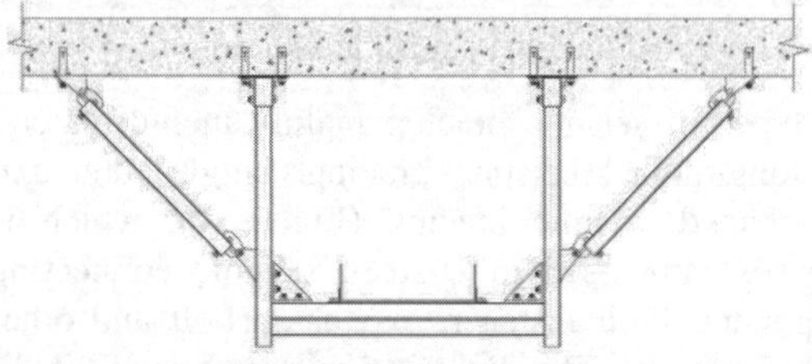

Figure 2. Construction of seismic bracing.

1.2 Seismic code requirements

China's code for seismic design of mechanical and electrical equipment (GB50981 2014) and specification of seismic supports for mechanical and electrical components (CJ/T476 2015) have made corresponding provisions for design and calculation of seismic bracings. The spacing of lateral and longitudinal seismic bracings shall be calculated by Equation (1), and comprehensive coefficient of horizontal seismic force can be calculated by Equation (2). Seismic bracings should be checked based on the load to satisfy the requirements through spacing adjustment.

$$l = \frac{l_0}{\alpha_{Ek} \cdot k} \tag{1}$$

$$\alpha_{Ek} = \gamma\eta\zeta_1\zeta_2\alpha_{\max} \tag{2}$$

Where l is the spacing of lateral and longitudinal seismic bracings, l_0 is the largest spacing of seismic bracings, α_{Ek} is comprehensive coefficient of horizontal seismic force, k is angle adjustment coefficient, γ is nonstructural function coefficient, η is nonstructural category coefficient, ζ_1 is condition factor, ζ_2 is location factor, $\alpha_{\max}$ is the maximum of earthquake influence coefficient.

1.3 State-of-the-art research

Recently, a few patents have been initiated to improve the construction of seismic bracings. A connector was designed for connecting a seismic brace to a support rod, such as for a cable tray, to inhibit movement of the rod during seismic activity (Rinderer 2002). A bracing apparatus was designed for bracing a flexible pipe extending between a fire sprinkler header pipe and a sprinkler and for holding the sprinkler in a desired location (Jensen & Jensen 2008). A sway brace fitting clamp was used for lateral bracing of sprinkler or other types of pipe (designated the service pipe) to a structure, to prevent movement of the service pipe perpendicular to the axis of the pipe relative to the structure to which it is attached (Allen et al. 2010). An apparatus was disclosed which provides connection points on a piping system to facilitate motion restraint using external motion-restraining systems (McMahon 2012).

2 SHAKING TABLE TEST

2.1 Test setup and specimens

The shaking table test of seismic bracings used in piping systems is carried out at the Key Laboratory of Earthquake Engineering and Engineering Vibration of China Earthquake Administration. The 5m×5m square shake table is characterized by three DOF along the two horizontal and one vertical directions. The maximum payload of the shake table is 300kN with a frequency ranging between 0.4 and 40 Hz, acceleration peak equal to 1g, velocity peak equal to 0.6m/s, and total displacement equal to 160mm (±80mm).

The tested bracings (Figure 3) were installed in three steel single-story framed structures with reinforced concrete floor which were designed to simulate the floor response. Considering the shaking table's size limit, the whole test model with two 6m length pipes was placed along the table diagonal (Figure 4), the total weight of which was 61.6kN.

Figure 3. Seismic bracing.

Figure 4. Test model.

2.2 Test input and program

According to the code for seismic design of buildings GB50011-2010 in China, five accelerograms were used as input in the test, consisting of three ground motions– the 1995 Kobe ground motion, the 1999 Chichi ground motion and the 2008 Wolong ground motion, and two sine resonance waves. The test input levels range from PGA=0.056g to 1.0g along two horizontal directions, in order to compose diagonal direction input from PGA=0.08g to 1.4g as floor response levels (Table 1).

Table 1. Test program.

Test ID	Input PGA(g)	Composed PGA(g)
1	0.056	0.08
2	0.16	0.23
3	0.35	0.50
4	0.64	0.90
5	1.00	1.40

2.3 Test results and analysis

The white noise frequency sweep was conducted before and after each input level, obtaining the natural frequency and damping ratio of pipes (Table 2).

Table 2. Natural frequency and damping ratio of pipes.

Composed PGA(g)	Natural frequency(Hz)		Damping ratio(%)	
	lateral	longitudinal	lateral	longitudinal
Pre-test	9.47	12.85	2.17	3.13
0.50	9.07	11.27	3.30	3.63
0.90	8.85	10.64	4.14	3.96
1.40	8.65	10.15	4.72	4.16

After the input of PGA=0.50g, the movement of structure was basically consistent with the shaking table, behaving as rigid motion. In the input of PGA=0.90g, the amplitude of structure increased obviously, while seismic bracings shake slightly due to bolt looseness. In the input of 1.40g, as seismic bracings shake violently with frames, one of the supporting U-steel rotated about 45 degrees along the vertical axis. Nevertheless, at the end of the test, neither seismic bracings nor piping systems occurred obvious damage, which means good seismic performance of bracings.

With the increase of input PGA, the acceleration amplification coefficient of pipes (the ratio of peak acceleration between pipes and floor) decreased gradually due to the degradation of model stiffness and the increase of damping ratio. The lateral acceleration amplification coefficient of pipes is always larger than the longitudinal one, with the maximum of which is 1.455 (Figure 5). The maximum displacement of seismic bracings relative to the floor is 68mm, while the strain of bracings is 58.74E-6, calculating the stress is 121MPa, less than the allowable stress 160MPa, which means bracings have no stress damage.

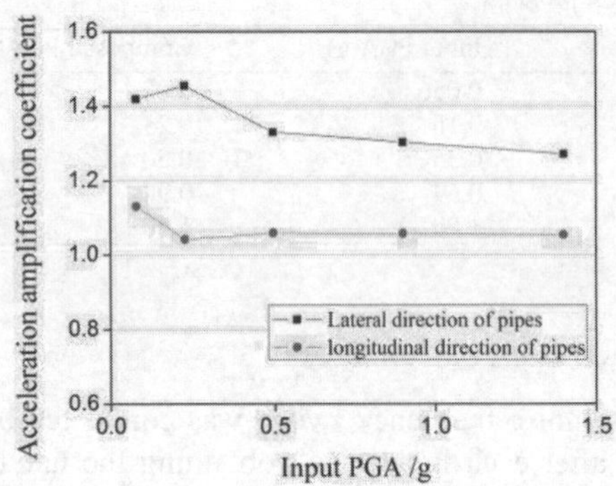

Figure 5. Acceleration amplification coefficient of pipes.

3 CONCLUSIONS

According to the shaking table test results of seismic bracings used in piping systems, the performance of bracings is good enough to satisfy the seismic requirements. Nevertheless, the problem of bolt looseness is easy to occur in the seismic connection parts, which may further cause the rotation or shift of bracings under continued earthquakes or aftershocks, therefore technical measures need to be improved and innovated to enhance weak connections. The use of real-time monitoring in bracings may be an effective way in condition-based maintenance. Meanwhile, the test results show that the longitudinal acceleration response of pipes is basically identical to the floor response. If cost saving is taken into account, designers and engineers can give priority to choose the type of lateral seismic bracings, on the basis of meeting seismic requirements.

ACKNOWLEDGEMENT

The authors appreciate the financial support from the Earthquake Scientific Research Funds Program (No. 201508023).

REFERENCES

Allen, P.B. Ambrogio, N.J. Osborn, E.C. et al. 2010. Lateral seismic brace: EP, EP2235416 A2.

CJ/T476. 2015. Specification of seismic supports for mechanical and electrical components. (in Chinese)

Cosenza, E. Sarno, L.D. Maddaloni, G. et al. 2014. Shake table tests for the seismic fragility evaluation of hospital rooms, *Earthquake Engineering & Structural Dynamics* **44(1)**, 23–40.

GB50981. 2014. Code for seismic design of mechanical and electrical equipment. (in Chinese)

Jensen, E.T. & Jensen, R.H. 2008. Fire sprinkler flexible piping system, bracing apparatus therefor, and method of installing a fire sprinkler: US, US7373720 B1.

Mcmahon, T.W. 2012. Industrial-piping system attachment devices for seismic bracing and methods of use: WO, WO2012087355 A1.

Retamales, R. Mosqueda, G. Filiatrault, A. et al. 2011. Testing Protocol for Experimental Seismic Qualification of Distributed Nonstructural Systems, *Earthquake Spectra* **27(3)**, 835-856.

Rinderer, E.R. 2002. Seismic bracing connector: US, US6415560 B1.

Proceedings of the International Conference on Smart Infrastructure and Construction
ISBN 978-0-7277-6127-9

© The authors and ICE Publishing: All rights reserved, 2016
doi:10.1680/tfitsi.61279.457

Effects of joint rotational stiffness on structural responses of multi-story modular buildings

A.J. Styles[*1], F.J. Luo[1] , Y. Bai[1] and J.B. Murray-Parkes[2]

[1] *Department of Civil Engineering, Monash University, Clayton, Australia*
[2] *Brookfield Multiplex Engineering Innovations Group, Melbourne, Australia*
[*] *Corresponding Author*

ABSTRACT The overall structural performance of multi-story modular buildings is largely dominated by the joint mechanical properties of inter- and intra-module connections. This paper presents a numerical investigation on the effect of joint rotational stiffness on the structural responses of multi-storey steel modular buildings. A three-dimensional detailed finite element (FE) model is first developed for typical connections in modular buildings. Bolt pre-tension and contact behaviour are considered in such detailed FE modelling for the involved bolted connections. The resulting joint rotational stiffness is verified against data available from the literature for implementation in subsequent structural analysis. Therefore, in the full structural models, connections are modelled using spring elements which enable the definition of connection semi-rigidity and non-linearity. Based on the structural model, parametric studies are performed to investigate the effect of joint rotational stiffness of intra-module (e.g. beam-to-column connections) connections, joint rotational stiffness of inter-module connections (e.g. horizontal and vertical ties). The resulting overall structural performances are examined against ultimate limit state (ULS) and serviceability limit state (SLS) requirements as per relevant standards and design guidelines.

1 INTRODUCTION

In recent years there has been rapid development in, and increased demand on, the building industry. As such, many building and engineering firms have sought cost and time-efficient methods of construction, leading to the development of modular construction.

Modular construction serves a diverse range of industries, from residential and commercial housing and the health sector (Mena Report 2013, Lawson et al. 2014 & Lawson et al. 2012), to industries such as the resources (Sullivan 2011) and technology sectors (Worthen 2011). Given the keen interest in modular construction, a number of companies have started entering the market to produce and distribute modular systems to fit project needs such as CIMC (CIMC 2015) as steel framed modules, Timber Building Systems (Timber Building Systems 2015) as timber based modular systems, whilst other systems exist using precast concrete (Australian Concrete Construction 2012). In recent years there have been a number of case studies presented by industry including Atlantic Yards B2 by ARUP (Farnsworth 2014) and the SOHO Apartments complex in Darwin by Irwin Consult (Irwin Consult 2015). These are underscored by increased interest in the ability for modular construction to benefit from techniques developed by the manufacturing industry (Yu et al. 2013) and the application of these construction techniques to increasing numbers of multi-story buildings (Lawson et al. 2012).

In multi-story modular buildings, both intra- and inter-module connections play a key role in the stability and serviceability of the final structure. Even given the recent interest and development in modular construction, quantitative data concerning the connection performance and design considerations or requirements is still very limited. One of the few examples being the handbook compiled for the British construction market by Lawson et al. (Lawson et al. 2014) and a detailed study conducted by Hong et al. on the performance of a newly developed modular building system (Hong et al. 2011).

However, little to no investigation is provided to research into the effect of joint stiffness on the overall performance and serviceability of modular buildings. This paper therefore applies the joint modelling methodologies outlined by Shi et al. (2008) and Luo et al. (2015), for typical connections in modular buildings and further implement the joint behaviours into full structural analysis. A parametric FEA study on this basis for steel framed modules is also presented in this paper.

2 STRUCTURAL FORM AND DESIGN REQUIREMENTS

The authors envisage three classes of modular construction as follows.

1. Self-supporting – Wherein the frame of the modules, both locally and globally, provides the requisite stiffness and strength to ensure stability and serviceability of the structure (see Figure 1a).
2. Externally braced – This class of construction incorporates external elements such as a core or shear wall (usually concrete) to provide additional resistance to lateral actions (see Figure 1b).
3. Externally framed structures – Resistance to lateral actions is provided primarily by an in-situ external frame into which modules are "slotted".

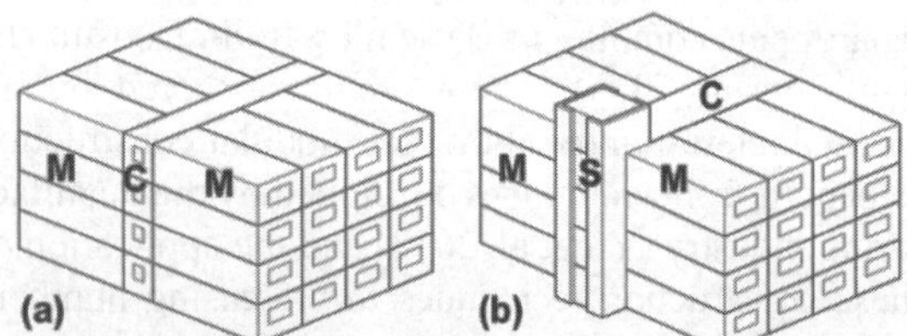

Figure 1. Forms of modular buildings (after Lawson et al. 2014): (a) Self-supporting and (b) with stabilizing core. Note: M = Module C= Corridor S = Stabilizing core

The deflection limits adopted for the serviceability limit state are summarised below.

Table 1. Deflection limits adopted for serviceability limit state

Element controlled	Phenomena controlled	Deflection limit
Beam deflection	d/L	L/250 (AS4100)
Total building drift	Δ/H	H/600 (Griffis, 1993)
Interstory drift	$(\delta_n - \delta_{n-1})/h$	h/500 (Griffis, 1993)

Note: d = beam sagging (deflection), L = beam length (spanning between supports), H = building height, h = storey height, δ = lateral displacement at floor n, Δ = total lateral displacement at roof,

When considering ultimate limit states (ULS), the design capacity of members shall not be exceeded by the design structural actions i.e. $S^* \leq \phi R$. Where ϕ is an appropriate capacity reduction factor as per AS4100 and S* is determined for appropriate ULS as per AS1170.0.

3 FE MODELLING

All models have been developed with the ANSYS software package using ANSYS Parametric Design Language (APDL) to facilitate the conduction of parametric studies with variation of the structural loadings and dimensions.

3.1 *Detailed connection modelling*

Detailed models of envisaged connections were first developed to determine the load-displacement (axial and transverse stiffness) and moment-rotation (rotational stiffness) curves needed for the following full structural model. Figure 2 illustrates the three major categories.

The beam-to-column connection is a typical double-cleat angle connection assembling a 200UB25.4 (universal beam) and a 300×200×6. RHS column (Figure 2a). The connection geometry was detailed as per section 4 and Appendix G of *Joints in Steel Connection: Simple Connection* (JSC) design guide (SCI and BCSA 2002). The vertical tie connection is a typical column splice connection using 370×410×25 end plates welded to the column end (Fig. 2b) detailed as per Table H.38 of the JSC (SCI and BCSA 2002). The horizontal tie connection (Fig. 2c) connects adjacent columns by side plates using bolts and nuts. These side plates are predrilled with bolt holes and welded to the face of the column. The connection geometry was determined as per chapter 8 of AS4100. Grade 350 steel and the class 10.9 M24 bolts are considered in the modelling. All dimensions in mm unless noted otherwise.

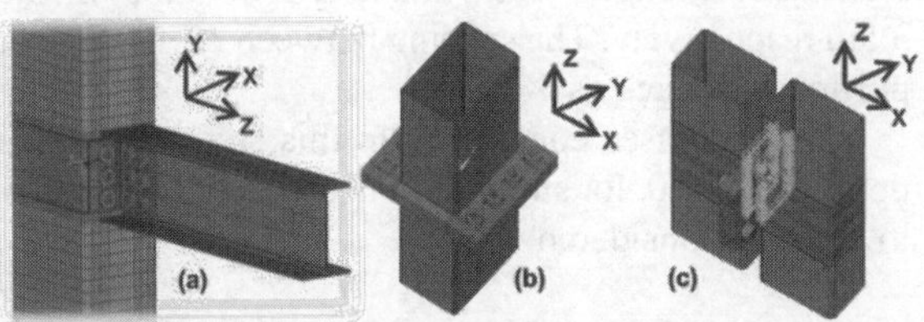

Figure 2. Three major types of connections, respectively: (a) beam-to-column; (b) vertical tie; (c) horizontal tie

In the FE models, all connecting elements were meshed with SOLID45 8-node structural elements. The bolt torque was simulated using PREST179 assigned in the mid-section of the bolt shank. Contact behaviours were defined using the CONTA170 contact and TARGE174 to simulate the interaction between connecting elements. Welding was simulated using the CPINTF node coupling technique.

Figure 3 illustrates typical results from the detailed modelling of the horizontal tie. The contour plot of stress state ratio on the deformed shape of the connection was taken from the applied load marked in the figure, showing the stress distribution and critical locations. The curves are the load-displacement (or moment-rotation) behaviour of the connection to be input to the full structural model. For vertical tie connection and beam-to-column connection, these curves are summarised in Figure 4.

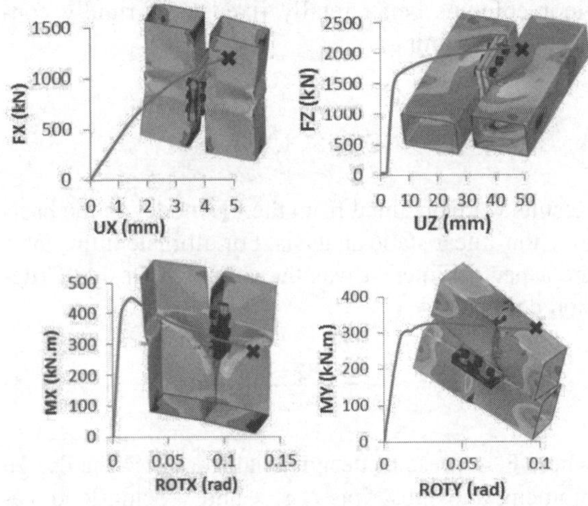

Figure 3. Sample set of results from FE modelling of horizontal tie connection (HZ) – Axes for FEA as in figure 2

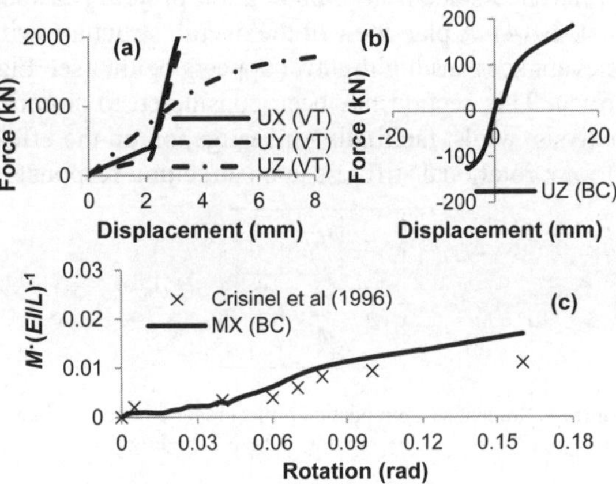

Figure 4. (a) load-displacement behaviour for vertical tie (VT) connection; (b) load-displacement behaviour for beam-to-column connection (BC); (c) moment-rotation behaviour of beam-column connection (BC) (with moment scaled by beam line stiffness)

The FE model exhibits similar in-plane moment-rotation behaviours to the experimental results of a typical double web-cleat I-beam to I-column connection (Crisinel et al. 1996, Figure 4c), with clearly four phases including friction and sliding behaviour (of bolted connection), beam in contact with column and finally the excessive deformation of web cleats.

The initial stiffness for beam-column connection was defined as per Eurocode 3 (CEN, 2005) namely the secant stiffness at 2/3 of the ultimate moment. For the moment-rotation behaviour obtained from both FEA and Crisinel et al. (1996) in Figure 4c, the initial stiffness is found approximately 26% of the EC3 lower limit of semi-rigid connection (based on the beam length used in the following structural analysis).

In cases of bolted shear connection, such as UZ for horizontal tie in Figure 3, UX and UY for vertical ties in Figure 4a and UZ for beam-column connections in Figure 4b, the typical initial friction-slip behaviour can be simulated. For the present configuration of vertical tie, a conservative value of tension capacity is suggested to be 1060kN based on failure of Class 8.8 bolt with yielding of S275 end-plate (SCI and BCSA 2002), indicating that the FEA value (1600kN with Class 10.9 bolts and Grade 350 steel plate) is within a reasonable range.

3.2 *Full structure model*

The structural model constructed herein is based upon

a reference student accommodation project (Lawson et al. 2014). A plan view of the overall structure, with relevant sections highlighted, appears below (see Figure 5). This section has been considered to simplify analysis, whilst facilitating investigation on the effect of joint rotational stiffness upon structural response.

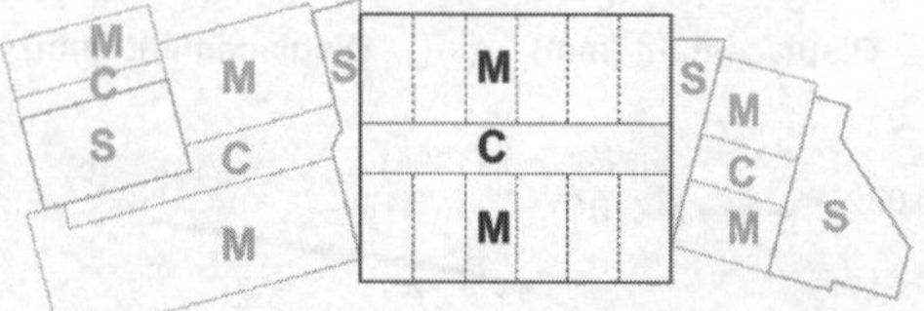

Figure 5. Student accommodation project reported upon by Lawson et al. (2014) – Section serving as basis is highlighted

Figure 6 illustrates the FEA model of a typical module and structure used within this study. In order to represent structural members BEAM188 elements are used. Non-linear spring elements (COMBIN39) are used to represent connections both between and within modules, which has been proved effective in modelling structural frames with non-linear semi-rigid connections in the previous studies (Nannan & Lai, 2010 & Bayo et al. 2006).

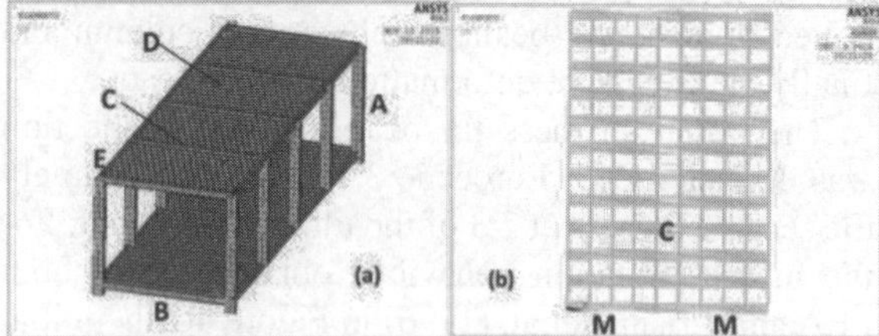

Figure 6 FEA models of (a) a typical module and (b) a complete modular structure. Note: For (a), A = module columns, B = Floor edge beam, C = intermediate beam location, D = roof/floor purlin location, E = roof edge beam; For (b) C = corridor, M = module

The modules considered within this study consist of 300×200×6.0 RHS columns, floor edge members of 200UB25.4, floor intermediate primary beams of 180UB22.2, roof edge members of 200UB22.3, roof intermediate primary members of 180UB18.1 and roof and floor purlins consisting of C15024 C-sections. All purlin-to-edge beam connections were assumed to be fully rigid (i.e. coupled using CPINTF). For the non-linear joint stiffness, the original moment-rotation curve was simplified as a bilinear curve, according to EC3 using the results from section 3.1 as a basis.

The final full structure model consisted of an 11 storey building containing a corridor, with 11 modules either side of the corridor. Spacing between the intermediate beams and columns was taken as 3m, with a total module length, width and height of 12m, 4m and 3.5m respectively. The spacing between roof and floor purlins was taken as 300mm.

The load cases considered in this study are based upon AS1170.0, for strength limit states (ULS) the following are considered.

$$E = 1.2G + 1.5Q \tag{1}$$

$$E = 1.2G + W_u + 0.4Q \tag{2}$$

$$E = 0.9G + W_u \tag{3}$$

And for serviceability (SLS), although these may not necessarily be the case for all structures.

$$E = G + 0.4Q + W_s \tag{4}$$

These load cases were chosen as structural design within Australia is largely governed by wind given the low severity of seismic events, with G, Q and W being the dead, live and wind loads respectively. Subscripts s and u being serviceability and ultimate loads respectively. All wind loads were determined as per AS1170.2 for a building in region A, typical of the vast majority of the Australian land mass.

When applying these loads to the FE structural model, the design pressures in kPa were turned into equivalent point loads which could conceivably apply at element nodes of members. Boundary conditions within the model consisted of the base of the ground floor columns being totally fixed as if rigidly connected to a footing.

4 RESULTS AND DISCUSSION

Results were obtained from the FE model on the basis of a non-linear static analysis. For ultimate limit states the aspect of interest was the so-called Design Criterion defined as

$$\zeta = \frac{F_c^*}{\phi F_s} + \frac{M_x^*}{\phi M_{sx}} + \frac{M_y^*}{\phi M_{sy}} \tag{5}$$

where F* represents design axial force, M* the design moment, and subscripts c, s, x and y being load, capacity, in-plane and out-of-plane respectively. The maximum design criterion was calculated for each trial and the following plots produced.

From the models, the interstory drift parameter for levels 3 to 4 (for an 11 storey building) was identified as the critical level for determining serviceability by the criteria outlined in Table 1, Figures 9 & 10 demonstrate the effect of varying connection stiffness.

The 3.7× and 183.8× factors in Figures 7 & 9 represent the lower and upper limits of a semi-rigid beam-to-column connection as per EC3.

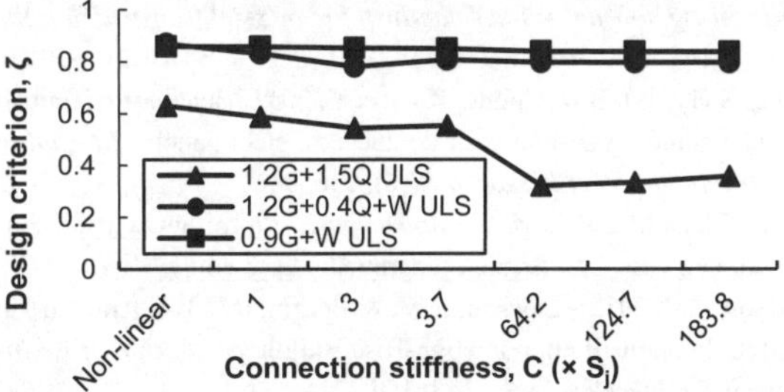

Figure 7. Parametric study – Varying intra-module connection stiffness S_j = 140.4kNm/rad

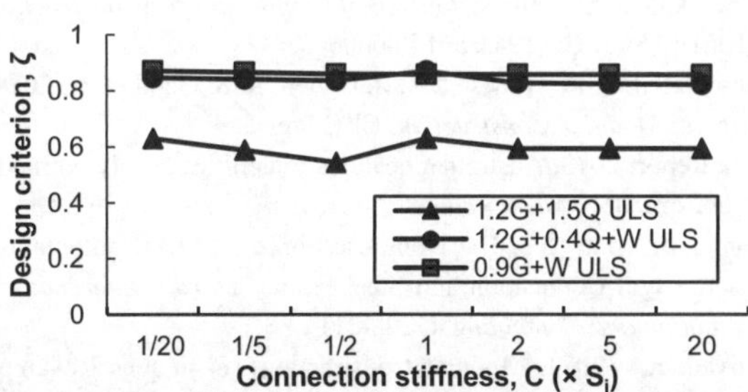

Figure 8. Parametric study – Varying inter-module connection stiffness S_j = 280 kNm/rad

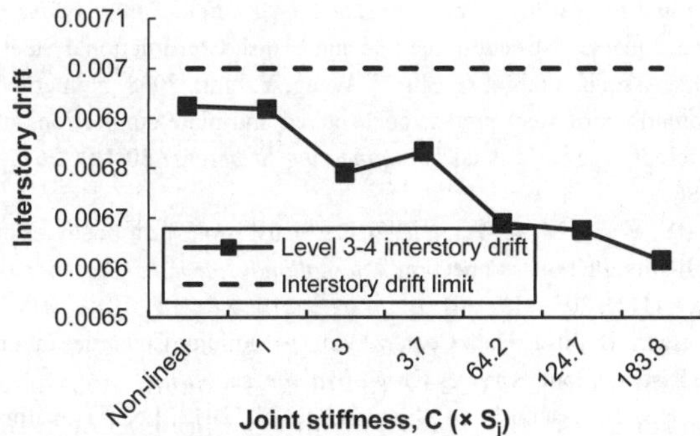

Figure 9. Parametric study (SLS) – Varying intra-module connection stiffness S_j = 140.4 kNm/rad

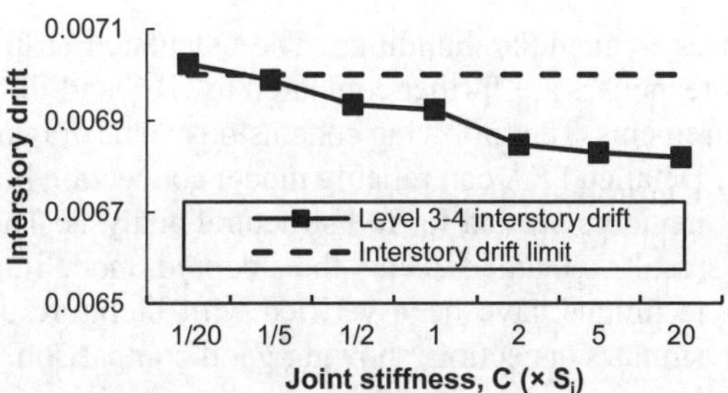

Figure 10. Parametric study (SLS) – Varying inter-module connection stiffness S_j = 280 kNm/rad

Clearly there is an improvement in serviceability criteria with increasing stiffness of inter- and intra-module connection stiffness (Figures 9 & 10 respectively), with the latter playing a seemingly greater role in the resistance to lateral actions. This would suggest that an efficient way to improve resistance to lateral action would be to add bracing elements to the modules, as this would essentially increase the overall stiffness. This lends itself to further investigation.

For the ultimate (strength) limit states (Figures 7 & 8), it is evident that for the load case presented in Equation (3) there is little variance in the maximum design criterion with increasing stiffness (either intra- or inter-module, Figures 7 and 8 respectively). However, there is moderate improvement for the load case as in Equation (2) with increasing stiffness in both parametric studies. The case presented in Equation (1) shows the greatest variance with connection stiffness. From the intra-module parametric study it is evident that this case is greatly improved by increased connection stiffness, this is likely due to improved frame action transferring loads more efficiently to the columns (the higher capacity members). However, for the inter-module study the opposite effect is seen, this may arise due to improved lateral transfer of loads, without improved frame action, resulting in more concentrated stress on certain modules within the building. Interestingly, in all cases increasing the stiffness from 3× to the lower EC3 limit for the intra-module study results in decreased performance.

5 CONCLUSIONS

A numerical study is conducted to investigate the effects of connection stiffness on the structural responses of modular buildings. The calculated structural responses are further compared to ULS and SLS requirements. The following conclusion can be drawn:

1) Detailed FEA can reliably model connection behaviours needed for full structural analysis. The results obtained using the adopted modelling technique have been verified with literature of similar connection, showing good comparison.
2) Increased intra-module connection stiffness results in greater performance increase (i.e., reduction of the corresponding structural responses in design criteria) under design load combinations.
3) Intra-module connection stiffness plays the greatest role in the overall structural resistance to lateral actions under SLS wind loads

This paper has only accounted for the role of intra- and inter-module lateral connection stiffness, a number of factors still require study, such as stiffness of vertical tie connections, role of external bracing (i.e. structural core), membrane effects and the number of stories. The approach outlined herein has revealed much about the importance of connection semi-rigidity to the response of modular buildings, and could form the basis for many studies where large scale testing is not feasible.

ACKNOWLEDGEMENTS

The authors would like to acknowledge the financial support of the industry partners of the Modular Construction Codes Board (MCCB) and the State Government of Victoria through the Manufacturing Productivity Networks program.

REFERENCES

Standards Australia. *AS4100-1998 Steel structures*, SAI Global.

Standards Australia. *AS/NZS 1170.0:2002 Structural design actions – General principles*, SAI Global.

Standards Australia. *AS/NZS 1170.2:2011 Structural design actions – Wind actions*, SAI Global.

Australian Concrete Construction 2012. Precast modular construction coming to Australia, *Australian Concrete Construction* 25(3).

Bayo, E., Cabrero J.M., & Gil, B. 2006. An effective component-based method to model semi-rigid connections for the global analysis of steel and composite structures, *Engineering Structures* **28**, 97–108.

CEN (European Committee for Standardization) 2005, *Eurocode 3: Design of Steel Structures – Part1.1: General Rules and Rules for Buildings*, CEN, Brussels, Belgium.

CIMC 2015. *CIMC Modular Building Systems*, http://www.cimc-mbs.com/.

Crisinel, M. Ren, P. & Carretero, A. 1996. Practical Design Method for Semi-Rigid Composite Joints with Double Web Cleat Connections, *IABSE REPORTS*, 185–196.

Farnsworth, D. 2014. Modular Tall Building Design at Atlantic Yards B2, *CTBUH 2014 Shanghai Conference Proceedings.*

Griffis, L.G. 1993. Serviceability Limit States Under Wind Load, *Engineering Journal - American Institute of Steel Construction Inc* **30(1)**, 1-16

Hong, S.G. Cho, B.H. Chung, K.-S. et al. 2011. Behavior of framed modular building system with double skin steel panels, *Journal of Constructional Steel Research* **67(6)**, 936–946.

Irwin Consult 2014. *Soho Apartments*, <http://www.irwinconsult.com.au/case_studies/soho-apartments-case-study/>

Lawson et al. 2012 - Lawson, R.M. & Bergin, R. 2014. Application of Modular Construction in High-Rise Buildings, *Journal of Architectural Engineering* **18(2)**, 148–154.

Luo, F.J. Bai, .Y., Yang, X. et al. 2015. Member Capacity of Pultruded GFRP Tubular Profile with Bolted Sleeve Joints for Assembly of Latticed Structures, *Journal of Composites for Construction* 10.1061/(ASCE)CC.1943-5614.0000643.

Lawson et al. 2014 - Lawson, R.M. Ogden, R. & Goodier, C. 2014. *Design in Modular Construction*, CRC Press.

Mena Report 2013. Australian healthcare facilities to rely on modular construction, *Mena Report*.

Nannan, Z. & Lai, W. 2010. Nonlinear Finite Element Computation on Semi-rigid Connection and Steel Frame, *Third Conference on Information and Computing* **4**, 208-211

Sullivan, J.A. 2011. Modular Construction Techniques Revamps Global LNG Construction, *Natural Gas Week.*

Timber Building Systems 2015. *Timber Building Systems*, <http://timberbuildingsystems.com/> Accessed 26/10/2015.

SCI and BCSA 2002, *Joints in steel construction: Simple connections*, Steel Construction Institute and British Constructional Steelwork Association.Shi, G. Shi, Y. Wang, Y. et al. 2008. Numerical simulations of steel pretensioned bolted end-plate connections of different types and details, *Engineering Structures* **30(10)**, 2677–2686.

Shi, Y. Wang, M. & Wang, Y. 2011. Analysis on shear behavior of high-strength bolts connection, *International Journal of Steel Structures* **11(2)**, 203–213.

Worthen, B. 2011. Data Centers Boom – Modular Construction on the Rise as Cloud Services Grow, *Wall Street Journal.*

Yu, H. Al-Hussein, M. Al-Jibouri, S. et al.. 2013. Lean Transformation in a Modular Building Company, *Journal of Management in Engineering* **29(1)**, 103–111.

Proceedings of the International Conference on Smart Infrastructure and Construction
ISBN 978-0-7277-6127-9

© The authors and ICE Publishing: All rights reserved, 2016
doi:10.1680/tfitsi.61279.463

Stethoscope-like smart sensing for system identification of building frames

I. Takewaki* and K. Fujita

Dept. of Architecture and Architectural Eng., Kyoto University, Kyoto, Japan
* *Corresponding Author*

ABSTRACT System identification of building frames using stethoscope-like smart sensing is presented in this paper. For shear building models and shear-bending building models, the recorded data at floors just above and below a specific story are used for the system identification of that story. The proposed system identification method based on such smart sensing does not need the simultaneous sensing at multiple stories. Direct physical-parameter system identification proposed here is suitable also for damage detection of specific stories. Some experimental results and numerical simulation during a recent mega earthquake are shown for demonstrating the realism of the proposed method.

1 INTRODUCTION

Extensive investigation has been conducted on system identification (SI) of structures (Housner et al. 1997, Takewaki et al. 2011).

It is well recognized that the modal-parameter SI and physical-parameter SI are two major branches in SI. The former is suitable for identifying the overall mechanical properties of a structural system and exhibits stable characteristics. While the latter is important from different points of view, e.g. enhancement of reliability and robustness in active controlled structures or base-isolated structures, its development is limited due to the requirement of multiple measurements or the necessity of complicated manipulation (Udwadia et al. 1978, Takewaki and Nakamura 2000, 2005, Takewaki et al. 2011, Zhang and Johnson 2013). A mixed approach is often used in which physical parameters are identified from the modal parameters obtained by the modal-parameter SI. However, in view of inverse problem formulation, a sufficient number of modal parameters must be obtained for the unique and reliable identification of the physical parameters. This requirement is usually hard to satisfy.

If the discussion is restricted to multi-storied building structures, it is believed in general that the acceleration records at all the floors above a specific story are necessary to evaluate the story shear force which is absolutely necessary in the stiffness-damping evaluation. This instrumentation may usually be unrealistic in multi-storied buildings due to cost and sensor-management problems.

To overcome this difficulty, Udwadia et al. (1978) proposed a unique SI theory for a shear model. They clarified that unique identification of stiffnesses and damping coefficients is possible when acceleration records at the floors just above and below a target story are available. However, the applicability of their theory to actual records with noises had never been investigated. Furthermore their theory includes unrealistic manipulation to take the limit at infinite frequency in order to identify the viscous damping coefficients. In addition their technique has a difficulty that some of stiffness and damping parameters are derived recursively. Takewaki and Nakamura (2000) overcame this difficulty by introducing a Taylor series expansion technique in a rigorous mathematical way. They also extended the technique to the SI problem with noise (Takewaki and Nakamura 2005).

In this paper, a new method of SI of building frames using stethoscope-like smart sensing is presented. The stethoscope-like smart sensing is an innovative one enabling actual structural 'health monitoring'. The method is an extension of the work by Takewaki and Nakamura to more general cases (bending-type structures, micro-tremor input). An ARX (Auto-Regressive with eXogenous input) model is utilized for overcoming the noise issue. For shear building models and shear-bending building models, the recorded data at floors just above and below a specific story are used for SI of that story. Some experimental results and numerical simulation during a recent mega earthquake are shown for demonstrating the realism of the proposed method.

It is also shown that the proposed SI method based on such smart sensing does not need the simultaneous sensing at multiple stories and is suitable even for damage detection of specific stories.

2 FORMULATION

2.1 Shear Building Model

Applicability to actually recorded data is extremely important in SI and structural health monitoring.

Takewaki and Nakamura (2000, 2005) introduced an 'identification function' (IF).

$$f_i(\omega) = M_i \times \mathrm{Re}\left\{\frac{(\mathrm{i}\omega)^2 \ddot{U}_i(\omega)}{\ddot{U}_{i-1}(\omega) - \ddot{U}_i(\omega)}\right\} \quad (1)$$

where $\ddot{U}_i(\omega)$ is the Fourier transform of the acceleration at the i-th floor and M_i is the sum of the floor masses from the i-th floor to the top floor. It was made clear that the limit of this IF at zero frequency indicates the i-th story stiffness of a shear building. Figure 1 shows a concept of stethoscope-like smart sensing for story SI. It should be emphasized that the method does not need the simultaneous sensing at multiple floors and the stethoscope-like smart sensing can be repeated for the respective story. The function for a damping coefficient can also be expressed in the form of a modified IF.

The ARX model was introduced to overcome the noise issue (Takewaki and Nakamura 2009, Takewaki et al. 2011, Minami et al. 2013). A high-pass filter was further found to be an important tool. Takewaki and Nakamura (2009) developed a time-variant nonparametric identification method using ARX models for investigating the change of natural frequencies and modal damping ratios. The method has been applied successfully to an actual base-isolated building (Takewaki and Nakamura 2009) and has also been applied to high-rise buildings during the 2011 off the Pacific coast of Tohoku earthquake (Minami et al. 2013, Ikeda et al. 2014a).

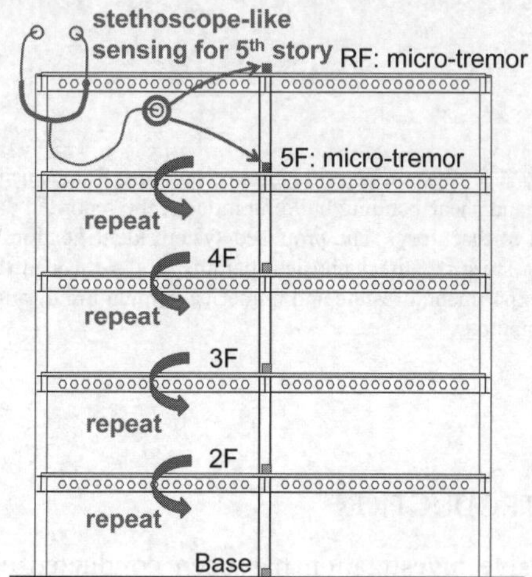

Figure 1. Stethoscope-like smart sensing for story system identification

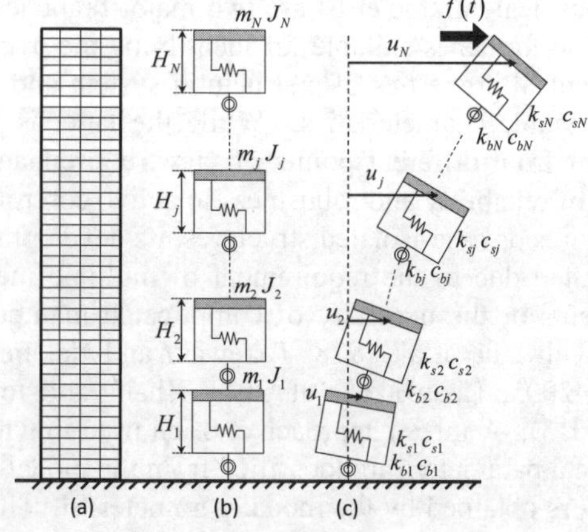

Figure 2. Shear-bending building model

2.2 Shear-bending Building Model

When the building becomes slender, a shear-bending building model becomes a more appropriate reduced model. Let H_j and H_i^t denote the story height of the j-th story and the floor height of the i-th floor from the ground. The bending and shear stiffnesses of the shear-bending building model as shown in Figure 2 can be derived (Kuwabara et al. 2013, Minami et al. 2013) with the stiffness ratio $R_j = k_{bj} / k_{sj}$.

$$k_{bj} = \frac{R_j + \dfrac{H_j}{\sum_{i=j}^{N} m_i} \sum_{i=j}^{N} \{m_i (H_i^t - H_{j-1}^t)\}}{\lim\limits_{\omega \to 0} \left[\mathrm{Re}\{F_j(\omega)\} \right] - \dfrac{H_j}{\sum_{i=j}^{N} m_i} \sum\limits_{m=1}^{j-1} \left[\dfrac{\sum_{i=m}^{N} \{m_i (H_i^t - H_{m-1}^t)\}}{k_{bm}} \right]} \tag{2}$$

$$k_{sj} = k_{bj} / R_j$$

where $F_j(\omega)$ is defined by

$$F_j(\omega) = \{(\ddot{U}_{j-1} / \ddot{U}_j) - 1\} \Big/ \left(-\omega^2 \sum_{i=j}^{N} m_i \right) \tag{3}$$

3 APPLICATION TO EARTHQUAKE RECORD

3.1 Shear Building Model

The formulation in Section 2.1 has been applied to actual earthquake records. Takewaki and Nakamura (2000) investigated the accuracy of this formulation by identifying the stiffness of the base-isolation story (Figure 3: left) and Nakamura et al. (2001) demonstrated the high accuracy and reliability of this formulation in the shaking table test of a scaled five-story steel frame (Figure 3: right).

Figure 3. Actual base-isolated building at Kyoto Univ. and shaking table test of a scaled steel frame

3.2 Shear-bending Building Model

Figure 4 shows a super high-rise building at Osaka bay area. An important recording was made in this building during the 2011 off the Pacific coast of Tohoku earthquake (Minami et al. 2011). It was reported that small non-structural damage was caused. Since the time varying identification is necessary in this case, the outline of the time-varying identification was explained in Ikeda et al. (2014a). The duration of 30s was used as the evaluation time and this process was repeated by moving the window by 1s. Figure 5(a) presents the top-story displacement and the time-varying fundamental natural period of this building (short-span direction) which was evaluated by the moving-window batch least-squares method (Ikeda et al. 2014a). The time-varying SI indicates the result by the moving-window batch least-squares method and the overall SI means the result which uses the batch least-squares method in the whole duration. The time dependency of the fundamental natural period may result from its amplitude dependence and reported slight damage to some non-structural elements (ceiling walls etc.). The accuracy of the employed method can be confirmed by the comparison with data from the record shown in Figure 5(b). The six points in Figure 5(b) have been plotted by investigating the period between zero-crossing points (Ikeda et al. 2015).

Furthermore, Figure 6 illustrates another super high-rise building at Shinjuku, Tokyo. Another important recording was made in this building during the 2011 off the Pacific coast of Tohoku earthquake (Minami et al. 2013). Figure 7 shows the top-story displacement and the time-varying fundamental natural period of this building (short-span direction) which was evaluated by the same method (Ikeda et al. 2014a). The fundamental natural period of this building has been evaluated by several other methods (Fourier analysis of recorded data and design document, etc.) and those values are also plotted in Figure 7 for comparison.

4 APPLICATION TO ACTUAL AMBIENT VIBRATION

The formulation in Section 2.1 can be applied to ambient vibration data. Figure 8 shows a five-story steel building at Kyoto University. The frame dimension is illustrated in Figure 9 and the measurement patterns are presented in Figure 10. Pattern A is aimed at identifying the 1st and 2nd story stiffnesses and Pattern B is introduced for identifying the

4th and 5th story stiffnesses. On the other hand, Pattern C is set for identifying the 3rd story stiffness. The measurement at the first and top floors is necessary in each measurement pattern in order to obtain the fundamental natural frequency from the transfer function using these sensors.

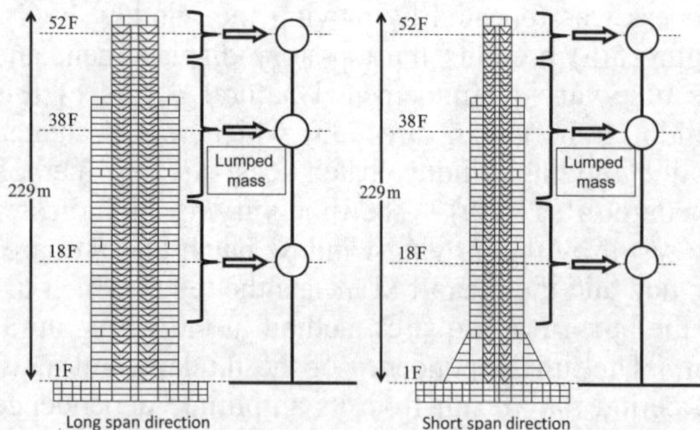

Figure 4. Super high-rise building at Osaka bay area (Minami et al. 2013)

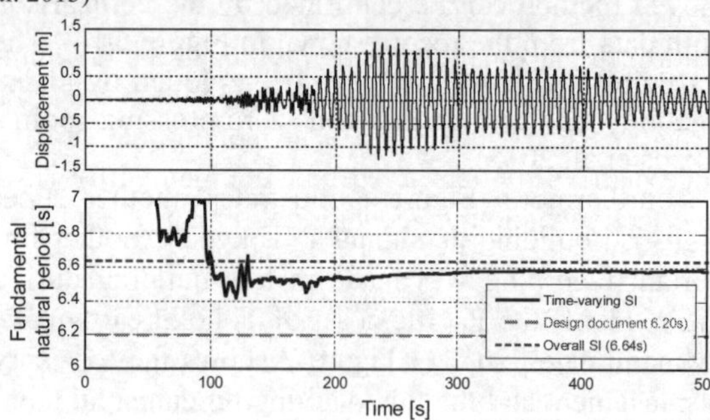

Figure 5(a). Top-story displacement and fundamental natural period of super high-rise building at Osaka bay area (short-span direction) (Ikeda et al. 2014a)

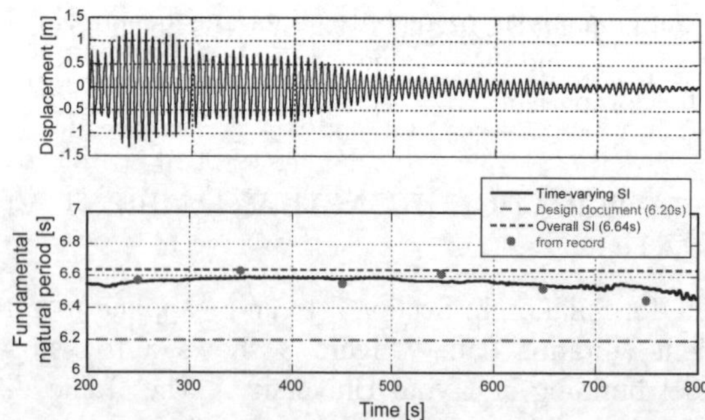

Figure 5(b). Top-story displacement and fundamental natural period of super high-rise building at Osaka bay area (short-span direction): Accuracy investigation through comparison with data from records (Ikeda et al. 2015)

As pointed out earlier, the simultaneous measurement at all floors is unnecessary in the proposed method.

Table 1 shows the summary of identification (short-span direction). It can be observed that the proposed method is almost reliable.

The formulation in Section 2.2 has been applied to the same micro-tremor vibration data. It has been confirmed that, while the shear stiffnesses can be identified relatively accurately, the identification of bending stiffnesses is often unstable. This property may result from the small dependence of the fundamental natural frequency on the bending stiffnesses. The enhancement of accuracy is desired.

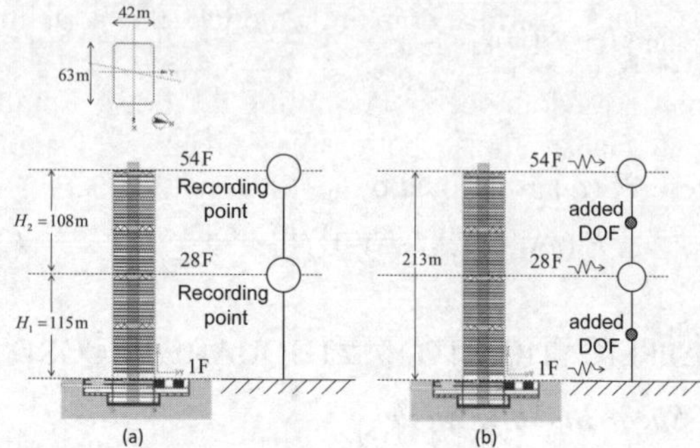

Figure 6. Super high-rise building at Shinjuku, Tokyo (Minami et al. 2013)

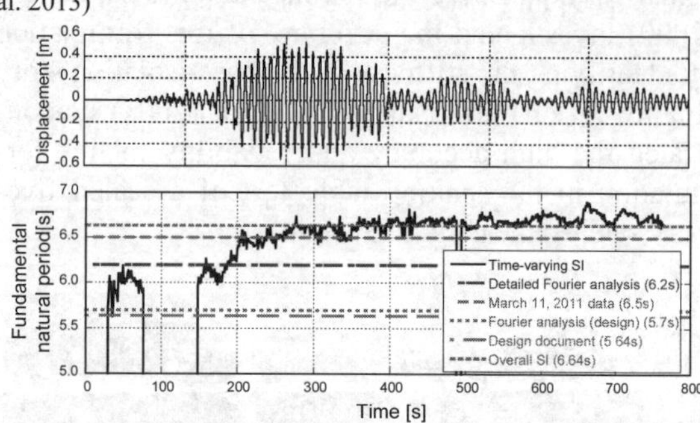

Figure 7. Top-story displacement and fundamental natural period of super high-rise building at Shinjuku, Tokyo (short-span direction) (Ikeda et al. 2014a)

Figure 8. Five-story steel building at Kyoto Univ.

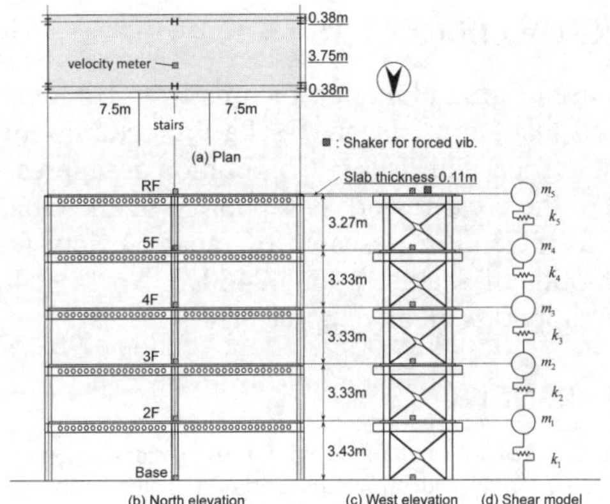

Figure 9. Frame dimension and shear building model

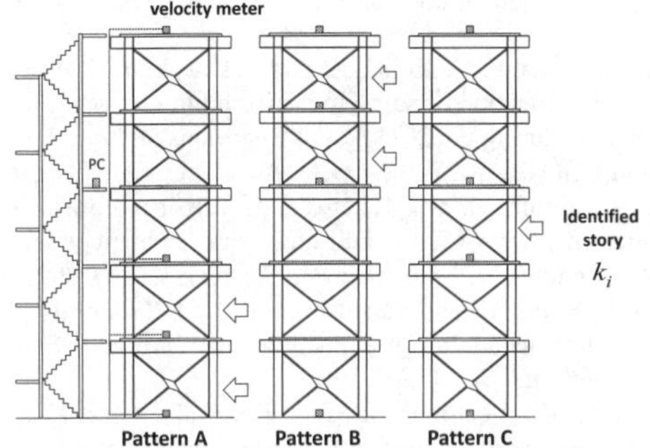

Figure 10. Location of velocity meters for three patterns of identification

Table 1. Summary of identification (short-span dir.)

		Static analysis	Forced vibration test	Ambient vib. data (Roof/ base)	Identified value
stiffness ($\times 10^4$ [kN/m])	1st story	10.82	-	-	**9.103**
	2nd story	9.83	-	-	**9.128**
	3rd story	9.56	-	-	**8.132**
	4th story	9.23	-	-	**6.252**
	5th story	8.49	-	-	**6.481**
natural circular freq. [rad/s]	1st	17.19	15.07	15.19	15.77
	2nd	48.75	46.83	46.89	42.65
	3rd	76.16	72.09	76.02	68.84
	4th	97.44	86.50	86.93	85.97
	5th	111.77	113.21	113.72	102.65
natural circular freq. [Hz]	1st	2.74	2.40	2.42	2.51
	2nd	7.76	7.46	7.46	6.79
	3rd	12.13	11.48	12.10	10.96
	4th	15.52	13.77	13.84	13.69
	5th	17.80	18.03	18.10	16.35

5 DAMAGE DETECTION

In damage detection, it is inevitable to prepare the data in the structurally healthy state. This treatment requires a tremendous amount of work before damage occurs. Unless we know the healthy state, it is not possible to know whether the state after an event is healthy or damaged. If restricted to a shear building model or a shear-bending building model, the recording at limited floors may be realistic (Kuwabara et al. 2013). This method is based on a pioneering nonparametric approach by Udwadia et al. (1978).

Figure 11 shows a schematic diagram for the damage detection in shear-bending building models (Kuwabara et al. 2013). On the other hand, some approaches using a lot of data and a great deal of repeated calculations have been proposed for shear building models (Zhang and Johnson 2013).

Figure 12 illustrates five examples of damage pattern. It was reported in Northridge earthquake (1994) and Hyogoken-Nanbu (Kobe) earthquake (1995) that not a few steel buildings exhibited brittle fracture behaviors at beam end connections due to the limited performance of the beam-end welding portion.

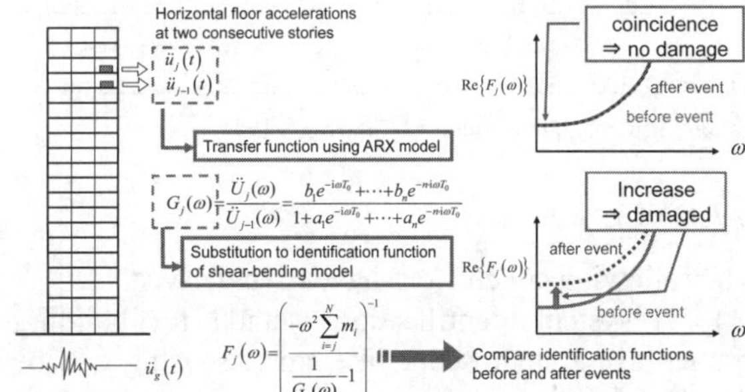

Figure 11. Schematic diagram for damage detection in shear-bending building model

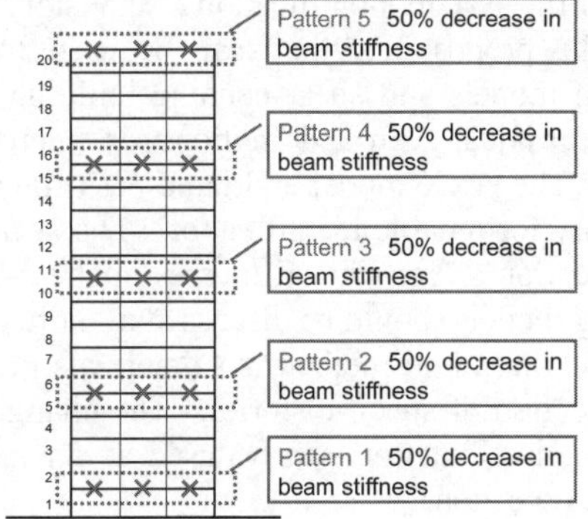

Figure 12. Examples of damage pattern

Figure 13 presents the identification functions $\mathrm{Re}[F_j(\omega)]$ defined by Eq.(3) before and after damage for pattern 3 in Figure 12. It can be observed that the damage of the 10-th story beam influences clearly the identification functions at the 10-th and 11-th stories. This demonstrates the validity and reliability of the proposed damage detection method.

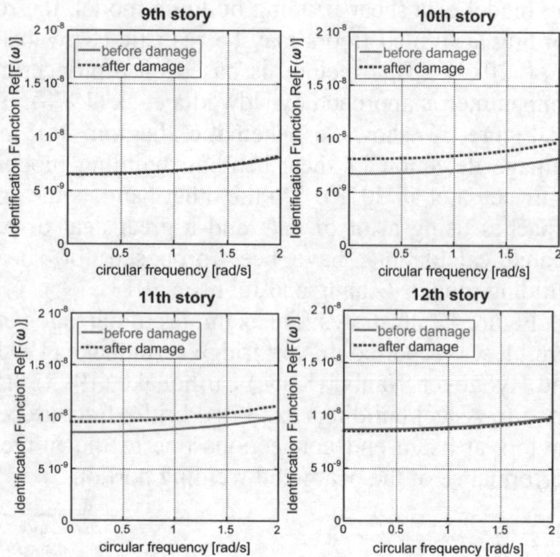

Figure 13. Identification functions before and after damage for pattern 3 (damage in beams): El Centro NS 1940

6 CONCLUSIONS

The following conclusions have been derived.

(1) A system identification method for building frames using stethoscope-like smart sensing can be developed which does not need the simultaneous sensing at multiple stories. The recorded data at floors just above and below a specific target story are used for the system identification of that story.

(2) The proposed method can be used for shear building models and shear-bending building models under earthquake ground motions or micro-tremor inputs. The ARX model and band-pass filtering are necessary for reliable identification. The accuracy of identification of bending stiffnesses in shear-bending building models should be discussed in more detail.

(3) The proposed method is suitable also for damage detection of specific stories. The change of the identification function can be used as an index for damage detection.

ACKNOWLEDGMENTS

The use of ground motion records from Building Research Institute of Japan, Osaka Prefectural Government Office and Taisei Corporation is appreciated. Part of the present work is supported by the Grant-in-Aid for Scientific Research of Japan Society for the Promotion of Science (No.24246095, No.15H04079). This support is greatly appreciated.

REFERENCES

Fujita, K., Ikeda, A. and Takewaki, I. (2015). Application of story-wise shear building identification method to actual ambient vibration, *Frontiers in Built Environment*,1: 2.

Housner G.W. et al. (1997). "Special issue, Structural control: past, present, and future." *J. Engng. Mech.*, ASCE, 123(9), 897-971.

Ikeda, A., Minami, Y., Fujita, K. and Takewaki, I. (2014a). "Smart system identification of super high-rise buildings using limited vibration data during the 2011 Tohoku earthquake." *Int. J. High-Rise Buildings,* 3(4), 255-271.

Ikeda, A., Fujita, K. and Takewaki, I. (2014b). "Story- wise system identification of shear building using ambient vibration data and ARX model." *Earthquakes and Structures*, 7(6), 1093-1118.

Ikeda, A., Fujita, K. and Takewaki, I. (2015). Reliability of system identification technique in super high-rise building, *Frontiers in Built Environment*, 1: 11.

Kuwabara, M., Yoshitomi, S. and Takewaki, I. (2013). "A new approach to system identification and damage detection of high-rise buildings." *Struct. Control Health Monitor.*, 20(5), 703-727.

Minami, Y., Yoshitomi, S. and Takewaki, I. (2013). "System identification of super high-rise buildings using limited vibration data during the 2011 Tohoku (Japan) earthquake." *Structural Control and Health Monitoring*, 20(11), 1317-1338.

Nakamura, M., Morita, K. and Takewaki, I. (2001). Structural Damage Detection on Five-story Steel Structure with Simulated Damage, Part2 Damage Detection Using Interstory-Stiffness Identification Method, No.01-253, pp224-227, The 44th Japan Automatic Control Society Conference.

Takewaki, I. and Nakamura, M. (2000). "Stiffness- damping simultaneous identification using limited earthquake records." *Earthquake Engineering Structural Dynamics*, 29(8), 1219-1238.

Takewaki, I. and Nakamura, M. (2005). "Stiffness- damping simultaneous identification under limited observation." *J. Engng. Mech.*, ASCE, 131(10), 1027-1035.

Takewaki, I. and Nakamura, M. (2009). "Temporal variation of modal properties of a base-isolated building during an earthquake." *J. Zhejiang University-SCIENCE A*, 11(1), 1-8.

Takewaki, I., Nakamura, M. and Yoshitomi, S. (2011). "*System Identification for Structural Health Monitoring*." WIT Press (UK) (ISBN-10: 1845646282).

Udwadia, F., Sharma, D. and Shah, C. (1978). "Uniqueness of damping and stiffness distributions in the identification of soil and structural systems." *J. Applied Mech.*, ASME, 45, 181-187.

Zhang, D. and Johnson, E. (2013). "Substructure identification for shear structures I: Substructure identification method." *Structural Control and Health Monitoring*, 20(5), 804-820.

Proceedings of the International Conference on Smart Infrastructure and Construction
ISBN 978-0-7277-6127-9

© The authors and ICE Publishing: All rights reserved, 2016
doi:10.1680/tfitsi.61279.469

Safety evaluation of complex civil structures using structural health monitoring

J. Teng[1, 2], W. Lu[*2] and Y. Cui[2]

[1] *Fujian University of Technology, Fujian, China*
[2] *Harbin Institute of Technology Shenzhen Graduate School, Shenzhen, China*
[*] *Corresponding Author*

ABSTRACT As the innovation developments of structures, especially the public large span structures, the complex problems regarding to the structural safety are always concerned and should be solved during their construction and service phase. The influencing factors to the safety of specific structure are various, such as the complex shape of structure, complex structure style with transfer stories and complex construction processes and so on. Three civil structures are represented to illustrate the complex problems in civil structures, the safety of which are tracked by using structural health monitoring. The first structure is Shenzhen Vanke Center, which is innovated structure style with cables, the complex problem is limitation control of displacements and stresses to assure the safety in its construction phase. The second structure is Shenzhen Bay Stadium, which is large span space structure, the complex problem is the limitation control of temperatures, stresses and displacements to decide the closing time and assure the safe closure process of two main parts of steel roof. The third structure is Zhuhai Opera House, which is located at Yeli Island and in wind effects district, the complex problem is the safety control of structure in its service phase when it is subjected with wind load. The structural health monitoring systems of these structures were implemented, where the measurements were used to evaluate the structural safety of each civil structure. As the real world structural health monitoring systems, the projects studied in the paper are examples for construction and service monitoring of complex civil structures.

1 INTRODUCTION

With the development of economy, upgrading of construction technologies and building materials, the investment volume of infrastructure grows rapidly in the world. Many public large span structures that have novel structural forms and large scales have been continuously emerging, especially in China (Teng et al. 2010; Zhang et al. 2015; Kuok & Yuen 2012). The characteristics of them are complex shape, structure style and construction processes. Once structural incidents and accidents occur, both the economy and human life loss will be inconceivable (Shen et al. 2013). So the structural health monitoring (SHM) system is very important and necessary for the large span structure (Chang et al. 2003), which helps to avoid the possible accidents, manage the schedule and progress of construction, control the construction quality, and assess the safety of structure in construction and service phase (Teng et al. 2014[b]).

Three civil structures are introduced in this paper, which are Shenzhen Vanke Center, Shenzhen Bay Stadium and Zhuhai Opera House. The relevant complex problems are discussed and analysed using the measurements from SHM to assure and control the safety of structure in its construction and service phase.

2 SHENZHEN VANKE CENTER

2.1 *Background knowledge*

Shenzhen Vanke Center is situated at the Shenzhen Dameisha Beach, which has a total construction area

137116 m^2, horizontal length 500 m. The four and five floors of the superstructure are supported by the giant tubes, solid web thick walls and columns in order to provide the large open space for the garden (Teng et al. 2015[a]). An effect drawing of Shenzhen Vanke Center is shown in Figure 1.

Figure 1. Effect drawing of Shenzhen Vanke Center.

This structure consists of the composite frame and cables system, in which the bottom floor is steel Q345, and the superstructure is a concrete frame with wide beams and flat columns. The main construction processes are comprised of the following aspects (1) Firstly, construction of the main vertical supporting members is performed. (2) Construction of the bottom floor and roof is then performed. (3) The cables are installed, and the multistage and sub-regional tensioning and repeated tensioning methods are adopted. (4) The sub-regional construction process of the superstructure is performed (Fu et al. 2012).

The cable tensioning and superstructure construction phase are two key phases during the whole construction process. The cable force, strains of the steel beams and concrete filled steel columns, degree of arching of the bottom floor and deformation of the vertical supporting members, which are the key factors for the safety construction and the final structural shape. A SHM system is established to control the safety and state of cables during the whole construction process, estimate the strains and displacements of the key structural locations and ensure safe construction.

For this paper, the local structure between tube 5# and tube 6# is selected as a standard unit to illustrate and analyse. The specific location is shown in Figure 1 (the region shaded with red dashed lines). The layout of measurement points of the standard unit is shown in Figure 2. The main construction process schedule for the selected standard unit is shown in Table 1.

a)

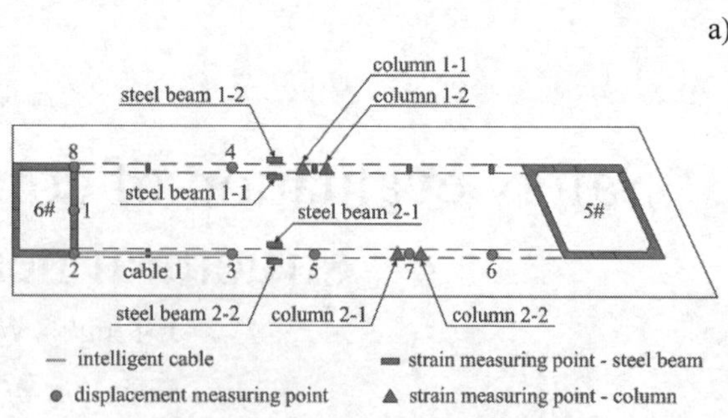

b)

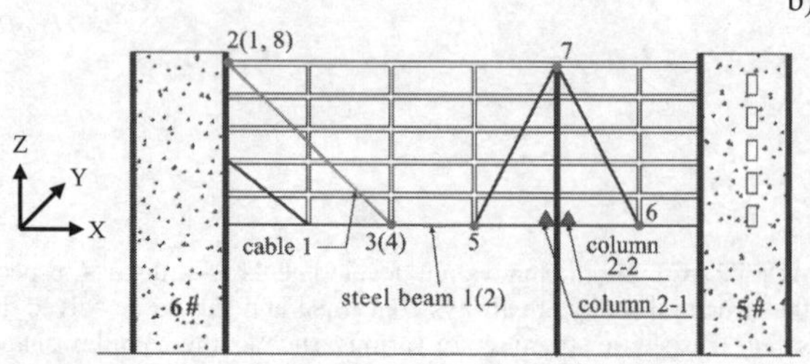

Figure 2. Layout of measurement points of the standard unit. a) Plan view. b) Vertical view.

Table 1. The main construction process schedule

Date	Process
4 April, 2008	cable tensioning phase
5 April, 2008	repeated cable tensioning phase
11 April, 2008	dismantle the full hall scaffold from 2nd floor to top floor
12 April, 2008	set a water tank at the second floor
4 May, 2008	complete the pouring work on the 4th floor slab
12 May, 2008	complete the pouring work on the 5th floor slab
27 May, 2008	complete the pouring work on the 6th floor slab

2.2 *Data analysis*

The monitoring results of displacements, stresses and cable force of some measurement points and cable 1 are respectively shown in Figures 3-6.

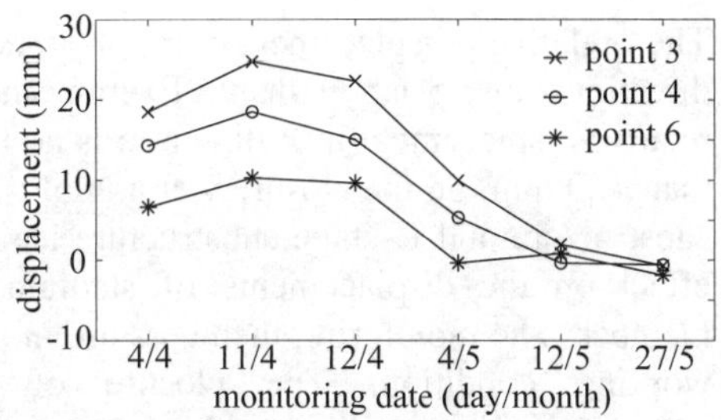

Figure 3. Z-axis displacement of points 3, 4 and 6 in construction phase.

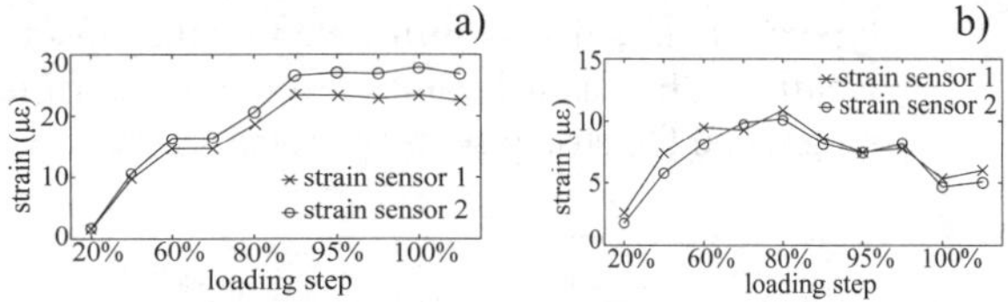

Figure 4. Strains of steel beam 1 and 2 in cable tensioning phase. a) Steel beam 1. b) Steel beam 2.

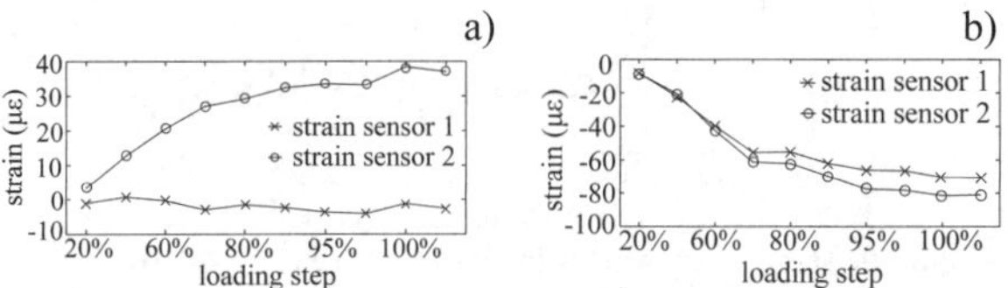

Figure 5. Strains of column 1 and 2 in cable tensioning phase. a) Column 1. b) Column 2.

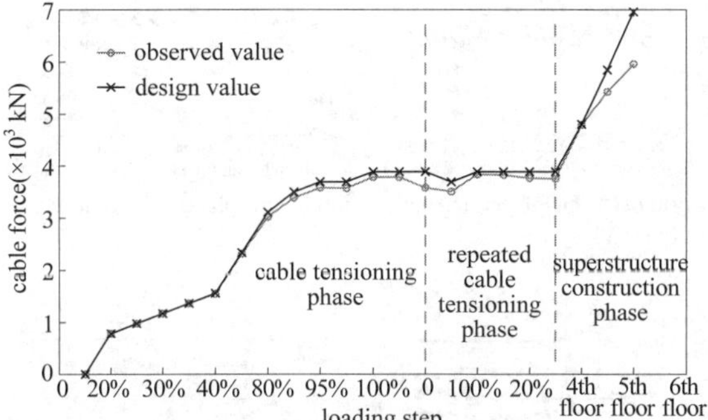

Figure 6. Contrast of cable force in different phases.

Through the analysis of the measurements, some conclusions can be drawn:

- During the superstructure construction phase, the bottom floor tended to recover the original horizontal state; the pre-arching of the bottom floor had basically returned to zero. This phenomenon coincides with the actual deformation. Therefore, the displacement in construction phase can be controlled effectively, the final state of structure can meet the design requirement.
- The stresses in the beams and columns are significantly lower than the yield strength of steel Q345, which is 345 MPa. Therefore, the beams and columns are well within the safe range of application.
- The intelligent bar can accurately monitor the cable force by comparing the measured value of cable force to the design value of cable force.

3 SHENZHEN BAY STADIUM

3.1 Background knowledge

Shenzhen Bay Stadium is constructed for the 26th Summer World University Games. It is situated at the Shenzhen Bay coastal recreation zone, the construction area is 307700 m^2, where the east-west length is about 730 m and the north-south length is about 480 m. The steel roof of Shenzhen Bay Stadium looks like "the spring cocoon", which consists of a single shell, double-layer space trusses and the vertical support system (Teng et al. 2014[a]). A effect drawing of Shenzhen Bay Stadium is shown in Figure 7.

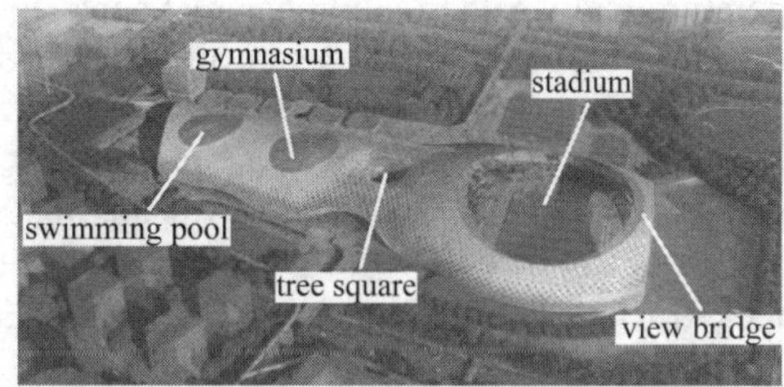

Figure 7. Effect drawing of Shenzhen Bay Stadium.

The main construction processes of the steel body are comprised of the following aspects (1) The vertical support system is installed. (2) The swimming pool and gymnasium are lifted overall. (3) Two substructure stages of the steel roof are installed respectively. (4) The closure of two substructure stages is performed.

The steel roof structure is divided into two substructure stages. I-stage substructure contains stadium, tree square, view bridge and conjoint tree columns. Π-stage substructure contains gymnasium, swimming pool and conjoint tree columns. Since these two substructure stages are installed respectively, it is necessary to weld the each substructure to form an integral structure. This process is called closure. In closure phase, the temperature, displacement and stress must be controlled to avoid the excessive residual stress and ensure the space configuration of structure. A SHM system is used to decide the closure time and assessing the safe of the closure process. The layout of measurement points of Shenzhen Bay Stadium is shown in Figure 8.

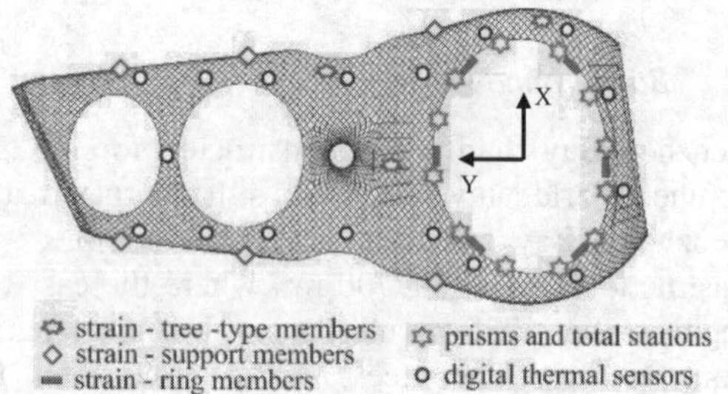

Figure 8. Layout of measurement points of Shenzhen Bay Stadium.

3.2 *Data analysis*

For the selection of closure temperature, the temperature and weather conditions of Shenzhen are taking into consideration to ensure that the closure temperature is closes to the average temperature as far as possible, where the average temperature of Shenzhen is about 23 °C ~30 °C. It means that the closure work can be performed when the temperature of steel structure lay in 23 °C ~30 °C. Otherwise, the closure work should be ceased. Furthermore, the change of temperature at night is relatively mild, the non-uniform temperature caused by the different irradiance could also be avoided. Therefore, the closure work is eventually conducted at night for this structure. The specific date of closure is 25 - 27 May, 2010. The results of some measurement points in closure phase are respectively shown in Figures 9-11.

Through the analysis of the measurements, some conclusions can be drawn (Teng et al. 2015b):

- During the whole closure phase, the range of temperature of the steel roof is 23 °C ~30 °C, which can meet the requirement. So the closure construction is performed safely.
- The relative displacements in *X*- and *Y*-directions are no larger than ±10 mm, and the relative displacement in *Z*-direction is no larger than ±20 mm. So the closure welding of I-stage substructure and Π-stage substructure has little effect on the displacements of stadium ring member. The monitoring members are in stable working condition. The closure of steel structure is completed smoothly.
- The change of stresses is less than 5 MPa, it indicated that the stress distribution of the tree-type column is hardly influenced by the closure welding. The change values are in a reasonable range, and the structure is in a security state.

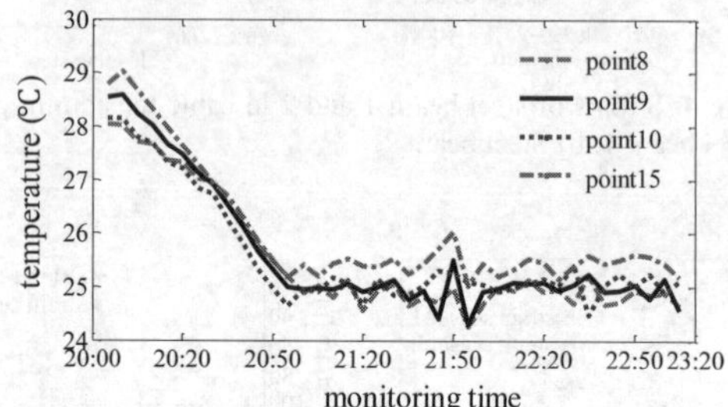

Figure 9. Temperature values in closure phase (25/5/2010).

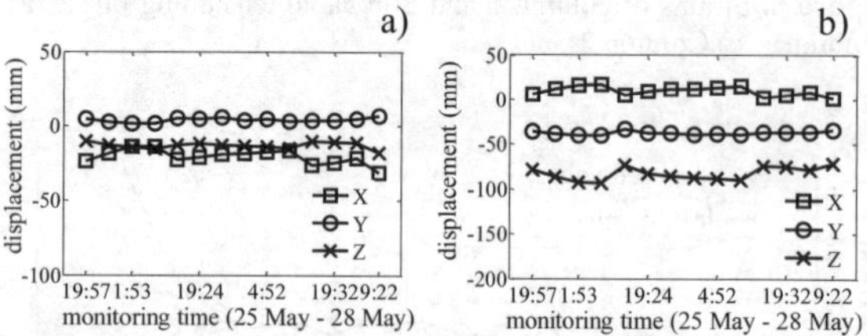

Figure 10. Displacement values in closure phase. a) Point 10. b) Point 11.

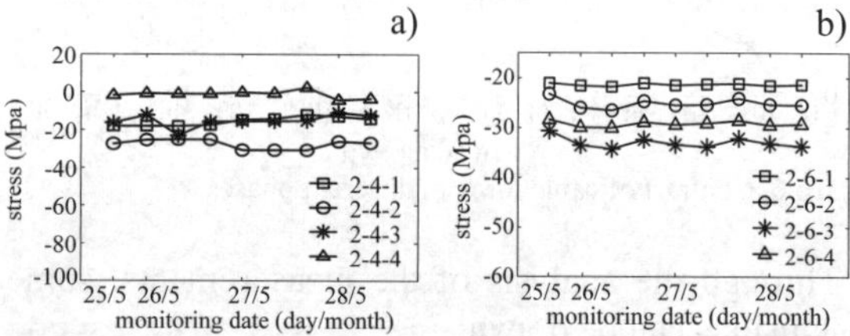

Figure 11. Stress values in closure phase. a) Point 2-4. b) Point 2-6.

4 ZHUHAI OPERA HOUSE

4.1 Background knowledge

Zhuhai Opera House is situated at the Yeli island which is the typhoon-affected area and the maximum wind speed exceeds the 40 m/s. This structure consists of grand theatre, little theatre and a shared lobby between two theatres, where the shape of theatres is shell, the height of shell are 90 m and 55.6 m respectively (Teng et al. 2015[c]). The physical map of Zhuhai Opera House is shown in Figure 12.

Figure 12. Zhuhai Opera House in construction phase.

Zhuhai Opera house is a typical wind load control structure and the unique shell-shape can form the narrow tuyere. On the other hand, the lowest and the maximum temperature in Zhuhai are about 5 °C and 36 °C, which is a larger temperature difference. So a SHM system is used to consider the influence of wind load and thermal effect on Zhuhai Opera House.

The SHM system of shell structure of little theatre includes the stress and temperature monitoring. The number of measurement points is 12. The layout of measurement points of Zhuhai Opera House is shown in Figure 13.

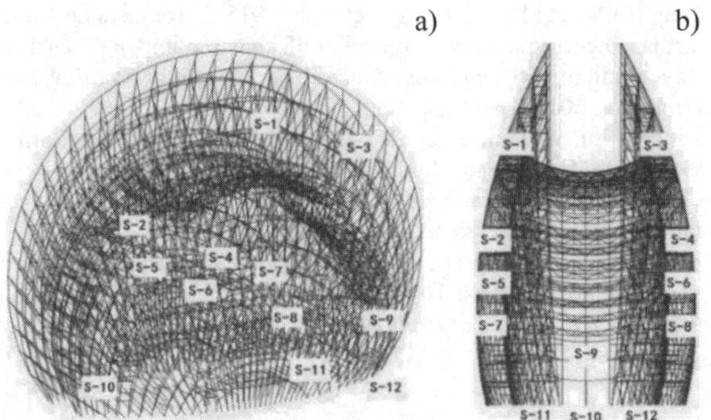

Figure 13. Layout of measurement points of Zhuhai Opera House. a) Front view. b) Side view.

The estimation method of the structural stress field includes the following steps: (1) The thermal value and partition of each region is determined. (2) Wind load is determined by the wind tunnel test and the measurements of wind speed; (3) The finite element model, which is subjected to the thermal effect and wind load, is analysed, then the distribution of the structural stress is obtained.

4.2 Data analysis

4.2.1 Region division and value acquisition of thermal action

According to the temperature fitting curve of different structural surface, height direction, and cumulative solar radiation, the region division based on the temperature is shown in Figure 14. On 25 June, 2014, the temperature values of all regions are shown in Table 2.

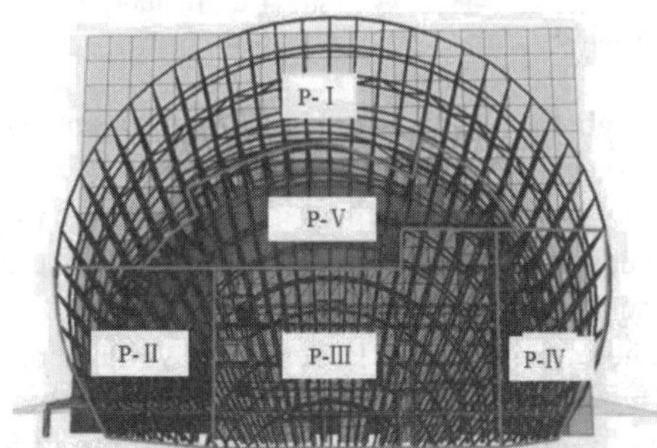

Figure 14. The temperature partitions of little theatre.

Table 2. Temperature values of all regions (°C)

Partitions	I	II	III	IV	V	Platform beam	Skylight 1	Skylight 2
Value	28.7	27	27.7	28.6	28.7	28.0	26.9	28.4

4.2.2 Region division and value acquisition of wind load

According to the analysis of wind tunnel test, the region division based on the wind load is 33. The wind pressure of each region can be calculated by the local wind speed monitoring system.

4.2.3 *Stress estimation and comparison*

The value of the thermal action and wind load of each designed region is estimated by the aforementioned method firstly, then these values and weight are applied to the little theater. Finally, the distribution of the structural stress can be obtained through the finite element analysis. Compared with the corresponding measured value of stress, the contrast of the simulated values and the measured values is shown in Figure 15. The interval of stress distribution of different members is shown in Figure 16.

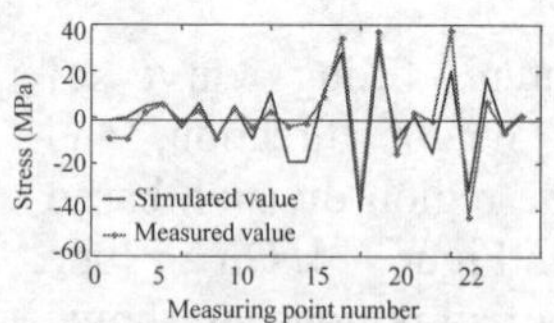

Figure 15. Contrast of the simulated values and the measured values.

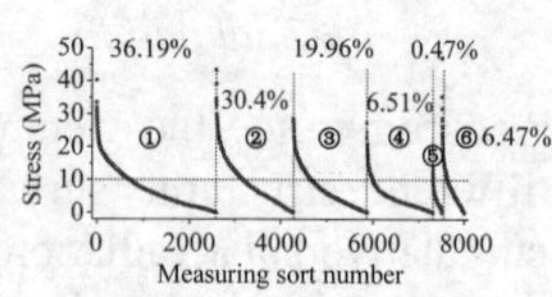

Figure 16. The interval of stress distribution of different members.

Through the analysis of the measurements, some conclusions can be drawn:

- The simulated stresses are close to the measured stresses, so the estimation method of the structural stress field is effective.
- The members with larger stress values are truss chords and ring bars, which should be paid more attention during the structural monitoring.

5 CONCLUSIONS

The SHM systems of Shenzhen Vanke Center, Shenzhen Bay Stadium and Zhuhai Opera House were implemented. Through the analysis of the measurements, it can be found that the SHM system is a very effective tool for providing the scientific reference information which helps to make decision more easily, quickly and safely, enhancing the structural safety and construction accuracy, tracking the performance of multi-type members and so on. The real world SHM systems in the paper are examples for construction and service monitoring of complex civil structures.

ACKNOWLEDGEMENT

This research is supported by NSFC (Grant No. 51308162), Shenzhen Knowledge Innovation Program - Fundamental Research (JCYJ20140417172417117), Natural Scientific Research Innovation Foundation in the Harbin Institute of Technology (HIT.NSRIF.2015085).

REFERENCES

Chang, P. C. Flatau, A. & Liu, S. C. 2003. Review paper: health monitoring of civil infrastructure, *Structural Health Monitoring*, **2**, 257-267.

Fu, X. Y. Gao, Y. Xiao, C. Z. et al. 2012. Horizontal skyscraper: innovative structural design of Shenzhen Vanke Center, *Journal of Structural Engineering*, **138**, 663-668.

Kuok, S. C. & Yuen, K. V. 2012. Structural health monitoring of Canton Tower using bayesian framework, *Smart Structures and Systems*, **10**, 375-391.

Shen, Y. B. Yang, P. C. Zhang, P. F. et al. 2013. Development of a multitype wireless sensor network for the large-scale structure of the National Stadium in China, *International Journal of Distributed Sensor Networks*, Article ID 709724, 16pages.

Teng, J. Zhu, Y. H. Lu, W. et al. 2010. The intelligent method and implementation of health monitoring system for large span structures, *Proceedings of the 12th International Conference on Engineering, Science, Construction, and Operations in Challenging Environments-Earth and Space*, United States, March, 2543-2552.

Teng, J. Lu, W. & Zhang, R. G. 2014[a]. Long-term performance monitoring of Shenzhen Bay Stadium, *Sixth World Conference On Structural Control And Monitoring*, Spain, July, 2037-2042.

Teng, J. Lu, W. & Lu, F. Z. 2014[b]. Structural health monitoring to multi-type members of Shenzhen Vanke Center in construction phase, *Sixth World Conference On Structural Control And Monitoring*, Spain, July, 2094-2100.

Teng, J. Lu, W. Wen, R. F. et al. 2015[a]. Instrumentation on structural health monitoring systems to real world structures, *Smart Structures and Systems*, **15**, 151-167.

Teng, J. Lu, W. Cui, Y. et al. 2015[b]. Temperature and displacement monitoring to steel roof construction of Shenzhen Bay Stadium, *International Journal of Structural Stability and Dynamics*, 2015: 1640020.

Teng, J. Lu, W. Cui, Y. et al. 2015[c]. Structural health monitoring system application to Zhuhai Opera House, *7th International Conference on Structural Health Monitoring of Intelligent Infrastructure SHMII 2015*, Italy, July, 1-7.

Zhang, Q. L. Yang, B. Liu, T. et al. 2015. Structural health monitoring of Shanghai Tower considering time-dependent effects, *International Journal of High-Rise Building*, **4**, 39-44.

Proceedings of the International Conference on Smart Infrastructure and Construction
ISBN 978-0-7277-6127-9

© The authors and ICE Publishing: All rights reserved, 2016
doi:10.1680/tfitsi.61279.475

Response enhancement of a 32-story building structure using two types of track nonlinear energy sinks

J. Wang*[1], X. Lu[2], N. E. Wierschem[3] and B. F. Spencer Jr[4]

[1] *Hunan University of Technology, Zhuzhou, China*
[2] *Tongji University, Shanghai, China*
[3] *University of Tennessee, Knoxville, U.S.A.*
[4] *University of Illinois, Urbana, U.S.A.*
* *Corresponding Author*

ABSTRACT Track nonlinear energy sinks (track NESs) and single-sided vibro-impact track nonlinear energy sinks (SSVI track NESs) are types of passive structural control devices that can rapidly mitigate the vibration of a primary structure. The smooth motion of the NES mass on a specially designed track provides a continuous nonlinear restoring force. Additionally, the use of an impact surface in the SSVI track NES increases the nonlinearity of the device and further enhances its robustness against changes in stiffness, input energy level, and damping of the device. While previous studies have limited their investigation to NESs on structures of relatively few degree-of-freedoms, in this work, a 32 degree-of-freedom model, based on an actual 32-story high-rise, is used as the primary structure. The mass ratio of the track NES attached to the top of this structure is set as 1%. Numerical optimizations are implemented to determine the properties of the track NES. Response comparison to a conventional tuned mass damper (TMD) shows that the track NES can reduce the structural responses to a similarly low level under both impulsive and seismic excitations. When the stiffness of either the primary structure or the device is shifted, the track NES has a strong advantage of robust performance compared to the TMD. The results also indicate that the SSVI track NES is robust against extensive changes in its viscous damping. Furthermore, the relative displacement of the control mass in the SSVI track NES system is reduced in comparison to the track NES and TMD systems.

1 INTRODUCTION

Nonlinear energy sinks (NESs), a type of passive structural control devices which exploit essential nonlinearities, have been proven to be effective in reducing the response of the primary structure of relatively few degree-of-freedoms (DOFs). In addition, the nonlinear restoring force provided by NESs enhances the robustness of the structure against the changes in structural stiffness and input energy level (McFarland et al. 2005; Quinn et al. 2012; Wierschem et al. 2012[a]; Wierschem et al. 2012[b]). A track NES utilizes specially designed tracks to direct the passive motion of the mass of the NES, which results in the addition of a nonlinear restoring force to the system (Figure 1 a) (Wang et al. 2015[a]; Wang et al. 2015[b]). A single-sided vibro-impact track NES (SSVI track NES) adds an impact surface to restrict the motion of the NES mass on one side (Figure 1 b) (Wang et al. In publication). In previous studies, the performance of various types of NESs has been investigated both numerically and experimentally, however, only structures with nine-DOF or fewer have been utilized (Wierschem et al. 2013; Luo et al. 2014). The track NES and the SSVI track NES, in particular, have only been examined on a small-scale two-DOF frame structure. Besides, a relatively high mass ratio of 5% is often used in those studies. Additionally, applications of NESs on more complex and realistic structures have rarely been considered.

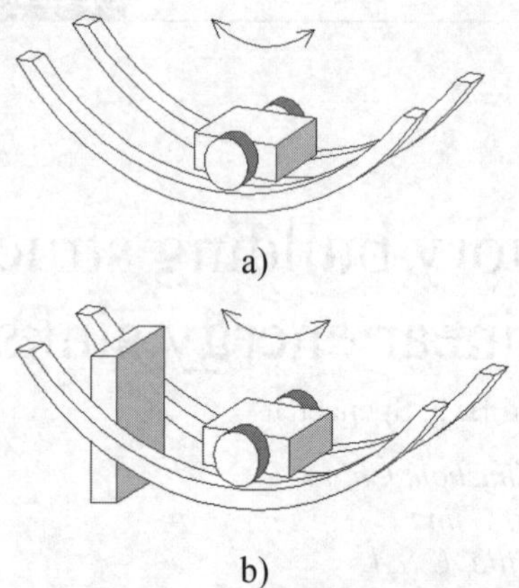

Figure 1. Conceptual model of a) track NES. b) SSVI track NES

In this work, a track NES and a SSVI track NES of a small mass ratio are attached to a condensed model of an actual 32-story building. The goal of this work is to investigate the performance of these nonlinear devices when incorporated with more realistic structures with a larger number of degrees of freedom.

2 PRIMARY STRUCTURE MODEL REDUCTION

The 32-story primary structure considered in this work is a 129 m high steel frame-steel core structure. The floor dimension is 48m×48m, and the height of each story is 4 m except for the first story of which the height is 5 m. The modal damping of this primary structure is set as 1% .

As the track NES studied in this paper is unidirectional, the primary structure therefore needs to be reduced to a 32-DOF lumped-mass model moving in only one direction. The model reduction utilizes the equivalent stiffness identification method in which the bending effect of tall buildings can be considered (Sun & Xu 1995). The consideration is necessary because high-rise buildings deform differently from shear frame structures of which the story drift and the shear force are only related to adjacent floors.

The reduced model has the same natural frequencies as the original structure. The first six natural frequencies of these models are 0.239 Hz, 0.729 Hz, 1.333 Hz, 1.944 Hz, 2.579 Hz, and 3.214 Hz.

3 OPTIMIZATION

The track NES and the SSVI track NES are attached to the top story of the primary structure. The SSVI track NES shares the same NES mass and track shape as the track NES, and as a result the equation of motions (EOMs) are the same for both NESs outside the impacts that occur with the SSVI track NES. In the case of the impacts with the SSVI track NES, the NES and the top story experience sudden velocity changes at every collision, which follows the law of conservation of momentum combined with a coefficient of restitution.

The numerical models of the systems with NES placed on the top story are constructed in the MATLAB environment. These models are used for optimization and response analysis. For both NESs, the NES mass is set as 1% of the mass of the primary structure. In the optimization of the track NES, the track shape and the NES damping are determined while in the optimization of the SSVI track NES, the previously determine track shape and damping are used and the clearance between the mass and the impact surface is determined.

3.1 *Equation of motion*

The EOM of the track NES is closely related to the shape of the track. As shown in Figure 2, the restoring force in the horizontal direction is a function of the normal force F_{Normal} and the slope of the track θ. The normal force F_{Normal} is a function of the track shape $h(u_{\text{N}})$, which is the vertical displacement of the NES relative to the primary structure, v_{N}, and a function of the relative horizontal displacement of the NES, u_{N}. The displacement of the primary structure is z. The mass of the NES is m_{N} and the vertical acceleration due to gravity is g.

Assuming the mass does not rotate and remains in contact with the track and the damping of the device is viscous damping, the EOM of the track NES (Eq. 1) can then be derived applying Lagrange's equation (Wang et al. 2015[a]).

$$m_N\ddot{u}_N + c_N\dot{u}_N + \left(\begin{array}{l} \left[h'(u_N)\right]^2 \cdot \ddot{u}_N + h'(u_N)g \\ +h'(u_N)h''(u_N)\cdot \dot{u}_N^2 \end{array} \right) \cdot m_N$$
$$= -m_N\left(\ddot{z} + \ddot{x}_g\right) \qquad (1)$$

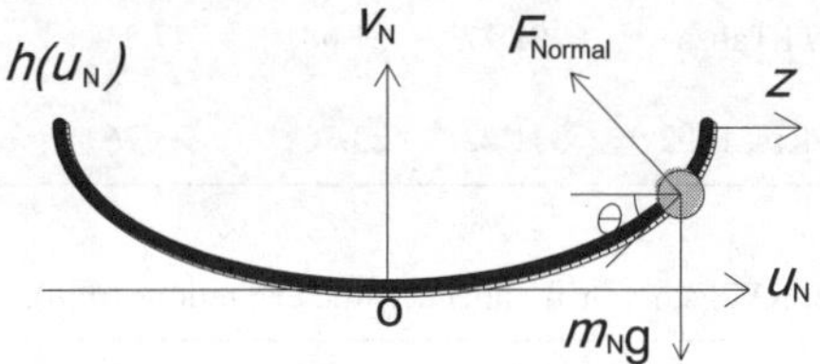

Figure 2. Free body diagram of track NES.

3.2 *Track NES optimization*

The optimization of the track NES consists of two general steps. The first step is to narrow down the ranges of the parameters to be optimized. The second step is to find the exact values of the parameters directly using the fmincon command from MATLAB function library. The first step is necessary to avoid a local extremum found in the second step. In the simulation, an initial velocity of 1 m/s is given to all stories and the NES mass and the objective is to lower the maximum root-mean-square (RMS) of story drift of the primary structure during 25 s.

The track shape after the optimization is $h(u_N) = 14.34 \times 10^{-3} u_N^4$ and the optimized NES damping is 14.90×10^4 N • m/s.

3.3 *SSVI Track NES optimization*

In the preliminary stage, the SSVI track NES is found to be highly effective even when the NES damping is largely deviated. To further explore this feature of the SSVI track NES, the NES damping is set as 10% of the optimized value in the optimization of the clearance.

The optimization of the SSVI track NES uses the same initial velocity and the same performance measure as the track NES. The optimized clearance is 2 m.

3.4 *TMD optimization*

To investigate the performance of the track NES and the SSVI track NES, they are compared with a conventional tuned mass damper (TMD) which is a linear passive energy absorber. The TMD has been optimized using the same method outlined in the previous section using the same mass ratio as the track NES. The optimized TMD stiffness is 16.78×10^5 N/ m and the TMD damping is 14.90×10^4 N • m/s.

4 RESPONSE ANALYSIS

4.1 *Performance effectiveness*

The systems in comparison are the track NES system, SSVI track NES system, the TMD system, and the locked system. For the locked system, the auxiliary mass of the NES is locked to the top story of the primary structure, and moves as a part of this story. All these systems are numerically tested by both impulsive and seismic excitations. The three earthquakes considered are records from the SAC Phase 2 Steel Project (SAC Steel Project: Technical Studies, 1997) and the information of these records is given in Table 1.

Figure 3 shows the displacements of the devices relative to the primary structure and the displacements of the top story when subjected to impulsive excitation. The dash line in the plot of device displacement represents the location of the impact surface in the SSVI track NES. All control devices perform effectively in reducing the structural responses, cutting the displacement in half within 15 s, which equates to less than four cycles of the fundamental period.

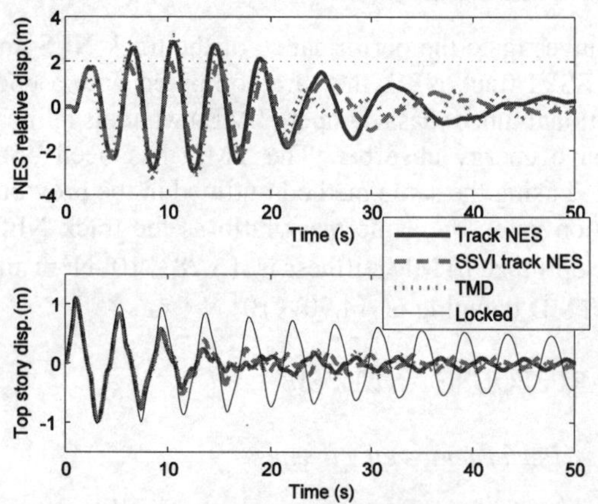

Figure 3. Responses under impulsive excitation.

Table 1 lists the reduction rates of the seismic RMS story drift of the track NES system, the SSVI track NES system, and the TMD system compared to the locked system. Table 2 lists their respective RMS story drifts. The reduction rate of the track NES system reaches as high as 51.68%. The SSVI track NES system reaches 47.33% and the TMD system reaches 49.32%. The top story displacements under the 1974 Tabas earthquake are shown in Figure 4. It can be concluded that the track NES and SSVI track NES exhibit a similar effectiveness as the TMD in normal circumstances where the structural stiffness and the device parameters remain unchanged.

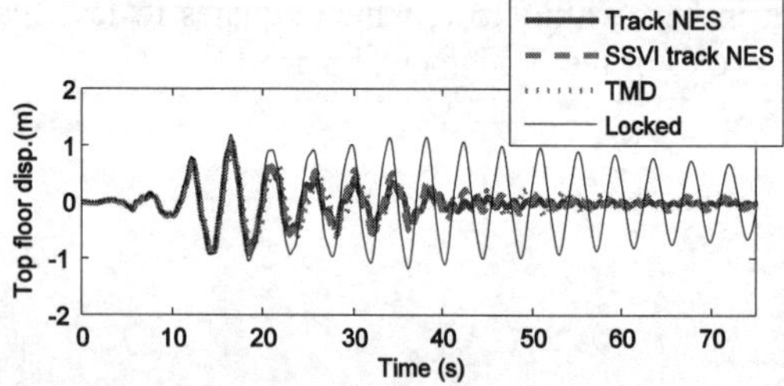

Figure 4. Responses under 1974 Tabas earthquake.

Table 1. Earthquake PGA (cm/s^2) and RMS story drift reduction rate (%).

Record	PGA	Track NES	SSVI track NES	TMD
Landers, 1992, Yermo	353.35	48.43	44.43	47.29
1974 Tabas	291.77	51.68	47.33	49.32
Landers, 1992	331.22	23.76	23.76	27.29

Table 2. RMS story drift under seismic excitations (mm).

Record	Locked	Track NES	SSVI track NES	TMD
Landers, 1992, Yermo	15.34	7.91	8.52	8.08
1974 Tabas	12.14	5.86	6.39	6.15
Landers, 1992	4.27	3.25	3.25	3.10

4.2 Performance robustness

This section studies the robustness of the devises against the changes in structure stiffness as well as the device stiffness, and extensive changes in the damping of the devices.

Structures in reality may experience stiffness decreases due to damages from severe excitations. Control devices themselves may differ from their original design due to manufacture inaccuracy and inappropriate installation or maintenance. Although those changes are usually within certain ranges, relatively big deviations are considered here for research purpose.

Figure 5 shows the top story displacements of the track NES system, the TMD system, and the locked system when either the stiffness of the primary structure or the device is changed. The performance of the SSVI track NES system is very similarly to the performance of the track NES system; therefore, it is not displayed in this figure for clarity. This figure can be compared with Figure 3. In the top plot (Figure 5 a) where the structure stiffness decreases by 50%, the TMD only reduces the response negligibly. The track NES is able to bring down the response more significantly, although it needs a longer time than in Figure 3. The distinction between the TMD and the track NES is more obvious in the bottom plot where the stiffness of the devices increases by 50% (Figure 5 b). The performance of the track NES system is very little affected by the increased device stiffness and it can quickly lower the magnitude of the response.

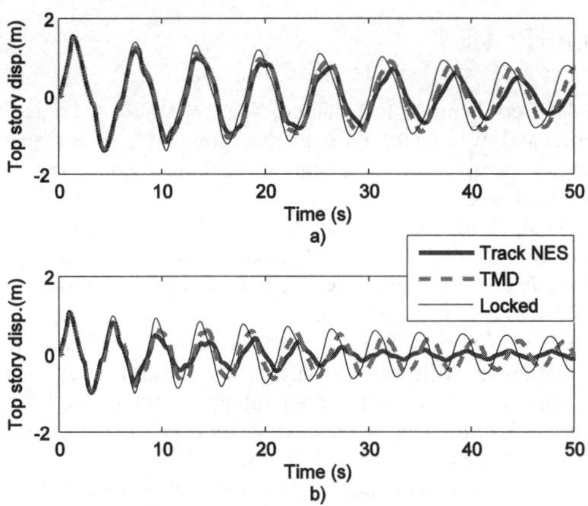

Figure 5. Responses under impulsive excitation when a) the structural stiffness decreases by 50%. b) the device stiffness increases by 50%.

Table 3 quantitatively compares the RMS story drift of the track NES system, TMD system, and the locked system in response to the impulsive excitation when primary structure's stiffness is unchanged, decreased by 50%, and device stiffness increased 50%. Although the effectiveness of both control systems deteriorates when either stiffness changes, the response reduction rates of the track NES are nearly three times the TMD, showing an improved robustness against stiffness changes.

Figure 6 shows the top story displacements of the track NES, the SSVI track NES, the TMD, and the locked systems when the damping of the devices is 50% and 10% of the optimal value. The structural displacements of the track NES system and the TMD system tend to be less stable as the device damping decreases. Beating phenomenon can be observed in these two systems where the magnitudes of the structural displacements oscillate, and at certain point, even exceed the locked system. The beating phenomenon happens due to the overlapping of two responses at very close frequencies and insufficient damping to dampen the overall responses. The TMD, which is linear and thus cannot exchange energy between modes, depends on the damping more than the track NES to be effective in reducing the structural responses. The SSVI track NES, on the other hand, shows little performance deterioration in response reduction. For all of the damping values investigated (10%, 50%, and 100% of the design damping value), the magnitude of the top story displacement in the SSVI track NES system monotonously goes down and the displacement oscillates at a low level, between ± 0.15 m, after 35 s. This behavior is indicative of the SSVI track NES's high level of robustness against extensive changes in the viscous damping of the device.

Table 3. RMS story drift and reduction rate under impulsive excitation with different stiffnesses

Case	Locked	Track NES		TMD	
	Story drift (mm)	Story Drift (mm)	Redu -ction (%)	Story Drift (mm)	Redu -ction (%)
Stiffness unchanged	10.28	5.86	43.03	5.96	41.99
Structure Stiffness decreased	16.05	13.76	14.28	15.17	5.53
Device stiffness increased	10.28	6.61	35.68	8.99	12.56

Figure 7 compares the relative displacement of the devices when the device damping is 10% of the optimal value. The stroke of the TMD is 5.03 m while the stroke of the track NES is 3.65 m and the stroke of the SSVI track NES is 2.65 m. The substantially lower stroke for the track NES and SSVI track NES is advantageous for practical purposes as smaller devices would likely be easier to incorporate into a real structure.

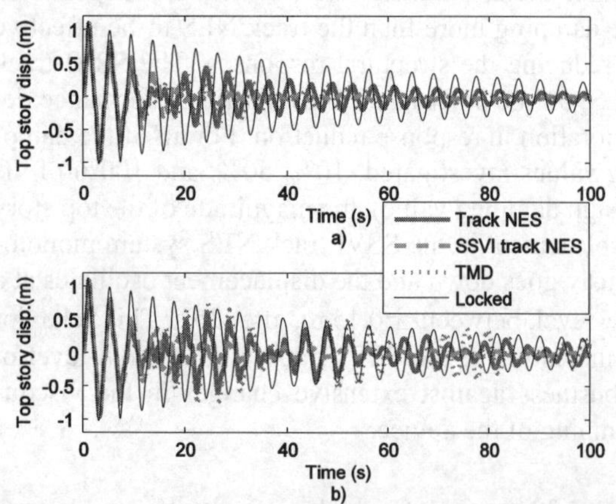

Figure 6. Responses under impulsive excitation when the damping of the device is a) 50% of the optimal value. b) 10% of the optimal value.

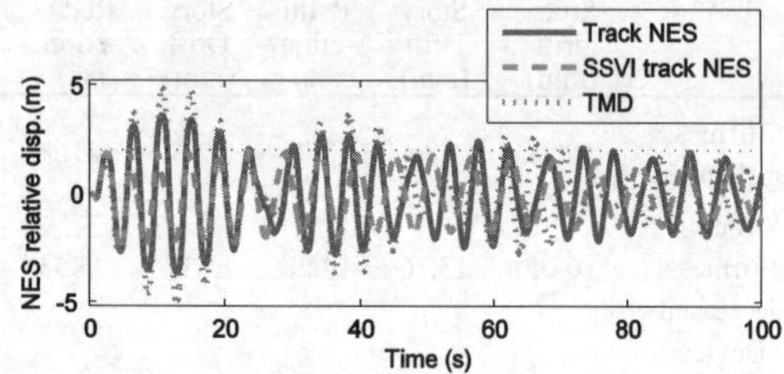

Figure 7. Relative device displacement under impulsive excitation when the damping of the device is 10% of the optimal value.

5 CONCLUSIONS

This paper numerically investigates the performance of a track NES and a SSVI track NES of a small mass ratio of 1% when attached to a 32-DOF high-rise building. These passive structural control devices can effectively mitigate the vibration of the primary structure not only at their optimal parameters, but also when changes in the stiffness and the devise damping are considered. The track NES differs from the TMD with its nonlinear restoring force and its ability to interact across a range of frequencies of the primary structure. By adding an impact surface to the track NES, the SSVI track NES possesses more nonlinearity and is therefore more robust against the uncertainty in the structure and the device itself. This study shows the potential of applying track NES and SSVI track NES to high-rise buildings when properly designed.

ACKNOWLEDGEMENT

The authors gratefully acknowledge the support from the National Natural Science Foundation of China (Grant No. 51261120377).

REFERENCES

Luo, J. Wierschem, N. E. Hubbard, S. A. et al. 2014. Large-scale experimental evaluation and numerical simulation of a system of nonlinear energy sinks for seismic mitigation. *Engineering Structures* 77: 34–48.

McFarland, D. M. Kerschen, G. Kowtko, J. J. et al. 2005. Experimental investigation of targeted energy transfers in strongly and nonlinearly coupled oscillators. *The Journal of the Acoustical Society of America* 118 (2): 791–99.

Quinn, D. D. Hubbard, S. Wierschem, N. E. et al. 2012. Equivalent modal damping, stiffening and energy exchanges in multi-degree-of-freedom systems with strongly nonlinear attachments. *Proceedings of the Institution of Mechanical Engineers, Part K: Journal of Multi-Body Dynamics* 226 (2): 122–46.

SAC Steel Project: Technical Studies. 1997. http://nisee.berkeley.edu/data/strong_motion/sacsteel/draftreport.html

Sun, H. & Xu, W. 1995. A parameter identification method for determining the equivalent rigidity coefficients of the simplified series multidegree-of-freedom system for a framed structure. Earthquake Engineering and Engineering Vibration 15 (2): 100–108 (in Chinese).

Wang, J. Wierschem, N. E. Spencer Jr., B. F. et al. 2015[a]. Track nonlinear energy sink for rapid response reduction in building structures." *Journal of Engineering Mechanics* 141 (1). 04014104.

Wang, J. Wierschem, N. E. Spencer Jr., B. F. et al. 2015[b]. Experimental study of track nonlinear energy sinks for dynamic response reduction. *Engineering Structures* 94 (July): 9–15.

Wang, J. Wierschem, N. E. Spencer Jr., B. F. et al. In publication. Numerical and experimental study of the performance of a single-sided vibro-impact track nonlinear energy sink. *Earthquake Engineering & Structural Dynamics.*

Wierschem, N. E. Luo, J. Al-Shudeifat, M. A. et al. 2012[a]. Simulation and testing of a 6-story structure incorporating a coupled two mass nonlinear energy sink. *Proceedings of the ASME 2012 International Design Engineering Technical Conferences & Computers and Information in Engineering Conference*. August 12-15, 2012, Chicago.

Wierschem, N. E. Quinn, D. D. Hubbard, S. A. et al. 2012[b]. Passive damping enhancement of a two-degree-of-freedom system through a strongly nonlinear two-degree-of-freedom attachment. *Journal of Sound and Vibration* 331 (25): 5393–5407.

Wierschem, N. E. Hubbard, S. A. Luo, J. et al. 2013. Experimental blast testing of a large 9-story structure equipped with a system of nonlinear energy sinks. *Proceedings of the ASME 2013a International Design Engineering Technical Conferences & Computers and Information in Engineering Conference.* August 4-7, 2013. Portland.

Proceedings of the International Conference on
Smart Infrastructure and Construction
ISBN 978-0-7277-6127-9

© The authors and ICE Publishing: All rights reserved, 2016
doi:10.1680/tfitsi.61279.481

Structural health monitoring system of railway bridge substructures utilizing soundness diagnosis index correlated with natural frequency

K. Abe*, T. Natori and Y. Kominato

Railway Technical Research Institute, Tokyo, Japan
** Corresponding Author*

ABSTRACT The soundness of structural stability of railway bridge substructures in a river deteriorates over time because the riverbed and overburden can be gradually eroded by river flow. Hence, it is necessary to diagnose the soundness of structural stability of the substructures over time. The soundness can be diagnosed by the size of natural frequency of a primary vibration mode of a pier that was loaded by an impact force. We proposed a new methodology to diagnose the soundness through a structural health monitoring approach. New soundness diagnosis indices based on microtremors and vibrations of piers during train passage were proposed. In order to confirm the applicability of the indices to the soundness diagnosis, a correlation between proposed indices and natural frequencies were investigated through model experiments and in situ measurements. In addition, a long term monitoring of actual railway bridge substructures were carried out using smart wireless sensors that can automatically measure the indices.

1 INTRODUCTION

The soundness of structural stability of railway bridge substructures in a river, which is designed by the bearing capacity of a foundation and stiffness of a pier body, deteriorates over time because the riverbed and overburden that support the substructures can be gradually eroded by river flow. Hence, it is necessary to diagnose the soundness of structural stability of such substructures over time.

In railway bridges across rivers in Japan, the soundness is diagnosed by the size of natural frequency of a primary vibration mode of a pier identified by measuring free vibration of the pier that was loaded by an impact force, which is called the impact and vibration test (**Figure 1**). The test was developed by Nishimura in 1986 (Shinoda et al., 2008). By using the test, deteriorated substructures can be identified because the unstable one has a low natural frequency. Impact and vibration test is widely used in railway companies in Japan. The threshold to judge the soundness through the size of natural frequency has already been established by using accumulated field data measured by the railway companies. Railway companies judge the soundness of the structural stability by comparing the measured natural frequency to the threshold.

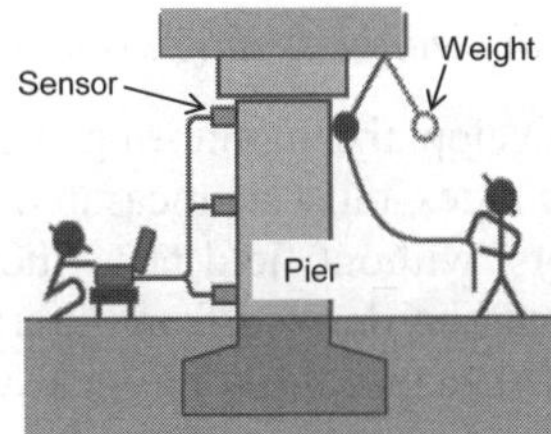

Figure 1. Schematic of impact and vibration test.

However, railway companies are seeking for a more efficient method to diagnose the soundness because the impact and vibration test requires a heavy weight for loading and considerable amount of labour regarding installation of weight and sensors at high place. In addition, with field test like impact and vibration test, it is impossible to diagnose the soundness constantly. Hence, a change in the soundness can be missed.

Accordingly, we proposed a new methodology to diagnose soundness through a structural health monitoring approach. With this method, the considerable amount of labour required in a field test is not necessary. In addition, soundness can be continually diagnosed. In order to develop such a monitoring method, we proposed new soundness diagnosis indices based on microtremors and vibrations of piers during train passage, i.e., area ratio of acceleration power spectrum of microtremors and ratio of amplitude of response acceleration of the pier during train passage, which will be explained in a later section.

In order to confirm the applicability of the indices to the soundness diagnosis, a correlation between proposed indices and natural frequencies introduced from impact and vibration test was investigated through model experiments and in situ measurements. Additionally, a long term monitoring of actual railway bridge substructures were carried out using smart wireless sensors that can automatically measure the indices.

2 SOUNDNESS DIAGNOSIS INDEX

2.1 *Outline of soundness diagnosis index*

In order to develop the monitoring method, soundness diagnosis index must be measured continuously through sensors, without field test. Such indices for substructures can be derived from microtremors and train vibrations because these physical values can be continuously measured by sensors such as accelerometers that are installed on the piers. Therefore, we focused on these physical values to propose the indices for the monitoring.

2.1.1 *Power spectrum area ratio of mincrotremors*

Samizo et al. (2007) reported that the spectrums of the microtremors measured by accelerometers installed on the top of the piers of bridges across a river were affected by the soundness of stability of the substructures, i.e., the spectrums of piers that lost an overburden were distributed in a low frequency range (1.0-5.0 Hz). Therefore, we proposed the power spec-

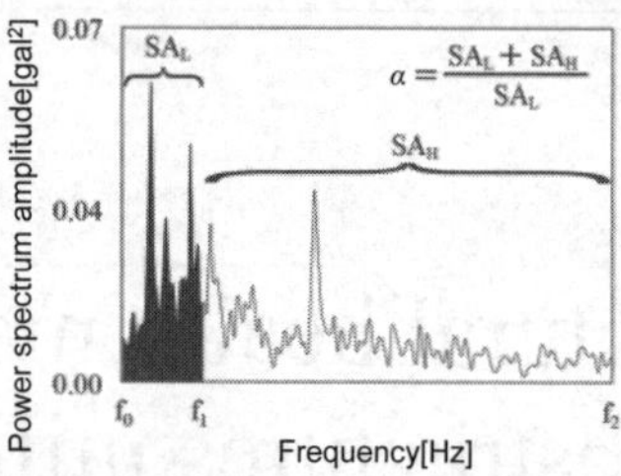

Figure 2. Schematic of power spectrum area ratio.

trum area ratio as one of the soundness diagnosis indices, which was described by the equation:

$$\alpha = \frac{SA_A}{SA_L} \quad (1)$$

where SA_A is the total area of the power spectrum of the microtremors at a specified range (f_0- f_2) and SA_L is the area of the power spectrum of the microtremors at the low frequency range (f_0-f_1) as shown in **Figure 2**. The power spectrum area ratio means the ratio of an energy at the low frequency range in that of the specified range.

2.1.2 *Train passing acceleration rate*

Another physical value that can be continuously measured by sensors is train vibration. When a train passes through a steel railway bridge, a bridge girder is loaded by a train running load. In that case, the bridge girder vibrates with a high frequency. Miyashita et al. (2007) and Yoshida et al. (2013) reported that the girders loaded by train running load had stress vibrations with high frequency (30.0-70.0 Hz) on web plates. Therefore, it is considered that the pier is loaded with high frequency by the girders through supports during train passage.

On the other hand, Shinoda et al. (2008) reported that general railway bridge piers had less than 20.0 Hz for natural frequency of the primary mode taken in the impact and vibration test. Hence, the ratio of a load frequency f to natural frequency f_n, f/f_n is considered to be larger than one.

When railway bridge pier is modelled by a single degree of freedom system (m: mass, k: spring constant, c: viscous damping coefficient) loaded by train running load F as shown in **Figure 3**, the acceleration resonance curve is described as shown in **Figure 4**. Therefore, when the ratio of the load frequency to the natural frequency f/f_n is larger than one, the ratio of an amplitude of a load wave to that of a response wave L_2 varies inversely with the f/f_n as shown in Figure 4.

By the way, the primary modes of the bridge piers are categorised as flexural vibration and rocking vibration modes. The categorisation is determined by the magnitude of a flexural stiffness of a pier body as shown in Figure 5. In general, the size of the flexural stiffness at the axial perpendicular direction is larger than that at the axial direction. Hence, the rocking vibration mode is dominant at the axial perpendicular direction. Then the size of natural frequency at the direction is greatly affected by change in the overburden thickness. Therefore, we proposed a train passing acceleration rate β, described by the following equation:

$$\beta = \frac{A_{TR}}{A_{LG}} \tag{2}$$

where, A_{TR} is a maximum amplitude of response acceleration wave of a pier loaded by train running loads at the axial perpendicular direction and A_{LG} is that at the axial direction. This index is a ratio of L_2 at the axial perpendicular direction to that at the axial direction. Hence, the amplitudes of load waves are cancelled. In addition, A_{TR} is greatly affected by change in overburden thickness compared with the A_{LG}. Hence, index β corresponds with the change in the size of natural frequency of primary mode.

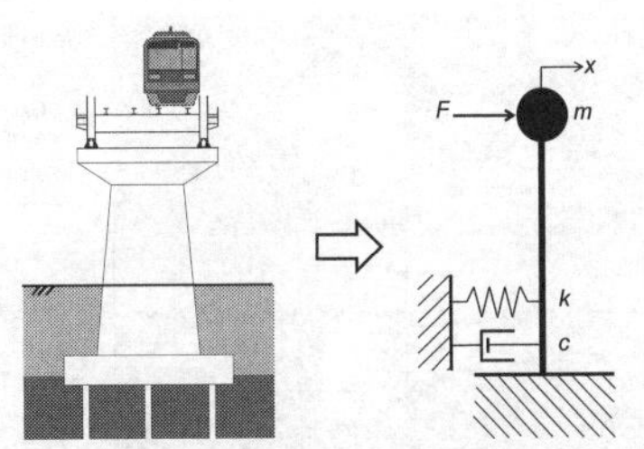

Figure 3. Schematic of single degree of freedom model.

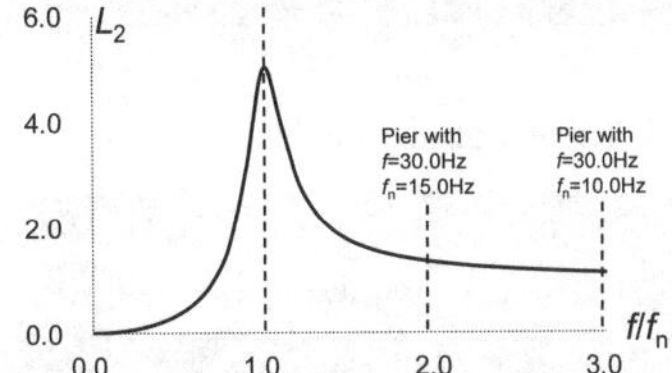

Figure 4. Schematic of acceleration resonance curve.

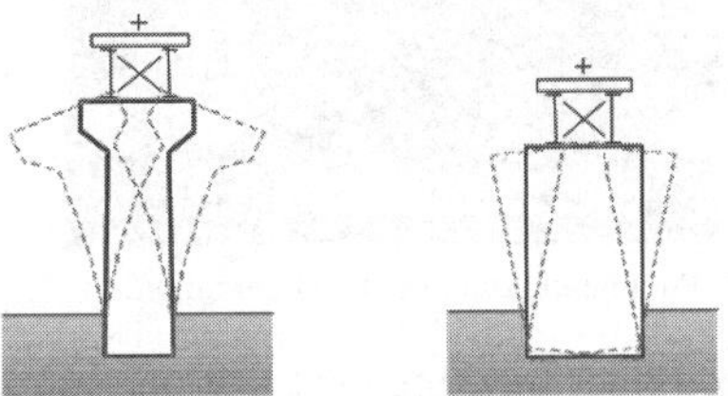
(a) Flexural vibration mode (b) Rocking vibration mode

Figure 5. Schematic of primary modes of bridge piers.

2.2 *Verification of indices by model experiments*

In order to verify the applicability of the proposed indices to the soundness diagnosis of piers, correlation between the indices and natural frequencies seen in impact and vibration test was investigated by model experiments. The target of the experiment was pier P12 (height: 16.7 m; crown width: 10.3 m; footing width: 11.0 m; overburden thickness: 4.35 m; foundation: 23 steel piles with 34.0 m high and 800 mm diameter; natural frequency taken in the impact and vibration test: 4.4 Hz) of a railway steel bridge, at which the in situ measurements and long term monitoring were carried out as shown in the next section.

The pier model was designed as a 1/40 scale model, referring to similitude shown in Kagawa (1979), i.e., the mass M and natural frequency f_n are described by the following equations:

$$M_m = \frac{1}{\lambda^2} M_p, \quad f_{nm} = \sqrt{\lambda} f_{np} \tag{3}$$

where λ is a scale factor, subscripts m and p indicates a model and a prototype, respectively. The schematic figures of the pier model are shown in Figure 6. The model consists of a 4.10 kg pier body made from a lightweight mortar and a 1.60kg aluminium girder plate fixed on the top of the pier body by 15.0 mm-tall 4 stainless screw supports. The piles were modelled by 9 polyethylene pipes with 12.0 mm diameter and 380 mm lengths. The ground was made by unsaturated siliceous sand with 90.0 % relative density.

The photograph of the experiment condition is shown in Figure. 7. Impact and vibration test, microtremor measurement test and vibration exciter test were carried out with 0.0 mm footing and overburden of following thickness: 300 mm, 200 mm, 50.0 mm, 0.0 mm, -50.0 mm and -100 mm.

In impact and vibration test, the response waves of the pier model were measured by hammering the top of the model at an axial perpendicular direction. In

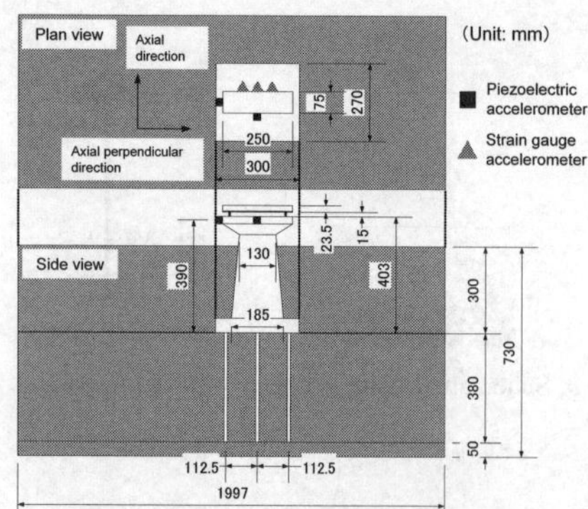

Figure 6. Schematic of pier model.

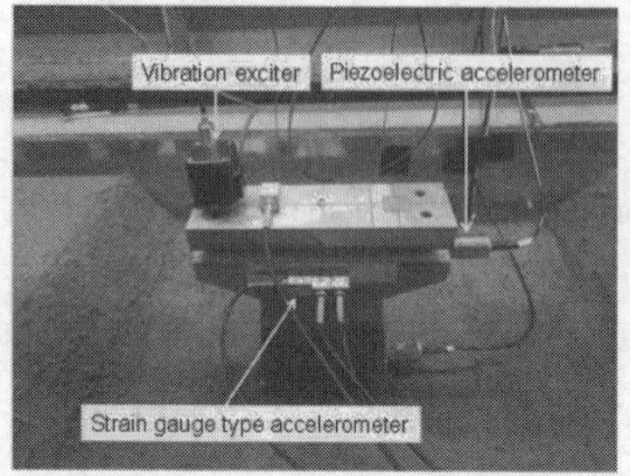

Figure 7. Photograph of experiment condition.

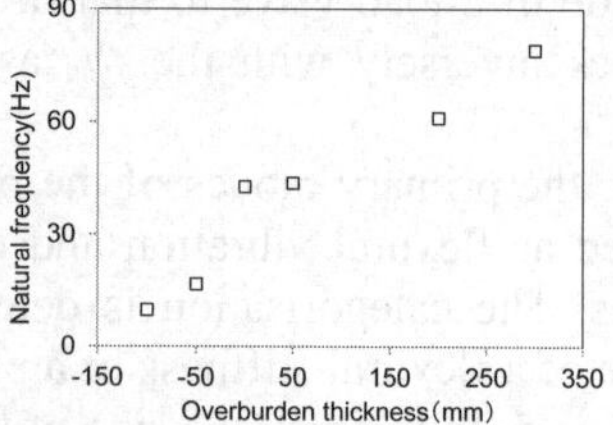

Figure 8. Relationship between natural frequency and overburden thickness.

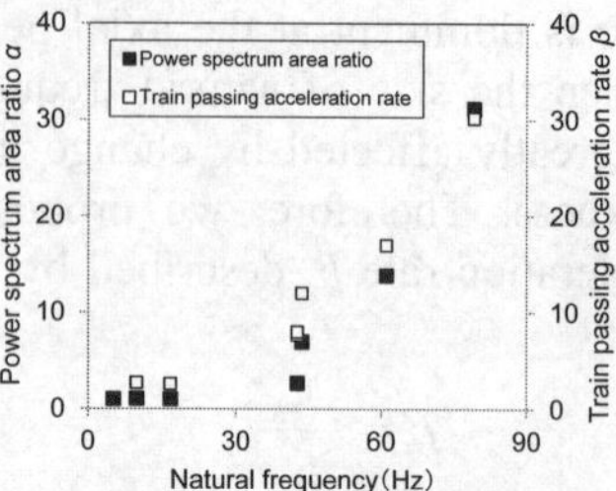

Figure 9. Relationship between soundness diagnosis indices and natural frequencies (model experiment).

the microtremor measurement test, the microtremors of the pier model were measured 5 times with 10.5 m/s wind from an electric fan positioned 1.0 m away from the model, and then the average of the power spectrum area ratio α was calculated. In the vibration exciter test, three sine sweep vibration tests were carried out with 20.0-150 Hz sweep frequencies, 2.17 Hz/s sweep velocity, 60.0 s sweep time and 7.5 m/s^2 acceleration amplitude by using a small exciter positioned at the edge of the top of the pier body, and then the maximum value of the train passing acceleration rate β in the range from 80.0 to 110 Hz, where the ratio of a load frequency f to natural frequency f_n, f/f_n was larger than one, was calculated.

The relationship between natural frequency and overburden thickness is shown in Figure 8. It was confirmed that reduction of natural frequencies corresponded with reduction of overburden thickness. The relationship between the power spectrum area ratio α and natural frequency is shown in Figure 9. The thresholds of the power spectrum area ratio α, i.e., f_0, f_1 and f_2, were set on 2.20 Hz, 42.7 Hz (natural frequency at 0.0 mm footing) and 100 Hz, respectively, covering all ranges of natural frequencies measured in the tests. It was confirmed that the power spectrum area ratio α was well correlated with natural frequency. The relationship between the train passing acceleration rates β and natural frequency is shown in Figure. 9. It was also confirmed that the train passing acceleration rate β was well correlated with natural frequency.

Hence, the correlation between the proposed indices and natural frequency was confirmed by the model experiments.

2.3 *Verification of indices by in situ measurements*

The schematic figure of double-track railway steel bridge targeted in the in situ measurements is shown in Figure 10. The bridge consists of 17 piers, in which P5-P13 are in a river. The superstructure of the bridge is a simple through truss girder. The foundation of P5 consists of 9 cast-in-place concrete piles with 1500 mm diameters. The foundations of P6-P9 are caisson foundations and those of P10-P13 consist of 23 steel pipe piles with 800mm diameters. The foundation of P11 consists of 23 steel pipe piles and 18 piles for reinforcing. Natural frequencies of individual piers measured in impact and vibration tests are shown below in Figure 10.

The microtremors and train vibrations were measured at 7 piers, i.e., P5-P6 and P9-P13 by using an accelerometer. Then the relationship between the

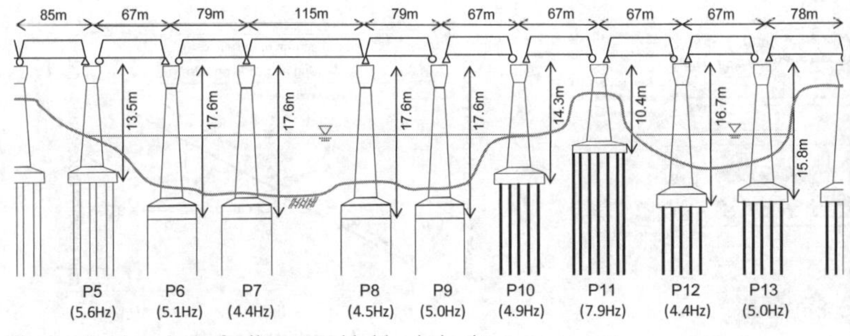

Figure 10. Plan view of railway steel bridge in in situ measurement.

Figure 12. Developed accelerometer

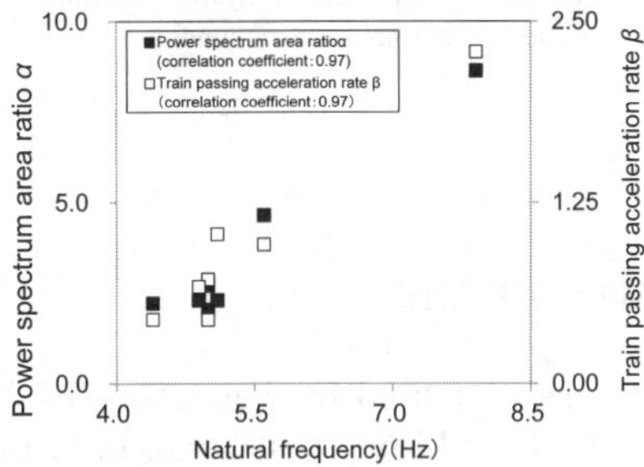

Figure 11. Relationship between soundness diagnosis indices and natural frequencies (in situ measurement).

proposed indices and natural frequencies were introduced. The thresholds of the power spectrum area ratio α, i.e., f_0, f_1 and f_2, were set on 1.0 Hz, 4.0 Hz (natural frequency of P12, which is minimum value among those of piers) and 20.0 Hz, respectively, covering all ranges of natural frequencies measured in the tests. The power spectrum area ratio α and the train passing acceleration rates β were measured three times, and then average of α and maximum value of β were calculated.

The relationship between the indices and the natural frequencies are shown in **Figure 11**. Indices were well correlated with natural frequencies as confirmed in the model experiments.

3 DEVELOPMENT OF SENSORS AND WIRELESS DATA TRANSMISSION SYSTEM

The developed sensor is shown in Figure 12. The size of the sensor is 200 mm high, 200 mm wide and 130 mm long. One directional microtremor and three directional train vibrations can be measured by MEMS accelerometers. The measured data can be recorded in a memory card and the calculated indices can be transmitted by wireless transmission system.

The battery of the sensor is 4 lithium batteries (voltage: 3.60 V; maximum electric current: 75.0 mA). The life span of the battery is two years, under following conditions:

1) Microtremors are measured five times per day, for 16.0 seconds with 500 Hz in sampling frequency from 2am to 4am every 30.0 minutes.
2) Train vibrations are measured three times per day for 20.0 seconds with 250 Hz in sampling frequency from 6am to 6:30am.

The data are transmitted to a gateway through multiple relays using a specified low power radio with 429 MHz band, and then sent to RTRI office from the gateway through a GPRS.

4 SOUNDNESS DIAGNOSIS OF PIERS BY USING MONITORING DATA

Long term monitoring was conducted on P5-P13 railway steel bridge shown in Figure 10. Figure 12 shows the sensors installed on the top of the pier using 8 anchor bolts.

The monitored data (five-point average) of P10-P12 are shown in Figure 13. The missing data were caused by the replacement of sensors. It was confirmed that the size of the indices well corresponded with that of natural frequencies, i.e., P11>P10=P12. In Figure 13, the predominant frequencies of the microtremors in 1.0-20.0 Hz band (five-point average) are also plotted. The proposed indices (variation coefficient: 0.12-0.18) were very uniform compared with the predominant frequencies (variation coefficient: 0.30-0.39).

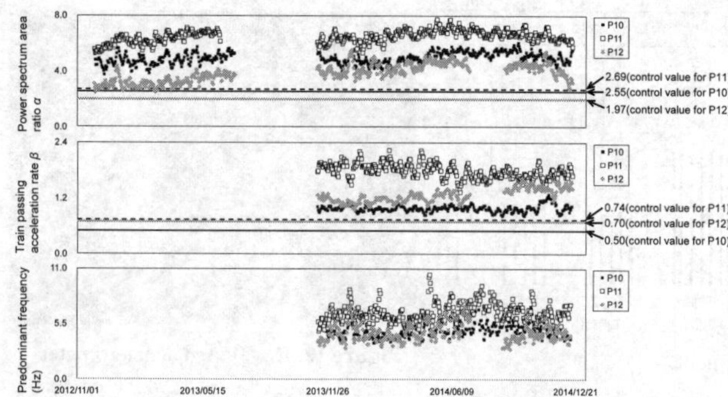

Figure 13. Monitored data of P10, P11 and P12.

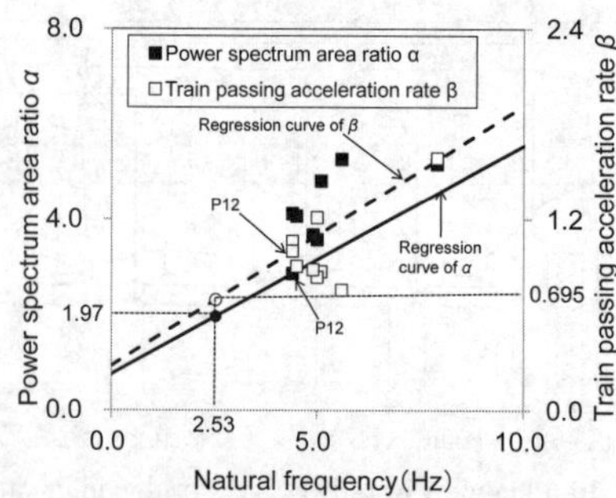

Figure 14. Relationship between minimum values of the monitored indices and natural frequencies.

The relationship between minimum values of the monitored indices and natural frequency is shown in Figure 14. The indices were correlated with natural frequencies (correlation coefficients of α and β are 0.66 and 0.70, respectively). The control values for the indices (=1.97 for α, 0.70 for β) can be calculated from the regression curves through the thresholds of natural frequencies (=2.53). The control values of the indices could be calculated if the number of piers with known natural frequencies were larger than two. Otherwise, the soundness can be diagnosed by the change in the indices.

5 CONCLUSIONS

The outcomes are summarized as follows:

- It was found from model experiments and in situ measurements that the proposed indices, the power spectrum area ratio of microtremors and the train passing acceleration rate were well correlated with natural frequency measured in impact and vibration test.
- The sensors and wireless transmission system were developed to monitor the proposed indices.
- Soundness diagnosis method using the monitoring data was proposed.
- The applicability of the indices to other bridge types and definition of the thresholds for the power spectrum area ratio of microtremors should be investigated in the future.

ACKNOWLEDGEMENT

This work was supported by a grant-in-aid for railway research of the Ministry of Land, Infrastructure, Transport and Tourism, Japan. In addition, authors wish to thank Tokyo Metro Co., Ltd., Japan, to the support for the research work.

REFERENCES

Kagawa, T. 1978. On the similitude in model vibration tests of earth-structures, *Proceedings of the Japan Society of Civil Engineers* **275**, 69-77. (In Japanese)

Miyashita, T., Ishii, H., Fujino, Y., Shoji, T. and Seki, M. 2007. Understanding of high-speed-train-induced local vibration of a railway steel bridge using laser measurement and its effect by train speed, *Dodoku Gakkai Ronbunshu A* **63(2)**, 277-296. (In Japanese)

Samizo, M., Watanabe, S., Fushiwaki, A. and Sugihara, T. 2007. Evaluation of the structural integrity of bridge pier foundations using microtremors in flood conditions, *Quarterly Report of RTRI* **3**, 153-157.

Shinoda, M., Haya, H. and Murata, S. 2008. Nondestructive evaluation of railway bridge substructures by percussion test, *Fourth International Conference on Sour and Erosion 2008*, 285-290.

Yoshida, Y., Kobayashi, Y. and Uchimura, T. 2014. Development of the monitoring system that operates with power generated from the bridge vibration, *Dodoku Gakkai Ronbunshu A1* **70(2)**, 282-294. (In Japanese)

Proceedings of the International Conference on Smart Infrastructure and Construction
ISBN 978-0-7277-6127-9

© The authors and ICE Publishing: All rights reserved, 2016
doi:10.1680/tfitsi.61279.487

Geo-material erosion and its effects on concrete tunnel lining

Ning Chen, Gang Ren and Chun-Qing Li

RMIT University, Civil Engineering Department, Australia

ABSTRACT Water seepage in tunnels often results in adverse effects both on the serviceability and stability of the tunnels. Inter alia, water tends to bring fines from the soil/rock matrix leaking into the tunnel, which will cause erosion in the surrounding geo-materials. Depending on the geological structure and geotechnical properties of the geo-materials surrounding the tunnel, the erosion may take the form of voids or suffosion which will weaken the stiffness of geo-materials around the tunnel lining. Consequently it will affect the stress distributions on the tunnel lining. In this paper, the geo-material erosion mechanisms were reviewed, and the stress redistributions affected by the erosion in the tunnel lining was analyzed. For this purpose, finite element modeling techniques was used for the analysis of additional stresses and bending moment on the lining as results of the erosion effects. The analytical results from finite element modeling showed that the water-leaking induced erosion will have significant effects on the stress distribution and bending movement in the tunnel lining and hence impact on the stability of the tunnel.

Keywords: Underground tunnels, erosion, rock failure mechanism, structural reliability, finite element analysis

1 INTRODUCTION

The need of traffic development requires the construction of tunnels in almost all kinds of soils or rocks in difficult ground conditions such as low coverage or high ground and water pressures.

Because of poor engineering properties, rapid deformation and large range of disturbances, soft rocks and soils have great impact on the stability of underground structures. Excessive deformation is likely to occur in the tunnel lining during construction, and sometimes, cracking or even collapse occurs. The ingress of groundwater through cracks in the tunnel lining will subsequently affect the tunnel stability and safety. Such as rebar corrosion and concrete deterioration. Furthermore, with time, above actions will accelerate crack propagation.

Inevitably, the ingress of groundwater water through cracks could disturb soil/rock behind tunnel lining, for example creating erosion voids if soil particles are eroded with water flow. A number of researches have been researched into those erosion voids affecting stress distributions on lining (Meguid and Dang 2009, Leung and Meguid 2011, Gao, Jiang et al. 2014, Wang, Huang et al. 2014). However little research work has been done looking into the mechanism of these erosion voids initiation. Also most of erosion phenomenon caused by groundwater movement are discussed for water retaining structure, with few covering underground structures. In this paper, a tunnel in difficult ground condition is taken as an example to study the underground erosion mechanisms (concentrated leak erosion and internal erosion) of surrounding geomaterial. And also the effect of the assumed erosions on tunnel lining is conducted by numerical analysis.

2 EROSION OF GEO-MATERIALS

An erosion problem in underground always has three components: the geo-material, the water flow and their interactions. The resistance of geo-material (soil or rock) is characterized by its erodibility; the water flow which causes erosion is quantified by its velocity; however the interaction of soil/rock and water will determine the type of erosion and the mechanisms.

This paper discusses two types of erosion: concentrated leak erosion and soil internal instability, which both are triggered by the ground water leaking through cracks of concrete tunnel lining.

2.1 *Concentrated Leak Erosion*

For a tunnel located below ground water table, when there is no through cracks in concrete lining, seepage through the lining would be minimal and can be ignored. The water imposes hydrostatic pressure around soil particles, therefore combining with electrical forces and forces at contacts between particles, whole matrix are stable.

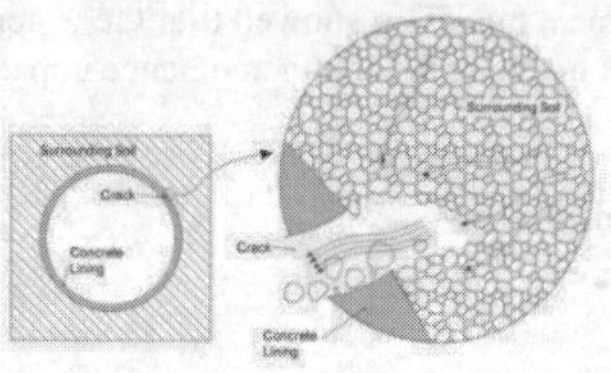

Figure 1. Initiation of concentrated leak erosion induced by crack

However as illustrated in Figure 1, when the condition of the concrete lining deterioration, cracks may develop in width and depth. Once the crack aperture is larger than soil particles' size, soil particle stability will be disrupted, potentially leading particles run off from matrix with the leaking water through cracks under the hydrostatic pressure. This kind of water inflow causes erosion behind lining and creating interconnected voids between interface of lining and soil.

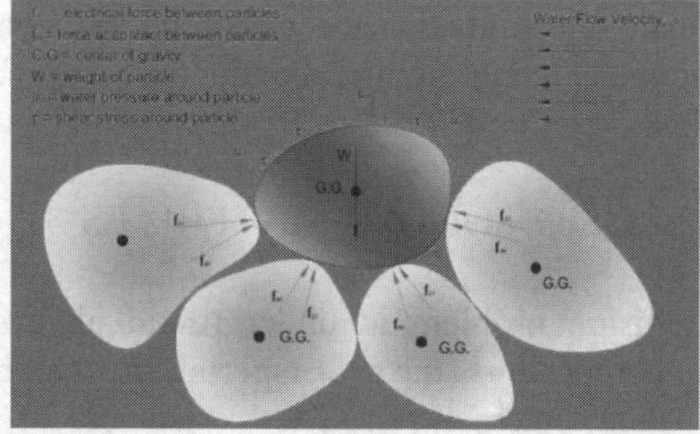

Figure 2. Free body diagram of soil particle when water flows

At beginning, the water starts flowing from soil to developed penetrating cracks, as shown in Figure 2, shear stress will be developed at the interface of soil particles and water. Based on Bernoulli's principle, with water flowing, pressure above soil particles is lower than below, this will develop uplift force to particles. As the water flow velocity increases due to cracks propagation and change in the ground water table, the shear force provided by flowing will be increased accordingly in terms of fluid mechanics. Once the shear stress over the threshold of erosion, erosion occurs. In terms of shear stress, this threshold is the critical shear stress τ_c . Briaud (2008) had summarized the measurements using the erosion function apparatus (EFA) as well as measurements published in the literatures. He found that for the soil mean grain size (D_{50}) smaller than 0.01mm (Clay to Silt), threshold decreases with grain size increase; The most erodible soils are fine sands with a mean grain size in the range of 0.1 to 0.5mm; For soil mean grain size larger than 0.5mm, which could be sorted as medium sand to gravel, the threshold increased when mean gran size increased.

Therefore, erodibility not only depends on soil types, but also depends on the properties of the water flowing over the soil. It can be defined as the relationship between the erosion rate $\dot{\varepsilon}$ and the shear stress τ, which governed by water flow at the soil – water interface (Eq 1). Massive testing results indicate that the critical shear stress and erosion rate are linearly (Eq 2).

$$\dot{\varepsilon} = f(\tau) \tag{1}$$

$$\dot{\varepsilon} = C_e(\tau - \tau_c) \tag{2}$$

C_e is obtained from testing which has small numbers in the order. So Wan and Fell (2004) use Erosion Rate Index to express C_e. Small number of I reflect rapid erosion rate and vice versa.

$$I = -\log(C_e) \tag{3}$$

Based on 20 years of erosion testing experience, Briaud (2008) provided a soil erosion categories as illustrated in Figure 3. This plot gives a macro cognitive of erodibility for different types of soil. However there are more researches conducted other methods to measure the threshold and erosion rate more accurately for a soil.

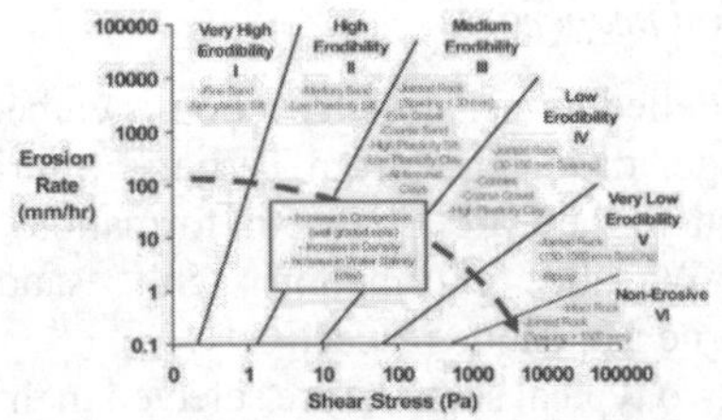

Figure 3. Erosion categories for geomaterial based on shear stress

The erosion function apparatus (EFA) was developed (Briaud 2008) for low cohesion soil to measure the erosion rate and critical shear stress. This has been widely used in North America to predict scour issue in river.

To evaluate piping phenomena which occurred in many hydraulic structures, Reddi (2000) designed tests studying internal and surface erosion, Wan and Fell (2004) further improved the tests and call it the Hole Erosion Test (HET) to measure the erosion rate and threshold for a soil. In this test, water flows through a 6mm diameter predrilled hole of a standard compacted soil sample in a confined tube, results of hole diameter increasement will be derived to indicate the soil erosion threshold as well as erosion rate at certain time (t).

$$\tau_t = \rho_w g s_t \frac{\emptyset_t}{4} \tag{4}$$

$$\dot{\varepsilon} = \frac{\rho_d}{2} \frac{d\emptyset_t}{dt} \tag{5}$$

Where,

ρ_w is density of the eroding fluid (kg/m^3)
g is acceleration due to gravity (9.8 m/s^2)
s_t is the hydraulic gradient across sample at time t
ρ_d is dry density of the soil sample
$\emptyset_t$ is the hole diameter at time t

2.2 *Soil internal instability*

The soil internal instability is induced by the soil movement internally, as indicated in Figure 3, the fine particles are removed through the voids between large particles by seepage flow. It contains two mechanism, suffusion and suffosion. In a long time, definition of suffusion and suffosion are confused, in the past, researchers just described both as "the transport of small particles from a soil". Until Moffat, Fannin et al. (2011) clarified that: migration of fines within its coarser particles without any loss of matrix integrity or change of in total volume, defined as *suffusion*; fines loss, leaving behind coarser particles, which yields a reduction in total volume and a consequent potential for collapse of the soil matrix, defined as *suffosion.*

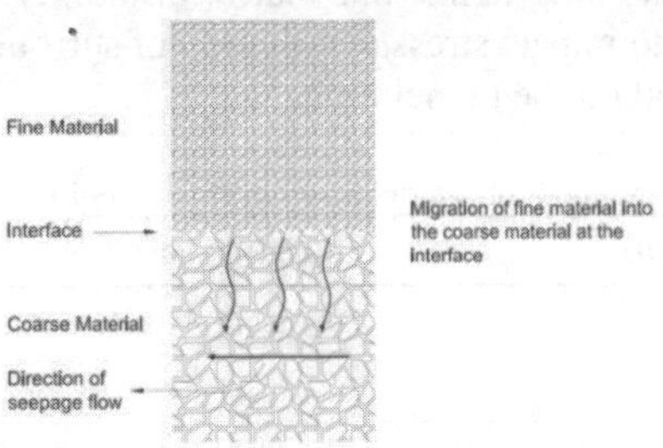

Figure 4. Diagram of soil internal instability

There are three semi-empirical criteria are widely used to determine the internal stability of soil, referring table 1(Moraci, Mandaglio et al. 2014).

In summary, all these three methods have one substance that the gapped soil are more potential to be eroded.

According to Bennet, Smith et al. (1992), in MURL project, from Flagstaff Station to Museum Station (Ch. 4191.450 – 4642.944 metres) tunnel through the geology where Quaternary Elizabeth Street formation (silty sandy clay) overlies the weathered surface of the Silurian Melbourne Formation (mudstone).

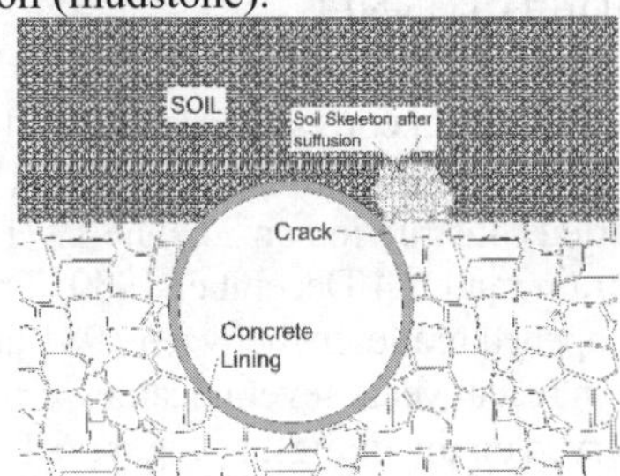

Figure 5. Internal instability mechanism behind tunnel lining

At the interface of soil and weathered rock, fine granular are mixed with gravels. For the purpose of discussion, this zone can be determined as critical internal instable zone. Fine particles runs into fractured rocks with seepage flow triggered by lining cracks, leaving behind an intact soil skeleton of coarser particles. Compared with no erosion stage, in terms of soil properties, the losing of fine particles will reduce the cohesion of the soil leading to possible voids in the matrix. Therefore, this erosion process will reduce the strength of the soil or collapse of matrix. The effects of internal instability on soil matrix have been studied by (Ke and Takahashi 2014).

In the long terms, this matrix instability will no doubt to impact stress distribution of surround geo-material on the tunnel lining.

Table 1. Methods of determine soil internal stability

Methods	**Criteria of unstable**
Kezdi's	$D_{15}/d_{85} \geq 4$
Sherard's	$D_{15}/d_{85} \geq 4$
Kenney and Lau (1985)	$H < 1.3F$ or $H < F$

$\boldsymbol{D_{15}}$ is the particle dia. of 15% by weight of coarser particles
$\boldsymbol{d_{85}}$ is the particle dia. of 85% by weight of finer particles
F is the value of "mass fraction smaller than" at any point on the grain-size distribution of a soil
H is the mass of fraction measured between particle diameters D and 4D

Figure 6. Photo of groundwater leaking observed in tunnel

3 STUDIED TUNNEL

The modelled tunnel corresponds to typical sections in the Melbourne Underground Rail Loop (MURL), which started construction on 22 June 1971, and the first test train ran on 4 December 1980. The whole loop was opened gradually between 1981 and 1985. However in recent years several leaking cracks have been observed on the linings, as Figure 6.The tunnel lining is composed by 450mm thick segments of reinforced concrete and minimum 50mm thick concrete grout. The overall thickness of the concrete lining is therefore 500mm (Baxter and Bennet 1981).

3.1 Soil properties

The modelled tunnel stretch section is embedded in three types of material from Tertiary, Silurian and Quaternary, they are Elizabeth formation (clayey sand), Werribee Formation (Silt sand) and Melbourne Mudstone respectively.

The two typical sections are selected in this study (a) Tunnel is completely buried in soil (Elizabeth Formation) where Concentrated Leak Erosion could occur once cracks formed on tunnel concrete lining. (b) Tunnel is embedded in both weathered rock and soil. In this scenario, as we discussed above, near the interaction, fines particles from soil could run into rock voids, leaving behind soil skeleton. This could lead the internal instability.

Table 2. Parameter values for the surround soil and tunnel concrete lining

	Element	Poisson ratio	E (kPa)	Weight (kN/m^2)
Soil	CPE4R	0.3	0.15e9	2000
Concrete Lining	B21	0.25	34.5e9	2500

Table 3. Parameter values for the surround soil and tunnel concrete lining

	Element	Poisson ratio	E (kPa)	Weight (kN/m^2)
Soil	CPE4R	0.3	0.15e9	2000
Soil skeleton	CPE4R	0.3	0.12e9	1600
Concrete Lining	B21	0.25	34.5e9	2500

4 NUMERICAL MODELLING

Recently, there has been some research carried out on the effects of erosion voids behind lining on tunnel performance (Meguid and Dang 2009, Wang, Huang et al. 2014). In this study, a two-dimensional finite element multistep simulation model for tunnel excavation and erosion is presented. The models takes into account results of two types of erosion: erosion voids behind lining and smear weaken area caused by suffusion.

The tunnel surround soil was modelled as homogenous 4-node bilinear, reduced integration with hourglass control element (CPE4R). The tunnel lining was simulated by B21 beam element. The interaction between lining and soil, also suffusion area and non-suffusion area were considered as no slipping. Both concrete lining and surrounding soil were assumed as an elastic material.

4.1 *Size of the domain*

The size of the model exactly based on MURL project. Tunnel is 23m under ground level, 5.95 internal diameters with 50mm thick lining.

Considered the boundary effects, the finite-element mesh extends to a depth of two times the tunnel diameter (D) below the tunnel spring line and laterally to a distance of 5D from the tunnel centerline and 8D along tunnel axis. Fig. 7 shows the finite element model used for this study.

4.2 *Model process*

The tunnelling lining installed following excavation underground, however in this study, we are more interested in stress from surrounding materials on lining rather than excavation effects, and also the tunnel is 40 years old, the disturbance of excavation would be very small. Therefore tunnel lining is considered installing before than excavation. The analysis process is demonstrated as following:

5 NUMERICAL RESULTS

The effect of erosion behind lining are analysed by comparing the stress distribution on lining for the case of no erosion. For the concentrated leak erosion, un-contact length of the void is assumed 3% of tunnel circumference and the location of the void to be consistent with the observation in the tunnel as illustrated in Figure 9.

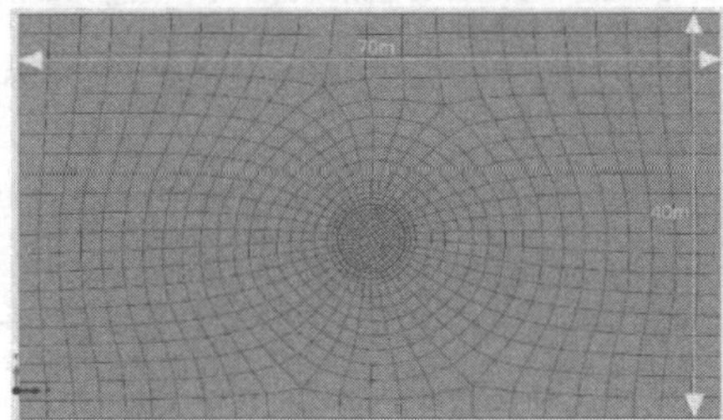

Figure 7. Numerical analysis domain

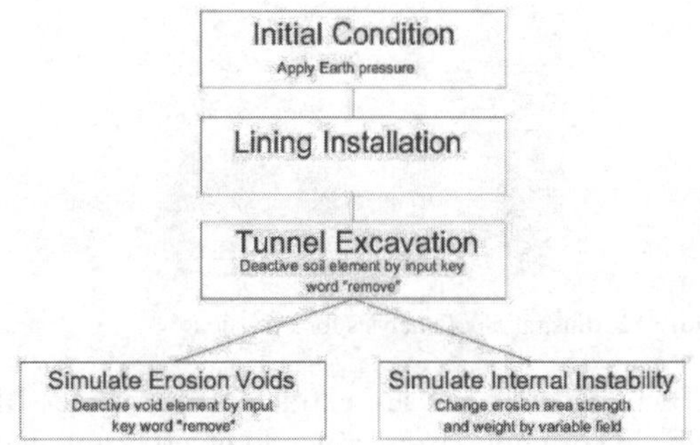

Figure 8. Numerical analysis steps

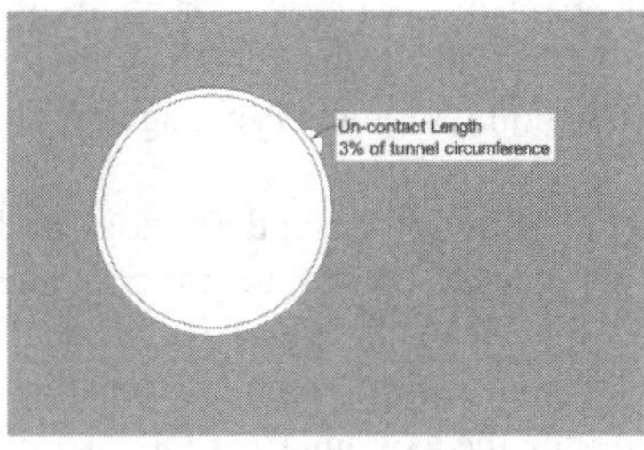

Figure 9. Illustration of analysis for concentrated leak erosion

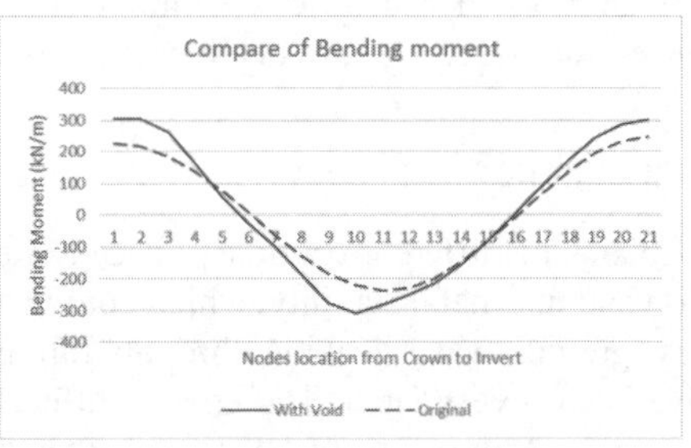

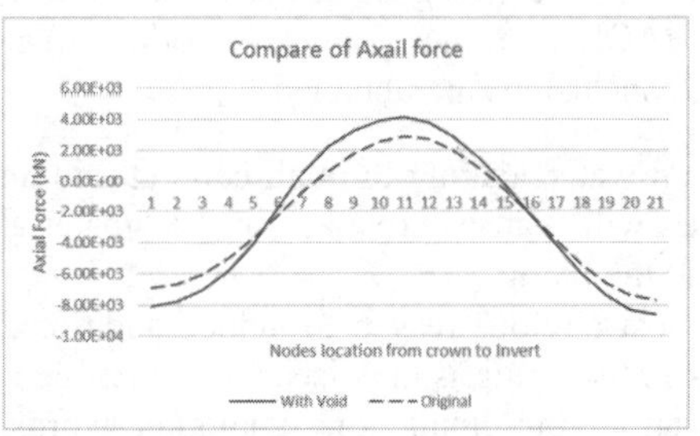

Figure 10 & Figure 11. Result of analysis for concentrated leak erosion

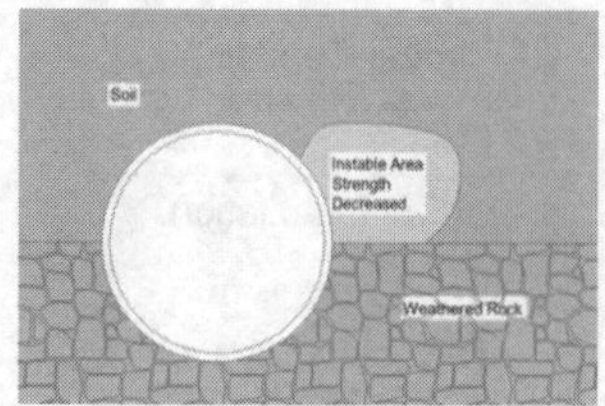

Figure 12. Illustration of analysis for suffusion

Bending moments are extracted from nodes of right side of lining, from crown to invert, as plotted in Figure 10. It indicates that comparing with original status (no erosion occurs), 3% of un-contact void will result in about 30% of bending moment increase. And it can be observed that the location of maximum bending moment is changed with existing voids.

Figure 11 shows the axial force at nodes from location of centre crown to invert. The results indicate a significant increase of the axial force in the liner for both compression and tension.

Considering the asymmetric of axial and bending moment distributed on tunnel concrete lining, it will lead to the increase of eccentricity at the section which causes the safety factor decline (Liu and Zhang 2014).

6 CONCLUSION

The mechanisms of different erosions, concentrate leak erosion and internal instability, which could be induced by ground water leaking through lining cracks have been investigated. The erosion effects on tunnel concrete lining have been investigated using ABAQUS finite element codes. The following conclusions are obtained:

- Ground water leaking through concrete lining crack, will cause erosion of geo-material behind the lining.
- The most obvious reason to induce additional stress in the lining is the disconnection of geo-material to the lining, or in another words, voids existing between lining and soil.
- With the increase of the length of disconnection between lining and geo-material, the effect of erosion voids will become more pronounced and therefore greatly affected stress distribution on lining, to reduced tunnel stability.
- The formation of voids in the geo-material behind the lining can be regarded as the extreme case of internal instability.

ACKNOWLEDGMENTS

Financial support from Metro Trains Melbourne, Australia and Australian Research Council under DP140101547 and LP150100413 is gratefully acknowledged.

REFERENCE

Baxter, D. A. and A. G. Bennet (1981). Aspects of Design and In-situ Testing for the MURL Rock Tunnels. *Fourth Australian Tunnelling Conference*. Melbourne, The Australasian Institute of Mining and Metallurgy**:** 41-60.

Bennet, A. G., N. B. Smith and J. L. Neilson (1992). *Tunnel.* The Seminar on Engineering Geology of Melbourne, Melbourne, Engineering Geology of Melbourne.

Briaud, J.-L. (2008). "Case Histories in Soil and Rock Erosion: Woodrow Wilson Bridge, Brazos River Meander, Normandy Cliffs, and New Orleans Levees." *Journal of Geotechnical and Geoenvironmental Engineering* **134**(10): 1425-1447.

Gao, Y., Y. Jiang and B. Li (2014). "Estimation of effect of voids on frequency response of mountain tunnel lining based on microtremor method." *Tunnelling and Underground Space Technology* **42**: 184-194.

Ke, L. and A. Takahashi (2014). "Experimental investigations on suffusion characteristics and its mechanical consequences on saturated cohesionless soil." Soils and Foundations **54**(4): 713-730.

Kenney, T. C. and D. Lau (1985). "Internal stability of granular filters." *Canadian Geotechnical Journal* **22**(2): 215-225.

Leung, C. and M. A. Meguid (2011). "An experimental study of the effect of local contact loss on the earth pressure distribution on existing tunnel linings." *Tunnelling and Underground Space Technology* **26**(1): 139-145.

Liu, J. and G. Zhang (2014). "The effect of hole behind lining on safety of tunnel." *Applied Mechanics and Materials* **477-478**: 600-603.

Meguid, M. A. and H. K. Dang (2009). "The effect of erosion voids on existing tunnel linings." *Tunnelling and Underground Space Technology* 2**4**(3): 278-286.

Moffat, R., R. J. Fannin and S. J. Garner (2011). "Spatial and temporal progression of internal erosion in cohesionless soil." *Canadian Geotechnical Journal* **48**(3): 399-412.

Moraci, N., M. C. Mandaglio and D. Ielo (2014). "Analysis of the internal stability of granular soils using different methods." *Canadian Geotechnical Journal* **51(9)**: 1063-1072.

Wan, C. and R. Fell (2004). "Investigation of Rate of Erosion of Soils in Embankment Dams." J*ournal of Geotechnical and Geoenvironmental Engineering* **130**(4): 373-380.

Wang, J., H. Huang, X. Xie and A. Bobet (2014). "Void-induced liner deformation and stress redistribution." *Tunnelling and Underground Space Technology* **40**: 263-276.

Proceedings of the International Conference on Smart Infrastructure and Construction
ISBN 978-0-7277-6127-9

© The authors and ICE Publishing: All rights reserved, 2016
doi:10.1680/tfitsi.61279.493

An event-adaptive control mechanism of the tunnel monitoring system based on the information granularity

Gang Li[1], Bin He [*1] and Hongwei Huang [2]

[1] *School of Electronics and Information Engineering, Tongji University, Shanghai, China*
[2] *Department of Geotechnical Engineering, Tongji University, Shanghai, China*
[*] *Corresponding Author*

ABSTRACT In applications of structural health monitoring of tunnels, wireless sensor network (WSN) is a useful technology for information extraction by using lots of different types of small-sized sensors. Existing tunnel monitoring systems have defects in understanding and using information, and there are few excellent data interpretation methods and adaptive feedback control mechanisms in WSNs. To improve the situation, an event-adaptive control mechanism of tunnel monitoring system based on the information granularity is presented in this work. The main idea in devising such a mechanism is to describe events quickly and then to adjust the monitoring system based on event description through the following two parts. Firstly, an information granularity model based on multi-sensor data is proposed. Its main process is implementing uncertain transforming between quantitative expressions and qualitative concepts through the fuzzy theory. Secondly, an event-adaptive control mechanism based on the information granularity model is presented. It is timely and effective to perceive the general condition of the tunnel by analyzing the information granularity evolved from large amounts of hybrid raw information. The information granularity is therefore used as the input of the event-adaptive control system which draws conclusions. Basing on these conclusions, this work realizes real-time feedback control of the tunnel monitoring system. The mechanism presented is validated on simulation experiments. The result shows that the information granularity can describe events with good accuracy and timeliness. And the event-adaptive control mechanism based on the information granularity can intelligently adjust the tunnel monitoring system with great efficiency and timeliness.

1 INTRODUCTION

Manual monitoring is gradually replaced by WSN in lots of applications including environment monitoring (He et al. 2014), traffic monitoring and military warfare (Rault et al. 2014). In structural health monitoring of civil engineering structures, WSN also plays an increasingly significant role (He et al. 2014). Some famous bridges (Kim et al. 2006) and tunnels (Hada et al. 2012) have deployed sensor nodes to perform monitoring tasks.

The existing tunnel monitoring systems mainly analyze hybrid sensor data to monitor the state of the tunnel after collecting and transmitting sensor data to a management center. Khan et al. (2015) provided a real-time monitoring system for underground mine and tunnels. The prototype has been deployed in four kinds of long and linear environments including underground mine, hallway, anechoic chamber and outdoor environment. These monitoring methods of directly analyzing large amounts of hybrid raw information have the high computational complexity and the poor real-timeliness (He et al. 2014), because they lack excellent data interpretation methods. Faced with large amounts of hybrid raw information, Campos et al. (2015) proposed a distributed autonomic inference machine. Using fuzzy logic, sensor nodes can perform self-management and contextualization tasks which fuse several sensor data into a more meaningful information. Though this kind of

WSN with data interpretation methods can express the event well, WSN still is hard to adjust itself through event-adaptive control. Mat et al. (2014) implemented WSN in a feedback loop to monitor temperature, humidity and moisture. Their control devices were activated by a threshold value.

Based on the situation referred above, this work presents an information granularity model to realize an information self-emerging mechanism. And then an event-adaptive control mechanism based on the information granularity is proposed to intelligently adjust the tunnel monitoring system.

2 INFORMATION GRANULARITY MODEL

Aiming at the low-level real-time character caused by the deficiency of excellent data interpretation methods in the tunnel monitoring system, this work completely changes the monitoring strategy that uniformly analyzing data after data collection. With the change of the event, a kind of information granularity is gradually formed in the information transmission process, which is termed information self-emerging mechanism. To express ideas clearly and effectively, this work adopts inclination sensor nodes, leakage sensor nodes and joint sensor nodes as examples. These sensor nodes are installed in the form of ring.

An excellent data interpretation method needs to fuse several sensor data into meaningful information quickly and accurately. Sensor nodes can perform contextualization tasks by using fuzzy logic. The fuzzy theory is therefore used in forming the information granularity. Fuzzy sets can be used to express several kinds of tunnel disasters.

Definition 1 Let X be a set of all monitoring rings, X={x}, termed the domain. It has several fuzzy sets (A, B, C,…) expressing the tunnel disaster. If there is a random number $\mu_A(x)$ with a steady tendency to fuzzy set A for any element x, $\mu_A(x)$ will be called membership degree x to A.

The tunnel disaster is divided into five grades from mild to severe, including Level 1, Level 2, Level 3, Level 4 and Level 5. And therefore, the fuzzy set A can be divided into five fuzzy subsets A1, A2, A3, A4 and A5. Depending on the type of sensor nodes, we divide the detectable tunnel disaster into segment tilting, water leaking and joint expanding, termed fuzzy set A, B and C.

According to the fuzzy theory, membership function of the fuzzy set must have the single peak feature. It is assumed that membership function obeys normal distribution $N(\mu,1)$. Regions of five grades of tunnel disaster can be determined by the expert experience method, termed $[z_{10},z_1]$, $[z_{20},z_2]$, $[z_{30},z_3]$, $[z_{40},z_4]$ and $[z_{50},z_5]$ separately. In the case of fuzzy subsets A1, its membership function $\mu_{A1}(x)$ can be calculated as shown in Formula (1). Membership functions of other fuzzy subsets follow the same calculation method. After the sensor value is assigned to the variable x in all membership functions, values of membership degree are calculated.

$$\mu_{A1}(x) = \frac{1}{\sqrt{2\pi}} \int_{z_{10}}^{z_1} \exp\left(-\frac{(z-x)^2}{2}\right) dz \qquad (1)$$

The information granularity is created like the bitmap, as shown in Figure 1. The information granularity also has the concept of resolution and pixel, termed G-resolution and G-pixel. G-resolution, which represents the size of the information granularity, is continually changing in process of data transmission. In G-resolution with $M \times N$, M is equal to $RN+1$, where RN denotes the number of monitoring rings that have completed data transmission at the moment. N is equal to $FS+1$, where FS denotes the number of fuzzy subsets at the moment. Every element of the information granularity is called G-pixel. G-pixel is a 1-byte data, of which value is in the range of 0 to 255 in decimal format.

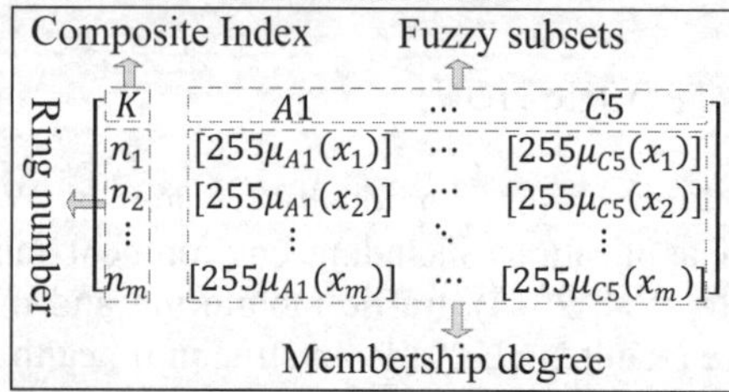

Figure 1. Information granularity

Since the information granularity comes into being with the data transmission, its description starts from the source node. The same type of sensor nodes in the monitoring ring with the source node compute the average of their sensor values, which is used to calculate the membership degree by Formula (1). And then the information granularity of the first monitoring ring is formed. If no tunnel disaster is monitored in the first monitoring ring, G-resolution will be 2×1. And if no tunnel disaster is monitored in *m* consecutive monitoring rings, G-resolution will be (*m*+1)×1. It can be seen that G-resolution with *M*×1represents no disaster occurs. The information granularity with this kind of G-resolution is considered to be a healthy information granularity. In the information granularity shown in Figure 1, ring number and fuzzy subsets increase gradually with the data transmission. Every value of membership degree, which is also a 1-byte data, is equal to $[255\mu_{XX}(x_X)]$ where the square bracket indicates rounding down. In G-resolution with *M*×*N*, smaller *N* means lesser tunnel disasters. Just like the diagnosis results of hospital, lesser tunnel disasters do not necessarily mean a healthier tunnel. The health of tunnel also depends on fuzzy subsets and their membership degree. And therefore, composite index *K* representing the general health of tunnel can be calculated as shown in Formula (2) where *n* denotes the number of fuzzy subsets. A severe tunnel disaster has a greater effect on the general health of tunnel. Because the tunnel disaster is divided into five grades, the coefficient of membership degree is set to 0.2*i* (*i*=1, 2,…, 5).

$$K = \left[\frac{1}{n \times m} \sum_{i=1}^{i=5} \sum_{j=1}^{j=m} \left\{ 0.2i \times \left([255\mu_{Ai}(x_j)] + [255\mu_{Bi}(x_j)] + [255\mu_{Ci}(x_j)] \right) \right\} \right] \quad (2)$$

The ever-changing information granularity is formed in the process of miscellaneous data transmission. It represents not only the general health of tunnel by the composite index *K* from the macro-perspective, but also the health of every monitoring ring by membership degree from the micro-perspective. And the composite index *K* can also indicate the priority of the information granularity, thus affecting the transmission mechanism.

3 EVENT-ADAPTIVE CONTROL MECHANISM

The above-mentioned information granularity representing the events is taken as the input of event-adaptive control mechanism. This kind of event-adaptive control mechanism based on the information granularity mainly controls two aspects, adjusting the collection cycle and the feedback cycle.

The feedback cycle is adjusted by the composite index *K*. A threshold value φ with 1 byte is set. When the information granularity enters the controller and meets $K>\varphi$, the event-adaptive control mechanism is triggered. It means that the event-adaptive control mechanism works only when the tunnel disaster reaches a certain level. This way can not only realize event-adaptive mechanism effectively, but also reduce the energy consumption caused by the transmission of feedback information.

The collection cycle, which is allocated according to ring, is adjusted by the health of every monitoring ring. The health of the n_kth monitoring ring K_k can be calculated as shown in Formula (3).

$$K_k = \left[\sum_{i=1}^{i=5} \left\{ 0.2i \times \left([255\mu_{Ai}(x_k)] + [255\mu_{Bi}(x_k)] + [255\mu_{Ci}(x_k)] \right) \right\} \right] \quad (3)$$

The collection cycle of the n_kth monitoring ring P_k can be calculated as shown in Formula (4)

$$P_k = \left(1 - \frac{K_k}{255}\right) P_k^* + \frac{K_k}{255} \left(P_{\min} + \frac{K_k (P_{\max} - P_{\min})}{255} \right) \quad (4)$$

where P_k^* indicates the last collection cycle of the n_kth monitoring ring, P_{min} and P_{max} indicates the lower and upper limits of the collection cycle separately.

The information granularity is also changing in the event-adaptive control mechanism. If K_k=0 is met, the row with ring number n_k will be deleted from the information granularity. And then P_k (k=1, 2,…, m, $K_k \neq 0$) are added into the last column of the information granularity. The new information granularity after the event-adaptive control mechanism has been formed, as shown in Figure 2 where *P* indicates the number of nonzero P_k. In G-resolution with *M*×*N*, *M* is changed to *P*+1 (*P*≠0), and *N* is changed to *FS*+2.

Collection cycle

$$\begin{bmatrix} K & A1 & \cdots & C5 & P \\ n_1 & [255\mu_{A1}(x_1)] & \cdots & [255\mu_{C5}(x_1)] & P_1 \\ n_2 & [255\mu_{A1}(x_2)] & \cdots & [255\mu_{C5}(x_2)] & P_2 \\ \vdots & \vdots & \ddots & \vdots & \vdots \\ n_m & [255\mu_{A1}(x_m)] & \cdots & [255\mu_{C5}(x_m)] & P_m \end{bmatrix}$$

Figure 2. Information granularity after the event-adaptive control mechanism

The schematic flow of the event-adaptive control mechanism based on the information granularity is shown in Figure 3. Firstly, the information granularity is gradually formed in the process of data forward transmission. In this stage, G-resolution will only be constantly increasing. Secondly, composite index K and threshold value φ are going to be compared. If $K \leq \varphi$ is met, the information granularity will be recorded and deleted so that adjusting the feedback cycle is realized. If $K > \varphi$ is met, the new collection cycle will be calculated as above. And then the new information granularity after the event-adaptive control mechanism will be formed. Thirdly, this new information granularity will be transmitted in the direction opposite to forward transmission. Every sensor nodes in the current monitoring ring will read their corresponding collection cycle P_k from the information granularity. If P_k is not equal to zero, these sensor nodes will adopt the new collection cycle P_k, and the row with ring number P_k will be deleted from the information granularity. Because P indicates the number of nonzero P_k, it becomes intuitive to see how many monitoring rings still do not adjust the collection cycle. In this stage, G-resolution will only be constantly decreasing. Before the information granularity arrives at the source node, the information granularity should be eliminated. If P is not equal to zero when the information granularity arrives at the source node, it means that there are P monitoring rings which are offline. According to the ring number of these P monitoring rings, corresponding sensor nodes should be checked.

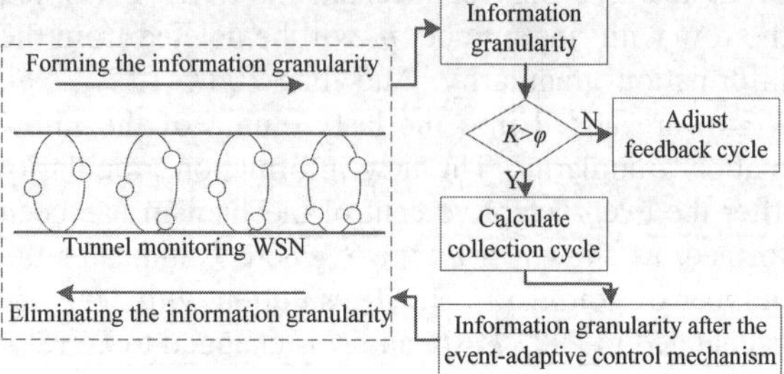

Figure 3. Event-adaptive control mechanism based on the information granularity

4 RESULT ANALYSIS

The information granularity is continually changing in the whole process. In the process of data transmission, there are two stages including forming the information granularity and eliminating the information granularity, as shown in Figure 3. It is assumed that data transmission time between the rings is invariable. The change of G-resolution in these two stages is recorded respectively, as shown in Figure 4. In the stage of forming the information granularity, M grows linearly, as shown in Figure 4(a), because normal or abnormal state of every monitoring ring will be recorded. With the rapid growth of fuzzy subsets, N grows within the upper limit. It can be seen that the initial value of M in Figure 4(b) is less than the final value of M in Figure 4(a), because the event-adaptive control mechanism has deleted data of normal monitoring rings. This way can not only meet the requirement of adjusting the collection cycle after recording the health status of tunnel, but also reduce the size of the information granularity to improve transmission performance. In the stage of eliminating the information granularity, M decreases irregularly, as shown in Figure 4(b), which means that only monitoring rings with events need to adjust the collection cycle. When M=0 is met, it indicates that all monitoring rings have finished adjusting the collection cycle.

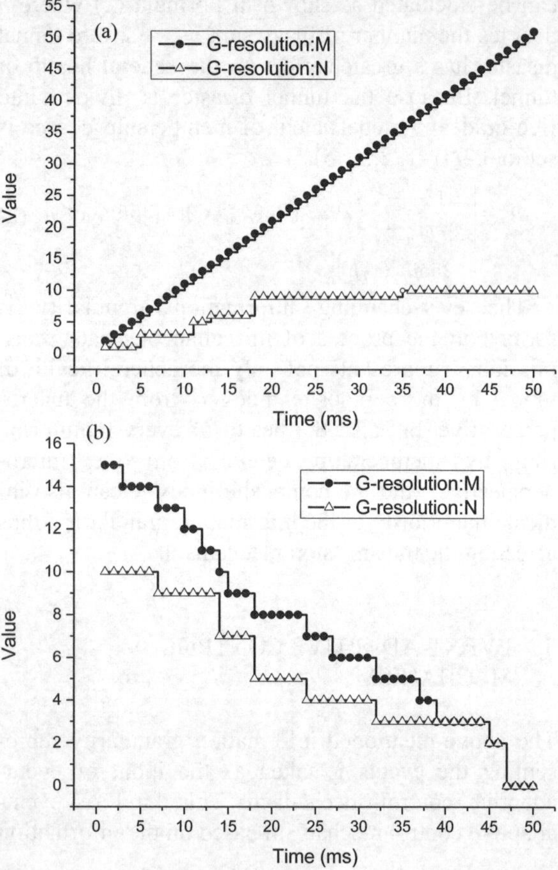

Figure 4. G-resolution in the two stages

And therefore transmission of the information granularity will be stopped to reduce energy consumption. *N* also irregularly decreases to 0.

Compared with the traditional monitoring systems, this monitoring method with the information self-emerging mechanism has the following three advantages. Firstly, real-time performance is good. As is stated above, the healthy degree of tunnel represented by the information granularity is updated in real time in the process of data transmission. Secondly, its analysis is multiscale. Meaning of the information granularity can be analyzed through G-pixel and G-resolution. G-pixel can represent not only the general health of tunnel by the composite index *K* from the macro-perspective, but also the health of every monitoring ring by membership degree from the micro-perspective. Thirdly, it has visual effects. Part of the meaning of the information granularity can be analyzed through its shape (G-resolution), as shown in Figure 5. Figures 5(a), 5(b), 5(c) and 5(d) are typical shapes of the information granularity in the forming stage. And Figure 5(e) is a typical shape of the information granularity in the eliminating stage. In Figure 5(a), $M>>N$ represents most of the monitoring rings still have good healthy degree in the late transmission stage. In Figure 5(b), $M<<N$ represents a lot of disasters have been monitored in the early transmission stage. In Figure 5(c), *M* and *N* are small relatively, which represents a few disasters have been monitored in the early transmission stage. In Figure 5(d), *M* and *N* are big relatively, which represents a lot of disasters have been monitored in the late transmission stage. Through comparative analysis of Figures 5(d) and 5(e), the size of the information granularity has been significantly reduced under meeting the requirement of adjusting the collection cycle, because only data of disasters have been retained completely.

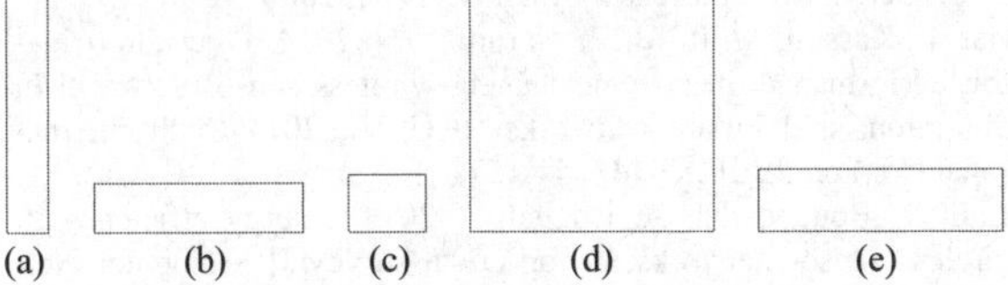

Figure 5. Shape of the information granularity (G-resolution)

To verify the effect of the event-adaptive control mechanism, the collection cycle and the feedback cycle are analyzed. The presented mechanism is compared with the traditional data collection with a fixed cycle, as shown in Figure 6. In Figure 6(a), the data collection with a fixed cycle is easy to lose valuable event data so that it is hard to restore the original appearance of the event. The presented mechanism can adjust the collection cycle by the event-adaptive control mechanism based on the information granularity, as shown in Figure 6(b). The collection cycle is inversely proportional to the degree of the events. This way can restore the trend of the events well by using less data.

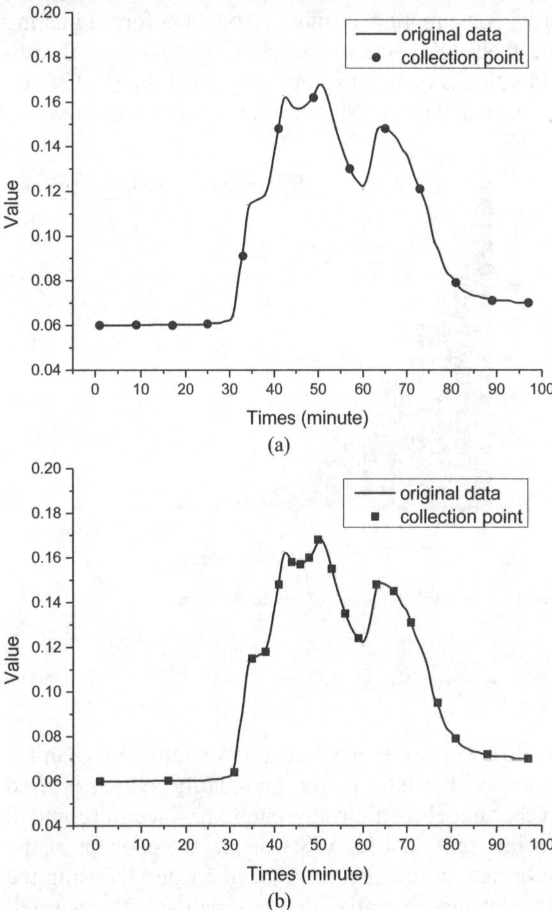

Figure 6. Two methods of data collection

The presented mechanism is also compared with the event-adaptive control mechanism without adjusting the feedback cycle. Data of a known disaster is selected as a test object. The number of feedback control is recorded by adjusting threshold value φ, as shown in Figure 7. The event-adaptive control mechanism without adjusting the feedback cycle (φ=0) has the most times of feedback control, because the feedback mechanism will be triggered every time finishing data collection. Transmitting feedback data inevitably leads to energy consumption. And therefore the number of feedback control should not be too many. Adjusting threshold value φ can adjust the sensitivity of event-adaptive mechanism to change the number of feedback control. The smaller φ causes a more frequent feedback control. Though the effect of restoring the event by the collection data will be better, energy consumption is more. And therefore adjusting the feedback cycle by selecting a reasonable threshold value φ can realize not only good data collection with variable cycle, but also low power consumption.

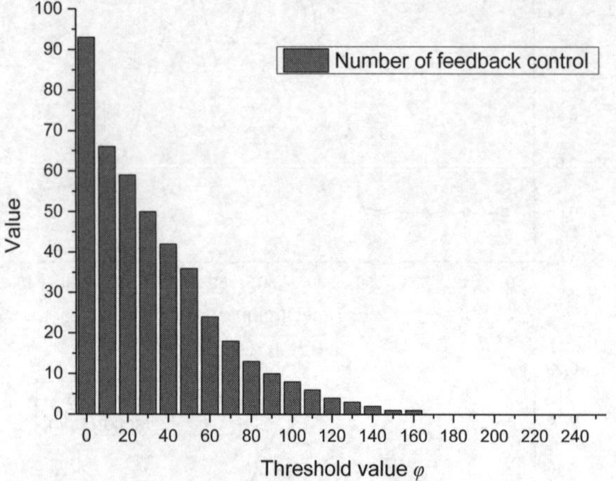

Figure 7. Recorded number of feedback control

5 CONCLUSIONS

In this work, the event-adaptive control mechanism based on the information granularity, which is used in the tunnel monitoring system, has been presented. To interpret data for WSN better, the concept of the information granularity has been created by using the fuzzy theory to realize the information self-emerging mechanism. The information granularity, which is created like the bitmap, has two important properties including G-resolution and G-pixel. Using analysis and simulation techniques, it has been proved that monitoring the tunnel through the ever-changing information granularity has three great advantages, including good real-time, multi-scale analysis and visual effects. The event-adaptive control mechanism based on the information granularity has been proposed. Through adjusting the collection cycle and the feedback cycle, the event-adaptive control mechanism can intelligently control the tunnel monitoring system according to the degree of the event.

ACKNOWLEDGEMENT

This work is supported by National Basic Research Program of China (973 Program: Grant No. 2011CB013803), International Research Cooperation Project of Shanghai Science and Technology Committee (Grant No. 15220721600), and Yangzi River Plan.

REFERENCES

Campos, N. G., Gomes, D. G., Delicato, F. C., Neto, A. J., Pirmez, L., & de Souza, J. N. 2015. Autonomic Context-Aware Wireless Sensor Networks. Journal of Sensors, **2015**.

Hada, A., Soga, K., Liu, R., & Wassell, I. J. 2012. Lagrangian heuristic method for the wireless sensor network design problem in railway structural health monitoring. Mechanical Systems and Signal Processing, **28**, 20-35.

He, B., Li, Y., Huang, H., & Tang, H. 2014. Spatial–temporal compression and recovery in a wireless sensor network in an underground tunnel environment. Knowledge and Information Systems, **41(2)**, 449-465.

He B, Li G. 2014. PUAR: performance and usage aware routing algorithm for long and linear wireless sensor networks[J]. International Journal of Distributed Sensor Networks, **2014**.

He, B., & Li, Y. 2014. Big data reduction and optimization in sensor monitoring network. Journal of Applied Mathematics, **2014**.

Khan, M. Y., Qaisar, S., Naeem, M., Aslam, A., Shahid, S., & Naqvi, I. 2015. Detection and self-healing of cluster breakages in mines and tunnels: an empirical investigation. Sensor Review, **35**, 3.

Kim, S., Pakzad, S., Culler, D., Demmel, J., Fenves, G., Glaser, S., & Turon, M. 2006. Wireless sensor networks for structural health monitoring[C]//Proceedings of the 4th international conference on Embedded networked sensor systems. ACM, **2006**, 427-428.

Mat, I., Kassim, M. R. M., & Harun, A. N. 2014. Precision irrigation performance measurement using wireless sensor network. In Ubiquitous and Future Networks (ICUFN), 2014 Sixth International Conf on. IEEE, **2014**, 154-157.

Rault T, Bouabdallah A, Challal Y. 2014, Energy efficiency in wireless sensor networks: A top-down survey[J]. Computer Networks, **67**, 104-122.

Proceedings of the International Conference on Smart Infrastructure and Construction
ISBN 978-0-7277-6127-9

© The authors and ICE Publishing: All rights reserved, 2016
doi:10.1680/tfitsi.61279.499

Strain monitoring and analysis of a steel pipe during horizontal directional crossing

Suzhen Li[*1,2] and Xiang Li[2]

[1]*State Key Laboratory of Disaster Reduction in Civil Engineering/Tongji University, Shanghai, China*
[2]*College of Civil Engineering/Tongji University, Shanghai, China*
[*]*Corresponding Author*

ABSTRACT Horizontal directional drilling (HDD) plays an increasingly important role in municipal pipeline construction in congested urban areas. Knowledge of the mechanical behaviors of a pipeline during HDD installation is significant for the design of an optimal construction scheme and the strategy making of life-cycle maintenance after installation. In combination with a practical project, the in-situ monitoring of a gas pipeline during HDD construction was conducted. Based on gyroscope position-detection system and fiber optic sensing technique, the three dimensional (3-D) spatial coordinates of the pipe position and the strain responses of the tested pipe segment traveling through the whole drilling track were measured. A theoretical model is presented to simulate the pipe pullback process and calculate the strain in the pipe during installation.The actual mechanical behavior of the pipe subjected to large bending deformation is particularly concerned. The case of the monitored HDD construction is examined to demonstrate the application of the model by comparing with the measurements.

1 INTRODUCTION

Horizontal directional drilling (HDD) is an important technology for the placement of municipal pipelines, including gas, water and sewer lines (Najafi2005, David 2005). The traditional open-cut methods in congested urban areas have been proved to be rather expensive and often associated with road damage, noise, traffic delays, and other disruptions of nearby commercial activities. Trenchless construction, which is regarded as an environmentally sound technology, has made a significant impact on the world-wide industries of underground utility installation over the past decade. In China, the HDD construction is a multi-billion dollar a year industry, occupying the fastest growing segment of the trenchless technology market (Ma and Najafi 2008).

However, the parallel development of standard engineering practice for the design, installation, and monitoring of trenchless crossings has not accompanied the rapid growth in the application and general acceptance of the construction method. Relatively little rational analysis has been done to examine the trenchless installation procedures, particularly regarding the prediction of the stresses developed in the pipeline product both during and after installation. The current design methodology largely relies on the experience and judgment of contractors, manufacturers and engineers.

To comply with limited underground space and reduce the influence of pipeline installation on above-ground traffic, large bending angle of the pipeline is usually desired. The responses of the pipe due to bending, which are caused as the pipe is forced through the curves in the bore hole, are key factors to ensure that the pipeline can be installed and operated without risk of damage. The layout of the bore path must be specific to keep the bending stress or strain within the allowable limits. To do that, knowledge of the relation between the bending radius and the mechanical behavior of the pipeline is critical. While some procedures and guidelines (GB 50424-2007, GB 50268-2008, David 2005a, David 2005b) for pipeline trenchless construction have provided the equations for calculating the bending stress, they are based on elastic assumptions and sometimes are too conservative to keep up withthe requirement of the resilience to limited underground space. With regard to horizontal directional drilling (HDD) installation, some researchers (Baumert and Allouche2002, Polak and Lasheen 2002, Cheng and Polak 2007, Yang et.al. 2014) have proposed several mechanical models for a pipe in HDD and focused their work mainly on the calculation of pulling loads. As many factors including the behavior of soil, pipe and drilling fluids are involved and some of them are difficult to determine in practice, the accuracy of the analytical results is hardly satisfying.

Using a gyroscope position-detection system and fiber optic sensing technique, in-situ monitoring of a gas pipeline during HDD construction is carried out in this work. A theoretical model is then presented to simulate the pipe pullback process and calculate the strain in the pipe during installation. The actual mechanical behavior of the pipe subjected to large bending deformation is particularly concerned.

2 HDD CONSTRUCTION MONITORING

Consider an engineering project for placing an underground gas transmission pipeline crossing beneath a small in-town river. Field tests have been conducted to investigate the performance of the pipe during the installation using the horizontal directional drilling (HDD) technique. The details of this project regarding the HDD monitoring of the pipeline can be found in the reference (Li 2013). Some important results and observations are briefly introduced here.

The electrical resistance welded (ERW) steel pipe was selected for gas transmission, with the full length of 520m and the maximum crossing depth of 21.4m. The other parameters of the pipe include: the diameter D=325mm; the thickness t=8mm; the gap between the drill hole and the pipe c=0.5D; the elastic modulus $E = 2.06\times10^{11}\,\mathrm{N/m^2}$.

Two sensing techniques were adopted to obtain the onsite measurements of the pipeline throughout the whole HDD procedure. The three dimensional (3-D) spatial coordinates of the overall pipeline were measured based on a gyroscope position-detection system. Figure 1 presents the measurements of the 3-D spatial coordinates of the underground pipeline and their projections on the XY, XZ, YZ planes. It can be seen that the pipeline was not restricted within a plane as most guidelines regarding trenchless construction require.

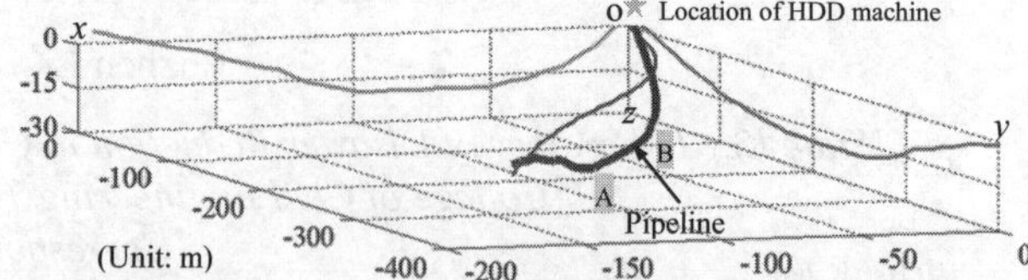

Figure 1. Spatial position of the underground pipeline.

The strain responses of the tested pipe segment traveling through the whole drilling track were obtained by using fiber Bragg grating (FBG) sensors. A 4m tested pipe segment is connected between the normal gas pipe and the drill rod, where the FBG sensors are attached to the inside wall. As shown in Figure 2, two circumferential strain sensors are attached at Section ① ; four longitudinal and two circumferential strain sensors are installed at Section ②.

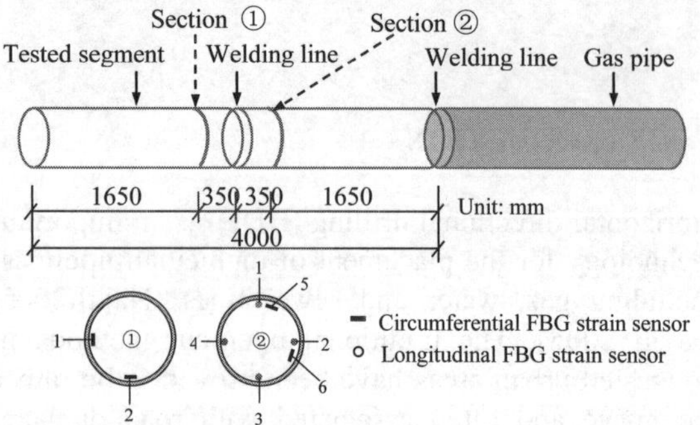

Figure 2. Sensor placement.

The important resultsachieved in the field test include: (1) the spatial 3-D coordinates of the constructed underground gas pipeline; (2) the strain responses of the tested pipe segment traveling through the whole drilling track during HDD; (3) the axial forces and moments of the tested pipe segment during HDD.

3 STRAIN ANALYSIS DUE TO LARGE DEFORMATION

The actual mechanical behavior of the pipe subjected to large bending deformation is first investigated, based on which the pullback force is calculated and evaluated in Section 4.

3.1 Mechanical Model

Consider a pipe segment (AB) in the drill hole subject to large deformation of the transit angle $2\psi_0$ and the maximum deflection $y_{\max}$, as shown in Figure 3(a). A simply supported beam is taken to represent the segment, with the corresponding parameters given in Figure 3(b). *R* is the reaction force with the vertical component *P*.

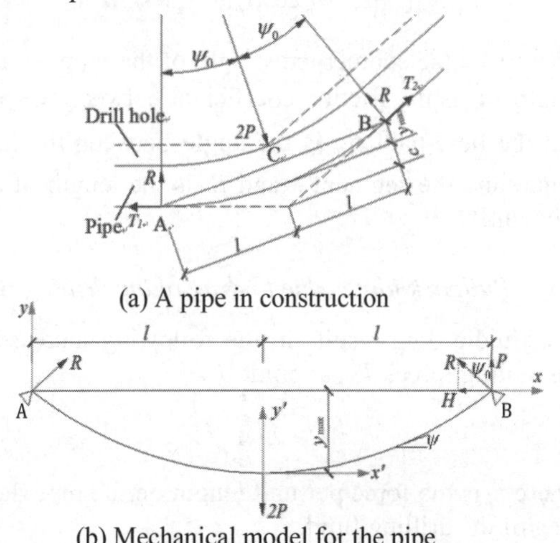

(a) A pipe in construction

(b) Mechanical model for the pipe

Figure 3.A pipe segment subject to large deformation

Based on plastic bending theory (Yu and Zhang 1996), the relation of the transit angle $2\psi_0$ and the maximum deflection $y_{\max}$ holds:

$$\frac{y_{\max}}{l}=\frac{\sqrt{2}\sin\psi_0\cos\phi_l-\cos\psi_0\cdot\Phi(\eta,\phi_l)}{\sqrt{2}\cos\psi_0\cos\phi_l+\sin\psi_0\cdot\Phi(\eta,\phi_l)} \tag{1}$$

in which,

$$\eta=\sin\left(\frac{\pi}{4}\right)=\frac{\sqrt{2}}{2} \tag{2}$$

$$\cos^2\phi_l=\sin\psi_0 \tag{3}$$

$$\Phi(\eta,\phi_l)=0.8472+\mathrm{F}(\eta,\phi_l)-2\,\mathrm{E}(\eta,\phi_l) \tag{4}$$

$$\mathrm{F}(\eta,\phi_l)=\int_0^{\phi_l}\frac{1}{\sqrt{1-\eta^2\sin^2\phi}}d\phi \tag{4-1}$$

$$\mathrm{E}(\eta,\phi_l)=\int_0^{\phi_l}\sqrt{1-\eta^2\sin^2\phi}\,d\phi \tag{4-2}$$

On the other hand, the following relation stands for the pipe segment subject to large deformation, as illustrated in Figure 3(a):

$$l=\frac{y_{\max}+c}{\tan y_0} \tag{5}$$

Here, *c* is the gap between the drill hole and the pipe. Normally, the inner-diameter of the drill hole is 1.5 times the outer-diameter (*D*) of the pipe, which makes it reasonable to set *c*=0.5*D*.

Based on the combination of Eq.(1) and Eq.(5), $y_{\max}$ and l can be determined. The reaction force *R* and the maximum strain responses $\varepsilon_{\max}$ are then derived as:

$$P=EI\cdot k^2 \tag{6}$$

$$R=\frac{P}{\cos\psi_0} \tag{7}$$

$$\varepsilon_{\max}=\frac{D}{2}k^2\left(l+y_{\max}\tan\psi_0\right) \tag{8}$$

with

$$k=\frac{\sqrt{\cos\psi_0}}{l}[-\sqrt{2}\cos\psi_0\cos\phi_l-\sin\psi_0\cdot\Phi(\eta,\phi_l)] \tag{9}$$

3.2 Parameter study

It can be seen that there is only one parameter (ψ_0) related to the strain of the pipe subject to large deformation, having nothing to do with the geometric and material properties of the pipe. Taking a pipe of 325mm diameter and 8mm thickness for example, the relations of the maximum strain and the parameter ψ_0 are presented in Figure 4. Clearly, the pipe strains are increased exponentially with the increase of ψ_0. This relation provides an important criterion for pipeline HDD construction. For one, it restricts the allowed values for transit angle of a pipeline during construction despite a wide range is desired regarding the limited underground space. For the other, it helps to assess the existing mechanical state of the pipeline after trenchless construction, which can be used for safety evaluation and life-cycle maintenance of the buried pipeline.

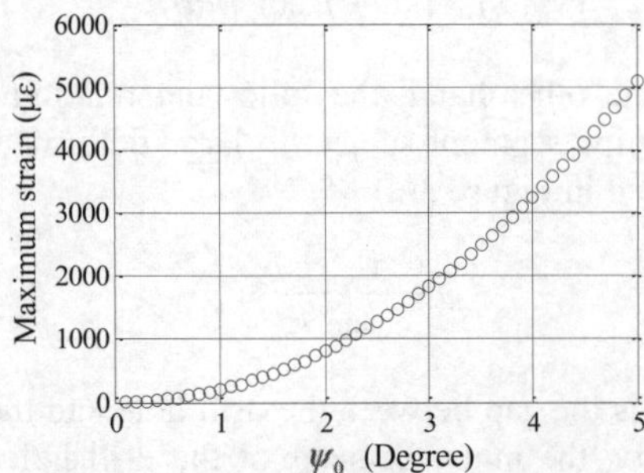

Figure 4. The relation of the maximum strain versus ψ_0

4 CALCULATION OF THE PULLBACK FORCE

To calculate the pullback force of the pipeline during HDD construction, the bore-path profile is approximated by fitting straight lines with defined angles of inclination, as shown in Figure 5. The total pulling force at point i can be written as:

$$T_i = T_{ig} + T_{is} + T_{id} + \sum_{j=1}^{i-1} T_{jif} \tag{10}$$

where $T_{ig}, T_{is}, T_{id}, T_{jif}$ are the components due to weight of the pipe outside the bore-path, weight of the pipe inside the bore-path, drag of the drilling fluid and large deformation, respectively. According to the reference (Polak and Lasheen 2002), they can be determined as follows.

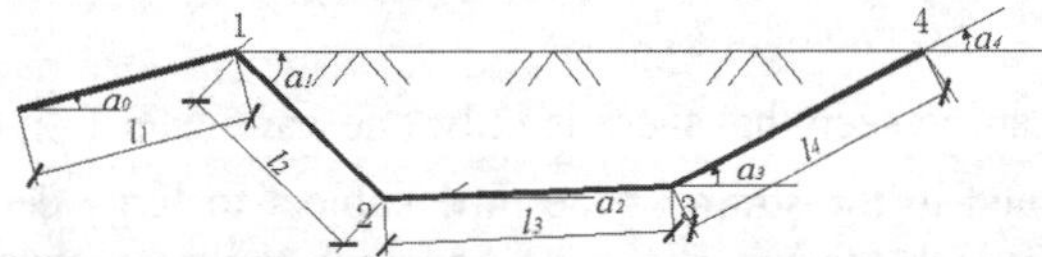

Figure 5. Profile of the bore-path

4.1 *Pullback forces due to weight of the pipe outside thebore-path*

The frictional force resulting from pipe weight outsidethe bore-path and the weight of the pipe create a component T_{ig} of the pulling force. At point '*i*' it is equal to:

$$T_{ig} = (w_p \mu_g \cos\alpha_0 + w_p \sin\alpha_0)(L - \sum_{k=1}^{i-1} L_k) \tag{11-1}$$

or

$$T_{ig} = C_0' \; (L - \sum_{k=1}^{i-1} L_k) \tag{11-2}$$

Where w_p is the weight of the pipe per unit length, μ_g is the friction coefficient between the pipe and the ground, L is the total length of the pipe, L_k are the lengths of segments of the bore-path and α_0 is the angle between the horizontal and the ground surface in front of the entry point.

4.2 *Pullback forces due to weight of the pipe inside the bore-path*

The component of the pulling force T_{is} results fromfriction due to the submerged weight of the pipe. At point '*i*' it is:

$$T_{is} = \sum_{k=1}^{i-1} (|L_k w \mu_b \cos\alpha_k| + L_k w \sin\alpha_k) \tag{12}$$

where w is the submerged weight of the pipe per unit length, μ_b is the friction coefficient between the pipe and the bore-path, α_k is the angle between the horizontal and the segment k and L_k is the length of this segment.

4.3 *Pullback forces due to drag of the drilling fluid*

The fluidic drag results in the following increase in the pulling forces T_{id}, at point '*i*':

$$T_{id} = f_d \sum_{k=1}^{i-1} L_k \tag{13}$$

where f_d is the force per unit length on the pipe dueto drag of the drilling fluid.

4.4 *Pullback forces due to large deformation*

Figure 3(a) shows a pipe segment (AB) that changes its direction from A to B. An equilibrating force exists and acts on the soil at the contact corner point C. The equation of equilibrium in the direction of AB is:

$$T_2 \cos\psi_0 = (T_1 + 2R\mu_b)\cos\psi_0 + 2P\mu_b + N_T\mu_b \tag{14}$$

Equilibrium in the direction normal to AB:

$$N_T = (T_1 + T_2)\sin\psi_0 \tag{15}$$

It can be derived for $\Delta T = T_2 - T_1$:

$$\Delta T = T_1\left(\frac{\cos\psi_0 + \mu_b \sin\psi_0}{\cos\psi_0 - \mu_b \sin\psi_0} - 1\right) + 4P\mu_b\left(\frac{1}{\cos\psi_0 - \mu_b \sin\psi_0}\right)$$

or $$DT = T_1' \, C_1(y_0) + P' \, C_2(y_0) \quad (16)$$

The axial force in the pipe at location '*j*' when the head is in location '*i*' is denoted as T_{ji}. The pulling force due to large deformation at point '*i*' is denoted as T_{if}. At the entry point, $T_{1f} = 0$. When the pipe reaches point 2, T_{2f} is due to the change of direction at point 1:

$$T_{2f} = T_{12}' \, C_1(y_{01}) + P_1' \, C_2(y_{01}) \quad (17)$$

where

$$T_{12} = (L - L_1)' \, C_0 \quad (18)$$

When the head of the pipe reaches point 3, the pulling force contains components due to large deformation at points1 and 2. For point 1:

$$T_{13f} = T_{13}' \, C_1(y_{01}) + P_1' \, C_2(y_{01}) \quad (19)$$

where

$$T_{13} = (L - L_1 - L_2)' \, C_0 \quad (20)$$

For point 2:

$$T_{23f} = T_{23}' \, C_1(y_{02}) + P_2' \, C_2(y_{02}) \quad (21)$$

where

$$T_{23} = T_{3g} + T_{2s} + T_{2d} + T_{13f} \quad (20)$$

For the pulling head at point 4 and further points, T_{jif} is calculated using the procedure analogous to the procedure for point 3.

5 COMPARISON AND DISCUSSION

5.1 *Strain responses due to large deformation*

It can be seen from Figure 1 that the pipeline was not restricted within a plane as most guidelines regarding HDD construction require. For each spatial position corresponding to the measuring point, the bending angle ($2\psi_0$) can be calculated by using the 3-D coordinates of the pipeline, and hence the strain responses are determined according to Figure 4. On the other hand, the strain measurements due to bending were obtained based on the records from the FBG sensors. Figure 6 presents the comparison of the analytic solutions and the measured strain responses. By and large, they reflect the similar tendency of the mechanical behavior of the pipeline during the whole HDD installation procedure. The larger strain responses due to bending can be observed at the location with the larger curvature (e.g. the location "A" and "B"). However, the strain responses from calculation and measurements display quantitative difference, which is largely due to three reasons: (1) measurement error exists in the gyroscope position-detection system; (2) the bending angle is determined based on the differential of a curve fitting by the spatial coordinates of the pipeline and inevitably leads to error amplification; (3) the proposed method is two dimensional whereas the profile of the actual pipeline is three dimensional.

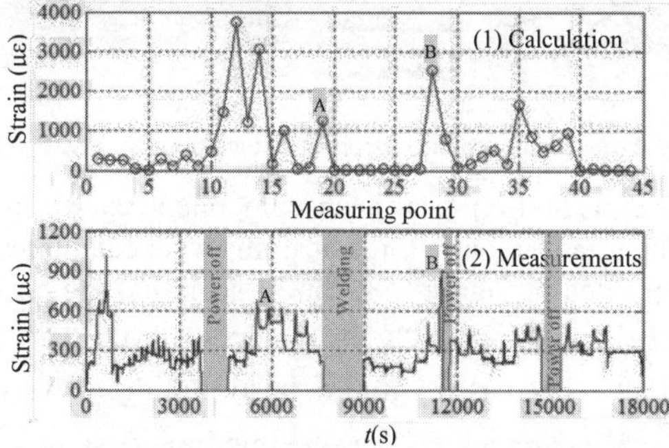

Figure 6. Strain responses of the pipeline due to bending

5.2 *Pullback force*

To calculate the pullback force of the pipeline during HDD construction, the bore-path profile in Figure 1 is approximated by fitting seven straight lines in a plane, as shown in Figure 7.

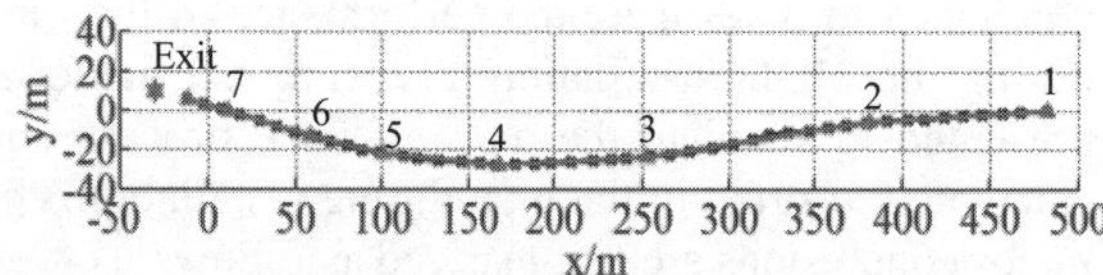

Figure 7. Profile of the bore-path

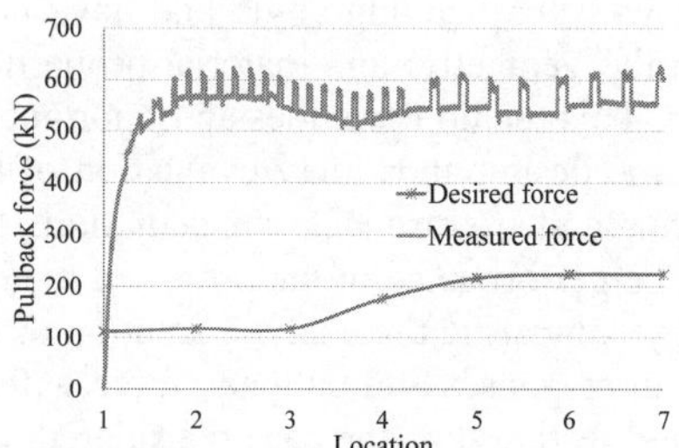

Figure 8. Measured and calculated pullback force

The desired pullback force at the given points is calculated based on Section 4 and displayed in Figure 8 by comparing with the measured forces. For safety, the pullback force applied in the HDD construction is much larger than the desired one. The safety factor is about 2.5.

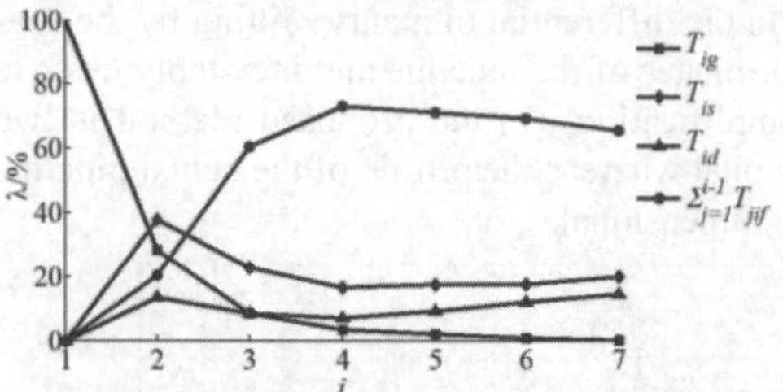

Figure 9. Proportion of each component of the pullback force

The proportion λ_i of each component contributed to the total pullback force at point '*i*' is calculated and presented in Figure 9. It can be seen that the proportion(l_{if}) of the component due to large deformation (T_{if}) increases with the head pulling forward and roughly maintain 65% at the middle-late stage of the construction, whereas l_{is} and l_{id} are around 20%.

6 CONCLUSIONS

Combined with an engineering project for placing an underground gas transmission pipeline crossing beneath a small in-town river, field tests have been conducted to investigate the performance of the pipe during the HDD installation. A theoretical model is presented to simulate the pipe pullback process and calculate the strain in the pipe during installation.The major conclusions are summarized as follows.

(1) The strain responses of the pipeline subjected to large bending deformation present an exponential relation with the bending angle and have nothing to do with the geometric and material properties of the pipeline. This relation provides an important criterion for scheme design and safety evaluation of the pipeline during and after trenchless installation.

(2) The performance of the proposed model is validated by comparing the analytic solution of the pipeline strain responses and pullback force with the data from field testing. For the case considered, the strain responses from calculation and measurements reflect the similar tendency of the mechanical behavior of the pipeline during the whole HDD installation procedure but present quantitative difference. For the pipeline which is not restricted within a plane during trenchless construction, a more sophisticated three dimensional model is desired.

ACKNOWLEDGEMENT

The authors are grateful to the Fok Ying-Tong Education Foundation, China (Grant No. 142004) and the State Key Laboratory of Disaster Reduction in Civil Engineering, China (SLDRCE14-B-19) for the support of this research.

REFERENCES

Ma Baosong and M. Najafi. 2008. Development and applications of trenchless technology in China, *Tunneling and Underground Space Technology*, Vol. 23,476-480

Michael E. Baumert and Erez N. Allouche. 2002. Methods for Estimating Pipe Pullback Loads for Horizontal Directional Drilling (HDD) Crossings, *Journal of infrastructure systems, ASCE*, Vol. 8(1), 12-19.

Cheng Elisabeth and Polak Maria Anna. 2007. Theoretical model for calculating pulling loads for pipes in horizontal directional drilling, *Tunneling and Underground Space Technology*, Vol. 22, 633-643

GB 50424-2007. *Code for construction of oil and gas transmission pipeline crossing engineering (in Chinese)*, National Standard of the People's Republic of China

GB 50268-2008. *Code for construction and acceptanceof water and sewerage pipeline works (in Chinese)*, National Standard of the People's Republic of China

Najafi, M. 2005. *Trenchless Technology-Pipeline and Utility Construction and Renewal*, McGraw-Hill, New York, USA

Polak Maria Anna and Lasheen Afdal. 2002. Mechanical modeling for pipes in horizontal directional drilling, *Tunneling and Underground Space Technology*, Vol. 16(1), S47-S55.

Suzhen LI. 2013. Construction Monitoring of a Municipal Gas Pipeline during Horizontal Directional Drilling, *Journal of Pipeline Systems Engineering and Practice,* Vol. 4(4), 04013005

Tongxi Yu and Liangchi Zhang.1996. *Plastic Bending: Theory and Applications*, World Scientific.

David A.Willoughby. 2005a. *Trenchless Technology Piping: Installation and Inspection*, McGraw-Hill, New York, USA

David A.Willoughby. 2005b. *Horizontal Directional Drilling: Utility and Pipeline Applications*, McGraw-Hill, New York, USA

C. J. Yang, W. D. Zhu, W. H. Zhang, X. H. Zhuand G. X. Ren. 2014. Determination of pipe pullback loads in horizontal directional drilling using an advanced computational dynamic model, *Journal of Engineering Mechanics, ASCE*, 04014060:1-12

Proceedings of the International Conference on Smart Infrastructure and Construction
ISBN 978-0-7277-6127-9

© The authors and ICE Publishing: All rights reserved, 2016
doi:10.1680/tfitsi.61279.505

Performance based design for the Crossrail Liverpool Street Station

H.L. Liew*, I. Farooq , Y.S. Hsu and A.S. O'Brien

Mott MacDonald, London, UK
* *Corresponding Author*

ABSTRACT Two out of three 42m deep shafts at Crossrail Liverpool Street Station required real time assessment of monitoring data and progressive modification to provide time and cost savings during construction. This paper describes this process which was based on the use of real time site monitoring data collected using ShapeAccelArray, inclinometers and piezometers. At the station's eastern end, Blomfield shaft involved an excavation into the Lambeth Group with a historically recorded groundwater pressure of more than 200kPa. Due to limited space, a depressurisation system with further contingency plans specified to deal with variations was devised. These contingency measures were required and progressively modified as the variability in the water pressures were observed as the excavation approached the final few stages. The designer worked collaboratively with the contractor and Crossrail in modifying the depressurisation design based on performance of each piezometer with additional site supports and monitoring points to complete the excavations just before the target date of Christmas 2014. On the western end, Moorgate shaft was delayed by 11 months prior to construction due to the prolonged time required to remove existing piles. To mitigate the delay, the designer pro-actively and progressively modified the design in response to observed behaviour, making use of real time monitoring data of the shaft and adjacent infrastructure as well as advanced numerical modelling. The outcome was that all previously specified temporary propping was omitted and construction stages for casting of ring beams were combined, leading to programme savings, which facilitated an early tunnel breakthrough.

1 INTRODUCTION

Deeper station excavations with limited footprints in an urban environment have become increasingly necessary as new tunnelling networks have to be dug deeper to avoid existing substructures. This creates a challenge for designers as excessive excavation-induced movements have to be designed out to meet external stakeholders concerns. Ideally, high quality site specific intrusive ground investigations should be available to characterize the ground and identify obstructions. However, they are often insufficient due to site constraints within a congested urban area. Therefore, the design produced tends to be conservative due to uncertainties in the ground conditions. Crossrail Liverpool street station is the deepest station on the Crossrail line, with two 42m deep shafts. Shaft wall displacements of less than 30 mm were needed to limit the risk of damage to adjacent structures. For excavations to depths of 42m this was less than 0.1 % of retained height which is usually the limit to economical design in the London environment. Wall movement limits below 0.1% retained height would have required significant and time consuming (to construct) propping systems.

The paper describes the use of real time monitoring data in the design of the two deep shafts at Crossrail Liverpool Street Station, i.e. Blomfield and Moorgate shafts. Allowance was made in the designs to progressively modify design as real time data became available, which provided significant time and cost savings whilst maintaining site safety, and minimizing the risk of damage to adjacent structures.

2 BLOMFIELD SHAFT

To the north, the Blomfield Shaft is bounded by the Circle, Hammersmith & City and Metropolitan Line, and to the south, it is bounded by 46 New Broad Street, Swedbank House and New Broad Street House (Figure 1).

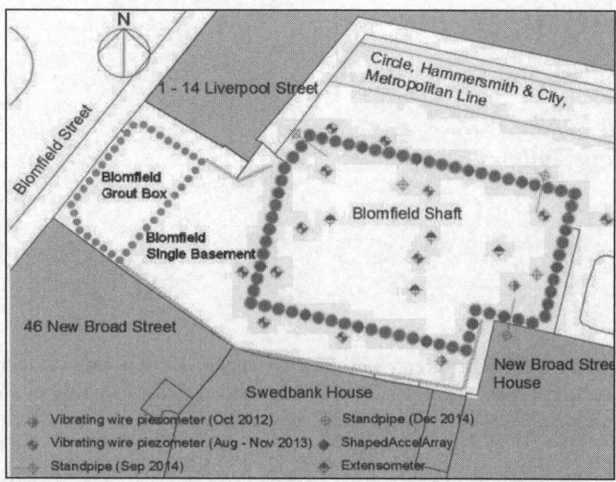

Figure 1. Blomfield Shaft site plan and instrument locations

The overall dimensions of the shaft are approximately 30m long and 20m wide. The perimeter of the shaft is formed of 1200mm diameter contiguous bored piles spaced at about 1.4m centre-to-centre (toe level at 61m ATD). Contiguous bored piles had to be used, rather than say diaphragm wall panels, because of the severe space restrictions. The top down excavation was supported by 6 levels of permanent reinforced concrete beams and walers as indicated in Figure 2. SCL tunneling took place less than 5m from the shaft, and these works included depressurization of the Lambeth Group sand/silt layers.

2.1 *Ground conditions at Blomfield Shaft*

The ground level at Blomfield Shaft location is approximately 112.5m ATD. The ground conditions consist of about 6.5m of superficial deposits (2.5m of Made Ground and 4m of Terrace Deposits), overlying London Clay, which extends from 106m ATD to about 76m ATD. The Harwich Formation is present at the base of the London Clay and is found to be highly variable. The underlying Lambeth Group is about 18m thick and overlies the Thanet Sands. Groundwater level is approximately 106m ATD, i.e. about 7m below street level. The clays are underdrained due to excessive pumping in the Lower Aquifer but during recent decades reduced pumping has led to a rise in ground water pressures (Figure 2).

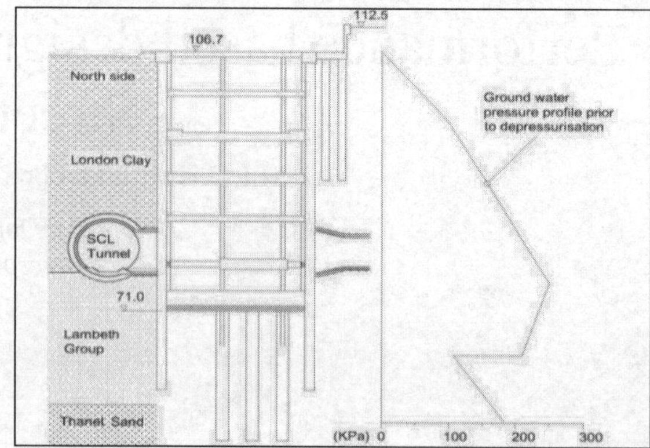

Figure 2. Blomfield Shaft north south section and ground water pressure profile

There were several significant sand-filled channels or sand/silt layers within the Lambeth Group which were hydraulically interconnected and pore water pressures up to 250kPa were measured in these layers. Hence, there was a risk of excavation instability depending on the size of the channels; their permeability; and their ability to recharge. Figure 3 shows the complexity and variability of the sand/silt layers in the Lambeth Group based on site specific ground investigation and careful logging of the exposed soils during construction. A fault zone was also anticipated to be present across the north-eastern corner of the shaft. This fault complicated the assessment of the hydrogeological regime, and likely changes in pore water pressures during excavations. There was also some uncertainty associated with the rise in groundwater pressures in sand channels, once the tunnel depressurization was switched off (which occurred part way through the shaft excavation).

2.2 *Key challenges*

The shaft was constructed top-down and the depth of excavation is approximately 42 m below street level. Although the excavation was predominantly within the London Clay, the critical challenge was excavating through 5m of the Lambeth Group to cast the base slab. To control the risks associated with high ground water pressures and permeable silt/sand layers, the excavation would need to be heavily propped and this would be counter-productive as the time required to install a prop itself may allow the clay to soften (due to water flow from adjacent sand/silt layers) especially when complex temporary propping is required. The contractor would need to treat the gaps in between piles when these permeable layers were encountered. A depressurisation system on both the retained and excavation sides to make the wall and prop forces more manageable whilst mitigating the ground water risks was also necessary. Installation of all the depressurisation wells in one stage, in particular the wells on the excavation side can also be counter-productive as they could become an obstruction and slow down the excavation process. Hence, a phased approach was adopted for the installation of the wells and the monitoring points with pre-specified trigger levels to ensure safe excavation.

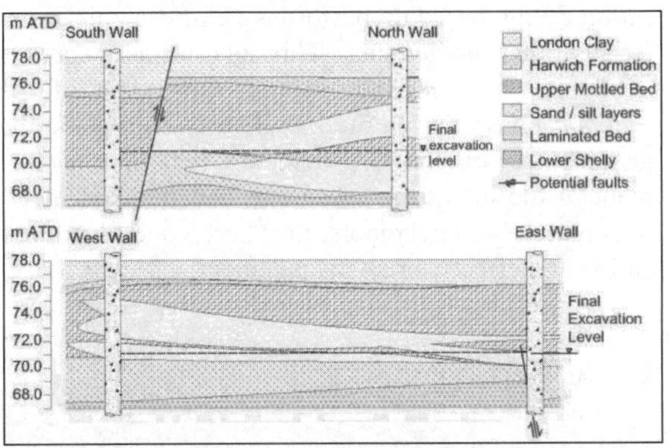

Figure 3. Blomfield shaft – sand/silt layers within the Lambeth Group

2.3 *Instrumentation and Monitoring*

The primary monitoring system (Figure 1) included 14 fast response piezometers and 5 open standpipes to monitor the performance of the depressurisation system, and 13 ShapedAccelArrays (SAAs) to monitor the deformation of the shaft. To monitor ground heave in the base of the excavation, 4 magnetic extensometer were installed.

To enable faster excavation, the installation of the piezometers was carried out in two different phases. The first phase involved 2 vertical vibrating wire piezometers targeting the sandy layers within the Lambeth Group. Apart from measuring ground water pressures, the purposes of these piezometers were also to gain a better understanding of the extent and nature of the sand/silt layers. They were rotary cored and carefully logged. This information was essential for the depressurisation design and also provided a set of base line readings for the piezometers which had not been influenced by adjacent dewatering/depressurization work carried out by the adjacent tunnelling contractor.

The second phase included 12 vibrating wire piezometers, and a further 5 open standpipes were installed when excavation approached the Lambeth Group. Three of the standpipes were installed behind the wall, via inclined drilling through the gaps in between piles when the excavation level reached 83m ATD (piezometers could not be installed behind the wall due to severe space constraints). The remaining two standpipes were installed when the excavation level reached 73m ATD. These additional standpipes were installed to improve the understanding of the distribution of ground water pressures in the Lambeth Group in order to confirm if the depressurisation system was adequate. The piezometers were checked for anomalies and the piezometer readings together with the other instrument readings were assessed on a daily basis by the design team. Full time design team input was essential to allow for rapid assessment of the monitoring data against a prescribed set of trigger levels.

2.4 *Progressive design modification*

The installation of depressurisation wells was carried out in two different phases: Phase I – 4 deep wells; and Phase II – 9 inclined and 2 vertical ejector wells. The purpose of the wells installed during the first phase (installed at 107m ATD) was to gain a better understanding of the Lambeth Group hydrogeology in terms of permeability and connectivity of the sandy and silty layers. The second phase wells were installed when excavation was at approximately 83m ATD, i.e. 7m above the Lambeth Group. A third phase of wells was envisaged. However, to provide cost and programme savings it was agreed that the third phase wells would be installed in response to monitoring data as excavation proceeded into the Lambeth Group. The contractor arranged for the necessary equipment to be available for quick deployment if monitoring data indicated that a trigger limit may be breached.

When excavation approached the Lambeth Group, i.e. at about 76m ATD, the trends of the piezometer readings inside the shaft (on the east side) indicated that the trigger levels were likely to be breached if no further actions were taken to reduce the ground water pressure in the Lambeth Group. It was considered that another 6 passive wells (capable of being converted into active wells) were required (Phase IIIa). However, this would have seriously delayed the construction of the shaft. Hence, the next stage was modified by dividing the excavation area into two zones. This was to allow for 3 wells to be installed in each zone from the current level whist allowing excavation to be continued in the other zone. This staged approach allowed the well installation to be carried out in parallel to shaft excavation.

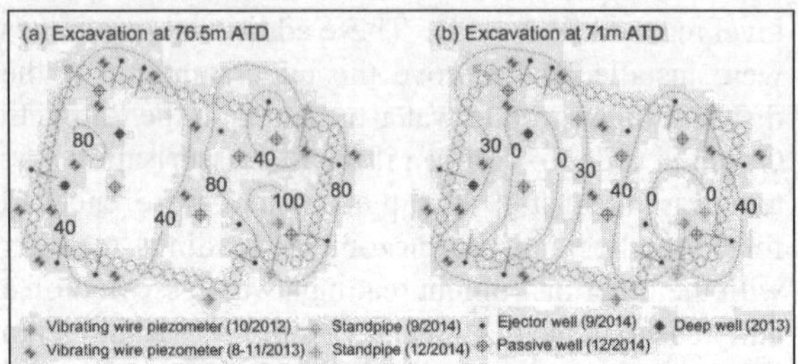

Figure 4. Blomfield Shaft – ground water pressure contours (kPa) before (a) and after (b) design modifications

When excavation approached the final stage, i.e. from 73m ATD to 71m ATD, further depressurisation measures (Phase IIIb) were considered necessary to allow for safe excavation. Additional trial holes to confirm ground water levels, and slit trenches down to 70m ATD with local sump pumps were installed and activated. Once the excavation had reached 71m ATD, a 300mm thick blinding slab was cast. To prevent structural failure of the blinding during the prolonged Christmas break, passive wells were extended to above the blinding and additional weep holes were created to relieve any excess pressures underneath the blinding slab. Figure 4 shows the ground water pressures before (excavation level at 76.5m ATD) and after (excavation level at 71m ATD) during the design modifications. Figure 5 shows the predicted and observed movements in one of the walls.

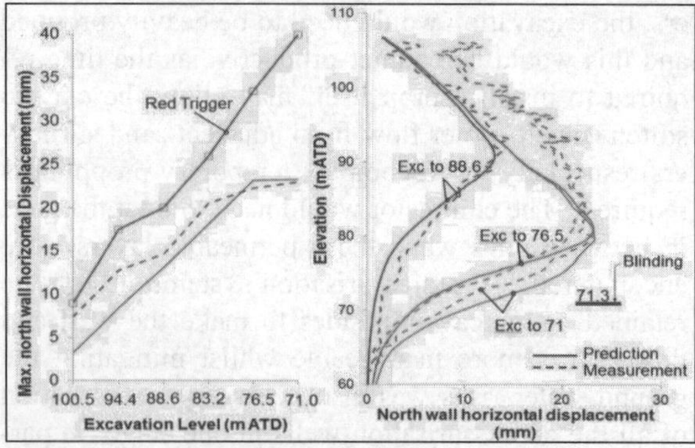

Figure 5. Blomfield Shaft wall displacements

The success of the implementation of these design changes in such a short time period was mainly due to the on-site presence of the designer, who worked collaboratively with the contractor and the client in modifying the construction sequence and depressurisation design based on performance of each piezometer and other instruments. Throughout this process, a geologist observed and mapped the extent of the sand/silt layers within the Lambeth Group whilst the geotechnical engineers provided a real time assessment of the monitoring data and had frequent meetings with key stakeholders, to ensure good communication and effective risk management.

3 MOORGATE SHAFT

Moorgate shaft is in close proximity to the Metropolitan Line to the North, the Northern Line tunnels to the east, listed buildings to the south and the existing Moorgate station ticket hall to the west (Figure 6).

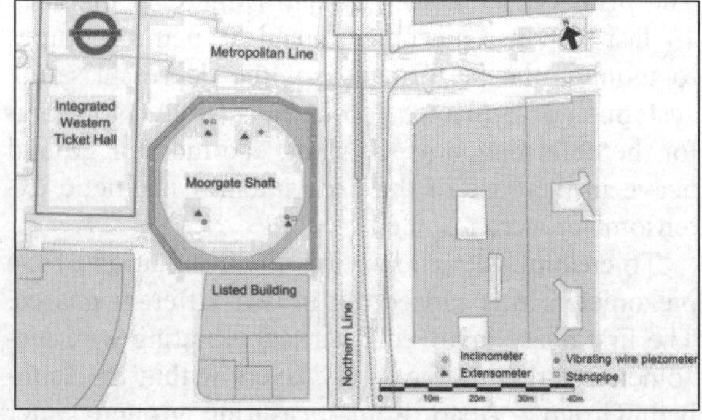

Figure 6. Moorgate Shaft site plan and instrument locations

The shaft is about 35m×35m in plan and extends approximately 42m below street level. The perimeter of the irregular shaped shaft is formed by a 1.2m thick diaphragm wall. The shaft was constructed top down and it was initially designed to be supported by 7-levels of reinforced concrete ring beams and 2-levels of temporary steel propping. With the Northern Line less than 5m from the shaft's east wall, two temporary cross walls (1.2m thick unreinforced panels) and a pair of slab strips at 80.5m ATD (3m wide by 1.5m deep) were included in the original shaft design; spanning in the east-west direction to limit wall deformation and protect nearby assets.

3.1 Ground conditions at Moorgate Shaft

The ground level at the Moorgate Shaft location is approximately 113m ATD, and the top of the diaphragm wall is between 106.5m ATD to 109.9m ATD. Figure 7 shows a typical geological section.

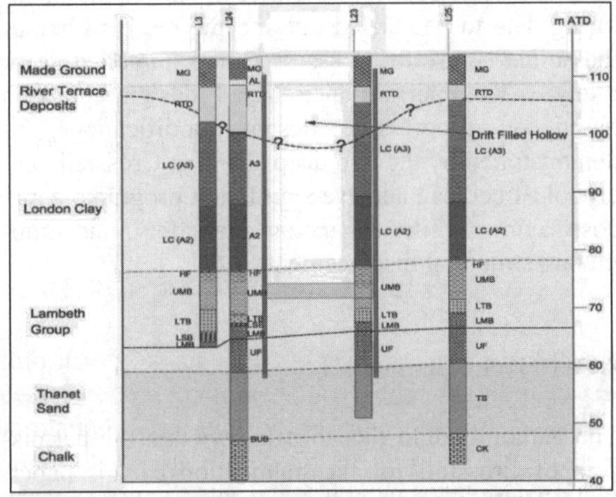

Figure 7. Moorgate Shaft geological section (north-south)

The ground conditions comprise Made Ground, Terrace Deposits, London Clay and Lambeth Group. The underlying Lambeth Group is 16m to 18m thick, and overlies the Thanet Sands. There were no significant sand layers identified in the Lambeth group. The large variation in thickness of Terrace Deposits (between 3m and 13m thick) and corresponding variation in London Clay thickness is due to the presence of a drift filled hollow, which is potentially associated with the course of the former River Walbrook and tributaries. This feature also raised some concerns about local variations in the London Clay properties.

3.2 Key challenges

The shaft was constructed top-down within the basement of a demolished building. The existing piles from the building (within the footprint of the shaft) were required to be removed prior to the installation of diaphragm walls. However, the removal of the piles took longer than anticipated and the shaft programme was delayed by 11 months. By the time the diaphragm walling was completed, meeting the project milestone for the shaft handover to the SCL tunnel contractor appeared unlikely.

3.3 Instrumentation and monitoring

To mitigate the delay, Mott MacDonald proposed an observation based verification process that made use of real time monitoring data for potential programme-saving measures including omission of temporary propping and combination of excavation stages for the construction of ring beams (Farooq et al, 2015). To ensure the success of the use of the verification process, a comprehensive monitoring strategy to monitor performance of the shaft was essential.

The primary instruments included 12 in-place inclinometers (IPI) and one SAA to monitor wall displacements, in particular adjacent to the Northern Line. Empty inclinometer tubes were also installed adjacent to each inclinometer to provide redundancy in the monitoring scheme. To verify inclinometer readings and to increase confidence in the measured wall displacements, mini-prism survey points were installed. Laser extensometer survey points were also installed to monitor convergence of the walls. Other instruments included 4 magnetic extensometers, 4 fast response vibrating wire piezometers and 2 Casagrande standpipes as indicated in Figure 6.

When the verification process commenced, the readings from both automated and manually read instruments were reviewed daily and anomalies were isolated. The potential influence on the monitoring data due to ambient temperature and other construction activities were considered. The processed data was then checked against a pre-agreed set of wall displacement trigger levels. Wall movements were predicted from FLAC 3D analyses, which utilized a bespoke non-linear small strain stiffness model, A* (Eadington & O'Brien, 2011). If the predictions differed from observations then the reasons for the discrepancies were sought and the FLAC model was modified if deemed appropriate. Hence the ground and structural model was refined progressively during the construction process. These refinements included (based on detailed site records and surveys): adjacent tunnelling and grouting activities, geological and topographical features, and stiffness of ring beams and cross wall. This process was repeated as excavation progressed. The calibrated model was then used to predict future movements and the implications of prop omission could be quantified.

3.4 *Progressive design modification*

Figure 8 shows the predicted and observed movements in two of the walls. The predicted results from the calibrated model were slightly larger than the measured movements but exhibited a similar overall displaced shape.

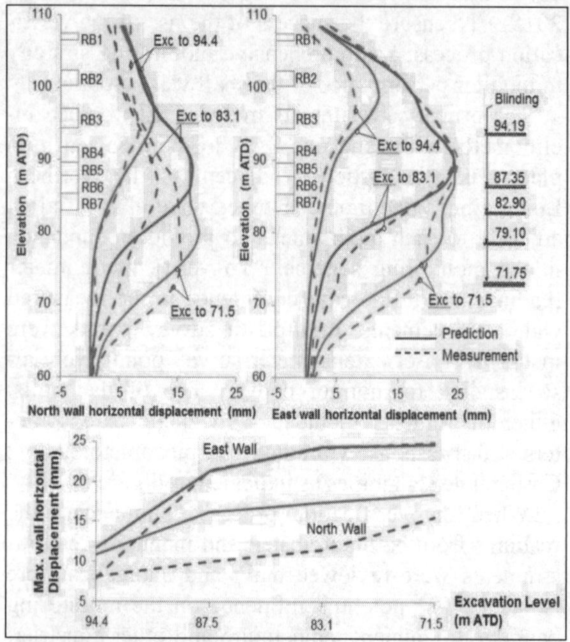

Figure 8. Moorgate Shaft wall displacements

Blinding layers (200mm thick) were cast at 4 levels (Figure 8) after the excavation level reached 94m ATD to minimize clay softening. At 71.5m ATD, a blinding slab (300mm thick) was cast to both minimize clay softening and to provide structural support.

Given the good match between the predictions and the observed shaft behaviour together with the effectiveness of the blinding slabs, all temporary steel propping including the slab strips were omitted and the number of construction stages for the lowest 4 ring beams were halved. This enabled the construction programme to be significantly accelerated. The shaft's base slab was completed ahead of programme, the shaft handover date to the SCL contractor was achieved and the potential 11 month delay was overcome.

4 CONCLUSIONS

Undertaking design modifications as excavation progresses, whilst ensuring full design assurance, requires a full collaboration between the designer, contractor and client. On-site presence by the designer is essential to enable real time assessment of the monitoring data to be carried out and the performance of the soil/structure to be assessed and modelled accurately. This performance based design approach based on progressive design modifications, as demonstrated by the two deep shafts at Crossrail Liverpool Street, can achieve significant programme and cost savings whilst enhancing site safety, and minimizing structural displacements.

ACKNOWLEDGEMENT

The authors would like to acknowledge the permission of Crossrail for the publication of this paper. They would also thank their colleagues in the Crossrail project team who contributed to the successful outcome of this project.

REFERENCES

Eadington, J. & O'Brien, A.S. 2011. Stiffness parameters for a deep tunnel – developing a robust parameter selection framework, *Proceedings of the 15th ECSMGE*, 531 – 536.

Farooq, I. Place, D. Steele, B. et al. 2015. Verification process speeds Crossrail's Moorgate Shaft, *Proc. Instn Civ. Engrs Civ. Engng*. Under review.

Powderham, A.J. 1998. The observation method – application through progressive modification, *Proceedings of Journal ASCE/BSCE* **13**, 87–110.

Proceedings of the International Conference on Smart Infrastructure and Construction
ISBN 978-0-7277-6127-9

© The authors and ICE Publishing: All rights reserved, 2016
doi:10.1680/tfitsi.61279.511

Bayesian updating of subsurface spatial correlation through monitoring of infrastructure and building developments

M.K. Lo and Y.F. Leung*

The Hong Kong Polytechnic University, Hong Kong
**Corresponding Author*

ABSTRACT As land space becomes increasingly sparse in urban areas, development of underground infrastructure becomes more important in major cities around the world. In these cases, uncertainties in subsurface ground conditions account for a significant proportion of construction problems and infrastructure underperformance. Recent advances in new sensing techniques open up opportunities to establish a holistic framework that duly considers spatial uncertainties of the subsurface environment, meanwhile processes sensing data to continuously update the predictions on system performance during construction and service life. Such potentials in promoting flexible design, 'observational method' in construction, and life cycle assessment of infrastructure have not yet been fully realised. This study introduces a preliminary framework for such purposes, by incorporating the Bayesian approach into recent findings of spatial correlations in geological profiles and geotechnical properties. The variables representing the subsurface spatial structure can be updated using the monitoring data obtained at various construction stages, through inverse analyses coupled with conditional random field simulations. This turns the subsurface environment into a self-aware system fed by multi-source sensing information, where the risks and uncertainties are quantified and gradually reduced as the construction progresses. The updating process can continue as long as the sensors remain active, throughout the life cycle of the infrastructure where disturbances may occur due to subsequent construction activities nearby. The proposed framework is illustrated with a case study of residual building development in London, and it can be extended to different infrastructure development pertaining to different urban settings.

1 INTRODUCTION

Field monitoring constitutes the backbone of proper construction control and safe operation of the infrastructure during its service life. While detailed planning and implementation of field monitoring are becoming a standard practice for major infrastructure projects, the interpretation of instrumentation data and the feedback mechanism to facilitate potential design revision is often incomplete. For example, the tragic failures at Heathrow Terminal 5 (1994) and Nicoll Highway, Singapore (2004) highlight the importance of timely review of monitoring data and proper interpretation when actual site conditions deviate from original design assumptions. This is particularly important in the field of geotechnical engineering as uncertainties inevitably exist regarding various geotechnical parameters, as discussed by Phoon & Kulhawy (1999).

The Bayesian updating approach provides an analytical framework to incorporate new data into revision of previous estimates. While the engineering profession is becoming familiar with these approaches (e.g. Wu 2011), a key component of geotechnical uncertainty, which is related to the spatial variability of geotechnical parameters, is usually omitted in discussions of the Bayesian approach. This paper presents the Bayesian framework that incorporates various sources of uncertainty, including spatial variability of soil parameters, and uncertainties related to model and observation errors. The approach is applied to a piled foundation case history in London, where predictions on foundation response and their corresponding uncertainties are quantified and continuously updated throughout the construction. This provides engineers and developers with the latest risk assessments as the construction progresses.

2 METHODOLOGY

The proposed approach aims to utilize monitoring data acquired during early stages of construction to enhance the knowledge in spatial correlations of underground soil properties. The approach includes modelling of the subsurface space as a conditional random field, coupled with the Bayesian approach to update associated spatial parameters to facilitate improved predictions of future infrastructure response.

2.1 *Soil media as conditional random field*

The general linear model for the spatial variability of soil properties can be represented by:

$$\boldsymbol{Z} = \mathbf{X}\boldsymbol{\beta} + \boldsymbol{e_1}$$
$$\boldsymbol{Z_{obs}} = \boldsymbol{Z} + \boldsymbol{e_2} \tag{1}$$

where $\boldsymbol{Z}$ and $\boldsymbol{Z_{obs}}$ represents the true and observed soil properties. $\mathbf{X}\boldsymbol{\beta}$ is the large scale trend, with $\mathbf{X}$ being the design matrix (containing information on spatial coordinates) and $\boldsymbol{\beta}$ the vector of trend coefficients. The residual $\boldsymbol{e_1}$ is the combination of smooth scale variation with variance σ_e^2 and correlation structure $\mathbf{C}$; and a natural nugget effect with variance σ_n^2. The residual $\boldsymbol{e_2}$ represents the measurement error during soil sampling process with variance σ_0^2. Therefore, the covariance matrix of $\boldsymbol{Z}$ and $\boldsymbol{Z_{obs}}$ are:

$$\mathrm{Cov}(\boldsymbol{Z}) = (\sigma_e^2 + \sigma_n^2)[s_1\mathbf{C} + (1 - s_1)\mathbf{I}]$$
$$\mathrm{Cov}(\boldsymbol{Z_{obs}}) = (\sigma_e^2 + \sigma_n^2 + \sigma_0^2)[s_2\mathbf{C} + (1 - s_2)\mathbf{I}]$$
$$\text{where } 0 \le s_1 = \sigma_e^2/(\sigma_e^2 + \sigma_n^2) \le 1$$
$$\text{and } 0 \le s_2 = \sigma_e^2/(\sigma_e^2 + \sigma_n^2 + \sigma_0^2) \le 1 \tag{2}$$

Typically, the spatial correlation structure, i.e. individual components of $\mathbf{C}$, is assumed to follow a certain functional form. In the current study, the Matérn function is adopted as a flexible model for $\mathbf{C}$ (Minasny & McBratney 2005):

$$\mathbf{C}(h_{ij}) = \frac{1}{2^{v-1}\Gamma(v)}\left(\frac{h_{ij}}{r}\right)^{v} K_v\left(\frac{h_{ij}}{r}\right) \tag{3}$$

where h_{ij} is the separation distance between sample points i and j, Γ is the gamma function, r is the range parameter, v is a smoothness parameter which ranges from 0 to infinity, and K_v is the modified Bessel function of the second kind with order v. Matérn function is a generalized function with its shape controlled by the smoothness parameter v. For example, it corresponds to the exponential function when $v = 0.5$, and is equivalent to the Gaussian function when v approaches infinity. The effective range r_{eff} is defined as the separation distance where the covariance of $\boldsymbol{Z}$ drops to $0.05(\sigma_e^2 + \sigma_n^2)$, i.e. beyond which the correlation between two locations becomes insignificant.

When soil samples $\boldsymbol{Z_{obs}}$ are obtained, the soil properties $\boldsymbol{Z_0}$ at unobserved locations $\mathbf{X_0}$ becomes conditional on $\boldsymbol{Z_{obs}}$, which means the soil properties near the sampling locations should be similar to $\boldsymbol{Z_{obs}}$. Assuming $\boldsymbol{Z_0}$ and $\boldsymbol{Z_{obs}}$ are both multivariate Gaussian, the conditional mean and conditional covariance of $\boldsymbol{Z_0}$ can be written as:

$$\mathrm{E}(\boldsymbol{Z_0}|\boldsymbol{Z_{obs}}) = \mathbf{X_0}\boldsymbol{\beta} + \boldsymbol{k}^T\mathbf{K}^{-1}(\boldsymbol{Z_{obs}} - \mathbf{X_{obs}}\boldsymbol{\beta})$$
$$\mathrm{Cov}(\boldsymbol{Z_0}|\boldsymbol{Z_{obs}}) = \mathbf{K_0} - \boldsymbol{k}^T\mathbf{K}^{-1}\boldsymbol{k} \tag{4}$$

where $\boldsymbol{k} = \mathrm{Cov}(\boldsymbol{Z_0}, \boldsymbol{Z_{obs}})$, $\mathbf{K} = \mathrm{Cov}(\boldsymbol{Z_{obs}})$, and $\mathbf{K_0} = \mathrm{Cov}(\boldsymbol{Z_0})$. The conditional mean and conditional covariance can be applied directly for the Monte Carlo Simulation of conditional random field.

2.2 *Bayesian updating of spatial variability*

Considering uncertainties associated with modelling and monitoring, the geotechnical model can be expressed as:

$$\boldsymbol{y} = g(\boldsymbol{\theta}) + \boldsymbol{\varepsilon} \tag{5}$$

where $\boldsymbol{y}$ is a $N \times 1$ vector representing the system performance, which may be observed through N various sensors. g represents the geotechnical model with spatial variability parameters $\boldsymbol{\theta}$, a $M \times 1$ vector (e.g. $\boldsymbol{\theta} = \{s_2, v, r\}$ for Matérn function). $\boldsymbol{\varepsilon}$ represents the uncertainties other than spatial variability, such as model errors arising from simplifying assumptions, and measurement errors of the sensors.

The uncertainty associated with the spatial variability parameters $\boldsymbol{\theta}$ can be quantified as a prior distribution, which is a multivariate Gaussian function $f(\boldsymbol{\theta})$, with $\boldsymbol{\mu_\theta}$ and $\mathbf{C_\theta}$ being the prior mean (initial guess) and prior covariance (initial uncertainty) of $\boldsymbol{\theta}$. Before construction, when only soil sample data is available, restricted maximum likelihood (REML) method can be applied to obtain $\boldsymbol{\mu_\theta}$ (Minasny & McBratney, 2005; Liu et al. 2016). $\mathbf{C_\theta}$ can be either evaluated explicitly or be assigned a reasonable value. In theory, the choice of prior distribution does not have significant impact of the Bayesian approach, as long as there are sufficient updating stages.

As the subsurface domain is modelled as a random field due to spatial variability, the prediction of $\boldsymbol{y}$ becomes a random vector. Assuming $\boldsymbol{y}$ follows a multivariate Gaussian distribution for a given set of spatial variability parameters $\boldsymbol{\theta}$, the likelihood function of the geotechnical model can be formulated as:

$$f(\boldsymbol{y}|\boldsymbol{\theta}) = \frac{1}{\sqrt{(2\pi)^{\mathrm{N}}|\mathbf{V}(\boldsymbol{\theta}) + \mathbf{E}|}} \exp\left\{-\frac{1}{2}[\boldsymbol{\theta} - \boldsymbol{\mu}(\boldsymbol{\theta})]^{\mathrm{T}}[\mathbf{V}(\boldsymbol{\theta}) + \mathbf{E}]^{-1}[\boldsymbol{\theta} - \mu(\boldsymbol{\theta})]\right\} \tag{6}$$

where $\boldsymbol{\mu}(\boldsymbol{\theta})$ and $\mathbf{V}(\boldsymbol{\theta})$ are the mean vector and covariance matrix of the geotechnical model $g(\boldsymbol{\theta})$, and $\mathbf{E}$ is the covariance matrix of $\boldsymbol{\varepsilon}$.

Once the observation $\boldsymbol{y}$ is obtained from monitoring, the distribution of $\boldsymbol{\theta}$ given a fixed $\boldsymbol{y}$ (the posterior distribution) can be derived using Bayes theorem:

$$f(\boldsymbol{\theta}|\boldsymbol{y}) = kf(\boldsymbol{y}|\boldsymbol{\theta})f(\boldsymbol{\theta}) \tag{7}$$

with k being an unknown normalizing constant. An objective function can be formed with the constant term dropped:

$$\begin{aligned} \mathrm{S}(\boldsymbol{\theta}) &= -2\ln f(\boldsymbol{\theta}|\boldsymbol{y}) \\ &= (\boldsymbol{\theta} - \boldsymbol{\mu_\theta})^{\mathrm{T}}\mathbf{C_\theta}^{-1}(\boldsymbol{\theta} - \boldsymbol{\mu_\theta}) + \ln|\mathbf{V}(\boldsymbol{\theta}) + \mathbf{E}| \\ &+ (\mu(\boldsymbol{\theta}) - \boldsymbol{y})^{\mathrm{T}}(\mathbf{V}(\boldsymbol{\theta}) + \mathbf{E})^{-1}(\mu(\boldsymbol{\theta}) - \boldsymbol{y}) \end{aligned} \tag{8}$$

$\mathrm{S}(\boldsymbol{\theta})$ can be minimized with respect to $\boldsymbol{\theta}$ to obtain $\boldsymbol{\theta_{MAP}}$, which is the maximum-a-posteriori estimate of $\boldsymbol{\theta}$. $\boldsymbol{\theta_{MAP}}$ then becomes the prior mean $\boldsymbol{\mu_\theta}$ for the next updating stage. The posterior covariance $\mathbf{C_{\theta|y}}$ can be evaluated by the Laplace's approximation:

$$\mathbf{C_{\theta|y}}^{-1} \approx \frac{1}{2}\frac{\partial^2}{\partial\theta_i\partial\theta_j}\mathrm{S}(\boldsymbol{\theta})\bigg|_{\theta=\theta_{MAP}} \tag{9}$$

with the partial derivative calculated by finite difference method. $\mathbf{C_{\theta|y}}$ then becomes the prior covariance $\mathbf{C_\theta}$ for the next updating stage.

2.3 *Conditional random field simulation, using Latin Hypercube Sampling with dependence*

Minimization process of $\mathrm{S}(\boldsymbol{\theta})$ involves repeated evaluations of $\boldsymbol{\mu}(\boldsymbol{\theta})$ and $\mathbf{V}(\boldsymbol{\theta})$. For each set of spatial variability parameters $\boldsymbol{\theta}$, hundreds of soil profiles can be generated by Monte Carlo Simulation to calculate the empirical mean $\hat{\boldsymbol{\mu}}(\boldsymbol{\theta})$ and empirical covariance $\hat{\mathbf{V}}(\boldsymbol{\theta})$. To reduce the number of simulations required, and to improve the accuracy of the Monte Carlo Estimator, Latin Hypercube Sampling with dependence (LHSD) is adopted. LHSD is essentially a multidimensional stratified sampling scheme, which ensures a uniform placement of random realizations in a multidimensional cube, while maintaining the correlation structure (Packham 2015). The conditional covariance matrix $\mathrm{Cov}(\boldsymbol{Z}|\boldsymbol{Z_{obs}})$ can be coupled with LHSD to effectively simulate conditional random field.

2.4 *Predictions with parameter uncertainty*

After updating spatial variability parameters $\boldsymbol{\theta}$ at each stage, predictions at the remaining stages could be performed based on the posterior distribution of $\boldsymbol{\theta}$, as shown in the following steps:

(1) Assume the posterior is multivariate Gaussian with mean $\boldsymbol{\theta_{MAP}}$ and covariance $\mathbf{C_{\theta|y}}$, sample 2^M sets of $\boldsymbol{\theta}$ by allowing two choices for each element in $\boldsymbol{\theta}$: $\theta_{MAP,i} \pm 1.96\sqrt{[\mathbf{C_{\theta|y}}]_{ii}}$.
(2) For each set of $\boldsymbol{\theta}$, simulate 100 soil profiles using LHSD, and evaluate the model output.
(3) From the 100×2^M model outputs, calculate the empirical mean as the predictor, and the 5th and 95th percentiles as the lower and upper bounds.

3 CASE STUDY

The proposed Bayesian updating approach for spatial correlation can be applied to various underground constructions, such as deep excavations or tunnelling projects. In this paper, the approach is illustrated using the data from various stages during the construction of a foundation project in central London, UK.

3.1 *Hyde Park Cavalry Barracks*

The Hyde Park Cavalry Barracks (HPCB) tower is a 90-m tall residual building at Knightsbridge, London. The tower is founded on a piled raft foundation, with a 1.52-m raft supported by 51 piles. The subsurface soil profile consists of fill soil, followed by a 58 m of London Clay overlying the Lambeth Group, which is in turn underlain by a thin layer of Thanet sand and Chalk bedrock. The groundwater level was approximately 4 m below the ground surface. The details of the superstructure and foundation have been described by Hooper (1973), and the modelling of the foundation has also been reported in Leung (2010).

Hooper (1973) presented undrained shear strength (c_u) data of the London Clay at the site, which is reproduced in Figure 1. It should be noted that the dataset was compiled using soil samples obtained from different boreholes across the site, but the exact locations of these boreholes were not reported. Therefore, in the current study, only the spatial variability of c_u in the vertical direction, and its influence on the inverse analyses, are investigated.

The piled raft foundation at HPCB was instrumented with settlement markers, pile load cells and earth pressure cells, which provide valuable data for the proposed inverse analyses and Bayesian updating approach. Figure 2 shows the locations of the instruments that are considered in the current study.

3.2 *Foundation modelling approach*

Substantial inverse analyses need to be performed for the proposed methodology, and an objective function (with efficient foundation analysis approach) needs to be defined. In the current study, the piled raft foundation is analysed using the approach described in detail by Leung (2010). Essentially, the raft is discretised into 4-node elements modelled using the finite element method, while the piles are represented by nodes and 0.5-m elements. The individual pile response is simulated using nonlinear load transfer relationships with the soil modulus varying with depth. Pile-to-pile interactions are modelled using the elastic solutions by Mindlin (1936), while for non-homogeneous soils, the average soil modulus representative of the two interacting pile elements is adopted as an approximation. The superstructure details are also available in Hooper (1973), and its stiffness is modelled in the current study using the matrix condensation method (Leung 2010).

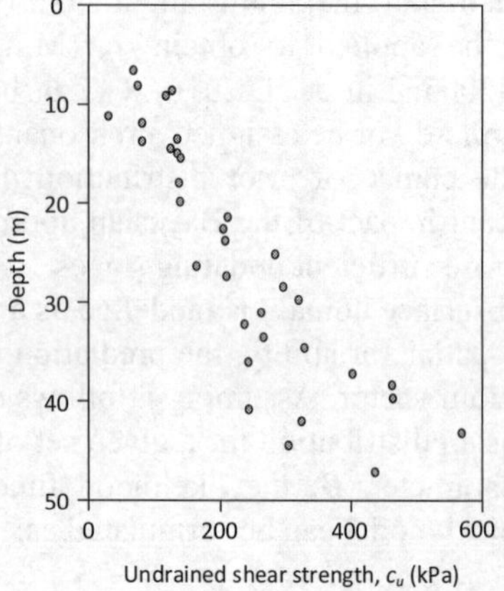

Figure 1. Soil strength data at HPCB (after Hooper 1973).

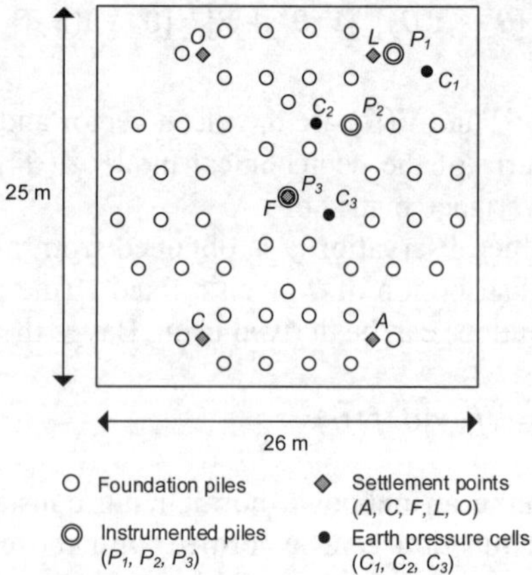

Figure 2. Instrumentation at HPCB tower (after Hooper 1973).

Apart from the c_u profile shown in Figure 1, in the current study, the initial shear modulus (G_0) of the London Clay is assumed to follow the G_0-c_u relationship proposed by Vardanega & Bolton (2011):

$$G_0 = 320.7c_u \tag{10}$$

In other words, the variation in G_0 is proportional to that of c_u in this study. For a different project with more test data of G_0, its variations can be used directly in the Bayesian approach. Since the 'short-term' behaviour during construction is the main concern, the Poisson's ratio of 0.45 is adopted for numerical modelling of undrained soil behaviour. Based on the monitoring data reported by Hooper (1973), four construction/development stages were selected to demonstrate the Bayesian updating procedures. They correspond to construction of the building up to the 3rd storey (Stage 1), 28th storey (Stage 2), structure completed (Stage 3) and building occupied (Stage 4).

3.3 Results

Before construction started (Stage 0), the spatial correlation features are evaluated with REML using available soil data (Figure 1). The associated parameters, s_2 and r_{eff}, are continuously updated making use of instrumentation data. Table 1 shows the changes in spatial correlation parameters at each updating stage, while Figure 3 shows the probability distribution of $\boldsymbol{\theta}$ at Stages 0 and 4 which illustrates the updating process. Both also show the reduction in variances of the parameters, representing improved knowledge in the correlation structure of the soil properties.

Table 1. Updating of spatial correlation parameters.

Stage	$\{s_2, r_{\text{eff}}\}$	$Var(s_2)$	$Var(r_{\text{eff}})$
0	{0.50, 1.30}	0.0056	0.038
1	{0.50, 1.25}	0.0048	0.037
2	{0.50, 1.25}	0.0047	0.036
3	{0.50, 1.00}	0.0041	0.035
4	{0.50, 1.00}	0.0040	0.034

The Bayesian procedures also allow the modelling uncertainties ($\boldsymbol{\varepsilon}$) at each stage to be evaluated, through differences between model prediction ($g(\boldsymbol{\theta})$) and the measured response ($\boldsymbol{y}$). The standard deviation of $\boldsymbol{\varepsilon}$ is assumed to be 30% of model prediction at Stage 0, and this is continuously updated at each subsequent stage. Also, at each stage, updated predictions are made on the subsequent foundation response using the updated $\{s_2, r_{\text{eff}}\}$ parameters and the corresponding $\boldsymbol{\varepsilon}$, as shown in Figure 4.

Figure 4 compares the prediction uncertainties with the updating procedures (grey area) and without (dotted line), i.e. if the variances of s_2, r_{eff} and $\boldsymbol{\varepsilon}$ remain unchanged as in the initial stage. It shows the improved confidence level using the updating framework: as construction progresses, the uncertainties in predictions are reduced. This is partly due to reduced variances in spatial correlation parameters, and partly due to reductions in $\boldsymbol{\varepsilon}$ in later stages. Modelling uncertainties may be higher in early stages potentially due to the uncertainties in seating loads for the load/pressure cells, and also because numerical models may not capture the complexities in the applied loads during excavations and construction of first few storeys of the building.

(a) Stage 0 (before construction)

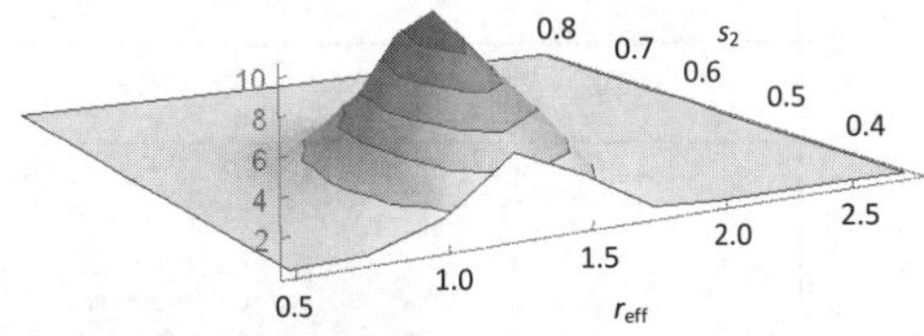

(b) Stage 4 (posterior distribution)

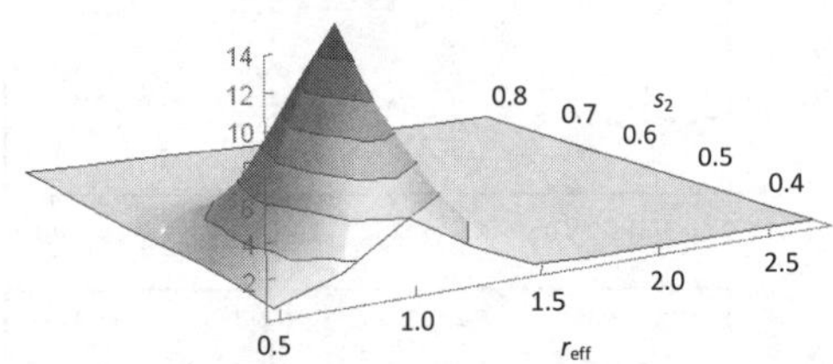

Figure 3. Probability distribution of $\{s_2, r_{\text{eff}}\}$ at Stages 0 and 4.

It should be noted that pile force estimates deviate more significantly from the measured values compared to estimates on settlements and contact pressures. $\boldsymbol{\varepsilon}$ was in fact higher than 30% for pile force estimates. This explains why the prediction variances are larger with the updating approach. The large discrepancy can be due to a number of potential reasons, including uncertainties in the superstructure and foundation stiffness, and perhaps more importantly, the variations of soil properties in the horizontal directions. In the current study, it is not possible to investigate effects of spatial variability in the horizontal directions, since the exact borehole locations were not reported in the literature. This, however, can be an important source of uncertainty regarding the differential settlements or force distributions within the piled foundation. The spatial correlation features can be significantly different between horizontal and vertical directions (e.g. Leung & Lo 2015), and a complete three-dimensional spatial correlation analysis will improve both the inverse analyses and model predictions under the current framework.

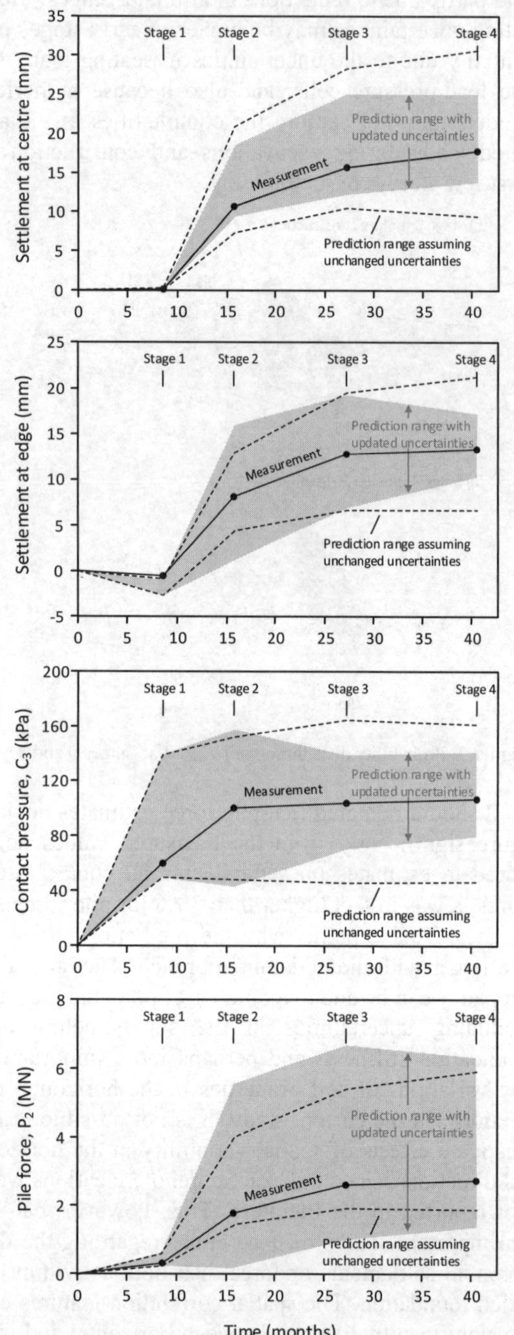

Figure 4. Updating of prediction uncertainties at various stages.

4 CONCLUSION

This paper presents the Bayesian framework to update spatial correlation parameters based on multi-source monitoring data during construction of infrastructure or building projects. The approach allows instrumentation data to be fully utilised for the quantification of project risks at various development stages. Additional work is being conducted to extend the approach to include anisotropy in soil variability (horizontal versus vertical directions). Also, the stochastic finite element method can be coupled with the present framework to improve the accuracies of inverse analyses in the updating process.

ACKNOWLEDGEMENT

The work presented in this paper is financially supported by the Research Grants Council of the Hong Kong Special Administrative Region, under Project No. 25201214.

REFERENCES

Hooper, J.A. 1973. Observations on the behaviour of a piled-raft foundation on London Clay, *Proc. Inst. Civ. Engrs.*, **55**, 855–877.

Ledesma, A., Gens, A. & Alonso, E.E. 1996. Parameter and variance estimation in geotechnical back analysis using prior information. *International Journal for Numerical and Analytical Methods in Geomechanics* **20**, 119-141.

Leung, Y.F. 2010. *Foundation optimisation and its application to pile reuse*, Ph.D. thesis, University of Cambridge.

Leung, Y.F. & Lo, M.K. 2015. Anisotropic spatial correlation of compression indices of marine clay deposits in Hong Kong. *Proc. Int. Conf. Soft Ground Engineering*.

Liu, W.F. Leung, Y.F. & Lo, M.K. 2016. An integrated framework for residual analysis in characterizing spatial variability of geological profiles, *Computers and Geotechnics,* Under review.

Minasny, B. & McBratney, A.B. 2005. The Matérn function as a general model for soil variograms, *Geoderma* **128** (3-4), 192–207.

Mindlin, R.D. 1936. Force at a point in the interior of a semi-infinite solid. *Physics* **7**, 195-202.

Packham, N. 2015. Combining Latin hypercube sampling with other variance reduction techniques. *Wilmott* **76**, 60-69.

Phoon, K.K. & Kulhawy, F.H. 1999. Characterization of geotechnical variability. *Canadian Geotechnical Journal* **36**(4): 612-624.

Vardanega, P.J. & Bolton, M.D. 2011. Predicting shear strength mobilization of London clay. *Proc. 15th European Conf. on Soil Mechanics and Geotechnical Engineering*, 487-492.

Wu, T.H. 2011. 2008 Peck Lecture: the observational method: case history and models. *Journal of Geotechnical and Geoenvironmental Engineering* **137**(10): 862-873.

Proceedings of the International Conference on Smart Infrastructure and Construction
ISBN 978-0-7277-6127-9

© The authors and ICE Publishing: All rights reserved, 2016
doi:10.1680/tfitsi.61279.517

The early detection method for embankment collapse using acceleration data

S. Ryuo*, A.Hada , N.Iwasawa, T.Kawamura, M.Nozue and K.Kawasaki

Railway Technical Research Institute, Tokyo, Japan
**Corresponding author*

ABSTRACT Various sensor data are obtained from the condition monitoring system. On the basis of the value of the sensor, an anomaly state is detected, in which case we focus on the individual one in most cases. In cases where a number of sensors are installed in the monitoring target, to consider the relationship of the sensors would be useful since they might interact with each other. The correlation coefficient generally represents the relationship between two sets of data; however, it is not appropriate if one data interact with the other data with a time lag. In this paper, the early detection method using the relationship of sensors is proposed, and the determination coefficient of time series model is utilized as the relationship of sensors. The relationship of sensors is calculated for a certain period of data sequence, and the next one is also calculated for the next data sequence, then the sequence of the relationship is derived. We define the change point of this sequence of the relationship as the detection time of the change of the state, and propose the evaluation criteria for calculating the change point of that. By applying the proposed method using the acceleration data of simulated embankment collapse caused by the heavy rainfall and the rise in the ground water level, we are able to detect the change of the state early.

1 INTRODUCTION

The railways in Japan are mostly constructed in soil embankment sections. Japan frequently suffers from natural disasters, and the landslides caused by heavy rainfall or a rise in ground water level occurs every year. The landslides along railway lines might result in a serious railway accident; hence in order to avert it, there is a rule of train operational control based on the critical amount of rainfall and the alarm given by disaster detecting systems (Shinomiya et al. 2003). The existing systems for detecting slides use optical fiber sensors and inclination sensors for embankment collapse, while they use a method of sensing cable disconnection for excavation collapse. Each detection system provide the ON/OFF switch, which emerges as an alarm, using the threshold value or disconnection, in which case the data before the collapse are often not stored in the system.

As for the monitoring of landslides, many systems of collecting the value of inclination sensors by using wireless networks are proposed (Uchimura et al. 2010 and Iwai et al. 2010). The features of such wireless network system are as follows; the multiple sensors are installed at the monitoring object, and they collect the data by using multi-hop communication. Next, the collected data is stored in the monitoring server, and then the data can be checked in chorological order. In the current situation, only the measurement data are displayed and the utilization of such data is hardly mentioned.

The continuous change of the state would exist before the state reaching an abnormal situation except for the mechanical failure. For the above disaster detecting system, it might be possible to detect faster the change of condition than the existing system if the data before the collapse are received, which would enable us to expeditiously work out the countermeasures against the collapse. Then getting and using effectively such data is desirable from the viewpoint of establishing a more safety railway system. In this research, we utilize the Wi-SUN network system, which is appropriate to the railway monitoring (Nakano & Yamada 2015), to create such a data collecting system.

2. WI-SUN MONITORING SYSTEM

The life of the most infrastructures in Japan will exceed fifty years in the next twenty years. The larger the number of infrastructures, the more difficult it is to maintenance them appropriately. Also, the abnormal state is more likely to occur by the deterioration of the infrastructure if its life is long. Then, to monitor their state, to foresee their danger and to respond quickly to a natural disaster are going to be requested, and the condition monitoring system becomes a way to solve these requests. The condition monitoring system in railways exists not only as the above disaster detecting system but also as the following systems: the level crossing obstruction detection using radar (Hisamitsu et al. 2008) and the platform door interference detection using the 3D infrared laser (Saito & Oyamada 2010). Methods of monitoring bridges and tunnels by using the wireless sensor networks which collect the data of sensors for inclination or displacement, and temperature and humidity have been researched (Bennett et al. 2010).

We currently have been undertaking researches of which purpose is to develop the basis of communication network using the Wi-SUN communication module and the M2M cloud platform. The Wi-SUN is a wireless communications standard developed by the National Institute of Information and Communication Technology (NICT). The Wi-SUN features a long communication distance of up to around 1km, low power dissipation and its data rate is 50 to 400kbps. In Japan, the 920MHz band is available.

The Wi-SUN network could provide stable and highly reliable communication since they rarely break down by radio wave interference and congestion. Also the 920MHz band is not easily influenced by the obstruction. The communication distance could extend by using multi-hop transmission. For all these reasons, the Wi-SUN network is suitable for the railway environment characterized by a number of rectilinear facilities (Nakano & Yamada 2015). It is required to find the problem with the application of the Wi-SUN network to the railway environment and to solve it so that the application of the Wi-SUN network may be possible. As for it, we have been testing it by utilizing it as the basic constitution of the condition monitoring system for the area along the railway track, the train bogie and the embankment collapse, and have been confirming their applicability and task (Iwasawa et al. 2015).

The possible use of the data before the landslides is mentioned in the above section. In this paper, a detection method of predictors as shown in the next section is developed by using the data which are obtained from the Wi-SUN network for the embankment collapse monitoring test. The developed method is available on the assumption that a number of sensors are installed on monitoring objects and they continue to measure and store the data such as acceleration and temperature before the occurrence of the incident which we want to detect.

3. DETECTION METHOD

3.1 Existing anomaly detection method

Many methods for anomaly detection using the sensor data from the condition monitoring system exist. Generally, the case where the sensor value exceeds the threshold set in advance is considered to be the occurrence of an anomaly state. The sufficient knowledge of the monitoring target is required to set the appropriate threshold. As for the study of anomaly detection, there are many approaches to use data mining techniques, such as clustering and classification. However, these approaches only focus on the behavior of an individual sensor data (Agrawal 2015).

The relationship between the sensors, however, is also valuable since the sensor data might interact with each other, which is described by the graphs where each node represents the sensors and each edge represents the relationship of any two sensors. In this graph, it is not essential to understand the characteristic of monitoring targets since we only focus on the change of the relationship between the data. As the condition monitoring system has become increasingly popular, it is required to construct the models for various monitoring targets; however it is hard to construct all models. Then we propose a widely useable model based on the relationship between the sensors.

The research of the anomaly detection from the time series of the graphs has been carried out in recent years (Ide & Kashima 2004). Most of their targets are the computer network systems, so the weight of the edge in the graph is given. In our research, the weight of the edge representing the strength of relationship of the sensors is defined and the evaluation criteria for the weighted graphs are proposed, which determine the predictor detection time.

3.2 Definition of the relationship between sensors

Let the set of sensors be V, and the set of times be $T=\{1, 2, ..., t\}$. We define the relationship of sensors at time t as a weighted graph $G_t = (V, A_t, W_t)$, where $A_t = \{(i, j) \mid \forall i \in V, \forall j \in V\}$ is a set of ordered pairs of sensors at time t called *arcs*, $W_t = \{w_{ij} \mid \forall i \in V, \forall j \in V\}$ is a set of weight of arcs at time t, and w_{ij} is the element of W_t, named *index value*. Depending on the definition of the relationship between sensor data, the index value changes.

The correlation coefficient is the well-known index which expresses the relationship between two data sequences; however this is not useful if one data interact with the other data with a time lag. In this paper, the determination coefficient derived from a vector autoregressive model (VAR model), which is one of the time series model, is utilized as the index value to express the strength of the relationship between sensors. The feature of the VAR model is that the present value is estimated by previous values.

3.3 Detection time of the change of the state

Let the length of data sequence be l. At the time t, the index values are calculated for all the combinations of the data sequences. The next index value is also calculated for the next data sequences by advancing the sequences by one term at the time $t+1$. Then the weight graphs are derived with each time as shown in Figure 1. The change point of the index value which determines the detection time of the change of state is equal to the change point of the graph structures. Therefore, to calculate the evaluation value of the weighted graph G_t is required.

First we propose the evaluation value for the graphs (V, A_t) which is obtained from the weighted graphs G_t whose weight of arcs is rounded so as to be zero or one by using the threshold. For example, in cases where the threshold is 0.7, the weight of arc is one if that is over 0.7, and is zero otherwise. The evaluation value of the sensor is determined by the number of tail of arcs, which we call for "(1) arc number", for a graph.

As the other method, the evaluation value is calculated from the weighted graphs. A set of weighted edge W_t also is described as the matrix. For any $t \in T$ and $n \in V$, we define the row vector $v_n^t = (w_{n1}, w_{n2}, ..., w_{nn})$. The evaluation value of sensor n for time t is obtained from the angle between v_n^t and v_n^{t-1}, called "(2) vector angle from before". This evaluation does not easily detect the change of the state in the situation in which the index value changes gradually. Then the other evaluation value is obtained from the angle between v_n^t and v_n^0 representing the state of before the experiment, called "(3) vector angle from initial".

The evaluation value of (1), (2) and (3) are calculated for each sensor and for each time, and the change of the value is regarded as the change of the state, i.e., the detection point of a predictor. In our research, (1), the combination of (1) and (2) and the combination of (1) and (3) are adopted for the test data. For each evaluation value, there is a possibility that many changes of the values are detected. Here, the one that is detected first is regarded as the detection time of a predictor.

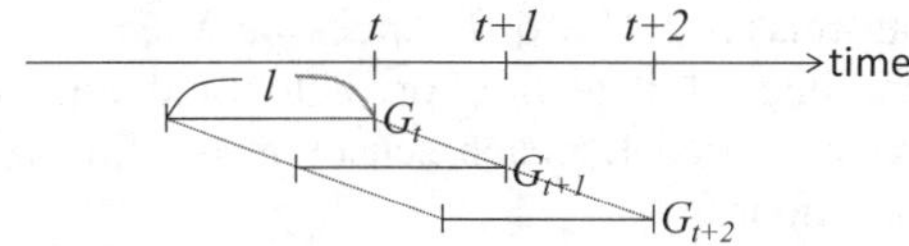

Figure1. The weight graphs at the time

4. EXPERIMENT

In this section, the embankment collapse monitoring test using Wi-SUN sensors is shown (Iwasawa et al. 2015). An embankment 6000mm × 4993mm × 2723mm in size is created in the large-scale rainfall simulator. 20 sensors measuring the 3D acceleration are used, the 16 of which are buried in the slope and the remaining 4 of which are buried in the slope shoulder (see Figure 2).

For this embankment, two situations are tested; heavy rainfall and the rise in ground water level,

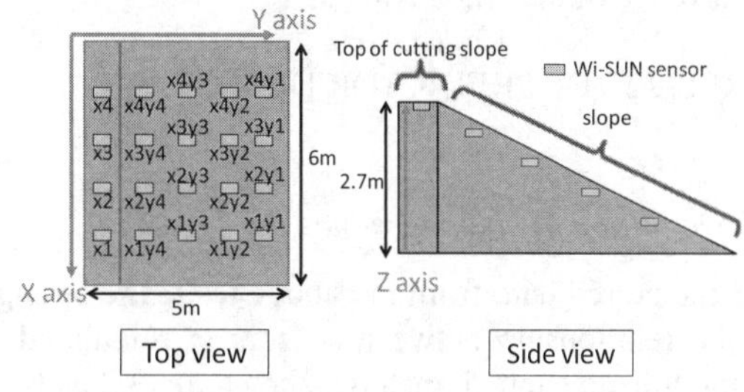

Figure 2. Deployment of the sensors

hereafter we call them the heavy rainfall test and the rising water level test. In the heavy rainfall test, we set an hourly rainfall of 30mm at the beginning of the test, and increase the rainfall by 5mm every nine hours until an hourly rainfall of 50mm. In the rising water level test, water is poured from the water supply hole in the back side of the simulator, and an hourly rainfall of 10mm is set after the rise in water level.

The Wi-SUN acceleration sensor has two kinds of transmission functions; the one is to send at a constant frequency named "constant send" and the other is to send when the difference of the acceleration exceeds the threshold which we name "threshold exceed send". A single or a hybrid transmission function can be given for the Wi-SUN sensor. In our test, the hybrid transmission function, with a constant transmission interval of two minutes, is given since the most required condition for the monitoring system is to detect the collapse of the embankment and to read the sign of a landslide.

The 3D acceleration data of the sensor node, which is the raw value of the sensors and whose range is -1000 to 1000 (mg), and the inclinations of y axis are shown in Figure 3 for the heavy rainfall test and in Figure4 for the rising water level test. In both figures, the horizontal axis shows the elapsed time from the beginning of the test. The selected sensor is the sensor which firstly sends the data by the threshold exceed send. We also attach the photos after the time of starting the threshold exceed send.

From the figures, it is understood that the "threshold exceed send" function behaved at the time of the collapse in both the tests. In both the figures, "CS" means "constant send" and "TES" means "threshold exceed send". From the photos, the timing of the first transmission of the data obtained by TES was the same as the timing of the visually confirmed slide occurrence of the embankment.

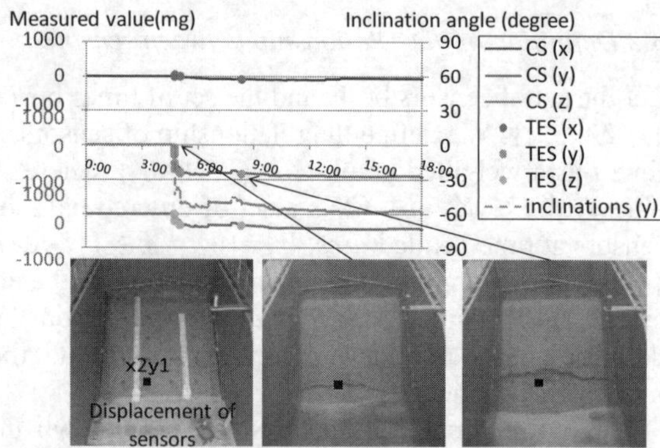

Figure 3. Measured value in the heavy rainfall test.

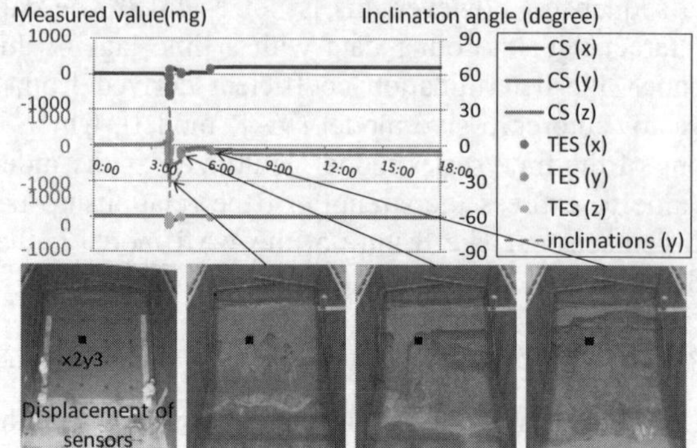

Figure 4. Measured value in the rising water level test.

5. RESULT OF PREDICTOR DETECTION

5.1 Definition of anomaly state

For the stored data from the above tests, the strength of the relationship between sensors is calculated by using the developed method shown in 3.2 subsections. 20 sensors are installed and there are 60 node since each of the 20 sensors measures three type of acceleration, that are x, y and z axis. The number of index values corresponding to the combination of any two sensors is 3540. The term of the data sequence is set at two hours which correspond to 60 data.

The existing system for detecting slides uses inclination sensors for embankment collapse and gives the warning when the inclination sensor of y axis exceeds 30 degrees (Shinomiya et al. 2003). From the 3D acceleration data, we can calculate the inclinations. Also the sensor detects the anomaly state for the embankment from the TES function as mentioned before section. Thus, we define the occurrence time of the anomaly state of the sensor as the earlier time when any of the three axis acceleration of the sensor exceeds 30 degrees or the sensor send the data by the TES transmission. The differences between the occurrence time of detection predictor and anomaly state are compared for each sensor in the following two subsections.

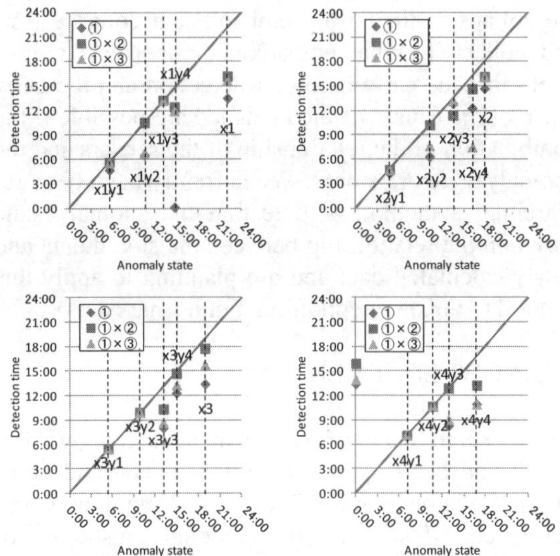

Figure 5. Detection time of anomaly and predictors of the heavy rainfall test.

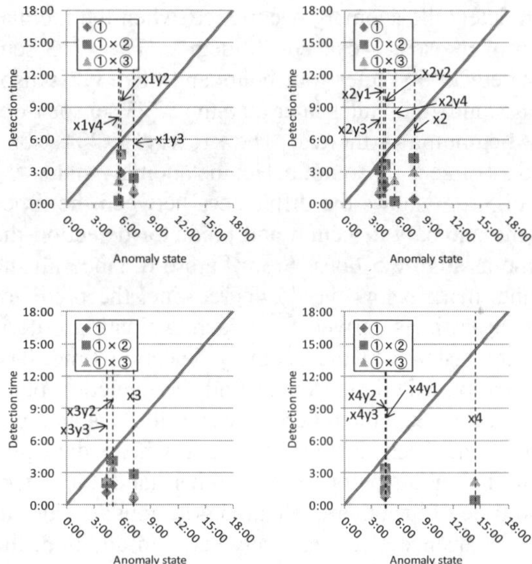

Figure 6. Detection time of anomaly and predictors of the rising water level test.

5.2 Result of the heavy rainfall test

In the heavy rainfall test, the collapse of the embankment starts from the bottom of the slope and has progressed gradually. 19 sensors are able to detect the anomaly occurrence. The x2y1 sensor detects first after four hours and three-quarters and the x3y1 sensor does second about three quarters of an hour later than the x2y1 sensor from the beginning of the test. After, the sensors detect the anomaly state in order from the bottom. Only the x4 sensor on the top of the cutting slope does not detect the anomaly state.

The differences between the detection time of the anomaly state and the predictor derived from the evaluation criterions of (1), the combination of (1) and (2) and the combination of (1) and (3) are compared in Figure 5. The sensor has three predictor detection times since one sensor has three axis acceleration data for the evaluation, and the minimum time of them is adopted. Figure 5 indicates the relationship between the predictor detection time and the occurrence time of the anomaly state. The former is plotted on the vertical axis and the latter is on the horizontal axis. Also in the figure, the 45 degree line is drawn which means that the proposed method detects the predictor earlier than the anomaly state in the case where the point is marked below the 45 degree line.

In Figure 5, the x1y3 and x3y2 sensors cannot detect the predictor faster than they detect the anomaly state. By means of the evaluation criteria(1), we are able to detect the predictor faster than the others. However, there are unreliable data such as the one detected 15 hours before the anomaly state by the x1y4 sensor. On the other hand, the predictor detection times obtained from other evaluation criteria are close to the occurrence time of the anomaly state. In addition, the methods could detect the predictor within ten minutes of the anomaly state although it could not detect the predictor earlier than the occurrence time of the anomaly state. Most of the predictor detection times are the same as or a few minutes faster than the anomaly state. In the case of the heavy rainfall test, the slope is excavated from the bottom and the sensor falls down suddenly when the sensor is in the excavated area, and then the anomaly state suddenly occurs without any sign. Therefore, it can be concluded that there is little difference between the detection time of anomaly state and the predictor.

5.3 Result of the rising water level test

In the rising water level test, the collapse of the embankment starts from the middle of the slope, this collapse affects the lower part of the slope, and finally the whole embankment settles down. 15 sensors can detect the anomaly occurrence, when the inclination of the sensor exceeds 30 degree. The x2y3 sensor detects first after four hours and the x3y2 sensor do second 20 minutes later than the x2y3 sensor from the beginning of the test. The x1, x1y1, x3y1, x3y4 and x4y1 sensors do not detect the anomaly state.

Figure 6 shows the differences between the times of the anomaly detection and predictor detection the same as Figure 5 does. From Figure 6, since all the points mark below the 45 degree line, the predictor detection times derived from each evaluation criterion are faster than the anomaly detection time. The evaluation criteria (1) could find the detection point quite faster than the others, however there is an unreliable one such as the x2 sensor whose detection point is 7 hours before. On the other hand, the detection time of the other evaluation criterions, is close to the occurrence of the anomaly state. In addition, the worst predictor detection point is 40 minutes before the anomaly state. The state had changed gradually for the rising water level test, and then the acceleration sensors had changed gradually before the anomaly state occurs. Therefore, the developed method enables the early detection of the predictor.

6. CONCLUSION

In this paper, we use the weighted graph to express the relationship of the sensor data obtained from the monitoring target, propose the evaluation criteria for the weighted graph so as to consider the change of the relationship, and define the change of the relationship as the detection time of the predictor for the sensor data. Using the acceleration data of embankment collapse tests assuming heavy rainfall and a rise in ground water level, the developed methods are able to detect the predictor state earlier than the anomaly state.

From the test results, the difference of the detection speed depends on the evaluation criteria. Then the reliability of the detection time of the predictor would improve with the use of a number of criteria. According to the patterns of the collapse of the embankment, the speed of predictor detection and the change of the value of the evaluation criterion are different. In the actual operation, no one knows what kind of collapse happens in advance. However, it is possible to link the patterns of predictors to those of the collapse with further embankment collapse tests and collected actual embankment data. Then, it is likely that the early predictor detection and the early grasp of the causes of the landslide are possible if the combination of the relationship of the sensors and the anomaly state are stored. We currently have been researching a method of detecting the anomaly state based on the relationship between the stored data and newly calculated data and are planning to apply this method to the other condition monitoring data.

ACKNOWLEDGEMENT

The research results have been achieved by "Research and Development on Fundamental and Utilization Technologies for Social Big Data", the Commissioned Research of National Institute of Information and Communications Technology (NICT), JAPAN.

REFERENCES

S.Agrawal and J.Agrawal. 2015. Survey on anomaly detection using data mining techniques, *Procedia Computer Science* **60**, 708-713.

P.J. Bennett, Y.Kobayashi, K.Soga. et al. 2010. Wireless sensor network for monitoring transport tunnels, *Proceedings of the ICE-Geotechnical Engineering* **163**, 147-156.

T. Ide and H.Kashia. 2004. Eigenspace-based anomaly detection in computer systems, *KDD 04 proceedings of the tenth ACM SIGKDD international conference on Knowledge discovery and data mining*, 440-449.

M.Iwai, D.Imai, M.Kobayashi. et al. 2010. iPicket:Slope Failure Detection System Using Wireless Sensor Nodes, *Mobile Computing and Ubiquitous Communications* **52**, 1-7.

N.Iwasawa, A.Hada, S.Ryuo. et al. 2015. The applicability for condition monitoring system in railway by using the Wi-SUN network (in Japanese), *ITS* **1**, 1-7.

T.Nakao and R.Yamada. 2015. The summary of Wi-SUN and its applications for railways (in Japanese), *Cybernetics* **20**, 41-46.

Y. Hisamitsu, K. Sekimoto, K. Nagata. et al. 2008. 3-D Laser radar level crossing obstacle detection system, *IHI Engineering Review* **41**, 51-57.

O.Saito and M. Oyamada. 2010. Development of 3-D safety sensor for platform door (in Japanese), *JR-East Technical Review* **33**, 39-42.

T.Shinomiya, S.Koshimizu and A.Kageyama. 2003. The development of landslide collapse detection system (in Japanese), *JR-East Technical Review* **2**, 40-45

T.Uchimura, I.Towhata, T.T.L.Anh. et al. 2010. Simple monitoring method for precaution of landslides watching tilting and water contents on slope surface, *Landslides* **7**, 351-357.

Proceedings of the International Conference on Smart Infrastructure and Construction
ISBN 978-0-7277-6127-9

© The authors and ICE Publishing: All rights reserved, 2016
doi:10.1680/tfitsi.61279.523

Study on mechanics performance of prefabricated trench block

Yun Zou*[1], Chengquan Wang[2], Lijun Zheng[1], Xin Cai[1], Xiaoping Feng[1]

[1]*Jiangnan University, Wuxi, China*
[2] *Zhejiang University, Hangzhou, China*
[*] *Corresponding Author*

ABSTRACT At present, the construction method for power cable trenches is casted in place in China. In this way, the construction quality and period are significantly influenced by weather and seasons. A kind of prefabricated common trench block is designed in this paper. 3D finite element models are established to analyze the mechanics performance under construction and utilizing conditions, respectively. Static analyses are made to simulate the block lifted in site. In utilizing condition, the block is buried underground. Dynamic analyses are also made to investigate the vibration characteristics when vehicle is passing by. Analysis results show that the performances of trench block can meet the requirements under the static and dynamic effects. The trench block will be used widely in municipal construction in future. The research might be valuable for the vibration study of underground pipelines under traffic loads.

KEYWORDS Prefabricated power cable trench; Lifting force; Buried pipeline; Vibration analysis; Finite element analysis

1 INTRODUCTION

With the acceleration of urban modernization, land resource becomes more valuable than before. Municipal infrastructure constructions are constrained by land resource, urban planning, landscape and environmental protection. The construction method for cast-in-situ pipe trench is influenced by weather and seasons and parts of pipe holes are jammed by construction waste to some extent. A prefabricated trench block is proposed and designed in this paper. Compared with traditional cast-in- situ construction, fabricated method is simple, convenient, high quality and fast. The static and dynamic performances of the fabricated trench block are investigated using finite element numerical analyses.

2 DESIGN OF PREFABRICATED POWER CABLE TRENCH

Three different sizes of blocks and different numbers of holes are designed for prefabricated trench, which are named DGG-9, DGG-12, DGG-15. Due to the length of the paper, only DGG-15 is discussed below as an example. Figure1 shows the cross section of DGG-15. With respect to materials, concrete is C30, while longitudinal reinforcement is HRB400 and Stirrups is HPB300. Concrete cover is 30mm. Diameter of stirrup is 10mm for three kinds of blocks. Longitudinal reinforcement Diameter for DGG-12, DGG-15 is 18mm, while the diameter is 16mm for DGG-9. The longitudinal length of the block is 6000 mm, and the height is 720mm.The rings are placed for lifting and the positions are shown in Figure2. Concrete cushion is designed as foundation for blocks with 150mm in thickness and 15Mpa in compressive strength.

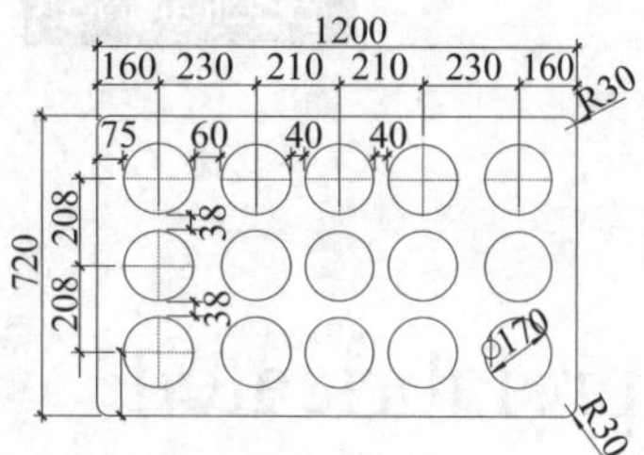

Figure 1. Cross section of DGG-15

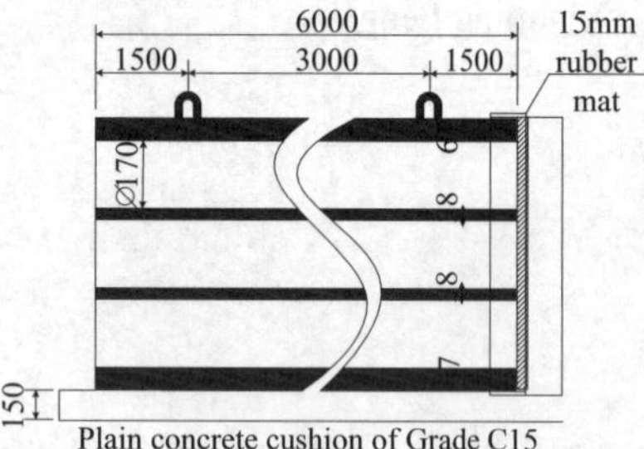

Figure 2. Longitudinal section of prefabricated trench block

Separate holes are arranged in order to place different pipeline dependently. At one end of trench block, a kind of guide protecting sleeve is designed specially for installing adjacent block conveniently and fast. Rubber strips are placed in the inner sides for sealing. Rings are embedded along the length of the block for hoisting in site.

3 STATIC ANALYSIS OF PREFABRICATED TRENCH BLOCK

3.1 Finite element analysis model

Finite element numerical simulation software ABAQUS is used to calculate the performance of the prefabricated blocks in this study. Properties including bending moment, concrete stress distribution, reinforcement stresses, rings stresses and deformation are investigated under different cases, in which block is hoisted and buried, respectively. Dynamic analyses are also made to simulate the vibration characteristics when vehicle is passing by.

Separate models are considered for reinforcement bar and concrete. Eight node linear hexahedral element (C3D8R) is selected to simulate concrete. Two node linear three-dimensional truss element (T3D2) is used for reinforcement bar. The model of the prefabricated trench block is shown in Figure 3. The model after element division is shown in Figure 4. Three kinds of nonlinear effects, including material, geometry and contact, are considered in the study. Newton-Raphson method is used for iteration calculation.

Figure 3. Finite element model of the prefabricated trench block

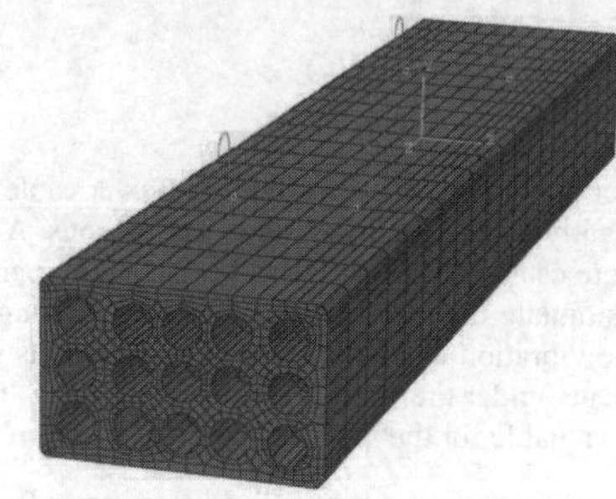

Figure 4. Mesh model of trench block

3.2 Results under lifting condition

Vertical displacements at the positions of rings are constrained to simulate the hoisting condition. Meanwhile amplification factor of 1.5 is taken into account to consider the gravity effect. The analytical results are shown in Figure 5- Figure8.

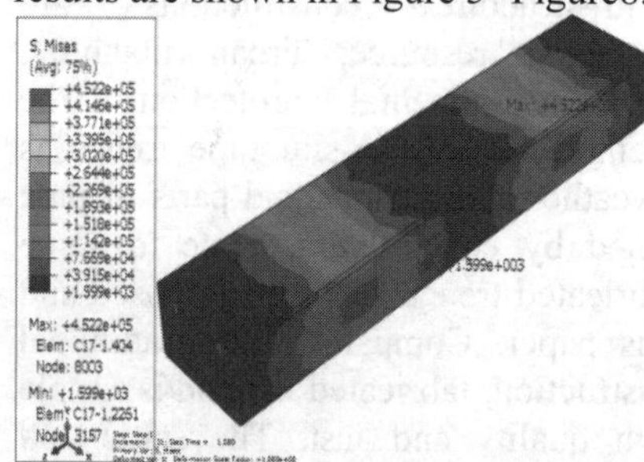

Figure 5. Concrete composite stress contour (Pa)

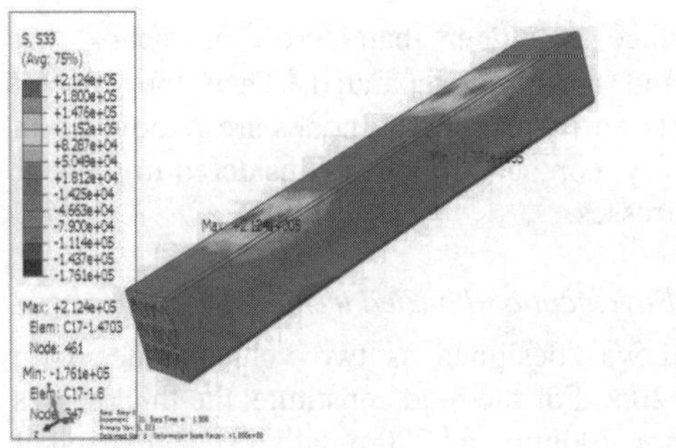

Figure 6. Concrete longitudinal stress contour (Pa)

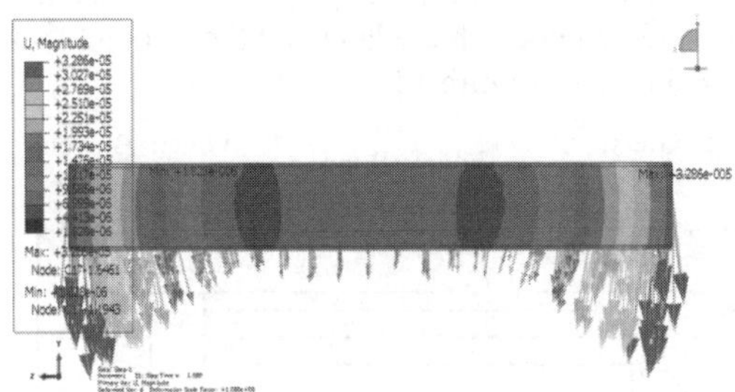

Figure 7. Vertical displacement contour (m)

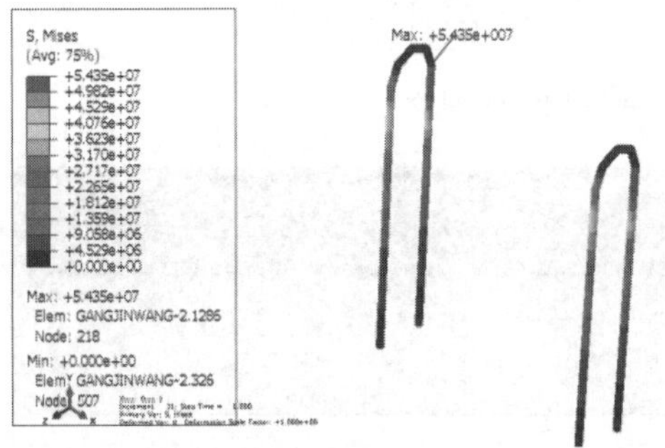

Figure 8. Rings reinforced stress contour (Pa)

In Figure 5, the maximum stress of the surface of block is 0.452MPa. The longitudinal stress contour of the pipe trench in Figure6 demonstrates that the maximum tensile stress is 0.212MPa, which is less than the tensile strength of concrete. From Figure 9, deformation downward occurs both in the middle and two ends of block due to gravity load. However the maximum displacement is only 0.033mm. In addition, bending moment in the middle of the section is 0.607kN.m and bending moment on ring's section is 12.99kN·m. The maximum tensile stress of local rings is 54.35MPa (Figure 8). To sum up， the performance of trench block under reasonable design meet the requirements and no damage occur in construction process.

3.3 Analytical results under vehicle lane

Considering that trench block is arranged under vehicle lane, it would be subject to soil pressure and vehicle loads. The vehicle loads is set to 10.5kN/m corresponding to I-grade road, which is transferred into pressure applied on top of the trench block. The analytical results loaded by soil pressure and vehicle are shown in Figure 9 and Figure 10.

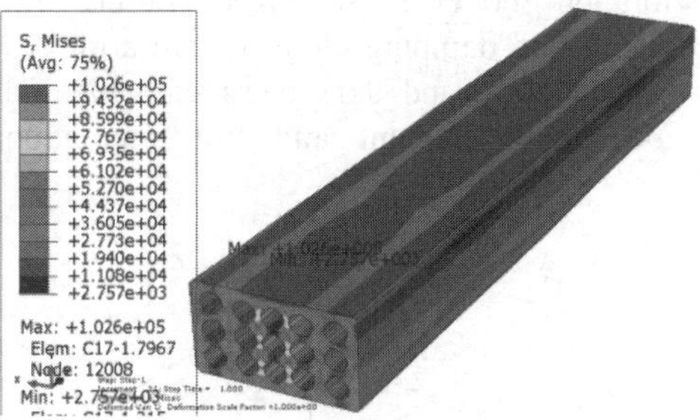

Figure 9. Concrete composite stress contour (pa)

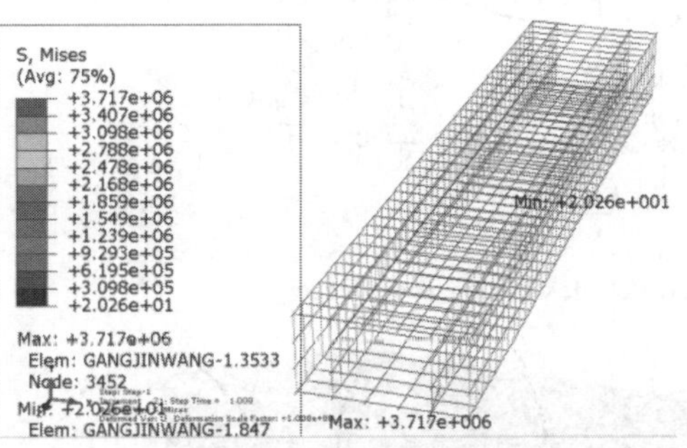

Figure 10. Stress contour of steel truss (Pa)

The results show that the stress and deformation on the block surface are much smaller than those in the wall of hole. The maximum stress of concrete is 0.102MPa, which appears in the middle of the block. The maximum stress of steel is 3.717MPa. Considering the results, it can be found that trench block is far less than damage. Even if the road is overloaded, there remains a large safety reservation.

4 DYNAMIC PERFORMANCE UNDER VEHICLE LOAD

4.1 Coupled vibration model of vehicle - Road - buried pipeline

4.1.1 Vehicle model

The vehicle consists of two kinds of loaded mass, which means one of them is with spring characteristic. The two kinds of loaded mass connected with each other, in which total of 7 degrees of freedom are considered including vertical, pitching, rolling for vehicle body as well as 4 vertical of freedom for each wheel (Figure 11). The vehicle vibration model is shown in Figure 12. Spring elements and damping elements are arranged between vehicle body and axle mass element and between axle mass element and tire simulation element.

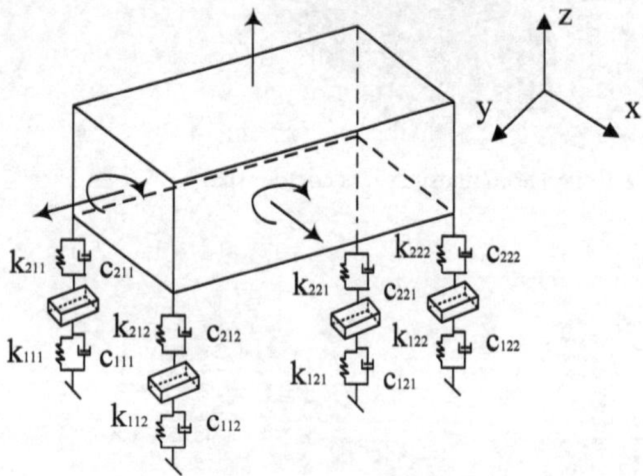

Figure 11. Vehicle vibration theory model

Figure 12. Vehicle vibration model

"Hard contact" is set between the wheels and the road surface. When the distance between wheels and road surface is bigger than zero， it means that wheels and road are separated. Then the contact constraints on corresponding nodes are removed. On the contrary, contact surface is considered to transfer normal stresses.

4.1.2 Pavement and buried trench block model

The road was designed as two vehicle lanes with width of 20m. For the road constitute, the thicknesses are 120mm, 200mm and 200m with respect to asphalt concrete, cement stabilized gravel and gravel soil, respectively. The trench block is buried under grassland with 500mm in depth and the distance between the edge of vehicle lane and the trench block is 3m as shown in Figure 13.

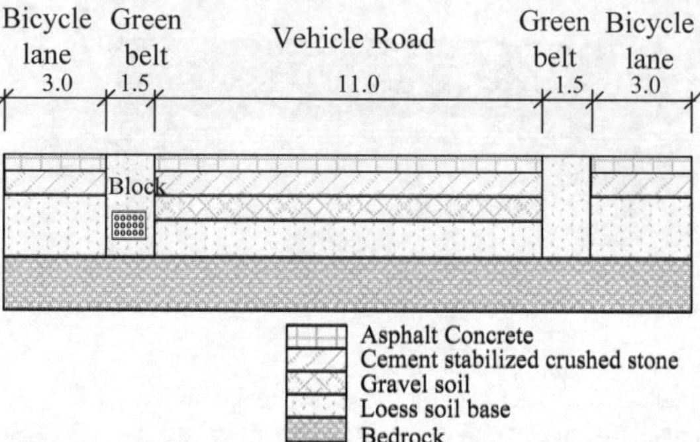

Figure 13. Vehicle lane model

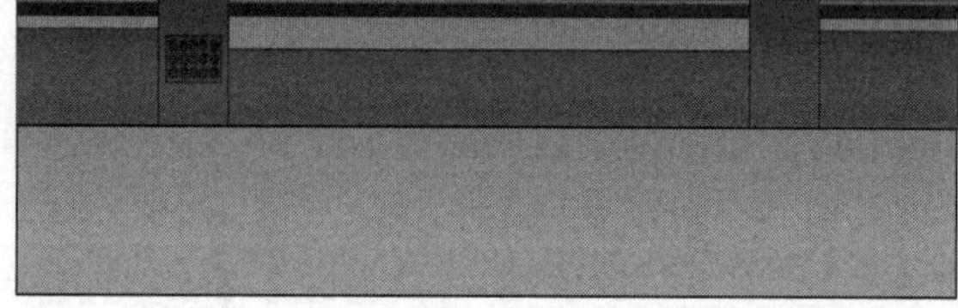

Figure 14. Finite element model of road

Assuming soil evenly distributed in the finite element model of the road and ideal elasticity method is used to simulate soil mass. The finite element model of road is shown in Figure 14, in which different colors represent corresponding soil layers. The parameters of the soil layer are shown in Table 1, which include oil density, dynamic shear modulus, Poisson's ratio and damping ratio.

Table 1. The parameters of vehicle - Road - buried pipeline mode

Site stratification	lift height h/cm	ρ (KN · m^{-3})	G (KN · m^{-2})	μ	ξ
asphalt concrete	10	24	1.8×10^7	0.16	0.06
cement stabilized gravel	20	22	2.6×10^6	0.20	0.12
gravel soil	50	20	3.8×10^5	0.26	0.12
soil	120	18	1.0×10^4	0.40	0.12

It is important and difficult to simulate wave propagation in soil for site vibration analysis. In reality, soil site is a semi-infinite body, while finite region must be intercepted from infinite body. In this case, artificial boundary is taken into account. In order to eliminate the reflection by artificial boundary, extended region with 10m loess is set up around the road model firstly and then 6 increasing damping layers are set up around extended region shown in Figure 15. The vibration wave in the soil is gradually absorbed by the artificial boundary.

The pavement and buried pipeline are simulated by using 3D solid element (C3D8R) in coupled vibration model of vehicle - Road - buried pipeline. The analysis model consists of 37388 nodes and 29746 elements. Elements for road and green belts are meshed from fine to coarse shown in Figure 16.

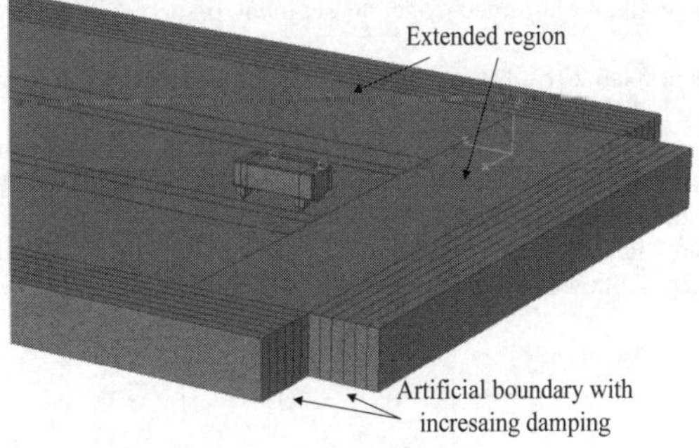

Figure 15. Non-reflective artificial boundary model

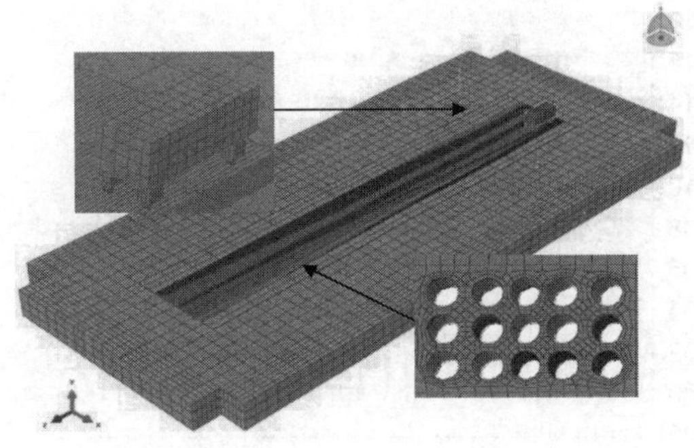

Figure 16. Coupled model of vehicle - Road - buried pipeline

4.2 *Finite element analysis under coupled vibration of vehicle - Road - buried pipeline*

Iterative calculation is used for dynamic analysis, while 40km/h is applied on vehicle body; meanwhile, displacement time-history curve of pavement roughness is applied on wheels to simulate the real pavement. From the displacement time-history curve of the front wheel beside grass side, the wheel moves forward with vertical vibration under the combination effect of the body, the axle and the pavement roughness. The peak acceleration value of buried trench block is 0.35 m/s^2 from the vibration acceleration time-history curve shown in Figure 17. The composite stress value of the buried trench block is reached to 0.15MPa under the action of vibration, which is 1.47 times of stress under static condition. From the above analysis, the stress value of the buried trench block is increased significantly under the action of vehicle dynamic load compared with static load. From the analysis of stress time-history curves of different direction of buried trench block, longitudinal stress response of buried pipe trench is the largest among lateral and vertical directions. Longitudinal stress and time-history curve at midpoint on trench block surface are shown in Figure 18. It could be drawn that vibration wave mainly transmits longitudinally along the trench.

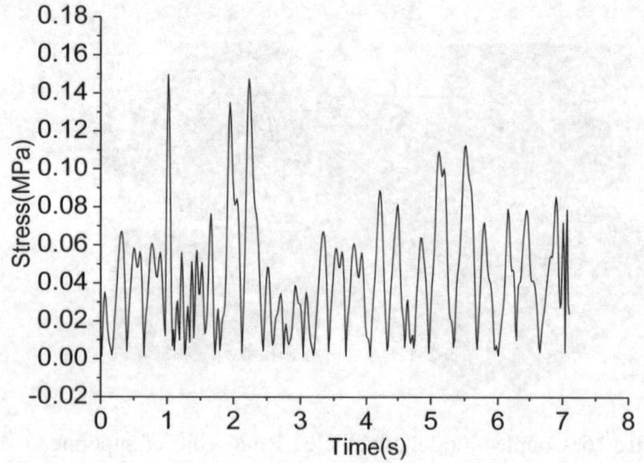

Figure 17. Vibration acceleration time-history curve

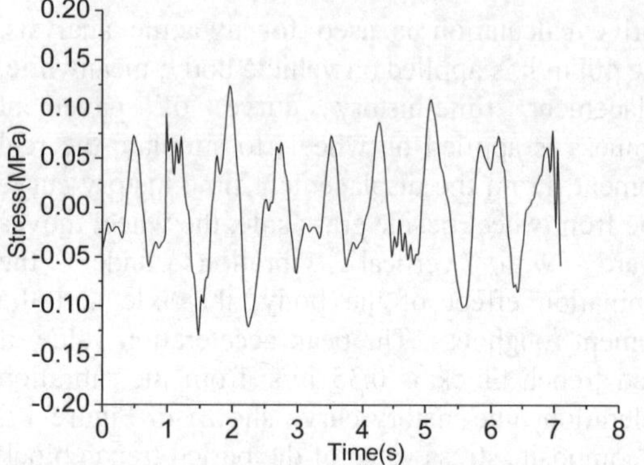

Figure 18. Longitudinal stress and time-history curve at midpoint on trench block surface

5 CONCLUSION

A kind of prefabricated trench block is presented in this paper. The stress and deformation are calculated by the finite element numerical simulation under the effect of hoisting condition, soil pressure, lane load and vehicle vibration load. From numerical simulation, conclusions are drawn as following:

- The performance of the prefabricated trench block could meet the requirements of the stress and deformation under construction and utilizing conditions, and there is enough safety reservation.
- The finite element model of the vehicle - Road - buried pipe trench could be used to simulate the action of vehicle vibration on the pavement as well as the propagation and attenuation of the vibration wave in soil.
- Compared with the action of static load, the stress value of buried trench block is obviously increased under dynamic load. Thus, dynamic analysis plays an important role for buried trench block design.

ACKNOWLEDGEMENT

This work was financially supported by National Natural Science Foundation of China through Grant 51378240 and 2015 Jiangsu province energy conservation and emissions reduction (building industry modernization) special guide capital projects.

REFERENCES

Boegly JR WJ & Griffith WL 1971. Underground utility tunnels.Mech Eng , 93(9):27-32.

Huck P.J., IyengarM .N., Makeig K.S., Chipps J. etal.1976. Combined Utility/Transportation Tunnel Systems-Economic, Technical and Institutional Feasibility .Mass.,United States.

Liyan Ju, Liang Wang & Xiong Zhang 2004. Subway tunnel composite fiber reinforced concrete segment of new technology [J].69-71.

Quanchun Xiao 2012. Design of shield tunnel segment [M], China Building Industry Press, Beijing.

Yiwen Yao, Lihua Jiang & Yiqun Fan 2012. Summary of underground space structure prefabricated assembly [J], City Bridge and flood control, 09:286-292+344.

Yonggang Jia, 2003.The mechanical properties of railway tunnel lining[D], Southwest Jiaotong University, Cheng Du.

SECTION B: ASSETS

Proceedings of the International Conference on
Smart Infrastructure and Construction
ISBN 978-0-7277-6127-9

© The authors and ICE Publishing: All rights reserved, 2016
doi:10.1680/tfitsi.61279.531

Bridge data – what do we collect and how do we use it?

J. Bennetts[1,2], P.J. Vardanega[*1] , C.A. Taylor[1] and S.R. Denton[2]

[1] *University of Bristol, Bristol, UK*
[2] *WSP | Parsons Brinckerhoff, Bristol, UK*
[*] *Corresponding Author*

ABSTRACT The organisations responsible for managing bridge assets in the United Kingdom collect large quantities of data on their bridges. A primary aspiration in the collection of asset data is that it can be processed into useful information that can inform decisions about future management of structures and enhance industry best practice. To enable this, bridge managers must take care to specify appropriate parameters to be recorded, in conjunction with a practical recording interval. In addition, the design of data collection and recording processes is key to ensuring that the data obtained can be transformed into useable information. This study draws on perspectives from a series of interviews with key agents involved in the management of bridges in the United Kingdom. The paper explores the nature of the data that is collected, and how this data is currently used.

1 BACKGROUND

With one of the oldest infrastructure networks in the world, the United Kingdom faces the large and growing challenge of maintaining and renewing its critical assets to allow them to continue delivering the social and economic benefits for which they were built (see also Thurlby 2013). Considering the scale of the investment that will be required in bridge assets in the coming years, it is important that we are able to understand their current and future condition and make informed decisions on what work to do, and when.

Understanding the current condition of bridge assets represents a significant challenge, with established practice being for periodic visual inspection of the structure by an experienced person. A balance has to be struck between the desire to have regular monitoring of the assets' condition and the cost and disruption to the network involved in carrying out an inspection; consequently thorough, touching distance, Principal Inspections (PI) are typically carried out at 6yr intervals (Highways Agency 2007). The recording of extant condition defects at a bridge is subject to the interpretation of the individual bridge inspector and their consideration of the defect type, extent and severity. Furthermore, visual inspections are often undertaken in non-ideal environmental and lighting conditions. Consequently, it is unsurprising that several studies have shown that there is considerable variation in the recoding of defects between inspectors and between individual inspections (Moore *et al.* 2001; Lea & Middleton 2002). Various technological solutions, and particularly Structural Health Monitoring (SHM) systems, have been proposed to supplement or replace visual inspections as a source of bridge condition data (McRobbie *et al.* 2015). These systems can offer dramatically improved data collection intervals, more objective and repeatable data, reduced network disruption, and measurement of variables that is not possible with visual inspection (e.g., Hoult *et al.* 2009).

Several approaches have been proposed to optimise spending on the management of infrastructure assets and to address the inherent uncertainties in the decision making process. Many authors have proposed systems for predicting the future condition of bridges based on imperfect current data (e.g., Enright & Frangopol 1999), and such processes are reported to be in use by bridge owners internationally (Mirzaei *et al.* 2012). Others propose decision support tools which consider evidence for current performance, such as inspection data and historic failures, and explicitly present the uncertainties to give an overview of current performance which could be used to inform future management (Hall *et al.* 2004).

The ownership of bridge assets in the UK is split based on: transport mode, strategic importance, and location. The management of these assets is often then further delegated to contractors, with specialist sub-contractors and consultants frequently picking up more complex work, load-rating assessments and renewal designs. The consequence of this is that asset data collection and decision making processes across the bridge stock are highly heterogeneous, with no clear view of current practice available in standards or the literature.

This paper presents a narrative around the management of the UK's bridge structures, focused on the collection and use of bridge condition data. The work has been built from a series of semi-structured interviews with key individuals in bridge management organisations around the UK.

2 RESEARCH METHODOLOGY

The research was designed as a cross-sectional series of semi-structured interviews with individuals in UK bridge management. In selecting the participants for such a study, it is important that the respondents are representative of the main population (e.g., Oppenheim 1992). Therefore, the participants interviewed were selected to be representative of the range of agents in UK bridge management, including individuals responsible for setting policy in major organisations, as well as those inspecting and making decisions on individual structures. Particular care was taken in ensuring that the all transport modes, levels of authority (i.e. strategic, city region and local authority) and elements of the supply chain were included.

In total, 9 interviews were conducted, with 11 participants who collectively have nearly 300 years' experience in the sector. Table 1 shows the details of the interviewees' organisational roles and the sectors in which they work. Throughout this paper, quotations from those interviewed are presented and are referenced using the notation shown in Table 1 (e.g., *C1*) printed in brackets following the quotation.

The interview approach was standardised using the same interview protocol for each interview. It explored key research questions and areas for enquiry. The interviews were recorded and transcribed and then analysed by coding against research questions and emerging themes in the transcripts (e.g., Saunders *et al.* 2009). Computer aided qualitative data analysis software (CAQDAS) was used to facilitate a thorough and auditable approach.

Table 1. Details of the interviewees' roles and sectors.

Ref.	Role	Sector	Scope
C1	Senior Policy Advisor	Highways	Strategic
C2	Structures Manager	Highways	Metropolitan Transport Authority
C3	Structures Manager	Highways	Local
C4	Structures Engineer	Highways	Local
C5	Structures Asset Manager	Rail	Strategic
C6	Structures Manager	Rapid transit	Metropolitan Transport Authority
C7	Regional Structures Specialist	Highways	Strategic
C8	Head of Engineering	Highways	Strategic, Concessionaire
C9	Assistant Head of Engineering	Highways	Strategic, Concessionaire
C10	Researcher	Highways	Local, Heritage
C11	Structures Watchman	Highways	Strategic, Service provider

3 DATA COLLECTED

The following sections set out the nature of bridge condition data which is collected in the UK.

3.1 *Visual Inspection*

Without exception, all of the organisations use visual inspection as their primary source of condition data, and many see it as driving the management of their structures. One participant said: *"inspections are, really, the foundation for everything we do" (C3)*. The majority of inspections record condition data as the nature, severity and extent of the defects, mostly using, or similar to, the County Surveyors' Society system (Sterritt 2002). The rail sector uses a similar process, but records defect risk in terms of consequence and likelihood. Inspections are typically also used to record maintenance actions, which may be tagged to specific defects and allocated indicative costs: *"we record suggested remedial works, indicative prices, that sort of thing" (C4).*

Recently, many organisations have begun to extend the inspection intervals for some structure beyond 6 years on a risk basis: *"the cycle is dependent on risk, so if you've got a brand new concrete or weathering steel structure you might want to look at it less frequently" (C5).*

An interesting feature of one inspection programme is that it has been aligned with the inspections for assessment required for an 18 year cycle of steady-state load-rating assessments such that *"Every 18 years you will get an engineer, doing an examination [whereas otherwise] ... our examiners are generally ex-trades[people]" (C5).*

Several participants noted the importance of ensuring the reliability of inspection data for example: *"... subsequently we obviously make the decisions on it, and if you're making it on the basis of unreliable data then that's clearly poor practice." (C1).*

Evidence for the variability of inspection data was noted, including an unpublished study where inspectors from 5 local authorities were each asked to inspect the same bridge, with marked variations between inspectors. Several respondents reported a lack of confidence in the quality of inspections delivered by their supply chain *"we are finding the quality of those inspections that we're getting done externally is ... inadequate" (C2),* consequently, some respondents reported that they are looking at changing the delivery of their inspection programmes: *"it may be that inspections are handled in-house or maybe with a contract that's separate from our service providers" (C1).*

3.2 Monitoring Inspections

If an element of a structure is deemed to require a higher level of data collection than the routine visual inspection process most of the organisations interviewed would implement a programme of monitoring inspections. The inspection periods are reviewed depending on the severity of the defect, on-going deterioration and the importance of the element *"it's a balance between keeping everything safe, and keeping an eye on everything and working within the resources we're given." (C3).*

3.3 Structural Health Monitoring

The deployment of structural health monitoring systems was generally limited to specific structures with particularly serious defects which are critical to the network: *"we have specific monitoring, so if we've got a specific problem we're concerned about and we want to gain information about it then we will ... have targeted monitoring, [that] definitely will help with what we need to do ... we're talking about a handful of cases" (C2).* Another interviewee similarly reported that: "*we have, probably a dozen sites where we have real-time monitoring. They're the stuff we're really worried about ... it's not very often, but we do do-it" (C5).* While another said that if they were to deploy SHM: *"it would be, very much, targeted" (C1).* Some interviewees responded that in terms of monitoring systems they have: "*none at the moment ... not any remote monitoring*" *(C11).*

The exception to this is for asset managers responsible for large and strategically important structures: *"Where do we start? We're monitoring wire breaks ... there's wind speed for bridge closure ... there's the weigh-in-motion system ..." (C8, C9).*

Some of the interviewees indicated their interest in potentially deploying structural health monitoring in the future: *"I am aware of ... remote monitoring as well" (C3*), while another interviewee stated that: *"we probably don't do as much as we should" (C2).* Others – when asked if there is monitoring they would like to do, but currently do not – noted that the condition of their structures does not currently warrant the use of monitoring systems: *"we've not really got anything that is of a serious concern, to say I really want that minute-by-minute" (C6).* Others noted the cost of monitoring systems as a deterrent: *"part of it would be cost, so, can we justify putting it in?" (C2)* it was also noted that managers need to ask themselves *"what is this monitoring really going to tell you?"(C5).*

3.4 *Recording of data*

The majority of the interviewees reported that bridge condition data is held in dedicated databases which typically hold inventory, inspection and maintenance data. These databases often also hold the results of load-rating assessments and risk assessments such as for scour or safety. The maturity of these tools varies, with a few organisations relying on spreadsheets for some aspects of their data management, while others have complex integrated IT solutions. Many participants mentioned either newly implemented or imminent IT solutions: *"we're in the process of rolling it [the new system] out ... it pulls all those databases together, so we've got one version of truth" (C5).* Another interviewee reported on developing a new system: *"well it's still in its infancy, I mean we've probably been running it for 3 or 4 years now and it's evolved slightly as well ... we've now got a refined approach ... we'll refine the process as well and keep reviewing it, and it'll become better and better and also we'll have more historical data to be able to verify against as well" (C2).*

4 USE OF DATA

Several participants linked the data that is collected and recorded and its use to inform management decisions: "*[the database] is just a repository for data, and perhaps some information, the knowledge is how you use it, and the wisdom is implementing that" (C1).* The use of the data varies across the organisations interviewed, however, generally it was possible to categorise it into: identification of need; informing assessment; analysis of trends; provision of an audit trail or use as a contractual tool.

4.1 *Identifying and Prioritising Need*

Identifying the need for maintenance interventions is the most common use for bridge condition data. *"So we get a great big long list [element by element, across all structures], so we can look at that and say those are the sorts of things we need to be looking at, and that's a first pass" (C2).* One interviewee reported that they rely on contractors to identify renewals: *"A lot of it relies on our service providers ... to identify need" (C1).*

Monitoring data too is used to identify needs and target interventions to resolve them: "*Take the example of acoustic emissions – we collate the data so we know where the highest instances of wire breaks is ... if we did get a cluster of wire breaks, then obviously when we went in to do our next intrusive inspection, then that [data] would feed into the selection of the panels for the intrusive inspection" (C9).*

4.2 *Informing Assessments*

Several candidates recognised the link between understanding the condition of their structures and assessing their capacity: *"There's interaction between the two sides, so it may be that an assessment triggers an additional inspection. Examination may trigger assessment [which is] more likely than assessment triggering an examination" (C5)* and one suggested change in condition can trigger a reassessment of load-rating: "*so it's as things change, or we're aware of some deterioration that effects the assessment, then we look at reassessing" (C3).* Monitoring data may also be used to verify structural analysis: *"as part of the assessment process, we do use strain gauges or whatever, so we can back analyse" (C5).*

4.3 *Analysing Trends*

All of the owners had some overview of the trends in their stock's condition with time and there was recognition that analysing trends is an opportunity for future development *"so we look at trends in condition ... but it's mainly used at a strategic level and obviously what we want to do is to be able to look at trends at an operational level as well ... looking to the future, there's a lot more opportunity to use the data in much smarter ways ... we're not probably very good at looking at trends, so it relies on individual's judgement to say whether we've got problems with particular types of structures" (C1).*

4.4 *Maintaining an Audit Trail*

One important use of data is to provide an evidence base to justify decisions to do work and what work to do: *"we have a finite resource; it's about justifying where's the best place to spend it" (C2).* Another organisation noted that it can be just as important to justify when work is not done: *"that priority score also helps us defend not doing something to politicians or the public" (C3).*

4.5 As a Contractual Tool

The management decisions for some structures are delegated from the asset owner to contractors who are given responsibility for maintaining the asset for a number of years. It is in the asset owners' interests to put measures in place to ensure that good decisions are made for the long-term performance of the structures, rather than short-term profit of the contractor. Two of the organisations interviewed noted contractual terms related to the condition of the assets: *"we have to hand it back in a condition which allows it to be operated for the remainder of its design life" (C9)* and one interviewee noted contractual terms that specifically use condition data *"on a fixed date at the end of the contract they have to hand back all structures with a BCI score of 90 or above" (C6).* A third organisation noted interest in using condition scores as a contractual tool for measuring service provider performance in the future.

5 DECISION MAKING

While the systems and processes by which management decisions are taken were found to vary considerably across the organisations, it was possible to identify some common themes.

5.1 Prioritisation Processes

All of the participants stated that they undertake some sort of prioritisation process to decide what work to do and when. For some organisations, this is quite a simple process: *"the priority is often very simple ... we've a high, medium or low priority" (C5)* other organisations have more quantitative approaches: *"we've got our own priority scoring system ... Which relates to the importance of the element, the severity of the defect, the size of the structure in terms of deck area, and cost" (C4).* One interviewee set out their prioritisation process as follows: *"we have an inspection programme, which highlights defects in structures, which generates what we call a risk score ... those highest risk scores go forward to a renewals programme and what we then try and do is, through Value Management, prioritise those renewals" (C7).*

It is worth noting that, while at least two organisations referred to a *"Value Management"* process, the mechanics of these processes had some significant differences, particularly in whether they are used to prioritise need, appraise solution options, or prioritise schemes put forward as the best solution to a need. The incorporation of costs into the processes also had significant differences, with some calculating a ratio of risk reduction per pound: *"effectively, we start off with the three risk categories and then we prioritise on that, and then we ... put the costs against each of those items there, and then we get a value ratio" (C2).* Others calculate the ratio of future anticipated savings in whole life cost over immediate cost and then combine that with risk scores. Some individuals reported processes that did not appear to consider cost.

5.2 Lifecycle Planning

Many of the interviewees consider the overall lifecycle of their assets. This may include deterioration modelling and whole life costing to inform planned preventative maintenance *" [the system] tries to predict the condition of different elements over the next 30/40 years, which gives us an indication of ... we don't have to do that now, we can do that in 5 years' time etcetera"(C4).* One of the organisations had the capability to review the costs and effects of different maintenance strategies for their whole asset stock *"the whole life costing's based on our lifecycle plans ... in terms of putting the programme together as a whole ... we will also do an absolute minimum scenario, see what does that look like, we'll run an optimised programme, what does that look like" (C2).*

Other interviewees noted frustrations in attempts to adopt whole life costing approaches *"We've tried in the past ... we used to have a system [which] I never got on with because it always came up with the same answer in my mind which was, 'the cheapest option today is the best'" (C5).*

5.3 Standard Asset Operating Policies

One approach to managing structures is to set out standard operating policies for different kinds of assets and components. For example prescribing that bearings are to be greased every *x* years, and then eventually replaced after *y* years. Alternatively, policy could set out standard interventions for common defects, and specify condition trigger levels for different intervention types. The benefits of this are a unified approach across an asset stock, and a move to planed preventative maintenance, with low downtime, rather than reactive maintenance. These approaches were noted to be taken by some organisations: *"so those maintenance manuals will have 'this area once every x years' so there's a rolling programme you take out every year" (C8)*. In the rail sector, it was reported that standard interventions are used as *"a starter for ten" (C5)*.

However, while noting an intention to develop policy in this area, some participants were more cautious about such approaches: *"you can make some broad assumptions about deteriorations but you've always got to look at the particular condition of those assets" (C1)*.

5.4 *Engineering Judgment*

All eleven individuals interviewed stressed the continued importance of engineering judgement in making bridge management decisions: "*Engineering judgement still rules the day" (C5)*. The two larger, strategic, organisations interviewed both mentioned peer review panels as key to their decision making processes: *"We have a peer review process to evaluate decisions ... where I have to pitch to my peers" (C5)*. A contracting reported discussing the work to be done with the client: *"the list I produce gets discussed at the monthly meetings, so it's pretty much pencilled in at that point which [schemes] are going to be focused on" (C11)*.

6 CONCLUDING REMARKS

The majority of organisations represented in this survey currently use a programme of visual inspections as their primary source of bridge condition data. The deployment of SHM systems is limited, except in targeted cases where there is a clearly articulated use for the data. Collected bridge condition data is used to inform decisions and, although this paper draws on a limited sample of stakeholder and practitioner views, the study does tend to confirm the perceived heterogeneity of approaches to the management of bridges, particularly in the decision making process.

ACKNOWLEDGEMENTS

The authors would like to express their thanks to the individuals who gave their time to participate in this research. This work was supported by the Systems Centre and the EPSRC funded Doctorate Centre in Systems at the University of Bristol and the sponsoring organisation, WSP | Parsons Brinckerhoff Ltd.

REFERENCES

Enright, M. & Frangopol, D. 1999, Condition prediction of deteriorating concrete bridges using Bayesian updating. *Journal of Structural Engineering (ASCE)*, **125(10)**, 1118–1125.

Hall, J.W., Le Masurier, J. W., Baker-Langman, E. A. et al. 2004, A decision-support methodology for performance-based asset management. *Civil Engineering and Environmental Systems*, **21(1)**, 51–75.

Highways Agency, 2007, BD63/07 Inspection of Highways Structures, Highways Agency, UK

Hoult, N.A., Bennett, P. J., Stoianov, I. et al. 2009, Wireless sensor networks: creating 'smart infrastructure'. *Proceedings of the Institution of Civil Engineers – Civil Engineering*, **162(3)**, 136-143.

Lea, F. C., and Middleton, C. R. (2002). Reliability of visual inspection of highway bridges. *Technical Rep. CUED/D-STRUCT/TR.201*, Dept. of Engineering, Univ. of Cambridge, Cambridge, U.K.

McRobbie, S., Wright, A. & Chan, A. 2015, Can technology improve routine visual bridge inspections ? *Proceedings of the Institution of Civil Engineers - Bridge Engineering*, **168(3)**, 197–207.

Mirzaei, Z. et al., 2012. IABMAS Overview of Existing Bridge Management Systems 2012, http://www.f.waseda.jp/akiyama617/rIABMAS/resources/IABMAS-BMC-BMS-Report-20120717.pdf [accessed 14/12/15].

Moore, M., Phares, B., Graybeal, B., Rolander, D., and Washer, G. (2001). Reliability of visual inspection for highway bridges, Vol. 1. *Rep. No. FHWA-RD-01-020*, U.S. DOT, Federal Highway Administration, Washington, DC.

Oppenheim, A.N. 1992, Questionnaire Design, Interviewing and Attitude Measurement New Ed., London, UK: Pinter.

Saunders, M., Lewis, P. & Thornhill, A. 2009, Analysing Qualitative Data. In Research Methods for Business Students, 6th Ed. Harlow, UK, pp. 544 – 594.

Sterritt, G. 2002, Bridge Condition Indicators, Volume 1, London: CSS Bridges Group.

Thurlby, R. 2013, Managing the asset time bomb: a system dynamics approach. *Proceedings of the Institution of Civil Engineers - Forensic Engineering*, **166(3)**, 134–142.

Proceedings of the International Conference on Smart Infrastructure and Construction
ISBN 978-0-7277-6127-9

© The authors and ICE Publishing: All rights reserved, 2016
doi:10.1680/tfitsi.61279.537

Harnessing BIM data in the management of project risks: the Bayesian risk-bearing capacity approach

Chen-Yu Chang

Bartlett School of Construction and Project Management, University College London, London, UK

ABSTRACT With the increasing proliferation of Building Information Modelling (BIM) worldwide, an emerging issue is how to better leverage the BIM data in decision making. This research demonstrates formally that the cost information attached to BIM can be utilised to inform risk management decisions by incorporating the newly developed risk-bearing capacity (RBC) approach into the Bayesian statistics framework. Under BIM, the deviations of outturn costs from planned costs can be systematically recorded and used to update the old "beliefs" that are normally formed by resorting to subjective probabilities. With the potential to integrate the data held by insurers, cost estimators and credit raters, this framework can greatly facilitate the effective use of enormous new data in improving risk management practices.

1 INTRODUCTION

In the OGC Gateway Review framework, the most important yet not fully resolved technical issue is how to efficiently incorporate risk impacts into decision making. With the advent of new technologies, the prevalent application of BIM to construction projects will allow a more systematic investigation on the causal links between cost performance and its determinants. When design and construction information is recorded in electronic forms, the trajectory of variances in construction cost and operating cost can be well preserved for each procurement stage, thereby enabling more advanced statistical techniques for improving the reliability of project risk analysis. This research aims to draw on the newly developed risk-bearing capacity (RBC) approach to demonstrate how the subjective risk estimates can be updated by accommodating new cost data generated by BIM to form a robust basis for risk analysis over different project stages.

2 THE RBC APPROACH

The concept of risk-bearing capacity (RBC) has a root in the economic concept of quasi rent (Klein, Crawford and Alchian 1978). By definition, quasi rent measures the return in excess of the minimum required by a contracting party to carry on with the transaction. Chang (2013) modifies this notion by interpreting it as a measure of the limit that contracting parties are willing to withstand during the construction process. As contract breakup is costly, any risk exposure over the risk-bearing capacity should be priced differently (Chang, 2013, Chang, 2014).

Consequently, a central focus of risk management should be placed on how to avoid contract breakup through the efficient use of feasible measures. In current practice, these decisions chiefly rely upon decision makers' experience and heuristic rules. Decisions on the use of these measures should be integrated so as to address the tradeoffs or compounding effects between them.

2.1 *How the RBC approach works?*

A hallmark of the current project risk management lies in the use of contingency funds for meeting the cash demand for unexpected losses. The determination of a contingency fund, or the risk allowance (RA) is critical in the risk management to all procurement stages. RA is referred to "the amount added to the base cost estimate for items that cannot be precisely predicted to arrive at the cost limit" (RICS, 2012). As shown in

Figure 1, where the random shock actually eventuates is crucial. w is a random variable with mean μ and standard deviation σ, i.e., $w \sim f(\mu,\sigma)$, where $f(w)$ indicates the probability density function of w. The RA is normally set against the expected loss μ to ensure that the client has resources to put out the "fire" in case it happens. There is an economic ground to argue that, instead of responding passively, the client should preempt a severe fire from occurring in the first place. For this purpose,

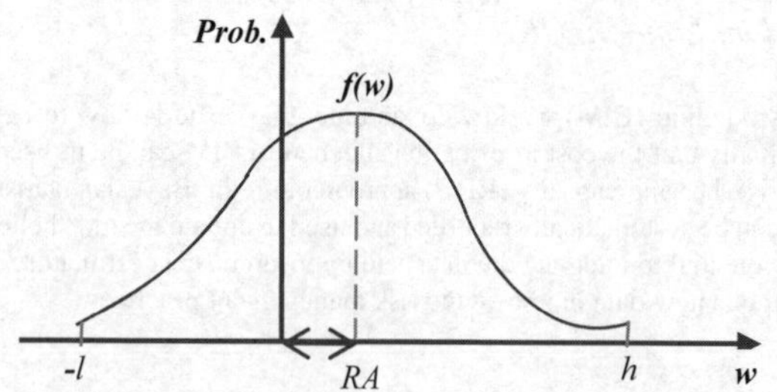

Figure 1. Relation of risk allowance and risk distribution.

The RBC of a project consists of two parts (Chang, 2013): the contractor's quasi rent and the client's quasi rent. When cost overruns occur, the contractor will shoulder most of the impact in accordance with the agreed-upon risk-sharing rules until his buffer (quasi rent) has exhausted. At this point, the contractor could either back out of the contract or seek to recoup some of the loss by holding up the owner in the negotiation of change orders. In practice, the latter dominates and thus requires further analysis.

Suppose the change order can yield a benefit v, but cost c. How the net benefit B (=v-c) is shared depends on the bargaining power of two parties. Chang and Qian (2015) derive a simple result using the Nash bargaining model,

$$b^o = \frac{1}{2}(B - \Delta Q) \qquad (1)$$

$$b^c = \frac{1}{2}(B + \Delta Q)$$

where b^o and b^c indicate the owner's and contractor's share of the surplus. From Eq.(1), it is evident that the owner takes a much smaller share, proportional to the quasi rent difference ΔQ (=Q^o-Q^c). It is worth noting that the benefit of a change order is relative to the "do nothing" option. In practice, change orders are more often triggered by design errors or unforeseen uncertainties than by the attempt to create value.

Eq.(1) should be employed as the prediction for the worst scenario for using one's bargaining power. In estimating risk-bearing capacity, two cases should be considered:

2.1.1 *No change order*

In this case, the contractor can only withstand the shock less than his quasi rent, which corresponds to the breakup probability A^n (=1-Φ(Q^c)) (see Fig.2(a)). Identifying A^c is crucial in cost estimation, because the shock over the contractor's risk-bearing capacity will result in an additional breakup cost (L). The average cost of shocks is

$$W^n = Ew = \int_{-l}^{Q^c} wf(w)dw + \int_{-l}^{Q^c} (L+w)f(w)dw \qquad (2)$$

The expected cost of the project should lie within the cost limit (C^T)

$$EC^n = C^B + W^n \le C^T \qquad (3)$$

2.1.2 *Change order*

If the owner needs to change requirements post contract, as explained previously, the breakup probability will reduce to A^c, which can translate into an extra risk-bearing capacity for the project (see Fig.2(b)).

$$W^c = Ew = \int_{-l}^{Q^c+Q^o} wf(w)dw + \int_{Q^c+Q^o}^{h} (L+w)f(w)dw \qquad (4)$$

Suppose the probability of change orders is denoted by π, the expected cost of the project becomes

$$EC^c = C^B + (1-\pi)W^n + \pi(W^c + (1+t)Q^o) \qquad (5)$$

In Eq.(5), the expropriation of the owner's quasi rent can increase the RBC, but at a high price. For instance, high priced change orders arising from holdup demands could prompt disputes and erode the mutual trust base. Thus, the cost of holdup is more than its nominal value, i.e, $(1+t)Q^o$, $t \ge 0$.

In Eq.(5), both π and t are parameters under the owner's control. First, the owner can choose to accept the holdup demand (t=0) or resolve the disputes through an agreed-upon mechanism (t>0). Second, the probability of change orders can be controlled through management of the design process.

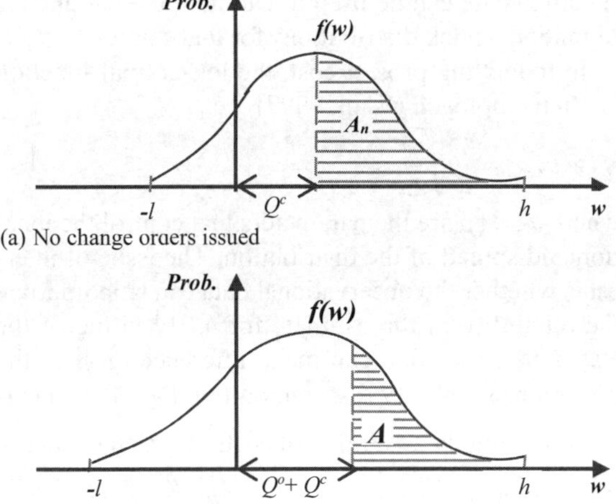

Figure 2. Breakup probabilities.

2.2 What the RBC approach can add?

2.2.1 Determining the feasibility of the project

At the stage of business justification, a central issue is seeking the best value-for-money design solution within the funding constraint. Under the RBC approach, the effectiveness of an insurance should be evaluated on the basis of its contribution to the increase in RBC against its cost (risk premium). Given the preliminary design information, the best estimate of the project's RBC is the contractor's profit included in cost estimation.

Suppose there is an externally imposed budget limit, C^T and the maximum risk allowance C^T-C^B cannot cover W^n, so an evaluation should be made to check if insurances can reduce the risk exposure to an acceptable level. For ease of notation, the premium of insurance coverage ($f(\alpha)$) is quoted in a percentage of base cost (C_1^B), α. If this protection is used to cover the upper end of the exposure, the whole distribution of w can be split into three areas (see Fig.3): first, in A_1, the outcome turns out to be favourable; in A_3, the exposure is zero as it is covered by the insurance policy; in A_2, the owner is still exposed to random shocks. All together, the risk exposure is changed to W^I

$$W^{c,I} = \int_{-l}^{h-\alpha C_1^B} wf(w)dw + f(\alpha) \tag{6}$$

The decision on insurances is seeking

$$\text{Min}_\alpha \left[C_1^B + W^{c,I}\right] \text{ (subject to } W^{c,I} \le C_T - C_1^B\text{)} \tag{7}$$

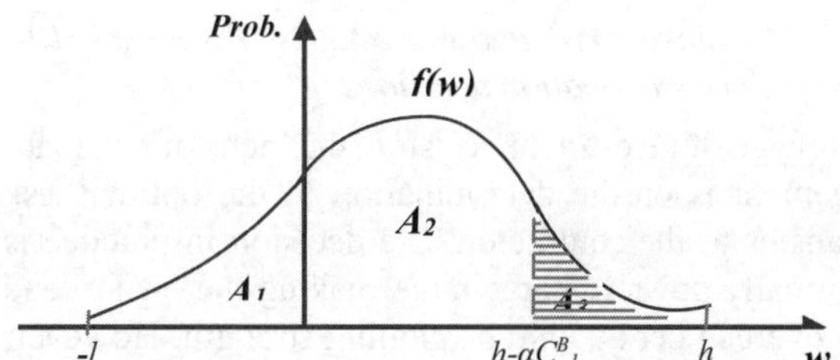

Figure 3. Effect of an insurance cover on the breakup probability.

2.2.2 How to incorporate risk impacts into decision making

In current practice, p.10 v.s. p.90 or p.5 v.s. p.95 are still widely used (Hillson, 2012). The purpose of contrasting two polar values is aimed at revealing the spread of risk outcomes, thereby informing procurement decisions. However, there is a case to argue that the whole distribution of possible outcomes should be considered. As demonstrated in Chang and Zhu (2014), using Monte Carlo simulation techniques together with off-the-self software @Risk, the RBC approach can be implemented to consider the effect of random shocks across its full range.

2.2.3 Linking risk analysis and risk responses (A)-procurement system selection

Once the procurement system is chosen, risks taking should be evaluated as organizational constraints and risk responses formulated accordingly. There are important tradeoffs to consider: whereas employing Design-Build, the system enabling the whole project reach the strongest RBC, can make the project more resilient to cost shocks, the owner will be exposed to a greater holdup threat. The benefit (B^c) of utilizing the owner's quasi rent as part of the risk-bearing capacity can be evaluated by subtracting Eq.(5) by Eq.(3) and differentiating with respect to π

$$B^c = \frac{d}{d\pi}(EC^n - EC^c) = W^n - W^c - (1+t)Q^o \\ = \int_{Q^c}^{Q^c+Q^o} (L-w)f(w)dw - (1+t)Q^o \tag{8}$$

Eq.(8) reveals that the benefit of reducing the probability of change orders by one percentage point amounts to the cost savings arising from a lower probability of contract breakup (first term in integral) net of the total cost of rent transfer ($(1+t)Q^o$).

2.2.4 *Linking risk analysis and risk responses (B)-risk allocation decisions*

At Stage of investment decision, the focus of risk management is on the determination of the optimal risk transfer to the contractor. The decision in practice is normally down to bargaining, making the owner tend to overuse her *ex ante* bargaining strength. However, the nominal favourable term of transaction may not transcend into a desirable outcome owing to the presence of bargaining power reversal (Chang and Ive, 2007). For ease of exposition, the allocation of risk is reduced to the choice of a single risk-sharing ratio (*b*) for the aggregate risk *w*, under which part of the risk exposure (*bw*) is borne by the contractor. This allocation will enable the contractor to withstand greater downside cost shocks, i.e., increasing the contractor's risk-bearing capacity to *Qc/b*. Given the optimal insurance cover obtained in Eq.(7), the owner's risk exposure after risk transfer can be revised to

$$W^{n,IR} = \int_{-l}^{Q_c/b} wf(w)dw + \int_{Q_c/b}^{h-\alpha^* C_{1,B}} (L+w)f(w)dw + f(\alpha^*) \quad (9)$$

Suppose, at this stage, the target cost is still adhered to, so the objective function becomes

$$\text{Min}_b \left[C_2^B + W^{n,IR}\right] (\text{subject to } W^{n,IR} \le C^T - C_2^B) \quad (10)$$

3 THE RBC APPROACH

The probabilistic model of the RBC approach (Eq.(9) & (10)) heavily depends upon the reliability of inputted parameters to maintain its predictive power. It is beneficial to accommodate the information of real cost data into the estimation of parameters.

As the RBC approach focuses on cost variance, it is useful to provide an account of how it could be measured. In each stage of project procurement, the estimation of cost variance could be different, so formally the outturn cost variance of stage *r* from project *i* ($\tilde{w}_{r,i}$) can be defined as

$$\tilde{w}_{r,i} = \tilde{C}_{r,i} - C^B_{r,i} \quad (11)$$

Where $\tilde{C}_{r,i}$ and $C^B_{r,i}$ represent the outturn cost and the base cost estimate at stage *r* in project *i*. BIM models can provide both cost information and the details of design solutions and procurement arrangement, whereby the causal relationship between cost variance and its determinants can be more thoroughly established (Fig.4) and the cost variance calculated for a specific stage can be used to calibrate the parameters of random shock distributions for that stage.

In modelling project cost, the log-normal function is often employed (Wall, 1997), i.e.,

$$f(\tilde{w}|\mu,\rho) = \frac{1}{\tilde{w}}\sqrt{\frac{\rho}{2\pi}}e^{-\frac{\rho}{2}(\ln \tilde{w}-\mu)^2} \quad (12)$$

where μ and ρ are the parameters that control the location and spread of the distribution. The issue of interest is whether the observational data can help improve the reliability of the estimate for ρ. By virtue of the Bayes theorem, one can make inference on ρ in the posterior probability distribution ($f(\rho|\tilde{w}_{r,1},\cdots,\tilde{w}_{r,u})$) after observing the evolution of costs at different stages.

$$f(\rho|\tilde{w}_{r,1},\cdots,\tilde{w}_{r,u}) = \frac{f(\rho)f(\tilde{w}_{r,1},\cdots,\tilde{w}_{r,u}|\rho)}{f(\tilde{w}_{r,1},\cdots,\tilde{w}_{r,u})} = \frac{f(\rho)f(\tilde{w}_{r,1},\cdots,\tilde{w}_{r,u}|\rho)}{\int_0^\infty f(\rho)f(\tilde{w}_{r,1},\cdots,\tilde{w}_{r,u}|\rho)d\rho} \quad (13)$$

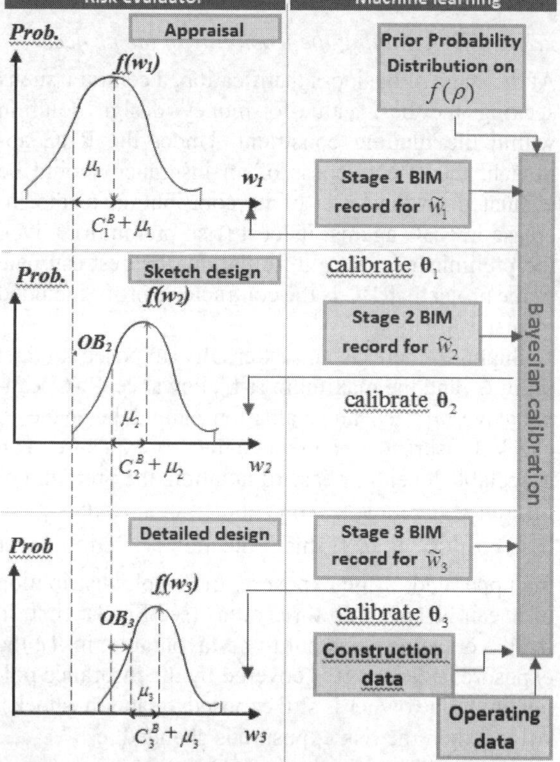

Figure 4. Integrating Risk Management Into Procurement Decisions.

In Eq.(12), the product of the prior on ρ ($f(\rho)$) and the conditional probability of $\tilde{w}_{r,1},\cdots,\tilde{w}_{r,u}$ for a given value of ρ ($f(\tilde{w}_{r,1},\cdots,\tilde{w}_{r,u}|\rho)$), is normalized by the marginal probability of the observed data given the model (prior and likelihood) (the integral in the denominator). The data collected ($\tilde{w}_{r,1},\cdots,\tilde{w}_{r,u}$) from u different projects can be incorporated into the prior knowledge of ρ through the likelihood function. Assume $\tilde{w}_{r,1},\cdots,\tilde{w}_{r,u}$ are identically and independently distributed from a lognormal process,

$$f(\tilde{w}_{r,1},\cdots,\tilde{w}_{r,u}|\rho)=f(\tilde{w}_{r,1}|\rho)\times\cdots\times f(\tilde{w}_{r,u}|\rho) \tag{14}$$

Thus, the likelihood function is

$$L(\rho|\tilde{w}_{r,1},\cdots,\tilde{w}_{r,u})\propto \rho^{\frac{u}{2}}\exp\left(-\frac{\rho}{2}\sum(\ln x-\mu)^2\right) \tag{15}$$

To make Eq.(12) analytically tractable, a popular solution is to search for a conjugate prior whereby the prior and posterior functions are in the same family of probability distribution. Fink (1997) suggests that a gamma distribution with hyperparameters and meets this condition, i.e.,

$$f(\rho|\alpha,\beta)=\begin{cases}\dfrac{\rho^{\alpha-1}\exp(\frac{-\rho}{\beta})}{\Gamma(\alpha)\beta^{\alpha}} & \text{where } \rho>0\\ 0 & \text{otherwise}\end{cases} \tag{16}$$

where $\Gamma(\alpha)(=(\alpha-1)!)$ indicates the gamma function. This prior can yield a posterior distribution with the same distributional form, but with a different set of hyperparameters (α', β'):

$$\alpha'=\alpha+\frac{u}{2} \tag{17}$$

$$\beta'=\beta+\frac{1}{2}\sum_{i=1}^{u}\left(\ln\tilde{w}_{r,i}-\mu\right)^2 \tag{18}$$

In doing so, posterior inference is reduced to the updating of hyperparameters of the prior. In circumstances where there is no historical data, the evaluator can estimate the hyperparameters on the basis of subjective evaluation first and then use the Bayesian statistics to improve the reliability of these estimates for next projects.

Although conjugate priors provide a convenient venue to solve the posterior distribution, the degree of fit of the assumed form to the observed project cost data should be scrutinized. In some complicated cases, the integration of the marginal probability in Eq.(12) could involve tremendous computational difficulty. Along with the development of Markov Chain Monte Carlo methodology, the Bayesian method becomes readily applicable to a wide range of practical problems. It is also well suited to the implementation of the new risk management framework developed in the current research.

4 CONCLUSIONS

Risks play a central role in construction procurement decisions. Currently, the focus of risk management seems more concerned with the setting of a contingency fund for unexpected bills than with the active management of risk. A full integration of risk management into the procurement process is an essential step to evolving risk management into an integral part of project procurement. The new risk management approach should provide a framework to link up the chain effect of design development on cost estimation, risk analysis and procurement decision. This research demonstrates that the RBC approach provides a flexible avenue to build this framework whereby new data can be accommodated to improve the reliability of estimation using the Bayesian statistics.

As suggested by the IRG report (Infrastructure Risk Group, 2013), the collection of data on outturn costs should be more systematically attempted. For instance, the integration of rich information held by insurers (e.g., Lloyds Register) and cost estimators (e.g., the Royal Institute of Quantity Surveyors) could be of enormous benefit to improve the effectiveness of project management in general and risk management in particular.

ACKNOWLEDGEMENT

The author is grateful for financial sponsorship from the CIOB Bowen Jenkins Legacy Research Fund.

REFERENCES

Buhlmann, H. 1985, "Premium calculation from top down." Astin Bulletin, 15(2).

Chan, W.-T. (2008). A source book in Chinese philosophy, Princeton University Press, Princeton.

Chang, C. (2014). "An Empirical Investigation of Procurement System Selection: A Transaction Cost Explanation." Journal of Construction Engineering and Management (under review).

Chang, C. Y. (2012). "Book Review- Adjudication in the Building Industry (3rd edn)." Construction Management and Economics, 30(7), 590-594.

Chang, C. Y. (2013). "Critiques of the existing techniques for the evaluation of renewable energy projects." International Journal of Project Management, 31(7), 1057-1067.

Chang, C. Y. (2013). "Understanding the hold-up problem in the management of megaprojects: The case of the Channel Tunnel Rail Link project." International Journal of Project Management 31(4), 628-637.

Chang, C. Y. (2014). "Principal-Agent Model of Risk Allocation in Construction Contracts and Its Critique." ASCE Journal of Construction Engineering and Management, 140(1), 04013032.

Chang, C. Y., and Ive, G. (2007). "The hold-up problem in the management of construction projects: A case study of the Channel Tunnel." International Journal of Project Management, 25(4), 394-404.

Chang, C. Y., and Ko, J. W. (2014). "A new advance in the application of Monte Carlo simulation to the risk analysis of construction projects." working paper, Bartlett School of Construction and Project Management, University College London.

Chang, C. Y., and Qian, Y. M. (2015). "An Econometric Analysis of Holdup Problems in Construction Projects." Journal of Construction Engineering and Management.

Chang, C., and Zhu, J. (2014). "A Risk-Bearing Capacity Approach to Optimal Risk Sharing Decisions in Construction Projects." working paper, Bartlett School of Construction and Project Management, UCL.

Chang, C.-Y. (2013). "When might a project company break up? The perspective of risk-bearing capacity." Construction Management and Economics, 31(12), 1186-1198.

Chapman, C., and Ward, S. (1994). "The efficient allocation of risk in contracts." Omega, 22(6), 537-552.

Chapman, C., and Ward, S. (2003). Project Risk Management: Processes, Techniques and Insights, Wiley, New Jersey, US.

Fink, D. (1997). "A compendium of conjugate priors." Department of Biology, Montana State Univeristy.

Fitzgerald, M., and Dwoskin, E. (2014). "Big Data Cuts Energy Use: Buildings find more information leads to efficiencies." Wall Street JournalNew York.

Gibbons, R. (2005). "Incentives between firms (and within)." Management Science, 50(1), 2-17.

Griffis, F., and Butler, F. M. (1988). "Case for cost-plus contracting." Journal of Construction Engineering and Management, 114(1), 83-94.

Grossman, S. J., and Hart, O. D. (1986). "The costs and benefits of ownership: A theory of vertical and lateral integration." The Journal of Political Economy, 94(4), 691-719.

Grout, P. A. (1984). "Investment and wages in the absence of binding contracts: A Nash bargaining approach." Econometrica: Journal of the Econometric Society, 52(2), 449-460.

Hillson, D. (2012). "Project Risk Management: Past, Present and Future Views From the Chair." Association for Project Management Risk Special Interest Group, Bucks, England.

HM Treasury (2007). "The Green Book - Appraisal and Evaluation in Central Government." TSO, London.

Infrastructure Risk Group (2013). "Managing Cost Risk & Uncertainty In Infrastructure Projects." I. UK, ed., HMSO, London.

Ive, G., and Chang, C. Y. (2007). "The principle of inconsistent trinity in the selection of procurement systems." Construction Management and Economics, 25(7), 677-690.

Klein, B., Crawford, R. G., and Alchian, A. A. (1978). "Vertical integration, appropriable rents, and the competitive contracting process." Journal of law and economics, 21(2), 297-326.

Kodukula, P., and Papudesu, C. (2006). Project valuation using real options: a practitioner's guide, J. Ross Publishing.

Markowitz, H. M. (1991). "Foundations of portfolio theory." The journal of finance, 46(2), 469-477.

Mishra, S., Khasnabis, S., and Dhingra, S. L. (2012). "A simulation approach for estimating value at risk in transportation infrastructure incestment decisions (Article in Press)." Research in Transportation Economics, 1-11.

Office of Government Commerce (2007). "Risk and value management- OGC Achieving Excellence in Construction Procurement Guide 4." HMSO, London.

Office of Government Commerce (2007). "Whole-life Costing and Cost Management: Achieving Excellence in Construction Procurement Guide No.7." OGC, London.

RICS (2012). "NRM 1 - Order of Cost Estimating and Cost Planning for Capital Building Works." Royal Institute of Chartered Surveyors, London.

RICS (2012). "NRM 2 - Detailed Measurement for Building Works ", Royal Institute of Chartered Surveyors, London.

Stukhart, G. (1984). "Contractual incentives." Journal of Construction Engineering and Management, 110(1), 34-42.

The Economist (2010). "It's a smart world." Special Issue: Smart SystemLondon.

Wall, D. M. (1997). "Distributions and correlations in Monte Carlo simulation." Construction Management & Economics, 15(3), 241-258.

Ward, S. C., Chapman, C. B., and Curtis, B. (1991). "On the allocation of risk in construction projects." International Journal of Project Management, 9(3), 140-147.

Williamson, O. E. (1996). The mechanisms of governance, Oxford University Press, USA.

Ye, S., and Tiong, R. L. K. (2000). "NPV-at-Risk method in infrastructure project investment evaluation." Journal of Construction Engineering and Management, 126(3), 227-233.

Zyphur, M. J., and Oswald, F. L. (2013). "Bayesian Probability and Statistics in Management Research A New Horizon." Journal of Management, 39(1), 5-13.

Proceedings of the International Conference on Smart Infrastructure and Construction
ISBN 978-0-7277-6127-9

© The authors and ICE Publishing: All rights reserved, 2016
doi:10.1680/tfitsi.61279.543

Modelling, management, and visualisation of structural performance monitoring data on BIM

J.M. Davila Delgado*, I. Brilakis and C. Middleton

University of Cambridge, Cambridge, United Kingdom
* *Corresponding Author*

ABSTRACT The use of systems to monitor the condition and structural performance of built-assets is becoming common practice. The data acquired by these systems could lead to reductions in construction, operational, and maintenance costs and improved performance and quality by enabling informed decision making. Nevertheless, the data as outputted by the systems is of little use and value. First, it needs to be processed and put into geometric context within the built-asset, which simplifies the interpretation and analysis of the data. This supports informed decision making that leads to effective actions. This paper presents an overview of an approach to model structural performance monitoring systems and to include and visualise sensor data on BIM models in a manner that facilitates decision making. This paper addresses aspects related to (1) interoperability and standards for data modelling; (2) processing and management of sensor data on BIM models; and (3) visualisation of sensor data directly on the BIM model. A precast concrete bridge, in Staffordshire, UK, has been used as case study. It has been installed with two types of fibre optic systems to monitor the manufacturing process of its pre-stressed beams. The paper shows that BIM provisions to support structural performance monitoring tasks are not sufficient yet. It also showcases that by including and visualising monitoring data directly on BIM models it gains geometrical context within the built asset. This facilitates its analysis and increases its value.

1 INTRODUCTION

Monitoring the condition and performance of built assets is becoming common practice across the entire range of types of projects in the construction industry. Systems for Structural Health Monitoring (SHM), an assessment of structural performance and damage identification (Farrar & Worden 2007), have been used mostly for critical infrastructure assets, in which large capital investments and large losses –due to failures and breakdowns– justify the investment on structural monitoring systems. The data acquired by these systems is the starting point to develop strategies to reduce operational and maintenance costs, and to improve performance and quality as well. These type of investments are now easier to substantiate given the advancements on sensing technologies that reduce fabrication and installation costs and increase reliability. In addition to that, the acquired data can be also used to validate structural designs and to devise more efficient designs and construction processes for future projects.

Structural monitoring systems generate data in raw form that is of little use as it is. This raw data needs to be organised –in such a way– that can be easily accessed, robustly exchanged (i.e. without any loss), and visualised, so that facilitates its analysis and aids decision making. The management and exchange of this data is part of the tasks addressed by the information technology approach known as Building Information Modelling (BIM), which intends to manage digital representations of all information related to a built asset during its entire life cycle. The BIM approach has being increasingly applied during the design and construction phases of the built assets' life-cycle. However, still not enough provisions (software solutions, standards, real-life demonstrations, etc.) exist that support tasks during the operational phase. Is in this phase that the largest reductions in cost can be achieved by applying the BIM approach; because this phase represents the largest share of the total life-cycle cost.

This paper presents an overview of an approach to model structural monitoring systems using a standard data model and to include and visualize structural performance monitoring data directly on BIM models. The paper addresses aspects related to (i) interoperability and standard data models and (ii) management and visualisation of monitoring data. To showcase the capabilities of the developed approach, a structural monitoring system installed in a pre-stressed concreted bridge in Staffordshire, UK, has been used as a case study. The monitoring system uses fibre optic sensors to measure strain in several elements of the bridge. The monitoring system has been modelled and monitoring data –acquired during the construction of the bridge– has been included and visualised directly on a BIM model.

2 INTEROPERABILTY AND STANDARD DATA MODELS

Ensuring that information is exchanged in an efficient and robust manner (without errors or data loss), between parties in the construction industry, is central to the successful adoption of the BIM approach for the entire life-cycle of built assets. Given the fragmented nature of the construction industry, this is one of the biggest challenges of the BIM approach. To achieve interoperability, open standard data models have been developed that define publicly available rules to format and exchange data.

The Open Geospatial Consortium (OGC) and buildingSmart are the two organisations that develop open standard data models for the Architecture, Engineering, and Construction (AEC) area. The Industry Foundation Classes (IFC) specification (Liebich et al. 2013), developed by buildingSmart, is the most widely used standard for the AEC area. It intends to provide data modelling specifications to describe information related to built assets during its entire life-cycle. In practice, it is mainly used to describe information related to buildings during the design and construction phases. OGC develops standards that focus on facility planning, asset and emergency management and navigation. For example, CityGML is used for 3D modelling of cities, IndoorGML for indoor navigation, and WaterML for representing data from water observations.

There are existing standards that describe monitoring system, as noted in literature (Hu et al. 2007; Lee 2007) However, they do not provide sufficient capabilities to fully describe structural monitoring systems (Davila Delgado et al. 2015). The IFC specification includes the entity called *IfcSensor*, which is used to describe predefined types of sensors commonly used for building services monitoring activities. For example, sensors that measure light levels, temperature, humidity, etc. or that detect movements or fire. User-defined entities can be used to describe sensors that are not predefined in the IFC specification. This is not sufficient because entities to model other devices such as sensor networks, processing units and directives for data storage and visualisation are not considered. Moreover, using user-defined entities is not a definitive and robust solution that should be used only in exceptional cases.

SensorML, developed by OGC, describes devices and processes related to generic and complex monitoring systems (Botts & Robin 2007). It can describe simulations, planning processes, monitoring and alert systems, and data storage and archiving systems; but it cannot describe the object being monitored, e.g. the built asset. SensorML is intended to describe complex monitoring systems with various types of sensors, large amounts of data in different formats and sources (Aloisio et al. 2006).

Due to the lack of current capabilities, informal approaches are being used to describe structural monitoring systems. Next, four examples of such approaches are presented. (1) The *IfcSensor* entity has been used to virtually describe structural sensors (Rio et al. 2013). In this case, smoke sensors with user-defined properties were used as proxies to model strain gauges. The IFC file had to be manually adjusted so that they could be used in different authoring tools. The paper proposes to include structural sensor types in the IFC specification. (2) Complex monitoring systems for buildings have been modelled combining IFC and SensorML (Liu & Akinci 2009). The building and the location of the sensors were modelled using the IFC specification and a detailed description of the monitoring system was modelled using SensorML. (3) Data collected by temperature sensors embedded in concrete structural elements has been included in user-defined sensor entities (Chen et al. 2014). The sensor data, stored in text files, is included in the BIM model and visualised using charts in the used authoring tool. The sensor data is then included in IFC files using user-defined entities; but to visualise the sensor data, the used authoring tool and a specific plug-in are needed. (4) Lastly, IFC models have been linked to external databases to store attributes and processes –for glass-pane manufacturing– that cannot be described with the IFC specification (Voss & Overend 2012). While this case does not refer to monitoring, it is a pertinent example of alternative approaches of data linkage stored in databases and IFC models.

These approaches may lead to inconsistencies and errors, therefore reducing the benefits of the BIM approach. More importantly, they indicate where the current lack of capabilities reside, i.e. lack of specific entities and attributes for modelling, and lack of directives for data management and visualisation. In this respect, an extension to the IFC specification is under development that seeks to address this lack of capabilities (Davila Delgado et al. 2016).

3 MODELLING, MANAGEMENT, AND VISUALISATION OF MONITORING DATA

This section describes an approach to model structural monitoring systems and to include and visualise sensor data directly on BIM models. The data modelling and visualisation requirements depend on the monitoring objectives. These are defined by the interests of the stakeholders, i.e. asset owners, asset managers, structural engineers, contractors, or researchers. The phase of the project, i.e. pre-construction, construction, operation, or decommission, also influences the monitoring objectives. Objectives are usually related to improve safety, performance, management, or to test sensing technologies. They can be categorised as follows: (i) model validation, (ii) anomaly detection, (iii) threshold check, (iv) damage detection, (v) performance record, (vi) problem investigation and diagnosis (Webb & Middleton 2013).

3.1 Modelling of structural monitoring systems

The first step to model a structural monitoring system is to define the purposes of the modelled system. Note that "structural monitoring system" refers to the physical set of devices installed on the built asset, while "modelled system" refers to the abstract description of the structural monitoring system. The most common purposes intend to aid with: (i) development of definitive designs of structural monitoring systems; (ii) deployment, maintenance, and operation; (iii) visualisation of monitoring data; and (iv) simulation of monitoring processes. Once the purposes of the modelled system have been defined, then the general monitoring objectives must be taken into account as well, so that the modelled system can effectively and efficiently provide the required data to achieve them. For example, a modelled system that visualises data from a structural monitoring system intended for threshold check would have different requirements than one intended for model validation.

Before to start developing the modelled system, a preliminary design of the structural monitoring system should be in place. For completeness, these are general steps to develop a preliminary design of a structural monitoring system: (1) Definition of structural behaviours to be monitored. E.g. to monitor strains caused on beams, in a bridge, caused by traffic loads. (2) Definition of the monitoring approach. The required devices to monitor the defined behaviours are selected in this step. For most cases, three types of devices are needed in a monitoring system i.e. sensors, a communication network, and processing unit. The structural elements to be monitored are also selected in this step. (3) Definition of the instrumentation approach. This step defines the physical installation process of the monitoring system.

Once a preliminary design of the structural monitoring system has been completed, then requirements for the modelled system can be defined. Requirements are related to (i) the level of detail of the modelled system, e.g. a modelled system for maintenance would require a higher level of detail regarding all the components of the system than one for visualisation; (ii) the types and frequency of recorded data; (iii) the users of the modelled system, e.g. an asset manager would require different data displayed in a different manner than a structural engineer; and (iv) retrieval of data, i.e. data stored directly on BIM models or referenced from databases.

The data modelling is done using an object oriented approach, i.e. that the system is modelled in a way that closely represents the real-life situation. Data is grouped in units called classes or entities, which have attributes that define their state. These classes represent physical entities, e.g. a sensor or a processing unit. Several modelled entities are first grouped into *sensor systems* and then into *monitoring assemblies*. The grouping is carried out depending on the types of structural elements that are installed on and on the structural behaviour to be monitored. The entities representing monitoring assemblies, devices, and single entities are linked to modelled structural assemblies and structural elements, as can be seen in Figure 2. The rectangles represent modelled entities and attribute sets, the dashed-rectangles represent aggregations of modelled entities, the lines represent simple associations in between them, and the indentations represent hierarchical levels. Lastly, attributes for each type of entity are defined as shown in Figure 3. These attributes should be defined based on the requirements of the modelled system, which could include physical quantities derived from the measured physical quantities.

Summarising, in this step of the process the framework of the modelled system is developed, which will be instantiated with the number of installed sensor systems and populated with the acquired –and inferred– data in the next step.

3.2 *Processing and management of sensor data*

In most cases, the data as outputted by the processing units cannot be directly used for the required analyses. The following steps are needed to include monitoring data into BIM models. (1) The raw data needs to be pre-processed into the correct physical quantity and units, and other phenomena that affect the measurements have to be corrected for. (2) The processing unit only provides the sensed measurements, but they do not include data concerning the location of the devices or its spatial and hierarchical relationship with the object being monitored and the structural monitoring system. For that reason, using the raw data outputted by the processing units and the design of the structural monitoring system, the modelled entities of every sensor system are instantiated, the data that belongs to every instance is identified, and their locations are derived. (3) Lastly, the attributes of the instantiated entities are populated. Note that depending on the requirements of the modelled system the raw data and the processed data may need to be included in the modelled system. Also, attributes that define functions to obtain –and to store (e.g. directly on the BIM model or in a database) – derived physical quantities are defined needed as well.

3.3 *Data visualisation*

In the previous step, the acquired data has been compiled into the required physical quantities and format. In this step, it is visualised to facilitate its analysis. Structural monitoring data is usually presented in tables and graphs. This approach proposes that, in addition, the data be visualised directly on the BIM model. This will clearly show its location and context within the built asset, which will provide more information for better analysis and decision making, and –ultimately– increase the value of the acquired data. In addition to that, the hierarchy, components, and functioning states of the modelled system can be visualised for operation and maintenance tasks. Visualisation for alert systems based on monitoring data can be developed as well. The visualisation approach should be based on the purposes of the modelled system. For different purposes, specific attribute sets for visualisation should be defined, as can is shown in Figure 2. E.g. an alert system would only need to indicate locations where anomalies have occurred, but probably not all the underlying data. More importantly, the monitoring data has to be linked to the corresponding visualisation attributes sets.

4 CASE STUDY

Structural monitoring systems, based on fibre optics, have been installed on two newly constructed bridges in Staffordshire, UK. The bridges are part of the Stafford Area Improvements Programme on the West Coast Main Line. A subset of the structural monitoring system installed in a pre-stressed concrete bridge is presented here as case study. The bridge is a single span structure that carries an electrified railway on ballasted formation. The bridge uses longitudinal precast pre-stressed beams (TY7 and TYE7) as main load bearing elements. The deck is simply supported on piled reinforced concrete abutments and wing walls. (Figure 1).

The installed structural monitoring systems intend to monitor various structural elements during manufacturing, construction, and operation. In this case study, only the monitoring of the manufacture of the pre-tensioned (pre-stressed) TY7 concrete beams is presented. The structural monitoring system employs distributed fibre optic sensors (DFOS), which are used to measure changes in the optical properties along the entire cable; and point fibre optic sensors (PFOS), which measure changes at discrete locations. These changes are due to changes in temperature and strain; therefore, strains in the cable –and in the structural elements they are attached to– can be deduced. Strain measurements have been taken during the following critical phases of the manufacturing process of the beams: before concrete casting, during the initial curing time, and immediately after the transfer of the pre-stressing force. A detailed description of the structural monitoring performed during manufacturing can be found in literature (Gibbons et al. 2015).

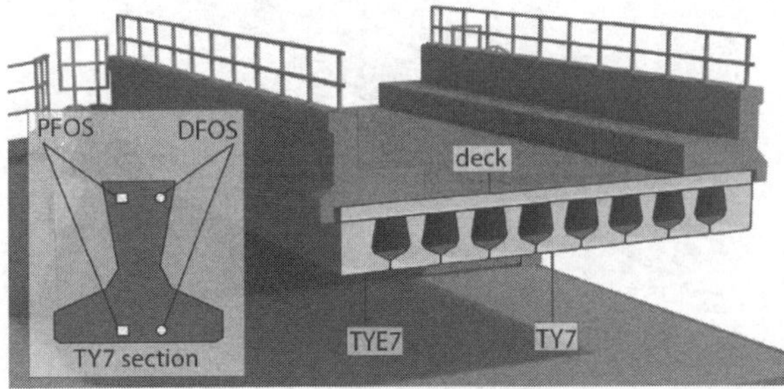

Figure 1. BIM model of a precast pre-stressed concrete bridge composed of TY7 and TYE7 beams; TY7 section showing the location of the PFOS and DFOS sensor devices.

The monitoring objectives for this phase were: (i) to quantify the elastic shortening in the concrete due to the transfer of the pre-stressing force, (ii) to provide baseline measurements for time-dependent losses, and (iii) compare the use of different fibre optic technologies, i.e. DFOS and PFOS. The purpose of the modelled system is to visualise the monitoring data and facilitate the comparison between systems.

The structural behaviour to be monitored is the axial deformation of the concrete beams due to the transfer of the pre-stressing force. For that, the pre-stressing strands, which run longitudinally along the top and the bottom of the TY7 beams, have been instrumented. DFOS have been attached along the entire length of the top and bottom strands in one side of the beam; while, PFOS, at 1 meter spacing, have been attached on the top (10 sensors) and the bottom (10 sensors) strands in the other side of the beam (Figure 1).

The modelled system required a low level of detail because it will be used only for visualisation purposes. For that reason only the sensors have been modelled. DFOS and PFOS generate different types of data with different recording frequencies. During casting of the concrete, readings from the DFOS were taken before and after pouring. Whereas, readings from PFOS were taken at every second during pouring. During the curing process, DFOS readings were taken every hour and every 10 min for the PFOS. Lastly, during the transfer of the pre-stressing force, readings for the PFOS were taken every second.

Figure 2 shows a diagram of the part of the modelled system considered in this case study. Note that for clarity only selected entities are presented in the diagram. The modelled system is composed of one type of monitoring assembly, which is associated with the structural assembly that groups the TY beams. One instance of the monitoring assembly is generated per each TY beam. Each monitoring assembly has a DFOS and a PFOS sensor system, which are associated with one pre-stressing strand of the TY beams. Each DFOS and PFOS sensor systems is composed of a sensor, a network communication element, and a connection element. Examples of the attributes of the monitoring assembly, the DFOS and PFOS sensor systems, and sensors are presented in Figure 3. Note that the sets of attributes presented in Figure 3 are only related to monitoring features. Attributes concerning geometrical shape, spatial positions, and visualisation are defined in another set of attributes.

Both types of sensor systems were used to measure strain and temperature given the linear varying relationship between changes in the wavelength of the backscattered (DFOS) or reflected (PFOS) light and strain in the fibre. The results as given by the processing units were converted to strain and corrected for temperature and separated into strain due to thermal and mechanical changes. This pre-processing of the data was carried out using established methods found in literature (Horiguchi et al. 1995; Kreuzer 2006).

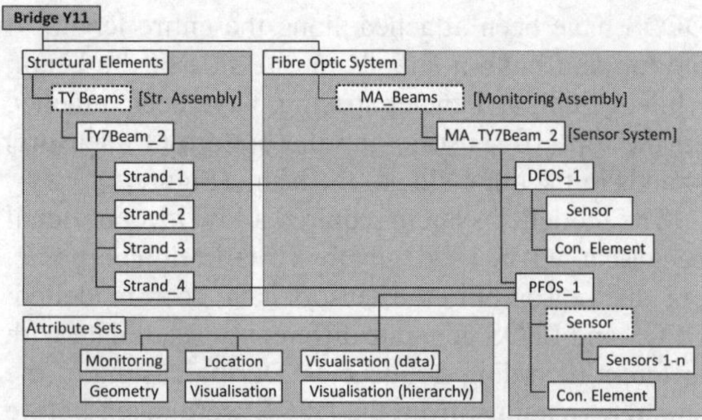

Figure 2. Diagram of the modelled system showing selected monitoring entities and attribute sets.

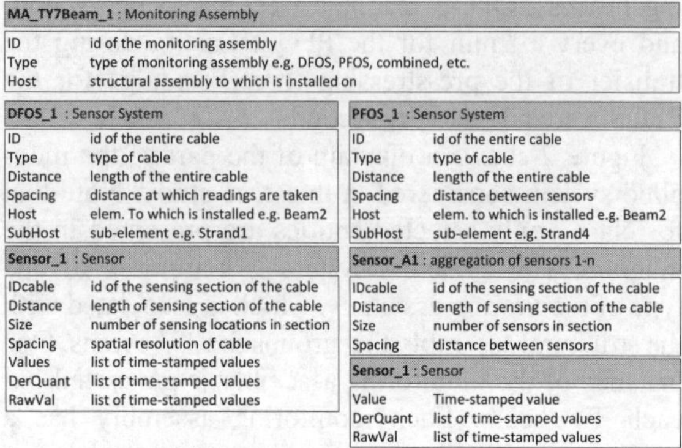

MA_TY7Beam_1 : Monitoring Assembly	
ID	id of the monitoring assembly
Type	type of monitoring assembly e.g. DFOS, PFOS, combined, etc.
Host	structural assembly to which is installed on

DFOS_1 : Sensor System	
ID	id of the entire cable
Type	type of cable
Distance	length of the entire cable
Spacing	distance at which readings are taken
Host	elem. To which is installed e.g. Beam2
SubHost	sub-element e.g. Strand1

Sensor_1 : Sensor	
IDcable	id of the sensing section of the cable
Distance	length of sensing section of the cable
Size	number of sensing locations in section
Spacing	spatial resolution of cable
Values	list of time-stamped values
DerQuant	list of time-stamped values
RawVal	list of time-stamped values

PFOS_1 : Sensor System	
ID	id of the entire cable
Type	type of cable
Distance	length of the entire cable
Spacing	distance between sensors
Host	elem. to which is installed e.g. Beam2
SubHost	sub-element e.g. Strand4

Sensor_A1 : aggregation of sensors 1-n	
IDcable	id of the sensing section of the cable
Distance	length of sensing section of the cable
Size	number of sensors in section
Spacing	distance between sensors

Sensor_1 : Sensor	
Value	Time-stamped value
DerQuant	list of time-stamped values
RawVal	list of time-stamped values

Figure 3. Examples of monitoring attribute sets for monitoring assemblies, sensor systems, and sensor entities.

The pre-processed data only includes the measurements and the relative distances at which they were taken, for DFOS, and a time stamp and a measurement for the PFOS. Using the order in which the values are listed and information from the design documentation the rest of the attributes were inferred. Then, all the required instances for the modelled entities are generated and their attributes are populated.

Figure 4 presents the development of the total micro-strain during curing of the concrete (2 hours and 7 days after casting) and after the des-tensioning of the strands as measured by the DFOS and PFOS sensor systems for one half of an internal beam. These values show the change in strain relative to the baseline values recorded before casting of the concrete. In the three cases a relatively uniform distribution of strain can be appreciated. After two hours of the casting, it can be seen an increase in tensile strain, which decreased after 7 days curing period. Compressive strain was introduced after the process of des-tensioning the strands. The strain measured at the bottom of the beam are generally higher than the ones recorded at the top of the beam. For a detailed explanation of the monitoring results and source data please refer to literature (Gibbons et al. 2015).

For clarity, a scaled-up and simpler geometry has been used to visualise both the DFOS and PFOS sensor systems. The values of the strain attributes of each sensor entity have been linked with the material attribute, in the monitoring visualisation attribute set, according to a predefined colour-coded scale. For the beam entities, the corresponding material attribute for visualisation of monitoring data has been adjusted as well. The modelling and visualisation of monitoring data has been done with additional software tool for Autodesk Revit, which is under development as part of this research project.

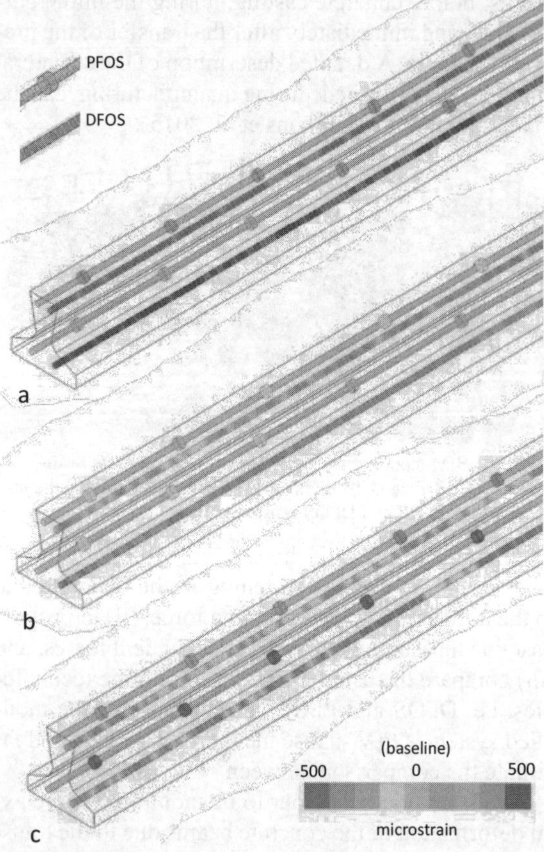

Figure 4. Change in strain relative to the measured baseline for (a) 2 hours after casting, (b) 7 days after casting, and (c) after de-tensioning.

5 CONCLUSIONS

Monitoring the structural performance of built assets is becoming common practice for various types of construction projects. Standard data models and software solutions are not sufficient yet to fully implement the BIM approach for structural monitoring tasks. The currently used informal approaches –or workarounds– are not the optimal solutions and indicate the lack of capabilities regarding specific entities, attributes, and directives for data management and visualisation.

This paper presents an approach to model structural monitoring systems and to manage and visualise the acquired data directly on BIM models. A case study has been presented, in which the strain changes during the manufacturing process of pre-stressed concrete beams were visualised on a BIM model. The required data model and attribute sets were developed, the acquired data was pre-processed, included, and visualised. This represents novel and useful features of the developed approach. Monitoring data is usually stored and presented –in tables and graphs– separately from the built-asset in different software solutions. By visualising the data directly on the BIM model, the acquired data gains geometric context within the built asset facilitating its analysis and increasing its value.

More importantly, the developed approach contributes to the advancement of the BIM approach for the operational phase by applying open data models standards. This ensures that modelled systems and acquired data can be used with different authoring tools and a robust exchange of information between the involved parties.

ACKNOWLEDGEMENT

The authors would like to acknowledge funding from the UK Engineering and Physical Sciences Research Council (EPSRC) and Innovate UK. This research was carried out under EPSRC (grant no. EP/L010917/1).

REFERENCES

Aloisio, G. et al., 2006. Globus monitoring and discovery service and sensorML for grid sensor networks. In *Proceedings of the Workshop on Enabling Technologies: Infrastructure for Collaborative Enterprises, WETICE*. pp. 201–206.

Botts, M. & Robin, A., 2007. OpenGIS Sensor Model Language (SensorML) Implementation Specification. *Design*, p.180.

Chen, J. et al., 2014. A Case Study of Embedding Real-time Infrastructure Sensor Data to BIM. In *Construction Research Congress 2014*. ASCE.

Davila Delgado, J.M., Brilakis, I. & Middleton, C.R., 2015. Open data model standards for structural performance monitoring of infrastructure assets. In J. Beetz, ed. *CIB W78 Conference 2015*. Eindhoven, The Netherlands, pp. 1–10.

Davila Delgado, J.M., Brilakis, I. & Middleton, C.R., 2016. Design and modelling of fibre optic systems to monitor reinforced concrete structural elements. Submitted to the ASCE Construction Research Congress (CRC) 2016, San Juan, Puerto Rico May 31 - June 2, pp. 1–10.

Farrar, C.R. & Worden, K., 2007. An introduction to structural health monitoring. *Philosophical Transactions of the Royal Society A: Mathematical, Physical and Engineering Sciences*, **365**(1851), pp.303–315.

Gibbons, N. et al., 2015. Monitoring the Early Age Behaviour of Prestressed Concrete Beams Using Fibre Optic Sensors. In *16th European Bridge Conference*. Edinburgh, United Kingdom, pp. 1–11.

Horiguchi, T., Shimizu, K. & Kurashima, T., 1995. Development of a distributed sensing technique using brillouin scattering: Optical fiber sensors. *Journal of lightwave technology*, **13**(7), pp.1296–1302.

Hu, P., Robinson, R. & Indulska, J., 2007. Sensor standards: Overview and experiences. In *Proceedings of the 2007 International Conference on Intelligent Sensors, Sensor Networks and Information Processing, ISSNIP*. pp. 485–490.

Kreuzer, M., 2006. Strain measurement with fiber Bragg grating sensors. *HBM, Darmstadt, S2338-1.0 e*.

Lee, K.L.K., 2007. Sensor standards harmonization-path to achiewving sensor interoperability. *2007 IEEE Autotestcon*.

Liebich, T. et al., 2013. Industry Foundation Classes Release 4 (IFC 4).

Liu, X. & Akinci, B., 2009. Requirements and Evaluation of Standards for Integration of Sensor Data with Building Information Models. In *Computing in Civil Engineering*. American Society of Civil Engineers, pp. 95–104.

Rio, J., Ferreira, B. & Poças Martins, J., 2013. Expansion of IFC model with structural sensors. *Informes de la Construcción*, **65**(530), pp.219–228.

Voss, E. & Overend, M., 2012. A Tool that Combines Building Information Modeling and Knowledge Based Engineering to Assess Façade Manufacturability. In *Proceedings of the Advanced Building Skins 2012 Conference*. Graz, Austria.

Webb, G. & Middleton, C.R., 2013. Structural Health Monitoring of Bridges. In M. Motavalli, C. Goksu, & B. Havranek, eds. *Proceedings of SMAR 2013 the 2nd Conference on Smart Monitoring, Assessment and Rehabilitation of Civil Structures*.

Proceedings of the International Conference on Smart Infrastructure and Construction
ISBN 978-0-7277-6127-9

© The authors and ICE Publishing: All rights reserved, 2016
doi:10.1680/tfitsi.61279.551

Green2.0: Socio-technical analytics of green buildings

Tamer El-Diraby[1], Thomas F. Krijnen[2] and Manos Papagelis*[3]

[1] *University of Toronto, Canada*
[2] *Eindhoven University of Technology, Netherlands*
[3] *University of California, Berkeley, United States*
[*] *Corresponding Author*

ABSTRACT Professionals and researchers of the AEC (Architectural, Engineering & Construction) industry, as well as, public policy makers are challenged by the increasing complexity and need to improve our understanding of the social, technical and business dimensions of green building projects. This typically requires close cooperation of the design team, the architects, the engineers, and the rest of the stakeholders at all project stages, but most importantly availability of new methods, tools and strategies that are enabled by emerging technologies. This paper builds around an online platform (*Green2.0*), that tries to leverage advancements in Building Information Models (BIM), energy-efficiency simulation tools and online social network analysis methods to enable a data-driven approach to building design, planning, construction and maintenance. The platform advances the current state of the art by providing an online integrated environment for (a) efficient storage, indexing, querying, 3D visualization and exploration of BIMs, (b) sharing BIMs and enabling online collaboration among the various stakeholders (c) interactive energy efficiency analysis of buildings by automatically linking IFC to external energy simulation libraries, and (d) interactive analysis of patterns of social interactions and collaboration networks of AEC professionals.

1 INTRODUCTION

Increasingly, we are noticing that green building research is a socio-technical process. This is because the decision of selecting energy/water saving measures, ultimately, rests on end-users. The more educated and engaged the users are, the better the chances that they will make informed decisions about greener options. To support this process, we have to match the advanced technical tools with socially-savvy tools that capture end-user needs and at the same time influence their attitude towards water and energy usage. Developing algorithms, tools and work processes that bridge the gulf between social and engineering aspects is therefore a major objective for the green agenda. This is quite a challenge given that the majority of analysis tools were developed by engineers for use by engineers.

In this paper, we introduce *Green2.0*, a web-platform that enables complex interactions between buildings and people, analysis of these interactions, as well as, interactive analysis of energy efficiency of green buildings (Papagelis et al. 2016). Fundamentally, it connects BIM to OpenStudio[1] to allow users to select different products from a catalog and study their green impact. At the same time, it allows participants (end-users or professionals) to comment and share views about design. Social network analysis (SNA) methods are then used to extract trends and visualize insights of these interactions. Green2.0 takes BIM from the realm of a software into the realm of collaborative platform for decision making. We aim to give people the controls of BIM software to study, choose, suggest and innovate new means to design, build and operate their facilities.

2 LITERATURE REVIEW AND NEED

Researchers have developed models to analyze the networked nature of project internal actors (see Di-

[1] https://openstudio.net

Marco et al. 2010; Pryke 2012). Others have considered the impact of project internal networks on the evolution of project scope (see Taylor and Levitt 2007; Wong et al. 2012). The most advanced approach is the proposal by Chinowsky and Galotti (2008) to model construction projects as social networks. Van Herzle (2004) found that inclusion of non-expert knowledge was beneficial to the planning process given that the diversity of perspectives (especially of those who are outside of the professional bubble) can (re)discover creative solutions. In fact, "citizen science" often results in superior solutions (Lakhani et al. 2009; Lakhani and Panetta 2007). Further, such solutions are by default, context-sensitive (Corburn 2003).

BIM technology has been developed and promoted as means to integrate all information of building designs. However, it is overly focused on the traditional design of facilities, i.e. not green-oriented. Designers and operators have to use an increasing set of heterogeneous software systems to complement the missing features in BIM, facing multitude of challenges in relation to interoperability and data integrity. With the increasing size and sophistication of BIM files and the increasingly iterative development cycles, the burdens of transferring data between software and the management of design changes is hindering fuller analysis.

Becerik-Gerber and Rice (2010) found that the top three BIM functions are visualization, clash detection, and creation of as-built models. While most professionals believed that sustainability analysis is of great importance, they didn't consider it to be a priority of the BIM agenda (Bynum et al. 2013). More alarming, researchers in green buildings found that BIM-based energy management is still an immature domain (Wang et al. 2013).

3 OBJECTIVE AND SCOPE

The current models and data structures for green aspects in BIM are lagging. Practitioners have been looking for incorporating green analysis (energy and water consumption) within BIM in an easy-to-use format. The solution is not just to expand BIM data standards to encapsulate all data related to green design, as this would just compound the data management tasks. Our approach and the main contribution of this research is to develop a middleware platform that serves as the bedrock upon which we can study and develop tools to enhance handling of the two challenges: how to engage users (both end-users and professionals) and harness their needs, and how to simplify energy analysis systems within a BIM environment—specifically:

Green-aware BIM: BIM models are large and complex—yet they currently have little focus on green-oriented issues throughout the building lifecycle - on accommodating alternative solutions during design; on the building operations phase; on engaging non-technical end users. Expanding BIM data standards to encapsulate semantics of green design would increase the complexity of existing data management tasks. Rather, our approach consists of establishing a middleware that can loosely integrate BIM and independent green analysis software and libraries, such as OpenStudio, without forcing a full merger.

Social-aware Analytics: Green2.0 embeds social commenting into BIM technology. This is coupled with analysis of the resulting discussion networks, which allows to understand the social dynamics between participating stakeholders and the semantics of their comments. In the era of the knowledge economy, these interaction networks constitute a rich source of data and can provide meaningful insights regarding design and operations plans. Indeed, this could provide the spark for a new realm in innovation democratization and bottom-up decision making.

4 HIGH-LEVEL ARCHITECTURE

In this section we describe the high-level architecture of our platform. Figure 1 illustrates the three components of the architecture and how they relate to each other, namely *Green2.0 MVC*, *Green2.0 BIM Management, and Green2.0 Modules.*

4.1 Green2.0 MVC

The main part of the Green2.0 core infrastructure is a web service that is based on a *Model-View-Controller* (MVC) web architecture. MVC is a popular software architectural pattern for implementing user interfaces. It divides a given software application into three interconnected parts, so as to separate internal representations of information from the ways

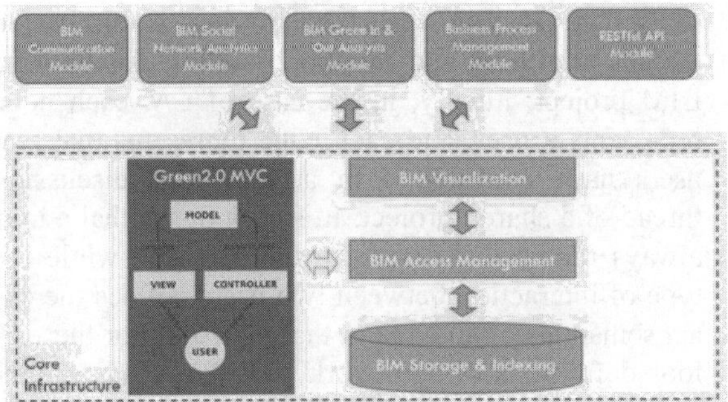

Figure 1. Green2.0 High-level architecture

that information is presented to or accepted from the users. This component is therefore responsible for managing all user interactions and domain-specific functionality. It is also responsible for integrating the BIM open source technologies, and facilitating the communication with various independent modules.

4.2 *Green2.0 BIM Management*

The most critical functionality of the Green2.0 platform's core infrastructure is the efficient management and visualization of BIM models. Towards this end, Green2.0 relies on a number of tightly-knit open source technologies:

BIM Storage & Indexing (*BIMServer*): The BIMServer (Beetz et al. 2010) enables to centralize the information of a building design project. The core of the software is based on the open standard IFC (Industry Foundation Classes) and therefore knows how to handle IFC data. The BIMserver is not a fileserver, but uses the model-driven architecture approach. This means that IFC data are interpreted by a core-object and stored in an underlying database (i.e., BerkeleyDB). The main advantage of this approach is the possibility to query, merge and filter the BIM-model and generate IFC files on the fly.

BIM Access Management (*Service Interfaces*): The *Service Interfaces* is a set of defined interfaces for interaction with BIMserver. These interfaces are defined as (heavily annotated) Java interfaces. All interfaces with namespace `org.buildingsmart.bimsie1` are implementations of the *BIM Service Interface Exchange* standard (BIMsie[2]). All calls in the `org.bimserver` namespace are BIMServer specific calls. Green2.0 uses a JavaScript Object Notation (JSON) interface to access the methods of the Service Interfaces. The JSON interface is mainly there to facilitate connecting to the BIMserver from web applications/web sites.

BIM Visualization (*BIMSurfer*): BIMSurfer[3] is an open source web-based viewer for the visualization of BIM models described as IFC models. It is based on WebGL (Web Graphics Library), a JavaScript API for rendering interactive 3D and 2D computer graphics within any compatible web browser without the use of plug-ins.

4.3 *Green2.0 Modules*

The Green2.0 high-level system architecture emphasizes separating the functionality of the system into independent, interchangeable modules, such that each contains everything necessary to execute only one aspect of the desired functionality. With modular programming, concerns are separated such that modules perform logically discrete functions, interacting through well-defined interfaces with the core architecture. Currently, Green2.0 consists of five modules.

4.3.1 *BIM Communication Module*

One of the main functionalities of the system is to provide means of online communication and collaboration of the various actors (engineers, owners, contractors, etc.,) around building design elements. These actors are coming from an Ontology that describe roles in AEC industry (see Zhang and El-Diraby 2012). Green2.0 supports online communication and collaboration through shared BIM models. In order to share a BIM online, it first needs to be uploaded by its owner in the system, typically as an IFC file. Then, the owner can share it by sending email invitations to known actors or by browsing the user database seeking for experts to join the project.

Once users have access to a shared BIM model, they can use the 3D building model visualization tool to navigate, explore, and select specific elements of the model (see Figure 2). Once an element is selected in the Tree View, the various element properties are listed that provide useful information to the expert. The collaboration is facilitated by means of a rich comment management tool that allows to submit, ed-

[2] https://buildingsmart.github.io/BIMSie/

[3] http://bimsurfer.org

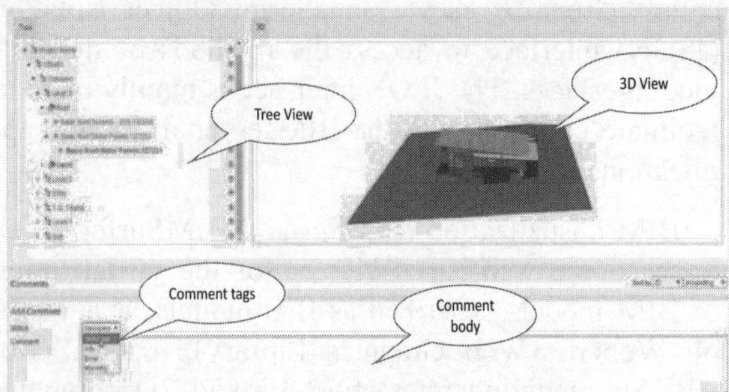

Figure 2. Snapshot of the Communication Module

it, delete, and filter comments about selected BIM elements. The functionality is similar to that found in an online discussion forum, with the exception that the discussion is domain-specific and thus domain-specific features are supported. For example, comments can only be of a specific type (info, error, warning, other). A user can navigate comments in chronological order or other semantic properties. Notifications are also available that inform actors for new dialogues or updated conversations.

4.3.2 BIM Social Network Analytics Module

Social interactions that occur among the various actors and BIM elements during collaboration processes consist valuable information for analysis. This module is responsible for the collection, storage, analysis and visualization of such data in a meaningful way. Revealing interesting patterns of communication can further enrich user experience and support decision making. The approach we follow is to define *discussion networks* based on interactions of actors and building elements and perform analysis on the underlying networks. These networks can be defined at many different levels of granularity. Aiming for a more flexible platform, we made the decision to define networks at three different levels of operation:

- Element-level Networks (*EN*)
- Project-level Networks (*PN*)
- Cross-project Networks (*CN*)

For each of the operational level above, a graph *G(V, E)* is defined comprising of a set of vertices *V* and a set of edges *E*. In the case of *EN*, each node represents a user and each edge represents that two users have contributed in a discussion thread about a specific building element. Accordingly, in the case of *PN*, each node represents a user and each edge represents that two users have contributed in discussion threads of at least one common building element of a BIM project. Finally, in the case of *CN*, each node represents a user, and each edge represents that two users have contributed in at least one discussion thread of a shared project. It is easy to see that a user always represents a node in the network, while the type of interaction between two users defines the exact semantics of an edge in that network. For the various definitions of a network (*EN*, *PN*, *CN*), a number of network insights are possible, based on network analysis. For each network, Green2.0 reports a number of important network structure measures, such as *network size*, *diameter*, *density* and *characteristic path length*. Note that due to the system's architecture, it is easy to plug-in more network measures in the future.

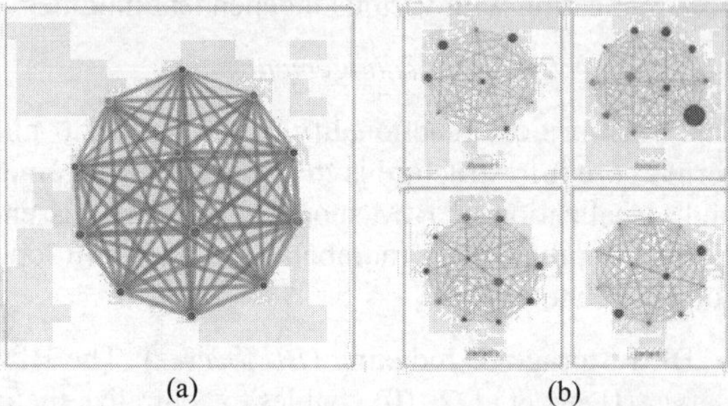

(a) (b)

Figure 3: Green2.0 SNA. (a) Example project-level network (*PN*), (b) Visualization of trending discussions (element-level networks)

This module is also responsible for the visualization of the various networks. Figure 3(a) shows an example *PN* network, while Figure 3(b) illustrates a number of *EN* networks about various elements of a specific project. Note that a user can easily depict trending discussions and navigate there directly by means of selecting the network.

4.3.3 BIM Green In & Out Analysis Module

The aim of the Green In & Out Module is to provide a comparative energy analysis of building models. For example, an engineer or an end-user might want to assess alternative window systems for the same building. In order to facilitate this, a product substitution functionality was developed. It allows to locally replace building products, such as a window, with a comparable product (from the product catalog).

To conduct the energy analysis (of each model being compared), OpenStudio is used. It is a cross-platform (Windows, Mac, and Linux) collection of software tools to support whole building energy modeling using EnergyPlus and advanced daylight analysis using Radiance. OpenStudio is an open source project which includes graphical interfaces along with a Software Development Kit (SDK). In addition, OpenStudio provides a rapid development mode and open application programming interface (API), which makes it highly extensible and customizable. It is rather simple for developers to either build on existing applications or create completely new ones to conduct customized building energy analysis. All of these aspects suggest OpenStudio as a suitable platform for initial targeting to support the data exchange needs of building energy modeling.

Recall that Green2.0 represents BIM models using the IFC standard. On the other hand, the OpenStudio API requires a specific file format, Open Studio Model (OSM), as input in order to run energy analysis of a model using tools such as Radiance (advanced daylight analysis tool) and EnergyPlus (whole building energy modeling). The first challenge therefore is to map the information represented in an IFC file to information that can be represented in an OSM file. A crucial difference between the two formats is that IFC files describe a building as a decomposition of individual components, which have one or more solid-volume geometrical representations and are enriched with semantic and relational information. An OSM file describes the building from the viewpoint of thermal zones and thin-walled space boundaries. Therefore, not only does the information need to be encoded differently, the geometrical information needs to undergo a translation process that flattens the solid-volume geometry for space bounding elements (such as walls, roof and floor slabs) into thin-walled thermal zone boundaries (see Krijnen et al. 2015). The geometrical conversion process sketched above requires a programming environment to efficiently and effectively manipulate IFC files and geometry in the Green In & Out Module. The choice has been made to develop this system in the Python programming language using the IfcOpenShell[4] and pythonOCC[5] modules. The former allows to efficiently parse IFC files and return geometrical definitions as Open Cascade[6] Boundary Representations (BReps). The latter allows BReps to be further manipulated so as to conform to the thin-walled thermal zone boundaries expected in OSM file. In addition to the geometrical definition, OpenStudio expects semantic information that pertains to the use of the building to deduce heating, cooling loads and functional constraints. These are to be defined by the user. In summary, the steps of conducting comparative energy analysis in Green2.0 are as follows:

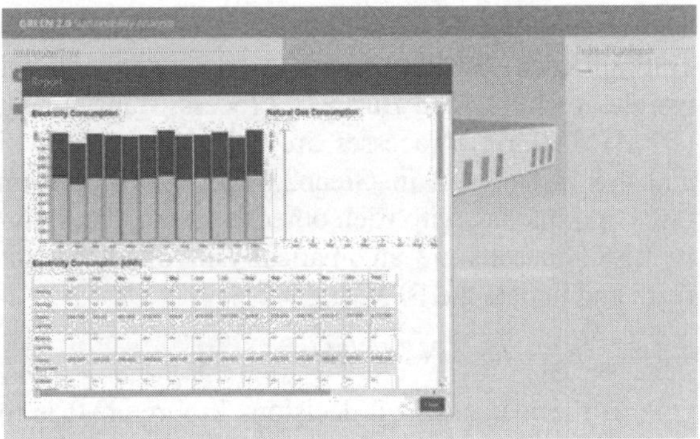

Figure 4. Green In & Out Module: Energy analysis results

- User can select any *targeted product* to study
- The user can choose a *replacement product* for the *targeted product*, from a predefined catalog
- The user must define the usage scenario of the facility (if not already done)
- IFC data of alternative designs are transliterated to OSM data
- OpenStudio is then invoked for energy analysis
- Results of the analysis are returned to user (see Figure 4)

4.3.4 *Business Process Management Module*

One of the objectives of the Green2.0 project is to improve corporate performance by optimizing business processes related to the building projects. To this end, we designed and developed a Business Process Management (BPM) module that operates on processes that become available in the Green2.0 platform and supports:

- Monitoring and exploration of business processes that evolve in Green2.0
- Offline analysis of BIM business processes

[4] http://ifcopenshell.org

[5] http://www.pythonocc.org

[6] http://www.opencascade.com

The above functionality becomes feasible by integrating Green2.0 with Activiti[7], an open source lightweight workflow and Business Process Management (BPM) platform. Processes are designed in Activiti and are instantiated in Green2.0. As users perform tasks and interact with each other in Green2.0, Activiti RESTful calls are automatically invoked that inform and update the BPM engine.

4.3.5 *RESTful API Module*

One of the major design decisions in Green2.0 is to allow third-party services to access Green2.0 resources. This is enabled through a RESTful API that is accessible via standard HTTP methods by a variety of HTTP clients, including browsers and mobile devices. Figure 5 illustrates a typical architecture for supporting RESTful APIs in Green2.0. Third-parties are accessing the Green2.0 REST API by submitting URL requests; the platform performs the necessary computation and compiles a REST answer to the request formatted as a JSON file. Through the API a number of resources become available to third-party services, clients and applications, such as, BIM users, projects, IFC elements, comments, information about the discussion networks, business processes.

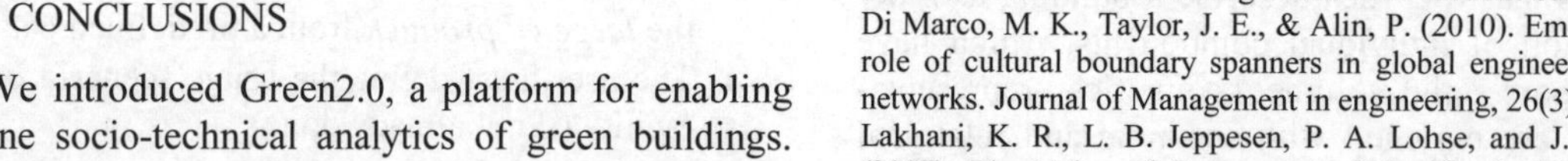

5 CONCLUSIONS

We introduced Green2.0, a platform for enabling online socio-technical analytics of green buildings. The platform has been successfully deployed at the University of Toronto and as a Canarie[8] research platform - a collection of platform services to be used by multiple independent research teams. Early phase users of the system include research and industry collaborators. Designing and developing Green2.0, we had to identify the scope of the system, investigate alternative system design and architecture concepts, explore data collection methods and assess the relevant emerging technologies. The premise of our work is that by engaging users in early design phases and by simplifying energy analysis within BIM environment it is likely to have a profound beneficial effect for both the AEC industry and the society at large. Our research describes a significant improvement over current practice and tries to advance the current state of the art in green building design towards sustainable development of cities.

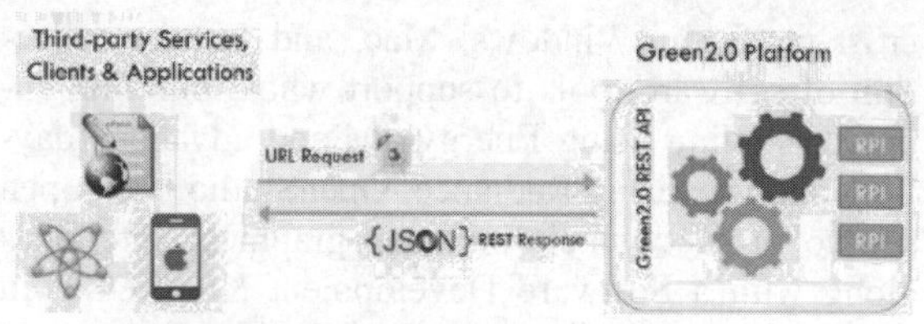

Figure 5. RESTful API Module Architecture

REFERENCES

Becerik-Gerber, B., & Rice, S. (2010). The perceived value of building information modeling in the US building industry. ITcon, 15, 185-201.

Beetz, J., van Berlo, L., de Laat, R. and van den Helm, P., 2010, November. BIMserver. org–An open source IFC model server. In Proceedings of the CIP W78 conference.

Bynum, P., Issa, R. R., and Olbina, S. (2013). "Building Information Modeling in Support of Sustainable Design and Construction." Journal of Constr. Engin. and Manag., 139(1), 24-34.

Chinowsky, P., Diekmann, J., & Galotti, V. (2008). Social network model of construction. ASCE Journal of Constr. Engin. & Manag.

Corburn, J., 2003. Bringing local knowledge into environmental decision making improving urban planning for communities at risk. Journal of Planning Educ. and Research, 22(4), pp.420-433.

Krijnen, T., Papagelis, M., & El-Diraby T. Green2.0: A web-based system fostering sustainable and socio-technical analysis and automatic conversion of building information models. Submitted.

Di Marco, M. K., Taylor, J. E., & Alin, P. (2010). Emergence and role of cultural boundary spanners in global engineering project networks. Journal of Management in engineering, 26(3), 123-132.

Lakhani, K. R., L. B. Jeppesen, P. A. Lohse, and J. A. Panetta (2007). The Value of Openness in Scientific Problem Solving. Harvard Business School Working Paper No. 07-050, 2007.

Lakhani, K. R., and J. A. Panetta. (2007). The Principles of Distributed Innovation. Innovations: Technology, Governance, Globalization, Vol. 2, No. 3.

Papagelis, M., Krijnen, T., Elshenawy, M., Konomi, T., Fang, R., El-Diraby, T. (2016). Green2.0: Enabling Complex Interactions Between Buildings and People. Proc. of ACM CSCW Companion.

Pryke, S. (2012). Social network analysis in construction. Wiley-Blackwell

Taylor , J., & Levitt , R. (2007). Innovation alignment and project network dynamics: an integrative model for change. Project management journal.

Van Herzele, A. (2004). "Local Knowledge in Action: Valuing Nonprofessional Reasoning in the Planning Process. Journal of Planning Education and Research, Vol. 24, No. 2.

Wong, K., Unsal, H., Taylor, J. E., & Levitt, R. E. (2010). Global dimension of robust project network design. Journal of construction engineering and management, 136(4), 442-451.

Zhang, J. and El-Diraby, T. E. (2012). "Social semantic portal for coordinating construction communication", J. of Computing in Civil Engineering, ASCE, Vol 26, No 1.

[7] http://activiti.org/

[8] www.canarie.ca

Proceedings of the International Conference on Smart Infrastructure and Construction
ISBN 978-0-7277-6127-9

© The authors and ICE Publishing: All rights reserved, 2016
doi:10.1680/tfitsi.61279.557

Information future-proofing assessment for infrastructure assets

T. Masood*[1,2], G. Yilmaz[1,2], D.C. McFarlane[1,2], A.K. Parlikad[1,2], K. Harwood[3,4] and R. Dunn[3]

[1] *Institute for Manufacturing, University of Cambridge, Cambridge, UK*
[2] *Centre for Smart Infrastructure and Construction (CSIC), University of Cambridge, Cambridge, UK*
[3] *Bridges & Structures, Hertfordshire County Council, Hertford, UK*
[4] *Bridge Engineering Midlands, Arup, The Arup Campus, Solihull UK*
* *Corresponding Author*

ABSTRACT Even though infrastructure assets serve society for long time, asset information does not last that long in all cases. This is primarily due to technological and organisational changes and disruptive events over long time. Therefore, information future-proofing considerations are required to be incorporated in order to sustain the assets. However, even though companies consider some aspects but largely mechanisms for information future-proofing assessments are not fully known. As part of the information future-proofing research conducted at CSIC ; what, why and how characteristics of information future-proofing were explored through a number of semi-structured interviews and workshops with leading companies dealing with infrastructure assets. This paper presents an information future-proofing assessment approach as an output of the CSIC future-proofing research. The information future-proofing assessment approach was applied in case studies of Hertfordshire County Council bridges and structures, University of Cambridge department building, Crossrail tunnels and pumps, and London Underground infrastructure.

1 INTRODUCTION

Information future-proofing is defined as "the process to ensure that required infrastructure asset information is retrievable (reusable) throughout or beyond the whole lifecycle of infrastructure assets when needed" (Masood, Cuthbert, McFarlane et al 2013). It is very important for key decision makers, future owners, operators, the environment and society. Hence, identifying through-life information requirements at earlier life cycle stages of infrastructure and ensuring availability of information at all stages by planning and taking appropriate actions for its collection, retention and reuse in long term is imperative (Masood, Cuthbert, McFarlane et al 2013).

Long-term availability of asset information for large-scale infrastructure e.g. during operation and maintenance phases or beyond is challenging. This is partly due to the nature of multi organisational management of infrastructure assets. Non-availability of critical information may cause severe impacts on decisions related to infrastructure operations e.g. in some cases resulting in closure of business or delays or disruptions in infrastructure operations. In another example, statutory retention of design, material and construction information is typically limited to a period of ten years and beyond this the data is often destroyed. Changing asset owners, loss of documentation, misplacement and obsolete file formats are some of the other factors that can impact on long term availability of critical infrastructure asset information.

Currently there is a major initiative underway, in the UK, to ensure the development of formalised infrastructure information via so called Building Information Modelling (BIM) (NBS 2015). Standards have been published to foster this agenda recently e.g. BS 1192-5, BS 1192-4, BS 1192-3, BS 1192-2, and ISO 55000 covering requirements and methodology for transfer of information relating to facilities, achieving BIM level 2 for CAPEX and OPEX, and general asset management requirements related to information retention (BSI 2015; BSI 2014a; BSI 2014b; BSI 2013; ISO 2014). These and other related initiatives are helping to ensure that adequate information retention becomes an organisational imperative.

With developments such as BIM as well as 3D design models, and electronic project management, long-term preservation and availability of such information has attracted much attention from stakeholders in the infrastructure sector. The industry has identified the above issues as being pivotal to be resolved for the ability to operate, service, maintain, and reconfigure their assets.

This paper presents an information future-proofing assessment approach for infrastructure assets, which is aimed at guiding life cycle stakeholders e.g. planners, designers, developers, managers and owners of major infrastructure assets to consider dependencies of decisions, information and enabling technologies, risks and impacts of information loss on the infrastructure assets in long-term and prepare to eliminate or mitigate the risks.

The information futureproofing assessment approach is tested in case studies of Hertfordshire County Council that manages a portfolio of more than 1200 bridges and structures, University of Cambridge department building, Crossrail tunnels and pumps, and London Underground infrastructure. The case studies illustrate usability and usefulness of the information future-proofing approach in industry, especially in improving the current risk assessment/management and stakeholder management processes.

The rest of the paper is structured as follows. Section 2 presents information futureproofing assessment approach. Case studies are included in section 3. Section 4 discusses the results while section 5 concludes the paper.

2 A STRUCTURED INFORMATION FUTURE-PROOFING ASSESSMENT APPROACH FOR INFRASTRUCTURE ASSETS

An information future-proofing assessment approach is developed based upon consideration of information retention requirements and assessment of risks of information loss in long-term. Following process is followed as part of the information future-proofing assessment approach (see Figure 1).

- Identify information retention requirements for long-term (information map);
- Identify key hazards leading to information loss in long-term; and
- Assess risks of information loss in long-term.

Some important considerations while applying the assessment approach are discussed in the following.

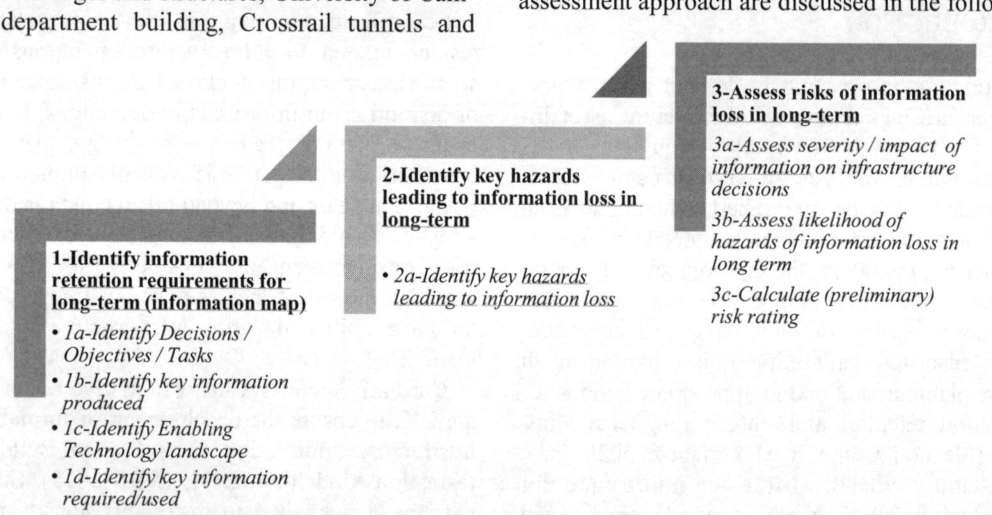

Figure 1. The information future-proofing approach.

2.1 *Information retention requirements for long-term (information map)*

An information map is developed to identify key decisions taken across infrastructure life cycle stages, key information types created and used for those decisions. This provides an understanding of the future-proofing challenges in terms of impacts of loosing related information, which could potentially affect respective decisions. The information map identifies key enabling technology landscape used to store key information and supports key decisions across infrastructure life cycle stages. This provides an understanding of the information retention requirements for long term and likelihood of loosing key information, which could potentially affect respective decisions.

2.2 *Hazards leading to information loss in long-term*

Hazards of information loss in long term are identified from extensive literature analysis as well as from semi structured interviews with industrial partners of the project e.g. London Underground, Crossrail, Arup, and Hertfordshire County Council. These are broadly categorised as technology (hardware and software) and organisational related.

2.3 *Risks of information loss in long-term*

Risks of information loss in long-term are identified as follows:

- Assessing severity/impact of information on infrastructure decisions (on a scale of 1-16);
- Assessing likelihood of hazards of information loss in long-term (on a scale of 1-5); and
- Calculating (preliminary) risk rating by multiplying severity and likelihood.

Case studies are presented in the following in which the assessment approach is applied.

3 CASE STUDIES

The information future-proofing assessment approach was piloted in following case studies: Hertfordshire County Council (bridges and structures), University of Cambridge (department building), Crossrail (tunnels and pumps) and London Underground.

The approach was piloted on Hertfordshire County Council (HCC) Bridges and Structures. The HCC has a portfolio of approximately 1200 bridges and structures. Out of these, a representative cross section of following seven bridges and structures were selected for further study (see Figure 2).

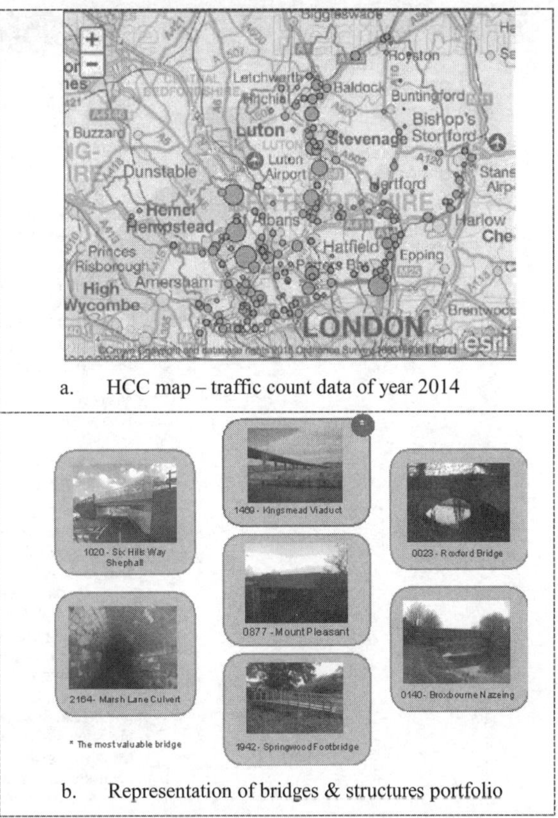

a. HCC map – traffic count data of year 2014

b. Representation of bridges & structures portfolio

Figure 2. Hertfordshire County Council – map and bridges & structures portfolio (Hertfordshire County Council 2015).

4 RESULTS AND DISCUSSION

The information future-proofing assessment approach provided a structured way of considering information future-proofing, and can be incorporated in routine assessment practices. Hazards of information loss in long term were identified from extensive literature analysis as well as from semi structured interviews with industrial partners of the project e.g. London Underground, Crossrail, Arup, Hertfordshire County Council and University of Cambridge (Masood, Yilmaz, McFarlane et al. 2015).

The hazards of information loss in long-term are broadly categorised as technology (hardware and software) and organisational related (see Figure 3) (Masood, Yilmaz, McFarlane et al. 2015). It is proposed that the hazards presented in Figure 3 be used as a minimum. However, additional hazards may be identified in each infrastructure asset where the information future-proofing assessment approach is applied.

Hardware

- The physical storage media may deteriorate in the long term period.
- The physical storage media may be obsolete in the long term period.
- May prevent accessing information if data storage company is not big enough to survive longer in the market.

Software

- May cause compatibility problems if new file format is not opened by older release of its software.
- May cause interoperability problems if different software is used for a same activity.
- May cause information loss if a file is converted into another file format.
- May cause interoperability problems if well established and widespread file formats are not preferred.
- May cause accessibility problems if common data environments are not used for information sharing.
- May cause conflicts when same information is stored in more than one database.
- May prevent retrieving information if information is sent and retrieved via emails.
- May cause loss of information if information is migrated from one software to another without any policy or standard.
- May cause information loss if information held in emails are not saved to secure accessible medium.
- May prevent access to information if information has authorization and authentication.
- May prevent accessibility if information is not viewed through different software platforms such as a hand held device.
- May cause ambiguity if files/folders do not have any version control or are not named according to naming conventions.
- May increase complexity if a new policy is adopted.

Organizational

- Lack of storage space when disposed of old data.
- Loss of device - portable storage media
- May cause information incompleteness if semantically related information are stored in different locations and spaces.
- May cause ambiguity if too much information type (Decisions, photos, drawings) are kept in a same file/folder.
- May prevent retrieving information if scanned papers are not are archived.
- Low resolution of scanned papers.
- Inadequate or no metadata.
- Unreliable information from unknown resource.
- May cause ambiguity if owner/facility manager of the physical asset changes.
- May cause ambiguity if there is unclarified ownership.
- Operator error
- May cause ambiguity if data stored in wrong directory, or with wrong file name, cannot be located later.

Figure 3. Identification of hazards of information loss in long term (Masood, Yilmaz, McFarlane et al. 2015).

An example of preliminary risk assessment of hardware related hazards is presented in Figure 4. Following hardware related hazards were considered against key information types identified to be retained for whole asset life:

- Physical storage media may deteriorate in long-term;
- Physical storage media may become obsolete in long-term; and
- May prevent information accessibility if data storage company go bust.

In order to conduct a meaningful assessment, the hazards of information loss need to be updated and contextualised.

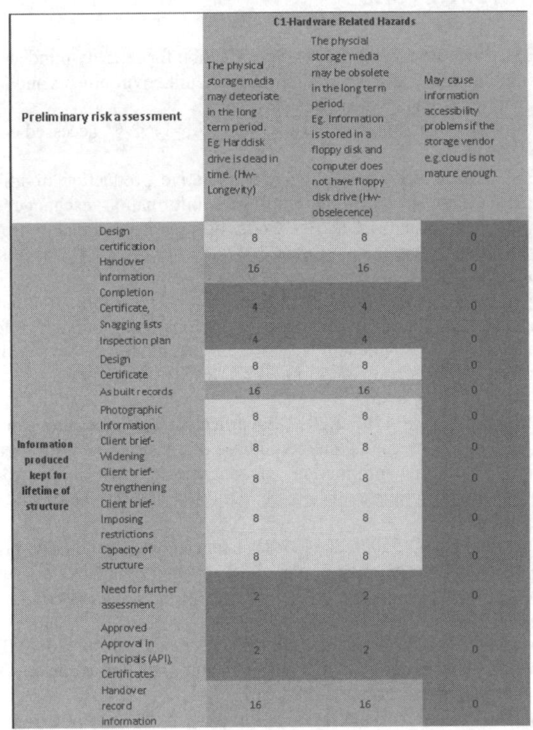

Preliminary risk assessment		C1-Hardware Related Hazards		
		The physical storage media may deteoriate in the long term period. Eg. Harddisk drive is dead in time. (Hw-Longevity)	The physcial storage media may be obsolete in the long term period. Eg. Information is stored in a floppy disk and computer does not have floppy disk drive (Hw-obselecence)	May cause information accessibility problems if the storage vendor e.g cloud is not mature enough.
Information produced kept for lifetime of structure	Design certification	8	8	0
	Handover information	16	16	0
	Completion Certificate, Snagging lists	4	4	0
	Inspection plan	4	4	0
	Design Certificate	8	8	0
	As built records	16	16	0
	Photographic Information	8	8	0
	Client brief-Widening	8	8	0
	Client brief-Strengthening	8	8	0
	Client brief-Imposing restrictions	8	8	0
	Capacity of structure	8	8	0
	Need for further assessment	2	2	0
	Approved Approval In Principals (API), Certificates	2	2	0
	Handover record information	16	16	0

Figure 4. Preliminary risk assessment - hardware related hazards.

The lessons learnt from the case studies suggest that the approach is usable in industry and can be useful in assessing the current state of information future-proofing for asset portfolios. The approach provides an assessment of information future-proofing, which can help in setting long term goals and priorities for infrastructure assets.

The outputs from the approach can guide infrastructure developers and managers with advisory input in their considerations of the potential impacts of information related future disruptive events and changes on the assets and how the assets can be most suitably prepared to respond or absorb them. This can include provision of guidelines to enable information future-proofing strategy based upon outcomes of the assessments. For example, by adopting open standards and open file formats; and having a clear information future-proofing strategy along with review cycles in place.

The information future-proofing is not a separate consideration but an extension to current practice. It is found from the case studies that the approach can be included in current practices e.g. risk assessment/management/registers, stakeholder management, future (upgrade) projects planning and building information modelling.

As part of the assessment approach, the information map of the HCC suggests that adopting information future-proofing strategies would be beneficial for following key information supporting key infrastructure decisions over the life time of bridges and structures.

Design certification produced at detailed design stage (has the designer signed off the design?) is useful for assessments every 10 years and ongoing reference. Handover information produced at as built stage (how will the structure be maintained?) is useful whenever maintenance is required following inspections. Completion certificate and snagging lists produced at as-built stage (certified completion and any final snags) are useful for audit trails. Inspection plan produced at inspection planning stage (where is the structure?) is useful for general inspections every two years and maintenance projects.

Structural maintenance works are benefited from following information: Photographic information produced at prioritisation of minor maintenance works (which structure should be maintained first?); Approved Approval In Principals and Certificates produced for acceptance of the design proposal; and Handover record information produced for handover of bridges between asset owners.

Structural reviews and technical approvals are benefited from following information: Client brief-Widening produced at as-built stage (change of use/modification – how will the bridge be widened?); Client brief-Strengthening produced at as-built stage (change of use/modification – how will the bridge be strengthened?); Client brief-Imposing restrictions produced at as-built stage (change of use/modification); Capacity of structure information produced for assessment (is the structure adequately designed to take the loads?); and Need for further assessment information for structural review whether the existing bridge is in the safety margins.

5 CONCLUSIONS

The following conclusions can be drawn from the research presented in this paper:

- Information future-proofing is a concern for asset managers.
- A meaningful approach for information future-proofing is not formally embedded into existing options appraisal processes but is needed for effective and long term infrastructure asset management.
- An information future-proofing assessment approach has been proposed along with two industrial case studies. The approach is helpful in formulating a strategy and priorities for wider asset portfolios.

It is recommended to assess the applicability of the information future-proofing approach to non-infrastructure asset portfolios.

ACKNOWLEDGEMENT

The authors would like to acknowledge the Centre for Smart Infrastructure & Construction, EPSRC (Grant EP/K000314/1), Innovate UK and the industrial partners, who collectively supported and funded this research: IBM, Crossrail, London Underground, Costain, Cementation Skanska, Redbite, Atkins, Halcrow (CH2M), Scottish Water, Arup, Lang O'Rourke, Hertfordshire County Council, The Woodhouse Partnership. The authors also acknowledge case study support from Hertfordshire County Council, Crossrail, London Underground, Costain, Boeing, National Archives and University of Cambridge (Institute for Manufacturing and Estates Management).

DATA STATEMENT

There is no real data to share for this paper.

REFERENCES

BSI. 2015. PAS 1192-5:2015 Specification for security-minded building information modelling, digital built environments and smart asset management, available from: http://shop.bsigroup.com/forms/PASs/PAS-1192-5/, accessed on 5/1/2016.

BSI. 2014a. BS 1192-4:2014 Collaborative production of information Part 4: Fulfilling employers information exchange requirements using COBie – Code of practice, available from: http://shop.bsigroup.com/forms/PASs/BS-1192-4-2014/, accessed on 22/2/2015.

BSI. 2014b. PAS 1192-3:2014 Specification for information management for the operational phase of assets using building information modelling, available from: http://shop.bsigroup.com/forms/PASs/PAS-1192-3-2014/, accessed on 18/2/2015.

BSI. 2013. PAS 1192-2:2013 Specification for information management for the capital/delivery phase of construction projects using building information management, available from: http://shop.bsigroup.com/forms/PASs/PAS-1192-2/, accessed on 27/4/2014.

ISO. 2014. ISO 55000:2014 Asset management – Overview, principles and terminology, pp 1-19, January, available from: https://www.iso.org/obp/ui/#iso:std:55088:en, accessed on 6/3/2015.

Masood, T. McFarlane, D.C. Parlikad, A.K.N. et al. 2015. Towards future-proofing of UK infrastructure, ICE Infrastructure Asset Management, accepted Jul 2015.

Masood, T. McFarlane, D.C. & Fielding, A. 2015. Assessing future-proofing of infrastructure, University of Cambridge, September.

Masood, T. McFarlane, D. Schooling, J. et al. 2014. The role of future proofing in the management of infrastructural assets, International Symposium for Next Generation Infrastructure, Vienna, 30 Sep - 1 Oct, pp. 147-152, available from http://discovery.ucl.ac.uk/1469416/1/Final%20Proceedings.pdf (accessed 3/8/2015).

Masood, T. Cuthbert, R. McFarlane, D.C. et al. 2013. Information future-proofing for large-scale infrastructure. In *Proceedings of the IET/IAM Asset Management Conference*, London, 27-28 November, DOI: 10.1049/cp.2013.1945

Masood, T. Yilmaz, G. McFarlane, D.C. et al. 2015. An information future-proofing approach for large-scale infrastructure, Department of Engineering, University of Cambridge, September.

NBS. 2015. NBS National BIM Report, National Building Specification, pp 1-40, available from: http://www.thenbs.com/pdfs/NBS-National-BIM-Report-2015.pdf, accessed on 13/1/2016.

Yilmaz, G. Masood, T. & McFarlane, D.C. 2015. Identifying and validating hazards in support of information future-proofing – case study of a building, In *Proceedings of IET/IAM Conference on Asset Management*, Institute of Engineering and Technology, London, November.

Proceedings of the International Conference on Smart Infrastructure and Construction
ISBN 978-0-7277-6127-9

© The authors and ICE Publishing: All rights reserved, 2016
doi:10.1680/tfitsi.61279.563

The Singapore cyber-civil-infrastructure project

I.F.C. Smith*and R. Pasquier

Swiss Federal Institute of Technology (EPFL), Lausanne, Switzerland
* *Corresponding Author*

ABSTRACT Sensor-data interpretation for asset management, when performed appropriately, has the potential to uncover unknown reserve capacity so that decision making related to repair, retrofit, improvement, extension and replacement results in timely, low-cost and sustainable solutions. Current proposals for data interpretation may lead to biased predictions of behavior due to unknown levels of systematic uncertainty and a lack of knowledge of their effects on correlations. This paper describes a project that is underway in Singapore at the Singapore-ETH Centre for Global Environmental Sustainability. The overall goal of the project is to build on more than fifteen years of research to determine which model-based data-interpretation method is the most reliable and robust within the context of decision making for civil-engineering infrastructure that contains sensors. Following this description, a summary of two studies, where simple beams are used to compare population-based approaches under various contexts of model uncertainty, is provided. Finally, the current plan for case-study-driven evaluation is outlined. It is concluded that this strategy has significant potential to add value, particularly for activities that focus on replacement avoidance for improved sustainability.

1 RESPONSIVE CITIES

The Singapore-ETH Centre for Global Economic Sustainability (SEC) began in 2010. Following a very successful first phase, it has recently been awarded a second phase of funding (2015-2020) for its "Future Cities" program.

One of the three "Scenarios" in this program is called "Responsive Cities" where the focus is on data interpretation. In a departure from the term "Smart City" which has predominately been driven by the computer-industry, the people and the infrastructure in a responsive city constructively react to disturbances, changes and needs. In a responsive city, inhabitants, designers, planners and engineers are the major focus, not the technology. All components, including people, work together to increase the livability and resilience of the city.

One of four projects in the Responsive-City Scenario is a project called "Cyber Civil Infrastructure". Funded with approximately $S2.4 million (£1.1 million), by the end of 2016 the project will have four PhDs and one postdoc in permanent residence in Singapore.

2 CYBER CIVIL INFRASTRUCTURE

In 2006, the National Science Foundation (USA) identified cyber physical systems as a strategic research area (Wolf, 2007). Efforts have concentrated on robotic surgery, nano-level manufacturing, deep-sea exploration and health-care monitoring. The construction sector has not yet received the support it deserves in spite of its growing participation in the World economy and its base-line contribution to the quality of life.

The general aim of the Singapore project is to study the application of cyber physical systems to civil infrastructure. The case of Singapore is an excellent example of how cities have increased their attractiveness through investment in civil infrastructure. Implementation of cyber civil infrastructure has the potential to ensure that these investments are used to optimize management, guide infrastructure modifications and improve future designs.

The research topic of cyber civil infrastructure includes studies of data-interpretation methods for civil infrastructure systems that contain sensors. These sensors produce data that need to be evaluated appropriately in order to obtain a greater understanding of real behaviour. The literature has many examples of full-scale case studies where real behaviour is uncovered to be significantly different from assumed design behavior (for example, Catbas et al, 2013). Furthermore, civil infrastructure systems of the future will need to perform in contexts and environments that today, can only be partially defined.

A good understanding of real behaviour supports decision making related to repair, improvement and extension of existing systems and ultimately, design of new systems. Savings due to sensor-data-informed decisions can exceed hundreds of millions of dollars, particularly when replacement of infrastructure, such as bridges, is avoided (Sweeney, 1990).

The main objective of this project is to determine which behaviour model-based data interpretation method is the most robust and reliable for the context of full-scale civil-engineering infrastructure. Specific objectives include 1) testing and extending a recently developed framework for data interpretation; 2) comparing Bayesian model updating with model falsification using criteria of fidelity, robustness to unrecognized errors and compatibility with engineering heuristic knowledge; 3) executing full-scale case studies to illustrate and confirm usefulness; and 4) developing appropriate methodologies for measurement system design (choice of sensors and placement). This paper will focus on the background research and plans related to Objective 2.

3 DATA INTERPRETATION METHODS

When sensors measure directly parameters of interest (for example, indoor temperature), data interpretation is usually straightforward and the required action is often clear (for example, a thermostat that triggers heating). When direct measurement of parameter of interest is not possible, data interpretation becomes more challenging. Low-cost approaches are signal-analysis techniques, such as correlation based methods (SOM, PCA) and wavelet analyses (for examples, see Catbas et al, 2013; Posenato et al, 2010; Laory et al, 2013). Use of these approaches may result in detection of anomalies with weak information related to location and cause, since there is no physical-principle-based behaviour model. These approaches are not very helpful for predictions and therefore, they cannot provide much support for decisions related to repair, improvement, extension and replacement of infrastructure.

Better decision support is provided through the application of identification methodologies that use behaviour models which are founded on principles of structural mechanics. In these applications, the word identification involves finding either a set or sets of model-parameter values that lead to reasonable explanations of measurement data. Identified models are then used for predicting future performance. This is especially useful when studying aspects such as new geometries, modified geometries, new reinforcing materials and changing environments. When managing civil-infrastructure assets, such aspects are usually more important than damage detection, which is a typical research driver in other engineering fields such as aeronautics and electrical engineering.

Model-based structural identification is essentially an inverse task of finding causes of observed effects and as a result, it is intrinsically ambiguous (Raphael and Smith, 2013). Therefore, population approaches (evaluation of many model instances) are attractive since the ambiguity of the task is explicitly represented. Bayesian updating (Beck and Katafygiotis, 1998) is the most widely studied population approach. Over the past ten years hundreds of papers have promoted the use of some form of Bayesian inference strategy. However, a recent article in Science warned of the dangers of using Bayesian approaches when key information (in this case, prior probabilities) is incorrect (Efron, 2013). More importantly, Efron noted that when in doubt, results obtained from Bayesian updating should be checked using other approaches.

Measurement data interpretation for identification involves uncertainty coming from many sources and having several statistical forms. The interpretation task is described as follows. Quantify unknown model-parameter values $\boldsymbol{\theta} = [\theta_1, \theta_2, \dots, \theta_N]$ using measurement data $\hat{y}_i$. Equation 1 relates measurement values, errors and model predictions.

$$\hat{y}_i + u_{\hat{y},i} = y_i = g(x_i, \boldsymbol{\theta}) + u_{g,i} \ \forall i = 1, \dots, n_m \quad (1)$$

where n_m is the number of measurements
i denotes the measurement location
$g(\boldsymbol{\theta})$ is a model class
$g(x_i, \boldsymbol{\theta})$ is the model-prediction value at degree of freedom x_i
y is the real value (unknown)
$u_{\hat{y}}, u_g$ are the measurement and modelling errors respectively

The function $g(\boldsymbol{\theta})$ may be either a statistical or a physical-model class. Civil infrastructure asset management (reserve-capacity estimation, repair, improvement, extension and replacement assessment) typically involves prediction through extrapolation, Therefore, the function, $g(\boldsymbol{\theta})$ is most useful when it is a physical-principles-based model class, such as a parameterized finite-element formulation based on structural mechanics. Often, on-site inspection guides the heuristic determination of $g(\boldsymbol{\theta})$ by experienced engineers.

Instead of using scalar values for errors, they can be described as distributions of uncertainty (U) as follows in equation (2):

$$\hat{y}_i + U_{\hat{y},i} = g(x_i, \boldsymbol{\theta}) + U_{g,i} \ , \forall i = 1, \dots, n_m \quad (2)$$

Rearranging Equation 2 to have the residual value (prediction minus measurement) on one side of the equation leads to the right side expressed in terms of modelling and measurement uncertainty. Traditionally, engineers use residual minimization methodologies which involve curve-fitting algorithms of various levels of sophistication. This is equivalent to assuming one of the following: 1) there is no uncertainty; 2) the measurement and modelling uncertainties cancel each other; 3) all uncertainties have zero means and there are enough measurements to correctly assume that the composite residual mean over all measurements is zero. None of these assumptions are correct for full-scale civil infrastructure, for example see Goulet et al (2014).

Two methods are compared next according to their ability to find $\boldsymbol{\theta} = [\theta_1, \theta_2, \dots, \theta_N]$ under several situations. Both methods involve sampling model instances and this creates a population of model predictions at each measurement location.

3.1 Bayesian Inference

In this approach, the following relationship is used to find probability distributions for parameter values:

$$p(\boldsymbol{\theta}|\hat{\boldsymbol{y}}) = \frac{p(\hat{\boldsymbol{y}}|\boldsymbol{\theta})p(\boldsymbol{\theta})}{p(\hat{\boldsymbol{y}})} \quad (3)$$

where $p(\boldsymbol{\theta}|\hat{\boldsymbol{y}})$ is the distribution function for a set of parameter values given a set of measurements $\hat{\boldsymbol{y}}$, $p(\hat{\boldsymbol{y}}|\boldsymbol{\theta})$ is called the likelihood function and it is the probability of a set of measured values given a set of parameter values, $p(\boldsymbol{\theta})$ is the probability prior to measurement and finally, $p(\hat{\boldsymbol{y}})$ is the probability of the set of measurements. The likelihood function in Equation 3 is typically assumed to have a Gaussian form

$$p(\hat{\boldsymbol{y}}|\boldsymbol{\theta}) = \left(\frac{1}{\sigma_\epsilon\sqrt{2\pi}}\right)^{n_m} exp\left(-\frac{1}{2\sigma_\epsilon{}^2}\sum_{i=1}^{n_m}\left(g(x_i, \boldsymbol{\theta}) - \hat{y}_i\right)^2\right) \quad (4)$$

where σ_ϵ is the standard deviation of the combined uncertainty distribution of the random variable $U_{\hat{y},i} - U_{g,i}$. The formulation in Equation 4 implicitly includes the assumption that this combined uncertainty function also has a Gaussian form and it has exactly the same value for all measurement locations. Furthermore, uncertainties are assumed to be independent and there is no systematic uncertainty. Although more sophisticated formulations are possible, they require knowledge of hard-to-determine factors such as covariance. This situation is further complicated by the knowledge that covariances in coupled systems can be dependent on the value of systematic uncertainty.

3.2 Error Domain Model Falsification

An alternative strategy to Bayesian updating is based on the observation that measurements are more useful when they falsify models rather than validate them

(Tarantola, 2005, 2006). This has led to methodologies for model falsification in areas such as environmental science (Beven, 2002) and civil engineering (Raphael and Smith, 1998). This approach has been applied successfully to full-scale bridges (Robert-Nicoud et al, 2008; Saitta et al, 2008; Goulet et al, 2013A; Pasquier et al, 2014) water supply networks (Goulet et al, 2013B; Moser, et al, 2015) and wind simulations around buildings (Vernay et al, 2015; Papdopoulou et al, 2015)

Using a target reliability of identification, upper and lower thresholds are defined in order to decide whether or not a set of model predictions is able to explain the measurements. If the residuals fall within the thresholds at *all* measurement locations, that combination of model parameters $\boldsymbol{\theta}$ is taken to be a candidate model instance.

For threshold bounds, $T_{i,low}$ and $T_{i,high}$, the following condition must be satisfied by all candidate model instances at all measurement locations.

$$\forall i: \quad T_{i,low} \leq g(x_i, \boldsymbol{\theta}) - \hat{y}_i \leq T_{i,high} \qquad (5)$$

If any residual is outside of the threshold at any measurement location, the model instance that is described by that combination of parameter values is falsified. Typically, more than a thousand combinations are tested in order to cover the parameter-value space. Thresholds are adjusted according to the number of measurements used in order to maintain a constant reliability of identification.

The methodology is called "error-domain" model falsification (EDMF) because uncertainties at the model-class level (modelling discrepancy) can be explicitly added to the effect of parametric uncertainty at measurement locations when determining the thresholds on the residual for use in Equation 5.

Thresholds are calculated for each measurement location, thereby allowing for spatial variations. Their values depend on a target reliability of identification and when there are a small number of measurements, the values are using the Sidak correction (Sidak, 1967). Also, there is no requirement for zero-mean uncertainty distributions; systematic uncertainties are represented explicitly. Finally for this approach, it is usually conservative to assume that correlations are zero regardless of their true value.

Application of Equation 5 provides a set of candidate-model instances that is defined by the measurements. Typically, ranges of parameter values are used to describe the results of falsification. Combinations that define these ranges are then used with model uncertainty to obtain model predictions.

Predictions may involve either interpolations, providing results of the same type and value range as the measurements that were used during identification. Alternatively, they may be extrapolations, where models are used to predict outside of the scope of measurements. Extrapolation is typically used for decisions related to important decisions concerning repair, improvement, extension and replacement.

4 INITIAL COMPARISONS

Two studies (Goulet and Smith, 2013; Pasquier and Smith, 2015) have involved comparisons of standard implementations of Bayesian inference with EDMF using simple beams. This created situations where the correct identification was known. Also, this has allowed comparisons to be carried out using identical input values.

When systematic errors are absent and the correct model class is known, Bayesian inference is better than EDMF. In the absence of unrecognized modelling errors, Bayesian inference provided results that were more precise and had equal fidelity as EDMF, see Figure 1.

Bayesian inference was not able to identify parameter values when incorrect assumptions of parameter independence were made, especially in the presence of systematic errors. Biased posterior distribution were proposed. EDMF is not as sensitive to correlations as Bayesian inference thus providing parameter value ranges that included the correct value.

Bayesian inference provided feasible values for parameters in wrong model classes. When the wrong model class was assumed, EDMF falsified the entire set of model instances, thus correctly declaring a null solution set (Goulet and Smith, 2013).

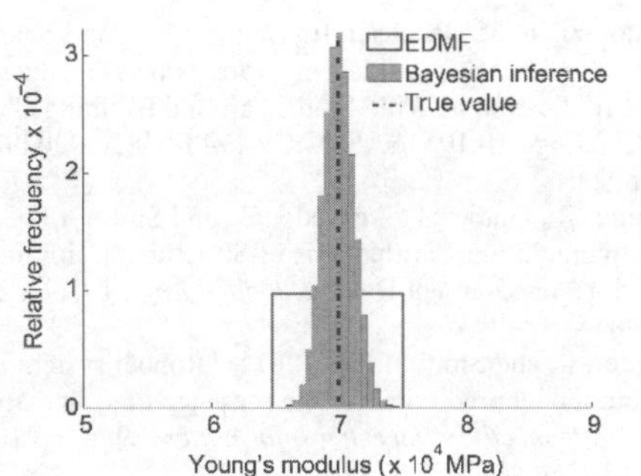

Figure 1. Parameter identification the beam Young's modulus using EDMF compared with Bayesian inference and the true value for 7 measurements.

The second study extended the previous work to include comparisons of predictions. Two types of predictions, interpolation and extrapolation, were compared for four model classes and three scenarios of modelling errors. A total of 30 comparisons were made. Bayesian inference did not provide accurate results in 16 cases while EDMF succeeded for the 30 cases (Pasquier and Smith, 2015).

Figure 2 gives an example of the results when an inaccurate model class is assumed.

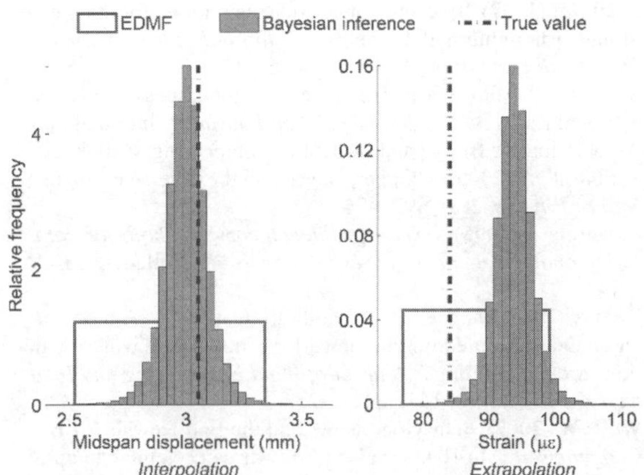

Figure 2. Comparison of EDMF and Bayesian inference with the true value involving interpolation and extrapolation.

As each parameter value compensates for each other to fit the measured value in Bayesian inference, interpolating with such models leads to accurate predictions. However, when extrapolating, since wrong parameter values and wrong modeling uncertainties are identified, the predictions are inaccurate. When interpolating, calculations may lead to accurate predictions even if the parameter identification is wrong. This can lead to overconfidence in model classes and thus, inaccurate subsequent extrapolations. Since systematic modeling uncertainty is included in both identification and prediction processes, EDMF is able to identify correctly the parameter value and to interpolate and extrapolate reliably.

Model-class selection was also studied using a metric called relative plausibility (Mackay, 2003). While EDMF was able to reject wrong model classes, Bayesian model class selection did not succeed consistently.

5 RESEARCH DIRECTIONS

Strengths and weaknesses of data-interpretation methods are most visible in the presence of systematic uncertainty. Systematic uncertainties are most likely outside of the laboratory when full-scale case studies are measured. Therefore, such case studies are an important aspect of this approach. It is planned to extend the comparisons described above to a range of data-interpretation challenges related to several types of full-scale civil infrastructure. Once the data interpretation methods that give good compromises between fidelity and precision in a range of situations are found, the strategy will be "inverted" to develop methods for measurement system design. This will include the optimal choice of sensors as well as their placement.

6 CONCLUSION AND OUTLOOK

This project has much potential to build on previous work in order to make progress in finding data-interpretation methods that are appropriate for the context of decision making for civil infrastructure. The ultimate goal is to create as many situations as possible where additional knowledge from sensors is used to avoid complete replacement of infrastructure. Therefore, this project will help create a future where infrastructure replacement occurs less often, thus contributing to more sustainable, cost-effective and safe asset management.

ACKNOWLEDGEMENTS

The work described in this paper was supported by the Swiss National Science Foundation under contracts 200020_144304 and 200020_155972 and by the Singapore-ETH Centre of Global Environmental Sustainability.

REFERENCES

Beven, K.J. (2002). "Towards a coherent philosophy for modelling the environment", *Proceedings of the Royal Society of London. Series A: Mathematical, Physical and Engineering Sciences,* 458:1–20.

Beck, J.L. and Katafygiotis, L.S. (1998). "Updating models and their uncertainties: Bayesian statistical framework", *Journal of Engineering Mechanics,* 124(4): 455–461.

Çatbas, F.N., Kijewski-Correa, T. and Aktan, A.E. (eds) (2013). *Structural Identification of Constructed Systems, American Society of Civil Engineers,* Reston, VA, USA, 248p.

Efron, B. (2013). 'Bayes' Theorem in the 21st Century', *Science,* 340, pp. 117–118

Goulet, J-A., Michel, C. and Smith, I.F.C. (2013A). 'Hybrid probabilities and error-domain structural identification using ambient vibration monitoring', *Mechanical Systems and Signal Processing* Vol. 37, No. 1–2, pp. 199–212.

Goulet, J-A., Coutu, S. and Smith, I.F.C. (2013B). 'Model falsification and sensor placement for leak detection in pressurized pipe networks', *Advanced Engineering Informatics,* Vol 27 No 2, 2013, pp. 261–269.

Goulet J-A. and Smith I.F.C. (2013). 'Structural identification with systematic errors and unknown uncertainty dependencies', *Computers and Structures,* Vol. 128, pp. 251–258.

Goulet, J-A., Texier, M., Michel, C., Smith, I.F.C. and Chouinard, L. (2014). 'Quantifying the Effects of Modeling Simplifications for Structural Identification of Bridges', *Journal of Bridge Engineering,* vol. 19, num. 1, pp. 59–71.

Laory, I., Trinh, T.N., Posenato, D. and Smith, I.F.C. (2013). 'Combined Model-Free Data-Interpretation Methodologies for Damage Detection during Continuous Monitoring of Structures', *Journal of Computing in Civil Engineering,* Vol. 27, No. 6, pp. 657–666.

Mackay, D. (2003) "Information theory, inference and learning algorithms" *Cambridge University Press.*

Moser, G., Paal, S. G. and Smith, I.F.C. (2015) "Performance comparison of reduced models for leak detection in water distribution networks" *Advanced Engineering Informatics*, Vol 29(3), pp 714-726.

Papadopoulou, M., Raphael, B., Smith, I.F.C. and Sekhar, C. (2015) "Optimal Sensor Placement for Time-Dependent Systems: Application to Wind Studies around Buildings." *J. Comput. Civ. Eng.*, 10.1061/(ASCE)CP.1943-5487.0000497 , 04015024

Pasquier, R., Goulet, J., Acevedo, C., and Smith, I.F.C. (2014) "Improving Fatigue Evaluations of Structures Using In-Service Behavior Measurement Data" *J. Bridge Eng.*, 19(11), 2014, 04014045

Pasquier, R. and Smith, I.F.C. (2015) "Robust system identification and model predictions in the presence of systematic uncertainty" *Advanced Engineering Informatics*, 29, 4, pp 1096-1109. doi:10.1016/j.aei.2015.07.007

Posenato, D., Kripakaran, P., Inaudi, D. and Smith, I.F.C.(2010). 'Methodologies for model-free data interpretation of civil engineering structures', *Computers and Structures,* Vol 88, No 7/8, pp. 467–482.

Raphael, B. and Smith, I. (1998). "Finding the right model for bridge diagnosis", *Artificial Intelligence in Structural Engineering.* LNAI, Springer, pp. 308–319.

Raphael, B. and Smith, I.F.C. (2013). *Engineering Informatics: Fundamentals of computer-aided engineering,* 2nd Edition, John Wiley & Sons, p 333.

Robert-Nicoud, Y., Raphael, B. and Smith, I.F.C. (2005) "System Identification through Model Composition and Stochastic Search" *J of Computing in Civil Engineering*, Vol 19, No 3, pp. 239--247

Saitta S., Kripakaran, P., Raphael, B. and Smith, I.F.C. (2008) "Improving System Identification Using Clustering" *Journal of Computing in Civil Engineering,* Vol 22, No 5, pp 292-302.

Sidak, Z. (1967) "Rectangular confidence regions for the means of multivariate normal distributions" *Journal of the American Statistical Association,* Vol 62, pp 626-633.

Sweeney, R A P (1990). "Update on Fatigue Issues at Canadian National Railways", *IABSE Workshop Lausanne*, International Association for Bridge and Structural Engineering, Zurich.

Tarantola, A (2006). "Popper, Bayes and the inverse problem", *Nature Physics,* 2(8): 492–494.

Tarantola A (2005). *Inverse problem theory: methods for data fitting and model parameter estimation.* Siam, Philadelphia, PA, USA.

Vernay, D.G., Raphael, B. & Smith, I.F.C. (2015) "A model-based data-interpretation framework for improving wind predictions around buildings" *Journal of Wind Engineering and Industrial Aerodynamics*, 145, pp.219-228

Wolf, W (2007). 'The Good News and the Bad News', *IEEE Computer* 40 (11) 104, see also http://varma.ece.cmu.edu/cps/

Proceedings of the International Conference on Smart Infrastructure and Construction
ISBN 978-0-7277-6127-9

© The authors and ICE Publishing: All rights reserved, 2016
doi:10.1680/tfitsi.61279.569

Big visual data analytics for damage classification in civil engineering

C.M. Yeum, S.J. Dyke*, J. Ramirez, and B. Benes

Purdue University, Indiana, USA
* *Corresponding Author*

ABSTRACT Visual data provide a wealth of information to better understand the world around us. A tremendous amount of visual data is collected in civil engineering applications through efforts such as scientific experiments, field surveys, resource management, and reconnaissance missions. Among these efforts, visual data generate crucial and abundant information evaluating the condition of a civil structure. As a typical example, during a disaster such as a natural catastrophe or industrial explosion, vast amounts of perishable image data are collected that may be used to generate new knowledge from the consequences of that event. However, not only does this process require time-consuming data collection by human engineers, it is also tedious and expensive to manually search through these data sets to find the most informative images. Autonomous collection, processing and analysis offer great potential to support structural evaluation. In this study, we propose a novel autonomous evaluation method to examine large volumes of images. Recent deep convolutional neural network (CNN) algorithms are applied to extract visual information from the collected images. Task-oriented engineering knowledge and experience are incorporated into the procedures to increase accuracy. The target application addressed in this study is post-disaster building damage evaluation. Illustration of the technique and capabilities for collapse classification is demonstrated using large-scale images gathered from past earthquake events.

1 INTRODUCTION

An astonishing amount of visual data is being collected worldwide through scientific experiments and field surveys in civil engineering. For example, during each natural disaster, vast amounts of perishable visual data are collected formally by teams of experts. That data is collected in order to generate new knowledge by learning from that event. With the availability of ubiquitous visual data sources, such as social media, news media and unmanned aerial vehicles for hire, large volumes of useful visual data are available for various purposes (Voigt et al. 2007; Yates & Paquette 2011; Computing Community Consortium 2013; Measure & American Red Cross 2015; Wang et al. 2015). Currently the major approach available to responders and researchers for the analysis of such data is tedious manual sorting and analysis of these photographs or videos. Only a small portion of the growing volumes of visual data collected are actually being used for research and extracting information for onsite decisions. Engineers must exploit the power of information technology to extract such information in an efficient manner.

Autonomous collection, processing and analysis of data offer great potential to aid the human decision maker. Whether he/she is conducting scientific research, or making a decision, the human operator and decision maker needs to quickly sift and sort visual data to identify and evaluate the important scenes. There is a compelling need to provide support to these human decision makers by giving them the analytic power to search, filter, prioritize, classify and annotate hundreds or thousands of images.

In recent years, powerful computer vision methods and machine learning algorithms have been established within computer science and engineering, and related disciplines. In some applications these have nearly reached human-level performance (Taigman et al. 2014; He et al. 2015). These methods have been considered for a broad range of applications, ranging from speech or text recognition to autonomous driving (Ciresan et al. 2012; Hannun et al. 2014; Chen et al. 2015). In civil engineering, vision-based remote or crowdsourcing structural inspection and construction management techniques have been researched and have achieved improvements in accuracy and efficiency (Golparvar-Fard et al. 2009; Jahanshahi et al. 2009; Ghosh et al. 2011; Zhu et al. 2011; German et al. 2012, 2013; Torok et al. 2013).

Previously developed techniques have been validated for specific damage types using a small quantity of images that were collected with the intension of using them for a specific purpose or application. However, in real circumstances during a disaster with realistic time and resource constraints, there is no guarantee that one may be able to collect favorable images for such a specific purpose due to the large uncertainty of locations, viewpoints, or contents. Thus, classification and filtering of the images will be able to support the decision-maker, particularly when time is limited. Furthermore, there is no assurance that these methods will be able to handle large-scale, complex, and unstructured images in such a way as to be tractable.

At this time such methods, when used in isolation, are still severely limited in their ability to extract useful information. Real-world visual data are quite diverse in nature (e.g. quality, resolution, subject, composition, illumination). And although these traditional computer vision methods are quite powerful, we still lack a good understanding of how to implement them in a domain-centric and task-oriented manner. Objects of interest to a civil engineer often need to be understood and analyzed in their spatial configuration as well as in their background context, and to extract meaningful information such analysis should be firmly based on engineering experience and knowledge.

In this paper, we develop and demonstrate a novel method for autonomous big data analytics that is intended to support decision-making in the field. The target application we addressed here is post-event building evaluation during a disaster. We implement task-oriented computer vision methods capable of detection, classification and evaluation of large volumes of visual data. A key factor in the method is that we incorporate prior knowledge from our target application, while we also investigate and utilize required optimal resolution of images for the scene classification and object detection actions. Recently developed deep convolutional neural networks (CNN) algorithm are applied for image classification and object detection, while ensuring success by integrating engineering domain knowledge into the procedure (Krizhevsky et al. 2012; Girshick et al. 2013; Russakovsky et al. 2014; Zhou et al. 2014; Simonyan & Zisserman 2014). Here we provide a feasible solution for analyzing a large-scale collection of real images from disasters. The proposed method can be expanded to incorporate new or existing damage detection methods for broad application in a range of disasters.

2 PROPOSED DAMAGE EVALUATION METHOD

An overview of the procedure developed here is shown in Figure 1. In step 1, images collected during a disaster are automatically filtered and prioritized based on available metadata from the images. Metadata acquired at the time of image acquisition includes geospatial and temporal (e.g. time/date, GPS) data, as well as information relevant to the event itself (e.g. previous building images, event intensity map). When available, these items can be incorporated into the overall process. The use of such metadata is beneficial for rapid access and flexible mining of a set of valuable images needed to explore their visual contents. In step 2, images are classified according to their content, particularly in terms of containing scenes associated with the target application. Scenes may be defined in terms of single or multiple objects and their spatial configuration. For instance, a scene of a building façade is composed of one or more objects and their spatial arrangement including an entrance at the bottom, and an array of windows or floor borders. Because scenes are typically recognized by low-dimensional features (e.g. general shape, colors, or compositions) and need not be interpreted using a detailed appearance of objects, they can readily be recognized in low resolution images. Thus, efficient and rapid computation is possible in this step. In step 3, specified target objects are identified and localized within the scenes classified in step 2. Here, the target objects encompass damage (e.g. spalling or cracking), and also the objects in which such damage is present (e.g. beam, column, wall). They may also include geometric or pattern features (e.g. window opening, stair-step cracking), which may be used subsequently for damage evaluation. Such object and scene categories must be designed and defined based on the applications for which they will be used. Lastly, in step 4, damage contained in those images is evaluated based on prior knowledge of the target applications. This step is performed to understand damage using its presence as well as its location on the object, appearance or surrounding objects. The steps proposed here are flexible depending on level of information provided with images.

A good example to illustrate the proposed concept is buckling detection after an earthquake. Buckling is detected by observing whether vertical rebar on a structural column is exposed and yielded. Such a domain-based definition of the problem provides good prior information to design the proposed technique. Instead of direct detection of rebar across a large collection of high-resolution images, the proposed method includes several steps: metadata filtering (with respect to location, date, or time), scene classification of indoor or outdoor building (or building façade), column/spalling object detection, and rebar detection. These steps are performed sequentially by gradually increasing the resolution of the images used for the relevant steps. Finally, we are able to evaluate the condition (e.g. bending or break) from the high resolution image containing the detected rebar. At each step, the number of images that are needed for processing with higher resolution decreases at each step. Thus, this approach is especially appropriate when time and resource constraints exist.

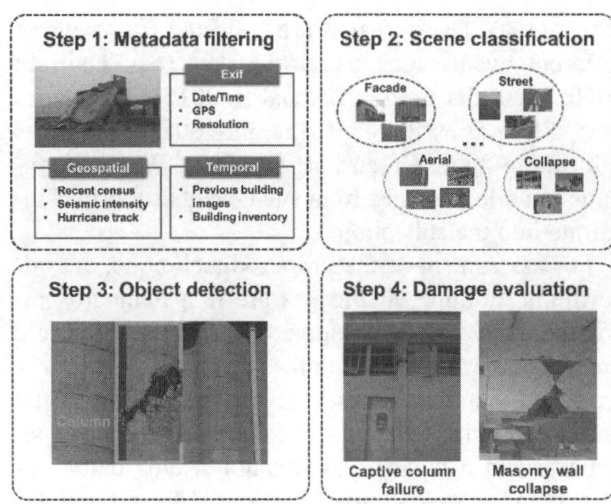

Figure 1. Overview of the proposed damage evaluation method

3 COLLAPSE CLASSIFICATION

As a pilot study, we use the case of collapse to demonstrate scene classification in the proposed method. The term collapse here refers to both significant damage/collapse of the building structure, as well as major damage/collapse of a single structural component. The reasons for selecting this damage case are that (1) collapse of buildings or their components represents a major mode of damage that is of interest in earthquake reconnaissance; (2) a large number of appropriate images are available for training and validation; and (3) there is no existing annotation image database for this situation. To perform this case study, a large annotation image collection is established and used for validation of the method.

We first introduce our post-event reconnaissance image data collection. We have gathered a collection of 67,000 color images acquired by various researchers and practitioners after past natural disasters including hurricane, tornado and seismic events (e.g., from datacenterhub.org at Purdue University, disaster responders, or Earthquake Engineering Research Institute (eeri.org)). Nearly all of these images preserve the original quality (resolution) as well as the basic information (e.g. date, time, and event), and a small portion of images have GPS information or a picture of a GPS navigator. However, no annotation was available for the visual contents of the images. At this time, the distribution across the types of events is earthquake (90%), hurricane (6%), tornado (3%), and others (1%). These images are collected from several different events such as earthquakes (e.g. Haiti in 2010, L'Aquila in 2009, Nepal in 2015), hurricanes (e.g. Florida in 2004, Texas in 2008), tornadoes (Florida in 2007; Greensburg in 2007). We will continue to collect images from such events to further integrate into the collection.

For assessment of the proposed technique, and algorithmic training, all images are first manually annotated using in-house annotation software. A single image is shown centered in the screen and annotators are asked to answer a yes or no question of "Does the image contain a collapse scene?". Based on our experience, such a binary classification yields better results than multiple choice questions. Manual annotation can be quite taxing for the human if there are several buttons/options and there is high potential for error.

Figure 2 shows several sample images used for collapse scene classification (from datacenterhub.org at Purdue University). For collapse classification, the dataset is composed of 1918 collapsed building and building components (b&bc) data as positive and 3427 other data as negative, which are composed of minor damage b&bc, irrelevant images, and undamaged b&bc in Figure 2. Such sampling of the negative dataset is designed to represent non-collapse image collection from a real earthquake reconnaissance scenario.

As mentioned in the introduction, the CNN algorithm will be implemented for collapse classification. In the last few years, CNNs have led to major breakthroughs in computer vision areas and have enabled the development of high-level abstractions using large-scale databases of general everyday objects. CNNs typically have one or more convolutional layers tied with weights and pooling layers to extract scale, translation and rotation tolerant features, and fully connected layers connected with these features classify image category or object(s). The parameters of CNNs are trained in advance of their implementation using large-scale training image data (Krizhevsky et al. 2012; Girshick et al. 2013; Russakovsky et al. 2014; Zhou et al. 2014; Simonyan & Zisserman 2014).

Figure 2. Sample images used for collapse scene classification: (a) collapse buildings and building components (b&bc), (b) minor b&bc, (c) irrelevant images, and (d) undamaged b&bc. (a) is assigned in a positive class and the others are in a negative class.

In this study, we used Vgg-f CNN architecture implemented in MatConvNet (Vedaldi & Lenc 2014), which comprises 8 learnable layers, 5 of which are convolutional, and the last 3 are fully-connected. Fast processing is ensured by the 4 pixel stride in the first convolutional layer. In the data augmentation process which produces a suitable set of input images for the CNN, a square region of input images are randomly cropped in original images followed by resizing them as 224 x 224 pixels. In each epoch, a batch at each iteration is assigned using randomly ordered pictures, and these data are augmented. Stochastic gradient descent with a batch of images are learned to optimize the parameters of the network.

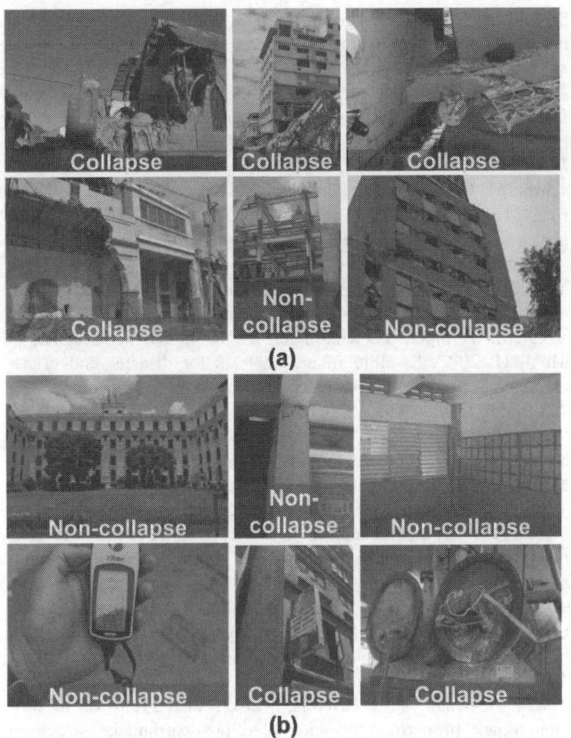

Figure 3. Examples of the collapse classification result: (a) classification of positive images and (b) classification of negative images. Note that the text labels indicate classification results, and blue and red colors are true and false classification, respectively (e.g. an image in (a) having a label of "collapse" in blue is true classification collapse building image)

A workstation having a Xeon E5-2609 CPU, 12 GB memory and two GPU, NVidia Titan X and Telsa k40, a total of 24 GB video memory is used for training and testing the algorithm. The MatConvNet library installed on Matlab 2014b is used for this demonstration (Vedaldi & Lenc 2014). To obtain these results, 80% of annotated images are used for training and the remainder are used for testing and assessment of the classifiers. Less than 20 epochs are required to reach convergence and this training processing required approximately 10 hours.

Finally, in this demonstration we obtain rates of 86.6% true positive (true collapse detection), 13.3% of false-positive, and 93.7% of false negative, respectively. The proposed collapse classification successfully attains a relatively high rate for true-positives. A sample of images showing the classification results are shown in Figure 3. Note that these rates will vary slightly depending on CNN architectures and their parameters. Overall the performance of this approach is quite successful. The method shows great promise for supporting decisions in the field and for enabling research using large volumes of image data.

4 CONCLUSION

A novel method for automated post-disaster image classification is proposed to perform processing and analyzing big visual data. The method is demonstrated on a specific example classification focused on collapse classification. However, the general method can be applied to other civil applications that use large-scale visual data. In the future we plan to incorporate and validate a broader array of damage evaluation methods for broader application.

ACKNOWLEDGEMENT

The authors would like to thank Dr. Santiago Pujol at Purdue University, Dr. Chungwook Sim at University of Nebraska, Dr. Alberto Pavese, and the EERI teams for providing access to large-volumes of post-earthquake reconnaissance images and Jongseong Choi at Purdue University for annotating large images.

REFERENCES

Computing Community Consortium (CCC). 2013. Computing for Disasters, *a workshop report funded by the National Science Foundation online at*: http://www.cra.org/ccc/files/docs/init/computingfordisasters.pdf

Ciresan, D.C., Meier U., & Schmidhuber, J. 2012. Transfer Learning For Latin And Chinese Characters With Deep Neural Networks, *The 2012 International Joint Conference on Neural Networks (IJCNN). IEEE.*

Chen, C., Seff, A., Kornhauser, A., & Xiao, J. 2015. DeepDriving: Learning Affordance for Direct Perception in Autonomous Driving. *arXiv preprint* arXiv:1505.00256.

German, S., Brilakis, I., & DesRoches, R. 2012. Rapid entropy-based detection and properties measurement of concrete spalling with machine vision for post-earthquake safety assessments, *Advanced Engineering Informatics*, **26(4)**, 846-858.

German, S., Jeon, J. S., Zhu, Z., Bearman, C., Brilakis, I., DesRoches, R., & Lowes, L. 2013. Machine vision-enhanced postearthquake inspection, *Journal of Computing in Civil Engineering*, **27(6)**, 622-634.

Ghosh, S., Huyck, C. K., Greene, M., Gill, S. P., Bevington, J., Svekla, W... & Eguchi, R. T. 2011. Crowdsourcing for rapid damage assessment: the global earth observation catastrophe assessment network (GEO-CAN). *Earthquake Spectra*, **27(S1)**, S179-S198.

Girshick, R., Donahue, J., Darrell & T., Malik, J. 2013. Region-based Convolutional Networks for Accurate Object Detection and Segmentation, *Pattern Analysis and Machine Intelligence, IEEE Transactions,* 99, 1-11.

Golparvar-Fard, M., Peña-Mora, F., & Savarese, S. 2009. D4AR–a 4-dimensional augmented reality model for automating construction progress monitoring data collection, processing and communication, *Journal of information technology in construction*, **14**, 129-153.

Hannun, A., Case, C., Casper, J., Catanzaro, B., Diamos, G., Elsen, E.,Prenger, R., Satheesh, S., Sengupta, S., Coates, A. & Ng, A. Y. 2014. DeepSpeech: Scaling up end-to-end speech recognition, *arXiv preprint* arXiv:1412.5567.

He, K., Zhang, X., Ren, S., & Sun, J. 2015. Delving deep into rectifiers: Surpassing human-level performance on imagenet classification, *arXiv preprint* arXiv:1502.01852.

Jahanshahi, M. R., Kelly, J. S., Masri, S. F., & Sukhatme, G. S. 2009. A survey and evaluation of promising approaches for automatic image-based defect detection of bridge structures, *Structure and Infrastructure Engineering*, **5(6)**, 455-486.

Karpathy, A., Toderici, G., Shetty, S., Leung, T., Sukthankar, Rahul, Fei-Fei, L. 2014. Large-Scale Video Classification with Convolutional Neural Networks. *IEEE Conference on Computer Vision and Pattern Recognition (CVPR).*

Measure & American Red Cross. 2015. Drones for disaster response and relief operations.

Russakovsky, O., Deng, J., Su, H., Krause, J., Satheesh S., Ma, S., Huang, Z., Karpathy, A., Khosla, A., Bernstein, M., Berg, A.C. & Fei-Fei, L. (* = equal contribution) 2014. ImageNet Large Scale Visual Recognition Challenge, arXiv:1409.0575.

Simonyan, K., & Zisserman, A. 2014. Very deep convolutional networks for large-scale image recognition. *arXiv preprint* arXiv:1409.1556.

Taigman, Y., Yang, M., Ranzato, M. A., & Wolf, L. 2014. Deepface: Closing the gap to human-level performance in face verification. *In Computer Vision and Pattern Recognition (CVPR), 2014 IEEE Conference on IEEE*, 1701-1708.

Torok, M. M., Golparvar-Fard, M., & Kochersberger, K. B. 2013. Image-based automated 3D crack detection for post-disaster building assessment, *Journal of Computing in Civil Engineering*, **28(5)**, A4014004.

Vedaldi, A., & Lenc, K. 2014. MatConvNet-convolutional neural networks for MATLAB. *arXiv preprint* arXiv:1412.4564.

Voigt, S., Kemper, T., Riedlinger, T., Kiefl, R., Scholte, K., & Mehl, H. 2007. Satellite image analysis for disaster and crisis-management support, *Geoscience and Remote Sensing, IEEE Transactions on*, **45(6)**, 1520-1528.

Wang, S., Purnell, N. & Bhattachary, S. 2015. Nepal aid workers helped by drones, crowdsourcing, *Wall Street Journal-Asia Ed.*, May 1.

Yates, D., & Paquette, S. 2011. Emergency knowledge management and social media technologies: A case study of the 2010 Haitian earthquake. *International Journal of Information Management*, **31(1)**, 6-13.

Zhou, B., Lapedriza, A., Xiao, J., Torralba, A., & Oliva, A. 2014. Learning deep features for scene recognition using places database, *In Advances in Neural Information Processing Systems*, 487-495.

Zhu, Z., German, S., & Brilakis, I. 2011. Visual retrieval of concrete crack properties for automated post-earthquake structural safety evaluation, *Automation in Construction*, **20(7)**, 874-883.

Proceedings of the International Conference on Smart Infrastructure and Construction
ISBN 978-0-7277-6127-9

© The authors and ICE Publishing: All rights reserved, 2016
doi:10.1680/tfitsi.61279.575

Failure assessment of underground tunnels

C.Q.Li[1], D.J.Robert[2*] , M.Mahmoodian[3] and J. Dauth[4]

[1,2,3] *RMIT University, Melbourne, Australia*
[4] *Metro Trains, Melbourne, Australia*
[*] *Corresponding Author*

ABSTRACT The increasing development of the world has demanded the construction of underground tunnels for needs such as transport, water supply, sewerage and telecommunication. As many of underground tunnels were constructed sometime in the last century or earlier, in most cases their condition has deteriorated mainly by weathering of rocks, corrosion of reinforcement and/or degradation of concrete. Consequently, the leaks have become a common problem in these assets and management of these aging infrastructures have become a pressing issue for asset managers. The current study identifies major failure mechanisms of underground tunnel and how such failures can be assessed. The study reveals that the ground water flow can be the major driver for commonly identified failure mechanisms of underground tunnels. Several failure mechanisms, extending from soil/rock failure to the concrete lining material failure, along with their corresponding failure criteria will be identified. The identification of such failure mechanisms can be helpful when characterising the limit state functions for the structural reliability assessments. An example is provided to demonstrate the proposed methodology for assessing the structural reliability of aging tunnels. The proposed failure assessment method can help the infrastructure managers to develop rehabilitation or replacement strategy for in-service tunnels with a view for optimum maintenance management of the tunnel assets.

1 INTRODUCTION

Construction of underground tunnels for needs such as transport, water supply, sewerage and telecommunication is an increasing worldwide demand. The reported statistics (Dimitrios, 2005) showed that many countries adopt underground tunnelling to facilitate their transportation systems due to limited above ground space and safe operational requirements. Furthermore, several of such systems are expected to be in operation in near future. Such tunnels are buried in ground with soil/rock which provides protection and support .However, these buried tunnels are often subjected to deterioration mainly due to weathering of rocks, corrosion of reinforcement and/or degradation of concrete. Consequently, the leaks have become a common problem in these assets and management of these aging infrastructures have become a pressing issue for asset managers. Such failures can cause significant consequences in terms of economic loss to asset managers, public safety, damage to property and also have an adverse effect on the overall performance of the tunnels.

The knowledge of the failure mechanisms by which the tunnels are failed is an essential to understand the in-service tunnel failures. The conditions of the rock/soil cover and the tunnel lining can significantly affect the failure modes/limit states of the tunnel. A tunnel can fail when the generated stress due to internal/external loading exceeds the nominal lining strength or when the stress intensity generated at a critical defect (such as lining cracks) exceeds the toughness of the lining material, or possibly as a combination of both.

The current study identifies major failure mechanisms of underground tunnels and clarifies how such failures can be assessed. Several failure mechanisms, extending from soil/rock failure to the concrete lining material failure, along with their corresponding failure criteria will be identified. The study would reveal that the ground water flow can be the major driver for commonly identified failure mechanisms of underground tunnels. The identification of failure mechanisms is necessary when characterising the limit state functions for the structural reliability assessments (Li and Mahmoodian 2013 and Mahmoodian and Li 2015). An example is provided to demonstrate the proposed methodology for assessing the structural reliability of aging tunnels.

2 POTENTIAL FAILURE MECHANISMS OF UNDERGROUND TUNNELS

The failure mechanisms of an underground tunnel can be multiple, and can be identified with a top-down approach; initiating from the soil/rock cover and extending towards the tunnel lining. Such approach has been conceptually summarized in Fig. 1. As noted in the figure, the main driver for most of the failure mechanisms is the water flow through the surrounding material. Secondarily, if the tunnel is driven in the rock, the process of weathering can also make an impact to the performance of the tunnel. However, the impact of weathering can only influence the tunnel in longer term compared to the consequences of the water flow which impacts more within short term stability. Due to deep embedment conditions, it can be assumed that most of underground tunnels would be under submerged conditions and hence the water flow can be of major concern when identifying the failure limit states of these tunnels.

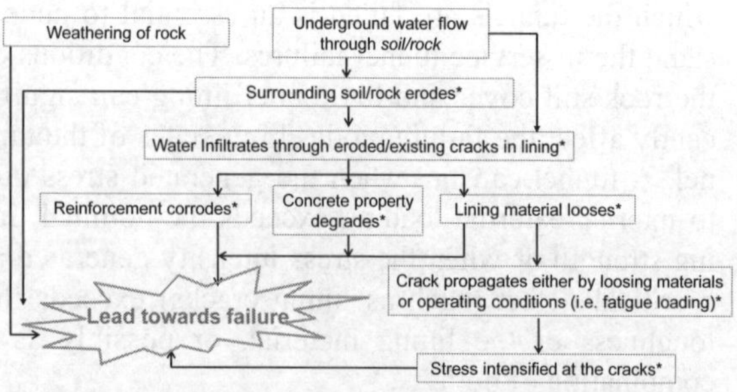

Figure 1. Possible failure mechanisms of underground tunnel

The water flow will first induce scour/erosion of the surrounding material, which can result a run off of fine particles from the soil/rock matrix. The erosion further causes to generate additional pathways for water flow (i.e. increases the permeability) due to particle run off. As a result of enhanced permeability in the rock, additional water percolates on the surface of the tunnel. If the outer surface of the tunnel lining is free of cracks, the seepage behind the lining would have a negligible impact as the water only imposes a normal stress (hydrostatics pressure) to the tunnel. However, if the lining possesses cracks (possible for aging tunnels) larger than the surrounding soil particles, it can lead to particle run off from the soil matrix with the leaking water through the cracks (Fig. 2). In addition, fine materials can be removed through the voids between large particles (i.e. suffusion) especially when the medium of embedment consists of stratification. These material flows can cause erosion behind the lining and could create interconnected voids (Fig. 2). Such formation of voids between the tunnel and surrounding rock can lead towards failure of the lining material during excessive loading. Therefore, a failure criterion can be induced using the level of erosion in the surrounding material. A widely accepted erosion function was proposed by Briadu (2008) as defined in Eq. 1. The current research work is focusing on how the erosion limits can be defined within limit state functionality.

$$\frac{\dot{Z}}{u} = \alpha\left(\frac{\tau-\tau_c}{\rho u^2}\right)^m \quad (1)$$

Where

$\dot{Z}$ is erosion rate (m/s);

u is water velocity (m/s);

τ is hydraulic shear stress;

τ_c is threshold or critical shear stress below which no erosion occurs (N/m^2)

ρ is mass density of water (kg/m^3)

α, m are parameters characterizing the soil being eroded.

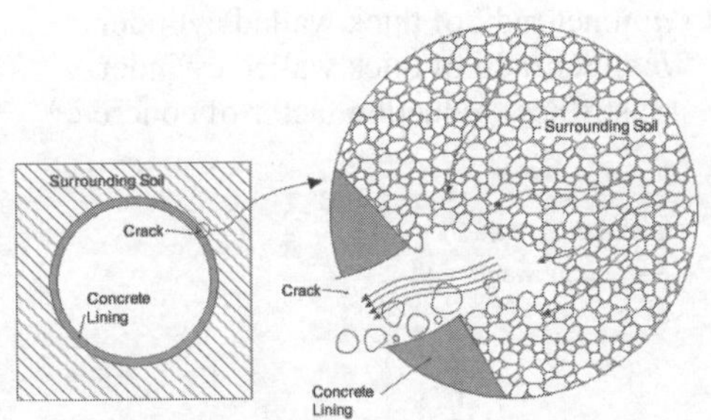

Figure 2. Erosion induced tunnel failure

Once the water percolates into the lining surface, it seeps into the tunnel lining due to the permeable (i.e. porosity) nature of concrete (intrinsic water permeability of concrete is 10^{-19} m²/s – 10^{-17} m²/s, Hoseini et al, 2009). The concrete porosity can significantly be affected by the externally applied loads which induce matrix cracks and extend the bond cracks through the cement paste until failure (Wong et al, 2009). These cracks increase the porosity of tunnel lining and create the porous path that is necessary for ground water to permeate into the lining (i.e. increases concrete permeability). The increased permeability can fail the tunnel in multiple ways; the increase in concrete porosity can result a reduction in concrete strength, leading towards a tunnel failure. In addition, continuing water flow could dissolve the material out of the concrete (Ekstrom, 2001). For example, the collected water samples from an underground tunnel in Australia depict that water contains high amounts of ca^{2+} than in the natural water (Fig. 3). Further, the increased permeability induced high water flows can bring ground water with aggressive agents into the lining. The presence of water in concrete (in forms of natural and/or contaminated water) causes corrosion of the rebars in concrete, which ultimately will cause failure to the tunnel lining. Thus, water flow in concrete can cause such diversified mechanisms of failure to the tunnel. Therefore, concrete permeability can be defined as an important parameter which significantly affects the failure of concrete. For a tunnel with a water-sealed concrete lining, Fernandez and Moon (2010) proposed a model to interpret water inflow rate through concrete lining, incorporating hydrostatic heads and hydraulic conductivities of lining and concrete (Eq. 2). Bieniawski (1990) classified tunnel failure (i.e. water flowing condition) when the flow rates exceeds 125L/min. The current research work is investigating the limits of concrete permeability which can be coupled with lining failure at microscopic level.

$$Q_L = \frac{2\pi K_l(h_{outer} - h_{inner})}{ln(\frac{(a+d)}{a})} \quad (2)$$

$$h_{outer} = \frac{h}{1 + C(K_l/K_m)} \quad (3)$$

Where

$$C = ln(\frac{2h}{a+d})/ln(\frac{a+d}{a}) \quad (4)$$

Where h_{outer} is the water head at the outer surface of the lining (using tunnel springline as datum level), h_{inner} is the water head at the inner surface of the lining, K_l is the hydraulic conductivity of the concrete lining, K_m is the hydraulic conductivity of the rock mass, Q_L is the water rate from the rock mass to the interface.

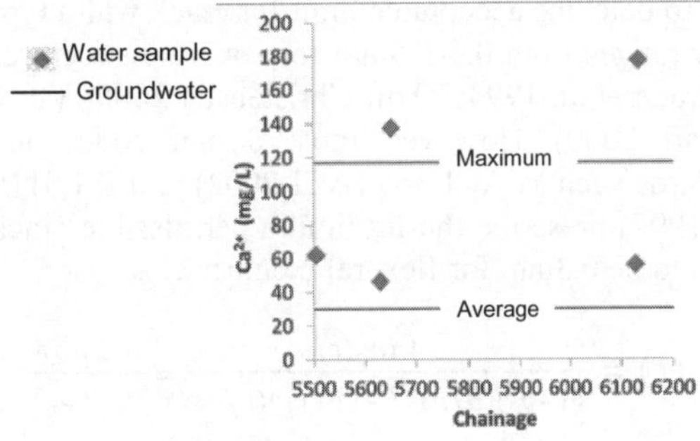

Figure 3. Analysis of the collected water samples from an underground tunnel for ca^{2+}

A major impact of water percolating through concrete is on the cracking of the tunnel lining which is induced by rebar corrosion. The corrosion of rebars in underground tunnels has been evidenced through the observation of collected water from an underground tunnel in Australia (Fig. 4). They revealed that the leaked water consisted with Fe^{2+} along with extremely low pH values. Considerable research has been undertaken through numerical and experimental studies to investigate the effect of corrosion induced cracking on the structure. Numerical investigations typically involve finite element (FE) methods. For example, Pantazopoulou & Papoulia (2001) investigated the effect of concrete cracking with the use of smeared cracking model and fracture mechanics using FE method. Similarly, Chen & Mahadevan, (2007) used FE method to examine the factors that affect corrosion induced cracking using smeared crack approach. In experimental investigations, several studies have been conducted to investigate the effect of corrosion induced cracking using accelerated corrosion technics (Alonso et al. 1998; Andrade et al. 1993; Liu & Weyers 1998). Liu and Weyers (1998) were among the pioneers to examine the corrosion induced cracking by proposing a model to predict surface cover cracking due to expansion of corrosion products. Several empirical models were also developed in the past to predict the corrosion induced cracking by means of deterministic models (Alonso et al. 1998) as well as probabilistic models (Thoft-Christensen, 2003). A corrosion induced crack width model has been analytically derived by Li et al (2005) using fracture mechanics and fundamental mechanics theory (Eq. 5). On the basis of investigations to date, the acceptable limit for crack width typically ranges from 0.1-0.5mm for concrete structures (Andrade et al. 1993; Thoft-Christensen 2000; Vu & Stewart 2000). However, most design codes and standards such as ACI 365 (ACI 2002) and BS 8110 (BS 1997) prescribe the maximum permissible crack width to be 0.3mm for flexural members.

$$w_c(t) = \frac{4\pi d_c(t)}{(1-v_c)(a/b)^{\sqrt{\varphi}}+(1+v_c)(b/a)^{\sqrt{\varphi}}} - \frac{2\pi b f_t}{E_{ef}}$$

Eq. 5

$w_c(t)$: crack width over time
$d_c(t)$: growth of rust band
v_c: Poisson's ratio
f_t: concrete tensile strength
E_{ef}: effective modulus of elasticity
a: inner radii of thick walled cylinder
b: outer radii of thick walled cylinder
φ: stiffness reduction factor of concrete

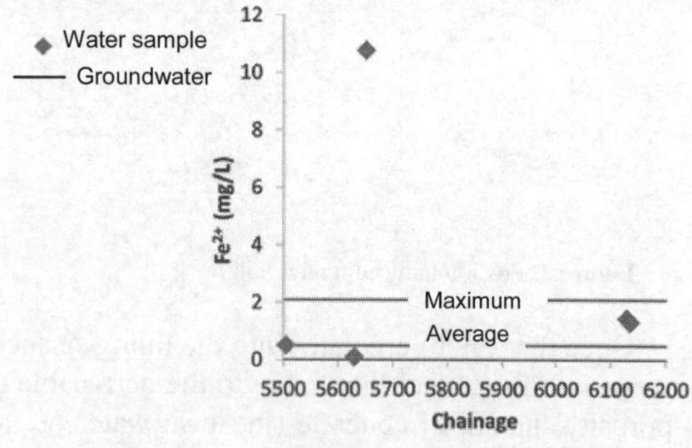

(a) *Water samples analyzed for Fe2+*

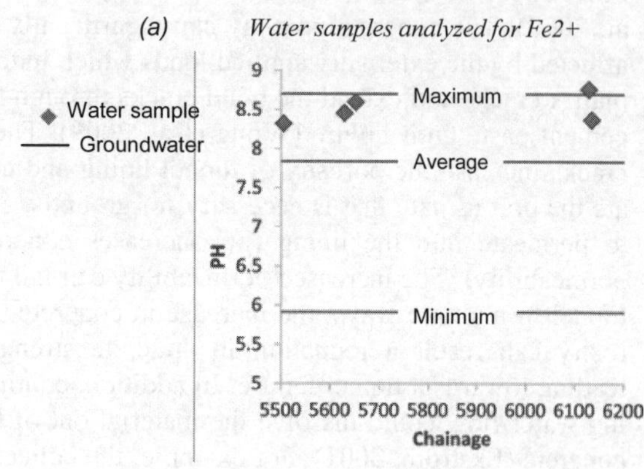

(b) *Water samples analyzed for pH*

Figure 4. Analysis of the collected water samples from an underground tunnel

3 RELIABILITY ASSESSMENT OF UNDERGROUND TUNNELS

The current study has been extended to demonstrate how the identified failure mechanisms can be used to assess the structural reliability of an underground tunnel. The failure mechanisms selected for this study are based on water flow rates and corrosion induced crack width. As obtained from previous studies, the limiting conditions of water flow and crack width can be 125L/min and 0.3mm respectively.

A tunnel can be defined as unserviceable in multiple modes due to different limit state violations. Therefore, the probability of a tunnel serviceability failure should be determined using the methods of systems failure analysis. In the current study, the two following serviceability failure modes (limit state functions) are considered for an underground tunnel failure:

Crack width limit state:
$$Z_1(W_{cr}, W_c, t) = W_{cr} - W_c(t) \quad \text{Eq. 6a}$$

Water flow limit state:
$$Z_2(Q_{cr}, Q_L, t) = Q_{cr} - Q_L(t) \quad \text{Eq. 6b}$$

where W_c *and* Q_L are crack width and water flow rate, respectively. W_{cr} and Q_{cr} are acceptable thresholds suggested by standards or practical manuals. Formulization of these two load actions (W_c *and* Q_L) for a reinforced concrete tunnel lining were presented in Equations 6a and 6b respectively.

As mentioned earlier, each failure mode happens when $Z_1 \& Z_2 \leq 0$. But to consider both two modes as a system, it is necessary to clarify the combination of the limit state functions. The two serviceability limit states (crack and water flow) are considered in a series configuration because violation of each of them will result in failure of the whole system. More discussion on system reliability analysis can be found in Melchers (1999) and Mahmoodian and Alani (2013).

Numerical method for estimation of the probability of failure

To estimate the probability of failure of a system, analytical or numerical methods can be used. In this study, the Monte Carlo simulation method is considered for estimation of the probability of failure. Monte Carlo simulation techniques involve sampling at random to simulate artificially a large number of experiments and to observe the results. To use this method in structural reliability analysis, a value for each random variable is selected randomly ($\hat{x_i}$) and the limit state function ($Z(\hat{x})$) is checked. If the limit state function is violated (i.e. $Z(\hat{x}) \leq 0$), the structure or the system has failed. The experiment is repeated many times, each time with randomly chosen variables. If N trials are conducted, the probability of failure then can be estimated by dividing the number of failures to the total number of iterations:

$$P_f \approx \frac{n(Z(\hat{x}) \leq 0)}{N} \quad \text{Eq. 7}$$

Details of the Monte Carlo method including sampling techniques can be found in Ditlevsen and Madesn (1996), Melchers (1999) and Rubinstein and Kroese (2008). The accuracy of Monte Carlo simulation results depends on the sample size generated and, in cases where the probability of failure is estimated, on value of the probability (the smaller the probability of failure, the larger the sample size needed to ensure the same accuracy, Melchers 1999).

Information regarding to the basic random variables have been gathered based on other literature and in some cases by assumptions (Tables 1 and 2). Considering the data showed in tables 1 and 2, Monte Carlo simulation method is performed in the current study for serviceability failure assessment of the tunnel lining. To consider uncertainties involved in the values of the parameters, a coefficient of variation for each random variable is assumed based on the experimental data. This will change the deterministic situation of the problem to a stochastic condition.

For the flow rate formulation (Equation 2), it is assumed that K_l and K_m are time dependent and follow a power low formulations:

$$K_l = a_1 t^{b_1} \quad \text{Eq. 8a}$$

$$K_m = a_2 t^{b_2} \quad \text{Eq. 8b}$$

where a_1, a_2, b_1 and b_2 are coefficient which will be treated as basic random variables. The statistical information of these four random variables can be estimated from experimental data.

Table 1 Values of basic variables for crack width limit state function

Basic variables	Mean	COV	Sources
C	31 mm	0.2	Li (2003), Mirza, et al. (1979)
D	12 mm	0.15	Li (2003), Mirza, et al. (1979)
d_0	12.5 μm	0.15	Liu and Weyers (1998)

f_t	5.725 MPa	0.2	Li (2003), Mirza ,et al. (1979)
i_{corr}	$0.1843Ln(t)+0.56$ µA/cm²	0.2	Assumption
E_{ef}	18.82 GPa	0.12	Li (2003), Mirza, et al (1979)
α_{rust}	0.57	-	Liu and Weyers (1998)
ρ_{rust}	3600 kg/m³	-	Liu and Weyers (1998)
ρ_{st}	7850 kg/m³	-	Liu and Weyers (1998)
υ_c	0.18	-	Li (2003)

Table 2 Values of basic variables for flow rate limit state function

Basic variables	Mean	COV
a	3 m	0.1
d	0.5 m	0.005
h	15 m	0.15
h_{inner}	12 m	0.15
a_1	2×10^{-7}	0.2
a_2	10^{-5}	0.2
b_1	0.9	0.2
b_2	10^{-2}	0.2

It is to be noted that values for the random variables for flow rate limit state function are assumed for the sake of illustration and implementation of the methodology to estimate the probability of serviceability failure of an underground tunnel.

By using the Monte Carlo technique, the probability of failure in time t for each failure mode ($p_{f,i}(t)$) is estimated. According to the theory of systems reliability, the probability of failure for a series system ($p_{f,s}$) can be estimated by (Thoft-Christensen and Baker 1982):

$$\max[p_{f,i}] \leq p_{f,s} \leq 1 - \prod_{i=1}^{n}[1 - p_{f,i}] \qquad \text{Eq. 9}$$

where $p_{f,i}$ is the probability of failure due to the ith failure mode of pipe (determined by Equation (5)) and n is the number of failure modes considered in the system.

Figure 5 shows the result of the probability of the tunnel serviceability failure (i.e., upper bound in Equation 9) considering the two failure modes. As a complete picture of failure assessment, the time for the tunnel to loose its serviceability, i.e., T_c, due to degradation, can be determined for a given acceptable probability of failure P_a. For example, using the graph for system failure in Figure 5, it can be obtained that $T_c = 23$ years for $P_a = 0.5$. If there is no intervention during the service period of (0, 23) years for the tunnel, such as maintenance and repairs, T_c represents the time for interventions or the end of service for the tunnel, based on the performance criteria of the two assumed failure modes. The information of T_c (i.e., time for interventions) is of significant practical importance to structural engineers and asset managers of tunnels in decision-making with regard to its repairs and/or rehabilitation which are usually dependent on the budget situation of the day. Therefore, when to intervene is the first question to decision-makers.

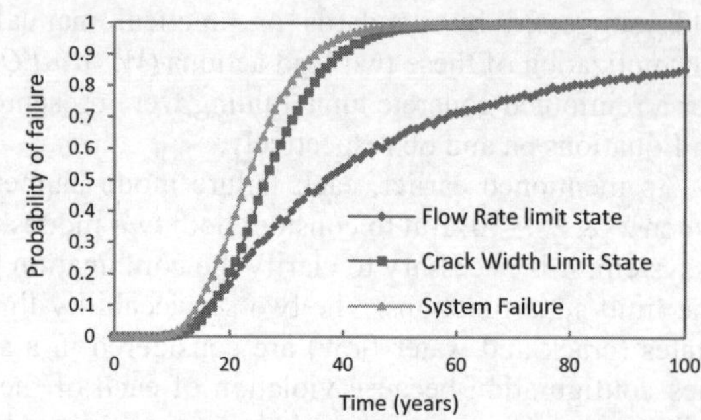

Figure 5 Probability of serviceability failure for individual failures and for system failure

Results from Figure 5 also confirm that considering multi-failure modes will result in higher probability of failure for the system compared to the individual failure modes. This shows the significance of a comprehensive multi-failure mode analysis for structures and infrastructures to prevent underestimating the probability of failure. It should be noted that the correlation between failure modes needs to be considered for more accurate results.

4 CONCLUSION

The current study identifies major serviceability failure mechanisms of underground tunnels and how such failures can be assessed. Several failure mechanisms, extending from soil/rock failure to the concrete lining material failure, along with their corresponding failure criteria were identified. The study revealed that the ground water flow can be the major driver for commonly identified failure mechanisms of under-ground tunnels.

The study has been extended to demonstrate a methodology to quantitatively assess the risk of serviceability failure of tunnels over a period of time and to predict their service life of aging infrastructure. The failure mechanisms selected for this study are based on water flow rates and corrosion induced crack width. The preliminary analysis showed that the performance prediction of the underground tunnels is mainly based on the accuracy of input data. Further, the results showed the importance of considering multi-failure mode analysis to estimate accurate probability of failure for an underground tunnel as a system. The proposed failure assessment method can help the infrastructure managers to develop rehabilitation or replacement strategy for in-service tunnels with a view for optimum maintenance management of the tunnel asset.

ACKNOWLEDGEMENT

Financial support from Metro Trains Melbourne, Australia and Australian Research Council under DP140101547 and LP150100413 are gratefully acknowledged.

REFERENCES

ACI Committee 365, 2002. Service life prediction – state of the art report, American Concrete Institute, Farmington Hills, Mich, 44.

Alonso, C. Andrade, C. Rodriguez, J. et al. 1998. Factors controlling cracking of concrete affected by reinforcement corrosion, Materials and Structures 31, 435-441.

Andrade, C. Alonso, C. & Molina, F.J. 1993. Cover cracking as a function of bar corrosion: Part I-Experimental test, Materials and Structures 26, 453-464.

Bennet, A.G. Smith, N.B. & Neilson, J.L. 1992. Tunneling. In: Peck, W.A. Neilson, J.L. Olds, R.J. & Seddon, K.D. eds. The Seminar on Engineering Geology of Melbourne, 1992 Melbourne. Engineering Geology of Melbourne, 321-344.

Bieniawski, Z.T. 1990. Tunnel design by rock mass classifications, DTIC Document.

Briaud, J.L. 2008. Case Histories in Soil and Rock Erosion: Woodrow Wilson Bridge, Brazos River Meander, Normandy Cliffs and New Orleans Levees, Journal of Geotechnical and Geoenvironmental Engineering 134, 1425-1447.

British Standards Institution, 1997. BS 8110:Structural use of concrete - Code of practice for design and construction - Part I:1997, London,UK.

Chen, D. & Mahadevan, S. 2008. Chloride-induced reinforcement corrosion and concrete cracking simulation, Cement and Concrete Composites 30, pp. 227-238.

Dimitrios K. 2005, Tunnelling and Tunnel Mechanics - A Rational Approach to Tunnelling, Springer, Austria.

Ditlevsen, O. & Madsen, H.O. 1996. Structural Reliability Methods, John Wiley and Sons.

Ekstrom, T. 2001. Leaching of concrete: experiments and modelling, Report TVBM 3090, 3090.

Fernandez, G. & Moon, J. 2010. Excavation-induced hydraulic conductivity reduction around a tunnel – Part 1: Guideline for estimate of ground water inflow rate, Tunneling and Underground Space Technology 25, 560-566.

Hoseini, M. Bindiganavile, V. & Banthia, N. 2009. The effect of mechanical stress on permeability of concrete: A review, Cement and Concrete Composites 31, 213-220.

Li, C. Lawanwisut, W. Zheng, J. et al. 2005. Crack width due to corroded bar in reinforced concrete structures, *International Journal of Materials & Structural Reliability* 3, 87-94

Li, C. Q., and M. Mahmoodian 2013. Risk Based Service Life Prediction of Underground Cast Iron Pipes Subjected to Corrosion, *Journal of Reliability Engineering & System Safety*, Volume 119, November, Pages 102–108

Liu, T. & Weyers, R.W. 1998. Modeling the Dynamic Corrosion Process in Chloride Contaminated Concrete Structures, Cement and Concrete Research 28, 365-379.

Mahmoodian, M. & Alani, A. 2013, Multi failure mode assessment of buried concrete pipes subjected to time dependent deterioration using system reliability analysis, Journal of failure analysis and prevention 13, 634-642.

Mahmoodian, M. and C. Q. Li 2015. Stochastic Failure Analysis of Defected Oil and Gas Pipelines In: *Handbook of Materials Failure Analysis With Case Studies from the Oil and Gas Industry,* Elsevier

Melchers, R.E. 1999. Structural Reliability Analysis and Prediction, John Wiley and Sons, Chichester.

Pantazopoulou, S. & Papoulia, K. 2001. Modeling cover-cracking due to reinforcement corrosion in RC structures, Journal of Engineering Mechanics 127, 342-351.

Rubinstein, R.Y. & Kroese, D.P. 2008. Simulation and the Monte Carlo Method, John Wiley and Sons.

Thoft-Christensen, P. & Baker, M.J. 1982. Structural Reliability Theory and Its Applications, Springer-Verlag Berlin, Heidelberg.

Thoft-Christensen, P. 2000, Modelling of the deterioration of reinforced concrete structures, Dept. of Building Technology and Structural Engineering, Aalborg University.

Thoft-Christensen, P. 2003. Corrosion and cracking of reinforced concrete, Dept. of Building Technology and Structural Engineering, Aalborg University.

Vu, K.A.T & Stewart, M.G. 2000. Structural reliability of concrete bridges including improved chloride-induced corrosion models, Structural Safety 22, 313-333.

Wong, H.S. Zobel, M. Buenfeld, N.R. & Zimmerman, R.W. 2009. Influence of the interfacial transition zone and microcracking on the diffusivity, permeability and sorptivity of cement-based materials after drying, Magazine of Concrete Research 61, 571-589.

Proceedings of the International Conference on Smart Infrastructure and Construction
ISBN 978-0-7277-6127-9

© The authors and ICE Publishing: All rights reserved, 2016
doi:10.1680/tfitsi.61279.583

Long-term behaviour of support elements in tunnelling - Results of investigations in several thirty year old tunnels

S. Lorenz*, R. Galler

Montanuniversität, Leoben, Austria
* *Corresponding Author*

ABSTRACT The technical lifetime of underground structures does not only depend on the long-term behaviour of individual support elements and material characteristics but also on various boundary conditions, which influence the entire long-term system behaviour of the structure. Latest by April 2019 all road tunnels with a traffic load of more than 20.000 vehicles per day must have two tubes due to the requirements of the directive 2004/54/EC of the European parliament. Thanks to the construction of second tunnel tubes and the corresponding cross passages sampling of support materials including membranes was possible. The paper presents outcomes regarding the long-term behaviour of support elements and membranes. To evaluate the stresses and the supporting effects of the tunnel shells flat-jack tests and different simulations were performed. The results presented can be used as a basis for future designs.

1 INTRODUCTION

Tunnels are an indispensable part of modern transport infrastructure. As multi-generation structures their long-term preservation as well as their technical service life is of central importance.

The current state of research in the field of durability of structural elements must be considered separately. In almost all areas a prediction of materials and structural components towards lifetime exists. System-wide investigations were rarely performed, so that there is no technical lifetime consideration of the outer lining in systems included.

1.1 Objective

The target of this work is, at first, to create a large database with the lifetime of different structural elements and materials, which may be the basis of durability considerations of tunnels. Secondly, the goal of these investigations is to optimise tunnel designs and their dimensioning regarding to the technical lifetime. For that matter, it is necessary to analyse the selected approaches of the loss of strength and stiffness of the outer lining.

1.2 Design method

In Austria tunnels are generally built with a double-shell construction. The outer lining serves as a temporary support during construction and usually consists of reinforced shotcrete, which is secured by further support elements. Between outer and inner lining a sheet membrane with a geotextile fleece is provided for sealing the tunnel. Finally an inner shell is built as a permanent structural system. The main tasks of the two shells are to bear the generated loads and keeping off mountain water.

2 PROJECT SCOPE

In the tunnels, mentioned in Table 1, sampling of support materials in cross passages was possible due to construction work required by the tunnel safety law (Directive 2004/54/EC, 2004).

Table 1. Evulated tunnel projects, investigation period 2010 – 2015.

Tunnel	Driving start of first tube	Geological conditions
Ganzsteintunnel	1976	phyllite, limestone
Katschbergtunnel	1970	gneiss, phyllite
Tanzenbergtunnel	1983	chlorite schist, amphibolite, marble
Roppener Tunnel	1987	dolomite
Tauerntunnel	1970	gneiss, phyllite
Pfändertunnel	1975	marl, sandstone
Bosrucktunnel	1980	dolomite, anhydrite
Gleinalmtunnel	1973	gneiss, amphibolite
Arlbergtunnel	1974	gneiss, phyllite

2.1 *Object of investigation*

In the process of investigation samples of

- shotcrete
- concrete of the inner lining
- rock bolts
- sheet membranes with geotextile fleece

as well as impacting mountain water were taken. Based on the concrete samples of outer and inner linings strength tests were carried out. Additionally to the strength tests on rock bolts metallographic analysis were performed. The changed material properties and defects of the samples of the membranes were also examined. To compare the quality of the samples investigations were also performed on reference materials.

In order to evaluate the load-bearing capacity of the inner lining stress compensation tests were executed. The stress within the concrete was determined by a local relaxation of the lining as a result of a saw cut and by the compensation of the deformation by a flat-jack. In FE-simulations the measuring procedure and the load transmission is simulated in different models.

Over the lifetime of a tunnel the linings are exposed to various environmental conditions. These impacts result from the rock mass, the construction conditions and the use of the tunnel as well as from aging effects of the materials.

3 INVESTIGATIONS

3.1 *Concrete tests*

For testing the shotcrete and inner lining concrete core samples were taken during construction of the cross passages and investigated in the laboratory. The uniaxial compressive strength (UCS), the shear parameters, the cohesion and the friction angle were determined. Furthermore investigations using scanning electron microscope (SEM) were performed to investigate the structure of shotcrete in more detail. For the analysis, the measured strength was compared with the documented and required strength during the construction phase.

The following Table 2 shows the calculated mean values of the strength tests of several individual samples. As one can see the measured shotcrete strengths are in general significantly higher than the strengths considered in the design requirements. The investigated shotcrete samples therefore indicate that slow-hardening cements (high C2S content), which can reach a post-hardening up to 50 % of the 28-day strength, were used. This is due to an increased formation of the CSH phases by extending the second hydration step, which leads to a more stable matrix (Benedix, 2011).

By the use of a scanning electron microscope these CSH phases and numerous clinker remains were observed which might be explained by the fact that the water-cement ratio was not constant when the shotcrete was manually applied.

Table 2. Values of different parameters obtained as a result of testing outer and inner lining concrete.

	outer lining		inner lining	
Tunnel	E [GPa]	UCS [MPa]	E [GPa]	UCS [MPa]
Ganzsteintunnel	29,7	44,7	32,3	34,1
Katschbergtunnel	23,9	43,7	33,8	36,2
Tanzenbergtunnel	33,2	69,5	39,2	36,2
Gleinalmtunnel	-	82,2	37,3	36,3
Arlbergtunnel	21,8	44,7	-	47,7
Roppener Tunnel	-	-	40	40,1
Bosrucktunnel	-	-	46	33,1

- not measured, E-modulus, UCS Uniaxial Compressive Strength

3.2 *Membrane test*

The re-extracted sheet membranes and geotextile fleeces have been used for about 30 years and were examined for changes of typical polymer properties. The membranes were removed from different cross passages of the investigated tunnels in order to be able to record different impacts. For evaluation of their conditions, chemical, mechanical and thermo-mechanical tests as well as reference measurements with a new membrane were executed.

Assuming that within a tunnel a uniform sheet membrane was used, the results clearly indicate that different aging mechanisms in various cross passages occurred. The tensile tests show some significant changes between measured parameters of different cross passages. The variation of tensile test results, in Figure 1, correlate with the determination of the de-hydrochlorination (DHC) temperature and the infra-red spectroscopy.

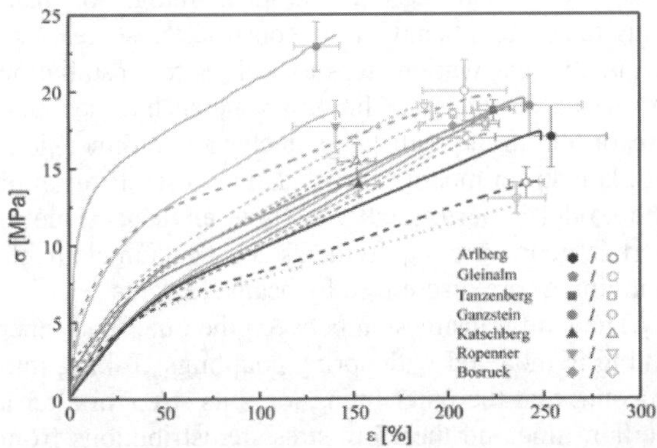

Figure 1. Summary of the stress-strain diagrams of the investigated membranes.

Individual samples show great embrittlement of the material. Changes in the DHC temperature point towards a higher thermal impact, while differences in the composition of plasticiser indicate a migration of plasticiser, eventually in combination with environmental conditions. In practice, a complex interaction of both mechanisms might be responsible for the observed changes in materials.

3.3 *Rock bolt tests*

As part of building the cross passages and the combined demolition of the inner lining of the existing tube, rock bolts were removed. Mechanical, chemical and metallographic analyses were performed on the rock bolts. 15 different rock bolts, each between 20 to 40 years old, were tested for their strength (Figure 2). Initial local corrosive attacks can be observed on the surfaces. However, it is currently assumed that the samples were not significantly affected by the external corrosive attacks due to the observed strength and hardness values as well as the structural examination.

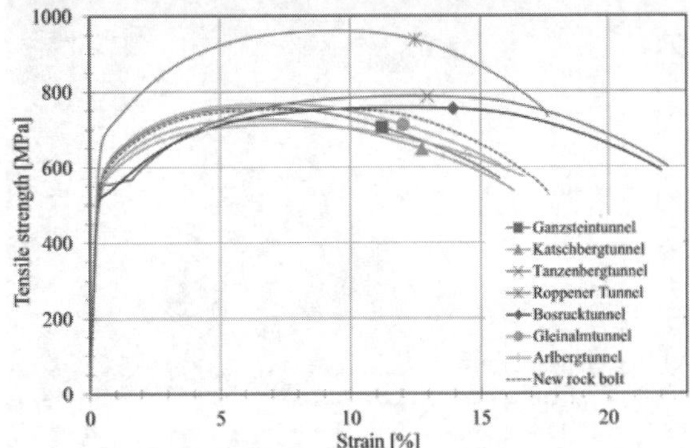

Figure 2. Stress-strain curves of the tested rock bolts.

Because of the parameters and structures of the 15 rock bolts it becomes clear that there are at least four different ferritic-pearlitic steels with a quality of standard construction steel or grain-refined construction steel. It is not possible to accurately classify the steel types based on the parameters of the tensile tests. In order to better assess the quality of the rock bolts they were compared to new rock bolts made of low-alloy carbon steel.

3.4 *Flat-jack tests*

In order to evaluate the load transmission from the outer lining onto the inner lining, stress compensation measurements, so-called flat jack tests, and numerical simulations were executed.

The test is based on the consideration that displacements which are caused by a local stress unload can be compensated by the compressive stress of a flat-jack. The applied compensation pressure can be read off of the pressure gauge. Taking the shape and geometry of the flat-jack into account and converting them including correction factors the compensation stress can be calculated. The calculated stress is assumed to be the tangential stress of the inner lining. The test is based on the principle of linear elastic deformation. In a numerical simulation the flat-jack test is simulated by a three-dimensional FE-model. This makes it possible to evaluate the measurement method, which in return provides information about the suitability of the used method for stress determination. Figure 3 shows the measuring method and the related FE-model.

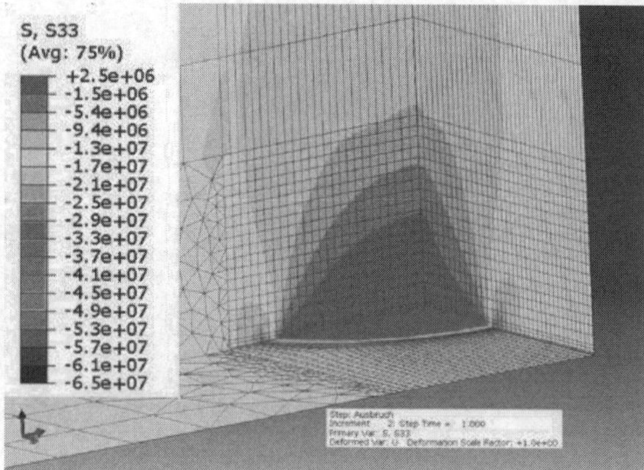

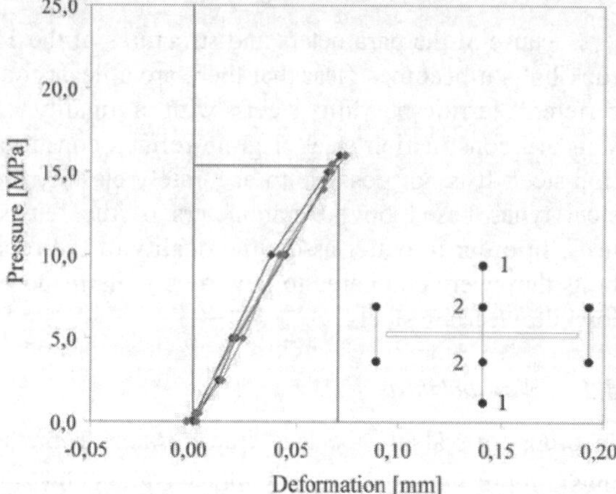

Figure 3. Numerical simulation and corresponding evaluation of stress compensation by flat-jack testing.

For the verification of the performed in-situ measurements, the documented displacement values of the test were compared with the displacements calculated by the numerical simulation. The results show that the test is appropriate for the stress evaluation in the tunnel shell.

Measurements are performed on the sidewall about 3 m above the banquet and if possible on the roof of the tunnel measurements were performed. The examined cross sections (CS) were arranged in geologically interesting areas according to the geotechnical longitudinal intersections.

3.4.1 *Numerical simulation*

For the estimation and the calculation of the strain transmissions between the outer and inner lining are based on numeric simulations of the tunnel cross sections. The input parameters for the simulation originate from geotechnical report of the respective tunnel projects. The designed tunnel cross sections and coverings correspond to the monitored cross sections that were part of the flat-jack tests. For the purpose of comparison, the stresses at the same sidewall points and roof points, on which the measurements in the tunnel were executed, are calculated.

In the numerical simulations the outer and inner lining are simulated together in one FE-model, which makes it possible to determine the load transmission from the outer lining onto the inner lining. For mapping the system behavior, all construction stages, like sequential excavation steps as well as the installation of the outer and inner lining are taken into account. According to the realistic geometry a two dimensional plain strain model is created. The discretization of the model is performed with finite elements. Due to certain geometric requirements construction elements and linings are discretized by beam elements.

The load transmission between the outer and inner lining is modeled with spring couplings. Taking into account that the outer lining loses its strength after a certain time and therefore stress redistributions from the outer onto the inner lining takes place, the alpha value method is used.

By decreasing the alpha value, a decrease of the outer lining resistance and consequently the deterioration of the system is simulated. It is assumed that if alpha equals 0.01 (1%), the entire load is affecting the inner lining. The most relevant advantage of this simulation is that the stress distributing impact of the outer lining is considered in the calculation of the inner lining.

3.4.2 *Results*

Table 3 shows the comparison of the stress measurements by flat jack test at the inner lining and the corresponding numerical simulation.

Table 3. Results of the simulations and the flat-jack tests

Tunnel	Meas uring CS	Sidewall flat-jack test [MPa]	Sidewall numerical calculation [MPa]	Roof flat-jack test [MPa]	Roof numerical calculation [MPa]
Arlberg-tunnel	CS 1	t	0,27	t	0,11
	CS 2	t	0,45	2,02	1,05
Bosruck-tunnel	CS 1	14,24	35.33*	21,36	33.33*
	CS 2	0,64	0,67	0,17	0,57
	CS 3	4,54	4	0,59	5
	CS 4	0,87	0.24*	-	0.11*
Gleinalm-tunnel	CS 1	t	0,37	-	0,17
	CS 2	t	0,33	0,75	0,16
	CS 3	t	1,1	t	0,8
	CS 4	t	0,48	t	0,27
Meistern-tunnel	CS 1	0,14	2.91*	-	-
	CS 2	0,14	4,07	-	-
Pfänder-tunnel	CS 1	6,05	15.40*	-	-
	CS 2	0,72	0,28	-	-
	CS 3	10,64	15.60*	-	-

* calculated with swelling pressure, t no compressive stresses, - not measured

4 RESULTS AND DISCUSSION

The results from the tested concrete samples of outer and inner lining from different tunnels indicate no degradation of the concrete characteristics. No reduction in strength of the samples after 30 years under load could be recorded. In a few cases an increased strength was measured, which implies a longer on-going hydration. The analysis of rock bolts show that there are significant local corrosive attacks on the surface of the bolts. However, according to the observed strength and hardness as well as the examined structure it is assumed that the samples were not significantly affected by external corrosive attacks. The investigations of the sheet membranes based on the performed characterisation methods provide no clear evidence of substantial material aging. Measurements of membranes from the same tunnel at different cross passages sometimes show significant variations in material properties, which indicates different external influences in the cross passages. Furthermore the examination of specific samples under special attention to surface damage lead to the conclusion that there is quite a high probability for leaks in form of point-like holes or penetrations. The measured differences in composition of the membranes and rock bolts go back to differences in composition and production methods. The testing results from the stress compensation measurements indicate that in geotechnical good rock mass it is only possible to detect low stresses in the inner lining after inserting a cut. Only in areas with swelling rock higher stresses can be measured and calculated, because it is difficult to estimate the swelling potential of the rock mass, which leads to a higher deviation of the measured stresses and the simulation in these areas. This leads to the conclusion that the outer lining is still intact and that there is nearly no stress redistribution or induction into the inner lining. This conclusion is supported by 24 individual measurements at 15 tunnel cross sections from the above mentioned tunnel projects.

5 CONCLUSION

The main focus of the research was to study the long-time behaviour of support elements in tunnels. The results show no reduction of the technical lifetime regarding the strength of the support elements. The durability of the static system is not negatively influenced but due to the operation of the tunnels the surface of the inner lining concrete is attacked. Stress measurements of the inner lining indicate that the bearing capacity of the outer lining is still intact.

As a result of these extensive tests on support materials the existing dimensioning concepts for double lined tunnels can be reconsidered. New concepts, in which the outer lining is part of the permanent support elements, can be considered. This approach would be applicable for conventional as well as mechanical tunnelling.

The results show almost no deterioration of all used support materials. It was shown that the support function of the primary support is unaffected even after 30 years and most likely for many years more.

REFERENCES

Benedix, R. 2011, Bauchemie: Einführung in die Chemie für Bauingenieure und Architekten, Vieweg+Teubner Verlag / Springer Fachmedien Wiesbaden. Wiesbaden

Directive 2004/54/EC of the European Parliament and the of the council of 29 April 2004 on minimum safety requirements for tunnels in the Trans-European Road Network [2004] OJ L167/37

Proceedings of the International Conference on Smart Infrastructure and Construction
ISBN 978-0-7277-6127-9

© The authors and ICE Publishing: All rights reserved, 2016
doi:10.1680/tfitsi.61279.589

Grading-based deterioration models for future performance predictions of coastal flood defences

M. B. Mehrabani and Hua-Peng Chen*

Department of Engineering Science, University of Greenwich, Chatham Maritime, UK
* *Corresponding Author*

ABSTRACT Accurate evaluation of the current and future conditions of coastal defence structures is essential for appropriate practical maintenance strategies, and as the structures get aged, it is more challenging to implement maintenance and management plans to avoid structural performance failure. This paper presents a probabilistic method for predicting future performance deterioration of coastal flood defence structures based on condition grading inspection data and deterioration curves. In state-based deterioration modelling, the main task is to estimate transition probability matrixes. The deterioration process of the structure is modelled according to Markov chain process, and a reliability-based approach is used to estimate the probability of overtopping failure. In the study, visual inspection data according to the Condition Assessment Manual CAM (Environment Agency 2006) are used to obtain the condition grade of the coastal flood defences in order to develop transition probabilities through optimisation algorithms. The Monte Carlo simulations are then used to evaluate the future performance of the structure on the basis of the estimated transition probabilities. Results from the case study show that the proposed method can provide an effective predictive model for various situations in terms of available condition grading data. The proposed model also provides useful information on time-variant probability of failure of the coastal flood defences.

1 INTRODUCTION

Management and maintenance of coastal defence structures during the expected life cycle have become a real challenge for decision makers and engineers. Also, as coastal defence structures get aged, it becomes more challenging to implement maintenance plans to avoid structural failure. Therefore, condition inspection data are essential for assessing current state and forecasting deterioration of aging flood defence structures in order to keep the structures in an acceptable condition (Chen and Alani 2012, Chen and Alani 2013).

In developing condition-based deterioration models for coastal defence structures, the inspection data, often collected using discrete visual condition rating schemes, which need to be adopted to develop probabilistic deterioration models. However, existing deterioration models may not provide a reliable prediction of performance deterioration for a long period due to uncertainties.

To tackle the limitation, a time-variant state-based model associated with a transition probability needs to be developed on the basis of condition grade scheme for flood defences. Recently, some studies have been carried out on performance prediction of civil engineering structures based on condition-based deterioration curves such as bridges (Saydam et al. 2013), pavements (Ortiz-García et al. 2006), waste waters (Baik et al. 2006), and coastal defences (Nepal et al. 2015, Chen 2006).

In this study, a condition-based stochastic deterioration model of flood defence structures such as earth sea dyke presented. The performance deterioration model due to aging is adopted by utilising a Markov chain model. Crest level deterioration due to settlement is also considered by applying into different condition grades. Transition probability matrix of the condition-based stochastic process is estimated based on the deterioration curves provided for flood defence structures (Halcrow Group 2013), and then optimisation algorithm is used for calibration.

Finally, failure probability of the structure due to overtopping is calculated with consideration of probability distribution of condition grades over time.

2 FLOOD DEFENCE CONDITION RATING SYSTEM

Condition Assessment Manual (CAM) is a condition grade assessment criteria in the context of performance based asset management (Environment Agency 2006). Condition grades are used to offer a standardised approach to assess the deterioration of flood defence structures, and this assists decision makers in managing the maintenance strategies. In the visual inspection system based on CAM, each component is visually inspected by a trained inspector and classified into one of five condition grades. It starts from grade 1, which denotes very good condition, and finishes at grade 5 for very poor condition, indicating the damage level.

Environment Agency provides a series of asset deterioration curves for different types of flood defences which can be used to estimate the future condition grade of a coastal defence structure (Halcrow Group 2013). Deterioration curves are determined based on the mentioned condition grading system to quantify the residual life of an asset with and without maintenance plans. Although the provided deterioration curves are useful to determine the residual life of flood defence structures, the level of uncertainty is still considerable due to nature of deterioration process.

Hence, in order to translate the deterministic condition grades into probabilistic deterioration models, a state-based stochastic model is considered. In this study, the asset deterioration condition grades given by Environment Agency (Halcrow Group 2013) will be used to estimate the failure probability of a structure due to wave overtopping.

3 MARKOV VHAIN DETERIORATION MODEL

Markov chain is a discrete stochastic time-dependent process to estimate the future event as random outcomes based on the present condition of the system (Norris 1998). This is a common tool to predict the deterioration of structure components which helps to deal with deterioration uncertainties. A Markov chain is considered as a series of transitions between certain condition grades.

When the Markov chain is used to model deterioration of a system in state i, a fixed probability p_{ij} exists when a system changes from state i to state j during the period. p_{ij} is the transition probability that given the system in state i at time t, it will be in a state j at time $t+1$.

Estimating the transition probability matrix $P_{(n)}$ for a system is based on available inspection data. Deterioration of flood defence structures is represented through the elapsed time to increase in condition grade ranging from 1 to 5, and it is assumed to have been derived through expert judgment. Therefore, the transition probability matrix of Markov chains from grade i to j is represented by a 5×5 matrix.

Under assumption of Do-nothing condition, which means no maintenance is applied to the components or structure during the service period, the asset will gradually deteriorate and hence the corresponding condition grade either transit to a higher number or remain unchanged during the inspection period. One more condition applies to the process when it is used to simulate coastal flood defence deterioration, as $p_{55}=1$, indicating condition grade where the sea dyke has reached its worst condition and cannot deteriorate further. Also it is assumed that the condition to deteriorate by no more than one grade in one specific period of time, in order to simplify the model. Hence, the transition probability matrix P by assuming a homogenous condition can be expressed as

$$P = \begin{bmatrix} p_{11} & 1-p_{11} & 0 & 0 & 0 \\ 0 & p_{22} & 1-p_{22} & 0 & 0 \\ 0 & 0 & p_{33} & 1-p_{33} & 0 \\ 0 & 0 & 0 & p_{44} & 1-p_{44} \\ 0 & 0 & 0 & 0 & 1 \end{bmatrix} \quad (1)$$

$$and; \sum_{j=1}^{4} p_{ij} = 1 \qquad for\ i = 1,2,3,4$$

Each element in the transition probability matrix P represents the probability that the condition of the system or component concerned will transfer from state i to state j during a certain period of time t. The probability that a process in state i will be in state j after n transition is defined as n-step transition probability.

The n-step transition probability matrix $P_{(n)}$, and the probability of each condition state (or grade) after n transitions $C_{(n)}$ can be expressed as, respectively (Ortiz-garcía et al. 2006)

$$P_{(n)} = P^n \tag{2}$$

$$C_{(n)} = C_{(0)}.P^n \tag{3}$$

where $C_{(0)}$ is often the starting vector based on the initial condition of the asset or component concerned. $C_{(n)}$ indicates the current condition, and the sum of all its proportions should be equal to 1, and all entries should be non-negative.

To satisfy the homogeneity of Markov chain models, a zoning concept is applied. A zone is a certain period of time assumed to produce constant transition probabilities, and the period for the zone is based on expert judgment of inspection intervals. The transition probability matrix P for n transition is obtained by minimising the difference between the observed and the predicted condition state of the sea dyke, given as

$$\text{Minimise: } \sum_{t=T_s}^{t=T_e} \sum_{n=1}^{N} \left| O_{(t)} - E_{(t)} \right|$$

$$0 \le p_{ij} \le 1,\ for\ i,j \in \{1,2,3,4,5\} \qquad (4)$$

$$\sum_{j=1}^{5} p_{ij} = 1,\ for\ i \in \{1,2,3,4,5\}$$

where t is the age of the flood defence structure; T_s and T_e are start and end age for each zone of the structure, respectively; $O(t)$ is the observed condition at time t for n transitions. The expected value $E_{(t)}$ of a specific condition grade at time t for n transitions based on Markov chain model, where transpose of condition rating vector is $S = [1\ 2\ 3\ 4\ 5]$, can be expressed as

$$E_{(t)} = C_{(n)} \cdot S^T \tag{5}$$

where T is the age of the asset, and by assuming that the structure is newly constructed, the initial condition grade vector is $C_{(0)} = [1\ 0\ 0\ 0\ 0]$.

4 FAILURE ASSESSMENT OF A SEA DYKE

The identifying of an asset's venerable parts is a critical step to process failure probability assessment. The Markov chain process used to model the deterioration of an earth sea dyke based on condition grading. The procedure of failure probability analysis in this approach can be categorised into three steps: 1) identifying the transition probability matrix for the asset or component 2) computing time-variant Markov chain condition grade probabilities and 3) assessing the failure probabilities for the possible condition grades over time.

Therefore, the failure probabilities of a sea dyke are different for various condition grades. In this paper, the overtopping failure mechanism is used to determine the failure probability of a sea dyke. Wave overtopping is an important failure mechanism of a coastal defence structures and its failures may result to significant consequences.

Excessive wave overtopping may lead to slope failure of the earth dykes by eroding the dyke crest and landside slope and by deteriorating soil strength due to saturation of the earth dyke. Average wave overtopping discharge q can be calculated by using Pullen et al. (2007), given here as

$$\frac{q}{\sqrt{g \cdot H_{m0}{}^3}} = \frac{0.067}{\sqrt{\tan\alpha}} \cdot \varepsilon_{m-1,0} \cdot \exp\left(-4.3 \frac{R_c\,(G_i)}{\varepsilon_{m-1,0} \cdot H_{m0}}\right) \tag{6}$$

with a maximum of:

$$\frac{q}{\sqrt{g \cdot H_{m0}{}^3}} = 0.2 \cdot \exp\left(-2.3 \frac{R_c\,(G_i)}{H_{m0}}\right)$$

where H_{m0} is significant wave height at the toe of structure; g is the acceleration due to the gravity;

$\varepsilon_{m-1,0}$ is breaker parameter; α is the angle between overall structure slope and horizontal line; and $R_c(G_i)$ is crest freeboard of the structure above the still water level at grade $i \in \{1,2,3,4,5\}$. For the existing coastal defence structure with crest level loss at specific condition grade, the freeboard $R_c(G_i)$ can be calculated from

$$R_c(G_i) = R_c(0) - H_d(G_i) \quad (7)$$

where $R_c(0)$ is initial freeboard; and $H_d(G_i)$ is the loss of dyke crest height at condition grade i due to deterioration. The failure of a sea dyke is described when the overtopping discharge rate calculated from equation (6) exceeds from a predefined critical rate e.g. 2 l/s/m suggested by CIRIA (2007).

The crest level loss of the sea dyke $H_d(G_i)$ is estimated for five different deterioration levels. The surface condition of an earth dyke crest is divided into three main categories as slight, minor, and major deteriorated conditions (Long et al. 2013). In this classifying, the condition of structure is defined based on the geometrical degradation in crest. Table 1 shows the translating of quantitative assessment into visual inspection condition rating system for crest features of an earth sea dyke. Loss of height is the change in elevation associated with the crest feature.

Table 1. Crest features related to damage and failure of an earth sea dyke.

Condition Grade	Loss of height (cm)	Density distribution
1	0-3.5	Lognormal
2	0-5.0	Normal
3	5.0-13.0	Normal
4	13.0-30.0	Normal
5	30.0+	Lognormal

To translate the condition grades into probabilistic terms, it was assumed that condition grade deterioration over time was linear and the deterioration intensity was normally distributed. Figure 1 shows the probabilistic distribution of crest level loss for various condition grades. This study considers high inspection quality, where the correct rating could be expected 90% of the time by utilising e.g. Kinematic GPS techniques in a low level noise situation or high experienced inspectors (Long et al. 2013).

The distributions are normal with the exception of the first and last condition grades, which are assumed to be lognormal, as the crest height will not increase due to deterioration (condition grade 1), and more deterioration considers as the functional failure (condition grade 5). The mean and standard deviation of the distributions are calculated by the method proposed by Ang & Tang (2007). For example, the mean value (μ) for probabilistic distribution for condition grade 3 is 9.0 cm with the standard deviation of σ =2.45 assuming the inspector being wrong 10% of the time. This means the probability of obtaining a value of crest height loss between 5.0 and 13.0 cm is 90%.

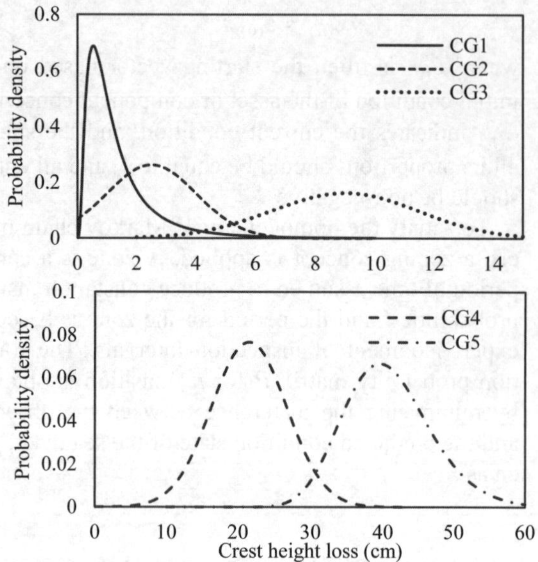

Figure 1. Probability density distribution associated with condition grade of an earth sea dyke crest.

The fragility curve of an earth sea dyke can be described as the failure probability conditional on a specific loading (Buijs et al 2007). The failure probability can be from different failure mechanisms, and in this study wave overtopping failure mechanism is considered. Water level including storm surge may be considered as loading, however it is suggested to consider combination of water level and wave heights at the same time for sea dykes (VanDerMeer et al. 2009). Therefore, by considering the decrease in crest level over time, the time-variant failure probability analysis can be conducted. The failure probability for different condition grades can be determined if the deterioration level at each condition grade is known.

5 NUMERICAL EXAMPLE

A numerical example is performed for an existing sea dyke described by Cong (2004). The structure has a crest height of MSL 5.50 m, a seaside slope of 1:4, a landside slope of 1:2, and crest width of 4.00 m. The sea slopes are protected by pitched stone revetment within 0.45 cm thickness to prevent wave attacks. The extreme height water level from combination of high tidal elevation and positive surge, is assumed at a level of MSL 3.19 m with a return period of 200 years. The critical overtopping discharge rate of 1 l/s/m is considered in the design of the sea dyke.

In this paper, the parameters relevant to the deterioration of crest level are considered. The initial grade probability vector for the crest of the earth sea dyke is assumed $C_{(0)} = [1\ 0\ 0\ 0\ 0]$. Estimation of the transition probabilities for a sea dyke requires adequate available inspection data history. The deterioration rate of typical earth sea dykes with progress of age is represented by five different condition grades (Halcrow Group 2013). The earth sea dyke deterioration rate over time matches to the situation is defined by the manual. Nonlinear optimisation based on equation (4) was performed to evaluate the transition probability. The estimated transition probability matrix for first stage of transition is expressed as

$$P = \begin{bmatrix} 0.815 & 0.185 & 0 & 0 & 0 \\ 0 & 0.867 & 0.133 & 0 & 0 \\ 0 & 0 & 0.891 & 0.109 & 0 \\ 0 & 0 & 0 & 0.680 & 0.320 \\ 0 & 0 & 0 & 0 & 1 \end{bmatrix} \tag{9}$$

The time-variant condition state probabilities of the asset are computed using equation (5) and presented in Figure 2. The results show how the flood defence structure deteriorates from one condition grade to another (CG1 to CG5) over time. From the results, the probability of CG1 vanishes at approximately 40 years. CG2 has the highest probability, compared with other condition grades within the range of 8 and 22 years. As expected, CG5 increase steadily with time, reaching approximately a probability of 80% at age of 60 years.

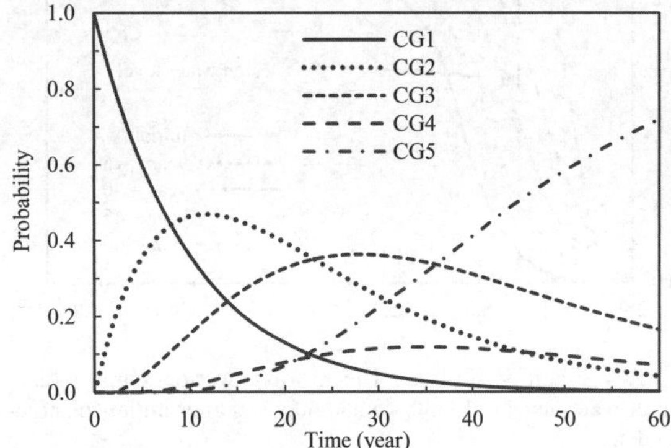

Figure 2. Time-variant Markov chain grade probabilities for an earth sea dyke.

The failure probability of the dyke in different condition grades has to be calculated. The time variant failure probability of five different deterioration levels (condition grades) can be assessed by applying time-variant Markov chain grade probability.

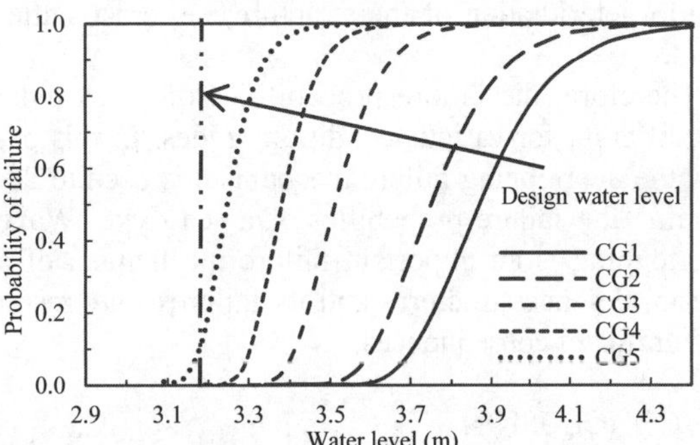

Figure 3. Fragility curve for 1.00 l/s/m wave overtopping discharge for different condition grades.

Figure 3 shows fragility curves of the earth sea dyke in different condition grades for 1 l/s/m allowable wave overtopping discharge associated with water level. The line illustrates design water level line that is used for safety assessment at MSL 3.19m. The probability of failure over time for the structure staying in condition grade 1 is not very critical, whereas the failure probability is very high for condition grade 5, as expected. The failure probability increases as condition grade increases, i.e. when structural performance of the earth flood defence deteriorates due to increase in settlement of the crest level.

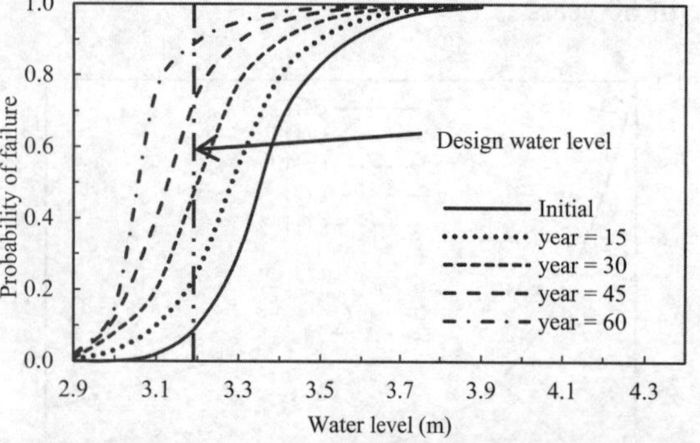

Figure 4. Fragility curves for 1 l/s/m wave overtopping as a function of water level at 15, 30, 45 and 60 years after initial construction date.

The overall failure probability of the structure can be assessed by summation of failure probability of all possible condition grades with consideration of their probability distributions. Figure 4 shows the overall failure probability of the earth dyke at initial time and after 15, 30, 45 and 60 years from the initial date due to wave overtopping. It is demonstrated that the overall probability of failure increases significantly due to deterioration of the structure, i.e. crest settlement.

Therefore, the failure probabilities of a sea dyke are different for various condition grades. In this paper, the overtopping failure mechanism is used to determine the failure probability of a sea dyke. Wave overtopping is an important failure mechanism of a coastal defence structures and its failures may result to significant consequences.

6 CONCLUSION

This paper presents a method for assessing the probability of overtopping failure over time. Here, visual inspection data according to CAM (Environment Agency 2006) are used to obtain the condition grade of the coastal flood defences in order to develop transition probability through optimisation algorithms. The possibility of different deterioration levels at a time is considered for different condition grades, and a Markov chain process is used to model the stochastic deterioration of the structure regarding the transition between the condition grades. An existing earth sea dyke is used to illustrate the method.

On the basis of the results from the case study of an earth sea dyke, it can be pointed out that the performance deterioration of flood defence structures can be predicted for overall condition grades utilising the proposed method. This can help to carry on more effective maintenance plans based on various condition grades. Also, the geometrical deteriorations such as crest level loss can be adopted into the condition grades, and then may take into account for failure probability analysis. The results will be used for generating specific fragility curves, and eventually determining the risk cost maintenance strategy.

ACKNOWLEDGEMENT

The authors are grateful for the financial support from the Institution of Civil Engineers through the R and D Enabling Fund (Project No. 1305). The findings and opinions expressed in this study are those of the authors alone and are unnecessarily the views of the sponsors.

REFERENCES

Ang, A. & Tang, W. 2007. *Probability Concepts in Engineering: Emphasis on Applications to Civil and Environmental Engineering*, 2nd Edition ed. John Wiley & Sons.

Baik, H.-S. Jeong, H. S. & Abraham, D. 2006. Estimating Transition Probabilities in Markov Chain-Based Deterioration Models for Management of Wastewater Systems, *Journal of Water Resources Planning and management*, **132**, 15–24.

Buijs, F.A. Hall, J.W. Sayers, B. & VanGelder, H.A.J.M. 2009. Time dependent reliability analysis of flood defences, *Reliability Engineering and System Safety*, **94**, 1942–1953.

Chen, H. P. 2006. Efficient methods for determining modal parameters of dynamic structures with large modifications, *Journal of sound and vibration,* **298(1)**, 462–470.

Chen, H. P. & Alani, A. M. 2012. Reliability and optimised maintenance for sea defences, *Proceedings of the ICE-Maritime Engineering*, **165(2)**, 51–64.

Chen, H. P. & Alani, A. M. 2013. Optimized maintenance strategy for concrete structures affected by cracking due to reinforcement corrosion, *ACI Structural Journal*, **110(2)**.

CIRIA, 2007, *The Rock Manual, The Use of Rock in Hydraulic Engineering*, Report C683, CIRIA(Construction Industry Research and Information Association), London.

Cong, M. V. 2004. *Safety Assessment of Sea Dikes in Vientam; A Case Study in Namadinh Province*, Ph.D. dissertation, Delft University, Netherland.

Environment Agency, 2006. *Condition Assessment Manual; managing flood risk*, Environment Agency, Bristol.

Halcrow Group, 2013. *Practical guidance on determining asset deterioration and the use of condition grade deterioration curves:Revision* 1, Environment Agency, Bristol.

Long, G. Smith, M. Mawdesley, M. & Taha, A. 2013. *Quantitative assessment methods for the monitoring and inspection of flood defences: new techniques flood defences: new techniques*, CIRIA, Classic House, London.

Nepal, J. Chen, H.-P. Gouldby, B. Simm, J. & Tarrant, O. 2015. State-Nased Stochastic Performance Deterioration Modelling of Flood Defence Assets, Torino, SHMU.

Norris, J. R. 1998. *Markov Chains, Cambridge Series in Statistical and Probabilistic Mathematics*, ISBN: 9780521633963, Cambridge.

Ortiz-García, J. Costello, S. & Snaith, M. 2006. Derivation of Transition Probability Matrices for Pavement Deterioration Modeling, *Journal of Transportation Engineering*, **132**, 141–161.

Pullen, T. Allsop, N.W.H. & Bruce, T. 2007. *EurOtop: Wave Overtopping of Sea Defences and Related Structure: Assessment Manual*, Environment Agency, Lodnon.

Saydam, D. Frangopol, D. & Dong, Y. 2013. Assessment of Risk Using Bridge Element Condition Ratings, *Journal of Infrastructure Systems*, **19**, 252–265.

VanDerMeer, J.W. Horst, W.L.A. & Velzen E.H. 2009.
Calculation of fragility curves for flood defence assets, *Flood Risk Management: Research and Practice*, 567–573.

Proceedings of the International Conference on Smart Infrastructure and Construction
ISBN 978-0-7277-6127-9

© The authors and ICE Publishing: All rights reserved, 2016
doi:10.1680/tfitsi.61279.597

Water infiltration estimation in tunnels: with reference to the Melbourne underground

Y.Q. Tan[1], J.V. Smith*[1] , C.Q. Li[1] and J. Dauth[2]

[1] RMIT University, Melbourne, Australia
[2] Metro Trains Melbourne, Melbourne, Australia
** Corresponding Author*

ABSTRACT Understanding the potential for water flow into a tunnel through the surrounding rock mass and the tunnel lining is essential for tunnel design and assessment. Typically, water penetrates through a lining along discrete paths such as joints and cracks. However, the hydraulic conductivity can be non-uniform through the lining for a range of reasons. A method describing the flow through a lining with progressively changing hydraulic conductivity is presented. When the hydraulic conductivity of the outer surface of the lining is higher than that of the inner surface, the water inflow rate can be several times higher compared to a homogenous lining with hydraulic conductivity equal to the inner surface. The inhomogeneity of the lining could also affect the water pressure on the tunnel, which could be an important factor for the stability study. A water seepage factor chart for engineering applications, incorporating inhomogeneous hydraulic conductivity, is presented.

1 INTRODUCTION

Tunnels are typically either unlined, lined allowing drainage or lined for water-sealing (Butscher, 2012). Although a tunnel lining may be intended to be water-sealed and impermeable it can deteriorate over time allowing seepage of water into the tunnel. The effect of water seeping into a water-sealed tunnel has been studied by Fernández (1994) with further development by Fernandez and Moon (2010). These studies assumed that the permeability of the lining, measured as hydraulic conductivity (m/s), is homogeneous. A more general case, presented here, is that the hydraulic conductivity of the concrete lining may be inhomogeneous. The hydraulic conductivity of the outer surface of the lining (in contact with surrounding rock and soil) could be different from the inner surface due to various reasons including the presence of steel reinforcement, crack closure and partial self-healing of the cracks in the concrete. A method to relate the observed inflow rate with hydraulic conductivity when the tunnel lining is inhomogeneous is presented in this study. The case of linear variation (increasing or decreasing) is considered. The results are illustrated by application to the Melbourne underground railway loop (MURL) which is a water-sealed tunnel system.

2 RESEARCH SIGNIFICANCE

Groundwater infiltration into tunnels is a common and critical problems all over the world (ITA, 1991). The condition when the rock mass and the lining are homogeneous and isotropic has been studied (Table 1).

Table 1 Hydraulic conductivity conditions in water inflow prediction models.

Type of Tunnel	Rock	Lining	Reference
Lined water sealing	Homogenous	Homogeneous	(Fernandez and Moon, 2010)
	Homogenous	Inhomogeneous	This paper

The rate of water inflow through a tunnel lining is very sensitive to the hydraulic conductivity of the lining material. Any inhomogeneity of the hydraulic conductivity through the lining can potentially cause significant water flow differences. Conversely, inferences of the hydraulic conductivity of a lining and surrounding rock mass, based on water inflow rate observations will differ between homogeneous and inhomogeneous hydraulic conductivity cases.

The hydraulic conductivity of sound concrete is very low (Neville, 2011). The main controlling factor of the bulk hydraulic conductivity of concrete is the presence of cracks (Reinhardt, 1997). The bulk hydraulic conductivity is usually estimated by the crack flow model developed by Snow (1965). Snow's model depends on the crack aperture (crack width) and crack density (abundance of cracks). However, one of the difficulties is that we can typically only observe the surface crack features or infer the internal continuity of cracks with geophysical methods (Fernandez and Moon, 2010). The crack aperture and crack density can vary through the concrete lining leading to a variation in the hydraulic conductivity through the lining. If the variation is systematic, such as a linear change, it can be readily incorporated into analytical models.

For the lining of the MURL tunnels (shown schematically in Figure 1), I beams were used as primary support for the surrounding rock mass of the tunnel, and steel reinforcement was placed in the in situ cast concrete close to the inner surface of the lining. The lining concrete will shrink after casting and it has been shown that reinforcement can reduce the crack width and increase the crack density in concrete (Bazant and Oh, 1983, Gilbert, 2001). Therefore, the outer surface could potentially have less abundant but wider cracks compared with the inner surface. The hydraulic conductivity would be affected accordingly.

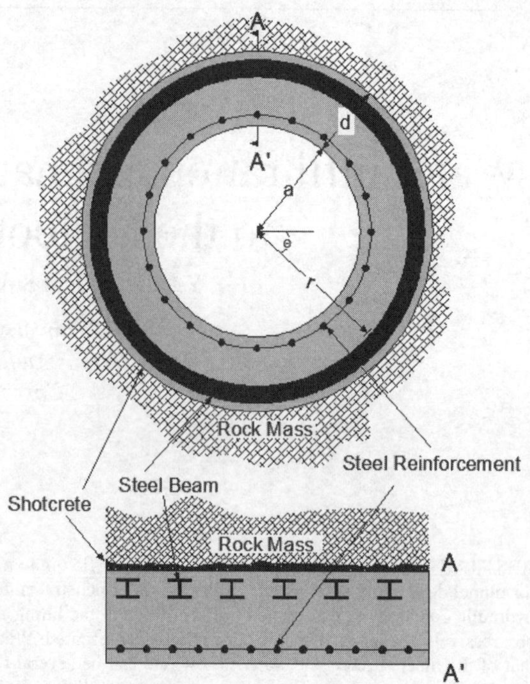

Figure 1. A schematic illustration of the MURL tunnel lining internal features. The circular reference coordinates are also shown.

Stress concentration around a tunnel can also result in changing of hydraulic conductivity by closing the cracks. According to Kirsch's equation the circumferential stress at the inner surface of the lining is greater than that of the outer surface. Joints or cracks in the lining are more likely to have a closure effect due to the confining stress closer to the inner surface of the tunnel (Bobet, 2001, Bobet and Nam, 2007). The hydraulic conductivity could potentially show a decrease from the outer to the inner surface of the lining as a result of the increased stress.

Inhomogeneity of the lining could be also caused by other source of crack width and crack density variation. A water flow study of inhomogeneous lining conditions is presented here.

3 ANALYSIS OF WATER FLOW THROUGH AN INHOMOGENEOUS LINING

The case in which the hydraulic conductivity of the lining changes linearly from the outer face of the lining to the inner face of the lining is discussed.

For an inhomogeneous lining, the water head loss through the lining is not a constant. The water head drop through the lining becomes non-linear since the hydraulic conductivity is non-constant (Figure 2).

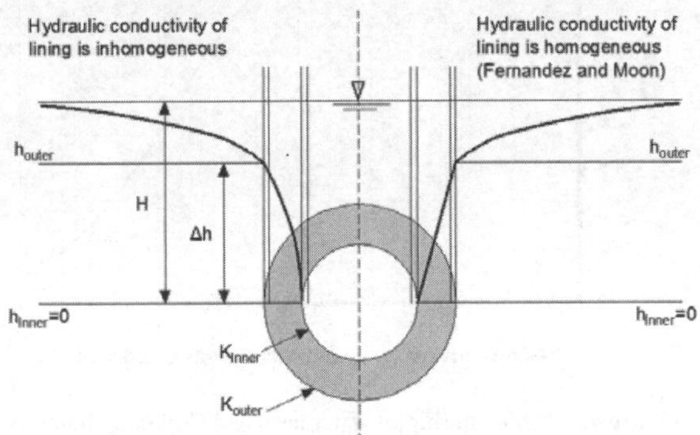

Figure 2. Head loss under both homogeneous (right) and inhomogeneous (left) conditions.

To quantify the inhomogeneity of the lining, an inhomogeneous hydraulic conductivity coefficient is defined as C_l. The coefficient is the ratio of the hydraulic coefficient between the outer lining and the inner lining ($C_l=K_{outer}/K_{inner}$).

When C_l is smaller than 1, it means the hydraulic conductivity of the outer surface of the lining is smaller than the inner surface ($K_{outer}/K_{inner}<1$). When C_l is equal to 1, it means the lining is homogeneous ($K_{outer}/K_{inner}=1$). When C_l is larger than 1, it means the hydraulic conductivity of the outer surface of the lining is larger than at the inner surface ($K_{outer}/K_{inner}>1$).

By defining the hydraulic conductivity changes through the lining, the water infiltration model can be written as Eq. 1 based on Darcy's law.

$$q = d\theta r K(r)\frac{dh}{dr} \tag{1}$$

Where q is the flow rate across a unit arc length of the tunnel; dθ is the unit central angle; K(r) is the hydraulic conductivity of the lining which is a function of r; h is the water head over the lining (Figure 1 and 2).

To determine Eq. 1, the boundary conditions need to be determined. For a lined tunnel with relatively low water infiltration velocity, the loss of velocity head is negligible compared with the total head. Therefore, the water head (total head) at the boundary of the rock and lining is assumed to be a constant. At the inner surface of the lining, there is no water pressure and the total head at the inner surface is equal to the elevation head.

Combining these boundary conditions of the tunnel, the water infiltration rate of the tunnel can be determined accordingly.

4 ANALYSIS OF AN ENGINEERING CASE

The Melbourne underground railway loop tunnels were constructed in the 1980s for public transportation. The deepest tunnel is approximately 35 m below surface. The water table is approximately 15 m above the spring line of this tunnel. The tunnels were designed as water-sealed tunnels to prevent water table draw down in the central business district. According to the design, the radius (a value, Figure 1) of the tunnel (from the tunnel centre to the inner surface of the lining) is approximately 3000 mm, and the lining thickness (d value, Figure 1) is approximately 570 mm.

Even though the concrete lining was designed to minimise water infiltration, water has been observed to flow into the tunnel leading to potential for deterioration of the lining. Seepage through cracks and joints has been observed in field inspections. The distribution of cracks in the lining has been recorded. The bulk hydraulic conductivity of the inner surface of the lining (K_{inner}) is approximately 1×10^{-7} m/s based on the crack density and crack width. The bulk hydraulic conductivity of the surrounding rock mass is approximately 1×10^{-6} m/s based on the previous geological surveys (Peck, 1992).

5 RESULTS

When the MURL parameters are modeled for conditions where the hydraulic conductivity through the lining is inhomogeneous (C_l not equal to 1) the water head distribution through the lining becomes nonlinear. This relationship leads to distinctive differences in water head behaviour through the concrete lining when the C_l value is equal to 0.01, 0.1, 1, 10 and 100 (Figure 3).

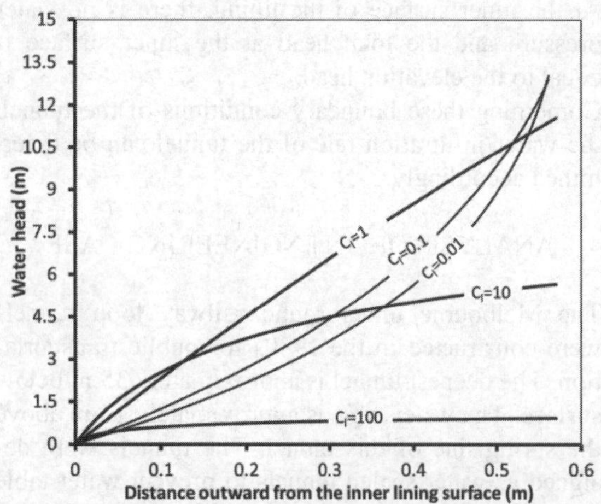

Figure 3. Relation of the water head at different lining position r-a with different C_l value when K_m/K_{inner} =10. The horizontal axis is the distance outward from the inner surface lining, equal to r-a. Parameters derived from MURL case study.

The water head at the outer surface (x=0.57, Figure 3) of the lining when C_l<1 is much larger than the value when C_l =10 and 100. This is because when C_l<1, the overall hydraulic conductivity is lower than the condition when C_l=10, 100. The lower hydraulic conductivity of the lining will build up more pressure on the outer surface of the lining than when C_l>1.

The inhomogeneous hydraulic conductivity coefficient C_l has an impact on the water head at the outer surface of the lining (Figure 4).

When C_l increases from 0.01 to 1, the water head at the outer surface of the lining does not show dramatic changes. When C_l increases from 1 to 10, the water head at the outer surface of the lining drops from 11.5 m to 5.6 m. In extreme circumstances, when C_l is 100, the water head drops to less than 1 m.

The water pressure on the lining is small when C_l increases. The air in the tunnel is more likely to flow into the lining, possibly causing a local unsaturated conditions. This is beyond the scope of this paper and will be discussed in the future.

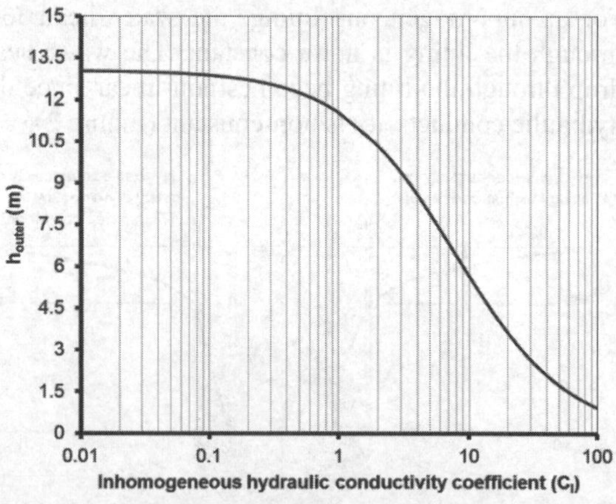

Figure 4. The relationship of water head and C_l at outer lining when K_m/K_{inner} =10. Parameters derived from MURL case study.

Variation of the C_l value will cause variable water inflow rate. An inhomogeneous hydraulic conductivity flow factor β was defined to describe the different flow rate.

The inhomogeneous hydraulic conductivity flow factor β is the ratio between water flow rate under different C_l value ($q_{cl=n}$) and the water flow rate of a homogeneous lining ($q_{cl=1}$) as in Eq. 2.

$$\beta = \frac{q_{cl=n}}{q_{cl=1}} \tag{2}$$

Generally speaking, the water flow rate does not increase linearly with the change of C_l. As the C_l value increases from 1 to 10, the water flow ratio shows an increase from approximate 0.25 to 1.9. In this range, the dominant factor controlling the water inflow is the change of the overall hydraulic conductivity. After the C_l value increases to 20, the water flow ratio shows a decrease. In this range, the decrease of water head will take over and become the dominant factor which could result in a decrease of the water inflow rate (Figure 5).

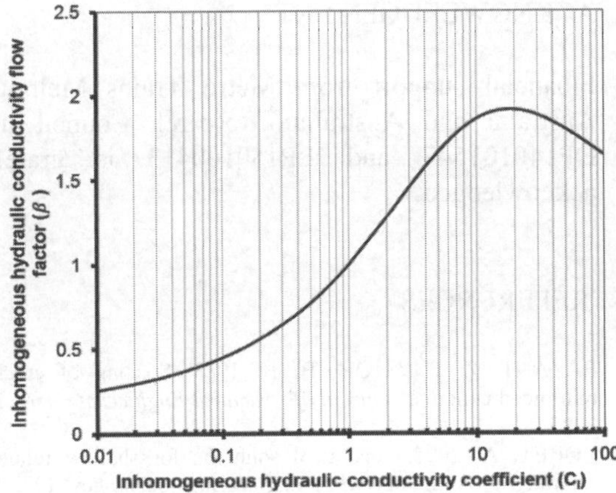

Figure 5. Relationship of inhomogeneous hydraulic conductivity flow factor β and inhomogeneous hydraulic conductivity coefficient C_l when K_m/K_{inner} =10. Parameters derived from MURL case study.

In the case of the MURL, some of the parameters are highly variable. The hydraulic conductivity of rock mass and the lining have a range of different values. A water seepage factor chart (Figure 6) is presented to illustrate the relationships of several different parameters.

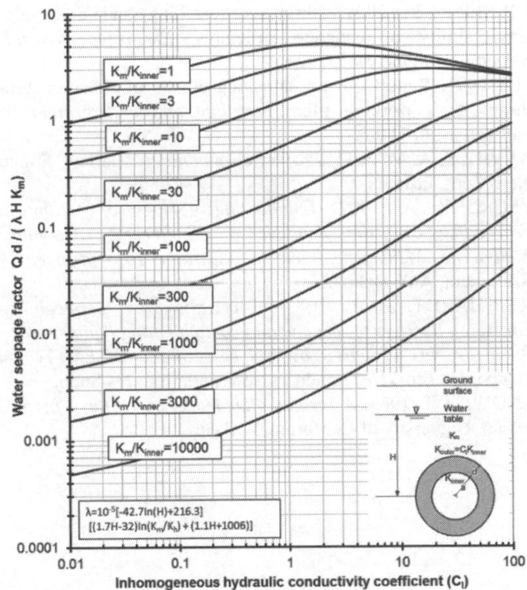

Figure 6. Tunnel water seepage factor chart incorporating inhomogeneous hydraulic conductivity of the lining.

In the water seepage chart (Figure 6), Q is the water inflow rate per metre length of the tunnel (m^3/s); H is the total water head above the spring line of the tunnel (m); K_m is the hydraulic conductivity of the surround rock mass (m/s); K_{inner} is the estimated hydraulic conductivity of the inner surface of the lining (m/s); C_l is inhomogeneous hydraulic conductivity coefficient equal to K_{outer}/K_{inner}; K_h is the equivalent hydraulic conductivity of the lining based on the lining if assumed to be homogeneous; d is the thickness of the lining (m); λ is a fitting coefficient.

The lining of a tunnel is usually intended to mitigate the water infiltration rate. The bulk hydraulic conductivity of the lining is considered to be lower than the rock mass in the water seepage chart. This chart can be used to understand the hydraulic conductivity condition inside the concrete lining if the water flow rate is known. The steps are shown as followed:

1) Measure the flow rate (Q) over a 1 m length of the tunnel.
2) Calculate the equivalent hydraulic conductivity of the lining based on the flow rate when the lining is assumed to be homogeneous (K_h).
3) Based on the original water head (H), lining thickness (d), fitting coefficient (λ) and the hydraulic conductivity of the rock mass (K_m). The y axial value can be determined.
4) To identify the value on the x axis (Figure 6), the crack width and density of the inner surface is used. According to the hydraulic conductivity of the cracked concrete model (Reinhardt, 1997), the hydraulic conductivity of the inner surface (K_{inner}) can be calculated using the observed crack width and crack density. And K_m/K_{inner} is known accordingly.
5) Find the correspondent inhomogeneous hydraulic conductivity coefficient (C_l) for different K_m/K_{inner} ratio according to the chart (Figure 6).

6 CONCLUSION

A water inflow solution for a linearly inhomogeneous tunnel lining has been presented. To describe the linear changes, an inhomogeneous hydraulic conductivity coefficient C_l has been defined. This coefficient is used to quantify the difference between the hydraulic conductivity at the inner and outer surfaces of the lining.

Using the parameters of the Melbourne underground rail loop, the water head distribution through the lining was calculated under different C_l values. As the C_l value is increased, the hydraulic gradient close to the inner surface of the lining increases. Correspondingly, the water head at the outer surface of the lining decreases. In general, if the hydraulic conductivity of the inner lining surface is applied to the entire lining then water pressure at the rock-lining interface would be over-estimated if the actual hydraulic conductivity at the interface were higher. This finding has implications for stability analyses as the water pressure in and behind the tunnel lining is an important contribution to a force equilibrium analysis. Since the hydraulic gradient through the lining can be calculated, the total water flow can be derived as a result. The water flow was defined using a factor called the inhomogeneous hydraulic conductivity flow factor β. Factor β defines the ratio between the water flow rate through an inhomogeneous lining ($q_{cl=n}$) compared to the water flow rate through a homogeneous lining ($q_{cl=1}$). One specific case was used to study the effect of the inhomogeneous hydraulic conductivity coefficient C_l. When the hydraulic conductivity of the outer surface of the lining is increased by a factor (C_l) of 10 and 100, the water inflow rate ratio approximated 1.9 and 1.6 respectively. This can be used to estimate the hydraulic conductivity conditions inside the concrete lining based on the measured water flow rate.

It is common that the underground conditions can be highly variable in many ways. The water table and the properties of the rock mass can be very different for different sections of a tunnel. A water flow factor chart relating several different parameters is presented. This chart can also be used to study the hydraulic conductivity variation through the concrete lining.

ACKNOWLEDGEMENT

Financial support from Metro Trains Melbourne, Australia and Australian Research Council under DP140101547 and LP150100413 is gratefully acknowledged.

REFERENCES

BAZANT, Z. P. & OH, B. H. 1983. Spacing of cracks in reinforced concrete. *Journal of structural Engineering,* 109, 2066-2085.

BOBET, A. 2001. Analytical solutions for shallow tunnels in saturated ground. *Journal of Engineering Mechanics,* 127, 1258-1266.

BOBET, A. & NAM, S. W. 2007. Stresses around Pressure Tunnels with Semi-permeable Liners. *Rock Mechanics and Rock Engineering,* 40, 287-315.

BUTSCHER, C. 2012. Steady-state groundwater inflow into a circular tunnel. *Tunnelling and Underground Space Technology,* 32, 158-167.

FERNÁNDEZ, G. 1994. Behavior of pressure tunnels and guidelines for liner design. *Journal of geotechnical engineering,* 120, 1768-1791.

FERNANDEZ, G. & MOON, J. 2010. Excavation-induced hydraulic conductivity reduction around a tunnel – Part 1: Guideline for estimate of ground water inflow rate. *Tunnelling and Underground Space Technology,* 25, 560-566.

GILBERT, R. 2001. Shrinkage, cracking and deflection-the serviceability of concrete structures. *Electronic Journal of Structural Engineering,* 1, 2-14.

ITA 1991. Report on the damaging effects of water on tunnels during their working life. *Tunnelling and underground space technology,* 6, 11-76.

NEVILLE, A. M. 2011. *Properties of concrete,* Harlow, England ; New York, Pearson.

PECK, W. A. 1992. *Engineering geology of Melbourne : proceedings of the Seminar on Engineering Geology of Melbourne, Melbourne, Victoria, Australia, 16 September 1992,* Rotterdam, A.A. Balkema.

REINHARDT, H. W. 1997. *Penetration and permeability of concrete : barriers to organic and contaminating liquids : state-of-the-art report prepared by members of the RILEM Technical Committee 146-TCF,* London ; New York, E & FN Spon.

SNOW, D. T. 1965. *A parallel plate model of fractured permeable media.* University of California, Berkeley.

Proceedings of the International Conference on Smart Infrastructure and Construction
ISBN 978-0-7277-6127-9

© The authors and ICE Publishing: All rights reserved, 2016
doi:10.1680/tfitsi.61279.603

Stochastic modelling of lifecycle delamination damage: evolution of composite blades of wind turbines

C. Zhang[1], H.P. Chen*[1] and T.L. Huang[1, 2]

[1] *University of Greenwich, Chatham Maritime, UK*
[2] *Central South University, Changsha, China*

* *Corresponding Author*

ABSTRACT This paper presents a stochastic approach to predict fatigue delamination development and failure probability for composite blades of wind turbines. On the basis of existing experiments, a Paris Law model for simulating the propagation of the fatigue crack growth is adopted to reproduce the crack evolution in fibre reinforced composite blades of wind turbines subjected to cyclically repeated loading. The failure probability for composite blades under cyclic loadings is estimated during the service life by stochastic deterioration modelling such as gamma process. The fatigue reliability of wind turbine blades is estimated by comparing the failure crack growth with its pre-determined critical crack length. A numerical example, where the parameters are obtained from the previous experimental studies, is investigated to simulate the fatigue damage process for wind turbine composite blades and the probability of fatigue failure by time-dependant reliability analysis. The results show that stochastic fatigue crack evolution modelling can provide a useful method for reliability analysis and optimum repair plan for composite blades of wind turbines.

1 INTRODUCTION

Energy crisis and global warming have led to increasing demand on clean and renewable energy. Wind farms have become one of most popular renewable power generators. Recently, wind turbines have gained a popularity as the most common offshore structures. Wind turbine blades play a key role to collect wind energy and they are also the most critical components taking up to 20% of total manufacturing costs (Liu et al. 2010). An efficient way to further improve the performance of wind turbines is to reduce the weight of the blades, thus, large wind turbine blades are usually manufactured by layered fibre reinforced polymer composite materials that have better properties and lighter weight.

Delamination is one of the most critical fatigue damage of wind turbine blades due to the cyclic loadings. During the designed lifetime (typically 20 years for wind turbines), the blades are suffered hundreds million times significant cyclic loadings and they are the most susceptible components damaged by fatigue (Zhou et al. 2014). It is vital to prepare maintenance strategies in advance to guarantee the safety and reliability of the structures in harsh marine conditions. Proper inspections and detections should be planned for these offshore structures in order to save the repairing caused by failures during the operation (Taylor et al. 2013). For monolithic fibre reinforced composite material blades, delamination fatigue cracks are typically divided into three modes, namely, opening for Mode I, sliding for Mode II and tearing for Mode III (Mandell & Cairns 2003).

Probabilistic analysis is a useful tool for simulating deteriorating composite materials affected by cyclic fatigue during the service life. The fatigue delamination growth of wind turbine blades can be modelled as a stochastic process with uncertainties. Considering the nature of cumulative growth of fatigue cracks, the gamma process model is an appropriate approach for performance deterioration since gamma process has been proved to be more versatile and increasingly used in stochastic deterioration modelling (Van Noortwijk 2009). Guida & Penta (2015) proved the gamma process to estimate the fatigue crack growth accurately and derive the distribution of the time to reach any crack size.

This paper focuses on the stochastic model of delamination fatigue damage of fibre reinforced composite material blades of wind turbines where the fatigue cracks depth is predicted by the Paris model. The parameters in this study are obtained by existing experiments for composite blades of wind turbines. The gamma process is used to estimate the cumulative crack propagation and the probability of fatigue failure because the modelling of the deterioration process considers uncertainties over the service life. Finally, a numerical example is presented to show the gamma process is accurate for stochastic modelling of fatigue analysis evolution. The failure probabilities in different cases for determining reliability are discussed for fatigue crack propagation to provide reliable prediction of fatigue damage accumulation and diagnoses of fatigue failure of fibre reinforced composite material blades of wind turbines.

2 DELAMINATION CRACK GROWTH

2.1 *Paris Law*

Delamination is a major fatigue damage weakening the physical properties of the original strength of the materials by cyclically loaded in long time. The prediction of fatigue crack propagation needs a proper model adopting the rate of fatigue crack growth with cyclic loadings. In order to find crack onset and crack growth process, a widely used method for constructing the relationship between the fatigue crack growth and the cycles of stress is based on a powerful equation known as Paris Law (Pugno et al. 2006).

Based on the empirical Paris–Erdogan crack growth model, the following linear elastic fracture mechanics model is used to describe crack propagation:

$$\frac{da}{dN} = C(\Delta G)^m \tag{1}$$

where a is the crack length; N is the number of load cycles; C and m are parameters related to material constants;

ΔG is strain energy release rate, defined as

$$\Delta G = \frac{\Delta G_{Imax}}{\Delta G_{IR}} \tag{2}$$

where ΔG_{Imax} is the strain energy release rate at maximum stress and ΔG_{IR} is the mode I delamination instantaneous resistance.

The crack growth propagation in Paris law can be divided into three stages, crack initiation stage, subcritical crack propagation stage and critical crack propagation stage, as shown in Figure 1 (Al-Khudairi et al. 2015). With fatigue crack propagation under repeated loading, the structures easily reach the first stage of the failure process. Thus, the life of a structure can be predicted by using the Paris Law model.

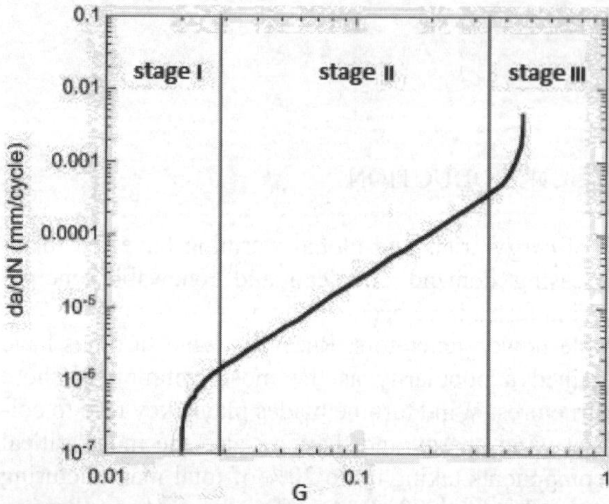

Figure 1 A schematic of the typical fatigue growth behaviour of cracks

The fibres in composite laminates increases the delamination resistance when the crack grows as fibres bridge connections. These effects such as applied load ΔG_{Imax} and instantaneous resistance value ΔG_{IR}

can be investigated by double cantilever beam (DCB) tests (Shivakumar et al. 2006). Therefore, determination of ΔG_{IR} for fatigue crack propagation is required.

According to the experimental data from double cantilever beam test by Al-Khudairi et al. (2015), the value of initiation fracture toughness for composite blades was taken as $\Delta G_{Ic} = 764 J/m^2$ in this study and the resulting resistance equation is

$$\Delta G_{IR} = \Delta G_{Ic} + 20.2(a - a_0)^{0.64} \tag{3}$$

where a_0 is the initial crack length

From the study by Shivakumar et al. (2006) and data from Al-Khudairi et al. (2015), the relation between maximum strain energy release rate and the number of loading cycles is expressed as:

$$\Delta G_{Imax} = \Delta G_{IC}(\log N + 1)^{-0.93} \tag{4}$$

Substituting the Equation (3) and (4) into Equation (1), the Paris Law for composite material in Mode I delamination is written as:

$$\frac{da}{dN} = C\left(\frac{\Delta G_{Imax}}{\Delta G_{IR}}\right)^m = C\left(\frac{\Delta G_{Ic}(\log N+1)^{-0.93}}{\Delta G_{Ic}+20.2(a-a_0)^{0.64}}\right)^m \tag{5}$$

2.2 *Gamma Process*

Gamma process is a stochastic process with independent non-negative having a gamma distribution with identical scale parameter and suitable to model gradual damage monotonically accumulating over time, such as wear, fatigue, corrosion, crack growth, erosion, etc. (Chen & Alani 2012; Van Noortwijk 2009). The gamma process was initially adopted as a proper model for deterioration occurring random in time (Abdel-Hameed 1975).

By using the crack length *a*, the probability density function $Ga(a|v,u)$ can be written as:

$$Ga(a|v,u) = \frac{u^v}{\Gamma(v)} x^{v-1} e^{-ua} I_{(0,\infty)}(a) \tag{6}$$

where *v* is shape parameter; *u* is scale parameter typically $0 < u < 10$ and $I_{(0,\infty)}(x)$ is defined as:

$$I_{(0,\infty)}(x) = \begin{cases} 1 & if\ x \in (0,\infty) \\ 0 & if\ x \notin (0,\infty) \end{cases} \tag{7}$$

the complete gamma function $\Gamma(v)$ is defined as

$$\Gamma(v) = \int_0^\infty x^{v-1} e^{-x} dx \tag{8}$$

and the incomplete gamma function $\Gamma(v,x)$ is

$$\Gamma(v,x) = \int_x^\infty x^{v-1} e^{-x} dv \tag{9}$$

where $v \geq 0$ and $x > 0$.

For the composite wind turbine blades, the fatigue failure is defined as experiencing *N* times loading at time *T*, where fatigue crack length *a* reaches the critical crack propagation stage. The critical crack length a_{cr} depends on the maintenance requirement, environmental conditions and the safety factor of the structures. The service life of the wind turbine blades can be predicted by summing the time period before reaching critical crack length. Therefore, the fatigue failure probability of the structure also increases if the maintenance for repairing is not undertaken in time. The equation for the failure probability can be calculated from Chen & Alani (2013), namely

$$F(t) = Pr\{t \geq T\} =$$

$$Pr\{a \geq a_{cr}\} = \int_{a_{cr}}^\infty F(a) da = \frac{\Gamma(v(t), ua_{cr})}{\Gamma(v(t))} \tag{10}$$

The shape function $v(t)$ can be obtained from the expected crack growth given in the previous section as

$$v(t) = nua(t) \tag{11}$$

where n is independent crack number; $a(t)$ is the crack length at *t* time; a_{cr} is critical crack length and the scale parameter *u* could be estimated from statistical estimation methods such as a maximum likelihood method and method of moments commonly (Van Noortwijk 2009);

The probability of failure per unit time at t_i is computed from

$$p_i = F(t_i) - F(t_{i-1}), For\ i = 1,2 \ldots T \tag{12}$$

When the fatigue crack length reaches the critical level, the probability of failure reaches unity and the structure fails. Before this stage, the requirement for maintenance becomes critical to reduce the risk of structural failure and to prevent the unacceptable possible loss. After repairing, the service time of composite blades can be extended.

3 NUMERICAL EXAMPLE

3.1 Fatigue Crack Growth

A fibre reinforced composite blade is used to examine the effectiveness of proposed approach. The parameters for simulating fatigue crack growth and reliability analysis related to Paris law and Gamma process are obtained from existing work. According to the experimental results, *C* and *m* for Mode I delamination at stage II crack growth in Paris equation is 4.47×10^{-2} and 5.27, respectively (Al-Khudairi et al. 2015). Marin et al. (2009) gave the relationship between the annual cyclic times and the years of operation for the blades, and stated that the annual cycle for a blades is approximately 14.26 million.

The initial crack length is 50mm at the beginning in double cantilever beam test for Mode I (Landry & LaPlante 2012). It is assumed the threshold is 100mm for structural failure. For wind turbines, the typical service life is from 20 to 30 years, and the service life of 25 years is used here. In Equation (10), the scale factor is taken as 6 and the crack number is taken as 4 in gamma process. Parameters relating to the crack development model and gamma process, are used to simulate the fatigue crack propagation and failure probability for composite blades.

Figure 2 shows the Paris law predictions for fibre reinforced composite material blades compared with steel fatigue predictions (Huang et al. 2014). The differences between the curves for two materials are significant. There is a sharp increase in steel where crack length approaches the critical length, but the crack length for composite increases steadily. From the simulation results for composite, the fatigue crack length curve by Paris Law is dramatically increasing at the beginning. Clearly, as the service of time increases, the crack length development slows down gradually. The blades fails after reaching the pre-defined critical crack length. For steel materials, the crack length grows slowly and gradually at beginning. As the service time increases, crack length becomes unstable and develops quickly. When the crack length reaches the threshold length, the crack becomes uncontrolled and steel material failures.

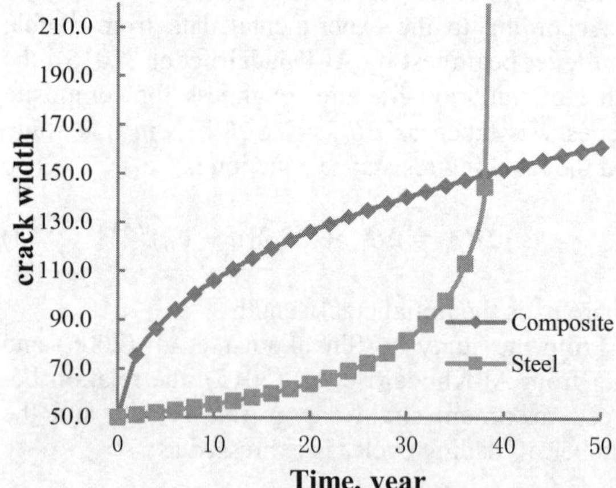

Figure 2 Fatigue crack growth of composite and steel

The reason for no obvious unstable crack for composite is explained by Al-Khudairi et al. (2015), Shivakumar et al. (2006) and O'Brien (1990). O'Brien (1990) suggested a linear relationship between G_{Imax} and $\log N$ data for $N < 10^6$ then obtained the maximum strain energy release rate G_{Imax} value from fitted equation at $N = 10^6$. This is confirmed by Shivakumar et al. (2006) to give the fitted Equation 4 and calculated the maximum strain energy release rate $G_{Imax} = 0.15G_{IC}$ when $N = 10^7$ cycles. The result shows if the applied strain energy release rate is less than 15% of Mode I delamination fracture toughness, the rate of crack length growth is very slow and the crack growth will remain unchanged until 10 million cycles.

3.2 Parameter Analysis

3.2.1 Failure Probability for Composite Blades

The deterioration of the performance for wind turbine blades during service life is modelled as a gamma process. As shown in Figure 3, the probability of failure curves for composite and steel are obviously different. For composite material, the crack length grows in a higher growth rate initially. Then, the probability of fatigue failure of composite reaches unity quickly. For metal material, the crack length grows slowly and the probability of fatigue failure also increases slowly.

With the service time increasing, the probability increases gradually until reaching approximately 20 years and then the curve has a rapid rise to a value of very close to unity.

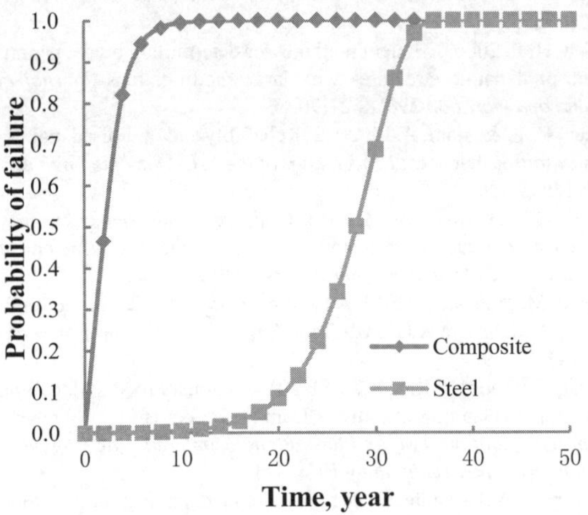

Figure 3 Failure probability for composite and steel

3.2.2 *Parameter Sensitivity Analysis*

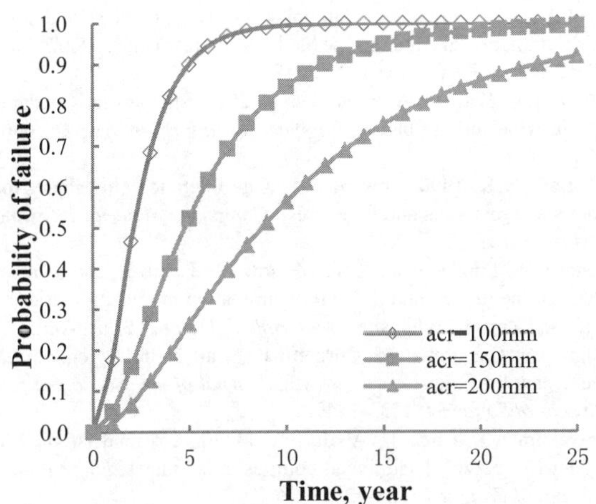

Figure 4 Failure probability for different critical crack lengths

The deterioration of the structural performance in terms of reliability (measured by the growth of the fatigue crack length) is modelled as a gamma process. The results of the lifetime distribution of time to failure probability are shown in Figure 4 for different acceptable fatigue crack length limits, i.e. a_{cr}=100mm, 150mm, 200mm, respectively. As expected, the probability of failure associated with the equivalent crack length depends on the given acceptable limit, with a higher probability of failure for a lower acceptable level at any given time. The probability of failure increases dramatically with time when the expected crack length approaches the given acceptable limit.

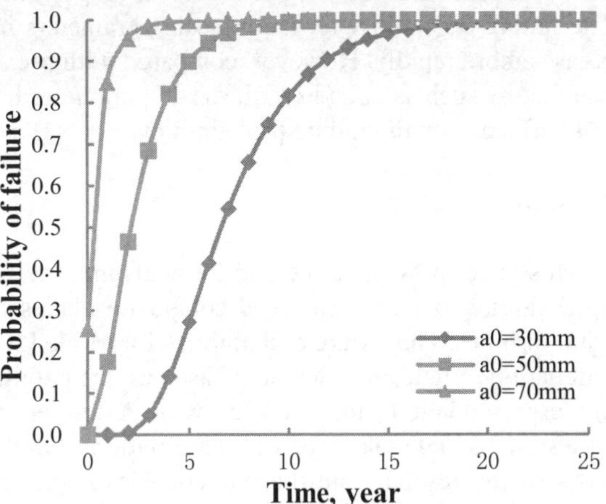

Figure 5 Failure probability for different initial crack lengths

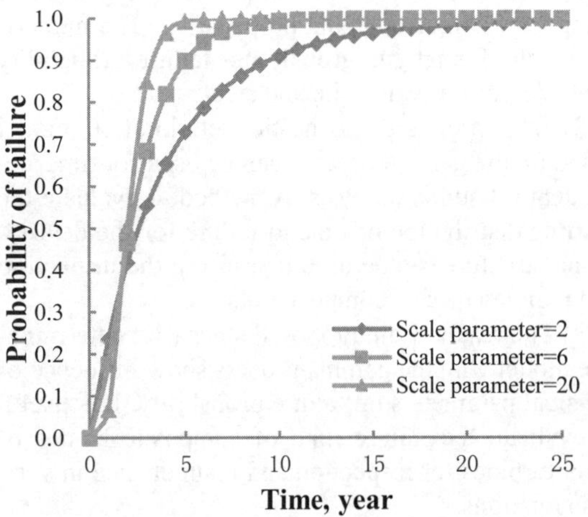

Figure 6 Failure probability for different shape parameters in gamma process

The results of the lifetime distribution of time to failure probability for fatigue crack growth are shown in Figure 5 for various initial crack lengths, i.e. a_0=30mm, 50mm, 70mm, respectively. As expected, the probability of failure for different initial fatigue crack lengths is significantly different, with a higher probability of failure for a higher initial crack length. For lower initial crack length, the results show that the time period before the sudden rise in probability of failure becomes much longer.

In Gamma process, the scale parameter u has influence on failure probability. From Figure 6, it is obvious that the smaller scale parameter makes the curve gentler and reaches the fatigue failure in longer time. The failure probability for larger scale parameters increases more rapidly. However, compared with previous factors such as a_0 and a_{cr}, the scale parameter has less influence on the failure probabilities.

4 CONCLUSIONS

This paper presents a method for analysing the fatigue failure of fibre reinforced composite blades of wind turbines. The failure probabilities for Mode I delamination fatigue are calculated based on the gamma process simulations for crack growth. A numerical case study is undertaken for various situations. On the basis of the results from the numerical example involving wind turbine composite blades, the following conclusions can be drawn:

1) The gamma process can predict the growth of composite fatigue cracking propagation. The numerical results for fatigue growth and failure probability are different for composite and steel.

2) The proposed stochastic deterioration model based on the gamma process can be used for time-dependent reliability analysis. A method to evaluate the lifetime distribution of time to failure for the deteriorating structure can be used to assist in the inspection and maintenance of composite blades.

3) The results from proposed stochastic determinative model with the gamma process show influence of physical parameters on failure probability. It is useful to evaluate the failure time of composite blades of wind turbines for inspection and maintenance in various situations.

Further efforts are needed to determine the optimum repair strategy by lifecycle cost analysis.

REFERENCES

Abdel-Hameed, M. 1975. A Gamma Wear Process. *Reliability, IEEE Transactions* **24**, 152-153.

Al-Khudairi, O. Hadavinia, H. Waggott, A. et al. 2015. Characterising mode I/mode II fatigue delamination growth in unidirectional fibre reinforced polymer laminates. *Materials & Design* **66**, 93-102.

Blanco, N. Gamstedt, E. K. Asp, L. E. et al. 2014. Mixed-mode delamination growth in carbon–fibre composite laminates under cyclic loading. *International journal of solids and structures* **41**, 4219-4235.

Chen, H. P. 2006. Efficient methods for determining modal parameters of dynamic structures with large modifications. *Journal of sound and vibration* **298**, 462-470.

Chen, H. P. & Alani, A. M. 2012. Reliability and optimised maintenance for sea defences. *Proceedings of the ICE-Maritime Engineering* **165**, 51-64.

Chen, H. P. & Alani, A. M. 2013. Optimized maintenance strategy for concrete structures affected by cracking due to reinforcement corrosion. *ACI Structural Journal* **110**, 229-238.

Guida, M. & Penta, F. 2015. A Gamma Process Model for the Analysis of Fatigue Crack Growth Data. *Engineering Fracture Mechanics* **142**, 21-49.

Huang, T. Zhou, H. Ren, W. et al. 2014. Gamma Process Modelling and Repair Planning of Fatigue Damaged Steel Bridge Members. *Proceedings of the Twelfth International Conference on Computational Structures Technology* **106**, 1-11.

Landry, B. & LaPlante, G. 2012. Modeling delamination growth in composites under fatigue loadings of varying amplitudes. *Composites Part B: Engineering* **42**, 533-541.

Liu, W. Tang, B. & Jiang, Y. 2010. Status and problems of wind turbine structural health monitoring techniques in China. *Renewable Energy* **35**, 1414-1418.

Mandell, J. F. Cairns, D. S. Samborsky, D. D et al. 2003. Prediction of delamination in wind turbine blade structural details. *ASME 2003 Wind Energy Symposium* **1**, 202-213

Marin, J. C. Barroso, A. Paris, F. et al. 2009. Study of fatigue damage in wind turbine blades. *Engineering failure analysis* **16**, 656-668.

O'Brien, T. K. 1990. Toward a damage tolerance philosophy for composite materials and structures. *Composite Materials: Testing and Design* **9**, 7-33.

Pegorin, F. Pingkarawat, K. & Mouritz, A. P. 2015. Comparative study of the mode I and mode II delamination fatigue properties of z-pinned aircraft composites. *Materials & Design* **65**, 139-146.

Pugno, N. Ciavarella, M. Cornetti, P. et al. 2006. A generalized Paris' law for fatigue crack growth. *Journal of the Mechanics and Physics of Solids* **54**, 1333-1349.

Shivakumar, K. Chen, H. Abali, F. et al. 2006. A total fatigue life model for mode I delaminated composite laminates. *International Journal of Fatigue* **28**, 33-42.

Taylor, S. G. Park, G. Farinholt, K. M. et al. 2013. Fatigue crack detection performance comparison in a composite wind turbine rotor blade. *Structural Health Monitoring* **12**, 252-262.

Van Noortwijk, J. M. 2009. A survey of the application of gamma processes in maintenance. *Reliability Engineering & System Safety* **94**, 2-21.

Zhou, H. F. Dou, H. Y. Qin, L. Z. et al. 2014. A review of full-scale structural testing of wind turbine blades. *Renewable and Sustainable Energy Reviews* **33**, 177-187.

Proceedings of the International Conference on Smart Infrastructure and Construction
ISBN 978-0-7277-6127-9

© The authors and ICE Publishing: All rights reserved, 2016
doi:10.1680/tfitsi.61279.609

Simulating the degradation and maintenance effects on an integrated urban transport infrastructure system

B. Zhao*[1], K. Soga[1] and E. Silva[2]

[1] *Department of Engineering, University of Cambridge, Cambridge, United Kingdom*
[2] *Department of Land Economy, University of Cambridge, Cambridge, United Kingdom*
* *Corresponding Author*

ABSTRACT Degradation is a major issue faced by transport infrastructures. Its negative effects and the associated maintenance costs are usually studied through degradation modelling. Current models for individual infrastructures are experiencing limitations in modern multi-modal transport networks. This paper starts the discussion towards a system-wide infrastructure condition prediction model, giving emphasis to network-level interconnections and the commonalities in the degradation of different infrastructures.

Existing transport infrastructure degradation models and techniques for large scale dynamics simulation are first reviewed. Based on these, a new model is proposed for the system-wide simulation, including key attributes, variables, inputs, data structures and dynamics. In Geographic Information System (GIS), infrastructure locations, conditions as well as the condition-influencing factors are stored in cell-based maps, which forms the basis for degradation simulation in Cellular Automata (CA). Key parameters that control the interaction are calibrated to match actual observations. Once calibrated, the model can predict infrastructure conditions under future scenarios. A case study is presented, which runs at low resolution to demonstrate the overall procedures. Options for enhancing the resolution, such as using High Performance Computing (HPC), are discussed.

Performance of transport infrastructures is not only a large concern of engineers and asset managers. It also has significant social implications. Infrastructure conditions projected by our model can be used to calculate traveller distributions within the network. Once fully developed, the model will provide an integrated platform for engineering, management and social study and show what interesting interactions may happen among different fields.

1 INTRODUCTION

Transport infrastructures consist of roads, railways, tunnels, bridges and etc. Subjected to nature erosion and human activities, these assets suffer inevitably from degradation. Degraded transport infrastructures cause many problems, mostly impairing traveller comforts and sometimes even creating safety threats (Greene & Ulm 2013). Maintenance on degradation defects are also costly (Esveld 2003; Jovanovic et al. 2011). As a result, various infrastructure degradation models have been developed to foresee degradation and to test and optimize maintenance schemes.

However, since individual infrastructure sections are analysed separately, existing degradation models cannot reflect the dynamic interactions within a highly interconnected transport network. For example, when Hammersmith Flyover in London was repeatedly closed during 2011-2015 for repair, huge amounts traffic was diverted to alternative routes, which would degrade faster than usual. Also, different infrastructures compete for maintenance funding. As a result, infrastructure conditions are no longer suitable to be analysed singly. System interactions need to be considered and can only be addressed effectively through a system wide traffic infrastructure condition model.

This paper presents such a system level infrastructure degradation and maintenance effects simulation model. It operates on a multimodal transport system, consisting of roads, railways and etc. Various condition influencing factors, such as traffic, climate and geotechnical properties, are included through Geographic Information System (GIS). Degradation is simulated using Cellular Automata (CA), where infrastructures are represented by cells and each cell will follow certain degradation rules. Key parameters in degradation rules are calibrated from actual infrastructure condition inspection data. Once calibrated, the model can predict infrastructure conditions under various future engineering, social and climate scenarios.

2 LITERATURE REVIEW

Two topics are reviewed for this study: existing transport infrastructure degradation models and techniques for large scale spatial dynamics simulation. The former gives insights into the degradation mechanisms while the latter is necessary for upscaling existing models to the network-level.

2.1 *Transport infrastructure degradation models*

Among the various forms of transport infrastructures, roads and railways are well-studied in terms of their degradation features. Undergrounds, although playing an important role in urban transportation, seems to be a blank page in degradation studies. Other transports, e.g. ferries, exist but the traffic volumes are usually small. Important aspects to learn from existing models are their structures, inputs, outputs and targets. These correspond directly to the dynamics, inputs, outputs and applicability of the system wide model.

Road degradation models are of great variety, from simple, deterministic models to complex, stochastic ones relying on many external factors. Eight models are reviewed and summarized in Table 1.

Railway consists of ballasts, sleepers, steel rails, switches and so on. These components do not coordinate in degradation. Consequently, it is crucial identify the targets of railway degradation models. Nine models are reviewed and summarized in Table 2.

2.2 *Large scale dynamic simulation techniques*

Above reviewed models offer insights to degradation mechanisms. However, these models only deal with a short section of infrastructure, which need to be scaled up to the transport network level. The study of urban progress usually do simulations on the city scale. The key techniques include GIS and CA. These can be adopted for building the proposed model.

2.2.1 *GIS for large scale spatial analysis*

GIS and its associated functions provide a powerful tool for spatial analysis. It enables the visualization, questioning, analysis and interpretation of geographic information to understand relationships, patterns and trends within the map environment (ESRI 2015; Ordnance Survey 2015a). In urban process studies, GIS is used to check input errors, overlay map layers, perform spatial inquiries (e.g., shortest distance) and visualize results (Aljoufie et al. 2013; Silva & Clarke 2002; Wu 2002). For transport studies, GIS has been used to evaluate social-economic and ecological impacts of new infrastructures, classify road hierarchies or build inventories (Ortega et al. 2014; Rodrigues et al. 2015; Duran-Fernandez & Santos 2014).

Recently, there is a great opportunity to implement GIS in traffic infrastructure degradation studies. First, digital information are available for infrastructure networks as well as site geology and climate, which can be easily manipulated in the GIS software (Ordnance Survey 2015b; British Geological Survey 2015; The Met Office 2015). Besides, infrastructure monitoring is becoming an increasingly automatic process. The up-to-date infrastructure conditions information is location-indexed and stored in asset management software that links well with GIS (PCIS 2014). These advantages facilitates the upgrading of traffic infrastructure degradation analysis to a network scale.

2.2.2 *CA for large scale dynamic simulation*

CA is a mathematical system consisting of many identical components, each simple, but capable of generating complex global behaviours (Wolfram 1984). Urban process studies use 2D CA, an array of square cells. Cell size vary between 20m to 1.5km depending on the desired resolution and computational cost.

In each step, cell is in a particular state. Globally this forms a spatial pattern. Simple models use binary

Table 1. Summary of road degradation models

Model	Nature	Region	Time	Inputs						Outputs										
				time	traffic	pave-ment	climate	mainte-nance	others	rut-ting	crack-ing	ravel-ling	pothol-ing	edge	rough-ness	tex-ture	com-bined	user	environ-ment	eco-nomic
ASSHTO 1993	Empirical	USA	1993		√	√			present condition								PSI*			
PARIS	Empirical	EU	1998	√	√					√	√	√			√					
HDM-4	Structured empirical	World	2000	√	√	√	√	√	Construction quality	√	√	√	√	√	√	√	√	√	√	√
MEPDG	M-E	USA	2004		√	√	√		initial quality	√	√				√					√
WLPPM/ LTPPM	M-E	UK	NA		√	√	√			√	√				√					
ADOT	Markov	Arizona, USA	1980	√				√	present condition						√					
HIPS	Markov	Nordic	1990	√				√	present condition	√	√				√			√		
HMEP	Markov	UK and abroad	2012	√			√	√	present condition, road hierarchy								CCI*			√

*PSI = Present Serviceability Index, CCI = Carriageway Condition Index

Table 2. Summary of railway degradation models

Model	Nature	Region	Time (Data)	Inputs								Outputs				
				time	traf-fic	rail-way	deterio-ration	inspec-tion	mainte-nance	environ-ment	others	rail	track	environ-ment	eco-nomic	others
INNOTRACK	Empirical	Europe	2009		√	√						√				
TCDD	Empirical	Turkey	2010		√	√				√			√			
TU Graz	Empirical	Austria	1997	√	√	√	√		√		Current condition		√		√	
MAINLINE	Empirical	Europe	2014	√	√	√	√		√		Current condition		√	√	√	intervention schedule
Melbourne tram	Markov	Melbourne Tram	2014	√			√				Current condition					
Nottingham Markov	Markov	UK	2013	√			√	√	√		Current condition		√			
Petri Net	Petri Net Monte Carlo	UK	2012	√			√	√	√		total life				√	time and number ofintervention
SNCF exponential	Stochastic	France, TGV/HSR	2011	√		√	√		√				√			
SNCF gamma	Gamma	France, TGV/HSR	2009	√			√	√	√				√			

states: rural/urban. While multiple states contain richer spatial information. Reasonable number of cell states depends on the specific problem, also the available computational power.

In terms of transition rules, models are either constrained or unconstrained, depending on whether the amount of change in one transition is fixed by external processes (Straatman et al. 2004). Most of the CA models adopt stochastic transition rules, either by using stochastic terms in transition rules, or defining the transitions as Markov processes (Balzter et al. 1998).

Calibration is about tuning parameters so that simulated results match the reality (Torrens 2011). It is crucial even for early urban CA models (White & Engelen 1993). Manual calibrations include trial & error (Barredo et al. 2003) and logistic regression (Wu 2002). Automatic methods are sweeping parameter space ("brute force") or heuristic algorithms (Silva & Clarke 2002; Al-Ahmadi et al. 2009). When observations are absent, parameters are calibrated from other models (Acevedo et al. 1995). Usually parameters are fixed after calibration, but it is possible to have dynamic calibration at each step (Silva & Clarke 2002).

Validation involves assessing the success of a simulation in achieving its intended goals (Torrens 2011), such as comparing the simulated land use with the actual map. Techniques for validation include pixel matching, spatial statistics analysis or structural measurements (Wu 2002; White & Engelen 2000).

3 METHODOLOGY

Joining together current degradation models and large scale spatial dynamics simulation, a framework is proposed for simulating the degradation of a transport infrastructure system. It consists of three parts: 1) Compiling spatial information using GIS. 2) Simulating degradation/maintenance condition variation. 3) Calibrating parameters.

3.1 GIS: adding a spatial dimension to degradation simulation

Degradation of transport infrastructures are ultimately related to condition influencing factors (traffic, climate and material properties, etc.). These various data can be organized effectively using GIS. Different layers of information are overlaid using a common projection. When enquiring an area, all layers of information at that location are displayed.

3.2 CA: cell-specific Markov degradation rule

Data compiled in GIS is read into CA cells. So each cell contains information on whether it represents road, railway, if so, what is the traffic and infrastructure conditions. Besides, cells also holds information of local climate and geology. Basically, what is presented is a cell-based map containing all crucial information of the transport network.

The adopted degradation mechanism is a cell-specific Markov degradation rule. Globally, the degradation of infrastructure cells follow a Markov process (HMEP 2012; Prescott & Andrews 2013). Global degradation probabilities (or Transition Probability Matrix, TPM) are obtained by cross tabulating infrastructure conditions at two observation years. For each cell, TPM differs slightly: sections in adverse climate degrade faster while sections not frequently used preserve their states longer than the average. Local conditions are quantified using a modification term. The better the local condition, the larger the term. Weighted average of the terms of nearby cells reflect neighbourhood influence. Stochastic simulation of network degradation is achieved by comparing the degradation threshold with a random number for each cell (Al-Ahmadi et al. 2009). If the random number is below the degradation threshold, cell remains in its current condition. Otherwise it degrades. Maintenance effects are simulated by forcing cell states to be up by one level (minor repair) or to the best possible state (reconstruction).

3.3 Calibration

Parameters need calibration are the weights of local influence factors in the "local modification term", as well as parameters associated with the neighbourhood influence. The most straightforward but computational intensive way of calibration is the "brute force" method. Heuristic optimization algorithms, such as Genetic Algorithm or Simulated Annealing (Al-Ahmadi et al. 2009), can be adopted. Multi-stage calibration can help to gradually narrow the ranges of the parameters (Silva & Clarke 2002).

4 CASE STUDY

Greater London (GL) is the capital of the UK. It has a sophisticated transport network, including roads, railways, undergrounds, overgrounds, trams and river buses. Managing such a big network is a challenging task, especially given that many infrastructures were built centuries ago and are suffering from significant aging problems (Department for Transport, 2015). In this case study, the transport network in the GL will be used as an example to illustrate the overall procedure.

4.1 Input information

Due to the time limit, only the railways and major roads in GL are investigated. Three condition influencing factors are traffic, temperature and superficial geology. As traffic and temperature are relatively large numbers, they are first normalised to [0, 1]. Geology conditions affect road and railway settlement. Being categorical data, it is converted into a numerical rating between [0, 1] based on expert judgement (0 represents the best geological condition). The infrastructure network and related inputs are overlaid as an integrated map in QGIS.

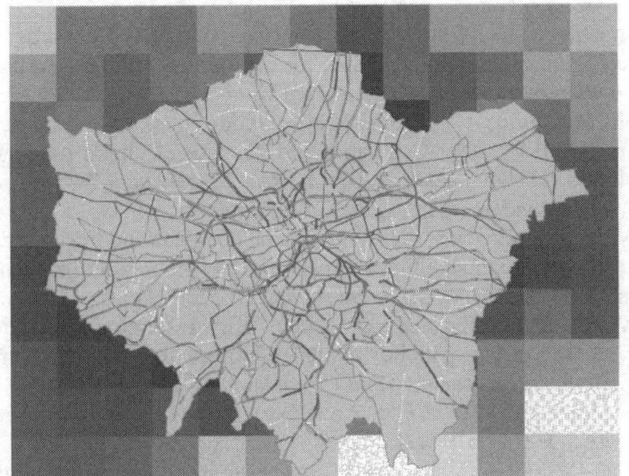

Figure 1. Input information (GL major roads, railways, geology and temperature on 5km grid)

4.2 Degradation simulation

In this study, binary cell states are used, i.e. the infrastructure condition is either "good" or "poor". It is assumed that all road and railway cells start with "good". The global degradation threshold is assumed as 1, i.e. no degradation would happen without considering local conditions. Local conditions serve to modify the degradation threshold, making it prone/difficult to degrade, depending on their weights. For weight=0, the local factor does not affect the degradation threshold. For weight larger/smaller than 0, the cell-specific degradation threshold will be higher/lower as the local factor increases. Degradation simulation is carried out in the CA software NetLogo. Weights associated with local factors are adjustable, allowing the infrastructures to degrade at varied speed.

4.3 Calibration

It is assumed that both types of infrastructures do not degrade over time. Though not realistic, this assumption facilitates testing the calibration procedure conceptually. Parameters to calibrate are the six weights (traffic, temperature and geology weights for road and railway cells). "Brute force" calibration method is used. Figure 2 shows the calibration results. The size of data points indicates the difference between the simulated infrastructure conditions and the target.

The calibration is performed on a desktop workstation. The model resolution (i.e., cell size) is 1km and the total parameter space consists of 250 parameter combinations. In the future, the desired resolution would be on meter level and the parameter space is also much larger. In face of the computational challenge for calibration, the model needs to be run on the High Performance Computing (HPC) system. Besides, a more intelligent calibration scheme, such as the Genetic Algorithm, needs to be deployed.

5 SUMMARY

Traditional degradation models for transport infrastructures cannot accommodate the system behaviours in modern transport network. A new model is proposed for upscaling traffic infrastructure degradation models to a system scale. It utilises techniques for large scale spatial dynamics simulation, namely the GIS and CA. A model framework is presented. From a simple case study, it is shown that the proposed framework is achievable. However, to improve the calculation speed, further computational improvements are necessary.

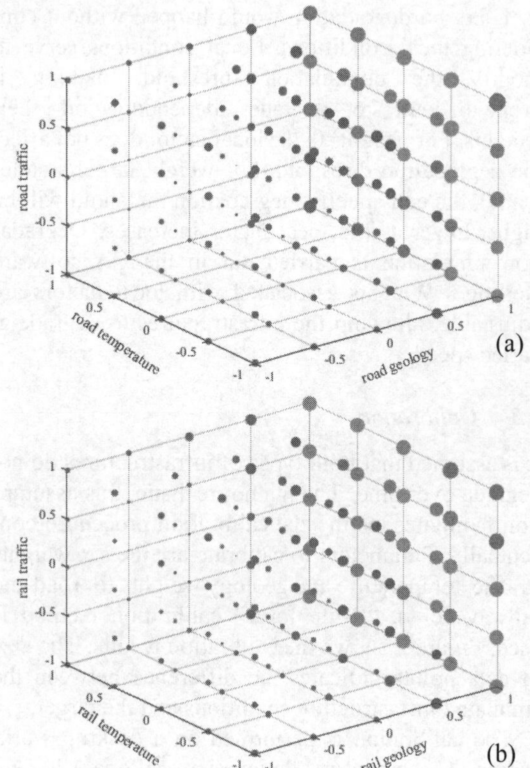

Figure 2. Calibration results, i.e., the weights of geology, temperature and traffic for the degradation of roads (a) and railway (b)

REFERENCES

Acevedo, M.F., Urban, D.L. & Ablan, M., 1995. Transition and gaps models of forest dynamics. *Ecological Applications*, 5(4), 1040–1055.

Al-Ahmadi, K. et al., 2009. Calibration of a fuzzy cellular automata model of urban dynamics in Saudi Arabia. *Ecological Complexity*, 6(2), 80–101.

Aljoufie, M. et al., 2013. A cellular automata-based land use and transport interaction model applied to Jeddah, Saudi Arabia. *Landscape and Urban Planning*, 112, 89–99.

Balzter, H., Braun, P.W. & Köhler, W., 1998. Cellular automata models for vegetation dynamics. *Ecological Modelling*, 107(2-3), 113–125.

Barredo, J.I. et al., 2003. Modelling dynamic spatial processes: simulation of urban future scenarios through cellular automata. *Landscape and Urban Planning*, 64(3), 145–160.

British Geological Survey, 2015. Digital Geology - Superficial theme.

Duran-Fernandez, R. & Santos, G., 2014. A GIS model of the National Road Network in Mexico. *Research in Transportation Economics*, 46, 36–54.

ESRI, 2015. ESRI: What is GIS? Available at: http://www.esri.com/what-is-gis [Accessed July 21, 2015].

Esveld, C., 2003. Recent developments in slab track. *European Railway Review*, 2, 81–85.

Greene, S. & Ulm, F., 2013. Pavement Roughness and Fuel Consumption. *CSHub @ MIT*, (August).

HMEP, 2012. *LIFECYCLE PLANNING TOOLKIT USER GUIDANCE DETERIORATION MODELS*, London.

Jovanovic, S., Evren, G. & Guler, H., 2011. Modelling railway track geometry deterioration. *Proceedings of the ICE - Transport*, 164(2), 65–75.

Ordnance Survey, 2015a. Ordnance Survey: Geographic Information System (GIS).

Ordnance Survey, 2015b. OS MASTERMAP INTEGRATED TRANSPORT NETWORK LAYER.

Ortega, E., Otero, I. & Mancebo, S., 2014. TITIM GIS-tool: A GIS-based decision support system for measuring the territorial impact of transport infrastructures. *Expert Systems with Applications*, 41(16), 7641–7652.

PCIS, 2014. The UK Pavement Management System (UKPMS).

Prescott, D. & Andrews, J., 2013. Modelling maintenance in railway infrastructure management. In *Annual Reliability and Maintainability Symposium*. Orlando, USA, 3–8.

Rodrigues, D.S., Ribeiro, P.J.G. & da Silva Nogueira, I.C., 2015. Safety classification using GIS in decision-making process to define priority road interventions. *Journal of Transport Geography*, 43, 101–110.

Silva, E. a. & Clarke, K.C., 2002. Calibration of the SLEUTH urban growth model for Lisbon and Porto, Portugal. *Computers, Environment and Urban Systems*, 26(6), 525–552.

Straatman, B., White, R. & Engelen, G., 2004. Towards an automatic calibration procedure for constrained cellular automata. *Computers, Environment and Urban Systems*, 28(1-2), 149–170.

The Met Office, U., 2015. UKCP09: Gridded observation data sets.

Torrens, P.M., 2011. Calibrating and Validating Cellular Automata Models of Urbanization. *Urban Remote Sensing: Monitoring, Synthesis and Modeling in the Urban Environment*, 335–345.

White, R. & Engelen, G., 1993. Cellular automata and fractal urban form: a cellular modelling approach to the evolution of urban land-use patterns. *Environment and Planning A*, 25(8), 1175–1199.

White, R. & Engelen, G., 2000. High-resolution integrated modelling of the spatial dynamics of urban and regional systems. *Computers, Environment and Urban Systems*, 24(5), 383–400.

Wolfram, S., 1984. Cellular automata as models of complexity. *Nature*, 311(4), 419–424.

Wu, F., 2002. Calibration of stochastic cellular automata: the application to rural-urban land conversions. *International Journal of Geographical Information Science*, 16(8), 795–818.

Proceedings of the International Conference on
Smart Infrastructure and Construction
ISBN 978-0-7277-6127-9

© The authors and ICE Publishing: All rights reserved, 2016
doi:10.1680/tfitsi.61279.615

Flexible street lighting solutions for traffic demand uncertainty

W. Fawcett*[1], I. Robles Urquilo[2], S. Perales[3], M. Hughes[1], and M. Rodriguez Fernandez [2]

[1] *Cambridge Architectural Research Ltd, Cambridge, UK*
[2] *Louis Berger International, Santander, Spain*
[3] *Cidro, Valladolid, Spain*
* *Corresponding Author*

ABSTRACT It is widely agreed that infrastructure projects should incorporate flexibility to mitigate the impact of future uncertainties. Rational decision-makers wish to compare the acquisition cost of flexibility with its expected benefits. The benefits are uncertain but estimates can often be made using Monte Carlo simulation, with multiple scenarios branching from a known initial state. These principles are applied to a case study of flexible street lighting. Traditional, non-flexible street lighting has fixed or limited scalability of light output, whereas flexible lighting with LED light sources uses less energy and also has the potential for variable light output. The cost-effectiveness of alternative degrees of flexibility are assessed in a street lighting renewal project in Spain.

1 FLEXIBILITY IN STREET LIGHTING

It is a well established proposition that infrastructure projects should incorporate flexibility to mitigate the impact of future uncertainties. The real options paradigm has been proposed as a way of turning the idea of flexibility into an operational mechanism and making it quantifiable (Ellingham & Fawcett, 2006; de Neufville & Scholtes, 2011). It provides a basis for rational decision-making by estimating the expected value of flexibility and comparing it with the acquisition cost, thus reducing the risk of over- or under-investing in flexibility.

The cost of acquiring flexibility is relatively easy to establish because it is part of current project budgeting. The service life benefits are harder to estimate because they occur in the future and the future is uncertain. Estimates can often be made using Monte Carlo simulation of multiple scenarios branching from a known initial state. This gives a probabilistic estimate of the value of flexibility.

These principles are applicable to street lighting infrastructure. Traditional, non-flexible street lighting installations are specified in accordance with codes of practice where the traffic intensity is a critical parameter, and have fixed or limited scalability of light output. In contrast, flexible lighting uses LED light sources that have a higher initial cost, but that have the potential for variable light output and are more energy-efficient. Flexible lighting with sensors linked to control software can adjust the light output to match the prevailing traffic conditions.

The decision between a flexible street lighting installation or a traditional non-flexible alternative is discussed in the context of a street lighting project in Spain.

2 PROBLEM DEFINITION

Street lighting installations in Spain must confirm to the Royal Decree on energy efficiency in external lighting installations (*see* Technical Note). This sets the illumination requirements for roads depending on the traffic speed and intensity classification, and is particularly aimed at the avoidance of over-lighting which wastes electricity. If a road's traffic speed and intensity increase and cross a classification threshold during the service life of a street lighting installation, it may be necessary to upgrade the installation.

The case study compares alternative renewal strategies for an existing street lighting installation that fails to meet current standards. The study uses a probabilistic life-cycle costing model that takes account of:

- initial traffic speed and intensity
- alternative strategies for upgrading the existing street lighting installation with traditional or flexible systems
- initial costs of the alternative strategies at current prices
- estimates for the future evolution of uncertain variables, viz. traffic demand and electricity price
- estimates for the service life electricity consumption/cost and running costs of the of the alternative strategies.

The LCC analysis uses the CILECCTA software for financial life-cycle costing (LCC) and environmental life-cycle assessment (LCA), developed in an EC-funded research project (CILECCTA 2013). Only the LCC component is active in the case study.

2.1 *Street lighting standards*

In the Spanish classification for streets and roads set out in the Royal Decree, urban roads fall into class A3. Within this class, the lighting requirements depend on threshold values for the average daily traffic intensity (IMD) (*see* Tables 1 and 2). For the case study road, the critical threshold is IMD = 7,000.

2.2 *Existing street lighting installation*

The case study is based on the street lighting in a small municipality in the province of Valladolid, Spain. It focuses on a section of road with 38 lamp posts that are currently fitted with energy inefficient luminaires. A lighting audit also showed that they do not meet current illumination standards. It is possible to improve both energy efficiency and lighting standards by replacing the luminaries on the existing lamp posts.

The current traffic intensity on the case study section of road is IMD = 5,218, below the 7,000 threshold. Before the recent economic crisis the traffic intensity was above this threshold, but it currently falls below the threshold; it may increase and cross the threshold again in the future.

Table 1. Traffic intensity and lighting standards for roads of type A3, from Royal Decree 1890/2008.

Traffic intensity	Lighting standard
IMD ≥ 25.000	ME1
IMD ≥ 15.000 and < 25.000	ME2
IMD ≥ 7.000 and < 15.000	ME3b
IMD < 7.000	ME4a / ME4b

Table 2. Illumination requirements for standards ME3b and ME4b, from Royal Decree 1890/2008.

	ME3b	ME4b
Luminance of the road suface under dry conditions		
Mean luminance, L_m (cd/m^2)	1	0,75
Global uniformity, U_o (minimum)	0,4	0,4
Longitudinal uniformity, U_L (minimum)	0,6	0,5
Diability glare		
Increase threshold, TI% (maximum)	15	15
Area lightning		
Enviromental relationship, SR (minimum)	0,5	0,5

2.3 *Alternative strategies*

Four alternative specifications for replacement luminaries were investigated, using sodium discharge and LED light sources.

A. New luminaries with 150W sodium discharge lamps. This has the lowest initial cost, but the lamps consume more energy than LEDs and have a shorter life. If the traffic intensity crosses the threshold the lamps have to be replaced with 250W lamps.

Table 3. Key data for the four alternatives being evaluated.

	Alternative A	Alternative B	Alternative C	Alternative D
Initial specification ME4b (< 7000 vehicles/day)	150W sodium discharge lamps	81W LED lamps	108W LED lamps regulated to 81W	108W LED lamps with variable output
Lamp service life (hours)	20,000	60,000	60,000	60,000
Sensors and controls	dusk to dawn	dusk to dawn	dusk to dawn	programmable, responsive
Hours of light per year (equivalent)	3200	2936	2936	varies with traffic – potential 65% saving
Initial cost of luminaires (€x1000)	18.3	23.9	25.3	26.2
Initial cost of sensors and controls (€x1000)	0.4	0.2	0.2	11.7
Annual electricity cost at €0.14/KWh (€x1000)	2.55	1.27	1.27	0.59 approx.*
Upgrade specification ME3b (≥ 7000 vehicles/day)	250W sodium discharge lamps	108W LED lamps	108W LED lamps regulated to 108W	108W LED lamps with variable output
Upgrade cost (€x1000)	15.6	23.3	4.5	nil
Annual electricity cost at €0.14/KWh (€x1000)	4.26	1.69	1.69	0.83 approx.*

* Varies with traffic level – 65% saving used in the base model

B. New luminaries with 81W LEDs. If the traffic intensity crosses the threshold LED assemblies have to be replaced with 108W versions. The LED output can be scaled down at the beginning and end of the night-time lighting period, reducing the effective number of hours of lighting per year.

C. New luminaries with 108W LEDs that are regulated to give 81W output to meet the requirements of the current traffic intensity. They are slightly more expensive than the 81W luminaires. If the traffic intensity crosses the threshold, the LEDs are regulated to full output.

D. New luminaries with 108W LEDs linked to sensors that continuously monitor traffic intensity and control the light output. It covers the lighting requirements across the traffic intensity threshold without upgrading. There is an initial cost for the sensors and controls, but they can give a substantial efficiency gain, reducing the effective number of hours of lighting per year.

Data for the four alternatives is given in Table 3.

2.4 *Uncertain variables*

For the case study evaluation, two uncertain variables were modelled.

First, the future price of electricity. This was modelled using a three-point estimate with low, mid and high estimates for the percentage annual change in electricity price of 94%, 103% and 112%, starting from the current electricity price of €0.14/KWh. This gives a long term rising trend, with moderate year-to-year variability. Alternative energy price scenarios can be investigated with sensitivity analysis.

Second, the future traffic intensity for the case study road. Official projections are set out in a three-point estimates with low, mid and high estimates for the annual percentage change in traffic intensity. The base values for the case study are 95%, 101.75% and 108% respectively. However, past experience from traffic surveys in the locality shows higher variability. Alternative values for traffic level uncertainty are investigated in the case study.

The light levels in alternative D are also dependent of the traffic level, because the sensors reduce the light intensity at quiet times. This feeds directly into electricity savings.

2.5 *Discount rate*

The discount rate is a critical factor for LCC. For public sector investment, like street lighting, it is often set by the government agency that provides funding. Government discount rates are usually lower than in the private sector, but they have varied widely in Spain since the financial crisis in 2007. The rate that will be applied to this project is not yet known; the case study uses a base value for the discount rate of 4% per year, supplemented by sensitivity analysis. This is a real rate and the LCC was carried out using real prices

3 LIFE-CYCLE EVALUATION

The alternatives strategies are evaluated using CILECCTA software for probabilistic LCC that can also model responsive changes triggered by the values of uncertain variables. This contrasts with typical deterministic LCC.

CILECCTA uses Monte Carlo simulation to estimate life-cycle costs when there are uncertain variables. In each MC run the values of the uncertain variables are sampled in each year from their defined probability distributions (three-point estimates in this case). All costs are discounted to the present.

In CILECCTA it is also possible to specify alternative states than the system can adopt, with defined criteria for transitioning between states. Here the alternative states are the lighting standards ME4b and ME3b, and the trigger is the IMD=7,000 threshold for traffic intensity (*see* Fig.1).

This feature of the CILECCTA software allows it to evaluate flexibility. The additional cost of alternative C over B can be seen as the acquisition cost of a real option which, when and if it is exercised, provides a low-cost upgrade. The reason for acquiring the option is that future traffic intensity is uncertain; it is not known at the time of acquisition whether or not the option will be exercised. The evaluation indicates whether the initial acquisition cost is balanced by probable service life savings. The flexibility in alternative D has a higher acquisition cost and provides a higher degree of flexibility or responsiveness to the future traffic intensity.

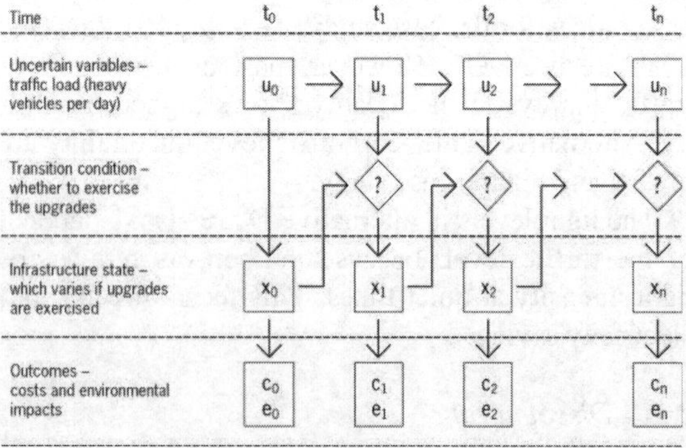

Figure 1. Diagram showing the time-steps in a Monte Carlo simulation run (from Fawcett et al, 2015).

The CILECCTA system produces charts showing the results of Monte Carlo runs for the alternatives under consideration, and also performs and generates charts for sensitivity analysis.

The LCC studies uses a twenty year study period.

4 COMPARISON OF ALTERNATIVE STRATEGIES

Using the base values for the variables in the model, a 1000-run Monte Carlo simulation produced the results shown as a percentile chart in Fig.2. The 70% confidence level is shown, because most decision makers are risk averse; the ranking of the alternatives at this confidence level may be a better guide for decision making than the average of the MC runs or the 50% (risk neutral) confidence level.

It is evident that alternative A using sodium discharge lamps performed worst. The lower initial cost was far more than outweighed by higher electricity consumption and more frequent lamp replacement during the service life.

Of the LED alternatives, B performed best in some cases – this was when the traffic level stayed below the upgrade threshold. When the threshold was crossed, as occurred in the majority of the MC runs, this alternative performed poorly.

The performance of alternatives C and D was very closely matched, with the increased initial cost of the sensors in D being balanced by reduced electricity consumption in the service life, and the avoidance of the upgrade cost which is incurred in C.

A revised evaluation was made for a greater range of uncertainty in the estimate of future traffic intensity, with other variables retaining their base data values. The revised values for the three-point estimate of annual traffic change were 75%, 101.75% and 125%. The percentile chart for a 1000-run Monte Carlo simulation with this revised data is shown in Fig.3.

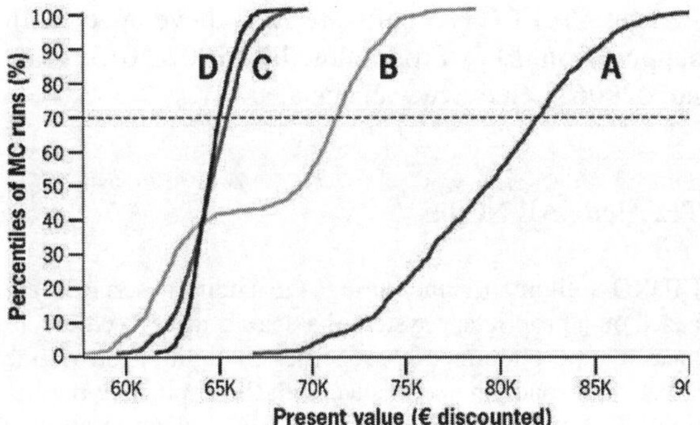

Figure 2. Percentile chart for 1000 Monte Carlo runs for the base input values. The 70% confidence level is marked.

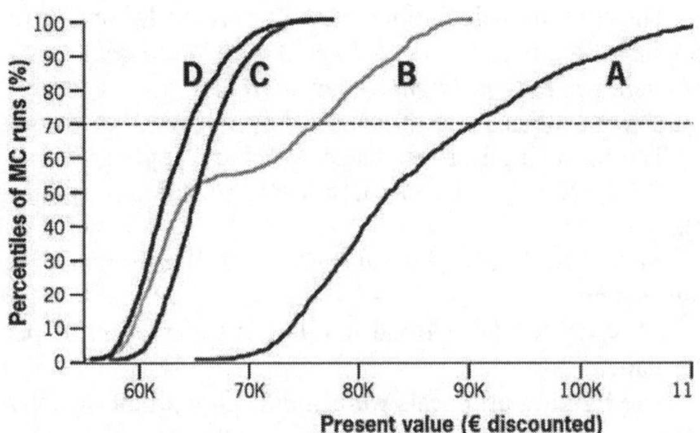

Figure 3. Percentile chart for 1000 Monte Carlo runs for higher volatility of future traffic intensity. The range of values is greater and alternative D is more highly favoured. The 70% confidence level is marked.

The spread of values is greater, and the performance of the most flexible of the alternatives, D, is enhanced, performing best across the whole range of MC runs.

A number of sensitivity analysis studies were carried out; two are shown.

First, sensitivity analysis for variations in the discount rate. Low discount rates allow initial investment to be be offset against service life savings, whereas with high discount rates evaluation is dominated by initial costs. The results (*see* Fig.4) show that alternative D with the highest initial cost performs best for low discount rates, because of the electricity savings in the service life; but it performs worst with high discount rates.

Second, sensitivity analysis for variations in the efficiency gain that is achieved with the sensors in alternative D. The results (*see* Fig.5) show that with a high efficiency gain alternative D performs best, but with low or zero efficiency gains it performs worst. The potential efficiency gain with sensors requires further investigation In this sensitivity analysis the discount rate was 4%.

5 CONCLUSIONS

In terms of methodology, the case study demonstrates that the cost-effectiveness of investing in flexibility can be evaluated using Monte Carlo simulation of uncertain variables, coupled with state transitions triggered by the values taken by the uncertain variables. This provides an evaluation tool for real options. The approach is embodied in the CILECCTA software.

In terms of the street lighting case study, the evaluation exercises indicate some provisional conclusions and point to the need for further investigations.

The immediate conclusion is that traditional, non-flexible sodium discharge lighting (alternative A) is decisively less cost-effective than LED lighting, despite the lower initial cost.

The data for the Monte Carlo runs (Figs. 2 and 3) and the sensitivity analysis (Figs.4 and 5) show that there is no fixed preference ranking for the three LED alternatives (B, C and D). Depending on the input data and the criterion for ranking (70% confidence level, 50% confidence level, or average value of MC runs) any of the three alternative could be the most attractive.

It is evident that high quality input data is vital for reliable decision-making. This requires site-specific surveys and data collection, and further research is required for the case study site. However, street lighting has a long service life and many factors will evolve in ways that cannot be accurately predicted. Hence the desire for flexible street lighting strategies.

The case study shows that flexibility is not a magical solution to all problems. It is a resource with valuable characteristics, but it must be rigorously evaluated with case-specific data.

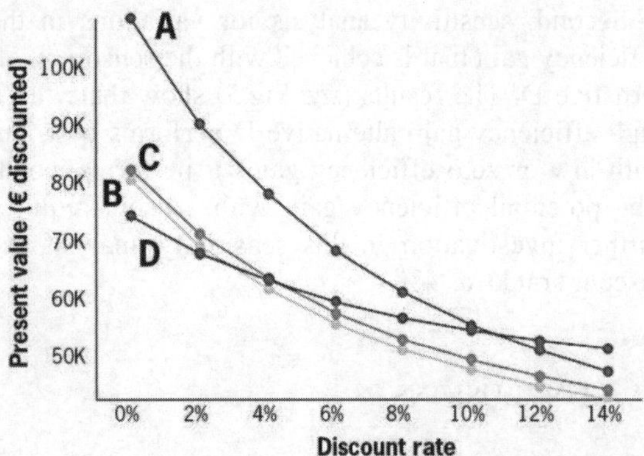

Figure 4. Sensitivity analysis chart for variations in the discount rate, with other variables at base data values. The y-axis is the expected value, ie. average of MC runs. Alternative D performs best for low discount rates and worst for high discount rates.

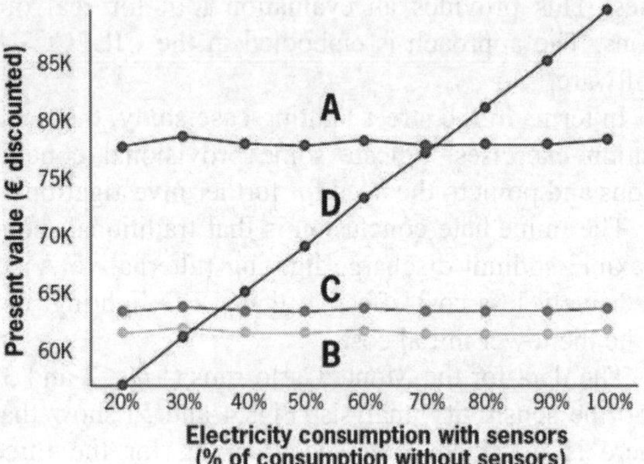

Figure 5. Sensitivity analysis chart for variations in the efficiency gain from the use of sensors, with other variables at base data values. The y-axis is the expected value, ie. average of MC runs. Alternative D performs best when the efficiency gain is high and and worst when it is low.

ACKNOWLEDGEMENTS

The Diputacion de Valladolid (Valladolid Province) commissioned CIDRO to assess public lighting in order to improve the energy efficiency in small municipalities.

Louis Berger IDC Transport department performs traffic studies and economic assessment for worldwide infrastructure development projects.

The CILECCTA software was developed with support from EC's Programme FP7/2007-2013 (grant no. 229061). See: www.cileccta.eu

TECHNICAL NOTE

CIDRO's Illumetric measuring and auditing services are based on a proprietary system that uses a survey vehicle to measure the photometric properties of lighting on public streets and roads in accordance with the applicable regulations (RD 1980/2008, EN 13201). This is done in a quick and accurate way, at road speed and without interrupting the traffic. An accurate inventory of the luminaries is also generated. The system is applicable to urban areas, bridges and tunnels.

The lighting calculations were performed following the Spanish Regulation *Royal Decree 1890/2008 on energy efficiency in external lighting installations* http://www.boe.es/buscar/doc.php?id=BOE-A-2008-18634

The following European standards were applied:

CEN EN 13201-1 Road lighting 1: Selection of lighting classes

CEN EN 13201-2 Road lighting 2: Performance requirements

CEN EN 13201-3 Road lighting 3: Calculation of performance

The lighting proposals were simulated with the software Dialux v4.12 using luminaire parameters from the manufacturer's photometric data (Philips):

Class ME4b (current traffic level): SGP670 FG P1H3V 1XSON-TPP150W R200, BGP 303 1xLED98-3S/740 DM

Class ME3b (upgraded traffic level): Visual IVF4 ST 250 33200 1950 E40 L, BGP 303 1xLED122-3S/740 DM

REFERENCES

CILECCTA. 2013. *Sustainability within the Construction Sector: CILECCTA – Life Cycle Costing and Assessment* (e-Handbook for EC-funded collaborative research project, 2009-13). cileccta.eu/sites/default/files/Cileccta_e-handbook.pdf

de Neufville, R. & Scholtes, S. 2011. *Flexibility in Engineering Design*, MIT Press, Cambridge, Mass.

Ellingham, I. & Fawcett, W. 2006. *New Generation Whole-life Costing: property and construction decision-making under uncertainty*, Taylor & Francis, Abingdon.

Fawcett, W. Robles, I. Krieg, H. Hughes, M. Mikalsen, L. & Ramón Ramos Gutiérrez, O. 2015. Cost and environmental evaluation of flexible strategies for a highway construction project under traffic growth uncertainty, *Journal of Infrastructure Systems* **21**, 05014006-1–14.

Proceedings of the International Conference on Smart Infrastructure and Construction
ISBN 978-0-7277-6127-9

© The authors and ICE Publishing: All rights reserved, 2016
doi:10.1680/tfitsi.61279.621

Future-proofing assessment of infrastructure assets

T. Masood*[1,2], D.C. McFarlane[1,2], J. Schooling[2], A. Fielding[3] and J. Downes[4]

[1] *Institute for Manufacturing, University of Cambridge, Cambridge, UK*
[2] *Centre for Smart Infrastructure and Construction (CSIC), Department of Engineering, University of Cambridge, Cambridge, UK*
[3] *Performance Management, Water Sector, Costain, Maidenhead, UK*
[4] *Engineering Governance & Services, London Underground, London, UK*
* *Corresponding Author*

ABSTRACT Infrastructure assets serve society for long time. Therefore, future-proofing considerations are required to be incorporated in order to sustain these assets. However, even though companies consider some aspects but largely key criteria and mechanisms for future-proofing assessments are not known. As part of the infrastructure future-proofing research conducted at CSIC ; what, why and how characteristics of future-proofing were explored through a number of workshops attended by leading companies dealing with infrastructure assets. This paper presents outputs of the CSIC future-proofing research that includes a framework for future-proofing considered asset management, future-proofing criteria, and a future-proofing assessment approach. This paper also presents case studies of Liverpool Wastewater Treatment Works Upgrade Project and London Underground Camden Town Capacity Upgrade Project, where the future-proofing assessment approach was applied.

1 INTRODUCTION

Infrastructure future-proofing is defined as "the process of making provision for future developments, needs or events that impact on particular infrastructure through its current planning, design, construction or asset management processes" (Masood, McFarlane, Parlikad et al 2015). There are two major dimensions of infrastructure future-proofing: (i) ensuring the ability of infrastructure to be resilient to unexpected or uncontrollable events e.g. extreme weather events; and (ii) ensuring the ability to adapt to required changes in structure and / or operations of the infrastructure in the future e.g. expansion of capacity, change in usage mode or volumes.

Most of the UK infrastructure 'requires attention' (e.g. in terms of flood management, waste minimisation and energy utilisation) or 'is at risk' (e.g. susceptibility of local transport to environmental change) (ICE 2014). The ICE State of the Nation: Infrastructure Report classified strategic transport and water infrastructure as 'adequate for now' but rail infrastructure has recently faced severe disruptions in service due to extreme weather conditions (ICE 2014). These comments imply that UK infrastructure would benefit from a systematic approach to how it should be maintained in the long term with future climate and other predictable and unpredictable changes in mind.

This paper presents outputs of CSIC future-proofing research: a framework to consider infrastructure future-proofing, future-proofing criteria and a future-proofing assessment approach, aimed at supporting key stakeholders of infrastructure assets in their asset management decision making. The outputs were informed by a series of workshops held by the Cambridge CSIC in 2014. The workshops, attended by leading companies dealing with infrastructure assets (listed in acknowledgements), focussed on examining what, why and how characteristics of future-proofing.

The rest of the paper is structured as follows. Section 2 presents a framework to consider infrastructure future-proofing, while section 3 presents an infrastructure future-proofing assessment approach. Case studies are included in section 4. Section 5 discusses the results while section 6 concludes the paper.

2 A STRUCTURED FRAMEWORK TO CONSIDER INFRASTRUCTURE FUTURE-PROOFING

It is proposed that, as a minimum, a systematic framework to future-proofing of infrastructure portfolios should be considered (see Figure 1) (Masood, McFarlane, Parlikad et al 2015). It is proposed that the following should be considered for infrastructure future-proofing:

1) Requirements analysis;
2) Current infrastructure management practice;
3) Future-proofing considerations;
4) Key issues related to a future-proofing strategy; and
5) Model for future-proofing-considered infrastructure management.

Considering these future-proofing elements can help in creating an overall infrastructure asset management plan. This is further developed as a future-proofing assessment approach, which is detailed in the following.

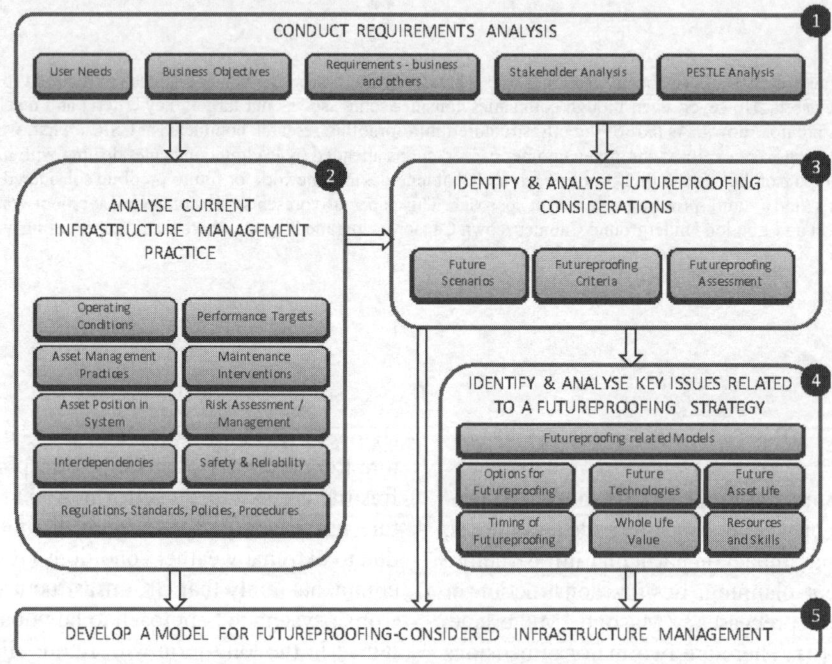

Figure 1. Framework for future-proofing of infrastructure portfolio (Masood, McFarlane, Parlikad et al 2015).

3 AN INFRASTRUCTURE FUTURE-PROOFING ASSESSMENT APPROACH

An infrastructure future-proofing assessment approach is developed based upon consideration of future scenarios for key infrastructure assets, key future-proofing criteria, an assessment mechanism and future-proofing gap analysis/visualisations. Following process is followed as part of the future-proofing assessment approach (see Figure 2) (Masood & McFarlane 2015; Masood, McFarlane, Fielding et al 2016).

1) Identify key infrastructure assets;
2) Identify and analyse future scenarios of changes and possible disruptions in infrastructure management;
3) Identify and contextualize the future-proofing criteria (including future scenarios);

4) Conduct future-proofing assessments; and
5) Conduct future-proofing gap analysis.

Some important considerations while applying the assessment approach are discussed in the following.

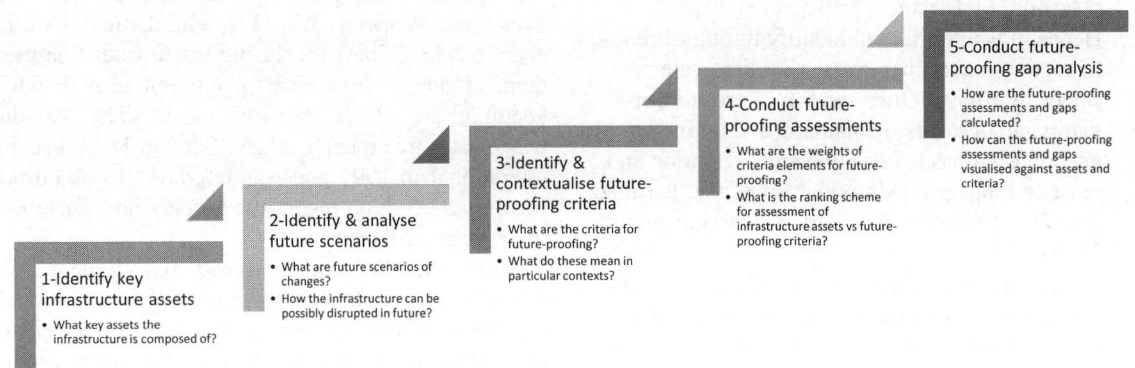

Figure 2. The infrastructure future-proofing assessment approach (Masood & McFarlane 2015; Masood, McFarlane, Fielding et al 2016).

3.1 Key infrastructure assets

Key infrastructure assets are identified in this stage, generally based upon their functional classification and importance in future-proofing considerations. For example, a typical over ground or underground railway infrastructure can be classified in following key assets: bridges & structures, tunnels, drainage, pumping systems, earth structures, premises, communications, signalling, control and information, electrical systems, power, mechanical systems, rolling stock, track, lifts and escalators, plant and equipment, fire systems, and automatic fare collection.

3.2 Future scenarios

A range of potential future changes including a number of future events can have impact on operating environments of infrastructures. Hence, possible future scenarios (e.g. wind storms, flood, snow) are identified to inform the future-proofing model.

3.3 Future-proofing criteria

It is crucial to understand and assess the fitness for the future of the infrastructure based upon the current infrastructure state, future scenarios (e.g. in the light of environmental change, future events or usage change), performance targets and a set of robust future-proofing criteria. This is in line with identifying and assessing specific risks as well as impacts of not future-proofing a particular infrastructure. This will help in identifying gaps and taking further actions to enable future-proofing of infrastructure as well as developing and analysing future business cases.

Following criteria are proposed as a means of assessing the ability of key infrastructure to cope with future developments (Masood, McFarlane, Parlikad et al 2015) (Masood, McFarlane, Schooling et al 2014):

- *Resilience* - the ability to withstand shocks and recover quickly.
- *Adaptability* - the ability of infrastructure to readily adapt or reconfigure if understanding of risks or requirements change over time.
- *Replace-ability* - the ability to be replaced during or at the end of infrastructure life or use, assuming the infrastructure has a finite life.
- *Reusability* - the ability of the infrastructure to be reused or extended at the end of its life. Even though extension is partially used in adaptability as well it is executed during operation phase there while in reusability, extension is meant to be at the end of asset life.
- *Operability* – the ability of infrastructure assets to be operated over its lifecycle. Information future-proofing is an important part of operability,

and it is especially important for decision makers, for a 'system of systems' view, for future owners, operators, the environment and society (Masood, McFarlane, Parlikad et al 2015). Hence, it is important to identify through-life information requirements at earlier life cycle stages of infrastructure and future-proof information at all stages by planning and taking appropriate actions for its collection, retention and reuse in long term (Masood, McFarlane, Parlikad et al 2015; Masood, Cuthbert, McFarlane et al 2013).

- *System-stability* - the ability of infrastructure assets to work for an overall balanced or positive effect, ensuring stability of a system or systems during or after future change(s). This could also mean that systems should work with rather against natural processes (McBain 2014).

3.4 Future-proofing assessments

The following process is followed for future-proofing assessments: assigning weights to criteria elements, rating future-proofing of infrastructure assets against each criterion (as actual and target), and calculating future-proofing scores (as actual and target).

3.5 Future-proofing gap analysis

Future-proofing gaps are calculated as differences between actual and target weighted future-proofing assessments of infrastructure assets against future-proofing criteria.

Two case studies are presented in the following where the assessment approach is applied.

4 CASE STUDIES

The infrastructure future-proofing assessment approach was tested in three case studies (of which two are presented here) as detailed below: a wastewater treatment upgrade project, an underground upgrade project, and a new underground development project. A number of focused workshops were conducted with relevant industrial directors, managers and engineers. The results will be discussed in the following section. However, further details of the approach and the case studies are presented in (Masood & McFarlane 2015; Masood, McFarlane, Fielding et al 2016).

4.1 Future-proofing assessment of Liverpool Wastewater Treatment Works Upgrade Project

The approach was piloted on Liverpool Wastewater Treatment Works (LWwTW) with United Utilities and Costain to meet increasing wastewater treatment demand due to long-term population growth while keeping the River Mersey clean. The existing wastewater treatment works at Sandon Dock became operational in 1991 and was upgraded to its current form in 2000. As a result, the Mersey now sustains a wide range of fish.

However, the works needed replacement. The LWwTW is a £200 million extension project to keep the Mersey clean for future. The recently completed new plant at Wellington Dock will serve around 600,000 residents, and will be able to cope with 11,000 litres of wastewater a second. An aerial overview of the Liverpool Wastewater Treatment Works is presented in Figure 3 (Costain 2014).

Figure 3. An aerial overview of the Liverpool Waste water Treatment Works Infrastructure (Costain 2014).

4.2 Future-proofing assessment of London Underground Camden Town Upgrade

A number of underground station capacity upgrade projects are underway in London. The assessment approach was applied on London Underground Camden Town Capacity Upgrade project. Together with Euston, Camden Town Station is one of the principal interchange stations on the section of the Northern Line. It is situated at the intersection of the Edgeware, High Barnet, Charing Cross and City Branches (see Figure 4) (London Underground 2015).

Figure 4. Overview of Camden Town Station (London Underground 2015).

The main objectives of the £200 million Camden Town Station Capacity Upgrade scheme include increasing station capacity, minimising passenger journey times and station congestion, and improving quality of access, interchange and ambience (London Underground 2015). It is also aimed to contribute to the achievement of the economic growth of London.

5 RESULTS AND DISCUSSION

The future-proofing assessments can be visualised as overall, criteria based and infrastructure asset based future-proofing gaps (see Figure 5). The actual and target future-proofing gaps are assessed and visualised on spider diagrams in both case studies.

The overall future-proofing gap analysis provides visualisation of sums of the gap values assessed from various criteria (C_1 - C_n) against various infrastructure assets (A_1-A_n). For example, Asset 3 has the biggest while Asset 1 has the lowest overall future-proofing gap as shown in Figure 5a. The results of such future-proofing assessments are considered useful for infrastructure stakeholders especially asset owners, overall planners, board members, and strategy directors.

The criteria based gap analysis provides visualisations of weighted sums of the future-proofing gaps as assessed for actual and target values. For example, Asset 2 has the biggest while Asset 1 has the lowest resilience gap (one of the future-proofing criteria) as shown in Figure 5b. The results of such future-proofing assessments are considered useful for infrastructure stakeholders especially asset owners, operators and maintainers.

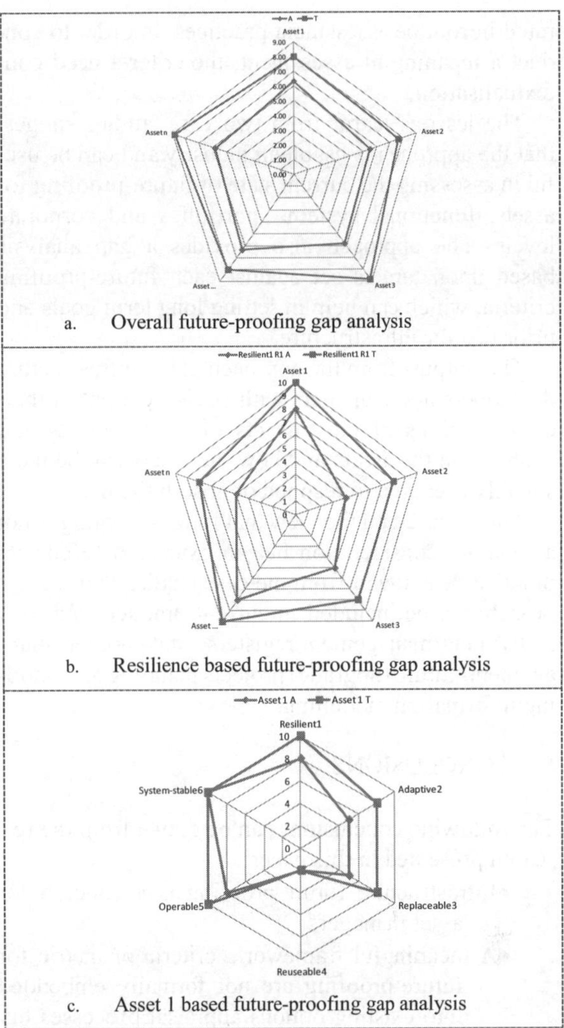

a. Overall future-proofing gap analysis

b. Resilience based future-proofing gap analysis

c. Asset 1 based future-proofing gap analysis

Figure 5. Example of future-proofing gap analysis (overall, criteria based and asset based using actual and target assessments).

The asset based gap analysis provides visualisation of weighted sums of the future-proofing gaps as assessed for actual and target values. For example, Asset 1 has the biggest future-proofing gap in Replace-ability while the lowest gap in Reusability and System-stability, as shown in Figure 5c. The results of such future-proofing assessments are considered useful for infrastructure stakeholders especially asset owners, operators and maintainers.

The assessment approach provides a structured way of considering future-proofing, and can be incorporated in routine assessment practices. In order to conduct a meaningful assessment, the criteria need contextualisation.

The lessons learnt from two case studies suggest that the approach is usable in industry and can be useful in assessing the current state of future-proofing for assets, functional systems, portfolios and corporate levels. The approach also provides a gap analysis based upon targets set against each future-proofing criteria, which can help in setting long term goals and priorities for infrastructure.

The outputs from the approach guide infrastructure developers and managers with advisory input in their considerations of the impacts of future events and changes on the assets and how the assets can be most suitably prepared to respond or absorb them.

This is based on the view that future-proofing is not a separate consideration but an extension to current practice. It is found from the case studies that the approach can be included in current practices e.g. risk assessment/management/registers, stakeholder management, future (upgrade) projects planning and building information modelling.

6 CONCLUSIONS

The following conclusions can be drawn from the research presented in this paper:

- Infrastructure future-proofing is a concern for asset managers.
- A meaningful framework, criteria or metric for future-proofing are not formally embedded into existing options appraisal processes but are needed for effective and long term infrastructure asset management.
- Key future-proofing criteria are based on resilience, adaptability, replace-ability, reusability, operability and system-stability concepts.
- An infrastructure future-proofing framework has been proposed in this paper.
- An infrastructure future-proofing assessment approach has been proposed along with two industrial case studies. The approach is helpful in formulating a strategy and priorities for wider asset portfolios.

It is recommended to assess the applicability of the future-proofing approach to non- infrastructure asset portfolios.

ACKNOWLEDGEMENT

The authors would like to acknowledge the Centre for Smart Infrastructure & Construction, EPSRC (Grant EP/K000314/1), Innovate UK and the industrial partners, which collectively funded this project. The authors are thankful to all the CSIC industrial partners involved in the future-proofing project especially London Underground, Costain, IBM, Crossrail, Cementation Skanska, Network Rail, Arup, Atkins, Halcrow/CH2M and Laing O' Rourke. The authors are also thankful to the speakers and delegates who attended the CSIC workshop(s) on infrastructure future-proofing.

DATA STATEMENT

There is no real data to share for this paper.

REFERENCES

ICE. 2014. The State of the Nation Infrastructure 2014, Institution of Civil Engineers, London, pp 1-28., available from https://www.ice.org.uk/ICEDevelopmentWebPortal/media/Documents/Media/Policy/State-of-the-Nation-Infrastructure-2014.pdf (accessed 6/1/2015).

London Underground. 2015. Concept Design Specification, Ref FSP-LU- B091-CAM-SPC-00088, London underground, London.

Masood, T. McFarlane, D.C. Parlikad, A.K.N. et al. 2015. Towards future-proofing of UK infrastructure, ICE Infrastructure Asset Management, accepted Jul 2015.

Masood, T. & McFarlane, D.C. 2015. CSIC Infrastructure future-proofing: approach and case studies, University of Cambridge, September.

Masood, T. McFarlane, D.C. Fielding, A. et al. 2016. Assessing future-proofing of infrastructure, journal paper draft, January.

Masood, T. McFarlane, D. Schooling, J. et al. 2014. The role of future proofing in the management of infrastructural assets, International Symposium for Next Generation Infrastructure, Vienna, 30 Sep - 1 Oct, pp. 147-152, available from http://discovery.ucl.ac.uk/1469416/1/Final%20Proceedings.pdf (accessed 3/8/2015).

Masood, T. Cuthbert, R. McFarlane, D.C. et al. 2013. Information future-proofing for large-scale infrastructure. In *Proceedings of the IET/IAM Asset Management Conference*, London, 27-28 November, DOI: 10.1049/cp.2013.1945.

McBain, W. 2014. Future-proofing infrastructure for disruptive events: Flooding, 1st CSIC Workshop on Infrastructure Future-proofing, Centre for Smart Infrastructure & Construction and Institute for Manufacturing, University of Cambridge, Cambridge, 23 January, pp 1-34.

Proceedings of the International Conference on Smart Infrastructure and Construction
ISBN 978-0-7277-6127-9

© The authors and ICE Publishing: All rights reserved, 2016
doi:10.1680/tfitsi.61279.627

Assessment of infrastructure resilience in developing Countries: a case study of water infrastructure in the 2015 Nepalese earthquake

H. Nazarnia[*1], A. Mostafavi[2] , N. Pradhananga[2], E. Ganapati[3]and R. R.Khanal[4]

[1]*Department of Civil and Environmental Engineering, Florida International University, Miami, USA*
[2]*OHL School of Construction, College of Engineering and Computing, Florida International University, Miami, USA*
[3] *School of International and Public Affairs, Florida International University, Miami, USA*
[4] *Department of Economic, Florida International University, Miami, USA*
[*]*Corresponding Author*

ABSTRACT Despite the emerging literature on resilient infrastructure systems, the number of studies related to developing communities is rather limited. The majority of the existing studies focus mainly on resilience of infrastructure networks in developed countries. Infrastructure networks in developed countries are less vulnerable to the impacts of catastrophic disasters due to the existence of established design codes and management processes and the availability of financial and technological resources. Catastrophic disasters usually have more extensive impacts on infrastructure systems in developing countires. The objective of this study is to investigate the resilience of infrastructure in developing countries using a case study of water system in Kathmandu Valley in the aftermath of the 2015 Nepalese Earthquake. First, a new systemic farmework for assessment of infrastructure resilience was developed. Second, data obtained from various sources including pre-disaster condition, post-disaster damage assessments, and interviews with different stakeholders were used in assessment of different components of resilience in the water system.The study investigated three dimensions of resilience in Kathmandu Valley's water system : (1) exposure ; (2) sensitivity ; and (3) adaptive capacity. Through a systemic analysis, various resileience characteristics such as coupling, response behaviors, and types of interdependencies that affect the resilience of the system were identified. The findings of the study highlight different factors that influenced the resilience of the water system in Kathmandu Valley. These results provide new insights regarding infratsructure resilience in the context of developing countries.

1 INTRODUCTION

The field of infrastructure resilience is an emerging area in science and engineering.Several studies (e.g., Rinaldi et al. 2001; O'Rourke 2007) have studied the determinants of resilience in infrastructure systems. The majority of the exiting studies in this area are related to infrastructure systems in developed countries. The extent and nature of resilience in Infrastructure systems in developed countries are different from the ones in developing counties due to the existence of established design codes and management processes, differences in social, economic, and political contexts, as well as the availability of financial and technological resources. There is a critical gap in the body of knowledge related to understanding the characteristics of resilient lifeline systems in developing countries. In addition,among different infrastructure sectors, water supply infrastructure plays a vital role in the resilience of communities in the face of natural disasters. A better understanding of the determinants of resilience in water supply infrastructure is essential in prioritizing the allocation of limited resources in developing countries to reduce the adverse impacts of natural disasters on communities. However, the existing studies related to the resilience of water systems in developing countries are rather limited. To this end, the objective of the study presented in this paper was to investigate the factors influencing the resilience of water systems in developing countries using a case study of the 2015 earthquake in Nepal.

2 RESEARCH FRAMEWORK

Lifeline Infrastructure systems are recognized as key elements in investigating the resilience of communities in the context of disasters (Cutter et al., 2003). The National Infrastructure Advisory Council (NIAC) defined infrastructure resilience as "the ability to reduce the magnitude and/or duration of disruptive events. The effectiveness of a resilient infrastructure or enterprise depends upon its ability to anticipate, absorb, adapt to, and/or rapidly recover from a potentially disruptive event" (NIAC, 2009).

In one stream of research, researchers have conceptualized resilience using four properties (known as 4Rs of resilience): robustness, redundancy, resourcefulness, and rapidity (Bruneau et al. 2003). In another stream of research, different studies have investigated the concepts involved in understanding system resilience. According to Gallopin (2006), the resilience of a system depends on: (1) the exposure of the system to hazard-related perturbations, (2) configuration of the system prior to a perturbation, (3) the transformation (a.k.a. sensitivity) of the system due to the perturbation, and (4) the adaptive capacity of the parts. The frameworks proposed in each of these streams of research are useful for investigation of resilience at different levels. For example, the 4Rs framework is appropriate for evaluation of resilience at facility or organizational level. On the other hand, the framework proposed by Gallopin (2006) is suitable for a system level analysis in which resilience is evaluated through the use of the concepts of exposure, sensitivity, and adaptive capacity. Hence, in this study, a framework for systemic assessment of infrastructure resilience was created to analyze the water supply system in Kathmandu Valley in 2015 Nepalese Earthquake. As shown in Figure1, the framework includes three dimensions of analysis (i.e., exposure, sensitivity, and adaptive capacity) consistent with the resilience model proposed by Gallopin (2006). Exposure is defined as the extent to which a system is subjected to perturbations induced by hazards. The exposure of a system can be understood based on the nature of hazards and value of economic and social resources at risk. Sensitivity of infrastructure systems is dependent on system condition (Mostafavi and Abraham 2012); dependencies with other infrastructure (Rinaldi et al. 2001), human-infrastructure coupling, and the preparedness of organizations managing and operating these systems. The third dimension of the framework investigates a system's adaptive capacity. The adaptive capacity of infrastructure systems depends on the social systems managing, operating, and utilizing the physical networks. Hence, infrastructure systems' adaptive capacity can be understood based on the analysis of the capacity of organizations to respond to hazard-induced perturbations as well as public's capacity to respond to service disruptions.

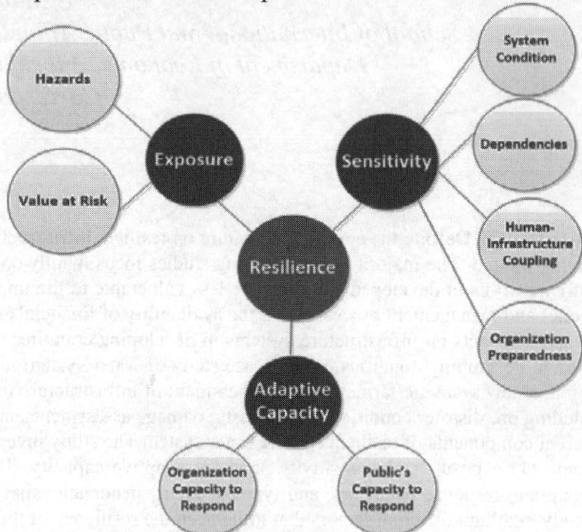

Figure 1. Framework for systemic assessment of resilience in infrastructure systems.

3 CASE STUDY

On April 25, 2015, Nepal witnessed one of the most destructive earthquakes in its history. This disaster claimed almost 8,500 lives, 22,000 people were injured, more than 800,000 houses were damaged or fully destroyed, and about 3 million inhabitants relocated. The earthquake affected 33 out of 75 districts in Nepal and various infrastructure sectors.

Among different sectors, water supply systems ranks second (after transportation systems) in terms of the value of damages caused by the earthquake. The extent of damages varies in different locations. Kathmandu Valley is among the districts that were severely impacted the earthquake. Kathmandu Valley is the most developed and fastest growing place in Nepal with population of 2.5 million. While other parts of Nepal are mainly rural and lack centralized water systems, Kathmandu Valley is an urbanized setting with an old water supply system. Hence, this study focused on the water supply system in Kathmandu Valley in order to investigate its resilience.

Data required for analysis of water system resilience in this study was obtained from four sources: (1) reports related to water system characteristics in Kathmandu Valley, (2) the post disaster need assessment (PDNA) report published by the government of Nepal in collaboration with international agencies such as the Asian Development Bank and Japan International Cooperative Agency (JICA), (3) the report published by the Earthquake Engineering Research Institute (EERI) Reconnaissance Team, and (4) field visits and interviews with different stakeholders. The primary method for collecting these data was in-depth interviews with elected (e.g., mayors, commissioners, and members of infrastructure agencies) and appointed public officials (e.g., public works managers and urban planners) at local and national levels, who were directly involved in water system operation, management, restoration, and response. The interviews were recorded (with permission of the interviewees) and transcribed in both Nepalese and English. The transcribed interviews were coded along with the secondary sources of information (e.g., PDNA and EERI reports) using NVIVO 11 software. The codes were refined through pattern analysis to summarize groups of codes into constructs, which will be explained in the following sections.

3.1 *Exposure of Water System*

Hazards: Kathmandu Valley is located in a seismic zone. Prior to the 2015 earthquake, the 1934 AD Bihar-Nepal Earthquake produced strong shaking in Kathmandu Valley. The seismic record of the region suggests that catastrophic earthquakes are expected approximately every 75 years (Dixit et al. 2000). In fact, the earthquake in April 2015 had an epicenter in the east part of the district of Lamjung and was not the expected seismic activity in Kathmandu Valley. If the epicenter was closer to the valley, the earthquake would have more severe damages.

Value at risk: Another factor affecting the exposure of Kathmandu Valley is its population growth, uncontrolled development, and poverty (Dixit et al. 2000). Nepal is urbanizing rapidly, and Kathmandu Valley as a major urban setting in the country has a population growth of approximately 7%. Population growth increases the demand for water supply and requires increased development of water supply system in Kathmandu valley. With increased development in water supply systems, there was more value of water utilities and facilities at risk. In addition, the population growth increased the adversity of the impacts due to water supply disruption on the people living in the region. Also, uncontrolled development led to improper connection of water mainlines to houses which caused damages and service disconnections due to the earthquake.

3.2 *Sensitivity of Water System*

System condition: Kathmandu Upatyaka Khanepani Limited (KUKL) operates and maintains the water supply and sewerage systems in most of Kathmandu Valley. Water services are provided by KUKL through six branch offices inside Kathmandu and four municipalities in Lalitpur and Bhaktapur. Kathmandu Valley's water supply system was built approximately 120 years ago. There are 2.7 million people and 200,000 connections in its service area. Fig. 2 depicts components of water system in Kathmandu Valley. The system is composed of eight subsystems, each with a different source and treatment plant (EERI 2015). There are about 45 water reservoirs supplying water to the valley. The water system source also includes 70 tube wells in the North part of the valley that provide about 30% of water in the region. The system was in a poor condition due to lack of periodic maintenance and rehabilitation causing water leakage in the system. The reason for the poor system condition can be attributed in part to insufficient management of operation and maintenance due to lack of technicians, lack of accurate information about registered users, and lack of maintenance funding (EERI 2015).

KUKL faces two major challenges in supplying water to the valley: a huge gap between supply and demand and ground water depletion. Due to the existing supply and demand gap, not all households have private connections and many use wells and taps in the community. In addition KUKL provides water through tankers to areas that do not have service connection. For households with private connection, the supply of water is limited to few hours during week requiring the households to store water in their houses. To this end, the use of in-house water tanks is ubiquitous in the valley. In addition, due to the uncertainty in the water supply, households dig wells in their property for water access. The increased use of ground water has led to decline of water tables causing challenges to KUKL for management of ground water. An immediate consequence of ground water depletion is that the connected wells and pumps will no longer be able to provide water.

In Kathmandu Valley, various modes of failure happened due to the 2015 earthquake. One of the eight water subsystems (in the South part) experienced damage due to a landslide in the Arniko Highway. The sub-system in the North part of the valley was disrupted due to power outage causing disruptions to the pumps extracting water from ground wells. The power outage lasted for two days during which the water supply was completely disrupted in Gonbagu area. Another major damage to the water system was house connection breaks. Due to loose connections, the seismic force led to connection breaks at the end point of the supply system causing increased water leakage. There were no damages reported for the storage tanks used for supplying water through water tankers. In fact, the storage tanks had full storage when the earthquake occurred. Thus, KUKL was able to mobilize tankers to supply water to different affected regions.

KUKL was able to restore the water supply system in 21 days after the earthquake. Despite the information presented above regarding water system disruptions, the complete extent of damages and service disruptions were not completely understood at the time of data collection for this study (five months after the earthquake). The reason for the lack of accurate information regarding the damages and service disruptions in the water system of Kathmandu Valley was the lack of service disruption reported by the customers due to: (1) damages to buildings; and (2) inconsistent quality of service prior to the earthquake.

System dependencies: The types of dependencies identified between water system and other infrastructure were geographic or physical. According to Rinaldi et al. (2001), physical dependencies exist when state of one infrastructure is dependent on the output of the other infrastructure. In the water supply system in Kathmandu Valley, two physical dependencies caused service disruptions. First, in the north

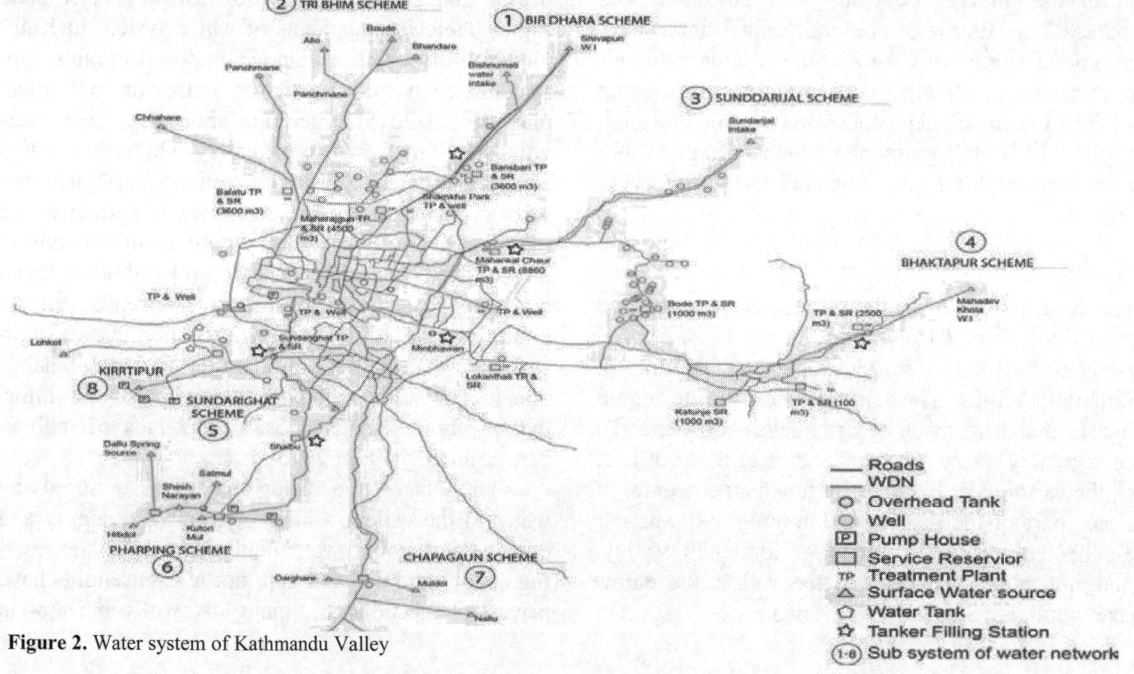

Figure 2. Water system of Kathmandu Valley

part of the valley, failures in the power supply system caused the pump stations to stop working, and hence, water supply from ground water sources was disrupted. The power supply system was restored after two days, and hence, water supply was also restored. The water supply system in other areas (such as Chitwan plant) also uses ground water sources. Fortunately, however, the power supply systems in those areas were not disrupted. The second physical dependency was between water supply and roads. This is a unique type of physical dependency that usually does not exist in infrastructure systems in developed countries. This dependency was unique to Kathmandu Valley since a considerable portion of water supply was delivered through water trucks. KUKL developed capacities over time to supply water by trucks in response to the supply and demand gap in the system. KUKL has six tanker stations for filling the water trucks, and fortunately, none of these stations were damaged by the earthquake. Also, the tankers had full storage capacity when the earthquake occurred. Hence, KUKL was able to deploy its truck fleets immediately after the earthquake; however, the closure of roads due to landslides or blockage due to building collapses caused difficulties for the water trucks to access certain areas.

The second type of dependencies identified in Kathmandu Valley's water supply system is geographic dependencies. Geographic dependencies exist when a local hazard can create state changes in different infrastructure (Rinaldi et al. 2001). In Kathmandu Valley, the majority of water utility lines delivering water from the source to plants or storage tanks were constructed along the major roads. Hence, landslides in roads caused breaks in the water trunks passing through the roads. One major incident in the aftermath of 2015 earthquake was the breakage of a 35 cm water trunk line due to a land slide along the Arniko highway causing service disruptions in Patan area. This incident caused one of the eight subsystems of KUKL to be disrupted for more than two weeks.

Human-infrastructure coupling is the extent to which the public is reliant on the services provided by a system. The supply-demand gap in the water system of Kathmandu Valley had reduced the public's reliance on the system for their water supply and storage. The use of on-site wells and purchase of water from private water trucks as well as on-site storage of water were the substitute solutions that the public had adopted to cope with the discontinuity of service in the water system. These substitutions, to some extent, reduced the human-infrastructure coupling, and hence, reduced the sensitivity of water system to the impacts of the earthquake. If the system had been able to supply 100% of the demand consistently, the impacts of the earthquake on the system and public would have been more deleterious.

Preparedness of organization: In the case of water supply system in Kathmandu Valley, KUKL did not have an established disaster management processes in place at the time of the earthquake in 2015. As mentioned earlier, the supply-demand disparity in the water system of Kathmandu Valley had created a chronic stress on the agency. Meeting the day-to-day water needs of the customers along with limitations in the agency's resources had reduced the capability of the agency to establish disaster management processes. Despite the lack of disaster management process, KUKL was able to respond to the service disruptions caused by the earthquake with the help of WASH Cluster. First, the agency prioritized the customers based on their urgency for receiving water supply. For example, hospitals and public buildings were prioritized for immediate service restoration. Another priority for KUKL was to provide water to government-established shelter camps.

The second component of KUKL's response activities included damage assessment. In the aftermath of the earthquake, KUKL did not know the extent of damages because a large portion of population had left Kathmandu Valley and many buildings were damaged. The KUKL's response was to deploy its personnel to facilities (e.g., treatment plants, storage tanks, reservoirs, pipelines, and pump stations) to collect information about the damages to water system component. The KUKL's capacity to monitor and assess the condition of underground utility conditions was very limited. Hence, the agency was collecting information about damages based on customers' complaints such as service disruptions or water leakage in the streets. However, there were many damages and leakages that were identified late since KUKL did not receive any complaints from the customers.

3.3 *Adaptive Capacity*

Two determinants of social system's adaptive capacity in infrastructure include: (1) adaptive capacity of administering agency; and (2) the adaptive capacity of general public. As mentioned earlier, the water supply system in Kathmandu Valley suffers from a significant supply-demand disparity. This supply-demand disparity created a chronic stress on the system, KUKL, and general public. The chronic stress caused both the agency and the public to develop adaptive capacity through enhancing redundancy. As for the agency, KUKL developed water trucking capacity to supply water to households during times of load shedding in the network. This additional capacity had not been developed for earthquake emergency management; as for KUKL, every day was an emergency situation to supply water to the people.

As for the general public, the chronic stress caused by water supply shortage caused the household to adopt in-house storage tanks. Almost every household in Kathmandu Valley had a storage tank to store water during the scheduled supply time and usage during load shedding periods. In addition, though it was illegal, many households had their own shallow wells as a backup source. Through these alternative solutions developed under chronic stress of water supply shortage, households built redundancy overtime. Hence, in the aftermath of the earthquake, service disruptions in KUKL water supply did not cause major problems to water access since households already had substitutions.

4 CONCLUSION

The findings of this study highlight the significant role of the social systems' adaptive capacity developed under chronic stressors (i.e., supply-demand gap) in enhancing the resilience of the water system. While the water system was very sensitive to hazards, its adaptive capacity reduced the negative impacts caused by service disruptions. In addition, the findings identify the extent of human-infrastructure coupling as an important component influencing the resilience of infrastructure systems. In the case of Kathmandu Valley, the coupling was not strong due to supply-demand disparity. Hence, the system disruptions did not have as extensive impacts. Finally, the findings of this study illuminate the type of dependencies between the water system and other infrastructures. For example, the KUKL's use of water trucks for water supply had created an emergent dependency between water and road infrastructure. These findings accentuate new dimensions of analysis in the emerging field of infrastructure resilience and also provide information for decision-makers in order to better understand the various factors influencing the resilience of infrastructure systems in the context of developing countries.

ACKNOWLEDGEMENTS

This material is based upon work supported by the National Science Foundation under Grant Number CMMI-1546738. Any opinions, findings, and conclusions or recommendations expressed in this material are those of the authors and do not necessarily reflect the views of the National Science Foundation.

REFERENCES

Bruneau, M., Chang, S. E., Eguchi, R. T., Lee, G. C., O'Rourke, T. D., Reinhorn, A. M.& von Winterfeldt, D. (2003). A framework to quantitatively assess and enhance the seismic resilience of communities. Earthquake spectra, 19(4), 733-752.

Rinaldi, S. M., Peerenboom, J. P., & Kelly, T. K. (2001). Identifying, understanding, and analyzing critical infrastructure interdependencies. Control Systems, IEEE, 21(6), 11-25.

Gallopín, G. C. (2006). Linkages between vulnerability, resilience, and adaptive capacity. Global environmental change, 16(3), 293-303.

O'Rourke, T. D. (2007). Critical infrastructure, interdependencies, and resilience. Bridge-Washington-National Academy of Engineering-, 37(1), 22.

Mostafavi, A., Abraham, D. M., & Lee, J. (2012). System-of-systems approach for assessment of financial innovations in infrastructure. Built Environment Project and Asset Management, 2(2), 250-265.

Dixit, A., Dwelley-Samant, L., Nakarmi, M., Pradhanang, S. B., & Tucker, B. (2000). The Kathmandu Valley earthquake risk management project: an evaluation. Asia Disaster Preparedness Centre: http://www. iitk. ac. in/nicee/wcee/article/0788. pdf.

Cutter, S. L., Boruff, B. J., & Shirley, W. L. (2003). Social vulnerability to environmental hazards*. Social science quarterly, 84(2), 242-261.

PDNA reports (2015). Retrieved on 20 December, 2015 from http://www.worldbank.org/content/dam/Worldbank/document/SAR/nepal/PDNA%20Volume%20A%20Final.pdf

KUKL Annual reports (2015). Retrieved 20 December, 2015 from http://www.kathmanduwater.org/reports/annual_report.php

EERI Reconnaissance Team Report (2015). Retrieved 20 December, 2015 from http://www.eqclearinghouse.org.

National Infrastructure Advisory Council (NIAC) (2009). Critical Infrastructure Resilience.
http://www.dhs.gov/xlibrary/assets/niac/niac_critical_infrastructure_resilience.pdf; Retrieved on Dec 1 2015.

Proceedings of the International Conference on Smart Infrastructure and Construction
ISBN 978-0-7277-6127-9

© The authors and ICE Publishing: All rights reserved, 2016
doi:10.1680/tfitsi.61279.633

Vulnerability analysis of hydrological infrastructure to flooding in coastal cities - A graph theory approach

R.I. Ogie*[1], T. Holderness[1], S. Dunn[2] and E. Turpin[1]

[1] *Smart Infrastructure Facility, University of Wollongong, Wollongong, Australia*
[2] *School of Civil Engineering & Geosciences, Newcastle University, Newcastle, UK*
* *Corresponding Author*

ABSTRACT Hydrological infrastructure such as pumps and floodgates are invaluable assets for mitigating flooding in coastal cities. These infrastructure components are often vulnerable to damage or failure due to the impact of flood waters, thus exacerbating the flood hazards and causing significant loss of life and destruction to property worth billions of dollars. Hence, there is a growing need worldwide to enhance the understanding of flood vulnerability and to develop key metrics for assessing it. This study proposes an approach for measuring the vulnerability of hydrological infrastructure to flood damage in coastal cities. In this approach, a hydrological infrastructure flood vulnerability index (HIFVI) is developed based on exposure, sensitivity and resilience of infrastructure assets to flooding. A graph-theoretic algorithm for implementing the proposed HIFVI is presented and applied to assess the flood vulnerability of floodgates in one of the most representative coastal cities - Jakarta, Indonesia. The application involves the construction of a graph-based spatio-topological network model of Jakarta's hydrological system, with floodgates represented as network nodes and waterways as edges. An analysis of the constructed network is carried out based on the underlying graph-theoretic algorithm to compute HIFVI for all nodes that represent floodgates. The results show that HIFVI can point to the most vulnerable hydrological infrastructure components and also highlight locations within coastal cities where additional infrastructure are required to improve resilience to flooding. These information are vital to decision makers when planning and prioritising infrastructure maintenance and resource allocation for flood preparedness in coastal cities.

1 INTRODUCTION

With the increasing frequency and intensity of rainfall and associated floods in coastal cities, there is a need to judiciously allocate limited resources for routine maintenance and upgrade of existing hydrological infrastructure (e.g. pumping stations, floodgates), in a manner that improves their resilience and minimises their failure during extreme flooding events (Sadoff et al. 2013). Ideally, such resource allocations and investment decisions should be effectively targeted at the most vulnerable components in the hydrological infrastructure system. By so doing, the failure of the hydrological infrastructure system and the resulting loss of life and property damage associated with flood inundation can be minimised (Hall et al. 2003). Though a quantitative assessment of vulnerability can point decision makers to the most vulnerable components in the hydrological infrastructure network, this is not a straightforward task that lends itself to a standardised process of finding suitable metrics (Balica et al. 2012). In the context of coastal cities situated in developing nations, this task is further complicated by the lack of sufficient data, potentially limiting the range of possible solutions (Brecht et al. 2012).

To address this issue, this study proposes a graph-based network approach for measuring hydrological infrastructure flood vulnerability index (HIFVI), using the concepts of exposure, sensitivity and resilience. The graph theory approach provides a rigorous mathematical basis for computationally reducing vulnerability to a single metric, using very little available data within the data-starved environment and allowing for further improvement from the initial results as additional data becomes available in the future (Bunn et al. 2000). In this approach, a graph-theoretic algorithm for implementing the proposed HIFVI will be developed and applied to assess and rank the flood vulnerability of Jakarta's floodgates, using the constructed spatio-topological network model of the city's hydrological system. The following section establishes the general equation for computing the flood vulnerability index of hydrological infrastructure components.

2 DERIVATION OF FLOOD VULNERABILITY INDEX FOR HYDROLOGICAL INFRASTRUCTURE COMPONENTS

Generally, vulnerability is determined based on three main factors: exposure, sensitivity (or susceptibility), and resilience (Balica et al. 2012). This can be represented mathematically using the general flood vulnerability index (FVI) formula (Eq. 1) (Balica et al. 2012).

$$FVI = \frac{E*S}{R} \tag{1}$$

The exposure of any given floodgate is determined by the length of all waterways that flow from upstream towards it (Balica et al. 2012). Given that the number of waterways that flow from upstream towards a given floodgate can range from 1 to n, the length, l, for each of these waterways can be summed to determine the exposure, E, of the floodgate. Mathematically, this can be represented as shown in Eq. 2.

$$E = \sum_{i=1}^{n} l_i \tag{2}$$

Susceptibility is a system characteristic, which determines the degree to which the system is affected by the impact of flood waters (Balica et al. 2012). In this study the capacity of the floodgate is used as a measure of susceptibility to flood damage. During intense flood events, a floodgate with lower capacity is considered more susceptible to failure or breakdown as compared to one with a greater capacity. Hence, given that C_g is the capacity of a given floodgate, susceptibility, S would decrease as C_g increases. This relationship can be represented mathematically as shown in Eq. 3.

$$S = \frac{1}{C_g} \tag{3}$$

Resilience can be derived as a function of redundancy (Chang & Shinozuka 2004). In this study, the resilience of a given floodgate, FG, in the hydrological infrastructure network is determined based on redundancy provided by connected upstream floodgates (Chang & Shinozuka 2004). Factors considered in measuring the redundancy provided by each connected upstream floodgate include capacity, c, geometric length, l (i.e. distance along flow path(s) to FG), and the upstream network configuration. The connected upstream floodgates with higher value of c and lower value of l contribute more to the resilience of FG. In terms of upstream network configuration, a connected upstream floodgate would contribute maximally to the resilience of FG if its location in the network allows it to divert floodwater from all the different channels flowing to FG. However, with additional number of channels, w, connecting the link between the two floodgates, the contribution of the upstream floodgate to the resilience of FG reduces accordingly. Hence, given that FG has m number of connected upstream floodgates, its total resilience, R can be estimated using Eq. 4.

$$R = R_s + \sum_{i=0}^{m} \frac{c_i}{l_i * w_i} \tag{4}$$

R_s is the structural resilience of the referent floodgate based on the physical property of its material, $\sum_{i=0}^{m} \frac{c_i}{l_i * w_i}$ is the total resilience contributed by the connected upstream floodgates, where i is an element in the set of connected upstream floodgates, which may be made up of 0 to m members. 0 member means that there are no connected upstream floodgates, in which case $\sum_{i=0}^{m} \frac{c_i}{l_i * w_i} = 0$ and $R = R_s$.

By substituting Eq. 2, 3, and 4 into Eq. 1, a general equation (Eq. 5) is obtained for estimating FVI (i.e. HIFVI) in the context of hydrological infrastructure (specifically floodgate) for flood mitigation.

$$HIFVI = \frac{\sum_{i=1}^{n} l_i}{C_g \left(R_s + \sum_{i=0}^{m} \frac{c_i}{l_i * w_i}\right)} \tag{5}$$

A graph-theoretic algorithm for applying Eq.5 to compute the HIFVI for the floodgates in a hydrological infrastructure network is shown below.

Begin

V = set of all nodes in the network, G.

A node, v represents either a junction or floodgate.

E = set of all edges in the network, G.

An edge, e is represented as $(v_s\ v_f)$, where v_s = start node and v_f = finish node.

F_g= set of all floodgates in the network, such that $F_g \in V$

For g such that $g \in F_g$:

Do

1. **Obtain the capacity, C_g of g**
   (Note: C_g is encoded as an attribute in network nodes).
2. **Compute total length of waterways, L_g connected to g.**
   --$L_g = 0$ (Initialisation)
   --Find V_g, set of all upstream nodes connected to g
   -- For e such that $e \in E$:
   if $(v_s \in V_g)$ and $(v_f \in V_g)$
   $L_g = L_g +$ (geometric length, l of e)
   --Return the value of L_g
3. **Compute F_c, the set of floodgates linked to g upstream.**
4. **Compute the total resilience, R_{fc} contributed by the upstream floodgates connected to g in four steps:**
   $R_{fc} = 0$ (Initialisation)
   -- For f such that $f \in Fc$:

   (i). Compute c, being the capacity of f.

   (ii). Compute the total number of additional waterways, W joining the link between each connected upstream floodgate and g (i.e. a measure of branchness factor).
   W = 1 (Initialisation)
   V_{fc}= set of all nodes in the shortest path between f and g.
   For p such that $p \in V_{fc}$:
   N_e= the number of inward edges to p
   if $N_e > 1$ (an indication of branchness)
   $W = W + N_e$
   Return the value of W

   (iii). Compute the total length of waterways, L between f and g.
   $L = 0$ (Initialisation)
   V_{fc}= set of all nodes in the shortest path between f and g.
   For e such that $e \in E$:
   if $(v_s \in V_{fc})$ and $(v_f \in V_{fc})$
   $L = L +$ (geometric length, l of e)
   Return the value of L

   (iv). Compute sum of the resilience contributed by all upstream floodgates connected to g.
   $R_{fc} = R_{fc} + \frac{c}{L*W}$
5. **Compute the total resilience, R of g.**
   -- $R_s = 1$ (structural resilience, R_s is assigned a constant value of 1 for all floodgates in the network.)
   -- $R = R_s + R_{fc}$
6. **Compute the flood vulnerability index, *HIFVI* of g.**
   -- $HIFVI = \frac{L_g}{C_g*R}$

End

3 A CASE STUDY APPLICATION: JAKARTA'S HYDROLOGICAL INFRASTRUCTURE NETWORK

Jakarta, the capital of Indonesia was selected for this study because it is one of the most exemplary coastal cities of developing nations that depend heavily on structural measures or hydrological infrastructure (e.g. floodgates, pumps, etc.) to mitigate the devastating impact of flooding on the people, property, economy, and environment (Li 2003). As a low-lying delta city served by thirteen rivers, Jakarta relies on a network of pumping stations and floodgates to control water flowing from surrounding hills and mountains, through the city to the Java Sea (Hartono et al. 2010). The frequent use of these ageing and poorly maintained hydrological infrastructure components during the annual monsoonal flooding (between November and March) exposes them to the damaging impacts of floodwaters, with possibility of breakdown or failure as a consequence (Turpin et al. 2013). Generally, the pumping stations are used to move out accumulating floodwater, particularly in low lying areas where drainage is difficult without pumping (Tingsanchali 2012). On the other hand, the action of closing a floodgate allows it to be used for diverting floodwater away from flooded areas located downstream of the floodgate. Because the acquired dataset for the pumping stations was incomplete at the time of this study, this application focuses on just the floodgates infrastructure in Jakarta.

The floodgate dataset, in addition to Jakarta's waterways (i.e. rivers, canals, and streams) were acquired and processed in readiness for network construction and subsequent vulnerability analysis. The data acquisition involved the use of ground survey, GPS locations and aerial imagery analysis to capture and record the names and locations of the different floodgates and waterways in Jakarta. The resulting waterways vector data is of line geometry type while the floodgates vector data made up of 30 records is of point geometry type. Using the topology toolset and GRASS plugin within the QGIS software, these datasets were processed to remove topological and locational errors introduced during survey and digitisation of mapped data. Furthermore, edges in the waterways dataset were programmatically split into separate line features where they self-intersected or intersected with floodgate infrastructure. This is to ensure that at the construction of the hydrological infrastructure network, junctions are created were they actually exist.

The graph-based spatio-topological network model of Jakarta's hydrological infrastructure was constructed using the PostGIS spatial database schema and coupled Python interface to the NetworkX graph analysis package developed by Newcastle University (Barr et al. 2012). This software was first extended to support the proposed graph type (i.e. multidigraph), which permits multiple edges between the same source and target nodes. Topology was encoded within the data using a system of unique node and edge primary keys. In the absence of high resolution and accurate elevation data to model flow direction of Jakarta's waterways, directionality was inferred by edge orientation assuming the general condition of water flowing from the mountains of Bogor to the south of Jakarta, and through the city to the Java Sea in the north. Where exceptions to this assumption existed based on actual field observations of water flow in the city of Jakarta, corrective adjustments of edge orientation were made by re-ordering (i.e. reversing) the geometric points in the linestrings.

The completed network comprised of 628 edges representing Jakarta's waterways, with a total geometric length of 1092 km. There were 560 nodes in the network, 30 of which represent floodgate infrastructure, and the remaining 464 representing network junctions (e.g. river confluences). Figure 1 highlights the locations of the floodgate infrastructure in the network.

Following the successful construction of the network model, the NetworkX and the Pandas Python libraries were used in implementing the underlying graph-theoretic algorithm, resulting in the computation of HIFVI for all 30 floodgate infrastructure in Jakarta. In this implementation, certain assumptions were made. For instance, in the absence of relevant data to determine structural resilience, R_s this parameter was assumed to be a constant value of 1 for each floodgate in the network. Similarly, in the absence of data for floodgate capacity, the number of gates in each floodgate was used as a proxy for capacity. The computed HIFVI values were stored in a PostGIS database table and accessible for visualisation using geographical information system software (e.g. QGIS).

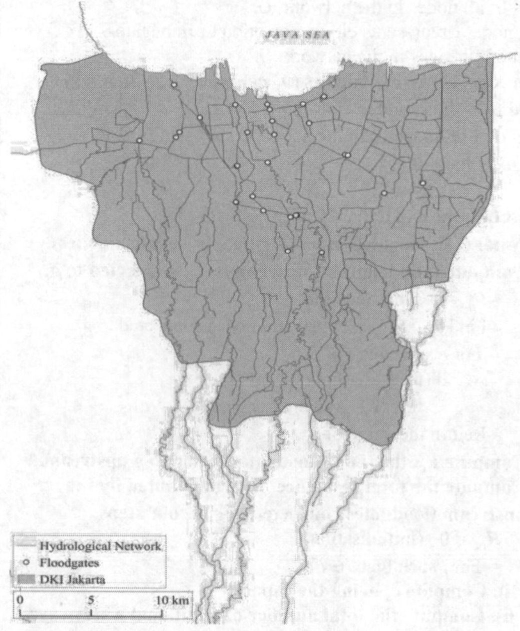

Figure 1. Jakarta's floodgate infrastructure network

4 RESULTS

The results of an application of the graph-theoretic algorithm to Jakarta's hydrological infrastructure are index values representing the degree to which each floodgate in the city is vulnerable to failure or damage due to the impact of flood waters. The index values (i.e. HIFVI) were normalised to give a number from 0 to 1, where 0 does not mean absence of vulnerability, but rather a representation of the lowest vulnerability and 1 indicates the highest in this dataset. This approach allows for a comparative assessment of infrastructure vulnerability to flood hazards (Balica et al. 2012).

To further characterise hydrological infrastructure based on computed HIFVI, index values were classified into 5 different levels of vulnerability as follows: 0-0.2 = very low vulnerability, 0.2-0.4 = low vulnerability, 04-0.6 = moderate vulnerability, 0.6-0.8 = high vulnerability, and 0.8-1 = very high vulnerability (see Table 1).

The results show that "Sunter C" topped the category of very high vulnerability floodgates, thereby ranking as the most vulnerable floodgate with computed HIFVI value of 1. On the other hand, "Sunter

Utara" had the lowest computed HIFVI value of 0, making it the least vulnerable to failure or damage arising from the impact of flood waters. Overall, Table 1 shows that 10% (i.e. three) of Jakarta's floodgates were classified as having very high vulnerability to failure or damage due to the impact of flood waters. Another 3.33% (i.e. one) of the floodgates was moderately vulnerable, but 0% (i.e. none) was classified as highly vulnerable based on computed HIFVI. Most floodgates (i.e. 17) came under the category of very low vulnerability, representing 56.67% of the entire sampled infrastructure. This is closely followed by another 30% (i.e. 9) classified as having low vulnerability to failure or damage due to the impact of flood waters. These results and their implications are discussed further in the subsequent section.

Table 1: Vulnerability ranking of Jakarta's floodgate infrastructure based on computed HIFVI

Name of floodgate	Susceptibility (1/no. of gates)	Resilience	Exposure	HIFVI	Ranking
Sunter C	1.00	1.45	199.89	1.000	VH
Ciliwung Lama	1.00	1.05	133.65	0.921	VH
Kebon Baru	1.00	1.00	122.59	0.890	VH
Muara Angke	0.50	1.43	203.95	0.518	M
Cakung Drainase	0.33	1.00	164.23	0.397	L
Karet 2	0.50	1.39	150.31	0.391	L
Pasar Ikan	0.25	1.51	307.40	0.370	L
Hailai	0.50	1.78	169.77	0.346	L
Istiqlal	0.33	1.13	160.47	0.343	L
Tangki	0.50	1.94	164.02	0.306	L
Jembatan Merah	0.25	1.10	163.09	0.268	L
Kali Cideng	0.33	1.50	157.59	0.253	L
Citra Land	0.33	1.75	153.79	0.212	L
Cengkareng Drain	0.25	1.00	101.23	0.182	VL
Pulogadung	0.17	1.00	143.50	0.172	VL
Ancol	0.20	1.52	176.82	0.168	VL
Pekapuran	0.20	1.64	170.12	0.149	VL
8	0.13	1.26	151.23	0.108	VL
Sogo	0.50	1.53	36.24	0.084	VL
Poglar	0.33	1.00	33.81	0.080	VL
Warung Pedok	0.50	1.00	12.81	0.045	VL
Manggarai	0.33	1.19	21.79	0.043	VL
Setia Budi	0.33	1.15	19.70	0.040	VL
Minangkabau	0.50	1.64	15.94	0.034	VL
Kampung Gusti	0.50	7.25	34.29	0.016	VL
Kalimati	0.50	1.00	3.04	0.009	VL
Honda	0.17	1.00	6.84	0.007	VL
Duri	0.33	1.00	3.09	0.006	VL
Karet	0.25	63.26	150.34	0.003	VL
Sunter Utara	0.25	1.00	0.92	0.000	VL

VH= Very High, M = Medium, L = Low, and VL = Very Low.

5 DISCUSSIONS

This study has proposed a new flood vulnerability index and an underlying graph-theoretic algorithm to comparatively assess and rank floodgates in coastal cities based on their exposure, susceptibility, and resilience to flooding. An application of the graph-theoretic algorithm to Jakarta's hydrological infrastructure produced index values that point to the most vulnerable floodgates in the network (see Table 1). "Sunter C" ranked as the most vulnerable floodgate in Jakarta, followed by "Ciliwung Lama", and "Kebon Baru" in that order. These three floodgates are characterised as having very high vulnerability and they represent the top 10% of Jakarta's floodgate infrastructure that are most likely to fail during a flood event. Hence, they should be prioritised during infrastructure maintenance and resource allocation for flood preparedness. To minimise their vulnerability to flood damage, limited resources can be judiciously spent on increasing their capacities by adding extra gate units where possible. No doubt, this outcome will be useful to coastal communities and external funding bodies who often require structured vulnerability assessment techniques that facilitate transparent and efficient decisions on where the limited resources allocated for flood mitigation should be invested.

Furthermore, because "Kebon Baru" does not currently have any upstream floodgate connected to it, its vulnerability can also be further minimised by improving its resilience through the installation of additional upstream floodgates. This way the pressure on "Kebon Baru" created by accumulating floodwaters can be controlled using the additional upstream floodgates, thereby reducing the probability of structural failure due to infrastructure fragility (Turpin et al. 2013). This demonstrates the usefulness of the adopted approach in highlighting locations where additional infrastructure may be required.

In addition, this approach to vulnerability assessment can be useful to decision makers who require justification for vulnerability attribution. For example, "Sunter C" ranked as the most vulnerable floodgate partly because of its huge exposure to 199.89km length of waterways as compared to very low vulnerability ranking floodgates like "Duri" and "Sunter Utara", which are only exposed to 3.09km and 0.92km length of waterways respectively. Another reason is because of its high susceptibility to flood damage, which can be attributed to the fact that it only has one gate unit compared to very low vulnerability ranking floodgates like "Sunter Utara", "Honda", and "8" which has 4, 6, and 8 gates respectively. Similarly, the very low vulnerability of 56.67% of Jakarta's floodgates is mainly due to their low exposure to flood waters when compared to other floodgates in the city. However, in the case of "Karet", it is its high resilience attained through redundancy provided by connected upstream floodgates that makes it rank as a very low vulnerability floodgate. No doubt, such detail of vulnerability attribution can leave clues as to what actions can be taken to minimise infrastructure vulnerability.

6 CONCLUSION

This paper has proposed a graph-based network approach for measuring hydrological infrastructure flood vulnerability index (HIFVI), using the concepts of exposure, sensitivity and resilience. An application of the proposed method produced HIFVI values for Jakarta's floodgates, demonstrating its usefulness in ranking and comparing the vulnerability of hydrological infrastructure components to flood damage in coastal cities. The results will facilitate transparent and efficient targeting of limited resources towards routine maintenance, future investments and upgrades to the flood control infrastructure within coastal cities situated in developing nations. Importantly, the method was found to be useful in highlighting locations where additional infrastructure may be required to improve resilience to flooding. This will enable coastal cities in developing nations plan for more resilient future and to improve the outcome of their structural response to flood hazards.

One limitation of this study is the absence of additional data to improve the quality and reliability of the technique. This issue can be addressed by taking advantage of the graph theory feature, which allows for incremental integration of additional data into the network model as they become available in the future (Bunn et al. 2000). Hence, future study will seek to improve the quality and reliability of the technique by introducing additional data related to hydrological infrastructure components (e.g. asset age, flood height capacity, maintenance and failure history). Moreover, the impact of flood waters on the hydrological infrastructure can be more accurately accounted for if additional data such as elevation, width, depth, roughness, and flow rate of river channels are available.

ACKNOWLEDGEMENT

The authors would like to acknowledge the support of this project by the Australian National Data Service (ANDS) through the National Collaborative Research Infrastructure Strategy Program, as well as the University of Wollongong Global Challenges Program.

REFERENCES

Balica, S. F. Wright, N. G. & van der Meulen, F. 2012. A flood vulnerability index for coastal cities and its use in assessing climate change impacts, *Natural Hazards* ***64***, 73-105.

Barr, S. Holderness, T. Alderson, D. et al. 2012. An open source relational database schema and system for the analysis of large scale spatial interdependent infrastructure networks, in *Proceedings of the 4th Open Source GIS Conference, Nottingham, UK* http://eprint.ncl.ac.uk/pub_details2.aspx?pub_id=196369.

Brecht, H. Dasgupta, S. Laplante, B. et al. 2012. Sea-level rise and storm surges: High stakes for a small number of developing countries, *The Journal of Environment & Development* **21**, 120-138.

Bunn, A. G. Urban, D. L. & Keitt, T. H. 2000. Landscape connectivity: a conservation application of graph theory, *Journal of environmental management* ***59***, 265-278.

Chang, S. E. & Shinozuka, M. 2004. Measuring improvements in the disaster resilience of communities, *Earthquake Spectra* **20**, 739-755.

Hall, J. W. Meadowcroft, I. C. Sayers, P. B. et al. 2003. Integrated flood risk management in England and Wales, *Natural Hazards Review* **4**, 126-135.

Li, H. 2003. Management of coastal mega-cities—a new challenge in the 21st century, *Marine Policy* **27**, 333-337.

Sadoff, C. Harshadeep, N. R. Blackmore, D. et al. 2013. Ten fundamental questions for water resources development in the Ganges: myths and realities, *Water policy* **15**, 147-64.

Tingsanchali, T. 2012. Urban flood disaster management, *Procedia engineering* **32**, 25-37.

Turpin, É. Bobbette, A. W. & Miller, M. 2013. Jakarta: Architecture+ Adaptation, Universitas Indonesia Press, Indonesia.

Proceedings of the International Conference on Smart Infrastructure and Construction
ISBN 978-0-7277-6127-9

© The authors and ICE Publishing: All rights reserved, 2016
doi:10.1680/tfitsi.61279.639

Conflation of data sources for the identification of residual ground related hazards and resilience planning for the Strategic Road Network of England

C.M. Power*[1], D. Wright[2], J. Mian[3] and D. Patterson[4]

[1] *Mott MacDonald, London, United Kingdom*
[2] *Atkins, Birmingham, United Kingdom*
[3] *Arup, Solihull, United Kingdom*
[4] *Highways England, Bristol, United Kingdom*
* *Corresponding Author*

ABSTRACT Highways England is responsible for the Strategic Road Network (SRN) of England, 4300 miles of Motorway and Trunk Road. Supporting this strategically vital transportation network are nearly 50,000 geotechnical assets. The generally good performance of the SRN during the poor winter of 2013/14 showed that the assets are still largely resilient. However, Highways England recognise that their geotechnical assets have been built over, or through, areas of ground related hazards and that measures designed to 'engineer out' these hazards are aging, and are subject to deterioration (as are the assets themselves). This paper summarises a suite of tasks undertaken by Highways England to increase their knowledge of the areas of residual geotechnical risk on the SRN. In line with the theme of the conference, this is very much a drive to transform the future of the SRN infrastructure through smarter information. It will describe analysis of the performance of geotechnical asset cohorts based on geological materials and slope morphology, allowing the hazard to the network posed by the construction of the assets themselves to be understood. Work to derive more value from geotechnical as-built and other report information will be described, as will work undertaken to determine the presence and condition of Special Geotechnical Measures (SGMs) in place to address ground related hazards. Pilot work undertaken to conflate these information sources to determine the areas of residual risk will be described. Finally, it will discuss how this increased knowledge will allow Highways England to improve the resilience of their network.

1 BACKGROUND

Highways England is the company responsible for the operation and maintenance of the Strategic Road Network (SRN) of England. The SRN comprises 4,300 miles of Motorway and Trunk Road, which carries a third of all road traffic and two thirds of all freight (Highways England, 2015a). The safe and reliable operation of the SRN is essential to the economy of the United Kingdom.

The SRN is a largely modern transport network, and as such the 50,000 geotechnical assets that support it (significant earthworks and sections at-grade) have mostly been built to a high standard, with designs having been undertaken based on modern geotechnical practice. However, Highways England recognise that the SRN has been built over, or through, areas of ground related hazards (of natural and man-made origin), and that measures put in place to 'engineer' out these hazards are aging, and are subject to ongoing deterioration. In order to improve the understanding of any residual ground related risk that may affect the safe and reliable operation of the SRN, Highways England have a programme of tasks in progress to better understand where vulnerabilities may lie, allowing targeting of works to increase the resilience of the network. This improved asset knowledge will also enable more effective asset management and can be used to accelerate the assessment of ground related risk in the development of schemes as part of the Capital Investment programme.

2 MANAGEMENT OF RESIDUAL GROUND RELATED RISK

Figure 1 sets out the framework for the programme of tasks being undertaken by Highways England, to improve and validate the knowledge of the residual risks posed to the SRN from the ground and buried elements constructed within the ground. The over-arching aim of these tasks is to improve the resilience of the geotechnical assets managed by Highways England, by understanding the presence, and location, of any un-addressed ground related hazards. Further, through understanding how these hazards, or the associated methods of mitigation, may be impacted by extreme weather events, particular vulnerabilities can be addressed.

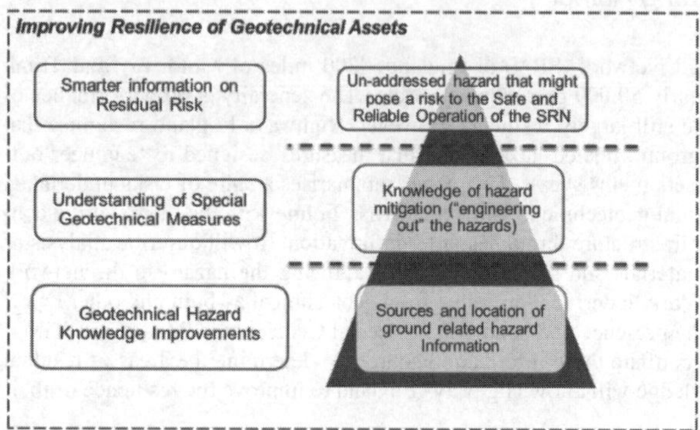

Figure 1. Management of residual ground related risk on the SRN

3 GEOTECHNICAL HAZARD KNOWLEDGE IMPROVEMENTS

The SRN faces ground related hazards from:

- Natural geological and hydro-geological hazards (for example dissolution features, see Figure 2)
- Man-made hazards in the ground (for example locations of historical mining)
- Hazards posed by the performance of the earthworks constructed to form the SRN (for example embankment failures undermining the highway infrastructure)

Figure 2. Sinkhole development on the SRN

To understand and assess these hazards, Highways England has undertaken, and continues to undertake a range of key activities:

- Development and support of an online Geotechnical Data Management system (HAGDMS), which is accessed by over 1000 users and holds a technical archive of over 200 mapping layers, 14,000 online geotechnical reports and regularly updated inspection records for 50,000 geotechnical assets (the Geotechnical Asset Database, GAD),
- An assessment of the performance of their geotechnical assets, based on cohorts of similar geological materials and slope morphology, to produce a slope hazard rating mapping layer available on HAGDMS (see Figure 3). This information allows proactive decisions to be made on the management of particular geotechnical assets, based on the performance of similar assets elsewhere on the network,
- Assessment of ground hazard information available from third party, national data providers to assess their relevance and suitability for use in relation to the SRN.

Whilst many third party datasets are very useful for providing context, they are often generalised to a national scale and can miss detail specific to the SRN. Often, a far greater density of information is available for the corridor of the SRN (which is narrow at a national scale), and much of this information is held within the HAGDMS. A recently completed Highways England task has concentrated on developing techniques to 'data mine' ground related hazard information from the technical archive and make it more readily available to users.

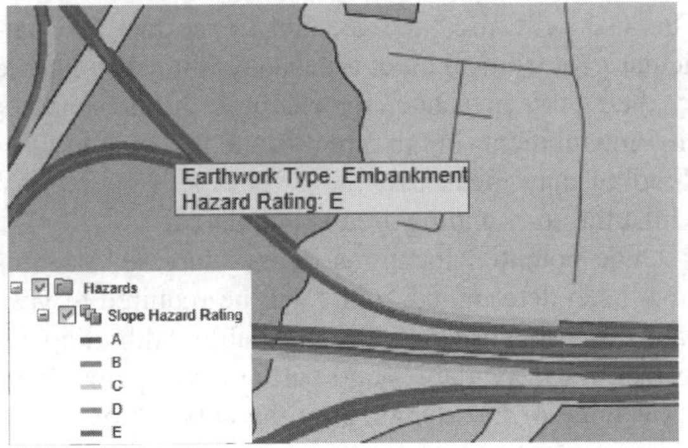

Figure 3. Slope Hazard Rating information available to users of the HAGDMS (E is the highest hazard rating)

4 UNDERSTANDING OF SPECIAL GEOTECHNICAL MEASURES

Work is also in progress to further the understanding of the type, location, purpose and performance of Special Geotechnical Measures (SGMs) on the network. These features are defined as measures over and above traditional earthworks construction required to, mitigate geotechnical risk associated with ground related hazards, or to remediate geotechnical defects that may have resulted from the presence of geo-hazards (examples include band drains and a basal drainage layer installed beneath an embankment during construction due to the presence of soft, compressible soils, or an area of slope failure on a pre-existing embankment reinstated using lime-stabilised arisings). Similar techniques implemented to facilitate widening or improvements are also classified as SGMs, albeit with a different purpose.

Whilst some of these features are visible, and their presence and extents are therefore captured as part of the routine earthworks inspections undertaken around the network, a significant proportion cannot be identified through inspections, and the presence of such features is being established through a process of data-mining over 14,000 technical reports, along with inventory and condition information, that are held by Highways England in order to extract relevant attributes.

In order to identify information relating to the presence of an SGM, a number of tasks have been completed;

- A hierarchy of terms has been developed to reflect the various types and sub-types of SGM utilised on the SRN. These terms are utilised as part of the data-mining to search the historical PDF reports and inspection records to identify those which contain relevant information. Highways England is liaising with other infrastructure owners on common nomenclature for this hierarchy through the Geotechnical Asset Owners Forum (GAOF),
- A data hierarchy has been developed for additional information relating to SGM's to be populated, where possible, for those locations identified by the data mining,
- A comparative analysis is undertaken between the locations identified from the historical reports and information from the visual inspections of the network. For visible SGMs, this allows confirmation of extents and avoidance of duplication.

This supplementary information will be added to the asset inventory, and correlated with current condition information, in order to enhance the information available to Highways England and those managing the network on their behalf.

Capturing the reasons for the inclusion of these SGMs in the construction of the network will also allow information on the location and type of measure to be correlated with the ground related hazards recorded by the Task described above. This assists in positively identifying existing hazard mitigation measures, and also raises awareness of locations where hazards exist with no recorded mitigation. The analysis will also highlight measures that have been introduced as a result of widening or other improvements to the network, rather than in response to the presence of a hazard.

Performance of SGMs will also be assessed through correlation with the recorded location and type of defects recorded in the GAD. Consideration will also be given to the design principles, standards and guidance that were in place at the time of SGM construction. Recognising that the condition of geotechnical assets is often influenced, or reflected, by the condition of other classes of asset (such as drainage and pavement), consideration will also be given to the use of data held by other disciplines in order to identify potential deterioration, along with the utilisation of third party data sets to provide supplementary information on performance trends of specific SGM's.

It is intended that this enhanced dataset, coupled with GIS mapping, will provide Managing Agents with locations of SGMs, or hazards, where current or potential performance issues may exist in order to supplement their knowledge and inform further development of risk-based asset management. The project will also develop a set of SGM data attributes to support future data collation and performance analysis.

Figure 4. Implementation of a Special Geotechnical Measure on the SRN

5 SMARTER INFORMATION ON RESIDUAL RISK

Whilst geotechnical assets of Highways England are largely robust, failures do occur (often with a causal linkage to another asset type, particularly drainage). Therefore, determination of areas of particular risk on the SRN is of high priority to Highways England. Through improved understanding of the type and geographical extent of ground related hazards (as outlined in Section 3), coupled with knowledge of Special Geotechnical Measures put in place to address these hazards (as outlined in Section 4), sections of the SRN where there may be unaddressed hazards can start to be determined, and further investigated.

An upcoming task that Highways England will be undertaking will seek to combine the developing knowledge base to begin to identify these locations. This task will also take account of the fact that particular geotechnical asset types may mitigate a hazard by their very presence, for example the presence of an embankment in an area prone to groundwater flooding may mean that the SRN is still operational whilst the surrounding area is inundated.

Once potential locations of un-addressed hazards have been determined, work will be required to validate these locations against available calibrating information (if available) and particularly against local knowledge of those managing the network on behalf of Highways England.

6 RESILIENCE TO SEVERE WEATHER EVENTS

Following the exceptional winter of 2013/2014, the Department for Transport review into the performance of the UK transport networks over this period (DfT, 2014) noted that:

"The SRN performed well during the winter of 2013/14, with most disruption from the weather being of relatively short duration with only minor delays."

Nonetheless, there is a recognised need to be able to anticipate the impact, reduce vulnerability and speed up recovery from geotechnical asset failures particularly during severe weather events and in anticipation of future climate projections. This will enable Highways England to be better prepared for the low frequency high impact events as well the more frequent weather events.

Resilience is defined by HM Government as "the ability of assets, networks and systems to anticipate, absorb, adapt to and/or rapidly recover from a disruptive event". Another important definition in this context is vulnerability "the degree of loss to each element should a hazard of a given severity occur" (Blaikie *et al.*, 1994). Shown in Figure 5 is an Adaption Framework for resilience infrastructure previously developed by the Highways Agency.

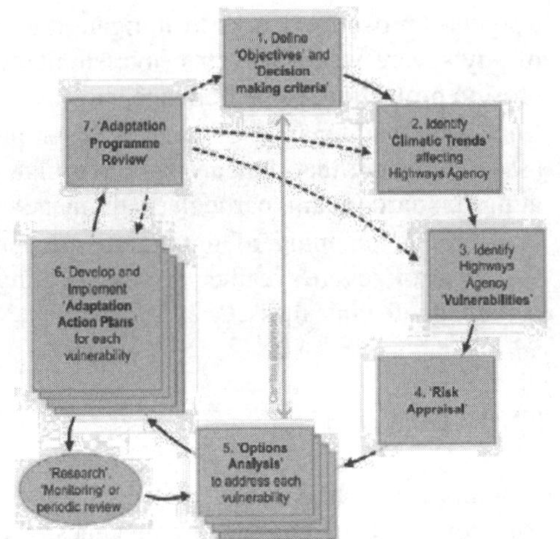

Figure 5. *Adaptation framework for resilient highway infrastructure (Highways Agency, 2011)*

Severe weather events are defined by Highways England as "any meteorological phenomena with the potential to endanger safe passage or cause disruption on the SRN".

Building on the improved use of data described in this paper, a prioritisation framework based on assessed vulnerability and risk for geotechnical assets in severe weather events will include the following components:

- It needs to align with the overall approach to severe weather resilience to avoid 'silo thinking' and recognise the interdependencies between asset groups, particularly drainage.
- An evaluation of 'tolerable risk' in terms of severe weather impacts on the SRN, which considers the criticality of the affected location, again in a cross asset context. This should recognise available diversionary routes, and the traffic affected.
- Definition of current and future exposure in terms of return periods of potential severe weather occurrences.
- Categorisation of geotechnical assets in terms of their 'coping capacity', i.e. different modes and levels of impact to severe weather events, considering inherent and man-made geohazards, and special geotechnical measures.

This prioritisation framework can be used both for targeted improvements (pre-event mitigation measures), and in event planning when severe weather events are forecast, such as re-routing of traffic.

Through the assessment of potential locations of un-addressed hazards from the ground outlined in this paper, this resilience prioritisation framework can be further targeted to the potentially most vulnerable sections of the SRN. These locations may be currently performing well, but may be vulnerable to future severe weather events.

7 IMPLEMENTATION

This paper sets out an ongoing body of work, much of it of a research and development nature. However, all the work is driven by a goal to produce improved ways of working that can be implemented by Highways England, and those working on their behalf, in the management of the SRN.

Improved knowledge about the types and locations of ground related hazards will be largely implemented through new datasets and functionality within the HAGDMS, making it readily available to over 1000 users, both within Highways England and their supply chain. Particularly important will be improved search facilities allowing users to target reference to particular hazard groups within the 14,000 reports available in the HE technical archive. Regular reviews of available third party datasets are valuable in ensuring that any national hazard datasets are assessed, and provided for users if appropriate. Highways England (and previously the Highways Agency), have for many years liaised with key third party providers (particularly the British Geological Survey, Environment Agency and Coal Authority) to provide feedback, and in many cases SRN specific data, to help improve the quality and applicability of available products.

As the knowledge of Special Geotechnical Measures develops the aim is for improvements to their management to be made possible. This might mean targeted inspection regimes for SGMs known to be deteriorating, improved proactive maintenance to prevent problems before they occur or perhaps targeted installation of monitoring to assess the performance of particular SGM locations or structures. The role of emerging and innovative monitoring techniques, a particular subject of this conference, may well have a significant role to play here, particularly where the costs of installed instrumentation can be significantly reduced (both in terms of initial installation cost and ongoing operation and maintenance costs).

As the understanding and validation of the particular locations of un-addressed ground-related hazards (and hence residual risk) develops, the real benefits of this body of work will become most apparent. Initial implementation will be through the entry of validated 'at risk' locations in the GAD, in accordance with the classification system set out in the HD41/15 Highways England Standard (Highways England, 2015b). From this, a range of improvements can be made, including:

- Improved Highways England geotechnical asset management policy and strategy
- Improved Geotechnical Asset Management Plans (GeoAMPs) prepared by Service Providers managing the SRN
- More targeted, risk-based inspection and monitoring of geotechnical assets
- Improved understanding of causal links to enhance the effectiveness of interventions and benefit whole-life costs
- Improved prioritisation of asset management interventions to proactively address areas of residual risk
- Improved understanding of the capability of assets to accommodate more onerous duties that may be imposed by future enhancement schemes
- Improved management of maintenance of other assets, particularly drainage, that significantly impact on the performance of the geotechnical assets

Greater understanding of the vulnerability of the geotechnical assets supporting the SRN to severe weather events (particularly locations of residual risk) will provide a further opportunity to strengthen the asset management processes of Highways England. As touched on in the previous section, this can be implemented in terms of:

- Targeted, proactive works to mitigate the impact of any severe weather events (inc. strengthening, erosion protection, slope drainage, etc)
- Enhanced maintenance of vulnerable or priority assets to ensure that they are kept at a high operating standard (again, particularly drainage)
- Pre-emptive planning of mitigation measures in the event of severe weather events affecting vulnerable geotechnical assets

ACKNOWLEDGEMENT

The authors would like to acknowledge considerable contributions of their combined colleagues, across a range of organisations, to the body of work described in this paper. The work described in the paper is funded through the Highways England Innovation Programme.

REFERENCES

Blaikie, P., Cannon, T., Davis, I. & Wisner, B. 1994. *At Risk: Natural Hazards, Peoples Vulnerability, and Disasters*. Routledge, London.

Department for Transport. 2014. *Transport Resilience Review: A review of the resilience of the transport network to extreme weather events*. Her Majesty's Stationery Office, London.

Highways Agency, 2011. *Climate Change Adaptation Framework*

Highways England, 2015a. Highways England website. https://www.gov.uk/government/organisations/highways-england/about

Highways England, 2015b. *HD41/15: Maintenance of Highway Geotechnical Hazards. Design Manual for Roads and Bridges*. Volume 4, Part 3. Her Majesty's Stationery Office, London.

Proceedings of the International Conference on Smart Infrastructure and Construction
ISBN 978-0-7277-6127-9

© The authors and ICE Publishing: All rights reserved, 2016
doi:10.1680/tfitsi.61279.645

Quantifying recovery for structures incorporating control

A. Wilbee*, F. Pena , S. J. Dyke, I. Bilionis, & P. Pandita

Purdue University, West Lafayette, Indiana, United States
* *Corresponding Author*

ABSTRACT Disasters, whether natural or manmade, strain the layered networks of our communities. When a crisis strikes, each facet of the community shatters and realigns, forming a new pattern of information from which critical decisions must be made. Though these decisions constitute the basis of the safety and continuing prosperity of the community, informing the potential paths forward requires intensive data processing. To choose among potential paths for rehabilitation, rapid methods for understanding and comparing options are needed. Over the past decade, researchers have been exploring this problem in terms of metrics for resilience. The frameworks proposed have provided several viewpoints from which to qualitatively and quantitatively understand the concept of resilience. To facilitate greater understanding of the application of resilience metrics to individual structures, this study seeks to apply and adapt these resilience frameworks to make decisions regarding the rehabilitation of a 20-story benchmark structure. Through the examination of the effects of decisions to control or rehabilitate a structure after a major event, the resulting performance and resilience of the structure is more clearly defined. Bayesian global optimization techniques are introduced to rapidly identify the design choices that will yield target performance levels.

1 INTRODUCTION

A community is built on a pattern of decisions that define its safety and productivity throughout its lifetime. These decisions range from the mundane to the extraordinary, but all are centered on improving community livelihood and sustainability. Perhaps the most critical choices are those made in response to disasters. In a disaster scenario a concentration of decisions are made that shape both the near-term and long-term future of the community and define its ability to return to, or even to exceed, its previous state.

This ability to bounce back after a disaster begins to shape the definition of a resilient community. Bruneau and Reinhorn complete it with the understanding that the resilience of a system is encapsulated in its ability "to reduce the chances of a shock, to absorb such a shock if it occurs … and to recover quickly after a shock" (2006). The resilience of the community as a whole is a multidimensional system, including technical, economic, organizational, and sociological components. Each of these components provides a certain level of resilience, and when considered together the contributions of the individual components set the resilience of the entire community (Bruneau et al., 2004).

Current research seeks to define these interactions in order to present a quantitative picture of resilience for communities. One of the groups making progress in this investigation is MCEER. In a 2004 paper on this topic, they introduced the concept of the quality of a system. As shown in Figure 1, quality is expressed as a time-varying function impacted by the health of the various system components and described by the robustness of the system, the redundancy in its elements, the resourcefulness of the users, and the rapidity of their response. The resilience of the society then becomes the time normalized integral of the quality curve (Bruneau et al., 2004).

Further work on this idea yielded the PEOPLES framework, which breaks a community into identifiable elements whose quality can be determined. Through the measurement of quality and uncertainty in these subsystems, a joint quality function can be formed which allows for the description of resilience for the entire community (Renschler et al., 2010). The principles of this framework were demonstrated in a case study to determine the resilience of a hospital network, performed by Cimellaro et al. By incorporating limited structural, social, and economic elements and simplifying the recovery portion of the quality curve, which is heavily impacted by societal aspects, the study resulted in quantitative measures of resilience (2010).

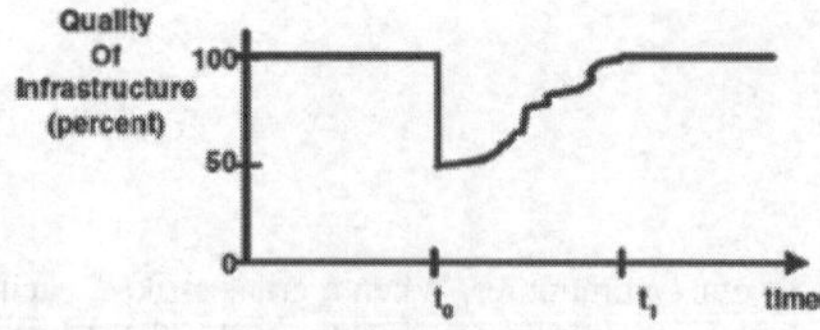

Figure 1. Quality Curve (Brunea et al., 2004)

Though this study demonstrates the potential for the framework, it also exhibits a weakness in characterizing the recovery portion of the quality curve. A variety of approximations have been proposed, however there is yet to be an accurate method for the determination of the recovery timeline. To propose an approach for tackling this issue, this study seeks to examine a component of the recovery path for a single structure through the study of post-disaster decision-making.

After a seismic event, decision makers have many choices. For example, they may have limited resources and need to consider carefully which structures should be rehabilitated and how should they be improved. Often the information and time which they have to work with is limited, particularly in the midst of a disaster. Informed decision making can be facilitated through the characterization of the structure at several points in its lifetime using probabilistic estimations of its behavior. Such a characterization may be accomplished through fragility curves.

Fragility is formally defined as the conditional probability of meeting or exceeding a given performance limit state at a certain hazard level (Taylor, 2007). The resulting curve serves as an important visualization tool as it is less costly than a full risk analysis, easy to understand, and provides an overall understanding of structural performance over a hazard range.

In recent years, researchers have worked to improve our use fragility curves through a variety of studies. For instance, Wen et al. provided a detailed analysis of the descriptive power of the fragility curve. His work is particularly helpful for identifying uncertainties inherent in the method, as well as for clarifying the importance of determining accurate limit states to describe the desired structural behavior (2004). Later, Jeong & Elnashai worked to establish the Parameterized Fragility Method, which creates a response database for the structure to reduce the time necessary to compute fragility functions (2007). This approach and other similar methods would be highly beneficial in rapid characterization of the structure in a post-disaster scenario. Additionally, Cha & Bai made use of fragility curves as an evaluation tool in reliability based design. After designing the structural control system using multi-objective genetic algorithms, they used the curves to perform a holistic evaluation of the response (2014).

Where these works use the fragility to evaluate a design, this study uses fragility as the design parameter. With pre-earthquake fragility as a baseline, a rehabilitation method can quickly be identified to return the structure to the approximate performance of its original state. The particular class of rehabilitation methods evaluated here are passive structural control strategies. Such strategies have been effective in protecting structures from natural hazards. Indeed, several examples of control strategies are in use, e.g. base isolation, tuned mass damping, and controllable viscous devices. Bayesian global optimization is applied as an efficient approach for establishing design options to meet target objectives. Thus, the evaluation of these strategies with respect to their rehabilitation ability results in an effective method to examine structural disaster recovery.

2 METHODOLOGY

At each period in the lifetime of a structure, fragility curves can be generated which detail the probability that certain damage states will be exceeded. These curves characterize the structure at discrete points in its lifetime, thus making it possible to form a comparison of the pre- and post-event states of the structure. Design of an appropriate passive control system for the structure is one potential avenue to minimize the differences between the fragility curves, thus returning the fragility of the structure to a reasonable semblance of its prior (undamaged) state. Generating these design options informs decision makers of viable options for rehabilitating the structure after an event has occurred, as shown in Figure 2.

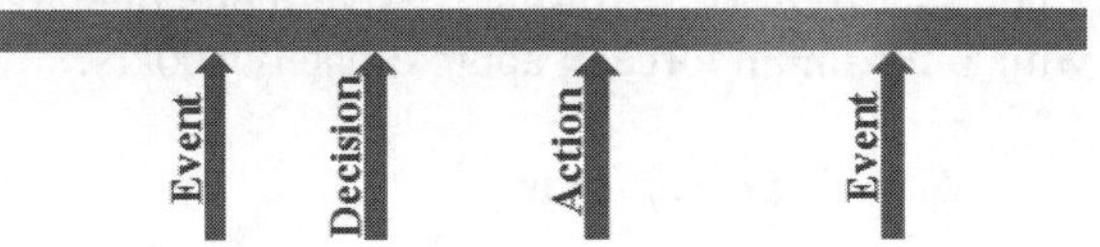

Figure 2. Reactionary Timeline for Disaster Scenario

However, making these decisions in a post-event scenario puts incredible stress and time constraints on the decision maker. By introducing uncertainty into the initial earthquake, the robustness of the controller design can be evaluated. Several alternative design/retrofit options can then be proposed immediately for that structure should an earthquake occur, thereby reducing reaction time, as shown in Figure 3.

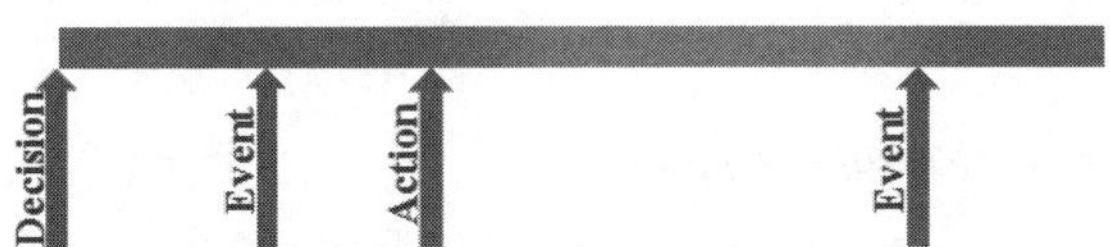

Figure 3. Planning Timeline for Disaster Scenario

The design of a retrofit option based on a targeted structural fragility is computationally expensive, as it involves the global minimization of a fragility-dependent loss function. In our problem the evaluation of the loss function at a given control input takes about 16 hours on a standard laptop. To deal with this issue, the efficient global optimization (EGO) strategy, as introduced by Jones, is adopted for this study (2001).

EGO starts by building a Gaussian process (GP) surrogate of the loss function based on an initial set of observed input-output pairs. In this case the pairs are the passive control strategies and the corresponding loss function evaluations (Sacks, et al., 1989; Rasmussen, 2006).

The Bayesian nature of the GP surrogate enables the calculation of the expected improvement (EI) the hypothetical simulation would yield on the currently observed minimum among the output set (Jones, et al., 1998). EGO proceeds by sequentially selecting the simulation that maximizes the EI. The function is evaluated at this chosen point and the new observation is added to the data set to obtain an updated surrogate model. With each iteration, the representation of the true objective function becomes more accurate and the optimization scheme approaches the global optimum. This greedy optimization technique provides fast and accurate results and requires fewer evaluations of the objective function than classical global optimization schemes, such as multi-start optimization and randomized search. The process continues until the value of the EI is below a pre-specified tolerance or if the computational budget is exhausted.

The implementation of EGO herein provides a feasible method to design a rehabilitation strategy within the time constraints of a post-disaster scenario.

3 CASE STUDY

A variety of methods to accomplish a return to original fragility can be conceived. In this study we consider two methods in particular, shown in Table 1. The first method involves the determination of a passive design that matches a single point on the original fragility curve of the structure with that of the rehabilitated structure. This guarantees a return in behavior at that specific hazard level, while leaving the structure to either increase or decrease in fragility at all other points. In this study, a point corresponding to a spectral acceleration of 0.334g is chosen, as it describes a relatively high level of hazard intensity. By choosing one point, it is ensured that the structure will be able to match its original fragility for larger amplitude earthquakes.

Table 1. Criteria for quantifying the rehabilitation strategy.

Criteria
1. Matching Point
$F_R(S_a^*\|Orig\ St) \approx F_R(S_a^*\|Rehab\ St)$
2. Minimizing Differences
$min\left\{\left[\int \left(F_R(S_a\|Rehab\ St) - F_R(S_a\|Orig\ St)\right)^2 dS_a\right]^{1/2}\right.$

The second method demonstrated in this study is to base the design on the minimization of the differences between the rehabilitated fragility curve and the original fragility curve. Though there is no guarantee that any specific fragility value will be matched, the overall fragility between the original and rehabilitated structures will be regulated, thus ensuring a general semblance of performance between the pre- and post-event states.

To understand the seismic fragility of a structure, it is necessary to model its response to a variety of input excitations. Here, the response of a 20 story benchmark structure to the SAC earthquakes is used to produce fragility curves. Several passive damping configurations exist through which the fragility of the structure can be modified. In this study we examine the achieved gains on an effective control configuration, which includes a total of 60 dampers distributed as: floors 1-7 have 4 dampers, floors 8-15 have 3 dampers, floors 16-18 have 2 dampers, and floors 19 and 20 have 1 damper, as in Wilbee et al. (2015).

3.1 20 Story Benchmark Model

A 20-story building, standing at 265 ft in height, was designed for the California region and used as the basis of a full-scale nonlinear benchmark problem (Ohtori et al., 2004). This steel structure has 5 - 20 ft bays in the N-S direction and 6 - 20 ft bays in the W-E direction. The perimeter of these bays forms the lateral resistance system, composed of steel moment-resisting frames (MRFs). One of the two N-S MRFs is modeled directly in the benchmark problem. More complete model details are available on the NEEShub database (https://nees.org/resources/2403).

The spectral acceleration of the structure is determined with respect to the first mode (0.261 Hz). Maximum inter-story drifts, normalized by the height of the floors, are recorded in each earthquake simulation to develop fragility curves.

For modeling the application of control forces, dampers are concentrated at a central location on each story, attached to the rigid floor systems both above and below.

3.2 MR Damper Model

The passive device used in this study is an MR damper operating in a passive mode. A nine parameter Bouc-Wen model is used to represent the damper to the benchmark structure (Yoshida & Dyke, 2004). To operate with the 20-story structure they established a damper model with a 1000kN capacity and a 10V control voltage, scaled from the identified model of a shear-mode prototype MR damper. Here, the parameters are chosen as α_a= 1.0872e5 N/cm, α_b= 4.9616e5 N/(cm-V), c_{0a}= 4.40 N-sec/cm, c_{0b}= 44.0 N-sec/(cm-V), A= 1.2, γ= 3 cm^{-1}, β= 3 cm^{-1}, and η= 50 sec^{-1}. For this study, the control voltage is a constant of 7.5V, ensuring passive mode operation with a maximum force of approximately 800kN.

3.3 Sample Earthquakes

The sample earthquakes used in this study are the SAC ground motions, which are time histories of synthetic earthquakes intended for case studies and trial applications (Woodward-Clyde, 1997). The earthquakes used were generated as being representative of the Los Angeles area (Friedman, et al. 2013; Ohtori et. al, 2004). The peak ground accelerations of these earthquakes range from 0.11g – to 1.33g, with spectral accelerations with respect to the first mode of the benchmark structure ranging from 0.19g – 0.538g. 60 SAC earthquakes were used in simulation to produce the fragility curves for this study.

3.4 Fragility Curves

According to Taylor, the formulation of a fragility curve is dependent primarily on two inputs, the structural response demand (D) and the hazard intensity (S) (2007; Wen et al., 2004). In this case, D corresponds to the maximum height normalized drift and S corresponds to the spectral acceleration of the structure with respect to the input earthquake. They are related according to *D*=*aS*^{b} where *a* and *b* are unknown coefficients determined by a logarithmic curve fit. With this information, the fragility can be calculated by

$$F_R = 1 - \Phi\left(\frac{\lambda_{CL} - \lambda_{D|S}}{\sqrt{{\beta_{D|S}}^2 + {\beta_{CL}}^2 + {\beta_M}^2}}\right) \quad (1)$$

where Φ is the standard normal distribution, λ_{CL} = ln(median drift capacity for a particular limit state), $\lambda_{D|S}$ = ln(calculated median demand drift given the spectral acceleration from the best fit power law line), $\beta_{D|S} = \sqrt{ln(1 + s^2)}$ [demand uncertainty], s^2 =

$\sum\left(ln(Y_i) - ln(Y_p)\right)^2/(n-2)$ [standard error], where Y_i and Y_p are the observed and power law predicted demand drifts, respectively, β_{CL} is the capacity uncertainty, and β_M is the modeling uncertainty. β_{CL} and β_M are taken as 0.3 here, due to the use of those values in similar studies (Wen et al., 2004; Taylor, 2007). In this study, the life safety limit state is examined (Cha et. al, 2014).

3.5 Results and Discussion

The preliminary design of control strategy for the structure is based on the response to synthetic earthquake #13 (EQ13), an earthquake of mid-level spectral acceleration. The control cases considered are defined according to the two criteria (Table 1). The transition from the reactionary analysis to the planning analysis is also demonstrated by changing the initial ground motion and verifying the robustness of the implemented structural control approaches.

The fragility curves shown in Figure 4 enable the visualization of the damage caused by the chosen ground motion. In this case, the translation of the initial into the damaged fragility curve exhibits a maximum increase of 20% in the probability of exceedance.

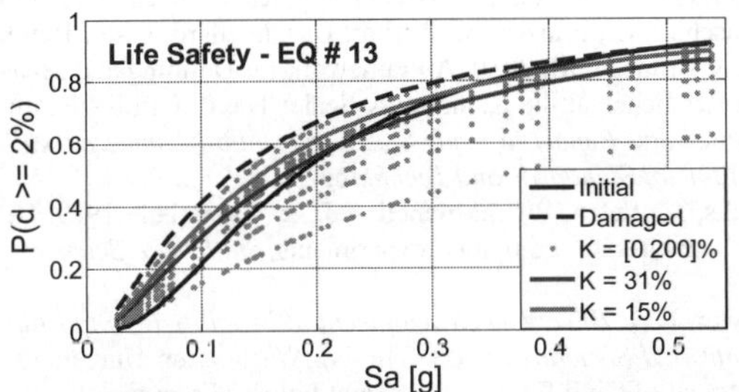

Figure 4. Variation of Fragility curves for Rehabilitated Structure

Using the first method, the point matching technique (Sa = 0.334g), simulations indicate the use of a gain (K) of 15% of the control force given by the damping strategy used in Wilbee et al. (2015). Identification of this design is performed using 10 iterations of trial and error. The results using the second method, the difference minimization result, are obtained through Bayesian global optimization (Locatelli, 1997). For this case, the control gain is 31%.

The robustness of the two control cases is examined by exchanging the initial ground motion (EQ13) with ground motions of different sizes ($S_a \in [0.03g - 0.54g]$). From Figure 5, it can be observed that both rehabilitation cases achieve good performance in terms of their respective criterion for both the small and medium earthquakes. However, EQ36, which produces the largest spectral acceleration of the set, exhibits almost complete limit state exceedance in both the damaged and rehabilitated states. The point matching criterion is summarized in Table 2. For each ground motion, the values on the left make reference to the fragility while those on the right reference the difference with respect to the initial structure. Similarly, the difference minimization criterion results are included in Table 3. The values on the left are loss functions and the ones on the right are normalized with respect to the initial fragility.

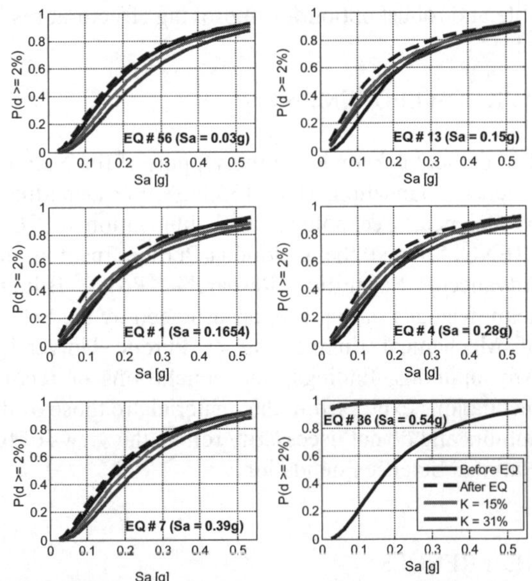

Figure 5. Variation of Fragility curves for Rehabilitated Structure during Different Events

4 CONCLUSION

Herein a method is proposed to establish passive control strategies to return a given structure to the performance of the pre-earthquake state. Bayesian global optimization is performed to identify the optimal retrofit strategy to meet a target fragility curve. The results lead to the classification of two scenarios. In the first, the event disturbing the structure is of mid to low range spectral acceleration, and the structure

Table 2. Results with Point Matching Criteria

K	Post EQ 13		Post EQ 1		Post EQ 4		Post EQ 7		Post EQ 36		Post EQ 56	
0%	0.82	-4.9%	0.80	-2.5%	0.82	-4.9%	0.79	-1.1%	0.99	-26.4%	0.79	-1.4%
15%	0.78	0.0%	0.76	3.5%	0.78	0.2%	0.75	4.4%	0.99	-26.4%	0.74	5.2%

Table 3. Results with Minimizing Difference Criteria

K	Post EQ 13		Post EQ 1		Post EQ 4		Post EQ 7		Post EQ 36		Post EQ 56	
0%	0.08	15.8%	0.07	13.7%	0.06	13.4%	0.02	3.5%	0.34	71.2%	0.02	3.4%
31%	0.04	7.7%	0.04	8.7%	0.04	7.3%	0.06	13.5%	0.34	70.9%	0.07	14.8%

suffers appreciable, but recoverable damage. In this case, the institution of a damping strategy for a single earthquake proves robust for both decision making schemes. In the second scenario, the initial earthquake damages the structure beyond repair. Thus, this study demonstrates that for situations in which rehabilitation is a possibility, control constitutes a viable and robust option for improving structural resilience.

ACKNOWLEDGEMENT

The authors acknowledge the support of the National Science Foundation, DGE-1333468, the Department of Science, Technology, and Innovation - COLCIENCIAS and the Colombia-Purdue Institute for Advanced Scientific Research. Prof. Bilionis acknowledges startup support provided by the School of Mechanical Engineering at Purdue University. Any opinions, findings, and conclusions or recommendations expressed in this material are those of the authors and do not necessarily reflect the views of the National Science Foundation.

REFERENCES

Bruneau, M., Chang, S., Eguchi, R., Lee, G., O'Rourke, T., Reinhorn, A., Shinozuka, M, Tierney, K., Wallace, W., & Winterfeldt, D. 2004. A Framework to Quantitatively Assess and Enhance the Seismic Resilience of Communities. *Proc., 13th World Conference on Earthquake Engineering.*

Bruneau, M, & Reinhorn, A. 2006. Overview of the Resilience Concept. *Proc., 8th National Conf. on Earthquake Engineering.*

Cha, Y.J., & Bai, J.-W. 2014. Seismic Fragility Estimates of Controlled High-Rise Buildings with Magnetorheological Dampers. *Proc., 10th U.S. National Conference on Earthquake Engineering.*

Cha, Y.-J., Agrawal, A.K., Phillips, B.M., & Spencer, B.F. 2014. Direct performance-based design with 200 kN MR dampers using multi-objective cost effective optimization for steel MRFs. *Engineering Structures*. **17**, 60-72.

Cimellaro, G.-P., Reinhorn, A., & Bruneau, M. 2010. Seismic Resilience of a Hospital System, *Structure and Infrastructure Engineer-ing.* **6:1-2**, 127-144.

Friedman, A.J., Dyke, S.J & Phillips, B. 2013. Over-driven Control for Large-scale Magnetorheological Dampers, *Smart Materials and Structures,* doi:10.1088/0964-1726/22/4/045001.

Jeong, S.-H. & Elnashai, A.S. 2007. Probabalistic Fragility Analysis Parameterized by Fundamental Response Quantities, *Engineering Structures*, **29**, 1238-1251.

Jones, D.R. 2001. A taxonomy of global optimization methods based on response surfaces, *J. of global optimization* **21:4**, 345-383.

Jones, D.R., Schonlau, M., & Welch, W.J. 1998. Efficient global optimization of expensive black-box functions, *J. of Global optimization* **13:4**, 455-492.

Locatelli, M. 1997. Bayesian algorithms for one-dimensional global optimization. *J. of Global Optimization* **10:1**, 57-76.

Ohtori, Y., Christenson, R.E., Spencer Jr, B. F., & Dyke, S. J. 2004. Benchmark control problems for seismically excited nonlinear buildings. *J. of Engineering Mechanics*, **130:4**, 366-385.

Rasmussen, C.E. 2006. Gaussian processes for machine learning.

Renschler, C., Fraizer, A., Arendt, L., Cimellaro, G.-P., Reinhorn, A., & Bruneau, M. 2010. A Framework for Defining and Measuring Resilience at the Community Scale: The PEOPLES Resilience Framework. *Report to the U.S. Department of Commerce National Institute of Standards and Technology.*

Sacks, J., Welch, W.J., Mitchell, T.J., & Wynn, H.P. 1989. Design and analysis of computer experiments. *Statistical Science*, 409-423.

Taylor, E.D. 2007. *The Development of Fragility Relationships for Controlled Structures* (M.S. Thesis), Washington University, Department of Civil Engineering, Saint Louis, Missouri.

Wen, Y. K., Ellingwood, B., & Bracci, J. M. 2004. Vulnerability function framework for consequence-based engineering. *MAE Center Report*, **4: 4**.

Wilbee, A.K., Pena, F., Condori, J., & Dyke, S.J. 2014. Fragility Analysis of Structures Incorporating Control Systems, *Proc.,6th Intl Conf. on Advances in Exper. Structural Engr. & 11th Intl Wksp on Advanced Materials and Smart Structures Technology.*

Woodward-Clyde Federal Services. 1997. Develop Suites of Time Histories: SAC Joint Venture Steel Project Phase 2. http://nisee.berkeley.edu/data/strong_motion/sacsteel/draftreport.html

Yoshida, O. & Dyke, S.J., 2004. "Seismic Control of a Nonlinear Benchmark Building Using Smart Dampers," *J. of Engineering Mechanics* Vol. 130, N. 4, pp. 386-392.

Proceedings of the International Conference on Smart Infrastructure and Construction
ISBN 978-0-7277-6127-9

© The authors and ICE Publishing: All rights reserved, 2016
doi:10.1680/tfitsi.61279.651

What should future design standards in the construction industry look like? The need for new value propositions

M. Angelino[*1,2], C. Taylor[1] and S.R. Denton[2]

[1] *University of Bristol, United Kingdom*
[2] *WSP Parsons Brinckerhoff, Bristol, United Kingdom*
[*] *Corresponding Author*

ABSTRACT Design standards play a fundamental role in the construction sector, in particular as a means by which the acceptability of designs can be verified and in enabling research outcomes to be translated for widespread practical application. Research has highlighted the need to address some usability issues with the structural design standards currently in use across Europe. Recognising the major changes that can be expected in the construction industry in coming years, it is worthwhile going beyond responding to these immediate issues and asking what future designs standards should look like to meet the emerging vision for the construction sector. From a study of the main changes affecting the construction industry and the current role of design standards and challenges in their development and use, the need for new value propositions for future design standards is established, particularly to meet needs, interests and capabilities of users of these documents. It is hoped that this paper will stimulate discussion, provide a better understanding of the research challenges in developing users-orientated design standards, and provoke interest among the research community to further explore this research area.

1 INTRODUCTION

Design standards occupy a key role in the construction industry. The scale of construction projects means that there are few opportunities to prototype designs, so design standards are used to verify the adequacy of designs to meet fundamental requirements for safety, serviceability, durability and robustness. Furthermore, design standards serve as a vital means for research outcomes to achieve widespread adoption within the sector and so provide a key means for the research community to achieve impact from their work. In recent years there has been an increasing interest in the role of standards in enabling or hindering the efficient delivery of construction projects in the UK (see Industry Standards Group report, 2012).

There exist some long-standing challenges in the development and use of design standards. For example, in 1970 it was observed that "*Like life in general our codes seem to get more and more complicated*" (IStructE, 2000). Forty years later, similar comments can still be heard in meetings and workshops with practicing structural engineers, clients and industry bodies. Despite such long-standing challenges and the impact of design standards on the working practices on many hundreds of thousands of structural engineers across Europe alone, research into design standards themselves and how they can best meet users' needs has been very limited (Angelino et al., 2014).

Set against this context, the construction industry is increasing recognising the potential of digital and smart technologies (building information modelling, automation, new sensor technologies, etc.), to positively impact how structures will be designed, built, managed, operated and dismantled. Recognising the major changes that can therefore be expected in the sector, it is worthwhile exploring what future designs standards should look like and the role they should play to meet this emerging vision for the future of construction.

The purpose of this paper is therefore to explore the potential need for new value propositions underpinning future design standards in the construction industry and to promote wider debate in an under-researched field. Literature review, open discussions, interviews and brainstorming sessions with practitioners, clients, industry bodies and standard writers, has been employed to explore this issue.

A "value proposition" is the statement of value that an organisation, product or service is going to deliver. It focuses on the value and benefits that will be experienced by customers. In the present context a "value proposition" therefore expresses key benefits which make design standards valuable to users.

2 THE CONSTRUCTION INDUSTRY

2.1 *Why the construction industry places particular reliance on design standards*

It is not rare to hear that the construction industry is considered different to other industries. Slaughter (1998) argued that constructed facilities differ from other manufacturing activities for (i) the physical scale of the components and the completed facility, which in turn do not allow full-scale prototypes, (ii) the complexity of facilities and the number of systems interacting with each other as well as with the environment, (iii) the longevity of use, (iv) the temporary alliance among independent organisations concentrated on a single specific project. In addition, construction is inherently a (v) site-specific project-based activity (Cox and Thompson, 1997), which is based on the (vi) coordination of specialised and differentiated tasks at the site level (Shirazi et al., 1996) (vii) often requiring local adjustment at the construction site (Dubois and Gadde, 2002). As a result, construction activities are (viii) discontinuous in their nature (Segerstedt and Olofsson, 2010). This in turn leads to discontinuity of demand for projects, (ix) uniqueness of each project in technical, financial and socio-political terms, and (x) complexity of each project in terms of the number of actors involved (Skaates et al., 2002). Vrijhoef et al. (2001) also argued that (xi) the actors involved in the design project organisation have no common and clear understanding of what should be designed. Lastly, Dubois and Gadde (2002) also note that (xii) the temporary nature of the construction project does not promote learning; consequently, the ability to form cognitive structures favouring learning is severely restricted. In addition, they argue that (xiii) too little effort appears to be devoted to transmitted knowledge and experience from one project to another and (xiv) learning takes place at an individual level rather than an industrial level. It can be argued that other fields may have similar features. However, the coexistence of all these aspects makes the construction industry *sui generis* and design standards so important (Angelino, 2016).

2.2 *Changes in the construction industry*

Advances in technology are transforming the way structures are designed, built, managed, operated and dismantled. Building information modelling, automation, crowd-sensing and crowd-sourcing, new sensor technologies, off-site construction, diagnostic tools and new materials are all changing civil and structural engineering. Based on interviews with industry leaders, Denton and Skinner (2014) identified eight key trends linked to advances in digital technology that are impacting the construction industry. These include how new ways of working will be unlocked, data will increasingly flow through and between projects, productivity will increase, and also that it is the opportunities associated with whole life asset management that offer the biggest prize for asset owners.

The vision for the UK construction sector in 2025 (HM Government, 2013) recognises the changes that the construction industry is going to face and the importance of adopting innovative technologies in sensors and data management to full understand assets performance. It is acknowledged that: "*This will result in smarter designs, requiring less material, reducing carbon and needing less labour for construction, whilst still ensuring full resilience of the assets*." Similarly, "The vision for civil engineering 2025" published by the American Society of Civil Engineers (2006) envisaged a future where the civil engineering enterprise is focused on fast-track development and deployment of technologies, which employ results from information technology and data management to significantly improve how facilities are designed, engineered, built, and maintained.

These changes would require design standards to evolve accordingly and defining clear value propositions would be helpful to support these changes.

3 DESIGN STANDARDS IN THE CONSTRUCTION INDUSTRY

3.1 Definition and scope

Generally speaking, standards are "*documents, established by consensus and approved by a recognised body, that provide for common and repeated use, rules, guidelines or characteristics for activities or their results, aimed at the achievement of the optimum degree of order in a given context*" (ISO, 2004). In the construction industry a variety of standards exists, including standards for health and safety, quality management, environmental issues, codes of practice and management. Design standards are a specific sub-category. They are defined as technical documents that give provisions (i.e. statements, instructions, recommendations and requirements) to satisfy fundamental requirements of safety, serviceability, durability and robustness of both new and existing structures. For the purpose of this paper the term "design standards" refers to both design and assessment standards. Product, material, test and execution standards are outside the scope of this discussion.

3.2 Current role

Design standards play a key role in construction projects and are expected to serve a variety of purposes as summarised in Table 1. This table has been derived from literature, discussions with industry experts and clients, and workshops with practicing structural engineers. The list is not exhaustive, but demonstrates the inherent complexity in developing and using design standards, particularly due to the diverse and differing needs and expectations of the users of these documents. Having clear value propositions for future design standards would help preserve the core purposes of current design standards, as well as navigate these sometimes competing demands.

3.3 Challenges with current design standards

A number of challenges exist with current design standards. The following summary of issues is based on feedback from practicing engineers, clients and standards makers, derived from meetings, workshops and interviews. Developing specific value propositions would be helpful to recognise and address them.

Increases in technical requirements – In recent decades the construction sector has been affected by a tremendous increase in technical requirements developed by organizations acting at international, regional (European) and national level.

'Systems' of provisions – The cross-references among standards stemming from different normative sources can cause overlapping (and sometimes conflicting) requirements or gaps, which may increase the risk of misapplication. This also causes problems in accessibility and navigation.

Table 1. Purposes of design standards

1.	Assist competent designers in verifying structural adequacy (WP)
2.	Define and disseminate best practice (Shapiro, 1997)
3.	Reflect and shape the exercise of practitioner judgment (Shapiro, 1997)
4.	Codify and share technical knowledge taking account of practitioner judgment (Shapiro, 1997)
5.	Contribute to the delineation of an appropriate discretionary or judgement space for technological practice (Shapiro, 1997)
6.	Provide a common understanding regarding the design of structures between owners, operators and users, designers, contractors and product manufacturers (Roberts, 2010)
7.	Be a common basis for research and development in the construction industry (Roberts, 2010)
8.	Embody the most up-to-date research
9.	Provide a comprehensive system of provisions relevant to design (WP)
10.	Provide a concise system of provisions for design (WP)
11.	Enable economical design of structures (WP)
12.	Aid common design situations (WP)
13.	Enable innovative design (WP)
14.	Support sustainable design (WP)
15.	Be a practical knowledge base for structural design which can be trusted by industry (WP)
16.	Give technical provisions which enable to strike the right balance between design costs, construction costs and maintenance costs (OD)
17.	Increase the competitiveness of civil engineering firms, contractors, designers and product manufacturers in their global activities (Roberts, 2010)
18.	Provide a framework for achieving economies, efficiencies and interoperability (BSI)
19.	Provide technical contents which are consistent with the regulatory and legal framework they exist in (OD)
20.	Support public policy objectives (BSI)
21.	Ensure consistency in design approaches (OD)
22.	Handle uncertainty and risk (OD)

OD = Open discussion with industry experts and clients

WP = Workshop with practicing structural engineers

More complicated technical provisions – There has been an increasing call for practicing structural engineers to apply more complicated technical provisions in order to meet societal demands, achieve greater consistency of structural reliability, or enable better economy and/or sustainability.

Unduly prescriptive provisions – There seems to be a general perception that design standards are unduly prescriptive. Over-prescription can inhibit innovation and the efficiency of construction projects. A move towards performance-based (or outcome-based) standards has been advocated as beneficial by many voices. However, for design standards this is not straightforward. It requires the performance requirements to be clearly defined, which for construction projects are often complex, multi-layered and interdependent. It would also demand major changes in aspects of procurement and the transfer of risk (liability) between parties for long term asset performance.

Inconsistent use of standards – The UK Industry Standards Group (2012) recognised that "*inconsistent approaches to the application of technical standards lead to inefficient, bespoke solutions that block innovation, add to whole life costs and fail to deliver the required performance and service improvements*". Similarly, Wilson, Grose and Rawlings (2015) have acknowledged that "*inefficient and inconsistent use of codes and standards can hamper effective delivery of infrastructure projects*".

'Soft' issues – Standards are the product of a socially constructed, multi-stakeholders process aimed at defining agreed technical solutions between all those likely to be influenced by them. A variety of different stakeholders are involved in the standardisation process, including designers, regulators, industry bodies, clients, contractors, professional institutions, research organisations, universities, learned societies, educators, software producers and lawyers. Standardisation should be a transparent and open process of cooperation; however, some authors suggest that this is not always the case. For instance, Weiss (1991) suggested that, while the stated goal of developing a standard may be adopted by most of the committee members, other – secondary – goals may also exist and may be in conflict. Allen (1992) argues that different stakeholders may have competing views and they might contribute negatively to the standardisation process, thus affecting the usability of standards and their success. Nethercot (2012) highlights the tensions that exist between the aspirations of practitioners for greater economy, simplicity and all-embracing provisions and the desire of the research community for technically advanced provisions; the latter can become an "*exercise in vanity rather than the guarantee of intellectual rigour*". The risk is that the standardisation process can become "*a political or economic power game although the topics discussed are mostly of a purely technical nature*" (Takahashi and Tojo, 1993).

Limited focus on users' needs - There is an urgent call to have more users-orientated design standards. However, research into how design standards can best meet users' needs appears to be very limited. The importance of considering users and how they use standards has been explicitly acknowledged at European level in the work on the second generation of Structural Eurocodes, where a major focus will be on improving their usability (CEN/TC 250, 2015). Likewise, in the US the purpose of the recent review of the ACI 318 Building Code has been "*to provide a more user-friendly backbone for design*" (Poston & Dolan, 2012). Similarly, a consultation has been recently carried out to explore the view of different stakeholders on how usability, structure and content of the DMRB should be improved (CIHT website, 2015).

Tensions – Some inherent tensions exist in the development of design standards. The recognition of tensions is longstanding, as demonstrated by the comments to the debates held at the Institution of Civil Engineers and the Institution of Structural Engineers in the early 1980s stemming from the question "*How should rules of structural design be codified?*" formulated by Moffatt and Dowling (1981) and from the statement "*Simple codes can stifle structural technology*" proposed by Sunley and Taylor (1982). Examples of tensions include:

- the need to avoid technical provisions that are unduly complex, whilst also not inhibiting experts from applying their knowledge and deploying advanced method of analysis;
- the attractiveness of stability versus the drive for the introduction of new approaches;
- the aspiration to address all users' needs whilst not making the standard more complex;

- the aspiration not to inhibit innovation, yet also provide clear provisions for common design situations and also help ensure long term performance and appropriate whole life cost.

4 VALUE PROPOSITIONS FOR DESIGN STANDARDS

The previous sections suggest the need to define clear value propositions for future design standards. As far as the authors are aware, the first explicit attempt to define value propositions for design standards in the construction industry has been made at European level for the development of the second generation of Structural Eurocodes. The ambition of the European Committee responsible for the development and maintenance of the Structural Eurocodes (CEN/TC 250) is "*to create a more user-orientated suite of design standards that are recognised as the most trusted and preferred in the world*". To attain this vision, one of the guidelines explicitly recognised the importance of identifying the main categories of users and a primary audience to target drafting efforts. For each of these categories specific statements of intent to meet their needs have been developed and unanimously agreed (CEN/TC 250, 2015).

Recognising this need for clear value propositions guiding future design standards and building upon the success of this European work, what extra factors should we bring to our thinking about standards to develop future design standards without extending their scope significantly?

Anderson et al. (2006) recognise that developing an effective value proposition requires a detailed understanding of the customer's requirements, preferences and – most importantly – of their priorities to avoid the pitfalls of the pure "benefits assertion" and "value presumption", and to deliver the greatest value to target customers. The value propositions for future design standards should: (i) retain and reinforce the core accomplishments and successes of current design standards; (ii) embrace currently recognised needs and overcome current challenges in developing and using design standards; (iii) take account of future changes anticipated in the construction industry.

It is not the aim of this paper to provide definitive value propositions, rather to provide themes that could potentially underpin the value propositions for design standards. These themes have been presented below in three groups.

Themes for established value propositions of design standards (existing core accomplishments)

1. **Ensure safety, serviceability, durability and robustness whilst providing economy**
2. **Assist competent designers in verifying structural adequacy**
3. **Codify and share technical knowledge**
4. **Ensure consistency in design approaches**

Themes for value propositions of design standards to address current recognised needs and challenges

5. **Support users' ability to form cognitive structures favouring critical application of design standards and learning**
6. **Improve clarity and understandability**
7. **Improve accessibility and ease of navigation**
8. **Provide appropriate freedom for innovation**

Themes for value propositions of future design standards to support expected changes in the construction sector

9. **Better enable performance data from monitoring to inform the design of structural modifications or rehabilitation schemes**
10. **Better enable structural verifications to be incorporated into digital models**
11. **Adapt the format of design standards to be better suited to digital working and able to be updated more rapidly**
12. **Present design assisted by testing more extensively, supporting innovation in modularisation and off site manufacture**

5 CONCLUSIONS

The construction industry is undergoing some profound changes. Digital and smart technologies are increasingly transforming the way structures are designed, built, managed, operated and dismantled. Coupled with that, future assets are expected to be more resilient, sustainable and adaptable.

Design and assessment standards play a key role in construction projects and will continue to do so. However, from discussions, interviews and brainstorming sessions with practitioners, clients, industry bodies and standard writers, it emerged that current design standards could perform better.

To guide the necessary evolution, the authors argue that defining clear value propositions would be helpful. These express key benefits, which make design standards valuable to users. It is proposed that these value propositions should not only address issues and needs that have already been recognised, but should also consider the profound changes that can be expected in the construction sector.

In an effort to stimulate debate, twelve themes that could potentially underpin future design standards have been proposed. These have been derived by drawing together themes for established value propositions of design standards, those responding to known challenges for design standards and those relevant to a future and 'smarter' construction industry.

ACKNOWLEDGEMENT

This work was supported by the Systems Centre of the University of Bristol, the EPSRC funded Industrial Doctorate Centre in Systems (Grant EP/G037353/1) and WSP Parsons Brinckerhoff.

REFERENCES

Allen, D.E. (1992) "The role of regulations and codes" in *Engineering Safety*, David Blockley, Ed., McGraw-Hill Book Co., London, UK, pp. 371-383

Anderson, J.C., Narus, J.A., van Rossum, W. (2006). Customer Value Propositions in Business Markets, *Harward Business Review*, 84 (3), pp. 91-99.

Angelino, M., Agarwal, J., Shave, J., Denton, S. (2014). The development of successful design standards: understanding the challenges. *37th IABSE Symposium*, Madrid.

Angelino, M. (2016), Doctoral thesis in preparation. University of Bristol, Bristol, UK

American Society of Civil Engineers (2006), The vision for civil engineering 2025, pp. 1-103, doi: 10.1061/9780784478868.001.

CEN/TC 250 (2015). *CEN/TC 250 Position paper on enhancing ease of use of the Structural Eurocodes*.

CIHT website (2015). http://www.ciht.org.uk/en/media-centre/news/index.cfm/highways-england-review-of-dmrb-and-tss

Cox, and Thompson (1997). 'Fit for purpose' contractual relations: determining a theoretical framework for construction projects, *European Journal of Purchasing & Supply Management*, Volume 3, Issue 3, pp. 127–135.

Denton, S.R. and Skinner, R (2014). *Digital Life - Digital Legacy, Realising the Digital Potential of Infrastructure Projects*. Parsons Brinckerhoff, ISBN 978-0-9933366-0-7

Dubois, A., and Gadde L.E. (2002). The construction industry as a loosely coupled system: implications for productivity and innovation, *Construction Management and Economics*, Volume 20, Issue 7, pp. 621–631.

HM Government (2013). Construction 2025 - Industrial Strategy: government and industry in partnership, UK.

Industry Standards Group, 2012. *Specifying Successful Standards: an Industry Enquiry into how Standards and Specifications can Enable the UK to Innovate, Lower Costs and Improve Whole Life Value of our Infrastructure Assets*. ICE, London, UK.

IStructE (2000). Only good code – old code! *The Structural Engineer*, 78(8).

ISO/IEC GUIDE 2:2004, Standardization and related activities - General vocabulary.

Moffatt, K.R., & Dowling, P.J. (1981). Discussion: How should rules of structural design be codified? *Proceedings Institution of Civil Engineers*, 70(1), 523-556.

Nethercot, D. (2012). Modern Codes of Practice: What is Their Effect, Their Value and Their Cost? *Structural Engineering International*, 22(2), 176-181

Poston, R.W., & Dolan, C.W. (2012). The Framework of the 2014 American Concrete Institute (ACI) 318 Structural Concrete Building Code. *Structural Engineering International*. 2, pp. 261-264

Roberts, J. (2010). The essential guide to Eurocodes transition. British Standards Institute, London

Segerstedt, A., and Olofsson, T. (2010). Supply chains in the construction industry, *Supply Chain Management: An International Journal*, Vol. 15 Iss: 5, pp. 347 – 353.

Skaates, M.A., Tikkanen, H. and Lindblom, J. (2002). Relationships and project marketing success, *Journal of Business & Industrial Marketing*, Vol. 17 Iss: 5, pp. 389 - 406

Shapiro, S. (1997). Degree of freedom: The Interaction of Standards of Practice and Engineering Judgment. *Science, Technology, & Human Values*, 22(3), pp. 286-316.

Shirazi, B.; Langford, D. A.; Rowlinson, S. M. (1996). Organizational structures in the construction industry, *Construction Management and Economics*, Volume 14, Number 3, pp. 199-212(14).

Slaughter, E.S. (1998). Models of construction Innovation. *Journal of construction engineering and management*, pp. 226-231.

Sunley, J.G., & Taylor R.G. (1982). Open discussion. Simple codes can stifle structural technology. *The Structural Engineer*, 60A(10), 320-333

Takahashi, S., & Tojo A., (1993). The SSI story What it is, and how it was stalled and eliminated in the International Standardization arena. *Computer Standards & Interfaces*, 15, 6, pp. 523–538.

Vrijhoef, R., Koskela, L. and Howell, G. (2001) Understanding construction supply chains: an alternative interpretation, *9th International Group for Lean Construction Conference*, National University of Singapore.

Weiss, Martin B.H (1991). *The Standards Development Process: A View from Political Theory*. Technical Report. School of Library and Information Science, University of Pittsburgh, Pittsburgh, PA .

Wilson, Grose and Rawlings, 2015. Improving infrastructure delivery though better use of standards. *Civil Engineering*. Volume 168 Issue CE1

Proceedings of the International Conference on Smart Infrastructure and Construction
ISBN 978-0-7277-6127-9

© The authors and ICE Publishing: All rights reserved, 2016
doi:10.1680/tfitsi.61279.657

Development of resource-efficient tunneling technologies – Results of the European research project DRAGON

Robert Galler

Head of Chair of Subsurface Engineering, Montanuniversität Leoben, Austria, Coordinator of the EU-Project

ABSTRACT The amount of different minerals in tunnel excavation material varies in a large range. Therefore DRAGON investigates in industrial clients who would be interested in raw materials extracted from underground construction sites. Parallel to the development of advanced online technologies for analysing the excavated materials DRAGON was looking for important information regarding requirements for raw materials used in industrial processes, which are grain size distribution, mineralogical composition, geochemistry as well as different water content and water absorption properties. Tunnel construction – not only with TBMs – is already a highly industrialized and technologically demanding process, with quite high penetration rates. So it was an important condition of both from the machine manufacturer as well as from the construction companies that the advance rate is not affected by new developments of DRAGON. Furthermore a goal was to assess the DRAGON project objectives with the wider strategic objectives of the EU policy and to highlight any areas where barriers may exist to guarantee the benefits of applying the DRAGON technology.

1 EUROPE 2020 STRATEGY

Europe 2020 is the European Union's ten-year growth strategy and is about delivering growth that is smart, through more effective investments in education, research and innovation and sustainable, thanks to a decisive move towards a low-carbon economy. Of relevance to the DRAGON project were the targets related to climate change and energy sustainability. The ambitious targets that were set by the EU have been adopted into the Europe 2020 strategy which are a reduction in EU greenhouse gas emissions of at least 20% below 1990 levels, 20% of EU energy consumption to come from renewable resources and a 20% reduction in primary energy use compared with projected levels, to be achieved by improving energy efficiency. Collectively these are known as the 20-20-20 targets. A resource-efficient Europe is one of seven flagship initiatives of the Europe 2020 Strategy and aims to support the shift towards a resource-efficient, low-carbon economy to achieve sustainable growth. This recognises that natural resources underpin our economy and our quality of life and that continuing our current patterns of resource use is not an option. Increasing resource efficiency is the key to secure growth and jobs for Europe and will potentially bring major economic opportunities, improve productivity, drive down costs and boost competitiveness. This flagship initiative has moved the extraction and use of natural resources into the centre of the political agenda of the European Commission. The strategy establishes resource efficiency as the guiding principle for EU policies on energy, transport, climate change, industry, commodities, agriculture, fisheries, biodiversity and regional development. The flagship initiative connects policies related to resources such as the Roadmap for a resource efficient Europe [EC 2011b] and the Raw Materials Initiative [EC 2009c].

2 OBJECTIVES OF DRAGON

The DRAGON project – Development of Resource-efficient and Advanced Underground Technologies – intends to improve resource efficiency in tunnelling and other underground construction processes by providing the excavated material as a raw material for the construction site as well as for other industrial sectors which are using primary mineral resources. Within DRAGON a system of automated chemical, physical and mineralogical online-analysis techniques all mounted on a bypass conveyor belt are developed. These analysing techniques are followed by a separation plant using recycling units like crushers, sieves, etc. (Figure 1). So the excavation material can be recycled and used directly onsite as construction material or by being transported to the particular industry sector. One aim of the project is to place the whole process, starting from the characterisation of the excavated material to its classification and processing, completely underground.

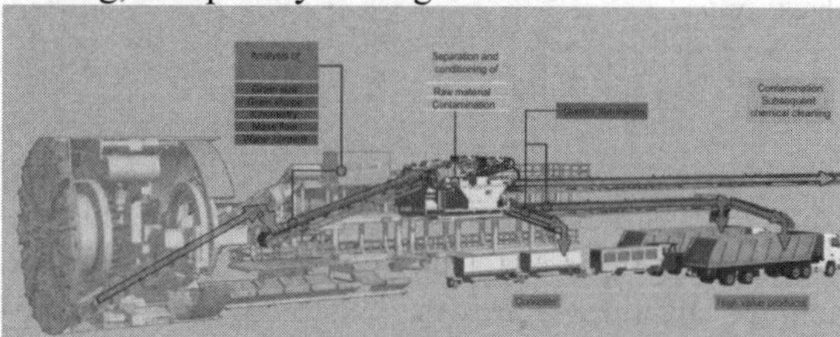

Figure 1. The concept of the DRAGON consortium

The results of the project will help to reduce or even eliminate any material deposits, thus aiming the production of zero waste originated from underground construction sites; furthermore environmental impacts like production of noise and dust will significantly be reduced by recycling the excavated material onsite. The DRAGON project applies the methodology of Life Cycle Assessment (LCA) according to ISO 14040/14044 (ISO 14040/14044, 2006) in order to include life cycle thinking into the project. Additionally the mass flows are analyzed by the Mass Flow Analysis (Brunner & Rechberger 2004) according to Baccini and Brunner (Baccini & Brunner 1991). The reason for applying these methods is to compare different scenarios of recycling or disposal of excavation material. This allows observation of the whole system in order to avoid any problem shifting into other parts of the environmental system. Substitution effects caused by replacement of primary material with excavation materials are identified and quantified. Considering these aspects, underground constructions in the future will much more contribute to resource efficiency and the reduction of CO2- emissions compared to today's state of the art.

3 OVERALL STRATEGY AND GENERAL DESCRIPTION

The current legal framework for the utilisation of excavation material is not yet completely satisfying and requires a change of rules where geochemical, geotechnical, mineralogical parameters are clearly specified for being allowed to reuse the excavated material as mineral product. The declared goal is to reach zero waste from underground construction sites, at least when rock properties and the demand allow it.

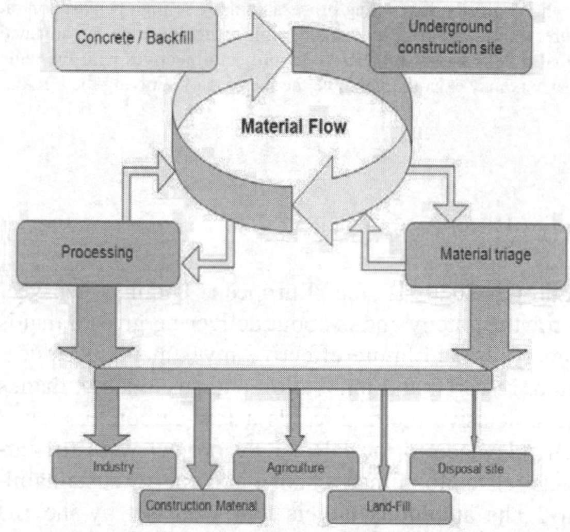

Figure 2. Material flow using new analyzing, separating and recycling techniques to reach zero waste

A legally binding rule that material extracted in tunnelling is preferred for use as long as its overall suitability can be demonstrated would not only create new raw materials potential. Companies in the mineral raw materials industry would obtain already crushed material at favourable rates, save their existing quarries and thus extend the lifetime of their companies. Following the results of the studies of Kündig et.al. (Kündig 1997) such a legally binding rule will get significantly important at least for European countries as mineable ressources get dramatically less in future (Figure 3).

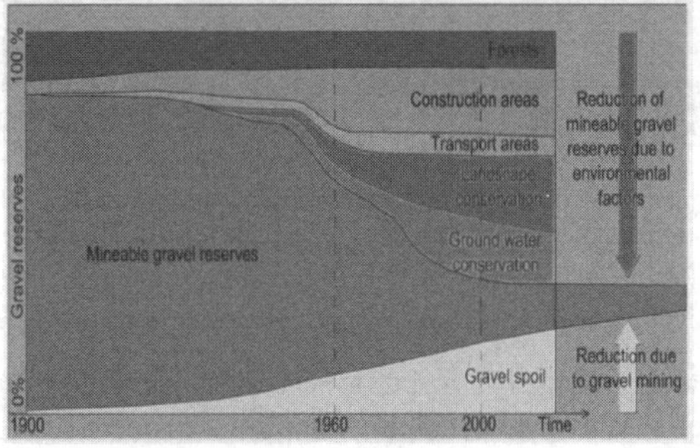

Figure 3. The conflicting utilisation claims of the resource gravel (according to Kündig et al., 1997, and Jäckli & Schindler, 1986, modified) (Geologische Bundesanstalt 2004)

To go a step ahead Austrian engineers together with the Federal Ministry of Agriculture, Forestry, Environment and Water Management wrote a first guideline called Verwendung von Tunnelausbruch published by the ÖBV – Österreichische Bautechnik Vereinigung (ÖBV, 2015)

4 IMPACT ON SOCIETY, ECONOMICAL AND ENVIRONMENTAL IMPACTS

The DRAGON prototype could also be of interest for mining companies which focus on resources of critical raw materials like gold, platinum, rare earths and others. Extracting the minerals containing such elements in a selective manner could facilitate the separation from dead rock and could allow the exploitation of minerals deposits which are currently not attractive enough due to their relatively low concentration of valuable elements. One of the main objectives of the DRAGON project is to contribute to the natural resource conservation within the European Union. Depending on the geological composition of the material it is possible to recycle up to 100% of the excavated material. The reduction of transport ways, the diminution of pollutants as well as the recycling of the excavation material possess a large environmental protection potential. The main expected outcome of the LCA (life cycle analysis) is to provide scientific evidence that the recycling of excavated tunneling material will result in more resource-efficient and more closed-loop related systems (even in the industry-related economy) in Europe.

The DRAGON consortium aims to establish a close relationship to the external surroundings (national environmental authorities; standardisation bodies etc.) of the project in order to implement and integrate the project results and findings quite smoothly within the specific regional, national and/or international environment. Main target of the DRAGON project is to act as Best Practise case how underground excavation material can be recycled as valuable material in diverse industrial processes and sectors. By finalizing the research project the DRAGON Consortium will get in close contact with diverse national & European stakeholders and will try to influence the European and national directives in order to guarantee the use of underground extraction material as new valuable input material for other industrial processes and industries.

5 TOPICS TO BE CONSIDERED

The dependencies of the individual fields of interest which have to be considered in order to recycle excavation material from underground construction sites are shown in Figure 4. The intention to continuously analyse the tunnel excavation material will result in an objective geological documentation of a previously unknown extent, which offers the potential to avoid disputes about the actually encountered conditions.

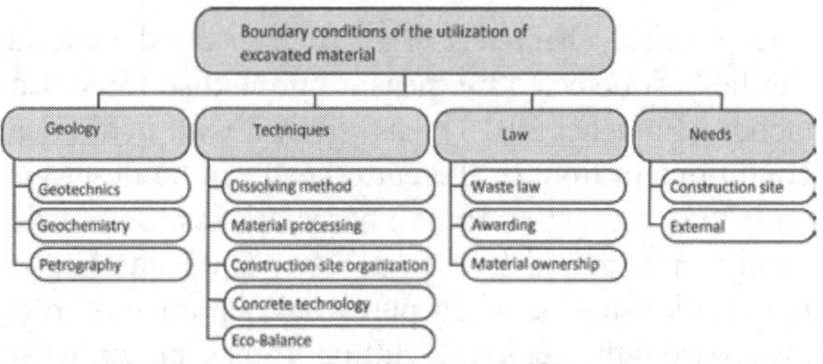

Figure 4. Topics to be considered for utilization of excavation materials from underground construction sites

6 TUNNELLING DATA BASE

DRAGON is providing a prototype of a tunnelling database. Within this database, inputs from the geological forecast are compared with data coming online from the face of the tunnel works. Furthermore, potential clients of mineral products are connected to the database. Thus the database represents a link between the actual developments on site which are compared to the prognosis and forms the basis for the material management, which enables interactive trading and exchange of material among an extended circle of interested parties.

7 A SHORT INSIGHT INTO SOME OF THE TECHNICAL DEVELOPMENTS WITHIN THE PROJECT

The main technological developments within the DRAGON project include fully automatic processes for:

1) fast detection of usable materials,
2) Recycling the materials immediately on the tunnel boring machine.

Conveyor belts offer great opportunities to automate analyses, classification and processing of the excavation material to ensure recycling in a proper way. Newly developed measuring gear for online-measurements need to be installed directly onto the hauling installations (Müller et al. 2011).The main parameters of the material concerning its suitability as a resource are Chemistry and impurities, grain size distribution, water content, grain shape and mass flow.

7.1 DCLM – Disc Cutter Load Monitoring

The physical characterisation of excavated material can be best detected by measuring the disc force. Entacher (Entacher 2013) has shown that geological conditions in front of the cutterhead can be described with the disc cutter force characteristics. These developments should lead to a better understanding of the relationship between pure rock parameters, rock mass strength, geology, cutting forces cutter wear. For this purpose, Herrenknecht AG and the Montanuniversität Leoben developed various methodologies for measuring the cutter force on discs while the machine is boring.

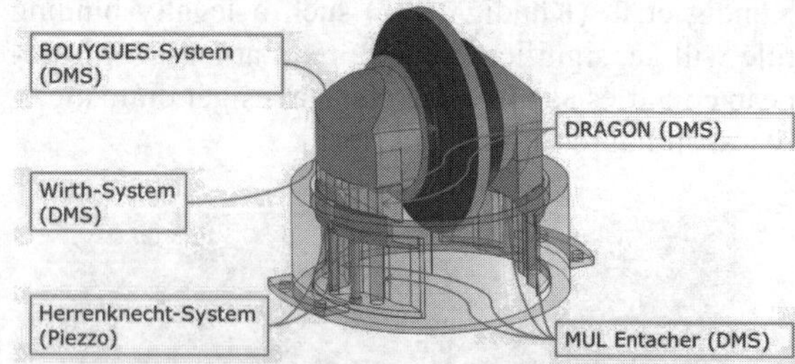

Figure 5. Overview of the available disc cutter measurement systems, example: wedge lock system of HERRENKNECHT

7.2 Automated sampling from the conveyor belt

The sampling has to be done very close to the tunnel face. With automated sampling procedures new standards are set in underground constructions. The sampling system has to be available throughout the whole construction time with only short breaks for cleaning and maintenance during TBM standstill times.

7.3 Automated measuring of grain size and grain shape

For various reuses the grain size distribution is an important input data. Therefore DRAGON uses a technology which determines the particle size distribution of the material fully automated. The excavated material is analysed photo-optical and the resulting image is evaluated digitally. The main advantages of photo-optical particle analysis to be used underground are the rapid detection of particle size, particle shape and particle number, the time savings in the measurements developed by Haver & Boecker, compared to a sieve analysis and the high level of automation.

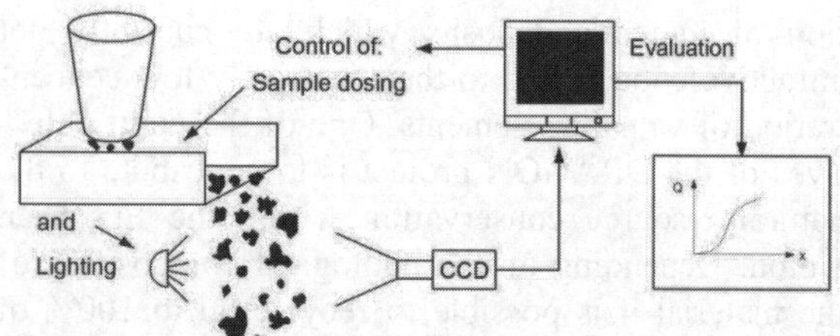

Figure 6. Photo-optical System for determining the grain size and shape

Figure 7. Photo-optical system mounted along the prototype

7.4 *Material analysis techniques by X-ray fluorescence analysis*

The X-ray fluorescence spectroscopy is one of several possible methods to find out the exact elementary composition qualitatively and quantitatively. The results are converted to the oxides of the single elements and give an overview of the percentage of each component in the sample. This analyzing technology is installed immediately behind the particle size analyser.

Figure 8. Online x-ray chemical element analyser OXEA (by INDUTECH)

8 MATERIALS MANAGEMENT

One of the key points for the successful use and recycling of tunnel excavation material is the management of materials. This includes the aspects of material classification, material separation, material transport, material intermediate storage, material processing and material storage. Each of these points is decisive for the success of recycling, since only correct classification and separation enables the quantity of material that is finally tipped to landfill to be minimised.

8.1 *Material processing*

In addition to the material demand on site, the suitability of the lithology encountered as a construction material is the most decisive factor. If there is no internal demand on the site, then preparation on site will be inappropriate. If, however, the excavated material is suitable for recycling in the construction works, the normal practice has to install a stationary processing plant on the site facilities area. However, the disadvantage here is that if the excavated material is to be used for support measures for tunnel construction, then the material has to leave the tunnel first in order to be processed above ground and then transported back into the tunnel as concrete. In case of long tunnels this causes increased electrical power consumption, complicated logistics, noise and dust nuisance as well as an increased need for space in front of the portal. DRAGON therefore pushes techniques for performing all required analysing and processing tasks underground directly behind the tunnel face.

8.2 *Material storage*

Material storage, whether intermediate or final, is a critical point on tunnel projects due to the large quantities of excavated material preceded in tunnelling. However, it is necessary to think about any future use of the material under changed economic conditions or raw materials policy. The material should be stored separately according to type, so that the exact tipping sequence is known, which means to create a secondary mineral deposit. This will be assisted by the detailed and complete recording of the entire material flow in the excavated material database. Thus it is ensured that raw materials are correctly sorted and can be extracted economically by future generations.

9 CONCLUSION

Using the analyses techniques developed in DRAGON allows an immediate analyses and recognition of valuable materials close to the tunnel face and thus forces the direct utilization of the excavated material. The DRAGON Consortium provides sufficient experience and scientific excellence in order to guarantee the successful implementation of the DRAGON project on a European level. The rate of utilization of excavated materials from tunnelling will be significantly increased. The goal of DRAGON is to achieve valuable minerals which could be used for construction materials, steel production, ceramics, electronics, pigments and others. Regarding the scientific impact DRAGON aims, in a multidisciplinary way, the optimization of the research outcome and commercial exploitation of its results by mobilising the critical mass of scientific knowledge as well as eco-environmental application skills. Most of this expertise is already based on previous funded national projects and initiatives, in which some of the partners have been involved dealing with the recycling of tunnelling excavation material and the development of an innovative separation technology for bentonite. The new and innovative technologies within DRAGON are the control of a tunnel boring machine by the fragmentation of the excavated rock which leads to an optimisation of the machine operation. Such a concept has never been realised before. However, previous fundamental research work at the Montanuniversität Leoben has shown that such a concept is feasible. Scientific challenges in the DRAGON approach are the combination of optical particle size measurements on a continuous sample and its evaluation with respect to the natural breakage characteristics. In addition the combination with a Laser based spectrometer has never been tested and realised before. Therefore the connected scientific impact embraces technologies which allow online-analyses considering grain size, grain shape, etc. under outstanding underground conditions as well as which allow separation processes and which facilitate recycling technologies of underground excavation materials under very limited space conditions. The exploitation of minerals is usually affected by conflicts between the economic interests of the extractive industries or the construction sector and environmental protection concerns, but also declining sizes of natural stocks have an influence. To overcome such problems new ways of making minerals available on a regional and local level through tunnel excavation projects is a possible solution. The excavation close to the users is very important as it will help to reduce the consumption of fossile fuels for production, processing and transport.

ACKNOWLEDGEMENT

This project has received funding from the European Union's Seventh Programme for research, technological development and demonstration under grant agreement No 308389.

REFERENCES

ISO 14040, 2006, *Environmental management - Life c cle assessment - Principles and framework / ISO 14044: Environmental management - Life cycle assessment - Requirements and guidelines*.

Brunner, P.H., Rechberger, H. 2004, *Practical Handbook of Material Flow Analysis*, CRC Press LLC, Boca Raton, Florida.

Baccini,P., Brunner, P.H. 1991, *Metabolism of the A throsphere*, Springer, Berlin, Heidelberg, New York.

Kündig, R. 1997, *Die mineralischen Rohstoffe der Schweiz*, Schweizerische Geotechnische Kommission, Zürich.

Geologische Bundesanstalt 2004, page 31. Translated into English

ÖBV, 2015, *Verwendung von Tunnelausbruch*, Richtlinie published by Österreichische Bautechnik Vereinigung.

Müller, S. B., Zwicky, C. N., Blahous, L. 2011, *Kalibrierung eines NIR-Online-Analysators mittels Bohrmehlproben zur Optimierung des Zement- Rohmaterials*, Berg- und Hüttenmännische Monatshefte, 156, Heft 6, 225-226.

Entacher, M. 2013, *Measurement and interpretation of disc cutting forces in mechanized tunnelling*, Dissertation, Lehrstuhl für Subsurface Engineering, Montanuniversität, Leoben.

Proceedings of the International Conference on
Smart Infrastructure and Construction
ISBN 978-0-7277-6127-9

© The authors and ICE Publishing: All rights reserved, 2016
doi:10.1680/tfitsi.61279.663

Cyber physical simulation of energy smart communities

Ali Ghofrani, Farbod Farzan, Jaimie Swartz, Khashayar Mahani, Nivetha Balsami, Pooria Ansari and Mohsen A Jafari*

Department of Industrial and Systems Engineering, Rutgers University, Piscataway, USA
* Corresponding Author

ABSTRACT Net-Zero buildings and communities are being established in the United States and all around the world. There is already a 2025-mandate in California on Net-Zero buildings and communities. Advanced operational controls and maintenance plans supported by data and technology are essential for ensuring sustained Net-Zero energy over the lifetime of a building or a community. In this work we describe a new cyber-physical test bed that is being developed at the Rutgers Laboratory for Energy Smart Systems (LESS) as part of the Center for Advanced Infrastructure and Transportation. This system is capable of a combined virtual and physical simulation of a community with user-defined buildings of different types, charging stations for electric vehicles, energy storage facilities, and renewable generation facilities, such as community based wind turbines and solar farms, and an infrastructure that supports power grid connection whenever necessary. Each building is equipped with advanced controls at zonal and whole building levels. There is also inter-building communication for the purpose of ensuring community level energy efficiency measures. Power generation farms, building level solar installations, and energy storage will be virtual and will use TRNSYS, renewable simulation software that allows for near real time control and monitoring of these systems. Buildings will be simulated using US DOE EnergyPlus with a modeling resolution that allows fast controls and monitoring of building energy assets. Electric vehicles will be digitally simulated and there will be high-level simulations of other transportation assets. The next generation of the test bed will extend to combined virtual and physical buildings, and will also include simulations of water efficiency measures and supporting technology simulations.

1 INTRODUCTION

Cyber physical systems are an integration of sensors, network components and controllers which gather data from environment and perform operations to control the environment. It is believed that Cyber Physical Systems will change the way we interact with the physical world, just like the advent of the Internet has changed the way we communicate with people [1]. The processing of physical world information autonomously in a cyber world is what differentiates Cyber Physical Systems from traditional Industrial Control Systems [2, 3]. [11] presents the scope of Cyber Physical systems in development of net zero energy buildings and optimal harvesting of wind and solar power. A net zero community is one that uses energy efficiently and switches over to renewable energy to meet the remaining demands. It is estimated that by 2025 nearly %62 of all commercial buildings will be net zero [4]. Such a realization cannot be possible by physical monitoring of a few parameters in the community. Using Cyber Physical Systems to model net zero communities enables monitoring and autonomous balancing between usage of the two energy sources. The major requirements of building communities like optimum energy usage, maintaining occupant comfort and efficient usage of renewable energy can therefore be efficiently realized using Cyber-Physical aspects [5]. The ultimate goal of a smart community is to provide better comfort while maintaining sustainability and energy efficiency. A smart community replaces the existing traditional grids with smart grids. Traditional grid falls short when it comes to communication capabilities. Smart grid, however, has a powerful system of sensing and computing components that facilitates communications between different parts the grid [6]. In fact, processing and analyzing large amounts of data is the real power of smart grids which can result in making smarter decisions by both energy consumers and suppliers [7]. Integrating different types of communication networks in smart grids and simulating their impact on electric power systems is a challenging topic in smart energy community studies [8]. Experimental studies of smart grids in large scales cannot be justified economically. On the other hand, development of cyber physical simulation of a smart energy community using real time data is an effective way to improve the efficiency and reliability of communications in these power systems which also is economically feasible [9,10].

In this work we present a cyber physical testbed that is able to address the needs in energy community assessments. Some of the external inputs, such as weather occupancy and prices of utilities will be real and obtained through physical sensors and/or from publically available sources. Building occupancy and use of transportation assets will be physically obtained real time through sensors and real time mobility data. Embedded controls for buildings and other assets will allow for experimentation with different controls at various levels. This test bed will also be instrumental in the assessment and evaluation of energy policies and regulations at community levels, and will also be used for valuation of major investments in a community.

2 SMART COMMUNITY DESCRIPTION

This community is composed of multiple commercial and residential building types, onsite generation, energy storage, geothermal, CHP, district cooling and heating systems, electric vehicles, data server, central and distribute control, energy storage, etc., which also has interaction with grid. Sixteen US DOE reference buildings along with real building models enables the testbed to evaluate the interaction between any combination of buildings, which covers a broad range of options, and other components in the smart energy system. Numerous types of wind turbines and solar panels are available in the testbed for the purpose of assessment of impact of type and design of renewable energy and its long term and short term cost and benefit analysis. In this testbed, not only the community level evaluation is available, also building and zone level simulation is carried out. Besides, the testbed has the ability of real-time simulation with real-time data sources.

Some of the tools used in this testbed are Energy Plus, TRNSYS, Matlab, Ptolemy II, and BCVTB. Energy Plus and TRNSYS are mostly implemented for building energy and control simulation. Renewable energy and storage systems are modelled in TRNSYS and C. The simulation of each component across the testbed is synced by the Building Controls Virtual Test Bed (BCVTB) developed by the Lawrence Berkeley National Laboratory. BCVTB allows the coupling of different simulation programs for co-simulation in real-time.

The framework of the testbed and architecture of the work can be seen in Figures 1 to 3.

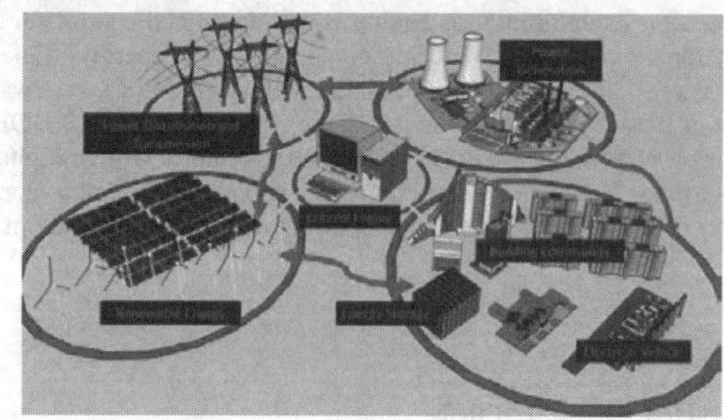

Figure 1. A smart community composed of residential and commercial buildings along with renewable sources and grid.

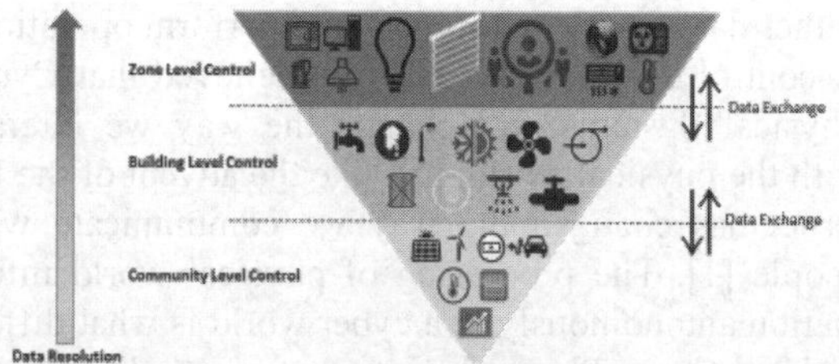

Figure 2. Hierarchy of the components, and data exchange in the community.

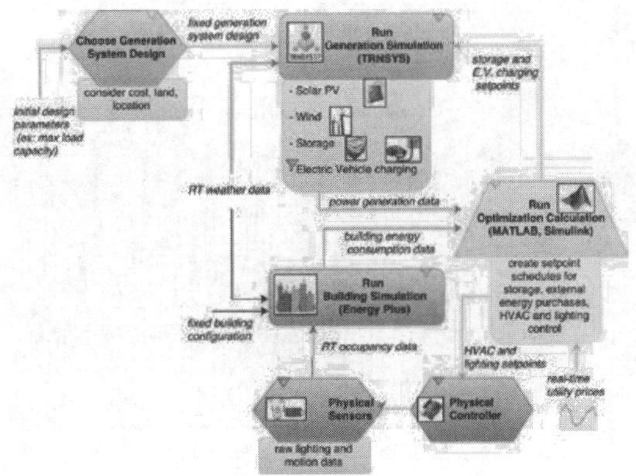

Figure 3. Testbed template.

The system is initially setup when initial design parameters are given. These parameters define the community, including the geometry of buildings, their average loads, and any land or budget constraints. These parameters are used to choose the best generation system design as well as defining the configuration of the buildings.

Initially the system uses deterministic building and generation models based on both historical data and initial parameters. On the other hand, the processor uses received real-time data to improve forecast models for community operation optimization. This process is done according to constraints defined by the required comfort of the building occupants, energy demand, and external utility energy prices. The processor will generate optimal lighting and temperature setpoint schedules for the building simulation, as well as optimal storage and external energy purchase setpoint schedules for the renewable generation simulation. This Optimization Iteration Process (OIP), comprised of simulating, analyzing, and sending control signals back, enables the system to approach its optimal settings. Less energy will be consumed and less energy will be wasted, resulting on the purchase of less external energy. When no external energy is purchased, a self-sustaining net-zero community has been achieved, which is the ultimate objective of the smart community.

One innovation of this work is that the test bed will include simulations of both virtual, or cyber, components as well as real-world, or physical, components. In earlier phases of the work the cyber aspect is the focus. This allows more flexibility when establishing the test bed itself and quickly collecting data. One challenge in the cyber part is choosing parameters and specification of the buildings and generation system that are realistic and scale up well. It may be easy to determine the behavior of a few solar panels, but doing so for a virtual farm of thousands of solar panels is a necessary challenge. Addressing these scaling, realistic, and complexity concerns is essential for creating a robust system that can be applied to any real-world community.

For the work's physical aspect, Rutgers University is the first test bed location. Physical sensors and meters are available throughout several buildings at the university to collect data. This data, once analyzed and converted, will contribute greatly to increasing the accuracy of the processor's forecast models that result in more realistic and applicative models. Generated control signals in the processor will be sent not only to the virtual building simulation but also has the capability to be applied on physical controllers that will assign the building operation schedules.

3 ILLUSTRATIVE EXAMPLE

A community of 8 buildings is built that utilizes onsite generation along with energy storage system. The buildings are US DOE reference buildings described in table 1.

Building type	Number of buildings	Area	Number of Thermal Zones
Medium Office	2	4,982 m^2	15
Midrise Apartment	4	3,135 m^2	27
Primary School	1	6,871 m^2	25
Service Restaurant	1	511 m^2	3

Table 1. Building simulation description.

Template of the community is also shown in figure 4.

Three communities are modelled in three different climate zones in Illinois, New Jersey, and Texas to evaluate the impact of climate on the combined onsite generation and storage behavior. It is assumed that the elevation and design of the renewable energy site and storage remains constant in the three scenarios. Also, all the buildings have the capability of communication to adjust their operation with different peak demands and availability of onsite generation. The timestep for the simulation is 15 minutes.

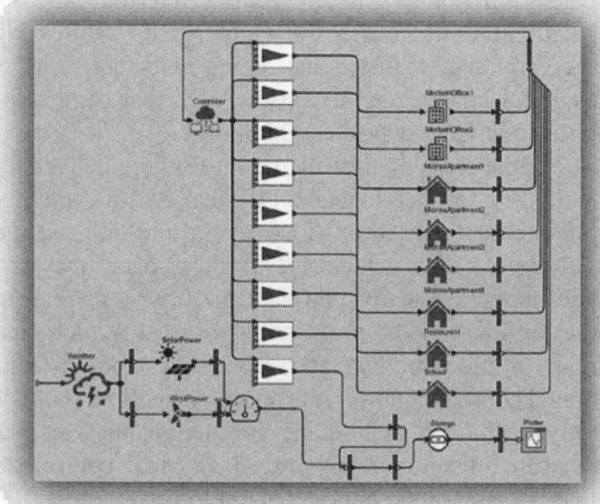

Figure 4. Schematic of the community.

Over the simulation, the human occupancy patterns and plug loads are not considered as constant; therefore, perturbation and stochasticity are applied to the behavior of the community.

Figure 5 compares three communities in different climate zones.

Figures 6 shows the impact of climate zone on the amount of energy that should be purchased from the grid due to unavailability of onsite generation. This can be defined as Equation 1.

$$\mathit{Purchased\ Energy\ Ratio} = \frac{\mathit{Purchased\ Power}}{\mathit{Maximum\ Demand\ in\ a\ Year}} \qquad \mathit{Equation}\ (1)$$

Figure 6 presents different storage patterns given that storage is available in the community.

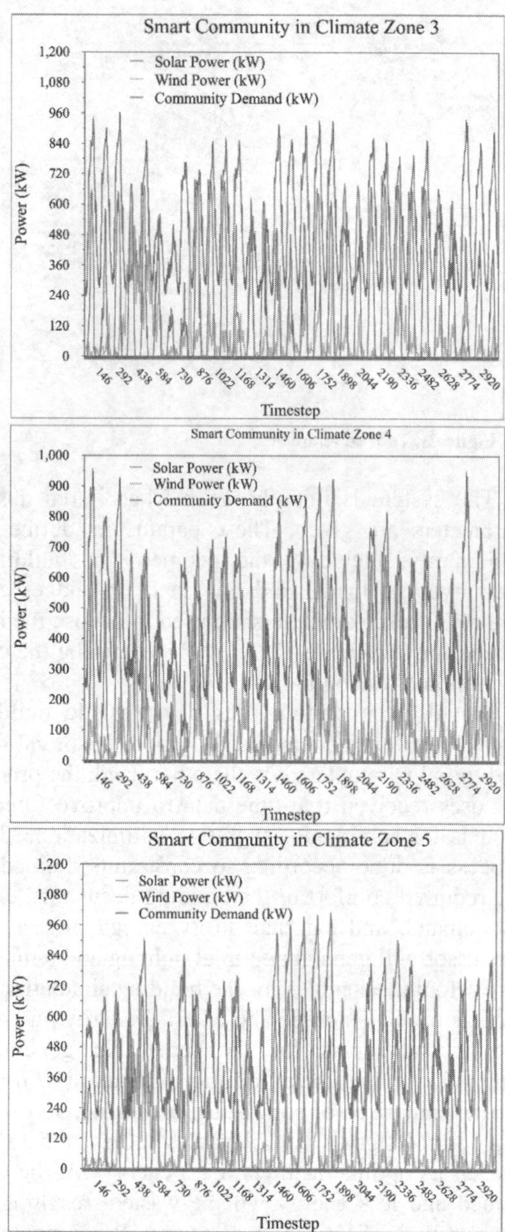

Figure 4. Community consumption and generation in December for three different climate zones.

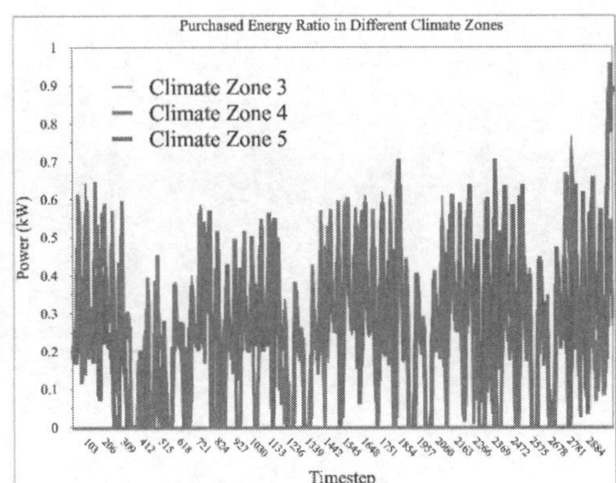

Figure 5. Purchased energy ratio.

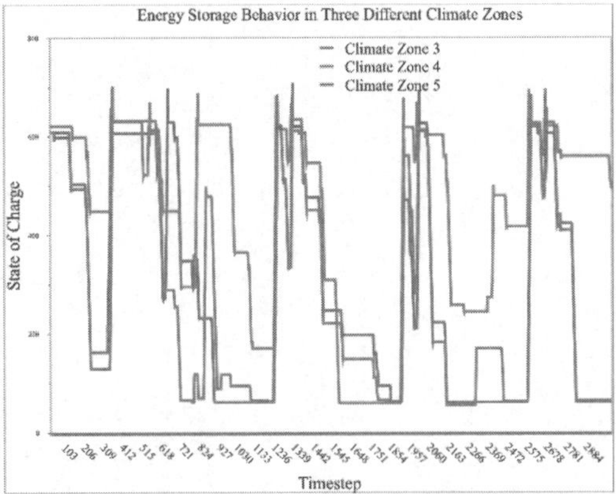

Figure 6. Storage pattern difference under different climate zone.

4 CONCLUSION

In data analysis for energy systems, demand response programs evaluation, energy community controls, etc., lack of data and a reliable assessment is one of the most important challenges. The current testbed has this ability to produce any scenario to develop data and make a highly reliable assessment. On the other hand, utilization of real sensor data and communication with physical world, dramatically decreases the evaluation costs and design risks for energy communities. Also, this can be implemented in decision making systems in grids to obtain a good estimation of the demand side behavior and micro grid participation.

ACKNOWLEDGEMENT

This publication was supported by a grant from the U.S. Department of Transportation, Office of the Secretary of Transportation (OST), Office of the Assistant Secretary for Research and Technology under Grant no. DTRT12-G-UTC16.

REFERENCES

S. Blumsack, A. Fernandez, "Ready or not, here comes the smart grid!" Energy, Volume 37. 61-68, 2014[7]

V. Gungor, D. Sahin, T. Kocak, S. Ergut, C. Buccella, C. Cecati, G Hancke, "Smart Grid Technologies: Communication Technologies and Standards," IEEE Transactions on Industrial Informatics, vol. 7, no. 4, 529-539, 2011. [6]

F. Guo, L. Herrera, R. Murawski, E. Inoa, C. Wang, P. Beauchamo, E. Ekici, J. Wang, "Comprehensive Real-Time Simulation of the Smart Grid," in IEEE Transactions on Industry Applications, vol. 49, no. 2, 899-908, 2013. [10]

L. Gurgen, O. Gunalp, Y. Benazzouz, and M. Gallissot, "Self-aware cyber-physical systems and applications in smart buildings and cities," Design, Automation Test in Europe Conference Exhibition, 1149–1154, March 2013. [3]

A. Hahn, A. Ashok, S. Sridhar and M. Govindarasu, "Cyber-Physical Security Testbeds: Architecture, Application, and Evaluation for Smart Grid," in IEEE Transactions on Smart Grid, vol. 4, no. 2, 847-855, June 2013. [9]

K. Managan, "Net zero communities: One building at a time," 2012. [4]

R. Rajkumar, "A cyber–physical future," Proceedings of the IEEE (Volume:100) 2012. [1]

"Report: Cyber-physical systems summit," [11]

Z. Y. Wang Jinlong, Zhao Qianchuan, "An effective framework to simulate the cyber-physical systems with application to the building and energy saving," Control Conference (CCC), 2013 32nd Chinese, 2013. [2]

A. Stefanov and C. C. Liu, "ICT modeling for integrated simulation of cyber-physical power systems," Innovative Smart Grid Technologies (ISGT Europe), 3rd IEEE PES International Conference and Exhibition on, Berlin, 1-8, 2012[8]

E. Widl, B. Delinchant, S. Kubler, D. Li, W. Muller, V. Norrefeldt, T. S. Nouidui, S. Stratbucker, M. Wetter, F. Wurtz, and W. Zuo, "Novel simulation concepts for buildings and community energy systems based on the functional mock-up interface specification," in Modeling and Simulation of Cyber-Physical Energy Systems (MSCPES), 2014 Workshop on, 1–6, 2014. [5]

Proceedings of the International Conference on Smart Infrastructure and Construction
ISBN 978-0-7277-6127-9

© The authors and ICE Publishing: All rights reserved, 2016
doi:10.1680/tfitsi.61279.669

Sustainable asset management for utility streetworks

A. Hojjati*, I. Jefferson, N. Metje and C.D.F. Rogers
Department of Civil Engineering, School of Engineering, University of Birmingham, Birmingham, UK
* *Corresponding Author*

ABSTRACT Utility infrastructures are one of the critical elements of urban environments. However, utility installation and maintenance operations are costly, both in terms of direct construction costs (estimated to be around £1.5 billion in the UK in 2006) and in terms of indirect and social costs, which adversely impact the UK economy. These costs are significantly increased when the considerable environmental costs are considered. These adverse impacts are mainly due to traffic congestion, both in terms of energy wasted and vehicle emissions. It is now established that the true total cost of any activity can only be measured by considering all aspects encapsulated by the three 'pillars of sustainability', i.e. taking account of social and environmental impacts along with economic (both direct and indirect) costs. After a critical review of the existing sustainability assessment tools, it is proposed that an existing tool should be adapted to provide a holistic, robust sustainability costing framework specifically for utility streetworks. This paper discusses key sustainability assessment indicators and provides recommendations for developing a value-based asset management framework for utility streetworks.

1 INTRODUCTION

Conventional development in industrialised and developing countries has caused human beings to face global warming and adverse climate changes. Additionally, increases in global population, the rate of consumption of natural resources without known replacements, and production of waste and emissions are increasing the pressure on the planet and its ability to supply resources. Consequently, societies now encounter serious problems such as flooding, forest fires, droughts, loss of biodiversity, and the negative impacts of urbanization on individuals (Butchart et al. 2010).

In recent years, however, the sustainability agenda has been introduced into world development plans, and consequently adopted for the built environment. The most common general assessment of sustainability, the Triple Bottom Line (TBL), emphasizes an enhancement in the three pillars of sustainability: economic, social and environmental. Several studies have investigated the impacts of economic developments on the environment and society within the context of decision-making (Ariaratnam et al. 2013). It is generally easier to quantify the economic and environmental elements of the 'TBL' sustainability than the social aspects, yet it is essential to consider all three pillars holistically (Hayes et al. 2012). This supports the argument underpinning 'Pareto Optimality (efficiency)' which states that in decision-making, performance of different options must be examined to establish the 'decision-frontier', where it is almost impossible to enhance one objective without having a negative effect on the performance of at least one other (Elghali et al. 2008) – i.e., it is normally difficult to improve one of the three pillars of sustainability without having any impact (often a negative effect) on the other two pillars (Hunt et al. 2008).

Holt at al. (2010) state that Sustainable Development is no longer a fringe activity, and over the past few decades has been recognized by governments and industry as a core activity. Academics and industry professionals, and even the aware public, use the term 'Sustainable Development', yet no fixed definition can truly explain the concept (Elghali et al. 2008). Nevertheless, as infrastructure systems act as an interface between citizens and the natural environment, all infrastructure projects can be considered as sustainable development projects.

This paper discusses sustainability assessments within the context of utility infrastructure operations (placement, maintenance, renewal), focussing on the challenges and opportunities of sustainability costing of utility streetworks, and concludes with recommendations for an asset-management framework.

2 SUSTAINABILITY COSTS OF UTILITY STREETWORKS

As the UK's utility infrastructure has been greatly expanded during the last 200 years, a vast amount of ground has been dug and different utility services have been placed underground. From the gas pipes installed to power streetlights, shops and homes in 1807 and sewer networks in 1866, to 20th century communication cables, maintenance, renewal and upgrading has been a constant feature (Rogers and Hunt, 2006), and one usually achieved by digging up the ground at significant economic cost. In response, road occupation charging schemes have been trialled (e.g. £1000 per day charged to utility companies for upgrading works by the Camden and Middlesbrough Borough Councils, Balance et al. 2002). The direct costs of utility streetworks excavations is ~£1.5 billion a year in the UK (McMahon et al. 2006), while the 'collateral damage' includes interrupting traffic flow, damaging tree roots, waste production, leakage and soil and air pollution (Kolator 1998).

Utility services are an essential part of well-functioning urban environments; indeed, quality of life in the 21st century is highly dependent on an invisible utility system (Rogers and Hunt, 2006), and as ever more people live in cities (estimated to reach ~70% of the world's population by 2050), these services will gain in importance (Sterling et al. 2012). However, McMahon et al. (2006) estimated that the social costs of traffic congestion in the UK in 2005 were as high as £5.5 billion per annum, with ~5% (~£275 million) attributed to utility streetworks. To this must be added the very considerable environmental costs due to traffic congestion, both in terms of energy wasted and vehicle emissions. Inaccurate location of pipes and cables lengthens streetworks operations and exacerbates congestion, e.g. via major delays due to repairs caused by third party utility damage, the annual direct cost to utility companies being estimated at ~ £150 million (McMahon et al. 2006).

3 ALTERNATIVE WORKING PRACTICES FOR UTILTY STREETWORKS

The combined pressures of population growth – an additional 10 million UK citizens by 2065 – new housing and urbanisation will see urban areas grow and densify, and demand for new buried utility infrastructure, while maintaining the existing, will grow concomitantly. Traditional methods of utility placement, i.e. open-cut trenching, are becoming unsustainable in terms of their disruption to city systems, and their (social, economic and environmental) costs will progressively become prohibitive. Thus, while trenching is now often considered to be the cheapest and most convenient option (for asset owners), and least risky construction method (all assets are exposed), growing awareness of the physical damage they cause (to roads, adjacent pipes) necessitates the use of alternatives such as Trenchless Technologies (TT) and Multi-Utility Tunnels (MUTs; Figure 1).

Figure 1. Combined use of MUTs and TT (Rogers and Hunt 2006)

3.1 *Comparison of utility placement techniques from a sustainability viewpoint*

Multi-Utility Tunnels (or Common Service Tunnels) have been advocated as the least expensive method of utility placement when a range of long-term costs were considered across all three pillars of sustainability (Hunt et al. 2014). However, traditional, open-cut methods of utility placement and maintenance are widely considered to be the cheapest short-term option from an economic point of view. For example, Laistner (1997) makes the point that trenching is less expensive than MUTs and trenchless technologies in terms of machinery and labour. It is therefore the length of view taken, as well as the misalignment between who pays and who benefits, that is important, even though it is the same citizens who benefit from both the utility services and the city systems that are disrupted (Hayes et al. 2012).

Rogers and Hunt (2006) derived a costing model for sustainability of MUTs compared with open-cut and TT. The model assesses both short-term and long-term costs via a simple credit system, suggesting that MUTs (by gaining +30 credits) have significant long-term advantages over TTs (with +8 to +18 credits) and trenching (with +6 credits). Although short-term economic costs of implementation of MUTs – e.g. 'cost of machinery' (e.g. excavators, barriers and traffic signals) for initial utility installation, 'cost of labour' for the construction works, 'cost of materials' needed for construction – will be high, the long-term economic advantages are significantly improved. Examples of these long-term economic advantages include, but are not limited to, 'quality of installation' causing long-term economic savings, 'leakage detection' (including gas leakage) and repairs (this is facilitated by sensors within MUTs), and 'preventing the costs of repairing and resurfacing of roads' due to excavations for the purpose of maintenance and renewal (road lives are not reduced).

Social costs of utility placement in the short-term by trenching were shown to be lower than for MUTs due to the existence of good and well-practiced management systems for trenching methods and the more extensive works required to introduce an MUT, whereas in the long-term MUTs offer an ideal socially-efficient system (Rogers and Hunt 2006) compared to open-cut and TTs. This is because the utility maintenance, upgrading, renewal or expansion will be carried out within the MUT (except for shallow, surface-opening MUTs), therefore, risks and disruptions to both public and workers can be minimised.

Environmental costs of utility placement vary between different methods of installation. Although short-term costs such as 'energy use' and 'materials' are less for TTs than MUTs, these would be improved in the long-term with MUTs. Examples of environmental damage – e.g. leakage from sewer pipes and air pollution resulting from 'delayed vehicles' (Kolator, 1998) – could be controlled through the use of MUTs. Similarly, ecological issues were considered; e.g. the damage to tree roots and also vice versa (i.e. damage from tree roots to utilities) is common with trenching methods, whereas these costs can be prevented or at least minimised by the implementation of MUTs and trenchless technologies.

4 SUSTAINABILITY ASSESSMENT TOOLS

In general, there are many tools and evaluation methods available for sustainability assessment. This is equally true for the field of civil engineering and for different types of construction projects. It is not easy to choose the most appropriate tool/assessment framework for use in a particular project. Many tools are commonly rejected because it is thought that using them might take too long or they might not be the right tool for a particular purpose. As stated in PETUS (2005), a tool could provide guidance or consist of a procedure, method, assessment or evaluation of a set of indicators based on a defined benchmark in order to accomplish an objective or to achieve a result.

In order to consider sustainability impacts and ultimately measure those impacts/costs, many tools have been developed by companies and organisations around the world such as BREEAM, LEED, DGNB and SBAT for buildings, and CEEQUAL, SPeAR® HalSTAR and EnvISIon for infrastructures, to name but a few. Rating systems such as BE^{2}ST-In-Highways, GreenLITES, Greenroads and I-Last have been designed particularly for the transportation infrastructure industry. Examples of other new sustainability assessment and rating tools developed for the construction industry in general are ATHENA (Canada), Estidama (UAE), and QSAS/GSAS (Qatar), which are likely to reflect the context and vision of their respective countries. Pearce et al. (2012) indicate that in the UK there are a range of assessment tools from voluntary rating schemes such as BREEAM and CEEQUAL through to a number of public and private sector 'Sustainability toolkits', and also statutory processes such as Sustainability Appraisal (SA). Most of the tools are either developed with a context in mind, which is normally very prescriptive and hence they do not bring much value, or they are so general that they do not include all of the aspects that are important in any particular context.

Surveys of sustainability assessments and costing of utility streetworks (Jung 2012; Hayes et al. 2012; Hunt et al. 2014) have indicated that in some cases monetary assessments of impacts could be suitable, but where this is not appropriate other measures should be employed. Where such monetary assignment is not possible, or would be unreliable, then a means of assessment should aim to determine value and its associated benefit. Only then can a direct cost associated with delivery of the benefit be estimated with any degree of certainty, allowing an assessment of the sustainability value-to-cost ratio to be made.

Methods to establish direct and indirect economic costs of streetworks are available (Hunt et al. 2014). However, building on recent research at the University of Birmingham a new method that captures the total costs and impacts is being developed. After an extensive review of the options, the suggested way forward is to adapt an existing assessment tool to provide a holistic, robust sustainability costing framework specifically for utility streetworks.

4.1 Assessment indicators

Although there are many sustainability and sustainable development indicator sets, or sets of criteria for sustainability assessment, there is not any standard or complete set of indicators available for utility streetworks sustainability assessment. It has been stated indicators associated with the United Nations Millennium Development Goals (UN 2008), either Global or National, and many other indicator sets developed for building rating systems are not directly related and appropriate for civil infrastructure (MacAskill and Guthrie 2013), and this is more acute still for the specificity of streetworks operations.

As described by Fenner and Ainger (2014), there are different types of general indicators. Some of them measure impacts or project outcomes, while others measure inputs, and there are also indicators that focus on the process, i.e. how the project is carried out rather than relating to specific inputs or objectives. MacAskill and Guthrie (2013) created a summary of sustainability indicators and themes under headline indicator categories collected from different infrastructure assessments and research including CIRIA's guidelines (Berry et al. 2011), AGIC (2010), BREEAM (2011) and CEEQUAL (2011). It is not, however, a comprehensive collection of all existing tools and indicators and none of the tools in that summary covered all the indicator themes.

Building on the growing literature in this field and avoiding unnecessary tool development, an existing well-established tool is being adapted to specifically deal with utility placement and maintenance works. However, this raises the question of how and to what extent these tools or methods should be modified to satisfy the requirements of utility streetworks projects while complying with their original design thinking. As a result, a new indicator system including headline costs, indicators and sub-indicators, has been developed to match the requirements and nature of the streetworks projects in particular (Figure 2).

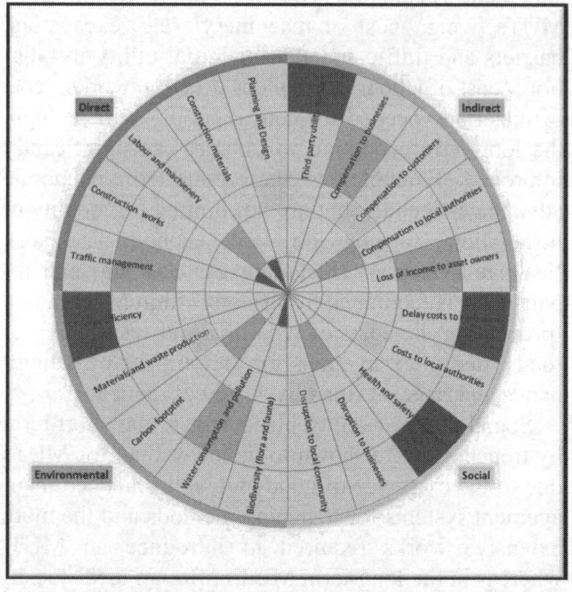

Figure 2. Rose diagram containing the set of bespoke indicators for utility streetworks

The indicator sets are divided into four main categories of impacts – Direct Economic, Indirect Economic, Social and Environmental – to create a complete picture of the total costs of utility streetworks. A summary of the headline costs with associated main indicators is shown in Table 1.

The indicator sets were initially developed from the literature and they have been reviewed based on initial discussions between the authors and a group of experts from the fields of sustainability and infrastructure. These initial indicator sets were then revised following three expert panel discussions to refine the final set.

As an example, 'Third Party Property Interference and Damage' as an initial indicator within the 'Construction Indirect Economic Impact' category was changed to 'Third Party Utility Damage' to crystallise the idea in terms of the utility streetworks context and highlight the potential seriousness of 'utility strikes' as an indirect cost of utility streetworks. Deeper research into its causes and costs (e.g. see Metje et al. 2015) will aid in further refinement.

Table 1- Sustainability indicators for utility streetworks costs and impacts

Headline Indicator	Indicator Category
Construction Direct Economic Impact	Planning and Design
	Labour and machinery
	Construction materials
	Construction works
	Traffic management
Maintenance Direct Economic Impact	Planned maintenance
	Monitoring
	Access
	Emergency repairs
	Decommissioning
Construction Indirect Economic Impact	Third Party utility damage
	Compensation to businesses for loss of profit
	Compensation to customers for interruptions to services
	Loss of income to asset owners or utilities
	Compensation to local authorities for damage to their assets
Maintenance Indirect Economic Impact	Goodwill
	Required Training (upskill)
	Insurance
	Loss of business to competitors
	Lost Opportunity Cost
Construction Social Impact	Delay costs to road users
	Disruption to businesses
	Disruption to local community
	Health and Safety (nuisance)
	Costs to local authorities
Maintenance Social Impact	Delay costs to road users
	Disruption to businesses
	Disruption to local community
	Health and Safety (nuisance)
	Costs to local authorities
Construction Environmental Impact	Energy efficiency
	Materials and waste production
	Carbon footprint
	Water consumption and pollution
	Biodiversity (flora and fauna)
Maintenance Environmental Impact	Energy efficiency
	Materials and waste production
	Carbon footprint
	Water consumption and pollution
	Biodiversity (flora and fauna)

4.2 *Advantages and limitations of the tool*

Unlike other sustainability assessment tools such as BREEAM and CEEQUAL, which apply weighting factors for the different categories or impacts, the indicators and sub-indicators in the tool under development are not weighted. This is an advantage because it prevents subjective bias in assigning weighting factors and maintains the tool's flexibility when applied to different projects. Furthermore, it is not a reward driven tool, which tends to introduce an in-built bias to the system (Holt et al. 2010). Moreover modification of an existing assessment tool addresses the concern over tool fatigue, which is routinely raised as an issue when the development of an entirely new assessment method is proposed.

5 THE WAY FORWARD: A VALUE-BASED ASSET MANAGEMENT FRAMEWORK

Due to the diverse nature of utility infrastructure projects, and the very large number of streetworks carried out annually – more than 480,000km of underground utilities, including water, wastewater, gas, electricity and telecommunication, are laid around the world each year (Najafi 2005) – a robust and comprehensive, yet simple and easy to use, assessment tool is required. This must be able to evaluate both the benefits and adverse impacts of the economic, social and environmental aspects associated with streetworks, and accommodate the variety of contexts in which the works are carried out. Given the complexity and bias that are often associated with cost based (monetary) assessment, the most appropriate approach is now considered to use 'value'. This enables the assessor to move away from simple, narrow, often misleading assessments deriving from a single monetary measure. Thus, further research should form the core of an assessment framework that informs decision-making in an environment where competing private and public financial interests interact with peoples' daily lives, as well as providing a basis on which to support investment decisions for streetworks projects, by enabling a more holistic view to be taken of the overall value of the works.

Oversimplification of the scoring system of some well-established assessment tools (e.g. SPeAR® and CEEQUAL) increases the potential of the tool to be misused. To avoid this, a detailed and rigorous analysis of the costs (impacts) as well as benefits of the chosen utility placement/rehabilitation/maintenance method is planned to identify the value of the proposed works. This will be in the form of a Whole Life Costing (WLC) exercise (considering short- and long-term impacts) within a broader asset management framework to realise the value of a particular streetworks option to both asset owners (utility companies) and society and the environment as a whole.

6 CONCLUSIONS

A tool for sustainability assessment of utility streetworks is much needed to help the process of decision-making for the selection of the best engineering solution in any given case for this important context. Consideration of 'value' as well as 'costs/impacts' is now considered the best approach. A tool and set of indicators were developed. The indicators are not weighted to avoid bias. Ultimately, time will be included in the assessment framework to determine whole life costs and provide better assessments of short-term and long-term benefits.

ACKNOWLEDGEMENT

The authors gratefully acknowledge the financial support provided by the UK's Engineering and Physical Sciences Research Council under the Assessing The Underworld (ATU) project grant EP/K021699.

REFERENCES

AGIC 2010. Climate Change Adaptation Guideline [online]. http://www.agic.net.au [Accessed 01/08/2015]

Ariaratnam, S.T., Piratla, K., Cohen, A. & Olson, M. 2013. Quantification of Sustainability Index for Underground Utility Infrastructure Projects, *Journal of Construction Engineering and Management* **139: (12).**

Balance, T., Reid, S. & Chalmers, L. 2002. Lane Rental Charging: A Way Forward, *Stone and Webster Consultants, London* **28.**

Berry, C. & McCarthy, S., 2011. *Guide to sustainable procurement in construction*, CIRIA, London.

BREEAM 2011. Building Research Establishment Environmental Assessment Method [online]. http://www.breeam.org [Accessed 21/09/2015]

Butchart, S.H., Walpole, M., Collen, B., van Strien, A., Scharlemann, J.P., Almond, R.E., Baillie, J.E., Bomhard, B., Brown, C. & Bruno, J. 2010. Global biodiversity: indicators of recent declines, *Science* **328: (5982)**, 1164-1168.

CEEQUAL 2011. Version 4.1 Assessment Manual for Term Contracts Part 1: Maintenance. [online]. www.ceequal.com/downloads.html London, UK CEEQUAL [Accessed 17/08/2015]

Elghali, L., Clift, R., Begg, K. & McLaren, S. 2008. Decision support methodology for complex contexts, *ICE Proceedings – Engineering Sustainability* **161: (1)**, 7-22.

Fenner, R. & Ainger, C.M. 2014. *Sustainable Infrastructure: Principles Into Practice*. ICE Publications, London.

Hayes, R., Chapman, D.N., Metje, N. & Rogers, C.D.F. 2012. Sustainability assessment of UK streetworks, *ICE Proceedings – Municipal Engineer* **165: (4)**, 193-204.

Holt, D.G.A., Jefferson, I., Braithwaite, P.A. & Chapman, D.N. 2010. Embedding sustainability into geotechnics Part A: methodology, *ICE Proc. – Engineering Sustainability* **163: (3)**, 127-135.

Hunt, D.V., Lombardi, D.R., Rogers, C.D. & Jefferson, I. 2008. Application of sustainability indicators in decision-making processes for urban regeneration projects, *Proceedings of the Institution of Civil Engineers-Engineering Sustainability* **161: (1)**, 77-91.

Hunt, D.V.L., Nash, D. & Rogers, C.D.F. 2014. Sustainable utility placement via Multi-Utility Tunnels, *Tunnelling and Underground Space Technology* **39**, 15-26.

Jung, Y.J. 2012. Evaluation of Subsurface Utility Engineering for Highway Projects: Benefit-Cost Analysis, *Tunnelling and Underground Space Technology* **27**, 111-122

Kolator, R. 1998. Economic comparison between conventional trench method and trenchless technology in an urban environment, *Proceedings of the 16th International No-Dig* ISTT, Lausanne.

Laistner, A. 1997. Utility Tunnels long-term investment or short-term expense? The new economic feasibility of an old idea, *No Dig International* **15**, 10-11.

MacAskill, K. & Guthrie, P. 2013. Risk-based approaches to sustainability in civil engineering, *ICE Proceedings – Engineering Sustainability* **166: (4)**, 181-190.

McMahon, W., Burtwell, M. & Evans, M. 2006. Minimising street works disruption: the real costs of street works to the utility industry and society, *UK Water Industry Research*, London.

Pearce, O.J.D., Murry, N.J.A. & Broyd, T.W. 2012. Halstar: systems engineering for sustainable development, *ICE Proceedings – Engineering Sustainability* **165: (2)**, 129-140.

Metje, N., Ahmad, B. & Crossland, S. M. 2015. Causes, impacts and costs of strikes on buried utility assets, *ICE Proceedings – Municipal Engineer* **168**, 165-174.

Najafi, M. 2005. *Trenchless Technology*, McGraw-Hill, NY, USA.

PETUS 2005. Practical Evaluation Tools for Urban Sustainability [online]. http://www.petus.eu.com/ [Accessed 07/11/2015]

Rogers, C. & Hunt, D. 2006. Sustainable utility infrastructure via multi-utility tunnels, *Proc. of the Canadian Society of Civil Engineering 2006 conference,* Towards a sustainable future, Calgary.

Sterling, R., Admiraal, H., Bobylev, N., Parker, H., Godard, J.-P., Vähäaho, I., Rogers, C.D., Shi, X. & Hanamura, T. 2012. Sustainability issues for underground space in urban areas, *ICE Proceedings – Urban Design and Planning* **165: (4)**, 241-254.

UN 2008. The Millennium Development Goals Report, United Nations Publications.

Proceedings of the International Conference on
Smart Infrastructure and Construction
ISBN 978-0-7277-6127-9

© The authors and ICE Publishing: All rights reserved, 2016
doi:10.1680/tfitsi.61279.675

Risk-cost optimised maintenance strategy for tunnel structures

Chun-Qing Li[1], Hassan Baji*[1], Mojtaba Mahmoodian[1] and Wei Yang[2]

[1]*School of Civil, Environmental and Chemical Engineering, RMIT University, Melbourne, Australia*
[2]*School of Civil Engineering and Architecture, Wuhan University of Technology, Wuhan, China*
**Corresponding Author*

ABSTRACT Due to limited maintenance budget, effectively spending the available funds for maintaining infrastructures is increasingly sought by asset managers. Tunnel is an essential infrastructure that plays a pivotal role in transportation network, economy, prosperity, social well-being, quality of life and the health of its population. In the light of considerable research that has been or is being undertaken on "aboveground" infrastructure, e.g. bridges, this threat cannot be more apparent for underground infrastructure such as tunnels. The situation has been exacerbated due to more unknowns and uncertainties relating to the factors such as underground water and soil/rock that affect the operation of tunnel infrastructure. In an ageing tunnel system, various potential deficiencies such as seepage, spalling, crack, delamination, steel corrosion, drainage, convergence and settlement of the lining structure can cause catastrophic life safety and economic consequences. Most collapses of tunnel structures in the world are related to tunnel deterioration with catastrophic consequences. Through an effective maintenance plan, the catastrophic failures of tunnels can be prevented. This research aims to develop a maintenance strategy for concrete tunnels which determines when (maintenance intervention times), where (segments of tunnel network) and what (failure mode of tunnel structure) maintenance actions need to be taken to ensure the safe and serviceable operation of tunnel with the intention to minimise the risk. The mathematical formulation of the proposed maintenance strategy, which is based on risk optimisation, is provided in a generic format. Application of the proposal to tunnel structures is presented in a numerical example.

1 INTRODUCTION

As a vital infrastructure, tunnels are important elements of transport network and play a pivotal role in national economy. To ensure the safe and serviceable operation of tunnel structure, maintenance including repairs, strengthening, and instalment is essential. The problem is how to determine when, where and what to maintain at minimal risk and effective cost. Various frameworks have been proposed to formulate strategies for inspection, maintenance and decision-making for deteriorated structures, using reliability-based optimization (Thoft-Christensen and Sorensen 1987, Sommer et al. 1993, Mori and Ellingwood 1994, Barone et al. 2013). However, maintenance strategy considering different failure modes during the service life has not been fully considered in a systematic manner. Furthermore, few studies on maintenance of underground structures such as tunnels can be found in the literature (Yuan et al. 2013).

The intention of this paper is to formulate a maintenance strategy based on risk cost optimization of a tunnel structure during its whole service life. A time-dependent reliability method is employed to determine the risk of attaining the limit state in each phase. To facilitate practical application of the formulated maintenance strategy, an algorithm is developed and programmed. An example is given to illustrate the application of the proposed maintenance strategy to an existing tunnel.

2 FORMULATION OF MAINTENANCE STRATEGY

A structure consists of many components, some of which are redundant. Thus the failure of a structure should be modelled as a combination of series system for non-redundant components and parallel system for redundant components. Similarly, a component can fail in many modes, some of which reach the ultimate limit state and some reach the serviceability limit state. Thus component failure should also be modelled as a combination of series system for ultimate failures and parallel system for serviceability failures. This concept can be logically illustrated in Figure1.

Since failure is not only random but also time-variant, a time-dependent (when) reliability method should be used to determine the probability of the basic failure, i.e., a component (where) failure by a mode (what), resulting in when, where and what maintenance actions are required for the tunnel. The rationale for the proposed maintenance strategy is that, whilst keeping the probability of ultimate failure under control (to ensure safety), only when the probability of serviceability failure is greater than an accepted limit, would the maintenance be warranted.

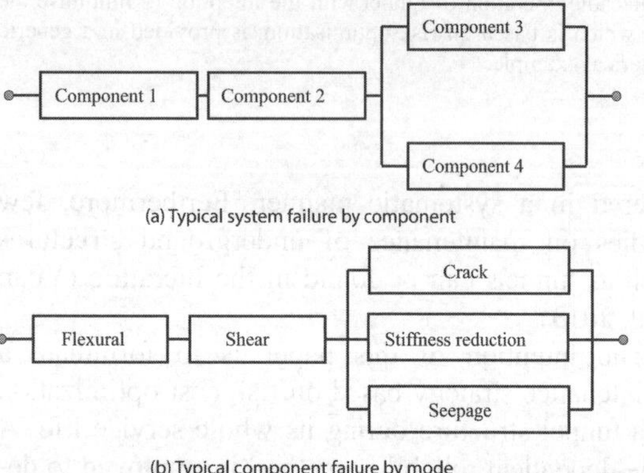

Figure 1. System reliability.

The merit of this rational is to minimise, if not eliminate, inspections for possible failures without compromising the safety of the infrastructure. The value of risk function is based on the system probability of failure (Figure 1) and action takes place on the most influential component for a given mode as schematically shown in Figure 2. The problem can then be formulated mathematically as follows:

$$Minimize: Risk = \sum_{i=1}^{N_r}\sum_{j=1}^{N_m}\sum_{k=1}^{N_c} C_{jk}\left(t_{i,j}\right) \times p_{sys}\left(t_{i,j}\right)$$

$$Subject\ to: \quad p_u\left(t_{i,j}\right) \le p_{u,a};$$

$$p_s\left(t_{i,j}\right) \ge p_{s,a}; \qquad (1)$$

$$0 \le t_{i,j} \le t_L$$

where, $t_{i,j}$ is the maintenance time sequence with i refereeing to time and j to failure mode. N_r is number of maintenance, N_m is number of failure modes and N_c is number of component in the system. C_{kj} is the cost (including interest rate) of repair for k^{th} component due to j^{th} failure mode. p_s and $p_{s,a}$ are the probability and acceptable probability of serviceability failure, p_u and $p_{u,a}$ are the probability and the acceptable probability of ultimate failure and p_{sys} is the probability of failure for the system. t_L is the time for strengthening or lifetime of the structure. The design variables in this optimization are the time of maintenance for each failure mode. For simplicity, interdependence between failure modes is not included in Equation (1) to achieve effective practical applications as will be shown in the example.

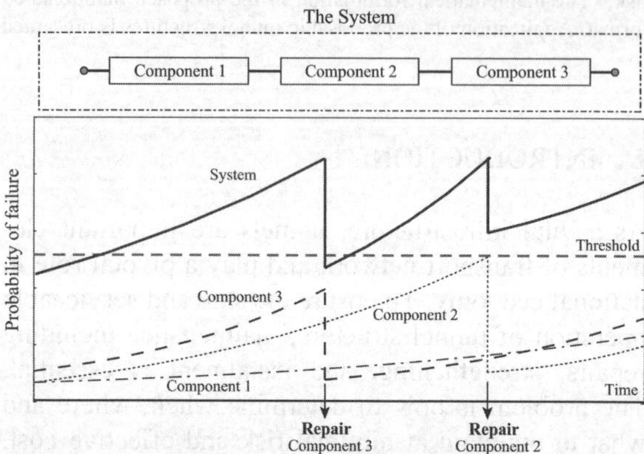

Figure 2. Maintenance strategy for a given mode.

In Equation (1) using annual discount rate of money all the cost terms are converted to a present value cost.

3 RELIABILITY ANALYSIS

3.1 *Modelling of structural response*

As may be appreciated, the structural response is a very random phenomenon, depending on many factors, such as material properties, geometry, stress conditions, defects and so on. It is justifiable to model the structural response as a stochastic process, expressed in terms of primary contributing factors, which are treated as basic random variables. It follows that the structural response is the function of basic random variables as well as time and can be expressed as:

$$S(t) = f(x_1, x_2, x_3, ..., x_n, t) \tag{2}$$

where x_i variables are the basic random variables, the probabilistic information of which are (presumed) available. With this treatment, the statistics of *S(t)* can be obtained using the technique of Monte Carlo simulation.

3.2 *Time dependent reliability*

In assessing the risk of failures for a structure, a performance criterion should be established for the structure. In the theory of structural reliability, this criterion is expressed in the form of a limit state function as follows:

$$g(S, R, t) = R(t) - S(t) \tag{3}$$

where *S(t)* is the structural response (or load effect), *R(t)* is an acceptable limit for structural response (or structural resistance) and *t* is time. With the limit state function of Equation (3), the probability of structural failure, p_f, can be determined by (Melchers 1999),

$$p_f(t) = \Pr(g < 0) = \Pr\left(R(t) < S(t)\right) \tag{4}$$

where *Pr* denotes the probability of an event. Equation (4) represents a typical upcrossing problem, which can be dealt with using time-dependent reliability methods. An analytical solution to this problem has been derived in Li and Melchers (1993) when $S(t)$ is a Gaussian process and the threshold *R* is deterministic. For the sake of brevity the reader is referred to mentioned reference.

4 OPTIMISATION ALGORITHM

Although each term in Equation (1) has been determined individually, the optimization itself is very computationally involved and complex. The computational procedure of optimization can be summarised as follows:

- For a given lifetime t_L, identify a specific number of failure modes (as inputs) for each component of the tunnel and categorize failure modes as either serviceability or ultimate limit states;
- Use Equation (2) to establish stochastic models for each failure and apply them to each component; and use Equation (4) to determine the time-variant probability for each failure mode and each component;
- Use system reliability methods to determine the probability of system failure;
- Determine maintenance intervention time for all failure modes (satisfying all the constraints)and find critical components to be repaired in each time and reset the deterioration process for these components;
- Repeat the last two steps until all maintenance actions are executed;
- Substitute results of steps 3 – 5 into Equation(1) to find the risk function and
- Repeat steps 3 – 6 in search of sequence of time which minimises the risk function;

The search for optimum time t_{ij} with given constraints will be performed using smart algorithms, e.g., the evolutionary algorithms (Deb 2001).

5 NUMERICAL EXAMPLE

5.1 *Case study*

Burnley tunnel, which is one of the Melbourne Underground Rail Tunnel (MURL) is considered as a case study. This tunnel comprises of five stations i.e. four segments. A cross sectional view of the tunnel is shown in Figure 3. It is assumed that all segments are connected in series. Three failure modes that are strength, seepage and crack are considered.

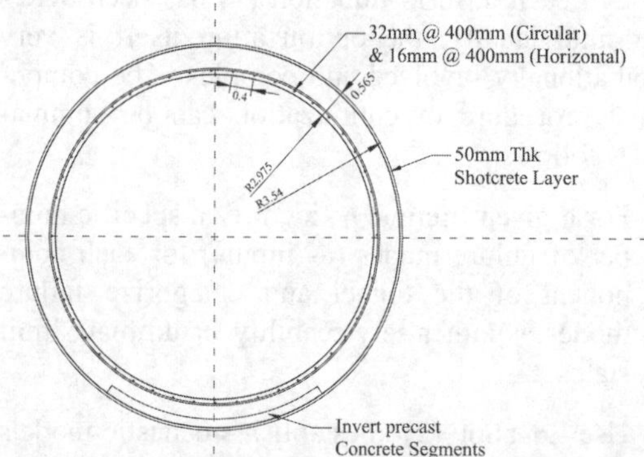

Figure 3. Cross section of Burnley tunnel

5.2 *Strength model*

Cross section of the tunnel is subject to interaction of axial force and bending moment interaction. Due to corrosion process, the rebar area is decreasing over time. On the other hand, the bond between steel reinforcement and concrete is also deteriorating over time. Therefore, the overall strength of the tunnel strength is degrading with time. In this study, a limit state based on degradation of flexural strength of tunnel cross section, shown in Equation (5), is used to quantify the structural performance of the tunnel.

$$g = \varphi_S(t)R_0 - R_a \tag{5}$$

where, $\varphi_s(t)$ is the deterioration function for strength, R_0 is the original structural strength and R_a is the minimum acceptable strength. Li (2003) showed that it may be appropriate to take the acceptable limit for strength deterioration as 60% of the original strength. A model proposed by Azad and Al-Osta (2014), shown in Equation (6), is employed for $\varphi_S(t)$ function.

$$\varphi_S(t) = 1.323\frac{(e/h)^{0.14}\left(\frac{D'}{D}\right)^{1.48}}{(i_{corr}t)^{0.192}} \leq 1.0 \tag{6}$$

where, $D' = D - 2\times 0.003185 i_{corr} t$ is diameter of corroded reinforcing bar, D is diameter of un-corroded reinforcing bar, e is the load eccentricity, h is depth of cross section across axis of bending moment, i_{corr} is the corrosion current density in mA/cm^2 and t is corrosion duration in days. A corrosion rate mode proposed by Li (2003), which is shown in Equation (7), is used in the reliability analysis.

$$i_{corr} = 0.3683\ln(t) + 1.1305 \tag{7}$$

In Equation (7), t is corrosion time in year. Considering rebar diameter of 32 mm and normalised eccentricity of 0.30 for the cross section of the tunnel lining, the reduction in strength can be represented as a function of time, as shown in Figure 4.

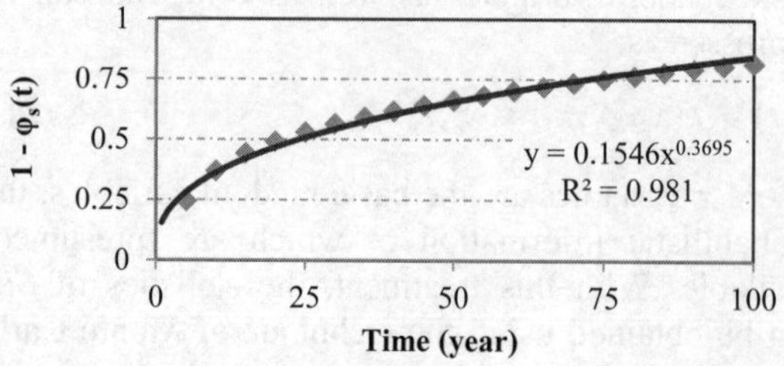

Figure 4. Time-dependent strength model.

Simulation is used for calculating the coefficient of variation for the strength deterioration model. Following Al-Osta et al. (2015) it is assumed that all these variables in Equation (6) are normally distributed and coefficients of variation of these variables are 0.20, 0.02 and 0.01, respectively. The simulation showed that the coefficient of variation of the strength deterioration process is about 0.25.

5.3 *Seepage model*

Tunnel lining structure is required to be watertight. The limit state used for seepage in this study, shown in Equation (8), is based on theoretical model developed by Murata et al. (2004). The original model accounts for both Darcy seepage flow and dissuasive seepage flow. As the water head pressure outside the lining is relatively low, only Darcy seepage flow is considered in this research.

$$g = D_a - \sqrt{2PK_s t} \tag{8}$$

where, P is the underground water pressure outside the tunnel lining, K_s is the seepage coefficient for concrete tunnel lining and D_a is the acceptable seepage penetration in the concrete tunnel lining. According to the available hydrological data the water head outside the tunnel lining is around 30 m. Experimental data reported in the literature e.g. Aldea et al. (1999) shows that for cracked concrete this coefficient can be as high as $10^{-6}\ m/s$. Water seepage penetration through the concrete lining as a function of time is shown in Figure 5. In the absence of probabilistic models for the external water head and seepage coefficient variables, it is assumed that both these variables follow the lognormal distribution with coefficient of variation of 0.30 and 0.10, respectively. Simulation of the seepage depth penetration showed that this deterioration process follows a normal distribution with coefficient of variation of 0.32.

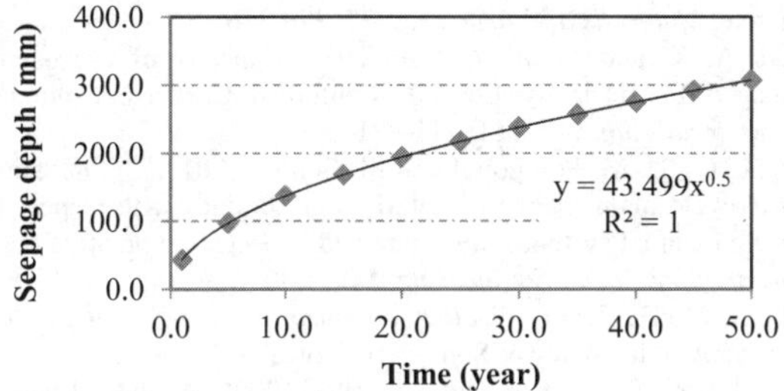

Figure 5. Time-dependent seepage model.

5.4 *Crack model*

Corrosion-induced crack width is expressed as a function of time and is compared with the acceptable crack width which is taken from guidelines in design codes and standards. In this study, an analytical model developed by Li et al. (2006), shown in Equation (9), is used for estimating the crack width as a function of time.

$$g = w_a - \left[\frac{4\pi d_s(t)}{(1-\nu_c)(a/b)^{\sqrt{\alpha}} + (1+\nu_c)(b/a)^{\sqrt{\alpha}}} - \frac{2\pi b f_t}{E_{ef}} \right] \quad (9)$$

where, a and b are the inner and outer radii of the thick wall cylinder (refer to Li et al study), ν_c is Poisson's ratio of concrete, E_{ef} is the effective modulus of elasticity of concrete and f_t is the tensile strength of concrete. Thickness of corrosion product, $d_s(t)$, is changing with time and α is a material characteristic value which also changes with time and can be determined as detailed in Li et al. (2006). For the rebar size of 32 mm and the cover of 60 mm, the crack width as a function of time is shown in Figure 6.

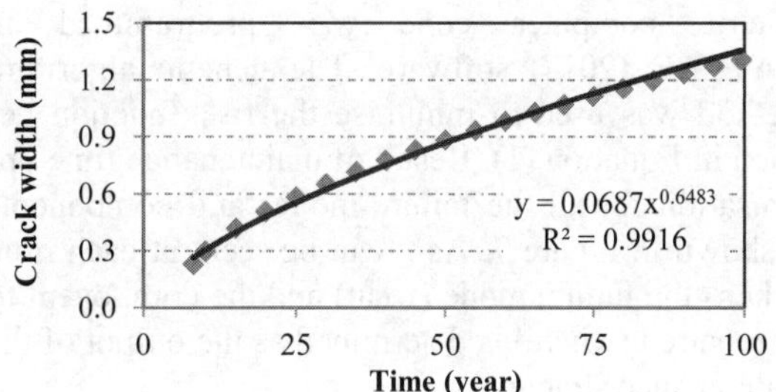

Figure 6. Time-dependent crack model.

Considering some probability density functions for the main random variables and using simulation technique, it is shown that the coefficient of variation of time-dependent function of crack is about 0.15 (Li et al. 2007). It is also assumed that the acceptable crack width, w_a, is 0.45 mm.

5.5 *Basic input*

It is assumed that cost of repair and maintenance for all components of the system is equal, with relative cost of repair for strength limit state being 3 times the cost of repair for crack and two times that of seepage repair. The acceptable probability of failure for strength limit state is taken as 0.05, while for serviceability limit states of crack and seepage an acceptable limit of 0.25 is used. The maintenance strategy is designed for a 50 years lifetime. The interest rate is assumed to be 5%. In this example, different deterioration models were assigned to each component. Otherwise, the optimisation results in all the components fixed at the same time. Table 1 summarises the assumption for deterioration of each component for different failure modes. It should be noted that the deterioration function for the strength limit state is $1-\varphi_S(t)$.

Table 1. Deterioration models for components.

Component	Failure mode		
	Strength	Seepage	Crack
1	$0.10t^{0.37}$	$40.0t^{0.50}$	$0.057t^{0.65}$
2	$0.11t^{0.37}$	$42.5t^{0.50}$	$0.060t^{0.65}$
3	$0.12t^{0.37}$	$45.0t^{0.50}$	$0.063t^{0.65}$
4	$0.14t^{0.37}$	$47.5t^{0.50}$	$0.065t^{0.65}$

5.6 *Optimum maintenance time results*

In order to find the optimised maintenance times, a generic computer code was programmed in MATLAB (2014) software. The genetic algorithm method was used to minimise the risk function defined in Equation (1). Result of maintenance time optimisation for all the failure modes and components is shown in Figure 7. As it can be seen, at each time (when) the failure mode (what) and the component to be repaired (where) is determined as the output of the maintenance strategy.

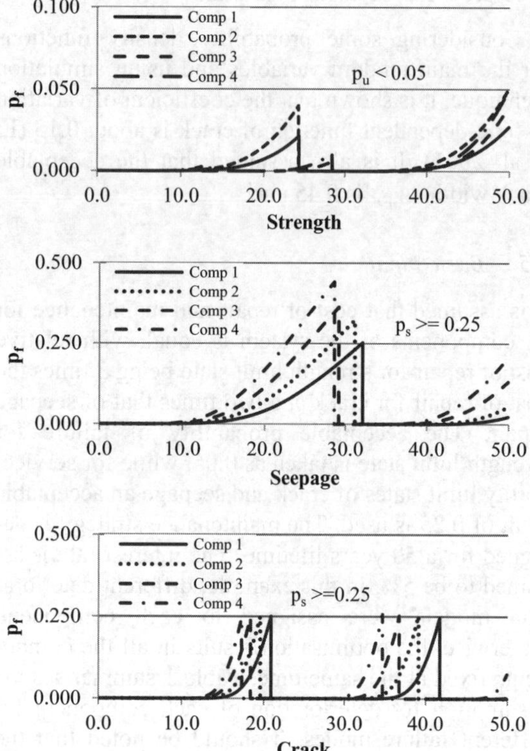

Figure 7. Optimised maintenance times.

6 CONCLUSION

A novel risk-cost optimisation maintenance strategy considering different failure modes and system components is proposed. Implementation of the method through an example confirms applicability of the method in maintenance of tunnel structures.

ACKNOWLEDGEMENT

Financial support from Metro Trains Melbourne, Australia and Australian Research Council under DP140101547 and LP150100413 is gratefully acknowledged.

REFERENCES

Al-Osta, M. A., A. K. Azad and S. Z. Selim 2015. Probability of Failure of Corroding Reinforced Concrete Columns under Eccentric Loading, *Arabian Journal for Science and Engineering*, **40**, 1-8.

Aldea, C.-M., S. Shah and A. Karr 1999. Permeability of cracked concrete, *Materials and structures*, **32**, 370-376.

Azad, A. K. and M. A. Al-Osta 2014. Capacity of Corrosion-Damaged Eccentrically Loaded Reinforced Concrete Columns, *ACI Materials Journal*, **111**, 711-721.

Barone, G., D. M. Frangopol and M. Soliman 2013. Optimization of life-cycle maintenance of deteriorating bridges with respect to expected annual system failure rate and expected cumulative cost, *Journal of Structural Engineering*, **140**, 1-13.

Deb, K. 2001. *Multi-objective optimization using evolutionary algorithms,* John Wiley & Sons, New York.

Li, C.-Q., R. E. Melchers and J.-J. Zheng 2006. Analytical model for corrosion-induced crack width in reinforced concrete structures, *ACI Structural Journal*, **103**, 479-487.

Li, C. 2003. Life cycle modeling of corrosion affected concrete structures-initiation, *Journal of Materials in Civil Engineering*, **15**, 594-601.

Li, C. and R. Melchers 1993. Outcrossings from convex polyhedrons for nonstationary Gaussian processes, *Journal of Engineering Mechanics*, **119**, 2354-2361.

Li, C. Q., R. Ian Mackie and W. Lawanwisut 2007. A Risk-Cost Optimized Maintenance Strategy for Corrosion-Affected Concrete Structures, *Computer-Aided Civil and Infrastructure Engineering*, **22**, 335-346.

MathWorks 2014. *Global optimization toolbox 3.2.2 user's guide*, Natick, MA.

Melchers, R. E. 1999. *Structural reliability analysis and prediction,* John Wiley & Sons, New York.

Mori, Y. and B. R. Ellingwood 1994. Maintaining reliability of concrete structures. II: Optimum inspection/repair, *Journal of Structural Engineering*, **120**, 846-862.

Murata, J., Y. Ogihara, S. Koshikawa et al. 2004. Study on watertightness of concrete, *ACI Materials Journal*, **101**, 107-116.

Sommer, A. M., A. S. Nowak and P. Thoft-Christensen 1993. Probability-based bridge inspection strategy, *Journal of Structural Engineering*, **119**, 3520-3536.

Thoft-Christensen, P. and J. D. Sorensen 1987. Optimal strategy for inspection and repair of structural systems, *Civil Engineering Systems*, **4**, 94-100.

Yuan, Y., X. Jiang and X. Liu 2013. Predictive maintenance of shield tunnels, *Tunnelling and Underground Space Technology*, **38**, 69-86.

Proceedings of the International Conference on Smart Infrastructure and Construction
ISBN 978-0-7277-6127-9

© The authors and ICE Publishing: All rights reserved, 2016
doi:10.1680/tfitsi.61279.681

Testing a PPP Performance Evaluation Framework

J. Liu*[1], P.E.D Love[2], J. Smith[3] and C.P. Sing[4]

[1,2] *Curtin University, Perth, Australia*
[3,5] *Bond University, Gold Coast, Australia*
[4] *City University of Hong Kong, Hong Kong*
* *Corresponding Author*

ABSTRACT Public-Private Partnerships (PPPs) have-become an integral part of infrastructure procurement strategy in many governments across the world. However, the use of PPPs has been being plagued with controversy, especially in the UK and Australia, as some procured ~~PPP~~ projects have experienced significant time and cost overruns and poor or less than optimal operational performance. A perspective that has been raised is that the unsatisfactory delivery of PPPs over the past decade has resulted from the absence of an ineffective and incomplete performance evaluation. Typically evaluation has focused been *ex-post* measuring construction deliverables and thus has ignored the projects' inception phses. With this in mind, an innovative evaluation framework is presented and empirically tested using a case study and Confirmatory Factor Analysis to evaluate the outputs of the early formative stages of PPPs (i.e., initiation, planning and procurement). The developed evaluation framework can provide governments with a reliable tool for measuring and managing their PPP projects.

1 INTRODUCTION

A plethora of factors contribute to the successful delivery of a PPP project. Problematic issues related to time and cost overruns in PPPs over the past decade have resulted from an incomplete and ineffective performance evaluation throughout their life-cycle (Liu *et al.*, 2015a). In fact, most PPPs have not undergone a comprehensive evaluation in terms of what had been delivered (Regan *et al.*, 2011).

Performance evaluation is essential for business success, particularly at the corporate and project levels (Love and Holt, 2000). Liu *et al.* (2015b) have suggested that effective and efficient project evaluation is a critical success factor (CSFs) for PPPs. Essentially, evaluating project performance is a core activity of the contract management of PPPs in most developed countries (European Investment Bank – EIB, 2011). Despite the importance evaluation, limited research was undertaken to explore this critical issue (Liu *et al.*, 2014). Against this contextual backdrop, this paper empirically develops and tests an innovative life cycle PPP performance measurement framework (PMF).

2 RESEARCH OF PPPs

PPPs have been perceived to be a time and cost efficient procurement approach for infrastructure projects. As a result, there has been an inordinate amount of PPP research. Liu *et al.* (2015) have identified that there are six key research areas in the normative literature of PPPs, including governments' roles and responsibilities, selection of concessionaire, risk identification and allocation, cost and time efficiencies of different contracts and project finance.

Although PPPs have attracted intensive attention of researchers, there have been limited studies that have empirically examined the performance of PPPs (Love *et al.*, 2015). Therefore, this study bridges this knowledge gap and contributes to the normative literature with an introduction of a Performance Management Framework (PMF) that provides governments with a reliable tool for measuring the outputs of inception phases of their projects.

3 METHODOLOGY

A detailed case study relying on an array of documentary sources and semi-structured interviews was conducted to develop hypotheses for conceptualising a PMF. The case used in this paper was chosen by communicating with a project director, who was experienced in delivering infrastructure PPPs, during the initial stage of the study. Thus, the findings of the recommended case study can be assumed to be reliable and significant.

Following the case study, a conceptual PMF was developed, constituting five performance measurement perspectives and a series of relevant key performance indicators (KPIs). To empirically test the feasibility of the developed framework, a questionnaire was designed to solicit PPP practitioners' views and insights about the PMF. The hypotheses to be tested using the survey are indicated as follows:

$F^1 - H_0$: The measurement perspectives are not significant for measuring PPPs.

$F^1 - H_1$: All measurement perspectives are significant for measuring PPPs.

$F^2 - H_0$: The conceptually-derived KPIs are not significant for measuring PPPs.

$F^2 - H_1$: The majority of the conceptually-derived KPIs are significant for measuring PPPs.

A *Likert* scale ranging from *1* (strongly disagree) to *5* (strongly agree) and purposive sampling were applied to the survey. Purposive sampling is useful when the researchers require the expertise of individuals who are specialised in particular fields to deal with a topic that integrates with a high level of uncertainties (Foreman, 1991). Respondents who participated in this study had to be knowledgeable about multiple aspects of PPPs, many of which are considered to be dynamic and uncertain projects (Yuan *et al.*, 2009). Thus, purposive sampling was ideal for this type of study and the target population were senior management personnel who had been involved in the procurement of social PPPs (Hodge, 2004).

The developed questionnaire was distributed to the selected respondents via *SurveyMonkey*. The collected data was analysed by applying Confirmatory Factor Analysis (CFA), which is a multivariate process to statistically examine how well the variables being measured represent their constructs. CFA relies on pre-constructed theory and it is used for confirming the theoretical relationships, rather than exploring the linkages between the items (Schreiber *et al.*, 2006).

4 CASE STUDY OF A SOCIAL PPP PROJECT

4.1 *Description of the Case Project*

The selected PPP project was initiated with an aim of delivering a new facility to replace an existing regional prison, which was built in the 1980s with 100 beds. The business case of Project-P for a redevelopment of a new prison was approved in 2009 by the State Treasury, and comprised: (1) 200-bed male medium security prison with the capacity to accommodate 20 maximum security prisoners; (2) 60-bed male minimum security sector; (3) 40-bed male open minimum security sector; and (4) 50-bed female maximum, medium and minimum sector including a 6-bed unit for women with children.

The procurement arrangement that the selected PPP undertook was a Design, Build, Finance and Maintenance (DBFM) model. The State Government expected value for money (V*f*M) from the private sector, which involved asset durability, efficiency and productivity, innovation in design, lower life-cycle costs and quality outputs.

4.2 *Delivery Process of Project-P*

After an evaluation of the Expression of Interests (EOI) submissions, the PPP project proceeded to the Request for Proposal (RFP), whereby the short-listed respondents of were requested to submit fully-costed and complete proposals in 2012. The government then evaluated the received proposals against a range of criteria, which included organisational structure, stakeholder relationship management, design solution and management, project management, delivery of facility maintenance (FM) services.

Table 1 reports the delivery process and timeframe of Project-P.

Table 1: Project-P delivery process and timeframe

Phases of Delivery Process	Timeframe
Initiation and Planning	
Business case study	2009 - 2011
Definition and procurement option	
Invitation for EOI	
Evaluation of the EOIs	
Procurement	
Release of RFP	Jan-Dec, 2011
Tendering & final negotiation	
Contract and financial close	
Design and Construction	
Commencement of construction	2013 – 2015
Stage 1 & 2 works completion	
Facility Maintenance	
Operation and FM	Since mid. 2015
Handover	

Source: 'Project Summary' of Project-P, p.12

4.3 *Practice in Performance Evaluation*

The aim of this paper is to develop a PMF for evaluating the life cycle of PPP, which were handled by the government. Hence, interviews with the Procurement Director, who was responsible for overseeing such aforementioned phases of Project-P, were conducted. The questions used for the interview focused on how Project-P was evaluated throughout its early stages. These included:

- How did you measure the deliverables of the inception stages of the project?
- What do you consider to be limitations in the current approach you used to measure the project's early-stage outputs?

It was identified from the interviews that the approach adopted to evaluate Project-P during its formative stages, for example, focused on the reviews of the business case and tender decision. As the Director stated:

> "In the pre-contracting phase, we had reviews for the business case development and the decision on the tendering to examine whether they are appropriate. Before these reviews, we also conducted a V*f*M assessment for the project."

This point of view was supported by the 'Project Summary', which stated that the evaluation approach of the inception phases of Project-P incorporated three parts: (1) a V*f*M assessment; (2) a review of the business case; and (3) a review of the appropriateness of the tendering decision (Figure 1).

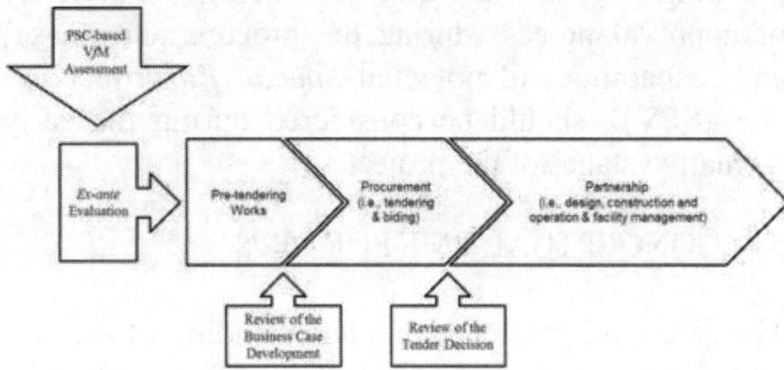

Figure 1: Performance evaluation in pre-contract of Project-P

4.4 *Limitations of Current Evaluation*

The Director identified that there were some limitations with the existing evaluation approach that was being used and stated:

> "Performance evaluation is used for improvement. In the case of this project, improving is first about the efficiency of the Procurement. It sometimes took us a little bit long to pass an approval process. Competition is an aspect we expected in the Procurement phase. But this is missed in the project's performance evaluation and it might be necessary for us to have a mechanism to examine whether the competitiveness of bidding can achieve the level we expected."

The Director also considered that the management of evaluation information in Project-P needed to be improved and an effective and efficient internal process would be required to capture the lessons learned from the evaluation results. Additionally, as outlined in the 'Project Summary', the V*f*M assessment of Project-P still relied on the *Public Sector Comparator* (PSC). The PSC has been criticised by both academia and practitioners owing to its pure focus on financial benefits. The Director acknowledged that the V*f*M assessment of Project-P should be ameliorated, and stated:

> "I think PSC gave us good results but we will have to continue refining it, of course, because it is true that PSC is a quantitative estimation and does not embrace important qualitative issues of PPPs."

To solve this problematic issue, the interview respondent tended to suggest that a process-based measurement with the measures capable of capturing the key stakeholders' expectations and strategic goals would be a promising way for future amelioration. The Director, for example, suggested other aspects, such as 'V*f*M for non-financial benefits', 'project planning', 'competitiveness of bidding', 'efficiency of approval process during the procurement phase' and 'capabilities of potential *Special Purpose Vehicles* (SPV)', should be considered during the early formative stages of the project.

5 CONCEPTUAL PMF FOR PPPs

The case study provided an understanding of the current practice used to evaluate the performance of a PPP used to deliver a social infrastructure project. It was identified that the inception phases of a PPP evaluation (e.g., Project-P) possess a range of 'gaps', involving an incomplete V*f*M assessment, incomprehensive measurement for the essential deliverables and ineffective and inefficient internal process used to supporting learning.

As suggested from the interviews, a process-based performance evaluation with learning mechanisms would be useful for government to improving that the performance their infrastructure assets delivered by PPPs. This view is supported by the research conducted by Liu *et al.* (2015a), who have suggested that process-based measurement is robust to effectively evaluate the performance of PPPs over their life cycles as it can capture all essential works required to deliver them. A process-based PMF is conceptualised with an aim of measuring the inception stages of PPPs (Figure 1).

Key Performance Indicators (KPIs) form the heart of performance measurement systems (PMSs) (Neely *et al.*, 2005). Hence, the *Performance Prism* developed by Neely *et al.* (2001) is applied to derive performance measures for the proposed PMF. Neely *et al.* (2002) has stated that the Performance Prism is a stakeholder-focused framework that can shed light on organisational measurement within a multiple stakeholder environment. In addition, Liu *et al.* (2015c) have demonstrated that Performance Prism is suitable for deriving measures for the PMSs devised for PPPs as it can accommodate their inherent complexities and uncertainties that result from a highly sophisticated stakeholder network.

The Performance Prism encompasses *five* measurement perspectives: (1) Stakeholder Satisfaction; (2) Strategies (i.e., project strategic goal); (3) Processes (i.e., project delivery process); (4) Capabilities (i.e., organisations' capabilities in delivering the project); and (5) Stakeholder Contribution (Neely *et al.*, 2001). Noteworthy, a sequence of phase-based KPIs that address the stakeholders' needs and expectations can be developed for the proposed PMF (Figure 2).

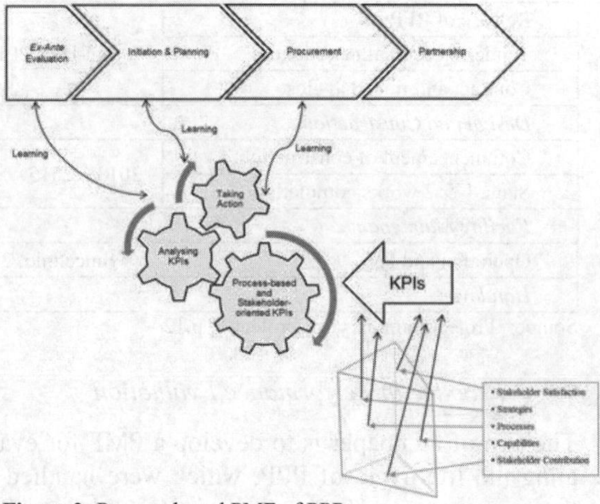

Figure 2: Process-based PMF of PPPs

6 FEASIBILITY TESTING

A questionnaire was designed to solicit senior practitioners' perspectives about the hypotheses derived from the case study. A pilot survey was conducted to test the reliability of the developed research instrument. It was distributed to 28 experienced PPP practitioners. A total of 25 responses were received, equating to a responsive rate of 89%.

Throughout the data collection period, a total of 368 questionnaires were distributed to practitioners who were knowledgeable of about issues associated with both the public and private sectors involvement with the delivery of PPPs. A total of 141 responses were received; 6 responses had to be discarded because of incompleteness. Data completeness is a prerequisite of CFA. Hence, a total of 135 valid datasets were entered into SPSS for related analysis.

Cronbach's α and the corrected item-total statistics were applied prior to conducting the CFA. All values generated from these two tests exceed 0.70 and 0.30, which indicate a high degree of internal consistency and reliability of the observed items (Nunnally, 1978; Nunnally and Bernstein, 1994).

After conducting the reliability tests, the CFA was undertaken. Essentially, CFA is a theory-driven factor analysis technique and the theoretical linkages drive its formulation between the observed and unobserved variables (Schreiber *et al.*, 2006). Within the configuration of the developed PMF (Figure 2), the five measurement perspectives and their relevant KPIs are viewed as the observed variables, while the outputs/deliverables of each PPP phase are considered to be the unobserved variables.

Initially, the CFA-hypothesised model was constructed to estimate a covariance matrix of the survey population in comparison with an observed covariance matrix. Accordingly, the formulated model was applied to examine whether the observed items (five performance measurement perspectives and their KPIs) were significant or could significantly contribute to the performance of a PPP. Items with comparatively low 'factor loadings' were eliminated to allow modifications to the initially hypothesised model to develop an optimal model.

The hypothesised model captures the features of the developed PMF, whereby the process-based KPIs are proposed according to the five measurement perspectives that were assumed to be causally significant to the performance of a PPP project.

The optimal models developed indicated that the coefficients of the five measurement perspectives are 0.78, 0.82, 0.77, 0.75 and 0.76, respectively, all of which are significant under 5% significance level. These suggest that such proposed perspectives are significant for measuring the performance of PPPs during their formative stages.

Apart from the findings relating to the five measurement perspectives, a sequence of critical implications can be drawn from the generated empirical evidence regarding the KPIs. Under the pre-contract phases (i.e., Phase 1: Initiation and Planning; Phase 2: Procurement), the coefficients of most of the derived KPIs are larger than 0.50 and are significant (under 5% significance level), which means that such observed KPIs are significant for the performance measurement of PPPs (Yuan *et al.*, 2012).

Most respondents acknowledged that the PPP market in Australia to be sophisticated and mature. Essentially, PPPs have become an integral part of both the federal and state governments' procurement strategies in Australia (Regan *et al.*, 2011). Australian PPP projects are procured by following strict well-developed guideline and processes (Infrastructure Australia, 2008). As a result, state governments and an array of private-sector entities within Australia have become experienced in procuring and delivering infrastructure projects via PPPs. The public and private sectors are familiar with solving process issues such as financing options and designing and determining an appropriate concession period, and/or how to organise and govern well the tendering and efficiently achieve financial close. This may explain why the KPIs relating to PPP project's finance option concession period and financial close, and the government's governance ability for the procurement phase were considered to be insignificant by the respondents of this survey.

After eliminating the four insignificant indicators an optimal model was constructed. The estimates of the optimal model are larger than 0.50 and significantly correlated to the performance of PPPs at a 5% significance level.

To examine the model, three Goodness-of-Fit Indexes (GFIs) were used: (1) *Chi-squared* statistic; (2) comparative fit index (CFI); and (3) root mean square error of approximation (RMSEA). Such GFIs are widely used for indicating how well the constructed structural model fits a set of observations (Sanders *et al.*, 2006). Table 2 identifies the benchmark values of the aforementioned GFIs. In other words, the constructed structural model is deemed as a 'fitted' model if its GFIs are within the intervals of the benchmark values.

Table 2: Benchmark values of the GFIs

Goodness-of-fit Indexes	Benchmark Values
Chi-squared	$1 \leq x^2/Df \leq 5$
CFI	≥ 0.90
RMSEA	$0.05 \leq$ good model fit ≤ 0.1

The empirical evidence relating to the three GFIs of the constructed optimal model are summarised as follows: 2.32 (*Chi-squared* statistic), 0.92 (CFI) and 0.076 (RMSEA). These indicate a good model fit. On the basis of the results derived above, it can be found that all of the five proposed measurement perspectives are significant for the performance measurement of PPPs, while 28 indicators out of the 32 proposed KPIs 'passed' the quantitative tests that relied on CFA. Therefore, the null hypotheses proposed above are rejected. The empirical findings confirmed the main proposition, that the stakeholder-oriented measurement perspectives and their relevant KPIs are significant for future PPP performance evaluation.

7 CONCLUSION

This paper proposed and empirically tested a life cycle PMF for evaluating the outputs of PPPs with particular emphasis being placed on the formative phases, such as initiation, planning and procurement. A detailed case study was used to develop the conceptual framework. Then CFA was performed to testing its validity using a questionnaire survey. Four insignificant KPIs were eliminated, and it was identified that the main components of the developed PMF (five measurement perspectives and 28 process-based KPIs) were feasible and applicable for evaluating social infrastructure PPPs.

The research has contributes to the literature by filling the knowledge gap of PPP performance evaluation. As the PMF was developed from a 'real-world' project and validated by experienced practitioners, this paper provides governments that will embark on PPPs with a robust tool and conceptual foundation to design effective and efficient PMSs for effectively and efficiently evaluate their future projects.

ACKNOWLEDGEMENT

The authors would like to thank those participants who participated in this study. The authors would also like to acknowledge the financial support provided by the Australian Research Council (LP120100347).

REFERENCES

EIB 2011. The guide to guidance: how to prepare, procure and deliver PPP projects, Luxembourg

Foreman, E.K. 1991. *Survey sampling principles*, Marcel Dekker, NY, USA

Hodge, G.A. 2004. The risky business of public-private partnerships. *Australian Journal of Public Administration*, **63**, 37-49.

Liu, J., Love, P.E.D., Smith, J., Regan, M. & Sutrisna, M. 2014. Public-private partnerships: A review of theory and practice of performance measurement. *International Journal of Productivity and Performance Management*, **63**, 499-512.

Liu, J., Love, P.E.D, Davis, P.R., Smith, J. & Regan, M. 2015a. Conceptual framework for the performance measurement of public-private partnerships, *Journal of Infrastructure Systems*, **21**, 04014023.

Liu, J., Love, P.E.D., Smith, J., Regan, M. & Davis, P.R. 2015b. Life cycle critical success factors for public-private partnership infrastructure projects, *Journal of Management in Engineering*, **31**, 04014073

Liu, J., Love, P.E.D., Smith, J., Regan, M. & Palaneeswaran, E. 2015c. Review of performance measurement: implications for public-private partnerships, *Built Environment Project and Asset Management*, **5**, 35-51.

Neely, A., Adams, C. & Crowe, P. 2001. The performance prism in practice. *Measuring Business Excellence*, **5**, 6-12.

Neely, A., Gregory, M., & Platts, K. 2005. Performance measurement system design: a literature review and research agenda. *International Journal of Operations & Production Management*, **25**, 1228-63.

Nunnally, J.C. 1978. *Psychometric Theory*. 2nd ed., McGraw-Hill, NY, USA

Nunnally, J.C. and Bernstein, I.H. 1994. *Psychometric Theory*, 3rd ed., McGraw-Hill, NY, USA

Regan, M., Smith, J. & Love, P.E.D. 2011. Infrastructure procurement: Learning from public-private experience 'down under', *Environment Planning C: Government and Policy*, **29**, 363-378.

Schreiber, J.B., Nora, A., Stage, F.K., Barlow, E.A., and King, J. 2006. Reporting structural equation modeling and confirmatory factor analysis results: A review, *The Journal of Educational Research*, **99**, 323-338.

Yuan, J., Zeng, A.Y., Skibniewski, M.J., & Li, Q. 2009. Selection of performance objectives and key performance indicators in public-private partnership projects to achieve value for money. *Construction Management and Economics*, **27**, 253-270.

Yuan, J., Wang, C., Skibniewski, M.J., and Li, Q. 2012. Developing key performance indicators for Public-Private Partnership projects: Questionnaire survey and analysis, *Journal of Management in Engineering*, **28**, 252-264.

Proceedings of the International Conference on Smart Infrastructure and Construction
ISBN 978-0-7277-6127-9

© The authors and ICE Publishing: All rights reserved, 2016
doi:10.1680/tfitsi.61279.687

Consequence-based management of North American railroad bridge networks enabled by wireless smart sensors

F. Moreu*[1], B.F. Spencer, Jr.[2], D.A. Foutch[2] and S. Scola[3]

[1] *Department of Civil Engineering, University of New Mexico, USA*
[2] *Newmark Structural Engineering Laboratory, University of Illinois at Urbana-Champaign, USA*
[3] *Bridges and Structures, Canadian National Railway, USA*
* *Corresponding Author*

ABSTRACT North American railroads are exploring means to improve the management of their bridge networks to increase overall profitability. Current maintenance, repair, and replacement (MRR) decisions are informed by bridge inspections and ratings, which recommend observing the response of bridges under trains. However, an objective relationship between bridge responses, bridge service state condition, and the associated impact to railroad operations has yet to be established. If the consequences of MRR decisions could be better determined, then the railroads could more cost-effectively allocate their limited resources. This paper develops an approach for consequence-based management of railroad bridge networks, adopted from the field of seismic risk assessment, for making MRR decisions on a network-wide basis. The proposed framework employs fragility curves to relate service condition limit-states to transverse displacement caused by traffic. The operational costs associated with these service conditions can be used to estimate the total costs of a given MRR policy. In this way, optimum MRR decisions minimize the total network costs. Additionally, measured bridge data can be used to update periodically the fragilities. This framework provides a consistent approach for the prioritization of railroad bridge MRR decisions.

1 INTRODUCTION

40% of the nation's freight tonnage transported by train (AAR, 2014). Furthermore, North American railroads expect to exceed their capacities over the next 20 years at many locations within their network and need to prepare their infrastructure accordingly. Capacity is the ability of a given railroad to move a volume of traffic over a specific line under a given Level of Service (LOS) (Lay and Barkan 2009), and it is affected by the maximum operating speed allowed. Railroads in North America have different track classes corresponding to different capacities. Higher track class corresponds to higher speed and higher capacity (FRA, 2015). Railroads in North America have doubled capital investments in the last few decades to meet capacity demands (Berman, 2012). This investment, combined with technology innovations in freight cars and locomotives, has resulted in a doubling of the average tons of freight per train loading (Weatherford, 2008). As a result, freight costs per ton-mile have been reduced by roughly 50%, portending that freight carried by North American railroads will increase significantly in the future (Thompson, 2010). Cambridge Systematics, Inc. (2007) estimated the cost of infrastructure investment needed to match the 2007-2035 projected demand at $148 billion (in 2007 dollars).

Bridges are a critical component of railroad infrastructure, with an average of one bridge for every 1.4 miles of track. Railroad companies need to continuously assess the structural condition (safety) of their bridges to ensure the operational performance of rail networks (Byers and Otter 2006). Of the 100,000 bridges, a significant portion is approximately 100 years old (Cambridge Systematics, 2007). In particular, the US Department of Transportation reported that more than half were built before 1920 (AREMA, 2003). According to Unsworth (2010), the weight/car has augmented rapidly in the last decades and the bridge capacities are being exceeded in old bridges.

During the last five years, Class I railroads have consistently invested approximately \$500M annually in Maintenance, Repair, and Replacement (MMR) only for bridges, even though this has been a period of general economic recession (AAR, 2015). To determine which bridges to include in each MRR decision at the network level, railroads use bridge inspection reports (AREMA, 2008, 2014). These inspections have been federally required annually since 2010 (FRA, 2010). Because funds are limited and capacity demands are growing, railroads need to develop MRR strategies that enable safe and cost-effective operations for the increasing demands.

One of the key problems to determine optimal policies for MRR decisions is the ability to assess the condition of the bridge. Bridge malfunctions may not always be captured by regular visual inspections. In this context, bridge response to revenue service traffic is believed to be a proxy for bridge health (i.e., if the bridge is not moving while the trains are crossing, then it is assumed to be in good shape). Indeed, a top research priority of the railroad bridge structural engineering community in North America is determining bridge displacements under revenue service traffic (Moreu and LaFave, 2012). However, measuring bridges responses under traffic is currently limited to subjective observations. As a result, when bridge response data is employed in MRR decisions, it is typically only qualitative. An objective relation for the serviceability of a given bridge is lacking to inform railroad bridge managers about their MRR decisions.

The cost to the railroad to maintain the bridge network is comprised of: (i) the operational costs (OC) and (ii) the MRR costs. Operational Costs (OC), as defined herein, have two components. The first is the Operational Expense (OE), or the expense beyond MRR investments to maintain the bridge network to meet operational needs. The second component is the Lost Revenue (LR) to the railroad associated with not doing MRR on specific bridges. Railroads decrease the speed of trains over bridges of poor condition, assuming the associated expenses related to traffic delay. The Total Network Cost (TNC) is the cost of MRR, plus the OC, which is uncertain. Thus, the goal is to choose MRR policies that will minimize the expected value of the TNC. Figure 1 shows the relationship between MRR investments and the TNC over a specified period of time. While the MRR costs are deterministic, the OC are uncertain. An optimal MRR policy minimizes the total costs to the network (TNC).

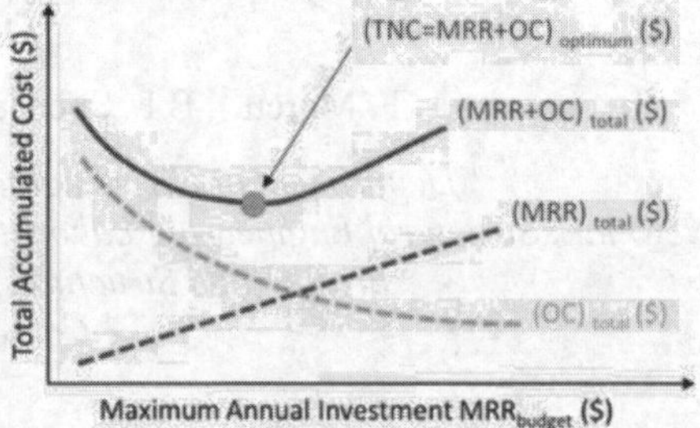

Figure 1. Minimization of Total Network Costs (TNC).

This research proposes a framework for the consequence-based management of railroad bridges for making network-wide MRR decisions. The goal established here is to minimize the expected value of the total network cost. Critical to the framework is the ability to assess bridge service condition. The proposed framework employs fragility curves to this end, which relate service condition limit-states to bridge transverse displacement under revenue service traffic. The operational costs associated with these service conditions can be used to estimate the total costs of a given MRR policy. In this way, MRR decisions can be prioritized, minimizing the total network costs to operations. This framework provides a consistent approach for intelligent management of railroad bridges, and more specifically, for the prioritization of railroad bridge MRR decisions. Using this framework, the rail owner can identify the most efficient use of a limited budget while maintaining safe railroad operations.

2 FRAMEWORK FORMULATION

This section describes the proposed consequence-based approach for the management of railroad bridge networks, using as an example timber railroad bridges in North America. The first component is the hazard, assumed as the maximum total transverse displacement of bridges, measured under a loaded train running at the maximum allowable speed for their track class. The second component of the framework is the inventory. The bridges in the inventory have already been identified by the railroad to need MRR decisions. The framework will inform how to prioritize MRR decisions of bridges within this inventory. The third component are fragility curves elaborated assuming that bridges within the inventory (component two) have similar structural properties and that the serviceability of each bridge is independent from the bridge location within the network. The fourth component uses the maximum measurement of transverse displacement for a given bridge under trains, following indications from the railroad. The fifth element calculates the operational costs per year assuming operational expenses of unplanned engineering work provided by the railroad as well as lost revenue related to delay or interruptions to traffic. The sixth component assumes that the operational costs related to the conditions of the bridge are the only variables in the decision making. The following section describes each of the six components in detail.

3 FRAMEWORK COMPONENTS

3.1 *Hazard*

The first component is the framework is the hazard. In the context of earthquake engineering, the hazard is characterized by some measure of the magnitude of an earthquake (i.e., peak ground acceleration or PGA). Fragility relations are then used to relate the PGA to the likelihood of a structure being in a certain damage state after the event. In the case of the railroad bridges, the train is the primary "hazard" or loading to the bridge. Moreu et al. (2014) found that maximum bridge transverse displacements can provide an important indication of the service level (or state) of railroad bridges (see Figure 2). Their study identified that large transverse displacements dictated installing additional bracing after field observations to increase safety and avoid lowering maximum allowable speed. This framework uses this preliminary field data to inform relations between transverse displacements and service limit states. This group of bridges and data sets is the foundational source of data informing the proposed framework. The maximum bridge transverse displacement under revenue service traffic will be considered to represent the hazard.

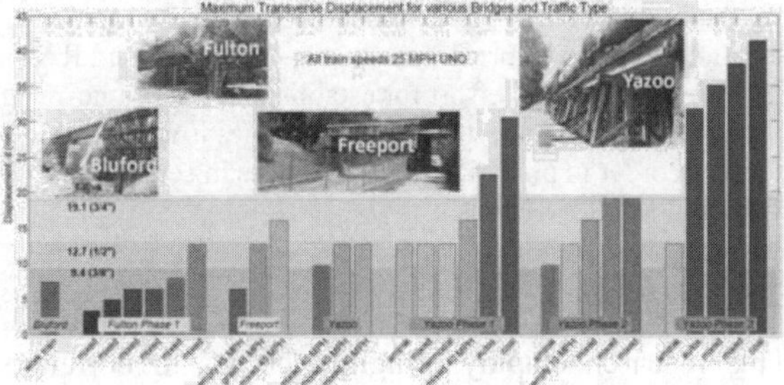

Figure 2. Figure caption.

3.2 *Inventory*

The second component is the inventory, which corresponds to the population of bridges owned by the railroad for which the MRR policies are being developed, and their current structural condition. The bridges in the inventory have already been identified by the railroad to need MRR decisions. This framework assumes that the bridges being monitored share similar structural properties and operational concerns, so that the measurement of the hazard of different components of the inventory can be used for relative comparisons within the inventory. Figure 3 shows a classification of North American railroad bridges showing percentages by length and material type (FRA, 2008).

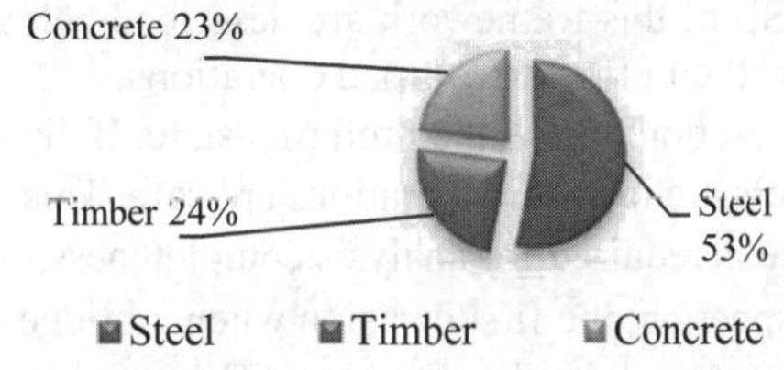

Figure 3. Railroad bridge population (a) from the most recent FRA survey (FRA, 2008)

Moreu et al. (2012) proposed a classification of the eleven railroad bridges types based on superstructure materials and structural type, developed in accordance with past railroad bridge classification efforts (AREMA, 2008; Sorgenfrei and Marianos, 2000; International Heavy Haul Association (IHHA), 2009; Parsons Brinckerhoff Quade & Douglas, Inc.,

1980; ENSCO, 1994). The current information of each bridge is provided by the railroad company owning the bridges, based on the most recent bridge annual inspection required by bridge safety standards and the FRA (2010). In this initial effort for establishing this framework, the relative importance of the bridges is assumed equal, in order to prioritize the differences in serviceability to inform MRR decisions.

3.3 *Fragility*

This research employs fragility curves to correlate bridge service condition to bridge transverse displacements under revenue service traffic. Fragility curves are a statistical tool representing the probability of exceeding a given performance (or damage) state as a function of an engineering demand parameter. In this paper, service limit-states (SL) represent the consequences to rail operations associated with bridge displacement. This framework proposes five different SL of railroad bridge serviceability using bridge performance under trains. Freight trains can be conservatively assumed to have the same weight, whereas their interaction during train crossing is different depending on multiple factors, such as the train speed and geometry and condition of both track and bridge (Hussain et al. 1980; FRA 2005; FRA 2012; Wolf, 2005; Watco, 2012). Based on past work by Moreu et al. (2014, 2015), loaded freight trains (of near hundred loaded cars) produce comparable maximum transverse displacements of individual timber railroad bridges if running at similar speed and direction. The SL of this framework are described below, followed by their effect to railroad operations:

- SL_0 – No Action: this is the preferred state. If displacements are low, rail operations are safe. This limit state is required for analysis completeness.
- SL_1 – Inspection: the first decision when a bridge moves excessively under regular traffic.
- SL_2 – Temporary Slow Order (TSO): the speed of trains is reduced with a TSO (plus maintenance).
- SL_3 – Permanent Slow Order (PSO): if the TSO does not address the serviceability of the bridge, a PSO is ordered to secure safe railroad operations, permanently slowing traffic over the bridge.
- SL_4 – Track Outage (TO): when the bridge condition is not safe for train crossing, the bridge is put out of service.

Figure 4 provides a conceptual representation of the variability between bridge displacements and the SLs.

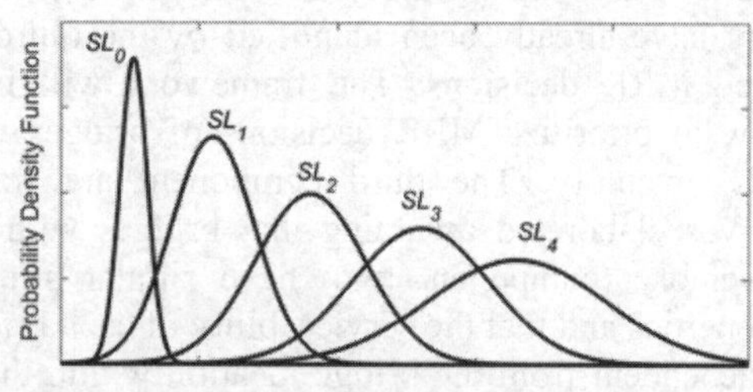

Figure 4. Probability density function of transverse displacements for each SL.

In the context of this research, the fragility function $F_{k,j}(d)$ is defined as the probability of being in service limit-state $SL \leq k$, given that the maximum displacement of the bridge is *d* for track class $Z = j$, i.e.,

$$F_{k,j}(d) = P(SL_k | D = d, Z_j)$$

where

D = random variable representing the maximum measured transverse displacement of the bridge,

d = realization of the random variable D,

SL_k = service limit-state, with $k = 1, 2, 3, 4$,

Z_j = track class, with $j = 1, 2, 3, 4, 5$.

A fragility curve for a particular railroad bridge SL_k is obtained by computing the conditional probabilities of a given SL_k being exceeded (Figure 5).

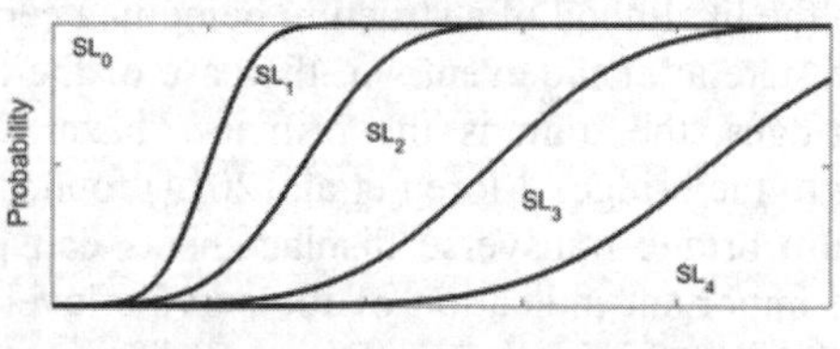

Figure 5. Conceptual depiction of railroad bridges fragility curves $F_k(d)$ for different SL_k for one specific track class Z_j.

3.4 *Campaign Monitoring Data*

Current advances in sensing technology now permit railroad managers to collect bridge transverse displacements under trains using wireless smart sensors in almost real-time, inexpensively and effectively. Wireless Smart Sensors (WSS) can provide reference-free displacements of multiple bridges with moderate effort. These measurements can inform of bridge condition and provide evidence to inform prioritizing or delaying MRR decisions (Moreu et al. 2015).

3.5 *Annual Operational Costs*

The fifth component of the framework is the relation between limit-states and operational costs to the owner for a given bridge. Service limit-states are related to operational expenses depending on both the bridge type and the LOS. For each bridge, the annual operational costs OC for each limit state has two components: operational expenses, or the bridge engineering expense (i.e., the cost of MRR) OE and lost revenue expense LR (e.g., caused by slow orders or by track outages). The total operational cost OC for a bridge is

$$OC = OE + LR \tag{2}$$

This paper estimates expenses assuming that the service limit states are mutually exclusive, as is done in seismic risk assessment (Shinozuka, 2000). Thus, the annual expected operational costs for one bridge can be calculated as:

$$< OC^n >= \sum_{k=1}^{K} P\left(SL = k \middle| D = d_{measured}, Z = j\right) \cdot OC_k^n \tag{3}$$

where

$< OC^n >$ = annual expected operational,

SL_k = service limit-state, with $k = 1, 2, 3, 4$,

OC_k^n = total expense for a bridge n in the kth SL.

3.6 *Annual Operational Costs*

This component of the framework permits railroads to minimize total network cost at the network level. Using the information from the prior components, operational costs are calculated from multiple MRR policies for a given population of bridges. The proposed framework minimizes operational costs for a given MRR policy, improving MRR budget decisions within the network. Consequence-based management can provide savings by quantifying the costs associated to service levels based on performance measurements.

4 CONCLUSIONS

This paper develops a consequence-based framework that prioritizes MRR decisions of railroad bridge networks, estimating the operational costs of bridge SL given bridge responses. The goal is to minimize the expected value of the total network cost. Critical to the framework is the ability to assess bridge service condition. Railroads can collect objective performance information of the bridge service condition under revenue service traffic using WSS. Fragility curves relate the measured bridge performance with a SL and also calculate the operational costs associated for each specific bridge and location. The railroad can prioritize the upgrading of their railroad bridge networks finding the optimal MRRs that minimized operational costs.

Additional future work includes measuring both track and bridge responses under different service limit-states. Finally, using evidence of transverse displacements of changes of bridge serviceability can assist to determine and include limit(s) on transverse displacements in their assessment practice and/or the AREMA manual, in addition to the current AREMA limit on normalized vertical displacements under trains. This framework provides an intelligent use of bridge response information to inform consequence-based management of railroad bridge networks, minimizing railroad bridge total network costs.

ACKNOWLEDGEMENT

The financial support from the following sources in supporting this research is gratefully acknowledged: the American Association of Railroads (AAR) Technology Scanning Program; the O. H. Ammann Research Fellowship of the Structural Engineering Institute (SEI) - American Society of Civil Engineers (ASCE); the Talentia Fellowship (Junta de Andalucía, Spain); the Illinois Graduate College Dissertation Travel Committee at the University of Illinois at Urbana-Champaign (UIUC); the Federal Railroad Administration (FRA) ; the Foreign Language and Area Studies (FLAS) Fellowships program (from the Department of Education of the United States); the Center for Global Studies and the Center for East Asian and Pacific Studies at the University of Illinois. The authors thank CN railways for their support in this research.

REFERENCES

Ang, A. H. S., & Tang, W. H. (2007). Probability concepts in engineering: emphasis on applications in civil & environmental engineering (Vol. 1). New York: Wiley.

AREMA (2003); Practical Guide to Railway Engineering.

AREMA (2008). "AREMA Bridge Inspection Handbook"; AREMA Committee 10, Structures, Maintenance, & Construction.

American Railway Engineering and Maintenance-of-Way Association (AREMA). (2014). Manual for Railway Engineering; Volume 2: Structures; Chapter 10: Structures, Maintenance and Construction; April.

AAR (2014). "A short history of U.S. freight railroads." www.aar.org/BackgroundPapers/A%20Short%20History%20of%22US%20Freight%20Railroads.pdf (Jun. 18, 2014).

AAR (2015). <http://freightrailworks.org/> (March 10, 2015).

Berman, Jeff (2012). "Class I railroads are on track to spend $13 billion in 2012 capital expenditures, says AAR". Logistics Management. January 30th.

Byers, W. G. and Otter, D. (2006). "Reducing the Stress State of Railway Bridges with Research: Researchers at TTCI Stay on Top of Railway Bridge Research to Ensure Safety, Cost Effectiveness and Maximum Life Cycle of Materials." Railway Track and Structures: RT & S, 1953, Chicago: Simmons Boardman Pub. Corp., February.

Cambridge Systematics, Inc. (2007) "National Rail Freight Infrastructure Capacity and Investment Study."

ENSCO Inc., Applied Technology & Engineering Division (1994), "Overview of Railroad Bridges and Assessment of Methods to Monitor Railroad Bridge Integrity"; U.S. Department of Transportation, Federal Railroad Administration; Report number DOT/FRA/ORD-94/20; June.

FRA (2005); "Safe placement of train cars: a report"; U.S. Department of Transportation Federal Railroad Administration (FRA); report to the Senate Committee on Commerce, Science and Transportation and the House of Committee on Transportation and Infrastructure; June.

FRA (2008), "Railroad Bridge Integrity Working Group Upgrade"; Railroad Safety Advisory Committee (RSAC), Railroad Bridge Working Group; Presentations (Railroad Bridge Working Group Report: Final Report and Recommendations); September 10th.

FRA (2010). "Bridge Safety Standards." DOT 49 CFR Parts 213 and 237, RIN 2130-AC04, Federal Register / Vol. 75, No. 135 / Thursday, July 15, 2010/Rules and Regulations. Pp. 41281-41309. [http://www.fra.dot.gov/downloads/BridgeSafetyStandardsFinalRle.pdf] (August 23rd, 2011).

FRA (2012); "Train Accidents Cause Codes"; Office of Safety Analysis (Appendix C of the FRA Guide for Preparing Accident/Incident Reports), p. 2 (downloaded November 19, 2012). [http://safetydata.fra.dot.gov/OfficeofSafety/publicsite/downloads/appendixC-TrainaccidentCauseCodes.aspx?State=0]

Federal Railroad Administration (FRA) (2015); Track Safety Standards; Title 49: Transportation, Part 213, (downloaded March 6, 2015) [http://www.ecfr.gov/cgi-bin/text-idx?c=ecfr&rgn=div5&view=text&node=49:4.1.1.1.8& idno=49#49:4.1.1.1.8.3.5.3]

Hussain, S.M.A., Garg, V.K., Singh, S.P. (1980); "Harmonic Roll Response of a Railroad Freight Car"; ASME Journal of Engineering for Industry, Volume 2, Issue 3, p. 282-288 (August).

IHHA (2009), "Guidelines to Best Practices for Heavy Haul Railway Operations. Infrastructure Construction and Maintenance Issues"; D&F; Scott Publishing, Inc., International Heavy Haul Association, 656 pp.

Lai, Y. C. R., & Barkan, C. P. (2009). Enhanced parametric railway capacity evaluation tool. Transportation Research Record: Journal of the Transportation Research Board, 2117(1), 33-40.

Moreu, F. and LaFave, J.M. (2012); "Current Research Topics: Railroad Bridges and Structural Engineering." NSEL Report No. 032, University of Illinois at Urbana-Champaign. [http://hdl.handle.net/2142/34749]

Moreu, F.; LaFave, J.; Spencer, B. (2012);"Structural health monitoring of railroad bridges – research needs and preliminary results"; Structures 2012 Congress (ASCE-SEI 2012), Chicago, IL. [http://dx.doi.org/10.1061/41016(314)36]

Moreu, F., Jo, H., Li, J., Kim, R.E., Cho, S., Kimmle, A., Scola, S., Le, H., Spencer Jr., B.F., and LaFave, J.M. (2014); "Dynamic Assessment of Timber Railroad Bridges using Displacements", Journal of Bridge Engineering.
[http://dx.doi.org/10.1061/(ASCE)BE.1943-5592.0000726]

Moreu, F.; Jo, H.; Li, J.; Kim, R.E.; Scola, S.; Spencer, Jr., B. F.; and LaFave, J. M. (2015); "Reference-Free Displacement Estimation and Assessment for Railroad Bridges using Wireless Smart Sensors"; Journal of Bridge Engineering.
[http://dx.doi.org/10.1061/(ASCE)BE.1943-5592.0000805]

Parsons Brinckerhoff Quade & Douglas, Inc. (1980); "Track and Bridge Maintenance Research Requirements"; U. S. Department of Transportation, Federal Railroad Administration; Report Number FRA/ORD-80/11; March.

Shinozuka, M., Feng, M. Q., Lee, J., & Naganuma, T. (2000). Statistical analysis of fragility curves. Journal of Engineering Mechanics, 126(12), 1224-1231.

Sorgenfrei, D.F. and Marianos, Jr., W.N. (2000); "Railroad Bridges"; Bridge Engineering Handbook; Ed. Wai-Fah Chen and Lian Duan; Boca Raton: CRC Press.

Thompson, Louis. (2010). "A vision for railways in 2050." International Transport Forum 2010.

Unsworth, J. F. (2010). Design of modern steel railway bridges. CRC Press.

Watco (2012); "System Special Instructions"; Effective January 15, 2012 (downloaded November 19, 2012).
[http://www.watcocompanies.com/safety/pdfs/2012%20Watco%20SSI%20Draft%20Final%2012-19-11.pdf]

Weatherford, Brian A., Willis, Henry H., Ortiz, David (2008). "The State of U.S. Railroads, a Review of Capacity and Performance Data." RAND Corporation.

Wolf, G. (2005); "It takes three to Rock and Roll (Causes and Prevention of Harmonic Rock and Roll"; Rail Sciences, Inc. (February).

Proceedings of the International Conference on
Smart Infrastructure and Construction
ISBN 978-0-7277-6127-9

© The authors and ICE Publishing: All rights reserved, 2016
doi:10.1680/tfitsi.61279.693

An explorative study to unveil the interdependency of public housing related infrastructure assets in Hong Kong

S. Thomas Ng, Chen Zhong and Frank J. Xu

The University of Hong Kong, Pokfulam, Hong Kong
* *Corresponding Author*

ABSTRACT As a typical sector of social infrastructure, public housing in many regions is playing an important role in underpinning the well-being, social cohesion and economic growth, offering accommodations to large percentage of their population. However, management of such infrastructure assets faces great challenges across their different lifecycle stages – a small incident may exert a huge negative social impact or lead to a catastrophic economic loss due to the internal interdependency between public housing buildings and facilities and the external interdependency of public housing with other urban infrastructures. Whilst most available literatures on public housing now focus on discussions of higher strategies like privatization and decentralization, this paper commences the research of public housing management from infrastructure asset management (IAM) perspective to reduce the lifecycle and social cost and improve the transparency of public housing programs. Taking public housing in Hong Kong as cases, this study aims to examine the interdependency among public housing related infrastructure assets from multiple perspectives: from physical, functional and informational interactions to management practices. Semi-structured interviews are conducted with relevant stakeholders to chart out the interrelationships among different public housing related infrastructure assets and their management processes. A dependency matrix is developed to illustrate the interdependencies of different infrastructure assets. causal loop diagrams (CLD) are constructed to identify complicated interactions and feedback loops among interwoven public housing related infrastructure maintenance management processes. The matrix and the diagrams can be used to develop decision support tools for the management of public housing related infrastructure assets

1 INTRODUCTION

Hong Kong is a famous metropolitan city as well as an important financial center with one of the highest population density all over the world. The city has large quantities of urban infrastructure assets like highways, drinking water networks, gas pipes, public housing and telecommunication networks, etc. Like other high-density and advanced megacities, Hong Kong has the urgent needs to improve its urban management efficiency through cross-sector city infrastructure asset management (IAM) practices. In Hong Kong, public housing together with its related infrastructure assets serves over 2 million residents – around 30% of the territory's population. However, it is encountering unique significant challenges in managing public housing related infrastructure assets, i.e. highly intensive and tightly coupled urban infrastructure facilities; large quantities of public buildings built in different eras; diverse public and private ownerships, operators and service providers; relatively isolated management systems; fragmented and inaccurate records of assets; etc. Due to limited land resources, expensive labor and the increasing needs of a large number of residents, Hong Kong has to improve the management efficiency of its existing public housing related infrastructure assets through efficient maintenance or refurbishment programs instead of by constructing new building facilities. Re-engineering the existing practices of Hong Kong to manage the public housing infrastructure urban assets motivates the in-depth study in this paper.

2 INTERDEPENDENCY

In general, interdependency describes a bidirectional relationship between states of a pair of infrastructures. In other words, infrastructure A is dependent on infrastructure B while infrastructure B is dependent on infrastructure A as well, then infrastructure A and infrastructure B can be considered to be interdependent. With interdependencies among different infrastructure components existing, the overall complexity of the urban infrastructure networks will increase remarkably (Rinaldi *et al.* 2001). There are several taxonomies to describe interdependency types. For example, Rinaldi *et al.* (2001) presented four categories of interdependencies, i.e. physical, cyber, geographic and logical interdependency; and Dudenhoeffer and Permann describe interdependency in five dimensions: physical, informational, geospatial, policy / procedural and societal dimension (Pederson *et al.* 2006). As a start, this study constructs a dependency matrix basing on the classification of Rinaldi *et al.*

3 INFRASTRUCTURE ASSET MANAGEMENT

IAM is a board process, providing the public, prospective users and owners with a required level of services through provision and maintenance of physical infrastructures and associated facilities at (Uddin *et al.* 2013). Generally, the whole lifecycle asset management consists of eight processes / stages namely: requirements definition; asset planning; asset creation; operations and maintenance; asset monitoring; renewal/rehabilitation; disposal; sustainable financial management (ALGENZ & IPWEA 2011). Lowest long-term cost, instead of short-term savings, is the crucial philosophy of lifecycle asset management. Since interdependency is a salient feature of infrastructure networks, an integrated inter-organization asset management approach becomes a key factor to succeed (ALGENZ & IPWEA 2011); smooth communication and highly efficient information sharing are required enablers for best inter-organizational asset management practices. The ability of information sharing and exchanging will greatly affect the implementation of management strategies in terms of efficiency and cost-effectiveness (Halfawy *et al.* 2006). Recently, increasing use of such information and communication technology as internet of things and big data across various infrastructure sectors of built environment has presented remarkable potentials for improving the efficiency, accuracy, reliability, effectiveness, safety and resilience of inter-organizational asset management (Bessis & Dobre 2014).

Hong Kong ranks the fifth in the recently released Global Cities Index (GCI) which describes current performance of a city, but it is not in the 'Global Elite' list which includes the top twenty-five cities with extraordinary current and future performance (A.T. Kearney 2015). Poor performance of Hong Kong in Global Cities Outlook (GCO) implies that Hong Kong needs to optimize its sustainable development further, while improving the sustainability and performance of its urban infrastructure is one of the critical factors (Labuschagne & Brent 2005). Among the sixteen global elite cities, ten of them have joined 100 Resilient Cities initiative pioneered by the Rockefeller Foundation (www.100resilientcities.org) to assist cities improving resilience against natural disasters and man-made disruptions.

It seems that Hong Kong has lagged behind its peers in terms of urban resilience capacity. Resilience describes the ability of a system to react and recover quickly when it suffers from disturbances and then preserve its dynamic stability (Johnsen 2010). Complex and correlative systems with more incidents have a stronger demand for resilience (Johnsen 2010). Hong Kong urgently needs to improve its resilience towards greater sustainability and livability. The significance of the demands is highlighted in its three-fold city management strategy – sustainable development management, asset management and hazard management, as "*a smart city must also be a safe one. Whilst being among the safest cities in the world, Hong Kong could still be vulnerable to natural and weather-related disasters*" (Chan 2014); "*hazard management is usually less emphasized in the community*" (Pun *et al.* 2015); and city resilience and asset management have become two components of its ongoing 'Smart City at Kowloon East' initiative (Hon 2015). ISO 55000 series is the first set of international standards for asset management (ISO 2014). Currently, several governmental departments of Hong Kong have begun to adopt the standard, but only at a single department level instead of at the inter-organizational level.

4 CASE STUDY

Whilst most available literatures on public housing now focus on the discussion of higher strategy like privatization and decentralization, it may be a good different starting point to commence the research of public housing management from IAM perspective to reduce the whole lifecycle and social cost and improve the transparency of public housing programs. Therefore, typical public housing projects will be taken as case studies in this research. Semi-structured interviews are conducted with relevant stakeholders to help chart out the existing management processes of public housing related infrastructure assets and then identify the critical barriers and solutions for continuous improvement. A dependency matrix is created to show the interdependencies of a range of public housing related physical infrastructure assets. Causal loop diagrams (CLD) are constructed to provide a graphical illustration of the causal interdependencies among various variables of interwoven asset management processes.

4.1 Semi-structured Interviews

Semi-structured interview is a standard interviewing method between structured interview which has restricted questions and unstructured interview which has few boundaries (Bernard 1988). Semi-structured interview is based on the use of an interview guide which includes a written list of questions and topics prepared ahead of time but also allow interviewees to come up with any new ideas during the interview. Semi-structured interview is flexible as well as reliable (Cooper & Fairburn 1987). It is better to use semi-structured interview when there might be only one chance to interview somebody and wish to get some fresh perspectives and understanding on the topic from respondents (Cohen & Crabtree 2006).

In this research, experts from the governments or utility companies who are managing the infrastructure facilities were invited to participate in the semi-interviews. The interviews were conducted in two rounds. First, a basic understanding on the practices of Hong Kong's IAM was obtained for unveiling the existing approaches of different organizations as well as their weaknesses and bottlenecks. The second round interview focused on the process and relationships among various stakeholders within maintenance of public housing. Information obtained from the interviews is used as inputs for constructing causal loop diagram and modeling interdependency of different infrastructure assets.

Five semi-structured interviews were conducted with four being face-to-face and one being a telephone interview. The first interview was conducted with the Estates Office of a local university in Hong Kong to learn the university's asset management system and the problems the office faced. The second interview was conducted with a chief engineer from a government department of Hong Kong Special Administrative Region (HKSAR); and the third one was with a research manager of construction related government organization. The second and third interviews mainly focused on reviewing existing IAM practices within utilities and the general construction industry respectively. Also, the weaknesses of existing practices in a dynamic environment had been discussed during the interviews. Combined with the findings from the desktop study and literature reviews on the IAM practices, general characteristics of IAM practices in Hong Kong can be summarized as follows:

- No professional regulations or guidelines for IAM
- No complete asset register taxonomy and depository
- More corrective maintenance than preventive maintenance
- Fragmented data record on existing assets
- Missing or inaccurate records of some old facilities
- Lack of open mechanisms for accessing asset information
- Elaborate division of labor on asset management within government's working departments
- Paper-based, file-based or e-mail based information sharing methods between working departments
- Relatively slow adoption of new computerized approaches
- Use geographic boundary (e.g. last manhole) to divide management allocation without sense of interdependency and collaboration

The fourth interview was conducted with two officers from public housing related department of HKSAR at the same time. One officer is a senior maintenance surveyor and another one is a maintenance surveyor. The two officers gave both macroscopic and microcosmic introductions on the development and implementation of maintenance plan. Generally speaking, there are three different ways to develop maintenance plan, from budget, program and responsive action aspects respectively. Detailed explanations are listed in Section 4.3.

Figure 1 presents the basic workflow of a public housing project according to the input of officers from the fourth interview. The department has two divisions, one responsible for the planning, design and construction (referred as 'HADCD' hereafter) of public housing and the other responsible for the management maintenance of the built public housing (referred as 'HAEMD' hereafter). At the moment, the separate administrations for the same public housing asset at different lifecycle stages could lead to unintended problems especially in handover stage.

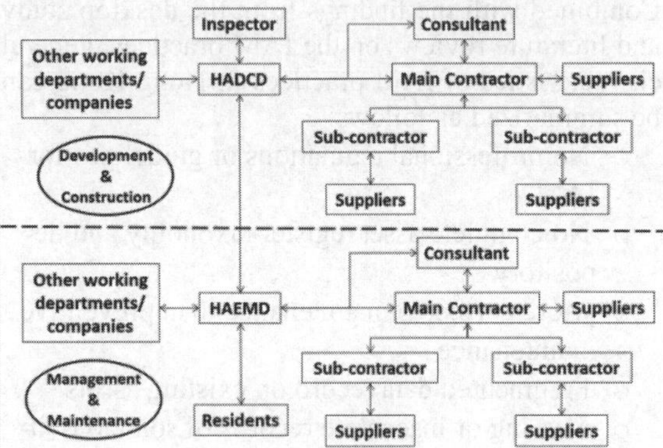

Figure 1. Flow diagram of interactions among different stakeholders for a public housing project

The fifth interview was a telephone interview with the same maintenance surveyor in the fourth interview, to make supplementary details to the whole maintenance process. Based on the type of maintenance, it can be divided into major project and minor project (or called in-house project). For major projects, consultants will be employed to take charge of design work and supervision to contractors. For minor projects, contractors are responsible for design and built and they are directly supervised by HAEMD.

Figure 2 shows a more detailed maintenance process flow diagram of public housing maintenance and management based on the input from the fifth interview.

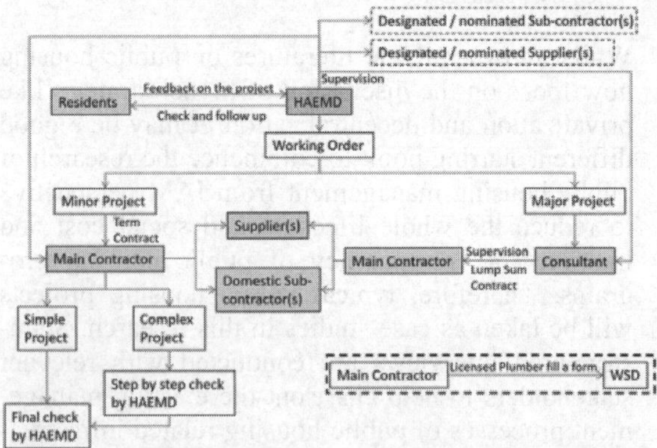

Figure 2. Maintenance process flow diagram for the maintenance & management of public housing

4.2 Dependency Matrix

Public housing related infrastructure assets mainly include buildings, utilities and transportation nearby. Typical utilities for a housing facility include water supply pipes, drainage pipes, gas pipelines, and electricity network and telecommunication cables. Infrastructure assets of transportation constitutes ground transportation like roads, bridges, tunnels and mass transit like bus transit and subways (Uddin *et al.* 2013). In terms of public housing related transportation in Hong Kong, road, bus and Mass Transit Railway (MTR) can be good representatives. Based on the desktop study, literature review and semi-structured interviews, a dependency matrix (Table 1) is developed on public housing related infrastructure assets to illustrate the level of dependency as well as types of interdependency.

Four types of interdependencies are considered: physical interdependency occurs "*if the state of each is dependent on the material output(s) of the other*"; cyber interdependency exists in an infrastructure "*if its state depends on information transmitted through the information infrastructure*"; geographic interdependency occurs "*when elements of multiple infrastructures are in close spatial proximity*" and logical interdependency occurs "*if the state of each depends on the state of the other via a mechanism that is not a physical, cyber, or geographic connection*" (Rinaldi *et al.* 2001).

For example, element "Electricity" from sector A is dependent on element "Telecom" from sector B via cyber and geographic connection, with the level of dependency between them is medium, but not vice versa. Element "Telecom" from sector A is dependent on element "Electricity" from sector B via physical, cyber and geographic connection, while the level of dependency between them is high.

Table 1. Dependency matrix of public housing related infrastructure assets

Sector A	B Element	Building	Utilities					Transportation		
		Building	Water supply	Drainage	Gas	Electricity	Telecom	Road	Bus	MTR
Building	Building	/	H; Lo	H; Lo	H; Lo	H; Lo	H; Lo	M; Lo	L; Lo	L; Lo
Utilities	Water supply	/	/	L; Ge	L; Ge	M; Ge	L; Ge	L; Ge	/	/
	Drainage	/	L; Ge	/	L; Ge	M; Ge	L; Ge	L; Ge	/	/
	Gas	/	L; Ge	L; Ge	/	M; Ge	L; Ge	L; Ge	/	/
	Electricity	/	L; Ge	L; Ge	L; Ge	/	M; Cy, Ge	L; Ge	/	/
	Telecom	/	L; Ge	L; Ge	L; Ge	H; Ph, Cy, Ge	/	L; Ge	/	/
Transportation	Road	L; Lo	L; Ge	L; Ge	L; Ge	L; Ge	L; Ge	/	M; Lo	M; Lo
	Bus	L; Lo	/	/	/	/	/	H; Lo	/	M; Lo
	MTR	L; Lo	/	/	/	L; Ge	L; Ge	M; Lo	M; Lo	/

Level of dependency: H – High M – Medium L – Low
Type of interdependency: Ph – Physical Cy – Cyber Ge – Geographic Lo – Logical

4.3 Causal Loop Diagram

Causal loop diagram (CLD) is one of several important diagramming tools for representing the feedback structure of systems when using a system dynamic model to capture the dynamics of a system; it is a graphical representation to elicit the casual relationships among variables within a system (Rehan *et al.* 2014). System dynamic model is a mathematical representation of the causal loop diagram (Rehan *et al.* 2014). The variables and qualitative relationships between variables within critical interactions among specified stakeholders and systems that are essential to sketch a causal loop diagram will be derived from the semi-structured interviews, case study and literature reviews.

There are two types of causal link linking variables in a cause-effect relationship: positive link and negative link. With a cause increasing, "*the effect increases above what it would otherwise have been*" and vice versa when there is a positive link (Sterman 2000). With a cause increasing, "*the effect decreases below what it would otherwise have been*" and vice versa when there is a negative link (Sterman 2000). Usually, plus sign (+) represents positive link while minus sign (-) for negative link. More than two links may form a feedback loop and there are also two loop types, one is reinforcing (or positive) loop and another is balancing (or negative) loop. Reinforcing loop indicates that a change in the initial component induces changes in subsequent components, which enhances the initial change.

On the contrary, balancing loop means a change in the initial component induces changes in subsequent components, which counteracts the initial change. Clockwise loop identifiers are for clockwise loops (and vice versa), with denoting by an R (or +) for reinforcing loops or a B (or -) for balancing loops (Sterman 2000). There are a number of software systems available for developing causal loop diagram; Vensim (http://vensim.com/) is applied in this paper to construct the CLD; and Vensim is well suitable for system dynamic modeling and simulation of real and complex systems.

Major variables associated with the maintenance management of public housing are identified through the fourth and fifth interview conducted to officers of public housing related department.

Development of maintenance plans for public housing is usually considered three perspectives: budget, program and response. "Budget" is based on a calculated baseline and adjusted by the exact capital works. "Program" includes both scheduled maintenance and condition driven maintenance. "Response" is mainly based on requests from residents, which may overlap with condition driven maintenance within "Program" sometimes. By contrast, the problems of condition driven maintenance are more serious, like large amount of concrete spalling.

Figure 3 demonstrates the causal loop diagram for maintenance development of public housing, capturing the dynamics of the maintenance process flow shown in Figure 2. Elements from the community are also taken into account, like residents' satisfactory. All the variables in the developed causal loop diagram are quantified with the corresponding units. Some variables can have more than one measurement. Taking "requests from residents" as an example, both "numbers of requests" and "grades of requests" are used for quantification. Suppose that there are one hundred requests from residents living in a public housing estate within one year. Only 10% deterioration from those requests is serious while the rest is general or slight. However, as for some descriptive variables with uncertain qualitative characteristics, like "quality of life", further description and clarification is needed.

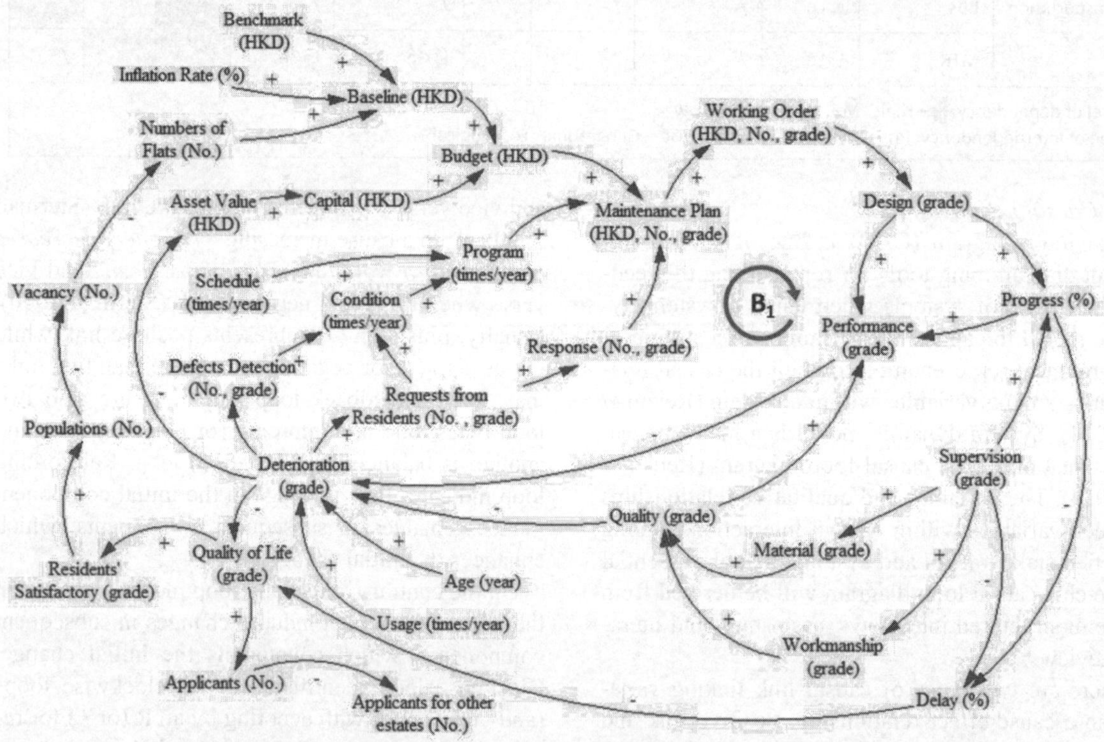

Figure 3. Causal loop diagram for maintenance plan development and implement of public housing

Eleven feedback loops can be identified in the sketched causal loop diagram, with seven reinforcing loops and four balancing loops. Balancing loop, **B_1** (Figure 4), could be directly identified from Figure 3. Starting from increment on the grade of deterioration, the number or grade of requests from residents will increase and then cause the increase on the number or grade of response. Increasing response induces increment on the number or grade of maintenance plan, following by the same trend of the number or grade of working order. The grade of design (e.g. quality, scale) will hence increase. The grade of performance for a construction after maintenance increases, which restrain the initial increment of the grade of deterioration.

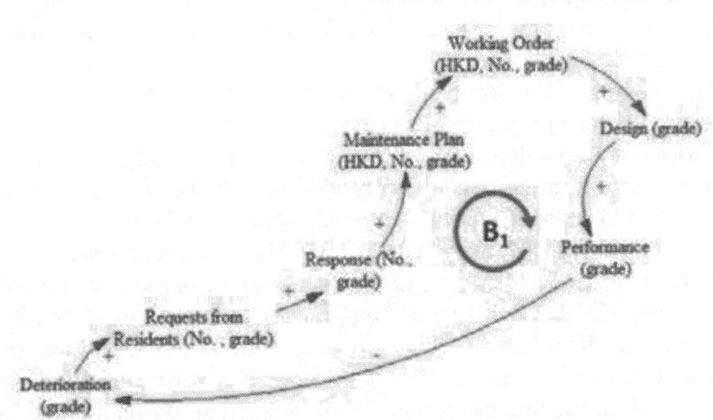

Figure 4. Balancing loop B_1

A reinforcing loop, **R_2** (Figure 5), extracted from the sketched causal loop diagram, is formed due to connection among deterioration, asset value, capital, budget, maintenance plan, working order, design and performance. If the grade of deterioration increases, the asset value will decrease, following by decrease on the capital. Decreasing capital induces decrement on budget, which leads to consequent decrement on the cost of maintenance plan. Then the cost of working order will decrease. Lower spend on working order causes decrease on the grade of design. The grade of performance for a construction after maintenance will hence decrease (lower than that of previous design quality or scale), which will enhance the increasing trend of the grade of deterioration.

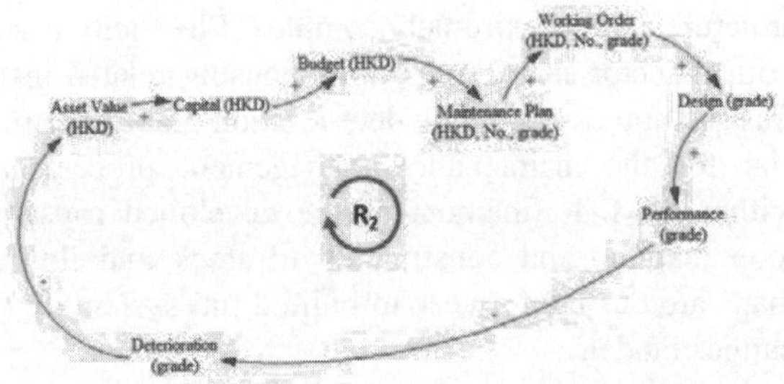

Figure 5. Reinforcing loop R_2

Except for essential variables for implement of maintenance plan like maintenance plan, working order and design, deterioration should be the critical variable in the sketched causal loop diagram since it is involved in ten feedback loops out of total eleven feedback loops.

Together with stock and flow maps which are to be developed once relevant data are collected, the developed causal loop diagram can be further developed into a system dynamic model to describe the structure and dynamics of public housing related IAM process. After thorough test, verification and validation, it is envisaged that the final full system dynamics model shall be helpful to monitor and predict the dynamics behaviors of the inter-sector IAM, as well as to contribute to improve city resilience.

5 CONCLUSIONS

This paper aims to unveil the interdependency of public housing related infrastructure assets in Hong Kong. Semi-structured interviews are conducted with relevant stakeholders to help identify current characteristics of IAM in Hong Kong and chart out existing management processes. A dependency matrix is developed to illustrate the interdependencies of infrastructure assets while causal loop diagrams are constructed to provide causal interdependencies among various variables of interwoven processes of public housing maintenance.

It can be drawn from the developed dependency matrix and causal loop diagrams that, as a technical system of systems the structure of the public housing related infrastructural assets and related infrastructure asset is extremely complex. Electricity is a critical sector among the public housing related infrastructural assets, while deterioration grade is crucial for the maintenance management processes within IAM. Refinement of the developed causal loop diagram and construction of stock and flow maps are our future work to build a full system dynamics model.

A well-developed system dynamic model can help capture, simulate and predict the dynamics behaviors of public housing related infrastructure assets and its management process. In long-term, the research findings will help clarify the requirements and enable the establishment of a smart, systematic inter-network IAM system for achieving the efficiency, effectiveness, transparency, accuracy and safety of the management of the public housing related infrastructure assets in Hong Kong. The study will also shed light on the public housing management of other cities for continuous improvement.

ACKNOWLEGMENT

The authors would like to thank the Research Grants Council of the HKSAR Government for funding this project under the General Research Fund (Project No.: 17202215). The financial support of The University of Hong Kong through the CRCG Small Project Fund (Project No.: 201409176107) is also gratefully acknowledged.

REFERENCES

ALGENZ & IPWEA 2011. *International Infrastructure Management Manual*, Association of Local Government Engineers of New Zealand. National Asset Management Steering Group and Institute of Public Works Engineering Australia, Wellington, New Zealand.

A.T. Kearney 2015. Global Cities 2015: The Race Accelerates, retrieved from https://www.atkearney.com/research-studies/global-cities-index/2015 on 15/12/22015.

Bernard, H. R. 1988. Research methods in cultural anthropology, Sage, Newbury Park, CA, 117.

Bessis, N. & Dobre, C. (Eds.) 2014. Big Data and Internet of Things: A Roadmap for Smart Environments, Springer International Publishing, Switzerland.

Chan, C.M. 2014. Smart cities, smart citizens, *Hong Kong: Annual Conference 2014 Hong Kong: Our Smart City in the Next 30 Years*, HKIS, 26-30.

Cohen, D., & Crabtree, B. 2006. Qualitative research guidelines project, retrieved from http://www.qualres.org/HomeSemi-3629.html on 18/12/2015.

Cooper, Z., & Fairburn, C. 1987. The eating disorder examination: A semi- structured interview for the assessment of the specific psychopathology of eating disorders, *International Journal of Eating Disorders*, *6*(1), 1-8.

Halfawy, M. R., Vanier, D. J., & Froese, T. M. 2006. Standard data models for interoperability of municipal infrastructure asset management systems, *Canadian Journal of Civil Engineering, 33*(12), 1459-1469.

Hon, C.K. 2015. Smart City at Kowloon East, presentation at the *First International Construction Innovation Conference: Explore, Exchange, Excel*, 16/12/2015, Hong Kong.SO 2014. *ISO 55000:2014 Asset management – Overview, principles and terminology*, International Organization for Standardization, Geneva, Switzerland.

Johnsen, S. 2010. Resilience in Risk Analysis and Risk Assessment. In T.

Labuschagne, C., & Brent, A. C. 2005. Sustainable project life cycle management: the need to integrate life cycles in the manufacturing sector, *International Journal of Project Management*, *23*(2), 159-168.

Moore and S. Shenoi eds. *Critical Infrastructure Protection IV*, Springer Berlin Heidelberg, 215-227.

Pederson, P., Dudenhoeffer, D. & Hartley, S. *et al.* 2006. Critical infrastructure interdependency modeling: a survey of US and international research, *Idaho National Laboratory*, 1-20.

Pun, W.K., Sun, H.W. & Yam, C.F. 2015. Emergency preparedness of disasters for safe and sustainable development. *Proceedings of the ICE HKA Annual Conference 2015*, HKICE, 218-230.

Rehan, R., Knight, M. A., Unger, A. J. A. *et al.* 2014. Financially sustainable management strategies for urban wastewater collection infrastructure–development of a system dynamics model, *Tunnelling and Underground Space Technology 39*, 116-129.

Rinaldi, S. M., Peerenboom, J. P., & Kelly, T. K. 2001. Identifying, understanding, and analyzing critical infrastructure interdependencies, *Control Systems, IEEE*, *21*(6), 11-25.

Sterman, J. D. 2000. *Business dynamics: systems thinking and modeling for a complex world* (Vol. 19), Irwin/McGraw-Hill, Boston.

Uddin, W., Hudson, W., & Haas, R. 2013. *Public infrastructure asset management*, McGraw Hill Professional.

Proceedings of the International Conference on Smart Infrastructure and Construction
ISBN 978-0-7277-6127-9

© The authors and ICE Publishing: All rights reserved, 2016
doi:10.1680/tfitsi.61279.701

Whole life costing of infrastructure investment: economic and social infrastructure projects in Australia

M. Regan[1], P.Love[2] and J. Smith*[3]

[1 3]*Bond University, Gold Coast, Australia*
[2] *Curtin University, Perth, Australia*
* *Corresponding Author*

ABSTRACT Public private partnerships (PPPs) commissioned by the Victorian State Government in Australia require life cycle costing of the public sector comparator, a risk-weighted and discounted model of traditional procurement as a benchmark for comparison of private bids. Value for money is a term used to measure procurement outcomes and is used widely in government as a proxy for *ex ante* assessment of the quantitative and qualitative performance of a specific procurement option. Value for money is also used by governments to select projects for delivery by public private partnerships. A public sector comparator for conventional delivery is used as a pricing benchmark (the quantitative evaluation) as well as an appraisal of the qualitative features of private proposals using project-specific criteria. A number of studies over the past decade show that public private partnerships may deliver value for money for projects in the built environment. However, less is known about the drivers of value for money. This study examines 13 PPPs delivered by the governments of the Northern Territory, New South Wales, and Victoria in Australia between 2005 and 2015. The major driver of value for money in PPPs is risk transfer of which life cycle costing and operating costs are a major component. These values are captured in the public sector comparator. The study also confirms that the qualitative attributes of bids received in a competitive market environment are contributing to greater asset utilisation, design and construction innovation and the delivery of improved services to or on behalf of government.

1 INTRODUCTION

The Life cycle costing (LCC) is a quantitative measure used widely by government as selection criteria for the procurement of public works including buildings, civil works, complex engineering systems, plant and equipment, and services. LCC is the aggregate cost of procuring, installing, maintaining, refurbishing, disposing and operating costs directly attributable to owning or using an asset over its economic or service life (New South Wales Treasury, 2014). LCC informs the design and procurement decisions, the specification and scope of works, benchmarking studies, bidder selection criteria, and is used to identify sustainability principles for new construction (National Audit Office [NAO], 2005). LCC can be used with most methods of project delivery including traditional contracting, it is an alternative to lowest cost bid evaluation under which the principal determinant is lowest initial acquisition cost. LCC often shows that a project with a higher initial cost may offer financial benefits over its life cycle that result in a lower overall procurement cost to the owner. In recent years, LCC has also been adopted along with qualitative performance measures to calculate value for money (VFM) in major project procurement.

LCC is suitable for use with both input and output-specified projects and the comparison of different design and construction approaches to complex and non-standard buildings and structures. LCC is used with output specification projects such as the *build operate transfer* family of contracts which includes public private partnerships (PPPs) and long-term outsourcing contracts particularly those delivering services in waste management and recycling, water supplies and desalination, transport equipment (rolling stock, signalling and electronic systems), information and communication technologies. Under these contracts, the private contractor meets the cost of financing assets required to deliver the service, operating costs and LCC. PPP contracts create an incentive for the contractor to invest in up-front design, innovation and technology with the objective of optimising LCC over service lives of 30 years or longer. A number of long-term studies have established that PPP contracts deliver projects with greater time and cost certainty than alternative procurement methods (National Audit Office 2003; Allen Consulting-University of Melbourne 2007). This study examines 13 PPP projects in the States of Victoria and NSW as well as the Northern Territory with collective capital costs of A$8.027 billion and forecast LCC of A$3.91 billion, a significant part of which was transferred to private contractors (see Table 1 for projects and costs).

The application of LCC is explained under international standards and was introduced in by governments in the early 1990s for most forms of procurement (ISO 15686-5:2008, Building and Constructed Assets, Service-life planning, Part 5: Life-cycle costing; AS/NZS 4536: 1999 (confirmed 2014) Life cycle costing – An application guide). LCC is mandatory under sustainability and asset management policies in a number of OECD countries and for the procurement of PPPs and major defence projects. The United States, Japan, Switzerland and Norway apply mandatory LCC as part of sustainable procurement policies, and Sweden, Britain, Denmark, Germany, The Netherlands, France, Korea, Austria, New Zealand and Australia conduct various derived methods of LCC for procurement of energy-efficient buildings, plant and equipment above specific cost thresholds (IISD 2009; Australian Government 2014, New South Wales Transport 2014).

Table 1. Life Cycle Costing for Major Projects

Projects Values in AU$m	**TDC**	**LCC**	**Final Year**
	a	*b*	
Hopkins Correctional Centre	181	158	2010
Victorian Schools Project	142		2009
NSW Schools Project 1	142	75	2005
NSW Schools Project 2	194	118	2007
Bendigo Hospital	422	249	2013
Darwin Correctional Facility	801	169	2011
Victorian Cancer Centre	779	183	2012
Ravenhall Prison Project	539	285	2015
Peninsula Link Project	680	43	2010
Barwon Biosciences Project	153	28	2007
Royal Children's Hospital	683		2008
Victorian Desalination Project	3,272	2,602	2009
Biosolids Project	39		2007
	8,027	3,910	

a Total development cost disclosed in the public sector comparator

b Designated lifecycle costs including repairs and maintenance, depreciation, energy, capital replacement, service costs.

Source: Project Summaries prepared by Northern Territory, Victorian and NSW

LCC can play an important role in assisting public and private organisations to select an optimal procurement method, to improve value for money outcomes, and provides a guide for future budgeting of operating, refurbishment, utilities and disposal expenditures (NAO 2005, 27). The NAO estimates that the potential for cost savings across the central and local government estate is through the better use of LCC to reduce energy, cleaning and security costs and improve environmental sustainability. LCC savings across the estate were in the order of £700m annually (NAO 2007). LCC assumes importance with infrastructure procurement because it involves capital intensive assets, high sunk costs, is generally site and use specific, and operates over service intervals of 25 years or longer. Studies suggest that nominal LCC for non-standard buildings over 20 years is around five times the initial capital cost of the building increasing to a multiple of 40 or more for complex assets such a fully operational hospital (NAO 2005,

27). Discounted cash flow is used to determine the present value of future costs outgoings with the discount rate in government either social opportunity cost, social time preference or the marginal cost of public debt (HM Treasury 2013). In the private sector, the rate is generally the weighted average cost of capital. LCC is widely used in benefit-cost analysis, cost effectiveness analysis and other methods for to determine the feasibility of projects and services.

LCC analysis has been the subject of an international standard for over 25 years, yet it is not common practice in government and industry. Studies by the National Audit Offices in the United Kingdom and Australia indicate that life cycle costing is being implemented slowly by government estate with many agencies failing to apply the analysis to all stages of the procurement cycle despite top-level department encouragement to do so throughout an asset's operational life cycle (Australian National Audit Office [ANAO] 2015, 1998; Victorian Auditor General's Office [VAGO] 2009; Pearson 2011).

LCC is an *ex ante* guide to estimating the full cost of ownership and operation of assets over the economic life of the undertaking. However, it is at best an approximation of future costs and vulnerable to the hazards of financial forecasting over periods of 20 years or more (Office of Government Commerce 1999, Barringer and Weber 1996). For governments, the focus of future financial planning is the forward estimates generally for period of three to five years. The assumptions to be made for long-term asset lifecycles are subject to wide margins of error. For many private firms, the business planning cycle is typically five years with detailed financial plans limited to three years and the remaining two years are broad estimates at best. Unlike government, private firms are subject to market forces, takeovers, mergers and acquisitions, which generally result in new management positions and revised business plans. For example, among the 23 infrastructure services companies listed on the Australian Stock Exchange in 1997, only one had retained its original structure, management and majority institutional ownership four years later.

Long-term forecasts are also susceptible to exogenous change in the form of new environmental and taxation laws, changed regulatory and compliance requirements, and systematic risks in the form of interest rate and refinancing risk. The changes may also be endogenous and take the form of technology obsolescence, change of use, poor management standards, optimistic revenue forecasts, new operating systems, and a decline in capital and labour productivity. Change may adversely affect operating costs and return on investment, a particular problem for private companies subject to competitive market forces.

A major impediment to LCC in the private sector is the short-term nature of the interest of construction firms, investors, clients and developers in the assets that they commission. With the exception of institutional investors, little reliance is placed on life cycle investment economics. The position is generally different for government with most assets having the character of public goods, which remain in use until service delivery is no longer required.

PPPs are the exception with the private operator required to deliver services to specification over the term of the contract and are one of many procurement options used by government to deliver public goods and infrastructure services. PPP policies were first introduced in the United Kingdom in 2001 although the methodology builds on earlier build operate transfer models employed in the late 1980s for the outsourcing of government services in waste management and recycling, water supplies and desalination, transport equipment (rolling stock, signalling and electronic systems), information and communication technologies. Under most policy frameworks, a private firm or consortium will bid for a project in a competitive bidding process to finance and construct assets that deliver public goods in response to a government's output specification. The successful bidder carries most of the risk of the undertaking and either applies a charge for services or government may pay a unitary or availability payment over the term of the contract. PPPs are long-term contracts of 20 years or longer and are commonly used to deliver services in health, education, roads and public transport, ports and airports, water and sanitary services, energy, and corrective services. PPPs are an important procurement option for government particularly for large-scale projects involving complexity, design and construction innovation, or new technologies. However, they account for a relatively small share of average global public infrastructure spending at around 5% gross domestic product (GDP) with wide variation between countries (International Monetary Fund 2015).

In the period 2011-15, PPPs accounted for 13% of core infrastructure spending in the United Kingdom, around 5% in Canada and India, and 4% in Australia (International Monetary Fund 2015, 10).

Most countries implement PPPs using enabling legislation and policies are designed for compatibility with domestic institutions and public procurement rules at central, regional and municipal government levels. Most countries have adopted value for money principles for their evaluation criteria but few actually use a public sector comparator (PSC) or other benchmarking techniques when choosing between procurement methods or the analysis of bids.

The economics of PPP transactions are framed by long-term incomplete contracts, high-powered incentives based on payment-for-service arrangements, and the transfer of project risks to the private party. The PPP contract effectively connects build quality, design and construction innovation and LCC to achieve beneficial procurement outcomes for government. A number of studies since the early 2000s have indicated that PPPs are delivering cost savings to government although the results show wide variation between jurisdictions and project types (H.M. Treasury 2013; National Audit Office 2003, 2009, 2011; Allen 2003; Fitzgerald 2004; Allen Consulting-University of Melbourne 2007).

LCC is a trade-off between capital and operating costs. When LCC risk is transferred to a private party under a long-term incomplete contract, the trade-off creates a high-powered incentive for the contractor to build a quality asset, invest time and effort in the early stages of the project and undertake innovative and sustainable approaches that reduce LCC risk and maximise the return on investment (NAO 2007,19; Adamson 2004). For this reason, PPPs are the preferred procurement option for government with school, hospital and corrective services projects (NAO 2005).

2 CASE STUDY SELECTION

The case studies examine the impact of LCC on project economics and the extent to which risk transfer is a significant factor in cost savings to government offered by the public private partnership procurement method. The case studies were selected from the 135 contained in the *Infrastructure Australia* database at 31 October 2015. Most PPPs in Australia are social infrastructure projects including hospital and school buildings, and corrective services establishments. The sample was selected with the following criteria:

- Projects were delivered under uniform national PPP policy introduced between 2001-2003
- Full financial information exists for the projects including the project's risk allocation schedule and the public sector comparator
- Projects were commissioned in the last 10 years
- A majority of projects were social infrastructure where government makes an availability payment to the private party for specified services to be delivered over the life of the contract
- A dispersion of projects across seven industry sectors including corrective services, education, health services, motorways, water supply and waste management, and public buildings
- The geographic dispersion of projects across no less than three states or territories.

Thirteen projects were selected for the study that met the selection criteria. The major difficulty was full access to project summaries, which are made available on the internet in Victoria and the Northern Territory. Detailed project summaries including a copy of the public sector comparator are not readily accessible in the remaining states and territories.

Notwithstanding differences in project size and particular industry applications, the case study selection criteria should eliminate selection bias. The qualifying projects are shown in Table 1. The sample includes three projects each for health services, education and corrective services, and one each for roads, environmental management, water resources and science.

3 THE ANALYSIS

3.1 *Payment Arrangements*

Nine of the 13 projects delivered social infrastructure services although all of the projects were based on an availability payment regime with government carrying market or patronage risk. In the case of the Desalination Project, the payment mechanism includes an off-take agreement whereby the government pays for a minimum quantity of output whether taken or not. Eight projects were commissioned in the period 2007-2010 during which time project finance was difficult to raise and at higher cost than any time in the preceding decade (See Table 1). For each of these projects, the credit rating of the sponsoring government at the time of commissioning was Standard and Poor's AAA level.

3.2 *Risk Allocation*

PPP projects in Australia are selected and implemented in compliance with the national policy framework although the states and territories may vary requirements in response to local conditions. The private party will be delivering core custodial services in each of the corrective services projects and the Biosolids project. In each of the other projects, government delivers the core services of teaching, health services, and medical care while the private party manages non-core services such as building maintenance, catering, repairs and maintenance, and meets asset operating costs over the term of the contract.

The uniform national policy has resulted in the adoption of common risk allocation practices by implementing agencies which is evident in the case studies. Government is responsible for land tenure and planning approval risk, asset utilisation (patronage risk) and utilities volume and pricing. The private party will typically carry design, construction, financial, commissioning, fitness for purpose and operational risks of which the most important is life cycle costs. The *ex ante* capital cost of the 13 projects disclosed in the PSC was $8,027m, life cycle costs were estimated at $4,233m (49% of capital cost) and operating costs a further $3,119m (around 92% of capital cost discounted over the project life cycle at a weighted average discount rate of 6.4% per annum). In nominal terms, LCC represented 68% of capital cost over the first 20 years of projects and LCC 92%. Both LCC and operating costs in nominal terms represented a multiplier of 1.6x capital cost.

3.3 *Risk Transfer*

Not all project risk in PPPs is transferred to the private party. A survey of risk allocation practices across the sample indicates that in the majority of projects, government retains risks associated with land title, planning approval, pre-existing contamination, utilities volume and pricing, change in law adversely affecting the project, demand risk, force majeure and for all but two projects, obtaining and meeting the cost of asset insurance. The PSC includes risk retained by government and for the sample of projects, risk transfer was valued at $1,637m. The difference between LCC and risk transferred can be partly explained with a survey of industry applications. Where full or partial responsibility for service delivery is allocated to the private party, higher levels of transferred risk are evident (John Hopkins (28%) and Ravenhall (38%) corrective services projects, Bendigo Hospital (34%), Biosolids project (31%) and Victorian Desalination project (24%) are above the weighted average of 22% for the sample. Bendigo Hospital is the exception for this class of projects largely as a result of the transfer of both hard and soft facilities management which was valued at $502m, a multiple of 1.2x capital cost.

3.4 *Value for Money*

The value for money performance of PPPs is measured using qualitative and quantitative measures. The primary analysis is quantitative which is a benchmarking exercise using the PSC, a life cycle and risk weighted costing for delivery of the project using a traditional method of procurement. The value for money saving for government from the 13 projects is based on the PSC adjusted for the value of risk retained by government and includes life cycle cost and operating risk. In Australia, the PSC is used for all PPP projects. A further assessment of VFM takes place during the bid selection phase of the project in which bidders may demonstrate other benefits to government such as improved services, better asset utilisation, design and construction innovation and new technology. To win the project, a bidder must offer a cost saving against the PSC and deliver a better outcome for government under the qualitative test (Infrastructure Australia, 2008).

The cost to government would have been $16,540m and the value of the winning bids $14,510m, a cost saving of $2,029m or an average 12.3%. Risk transfer is the single most important driver of value for money with economic and social infrastructure projects, accounting for an average 58% of project risks with the sample.

4 CONCLUSIONS

Projects with the highest ratio of life cycle to capital costs (John Hopkins, Desalination, Bendigo Hospital and Ravenhall) also disclosed higher levels of risk transfer to the private parties and greater private participation in the delivery of core services in corrections, hospitals and water supplies. Analysis suggests the ratio was more likely to be determined by differences in project requirements, the specification and scope of works than by differences in industry sectors or locational factors. The projects with highest ratio of life cycle and facilities management to capital costs (John Hopkins, Desalination, Bendigo Hospital, Ravenhall, and Biosolids) also disclosed higher levels of risk transfer under the contract and collectively accounted for 92% of the sample's aggregate cost savings to government.

The case studies do not establish a correlation between LCC, value for money outcomes to government or particular industry applications. However, the analysis does identify the proposition that LCC is an important analytical tool for government procurement, it informs risk allocation decision-making and confirms the role that risk transfer plays with improved value for money outcomes for government.

The advantage of the PPP procurement method is the requirement for full costing of the project over its effective life. Assets are transferred to government at the conclusion of the contract term, which is between 20 and 30 years for case study projects. Notwithstanding the dangers of long-term forecasting and the limited scope for both government and private parties to make long-term commitments for future outlays, LCC and facilities costs over the project life-cycle informs the initial procurement decision, identifies optimal risk allocation strategies, and provides a benchmark for improved bidder selection methods. A major achievement of modern government would be to strictly require the use of LCC across all forms of procurement in the built environment.

REFERENCES

Australian National Audit Office 1998, *Life-cycle costing in the Department of Defence*, Department of Defence, Canberra.

Australian National Audit Office 2015, *Major Projects Report 2013-14*, Defence Materiel Organisation, Canberra.

Adamson, D. 2004, Whole-life Costing – A Client's perspective, University of Cambridge, United Kingdom.

Allen Consulting-University of Melbourne 2007, *Performance of Public Private Partnerships and Traditional Procurement in Australia*, Final Report to Infrastructure Partnerships Australia, 30 Nov.

Barringer, H.P. Weber, D.P. 1996, *Life cycle cost tutorial*, Fifth International Conference on Plant Reliability, 2-4th December 1996, Houston, pp. 1-58.

H.M. Treasury 2013, *The Green Book, Appraisal and Evaluation in Central Government*, H M Stationery Office, London.

Infrastructure Australia 2008, *National Public Private Partnership Policy Framework*, Canberra.

International Monetary Fund 2015, *Making Public Investment More Efficient*, Staff Report, Washington.

National Audit Office 2003, *PFI: Construction Performance*, Report by the Comptroller and Attorney General, HC371, Session 2002-03, 5 February.

National Audit Office 2005, *Improving Public Services Through Better Construction*, Part 1, Report by the Comptroller and Auditor General, HC364-1, Session 2004-05, 15 June.

National Audit Office 2007, *Building for the future: Sustainable construction and refurbishment of the government estate*, Report by the Comptroller & Auditor-General, HC 324, Session 2006-07, April 20.

New South Wales Treasury 2014, *Life Cycle Costing, Management Standard*, Transport for New South Wales, Sydney, June.

Office of Government Commerce 1999, *Life-cycle costing*, Guidance, London.

Pearson, D. 2011, Managing Infrastructure Assets One Year On – Lessons Learnt, Presentation by the Auditor General of Victoria, 4 November and viewed on 12 October 2011 at www.audit.vic.gov.au /presentations/AAMCOG-4-November-2011.pdf.

Victorian Auditor General's Office (VAGO) 2009, *Audit Summary of Vehicle Fleet Management*, Performance Review, 25 Nov. 2009.

Proceedings of the International Conference on Smart Infrastructure and Construction
ISBN 978-0-7277-6127-9

© The authors and ICE Publishing: All rights reserved, 2016
doi:10.1680/tfitsi.61279.707

Critical risk factors in construction projects: perspective of contractors in South Africa

Berenger Y. Renault* and Justus N. Agumba

University of Johannesburg, Johannesburg, South Africa
**Corresponding author*

ABSTRACT The significant impact of construction projects on a nation has been characterised in literature in terms of infrastructure development and job creation. These projects are nevertheless associated with various risks that need to be managed to achieve set goals. Despite the extensive research on risk management in the construction industry, there are still significant risk factors that hinder projects from being achieved. In order to manage the project for a successful delivery, the identification of these risk factors is of utmost importance. Therefore, this study seeks to investigate risk factors from contractors' perspective since contractors are the key players in the success of a project. To achieve the purpose of this study, questionnaires were distributed to contractors in the construction sector in Gauteng (South Africa). Results revealed that supply of faulty materials, poor communication between involved parties, financial failure of the contractor, working at dangerous areas and closure were the five significant risk factors in construction projects. These significant factors are from four major categories of physical, management, financial and political group risks. The results of this study would considerably boost the understanding of risk factors and also assist contractors in handling various risks faced in construction projects.

Keywords: Construction, Contractor, Risk Factors, South Africa.

1 INTRODUCTION

The turbulent impact that the construction industry makes in terms of infrastructure development and job creations among economies have become overwhelmed in literature. The significant impact that is associated with its benefits is usually attached to critical risks that must be managed before achieving a successful delivering of the project. In order to meet the targeted objectives of project success (time, cost and quality), effective management tools must be put in place as risk may appear in many ways and could result in increased cost and time, decreased quality and many more failures. One of the major reasons for this situation is not handling the risks, which is about thinking ahead, simulating and searching for better solutions (keçi & mustafaraj 2013). Thus, the project can be achieved successfully by considering the risks where it normally tends to give positive and negative effect on the project (Ayyub 2003).

In recent years, some exhaustive studies and development have concentrated on project risk management. Project risk management is acknowledged as one of the most critical procedures and capability areas in the field of project management (Mahendra et al. 2014). One of the areas of project management is risk management. This is undoubtedly the most difficult aspect of project management (Mahendra et al. 2014). Managing risks in construction projects have been accepted as a very important process so as to meet project goals. Hence, risk management can be defined as a complete set of activities and actions aimed at dealing with any risk to maintain control over the entire (Van Well-Stan 2004). To achieve the set objectives, a proper risk management is indispensable. For this reason, the identification of risk factors affecting risk management is a crucial step in the risk management process; as if risks are not identified it will be almost impossible to respond to them. Moreover, it is well recognized that in construction projects, contractors are the key players in carried out construction works and are directly involved in the physical phase of the project. They are required to control the risks that occur during construction activities to ensure the effective completion of projects (Tang 2012). Therefore, this study focused on examining risk factors from the perspectives of contractors involved in construction projects in Gauteng (South Africa).

1.1 Research Objectives

The main objective of this study was to examine contractors' perception of risk factors associated with construction projects in Gauteng, South Africa.

2 LITERATURE REVIEW

2.1 Risks in Construction Projects

The opinion that the construction industry is the most exposed to threats (risks and uncertainty) is an agreement among authors due to the nature of its activities (Akintoye & Macleod 1997; Dey 2001).

Still, we found diverse tactics in the literature regarding the factors and characteristics of projects that expose the CI to numerous risks. Dey (2001), for example, indicate the following: the complexity of planning and design, changes in the environment, resource availability, and the presence of various interest groups, climate change, economic insecurity and political and regulatory statutes. In succession, Zoo et al. (2007), made reference to long, complex environment, complicated process, and the need for investment-intensive, dynamic organizational structures, technological and organizational complexity and the diverse interests of stakeholders. Ghani (2009), Points out as factors and essential features high life cycle design, size, complexity, location, the different parties implicated and familiarity with the performer's work to be done. Zeng and Smith (2007), found a persistent change of environment, direct exposure to hazards, the high pressure involved in the compliance of costs and deadlines, and increasing the complexity of construction techniques. Furthermore, Shen (1997), emphasize how the main features: A large number of people with different interests and abilities necessary to coordinate a wide diversity of interrelated activities. Likewise, in the study, conducted by Chapman & Ward (2003), the changeability in the performance objectives of cost, time and quality, the ambiguity related to various aspects such as lack of clarity owing to the behaviour of participants involved, as well as the lack of evidence and detail, are listed among the critical factors.

2.2 Risk Management Process (RMP)

Risk management systems are used to ensure the control of risks in the business process. In this study, the simplest possible approach to describing the risk management process is adopted due to the context of the construction sector. There is no common definition of the scope of risk analysis, risk management or the risk process in the literature, as each one has its own twist (Chapman & Ward 2003; Ingvarsson & Roos 2003). The risk management process in this study consists of the steps. It comprises the risk analysis followed by the risk response. Risk analysis includes risk identification and assessment, as depicted in Figure 1.

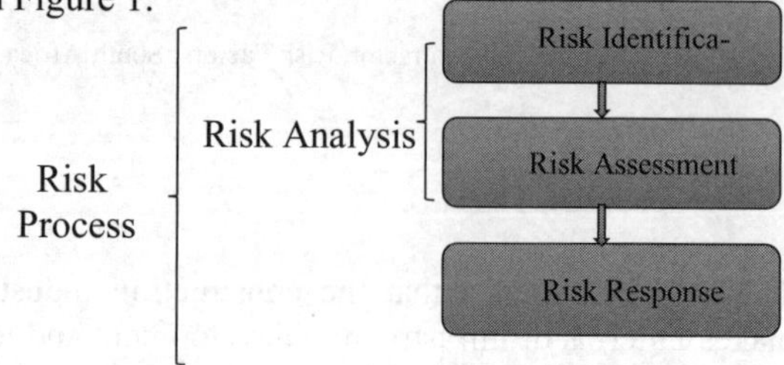

Figure 1: Risk Management Process (Adapted from Simu 2006)

The first step of the RMP is risk identification. This is probably the most important and time-consuming step, because if risks are incorrectly identified, incorrect assessments and responses will follow (Simu 2006). Several techniques are available for identifying risks; the most known in construction are brainstorming, interviews, Expert opinion, questionnaire, checklist, Delphi technique, Expert systems, past experience and documentation review (PMI 2013; Khalafallah 2002).

The second step is assessing the risks. It can be assessed based on the possibility of risk occurrence and severity of its impact (Lester 2007) by developing risk matrix. It aims at assessing the risk to evaluate the effect of each risk on the project. Risk assessment is conducted in various ways. There are tools and techniques that have been developed to consider probabilities and consequences, using historical data, statistical data or estimated judgment translated to numerical information (Aven 2003; Grey 1995). Common are the estimation of probability and consequence and the usage of software tools to manage the data. Scoring techniques (Grey 1995) are developed checklists that include the judgment of both probability and consequence of a risk breakdown. This is a common technique for risk assessment in construction projects that is widely used due to its simple approach.

In the risk response step, actions are taken to control the risks analyzed in the first two steps. In this study, the response step includes both the planned and the monitoring responses. There are four different ways of responding to risks in a construction project, namely, risk avoidance, risk reduction, risk retention and risk transfer (Ahmed et al. 2001; Enshassi & Mayer 2001). These are briefly discussed below:

***Risk avoidance*:** this method is at times referred to as risk elimination. Risk avoidance in construction is not recognized to be impractical as it may lead to projects not going ahead, a contractor not placing a bid or the owner not proceeding with project funding are two examples of totally eliminating the risks. There are some ways through which risks can be avoided, e.g. tendering a very high bid; placing conditions on the bid; pre-contract negotiations as to which party takes certain risks, and not biding on the high-risk portion of the contract (Flanagan & Norman 1993).

***Risk Reduction*:** this is essentially reducing the probability and consequences of an adverse risk event. In the extreme case, this can lead to eliminating entirely, as seen in "risk avoidance." However, in reduction, it is not sufficient to consider only the resultant expected value, because, if the potential impact is above a certain level, the risk remains unacceptable. In this case, one of the other approaches will have to be adopted (Piney 2007).

***Risk Retention*:** is a process by which risks are retained in a situation where they can neither be transferred nor avoided. Potts (2008), Stated that in this situation the risk must be controlled to reduce the effect of its occurrence. Two methods are available: active retention (also known as self-insurance) and passive retention (also referred to as non-insurance). The first one deliberate management strategy after a conscious evaluation of the possible losses of alternative ways of handling risks. The later, however, occurs through negligence, ignorance or absence of a decision, e.g. a risk has not been identified and handling the consequences of that risk must be borne by the contractor performing the work.

***Risk Transfer*:** risk transfer is concerned with transferring risks to another party. In other words, if a risk can be managed by a party who is well positioned or has a greater capacity to deal with it, it is better to transfer it to the party in question. (Potts 2008) Indicates that risk should be shifted to those who know how to manage it. For a construction project, an insurance premium would not relieve all risks, although it gives some benefits as a potential loss is covered by fixed costs (Tummala & Burchett 1999).

2.3 Risk identification factors

Some studies have identified risk factors for construction projects. In a survey conducted by (Abu Mussa 2005), a total of forty-four risk factors were identified and categorized into nine groups namely Physical, Environmental, Design, Logistics, Financial, Legal, Construction, Political, and Management group. It was revealed that financial failure of the contractor, working in the hot environment, closure, defective design and delayed payments on the contract were the most important risk factors. The findings show that there are some risk factors contractors could not allocate them to the party that should bear these risk's consequences. Wong and Cheung (2005), also stressed that the most significant risk occurred in design and built include time and cost overrun. The main reason for these risks is an employer or government delay, lack of information from the employer, the difficulty of following instructions, conflict of interest and variation to changes. Ibrahim et al. (2006), Opined that construction projects are attributed to financial, technical, politics, act of God and social risks that may influence the projected profit. Therefore, for this study, a thorough review of existing literature was performed to identify common risk factor that may stand in front of construction projects. The literature identified forty-four factors categorized into nine groups as mentioned early. These group risk factors are illustrated in Table 1.

3 METHODOLOGY

Both secondary and primary data were employed to examine contractors' perception of the risk factors. The secondary data was gathered through a comprehensive related literature review. The primary data, on the other hand, was obtained from a well-organized questionnaire that was distributed to construction professionals in Gauteng, who are presently practicing or have worked on completed or proceeding construction projects. The target population of this research was Project Manager, Construction Manager, Architect, Quantity Surveyors and Contractors, registered with the various professional bodies. A five-point Likert scale was used to examine the impact each identified risk factor. The adopted scale was as follows: 1- No impact, 2-Law impact, 3-Medium impact, 4-High impact, 5-Very high impact. Data collected were analyzed statistically using the Mean Item Score (MIS). The indices were used to determine the relative impact and ranking of each item. The ranking made it possible to cross compare the relative importance of the items as perceived by the respondents. The similar approach has been used by some researchers to analyze the data gathered from questionnaire survey (Le-Hoai et al. 2008). The computation of the relative mean item score (MIS) was calculated from the total of all weighted responses and then relating it to the total responses on a particular aspect. This was based on the principle that respondents' scores on all the selected criteria, considered together, are the empirically determined indices of relative importance. The index of MIS of a particular factor is the sum of the respondents' actual scores (on the 5-point scale) given by all the respondents' as a proportion of the sum of all maximum possible scores on the 5-point scale that all the respondents could give to that criterion. Weighting was assigned to each responses ranging from one to five for the responses of 'No impact risk' to 'Very high impact. The mean item score (MIS) was calculated for each item as follows;

$$MIS = \frac{1n_1 + 2n_2 + 3n_3 + 4n_4 + 5n_5}{\sum N} \quad \text{.......... Equation 1.0}$$

Where: n_1 = Number of respondents for 'No impact', n_2 = Number of respondents for 'Law impact', n_3 = Number of respondents for 'Medium impact', n_4 = Number of respondents for 'High impact', n_5 = Number of respondents for 'Very high impact', N = Total number of respondents. After mathematical computations, the criteria were then ranked in descending order of their mean item score.

4 FINDINGS AND DISCUSSIONS

4.1 *Demographic Results*

The questionnaires were completed by top management in the organizations (Table 1). A total of 50 questionnaires were sent out, 44 were returned and used which represent 88% the overall sample, which formed the basis of this study. Table 2 below presents the responses from the participants as well as their position. It is shown that of the 44 responses, the majority of respondents were construction managers 15 (34.1%), 11 (25%) were quantity surveyors, 9 (20.5%) were project managers, 4 (9.1%) were named as others while 3 (6.8%) were directors, and 2 (4.5%) were architects.

Table 2: Position of the Respondent

Respondent Position	Frequency	Percentage %
Director	3	6.8
Project Manager	9	20.5
Construction Manager	15	34.1
Architect	2	4.5
Quantity Surveyor	11	25.0
Other	4	9.1
Total	44	100

Figure 2 below illustrates the working experience of the respondents in the industry, it is shown that 48% had working experience that ranged from 1-5 years, 25% had working experience that ranged from range 6-10 years, 14% had working experience that ranged between 11-15 years, 5% had experience that ranged from 16-20 years and 8% had more than 20 years of working experience. On the basis of their function, education, work experience and professional background, it can be concluded that the respondents have sufficient knowledge of construction activities. The response rate for completed questionnaires is shown in figure 2 below.

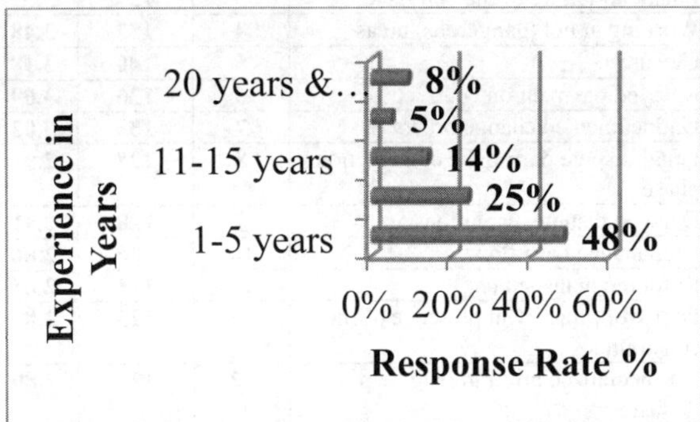

Figure 2: Experience of the organization in construction (years)

Figure 3 below represents the number of construction projects executed in the last five years by contractor respondents. The results revealed that 32% of the respondents were involved in 3-4 projects, 26% were involved in more than 8 projects, 18% were involved in 1-2 projects, 17% were involved in 5-6 projects, and 5% of the respondents were involved in 7-8 projects, and 2% were not involved in any construction projects during the last five years.

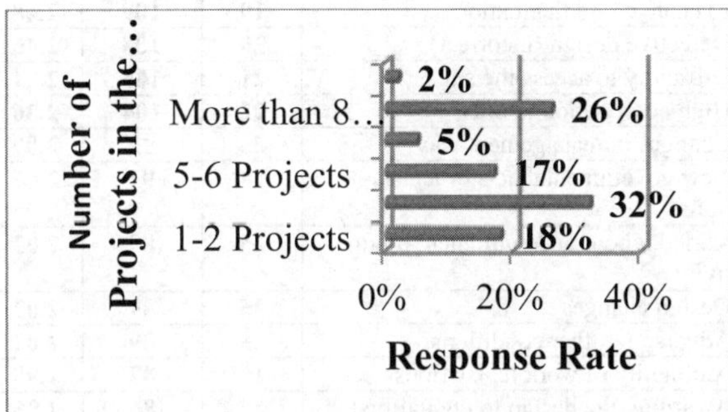

Figure 3: Number of executed projects in the last five years

4.2 *Ranking Study of Risk Factor*

Results from the study revealed that supplies of defective materials, poor communication between involved parties, financial failure of the contractor, working at dangerous areas and closure are the five most significant risk factors in construction projects. These five risk factors are from four major categories namely physical, management, financial and political group risks. Of the five most significant risk factors, supplies of defective materials, poor communication between involved parties and financial failure of the contractor were perceived by respondents as the most three dominant risk factors ranked 1 and 2. These results draw the contractor's attention to the appropriateness of materials that contribute 70% of the total value of the project (Enshassi et al. 2003). Hence, any problems related to construction materials would affect the project (Enshassi et al. 2003). These findings are in agreement with the studies of Ahmed et al. (1999) and the findings of National Audit Office (2001), which discovered the risks defective materials as very important risks. The results further emphasize the importance of communication in early stages of the project as poor communication between involved parties results in a waste of time and thus affecting the budget. These results concord with the findings of Hoezen et al. (2006), where it was found that making adjustments in later stages of the building process, as a result of poor communication usually cost extra money. Contractors are advised to communicate at early stages of the project, as early and or improved communication would undoubtedly lead to fewer delays and lower expenses.

Another important risk factor is the financial failure of the contractor. This can significantly affect the procurement of material, therefore, delaying the project from being delivered in due time. These results are in line with the studies of Hallaq (2003) that concluded that more than 80% of financial contract failures were caused by financials factors such as depending on banks and paying high, low margin of profit due to competition, award contract to the lowest price, lack of capital and cash flow management. Consequently, Contractors are recommended to have enough cash to lessen financial problems (Enshassi et al. 2003).

Working at hot (dangerous) areas and closure in the political group came as the fourth and fifth most important risk factors. It is evident that working at dangerous areas risk is perceived as a significant risk, contractors cannot be imposed to work in such conditions. On the other hand, the closure could be the result of material unavailability and inflation due to monopoly.

Table 1. Risk Factors revealed by existing Literature.

Risk Group	Risk Factor
Physical	Occurrence of accidents due to poor safety procedures
	Supplies of defective materials
	Varied labor and equipment productivity
Environmental	Environmental factors
	Difficulty to access the site
	Adverse weather conditions
Design	Defective design
	Not coordinated design
	Inaccurate quantities
	Lack of consistency
	Rush design
	Awarding the design to unqualified designers
Logistics	Unavailable labor, materials, and equipment
	Undefined scope of working
	High competition in bids
	Inaccurate project program
	Poor communications
Financial	Inflation
	Delayed payments on contract
	Financial failure of the contractor
	Unmanaged cash flow
	Exchange rate fluctuation
	Monopolizing of materials
Legal	Difficulty to get permit
	Ambiguity of work legislations
	Legal disputes during the construction phase
	Delayed disputes resolutions
	No specialized arbitrators to help settle fast
Construction	Rush bidding
	Gaps btwn the Implementation & specifications
	Undocumented change orders
	Poor work quality in the presence of time constraints
	Design changes
	Actual quantities differ from contract quantities
Political	New governmental acts or legislations
	Unstable security circumstances
Political continue	Closure
	Segmentation of Gauteng
Management	Vague planning due to project complexity
	Resource management
	Changes in management ways
	Information unavailability
	Poor communication (involved parties)

Table 3: Construction Risk Factors ranking

Risk Factors	Rank	Weight	MIS
Supplies of defective materials	**1**	**178**	**4.05**
Poor communication between involved parties	**2**	**156**	**3.55**
Financial failure of the contractor	**3**	**155**	**3.52**
Working at hot (dangerous) areas	**4**	**153**	**3.48**
Closure	**5**	**140**	**3.18**
Delayed payment on contract	**6**	**136**	**3.09**
Undocumented change orders	**7**	**133**	**3.02**
Legal dispute during the construction phase	**8**	**128**	**2.91**
Delayed dispute resolutions	**9**	**128**	**2.91**
Unmanaged cash flow	**10**	**126**	**2.86**
Resource management	**11**	**125**	**2.84**
Poor work quality in presence of time constraints	**12**	**123**	**2.80**
No specialized arbitrators to help settle fast	**12**	**123**	**2.80**
Unavailable labour, materials and equipment	**13**	**122**	**2.77**
Poor communication between the home and field offices	**13**	**122**	**2.77**
Gaps between the implementation and the specifications	**14**	**121**	**2.75**
Segmentation of Gauteng	**15**	**120**	**2.73**
Unstable security circumstances	**15**	**120**	**2.73**
Monopolising of materials	**16**	**117**	**2.66**
Occurrence of accidents due to poor safety procedures	**16**	**117**	**2.66**
Vague planning due to project complexity	**17**	**116**	**2.64**
Inflation	**18**	**111**	**2.52**
Exchange rate fluctuation	**19**	**109**	**2.48**
Defective design (incorrect)	**20**	**108**	**2.45**
Difficulty to access the site	**21**	**106**	**2.41**
High competition in bids	**22**	**104**	**2.36**
Changes in management ways	**23**	**92**	**2.09**
New governmental acts or legislations	**24**	**91**	**2.07**
Varied labour and equipment productivity	**25**	**89**	**2.02**
Design changes	**25**	**89**	**2.02**
Adverse weather conditions	**25**	**89**	**2.02**
Ambiguity of work legislations	**26**	**87**	**1.98**
Awarding the design to unqualified designers	**27**	**86**	**1.95**
Actual quantities differ from the contract quantities	**28**	**83**	**1.89**
Environmental factors	**28**	**83**	**1.89**
Undefined scope of working	**29**	**82**	**1.86**
Not coordinated design	**30**	**80**	**1.82**
Lack of consistency	**31**	**79**	**1.80**
Information unavailability	**31**	**79**	**1.80**
Inaccurate project programme	**31**	**79**	**1.80**
Difficulty to get permit	**32**	**78**	**1.77**
Rush bidding	**34**	**74**	**1.68**

Inaccurate quantities	35	55	1.25
Rush design	36	53	1.20

5 CONCLUSIONS

This study has proved that risks factors are the key elements that need to be considered to achieve successfully the fundamental elements of a project (time, cost and quality). Forty-four risk factors were revealed through a detailed literature review that was then categorized into nine groups namely physical, environmental, design, logistics, financial, legal, management, political, and construction. The five most critical risk factors as perceived by respondents are a supply of defective materials, poor communication between involved parties, financial failure of the contractor, working at dangerous areas and closure. These five critical risk factors are from four different categories of risk, i.e., physical, management, financial and political group risk. These findings will strengthen the contractors' evaluation of the risk factors.

To reduce the probability of failure of construction projects, contractors are recommended to take into consideration the importance of handling risk factors associated with construction projects. Contractors should have an adequate project planning that would allow them to foresee these risks factors. Moreover, risk should be taken into account by adding a risk premium to quotation, time estimation and this has to be supported by organizations such as the Construction Industry Development Board (CIDB), the Association of South African Quantity Surveyors (ASAQS), the Chartered Institute of Building (CIOB), the South African Federation of Civil Engineers Contractors (SAFCEC) and other organizations involved in the construction sector. Additionally, contracting firms should provide training programs for their personnel to properly apply management principles as it is the duty of organizations to provide such training.

REFERENCES

Abu Mousa, J.H.E. (2005). Risk Management in Construction Projects: from Contractors and Owners perspectives. Master of Science in Construction Management. The Islamic University of Gaza, Palestine.

Ahmed, S.M., Ahmad, R., & De Saram, D.D (1999). Risk management trends in the Hong Kong construction industry: a comparison of contractors and owners perception. *Engineering, Construction and Architectural Management*, 6(3): 225-234.

Ahmed S., Azhar S., and Ahmed I., (2001). Evaluation of Florida General Contractors' Risk Management Practices, Florida International University.

Akintoye, A.S., & Macleod, M. (1997). Risk Analysis and Management in Construction, *International Journal of Project Management*, 15 (1): 31- 38.

Aven, T. (2003). Foundations of risk analysis. Chichester: John Wiley & Sons Ltd.

Ayyub, B.M. (2003). Risk Analysis in Engineering and Economic. Chapman & Hall/CCRC, 2003.pp.35.

Chapman, C., Ward, S. (2003). *Project Risk Management: Processes, Techniques and Insights*, 2nd ed. West Sussex

Dey, P.K. (2001). Decision support system for risk management: a case study, *Journal of Management Decision*, 39(8): 634-649.

Enshassi A. & Mayer P., (2001). Managing risks in construction projects, 18th Internationales Deutsches Project Management Forum, Ludwig burg, Germany.

Enshassi, I., A., Lisk, R., Sawalhi, I., & Radwan. (2003). Contributors to construction delays in Palestine, *the Journal of American Institute of Constructors, 27(2): 45–53.*

Flanagan R. & Norman G., (1993). Risk Management and Construction, 2nd Edition. Blackwell Science.

Ghani J.A (2009). *Construction Risk Management*. Punjab Information Technology Board. Available from http://www.pitb.gov.pk/downloads.aspx. [Accessed: 27/08/2015]

Grey, S. (1995). Risk assessment for project management. Chichester: John Wiley & Sons Ltd.

Hallaq, K. (2003). Causes of contractors' failure in Gaza Strip; Master dissertation, Islamic University of Gaza Strip.

Hoezen, M.E.L, Reymen, I.M.M.J, & Dewulf, G.P.M.R. (2006).The problem of communication in construction. University of Twente, Enschede, the Netherlands.

Ibrahim, A.D, Price, A. D. F, Dainty, A.R.J, Engineering, Road, A., and Adibrahimlboroacuk, E. (2006). "The analysis and allocation of risks in public private partnerships in infrastructure projects in Nigeria," 11(3): 149–163

Ingvarsson, J., & Roos, A. (2003). Risk Analyst: Metodbeskrivning för beställare-utförare- gran skare. Stockholm: Svenska brandförsvarsföreningen

Keçi, J. and Mustafaraj, E. (2013)."Practices, Barriers and Challenges of Risk Management Implementation in Albanian Construction Industry" Athens: ATINER'S Conference Paper Series, No: CIV2013-0639.

Khalafallah, A.M.G.E.I. (2002). Estimating cost contingencies of residential building project using belief networks. Master of Science Thesis, Cairo University Giza, Egypt.

Le-Hoai, L., Lee, Y. D., and Lee, J. Y. (2008). Delay and cost overruns in Vietnam large construction projects: A comparison with other selected countries, *KSCE Journal of Civil Engineering,* 12 (6): 367– 377.

Lester, A. (2007). Project Management Planning and Control, Fifth ed. Elsevier Ltd.

Mahendra, P.A, Pitroda, J.R., and Bhavsar, J.J. (2014). Risk Assessment in Residential Construction Projects by SPSS. *International Journal of Engineering Sciences and research technology*, 3(5): 498-504.

National Audit Office. (2001). Modernising Construction, NAO, UK.

Piney, C. (2002). Risk response planning: Selecting the right strategy, the 5th European Project Management Conference, PMI Europe, France.

Potts, K., (2008). Construction cost management, learning from case studies. Abingdon: Taylor Francis

Project Management Institute (PMI). (2013). A Guide to the Project Management Body of Knowledge (PMBOK® Guide) – Fifth (5) Edition

Shen, L.Y. (1997). Project Risk Management in Hong Kong, *International Journal of Project Management*, 15(2), 101-105

Simu, K. (2006). Risk Management in Small Construction Projects. Licentiate Thesis, Luleå University of Technology.

Tang, w., Qiang, M., and Duffield, C. (2012). Risk management in the Chinese construction industry, *Journal of Construction Engineering and Management*, no. May 2012, pp. 944–956, 2007.

Tummala, V., & Burchett J., (1999). Applying a risk management process (RPM) to manage cost risk for an EHV transmission line project, *International Journal of Project Management,* 17, 223-235.

Van Well-Stam, D., Lindenaa, F., Van Kinderen, S., & Van den Bunt, B. (2004). Project Risk Management: An essential tool for managing and controlling projects. London: Kogan Page

Wong, P.S.P., and Cheung, S.O. (2005). "No Structural Equation Model of Trust and Partnering Success," *Journal of Management in Engineering*, 21(2): 70 – 80.

Zeng, J.A.N.M., & Smith, N.J. (2007). Application of fuzzy based decision-making methodology to construction project risk assessment, *International Journal of Project Management*, 25, 589-600.

Zou, P.X.W., Zhang, G., & Wang, J. (2007). Understanding the key risks in construction projects in China, *International Journal of Project Management*, 25, 601-614.

Proceedings of the International Conference on Smart Infrastructure and Construction
ISBN 978-0-7277-6127-9

© The authors and ICE Publishing: All rights reserved, 2016
doi:10.1680/tfitsi.61279.715

Value based maintenance prioritization for a portfolio of bridges

R. Srinivasan* and A.K. Parlikad

University of Cambridge, Cambridge, UK
* *Corresponding Author*

ABSTRACT One of the main challenges in infrastructure asset management is the ability to prioritise maintenance activity on portfolio of assets. These maintenance activities consist of reactive repair work and periodic routine maintenance. However, with limited budget available, it becomes quite complex to delay or postpone maintenance activity as it might incur higher impact during the later stages of the life of the asset. Moreover, infrastructure assets are network of assets and consist of multiple stakeholders and longer service life. Therefore, it is essential to understand the value provided by infrastructure assets and use it to prioritise maintenance activities. In this paper, a structured methodology for identifying and assessing the value of bridges is presented, which is then used to assess the impact of maintenance activity on a particular bridge. A case example involving a portfolio of bridges is presented to illustrate the proposed approach. The essential consideration of value is expected to allow organisations in evaluating the impact of short term decisions on long term impact of cost, risk and performance.

1 INTRODUCTION

Management of infrastructure assets such as bridges and road networks are becoming critical as the number of deteriorated assets increases continuously. Decrease in public spending, constraints in budget, and demand for better service further augments the problem of balancing between spending and the risk associated with the postponement of maintenance activities. Traditional methods for prioritizing and planning maintenance activities are predominantly based on cost and therefore are not suitable to understand the value of the asset and the assessment of associated risks. There is increasing interest in the need for value based maintenance as proposed by ISO 55000 standard. However, there is a lack of systematic approach to identify value and to utilize this value to drive asset management decisions.

Existing research on infrastructure asset management focusses on solving problems specific to an asset class such as pavements, roads or bridges. Maintenance modelling, optimization and decision making have gained considerable interest for infrastructure asset management. Chen et al formulate and solve a multi-objective optimization problem for selecting the best strategy for road maintenance (Chen et al, 2015). Buttlar and Paulino integrated pavement crack prediction model with asset management and vehicle-infrastructure interaction models (Buttlar and Paulino, 2015). From a bridge perspective, existing works have focused on modelling deterioration and integration with maintenance planning (Frangopol,

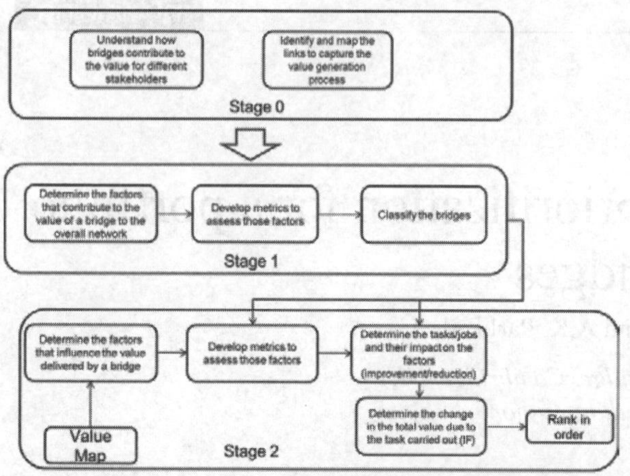

Figure 1. Approach.

and Bocchini, 2012; Furuta et al, 2015). One of the major focus areas in bridge asset management is the modelling of deterioration. Various techniques have been proposed in the literature focusing either on a bridge type or element, such as reinforced concrete bridges and steel bridges. For example, Ghodoosi et al propose a system level deterioration model for reinforced concrete bridge decks (Ghodoosi et al, 2014). Others, such as Wang et al propose a dynamic Bayesian networks approach to predict the condition deterioration of bridge elements (Wang at al., 2012). Analysis of the literature reveals that there has been less focus on portfolio management of bridges and also on prioritising maintenance activities for different bridges. Additionally, most of the existing models use cost as the objective function. However, infrastructure assets have multiple stakeholders with different objectives and there is a need to consider value as a decision making criteria.

Typical infrastructure management requires consideration of portfolio of a various asset types. The assets within the portfolio are in different life cycle stages and their maintenance requirements vary. From operational point of view, the decision to carry out maintenance activities on these assets is generally done annually for the next calendar year. However, due to budget constraints not all planned activities can be carried out. This increases the risk and cost for the subsequent years. Therefore, there is a need to prioritise the maintenance activities on specific assets based on their value to network. In this paper, we present a value based approach to prioritise maintenance activities for a portfolio of bridge assets.

2 APPROACH

Fig:1 illustrates the methodology adopted to prioritise the maintenance activities for a portfolio of bridges. Stage 0 is the process of identifying the various factors of the bridge that contribute to the value for different stakeholders. In order to understand this value generated by a bridge, a value based process as proposed in (Srinivasan and Parlikad, 2015) is used. This consists of a three stage process, namely, value identification, value mapping and value quantification. The output of this stage is a map linking the value generation factors in a systematic manner.

The next stage involves classifying the bridges based on their value to different stakeholders. The first step in this stage is to determine the key factors that contribute to the value of a bridge to the overall network. These factors are identified from the value map generated in the previous stage. The next step is to develop metrics to assess those factors, which can then be used to classify the bridges in the portfolio. In this stage each bridge in the portfolio is assigned a classification reflecting the value of a bridge.

Stage 2 involves prioritising maintenance activity based on the value of a bridge. The first step in this stage is to identify the key factors that influence the value delivered by a bridge. The next step is to develop metrics to assess the factors. Using these metrics, then it is possible to determine the impact of the scheduled maintenance activity on the bridge. This impact can then be used to assess the change in risks and will allow the ranking of various maintenance activities associated with all bridges. In the next section, a case example involving a portfolio of bridges is presented and the proposed methodology is implemented.

3 CASE EXAMPLE

To demonstrate the proposed approach, we present a case example of a local county council, which is responsible for the up keeping of all the bridges under

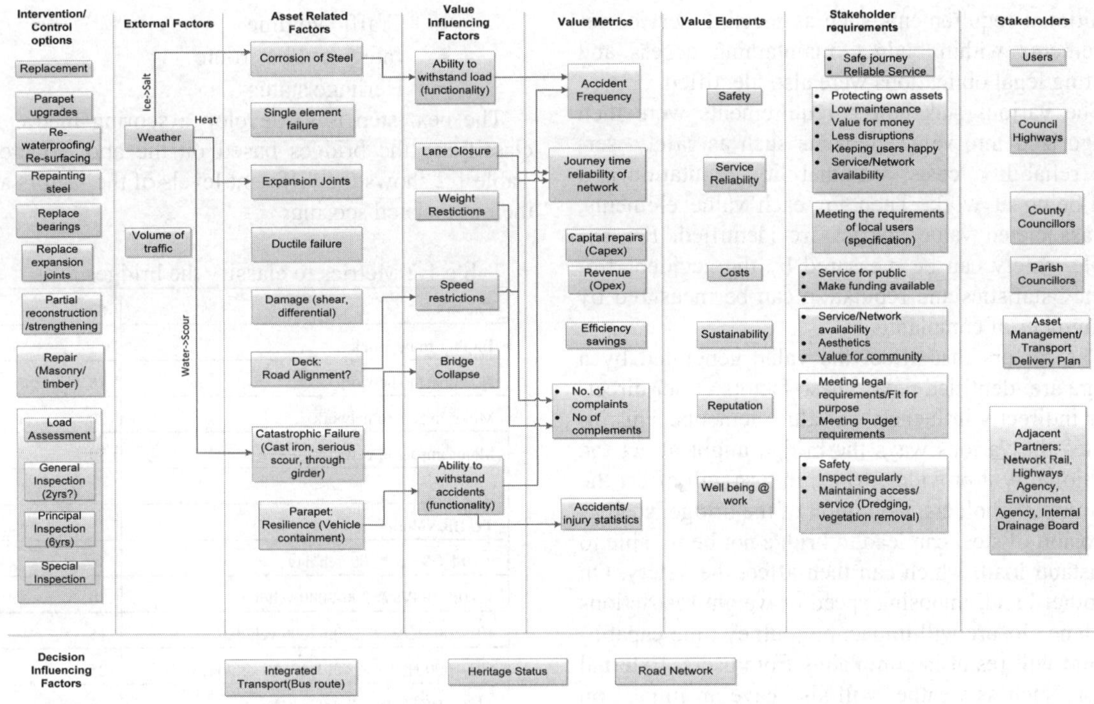

Figure 2: Value map for a bridge

its jurisdiction. The council has 1500 bridges under its care and the bridges are all in different deterioration stages. Currently, the council has a limited budget and the planned maintenance activities on bridges are prioritized based on a scoring system. However, the current approach of prioritizing does not take into account the importance of the bridge and the impact of proposed work on the bridge. For example, a low priority bridge such as a farm access bridge might come higher up the order. The bridge portfolio manager manually decides which bridge work will be carried out. In order to alleviate this problem, the proposed methodology is applied and is presented in the subsequent sub-sections. For this case example, a selection of five bridges and their associated tasks are chosen as shown in table 1.1

Table 1.1 Bridge tasks for prioritisation

Bridge	Tasks	Cost of tasks (M£)
Bridge 1	Stone work refurbishment	1
Bridge 2	Strengthening	0.6
Bridge 3	Edge beam replacement	0.4
Bridge 4	Re-deck	0.2
Bridge 5	Parapet upgrade/refurbishment	0.4

3.1 *Stage 0: Value Identification and mapping*

In this stage the value generation process is captured in the form of a map, as shown in fig:2. In this process, the various stakeholders associated with bridges and their requirements are identified. The critical external stakeholders are the users, politicians such as county councilors and external partners where the bridge interfaces with other organisations such as railways. The internal stakeholders are the council's highways department and the internal funding committee. The key requirements of the various stakeholders consisted of safety and service reliability.

Additional requirements such as meeting service requirements within budget, maintaining access and meeting legal obligations were also identified.

The various stakeholder requirements were then categorized into value elements such as safety, service reliability, costs, sustainability, reputation and well-being at work. Then for each value elements, the associated value metrics are identified. For example, safety can be measured by the accident frequency statistics and reputation can be measured by the number of complaints.

The factors that affect the value generated by a bridge are identified next. These factors either directly or indirectly influence the value elements. This indicates the various ways the bridge might affect the functionality it provides, which in turn will affect the various stakeholders. Condition of the bridge, such as corrosion of steel can lead to bridge not being able to withstand load, which can then affect the safety. On the other hand, imposing speed or weight restrictions and lane closure will impact on journey time capability and will result in complaints from users. External factors such as weather will also have an impact on the bridge. For example, using salt to treat road ice will cause corrosion. Excessive heat will cause issues in expansion joints.

Next, the various intervention and control options are identified. These include replacement, partial replacement, repair of masonry and inspection options. Additionally, various decision influencing factors are identified. These are factors that influence the decisions as they indicate the criticality of the bridge. The key decision influencing factors are the road network or the location of the bridge in the network, heritage status and whether the bridge is in an existing bus route.

3.2 Stage1: Classification of bridges according to value

The next stage is to classify the various bridges based on the value it provides to different stakeholders. The first step in this stage is to determine the various factors of a bridge that influence the decisions. This is obtained from the decision influencing factors identified from the value map. For this particular example, the various factors that influence the decisions are:

- Impact to network
- Traffic volume
- Integrated bus route
- Heritage status

The next step is to develop a scoring metric for classifying the bridges based on the above factors. Table 1.2 shows the different levels of the factors and their associated scoring.

Table 1.2 Metrics to classify the bridges

Factors	**Score**
Impact to network	
No impact to network	10
Minor impact to network	20
Major impact to network	30
Traffic volume	
0-10 HGVs & <200 veh/day	5
11-100 HGVs & 200-2000 veh/day	10
101-500 HGVs & 2000-7000 veh/day	15
501-1000 HGVs & 7001-12500 veh/day	25
>1000 HGVs & >12500 veh/day	30
Integrated bus route	
No bus route and or not strategically important	0
Bus route or strategically important	10
Heritage status	
No heritage or local interest	0
Not listed but local interest	5
Listed or heritage structure	10

From the table, it is possible to assign scores for each individual bridge for each of the influencing factors. In order to classify the bridges, the individual scores of each factor are added together. This sum is then classified into three levels to identify the value of the bridge.

- High value bridges: Scores>60
- Medium value bridges: Score 40-60
- Low value bridges: Score<40

For example, Bridge 2 is a railway bridge on a high-priority road network. The scores assigned are as:

- Major impact to network – 30
- Traffic: 501-1000 HGVs & 7001-12500 veh/day – 25
- Integrated bus route – 10
- No heritage status – 0

Therefore the total score obtained is 65, which implies that the bridge is of high value.

3.3 *Stage 2: Prioritisation of maintenance activities*

The next stage is to prioritise the maintenance activities on the bridges based on the value provided by each bridge. The first step in this stage is to determine the various factors that influence the value of a bridge. These factors are identified in the value map, marked as value drivers. The critical factors that influence the value are:

- Safety
- Service
- Cost
- Sustainability
- Reputation

However, in this particular example, the council wanted to utilize only safety, service and cost to assess the impact of maintenance on value.

The next step is to develop the scoring mechanism for each of the value drivers and this is shown in table 1.3.

Table 1.3 Metrics to assess the value drivers

Factors	Score
Safety	
No impact to safety	0
Minor impact to safety	30
Major impact to safety	50
Service	
No service disruption	0
Less impact on service	20
Minor impact on service	30
Major impact on service	50
Cost (£)	
> 2M	50
1M> <2M	60
0.5M> <1M	80
0.1M> <0.5M	90
<0.1M	100

From the above table, it is possible to assign a score for the bridge depending upon the condition and its consequent impact on safety, service and cost of scheme. For the purposes of prioritising maintenance tasks, the score for safety and service can be calculated before and after the maintenance activities. This will indicate the reduction or improvement in the value drivers due to maintenance. The resulting impact, the difference between the sums before and after maintenance indicates the change in risk and will also depend on the value of bridge (from the previous stage). The resulting change in impact is weighted by the value of the bridge as shown in table 1.4.

Table 1.4 Score for assessing maintenance impact based on the value of the bridge

Change in impact	Bridge Value		
	High	Medium	Low
>60	100	80	60
>40 and <60	80	60	40
<40	60	40	30

Based on the impact of the maintenance activity and the value of bridge, a score is assigned. This indicates the value of carrying out a maintenance activity on a bridge. High impact maintenance job on a high value bridge will result in a high score, whereas a low impact job on a low value bridge will yield a very low score.

The next step is to weigh the maintenance impact with the cost score. This is done by combining the value obtained from table 1.3 with the cost score. The resulting sum (score out of 200) is then used to prioritise which bridge maintenance activity needs to be carried out in the next year. For example, Bridge 2, the following scores are obtained:

Before maintenance

- Major impact to safety – 50
- Minor impact on service – 30

After maintenance

- No impact on safety - 0
- Less impact on service – 20

Therefore the change in impact is 60 (80-20) and from table 1.3, the overall impact is 100 given that bridge 2 is high value. The cost of works is between £0.5 million to £1.0 million and the score obtained is 80. Hence, the overall value score then is 180.

The resulting scores are sorted in ascending order and high scores indicate that the maintenance activities on those bridges will yield the maximum value. Table 1.5 shows the final output of the proposed approach. An excel based tool was developed as part of the case study and was validated with a set of examples.

Table 1.5 Prioritised list of jobs for each bridge

Bridge	Tasks	Cost of tasks (M£)
Bridge 2	Stone work refurbishment	0.6
Bridge 4	Strengthening	0.2
Bridge 3	Edge beam replacement	0.4
Bridge 1	Re-deck	1
Bridge 5	Parapet upgrade/refurbishment	0.4

4 CONCLUSIONS

In this paper, a systematic approach is proposed to prioritise maintenance activity for a portfolio of bridges using a value based approach. The proposed approach is illustrated with a case example. The application of value based decision making principle allows infrastructure managers to objectively plan maintenance activities and also enables them in better management of budget. Additionally, infrastructure managers can utilize the proposed approach to understand budge requirements for future planning purposes.

The current approach fails to capture the impact of postponing the maintenance activities on the value. The next logical approach is to develop a portfolio optimization method based on value, which will enable asset management decision makers to better understand the implication of future risks and performance. Furthermore, such modelling approach will enable better understanding of budgetary constraints and will allow organisations and governments to allocate funds for long-term management.

REFERENCES

Chen, L., Henning, T.F., Raith, A. and Shamseldin, A.Y., 2015. Multiobjective Optimization for Maintenance Decision Making in Infrastructure Asset Management. Journal of Management in Engineering, p.04015015.

Buttlar, W.G. and Paulino, G.H., 2015. Integration of Pavement Cracking Prediction Model with Asset Management and Vehicle-Infrastructure Interaction Models (No. NEXTRANS Project No. 073IY03).

Frangopol, D.M. and Bocchini, P., 2012. Bridge network performance, maintenance and optimisation under uncertainty: accomplishments and challenges. Structure and Infrastructure Engineering, 8(4), pp.341-356.

Furuta, H., Fujikawa, H., Kagawa, Y., Nakatsu, K. and Ishibashi, K., 2015, May. Multi-stage Optimization for Flexible Bridge Maintenance and Management Planning against Modification of Deterioration Curve. In IABSE Symposium Report (Vol. 104, No. 6, pp. 1-8). International Association for Bridge and Structural Engineering.

Ghodoosi, F., Bagchi, A. and Zayed, T., 2014. System-Level Deterioration Model for Reinforced Concrete Bridge Decks. Journal of Bridge Engineering, 20(5), p.04014081.

Wang, R., Ma, L., Yan, C. and Mathew, J., 2012, June. Condition deterioration prediction of bridge elements using Dynamic Bayesian Networks (DBNs). In Quality, Reliability, Risk, Maintenance, and Safety Engineering (ICQR2MSE), 2012 International Conference on (pp. 566-571). IEEE.

Srinivasan, R. and Parlikad, A.K., 2015, A process for value based asset management decision making. In World Congress in Engineering Asset Management (WCEAM), Finland

Proceedings of the International Conference on Smart Infrastructure and Construction
ISBN 978-0-7277-6127-9

© The authors and ICE Publishing: All rights reserved, 2016
doi:10.1680/tfitsi.61279.721

How smart sensoring improves tunnel resilience: from theoretical model to future application

D.M. Zhang*[1], H.W. Huang[1], Q.F., Hu[2] and Y.J. Zhang[1]

[1] *Key Laboratory of Geotechnical and Underground Engineering of Minister of Education, and Department of Geotechnical Engineering, Tongji University, Shanghai, China.*
[2] *Shanghai Institute of Disaster Prevention and Relief, Shanghai, China*
* *Corresponding Author*

ABSTRACT Preventive maintenance has gained more and more attentions since tunnel performance would inevitably degrade against time. It is generally accepted that smart sensoring could in some way assist in the decision for preventive maintenance. However, the timing and cost-benefit when using smart sensoring is quite vague. With regard to this circumstance, applying resilience analysis for tunnels could evaluate the effectiveness of smart sensoring rigorously. The resilience is explained conceptually as the ability of a tunnel to absorb the disruption and the ability to recover to the acceptable performance level. Using the framework of resilience model proposed by the authors recently, this paper illustrates explicitly the timing and cost-benefit in using the smart sensoring to improve the tunnel resilience. It has been derived that if the response time to disruption when applying smart sensoring were n times faster than the time using the traditional monitoring technique, the loss of tunnel resilience could be n^2 times less than the loss in traditional way. The merit of using smart sensoring for tunnel resilience is thus numerically appreciated. Furtherly, preliminary study on resilience-based strategies for two types of repair works for tunnel is presented. One is the repair for disrupted tunnel subjected to unexpected extreme disruption and the other is repair for preventive maintenance under the condition of degradation of tunnel performance in long-term. The time duration and cost-benefit have been included in this design where the multi-objective optimization is applied.

1 INTRODUCTION

The numbers of the operated metro and road tunnels are increased incredibly in the world and China as well. Engineers are facing the ever growing pressures for maintenance and repair of tunnels under operation. The system resilience could reflect its ability to absorb the disruption caused by hazards and the subsequent ability to rapidly recover the performance to its normal level (Ayyub, 2014; Francis and Bekera, 2014). This resilience concept could potentially offer a possibility to assess the timing and repair strategy for preventive maintenance of structures (Wang and Ellingwood, 2015). Hence, needless to say, it is greatly welcomed if a tunnel structure could have a strong resilient ability to remain its function at a high level. But, how to make a tunnel more resilient under the current technology and facilities? It is not clear at present.

The smart sensoring technique, e.g., wireless sensoring network (WSN), nowadays is becoming an effective way to implement a real-time monitoring on the structural health state (Huang, et al., 2013). It might be generally accepted that smart sensoring could in some way assist in the decision for preventive maintenance. But quite often, in view of the additional cost during the long-term monitoring before a real disruption happens to the tunnel structures, the benefit of real-time monitoring usually is not well appreciated by the decision makers.

The authors (Huang and Zhang, 2016) have presented a resilience model for shield tunnel linings under extreme surcharge. This model has been applied to a real tunnel disruption case in Shanghai. From the case study, the effect of real-time monitoring on the tunnel resilient ability has been firstly discussed but without a rigorous derivation. Thus, this paper tries to rigorously derive the effect of real-time monitoring on resilient ability of tunnels. Based on the real-time monitoring technique, preliminary study on resilience-based design of repair strategy for two types of repair is discussed at the end. Before that, the resilience analysis model for tunnels is briefly reviewed.

2 RESILIENCE OF SHIELD TUNNEL

Before a detailed description of the tunnel resilience, the index for tunnel performance should be first specified. An index should be easily measured in site and significantly reflecting the structural response. The tunnel horizontal convergence ΔD is adopted in this paper (shown in Fig. 1), which is probably the widely used index both in practices (JTG/D70-2004, 2004) and researches (Mair, 2008).

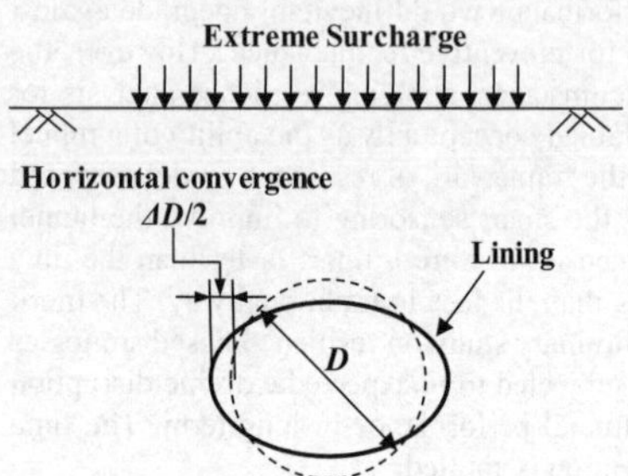

Figure 1 Performance index of lining convergence

Eq. 1 is denoted as the performance index $Q_n(t)$ by a normalization transformation form with ΔD, where ΔD_0 is the initial convergence deflection once the tunnel is built and $\Delta D(t)$ is the convergence at time t which can consider the degradation effect with time.

$$Q_n(t) = \frac{\Delta D_0}{\Delta D(t)} \tag{1}$$

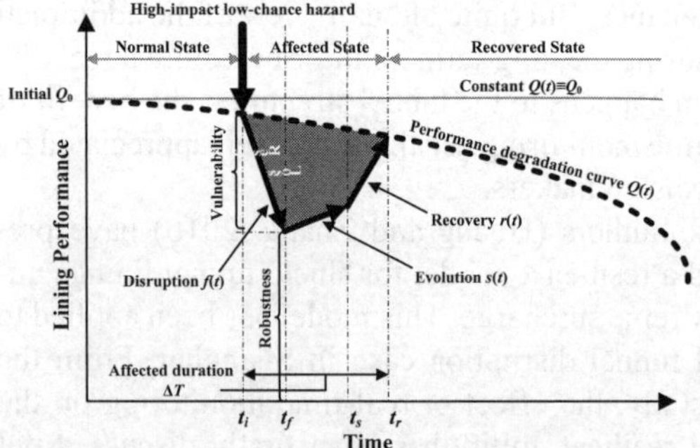

Figure 2 Definition of convergence resilience for tunnels

The resilience is explained conceptually as the ability of a system to absorb the disruption caused by hazards and the ability to recover to an acceptable performance level (Ayyub, 2014). Fig. 2 has illustrated the detailed frame of lifetime performance evolution for tunnel convergence. In general, there are three stages including before surcharge, after surcharge and after recovery. The resilience assessment locates in the second stage, i.e., after surcharge.

Once the surcharge is loaded at time t_i, the tunnel convergence will make a response to this action. The performance will then experience a decrease described by function $f(t)$. The residual performance f_d after this response (at time t_f) stands for the robustness of the tunnel lining. Due to the time cost for decision-making process, the performance will experience a relative stable evolution illustrated by function $s(t)$. Then, once the recovery measures are implemented at time t_s, the recovery will take place until the time t_r reaching to an acceptable level of performance. Hence, the resilience metric can be visually explained by the ratio of the area for the performance evolution function, i.e., $f(t)$, $s(t)$ and $r(t)$, over the area of normal performance function $Q_n(t)$. Mathematically, the resilience index is calculated by following equation:

$$\mathrm{Re} = \frac{t_f - t_i}{t_r - t_i} F + \frac{t_s - t_f}{t_r - t_i} S + \frac{t_r - t_s}{t_r - t_i} R \tag{2a}$$

$$F = \frac{\int_{t_i}^{t_f} f dt}{\int_{t_i}^{t_f} Q dt} \tag{2b}$$

$$S = \frac{\int_{t_f}^{t_s} s dt}{\int_{t_f}^{t_s} Q dt} \tag{2c}$$

$$R = \frac{\int_{t_s}^{t_r} r dt}{\int_{t_s}^{t_r} Q dt} \tag{2d}$$

Several dimensions can be covered in the above metric of the resilience, including degradation, robustness, vulnerability, rapidity and recovery. Details are summarized in Table 1.

Table 1. Resilience dimension and its property

Dimension	Symbol	Property
Degradation	f_i	Degraded performance $Q_n(t)$ at t_i
Robustness	f_d	Residual performance at t_f
Vulnerability	$f_l = f_i - f_d$	Performance loss at t_f
Rapidity	$\Delta T = t_r - t_s$	Speed of recovery
Recovery	f_r	Recovered performance

3 EFFECT OF SMART SENSORING ON TUNNEL RESILIENCE

The rapidity is a crucial dimension in assessing the tunnel resilience. The smart sensoring could increase the response time of disruption of tunnels and further increase the rapidity of recovery. If there is an ideal tunnel structure that its performance do not degrade with time t as shown in Fig. 3. In other words, the performance Q is always equal to unit. However, once the tunnel is unfortunately disrupted by extreme hazard at time t_i, the tunnel performance has been reduced to f_d ($f_d<1$) through a period of time ΔT_1. After implementing the repair measures on the disrupted tunnels, the performance has been recovered to normal state (i.e., $Q=1$) through a period time of ΔT_2. By applying the above mentioned resilience metric, the calculated resilience index Re_1 could be expressed as below:

$$\mathrm{Re}_1 = \frac{1+f_d}{2} \tag{3a}$$

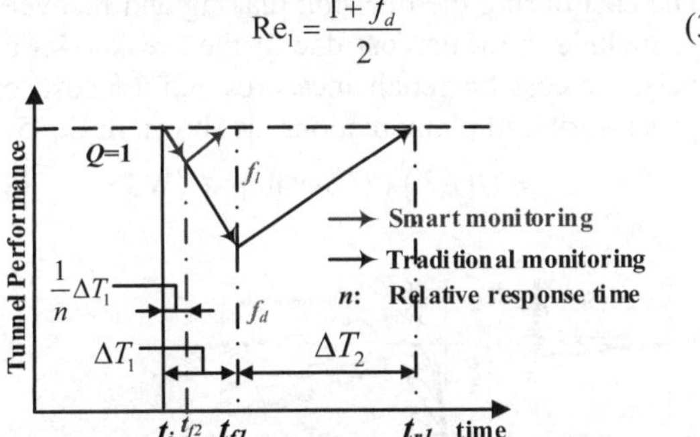

Figure 3 Difference of performance transition curves between smart sensoring and traditional monitoring

In this benchmark problem, if the smart sensoring technique could be used before the disruption happens, the reduction of performance could be captured once it is being reduced. Thus, suppose the time period for tunnel response time in this case of applying smart sensoring equal to $1/n$ times ΔT_1, as shown in blue arrow line in Fig. 3. By applying the same repair measures, the recovery duration could also be $1/n$ times ΔT_2 on the basis of geometric laws. In other words, by applying the same repair measures, the rapidity of recovery by using smart sensoring could be n times faster than the traditional monitoring system. It could be derived further that the area of performance loss in Fig. 3 for the smart sensoring case could be n^2 times smaller than that for the traditional sensoring case. The resilience index Re_2 for the smart sensoring case could be generally n^2 times larger than the Re_1 for traditional sensoring case and is represented as below:

$$\mathrm{Re}_2 = 1 - \frac{1}{2n^2}(1-f_d) \tag{3b}$$

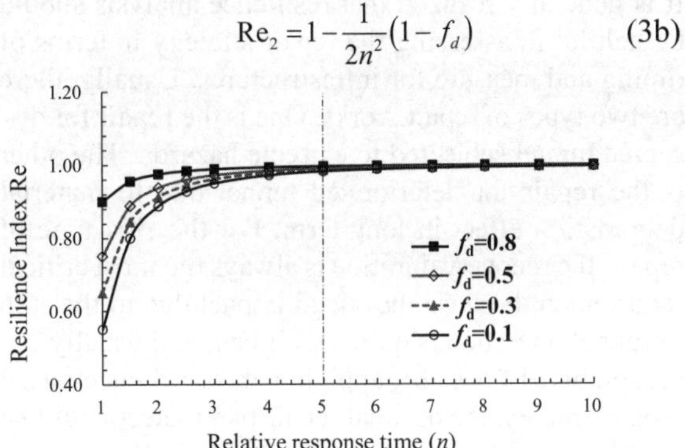

Figure 4 Effect of rapidity of response time by using smart sensoring on the tunnel resilience

Given the robustness performance f_d after the disruption at the level of 0.8, 0.5, 0.3 and 0.1, by applying Eq. 3b, the calculated index Re_2 for smart sensoring case could be plotted against the relative response time coefficient n. It is clear that the coefficient n could greatly affect the results of Re_2. If n is larger than 5, the resilience could be incredibly high and almost equal to unit. That is to say, the tunnel performance could be strongly resilient, regardless of the vulnerability under disruption. This is the reason that the smart sensoring could improve the tunnel resilience even with the same repair or rehabilitation techniques.

Table 2 Comparison of resilience dimension between smart sensoring and traditional sensoring in this benchmark problem

Dimension	Effect of smart sensoring
Response duration	$1/n$
Vulnerability	$1/n$
Robustness	$A - B/n$
Rapidity	$1/n$
Resilience	$C - D/n^2$
Resilience loss	$1/n^2$

Note: Parameter A, B, C, and D is constants when the case is specific and could be calculated by Eq. 2.

Table 2 has summarized the overall effects of smart sensoring on tunnel resilience expressed by using relative response time coefficient n. Since the loss of resilience is related to second order of n, i.e., $\Omega(n^2)$, it thus could clearly indicate the great effect of smart sensoring on the tunnel resilience.

4 RESILIENCE-BASED REPAIR STRATEGY

It is generally realized that resilience analysis should be helpful in assessing the repair strategy in terms of timing and measure for infrastructures. Usually, there are two types of repair works. One is the repair for disrupted tunnel subjected to extreme hazards. The other is the repair for deteriorated tunnel due to material degradation effect in long-term. For the first type of repair, the recovery duration is always the most critical requirement because the social impact due to the stop of tunnel operation is quite significant and usually unacceptable. Hence, by applying the resilience-based repair strategy, the optimal repair parameters could be found to minimize the recovery duration. For the second type of repair, the cost might be the most critical issue compared to the time duration in long-term. By applying the resilience-based repair strategy, the optimal timing to do the recovery could be found to minimize the recovery cost. Hence, the objectives of the resilience-based repair strategy for these two types of repair is different, which is discussed in detail as below:

4.1 *Scenario 1: Hazard-caused disruption case*

For the first scenario, the disruption due to hazard occurs before any notification or preparation. Hence, after discovering the disruption, the response time in decision making and recovery usually is limited. The stop of operation could trigger the community instability and social risk. Hence, there should be a clear deadline of the recovery, saying $\boldsymbol{T_{\max}}$. Apart from the duration, the disrupted performance f_s after the decision making and the recovered performance f_r after the recovery should be larger than a minimum requirement, saying $\boldsymbol{F_d}$ and $\boldsymbol{F_r}$, respectively. As for the overall performance evaluation, the resilience of disrupted tunnel under such a type of hazard should be larger than a minimum resilience index $\mathbf{Re_{min}}$. Subjected to all these conditions, by varying the duration time $\boldsymbol{t_s}$ and $\boldsymbol{t_r}$ and the recovery parameter vector $\boldsymbol{A_r}$, the objectives of the resilience-based repair strategy for scenario 1 is essentially an optimization that maximizes the index Re and f_r, while minimizes the cost C and total time duration ΔT. The optimization algorithm is described in Fig. 5.

Find:	(t_s, t_r, A_r)
Subject to:	$t_i \leqslant t_r \leqslant T_{\max}$, $f_s \geq F_d$, $f_r \geq F_r$, $\mathrm{Re} \geq \mathbf{Re_{min}}$
Objectives:	Maximizing resilience index Re;
	Maximizing recovered performance f_r
	Minimizing cost C
	Minimizing time duration ΔT

Figure 5 Optimization algorithm of repair strategy for scenario 1.

The recovery parameter vector includes two sets of parameters. One is the parameters for repair measures $a_{r,1}$, including the type of measure, the intensity of measure, etc,. The other is the parameters for smart sensoring frequency $a_{r,2}$. Because different frequency could result in different cost and the final recovered performance.

$$A_r = \left[a_{r,1}, a_{r,2} \right] \tag{4}$$

The cost during the decision making and recovery stage include the time cost due to the breakdown of tunnels, the cost for repair measures and the cost for smart sensoring implementations, as shown in Eq. 5.

$$C_{\mathrm{tol}} = C(\Delta T) + C(\mathrm{repair}) + C(\mathrm{WSN}) \tag{5}$$

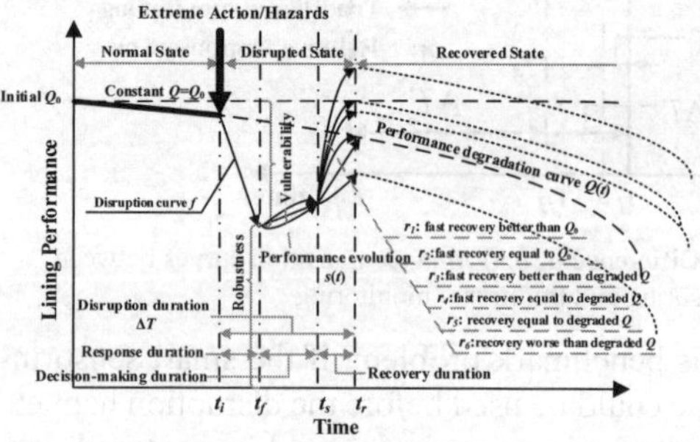

Figure 6 Transition curves with different recovery measures for a disrupted tunnel.

A graphical explanation of the above optimization algorithm is shown in Fig. 6. The performance disruption curve f is determined in the case of scenario 1. The varied parameter is the time t_s and t_r in the horizontal axis and the performance f_s and f_r. However, the performance f_s and f_r is determined by many factors, including the structural properties of tunnel lining, the repair measures and the smart sensoring strategy in terms of devices and frequency. The effect of smart sensoring on the transition curves has been mentioned in previous session. From the graphical point of view, the less the performance area loss has, the higher the resilience index could obtain. But, the overall optimization results from the above algorithm might not be the one with fast recovery better than initial performance Q_0 since the time duration and cost effect have been included.

4.2 Scenario 2: Degradation-caused case

For the second type of repair, apart from the tunnel operational performance, the overall cost during the long-term operation management is prior to the time duration in the recovery stage since engineers could well prepare to do the rehabilitation works. Figure 7 shows an example of this type of repair. In this case, it differs from the previous discussed performance transition curves. The natural degradation curve is just the disruption curve as shown in Fig. 2. It is widely accepted that the tunnel performance could be deteriorated due to time-dependent factors on material in long-term. Hence, engineers have to do the preventive maintenance for degraded tunnels. However, the question of the preventive maintenance is to find a best timing for conducting the rehabilitation or repair works. As shown in Fig. 7, it needs to be decided for engineers whether we should repair the degraded tunnel at the 1st year of operation, 5th year or something later than that. It matters with the robust tunnel performance, sensoring frequency, maintenance cost and maintenance time duration.

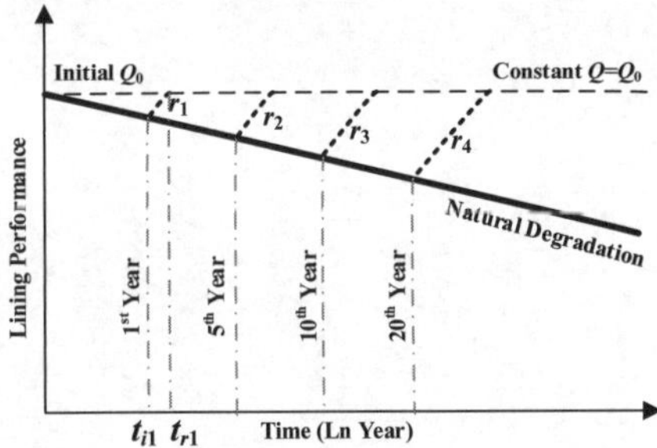

Figure 7 Transition curves with different repair measures for a naturally degraded tunnel

The question of repair timing could be partially answered by doing the following optimization analysis. By varying the time t_i that starts to do the repair work, time t_r that the repair work is finished and varying the repair parameter vector A_r (mentioned previously), the objectives of this optimization is to maximize the resilience index Re and recovered performance f_r and minimize the overall cost C and time duration in recovery ΔT. The condition is that 1) the degraded performance f_i and recovered performance f_r should be larger than the required minimum $\boldsymbol{F_i}$ and $\boldsymbol{F_r}$; 2) the cost should be limited within the maximum acceptable level of $\boldsymbol{C_{max}}$; and 3) the overall resilience index should be larger than the minimum requirement $\mathbf{Re_{min}}$.

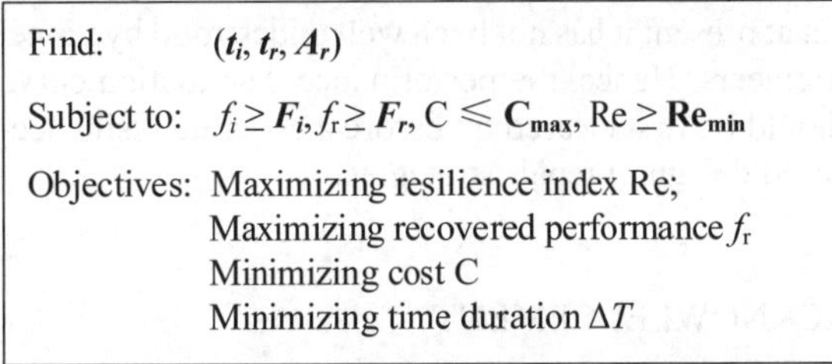

Find: (t_i, t_r, A_r)

Subject to: $f_i \geq F_i, f_r \geq F_r$, $\mathrm{C} \leqslant \mathbf{C_{max}}$, $\mathrm{Re} \geq \mathbf{Re_{min}}$

Objectives: Maximizing resilience index Re;
Maximizing recovered performance f_r
Minimizing cost C
Minimizing time duration ΔT

Figure 8 Optimization algorithm of repair strategy for scenario 2.

These two types of optimization inevitably involve the multi-objective optimization method. The Pareto front is formed due to the multi-objectives since all the objectives hardly could be optimized simultaneously (Juang, et al., 2013; Gong, et al., 2014).

5 CONCLUSION

As the mileage of operated tunnels in cities has been boost up these days, the preventive maintenance and emergency response to the tunnel disruption due to hazards is becoming more and more important. The presented resilience analysis coupled with the implementation of smart sensoring technique could in some way assist the decision maker in a scientific manner to propose a repair strategy for "unhealthy" tunnels. Some of the conclusion could be drawn from this paper as below:

1. By applying the smart sensoring technique in the structural health monitoring system, if the response time could be n times faster than that using traditional technique, the tunnel resilience loss is n^2 times less than the loss for traditional monitoring. This is how the smart sensoring technique to improve the structural resilience.

2. Coupling with the smart sensoring technique, the resilience model could be applied into the design of two types of repair works. The first is the repair for unexpected disruption of tunnel caused by hazard. The

resilience-based design could obtain an optimal repair parameters in terms of specific measure, volume. The second is the repair for preventive maintenance for naturally degraded tunnel due to material time effect in long-term. The resilience-based design could obtain an optimal timing to start the repair.

It should be noted that this is the preliminary study on the resilience-based repair strategies in preventive maintenance. The performance degradation curve in these two types of repair works plays an important role, but at present it has not been well understood by tunnel engineers. Hence, the performance degradation curve should be first cleared up before a concrete resilience-based design of repair strategies.

ACKNOWLEDGEMENT

This study is substantially supported by the Natural Science Foundation Committee Program (51538009, 51278381) and international Research Cooperation Project of Shanghai Science and Technology Committee (15220721600). Hereby, the authors are grateful to these programs.

REFERENCES

Ayyub, B. M. (2014). "Systems resilience for multihazard environments: definition, metrics, and valuation for decision making." *Risk Analysis*, 34(2), 340-355.

Francis, R., and Bekera, B. (2014). "A metric and frameworks for resilience analysis of engineered and infrastructure systems." *Reliab. Eng. Syst. Saf.*, 121, 90-103.

Gong, W., Wang, L., Juang, C. H., Zhang, J., and Huang, H. (2014). "Robust geotechnical design of shield-driven tunnels." *Computers and Geotechnics*, 56, 191-201.

Huang, H.-w., and Zhang, D.-m. (2016). "Resilience analysis of shield tunnel lining under extreme surcharge: Characterization and field application." *Tunnelling and Underground Space Technology*, 51, 301-312.

Huang, H. W., Xu, R., and Zhang, W. (2013). "Comparative Performance Test of an Inclinometer Wireless Smart Sensor Prototype for Subway Tunnel." International Journal of Architecture, Engineering and Construction, 2(1), 25-34.

JTG/D70-2004 (2004). "Code for design of road tunnel." Ministry of Transportation of the People's Republic of China, Beijing.

Juang, C. H., Wang, L., Liu, Z. F., Ravichandran, N., Huang, H. W., and Zhang, J. (2013). "Robust Geotechnical Design of Drilled Shafts in Sand: New Design Perspective." *Journal of Geotechnical and Geoenvironmental Engineering*, 139(12), 2007-2019.

Mair, R. J. (2008). "Tunnelling and geotechnics: new horizons." *Geotechnique*, 58(9), 695-736.

Wang, N. Y., and Ellingwood, B. R. "Disaggregating Community Resilience Objectives to Achieve Building Performance Goals." Proc., 12th International Conference on Applications of Statistics and Probability in Civil Engineering (ICASP12).

SECTION C:
CITIES AND URBAN INFRASTRUCTURE

Proceedings of the International Conference on Smart Infrastructure and Construction
ISBN 978-0-7277-6127-9

© The authors and ICE Publishing: All rights reserved, 2016
doi:10.1680/tfitsi.61279.729

A dependency network description of building information models

K. Al Sayed*[1], M. Bew[2], and A. Penn[1]

[1] *University College London, London, UK*
[2] *The Department for Business, Innovation & Skills (BIS), London, UK*
* *Corresponding Author*

ABSTRACT. The pervasive deployment of "smart building" projects world-wide is driving innovation on many fronts including; technology, telematics, engineering and entrepreneurship. This paper focuses on the technical and engineering perspectives of BIM, by extending building morphology studies as to respond to the challenges posed by Big Data, and smart infrastructure. The proposed framework incorporates theoretical and modelling descriptions to verify how network-based models can act as the backbone skeletal representation of building complexity, and yet relate to environmental performance and smart infrastructure. The paper provides some empirical basis to support data information models through building dependency networks as to represent the relationships between different existing and smart infrastructure components. These dependency networks are thought to inform decisions on how to represent building data sets in response to different social and environmental performance requirements, feeding that into void and solid descriptions of data maturity models. It is concluded that network-based models are fundamental to comprehend and represent the complexity of buildings and inform architectural design and public policy practices, in the design and operation phases of infrastructure projects..

1 INTRODUCTION

There is a vast amount of data that are made available through technology. Yet, there is no comprehensive regulatory framework by which different types of data can be grouped and organised in response to performance requirements. On a building scale, Building Information modelling (BIM) schemes often account for the solid built "atomic" elements and their associated supply and operational infrastructure. Where there are "abstract" void descriptions (Jeong & Ban, 2011), they need to be organised and systemised to relate to social, cognitive and behavioural performance criteria (Schultz & Bhatt, 2011). One could argue that, with a structured and semantic data (LOD) the best possible "fidelity" of any output would be proportional to the lowest quality data. There is therefore a need for structuring information about the built form in such a way as to improve on delivering performance indicators. A proposition for a network description of built spaces that is perhaps associated or complemented by a shape description might hence be sensible in this context. A combined spatial and shape descriptions of the built environment that are compatible with and complementary to energy and lighting performance requirements would enable forecasting user behaviour and comfort during the design stage. The key issue is to really outline the set of performance requirements for buildings, hence find the reduced set of variables and parameters that are essential for analysing and forecasting the performance of built assets. There is also a need to identify a priority structure for different performance criteria depending on what is essential for a building to function and what would improve the comfort of the built environment.

This paper aims to outline key aspects of the nature of building dependencies, in an effort to build a dependency network description of the variables that make up their complexity. At essence, we plan to model the network of relationships that characterize; how the physical infrastructure and its configurations relate to different types of performance criteria and how a smart infrastructure corresponds to performance requirements.

With this objective, this paper addresses how network-based descriptions of built environment might be incorporated in BIM frameworks, through establishing a relationship between the configurations of built form and social structures, as well as environmental performance. For that, a methodology for visualising dependency networks is introduced, along with some propositions on how to integrate frameworks and incorporate empirical models of dependency networks as to inform data information models and public policies.

2 A DATA MATURITY MODEL FOR BIM

Construction contributes with 90BN to the UK economy (6.7% of the national GDP) (2013). About 10% of the UK population works in the construction sector. Construction 2025 set targets of 33% lower costs, 50% faster delivery, 50% lower emissions, 50% improvement in exports. In the current government strategy, and due to funding restrictions, the delivery of BIM level'2, level'3 and level'4 strategies needed to be separated (Bew & Underwood, 2009). Ideally, this separation should have been avoided, particularly in what concerns the link to human behaviour in built assets and the cultural aspects of smart buildings and smart cities, but the need to effectively communicate with the 3M people involved in the industry a managed migration was seen as essential.

To complement the vision for data analytics, there is a need to attend to the value of the social performance embedded in the description of building layouts, e.g. how room spaces connect through adjacency relationships, and how these adjacencies influence social interactions on the long term. It might be argued that any improvements made on the performance of building layouts, would have positive impact on the social and economic performance of buildings during the operational phase.

In the model proposed by Mark Bew[1] for BIM level 3 strategy, the design and operation of buildings needed to account for dependency analytics, in order to better outline the performance requirements of infrastructure. From these requirements stems the relationship between existing building infrastructure and the smart systems that are designed to improve its function. It is usually argued that performance requirements might be mined from smart building projects. However, in order not to be limited to current descriptions of smart infrastructure projects, there is a need to go beyond the smart layer to reveal intelligent relational descriptions in the physical built environment, and perhaps expose how the smart layers might be integrated with building infrastructure to improve its overall performance. In this context, a network description of building layouts might be used, but need to also have a complementary description of solid surfaces that envelope spaces. This is mainly to do with the impact of data fidelity when dealing with more than one performance criteria. For example, if we have a complete data model of the built environment, where BIM values are filled in along with the spatial attributes of each room space (e.g. network configurations (space syntax), shape proportions), the values for each component will fall into the same attribute and entity positions with different provenance, to make a data maturity model of the built environment (figure 1). The analytics devised to measure performance will need to be adapted to provide a tolerance of error value so the user can interpret potential uses of data analytics.

For the purpose of building a universal and integrative framework that brings together BIM and social performance indicators, there is a need to outline an extended network-based representation, accounting for the dependencies between different layout attributes and the temporal, operational and economic dimensions. For BIM models and tools, a network-based description of building space that accounts for the shape parameters of each room might perhaps offer the inverted void description of buildings.

---

[1] See Digital Built Britain plan; https://www.gov.uk/government/uploads/system/uploads/attachment_data/file/410096/bis-15-155-digital-built-britain-level-3-strategy.pdf [accessed 23 April 2015]

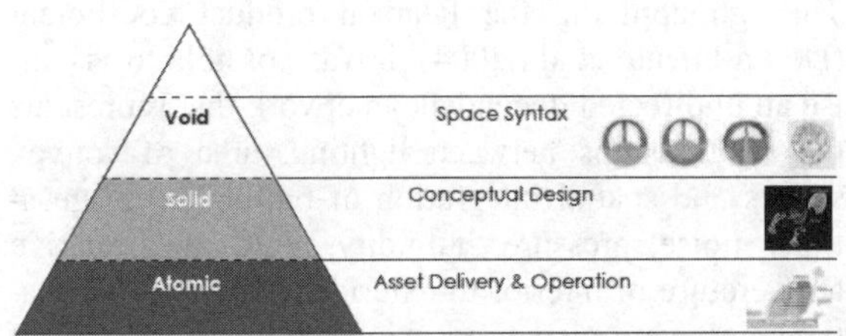

Figure 1 A data maturity model for Building Information Modelling; accounting for void and solid descriptions of the built environment (Source of GLA data: Foster and Partners, 2007).

3 RESEARCH ON DEPENDENCY ANALYTICS IN BUILDINGS

There are multiple performance criteria in buildings; some are intended and some are a by-product of their size, shape and configurations (Al_Sayed, 2014a). It is possible perhaps to describe buildings as organised complex systems, but this description is restricted and incorporates limited dynamics; in that the dynamics are mostly affiliated with the way the smart grid and supply networks operate, and with human occupation and behaviour in facilities, and perhaps with changes on furniture and temporary structures. This is less the case with the physical structure of building; unless the building incorporates dynamic components in its structure.

It is perhaps useful to start from the implicit dependencies in the void descriptions of buildings, and move further to explain how the shape, configurations and size of spaces in buildings might have many implications on different performance criteria; such as sensed social behaviour (Sailer & Penn, 2007), social media (Conroy Dalton et al., 2013), Behavioural psychology, wayfinding and cognition (Kuliga et al., 2013; Orellana & Al_Sayed, 2013), morphological and typological parameters (Shayesteh & Steadman, 2005; Steadman, 2014), and energy performance (Steadman et al., 1991; Batty et al., 2008; Salat, 2009). It is then important to acknowledge dependencies between the atomic void and solid elements of buildings and different utility networks that supply buildings with water, gas and electricity.

Interdependencies between shapes and configurations in buildings can be described discursively, through relating the network structure of spaces in a building to the shape proportions and size (Al_Sayed, 2014a; 2014b). These basic dependencies might have many implications on the social and energy performance of buildings, hence the need to distinguish between core dependencies that characterise other more specific dependencies. An understanding of these fundamental compositions and performance criteria of void descriptions is much needed to complement the solid descriptions of buildings.

4 USING PARTIAL CORRELATIONS FOR CAUSAL INFERENCE IN SPATIAL DATASETS

Previous sections have discussed how dependence between pairs of variables was investigated separately in the literature. For the purpose of representing relationships between larger groups of variables in the built environment, there needs to be a methodological intervention that explains the sequence and structure of interactions between performance variables and the affordances of the physical infrastructure. To reveal networks of dependencies between different data sets in buildings, a methodological framework was adapted from biomedical research (De La Fuente et al., 2004)[2] to outline the relationships between different spatial components in architectural layouts. The Pearson product[3] moment correlation coefficient was used in measuring associations between continuous random variables. For this purpose, a partial correlation coefficient was used to reveal dependencies and identify independence between built environment data sets. A partial correlation coefficient[4] quantifies the correlation between two variables (e.g. temperature x and humidity y) when conditioning on one z or several other varia-

[2] Please refer to this paper for further details about the algorithms. The associated software was used to calculate the Pearson coefficients.

[3] As an alternative, Spearman rank correlation could be used for this analysis since it does not depend on normality and linearity of interactions, thus can be useful for a variable like Choice (Betweenness Centrality) in street networks which follows an exponential distribution.

[4] See Appendix

bles (z1, z2, z3, … zi)[5]. If a correlation between two variables yields a zero partial correlation (or a correlation not significantly different from zero), the algorithm removes that edge (representing a relationship between two variables) from the correlation network. The recursive application of this algorithm on all possible edges results in a network that represents putative direct interactions (a second-order UDG approximation graph). In this study, we propose to use a 0 to 2nd –order correlation coefficient to interpret relationships between spatial components in buildings. The application of partial correlation coefficients to represent dependencies between spatial variables in the built environment can reveal some interesting patterns that might help understanding different types of social, configurational, functional and environmental performance and link it to existing and smart infrastructure.

4.1 Revealing dependency networks in buildings

This section will demonstrate the possibility of applying graph theoretic models of dependency networks to represent relationships between building data sets (configurations and room size), and environmental datasets that are collected from 7 sensors reporting a set of environmental qualities[6] of a 6th form school building[7]. In the context of buildings, social performance variables can be inferred from building configurations using convex representations[8] of space (Hillier & Hanson, 1984). The topological connections between different convex spaces might be represented by an adjacency graph (figure 2). Spaces with high connections might have more accessibility and afford higher likelihood of people moving through them to reach others. Hypothetically, the accessibility of a convex space along with its physical area might have implications on the sensed environmental and comfort qualities of the environment.

[5] A partial correlation coefficient between *rxy.z* is the correlation between the parts of x and y that are uncorrelated with z. To obtain these parts of *x* and *y*, they are both regressed on *z*. The residuals of the regression are then the parts of *x* and *y* that are uncorrelated with *z*.

[6] Environmental performance is calculated based on average sensed values during normal workday operational hours.

[7] The data belongs to Mark Bew, EC Strategies.

[8] Fewest and fattest spaces in a layout

Through applying the Pearson product coefficient (De La Fuente et al., 2004), it was possible to visualise an undirected dependency network that represents the relationships between lighting, area of convex spaces and spatial integration of building configurations, noise, pressure, humidity, VOC, and relative temperature of interior to exterior (table 1). The relationships were visualised using the energy model of Kamada Kawai (separate components) in figure 3, revealing that temperature, pressure and noise are strongly related. Integration, humidity and VOC form another cluster; where humidity seems to bear a strong connection to noise. The analysis yields negative correlations between the physical area of building spaces, and integration, pressure, temperature, and VOC. The analysis also yields that light bears significant positive correlations with integration and noise, and less significant with pressure and temperature. It is not clear whether these performance criteria do actually relate to each other in reality. Due to the limited number of observations and issues with accuracy of the data being generated at present, the results of this approach should be seen as an initial estimate of the real underlying network, enabling us to develop new hypotheses for interactions between configurations, physical characteristics of building components, and environmental performance.

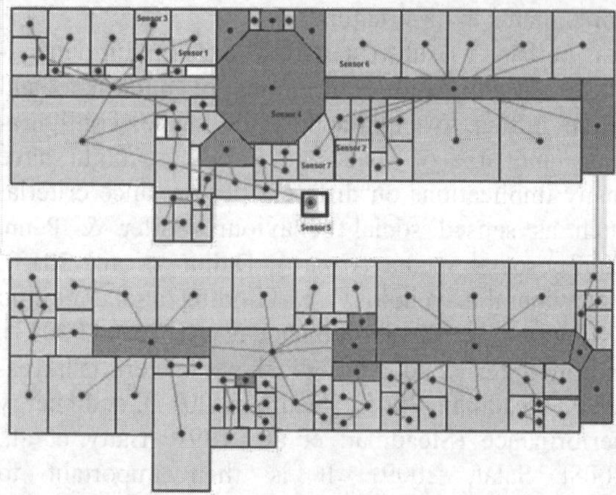

Figure 2 A topological network description[9] of a 6th Form school building, with the locations of sensors identified. Darker colours indicate higher levels of centrality closeness.

[9] The topological network was visualized using DepthmapX, UCL.

Table 1 zeroth[10] order Pearson correlation matrix for school data

	R_Temp	VOC	Light	Noise	Humidity	Pressure	Integ	Area
R_Temp	1.00	-0.49	0.21	0.62	-0.20	0.96	-0.52	-0.46
VOC	-0.49	1.00	0.13	0.21	0.42	-0.47	0.66	0.07
Light	0.21	0.13	1.00	0.36	0.13	0.21	0.41	0.00
Noise	0.62	0.21	0.36	1.00	0.43	0.66	0.11	-0.74
Humidity	-0.20	0.42	0.13	0.43	1.00	0.00	0.80	-0.42
Pressure	0.96	-0.47	0.21	0.66	0.00	1.00	-0.35	-0.61
Integ	-0.52	0.66	0.41	0.11	0.80	-0.35	1.00	-0.03
Area	-0.46	0.07	0.00	-0.74	-0.42	-0.61	-0.03	1.00

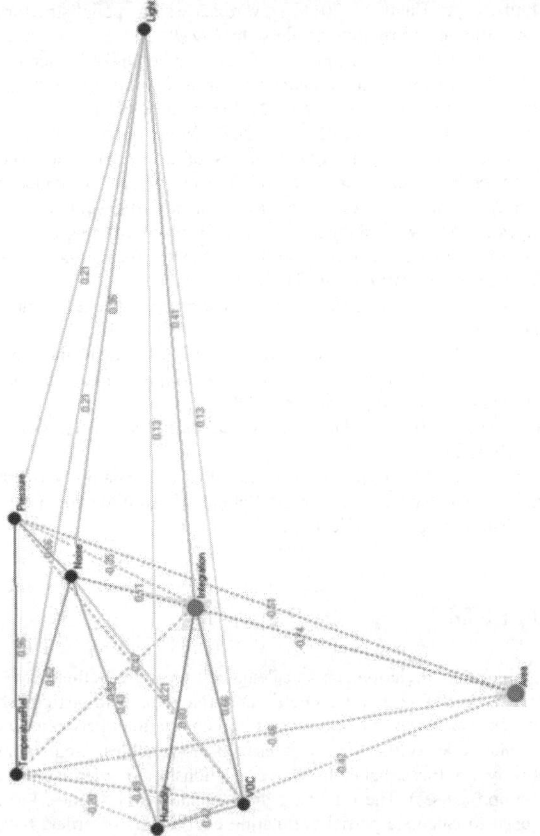

Figure 3 Dependency network[11] of the school building dataset (see figure 2), revealing relationships between eight variables. Darker edges indicate higher values for zero-order partial correlation coefficient between each two variables. The green coloured nodes represent the physical and configurational variables of rooms, the rest of the nodes represent environmental variables.

[10] The order of the partial correlation coefficient is determined by the number of variables it is conditioned on. The zero-order For example, *rxy.z* is a first-order partial correlation coefficient, because it is conditioned solely on one variable (*z*).

[11] The dependency network was visualized using PAJEK software (De Nooy et al., 2005).

5 CONCLUSION

This paper introduced a theoretical framework on how to address the use of social performance analysis (using space syntax) in data maturity models. The paper has also demonstrated a method on how to empirically represent dependencies between different building data sets by adapting a novel Pearson Correlation technique -used previously in biomedical research (De La Fuente et al., 2004)– and exploring its application in the context of buildings. Using this method, it was possible to derive dependency network representations from partial correlation coefficients.

There are nontrivial benefits for dependency network representations in the context of smart buildings; some are to do with testing the degree of fitness between artificial smart systems and existing infrastructure, whilst others are to do with outlining redundancies, disruptions, and cascading effects in building systems. On a building scale, complexity might also have some underlying universal principles; in how spatial structures relate to shape and size of spaces, and in how a combined description of shapes and configurations bears a relationship to energy consumption, carbon emissions, lighting, and noise.

At this stage, it is important to raise some caveats with regards to the interpretation of our findings, considering the small data set we had for buildings and the variance in environmental performance measures that have much to do with the operation of buildings alongside many other factors. It is also important to recognise that, whilst partial correlation coefficients do not necessarily indicate causality, their ability to exclude weak correlations legitimises their use as indicators for causal inference, hence their use makes it possible to rule out primary from secondary datasets. There is a need to emphasise here that "spurious" correlation models of building relationships must not be explained as matters of causality (Simon, 1957), since many different causal relationships can be mapped onto a correlation. Therefore, the application of zero-order to 2nd order correlation networks in the context of buildings need to be cautiously interpreted. Pearson correlations might fail in some occasions to correctly identify a system of significant relationships between different variables, and might on other occasions coincidentally show unrealistic correlations between variables that don't have any shared performance requirements. Despite these deficiencies, the method can be used to develop, with reasonable degree of confidence, plausible hypotheses of interactions between physical, configurational and performance variables, whilst also revealing correspondence between existing and smart infrastructure. The use of dependency networks will therefore be very helpful in building an empirical basis for building information models, and in structuring performance data to enable better predictions about design and operation of buildings.

REFERENCES

Al_Sayed, K. 2014a. On the evolution of thoughts, shapes and spatial structure in architectural design. In *Design Computing and Cognition* DCC'12. J.S. Gero (ed), pp. 393-411. © Springer 2014.

Al_Sayed K. 2014b. How designs evolve, *The Journal of Space Syntax*. Special Issue on Models and Diagrams in Architectural Design, In D. Koch and P. Miranda Carranza (ed), 5(1): 68-90.

Al_Sayed, K., Bew, M., Penn, A., et al. 2015. Modelling dependency networks to inform data structures in BIM and smart cities. In *Proceedings of the 10th International Space Syntax Symposium*. Edited by K. Karimi et al, London, UK.

Bew, M. & Underwood, J. 2009. Delivering BIM to the UK Market. *Handbook of Research on Building Information Modeling and Construction Informatics*: Concepts and Technologies, Hershey, PA: IGI Global, 30-64.

Conroy Dalton, R., Kuliga, S. F. & Hoelscher, C. 2013 POE 2.0: POE 2.0: exploring the potential of social media for capturing unsolicited post occupancy evaluations. *Intelligent Buildings International*. 5(3), 162-180.

De La Fuente, A., Bing, N., Hoeschele, I., et al. 2004. Discovery of meaningful associations in genomic data using partial correlation coefficients. *Bioinformatics*, 20(18), 3565-3574.

De Nooy, W., Mrvar, A., & Batagelj, V. 2005 Exploratory Social Network Analysis with Pajek, *Structural Analysis in the Social Sciences* 27, Cambridge University Press.

Department for Business, Innovation & Skills. 2015. Digital Built Britain, Level 3 Building Information Modelling - Strategic Plan. [Online], Available: https://www.gov.uk/government/uploads/system/uploads/attachment_data/file/410096/bis-15-155-digital-built-britain-level-3-strategy.pdf, [Accessed: 23rd April 2015]

Hillier, B. & Hanson, J. 1984. *The social logic of space*. Cambridge: Cambridge university press.

Jeong, S. K. & Ban, Y. U. 2011. Developing a topological information extraction model for space syntax analysis. *Building and Environment*, 46(12), 2442-2453.

Kuliga, S., Conroy Dalton, R. & Hölscher, C. 2013. Aesthetic and Emotional Appraisal of the Seattle Public Library and its relation to spatial configuration. In*: 9th International Space Syntax Symposium*, Seoul, South Korea

Orellana, N. & Al_Sayed, K. 2013. On spatial wayfinding: Agent and human navigation in virtual and real worlds, In *Proceedings of the 9th International Space Syntax Symposium*, Edited by Y O Kim, H T Park, K W Seo, Seoul, Korea.

Sailer, K., & Penn, A. 2007. The performance of space–exploring social and spatial phenomena of interaction patterns in an organisation. In: *Proceedings of the Architecture and Phenomenology Conference*. Faculty of Architecture and Town Planning, The Technion, Israel Institute of Technology: Haifa, Israel.

Salat, S. 2009. Energy loads, CO2 emissions and building stocks: morphologies, typologies, energy systems and behaviour. Building Research & Information, 37(5-6), 598-609.

Schultz, C., & Bhatt, M. 2011. Toward accessing spatial structure from building information models. In *UDMS 2011: 28th Urban Data Management Symposium*, Delft, The Netherlands, September 28-30, 2011. Urban Data Management Society; OTB Research Institute for the Built Environment; Delft University of Technology.

Schultz, C., Bhatt, M. & Mora, R. 2013. Mindyourspace - a tool for evidence-based qualitative analyses of user experience and navigation behavior in the built environment. In edra44providence - 44th *Environmental Design Research Association Conference*.

Shayesteh, H. & Steadman, P. 2005. Typo-morphological approach to housing transformation in Tehran. In *Proceedings of the 5th Space Syntax Symposium*, TU Delft.

Simon, H. A., 1957. *Models of man; social and rational*. Wiley. New York.

Steadman P., Brown F. & Rickaby P. 1991. Studies in the morphology of the English building stock. *Environment and Planning B: Planning and Design*. 18(1) 85 – 98

Steadman, P. 2014. *Building Types and Built Forms*. Troubador Publishing Ltd.

Varoudis T. 2012 "DepthmapX - Multi-Platform Spatial Network Analysis Software", The Bartlett School of Architecture, UCL, http://varoudis.github.io/depthmapX/

APPENDIX

A partial correlation can be calculated to any pre-defined order (table 2). Partial correlation coefficients can be used to distinguish between causal type of correlations and correlations between two variables that originate via intermediate variables (sequential pathways) or those that embed direct relationship to other variables (common causes). The following three Equations (1,2, and 3) can be used to calculate partial correlation coefficients of orders 0–2. Similar type of equations can also be used to calculate higher order partial correlation coefficients.

Table 2. Different orders for the partial correlations.

0th order correlation	$r_{xy} = \frac{cov(xy)}{\sqrt{var(x)var(y)}}$	(1)
1st order correlation	$r_{xy.z} = \frac{r_{xy} - r_{xz}r_{yz}}{\sqrt{(1-r_{xz}^2)(1-r_{yz}^2)}}$	(2)
2nd order correlation	$r_{xy.zq} = \frac{r_{xy.z} - r_{xq.z}r_{yq.z}}{\sqrt{(1-r_{xq.z}^2)(1-r_{yq.z}^2)}}$	(3)

Proceedings of the International Conference on Smart Infrastructure and Construction
ISBN 978-0-7277-6127-9

© The authors and ICE Publishing: All rights reserved, 2016
doi:10.1680/tfitsi.61279.735

An operational strategy to increase average door-to-door speeds of metro systems in megalopoleis

Marcelo Blumenfeld*, Clive Roberts and Felix Schmid

Birmingham Centre for Railway Research and Education, Birmingham, UK
* *Corresponding author*

ABSTRACT. The increase in average speeds with the predominance of the automobile has led to a geographical expansion of megalopoleis in the form of longer distances travelled. This paradigm has emphasised the systemic shortcomings of public transport systems. Current metro infrastructures are inherently hindered by a paradox between the time to access stations and the average speed on the line, which prevents them offering sufficient door-to-door speeds to compete with the car. The solution proposed in this paper comprises an operational strategy where autonomous vehicles stop in different patterns at stations along a line. It is proposed that vehicles will travel in platoons and are controlled by vehicle-to-vehicle communication algorithms similarly to automated highways. Simulations show that this strategy can reduce the time to access stations by 50% while increasing average speed on the line by 65% and reduce door-to-door journey times by approximately 42% compared to conventional metro operations. In addition, capacity is also increased by 30% within the conventional platform lengths.

1 TRAVEL TIME BUDGETS AND THE GROWTH OF CITIES

It is widely agreed that the limits to urban growth are intrinsically related to the speed of movement, under stable budgets of time and cost. Several studies show historical and geographical stability on travel time expenditures in cities of approximately 1.2 hours a day on average (Zahavi and Ryan, 1980; Zahavi and Talvitie, 1980; Laube et al, 1999; Schäfer and Victor, 2000; Zahavi, 1974; Bieber et al, 1994; Schäfer, 2000; van Wee et al, 2006). Consequently, the average distance travelled across a whole population will be a product of the travel time budget and the absolute speed of the transport network (Laube et al, 1999). It logically follows that under a constant time budget, people will try to maximise the speed that their budget can afford so they can increase their area of potential exploration. Based on these premises, Marchetti (1994) suggests an "anthropological invariant" relationship between travel speeds and urban scale. His research shows evidence that, throughout history, cities have always been one-hour wide.

It can be said, therefore, that transport system capability has been the main driver of urban growth. Just as the railway transformed the city into the metropolis, the car was responsible for the birth of the megalopolis. The late nineteenth century, with the advent of the railways, witnessed the emergence of a city that was based on quite different spatial, social and economic relations than the walking city that preceded it (Schaeffer and Sclar, 1980). Subsequently, once the automobile became widely affordable in the twentieth century, cities grew even further to accommodate even longer distances covered by the car. As cars within cities have an average speed of 6 or 7 times greater than a pedestrian, they expand daily connected space 6 or 7 times in linear terms, or about 50 times in area (Marchetti, 1994).

Nowadays, the average area of the 75 cities with more than 5 million inhabitants is 2,241 km², which results in an average radius of 26.7 km were they perfectly round (Cox, 2015). The dominance of the car in the twentieth century has resulted in a geographical sprawl that makes public transport systems such as the metro less efficient and attractive than private modes. Because of their size, metropolitan areas generally tend to impose longer journey times than their smaller counterparts (Knox, 2014; Schwanen and Pereira, 2013; McKenzie and Rapino, 2011; TUC, 2014), and because of their operations, metro systems cannot offer the same journey times as private modes for similar distances (US FHWA, 2011; TUC, 2014; Infocidade 2007)

2 THE COVERAGE PARADOX

When distances increase and the cost budget permits, travellers will look for higher speeds. Once public transport speeds fall short of private cars, it reinforces the vicious cycle of urban sprawl that makes the car even more attractive. In addition, travel time savings are converted into more journeys (Zahavi, 1974; Schäfer, 2000), so urban dwellers enjoy socio-economic advantages from greater speeds, in the form of more access to opportunities.

Although metro systems can achieve higher top speeds than permitted on roads, their main disadvantage lies on the coverage paradox. Since the journey by car involves mainly a motorised component and thus enables a higher average speed, the distances it permits travelling are greater for the same time budget. The same journey by metro involves at least two components: namely one of individual access to stations, where distance is the critical factor; and in-vehicle time, where speed is the critical factor. Hence, if stations are far apart, a metro will have a higher average speed, but access time increases as well. Conversely, when stations are close together, access is fast but the average speed of the metro is reduced and in-vehicle time increases.

Findings from Gyimesi et al (2011) from 8,000 drivers in 20 of the world´s largest cities show that the average distance of commuters is 19.7 km, covered in 33 minutes. This trip simulated in a metro system (Figure 1) shows that the minimum travel time achievable over the same distance is approximately 47 minutes, or 45% more. The model takes into account the access time (T_a) to walk to and from the station at half the distance between stations at a pace of 80 m/min; an average waiting time of 1 minute at half the headway (T_w); an in-vehicle time (T_v) spent travelling between stations; and a constant average time of entry and exit from payment to the platform (T_e) set at 2 minutes each. Total journey time (T_t), thus, is calculated by:

$$T_t = 2T_a + T_w + T_v + 2T_e$$

This model does not include line transfers that would increase the journey time even further. For this specific simulation, we have used the design parameter used by Siemens (2012) for the new Thameslink fleet in London; these provide maximum speed, acceleration and braking rates of V=90 km/h, α=1.3 m/s² and β=1.2 m/s² respectively. It logically follows that in-vehicle time is calculated by multiplying the station-to-station travel time (T_s) by the number of stations (N_s), which is found by dividing journey length by the distance between stations (S):

$$Tv = \frac{Ts * Ns}{60}$$

As suggested by Vuchic (2004), station-to-station travel time (T_s) is calculated by :

$$Ts = \frac{3.6 * S}{V} + Tl$$

Where S is the distance between stations in metres, V is the maximum speed in km/h and T_l is the incremental time loss per stopping at one station. This incremental time, subsequently, is calculated by:

$$Tl = \frac{V}{7.2} * \left(\frac{1}{\alpha} + \frac{1}{\beta}\right) + Ts$$

Where α and β are the acceleration and braking rates respectively, and T_s is the dwell time at the station.

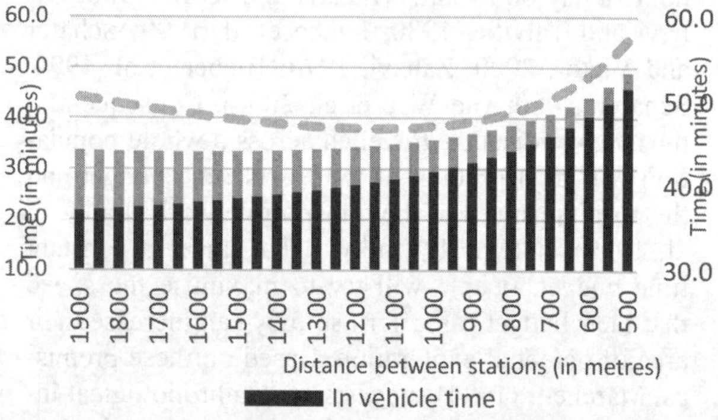

Figure 1. The coverage paradox, illustrating in-vehicle and access time (left hand side) and door-to-door travel time (right hand side) according to the distance between stations for a 19.7km trip. Scales have been adapted to increase clarity.

However, simply increasing top speeds in metros does not appear to solve the problem. A 40% increase in the maximum speed to 35 m/s (126 km/h) would only result in an overall reduction of 1.6% in travel time for the same 19.7 km journey when stations are kept at 1,000 m apart. Increasing interstation distances would only reduce door-to-door journey time by 3.7%. This is because most metro systems stop at all stations along the line, which inevitably reduces the average speed to approximately a third of the top speed.

The solution seems to be reliant on smarter use of infrastructure than on its further expansion. In face of the systemic shortcomings of mass transport systems, several studies have proposed strategies to increase average in-vehicle speeds without increasing the distance between stations (Furth and Day, 1985; Fu, 2003; Vuchic, 2004). Nonetheless, while strategies for accelerated operations can increase efficiency in service patterns and/or journey times, they are counterbalanced by their inability to optimise metro journeys at the systems level. Such irregular stopping patterns provide uneven levels of accessibility for users in different stations. Even with these solutions, door-to-door journey times still exceed the anthropological invariant of 1.2 hours.

3 THE OPERATIONAL CONCEPT

The high level objectives of this concept focus on a system that reduces door-to-door journey times, operating in a way that all users can benefit from it, while maintaining or increasing its capacity. Gyimesi et al (2011) showed that drivers in 20 of the largest cities in the world take an average of 33 minutes to cover 19.7 km. It logically follows that the system should cover the same distance in at least the same amount of time, including access, entry, waiting, and exit times, with an average door-to-door speed of 35.8 km/h.

Current systems, according to the model used in this research, cannot achieve the desired door-do-door journey time. From our experiments, we have found that interstation distances are critical because non-motorised speeds are limited by human ability. For this reason, this research considers all access trips to the stations as made by walking at 84 metres per minute, based on observations by Pachi and Ji (2005). Consequently, leaving all other trip components constant, halving the distance between stations to 500 m would result in a 6 minute saving in journey times. That alone accounts for a 12% reduction of the 47 minutes previously calculated in the model and considerably more for shorter distances.

However, the model also shows that small interstation distances can severely affect average speeds of the metro system by adding extra stops to the line. With stations spaced 500 m between each other, in-vehicle speeds drop to 25.7 km/h, in-vehicle time increases to 46 minutes, and total door-to-door time exceeds 57 minutes each way, thus exceeding the suggested travel time budget. Therefore, the system must also offer such accelerated operation to all passengers, as opposed to the strategies previously described. All services must attend all stations, and connect all origin-destination pairs available. In addition, overall infrastructure needs cannot be increased by the additional stations on the line.

Firstly, the operational strategy assigns different services to vehicles (carriages of a train). Each service stops at stations along the line in different patterns (P_x). The main advantage of this model is that every station is eventually served by all types of services and thus every passenger will experience reduced journey times. If a line is either circular or operated as an infinite loop, the mathematical model shows that if the number of stations (N_s) is not divisible by the number of patterns, then every vehicle will eventually stop at every station, taking a number of 'laps' equal to its pattern to attend all stations. For example, a service that stops every 3 stations (P_3) will take 3 laps to attend all stations on the line, regardless of the total number of the stations.

Secondly, because of the different patterns and service provision to every station, the distance between stations can be reduced to a minimum with no impact on the average in-vehicle speed. Considering the acceleration and braking rates proposed for Thameslink by Siemens (2012), the minimum distance between stations is approximately 500 m, which reduces access time by a half, to 3.1 minutes. Because of the shorter distance, average speed and overall time is increased for the local service. Thus, a system with patterns (P_x) of 1, 2, 3, 5 and 7 in the same 19.7 km trip, can increase in-vehicle average speeds by 65% and, keeping other trip components constant, theoretically reduce door-to-door journey times by approximately 42% (Table 1). Similarly to the model previously used to simulate the coverage paradox, the model at this point does not include transfer times for line changes.

This theoretical simulation considers a network coverage equal to the line, where stations are on average 1000 m and 500 m apart throughout the urban area, hence the access distances of half the distance between stations.

Table 1. Door-to-door time in minutes and speed in km/h of different patterns with stations 500 m apart for a 19.7 km journey

Px	In vehicle time (mins)	Accesss time (mins)	Wait time (mins)	System access (mins)	Door-to-door time (mins)	Door-to-door average speed (km/h)
P1	46.7	3.1	1	2	52.9	20.4
P2	30.0	3.1	1	2	41.2	28.7
P3	24.4	3.1	1	2	35.6	33.2
P5	20.0	3.1	1	2	31.2	37.9
P7	18.1	3.1	1	2	29.3	40.3

Thirdly, in order to allow different vehicles to stop in different patterns without significantly affecting service, it is necessary to reconfigure operational strategies. If each pattern is assigned to one train in a conventional service, headway between two trains of the same pattern in stations increases and could surpass $H_{min}(P_x+1)$ at certain points of the line, where H_{min} is the headway between trains and P_x the number of patterns. Such increases in time and variability in headways considerably affects capacity and influences journey times.

The solution proposed, therefore, is one of autonomous vehicles platooning together such as in automated highways (Tank and Linnartz, 1997; Gehring and Tritz, 1997; Robinson et al, 2010; Coelingh and Solyom, 2012; Fernandes and Nunes, 2015). In this concept, one pattern is assigned to each vehicle, and a number of vehicles of different patterns are virtually coupled to form a platoon that simulates a train. Each train is composed of vehicles of all patterns. Since all trains contain vehicles of all patterns, this strategy guarantees that every station is serviced at the minimum headway between trains.

Such a strategy requires stations to be located off the main line so headways can be reduced and capacity increased (Figure 2). Vehicles that are supposed to skip a certain stop will continue on the main line while those stopping will move to the loop and start braking. Consequently, the length of the loops must be equal to acceleration and braking distances. The platoon, therefore, is variable in length once vehicles leave the main line to attend the next station and other vehicles join after attending the previous station.

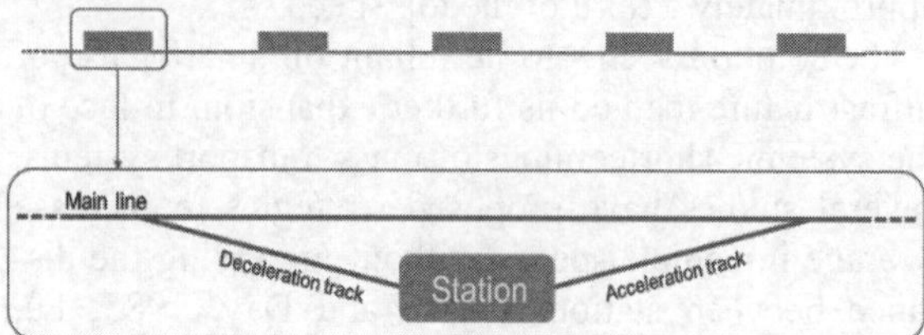

Figure 2. Illustration of off-line stations

Moreover, to avoid conflict and delays, vehicles of different patterns need different platforms to operate. Considering the distribution of passengers in the different patterns, the number of platforms is the number of patterns plus an emergency exit in case of faults in the point switches. The most efficient layout to reduce land use and permit retrofitting of legacy stations is to distribute platforms linearly (Figure 3). The platform length for each pattern is found in simulations according to vehicle length and the largest possible number of vehicles of a pattern in a platoon at a certain time, multiplied by the number of patterns plus one emergency unit.

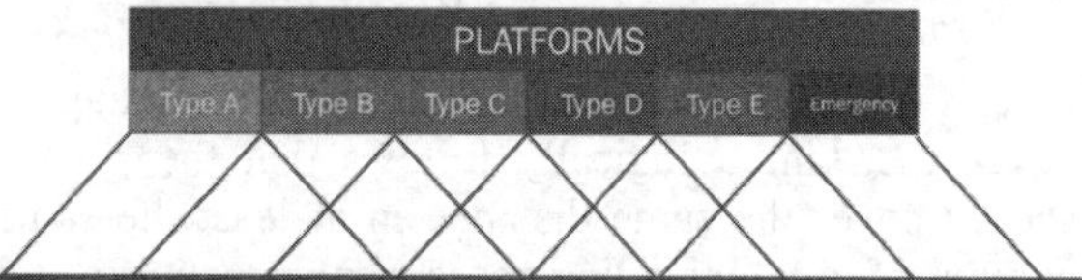

Figure 3. Illustration of off-line stations

Finally, as vehicles stop off the main line, headway calculation need not include acceleration and braking and dwell times of the preceding trains. Headway between trains, in this case, can be as short as the general dwell time. They must be at least equal to the dwell time, however, otherwise it will lead to progressive queuing and subsequently leading to a complete stop of the system as the queue reaches the main line. However, headways must also include the time for accelerating vehicles to access the main line, which is calculated by dividing the maximum platoon length by the speed on the main line. That way, the vehicles leaving the station join the approaching platoon on its front.

It logically follows that headways are reduced when dwell times and platoons are shortened. This dependency creates a set of trade-offs for vehicle design specifications, as dwell times depend on the number of passengers and the number of doors in each vehicle. Assuming that all vehicles are the same width, longer vehicles will carry more passengers, however they will require longer dwell times and also increase platoon length. Plus, longer vehicles and longer platoons require longer platforms, which increase infrastructure costs and transfer times on platforms.

Simulations show this operational strategy significantly increases the theoretical capacity of the system within the same or even shorter platform length. As Figure 5 shows, platoons of 7 vehicles can achieve the same theoretical capacity as the Yamanote line of the Tokyo metro, the busiest in the world, at 100,000 passengers per hour per direction (Parkinson and Fisher 1996). For this, vehicles were set at 12 metres long with a carrying capacity of 110 passengers each, and headway as the headway plus the necessary time for the platoon to pass the transition point. Dwell time (T_d) is set at 20 s and headway (H_{min}) is calculated by

$$H_{min}=T_d+L_p/V$$

where L_p is the length of the platoon and V is the speed on the main line set at 25 m/s.

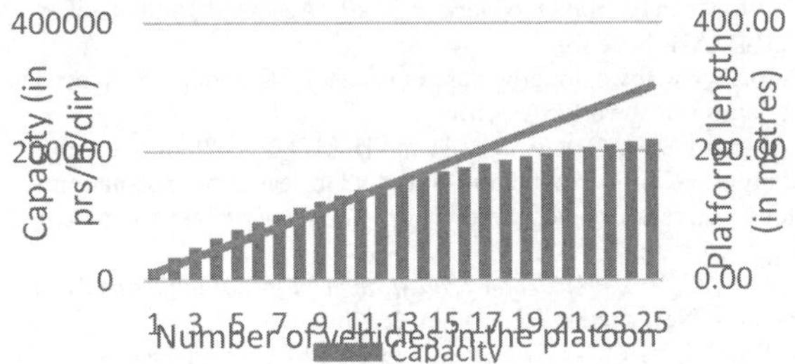

Figure 4. Capacity simulations of the system with different N with L_v=12 m

In this simulation, it is also possible to see the trade-off of longer platoons influencing headways and platform length. Although theoretical capacities can exceed 200,000 passengers, they come at a cost of longer platforms and longer platoons, which not only increases headways but also transfer times. Figure 10 also illustrates the optimal point in the curve at which capacity exceeds the linearity of platform length. Calculating these differences, Figure 11 shows that the most efficient operation arises when platoons are between 8 and 12 vehicles long, and the largest excess is when platoons are 10 vehicles long. In this case, considering a stationary distance between vehicles of 0.5 metres, platforms need to be 125 m long, similar to the average length of current platforms. Moreover, these excesses are also maintained when the vehicle length is changed, suggesting an optimal operational concept for different capacity requirements.

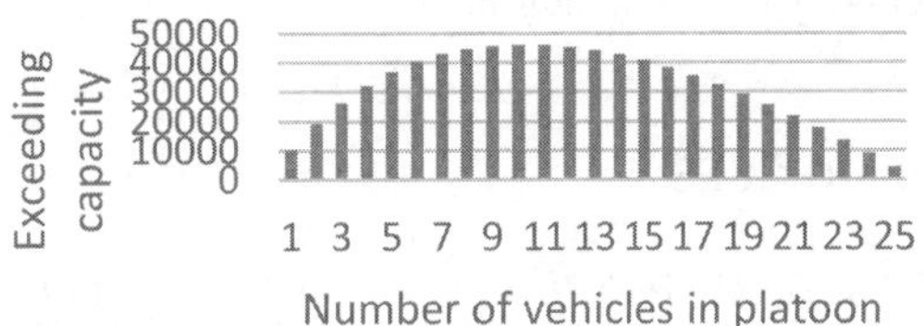

Figure 5. Exceeding capacity of the system with different N under constant L_v=12 m

4 CONCLUSION

In this paper, we have presented an operational concept that solves the coverage paradox and is able to offer higher door-to-door speeds than the automobile. It allows shorter interstation distances which reduces the critical access time, but prevents this from increasing in-vchicle times with a strategy that assigns different stopping patterns for the vehicles. Consequently, as the number of stops between origin and destination stations are reduced, in-vehicle speeds are significantly higher. This is made possible when stations are located off the main line, so vehicles attending the station do not interfere with the journey of those not stopping at that location on the line at that moment. To avoid this from increasing headways, trains must be composed of vehicles of all patterns that egress and return to the main line in different points at different times.

The solution suggested is the application of platooning algorithms discussed in automated highways literature. Trains are composed of autonomous vehicles of different patterns virtually coupled and travelling at a safety distance apart. That allows the system to operate different services at shorter headways at all stations, thus waiting time is also minimised. Headways, as the model shows, can be as short as the dwell time. Platforms are divided in sections for each pattern in order to facilitate accessibility and enable retrofitting of existing networks.

Our simulations show that this model can increase in-vehicle average speeds by 65% and, keeping other trip components constant, theoretically reduce door-to-door journey times by approximately 42%. Because headways are also reduced, a system with 7-vehicle platoons already exceed the theoretical capacity of the busiest metro line in operation, while 10-vehicle platoons provide the optimal capacity per necessary platform length.

REFERENCES

BIEBER, A. , M. MASSOT and J. ORFEUI. 1994. Prospects for daily urban mobility. *Transport Reviews*. **14** (4). pp. 321–339.

COELINGH, E. and S. SOLYOM. 2012, *All aboard the Robotic Road Train*. IEEE Spectrum [online]. [accessed 28 September 2015]. Available at http://spectrum.ieee.org/transportation/advanced-cars/all-aboard-the-robotic-road-train

COX, W. 2015. *Demographia world urban areas* [online]. [Accessed 15 September 2015]. Available at http://demographia.com/db-worldua.pdf.

FERNANDES, P. and U. NUNES. 2015. Multiplatooning leaders positioning and cooperative behaviour algorithm of communicant automated vehicles for high traffic capacity. *IEEE Transactions on Intelligent Transportation Systems*. **16**(3). pp. 1172-1187.

FU, L., Q. LIU and P. CALAMAI. 2003. Real-time optimization model for dynamic scheduling of transit operations. *Transportation Research Record*. **1857**. pp. 48-55.

FURTH, P. and F. DAY. 1985. Transit Routing and Scheduling Strategies for Heavy Demand Corridors. *Transportation Research Record*. **1011**. pp. 23-26.

GEHRING, O. and H. FRITZ. 1997. *Practical results of a longitudinal control concept for truck platooning with vehicle to vehicle communication*. IEEE Conference on Intelligent Transportation System. 9-12 November, 1997. Boston, Massachusetts, USA.

GYIMESI, K., C. VINCENT and N. LAMBA. 2011. *Frustration rising: IBM Commuter Pain Survey*. Final Report. IBM.

INFOCIDADE. 2007. Tempo de Viagem por Tipo: Região Metropolitana de São Paulo [online]. [Accessed 15 September 2015]. Available from: http://infocidade.prefeitura.sp.gov.br/htmls/12_tempo_de_viagem_por_tipo_2007_252.html

KNOX, P. 2014. *Atlas of cities*. New Jersey, USA: Princeton University Press.

LAUBE, F. B., J.R. KENWORTHY and M. E. ZEIBOTS. 1999. Towards a science of cities: city observation and a formulation of a city theory. Workshop paper: How can cities become more sustainable?. Tequisquiapan, Queretaro, Mexico.

MARCHETTI, C., 1994: Anthropological Invariants in Travel Behaviour. Technological Forecasting and Social Change. 47. pp.75-88. Internal Publication, International Institute for Applied Systems Analysis, Laxenburg, Austria.

MCKENZIE, B. AND M. RAPINO. 2011. Commuting in the United States: 2009. *American Community Survey Reports*. ACS-15. U.S. Census Bureau, Washington, DC.

PACHI, A. and T. JI. 2005. Frequency and velocity of people walking. *The Structural Engineer*. **83**(3). pp. 36-40.

PARKINSON, T. and I. FISHER. 1996. *Rail Transit Capacity*. TCRP Report 13. Transit Cooperative Research Program. Transportation Research Board.

ROBINSON, T., E. CHAN and E. COELINGH. 2010. *Operating Platoons on Public Motorways: An Introduction to the SARTRE Platooning Programme*. Proceedings of the 17th ITS World Congress. 25-29 October, 2010, Busan, Korea.

SCHAEFFER, K. H. and E. SCLAR. 1980. *Access for all*. Columbia University Press.

SCHAFER, A. 2000. Regularities in Travel Demand: An International Perspective. *Journal of Transportation and Statistics*. **3**(3). pp. 1-31.

SCHAFER, A. and D. G. VICTOR. 2000. The future mobility of the world population. Transportation Research Part A. 34. pp. 171-205.

SCHWANEN, T. and R. H. M. PEREIRA. 2013. Tempo de deslocamento casa-trabalho no Brasil (1992-2009): diferenças entre regiões metropolitanas, níveis de renda e sexo. Working paper. IPEA ISSN 1415-4765.

SIEMENS. 2012. *Siemens to supply new metro for Munich*. InnoTrans 2012, 18-21 September 2012, Berlin.

TANK, T. and J. M. G. LINNARTZ. 1997. Vehicle-to-vehicle communication for AVCS platooning. *IEEE Transactions on Vehicular Technology*. **46**(2). pp. 528-536.

TRADE UNION CENTRE. 2014. *Commute times increase as the UK's transport system gets more crowded* [online]. [Accessed 15 September 2015]. Available at https://www.tuc.org.uk/economic-issues/labour-market-and-economic-reports/industrial-issues/transport-policy/commute-times.

UNITED STATES FEDERAL HIGHWAY ADMINISTRATION. 2011. Means of travel to work [online]. [Accessed 15 September 2014]. Available from: http://www.fhwa.dot.gov/planning/census_issues/ctpp/data_products/journey_to_work/jtw4.cfm

VAN WEE, B., P. RIETVELD and H. MEURS. 2006. Is average daily travel time expenditure constant? In search of explanations for an increase in average travel time. *Journal of Transport Geography*. **14**. pp. 109-122

VUCHIC, V. 2004. *Urban Transit: operations, planning and economics*. New Jersey: John Wiley & Sons.

ZAHAVI, Y. 1974. *Traveltime budgets and mobility in urban areas*. Final report. FHWA PL 8183. US Department of Transport.

ZAHAVI, Y. and J.M. RYAN. 1980. Stability of travel components over time. *Transportation Research Record*. **750**. pp. 19–26.

ZAHAVI, Y. and A. TALVITIE. 1980. Regularities in travel time and money expenditures. *Transportation Research Record*. **750**. pp. 13–19.

Proceedings of the International Conference on Smart Infrastructure and Construction
ISBN 978-0-7277-6127-9

© The authors and ICE Publishing: All rights reserved, 2016
doi:10.1680/tfitsi.61279.741

Understanding traveller decision making – a crowd sourced big data analysis of the London Travel Demand Survey

G. Casey*, K. Soga, E. Silva and P. Guthrie

University of Cambridge, Cambridge, UK
* *Corresponding Author*

ABSTRACT Engineers have begun to take interest in the interface between the structures they build and the people that use them. This has resulted in a need to better understand behavioural decision making at the micro scale. An analysis was carried out on the London Travel Demand Survey in order to see what behavioural rules could be derived from the survey data. This analysis attempted to relate the financial cost and time cost of a journey to the socio-economic status of a traveller in order to see what relationship existed. Relating this empirical exhibited data to real-world data is challenging. Crowd sourced big-data avenues were explored in order to derive realistic information that considers factors such as traffic congestion, service time tables and the underlying complexity of the London transport network. It was hypothesised that the socio-economic status of the traveller, the distance and the motivation of the journey would impact on traveller decision making. However, no statistically significant results were found. It is concluded that the use of crude small sample size survey data to analyse fine grained crowd-sourced data is inappropriate and highlights a critical need to find other means of recording actual traveller decisions in order to understand the decision making process better. Such insights are critical if better strategic decisions are to be made on transport infrastructure.

1 INTRODUCTION

The purview of civil engineering has broadened in recent years. Historically the civil engineer was responsible for the design, construction, maintenance and other hard quantifiable roles required to build our infrastructure. However, the challenges of climate change and economic uncertainty have led to a branching out into the interface between the structures created and the people that use them. This relationship is very apparent in transport infrastructure projects where civil engineers are attempting to understand the decision making process of its potential users in an attempt to create more suitable, more sustainable and more appropriate engineering solutions.

The wider aim of this research is to aid strategic decision making specifically within the realms of lower carbon transport projects. This can only be done by understanding the specific nature of an environment and the potential transport solutions that exist. Our current challenges require tailored and context specific solutions.

This paper builds upon previous work which is trying to understand why transport infrastructure projects suffer from large ridership forecasting uncertainty. A study of 27 rail projects by Flyvbjerg found that ridership was overestimated in 90% of the projects and that for 67% of these projects the overestimation was greater than 66% (Flyvbjerg *et al*, 2005).

Such overestimations pose great challenges to strategic planners who are required to make decisions in the face of great uncertainty. The ridership, or load factor of a transport mode can be a dominant factor in its carbon credentials (UIC, 2011; Saxe *et al*, 2015). A train that will run at 85% of load factor is a very different carbon story to a train that runs at 40-50% load factor.

Ultimately the load factor of a large infrastructure project such as a high speed rail line is the macro output of many individual and local decisions. The small scale decisions of hundreds, thousands and millions of individual, distinct and highly autonomous travellers all interact to bring about the macro behaviours that we have historically struggled to forecast.

A new method has been developed in an attempt to connect this micro scale to the macro scale (Casey *et al*, 2015). This method employs an agent based model in an attempt to better understand the distinct and heterogeneous nature of decision makers. This bottom up method then uses recent computational advances in order to find the collective macro output of these small and local decisions. This agent based model is informed with crowd sourced real-time big-data in order to better reflect dynamic and real-world conditions.

The assessment of the behavioural rules in this model is the subject of this paper. A survey is used as an empirical data source in order to assess the predictive ability of the proposed new method.

2 UNDERSTANDING TRAVEL DECISION MAKING

2.1 *Model*

To understand traveller decision-making this model considers 3 factors, financial cost, time cost and a metric of socio-economic status. Modern transport infrastructure offers a range of different options in order to achieve the same mobility. The complexity of the transport network in the Greater London Area poses a great challenge for travellers and in turn modellers who wish to attempt and predict their behaviour. This complexity exists in two dimensions - spatially and temporally. The spatial dimension explains the intricacy of the differing services and how they lead to differing journey times that do not always relate linearly to distance. Train, tram, underground, bus services, taxis and so on all provide mobility. The spatial complexity is clearly shown in a map visualizing all the different transport corridors available.

The temporal dimension explains how the exact same journey may vary in terms of financial and time cost. For example, car journeys of the same origin and destination can have highly variable time costs as a result of changing traffic conditions and public transport journey times are dependent on timetables and scheduling. The end result is that there is no fixed data input for journey decision making. In order to consider real world decision making we must find ways of giving the model real world temporally and spatially dynamic inputs.

The traveller decision making model considers the many different options that are available from an origin to a destination and for each of these we consider its financial and time costs and depending on our socioeconomic status we will make a modal decision. The socio-economic metric enables the model to consider how different people will value their time and financial factors with differing importance. This is represented in Figure 1.

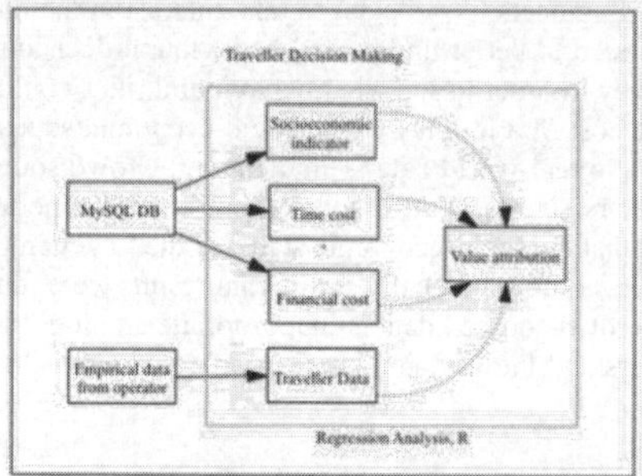

Figure 1 Traveller decision making model

As an illustration, to go from Heathrow to central London you have various options, Heathrow Express, London Underground, a taxi or driving a personal car. Each of these has a different financial cost and time cost. This model therefore infers that those with a higher socio-economic status will value their time higher than the financial cost more so than someone of lower socio-economic status.

In order to assess how reasonable this is, an analysis was carried out on the London Travel Demand Survey (LTDS) of 2013.

3 DATA SOURCES

3.1 *The London Travel Demand Survey*

The LTDS is a Transport for London (TfL, 2014) survey, which contains detailed travel information for

18,931 driving journeys, 11,000 cycling journeys, 10,386 transit journeys and 13,574 walking journeys.

Thus, there is a dataset that we can perform a multi-variable regression analysis on. This will enable us to relate journey time (which is related to mode choice) with financial cost and the socioeconomic status of a traveller.

3.2 Real-time big-data sources

The Google Directions API provides travel information for a range of travel modes between two points using a HTTP request. Google use a modified Djikstra's algorithm to compute the shortest path in a complex network in real-time. This service offers results for a range of transport modes including driving and multiple forms of public transport.

3.3 Driving

Finding the lowest time cost route in the Greater London area is spatially and temporally complex. Firstly, the range of roads available offers a multitude of different possibly paths. These roads can be viewed as a weighted graph where the links and nodes represent roads and junctions respectively. The weight of each individual link will be intrinsic to the road itself and will consider factors such as speed limit, lane count and road surface. Such information can be sourced via the Ordnance Survey in the UK or, more recently, online map providers such as Google, OpenStreetMap and Apple. This research made use of Google's map infrastructure. Using a combination of satellite imagery, street view data and other crowd sourced data they have created an underlying map with very fine resolution.

Beyond the spatial complexity there is the challenge of quantifying the impact of users on this hard infrastructure. Chronic traffic issues plague most cities. As a result of this journey times and lowest cost journey routes can vary considerably. The advent of GPS enabled mobile phones has allowed for real-time location data to be used to derive real world traffic conditions. Google gathers anonymised location data from users and from this traffic flow models can be created (Google, 2009). The density of cities and very high uptake of smart phones has made this method particularly appropriate in dense urban areas.

As an illustration why this is important, consider the same journey but carried out at differing times of the day. Figure 2 shows the total journey time for an example origin and destination but for different times of the day. The temporally static value of around 50mins illustrates what a conventional shortest time cost path would output for the known infrastructure. However, as is clear this is a distortion of reality when compared with the highly dynamic journey times that are derived when considering crowed sourced mobile phone data. The morning rush hour is significantly peaked and equates to nearly double the journey time of that with no traffic. The evening rush hour is less pronounced as work end times are more staggered and results in a longer but less intense increase in journey times.

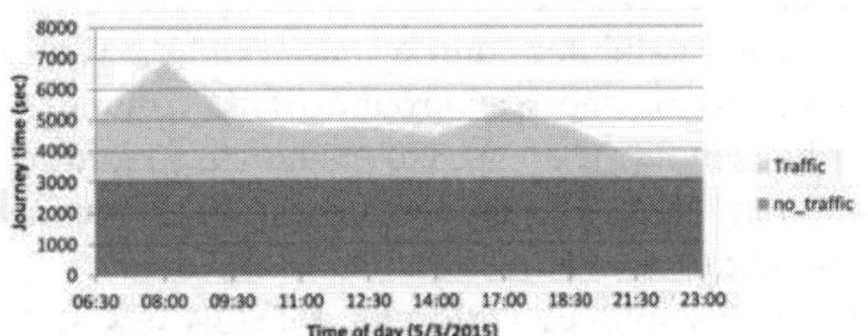

Figure 2 Downe to London St Pancras Journey times at different times of day (Casey et al, 2015)

Combining spatial data with this real-time empirically sourced traffic data enables a shortest time cost path that reflects real world conditions possible. The ability to quantify journey times in close to real world conditions has opened up new avenues to help understand actual traveller behaviour rather than idealised traveller behaviour as has been traditionally done.

3.4 Public Transport

Transport for London is responsible for Buses, Cycling, River services, Dial-a-Ride, Taxi & private hire, Underground, Overground, Docklands Light Railway, Tram, Emirates Air Line and TfL Rail. These services are heavily integrated and vary spatially and temporally. Travellers use them in different ways and may use a combination of differing services in one journey.

Systems such as the Underground can operate at close to maximum capacity for prolonged periods and as such are constantly dynamically reacting to new information. Similarly to driving, the algorithm is required to have up to date information on the services themselves, the access points of the stations, the interchange times between different platforms and so on. TfL publish real-time feeds and encourage software developers to use the data in a way which can inform traveller decision making. These feeds can be harnessed in a similar way to the phone location data and enable real world shortest paths to be derived. TfL share the same data with Google via the General Transit Feed Specification (GTFS) and thus allow Google to present real-time informed travel information.

4 METHODOLOGY

Two different data sources can now be combined and an attempt to relate the two. There is an empirical data source, the LTDS survey carried out by TfL. This contains actual exhibited journeys as well as a wealth of information on the background of the traveller and their household. There is also the idealized journey information that has been sourced using a combination of empirical data sources and real-time data.

For every journey within the LTDS we wish to attribute a realistic time and financial cost. As has been illustrated, assuming a fixed time cost for a journey distorts the real world reality, thus the time cost must be estimated based upon when the journey was carried out.

In order to assess the likely time and financial inputs walking/cycling journeys are deemed time independent and driving/transit journeys are deemed time dependent. By associating the survey journeys with matching crowd-sourced big data from Google we can assess the relationship between these identified factors. This crowd-sourced data then enables us to find reasonable financial and time costs for the traveller. From the LTDS, we are given the wage grade of the traveller which is used a socio-economic metric. Using coefficients from AA and accessing the method of payment (from the LTDS) we can estimate the financial cost for their journey.

4.1 Time Dependent

The LTDS was imported into a MySQL database for ease of manipulation. All driving journeys were exported. These driving journeys were then filtered by day of week and time. This was transformed into a schedule per weekday, Saturday, Sunday and per 2-hour time intervals. Thus, for a survey journey that occurred at 13:30 on Saturday 14th September in 2013 the data was harvested at 13:00 on Saturday 15th August. Thus attempting to bridge the gap between real-time crowd sourced data and that data which was harvested via surveys.

This schedule was then exported in the format shown in Figure 3. The trip_id is used to then relate this journey back to its parent data in the LTDS. Time gives the sample time in Unix time (seconds since 1st September 1970) and the the remaining coordinates give the origin and destination locations. These were converted from British Easting/Northings to WGS84, as used by Google.

trip_id	Time	O_lat	O_lng	D_lat	D_lng
130192110105	1434326400	51.51468471	-0.162015601	51.63822958	-0.178644361

Figure 3 Example driving request tabulated

There were 41 of these files, each for a different day and 2-hour time slot. These were then batch processed and sent at the appropriate time to Google's API using a Python script. Similarly to driving, public transport is highly time dependent. The same method is used to filter and batch these requests as driving.

4.2 Time Independent

Cycling and walking were deemed time independent, i.e. that they are not impacted by congestion. This simplifies their scheduling and allows them to be sent in bulk at one time.

4.3 Pricing

One further metric must be derived from the journey results – financial price. For driving, cycling and walking distance related coefficients can be applied to estimate the financial cost for that journey, dependent on its mode. Driving coefficients were provided by the AA and considered aspects such as road tax, insurance, capital cost and so on. This equated to 31p per km (AA, 2014). A cycling pricing methodology by Arundell followed a directly comparable method to that of driving and also provided a co-efficient related to distance, equating to 24p per km (Arundell, 2008). Walking had no estimates in the literature and so an estimation of 10p per km was used, justified based upon the extra calorific demand calculation used as per Arundell. Public transport was considerably more complex. Pricing structures are not simply dependent on distance but service provider, travel zones, payment method, timing and political boundaries. Pricing boundaries were harvested from TfL and tabulated. Underground stations were allocated a travel zone by referencing the Google results to the TfL boundaries. MySQL string matching was performed to overcome small spelling differences (for example Kings Cross vs King's Cross). Bus and tram journeys operated by TfL were much simpler to price due to their static price and cashless systems. Finally, the total journey cost was calculated depending on the sum of its constituent legs with consideration made to price capping.

5 ANALYSIS

The Google results give the key outputs from the journey, such as total distance, total time as well as fine grained information on the exact constitutional parts of the route. Google returns the journey information in JavaScript Object Notation (JSON) format. This JSON is parsed and transferred to a MySQL database where it can be manipulated. The first step is to pair the Google results to its parent in the LTDS. Once this has been done we are presented with a database which contains journey time, journey cost, journey type and traveller data (such as household income and method of travel payment).

This collective dataset is then inputted to the statistical software R in order to establish a link between the differing factors and ultimately assess if differing socio-economic groups value financial and time cost differently. A multi-variable regression analysis is carried out to relate time cost (duration) in terms of financial cost and household income.

5.1 *Coefficient of Determination*

The coefficient of determination (R^2) indicates how well the data correlates. For each of the different modes, a different R^2 value was found and displayed in Table 1.

Table 1. Coefficient of determination by travel mode

Travel Mode	R^2
Driving	0.8544
Cycling	0.9835
Walking	0.99
Public Transport	0.3604

It is clear that cycling, driving and walking exhibit strong correlations. This is unsurprising as the financial cost is derived from the distance of the journey. The distance of the journey in turn has a strong relationship to the duration. However, it is also clear that for public transport the correlation is much less strong at 0.36. This may be explained by the non-linear pricing structures that are more related to supply and demand than they are to distance or duration of travel.

5.2 *Socio-economic Metric*

The hypothesis is that those of different socio-economic groups will value the time cost and financial costs differently and thus exhibit different modal choice decisions. For example, the Heathrow Express versus the Piccadilly Line on the London Underground. The Heathrow Express service offers a large journey time saving for many journeys but at a significant cost penalty, thus it tends to attract business related travel primarily.

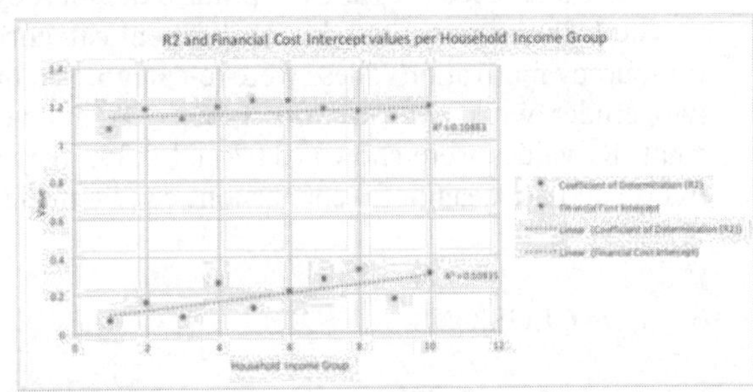

Figure 4 R^2 and financial cost intercept value per household income group

To assess this hypothesis, each household income group was considered separately and their multi-variable regression results compared. As is clear in Figure 4 the coefficient of determination value was low for all socio-economic groups and the intercept (which indicates the relative importance of the financial journey cost) remained fairly constant. The first concern is that the R^2 values are too low and show little statistical significance. Despite this, the financial cost intercept values show little variation in value attribution between different income groups.

5.3 *Distance Metric*

The value a traveller attributes to time and financial cost is dynamic and will change with varying elastically dependent on the distance travelled (DfT, 2013). For example, it is generally held that the same person is unlikely to spend 5% more on a short journey to save 5% time but is quite likely to spend 5% more to save 5% time on a longer journey. The journeys were categorized into 5 distance categories and each set was analysed separately. The exhibited R^2 values were extremely low and the intercepts for income and financial cost were extremely volatile. No conclusions could be made.

5.4 *Journey Motivation*

This hypothesis considers how a traveller may behave differently depending on their role at the time. A clear example being the difference between work related travel and non-work related travel. Work related travel may exhibit a higher premium on time savings as the traveller may not directly incur the cost or their employer may value their time at a higher level. Conversely, non-work travel will have personal financial costs and a reduced pressure to optimise the journey.

The LTDS survey provides 21 different categories for journey motivation. These were re-categorised into two cruder work and no work groups. Again very small R^2 values were shown (0.07 and 0.355 respectively) and no meaningful conclusions can be made.

6 CONCLUSIONS

It is clear that traveller decision making is complex and considers many factors that were not captured in this analysis. Despite using crowd-sourced big data sources in order to better reflect real world conditions the sample size with which to compare this to was very small. The survey contained around 43,000 journeys which is small when compared to the 1.3 billion passengers TfL carry every year. Attempting to tease out the differences at the small scale resulted in very little in the way of significant results. As the time and cost of completing such travel surveys becomes ever more scrutinised and the sample size continues to drop in proportion to ridership numbers other alternatives must be found. Appropriate engineering solutions require tailored and context specific solutions. These require data sets that can permit the teasing out of the subtle behavioural changes that have been long hypothesised, generally held and unsuccessfully quantified here. The 3 hypothesises assessed here are widely held and yet their macro impacts were not detectable.

Mobile phone related information offers the scope and more importantly sample size to avoid the issues identified here. Manageable privacy concerns should not be a permanent blocker to their use.

ACKNOWLEDGEMENTS

This research made use of the London Travel Demand Survey, provided by Transport for London (TfL). It also made use of the Directions API by Google. This research was funded by an EPSRC I-case studentship in collaboration with Arup.

REFERENCES

Flyvbjerg, B. Skamris Holm M. & Buhl, S. 2005. How (In)accurate are demand forecasts in public works projects? Journal of the American Planning Association. Vol. 71, No.2.

International Union of Railways (UIC) 2011. Carbon footprint of High Speed Rail. Paris, France.

Saxe, S. Casey, G. Guthrie, P. Soga, K. Cruickshank, H. 2015. GHG considerations in rail infrastructure in the UK. ICE: Engineering Sustainability

Casey, G. Silva, E. Soga, K. Guthrie, P. 2015 (in review). An ABM supported by real-time big data: Case of HSR & Aviation. ICE: Transport

Transport for London (TfL). 2014. London Travel Demand Survey. London, UK.

Google. 2009. The bright side of sitting in traffic: Crowdsourcing road congestion data.

The Automobile Association (AA), 2014. Motoring costs 2014.

Arundell, L. 2008. The cost of cycling. Thinking on two wheels cycling conference. Adelaide, Australia.

Department for Transport (DfT). 2013. Valuation of travel time savings for business travellers.

Proceedings of the International Conference on Smart Infrastructure and Construction
ISBN 978-0-7277-6127-9

© The authors and ICE Publishing: All rights reserved, 2016
doi:10.1680/tfitsi.61279.747

Employing a system-of-systems approach to water management in the shale gas value chain

Gregory J. Fitch*[1], Ibrahim Odeh[1], and Amr Kandil[2]

[1]*Columbia University, New York, NY, USA*
[2]*Purdue University, West Lafayette, IN, USA*
**Corresponding Author*

ABSTRACT: This paper presents a model for managing resources impacted by the nexus between water and energy. The current research recognizes the problem of implementing long-term resource management policies related to water impacted from unconventional natural gas drilling. The problem stems in-part from the fact that the various stakeholders of these two "systems" fundamentally operate at different levels, use different languages and are built from different frames of reference. In these terms the current research develops a conceptual model that distinguishes sustainable resource management of water and energy as a system-of-systems problem. The model investigates forecasted energy demands with the purpose of creating a navigation aid for sustainable resource management. As a first step in the development of this model, this paper presents a system lexicon populated with the information that is already known about water resource management and what is being forecasted about future energy markets. The research presented in this paper contributes to the existing body of knowledge by creating a road map where horizontal thinking across "stovepipe" columns of knowledge bridges the understanding and communication gap between these two systems. This navigation aid is aimed at assisting policymakers to make wiser decisions related to the interdependencies between energy demands and water resource management.

1 INTRODUCTION

The nexus between energy and water is far-reaching and requires industry leaders and concerned governments to strategically manage these interdependent resources in a joint fashion. A strategic goal of the U.S. Department of Energy (D.O.E.) is to promote America's energy security through reliable, clean and affordable energy. At the same time, the D.O.E. recognizes that energy and water are essential and interdependent resources. To maintain reliable energy and water supplies these two resources must be managed together as demand grows and limitations increase (U.S. Department of Energy, 2006).

The current research builds upon this concept by proposing a methodology for developing simulation models within the framework of system-of-systems (SOS) engineering. While the term "system-of-systems" has no widely accepted definition, the current research is consistent with Jamshidi (2008) that defines SOS as "large–scale integrated systems that are heterogeneous and independently operable on their own, but are networked together for a common goal."

To begin to develop the simulation model, this research constructs a system lexicon to identify and better understand the agents that are necessary for the long-term planning and financing of infrastructure required for sustainable water resource management (WRM) while reliable sources of energy are secured. The lexicon includes the agents representing the constituent components and takes into consideration the exogenous drivers and a significant amount of uncertainties related to energy markets. The lexicon is used to identify the appropriate level of abstraction that is required in order to define the problem statement and construct the proposed simulation model.

2 SOS AND PUBLIC POLICY

In DeLaurentis et al. (2004) a navigation aid in the form of a lexicon is presented to analyze problems of increasing complexity that face decisions-makers within government and private industry. Many of these problems of increasing complexity that face public policy decision-makers are recognized as SOS with the effective analysis becoming unmanageable within the "stovepipe" context that characterizes organizations in the research and development communities.

The current research recognizes that a lexicon that describes the hierarchy of components within WRM that is related to energy markets would serve as a valuable roadmap. The objective of using such a lexicon from the SOS engineering perspective is not prediction but is instead gaining a greater understanding of the essence of the problem. The authors recognize that the genesis for the "stovepipe" environment within the current problem of sustainable WRM related to energy markets stems from apparently opposing driving factors for each agent. Specifically, the service provider (SP) agents represent constituent components within energy markets and its primary driving factor is profitability. In contrast, the infrastructure provider (IP) agents represent sustainable WRM and its primary driving factors are public health and environmental conservation.

Consistent with DeLaurentis et al. (2004) construction of a lexicon for sustainable WRM related to energy markets begins with establishing a hierarchy of components. The framework of the hierarchy is shown in Table 1 and consists of two major structures: (1) Categories of systems and (2) levels of the hierarchy. The categories highlight the important distinguishing trait of a SOS problem, which is heterogeneous mix of engineered and responsive systems existing in a single problem (DeLaurentis et al., 2004).

Again, consistent with the methodology outlined in DeLaurentis et al. (2004) and to avoid confusion the lexicon uses Greek symbols at each hierarchy. Delta (δ), Gamma (γ), Beta (β) and Alpha (α) indicate the relative position. The collection and interdependencies of constituent components at the α level determine the construct of the constituent components at the next higher β level. In a similar fashion, the collective components at the γ level are comprised of the lower constituents at the β level. Hence, the highest δ level, which in the current research represents the sustainable WRM within and related to energy markets, takes into consideration all constituent components at each lower γ, β and α level.

Table 1. Categories and Levels for achieving energy security and sustainable WRM.

Categories	**Descriptions**
Policies	The external forces from decision-makers that impact energy production and WRM.
Economics / Conservation	Forces that direct SP / IP agents to react in a responsive manner within a responsible market economy.
Operations	The application of the policies for energy production and WRM.
Resources	The constituent components of the SP and IP agents.
Levels	**Descriptions**
Delta (δ)	Collection of the lower levels γ, β and α, organized into a SOS.
Gamma (γ)	Collection of the lower levels β and α levels, organized into two heterogeneous systems: (1) Energy Markets and (2) WRM.
Beta (β)	Energy Production and WRM where policy is implemented and includes the collection of the lower α level constituent components.
Alpha (α)	The base level of entities for which further decomposition will not take place beyond its constituents.

3 CASE STUDY

In order to develop the simulation model, this paper presents a system lexicon that is populated with the information that is already known about WRM and what is being forecasted about future energy markets. The current research is aimed at developing best management practices (BMP) for policy planners and decision-makers by identifying the appropriate level of abstraction within the system lexicon.

Once the appropriate level of abstraction has been determined and the problem statement defined, the agent based simulation model can be constructed.

A review of existing literature related to energy markets and WRM reveals a growing conundrum between unconventional natural gas (UNG) production and BMPs related to WRM. As the current research demonstrates the UNG industry is still relatively new but is anticipated to exist for several decades. For these reasons the case study presented in this paper is focused on populating a system lexicon and identifying the appropriate level of abstraction specifically related to the agents within UNG production in the Marcellus Shale Formation in the U.S. Commonwealth of Pennsylvania.

3.1 Forecast Energy Demands

A review of existing literature suggests that UNG production within the Marcellus Shale Formation will continue for at least the next two decades. The U.S. Energy Information Administration Annual Energy Outlook 2014 (AEO2014) forecasts that total natural gas consumption in the U.S. will grow from 724.9 billion cubic meters (bcm) in 2012 to 894.8 bcm by 2040. With the exception of residential use, natural gas use will increase in all of end-user sectors. In the case of the residential sector, natural gas use is forecast to decline as a result of population shifts to warmer regions of the country as well as improvements in appliance efficiency.

For the Marcellus Shale Formation, natural gas production is forecast to increase from 53.8 bcm in 2012 to peak production of 141.6 bcm between 2022 through 2024. During these peak years, natural gas from the Marcellus Shale Formation is forecast to provide up to 39% of market demand east of the Mississippi River. Between 2016 and 2040, natural gas production from the Marcellus Shale Formation will exceed market demand for the Northeastern and Mid-Atlantic regions of the United States, requiring other and more distant markets to purchase the surplus gas. During the peak years 2022 – 2025, total production from the Marcellus Shale Formation will exceed market demand in the Northeastern and Mid-Atlantic Regions by 28.3 bcm. Current forecasts of gas production from the Marcellus Shale Formation show a continuation of a sustained rate before declining to 130.3 bcm in 2040. Figure 1 graphically shows the AEO2014 forecasts in percentages of gas production from the Marcellus Shale Formation against total U.S. gas consumption east of the Mississippi River between 2010 and 2040. (USAEO, 2014)

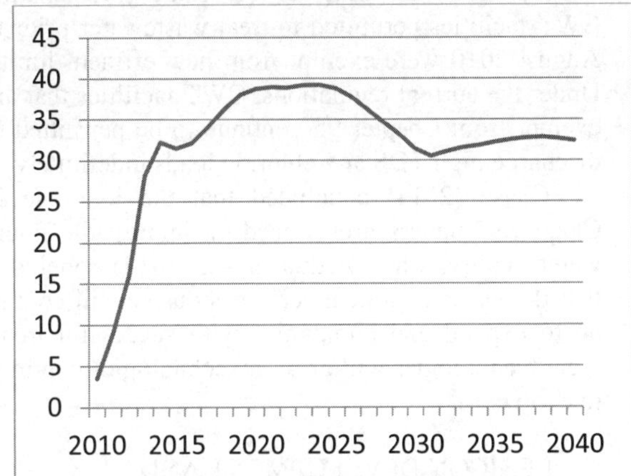

Figure 1. Forecast Marcellus Shale Gas Production.

3.2 WRM within UNG Production

UNG, also known as hydraulic fracturing (fracking) used in gas production from the Marcellus Shale Formation requires extensive amounts of water and its treatment and disposal are of particular concern. In Chase (2014) wastewater associated with UNG production from the Marcellus Shale Formation is characterized uniquely different than wastewater produced from conventional vertical drilling. Chase (2014) recognized that Central Waste Treatment (CWT) facilities that were historically permitted to treat wastewater from oil and gas wells from conventional vertical drill wells were not able to meet the increased loads from UNG production. Beginning in the late 2000s, with the boom in UNG production, drilling companies overwhelmed these existing CWTs and began sending their wastewater to municipally owned wastewater treatment facilities that were not designed or permitted to treat wastewater from UNG production. This ineffective treatment process resulted in elevated toxicity levels to aquatic life and created a potential impact on drinking water supplies (Chase 2014).

In response, the Pennsylvania Department of Environmental Protection (DEP) in August 2010 added new effluent limits under 25 Pa. Code § 95 with TDS and chloride at 500 mg/L and 250 mg/L respectively. The improved standards effectively eliminated municipal facilities from being an alternative source for treating wastewater from UNG production; however, under 25 Pa. Code § 95 existing CWT facilities permitted to treat wastewater prior to August 2010 were exempt from new effluent limits. Under the current regulations, CWT facilities that are exempt from Chapter 95 continue to be permitted to discharge high TDS and chloride loads indefinitely.

Chase (2014) concluded that the loophole in Chapter 95 underscores a need for improved effluent water quality, while Arthur et Al. (2014) concluded that the water treatment service industry will continue to expand and companies with successful treatment technologies will grow as technologies continue to evolve.

4 LEXICON DEVELOPMENT AND ABSTRACTION PHASE

The research presented in this paper demonstrates that SP agents will increasingly be required to meet current and forecast demands for UNG production from the Marcellus Shale Formation through 2040. The investigation also indicates that UNG production will result in indeterminate amounts of water requiring treatment as the various processes and technologies continue to evolve in this relatively new industry. The following sections present: (a) the development of a system lexicon that is populated with the agents required for sustainable WRM within the UNG value chain; (b) an examination of the taxonomy of the lexicon for purposes of identifying the appropriate level of abstraction required for developing a simulation model; and (c) defining the problem statement at the appropriate level of abstraction.

3.3 Lexicon for UNG Production and WRM

The lexicon shown in figure 2 is used to determine the appropriate level of abstraction for developing a simulation model for sustainable WRM within UNG value chain. The lexicon is populated with the agents related to both UNG production within the Marcellus Shale Formation and WRM.

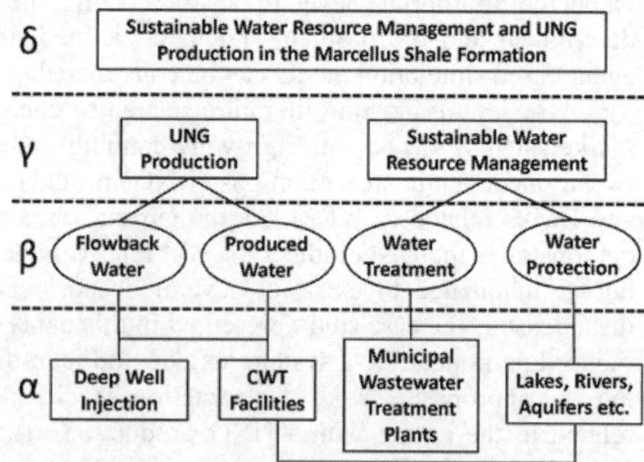

Figure 2. Lexicon for UNG Production and WRM

It's important to note that the current research does not consider all constituent components at every hierarchy of the current SOS problem. Most notably, at the β level, the current research focuses on flowback and produced water that are resultant byproducts from UNG production. Strictly for brevity, the β level excludes a significant source for controversial public policy related to the potential and unintentional contamination of aquifers from hydraulic fracturing fluid.

Nevertheless, the current research aims at investigating sustainable WRM within the bounds of flowback and produced water and presents a methodology for constructing models for simulating the relationships between water and energy markets within the framework of SOS engineering. To achieve both goals, it is necessary to recognize decisions made at the δ level can result in disruptive and unintended consequences at lower levels γ, β and α; where policy is implemented at the operations level and impacts lower level constituent components.

For UNG production in the Marcellus Shale Formation, the Pennsylvania State Government strived to maintain sound public policy at the δ level by enacting new effluent limits for CWTs under 25 Pa. Code § 95. At the same time, the Pennsylvania state government benefited fiscally from UNG production through additional funds provided by impact fees that drilling companies are required to pay for each well drilled at the γ level. As of March 2015, the impact fee has provided Pennsylvania $210 million per year, with most of the money going back to communities hosting drillers (NPR, 2015).

The dynamics between the Pennsylvania state government at the δ level and the UNG production industry at the γ level continues to evolve and will become more apparent in the near term. In March 2015, as UNG production in the Marcellus Shale Formation continued to expand at the γ level, the Pennsylvania state government debated at the δ level a proposal for a new severance tax on the state's natural gas industry. During the same period, the loophole in Chapter 95 became clearer to the UNG and CWT industries at the γ and β levels while the DEP moved to discourage and, in some cases, ban the use of open waste pits in shale gas operations in favor of closed tanks at the α level (Pittsburgh Post-Gazette, 2015).

The economic and environmental factors that will drive the direction of the UNG industry, and in turn drive the CWT industry at the α level will become more apparent and will continuously be impacted as these and similar government regulations and policies are established. As a result, the CWT procurement process that is most dynamic and best able to respond to future changes in the UNG industry is most likely to be successful.

4.2 Level of Abstraction and Problem Defined

Historically, government agencies and authorities issued request for proposals from IP agents for specific platform systems. Increasingly, government owners, however, are soliciting proposals for wider scopes of services in order to meet broader sets of capabilities and over significantly longer time spans. Hence a system engineering (SE) approach is required that first identifies and then analyzes the problem to validate the existing systems and determine the systems that have yet-to-be procured in order to provide these capabilities. The core activity within SE is the process of developing the system or systems required to achieve a particular set of goals on a concurrent basis with managing current and future constraints.

In the work of DeLaurentis et al. (2010) a three-dimensional taxonomy is presented that is used to direct the SE process and is used as a means of classifying entities according to their relationships. Consistent with DeLaurentes et al. (2010), Figure 3 is a representation of the three dimensional taxonomy that is used to examine the relationships between systems in terms of the system type, control and connectivity.

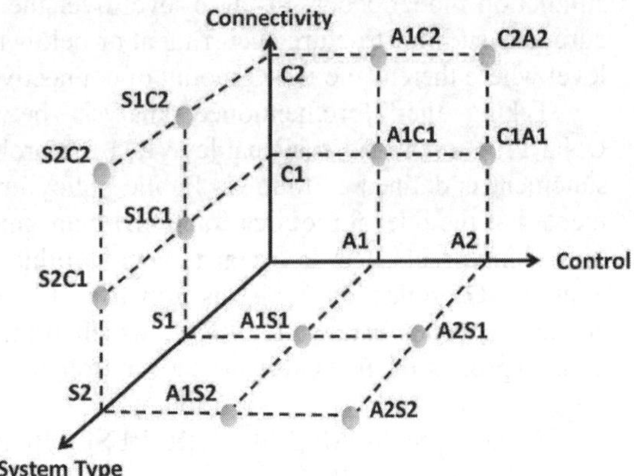

Figure 3. Taxonomy: Key Dimensions and Relative Locations within SOS Applications.

The current research uses the taxonomy in Figure 3 to identify the appropriate level of abstraction in the SE problem at hand by using multiple views pointed toward the operational objective: sustainable WRM and UNG production. It is noted that along the system type axis, infrastructure is procured at the γ level and includes independently operating monolithic infrastructure systems: SP agents required for UNG production and IP agents required for sustainable WRM. These SP agents include the systems of UNG drill wells and related infrastructure. The IP agents include the system of the infrastructure that is required for transporting and treating wastewater produced from UNG production.

Analyses of both monolithic systems indicate that currently there exists a low level of control occurring at or below the γ level. Energy security as well as exogenous economic drivers is causing UNG production to ramp up and continue through 2040. A significant amount of uncertainties and a low level of control also exist and are related to the anticipated quantities and variants of produced wastewater. Low levels of connectivity also exist and are related to the procurement of infrastructure at and below the γ level between the SP and IP agents.

The DEP and public policymakers within the Pennsylvania state government best represent the agents at the δ level. Furthermore, the current research demonstrates that an evolving level of control exists at the δ level. Hence, the current research determines that the appropriate level of abstraction for a simulation model occurs at the δ level over the procurement of infrastructure occurring at or below the γ level where there is the least amount of connectivity.

Taking the aforementioned analysis between UNG Production and sustainable WRM the problem statement is defined as follows: Public policy implemented at the δ level requires fractional management on a concurrent basis between two monolithic systems: UNG production, which is primarily driven by economics; and sustainable WRM, which primarily aims to protect public health and the environment.

5 CONCLUSION AND FUTURE RESEARCH

The current research begins to create a methodology for examining the relationships between WRM and energy markets. A case study is presented within SOS framework that facilitates a clearer examination of the growing conundrum for treating water impacted and produced from UNG production from the Marcellus Shale Formation. To better understand how to achieve sustainable WRM on a concurrent basis with UNG production, the current research concludes that the proper level of abstraction for constructing a simulation model occurs at the δ level where it naturally reflects the structure of these real systems. The current research also concludes that there is currently a significant amount of uncertainty and a low level of control that exists for the anticipated quantities and variants of wastewater produced from UNG production.

As a result of these findings, the authors conclude that in order to construct the simulation model, future research should be conducted to identify the specific stochastic processes that are best suited for modeling the fractional management of these two monolithic systems. This future research should specifically include simulation runs for the anticipated quantities and variants of wastewater produced from UNG production through 2040. A theoretical integrated project delivery structure should be developed in order to facilitate collaboration between these systems: UNG drilling industry, which is driven primarily by profit, and sustainable WRM, which primarily aims to protect public health and the environment.

ACKNOWLEDGEMENT

The authors express their appreciation to Professor William Ibbs of U.C. Berkeley's Department of Civil and Environmental Engineering; Greg Hamm, Ph.D. and Senior Economist with NERA Economic Consulting; and Michael Long of Black & Veatch for their review and consultation on the development of this manuscript.

REFERENCES

Annual Energy Outlook 2014 with Projections to 2040. (2014) U.S. Energy Information Administration, Washington, DC

Arthur, J.D., Bohm, B., and Zampogna, D. 2014 Evolution of Best Management Practices and Water Treatment in High Volume hydraulic Fracturing Operations *Shale Energy Engineer2014,* 33-44

Chase, E. H. 2014, Regulation of EDS and Chloride from Oil and Gas Wastewater in Pennsylvania, *Shale Energy Engineering2014,* 95-105

DeLaurentis, D. A. and Callaway, R.K. (2004). "A System-of-Systems Perspective for Public Policy Decisions." *Review of Policy Research,* Vol. 21, Issue 6, 1 – 9.

DeLaurentis, D. A., Crossley, W. A. and Mane, M. (2010). "Taxonomy to Guide System-of-Systems Decision-Making in Air Transportation Problems." *Journal of Aircraft,* Vol. 48, No.3 6, 760 – 770.

Jamshidi, M. 2008, Introduction to system of systems, *System of systems engineering*, Wiley, New York, 1 – 20.

National Public Radio https://stateimpact.npr.org/pennsylvania/2015/03/03/wolfs-budget-plan-calls-for-new-tax-on-natural-gas-industry/

Pittsburgh Post-Gazette, http://powersource.post-gazette.com/powersource/policy-powersource/2015/03/09/Pennsylvania-DEP-moves-to-ban-open-waste-pits-in-shale-operations/stories/201503090133

Proceedings of the International Conference on Smart Infrastructure and Construction
ISBN 978-0-7277-6127-9

© The authors and ICE Publishing: All rights reserved, 2016
doi:10.1680/tfitsi.61279.753

Seismic evaluation of bridge network of urban city considering the post-disaster traffic capacity

Anxin Guo*, Zhenliang Liu and Hui Li

School of Civil Engineering, Harbin Institute of Technology, Harbin, China

**Corresponding author*

ABSTRACT In an urban city, transportation network system, which is mainly composed of highway bridges and roads, is a crucial linkage for the development of the economy and society. However, it is usually subjected to natural or man-made hazards, of which earthquake is one of the most common. After a destructive earthquake, the components of transportation network systems will be significantly impaired, leading to huge loss to the economy and society. Especially, the failure or damage of highway bridges in the transportation network system would hinder the efforts of relief and cost large amount money for the repair and reconstruction. Therefore, it is important to evaluate the seismic traffic capacity risk of highway bridge networks. This paper presents a methodology for the regional seismic risk assessment of highway bridge networks considering the post-disaster evacuation capacity of transportation system. Firstly, the framework of the proposed method is introduced based on the quantification of importance and vulnerability of individual bridge in the network system. Then, the approach to evaluate the importance of the bridge is established through the topological structure of the network and traffic flow of bridges carried. Afterward, the vulnerability of highway bridge networks is quantified by using the individual bridge fragility based on which the vulnerability of links is measured. Finally, a numerical simulation is conducted to validate the effectiveness of the proposed methodology based on the highway bridge network in San Jose, California, USA. The simulation results indicates that the methodology that integrates the analysis of importance and vulnerability is rational.

Keywords: Highway bridge networks; Importance analysis; Vulnerability analysis; Seismic risk assessment.

1.INTRODUCTION

Transportation network system, which is mainly composed of bridges and roads, is a crucial linkage for the development of the economy and society. Earthquake is a common hazard threatening the safety of civil infrastructures. In the earthquake-prone region, transportation network systems are also vulnerable to the seismic-induced damage. After a destructive earthquake, the components of transportation network systems will be significantly impaired. Especially, the failure or damage of highway bridges in the transportation network system would hinder the efforts of evacuation, emergency responses of post-disaster rescue and relief, and also cost large amount money for the repair and reconstruction.

In the evaluation of the hazard risk of thebridge networks, the concerned consequences of the hazard play an important role in the performance assessment process. Generally, social, environmental and economic losses, which are also covered and referred by "sustainability", are always selected as the performance measure for the analysis. However, the post-earthquake traffic capacity of highway bridge networks, which strongly affects the post-earthquake evacuation and relief, is seldom addressed by researchers and will begin herein. In order to evaluate the post-earthquake capacity of highway bridge networks, it is crucial to develop a methodology that integrates the analysis of network configuration, traffic flow and vulnerability of bridges in the network. In recent years, the seismic performance assessment of transportation network system has widely investigated by some researchers (Padgett et al., 2010; Saydam et al., 2013). In those works, it is usually assumed that damage of the roads was ignored, and bridges in the transportation network system were the main components vulnerable to earthquakes (Kiremidjian et al., 2007). Liu and Frangopol (2005) developed a computational procedure for the time-dependent and probability-based reliability analysis of bridge networks. Shiraki et al. (2007) investigated the travel delays of transportation network system caused by the bridge damage in earthquake scenario; Bocchini and Frangopol (2011) proposed an approach to combine structural fragility analysis, network flow and random field theory for the quantitative assessment of bridge network; Dong et al. (2014) develop a novel methodology in terms of time-variant risk accounting for the associated consequences of society, environment and economy to assess the seismic risk of transportation networks. This paper presents a methodology for the regional seismic risk assessment of highway bridge networks considering the post-disaster evacuation capacity of transportation system. Firstly, the framework of the proposed method is introduced in Section 2.1 based on the quantification of importance and vulnerability of individual bridges, traffic flow and traffic carrying capacity of damaged bridge network system. In Section 2.2, the importance assessment of bridges is analyzed to evaluate the comprehensive importance of bridges. Afterward, Section 2.3 presents the approaches for vulnerability analysis of links, traffic flow and traffic carrying capacity analysis of highway bridge networks. Integrating the topological properties and traffic analysis in emergency condition, the importance of individual bridge is identified and the seismic performance of highway bridge network in an earthquake scenario is evaluated and presented. Finally, a numerical simulation is conducted to validate the effectiveness of the proposed methodology based on the highway bridge network in San Jose of California.

2. METHODOLOGY FOR THE SEISMIC RISK ASSESSMENT OF HBNs

2.1 The Methodology Flowchart

After an earthquake, the emergency responses and the post-disaster evacuation are depended upon the damage level of bridges, the connectivity of the bridge networks and the traffic demand in the emergency condition. Considering the above factors, the methodology for the seismic performance assessment of highway bridge networks is summarized and presented as the flow chart in **Fig. 1**. It can be seen from the figure that the first step of the methodology is to identify the bridge inventory (e.g., location, bridge type and specifications, operation conditions, built year) and the information of bridge links (e.g., daily traffic flow). With those basic data, the evaluation procedure of the seismic performance of highway bridge networks can be divided into two branches, which can be named as the importance analysis and vulnerability analysis of HBNs, and from which the importance indexes and post-earthquake traffic capacity of links can be measured. In fact, links and bridges of HBNs must be analyzed individually to estimate their vulnerability in case of earthquakes, but the overall performance depends on the integration of the vulnerability and importance.

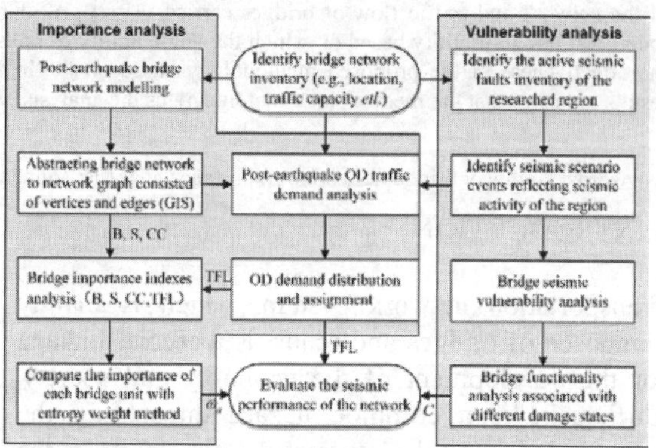

Fig. 1. Flow chart of the proposed seismic risk assessment methodology

Betweenness, substitutability and clustering coefficient measures (B, S, CC) are chosen as the topology properties, and traffic flow of links (TFL) is used to measure the dynamic importance of links. The two measures are integrated to measure the comprehensive importance of bridges which is eventually combined with functionality measure to assess the seismic performance of bridges. At last, the comprehensive importance of bridges or links can be determined with the entropy weight method.

As bridges are assumed to be the only components vulnerable to earthquakes, the vulnerability of bridges on the same link can be integrated to determine the vulnerability of the link. A set of seismic scenarios of the region where a HBN located should be identified to capture the seismic intensity. For a given earthquake scenario, fragility matrix analysis method can be employed to obtain the damage state probability of each bridge, based on which the damage states of links is determined. The damages states of links can impair the traffic capacity and speed, which can be determined by the relationship of performance and damage states. Then the expected traffic capacity of links can be quantified by summarizing the multiplication of the residual capacity and the probability of each damage state. Combined with the importance analysis, the seismic performance indexes of HBNs can be determined and compared with each other.

2.2 Importance assessment of bridges

The entropy weight method is a scientific and reasonable method to combine the subjective factors with the objective factors together to establish the importance index. Firstly, the above topology index and the traffic flow is standardized into an initial matrix:

$$R' = (r'_{ij})_{m\times n} \tag{1}$$

where r'_{ij} is the *i*-th importance index of bridge *j*, such as topology properties and traffic flow of links used in this paper; m is the total number of bridge importance indexes; n is the total number of bridges. In general, the indexes can be divided into two kinds. One is the kind of indexes whose importance increases with values, and the other one is the kind of indexes whose importance decreases with values.

Let P_{ij} be the proportion of index i in the importance indexes of bridge j:

$$P_{ij} = r_{ij} / \sum_{j=1}^{n} r_{ij} \tag{2}$$

The entropy of S_{ij} can be calculated by (Hsu and Lin, 2006):

$$S_i = -\frac{1}{\ln(n)}\sum_{j=1}^{n} P_{ij}\ln(P_{ij}) \tag{3}$$

It can be assumed that the topology properties and the traffic flow are equally important and the roles the three topology indexes play are the same. Therefore, the subjective weight of each index can be computed with analytic hierarchy process or AHP.

According to the fuzzy set theory proposed by Zadeh (1965), the importance indexes can be expressed in the form of distance:

$$L_{pj}(\mu, a) = \left[\sum_{i=1}^{m} \mu_j^p \cdot r_{ja}^p\right]^{\frac{1}{p}} \tag{4}$$

In general, when $p=1$, $L_{1j}(\mu, a)$ is called hamming distance, and when $p=2$, $L_{2j}(\mu, a)$ is called Euclidean distance. For the purpose that the importance indexes can be distinguishable easily, $L_{1j}(\mu, a)$ is used. Finally, the final importance index of link a is shown as:

$$\omega_a = \frac{n\sum_{j=1}^{m} \mu_j \cdot r_{ja}}{\sum_{a=1}^{n}\sum_{j=1}^{m} \mu_j \cdot r_{ja}} \tag{5}$$

2.3 Vulnerability assessment of the bridge

The damage and failure of the bridge components induce the function loss of the links, including the traffic flow capacity and the flow speed. Table. 1 listed the traffic flow of links associated with different damage states of the bridge components according to reference (Chang et al., 2000). Since there may be more than one bridge segments in a link, the traffic flow of a link is impaired by the damage of all the bridge segments it contains. While the bridges in a link are in series, and the post-earthquake performance of the link can be determined in terms of link damage index, expressed as $\beta(a)$, based on BDI_i^k of bridge i in the concerned link as follows:

$$\beta(a)=\begin{cases}\sqrt{\sum_{i=1}^{n}\left[BDI_i^k\right]^2} & \text{All } \mathrm{BDI}_i^k<1\\ \infty & \text{At least one } \mathrm{BDI}_i^k=1\end{cases} \tag{6}$$

where BDI_i^k is the BDI of bridge i at the damage state k, n is the total number of bridges located on link a. Table 4 shows the corresponding relationship between β and qualitative performance indexes.

Table. 1 Link damage index, damaged capacity and damaged speed

DS	LDI or β	DC	FS
None	$0.0\le\beta<0.5$	0.00	0.00
Slight	$0.5\le\beta<1.0$	0.00	0.25
Moderat	$1.0\le\beta<1.5$	0.25	0.50
Excessiv	$1.5\le\beta<\infty$	0.50	0.50
Complete	∞	1.00	1.00

The post-earthquake network traffic capacity considering the importance of the links can be calculate by:

$$C_{N_d}=\sum_{a=1}^{m}\omega_a\sum_{i=0}^{4}P_{LDI_i|IM}^{a}\left(1-R_{\mathrm{dc},i}^{a}\right)C_0^a \tag{7}$$

where $P_{LDI_i|IM}^{a}$ is the conditional probability of link a being in the none, slight, moderate, major and complete damage states, quantified with $\beta(a)$, subjected to the earthquake with intensity IM; $R_{dc,i}^{a}$ is the reduction factor of the traffic carrying capacity of the corresponding link at the i-level damage state.

3. CASE STUDY

The methodology proposed in this paper is herein applied to a HBN located in San Jose, California as a numerical case study. The damaged capacity assessment of the HBN is implemented in the MATLAB 2014 environment on a Dell computer (Intel Core i5 3.2 GHz and 4 GB RAM).

It is assumed that the region is subjected to evenly distributed earthquakes. Fig. 2 shows the schematic layout of the HBN and seismic hazards. The network HBN connects the cities of San Jose with three interstate highways and six state highways by a total of 20 nodes vertices and 33 edges links which contain 75 bridges, the roads other than interstate or state roads are not considered. Each vertice corresponds to a traffic analysis zone while each edge corresponds to a highway segment or link. It is clear that there is in total 190 OD sets in this network. The detailed link information is described in another literature with some modifications, including the limit speed limit, physical length, traffic capacity etc.

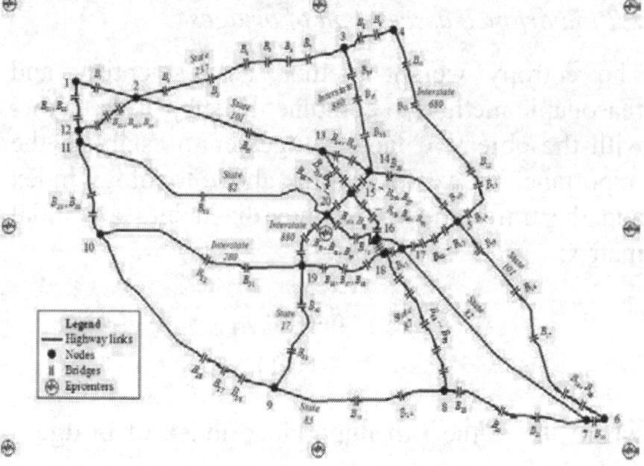

Fig. 3. Schematic layout of the HBN in the case study

The comprehensive importance of each link is measured by means of the entropy weight method in section 2.2, as shown in Fig. 3. It can be seen that the most important link is link 21 which connects the vertices 11 and 12. The effects caused by the damage of this link may therefore be more severe than that by others.

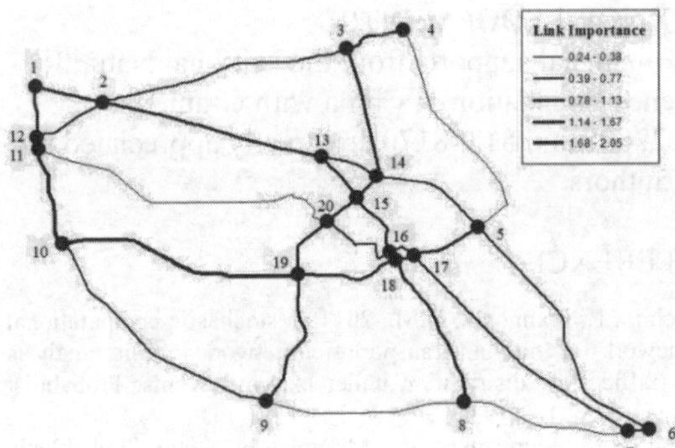

Fig. 4. Link importance for the whole HBN

6.3 Vulnerability assessment

Vulnerability assessment of the HBN system relates to determination of the damage states and comprehensive importance of the links in the system and can be identified and evaluated in terms of the susceptibility of vulnerable bridges to possible seismic hazards. The seismic hazards depend on the seismicity of the faults located in the region, the distance between the sites of the bridges and those earthquake sources and other ground effects etc. In this paper, the earthquakes are characterized by magnitude and distance between the epicenter and bridge site.

The seismic vulnerability of the bridges, links and whole HBN is assessed considering eight groups of earthquakes with nine epicenters and depth 10 km, have magnitudes ranging from 5.5 to 9.0 respectively. They are employed to perform fragility analyses for the bridges using HAZUS automatically. The damage states of bridges can be converted to the links' capacity and speed limit based on the TFL and fragilities relationship described in section 5. The vulnerability assessment process is repeated for all the bridges with all the specific earthquakes located in this region.

To evaluate expected damage for different scenario events, HAZUS was used to perform the vulnerability analyses of the bridges for earthquakes with the nine epicenters in Fig. 3 with magnitudes 5.5 to 9.0 and then can be translated into the vulnerability matrices of links. As illustrative of the probability of damage states, Fig. 9 illustrates the damage probability distribution of the bridges 1~20 in the HBN region due to the M 7.5 earthquake event. According to section 4.2, the expected LDI for earthquakes with magnitude 7.5 is calculated based on the damage states of the bridges, as shown in Fig. 4.

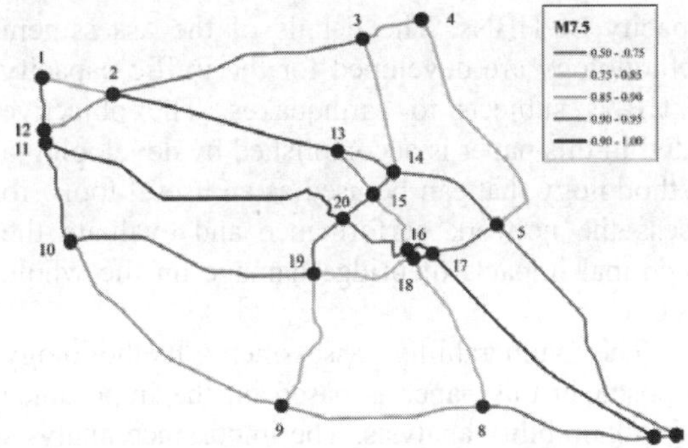

Fig. 5. Expected LDI or β for M7.5 Earthquake

As indicated previously, the traffic capacity sustainability of the HBN is evaluated in terms of GNC metrics. As a comparison, the transportation cost indexes or TSTT (Martin and McGuckin, 1998) and the traffic capacity sustainability of the HBN is evaluated with the assumption that the links are equally important or $\alpha = 1$. The results for the illustrative HBN are collected in Fig. 5.

It is worth noting that the results from the weight model and no-weight model differ significantly immediately after the earthquake. The weighted model shows that the traffic capacity of the illustrative HBN is in accordance with TSTT, while the no-weighted model will underestimate the seismic performance of HBNs. The GNC can be used to provide the means of comparison of the traffic capacity of HBNs located in different regions in terms of seismic risk.

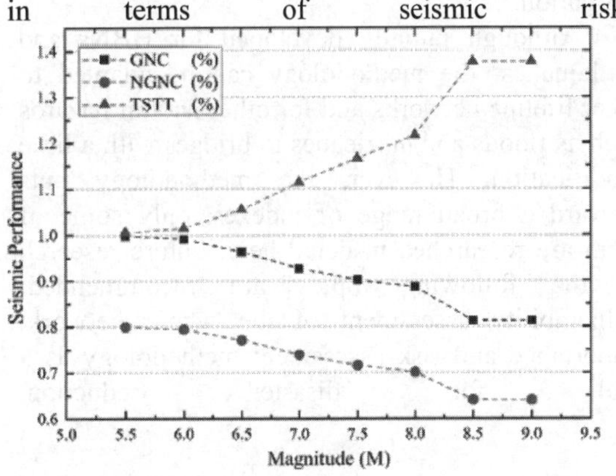

Fig. 6. Results of the numerical example

4. CONCLUSIONS

The objective of this paper is to develop a regional seismic risk assessment methodology for the traffic capacity of HBNs. The details of the assessment methodology are developed for the traffic capacity of HBNs subjects to earthquakes. The objective stated in this paper is accomplished by developing a methodology that can be used as an useful tool to assess the network performance and evaluate the functional impacts of bridge damage on the whole system.

The vulnerability assessment methodology proposed in this paper is based on the importance and vulnerability analysis. The importance analysis approach employs the topology and TFL as the two independent importance parameters. However, the estimation of the importance is performed by the entropy weight method that take into account both subjective and objective factors of the considered links. Other types of importance indexes can be included in the methodology to develop a more comprehensive approach. For instance, links can be classified into subsets based on the population of TAZ they connect. Nevertheless, the proposed importance analysis methodology can be adopted for practical applications due to its simplicity. For example, the proposed importance analysis framework can be used as an intuitive tool for decision of post-emergency response management. Furthermore, with the framework presented in this paper, the pre-earthquake retrofitting prioritization of bridges can serve as a means of seismic risk mitigation.

Although initially developed for HBNs and earthquakes, the methodology can be adapted to other lifeline networks and for other natural hazards such as floods and hurricanes to bridges with a little modification. However, the methodology puts forward a broad range of indexes, only some of them are researched in detail here. Future research in the following topics are recommended: vulnerability assessment of the whole network, importance and risk assessment methodology as a tool for disaster reduction.

ACKNOWLEDGEMENTS

The financial supports from the National Natural Science Foundation of China with Grant No. 51222808 and 51308170 are greatly appreciated by the authors.

REFERENCES

Bocchini, P., Frangopol, D.M., 2011. A stochastic computational framework for the joint transportation network fragility analysis and traffic flow distribution under extreme events. Probabilist Eng Mech 26, 182-193.

Chang, S.E., Shinozuka, M., Moore, J.E., 2000. Probablistic earthquake scenarios: extending risk analysis methodologies to spatially distributed systems. Earthq Spectra 16, 557-572.

Dong, Y., Frangopol, D.M., Saydam, D., 2014. Sustainability of highway bridge networks under seismic hazard. J Earthq Eng 18, 41-66.

Hsu, T.H., Lin, L.Z., 2006. QFD with fuzzy and entropy weight for evaluating retail customer values. Total Qual Manag Bus 17, 935-958.

Kiremidjian, A., Moore, J., Fan, Y.Y., Yazlali, O., Basoz, N., Williams, M., 2007. Seismic risk assessment of transportation network systems. J Earthq Eng 11, 371-382.

Liu, M., Frangopol, D.M., 2005. Time-dependent bridge network reliability: Novel approach. J Struct Eng-Asce 131, 329-337.

Martin, W.A., McGuckin, N.A., 1998. Travel estimation techniques for urban planning. National Academy Press Washington, DC.

Padgett, J.E., Desroches, R., Nilsson, E., 2010. Regional Seismic Risk Assessment of Bridge Network in Charleston, South Carolina. J Earthq Eng 14, 918-933.

Saydam, D., Bocchini, P., Frangopol, D.M., 2013. Time-dependent risk associated with deterioration of highway bridge networks. Eng Struct 54, 221-233.

Shiraki, N., Shinozuka, M., Moore, J.E., Chang, S.E., Kameda, H., Tanaka, S., 2007. System Risk Curves: Probabilistic Performance Scenarios for Highway Networks Subject to Earthquake Damage. J Infrastruct Syst 13, 43-54.

Zadeh, L.A., 1965. Fuzzy sets. Information and control 8, 338-353.

Proceedings of the International Conference on Smart Infrastructure and Construction
ISBN 978-0-7277-6127-9

© The authors and ICE Publishing: All rights reserved, 2016
doi:10.1680/tfitsi.61279.759

Assessing the discrepancies between recorded and commonly assumed journey times in London

T. Hillel[*], P. Guthrie, M. Elshafie and Y. Jin

University of Cambridge, UK
[*] *Corresponding Author*

ABSTRACT Transport models for infrastructure investment and operations planning make use of generalised trip cost to predict travel choice decisions. In cities, the most important factors in the generalised cost is trip duration. When calibrating such models to achieve simulation fidelity, observed data such as the choice of destination and means of travel recorded in travel surveys are used in estimating model parameters. Ideally, observed travel durations should also be used in the model estimation. However, in the past it was infeasible to record the actual trip durations to any degree of accuracy in travel surveys. Trip durations derived from a transport network model were commonly assumed to be sufficiently representative. Increasing availability of better recorded trip durations from travel surveys and better modelled trip durations from online mapping present the promise of significant improvements in the fidelity of transport models. As a preamble to adopting such data, we investigate how the best developed recording of actual trip durations from the London Travel Demand Survey compares with the most advanced trip duration modelling from Google Map travel directions API. We find clear discrepancies between the two, with the discrepancies varying systematically for different means and purposes of travel. The magnitude of the discrepancies is greater than can be attributed to randomness or noise. The systematic nature of the discrepancies suggests that transport network modelling even in its advanced form still has a long way to go to represent the observed patterns of behaviour, particularly for non-commuting journeys which account for about 80% of all trips made in cities. Since the discrepancies may create a systematic bias in the model parameters, it is of critical importance to understand them better in future analysis.

1 INTRODUCTION

Transport models for infrastructure investment and operations planning use discrete choice models to predict travel on the transport network, based on generalised travel costs (Train 2009; Prato 2009; TfL 2014). In order to obtain sound predictions, it is essential to have good measurements of the generalised travel costs. In cities, the duration of travel is usually the greatest influence on generalised cost.

The *commonly assumed* definition of trip duration, used in both research and industry, is the time taken to complete the optimal route between two points on the network, as predicted by a transport network model. This includes timetable information on public transport services, and either observed or predicted congestion delays on roads. This common assumption has been made by convention because it is difficult to record systematically the actual trip durations in travel surveys, where all travel within a day or week need to be recorded.

Increasing availability of better recorded trip durations from travel surveys and better modelled trip durations from online mapping present the promise of significant improvements in both data sources. In this paper we investigate how the best developed recording of actual trip durations from the London Travel Demand Survey (LTDS) compares with the most advanced trip duration modelling from Google Maps API. The analysis is for passenger travel only.

2 DEFINITIONS OF TRIP DURATION

There are four alternative definitions of trip duration:

1. *Ideal duration* (t_i): time to complete a trip as predicted by a transport model, free of effects of traffic congestion or delays.

2. *Commonly assumed duration (t_c)*: time to complete a trip as predicted by a transport model given predicted/observed traffic conditions, congestion and delays.
3. *Expected duration (t_e)*: time the passenger expects to take.
4. *Recorded duration (t_r)*: time the passenger records a trip as having taken.

Their characteristics are summarised in Table 1.

Table 1. Summary of trip duration definitions used in this study.

	Name	Includes congestion/ disruption	Modelled	Measured	Pertinence
t_i	Ideal	✗	✓	n/a	Low
t_c	Commonly assumed	✓	✓	n/a	Mid
t_e	Expected	?	✗	✗	High
t_r	Recorded	✓	✗	✓	High

Both the ideal and optimal durations are theoretical values which represent minimum journey times and are computed using a model. The expected and recorded durations are real world values that need to be observed or measured rather than computed.

Galotti & Bathelemy (2014) analyse the theoretical efficiency of the British public transport network by comparing the ideal route and duration for multiple journeys *(t_i)*. This value is independent of the conditions of the transport network, and so is of low relevance for real passenger journeys.

As discussed, trip assignment within state-of-the-art transport models uses the commonly assumed duration *(t_c)*. This is of greater relevance to real world journeys than the ideal duration *(t_i)*, as it is dependent on the network conditions. However, it still represents an idealised case, where the passenger takes the optimal route and travels as quickly as possible.

In reality, passengers make decisions based on their expected duration of a trip *(t_e)*. It is not possible to model the expected duration of a trip directly as it is highly dependent on a passenger's individual experience at that time. Instead this study investigates the recorded trip duration *(t_r)*, which is how long a passenger reports a trip to have taken. This is likely to have a strong relationship with the expected duration of repeating a similar trip in the future.

Each of these definitions of trip duration is separate from the duration a passenger actually takes.

3 METHODOLOGY

This study assesses the discrepancies between recorded trip durations *(t_r)* taken from the London Transport Demand Survey (LTDS) and their corresponding commonly assumed trip duration *(t_c)*, generated using the Google Maps Directions Application Programming Interface (API).

3.1 LTDS

The data source for completed journeys for this study is the LTDS, a continuous survey carried out by TfL of a sample of households within London's orbital motorway, the M25 (TfL 2011).

Each household is surveyed on one day of the year, listing all of the members of the household, all of the vehicles that the household owns or has access to, and the estimated total household income. Each household member over 5 years of age then completes a trip diary, giving details of all of the trips made on the survey date. Details include the trip start-point, end-point, start time, trip duration *(t_R)*, means of travel and trip purpose.

This study uses data from the 2013/14 survey year, which contains 44,981 trips made by 18,877 individuals.

3.2 Google Maps API

The data source for generating optimal journey times is the Google Directions API. It generates more than one route for any origin-destination pair. In line with the modelling convention, we retrieve the optimal route as the commonly assumed trip duration *(t_c)*.

Google's representation of London's transport network is commonly considered fine grained and accurate. On the network, Google generates real-time traffic routeing using crowd-sourced movements data. Google also receives up-to-date public transport timetable and delay information from TfL and Network Rail. It is reasonable to consider this dataset to the most advanced estimation of trip durations.

Using this information, the Google Maps API can return an optimal route and the commonly assumed trip duration (t_c) calculated using a modified Dijkstra's algorithm (Dijkstra 1959; Casey et al. 2015).

3.3 *Processing the data*

The trips from the LTDS are sorted into the same trip classes as used in London's transport policy model LTS (Table 2). For each trip in the LTDS, an optimal route and duration is obtained from the Google Maps API. The trip requests to the Google Directions API are performed in time bracketed groups, according to their departure time and day of the week from the LTDS, for each means of travel:

- **Driving:** Trips sorted by weekday, Saturday, or Sunday departure. Within each day, trips sorted into groups of departure time within two hour intervals.
- **Transit (public transport):** Trips sorted into weekday or weekend departure. Within each day, trips sorted into day and night departure trips.
- **Cycling and walking:** No time bracketing used, as walking and cycling durations returned by Google are time independent.

Table 2. LTS model trip classes.

Time periods (weekday)	Morning peak:	07:00-10:00
	Inter-peak:	10:00-16:00
	Evening peak:	16:00-19:00
Means of travel	Walking	
	Cycling	
	Transit	
	Driving	
Trip purposes	Home-based work	
	Home-based education	
	Home-based other	
	Non-home-based work	
	Non-home-based-other	

4 RESULTS

4.1 *Scatter plots*

Figure 1 shows a scatter plot of t_c against t_r for all trips within the study. A bi-square linear regression, which is robust to outliers, is performed on the data. The regression line is well below the line $y = x$ which corresponds to $t_r = t_c$. This shows that the recorded trip durations tend to be substantially longer on average than the commonly assumed durations.

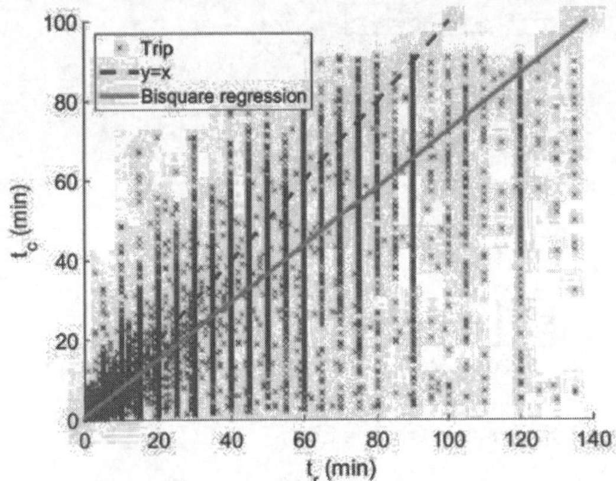

Figure 1. Scatter plot of all trips.

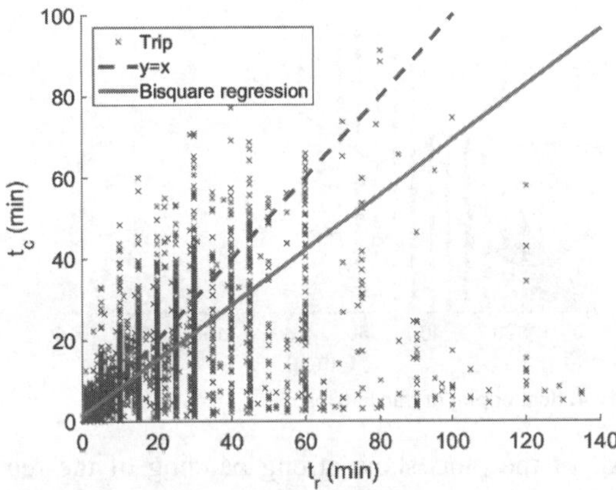

Figure 2. Scatter plot of walking trips.

Figures 2-5 show the scatter plots for each transport mode. Each plot contains trips for all trip purposes and trip departure periods. Each plot has different visual characteristics, which are shown numerically in Table 3. The bi-square regression gradient shows the average relationship between t_c and t_r and the Pearson correlation coefficient demonstrates the spread of the data. These values are also calculated for each journey purpose and departure period. There is wide variation in both the gradient of the linear regression and the value of the cross-correlation coefficient for each trip class.

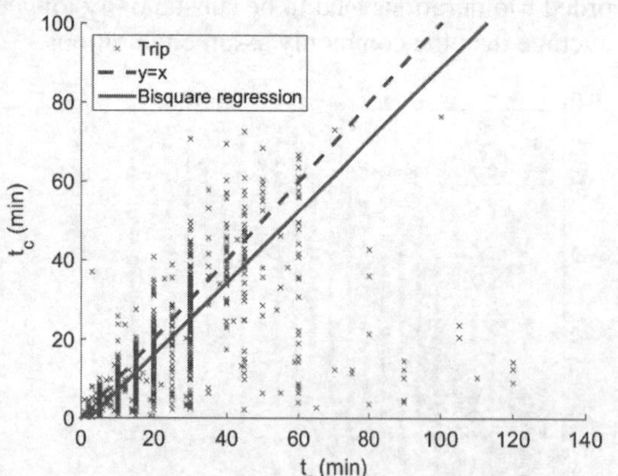

Figure 3. Scatter plot of cycling trips.

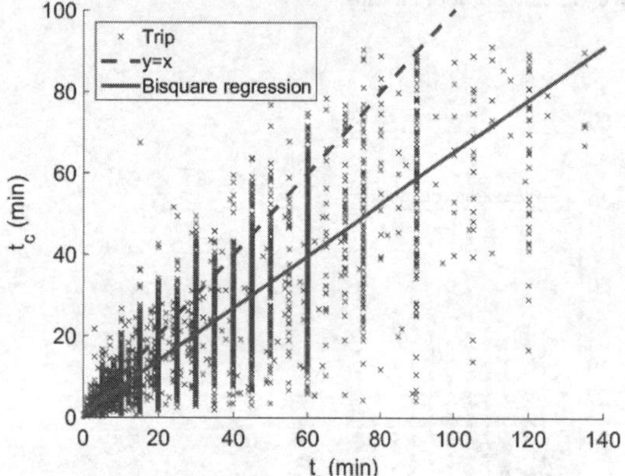

Figure 4. Scatter plot of transit trips.

All of the plots show strong banding of the recorded duration *(t_r)*. This relates to the fact that the recorded duration is a measure of how long a passenger perceives a journey to have taken. Below 60 minutes, the bands occur at 5 minute intervals, demonstrating that for short journeys the resolution of perceived duration is ±2.5 minutes, i.e. the trip durations are rounded to the nearest 5 minutes. For all of the plots, the strongest band above 30 minutes is at 60 minutes.

The bands at 55 minutes and 65 minutes are also much weaker than the other bands. This suggests that for the majority of the population, there is a tendency to round trip durations to 60 minutes. Above 60 minutes, the plot for all trips shows the strongest bands at 75 minutes, 90 minutes, and 120 minutes, showing the resolution for the majority of the population reduces to 15 minute and then 30 minute intervals.

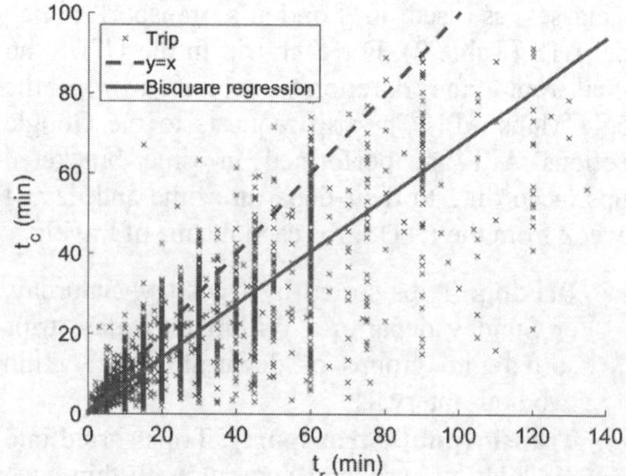

Figure 5. Scatter plot of driving trips

Table 3. Linear regression gradient, and correlation coefficient for each trip mode, trip purpose, and departure time period.

Category	Class	Gradient	Correlation
All	All	0.718	0.830
Transport mode	Walking	0.679	0.548
	Cycling	0.908	0.588
	Transit	0.679	0.751
	Driving	0.647	0.837
Purpose	Home-based work	0.752	0.840
	Home-based education	0.629	0.849
	Home-based other	0.681	0.822
	Non-home-based work	0.573	0.760
	Non-home-based other	0.655	0.712
Period	AM Peak	0.766	0.866
	Inter peak	0.669	0.797
	PM peak	0.694	0.848
	Other	0.736	0.820

4.2 *Probability distributions*

In order to create the probability distributions, a dimensionless ratio of recorded duration *(t_r)* to commonly assumed duration *(t_c)* is defined:

$$d = t_r/t_c \tag{1}$$

The ratio of two values is not a symmetrical operation, and as such the distribution of the ratios show heavy positive skew. This is shown in Figure 6, which plots smoothed kernel distributions of the ratio for each transport mode. The line $d = 1$ corresponding to $t_r = t_c$ is given for reference. To deal with the heavy skew, the natural logarithm of the ratio is taken to provide a symmetrical operation. This gives the following formula for the log-ratio *(r)*:

$$r = \ln(d) = \ln(t_r/t_c) \quad (2)$$

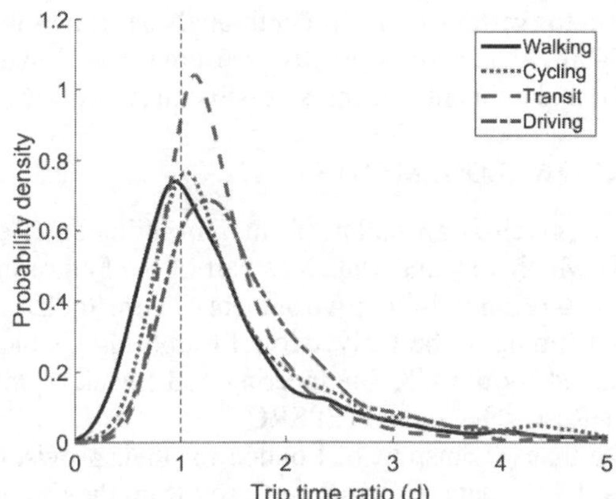

Figure 6. Skewed probability distributions by means of travel.

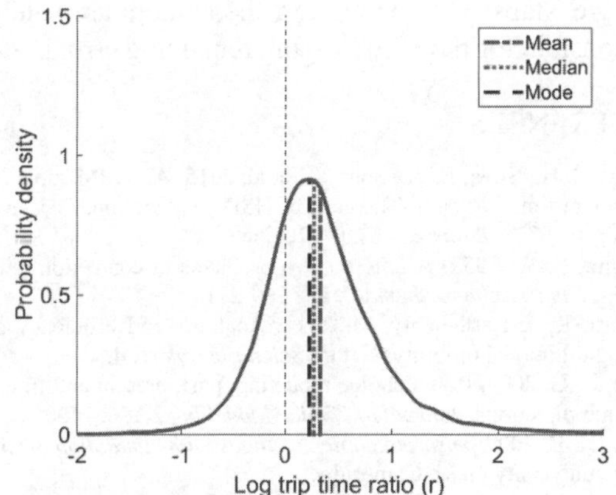

Figure 7. Probability distribution for all trips.

Figure 7 shows the smoothed kernel distribution plot of all trips combined. Here the line $r = 0$ corresponds to $t_r = t_c$. The mean, median, and mode are all to the right of this line, once again showing that the recorded durations *(t_r)* are on average significantly higher than the commonly assumed durations *(t_c)*.

Smoothed kernel distributions of the log-ratio *(r)* are generated for each trip class. The sample geometric mean and standard deviation of the ratios *(d)* can be calculated directly from the log-ratio *(r)*, using the following formulae:

$$\mu_g = \left(\prod_{i=1}^{n} d_i\right)^{1/n} = \exp\left[\frac{1}{n}\sum_{i=1}^{n} \ln d_i\right] \quad (3)$$

$$s_g = \exp\sqrt{\frac{\sum_{i=1}^{n}\left(\ln\frac{d_i}{\mu_g}\right)^2}{n-1}} \quad (4)$$

where:

$$r_i = \ln d_i$$

These values are given in Table 4 for all of the primary trip classes, alongside a calculation of the Pearson's moment coefficient of skewness of the log-ratios.

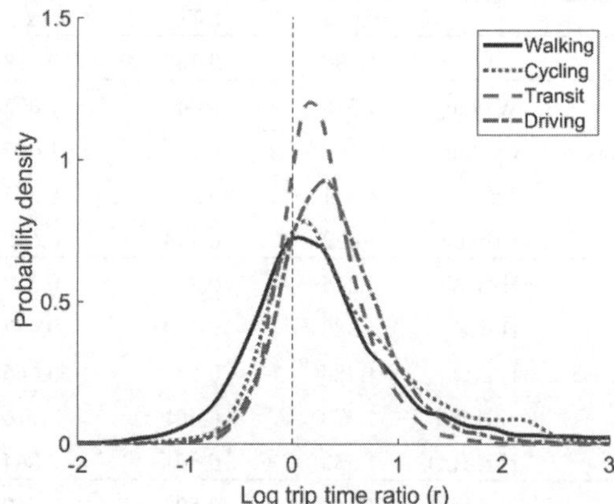

Figure 8. Probability distributions for each means of travel.

Figure 8 shows the smoothed kernel distribution plots for each means of travel. Walking has the modal value closest to the $r = 0$ line. However, it has high positive skewness and variance. Cycling trips show a very similar distribution to walking. Transit trips have the lowest variance and skewness, reflecting their constrained nature (transit trips are constrained to train lines/bus routes, which generally run to a fixed schedule). The value of transit trips is also lower than that for driving trips.

Table 4 also gives the statistical properties of the distributions for each trip purpose and departure time. As with choice of the means of travel, the distributions for trip purpose are distinct with clear differences. Home-based work (commuting) trips have the geometric mean closest to the origin, as well as the lowest variance and skew. These trips are repeated regularly, and as such there is a high incentive for passengers to research and select the quickest route. Home-based trips tend to show lower variance and skewness to non-home-based trips.

The distributions for each departure period are relatively closely matched compared to those for different means of travel and trip purpose, as shown by their similar geometric mean, standard deviation and skewness.

Overall the distribution of the log ratios varies significantly for each trip class. This is indicated with the properties shown in Table 4.

Table 4. Geometric mean, standard deviation and skewness of the log-ratio.

Category	Class	Geometric mean	Geometric S.D.	Skewness
All	All	1.390	0.568	0.918
Means of travel	Walking	1.328	0.741	1.002
	Cycling	1.608	0.719	1.047
	Transit	1.311	0.389	0.575
	Driving	1.472	0.494	0.648
Purpose	H.B.W.	1.275	0.396	0.527
	H.B.E	1.380	0.519	0.477
	H.B.O.	1.384	0.527	0.666
	N.H.B.W	1.473	0.601	0.716
	N.H.B.O.	1.538	0.801	0.864
Period	AM Peak	1.363	0.502	0.560
	Inter peak	1.431	0.623	0.966
	PM peak	1.421	0.534	0.689
	Other	1.355	0.562	1.012

5 CONCLUSIONS

There are clear discrepancies between the commonly assumed trip durations such as used in transport models and trip durations as recorded by the passenger as reflected in the survey data. Crucially, as is shown by the geometric mean and skewness of the data, the discrepancies are non-uniform across the modes of travel and trip classes. The patterns of variation in the duration of actual trips compared to the commonly assumed duration for different classes of trip is not captured in the generalised costs calculated by current transport models, which may have significant implications regarding the assumptions made for model calibration, validation and predictions.

The analysis carried out in this study is subject to imprecisions inherent in both the recording by the surveyed travellers and in the derivations of the Google based travel times, but the discrepancies are both greater in magnitude and more systematic than can be attributed to randomness or noise. This would appear to warrant more in-depth analysis. Emerging availability of more directly sensored travel data would make this increasingly feasible in future work.

ACKNOWLEDGEMENTS

This research is an update from Tim Hillel's MRes thesis which was undertaken as part of the Future Infrastructure and Built Environment Centre for Doctoral Training at the University of Cambridge, which is funded by the UK Engineering and Physical Sciences Research Council (EPSRC).

We thank Transport of London for their provision of the LTDS data, and Gerard Casey from the Centre for Sustainable Development at the University of Cambridge for assistance in data retrieval from the Google Maps API. However, the authors are solely responsible for the views or any remaining errors.

REFERENCES

Casey, G. H., Silva, E. A., Soga, K. et al. 2015. An ABM supported by real-time big data: The case of HSR vs. Aviation, *Proceedings of the ICE - Transport* Under Review.

Dijkstra, E. W. 1959. A note on two problems in connexion with graphs, *Numerische mathematik* **1.1**, 269-271.

Gallotti, R. & Barthelemy, M. 2014. Anatomy and efficiency of urban multimodal mobility, Nature *Scientific reports* **4**.

Prato, C. G. 2009. Route choice modeling: past, present and future research directions, *Journal of Choice Modelling* **2.1**, 65–100.

Train, K. E. 2009. *Discrete choice methods with simulation*. Cambridge university press, Cambridge.

Transport for London 2014. *The London Transportation Studies Model (LTS)*. Transport for London, London.

Transport for London 2011. *Travel in London, Supplementary Report: London Travel Demand Survey (LTDS)*. Transport for London, London.

Proceedings of the International Conference on Smart Infrastructure and Construction
ISBN 978-0-7277-6127-9

© The authors and ICE Publishing: All rights reserved, 2016
doi:10.1680/tfitsi.61279.765

A methodology to estimate losses in level of service for urban infrastructure networks

C. Kielhauser*, N. Lethanh and B.T. Adey

ETH Zurich, Zurich, Switzerland
* *Corresponding Author*

ABSTRACT Interventions on infrastructure networks in cities cause disruptions to the service provided by the network that requires the intervention. They also cause disruptions to the service provided by other networks that have to be at least partially shut down so that the intervention can be executed. Due to these effects, there is substantial benefit to be obtained by grouping interventions on all networks that are spatially close to one another, i.e. intervention programs for spatially close networks should be developed together. This benefit is principally due to reduced interruption to services and reduced costs of intervention. The challenge of determining such combined intervention programs is, that it is difficult to quantify the value of lost service, which depends on how different stakeholders value the services as well as how long and in which way the services are interrupted. This paper presents a methodology to estimate the loss of service on five infrastructure networks due the deterioration and due to the execution of both preventive and corrective interventions, and a way to formulate an optimisation equation in order to construct an optimal combined intervention program. The advantages and disadvantages as well as next steps in the research are discussed.

1 INTRODUCTION

Infrastructure networks, such as electricity, gas, road, sewer and water distribution networks, are built to provide services to the public. Since these networks deteriorate over time, and thus are less able to provide the desired levels of services, interventions must be executed to minimise the loss of service to the public, due to the execution of interventions, taking into consideration the possible synergies between networks (Kielhauser et al., 2015). The development of such an optimal intervention program requires accurate modelling of the loss of service in each type of network if interventions are executed or if failures occur. It is mathematically advantageous to model the loss of service of all network in the same way. This paper presents a methodology to do just that. The presented methodology is based on an electrical power analogy, and can be used to estimate the loss of service during and between interventions.

2 GENERAL CONCEPT

In this paper, a framework is presented to be used to estimate the loss of service for electricity distribution, gas distribution, road, sewer, and water distribution networks.

For the electricity distribution network, the demanded (and thus also billed) service is electric energy E_{el}, or more precise: the ability to use an electric power $P_{el}(t)$ for a certain time t. This electric power is produced by an electric current I_{el} consisting of a charge of Q_{el} coulomb every t seconds passing through an electric potential (=voltage) difference of V_{el}. This gives the basic equation for electric energy:

$$\int_t V_{el} \cdot \frac{Q_{el}}{t} dt = \int_t V_{el} \cdot I_{el}(t) dt = \int_t P_{el}(t) dt = E_{el} \quad (1)$$

Note that the energy is provided by the electrons carrying the charge passing through the voltage difference over time.

For the gas network, the demanded and billed service is also energy. However, contrary to electricity, the energy is stored chemically in the gas molecule instead of the electron. Nevertheless, the equation is similar: The gas energy is produced by a mass flow I_{gs} of Q_{gs} gas molecules every t seconds exiting at the consumer at a certain pressure difference V_{gs} (which is responsible for the flow) for a certain amount of time:

$$\int_t V_{gs} \cdot \frac{Q_{gs}}{t} dt = \int_t V_{gs} \cdot I_{gs}(t) dt = \int_t P_{gs}(t) dt = E_{gs} \quad (2)$$

It might seem unusual to call the product of gas flow and pressure "gas power" $P_{gs}(t)$, but this is done so here for consistency. The similarity of both networks and services is also demonstrated by the way, gas is billed (either in kWh or therms), with both being energy units. Note that the variable naming differs in literature, with the variable P sometimes also used for pressure, which is in this paper expressed as V_{gs}; the variable V sometimes is used for volume, and the variable Q sometimes is also used for flow, which is in this paper expressed as $I_{gs} = Q_{gs} / t$.

For the road network, the demanded service is mobility (Litman, 2011), i.e. the product of travel distance and transported good. This service also exhibits properties that are similar to the electricity and gas distribution networks: Mobility E_{rd} is produced by (the ability of providing) a traffic flow I_{rd} consisting of a load of Q_{rd} transported units (persons, vehicles, tons, etc.) every t seconds being able to travel a certain distance V_{rd}.

$$\int_t V_{rd} \cdot \frac{Q_{rd}}{t} dt = \int_t V_{rd} \cdot I_{rd}(t) dt = \int_t P_{rd}(t) dt = E_{rd} \quad (3)$$

However, the service is partially billed differently, with mobility directly billed via toll collection (depending on travel distance and vehicle type) or otherwise financed by taxes.

For the sewer network, the service provided is a mass flow at a certain slope, i.e. with the given sewer pipes leading from the customers (=sources) to the wastewater treatment plant (WWTP), the ability to remove a certain mass flow of wastewater. In other words, the service provided is a mass flow I_{sw} consisting of Q_{sw} cubic meters every t seconds for a certain (geodetic) potential difference to the WWTP, V_{sw}. Again, the equation takes the same form as before. The resulting integral, the service, is referred to here "elimination" E_{sw}, the product of mass flow and potential "sewage removal power" $P_{sw}(t)$.

$$\int_t V_{sw} \cdot \frac{Q_{sw}}{t} dt = \int_t V_{sw} \cdot I_{sw}(t) dt = \int_t P_{sw}(t) dt = E_{sw} \quad (4)$$

For sewer networks, the billing is based on the maximal possible removal power, i.e. the possibility to safely dispose of the occurring amount of wastewater. For water distribution network, the service provided is a mass flow at a certain pressure. The equation is similar to the equations before: The service is provided as a mass flow I_{wa} consisting of Q_{wa} cubic meters every t seconds at a certain pressure (potential), V_{wa}. The product of pressure and mass flow is called "hydraulic power" $P_{wa}(t)$, a term that is commonly seen in mechanical engineering, but rarely in other engineering fields.

$$\int_t V_{wa} \cdot \frac{Q_{wa}}{t} dt = \int_t V_{wa} \cdot I_{wa}(t) dt = \int_t P_{wa}(t) dt = E_{wa} \quad (5)$$

However, for water networks, only the amount in m^3 is billed for private consumers. For larger consumers, the pressure delivered is also included in the billing.

As can be seen, the general equation describing the provided services on these five networks has the same structure, with only differences in the way of billing the service, but not the way service is provided nor in the formulation of the basic equations.

3 DEFINITION OF THE LOSS OF SERVICE POWER

From the equations in section 2, it can be seen that the service provided is an integral of power over time. Therefore, the loss of service should also be measured by a power integral, as the service provided changes over time and thus also the possible loss of service may be. In general, loss of service occurs during the transportation of the service over the network and is related to the condition of the network objects.

For the electricity network, this loss of service is caused by the resistance of the conductors. This leads the electricity producer to increase the produced input power $P_{el,in}$ such that the lost power due to resistance $P_{el,los}$ is overcome and the consumer is provided the demanded output power $P_{el,out}$. Written as equation:

$$P_{el,in}(t) = P_{el,out}(t) + P_{el,los}(t) \quad (6)$$

For the water and gas distribution networks, the loss of service is related to two things: 1) friction along the pipe walls and internal friction, and 2) leaks in the networks. To counteract this, the producer has to increase the produced hydraulic/gas power P_{in} to overcome the lost power due to flow through leaks P_{le} and the lost power due to friction P_{fr} to provide the demanded output power P_{out} to the consumer. As both P_{le} and P_{fr} contribute to the loss of service, they can be combined to P_{los} :

$$P_{gs/wa,in}(t) = P_{gs/wa,out}(t) + P_{gs/wa,le}(t) + P_{gs/wa,fr}(t) = P_{gs/wa,out}(t) + P_{gs/wa,los}(t) \quad (7)$$

For the road network, there is no distinct "producer" or "consumer" of mobility, every road user acts as both. However, there are producer and consumer nodes, i.e. the origins/destinations of each trip - with producer nodes $P_{rd,in}$ being nodes with more trip origins than trip destinations and vice versa ($P_{rd,out}$). When looking at the unit of "mobility power" i.e. mobility per time, this corresponds to $[Veh.\cdot m \cdot s^{-1}]$, i.e. the number of vehicles going at a certain speed. Therefore, the loss of service can be related to the loss of speed (due to road condition), certain vehicles occur on their trip, also being in the same unit. Therefore, this loss can be added to the equation, thus obtaining:

$$P_{rd,in}(t) = P_{rd,out}(t) + P_{rd,los}(t) \quad (8)$$

For the sewer network, the loss of service is related to increased roughness up until clogging of the sewer pipes. The producer (the WWTP, in this case the "out" node) cannot counteract this and only can accept the arriving sewage (power) $P_{sw,out}$, so the consumer (who is, in this network, the source, i.e. the "in" node) cannot dispose over the full sewage removal power $P_{sw,in}$, but is reduced by the lost removal power due to friction and clogging $P_{sw,los}$:

$$P_{sw,in}(t) - P_{sw,los}(t) = P_{sw,out}(t) \quad (9)$$

which can be rewritten as

$$P_{sw,in}(t) = P_{sw,out}(t) + P_{sw,los}(t) \quad (10)$$

As can be seen, all the equations describing the loss of service conveniently have the same structure, however the units differ from one network to the other.

The next step in building an optimisation routine is the valuation of the loss of service in order to compare it to the costs caused by an intervention that prevents that loss of service.

For the electricity distribution network, this is straightforward, as the power producer bills energy provided to the consumer, and incurs costs for producing this energy, but also production costs for producing the lost power due to resistance. With a deteriorating electricity distribution network, the resistance increases, and so does the lost power. This leads to more production costs without additional income, as only the delivered energy can be billed. Additionally, if a consumer is shut off from service due to a conductor being inoperable, no energy is delivered to him so nothing can be billed, which reduces the income of the producer further. To counteract this, the producer can choose to execute a maintenance intervention in order to keep the network in a good state. The optimal intervention program would then be the set of interventions that balances out (i.e. minimises) the cost of interventions versus the cost of increased power production/ reduced income due to power outage.

For the gas distribution network, the same principle can be used, only with additionally accounting for the losses due to leaks, which can be included in the optimisation by defining the leak as a customer, who does not pay the gas bill.

For the water distribution network however, the service provided does not directly correspond to the service paid for, except for the case of large industries. Nevertheless, the producer still incurs production costs for hydraulic power, and also the power loss due to leakage and increased friction as the pipe condition deteriorates. Although the consumption is only measured in the amount of water, pressure is needed to provide it, and so the hydraulic power at the consumer in order to provide this amount can be calculated, as a water network is, from a physical point of view, a potential field. With the consumed hydraulic power now known, these costs can be compared again with the costs of interventions.

For the road network, the valuation of service and the loss thereof is more complex. As stated in section 3, there is no "producer" nor "consumer". However, the loss of speed times the number of vehicles (mobility power) can be translated to additional travel time and the cost associated with it. Reducing travel time (costs) by a set of interventions can then be balanced out against the costs for these interventions.

For the sewer network, the direct valuation of service is hardly possible, as a reduction in service is only visible if the service is heavily reduced, i.e. if the flow is reduced to (almost) zero. This means that costs for interventions have to be balanced out against the loss in income due to disconnection of a consumer, but are not related to deteriorating performance of the network until it ceases to serve a specific customer.

4 CALCULATION OF THE LOSS OF SERVICE POWER

With the loss of service defined in terms of power (section 3), the power state and the loss of power can be estimated for each network. To facilitate reading, the index (t) for each time point has been omitted. Following basic physical relations, the general form of the power equation is

$$P_m = f(\gamma_{n,m}, \sigma_{n,m}, I_{n,m}) \quad (11)$$

with $\gamma_{n,m}$... conductivity term for object $o_{n,m}$, $\sigma_{n,m}$... object slimness, $I_{n,m}$... flow through object $o_{n,m}$, P_m... power in network m. The use of the conductivity (which is the inverse of the resistivity) is due to the fact that an object out of service has infinite resistivity (=zero conductivity), with the infinite being mathematically more difficult to handle than the zero. The object slimness takes into account the length and the cross-section of the object, which also influence the flow and thus the power.

For example, in electricity distribution networks the respective variables are: P_{el}... electric power [W], $\gamma_{n,el}$... electric conductivity $[\Omega^{-1}m^{-1}]$ of object $o_{n,el}$ in electricity network, $\sigma_{n,el}$... conductor slimness $l \cdot A^{-1}[m^{-1}]$, $I_{n,el}$... electric current [A]. Note that the variable naming in this paper differs from the conventions in electrical engineering, where σ is usually being used as the variable for conductivity. However, to be consistent within this paper, the above-presented notation is used. The power lost in the network can be obtained by solving the following equation (Kirchhoff, 1845):

$$\mathbf{G} \cdot \vec{\varphi} = \vec{I} \quad (12)$$

with $\mathbf{G}$... conductance matrix according to (Kirchhoff, 1845), $\vec{\varphi}$... Voltage vector, and $\vec{I}$... current vector. The conductance is simply the ratio of conductivity to slimness $\gamma_n \cdot \sigma_n^{-1}$. With at least voltage or current in-/outflow known at all nodes, the equation is complete and all missing currents/potentials, including currents from node to node can be calculated. The loss of power (following ohm's law) at each object $o_{n,el}$ is then:

$$P_{el\,los,o_{n,el}} = (\varphi_2 - \varphi_1)^2 \cdot \frac{\gamma_{n,el}}{\sigma_{n,el}} = I_{1\to 2}^2 \cdot \frac{\sigma_{n,el}}{\gamma_{n,el}} \quad (13)$$

with φ_1, φ_2... voltage at start/end of object $o_{n,el}$ and $I_{1\to 2}$... current flowing from start to end of object $o_{n,el}$. Summing the loss of service over the network gives the total loss of power for electricity:

$$P_{el\,los} = \sum_n P_{el\,los,o_{n,el}} \quad (14)$$

For all other networks, the principle is the same - however with methodologies different from Eq. 12 to calculate the potentials in and the currents flowing between nodes *i* and *j*. For water distribution and sewer networks, the Darcy-Weisbach equations (Brown, 2002) can be used, for gas networks, the Newton-nodal method (Li et al., 2003) can be used, while for roads, traffic simulations can be used. With the potential and/or the currents known, they can be inserted in Eq. 13 to obtain the loss of service power.

5 CHANGING CONDUCTIVITY OVER TIME

To be able to incorporate the changing condition of the network, the conductivity of each object has to be made time dependent. A schematic of this is shown in Fig.1.

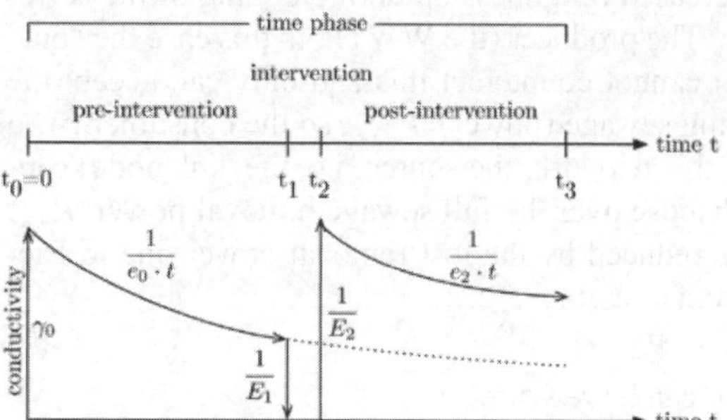

Figure 1. Conductivity over time

At $t_0 = 0$, the object has a base conductivity γ_0. As deterioration of the objet occurs, the conductivity decreases over time with a function $1/(e_0 \cdot t)$. If an intervention is executed, the conductivity will become 0 between t_1 and t_2, i.e. during the execution of the intervention. Afterwards, the conductivity will be restored to a value $1/E_2$ (which may be higher or lower than γ_0, depending on the type of intervention) and then start decreasing again with a function $1/(e_2 \cdot t)$. If no intervention is executed, the conductivity follows the function $1/(e_0 \cdot t)$ along the dashed line. Mathematically:

$$\gamma_{n,m}(t) = \frac{\gamma_{0,n,m}}{1+\gamma_{0,n,m} \cdot \left(\int_0^t \left(e_{0,n,m} + \delta_{n,m}\left(e_{2,n,m} \cdot \langle t-t_2\rangle^0 + E_{1,n,m} \cdot \langle t-t_{E_1}\rangle^{-1} + E_{2,n,m} \cdot \langle t-t_2\rangle^{-1}\right)\right)dt\right)} \quad (15)$$

with $\gamma_0 \ldots$ base conductivity of object in perfect condition, and $\delta_{n,m} \ldots$ binary variable indicating if an intervention is executed. With this formulation, different types of interventions can be accounted for. The angled brackets denote singular functions as defined in (Mang et al. 2013)

With the conductivity now defined in general, the last step is to transfer it to the investigated networks. For the electricity distribution network, the conductivity is simply the material-inherent conductivity that deteriorates over time due to oxidation and corrosion, the slimness is the ratio of conductor length to conductor diameter. For the hydraulic networks (sewer, water distribution), the conductance is the pressure head loss, i.e. the product of conductivity and slimness, with the slimness being the ratio of length to hydraulic diameter. This gives the hydraulic conductivity as $\gamma_{n,hyd} = 8 \cdot f_D \cdot \rho \cdot I^2 \cdot \pi^2 \cdot D^{-4}$ with $f_D \ldots$ Darcy friction factor, $\rho \ldots$ fluid density, and $D \ldots$ hydraulic diameter, with only the Darcy friction factor being dependent on deterioration and improvement due to interventions. For the gas network, the equation is similar, but adjusted to take into account the temperature and compressibility. This gives the gas conductivity as $\gamma_{n,gas} = f_M \cdot Z \cdot T \cdot \rho \cdot I^2 \cdot A^{-4}$ with $f_M \ldots$ Moody friction factor, $Z \ldots$ compressibility factor, $T \ldots$ gas temperature, $A \ldots$ pipe diameter. Again, only the friction factor is dependent on deterioration and improvement due to interventions. For the road network, the terms shall be explained further. The mobility power represents the ability of the system to transport vehicles, given by the mobility per second, i.e. desired number of vehicles that want to go at a certain speed $[Veh. \cdot m \cdot s^{-1}]$. Therefore, (see also Eq.3) the conductance is given in $[\mathrm{m}]$, and thus the conductivity (conductance divided by slimness $[\mathrm{m}/1]$) is given as $[\mathrm{m}/\mathrm{m}]$. This road conductivity expresses the ratio of the real length to the apparent length of the road due to the non-perfect condition. This apparent length represents the loss of speed one occurs due to non-optimal road condition, but instead of being expressed as reduction in speed, it is expressed as the additional length one has to travel at the desired speed. The slimness is given as length divided by the number of lanes, and thus carries the unit $[\mathrm{m}/1]$.

6 COST EQUATION

To be finally able to do an optimisation, the components of the cost equation, which were shown in sections 2 to 5 can be put together. This will be presented starting from the most general form of the cost equation to the detailed form for each network. The optimal intervention program can be found by:

$$min.: C_{LOS}(\delta_{n,m}) + C_{INT}(\delta_{n,m}) \quad (16)$$

with $C_{LOS} \ldots$ cost for service, including loss of service and $C_{INT} \ldots$ cost of interventions on all networks. The methodology for calculating the intervention cost is explained in Kielhauser et al. (2015) and will not be further explained here. The values taken over from the methodology are the costs C_{INT} and the decision vector $\delta_{n,m}$, indicating the objects n on network m for which an intervention is executed, which is the also the decision variable for this optimisation. C_{LOS} can be subdivided in:

$$\begin{aligned} C_{LOS}(\delta_{n,m}) &= C_{LOS,el}(\delta_{n,el}) + C_{LOS,gs}(\delta_{n,gs}) \\ &+ C_{LOS,rd}(\delta_{n,rd}) + C_{LOS,sw}(\delta_{n,sw}) + C_{LOS,wa}(\delta_{n,wa}) \end{aligned} \quad (17)$$

The single components are, in conjunction with section 3 further detailed in:

$$\begin{aligned} C_{LOS,el} &= \int_0^t \left(P_{el,in}(\gamma_{n,el}(t), \delta_{n,el})\right)dt \cdot c_{prod,el} \\ &- \int_0^t \left(P_{el,out}(\gamma_{n,el}(t), \delta_{n,el})\right)dt \cdot p_{sell,el} \end{aligned} \quad (18)$$

with $c_{prod,el} \ldots$ cost of electric energy production and $p_{sell,el} \ldots$ income per sold unit electric energy. Note that the loss of service doesn't appear directly in the equation, but is indirectly included via the relation given in Eq.6.

For the gas distribution network, the equation is similar:

$$\begin{aligned} C_{LOS,gs} &= \int_0^t \left(P_{gs,in}(\gamma_{n,gs}(t), \delta_{n,gs})\right)dt \cdot c_{prod,gs} \\ &- \int_0^t \left(P_{gs,out}(\gamma_{n,gs}(t), \delta_{n,gs})\right)dt \cdot p_{sell,gs} \end{aligned} \quad (19)$$

with $c_{prod,gs} \ldots$ cost of gas production and $p_{sell,gs} \ldots$ price at which gas is sold. Again, the loss of service is indirectly included via Eq. 7

For the road network, the cost is related to the additional travel time, as stated in section 4. The additional travel time, and such the costs, can be calculated from the ratio of the reduced mobility divided by the outflowing mobility:

$$C_{LOS,rd} = c_{travel} \cdot t_{travel,avg} \cdot \int_0^t \left(\frac{P_{rd,out}(\gamma_{n,rd}(t), \delta_{n,rd}) + P_{rd,los}(\gamma_{n,rd}(t), \delta_{n,rd})}{P_{rd,out}(\gamma_{n,rd}(t), \delta_{n,rd})} \right) dt$$
$$- \int_0^t \left(P_{rd,out}(\gamma_{n,rd}(t), \delta_{n,rd}) \right) dt \cdot p_{rp,rd} - p_{tax,rd} \quad (20)$$

with c_{travel} …travel time costs to the taxpaying public, $t_{travel,avg}$ …average travel time without disturbance, $p_{rp,rd}$ … income due to road pricing, $p_{tax,rd}$ … income due to road-dedicated taxes. Note that this cost equation relies on a simplification about the travel time. Implications of this will be discussed in section 7.

For the sewer network, the cost is related to the inability to dispose of the waste water via the sewer. This inability is expressed as $P_{los,sw}$ in Eq. 10 as it is the difference of what can be accepted at the WWTP ($P_{sw,out}$) and what is actually transported from the origins due to restrictions in the network ($P_{sw,in}$). Therefore, the costs can be calculated:

$$C_{los,sw} = \int_0^t \left(P_{sw,los}(\gamma_{n,sw}(t), \delta_{n,sw}) \right) dt \cdot c_{disp,sw} \quad (21)$$

with $c_{disp,sw}$ … cost for alternative disposal. Note that this calculation relies on an assumption about sewage disposal. Implications of this will be discussed in section 7.

In summary, the optimisation of the intervention program is done by choosing the interventions as decision variables in such a way, that the sum of costs for service (including income due to billing) C_{LOS} and costs for interventions C_{INT} is minimal.

7 DISCUSSION AND CONCLUSION

A methodology has been presented to estimate the loss of service on five infrastructure networks due the deterioration and due to the execution of both preventive and corrective interventions, and a way to formulate an optimisation equation in order to construct an optimal combined intervention program. In the methodology the same mathematical structure is used to model the provided services for all networks, although there are differences between the services billed and the services provided. The methodology is able to account for deterioration of the objects and the resulting changes in the network, as well as, the inoperability of objects during interventions and the better condition after the execution of interventions. The measure used for this is conductivity. It has been demonstrated, that the conductivity has a physical basis in four of five networks. For the road network, the proposed relation seems logical and is also coherent in terms of units.

For the cost equations, some assumptions had to be made. For all networks, fixed costs have been omitted. This is reasonable, as fixed costs do not change with network condition, and thus it is feasible that it might be possible to omit them when determining the optimal intervention program. Additionally, for travel costs the assumption has been made that the costs to the public are dependent on the additional travel time, expressed as a multiplication of the average travel time. This is a generalisation, that is reasonable for a comparing view over more networks, like in this paper, but for detailed analysis of the road network, better models like agent based models can be used. For the sewer network, the assumption has been made, that the occurring wastewater cannot be stored, but is removed by the network operator via sewage tank trucks at his expense. Although this is probably only happening for consumers with high-availability contracts, this is a reasonable assumption if the costs are reduced accordingly.

Future research in this area is to be focused on the validity of excluding fixed costs in the determination of the optimal intervention program, verifying the use of this model with real world assessments of the value of lost service, the inclusion of priority contracts and the development of improved cost/price models for sewer and road networks.

REFERENCES

Kielhauser, C. Adey, B.T. & Lethanh, N. 2015. A comparison of geographic intervention grouping methods for infrastructure intervention planning across multiple networks, *Proceedings of the 2015 ICSC Conference,* UBC, Vancouver, Canada.

Kirchhoff, G.R. 1845. Ueber den Durchgang eies elektrischen Stromes durch eine Ebene, insbesonders durch eine kreisförmige, *Annalen der Physik* **140(4)**, 497-514.

Li, Q. An, S & Gedra, T.W. 2003. Solving natural gas loadflow problems using electric loadflow techniques, *Proc. Of the Norcht American power Symposium*, Citeseer

Litman, T. 2011 *Measuring Transportation*, Victoria Transport Policy Institute, Victoria, Australia

Mang, H. A. & Hofstätter, G. 2013. *Festigkeitslehre.* Springer, Berlin

O'Brown, G. 2002. The history of the Darcy-Weisbach equation for pipe flow resistance, *Environmental and Water Resources History* **38(7)**, 34-43.

Proceedings of the International Conference on Smart Infrastructure and Construction
ISBN 978-0-7277-6127-9

© The authors and ICE Publishing: All rights reserved, 2016
doi:10.1680/tfitsi.61279.771

Monitoring city-scale land use development using new online sources – the case of Beijing

X. Rong *
Martin Centre for Architectural and Urban Studies, University of Cambridge, UK
* *Corresponding Author*

ABSTRACT In the on-going, fast urbanisation in developing countries, monitoring urban land use development has an immediate relevance to the pressing challenges in promoting citizens' quality of life, social inclusion, environmental sustainability and economic growth. In fact in all fast growing city regions in both poor and rich countries, there exists a significant gap between rapid urban development taking place on the ground and the statistical data, which often hampers effective analysis for urban development policies, regulations and infrastructure planning. However, administrative reporting of urban land and development permissions has become increasingly more timely, as part of the Open Government initiative. This has opened up new potentials for monitoring and analysing emerging patterns of urban land and floorspace development, if the new sources could be joined up with exiting statistics. This paper examines the potential and effectiveness in using such online administrative sources through the case of Beijing. The paper uses urban land plot transaction data collected from the websites of Beijing Municipal Bureau of Land and Resources to analyse Beijing's urban expansion between 2004 and 2013. It assesses the strength and weaknesses of the data by comparing it to published statistics. And it examines the growth pattern by ring of growth and the direct distance to the nearest station.

1 INTRODUCTION

The world is in its most rapid urbanisation phase in the history of humanity: World urban population has tripled to reach 3.15 billion between 1950 and 2005, and is expected to rise to more than 6 billion by 2050 (United Nation 2014). This is associated with urban land use expansion at a similar, and often faster, speed. Much of the growth is taking place in cities in developing countries, where inadequate planning has led to problems of informal settlements with inadequate infrastructure support. Monitoring urban land use development has an immediate relevance to the pressing planning objectives to enhance the citizens' quality of life, social inclusion, environmental sustainability and economic growth.

Rapid urban development tends to create a significant gap between what takes place on the ground and the statistical data for strategic policy analysis. Typically, the data sources lag behind for one or more years, and for urban land use the data has been really patchy, in spite of great efforts by statisticians and scholars. Beijing is a good example. Studies on land use development have to rely on satellite photos (e.g. Liu et al. 2000, Kuang et al. 2009) and confidential official data (e.g. Yan & Feng 2009). But neither is not in the public domain, and the satellite images are limited to land use categorisation only. This hampers effective policy analysis.

However, the increasingly timely reporting of administrative records for urban land provision by the government has opened up new opportunities to systematically monitor and analyse the emerging patterns of urban land use development. Currently, these data sources are rarely used, mainly because the online data is difficult to process and its quality is not well understood. The few studies are focused on land price (Gao et al., 2014) or affordable housing development patterns (Dang et al., 2014), rather than city's land use expansions. Little attention has been paid to the real strengths and weaknesses of the online data.

This paper aims to explore the potential of using the online administrative data for monitoring city-scale land use development through the case of Beijing. The online data is carefully verified by comparing to all publicly available statistics. It is then used in revealing the pattern of land supply, where little up to date knowledge exist.

2 DATA ASSEMBLY AND VERIFICATION

2.1 Data sources

The paper uses urban land plot provision data collected from Beijing Municipal Bureau of Land and Resources website, which has a good coverage for land development and associated building floor space information. It mostly starts in 2003 (i.e. covering most years after the reform of land provision), and is consistently updated. Three datasets, respectively on land plot transaction, land plot allocation and land plot acquisition, are published.

- Land plot transaction deals with urban land plots to be used for profitable purposes, such as residential, industrial and commercial. After acquiring the permission from the local government, the land plots go to open market for bidding. The winning bidder(s) pay the government a market transaction price and the previous user/the primary land developer a specified fee (based on the cost incurred for infrastructure provision on site).
- Land plot allocation deals with urban land plots to be used for non-profitable, public purposes, such as infrastructure, government and institutions, affordable housing, etc. The recipients do not need to pay a fee to the government, but in some cases they have to pay the previous user a small one-time fee for replacement cost and other expenses.
- Land plot acquisition deals with rural land, which is collectively owned. Because the land use right of such land is not transferable, the government usually has to first convert it into state-owned land through acquisition. This data is less relevant for our purposes here because much of the land acquired in this way will show up again in the land transaction and allocation data (i.e. after primary land development).

This paper therefore chooses to use the land plot transaction and allocation data. An additional incentive for doing so is that for those two types of data, there is a decade-long data series since 2003 with reasonably consistent formats[1]. By contrast, the land plot acquisition data, which is separately collected and published by each city district, has little consistency in content and format, and dates back to different years, typically 2009.

2.2 Data Assembly

We first collect all the items published on the website using a web-scraping software (LocoySpider). After cleaning the addresses (mainly by confirming the integrity of the addresses and deleting unnecessary details), we used a VBA program to geocode the addresses into longitude and latitude. By checking the longitude and latitude with the expected locations in the online maps, we clean the addresses that were mis-recognised by Google and controlled the level of accuracy in a semi-automatic way. To solve the offset issue of Google mapping in China, we transform the geo-coded data to correct locations for calculating distances to stations, etc.

The land use types from the data were categorized into seven groups, including housing, offices and institutions, manufacturing, R&D, retail, community services and infrastructure. The data is then processed and analysed using Arc GIS. The plots which have no information on either land area or planned building floorspace are deleted.

2.3 Data verification

The online land provision data is not collected for the purpose of statistics and not guaranteed in data quality. Therefore, it is necessary to examine the quality of the data in order to make the best use of it. This is

[1] The data for 2003 is only available in summary form at the level of urban district and counties, and is thus not included in later analysis.

done by examining how well it compares with the sources published in official statistics, including China Urban Construction Statistical Yearbooks (CUCSY), published reports of Beijing Municipal Bureau of Land and Resources (BMBLR) and Beijing statistical Yearbooks (BSY).

Table 1. Comparison of online data and statistics on land area change (with one year lag)

	(Unit: ha)	**Online data** Total of land area transacted and allocated from 2005 to 2010	**CUSCY** Change between 2006 and 2011
Land use types included in the online data	Residential Area	5288	4170
	Public Facilities	2868	2356
	Industrial Operations	3110	3142
	Storage	186	-104
	Transport	468	1459
	Other Municipal Utilities	459	216
Land use types not included in the online data	Roads and Plazas	0	3289
	Green Space	0	2662
	Other Specific Purposes	0	-26

The comparison on land area change shows that for the land use types that are included in the online data, the online data reports the same magnitude of changes as the official statistics, especially for residential, public facilities and manufacturing. It slightly overestimates the residential area and area for public facilities, and reports very accurately for industrial operation. This is partly because the online data demonstrates the growth, whereas the CUSCY data shows the net changes. But it also suggests that the online data may have a more comprehensive coverage than CUSCY data. On the other hand, the online data does relatively badly in land for storage, traffic and municipal utilities (**Table 1**), which is mainly caused by the exclusion of land acquisition data.

This is further confirmed by the comparison with the published reports of BMBLR. The online data has captured the main bulk of the land provided through transaction and allocation, although the online data is doing relatively poor for land allocation (**Table 2**). Because the official report only includes the total, it is currently impossible to speculate which kind plots are excluded either in the online data or in the reports.

Table 2. Comparison of online data and BMBLR reports on land provided through transaction and allocation (2009, 2011 and 2012)

		Land transaction		Land allocation	
		Plots	Area (ha)	Plots	Area (ha)
2009	Online total	541	1271	87	475
	Official report	677	1854	133	722
	Percentage	80%	69%	65%	66%
2011	Online total	604	2156	68	215
	Official report	351	2086	121	586
	Percentage	172%	103%	56%	37%
2012	Online total	402	1118	139	345
	Official report	/	1077	/	450
	Percentage	/	104%	/	77%

We further examine the coverage of online data in housing and related land use types by comparing to the real estate development data in BSY. In terms of land area, the online data is larger than BSY statistics in the majority of the years, which suggests that the online data has a better coverage possibly because it accounts for other types of developers than non-real estate companies (**Table 3**). In terms of land purchasing cost, the online data is lower than BSY statistics in almost all the years (**Table 3**). This is because the online data only covers the land transaction fee, whereas the BSY statistics also includes compensation fee and the administrative fee.

In terms of the floorspace, the planned floorspace in the online data is always lower than floorspace of newly started construction in the BSY statistics. This is contrary to the trend observed in land area of housing construction (**Table 4**). Our conjecture is that the online data has a better coverage on land, and the land recipients increase the building floorspace or Floor Area Ratio after they have obtained the land in later applications for planning commissions.

To summarise, the verification shows that the online data does not fully represent what is taking place on the ground. It is relatively weak in planned floorspace; it excludes the land acquisition data and does not cover all the land plots that are transacted and allocated; it by nature cannot fully reflect the actual construction on the ground. However, the land statistics has always been notoriously difficult. Despite these weaknesses, the verification shows that the online data has captured the main projects for housing and employment land, and is able to demonstrate the land use development pattern at the city scale.

Table 3. Comparison of online data and BYS statistics on real estate land development and land purchasing cost (2005-2012)

	Land transacted/ purchased (Unit: ha)			Land purchasing cost (Unit: 100 million Yuan)		
	Online data (housing only)	Online data (also including related land use types)	BSY	Online data (housing only)	Online data (also including related land use types)	BSY
2005	354.6	588.9	773.9	43.9	75.3	239.8
2006	805.4	1078.7	347.1	116.3	165.7	477.9
2007	514.0	712.5	521.8	252.1	313.4	644.7
2008	635.5	866.3	915.1	466.1	539.0	639.0
2009	607.4	863.3	625.0	517.7	695.0	587.7
2010	430.5	656.6	858.7	640.5	840.7	1292.7
2011	992.3	1293.4	507.0	799.7	1267.5	1301.2
2012	487.9	666.8	306.0	424.7	656.9	1102.7

Table 4. Comparison of online data and BYS statistics on housing building floorspace (2008-2011)

Unit: 10000 m^2	Total		Market housing		Non-market housing	
	Online data	BSY	Online data	BSY	Online data	BSY
2008	1453.7	2177.4	1148.5	1565.3	305.2	612.1
2009	1599.9	2212.6	1174.0	1380.3	425.9	832.3
2010	1128.7	2999.3	851.3	2063.4	277.4	935.9
2011	2330.3	4047.4	1479.4	2596.4	850.9	1451.0

3 PATTERNS OF LAND USE DEVELOPMENT IN BEIJING (2004-2013)

3.1 Overall growth

From the online data, in the decade between 2004 and 2013, for housing and employment, altogether 6,913 projects have acquired 21,995.4 hectares of land through both land transaction and land allocation, which are planned to provide 351.6 million m^2 of floorspace[2]. The land provided in this decade is equivalent to around 17.3% of Beijing's total built area in 2013 (1,268 km^2), which reveals the fast speed of city's growth. Similar to the majority of growing cities, the largest proportion of this growth comes from housing. The dominance of housing is firstly shown in the number of projects at 31%, and is further highlighted by its share in land area and planned building floor space, respectively 54% and 67% (**Table 5**).

Table 5. Summary of number of plots, land area and planned building floorspace by land use type (2004-2013)

	Number of plots		Land area (ha)		Planned building floor space (million m^2)	
Housing	2164	31%	11893.8	54%	233.6	67%
Offices and institutions	770	11%	736.0	3%	21.3	6%
Manufacturing	1078	16%	5049.3	23%	25.1	7%
R&D	373	5%	1610.8	7%	19.0	5%
Retail	1411	20%	1285.3	6%	30.1	9%
Community services	1135	16%	1420.4	7%	22.5	6%
Total	6931	100%	21995.4	100%	351.6	100%

3.2 Rings of growth

3.2.1 Defining the rings of growth

To examine the trend of centralising/decentralising, we divide Beijing into 6 concentric rings in line with the ring roads (**Figure 1**). The area inside Ring Road 2 is defined as the city centre. The wide ring roads not only act as physical boundaries, but also psychological ones e.g. when discussing property prices (see Tian et al, 2010).

3.2.2 FAR changes

As reported in Rong et al. (2015), the rings of growth analysis of both land area and planned floorspace demonstrates that housing and most types of em-

[2] Here include 97 plots which have land area information but no planned building floor space information, and 14 plots which have planned building floor space information but no land area information.

ployment are decentralising, but housing is decentralising at a faster speed. In particular, offices and institutions continue to grow in a centralising way (Figure 2 & 3). But this may well be due to the larger areas of outer rings. Therefore, we further analyse the FAR changes of housing and non-manufacturing employment, which take both the land area and the existing developments into consideration.

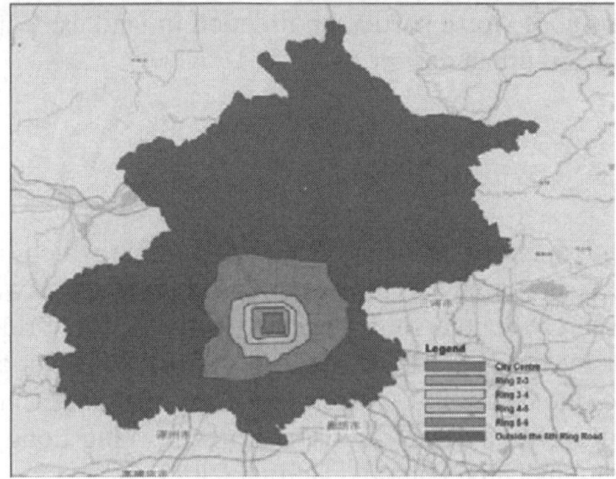

Figure 1. Rings of growth

The FAR of housing in 2000 is estimated using the number of residents by sub-district and the average housing floorspace per capita by district from the population census. The FAR of non-manufacturing employment in 2000 is estimated using the number of employed workers by sub-district from economic census, assuming each worker takes 20 m^2.

The FAR changes (**Table 6**) reconfirm the aforementioned pattern. For housing, Ring 5-6 has seen the largest increase of FAR (176%), Ring 3-4 comes the second (104%), followed by Ring 4-5 and Ring 2-3 (72% and 71% respectively). For employment as a whole, Ring 5-6 also sees the largest change of FAR, but only by 125%, slightly above Ring 3-4 (119%). And Ring 2-3 (75%) exceeds Ring 4-5 (51%) by a large margin. In particular, offices and institutions continue to develop faster within the 4th Ring Road, with FAR increases by 43% in both Ring 2-3 and Ring 3-4 followed by City Centre (20%). Retail sees the largest increase of FAR in Ring 3-4 (45%). R&D and community services are both decentralising. But they are developing at a smaller scale than housing, and the community services are decentralising at a slower rate than housing.

Table 6. Gross floor area ration (FAR) of total estimated stock in 2000 and of increment 2004-2013, and the percentage of change of FAR for housing and employment by ring of growth

Gross Floor/Area Ratio (FAR)	City Centre	Ring 2-3	Ring 3-4	Ring 4-5	Ring 5-6	Outside 6th	All areas
Housing Total estimated stock in 2000	0.515	0.241	0.193	0.158	0.031	0.007	0.017
Increment 2004-2013	0.165	0.171	0.201	0.114	0.055	0.003	0.014
% change	32%	71%	104%	72%	176%	50%	82%
Business floorspace Total estimated stock	0.387	0.144	0.081	0.077	0.016	0.003	0.009
BFS 2004-2013 – increment in FAR							
Office and institution	0.076	0.063	0.034	0.006	0.002	0.000	0.001
R&D	0.002	0.003	0.006	0.007	0.007	0.000	0.001
Retail	0.040	0.029	0.036	0.019	0.006	0.000	0.002
Community services	0.032	0.014	0.020	0.007	0.006	0.000	0.001
All	0.150	0.108	0.096	0.039	0.020	0.001	0.006
BFS - % change in FAR							
Office and institution	20%	43%	43%	8%	10%	2%	15%
R&D	1%	2%	7%	10%	44%	9%	13%
Retail	10%	20%	45%	24%	35%	9%	21%
Community services	8%	9%	25%	9%	35%	11%	15%
All increments	39%	75%	119%	51%	125%	30%	65%

3.3 Land use development and metro provision

The increasingly more decentralised pattern of growth and a continuing separation of jobs and housing may lead to problems of longer commuting hours and increased traffic congestion. But Beijing municipal government has been trying to mitigate this problem by extending metro provision. Therefore, we further examine the average distance weighted by planned floorspace to the nearest metro station. Here we excluded the area outside the 6th Ring Road, as people do not rely on metro to travel there.

Offices and institutions are closest to the metro station with the average of 792m, followed by retail (1212m). Community services and housing are at the same level with 1557m and 1618m respectively.

Manufacturing and R&D are much further away from the metro station, with 2309m and 2798m. For all the land use types, the distance to the nearest metro station increases from more central rings to more periphery rings, with the exception of Ring 3-4. And there is a clear leap of the distance between rings within the 4th Ring Road and outside. The long distance to the station of housing in Ring 5-6 (2,134m) and Ring 4-5 (1,575m) is worth special attention, as both areas have a large share of housing (38% and 18% respectively) (**Table** 7). As this is direct distance, the actual route distance would be even longer.

Table 7. Average distance (weighted by planned floorspace) to the nearest metro station (Unit: m) and share of planned floorspace by ring of growth within the 6th Ring Road

	City Centre	Ring 2-3	Ring 3-4	Ring 4-5	Ring 5-6	All within 6th
Housing	751	1,027	754	1,575	2,134	1,618
	4%	7%	12%	18%	38%	79%
Office and institution	438	647	622	1,287	1,652	792
	23%	28%	23%	10%	13%	97%
Manufacturing	978	1,154	697	1,987	2,669	2,309
	1%	1%	2%	20%	30%	54%
R&D	501	658	580	1,474	3,336	2,798
	1%	1%	4%	14%	62%	82%
Retail	610	754	539	1,569	1,631	1,212
	8%	9%	17%	23%	31%	88%
Community services	555	599	799	1,083	2,300	1,557
	9%	6%	13%	12%	41%	81%

4 CONCLUSIONS

Through a case study of Beijing, this paper is a first attempt to use the online land data to monitor the trend of city-scale land use development, with particular focus on housing and different types of employment land. Data verification demonstrates that the online data in general has a better coverage of land plots than official statistics, and in particular capture well the main projects. Thus it represents a major new source for continuous monitoring and policy assessment, complementing notoriously patchy traditional statistics.

The analysis of FAR changes by rings of urban growth confirms that between 2004 and 2013, housing is developing and decentralising at a much faster rate than non-manufacturing employment. Where housing is growing at the fastest rate (i.e. Ring 4-5 and Ring 5-6) the average distances to metro stations are considered inappropriately long (on average 1,575 and 2,134m respectively) for a city that aims to achieve good public transport accessibility for all, and this is worth particular attention in land use planning and urban design.

ACKNOWLEDGEMENTS

This paper is part of my PhD programme which has been supported by the Raymond and Helen Kwok Research Scholarship at Jesus College, Cambridge and by the Cambridge Overseas Trust. Thanks are also due to my PhD Supervisor Dr Ying Jin at Cambridge and my PhD Advisor Professor Ying Long at Tsinghua University and Beijing City Lab.

REFERENCES

Dang, Y., Liu, Z., & Zhang, W. 2014. Land-based interests and the spatial distribution of affordable housing development: The case of Beijing, China. *Habitat International*, *44*, 137-145.

Gao, J, Chen, J, & Su, X 2014. Influencing factors of land price in Nanjing Proper during 2001-2010. *Progress in Geography*, *33*(2), 211-221.

Kuang, W., Liu, J., Shao, Q., & Sun, C. 2009. Spatio-temporal patterns and driving forces of urban expansion in Beijing Central City since 1932. *Journal of Geo-Information Science*, *4*(005).

Liu, S. H., Wu, C. J., & Shen, H. Q. 2000. A GIS based model of urban land use growth in Beijing. *Acta Geographica Sinica-Chinese Edition*,*55*(4), 416-426.

Rong, X., Jin, Y., & Long, Y. 2015. Understanding Beijing's Urban Land Use Development from 2004–2013 Through Online Administrative Data Sources. In *Recent Developments in Chinese Urban Planning* (pp. 183-217). Springer International Publishing.

United Nations 2004. *World Urbanization Prospects 2014: Highlights.* United Nations Publications.

Yan, Y. T., & Feng, C. C. 2009. Research on spatial structure of urban land use intensity of Beijing. *China Land Science*, *3*(008), 12.

Proceedings of the International Conference on Smart Infrastructure and Construction
ISBN 978-0-7277-6127-9

© The authors and ICE Publishing: All rights reserved, 2016
doi:10.1680/tfitsi.61279.777

The spatial relationships between Beijing–Shanghai high-speed railway station and the cities

Chen Wu

Beijing Institute of Architecture Design- Central Design Group International, Beijing Engineering Research Center of Urban Design and Urban Renaissance, Beijing, China

Abstract: Over the past ten years, the high-speed railway construction of China has aroused the attention of the world, which has also exerted extensive impact upon the development pattern of Chinese urban space. This study attempts to elaborate a typology of relationships between the railway stations and their host cities.

Beijing-Shanghai High-speed Railway of China starts from Beijing South Station in the north and ends in Shanghai Hongqiao Railway Station with a full length of 1318km. The "Report in the Conception of Beijing-Shanghai High-speed Railway" had already been accomplished as early as 1990, which went through more than twenty years of winding and complicated process of argumentation. The project was officially initiated on April 8th, 2008 and opened to traffic on July 1st, 2011. The total project investment is RMB 220 billion (23.9 billion pounds). It proves to be not only the high-speed railway with the longest completed lines and highest standard all across the world, but also the construction project fulfilled with the largest investment scale since the founding of China. (The longest high-speed railway of China and even the world is Beijing-Guangzhou High-speed Railway with the overall length of 2240km.)The designed speed target value of Beijing-Shanghai High-speed Railway is: 350km/h with 380km/h as an alternative target. The current running speed is 300km/h. It takes 4 hours and 48 minutes at the soonest from Beijing to Shanghai; 24 passenger terminals are established along the line; it is able to realize the one-way transportation of more than 80 million passengers per year and the trains are designed based on the 3.5-minute minimum tracking intervals; the Beijing-Shanghai High-speed Railway will realize the passenger and cargo division with the original Beijing-Shanghai Railway after the completion. The former is the railway line for passenger traffic and the latter is for freight traffic.

Beijing-Shanghai High-speed Railway connects Beijing-Tianjin-Hebei economic belt and Shanghai-Nanjing-Hangzhou Yangtze River Delta economic belt. It penetrates through three major municipalities directly under the central government including Beijing, Tianjin and Shanghai as well as four provinces including Hebei, Shandong, Anhui and Jiangsu. The passing regions take up 6.5% of the entire national territorial area with 28% of the national population and 37% of national GDP (according to China City Stastical Yearbook of 2011.) The completion of Beijing-Shanghai High-speed Railway is able to not only provide the two-way passenger capacity for 160 million people and release the 100-million-ton freight capacity of the original Beijing-Shanghai Railway, but also reduce the entire traveling time to less than 5 hours. This inducts that Beijing-Shanghai High-speed Railway takes up an essential strategic status in the national economic development of China.

According to the statistics of January, 2015 made by China Railway, Beijing-Shanghai High-speed Railway transported more than 100 million passengers in 2014, which increased by 27% from the year earlier and took profits of about RMB 1200 million. It was the first time to realize profit. Beijing-Shanghai High-speed Railway transported 330 million passengers during the four years from 2011 to 2014.

1. ANALYSIS IN THE SPATIAL DEVELOPMENT OF CITIES AND TOWNS ALONG THE HIGH-SPEED RAILWAYS OF CHINA

The construction of Chinese high-speed railways has brought forward enormous opportunity for the development of regions and cities along the line. On the one hand, the cities along railway lines have grasped the social and economic benefits brought forward by the high-speed railways through not only improving urban transportation facilities in succession, but also vigorously promoting the development of urban service industry, the real estate exploitation and construction of new high-speed railway districts. On the other hand, however, quite a few cities are lack of correct understanding in the influencing mechanism of high-speed railways. They neglect their development level and the actual situation of specific development stage and cannot comprehensively think over their position in the district. Therefore, problems like over-fast urban spatial expansion speed come out, which is particularly presented by the various types of "high-speed railway new towns" widely forged by all cities. 15 new towns are planned to be constructed among the 24 Beijing-Shanghai High-speed Railway stations. These high-speed Railway new towns can be divided into two categories in general.

High-speed Railway New Towns

Since lots of high-speed railway stations are far away from the city center, most of them plans to construct high-speed railway new town.They are brand new large scaleurban planning projects, whichcovers several tens of kilo meters in general.They are provided with various urban elements including residence, livelihood, service and business, etc. The Dezhou new town covers 13 km^2, which core area is 4.6 km^2. The Nanjingnew town is centered by the high-speed railway station, which covers 164 km^2 in total with a population of 1.6 million.

High-speed Railway New District

Since the high-speed railway station is adjacent to the main city, only the districts in front of the high-speed railway stations are constructed. They are relatively in small scale covering several square kilometers, such as Suzhou Interurban Railway Station.

2. SPATIAL RELATIONSHIP BETWEEN BEIJING-SHANGHAI HIGH-SPEED RAILWAY STATIONS AND CITIES ALONG THE LINE

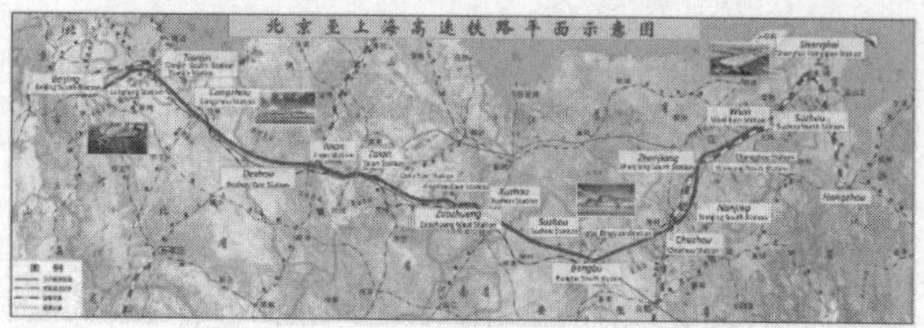

Figure 1. Distribution Diagram of Cities and Stations of Beijing-Shanghai High-speed Railway (Source: Institute of Economic Zoning of the Ministry of Railways)

2.1 General Development Features of Cities along Beijing-Shanghai High-speed Railway

Beijing-Shanghai High-speed Railway connects Beijing and Shanghai, the two major municipalities directly under the central government. The economic development level of cities along the line presents the characteristics of peaked ends and lower middle part: Beijing and Tianjin in the north end are located in Bohai Economic Circle and are in the middle of rapid economic development. They need to be linked with cities of the Yangtze River Delta economic belt relying on Beijing-Shanghai High-speed Railway, so as to generate an efficient complementary relationship; economic development of Hebei, Shandong and Anhui provinces of the middle part is relatively lagging. They are lack of motive force of development. The completion of Beijing-Shanghai High-speed Railway will provide opportunities for the urban development of the middle part; the south end is the relatively developed Yangtze River Delta, which is able to not only expand its central region scope, but economically complemented with cities of North China relying on Beijing-Shanghai High-speed Railway.

2.2 *Location Relation between Train Stations and Cities*

The location relation between train stations and cities mainly refers to the location of stations of Beijing-Shanghai High-speed Railway in the cities (whether the station is built within the built-up area and the distance between the station and downtown area) as well as land-use condition, population and economic development surround the stations, etc. This also reflects high-speed railways' role in urban economic and spatial development.

Table 1 indicates the spatial relationship of the 24 stations along Beijing-Shanghai High-speed Railway. 3 of them are located in the central area. 10 are in fringe area and suburbs. And the rest 11 stations are built in rural area.

Serial No.	Station	Location (Central Area-Fringe Area-Suburbs-Rural Area)	Distance from Downtown (km)	Serial No.	Station	Location (Central Area-Fringe Area-Suburbs-Rural Area)	Distance from Downtown (km)
1	Beijing South Railway Station	Central Area	4.4	13	Suzhou East Railway Station	Rural Area	24
2	Langfang Railway Station	Central Area	3.3	14	Bengbu South Railway Station	Suburbs	4.5
3	Tianjin West Railway Station (Interconnection Line)	Central Area	8.5	15	Dingyuan Railway Station	Rural Area	16
				16	Chuzhou South Railway Station	Rural Area	11
4	Tianjin South Railway Station	Fringe Area	13	17	Nanjing South Railway Station	Fringe Area	9.7
5	Cangzhou West Railway Station	Suburbs	6.8	18	Zhenjiang South Railway Station	Fringe Area	6.1
6	Dezhou East Railway Station	Rural Area	16	19	Danyang North Railway Station	Rural Area	9.2
7	Jinan West Railway Station	Fringe Area	9.5	20	Changzhou North Railway Station	Fringe Area	4.5
8	Taian Railway Station	Fringe Area	6.1	21	Wuxi East Railway Station	Fringe Area	15
9	Qufu East Railway Station	Rural Area	7.5	22	Suzhou North Railway Station	Fringe Area	15
10	Tengzhou East Railway Station	Rural Area	7.9	23	Kunshan South Railway Station	Fringe Area	4.8
11	Zaozhuang West Railway Station	Suburbs	5.9	24	Shanghai Hongqiao Railway Station	Fringe Area	16
12	Xuzhou East Railway Station	Suburbs	7.1				

Table 1: The Table of the Spatial Relationship between Stations of Beijing-Shanghai High-speed Railway and the Cities they located (Source: The data is measured based on Baidu Map)

According to table1, the distance between various stations and downtown areas range from 3km to 24km. Langfang Railway Station, Zaozhuang West Railway Station, Bengbu South Railway Station and Kunshan South Railway Station are those in the distance ranging from 0-4.9km. Although these four stations are located not far from downtown, all of the cities are featured by relatively small scale and lower development level, especially Bengbu South Railway Station and Zaozhuang West Railway Station are not established in the built-up areas. The relevant facilities of the four stations are lagging, which cannot meet the demand of the station development. Beijing South Railway Station is though located in downtown with easy transportation and intensive land usage.

2 stations are built within the distance range from 10 km to 14.9km, which are Tianjin South Railway Station and Chuzhou South Railway Station respectively. These two cities are in different development levels. Comparatively speaking, Chuzhou South Railway Station is surrounded by relatively worse environment and lagging facilities.

6 stations are built more than 15km away from downtown areas. They are Dezhou East Railway Station, Dingyuan Railway Station, Suzhou East Railway Station, Wuxi East Railway Station, Suzhou North Railway Station and Shanghai Hongqiao Railway Station. Cities that Dezhou East Railway Station, Dingyuan Railway Station and Suzhou East Railway Station are located in are featured by the lowest development level. The stations are constructed outside the built-up areas. The surrounding environment and supporting facilities are far worse than the other three stations; Wuxi East Railway Station and Suzhou North Railway Station are also located in the fringe area of the built-up area. There is still much room for development as for the surroundings; the built environment and supporting facilities of Shanghai Hongqiao Railway Station are comparatively better.

Most stations are established within the distance range from 5km to 9.9km. There are 12 in total, among which Cangzhou West Railway Station, Qufu East Railway Station, Tengzhou East Railway Station and Xuzhou East Railway Station are located not only outside the built-up areas, but in the west of Western Shandong. Districts with relatively weak economic foundation like mid-Anhui, etc, are in need of larger investment in the station infrastructure, which has brought forward certain adverse factors toward the development of the station. (Figure 2)

It is thus clear to see that the megalopolises like Beijing and Shanghai as well as the regional central cities have their high-speed railway stations constructed in city center or fringe area; the high-speed railway stations of local cities like Jinan, Xuzhou and Suzhou, etc, are usually established in fringe area and suburbs; by contrast, the second-class local cities like Qufu and Danyang, etc, usually select suburbs or rural area that are far from downtown to construct high-speed railway stations. Such relationship indicates the urban system under the impact of typical central location mode of China.

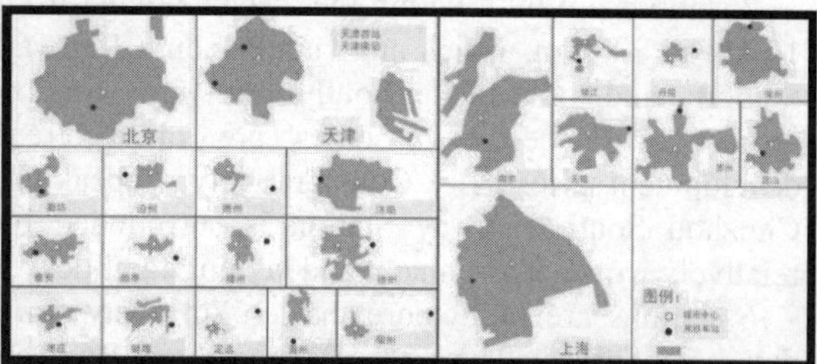

Figure 2. The Summary of the Relationships Between Stations on the Beijing-Shanghai HSR

Analysis of the Drivers of the Location Relationship

The urban center already possesses the development foundation such as high-density passenger flow, frequent economic activities as well as convenient infrastructure, which made it the ideal location for railway stations. Both passenger flow volume and economic development grade would surpass stations located in suburbs and rural area. However, the station site selection varies among different cities in account of a series of objective elements, namely the economic development level of the city, land-use condition and policy, etc. The following aspects shall be taken into consideration for the station site selection in accordance with the urbanized level of China, railway-related land policy as well as political policy, etc.

China is in the middle of rapid urbanization. The population and income of various cities are in the significant rise, which is manifested by not only the formation of megalopolis-belt surround Beijing and Shanghai, but also the rapid expansion of the small-and-medium cities in underdeveloped areas like Zaozhuang and Bengbu, etc.

It is comparatively convenient for Chinese government to acquire land for constructing high-speed railway stations at low price due to the public ownership of land and relevant laws. It is possible to gain high benefits with low input. This feature seems more obvious in suburbs or rural area than in downtown.

GDP index has been regarded as an essential indicator for the officials' performance appraisal under the special political system of China. The government is more inclined to construct high-way railway stations in suburbs since the infrastructure like high-speed railway is a vital driving force for urban development. Stations are built with the purpose of attracting more investment and business, so as to mobilize overall economic level and GDP level of the locality.

3. TYPICAL SPATIAL DEVELOPMENT PATTERN OF BEIJING-SHANGHAI HIGH-SPEED RAILWAY STATIONS AREAS

Based on the characteristics of the spatial development in regions locating Beijing-Shanghai High-speed Railway Stations and the overall consideration of the current development situation of various elements of Beijing-Shanghai High-speed Railway station areas as well as the difference in the development level among cities of various sizes and grades, the spatial development pattern of the 24 station areas of Beijing-Shanghai High-speed Railway can be divided into three typical patterns, namely megalopolis / metropolis "urban renewal type", megalopolis / metropolis "new town & new district type" and small-and-medium sized cities "new town & new district type" (Table 2)

Category	Megalopolis/metropolis "urban renewal type"	Megalopolis/metropolis "new city / new district"	Small and medium sized city "new city/ new district"
Station	Beijing South Railway Station, Tianjin West Railway Station	Shanghai Hongqiao Railway Station, Nanjing South Railway Station, Jinan West Railway Station, Tianjin South Railway Station, Wuxi East Railway Station, Suzhou North Railway Station, Xuzhou East Railway Station, Bengbu South Railway Station, Zhenjiang Railway Station, Changzhou North Railway Station, Kunshan South Railway Station	Cangzhou West Railway Station, Qufu East Railway Station, Tengzhou East Railway Station, Chuzhou East Railway Station, Danyang North Railway Station, Dezhou East Railway Station, Zaozhuang West Railway Station, Tai' an Railway Station, Suzhou East Railway Station and Langfang Railway Station
Urban Size	Megalopolis	Regional City / Local City Grade-I City	Local City Level II City
Station Area Level	Class 1	Class 1 / Class 2 / Class 3	Class 4
Location Characteristics	Downtown	Fringe of the Built-up Area or Suburbs	Urban Suburbs
Functional Characteristics	Developed traffic nodes and regional functions	Relatively developed traffic nodes and regional functions	Developed traffic node function, relatively weak regional function

Table 2. Spatial Development Patterns of Beijing-Shanghai High-speed Railway Station Areas

3.1 *Megalopolis / Metropolis "Urban Renewal" typed Station Areas*

The stations belonging to this type along the Beijing-Shanghai High-speed Railway line are merely Beijing South Railway Station and Tianjin West Railway Station. The larger cities' station areas with relatively higher economic aggregate are mostly located in suburbs and outer suburbs. This type of railway station improvement in the way of reconstructing original old stations, situating in the downtown area, provides opportunities for old town innovation and environment.

The "urban renewal" typed stations are located in the city center, which constitute a complex traffic junction integrating multiple transportation means including metro, bus, high-speed railway and intercity bus, etc. Such station areas are provided with high accessibility and relatively few available land resources. Thus land cost and land use efficiency turn to be the key factors for the station area development. Due to the impact of historical development, the stations are surrounded by abundant residential areas scaled by traditional streets with high land improvement cost, which contain the development of both the supporting facilities of the station areas and the tertiary industry.

Case Study: Beijing South Railway Station Area

Beijing South Railway Station is located in the intersection of Xicheng District, Dongcheng District and Fengtai District outside the Yongding Gate. This area is lagging in with regard to development. The core station area is enclosed by the Second South Ring Road, the Third South Ring Road, the East Majiabao Road and Kaiyang Road, which is provided with advantageous regional conditions. It is expanded and reconstructed based on the original railway station area and was officially opened to traffic at the same time of the opening of Beijing-Tianjin Inter-city High-speed Railway in August, 2008. As the opening of Beijing-Shanghai High-speed Railway and with the rapid increase in the passenger flow volume, the construction and development of Beijing South Railway Station have provided great opportunities for the realizing the revitalization of south-west Beijing, reorganizing and optimizing the interior structure of central urban area and strengthening the central status.

The new Beijing South Railway Station takes up 499,200 m^2 and a structure area of 420,000 m^2, which include 310,000 m^2 of main station structure area. The station contains five floors with two above the ground and three under the ground. The station is provided with relatively comprehensive functions. Shops of commerce and service industries including groceries, bookstores, restaurants and post-offices are also allocated to the ground waiting area other than the main transportation function. (Figure 3)

Figure 3. Live-scene of Beijing South Railway Station

3.2 Megalopolis / Metropolis "New City & New District" Typed Station Areas

The megalopolis / metropolis "new city & new district" typed station areas along Beijing-Shanghai High-speed Railway are relatively more in number, which are mainly scattered in the economically developed areas of Shandong, Jiangsu and Yangtze River Delta. Compared with the development pattern of "urban renewal" typed station areas, the station areas of this pattern are mainly featured by building the high-speed railway stations in the fringe urban areas or suburbs. High-speed railway stations have comparatively large impact upon the spatial structure of the cities. Current land utilization situation of the station areas is simple with ample available land resources. The station areas will be developed into new district or urban sub-centers and will directly promote the cities to develop toward multi-center spatial structure.

The station areas differ from one another with respect to the space-forming elements like comprehensive transportation, land development and landscape styles, etc, due to the difference in city size and economic level. This category of station areas can be further divided into the following three types in accordance with the relationship between station area and city size. Among the stations, Shanghai Hongqiao Railway Station, Nanjing South Railway Station and Jinan West Railway Station are stations of departure located in the fringe urban area. The stations are operated in the integrated air-rail transportation development pattern. The impact of these stations upon urban spatial structure is mainly manifested in the realization of regional aspiration of the cities through the development of corresponding stations areas. The rest stations like Tianjin South Railway Station, Wuxi East Railway Station and

Kunming South Railway Station are intermediate stations located in the suburbs of corresponding cities.

Typical Stations: Nanjing South Railway Station

Nanjing South Railway Station area is located in the main development axis of Nanjing situating the space joint between and traffic node of the main city and Dongshan Xinshi District. The urban spatial development pattern requires the district to not only suffice the function as traffic junction, but also become the joint that would both carry on the functional fill-over of the main city and promote urban functional reconstruction. The overall covered area of Nanjing South Railway Station is 480,000 m^2 with 48- m^2 planning area. The core area takes up nearly 6 km^2.
(Figure 4)

Figure 4. Birds Eye View of Nanjing South Railway Station

3.3 Small and Medium Sized City" New City& New District Type"

Stations of this type along the Beijing-Shanghai High-speed Railway are mainly distributed in the economically under-developed regions like Hebei, Shandong and Anhui, etc. Compared with megalopolis / metropolis "new city & new district" typed stations areas, this type of stations are located in the outer suburbs, which will not only induce the formation of high-speed railway new towns,but also become the major driving force for cities to develop toward multi-center spatial structures. However, it would take a relatively long period of time to accomplish the high-speed railway new town construction limited by the city size, economic development level and traffic conditions.

4. CONCLUSION

This paper summarizes three typical types of station areas along Beijing-Shanghai High-speed Railway based on city scale, urban functions, station status, station area location and functional characteristics. They differ in terms of development backgrounds, key influential factors, and development characteristics, and as a result face distinct challenges in spatial development and infrastructure construction. To ensure the long term success of Beijing-Shanghai High-speed Railway, it is of critical importance to understand the different starting points of the station areas, which will require distinct approches to land use planning, infrastructure investment and urban design.

REFERENCES

Bullock, Richard G.; Jin, Ying; Salzberg, Andrew. High-speed rail : the first three years taking the pulse of China's emerging program, 2012.02.01

Li Changbo, Chen Caiyuan. The Impact of High-speed Railway on the Cities along the Line, Seminar in Chinese Metropolis Transportation Planning, Collected Papers of 2010 Annual Meeting of China's Urban Transportation Planning and the 24th Academic Seminar, 2010.

Li Xiangguo. High-speed Railway Technology. Beijing: China Railway Publishing House, 2012:3-4

Li Xuewei. Introduction to High-speed Railway. Beijing: China Railway Publishing House, 2013:1-9

Lu Dongfu. The Study of the "12th Five-Year Plan" in Railway Development and Planning. Beijing: China Railway Publishing House, 2013:130-132.

Salzberg, Andrew; Bullock, Richard G.; Fang, Wanli; Jin, Ying, High-speed rail, regional economics, and urban development in China, 2013.01.01

Urban Transportation Institute of China Academy of Urban Planning and Design. The Impact of Railway Development and Railway Station Construction upon Urban Development, 2009

Wu Chen. Planning and Architectural Design Text of Beijing South Railway Station.

Wu Chen. Planning and Architectural Design Text of Beijing South Railway Station.

Zhou, Nanyan; Bullock, Richard G.; Jin, Ying; Lawrence, Martha B.; Ollivier, Gerald Paul, High-speed railways in China : an update on passenger profiles, 2016.01.01.

Proceedings of the International Conference on Smart Infrastructure and Construction
ISBN 978-0-7277-6127-9

© The authors and ICE Publishing: All rights reserved, 2016
doi:10.1680/tfitsi.61279

Author Index

Author Index

Author Index